Morphology and Evolution of Turtles

Vertebrate Paleobiology and Paleoanthropology Series

Edited by

Eric Delson
Vertebrate Paleontology, American Museum of Natural History,
New York, NY 10024, USA
delson@amnh.org

Eric J. Sargis
Anthropology, Yale University
New Haven, CT 06520, USA
eric.sargis@yale.edu

Focal topics for volumes in the series will include systematic paleontology of all vertebrates (from agnathans to humans), phylogeny reconstruction, functional morphology, Paleolithic archaeology, taphonomy, geochronology, historical biogeography, and biostratigraphy. Other fields (e.g., paleoclimatology, paleoecology, ancient DNA, total organismal community structure) may be considered if the volume theme emphasizes paleobiology (or archaeology). Fields such as modeling of physical processes, genetic methodology, nonvertebrates or neontology are out of our scope.

Volumes in the series may either be monographic treatments (including unpublished but fully revised dissertations) or edited collections, especially those focusing on problem-oriented issues, with multidisciplinary coverage where possible.

For other titles published in this series, go to
www.springer.com/series/6978

Morphology and Evolution of Turtles

Proceedings of the Gaffney Turtle Symposium (2009) in Honor of Eugene S. Gaffney

Edited by

Donald B. Brinkman

*Royal Tyrrell Museum, Midland Provincial Park, Highway 838
Drumheller, AB T0J 0Y0, Canada*

Patricia A. Holroyd

*Museum of Paleontology, University of California, 1101 Valley Life
Sciences Building, Berkeley, CA 94720-4780, USA*

James D. Gardner

*Royal Tyrrell Museum, Midland Provincial Park, Highway 838
Drumheller, AB T0J 0Y0, Canada*

Springer

Editors
Donald B. Brinkman
Royal Tyrrell Museum
Midland Provincial Park
Highway 838
Drumheller, AB T0J 0Y0
Canada

James D. Gardner
Royal Tyrrell Museum
Midland Provincial Park
Highway 838
Drumheller, AB T0J 0Y0
Canada

Patricia A. Holroyd
Museum of Paleontology
University of California
1101 Valley Life Sciences Building
Berkeley, CA 94720-4780
USA

ISSN 1877-9077 ISSN 1877-9085 (electronic)
ISBN 978-94-007-4308-3 ISBN 978-94-007-4309-0 (eBook)
DOI 10.1007/978-94-007-4309-0
Springer Dordrecht Heidelberg New York London

Library of Congress Control Number: 2012937464

Cover Illustration: "The Ghosts of Turtles Past" by Don Brinkman

Printed on acid-free paper

Springer is part of Springer Science+Business Media (www.springer.com)

Preface

With over 100 publications, many of monographic length, Gene Gaffney has brought research on fossil turtles to the forefront of paleontological study. His work set the stage for the current explosion of research on fossil turtles, which is showing exceptional potential for contributing to higher level concepts, such as the nature of evolutionary processes, paleobiogeography, and paleoecology. Upon Gene's retirement in 2007, the consensus among turtle researchers was that his valuable contribution to turtle research should be recognized. An informal survey of researchers at the 2007 Society of Vertebrate Paleontology meetings indicated that a dedicated meeting on fossil turtles and turtle relationships, together with a Festschrift publication, was the most appropriate way to honor Gene.

The idea of a stand-alone meeting dedicated to fossil turtles follows a tradition established by France Lapparent, who organized the first international meeting on fossil turtles in Paris, France, in 1983. Subsequent fossil turtle symposia were held at: the American Museum of Natural History, New York, USA, in 1987; the University of Alberta in Edmonton, Alberta, Canada (in conjunction with the 75th annual meeting of the American Society of Ichthyologists and Herpetologists), in 1995; and at the Zoological Institute of the Russian Academy of Sciences in St. Petersburg, Russia, in 2003. At the time of writing, plans are well underway for hosting the next symposium at the University of Tübingen, Germany, in 2012.

To emphasize that the meeting was being held in Gene's honor, it was formally named the "Gaffney Turtle Symposium", and quickly became known simply as "Gaffneyfest". The success of the meeting was proof of the high esteem in which Gene is valued by his peers. Held at the Royal Tyrrell Museum of Palaeontology in October 2009, the meeting lasted 2 full days, featured 37 talks and 16 posters, and was attended by approximately 100 people. In keeping with the desire to mark Gene's latest career milestone, the meeting was a gathering of several generations of researchers. One end of the spectrum was represented by 6 of the 13 attendees at the original 1984 Paris meeting and by 2 of Gene's classmates from university, all of whom made presentations. The other end of the generational spectrum was represented by an equal number of presentations from researchers who were either within the final stages of their graduate training or had graduated within the past year. And somewhere in the middle were presenters who had never known a time when some of Gene's classic papers, such as "The systematics of the North American Family Baenidae (Reptilia, Cryptodira)" (1972) and "Comparative cranial morphology of Recent and fossil turtles" (1979) were not available as standard references.

Many of the previous turtle meetings resulted in benchmark publications that have played a key role in the development of fossil turtle research, and in some cases, continue to be standard references. The meeting in Paris in 1984 resulted in the first volume of

Vertebrata Paleochelonica, a series that included, among other topics, reviews of fossil turtles from central Asia and Mongolia that provided western researchers with the first easily accessible overview of turtle research being undertaken in the Soviet Union. During the second meeting at the American Museum of Natural History in 1987, attendees received a compilation of translated papers on fossil turtles that fostered further development of an international perspective by turtle researchers. Thus, it was clear in the organizers' minds that to have a long-lasting impact, the Gaffney Turtle Symposium must be followed by a Festschrift volume. It was recognized early on by the editorial team that to be a fitting tribute, the Festschrift ideally should contain a mix of focused papers that would appeal to specialists along with papers having a broader scope that would appeal to a more general audience. For the latter, we solicited several review-style papers on subjects of current interest, such as the origins of turtles and development of the turtle shell. Although there are some obvious gaps in this volume (e.g., no comprehensive analyses of turtle relationships or paleobiogeography), this Festschrift contains a broad range of research and review papers, and we hope it will be a useful resource for years to come.

Don Brinkman got the ball rolling on this Festschrift project. Soon thereafter, Jim Gardner and Pat Holroyd stepped up to help the volume come to fruition, bringing with them extensive experience as editors for the *Journal of Vertebrate Paleontology*. Due to the enthusiastic response to our invitation to contribute to Gene's Festschrift, the editors realized the volume would have to take the form of a book, rather than a special issue of the Bulletin of the American Museum of Natural History as originally planned. At the suggestion of Walter Joyce, the editors approached Eric Delson and Eric Sargis with the idea of including this in Springer's *Vertebrate Paleobiology and Paleoanthropology* series. Pat Holroyd did the key work in drafting the proposal, which was positively received by Springer. Don Brinkman and Jim Gardner shared the workload in getting the manuscripts and other parts of the text ready for publication. Several years later, this volume is the final result of the editors' and authors' collective efforts.

GAFFNEY TURTLE SYMPOSIUM 09

Fig. 1 Logo for the Gaffney Turtle Symposium

Fig. 2 Traditional Chinese painting presented to Eugene Gaffney during the Gaffney Turtle Symposium in October 2009

Acknowledgments

Many of the papers included in this volume were presented at the Gaffney Turtle Symposium, held in October 2009 at the Royal Tyrrell Museum Palaeontology in honor of Gene Gaffney. This symposium was made possible through the financial support of The Cooperating Society of the Royal Tyrrell Museum, the American Museum of Natural History, and a Community Spirit Grant from the Department of Culture and Community Spirit of the Province of Alberta. The support of those organizations is greatly appreciated, especially because it helped defray travel expenses for several international participants who otherwise would not have been able to attend the symposium. The symposium also would not have been possible without the assistance of the organizing committee and staff of the Royal Tyrrell Museum. As well, the editors thank the participants of the symposium, whose enthusiastic discussions led to the development and refinement of many of the ideas presented in this volume.

The role of reviewers in the peer-review process is critical for a volume of this kind. As well as generously donating their time and energy, the reviewers for this volume maintained a high degree of professionalism, provided constructive and supportive comments, and returned their reviews promptly; all of this made the work of the editors much easier. For their efforts in reviewing manuscripts, the editors thank the following individuals:

David Brockhurst,
Ann Burke,
Edwin Cadina,
Bob Carroll,
Zoltán Csiki,
Igor Danilov,
Mike Everhart,
Andrew Farke,
Nadia Fröbisch,
Marcelo de la Fuente,
Emiliano Jiménez Fuentes,
Gene Gaffney,
Donald Henderson,
Ren Hirayama,
Howard Hutchison,
Walter Joyce,
Ben Kear,
Takuya Konishi,
France Lapparent de Broin,
Derek Larson,
Peter Meylan,
Andrew Milner,
Richard Moody,
Gary Morgan,
Jacqueline Moustakas,
Hiroshi Nagashima,
Jim Parham,
Larry Powell,
Adán Pérez-García,
Márton Rabi,
Olivier Rieppel,
Pedro Seyferth Romano,
Krister Smith,
Juliana Sterli,
Steven Sweetman,
Anthony Russell,
Michael Ryan,
Juliana Sterli,
Haiyan Tong,
Matthew Vickaryous,
Michael Wuttke,
Chang-Fu Zhou,
and several anonymous reviewers.

Three of the above-listed reviewers (Igor Danilov, Walter Joyce, and Jim Parham) deserve particular recognition, because they were called upon repeatedly and often at short notice to review manuscripts for this volume.

We are grateful to Eric Delson and Eric Sargis (a.k.a. "the Erics"), co-editors for Springer's Vertebrate Paleobiology and Paleoanthropology Book Series, for their help and guidance in shepherding this project from proposal to publication. Judith Terpos, Tamara Welschot, and other staff at Springer also provided invaluable help in producing this volume. The Cooperating Society of the Royal Tyrrell Museum provided funds to cover additional costs for printing some of the color figures in this volume.

Finally, we thank Gene Gaffney for his leadership and willingness to share information and ideas about fossil turtles, paleontology, and science throughout his more than thirty year career as a student of fossil turtles. His passionate interest in turtle evolution and the information they provide for studying the processes and patterns that have shaped the history of life, his openness in sharing unpublished information and ideas, and his willingness to provide access to undescribed specimens have led to the development of a uniquely collegiate atmosphere among fossil turtle researchers. We hope this book, and the symposium that preceded it, demonstrate the esteem and respect with which Gene and his work are held.

Drumheller, Alberta, Canada, 13 January 2012 — Donald B. Brinkman and James D. Gardner

Berkeley, California, USA, 13 January 2012 — Patricia A. Holroyd

Contents

Part IV Pleurodire Diversity and Biogeography

Contributors

Sergio A. K. Azevedo
Departamento de Geologia e Paleontologia, Universidade Federal do Rio de Janeiro, Museu Nacional, Quinta da Boa Vista, São Cristóvão, Rio de Janeiro 20940-040, Brazil
sazevedo@mn.ufrj.br

Christopher J. Bell
Jackson School of Geosciences, The University of Texas at Austin, Austin, TX 78712, USA
cjbell@mail.utexas.edu

Jonathan I. Bloch
Division of Vertebrate Paleontology, Florida Museum of Natural History, Dickinson Hall, University of Florida, Gainesville, FL 32611, USA
jbloch@flmnh.ufl.edu

Jason R. Bourque
Division of Vertebrate Paleontology, Florida Museum of Natural History, Dickinson Hall, University of Florida, Gainesville, FL 32611, USA
jbourque@flmnh.ufl.edu

Donald B. Brinkman
Royal Tyrrell Museum of Palaeontology, Box 7500, Drumheller, AB T0J 0Y0, Canada
don.brinkman@gov.ab.ca

Robert W. Burroughs
Jackson School of Geosciences, The University of Texas at Austin, Austin, TX 78712 USA
utexasrburroughs@gmail.com

Edwin A. Cadena
Division of Vertebrate Paleontology, Florida Museum of Natural History, Dickinson Hall, University of Florida, Gainesville, FL 32611, USA; Smithsonian Tropical Research Institute, Balboa, Ancon 0843-03092, Panama; North Carolina State University, Raleigh, NC USA
eacadena@ncsu.edu

Diógenes de Almeida Campos
Departamento Nacional de Produção Mineral, Museu de Ciências da Terra, Avenida Pasteur 404, Urca, Rio de Janeiro 22290-902, Brazil
diogenes.campos@dnpm.gov.br

Robert L. Carroll
Redpath Museum, McGill University, Montreal, QC H3A 2K6, Canada
robert.carroll@mcgill.ca

Sandra D. Chapman
Department of Palaeontology, The Natural History Museum, Cromwell Road, London SW7 5BD, UK
s.chapman@nhm.ac.uk

James M. Clark
George Washington University, Lisner Hall RM 340, 2023 G Street NW, Washington, DC 20052, USA
jclark@gwu.edu

Igor G. Danilov
Department of Herpetology, Zoological Institute of the Russian Academy of Sciences, Universitetskaya Emb. 1, St. Petersburg, Russia 199034
igordanilov@gmail.com

Massimo Delfino
Dipartimento di Scienze della Terra, Università di Torino, Via Valperga Caluso 35, 10125 Torino, Italy; Institut Català de Paleontologia Miquel Crusafont, Universitat Autònoma de Barcelona, Edifici ICP, Campus dela UAB s/n, Cerdanyola del Vallès, 08193 Barcelona, Spain

David A. Eberth
Royal Tyrrell Museum of Palaeontology, Box 7500, Drumheller, AB T0J 0Y0, Canada
david.eberth@gov.ab.ca

Sara J. ElShafie
Department of Organismal Biology and Anatomy, University of Chicago, 1027 East 57th Street, Chicago, IL 60637, USA
selshafie@uchicago.edu

David C. Evans
Department of Natural History (Palaeobiology), Royal Ontario Museum, 100 Queen's Park, Toronto, ON M5S 2C6, Canada
d.evans@utoronto.ca

Eugene S. Gaffney
Division of Paleontology, American Museum of Natural History, Central Park West at 79th Street, New York, NY 10024, USA
genegaffney373@comcast.net

Georgios L. Georgalis
School of Chemical Engineering, Aristotle University of Thessaloniki, 54124 Thessaloniki, Greece
dimetrodon82@yahoo.gr

Dean A. Hendrickson
Texas Natural Science Center, The University of Texas at Austin, Austin, TX 78705, USA
deanhend@mail.utexas.edu

Tsuyoshi Hibino
Shiramine Institute of Paleontology, Ishikawa, Japan
s-kyoiku@city.hakusan.lg.jp

Ren Hirayama
School of International Liberal Studies, Waseda University, Tokyo, Japan
renhirayama@waseda.jp

J. Howard Hutchison
Museum of Paleontology, University of California Berkeley, Berkeley, CA 94720, USA
howard.hutchison@gmail.com

Shinji Isaji
Natural History Museum and Institute, Chiba, Japan
isaji@chiba-muse.or.jp

Carlos A. Jaramillo
Smithsonian Tropical Research Institute, Balboa, Ancon 0843-03092, Panama
jaramilloc@si.edu

Steven E. Jasinski
Section of Paleontology and Geology, The State Museum of Pennsylvania, 300 North Street, Harrisburg, PA 17120-0024, USA
sej139@yahoo.com

Yoshie Kawashima-Ohya
Laboratory for Evolutionary Morphology, RIKEN Center for Developmental Biology, Kobe 650-0047, Japan
Present Address: Laboratory of Immunology, Department of Health Pharmacy, Yokohama College of Pharmacy, 601 Matano-cho, Totsuka-ku, Yokohama 245-0066, Japan
e-mail: y.kawashima@hamayaku.ac.jp

Benjamin P. Kear
Palaeobiology Programme, Department of Earth Sciences, Uppsala University, Villavägen 16, SE-752 36, Uppsala, Sweden
benjamin.kear@geo.uu.se

Alexander W. A. Kellner
Departamento de Geologia e Paleontologia, Universidade Federal do Rio de Janeiro, Museu Nacional, Quinta da Boa Vista, São Cristóvão, Rio de Janeiro 20940-040, Brazil
kellner@mn.ufrj.br

James L. Knight
South Carolina State Museum, 301 Gervais Street, Columbia, SC 29202, USA
jim.knight@scmuseum.org

Shigehiro Kuraku
Laboratory for Evolutionary Morphology, RIKEN Center for Developmental Biology, Kobe 650-0047, Japan
Present Address: Laboratory of Immunology, Department of Health Pharmacy, Yokohama College of Pharmacy, 601 Matano-cho, Totsuka-ku, Yokohama 245-0066, Japan
shigehiro.kuraku@uni-konstanz.de

Shigeru Kuratani
Laboratory for Evolutionary Morphology, RIKEN Center for Developmental Biology, Kobe 650-0047, Japan
saizo@cdb.riken.jp

Travis J. LaDuc
Texas Natural Science Center, The University of Texas at Austin, Austin, TX 78705, USA
travieso@mail.utexas.edu

Derek W. Larson
University of Alberta Laboratory for Vertebrate Paleontology, University of Alberta, CW 405 Biological Sciences Building, Edmonton, AB T6G 2E9, Canada
derek.larson@utoronto.ca

Nicholas R. Longrich
Department of Geology and Geophysics, Yale University, P.O. Box 208109, New Haven, CT 06520-8109, USA
longrich@gmail.com

Spencer G. Lucas
New Mexico Museum of Natural History and Science, 1801 Mountain Road NW, Albuquerque, NM 87104, USA
spencer.lucas@state.nm.us

Peter Meylan
Division of Paleontology, American Museum of Natural History, Central Park West at 79th Street, New York, NY 10024, USA; Natural Sciences, Eckerd College, 4200 54th Avenue S., St. Petersburg, FL 33711, USA
meylanpa@eckerd.edu

Andrew R. Milner
Department of Palaeontology, The Natural History Museum, Cromwell Road, London SW7 5BD, UK
andrew.milner@nhm.ac.uk

Christian Mitgutsch
Paläntologisches Institut und Museum, Universität Zürich, Karl Schmid-Strasse 4, 8006 Zürich, Switzerland
christian.mitgutsch@gmail.com

Tetsuto Miyashita
Department of Biological Sciences, University of Alberta, CW 405 Biological Sciences Centre, Edmonton, AB T6G 2E9, Canada
tetsuto@ualberta.ca

Richard T. J. Moody
School of Geology and Earth Sciences, Kingston University, Penrhyn Road, Kingston, Surrey KT1 2EE, UK
rtj.moody@virgin.net

Xabier Murelaga
Departamento de Estratigrafía y Paleontología, Facultad de Ciencia y Tecnología, Universidad del País Vasco/Euskal Herriko Unibertsitatea, Apartado 644, 48080, Bilbao, Spain
xabier.murelaga@ehu.es

Hiroshi Nagashima
Laboratory for Evolutionary Morphology, RIKEN Center for Developmental Biology, Kobe 650-0047, Japan
Present Address: Division of Gross Anatomy and Morphogenesis, Department of Regenerative and Transplant Medicine, Niigata University, Niigata 951-8510, Japan
azusagawakappa@gmail.com

Yuichi Narita
Laboratory for Evolutionary Morphology, RIKEN Center for Developmental Biology, Kobe 650-0047, Japan;
Present Address: Department of Food Business, Faculty of Health and Human Life, Nagoya Bunri University, 365 Maeda, Inazawa-cho, Inazawa 492-8520, Japan
narita.yuuichi@nagoya-bunri.ac.jp

Gustavo R. Oliveira
Departamento de Geologia e Paleontologia, Universidade Federal do Rio de Janeiro, Museu Nacional, Quinta da Boa Vista, São Cristóvão, Rio de Janeiro 20940-040, Brazil
gustavoliveira@gmail.com

Francisco Ortega
Grupo de Biología, Facultad de Ciencias, UNED, C/ Senda del Rey, 9, 28040 Madrid, Spain
fortega@ccia.uned.es

Rodrigo Pellegrini
New Jersey State Museum, PO Box 530, Trenton, NJ 08625-0530, USA
Rodrigo.Pellegrini@sos.state.nj.us

Adán Pérez-García
Departamento de Paleontología, Facultad de Ciencias Geológicas, Universidad Complutense de Madrid, Ciudad Universitaria, C/ José Antonio Novais, 2, 28040 Madrid, Spain
paleontologo@gmail.com

Márton Rabi
Department of Paleontology, Eötvös Loránd University, 1117 Pázmány Péter sétány 1/C, 96, Budapest, Hungary
iszkenderun@gmail.com

Olivier Rieppel
The Field Museum, 1400 South Lake Shore Drive, Chicago, IL 60605-2496, USA
orieppel@fieldmuseum.org

Pedro S. R. Romano
Departamento de Biologia Animal, Museu de Zoologia João Moojen, Universidade Federal de Viçosa, Campus Universitário, Viçosa, Minas Gerais 36570-000, Brazil
psrromano@gmail.com

Bruce M. Rothschild
Biodiversity Research Center and Natural History Museum, The University of Kansas, Lawrence, KS 66045, USA
brothschild@ku.edu

Michael J. Ryan
Cleveland Museum of Natural History, 1 Wade Oval Drive, University Circle, Cleveland, OH 44106, USA
mryan@cmnh.org

Marcelo R. Sánchez-Villagra
Paläntologisches Institut und Museum, Universität Zürich, Karl Schmid-Strasse 4, 8006 Zürich, Switzerland
m.sanchez@pim.uzh.ch

Torsten M. Scheyer
Paläntologisches Institut und Museum, Universität Zürich, Karl Schmid-Strasse 4, 8006 Zürich, Switzerland
tscheyer@pim.uzh.ch

Hans-Peter Schultze
Biodiversity Research Center and Natural History Museum, The University of Kansas, Lawrence, KS 66045, USA
hp1937@ku.edu

Paul C. Sereno
Department of Organismal Biology and Anatomy, University of Chicago, 1027 East 57th Street, Chicago, IL 60637, USA
dinosaur@uchicago.edu

Elizabeth T. Smith
Vertebrate Palaeontology Laboratory, University of New South Wales, PO Box 1, Kensington, NSW 2052, Australia; Australian Opal Centre, PO Box 229, Lightning Ridge, NSW 2834, Australia
elizabethtsmith@exemail.com.au

Robert M. Sullivan
Section of Paleontology and Geology, The State Museum of Pennsylvania, 300 North Street, Harrisburg, PA 17120-0024, USA
rsullivan@state.pa.us

Vladimir B. Sukhanov
Paleontological Institute of the Russian Academy of Sciences, Profsoyuznaya 123, Moscow, Russia 117997
sukhanovturtle@yandex.ru

Elena V. Syromyatnikova
Department of Herpetology, Zoological Institute of the Russian Academy of Sciences, Universitetskaya Emb. 1, St. Petersburg, Russia 199034
esyromyatnikova@gmail.com

Haiyan Tong
Division of Paleontology, American Museum of Natural History, Central Park West at 79th Street, New York, NY 10024, USA;, 30 Rue Carnot, 94270 Le Kremlin-Bicêtre, France

Katsuhisa Uchida
Laboratory for Evolutionary Morphology, RIKEN Center for Developmental Biology, Kobe 650-0047, Japan;
Present Address: Department of Marine Biology and Environmental Sciences, Faculty of Agriculture, University of Miyazaki, 1-1 Gakuen Kibanadai-nishi, Miyazaki 889-2192, Japan
k-uchida@cc.miyazaki-u.ac.jp

Dimitrios E. Velitzelos
Department of Historical Geology and Paleontology, National and Kapodistrian University of Athens, 15784 Athens, Greece
dvelitzel@geol.uoa.gr

Evangelos Velitzelos
Department of Historical Geology and Paleontology, National and Kapodistrian University of Athens, 15784 Athens, Greece
velitzel@geol.uoa.gr

Natasha S. Vitek
Yale University, PO 202411, New Haven, CT 06520, USA
nsvitek@utexas.edu

Mátyás Vremir
Department of Natural Sciences, Transylvanian Museum Society, 2-4, Strada Napoca, 400009 Cluj-Napoca, Romania
vremirmatyi@yahoo.co.uk

Cyril A. Walker
Department of Palaeontology, The Natural History Museum, Cromwell Road, London SW7 5BD, UK

Robert E. Weems
Paleo Quest, 14243 Murphy Terrace, Gainesville, VA 20155, USA
rweems@usgs.gov

Ingmar Werneburg
Paläntologisches Institut und Museum, Universität Zürich, Karl Schmid-Strasse 4, 8006 Zürich, Switzerland
ingmar_werneburg@yahoo.de

Xiao-Chun Wu
Earth Sciences Section, Canadian Museum of Nature, PO Box 3443, Station D, Ottawa, ON K1P 6P4, Canada
xcwu@mus-nature.ca

Xing Xu
Key laboratory of Evolutionary Systematics of Vertebrates, Institute of Vertebrate Paleontology and Paleoanthropology, Chinese Academy of Sciences, 42 Xi-Zhi-Men-Wai, 100044 Beijing, People's Republic of China
xu.xing@ivpp.ac.cn

Part I
Perspectives on the Life and Accomplishments of Eugene S. Gaffney

Chapter 1
Eugene S. Gaffney: A Professional Biography and Bibliography

Robert L. Carroll

Introduction

Eugene (Gene) Gaffney, long time Curator of Vertebrate Paleontology at the American Museum of Natural History, has made the most significant contributions to understanding the fossil record, evolution, and interrelationships of the Chelonia of any paleontologist. He has studied all the significant fossil and living turtles assembled in the major museums of the world, as well as collected specimens from localities throughout North America and as far distant as Lord Howe Island, off the coast of Australia. Gene was also instrumental in the conception, detailed planning and implementation of the current vertebrate paleontology gallery at the American Museum of Natural History. Its six fossil halls constitute the largest vertebrate fossil exhibition in the world. A significant aspect of this exhibition was the explicit integration of cladograms in association with the phylogenetic distribution of the specimens.

Gaffney's Accomplishments

Gene's first interest was in dinosaurs and primitive mammals, but through a visit to Princeton University he met Don Baird who became his advisor on a senior thesis based on a specimen of the side-necked turtle *Taphrosphys*. Don also introduced Gene to the then dean of fossil turtles, Rainer Zangerl, who became his co-author on the related genus *Bothremys*, whose type specimen had just been re-located by Gene in the Rutgers teaching collection (Gaffney and Zangerl 1968).

From Rutgers, Gene went on to Columbia University for his PhD and took advantage of its links with the American Museum of Natural History, which has been Gene's professional home ever since. Bobb Schaeffer instilled in Gene the importance of the highest standards in anatomical description and scientific objectivity. What is most important is Gene's thorough and highly detailed preparation, illustration and analysis of both fossil and living turtles at the broadest scale possible. Already in his PhD thesis he was investigating the highest taxonomic level of chelonian systematics, the cryptodire-pleurodire dichotomy (Gaffney 1969). Between 1972 and 1977, he published 10 major papers on the anatomy and systematics of turtles at the generic and family level including: "The Systematics of the North American Family Baenidae (Reptilia, Cryptodira)" and "An Illustrated Glossary of Turtle Skull Nomenclature" (both in 1972) and "A Revision of the Side-necked Turtle *Taphrosphys sulcatus* (Leidy) from the Cretaceous of New Jersey", "A Taxonomic Revision of the Jurassic Turtles *Portlandemys* and *Plesiochelys*", "*Solnhofia parsonsi*, a New Cryptodiran Turtle from the Late Jurassic of Europe", "A Phylogeny and Classification of the Higher Categories of Turtles", and "Phylogeny of the Chelydrid Turtles: a Study of Shared Derived Characters in the Skull" (all in 1975). "Cranial Morphology of the European Jurassic Turtles *Portlandemys* and *Plesiochelys*" was published in 1976 followed in 1977 by "The Side-necked Turtle Family Chelidae: a Theory of Relationships Using Shared Derived Characters" and "An Endocranial Cast of the Side-necked Turtle, *Bothremys*, with a New Reconstruction of the Palate". His later papers were closely linked with the development of phylogenetic systematics, as seen through the eyes of Karl Popper and discussed in a trio of papers by Platnick and Gaffney in 1977 and 1978.

Major integrative contributions in 1979 included "The Jurassic Turtles of North American" and "Comparative Cranial Morphology of Recent and Fossil Turtles." In the same year, he published his only paper (with McKenna) discussing a specific genus as a possible sister-taxon of

R. L. Carroll (✉)
Redpath Museum, McGill University, Montreal,
QC H3A 2K6, Canada
e-mail: robert.carroll@mcgill.ca

D. B. Brinkman et al. (eds.), *Morphology and Evolution of Turtles*, Vertebrate Paleobiology and Paleoanthropology,
DOI: 10.1007/978-94-007-4309-0_1, © Springer Science+Business Media Dordrecht 2013

turtles, followed in 1980 by a more general investigation, "Phylogenetic Relationships of the Major Groups of Amniotes." His paper "The Cranial Morphology of the Extinct Horned Turtle, *Meiolania platyceps*, from the Pleistocene of Lord Howe Island, Australia" (1983) and other later papers summarized results from his extensive field work off the coast of Australia (e.g., Gaffney 1996a, b). Other field work included collecting trips to the Grizzly Buttes localities in the Bridger Basin, the Lance Formation of Wyoming, the Cretaceous-Paleocene of Montana, the San Juan Basin in New Mexico, and the Lower Jurassic Kayenta beds of Arizona.

An exhaustive phylogeny of all turtles was undertaken with Peter Meylan in 1988 and in 1990 Gene published a 263 page monograph on the oldest turtle then known: "The Comparative Osteology of the Triassic Turtle, *Proganochelys*". Through the 1990s and early years of this century, Gene published a wide range of descriptive papers based on turtles from Australia, Africa, China, Europe, the Middle East, South America, and India.

So far, Gene's largest publication has been the 698 page monograph "Evolution of the Side-Necked Turtles: The families Bothremydidae, Euraxemydidae, and Araripemydidae", co-authored with Haiyan Tong and Peter Meylan (2006), with a core data base of 41 taxa.

One of his most recent publications (2010) was entitled "*Kayentachelys*, an Early Jurassic Cryptodire, and the Early History of Turtles", with Farish Jenkins as co-author. This paper discussed the problems of establishing the phylogenetic position of this genus, whether as a sister-taxon of all more advanced cryptodires or as being outside the common ancestry of cryptodires plus pleurodires. Gaffney and Jenkins (2010) demonstrated that very different phylogenetic positions can be hypothesized, even when dealing with the same specimen and using comparable means of phylogenetic analysis.They compared the results of an earlier paper on this genus by Gaffney et al. (1987) and ones by Joyce (2007) and Sterli and Joyce (2007). Gaffney and Jenkins (2010) pointed out that a general problem, the absence of adequate knowledge of well preserved skeletal remains of pre-Late Cretaceous pleurodires, is a major difficulty in establishing their phylogenetic position.This applies specifically to the skull. Joyce (2007) and Sterli and Joyce (2007) had depended heavily on shell characters, which are known to be much more homoplastic than those of the skull and primarily involve loss of primitive features. In contrast, cranial features support close affinities between *Kayentachelys* and other cryptodires.

Gaffney and Jenkins (2010, p. 365) wrote: "The frequent proliferation of contrary results of analysis of the same taxa shows that at least a moderate degree of subjectivity is involved, despite the use of PAUP... An unfortunate by-product of the computer algorithm is the trend to assemble large data sets, often containing a high percentage of unanalyzed and miscoded characters leading to results that are more phenetic than cladistic.Using large number of taxa also increased the likelihood of making a mistake in the homology assessment and distribution.As taxa are included which are farther and farther removed from each other, the morphology becomes harder to compare as differences increase, and the coding of characters becomes even more arbitrary." As the pioneer cladist Colin Patterson and David Johnson (1997, p. 361) said "This change of emphasis replaces our pernicious old black box, evolutionary systematics, with a new one, the [data] matrix."

All but two or three of Gaffney's papers (now totaling more than 100) have been focused on members of the Chelonia. This has been possible, and remains highly significant for understanding evolutionary processes, because all turtles, fossil and living, can be unquestionably united in a single taxonomic assemblage based on a long list of unique synapomorphies (autapomorphies) known only in this assemblage.The structure and modes of development of the carapace and plastron are the most striking examples, but many other aspects of the skeleton and its functions also support a unique evolutionary history.

Ironically, neither Gene's studies of all groups of living turtles and their known fossil record nor his application of phylogenetic analysis have enabled him to determine their specific relationships among other lineages of primitive amniotes. Gene has subsequently avoided active participation in the current arguments regarding the specific sister-taxa of chelonians. In the summary of his recent magnum opus on the evolution of side-necked turtles (2006, p. 570) it was stated: "We consider turtles to be the sister group of diapsids, not within diapsids or within pareiasaurs/procolophonids." Furthermore, "We do not consider the hypothesis that turtles are within diapsids (e.g., deBraga and Rieppel 1997) to have merit, but even accepting this would not alter relationships within turtles as analyzed here."

Nevertheless, Gene's research provides the basis for understanding not only the anatomy and phylogeny of all known turtle taxa, but also demonstrates the clear distinction of even the oldest recognized Late Triassic fossils from any of the known clades of earlier amniotes that have been proposed as the sister-group of chelonians over the past 120 years.

Quoting from Gene's autobiography (Gaffney 2012): "Although I was an active participant and minor contributor to the Clade Wars of the 1970s, I've never been much of a systematics theoretician, at least by AMNH standards.My real love in paleontology has always been the intricacies of morphology—systematics was simply the obvious thing to do with it."

References Not Included in " Bibliography of Eugene S. Gaffney"

deBraga, M., & Rieppel, O. (1997). Reptile phylogeny and the interrelationships of turtles. *Zoological Journal of the Linnean Society, 120*, 281–354.

Joyce, W. G. (2007). Phylogenetic relationships of Mesozoic turtles. *Bulletin of the Peabody Museum of Natural History, 48*, 3–102.

Patterson, C., & Johnson, G. (1997). The data, the matrix, and the message: comments on Begle's "Relationships of the Osmeroid fishes". *Systematic Biology, 46*, 358–365.

Sterli, J., & Joyce, W. G. (2007). The cranial anatomy of the Early Jurassic turtle *Kayentachelys aprix*. *Acta Palaeontologica Polonica, 52*, 675–694.

Bibliography of Eugene S. Gaffney

Burke, A. C., Anderson, M., Weld, A., & Gaffney, E. S. (1983). The reconstruction and casting of a large extinct turtle, *Meiolania*. *Curator, 26*, 5–26.

Burke, A. C., Gaffney, E. S., & Rich, T. H. (1983). Miocene turtles from Lake Tarkarooloo, South Australia. *Alcheringa, 7*, 151–154.

Colbert, E. H., & Gaffney, E. S. (1984). Bobb Schaeffer, a biographical sketch. *Journal of Vertebrate Paleontology, 4*, 285–291.

Dingus, L., Tedford, R., Gaffney, E. S., McKenna, M., Novacek, M., & Delson, E. (1994). *Mammals and their extinct relatives: A guide to the Lila Acheson Wallace Wing*. New York: American Museum of Natural History.

Dingus, L., Gaffney, E. S., Norell, M. A., & Sampson, S. D. (1995). *The halls of dinosaurs. A guide to saurischians and ornithischians*. New York: American Museum of Natural History.

Gaffney, E. S. (1967). Early evolution of pleurodires. *American Society of Ichthyologists and Herpetologists, Forty-seventh Annual Meeting, Abstracts, 7*

Gaffney, E. S. (1969). The North American Baenoidea and the cryptodire-pleurodire dichotomy (Ph.D. Dissertation, Columbia University)

Gaffney, E. S. (1971a). Chelonia. In McGraw-Hill (Ed.), *Yearbook of Science and Technology* (pp. 140–142). New York: McGraw-Hill Inc.

Gaffney, E. S. (1971b). Inadequacy of lower tetrapod fossils for determination of past continental relationships. *Geological Society of America, 3*(7), 576. (Abstracts.)

Gaffney, E. S. (1972a). The systematics of the North American family Baenidae (Reptilia, Cryptodira). *Bulletin of the American Museum of Natural History, 147*, 241–320.

Gaffney, E. S. (1972b). Field trip to the American Museum of Natural History. New York *Paleontological Society, Notes, 3*, 4–11.

Gaffney, E. S. (1972c). An illustrated glossary of turtle skull nomenclature. *American Museum Novitates, 2486*, 1–33.

Gaffney, E. S. (1974). (Review of) Vertebrate history: Problems in evolution. Barbara J. Stahl. New York, McGraw-Hill, 594 pp. *Systematic Zoology, 23*, 299

Gaffney, E. S. (1975a). A revision of the side-necked turtle *Taphrosphys sulcatus* (Leidy) from the Cretaceous of New Jersey. *American Museum Novitates, 2571*, 1–24.

Gaffney, E. S. (1975b). A taxonomic revision of the Jurassic turtles *Portlandemys* and *Plesiochelys*. *American Museum Novitates, 2574*, 1–19.

Gaffney, E. S. (1975c). *Solnhofia parsonsi*, a new cryptodiran turtle from the Late Jurassic of Europe. *American Museum Novitates, 2576*, 1–25.

Gaffney, E. S. (1975d). A phylogeny and classification of the higher categories of turtles. *Bulletin of the American Museum of Natural History, 155*, 389–436.

Gaffney, E. S. (1975e). Phylogeny of the chelydrid turtles: A study of shared derived characters in the skull. *Fieldiana: Geology, 33*, 157–178.

Gaffney, E. S. (1976a). Cranial morphology of the European Jurassic turtles *Portlandemys* and *Plesiochelys*. *Bulletin of the American Museum of Natural History, 157*, 489–543.

Gaffney, E. S. (1976b). The history of turtles in Australia. 25th *International Geological Congress, 25*(1), 325–326. (Abstracts.)

Gaffney, E. S. (1977a). The side-necked turtle family Chelidae: A theory of relationships using shared derived characters. *American Museum Novitates, 2620*, 1–28.

Gaffney, E. S. (1977b). An endocranial cast of the side-necked turtle, *Bothremys*, with a new reconstruction of the palate. *American Museum Novitates, 2639*, 1–12.

Gaffney, E. S. (1977c). Phylogeny as science, North American Paleontological Convention II. *Journal of Paleontology, 51*(suppl 2), 11

Gaffney, E. S. (1979a). The Jurassic turtles of North America. *Bulletin of the American Museum of Natural History, 162*, 91–136.

Gaffney, E. S. (1979b). Turtles from the age of dinosaurs. *Explorers Journal, 57*, 8–9.

Gaffney, E. S. (1979c). Tetrapod monophyly: A phylogenetic analysis. In J. H. Schwartz & H. B. Rollins (Eds.), *Models and methodologies in evolutionary theory* (pp. 92–105). *Bulletin of the Carnegie Museum of Natural History, 13*

Gaffney, E. S. (1979d). An introduction to the logic of phylogeny reconstruction. In J. Cracraft & N. Eldredge (Eds.), *Phylogenetic analysis and paleontology* (pp. 79–111). New York: Columbia University Press.

Gaffney, E. S. (1979e). Comparative cranial morphology of recent and fossil turtles. *Bulletin of the American Museum of Natural History, 164*, 65–375.

Gaffney, E. S. (1979f). Fossil chelid turtles of Australia. *American Museum Novitates, 2681*, 1–23.

Gaffney, E. S. (1979g). Description of a large trionychid turtle shell from the Eocene Bridger Formation of Wyoming. *Contributions to Geology, University of Wyoming, 17*, 53–57.

Gaffney, E. S. (1979h). Fossil trionychids of Australia. *Journal of Paleontology, 53*, 1354–1360.

Gaffney, E. S. (1980). Phylogenetic relationships of the major groups of amniotes. In A. L. Panchen (Ed.), *The terrestrial environment and the origin of land vertebrates* (pp. 593–610). London: Academic Press.

Gaffney, E. S. (1981a). A review of fossil turtles of Australia. *American Museum Novitates, 2720*, 1–38.

Gaffney, E. S. (1981b). Phylogeny and biogeography of meiolaniid turtles. In H. F. Recher & W. F. Ponder (Eds.), *Lord Howe Island* (pp. 26–29). *Occasional Reports, Australian Museum, 1*

Gaffney, E. S. (1982a). Cranial morphology of the baenid turtles. *American Museum Novitates, 2737*, 1–22.

Gaffney, E. S. (1982b). The lower jaws of baenid turtles. *American Museum Novitates, 2749*, 1–10.

Gaffney, E. S. (1982c). On the horns of a dilemma (Letters from the field). *Rotunda, 56*(4), 4–5.

Gaffney, E. S. (1982d). (Review of) Kemp, T. S.: Mammal-like reptiles and the origin of mammals. London: Academic Press. *Systematic Zoology, 31*, 526–527.

Gaffney, E. S. (1983a). The basicranial articulation of the Triassic turtle, *Proganochelys*. In A. G. J. Rhodin & K. Mayata (Eds.), *Advances in herpetology and evolutionary biology; Essays in honor of Ernest E. Williams* (pp. 190–194). Cambridge: Museum of Comparative Zoology, Special Publications.

Gaffney, E. S. (1983b). The cranial morphology of the extinct horned turtle, *Meiolania platyceps*, from the Pleistocene of Lord Howe Island, Australia. *Bulletin of the American Museum of Natural History, 175*, 326–479.

Gaffney, E. S. (1984a). Historical analysis of theories of chelonian relationship. *Systematic Zoology, 33*, 283–301.

Gaffney, E. S. (1984b). Progress towards a natural hierarchy of turtles. *Studia Geologica Salamanticensia, Vol. especial, 1*, 125–131.

Gaffney, E. S. (1985a). The cervical and caudal vertebrae of the cryptodiran turtle, *Meiolania platyceps*, from the Pleistocene of Lord Howe Island, Australia. *American Museum Novitates, 2805*, 1–22.

Gaffney, E. S. (1985b). The shell morphology of the Triassic turtle, *Proganochelys. Neues Jahrbuch für Geologie und Paläontologie Abhandlungen, 170*, 1–26.

Gaffney, E. S. (1985c). *Meiolania platyceps*. The Lord Howe Island horned turtles. In P.V. Rich & G.F. Van Tets (Eds.), *Kadimakara: Extinct vertebrates of Australia* (pp. 132–136). Lilydale: Pioneer Design Studio.

Gaffney, E. S. (1986). Triassic and Early Jurassic turtles. In K. Padian (Ed.), *The beginnings of the age of dinosaurs* (pp. 183–187). New York: Cambridge University Press.

Gaffney, E. S. (1988). A cladogram of the pleurodiran turtles. *Acta Zoologica Cracoviensia, 31*, 487–492.

Gaffney, E. S. (1990a). The comparative osteology of the Triassic turtle *Proganochelys. Bulletin of the American Museum of Natural History, 194*, 1–263.

Gaffney, E. S. (1990b). *Dinosaurs a golden guide*. Racine: Western Publishing Company, Inc.

Gaffney, E. S. (1991a). The fossil turtles of Australia. In P. Vickers-Rich, J. M. Monaghan, R. F. Baird & T. H. Rich (Eds.), *Vertebrate Palaeontology of Australasia* (pp. 704–720). Lilydale: Pioneer Design Studio.

Gaffney, E. S. (1991b). (Review of) D. B. Weishampel, P. Dodson, & H. Osmólska (Eds.), *The Dinosauria*, 1990, Berkeley: University of California Press, pp. 733). *Systematic Zoology, 40*, 378–379.

Gaffney, E. S. (1991c). Phylogeny and biogeography of the horned turtles. *Journal of Vertebrate Paleontology, 11*(suppl 3), 30A.

Gaffney, E. S. (1992). *Ninjemys*, a new name for "*Meiolania*" *oweni* (Woodward), a horned turtle from the Pleistocene of Queensland. *American Museum Novitates, 3049*, 1–10.

Gaffney, E. S. (1996a). The postcranial morphology of *Meiolania platyceps* and a review of the Meiolaniidae. *Bulletin of the American Museum Natural History, 229*, 1–165.

Gaffney, E. S. (1996b). Unique among vertebrates. *Natural History, 96*(6), 38–39.

Gaffney, E. S. (2000). Turtles of the Santana Formation. *International Geological Congress*, Rio de Janeiro, Brazil: CD-ROM

Gaffney, E. S. (2001). Phylogeny of turtles as seen in patterns of skull diversity. ICVM-6. *Journal of Morphology, 248*, 232–233.

Gaffney, E. S. (2003) Lord Howe Island Australia. *Natural History, 112*(4A), 44.

Gaffney, E. S. (2012). Autobiography (through May 2009). In D. B. Brinkman, P. A. Holroyd, & J. D. Gardner (Eds.), *Morphology and evolution of turtles*. Dordrecht: Springer.

Gaffney, E. S., Archer, M., & White, A. (1989). Chelid turtles from the Miocene freshwater limestones of Riversleigh Station, Northwestern Queensland, Australia. *American Museum Novitates, 2959*, 1–10.

Gaffney, E. S., Archer, M., & White, A. (1992). *Warkalania*, a new meiolaniid turtle from the Tertiary Reversleigh deposits of Queensland, Australia. *The Beagle, Records of the Northern Territory Museum of Arts and Sciences, 9*, 35–48.

Gaffney, E. S., Balouet, J. C., & De Broin, F. (1984). New occurrences of extinct meiolaniid turtles in New Caledonia. *American Museum Novitates, 2800*, 1–6.

Gaffney, E. S., Brinkman, D. B., Hutchison, H., & Meylan, P. A. (1995). Early history of the Eucryptodira. *Seventy-fifth Annual Meeting of the American Society of Ichthyologists and Herpetologists*, (University of Alberta, Edmonton, Alberta, Canada), Abstracts, 106.

Gaffney, E. S., Campbell, K. E., & Wood, R. C. (1998). Pelomedusoid side-necked turtles from late Miocene sediments in southwestern Amazonia. *American Museum Notivates, 3245*, 1–11.

Gaffney, E. S., Chatterjee, S., & Rudra, D. K. (2001). *Kurmademys*, a new side-necked turtle (Pelomedusoides: Bothremydidae) from the Late Cretaceous of India. *American Museum Novitates, 3321*, 1–16.

Gaffney, E. S., de Campos, D. A., & Hirayama, R. (2001). *Cearachelys*, a new side-necked turtle (Pelomedusoides: Bothremydidae) from the Early Cretaceous of Brazil. *American Museum Novitates, 3319*, 1–20.

Gaffney, E. S., DeBlieux, D. D., Simons, E. L., Sánchez-Villagra, M. R., & Meylan, P. A. (2002). Redescription of the skull of *Dacquemys* Williams, 1954, a podocnemidid side-necked turtle from the late Eocene of Egypt. *American Museum Novitates, 3372*, 1–16.

Gaffney, E. S., Dingus, L., & Smith, M. K. (1995). Why cladistics? *Natural History, 6*(95), 33–35.

Gaffney, E. S., & Forster, C. A. (2003). Side-necked turtle lower jaws (Podocnemididae, Bothremydidae) from the Late Cretaceous Maevarano Formation of Madagascar. *American Museum Novitates, 3397*, 1–13.

Gaffney, E. S., & Hiatt, R. (1971). A new baenid turtle from the Upper Cretaceous of Montana. *American Museum Novitates, 2443*, 1–9.

Gaffney, E. S., Hooks, G. E. III., & Schneider, V. P. (2009a). New material of North American side-necked turtles (Pleurodira: Bothremydidae). *American Museum Novitates, 3655*, 1–26.

Gaffney, E. S., Hutchison, J. H., Jenkins, F. A., Jr., & Meeker, L. J. (1987). Modern turtle origins: The oldest known cryptodire. *Science, 237*, 289–291.

Gaffney, E. S., & Jenkins, F. A., Jr. (2009). The cranial morphology of *Kayentachelys*, an Early Jurassic cryptodire, and the early history of turtles. *Gaffney Turtle Symposium (October 17–18, 2009, Royal Tyrrell Museum, Drumheller, Canada), Abstract Volume*, 59–63

Gaffney, E. S., & Jenkins, F. A., Jr. (2010). The cranial morphology of *Kayentachelys*, an Early Jurassic cryptodire, and the early history of turtles. *Acta Zoologica (Stockholm), 91*, 335–368.

Gaffney, E. S., & Kitching, J. W. (1994). The most ancient African turtle. *Nature, 369*, 55–58.

Gaffney, E. S., & Kitching, J. (1995). The morphology and relationships of *Australochelys*, an Early Jurassic turtle from South Africa. *American Museum Novitates, 3130*, 1–29.

Gaffney, E. S., Kool, L., Brinkman, D. B., Rich, T. H., & Vickers-Rich, P. (1998). *Otwayemys*, a new cryptodiran turtle from the Early Cretaceous of Australia. *American Museum Novitates, 3233*, 1–28.

Gaffney, E. S., Kool, L., Rich, T., Vickers-Rich, P., Pledge, N., Archer, M., et al. (1992). Early history of the cryptodiran turtles in Australia. *The Beagle, Records of the Northern Territory Museum of Arts and Sciences, 9*(1), 261.

Gaffney, E. S., & Krause, D. W. (2011). *Sokatra*, a new side-necked turtle (Late Cretaceous, Madagascar) and the diversification of the main groups of Pelomedusoides. *American Museum Novitates 3728*, 1–28.

Gaffney, E. S., Krause, D. W., & Zalmout, I. S. (2009b). *Kinkonychelys*, a new side-necked turtle (Pelomedusoides, Bothremydidae) from the late Cretaceous of Madagascar. *American Museum Novitates, 3662*, 1–25.

Gaffney, E. S., & McKenna, M. C. (1979). A Late Permian captorhinid from Rhodesia. *American Museum Novitates, 2688*, 1–15.

Gaffney, E. S., & McNamara, G. (1990). A meiolaniid turtle from the Pleistocene of Northern Queensland. *Proceedings of the De Vis Symposium, memoirs of the Queensland Museum, 28*, 107–113.

Gaffney, E. S., & Meeker, L. J. (1983). Skull morphology of the oldest turtles: A preliminary description of *Proganochelys quenstedti*. *Journal of Vertebrate Paleontology, 3*, 25–28.

Gaffney, E. S., & Meylan, P. A. (1988a). A phylogeny of turtles. In M. J. Benton (Ed.), *The Phylogeny and Classification of Tetrapods* (pp. 157–219). Oxford: Clarendon Press.

Gaffney, E. S., & Meylan, P. A. (1988). A new phylogeny of cryptodiran turtles. *Journal of Vertebrate Paleontology, 8*(suppl 3), 15A.

Gaffney, E. S., & Meylan, P. A. (1989). The early history of the side necked turtles. *Journal of Vertebrate Paleontology, 9*(suppl 3), 21A–22A.

Gaffney, E. S., & Meylan, P. A. (1990). A history of the cranial circulation in turtles. *Journal of Vertebrate Paleontology, 10*(suppl 3), 6A.

Gaffney, E. S., & Meylan, P. A. (1991). Primitive pelomedusid turtle. In J. G. Maisey (Ed.), *Santana fossils: An illustrated atlas* (pp. 335–339). Neptune: Tropical Fish Hobbyist Publications.

Gaffney, E. S., & Meylan, P. A. (1992a). The Transylvanian turtle, *Kallokibotion*, a primitive cryptodire of Cretaceous age. *American Museum Novitates, 3040*, 1–37.

Gaffney, E. S., & Meylan, P. A. (1992b). *Sinaspideretes* is not the oldest trionychid turtle. *Journal of Vertebrate Paleontology, 12*, 257–259.

Gaffney, E. S., & Meylan P. A. (1993). Relationships of primitive Eucryptodira. *Journal of Vertebrate Paleontology 13*(suppl 3), 36A.

Gaffney, E. S., Meylan, P. A., & Wood, R. C. (1997). Changing interpretations of the history of the side-necked turtles. *Journal of Vertebrate Paleontology, 17*(suppl 3), 47A–48A.

Gaffney, E. S., Meylan, P. A., Wood, R. C., Simons, E., & Campos, D. A. (2011). Evolution of the side-necked turtles: The family Podocnemididae. *Bulletin of the American Museum of Natural History, 350*, 1–236.

Gaffney, E. S., Meylan, P. A., & Wyss, A. (1991). A computer assisted analysis of the relationships of the higher categories of turtles. *Cladistics, 7*, 313–335.

Gaffney, E. S., Moody, R. T. J., & Walker, C. A. (2001). *Azabbaremys*, a new side-necked turtle (Pelomedusoides: Bothremydidae) from the Paleocene of Mali. *American Museum Novitates, 3320*, 1–16.

Gaffney, E. S., Rich, T. H., Vickers-Rich, P., Constantine, A., Vacca, R., & Kool, L. (2007). *Chubutemys*, a new eucryptodiran turtle from the Early Cretaceous of Argentina, and the relationships of the Meiolaniidae. *American Museum Novitates, 3599*, 1–35.

Gaffney, E. S., Roberts, E., Sissoko, F., Bouaré, M. L., Tapanila, L., & O'Leary, M. A. (2007). *Acleistochelys*, a new side-necked turtle (Pelomedusoides: Bothremydidae) from the Paleocene of Mali. *American Museum Novitates, 3549*, 1–24.

Gaffney, E. S., Sahni, A., Schleich, H. H., Singh, S. D., & Srivastava, R. (2003). *Sankuchemys*, a new side-necked turtle (Pelomedusoides: Bothremydidae) from the Late Cretaceous of India. *American Museum Novitates, 3405*, 1–10.

Gaffney, E. S., Scheyer, T. M., Johnson, K. G., Bocquentin, J., & Aguilera, O. A. (2008). Two new species of the side necked turtle genus, *Bairdemys* (Pleurodira: Podocnemididae) from the Miocene of Venezuela. *Paläontologische Zeitschrift, 82*, 209–229.

Gaffney, E. S., & Tong, H. (2003). *Phosphatochelys*, a new side-necked turtle (Pelomedusoides: Bothremydidae) from the Paleocene of Morocco. In L.G. Flynn (Ed.), *Vertebrate Fossils and Their Context: Contributions in Honor of Richard H. Tedford* (pp. 644–659). *Bulletin of the American Museum of Natural History, 279.*

Gaffney, E. S., & Tong, H. (2008). Redescription of the skull of *Ummulisani rutgersensis* Gaffney, Tong, and Meylan, 2006, a bothremydid side-necked turtle from the Eocene of Morocco. *American Museum Novitates, 3615*, 1–20.

Gaffney, E. S., Tong, H., & Buffetaut, E. (1998). *Foxemys*, a new side-necked turtle (Bothremydidae: Pelomedusoides) from the Late Cretaceous of France. *American Museum Novitates, 3251*, 1–19.

Gaffney, E. S., Tong, H., Chatterjee, S., Moody, R. T. J., & Hirayama, R. (1998). Evolution of the bothremydid turtles. *Journal of Vertebrate Paleontology, 18*(suppl 3), 44A.

Gaffney, E. S., Tong, H., & Meylan, P. A. (2002). *Galianemys*, a new side-necked turtle (Pelomedusoides: Bothremydidae) from the Late Cretaceous of Morocco. *American Museum Novitates, 3379*, 1–20.

Gaffney, E. S., Tong, H., & Meylan, P. A. (2006). Evolution of the side-necked turtles: The families Bothremydidae, Euraxemydidae, and Araripemydidae. *Bulletin of the American Museum of Natural History, 300*, 1–698.

Gaffney, E. S., & Wood, R. C. (2002). *Bairdemys*, a new side-necked turtle (Pelomedusoides: Podocnemididae) from the Miocene of the Caribbean. *American Museum Novitates, 3359*, 1–28.

Gaffney, E. S., Wood, R., & Sanchez-Villagra, M. (1996). Relationships of the *Shweboemys*-group of side necked turtles. *Journal of Vertebrate Paleontology, 16*(suppl 3), 36A.

Gaffney, E. S., & Ye, X. (1992). *Dracochelys*, a new cryptodiran turtle from the Early Cretaceous of China. *American Museum Novitates, 3048*, 1–13.

Gaffney, E. S., & Ye, X. (1993). A new cryptodiran turtle from the Early Cretaceous of China. Zigong Dinosaur Museum Newsletter, 1, 5–11. (In Chinese)

Gaffney, E. S., & Zangerl, R. (1968). A revision of the chelonian genus *Bothremys* (Pleurodira: Pelomedusidae). *Fieldiana: Geology, 16*, 193–239.

Lewy, Z., & Gaffney, E. S. (2005). First record of a possible chelonioid sea turtle from the Upper Campanian of southern Israel. *Israel Journal of Earth Sciences, 54*, 55–58.

MacPhee, R. D. E., Iturralde-Vinent, M. A., & Gaffney, E. S. (2003). Domo de Zaza, an early Miocene vertebrate locality in South-central Cuba, with notes on the tectonic evolution of Puerto Rico and the Mona Passage. *American Museum Novitates, 3394*, 1–42.

Maisey, J. G., Gaffney, E. S., Norell, M. A., Posen, M., & Dingus, L. (1996). *The Hall of Vertebrate Origins. A guide to fishes, amphibans, turtles, lizards, crocodiles, and pterosaurs*. New York: American Museum of Natural History.

Meylan, P. A., & Gaffney, E. S. (1989). The skeletal morphology of the Cretaceous cryptodiran turtle, *Adocus*, and the relationships of the Trionychoidea. *American Museum Novitates, 2941*, 1–60.

Meylan, P. A., & Gaffney, E. S. (1991). *Araripemys* Price, 1973. In J. G. Maisey (Ed.), *Santata fossils: An illustrated atlas* (pp. 326–334). Neptune: Tropical Fish Hobbyist Publications.

Meylan, P. A., Gaffney, E. S., & De Campos, D. A. (2009). *Caninemys*, a new side-necked turtle (Pelomedusoides: Podocnemididae) from the Miocene of Brazil. *American Museum Novitates, 3639*, 1–26.

Meylan, P. A., Gaffney, E. S., & Wood, R. C. (1995). The case for recognition of the Podocnemidae. *Seventy-Fifth Annual Meeting of the American Society of Ichthyologists and Herpetologists* (University of Alberta, Edmonton, Alberta, Canada), Abstracts, 143.

Norell, M. A., Gaffney, E. S., & Dingus, L. (1991). *Barosaurus* on Central Park West. *Natural History, 12*, 36–40.

Norell, M. A., Gaffney, E. S., & Dingus, L. (1995). *Discovering dinosaurs in the American Museum of Natural History*. New York: Alfred A. Knopf.

Olsson, R. K., & Gaffney, E. S. (1970). The Cretaceous-Tertiary datum in New Jersey. *Geological Society of America, 2*(1), 30. Abstracts.

Platnick, N. I., & Gaffney, E. S. (1977). (Review of) Systematics: A Popperian perspective. The logic of scientific discovery by Karl R. Popper. Conjectures and refutations: The growth of scientific knowledge by Karl R. Popper. *Systematic Zoology, 26*, 360–365.

Platnick, N. I., & Gaffney, E. S. (1978a). [Review of] Evolutionary biology: a Popperian perspective. The poverty of historicism by Karl R. Popper objective knowledge: an evolutionary approach by Karl R. Popper. Unended quest: An intellectual autobiography by Karl R. Popper. *Systematic Zoology, 27*, 137–141.

Platnick, N. I., & Gaffney, E. S. (1978b). (Review of) Systematics and the Popperian paradigm. The philosophy of Karl R. Popper, Paul A. Schillp (Eds.), Criticism and the growth of knowledge; Imre Lakatos, Alan Musgrave (Eds.), the philosophy of Karl R. Popper by Robert J. Ackermann. Karl Popper by Bryan Magee. *Systematic Zoology, 27,* 381–388.

Rueda, E. A. C., & Gaffney, E. S. (2005). *Notoemys zapatocaensis*, a new side-necked turtle (Pleurodira: Platychelyidae) from the Early Cretaceous of Columbia. *American Museum Novitates, 3470*, 1–19.

Sánchez-Villagra, M. R., Burnham, R. R. J., Campbell, D. C., Feldmann, R. M., Gaffney, E. S., Kay, R. F., et al. (2000). A new near-shore marine fauna and flora from the early Neogene of northwestern Venezuela. *Journal of Paleontology, 74*, 957–968.

Schleich, H. H., & Gaffney, E. S. (1994). New reptile material from the German Tertiary. 16. On *Chelydropsis murchisoni* (Bell, 1892) from the Middle Miocene locality of Unterwohlbach/South Germany. *Courier Forschungsinstitut Senckenberg, 173*, 197–213.

Siddall, M. E., & Gaffney, E. S. (2004). Observations on the leech *Placobdella ornata* feeding from bony tissues of turtles. *Journal of Parasitology, 90*, 1186–1188.

Singh, S. D., Sahni, A., Gaffney, E. S., & Schleich, H. H. (1998). Reptilia from the Intertrappean Beds of Bombay (India). *Veröffentlichungen aus dem Fuhlrott-Museum, 4*, 307–320.

Tong, H., & Gaffney, E. S. (2000). Description of the skull of *Polysternon provinciale* (Matheron, 1869), a side-necked turtle (Pelomedusoides: Bothremydidae) from the Late Cretaceous of Villeveyrac, France. *Oryctos, 3*, 9–18.

Tong, H., & Gaffney, E. S. (2001). Turtle fauna from the Cenomanian (Late Cretaceous) of Southern Morocco. *Bulletin de la Société d'etude des Sciences Naturelles d'Elbeuf, 2001*, 89–90.

Weldon, P. J., & Gaffney, E. S. (1998). An ancient integumentary gland in turtles. *Naturwissenschaften, 85*, 556–557.

Wood, R. C., Johnson-Gove, J., Gaffney, E. S., & Maley, K. F. (1996). Evolution and phylogeny of the leatherback turtles (Dermochelyidae), with descriptions of new fossil taxa. *Chelonian Conservation and Biology, 2*, 266–286.

Woodburne, M. O., Hunt, R. M., Jr., Gould, G. C., Gaffney, E. S., & Flynn, L .J. (2003). Richard H. Tedford: Field man, systematist, professor, and mentor. In L.G. Flynn (Ed.), *Vertebrate Fossils and Their Context: Contributions in Honor of Richard H. Tedford* (pp. 1–17). Bulletin of the American Museum of Natural History, 279.

Chapter 2
Autobiography (Through May 2009)

Eugene S. Gaffney

My research interests are the morphology and systematics of turtles, primarily as seen in the fossil record. Turtles are a good group for phylogenetic problems: they have a long history with a diverse fossil record and a good sampling of living taxa that allow for accurate identification of structures. I became an early proponent of phylogenetic systematics due to the influence of Gary Nelson and Bobb Schaeffer in the late 1960s, and I was very fortunate to be present during the development of cladistics at one of its primary centers, the American Museum of Natural History (AMNH). Although most of my publications consist of the documentation and analysis of morphology, I think that my main scientific achievement has been the increase of phylogenetic knowledge of turtles using cladistic methodology. I never had an overall plan, but in retrospect my research reveals an accidentally sensible pattern. After doing a general treatise on turtle skull morphology (Gaffney 1979), I concentrated on the earliest turtles (Triassic and Jurassic) and the largest group of turtles, the cryptodires. I have devoted the past decade to the other main clade, the pleurodires.

I was born in Jersey City, New Jersey, on August 12, 1942. We lived only a few miles from the AMNH, where my parents did take me when I was 5 or 6 years old, but I don't remember seeing the dinosaurs. After my father died in 1949, my mother remarried and we moved to Texas, where I spent most of my childhood. My stepfather was in the Air Force and we did a fair amount of traveling. It was in Texas that I discovered paleontology and collected my first fossils, invertebrates not vertebrates. When my Dad left the Air Force we returned to New Jersey and I entered high school in East Orange. For various reasons, my parents went from a good life in the military to poverty in a New Jersey slum. After high school, I attended Rutgers, the State University of New Jersey, where I received full scholarships. I first majored in biology, as this was the facet of paleontology I was most interested in. However, the very large biology department at Rutgers was very oriented to pre-med and getting as many graduates into medical school as possible. When I saw that the Department of Geology actually had a museum with a mastodon skeleton, a mosasaur skull, and a hind limb of a sauropod, I left the brand-new Biology building for the hundred-year old Geological Museum and never regretted it for a second.

Although Rutgers was and is a large school, the Geology Department was small at that time, and, with a faculty/student ratio (for majors) of close to 1:1, the faculty had a very personal relationship with the students. We did a lot of field work, possible even in New Jersey (Fig. 2.1), and I learned a lot. My interests in paleontology at this time were dinosaurs and primitive mammals. But as a senior, Steven K. Fox and Richard K. Olsson of the Rutgers Geology faculty took me to Princeton University where I met Don Baird. Don, who became my advisor on my senior honors thesis, persuaded me to do a description of the side-necked turtle, *Taphrosphys*, based on new material from New Jersey that he provided. Don was a great inspiration for me, and my first real contact with a vertebrate paleontologist. In 1965 he took me to Nova Scotia as his field assistant. We met and worked with Bob Carroll there, another life-long friendship I acquired.

While I was working on my senior thesis at Rutgers, Don told Rainer Zangerl (Chairman of the Dept of Vertebrate Paleontology and Curator of Fossil Reptiles at the Field Museum in Chicago) about this kid working on pleurodires. At this time Rainer was the world turtle expert (although rapidly changing to fishes), and he offered me a travel grant to Chicago to do a paper with him on the related pleurodire, *Bothremys*. I had just re-located the type skull of *Bothremys cooki* Leidy in the Rutgers teaching collection, along with about a dozen other fossil reptile types dating back to Leidy from the 1860s, including *Taphrosphys sulcatus*. So suddenly

E. S. Gaffney (✉)
Division of Paleontology, American Museum of Natural History, Central Park West at 79th Street, New York, NY 10024, USA
e-mail: genegaffney373@comcast.net

D. B. Brinkman et al. (eds.), *Morphology and Evolution of Turtles*, Vertebrate Paleobiology and Paleoanthropology,
DOI: 10.1007/978-94-007-4309-0_2, © Springer Science+Business Media Dordrecht 2013

Fig. 2.1 Gene in 1964 collecting (mostly mud) in the Cretaceous-Paleocene Greensands of Sewell, New Jersey, while an undergraduate at Rutgers University. Photo by J. Cunliffe

from a lowly undergrad I became a turtle expert going on my first museum trip to study collections for a genuine paper with a real honcho. I don't think I could have gotten this far if I tried sticking to mammal teeth or dinosaurs. This was heady stuff and I was on my way. My association with Rainer lasted a long time; I always valued his friendship.

After graduating from Rutgers in 1965, I was accepted to the Columbia University Department of Geology cognate field program in vertebrate paleontology with Edwin H. Colbert, Curator of Fossil Reptiles at the AMNH, as my advisor. In that year I moved into office space at the AMNH and have had an office there ever since. Although my original intention was to work on dinosaurs, Colbert turned out to be less than supportive, and I turned to a subject I knew something about, turtles. I wrote my paper with Rainer (Gaffney and Zangerl 1968). I also met Sam McDowell, a Research Associate in Herpetology at the AMNH. Sam is a gifted anatomist and became a good friend, filling me with his sometimes disorganized, but seemingly infinite, knowledge of turtle basicrania. Sam also inspired me to do dissections of recent turtles, using the nearly limitless corpses from the Bronx Zoo. But my real mentor (and close friend in later years) in grad school was Bobb Schaeffer, Curator of Fossil Fishes and Chair of the Department of Vertebrate Paleontology at the AMNH. Bobb gave the best course I've ever taken, on fossil fishes, but it was really a basic vertebrate anatomy course, turning us on to Edwin Stephen Goodrich and Sir Gavin de Beer. I was not the only one who Bobb inspired to accept only the highest standards in anatomical description and scientific objectivity. But Bobb was the one who helped me through the thesis, read it with care, and inspired me to do things I didn't think I could.

Although I had no particular interest in turtles, they kept appearing as excellent targets of opportunity. I don't know anymore where I first heard the story, but I used it when people would ask why I was only working on turtles. I wrote it up in a departmental review we had sometime in the 1980s and was very enamored of it, even if it's kind of silly. A famous geologist gave a public talk on continental drift and the history of the earth. After the talk an old lady came up to him and complained that he had it all wrong. She said the earth was supported on the backs of four giant turtles. The great man replied that that was impossible, what were the turtles standing on? The old lady said they were standing on four even larger turtles. So the professor thought a second and said, "Aha, but what are those turtles standing on?" And the old lady looked at him, shook her head a bit in disappointment and replied, "It's no use, Professor, it's turtles all the way down."

The AMNH has an excellent collection of turtles, including many turtle skulls, including baenids, supposed "Amphichelydia" or ancestral group turtles. Even before cladistics, it was apparent to me that using the basicranial criteria developed from Sam McDowell and my own work, baenids were cryptodires. My thesis included both a revision of the baenids as well as a redefinition of cryptodires and pleurodires, and was called: "The North American Baenoidea and the Cryptodire-Pleurodire Dichotomy." Colbert never actually read it, but both Bobb Schaeffer and Malcolm McKenna made major improvements to it. Colbert, however, gave me the best possible legacy, his job. Actually, it was Schaeffer who did that, although I subsequently learned that he asked the opinion of a number of people. There were no search committees in those days. Malcolm was very supportive, as were other curators in the

Fig. 2.2 After his PhD thesis defense in 1969, Gene celebrated with his close friends in the joint Columbia University-AMNH graduate program. From *left* to *right* (corresponding AMNH advisor indicated in parentheses): Gene Gaffney (E. H. Colbert, fossil reptiles), Niles Eldredge (N. Newell, fossil invertebrates), John Boylan (B. Schaeffer, fossil fishes), and Bob Hunt (M. McKenna and R. H. Tedford, both fossil mammals). Photo by H. Osborn

museum. Schaeffer told me to finish up quickly as Colbert was retiring soon, and I completed my grad career in 4 years (1965–1969; Fig. 2.2).

Although my thesis as submitted (not the published version) showed no influence from cladistics, the topic was a burning controversy during my last years as a grad student. Gareth Nelson, the new curator of recent fishes, brought cladistics to the vertebrate departments (entomology had an independent origin), around 1967. I saw a real opportunity with the turtles. Here was a group with many living representatives to provide excellent anatomical information, a relatively good fossil record, systematics largely uninfluenced by skull morphology, and a good collection of "primitive" taxa with lots of skulls in my own basement. My first cladistic paper was the baenids in 1972 when I renovated that part of the thesis. The cryptodire-pleurodire part I rewrote and published in 1975.

After I got the AMNH job in 1970, I went after anything called "Amphichelydia" to see if they were cryptodires or pleurodires. I first toured European museums in 1971 and started on Jurassic turtles primarily. While I was still a graduate student I was approached by Carl Gans to do a chapter on the turtle skull for his "Biology of the Reptilia" series. But it got out of hand, became too long for the series, and I ended up publishing it as an AMNH bulletin on turtle cranial morphology in 1979, illustrating and discussing all the living and extinct genera then known. My life-long interest in morphology always depended on high quality illustrations and illustrators. Over the years, first Lorraine Meeker, then Frank Ippolito supported me with superb talent and dedication.

Although I was an active participant and minor contributor to the Clade Wars of the 1970s, I've never been much of a systematics theoretician, at least by AMNH standards. My real love in paleontology has always been the intricacies of morphology—systematics was simply the obvious thing to do with it. In any case, the 1970s was mostly spent working up the material I could get easily, largely from the Jurassic of North America and Europe. I did much of my own acid preparation of European specimens where necessary.

One of the classic "Amphichelydia" was the horned turtles, the meiolaniids. At the AMNH we had one partial specimen collected from South America and described by George Gaylord Simpson in the 1930s, but most of the

Fig. 2.3 Gene with his wife Barbara and daughter Karen in 1980, while collecting Pleistocene *Meiolania* on Lord Howe Island, Australia. As recounted by his then PhD student Paul Sereno, "a tropical paradise isn't the usual setting for paleontological exploration and can make for family fun. The expedition was supported by food, beer and other luxuries, airlifted by Hercules aircraft to the short airstrip on Lord Howe Island courtesy of the Australian army." Photo by P. Sereno

known specimens were Australian. I was never much of a field man, but Australia really appealed to me, so in 1976 I went to Australia. I had heard about the place of course, from Dick Tedford, a fellow AMNH Department of Vertebrate Paleontology curator, who was famous for great discoveries in the Australian deserts from years of work there. I was still interested in pleurodires, which were big in Australia, and hoped to find some fossil chelids, at least in Australian collections. I loved Australia. Even the 1976 trip was a great adventure for me. I visited a lot of museums, saw a lot of people, and went into the Dead Center. I even slept in a tent, for a couple of days anyway. The roughing-it life is not mine, I'm more the Hilton-type. I studied meiolaniid skulls in the collections in Sydney and decided I wanted to go to Lord Howe Island where they came from, collect some more, and resolve the conflicting statements about the source. Eventually, I managed two ambitious field seasons on Lord Howe with field parties of up to 12 people, and a number of supporting trips. I visited Australia eight times, many of those with my wife, Barbara, and our daughter, Karen (Fig. 2.3). We also collected in Oligocene deposits at Lake Palankarinna, another Tedford turtle locality.

I think most of my contemporaries thought I was crazy wasting time and effort on the horned turtles, which were an extreme of evolution no matter what their relationships were. But they had not visited Lord Howe Island. Lord Howe was my kind of field work. It is fantastically beautiful, and no camping allowed. Meiolaniids and other Australian turtles formed the subjects of a number of my monographs

Fig. 2.4 Gene's usual habitat was in the turtle collections of institutions, in this case searching out podocnemidids in the depths of the Departamento Nacional de Produção Mineral, Rio de Janeiro, Brazil, in 1995. Photo by P. Meylan

Fig. 2.5 Gene with a specimen of the rare Late Triassic turtle *Palaeochersis*, La Rioja, Argentina, in 1997. Photo by B. Gaffney

Fig. 2.6 Gene and long-time turtle research partner Peter Meylan in the Jurassic turtle quarry near Solothurn, Switzerland, in 1986. Photo by C. Meyer

and other papers over the following years. In fact, the 1980s were characterized by work on two topics, the Australian turtles and the Triassic *Proganochelys*.

In 1979 I asked for and received permission to work on the best preserved Triassic turtles known, the collection of *Proganochelys* found in Trossingen, Germany, and housed in Stuttgart. This was possible only through the efforts of Dr. Rupert Wild, the Curator of Fossil Reptiles and Amphibians at the Staatliches Museum für Naturkunde. *Proganochelys* is the sister taxon to all other turtles and, thus, provides an important view of turtle evolution and an outgroup for the systematic analysis of other turtle groups. I visited the collection in 1979 and quickly saw that the

Fig. 2.7 Gene with colleagues at the Gaffney Turtle Symposium, in honor of his retirement from the AMNH, held at the Royal Tyrrell Museum of Palaeontology, Drumheller, Canada, in October 2009. From *left* to *right*: Hans-Peter Schultze, Gene Gaffney, Roger Wood, and Bob Carroll. The golden bust in the background is of Joseph B. Tyrrell, namesake of the museum. Photo by the Royal Tyrrell Museum of Palaeontology

three skeletons not only belonged to the most primitive turtle known, but their documentation was going to be a very big project, if done right. I began the work then, but decided to seek a National Science Foundation grant so I could bring scientific illustrators to Stuttgart and Berlin to figure the material, and for further preparation and casting. I got the grant and managed to publish my AMNH bulletin on *Proganochelys* in 1990. I emphasized the comparative morphology of *Proganochelys* by making comparisons of its entire osteology with cryptodires and pleurodires. During the 1980s I was also able to work on the oldest African turtle, *Australochelys,* and the earliest cryptodire, *Kayentachelys*.

Over the years I have done some less ambitious field work, nothing as extensive or prolonged as Lord Howe Island. During the 1980s, along with Frank Ippolito, I collected in the Bridger Basin at the classic Grizzly Buttes turtle localities of Hay, Granger, and other AMNH parties of 100 years ago. I've also collected in the Lance Formation of Wyoming, the Cretaceous-Paleocene of eastern Montana, the San Juan Basin of New Mexico, and the Lower Jurassic Kayenta Formation of Arizona. Besides the Pleistocene of Lord Howe, I went into the field in Australia to the Oligo-Miocene of the Tirari Desert and the Pleistocene of the Atherton Tableland.

During the 1990s I continued work on the cryptodires, particularly the extinct groups outside the living forms, the macrobaenids and sinemydids. Don Brinkman (Royal Tyrrell Museum) and Howard Hutchison (University of California Museum of Paleontology) shared their knowledge of these groups with me. When I started work on fossil turtles, there was almost no one who was devoting their entire careers to turtles, but by the 1980s, a number of excellent people had been attracted to the group.

Pleurodires, the other major group of turtles, got me when I wasn't looking. My first paper (Gaffney and Zangerl 1968) was on a pleurodire, but I did only a few papers on the group until the 1990s. Even then I thought I would only do a couple of short papers, then a couple of longer ones became necessary, then the thing ballooned into the largest monograph I've ever participated in, with Haiyan Tong and Peter Meylan as co-conspirators. In 1996, after 10 years involvement in the exhibition program, I was again able to devote most of my time to research. It made sense to intensify work on the pleurodires, as more and more material was turning up. Pleurodires, living and extinct, were very poorly known. At the beginning of the project most extinct taxa were based on undiagnosable rubble, and there were less than a dozen skull-based taxa. Now there are over 40, mostly new, skull-based taxa. Much of this material has been discovered in recent years, but many specimens were already in collections (Figs. 2.4, 2.5). All of this newly recognized diversity made its documentation a much bigger project than I had envisioned when it was begun.

For many years I taught in the Columbia University-AMNH vertebrate paleontology program (the same program I graduated from), teaching a graduate course called "The History of Reptiles and Amphibians." I have only had three graduate students, Paul Sereno (the only one who went in the field with me, if you call Lord Howe Island "the field"), Dan Chure, and Ron Coldiron. My first postdoctoral student, Peter Meylan (see Fig. 2.6), also became my closest scientific associate, coauthoring 10 papers, most of them substantial, and helping me when the transition to computer cladism became unavoidable in the late 1980s. By the early 1980s, cladistics was pretty much state-of-the-art for systematics, and toward the end of the decade, computer programs were devised that married cladistics and numerical taxonomy. Today, few students are even aware that cladistics existed for about 20 years before computerization.

As part of my responsibility for the fossil reptile collection, I have always been involved in exhibitions at the AMNH. Since 1970, I had strongly advocated renovation of the dinosaur exhibitions and incorporation of cladistics with no success. But the museum changed significantly in 1986 and for the following 10 years I was fortunate enough to be able to participate in a series of exhibitions that culminated in the complete renovation of the six fossil halls, the world's largest vertebrate fossil exhibition. In 1986 we brought a large collection of Chinese fossils to the AMNH for an exhibition that featured cladistic methodology, then in 1989 the administration accepted the proposal from Lowell Dingus and me for mounting the *Barosaurus* group now in the Rotunda. My involvement with the fossil hall renovation lasted until its completion in 1996. I worked on all of the halls, particularly in the incorporation of cladistics as the organizing theme. Thanks largely to the management skills and friendship of Lowell Dingus, the fossil hall exhibition program was immensely satisfying for me, and the highlight of my museum career. It was a great honor to be able to deal first-hand with that rich legacy of material collected by so many giants in paleontology.

After 37 years as a curator, I retired in 2007 as a Curator Emeritus (Fig. 2.7). I've written about 100 scientific papers. I continue working as I always have, just doing less of it.

Note added in proof The references for publications mentioned in this chapter can be found in the complete bibliography of Chapter 1.

Part II
The Origin of Turtles

Chapter 3
Problems of the Ancestry of Turtles

Robert L. Carroll

Abstract The unquestioned unity of the Chelonia provides a necessary basis for establishing their interrelationships and determining the evolutionary history within the group. On the other hand, the host of uniquely derived features of the oldest known turtles make it extremely difficult to establish their ancestry among more primitive amniotes. This is illustrated by the great diversity of taxa that continue to be proposed as putative sister-taxa of turtles without general acceptance of any. Nearly every major clade of early amniotes from the late Paleozoic and early Mesozoic has been proposed as a possible sister-taxon of turtles, from synapsids to anapsids and diapsids, including pelycosaurs, captorhinomorphs, procolophonids, pareiasaurs, aquatic placodonts and crocodiles, but none possess derived characters that could be synapomorphic with the unique skeletal structure and patterns of development of the chelonian skull, carapace or plastron, which had reached an essentially modern configuration by the Late Triassic. Numerous molecular biologists have attempted to establish the closest sister-group of turtles through analyses of a host of living species, but there is no way for them to preclude turtles from having evolved from one or another of the Paleozoic or early Mesozoic clades that have become extinct without leaving any other living descendants. On the other hand, recent studies of the genetic and molecular aspects of the development of the carapace and plastron imply unique patterns of evolutionary change that cannot be recognized in any of the other amniote lineages, living or dead. This, together with the retention of a skull without temporal fenestration implies a very early divergence from a lineage that probably retained an anapsid skull configuration. This problem may be resolved by more detailed study of the enigmatic genus *Eunotosaurus*, from the Late Permian of South Africa.

R. L. Carroll (✉)
Redpath Museum, McGill University, Montreal,
QC H3A 2K6, Canada
e-mail: robert.carroll@mcgill.ca

Keywords Captorhinomorphs • *Eunotosaurus* • Pareiasaurs • Procolophonids • Turtle origins

Introduction

Gene Gaffney has provided a thorough understanding of the skeletal anatomy of all adequately known turtles, fossil and living. This research has documented their distinction from all other amniotes, going back to the Late Triassic, demonstrating the difficulty of establishing sister-group relationships with any other major tetrapod lineage.

The absence of any plausible intermediate forms can be attributed to gaps in the known fossil record, but might also be the result of unique and relatively rapid changes in molecular aspects of their development that led to the unique position and configuration of their ribs and the formation of first the plastron and later the carapace, as well as other aspects of their skeletal anatomy.

Investigation of such molecular and developmental changes can only be carried out through studies of living turtles. A great deal of such research as been has been published in leading journals of molecular biology and development over the past 20 years, for example: *Evolution & Development*, *Journal of Morphology*, and *Development Genes and Evolution*. Unfortunately, very few vertebrate paleontologists have made use of this literature, which is key to understanding how turtles evolved a complex of unique structure, not known in any other taxa. However, before discussion of the molecular and developmental aspects of chelonian evolution, it is necessary to review previous attempts to discover their affinities among early amniotes based on the fossil record.

Unfortunately, the uniqueness of turtles, from their first appearance in the fossil record makes it difficult, if not impossible, to recognize synapomorphies with any other amniote groups, other than features common to many, if not all, other contemporary anapsids, diapsids, and parapsids.

D. B. Brinkman et al. (eds.), *Morphology and Evolution of Turtles*, Vertebrate Paleobiology and Paleoanthropology,
DOI: 10.1007/978-94-007-4309-0_3, © Springer Science+Business Media Dordrecht 2013

Ancestral Amniotes

The Reptilia are considered a monophyletic assemblage, but two groups have been recognized as sub-units. The Eureptilia includes the vast majority of species, encompassing the ancestral lineage and all living forms. The much more restricted Parareptilia includes fossils from the Late Permian through the Triassic, and possibly the turtles, but no other living taxa.

Three closely related lineages of early Eureptilia have long been recognized on the basis of openings in the temporal region of the skull (Figs. 3.1, 3.2). The most primitive of these are the anapsids, named for the absence of temporal fenestration—clearly a retention of the primitive state of their putative sister taxon, the anthracosaur labyrinthodonts. The oldest known amniote genera, *Hylonomus* and *Paleothyris,* are Late Pennsylvanian in age, from deposits ranging from 314-311 million years old (Carroll 1969). The second clade to appear in the fossil record is the Synapsida, or mammal-like reptiles, with a single temporal opening low in the cheek. The third, appearing close to the end of the Pennsylvanian, is the Diapsida (Reisz 1977), with both dorsal and lateral temporal openings. The appearances in the fossil record of these groups correspond with their probable sequences of divergence over time.

The terms Synapsida and Diapsida are recognized as formal taxonomic designations, as, for a long time, was the Anapsida. However, because the anapsid condition is actually a primitive character state, common to the putative antecedent of amniotes, it cannot be considered as an apomorphy uniting a particular assemblage of taxa, and so Anapsida has been dropped as a formal name. Genera including *Hylonomus, Paleothyris,* and other early amniotes grouped as the Captorhinomorpha, are now referred to as members of the Diapsida, because they share other features similar to those of other primitive genera, such as *Petrolacosaurus,* which are diapsids in terms of having two pairs of temporal openings. This makes it possible to state that the sister group of turtles is the Diapsida, as is common in recent papers (e.g., deBraga and Rieppel 1997; Li et al. 2008), even though the oldest possible sister taxa among amniotes are anatomically anapsid. This terminology is further complicated by the fact that there is no evidence from either modern turtles or any of their possible antecedents that they ever had temporal openings, either during their evolutionary history or during their early development.

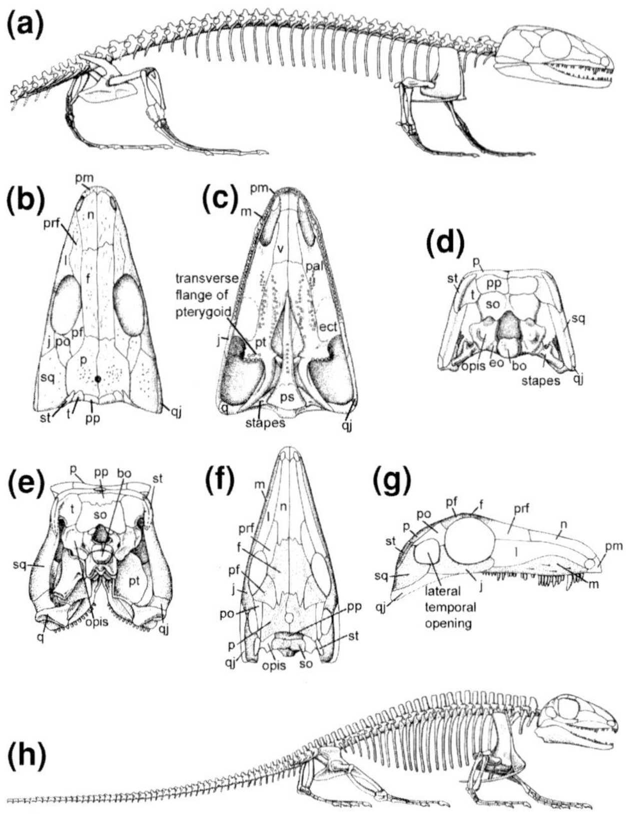

Fig. 3.1 Skulls and postcranial skeletons of genera that represent early members of the three major amniote lineages. **a–d** The anapsid *Paleothyris*, from the Late Pennsylvanian of Florence, Nova Scotia, Canada: **a** skeleton in lateral view, skull-trunk length approximately 12 cm; **b–d** skull in dorsal, palatal, and occipital views (from Carroll 1969). **e** Occiput of the early pelycosaur *Ophiacodon* (6 cm in height), from the Early Permian of Texas, USA (modified from Romer and Price 1940). **f, g** The early synapsid *Archaeothyris*, (10 cm in length) from the Middle Pennsylvanian of Florence, Nova Scotia, skull in dorsal and right lateral views (from Reisz 1972). **h** The early synapsid *Haptodus*, approximately 140 cm in length, from the Late Pennsylvanian of Kansas, USA (modified from Currie 1977). Abbreviations used in this and subsequent illustrations: *a* angular, *art* articular, *as* astragalus; *bo* basioccipital, *bs* basisphenoid, *co* coronoid, *CR* carapacial ridge, *d* dentary, *dep* dorsal process of epiplastron, *dsc* dorsal process of scapula, *ect* ectopterygoid, *ent* entoplastron; *eo, ex* exoccipital, *ep* epiplastron, *f* frontal, *fe* femur, *fi* fibula, *fm* foramen magnum, *gpep* gular projection of epiplastron, *hyo* hyoplastron, *hyp* hypoplastron, *il* ilium, *ipt* interpterygoid vacuity, *j* jugal, *ipt* interpterygoid vacuity, *l* lacrimal, *ldv* last dorsal vertebra, *m* maxilla, *meso* 1 first mesoplastron, *meso* 2 second mesoplastron, *n* nasal, *na* naris, *nc* neural canal, *nt* neural tube, *nu* nuchal, *opis* opisthotic, *p* parietal, *pal* palatine, *pf* postfrontal, *phyis* posterolateral process of hypoischium, *pm* premaxilla, *po* postorbital, *pp* postparietal, *pre* prearticular, *prf* prefrontal, *pro* prootic, *prp* posterior ramus of pterygoid, *ps* parasphenoid, *pt* pterygoid, *ptf* post-temporal fenestra, *q* quadrate, *qj* quadratojugal, *qrp* quadrate ramus of pterygoid, *sa* surangular, *so* supraoccipital, *sp* splenial, *sq* squamosal, *st* supratemporal, *sv*1 1st sacral vertebra, *t* tabular, *ti* tibia, *trpt* transverse flange of pterygoid, *ul* ulna, *v* vomer, *xi* xiphiplastron

Parareptilia

Ancestral anapsids, Diapsida, and Synapsida, all known from the Late Pennsylvanian, are sufficiently similar to one another that they almost certainly had a single common ancestor, late in the Carboniferous, forming the basis of the taxon Eureptilia, as used by deBraga and Rieppel (1997). This name was coined to distinguish them from a second

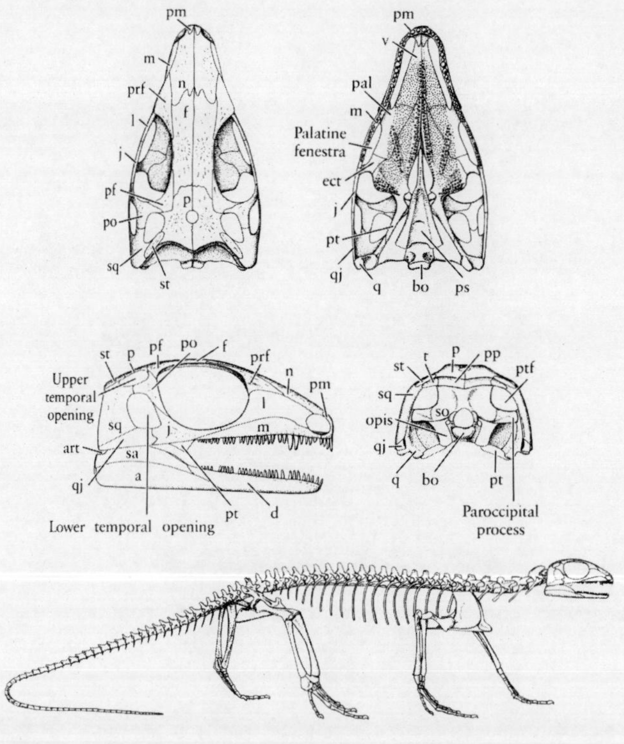

Fig. 3.2 Skull and skeleton of the earliest known diapsid, *Petrolacosaurus*, from the Late Permian of Kansas, USA (Modified from Reisz 1981). Skull and skeleton at different scales

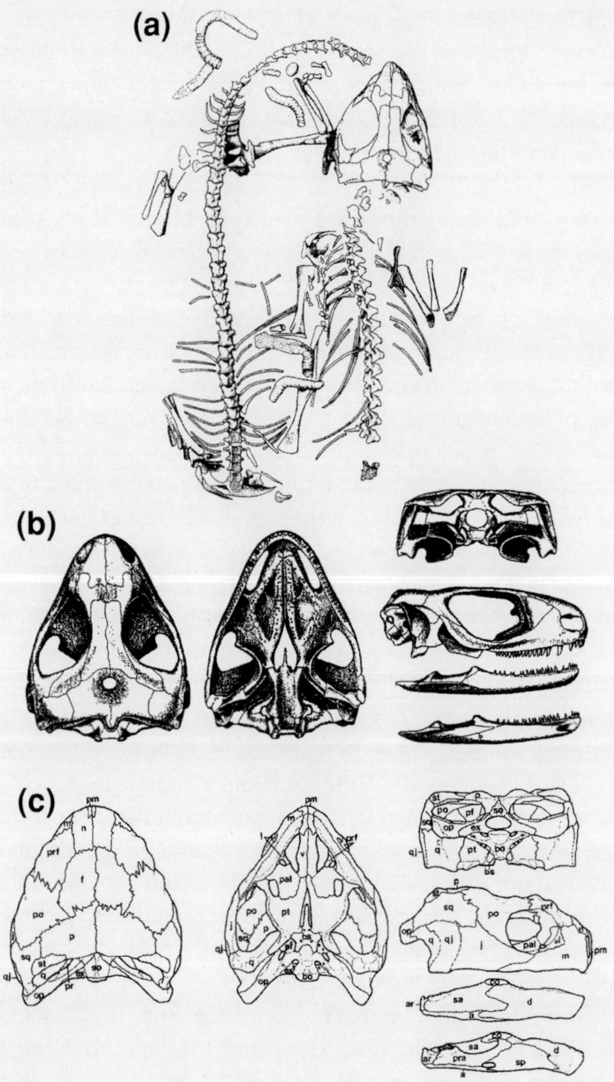

Fig. 3.3 Comparison of a procolophonid and a basal turtle. **a**, **b** The procolophonid *Owenetta,* from the Late Triassic of South Africa: **a** two incomplete skeletons in dorsal view; **b** skull in multiple views (modified from Reisz and Laurin 1991). **c** Several views of the skull of *Proganochelys*, approximately 12 cm in length, from the Late Triassic of Germany (modified from Gaffney and Meeker 1983)

group of primitive amniotes, the Parareptilia, a name initially coined by Olson (1947) to include a number of distinct taxa that had presumably achieved the status of amniotes, but were not closely related to the Eureptilia. The Parareptilia originally included the Diadectidae, a group known from the Late Pennsylvanian and Early Permian, together with several other distinct taxa from the mid-Permian and Triassic: the terrestrial Millerettidae, Lanthanosuchidae, Procolophonia, and Pareiasauria and the aquatic Mesosauria. The Diadectidae are now considered a remote sister taxon of amniotes. More recently, both the Procolophonia and the Pareiasauria have been proposed as sister taxa of turtles (Reisz and Laurin 1991; Lee 1993, 2001).

In 1991, Reisz and Laurin postulated that the procolophonid *Owenetta* (Fig. 3.3) was a plausible sister taxon of turtles. This was based on the following derived characters also recognized in chelonians: cultriform process of the parasphenoid greatly reduced in length; teeth on transverse flange of pterygoid lost; distinctly shaped anterodorsal expansion of the maxilla; prefrontal and palatine massively buttressed against each other; dorsal process of quadrate exposed laterally; slender stapes has lost dorsal process and foramen; anterior edge of splenial excluded from symphysis; dorsal surface of retroarticular process (formed by articular, angular and prearticular) is broad and concave; postparietal is reduced or lost. More notable are the presence of large post-temporal fenestrae and ventral emargination of the cheek, as in some turtles, but not in primitive anapsids or synapsids. In the postcranial skeleton, the absence of the entepicondylar foramen is a chelonian character, but the presence of 27 presacral vertebrae is not. *Owenetta* also lacks any trace of dermal armor, either dorsally or ventrally.

A much more extensive paper by Laurin and Reisz (1995) provided data matrices for a host of reptilian groups in an effort to reevaluate early amniote phylogeny. This included the erection of a new taxon, Testudinomorpha, defined as the last common ancestor of the Procolophonidae and Testudines and all its descendants. This new clade was supported by 17 synapomorphies, 14 of which were listed as unambiguous. Surprisingly, in a later paper co-authored by Rieppel and Reisz (1999) this hypothesis was replaced by one supporting a diapsid origin for turtles.

More lasting support for a chelonian ancestry from among the Parareptilia was provided by Michael Lee, who between 1993 and 2001 published a series of papers (Lee 1993, 1994, 1995, 1996, 1997, 2001; Lee et al. 1997) arguing for pareiasaurs (Fig. 3.4) as the sister taxon of turtles. His 1995 paper, "Historical burden in systematics and the interrelationships of 'parareptiles'", was based on a complex cladistic analysis that focused on turtles, pareiasaurs, and procolophonid-like taxa and incorporated 56 characters. This resulted in a single most parsimonious tree with 80 steps. Pareiasaurs were united with turtles based on 20 unambiguously derived characters absent in basal amniotes. Only one synapomorphy supported affinities with procolophonoids. Characters were chosen from throughout the skeleton: 31 from the skull; three from the mandible and dentition; five from the axial skeleton; seven from the shoulder girdle and forelimbs; eight from the pelvic girdle and hind limbs; and two in reference to dermal ossification. Evidence from most of the skeleton appeared to provide strong support for pareiasaur-chelonian affinities, but little specific attention was paid to the nature of the dermal ossification. Pareiasaurs do not have a ventral structure that is comparable to the plastron of turtles, and although the carapace appears superficially like that of turtles, its mode of development is highly divergent.

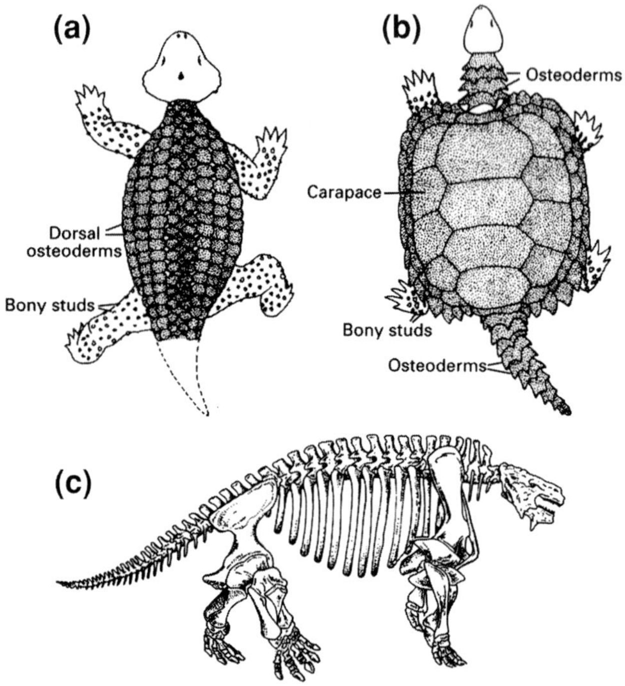

Fig. 3.4 Comparison of a pareiasaur and a basal turtle. **a** The pareiasaur *Anthodon* (about a meter in length) from the Late Permian of eastern and southern Africa and eastern Europe showing dorsal armor formed by a large number of small oestoderms (modified from Lee 1995). **b** *Proganochelys*, until recently the oldest known turtle, has extensive dorsal armor made up of a small number of very large plates extending above the should and pelvic girdle. It is about 1 m in length. Fossils are known from the Late Triassic of Germany (modified from Lee 1995). **c** The advanced pareiasaur *Scutosaurus* (2 m in length) from the Late Permian of Russia, in lateral view showing the ventrally oriented rib cage located medial to the shoulder girdle (redrawn from Carroll 1988)

Post-embryonic growth sequences of pareiasaurs demonstrate that their carapace formed by the growth and coalescence of large numbers of relatively small ossicles that form in the dermis, as do those of crocodiles, other archosaurs and some lepidosaurs, including lizards (Vickaryous and Hall 2008). In striking contrast, no turtles, fossil or living, form their carapace from osteoderms, but rather from mesodermal tissue induced by its close association with highly specialized ribs, which, uniquely among vertebrates extend laterally, toward the carapacial ridge, rather than ventrally (Gilbert et al. 2001).

Lee (1995, p. 2001) also stated that all Paleozoic outgroups of pareiasaurs had gastralia, but "there is little evidence that the plastral plates (of turtles) are homologous with gastralia". In fact, it is certain from embryonic studies of modern turtles and knowledge of the plastron gained from the oldest known turtle, *Odontochelys* (Li et al. 2008), that much of the turtle plastron is formed from gastralia, in addition to the dermal bones of the shoulder girdle, the clavicular blades, and the interclavicle. Among the Parareptilia, neither the procolophonoids nor the pareiasaurs possess any significant synapomorphies that link them to turtles. The procolophoids lack any evidence of a carapace or plastron, and the carapace of pareiasaurs developed in an entirely different way than does that of chelonians. In contrast with the hypothesis regarding the procolophonoid origin of turtles, which was permissive (no elements of their skeletal anatomy clearly preclude them from being ancestral to turtles), that of pareiasaurs would seem to be prohibited, on the basis of the antithetical nature of the carapace and its development.

In a later paper, Lee (2001) broadened his approach by including a large number of molecular characters, giving a matrix of 176 osteological, 40 soft anatomical, and 2903 molecular characters for 28 amniote taxa. This expanded analysis yielded a tree in which turtles grouped with anapsid parareptiles and fell outside a monophyletic Diapsida. This result corresponded basically with his previous studies. However a turtle-archosaur clade could not be statistically rejected. In the concluding remarks, he stated (Lee 2001, p. 10): … "the anapsid hypothesis of turtle affinities—and thus diapsid monophyly—remains not only tenable, but is the most supported arrangement when the morphological, molecular and fossil evidence is considered simultaneously". More specifically he stated (2001, p. 10): "Within parareptiles, the nearest relatives of turtles are pareiasaurs …", also agreeing with conclusions in his previous publications.

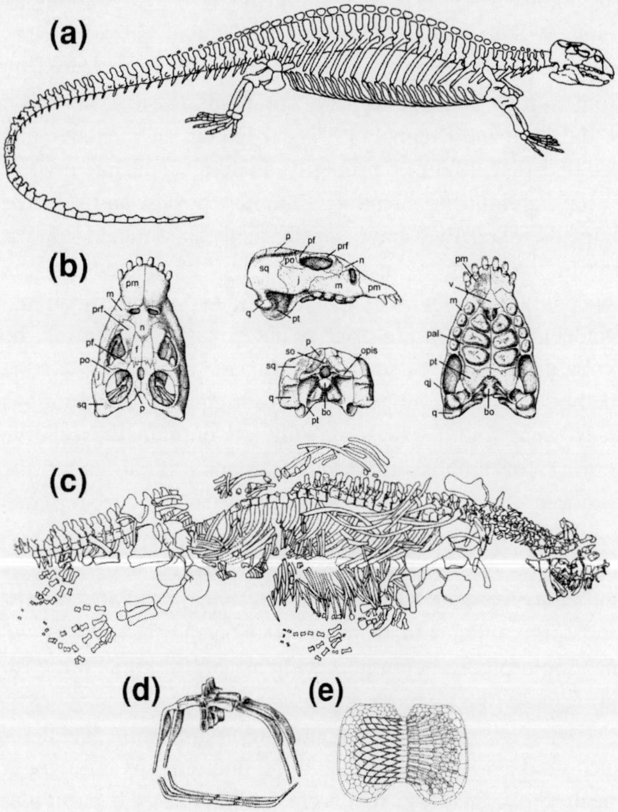

Fig. 3.5 Representatives of the diapsid assemblage Sauropterygia, specifically members of the Placodontia, alternatively assigned to the archosauromorphs and lepidosauromorphs. **a**, **b** *Placodus*, from the Early and Middle Triassic of Europe and Middle Triassic of southwest Asia: **a** skeleton, about 150 cm long, in lateral view; **b** skull in multiple views. **c** Dorsal view of skeleton of *Helvetiosaurus*, approximately 2 m in length, from the Middle Triassic of Europe. **d** Oblique view of ribs and gastralia of *Paraplacodus*, from the Middle Triassic of Europe. **e** Dorsal view of carapace of *Henodus* (approximately 1 m in width), from the Late Triassic of Europe. All images from Carroll (1988)

Eureptilia

Other paleontologists have proposed one or another group of eureptiles as the probable antecedent of turtles. Gardiner (1982) suggested an ancestry among the synapsids (or mammal-like reptiles), or more specifically, a sister-group relationship with an assemblage including the Crocodilia, which in his analysis were nested within Aves and Mammalia, on the basis of the configuration of the epipterygoid, development of the stapes, presence of a single protrusible penis, characteristics of the eye, and forward position of the larynx. However, this hypothesis has not been seriously discussed since.

On the other hand, early members of the Eureptilia, the captorhinomorphs and their immediate sister taxa, have repeatedly been suggested as likely sister taxa of turtles, primarily because they, like turtles, lack evidence of temporal openings and are sufficiently primitive to have given rise to any of the more advanced reptilian lineages (Carroll 1988). Gaffney and McKenna (1979, p. 1), in a description of *Protocaptorhinus* sp. from the Late Permian of Zimbabwe, stated: "The Captorhinidae is a monophyletic group (possibly including turtles) with these derived characters: downturned premaxilla, ectopterygoid and tabular absent, medial process of jugal". These character states are shared with ancestral chelonians but also occurred in some other early amniotes, and so do not provide unique evidence for close affinities.

In contrast, Gaffney's (1980) contribution to a volume honoring Alec Panchen, proposed a phylogeny in which Testudines were the sister taxon of a clade including synapsids and diapsids, based primarily on the absence of a Jacobson's organ among all chelonians. Subsequently, in a book chapter written with Peter Meylan, it was concluded that "turtles are considered to be the sister-group to the Diapsida alone" (Gaffney and Meylan 1988, p. 157). However, it was indicated in the introduction of the same paper that those authors included the anapsid captorhinids among the Diapsida. Otherwise, their paper was based primarily on a very extensive phylogenetic analysis of all turtles, supporting the sister-taxon relationship of cryptodires and pleurodires, traced back to the Late Triassic.

More detailed investigation of the possible relationships between turtles and diapsids (as defined by the presence of dorsal and lateral temporal openings) was carried out by deBraga and Rieppel (1997), and later by Rieppel and Reisz (1999), who expanded the search for plausible chelonian ancestors to a wider diversity of archosauromorphs and lepidosauromorphs. Among these groups are several lineages that have dorsal armor, which suggests the possibility of chelonian affinities.

These included several highly derived aquatic taxa, common in the Mesozoic, notably the Nothosauria, Plesiosauria, and Placodontia, collectively termed the Sauropterygia. The placodonts encompass several genera with dermal armor, the most conspicuous being *Henodus*, which had an extensive covering of dorsal scutes covered by epidermal plates, but not at all comparable to the larger plates of turtles (Fig. 3.5). Interestingly, this genus lacks the dorsal temporal opening, whereas the related *Placodus* has a dorsal opening, but lacks the lateral opening, demonstrating the possibility of the reduction of the temporal opening to zero, as in turtles. *Henodus* has a short trunk, but other placodonts, such as *Helveticosaurus,* have a very elongate trunk. All placodonts have well developed gastralia. Sauropterygians are known throughout the Triassic, primarily from Europe and questionably from southwest Asia.

deBraga and Rieppel (1997) carried out an extensive and detailed discussion of primarily anatomical characteristics of the Parareptilia, primitive amniotes, turtles, and selected diapsids. Their total phylogenetic analysis included 33 taxa within the Reptilia, distinguished by 168 characters. Six synapsid taxa, as well as the anamniotes Seymouriidae and Diadectomorpha, were used as outgroups to determine the probable polarity of character change. Depending on the method of rooting, either two or four equally parsimonious trees were found, but those differed only in the relative position of archosauromorph taxa and did not alter the position of turtles. The shorter of two trees (by one step) had a length of 771 steps and a consistency index of 0.507. Characters were not differentially weighted and reversals and convergences were treated as equally likely events. Forty-three clades were recognized on the basis of nested autapomorphies, beginning with all of Reptilia, but also citing individual genera, e.g. *Placodus.*

Characters indicated as autapomorphies were listed for each taxon, with all assumed to be homologous throughout the list, except those that were ambiguous at a particular node. Most recognized characters were cranial, but good coverage was provided for the postcranial skeleton. Surprisingly, very few characters applied to the most striking features of Testudines, namely the carapace and plastron, although osteoderms were mentioned in the autapomorphy list of Testudines (where reference was made to *body osteoderms,* and cited in the discussion for Lanthanosuchidae) but this character did not appear in that taxon's apomorphy list. Osteoderms were also listed as autapomorphies for *Placodus,* but not for Turtles + Sauropterygia, or for Sauropterygia. Nor was reference made to the placodont *Henodus,* which has a very complex carapace, except that it is formed by osteoderms rather than reflecting the unique feature of the chelonian shell.

In addition to the 168 morphological characters used by deBraga and Rieppel (1997) in their phylogenetic analysis, they also looked at five aspects of development that could be determined by study of growth stages in extinct taxa: changes in the configuration of the jugal during ontogeny; development of the interclavicle and clavicle; ontogeny of the carpus; development of the tarsale proximale; and presence of a hooked fifth metatarsal. Of these, the first, third and fourth support a turtle-lepidosaur sister-group relationship, but the other two lack sufficient evidence to support or refute those affinities.

The final conclusions drawn from deBraga and Rieppel's (1997) phylogenetic analysis were that Testudines may well be related to diapsids, indeed they may be nested within crown-group diapsids rather than within 'parareptiles'. More specifically, chelonians were found to be a sister group of the Sauropterygia, a group that includes the marine Triassic placodonts and the living Lepidosauriformes, including lizards and snakes. However, questions arise when we look at the number of autapomorphies listed by deBraga and Rieppel (1997) for selected taxa (Table 3.1).

Table 3.1 Number of autapomorphies of selected taxa from de Braga and Rieppel (1997)

Taxon	Autapomorphies
Parareptilia	6
Procolophoniformes	12
Pareiasauria	26
Anthodon	4
Owenettidae	4
Eureptilia (anapsids + diapsids)	5
Diapsida	7
Eosuchia	22
Lepidosauromorpha	15
Lepidosauriformes	3
Turtles + Sauropterygia	10
Testudines	41
Sauropterygia	11
Placodus	18
Archosauromorpha	7

Turtles (Testudines) stand out in having by far the highest number of autapomorphies, 41, with the next closest taxon being the Pareiasauria with 26, Eosuchia with 22, and *Placodus* with 18. This suggests that turtles show the greatest phylogenetic distance from all other clades of reptiles. On the other hand they share only 10 synapomorphies with their proposed sister taxon, the Sauropterygia. It should be noted that none of these characters involve molecular or developmental factors.

The problems associated with this methodology of establishing interrelationships were commented on by deBraga and Rieppel (1997, p. 325): "The present analysis, albeit represented by a larger data base than previous phylogenetic analyses involving Testudines, does not necessarily represent a more accurate *de facto* representation of turtle origins when compared to other recent interpretations (Laurin and Reisz 1995; Lee 1994, 1995, 1996). In fact, it can be argued that the phylogenetic hypothesis presented here is yet another indication of how poorly understood the issue of turtle origins is. The only significant, methodological advantage between this analysis and previous ones is in testing for congruence between developmental data and the morphological data set. The a priori assumption here is that the present phylogeny is congruent with the evolution or distribution of developmental data".

In addition, deBraga and Rieppel's (1997) analysis did not place sufficient attention on individual characters such as the anatomy and nature of development of the carapace and plastron of turtles, in contrast with what can be observed or readily assumed in putative sister taxa such as the sauropterygians *Henodus* and *Placodus.* As was pointed out in a later paper (Rieppel and Reisz 1999, p. 7). "As such, neural and costal plates are components of the turtle carapace and cannot be derived from a hypothetical ancestral condition by fusion of exoskeletal osteoderms. … the turtle carapace is unique i. e., autapomorphic for turtles, and morphologically very distinct even from its closest counterpart among other amniotes, which is the carapace of cyamodontoid placodonts".

The next paper to investigate a diapsid ancestry of turtles was that of Rieppel and Reisz (1999). This was written in response to a paper by Lee (1997) and incorporated changes in their data matrix in response to Lee's findings. They also added data involving the dermal armor, the homology of the "acromion", the astragalo-calcaneal complex, and the hooked fifth metatarsal, in order to evaluate the problems of establishing primary homology. Despite these changes, they still found sauropterygians to be the most probable sister taxon of turtles. But as noted in their abstract (Rieppel and Reisz 1999, p. 7) "The high Homoplasy Index raises concerns about the phylogenetic information content of various morphological characters in broad-scale phylogenetic analyses."

The most parsimonious tree generated by Rieppel and Reisz (1999) had a length of 793 steps and resulted in an unresolved trichotomy of turtles within archosauromorphs. But rearranging the branches to show a sister-group relationship with pareiasaurs required only five more steps, suggesting extremely weak support for either hypothesis, despite the great phylogenetic distance that deBraga and Rieppel (1997) had found between pareiasaurs and diapsids. Allowing PAUP to search for all trees that were five steps longer than the most parsimonious tree resulted in more than 1000 trees, with little or no resolution. The Homoplasy Index suggested rampant convergence.

In addition, further analysis of specific morphological characters unique to turtles showed the improbability of their affinities to pareiasaurs. The most significant was that of the nature of the carapace. Rieppel and Reisz (1999, p. 7) pointed out that "As such, the neural and costal plates are endoskeletal components of the turtle carapace and cannot be derived from a hypothetical ancestral condition by fusion of exoskeletal osteoderms … the turtle carapace is unique, i.e., autapomorphic for turtles, and morphologically very distinct even from its closest counterpart among other amniotes, which is the carapace of cyamodontoid placodonts."

Nevertheless, Rieppel and Reisz (1999) went on to suggest (page 16) "Triassic Sauropterygia, sister-group of Testudines, comprise two major clades, the Placodontia and the Eosauropterygia". They also stated (p. 17) that molecular data further supported diapsid affinities of turtles, and proposed an age of 207 ± 20.5 million years ago (~Middle Triassic) for the time of divergence.

Rieppel and Reisz (1999) also spoke of the question of the ecological setting of turtle origins in association with respiration and locomotion, the former of which also required major developmental and structural changes prior to the level of *Odontochelys* and *Proganochelys.* An aquatic origin would clearly support the relationship of turtles to sauropterygians. Because turtle ribs are embedded in the costal plates, respiration depends on volume changes of the thoraco-peritoneal cavity inside the rigid dermal armor. Expansion of this space for inhalation results from the contraction of the testocoracoideus muscle, which is apparently a homologue of the serratus ventralis muscle of other reptiles. Because lung ventilation in aquatic turtles is also supported by hydrostatic pressure and gravity it was argued that turtles probably evolved in an aquatic environment, although the limb proportions of *Proganochelys* suggest terrestrial locomotion.

Later, Rieppel (2001) added molecular data to the investigation of chelonian relationships, stating in his summary (p. 991): "Without exception, molecular data place turtles closer to archosaurs than to lepidosaurs". This is also the case for morphological characters that have been studied in living reptiles by de Beer (1937), Hofsten (1941), Remane (1959), Løvtrup (1977, 1985), Ax (1984), and Gardiner (1993), the last of whom also discussed whether turtles were more closely to crocodiles or lepidosauromorphs (e.g., *Sphenodon).*

In contrast, Rieppel (2001) emphasized the great difference in the embryological development of the chelonian carapace, based on research by Gilbert et al. (2001). He was specifically struck by the possibility that large scale changes in adult anatomy might be the result of comparatively rapid evolutionary changes in the mode of development. Rieppel (2001) documented the clear distinction between the development and configuration of the bones that make up the carapace of turtles and the osteoderms of any of the reptilian groups that had been suggested as their sister taxa. Instead of having a large number of small plates of similar size, he recognized that the large thecal plates make up the bulk of the carapace. Those above the neural arches were distinguished as neural plates, and those more lateral, which are closely integrated with the ribs, as costal plates. In most turtles the costal plates are succeeded laterally by the marginal plates. In addition, one can recognize an anterior nuchal plate, and one or two posterior pygal plates. No homologous elements have ever been recognized in any of the other groups of reptiles that have been hypothesized as sister taxa of turtles.

Not only are osteoderms much smaller and more numerous than the chelonian plates, but they form independently of the ribs and extend medial to the scapular blade, rather than dorsal to it. Of even greater significance is the fact that the ribs and the carapacial bone are preformed in cartilage, rather than directly in the dermis, as are the osteoderms of various other groups of reptiles. These elements cannot be homologous, and hence it is extremely unlikely that the chelonian carapace evolved from any group of reptiles that had previously evolved a carapace formed from osteoderms.

Rieppel (2001) also cited studies that identified inductive interaction triggered by the carapacial ridge as the probable cause of the deflection of rib growth (Yntema 1970; Burke 1989; Gilbert et al. 2001). Another unique feature of turtles is the anterior displacement of the neural arches and ribs relative to the centra by half a segment, as discovered by Goette in 1899. In summary Rieppel (2001, p. 991) stated: "The initial segmentation of the paraxial mesoderm is the same in turtles as in all other tetrapods, but further development of the structures derived from the somites (dermis, vertebrae and ribs) proceeds along a different trajectory in turtles compared to all other tetrapods".

These factors would seem to meet the criteria for the rejection of the sauropterygian hypothesis that deBraga and Rieppel (1997) stated on their page 325: "Therefore, in order to reject or fail to support the present phylogeny it would be necessary only to show that lepidosauromorphs share more developmental characteristics in common with archosauromorphs than with turtles. Lepidosauromorphs and archosauromorphs are equivalent in sharing the absence of the unique factors involved in the development of the carapace and plastron of turtles. This subject will be discussed more fully after consideration of molecular means to analyse the phylogenetic position of turtles.

Rieppel's later book chapter on this subject (2008) continued to support a diapsid origin for turtles, but recognized problems of accounting for the origin of the carapace and plastron from any known group of early amniotes. He began by stating that (p. 345): "The turtle carapace combines endoskeletal (ribs and vertebral neural arches) and exoskeletal (dermal) components in an entirely novel, indeed unique structure .... The development of the carapace involves a redirection of sclerotome cell migration to form the ribs in a more superficial position, i. e., with the dermal carapacial disk (Burke 1989; Gilbert et al. 2001, 2004). The profound structural changes that are correlated with the development of the turtle shell render the skeletal morphology of turtles difficult to compare with that of other reptiles. There are also no fossil intermediates known that link turtles with a more generalized group of reptiles".

Figure 3.6 shows the alternative hypotheses of turtle relationships current at the time of Rieppel's (2008) review.

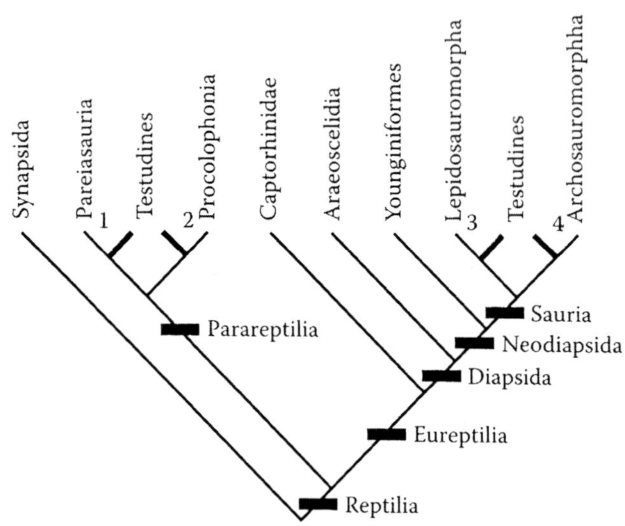

Fig. 3.6 Cladogram depicting four of the proposed sister-group relationships for turtles, all based on cladistic analyses. From left to right, turtles are shown as the sister taxon of pareiasaurs, procolophonids, lepidosauromorphs, or archosauromorphs (from Rieppel 2008). Turtles previously had been affiliated with synapsids or captorhinomorphs, providing a choice among six phylogenies

Molecular Analyses of Living Amniotes as a Means of Establishing Chelonian Relationships

A very different approach to the problem of identifying the sister taxon of turtles has been taken by molecular biologists who have concentrated entirely on living taxa. A sample of their results is listed in Table 3.2. On the basis of molecular data of a variety of taxa, most authors found turtles to be an out group of archosaurs and birds, with lepidosaurs (squamates) the out group of all three. This arrangement fits with the temporal sequence of first appearances of each group in the fossil record. One exception was the study by Cao et al. (2000), which gave equal support to the above-described arrangement and one in which turtles and crocodiles were sister taxa, with birds, squamates, and mammals as successively more distant out groups. None recognized a sister-group relationship between turtles and primitive amniotes, but this was only expressly excluded by Zardoya and Meyer (1998).

However, none of these authors acknowledged that it is impossible to test molecular affinities with ancestral anapsids such as procolophonids and pareiasaurs or with diapsid placodonts, all of which had been proposed by paleontologists to be closely related to turtles, because no species are living today that are direct descendants of these Permian or Triassic amniotes. Hence, neither their DNA nor other molecules can be sequenced and compared with living turtles.

Table 3.2 Sample of molecular studies that have considered the position of turtles relative to other extant groups

Authors	Molecules used	Chelonian relationships	Divergence time and other details
Hugal et al. (2007)	Long Nuclear Gene RAG-I	Turtles sister taxon of crocodiles + birds, with squamates an out group of all three. Supports diapsid origin	$\sim 270 \times 10^6$ years, 88 taxa studied
Kumazawa (2007)	Nearly complete mitochondrial DNA 9542 alignable sites	Turtles sister taxon of crocodiles + birds, with squamates an out group	Mid-Permian
Krenz et al. (2005)	RAG-1 2793 Nucleotides	Turtles sister taxon of crocodiles + birds. No further out group	Basically looking at classification of turtles per se
Iwabe et al. (2005)	Nuclear DNA-coded proteins	Turtles sister taxon of crocodiles + birds. Squamates an out group	Relationships with basal amniotes rejected
Cao et al. (2000)	Mitochondrial and nuclear genes	Turtles sister taxon of crocodiles + birds, squamates an out group, or ([(crocodiles, birds) squamates] turtles)	Study based on 3 turtles, 1 squamate, 1 crocodile, and 8 birds
Hedges and Poling (1999)	2 nuclear genes, mitochondrial DNA and 22 additional nuclear genes	Turtles sister taxon of crocodiles, squamates an out group	18 rRNA favors a bird + mammal grouping. Turtles diverged from crocodilians 207 ± 20.5 million years ago
Mannen and Steven (1999)	cDNA sequences	Turtles sister taxon of crocodiles	Turtles diverged after divergence of squamates

Another approach was taken by Hugall et al. (2007) who attempted to establish times of divergence between the various amniote clades through estimates of the rates of change over time in the nuclear gene RAG-1. Their study appeared to support archosauromorphs as the most likely sister taxon of turtles, with an inferred divergence date between 265 and 273 million years ago. They then stated that this relatively late date of turtle divergence refuted Gaffney and McKenna's (1979) suggestion that captorhinids were the nearest relatives of turtles. In fact, the genus that Gaffney and McKenna (1979) studied was discovered in Upper Permian beds, now dated as about 255 million years old. This is far later than Hugall et al. (2007) stated date for the initial appearance of captorhinids at ~300 million years ago. The latter date is approximately correct, but an earlier time of initial appearance of this group does not preclude a later time of divergence between captorhinids and turtles. In fact, the age of the African specimen described by Gaffney and McKenna (1979) is only about 35 million years older than the oldest known turtle, *Odontochelys,* dated at 220 million year old. Judging by the changes between *Odontochelys* and *Proganochelys* over their 20 million years of temporal separation, this is about the time interval that one might expect for the divergence of the oldest known turtle from a primitive, stem amniote. In fact, the age of the youngest known captorhinid is the same as the oldest known archosaur, as given by Hugall et al. (2007), suggesting that both clades are equally plausible as the sister taxon of turtles, using the criteria of timing of first occurrences. This, however, is not sufficient to establish the probability of either group having been the phylogenetic antecedent of turtles.

Organogenesis of Living Tetrapods as a Means of Establishing Chelonian Relationships

Another approach to establishing the ancestry of turtles was that of a recent study by Werneburg and Sanchez-Villagre (2009), which considered the relative timing of organogenesis among the major clades of reptiles (Aves, Crocodylia, Sphenodontia, and Squamata) based on a standard reference series of embryological events in amniotes. They used 104 developmental characters of external morphology that appeared prior to birth among 15 turtles and seven other amniote species and one salamander species, using Parsimov. Considering all specimens, this involved total of 1284 character states, and resulted in 24 topographically different phylogenies. All supported a basal position for turtles among the living clades within the Class Reptilia.

On the other hand, highly informative genetic data involving the unique mode of development of the carapace and plastron in living turtles provides a much more effective means for investigating the origin of the anatomical features that distinguish them from all other living amniotes.

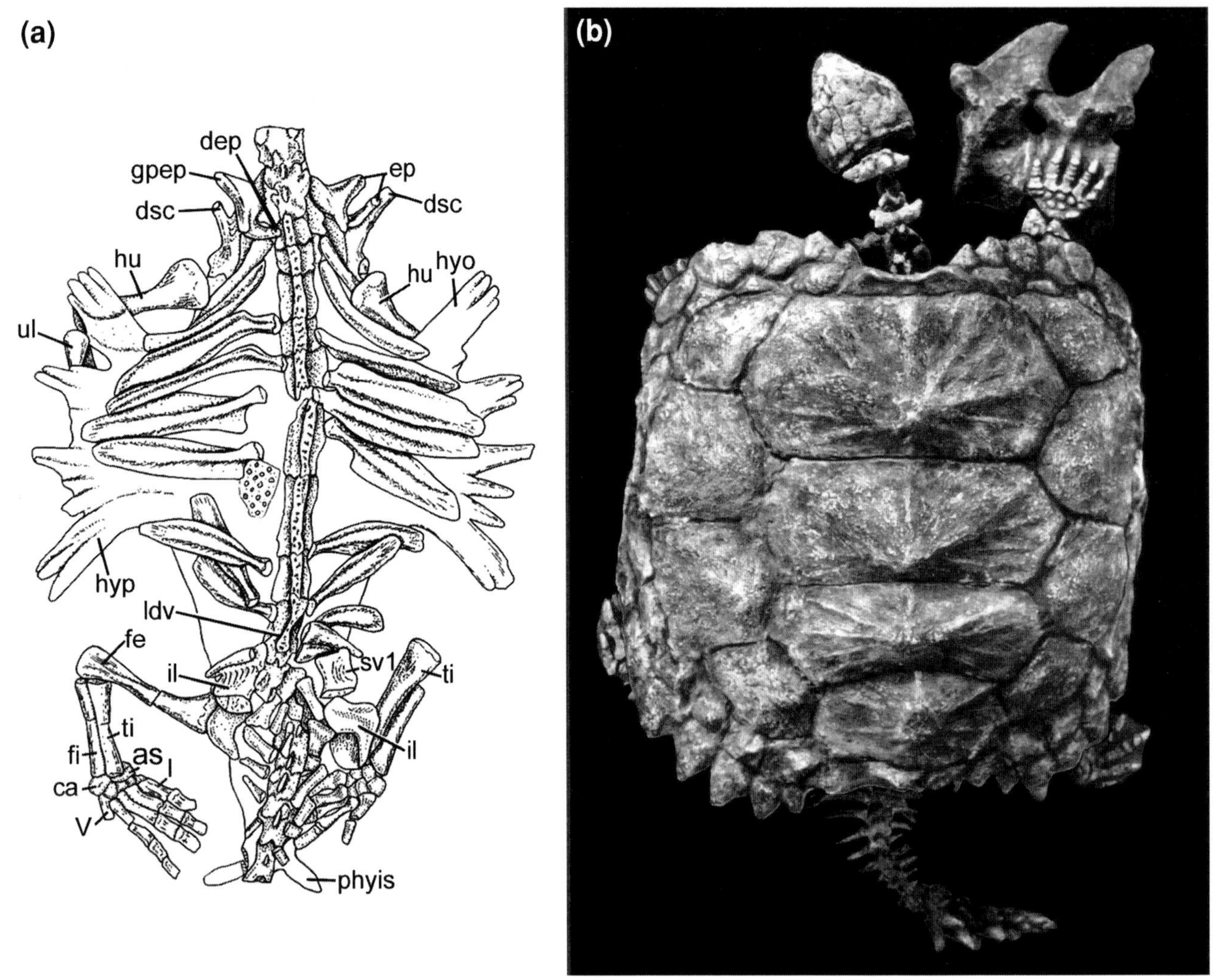

Fig. 3.7 Dorsal views of the skeletons of the oldest currently known turtles. **a** The oldest known turtle, *Odontochelys*, from the Late Triassic of China, lacking a carapace (from Li et al. 2008); see photograph (Fig. 3.10a) of same specimen. **b** The oldest known turtle having a fully developed carapace, *Proganochelys*, from the Late Triassic of Germany (from Gaffney 1990). See Fig. 3.10 for scale

Developmental Aspects of Chelonian Anatomy

Surprisingly, neither paleontologists who have used cladistic analyses to search for the sister-taxon of turtles nor the molecular biologists just cited, have made effective use of a wealth of new evidence that has become available over the past 15 years from the fields of molecular and developmental biology. This combined study has been termed Evolutionary-Developmental Biology, or EVO-DEVO, recognized by a distinct new journal, *Evolution & Development.* Its relevance to the origin of turtles is demonstrated by the cover of the May/June 2007 issue of that journal, adapted from Gene Gaffney's (1990) paper in the Bulletin of the American Museum of Natural History. Of specific interest are the many recent articles discussing the genetic, molecular, and developmental aspects of their most conspicuous attributes, the unique nature of the carapace and plastron.

Carapace

Despite the relatively earlier origin of the plastron, this discussion begins with the carapace because it is the canonical character of the Chelonia. As we have seen, there are a variety of amniotes that have evolved dermal armor that covers the dorsal surface of the body but in most cases, including pareiasaurs, placodonts, and some archosaurs, it consists of a mosaic of osteodermal plates superficial to the ribs and the girdles. None of these taxa had the structural details or manner of development comparable to those of

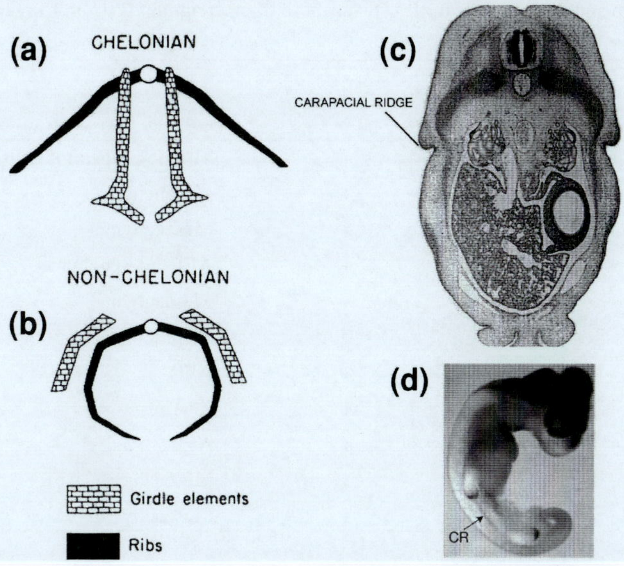

Fig. 3.8 Configuration of ribs and girdle elements in adult turtles versus non-turtles and position of carapacial ridge in turtle embryos. **a**, **b** Simplified, cross-sectional diagrams of adult tetrapods: **a** the derived turtle condition, in which the ribs extend laterally and dorsally above the shoulder girdle; **b** the primitive, non-chelonian condition, in which the ribs extend ventrally, around the trunk, and medial to the scapula. **c**, **d** Photographs of turtle embryos, showing position of the carapacial ridge, a developmental feature unique to turtles: **c** transverse section through the trunk; **d** lateral view of an entire turtle embryo, with carapacial ridge (CR) between the limb bud of the arm above and the rear limb bud below. First three images are from Burke (1989); last image is from Moustakas (2008). Images at different scales

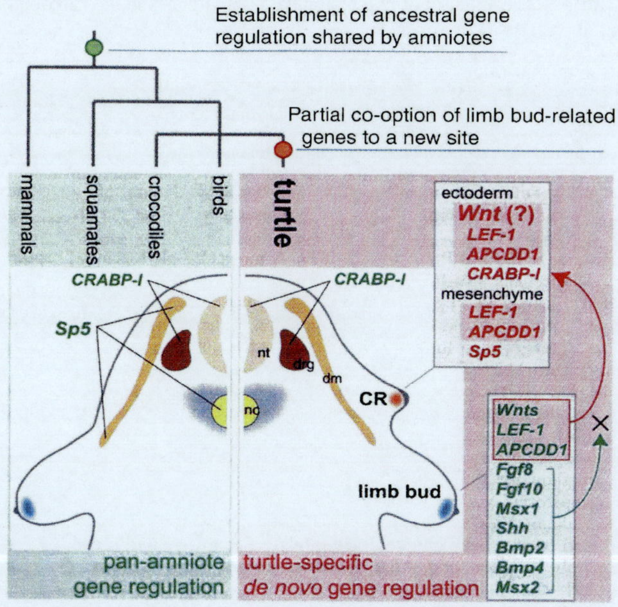

Fig. 3.9 Diagram of the cross-section through the trunk of an amniote other than a turtle on the left, and a turtle on the right showing the position of the carapacial ridge (CR) in relationship to a limb bud and the various genes, common to all amniotes (in areas of lighter shading) and those in darker shading that have been co-opted by the area of the carpacial ridge in turtles. See Table 3.3 for functions of genes in turtles versus other amniotes. Illustration from Kuraku et al. (2005)

any turtles, fossil or living, in which the carapace is closely integrated with the ribs, which lie *external* to the shoulder girdle (Burke 1989; Gilbert et al. 2001).

Although the extent of ossification of the carapace differs somewhat among modern turtles, its basic bony constituents remain constant from the Late Triassic to the present (Fig. 3.7). *Proganochelys* has a complete carapace and plastron essentially identical to those of modern forms except for the presence of additional supramarginal plates forming the periphery of the carapace, and an additional pair of mesoplastra in the plastron. The skeletal similarity of the carapace and plastron of living turtles is sufficiently similar to that of their most primitive known antecedents to support the assumption that their genetics and modes of development were also comparable. This assumption is further supported by comparable modes of development in both the living cryptodires (Kuraku et al. 2005) and pleurodires (Scheyer et al. 2008), whose time of divergence can be traced to the Late Triassic (Gaffney and Jenkins 2010).

The chelonian carapace is uniquely composed of the endochondral ribs and vertebrae associated with a specialized dermis. The ribs are found in a unique position, superficial and dorsal to the limb girdles. Beginning with *Proganochelys,* the carapace is solidly linked to the plastron by a bony bridge. Nothing of this nature occurs in other tetrapod groups.

The exceptional configuration of the carapace, particularly the position of the ribs, occurs in relationship to patterns of ontogenetic development that are seen in no other vertebrates. It is based on a unique embryonic feature, the carapacial ridge (CR) extending anteroposteriorly above the bases of the pectoral and pelvic limbs (Fig. 3.8). This ridge is a unique raised area composed of mesenchyme of the dermis and the overlying ectoderm that forms dorsal to the boundary between the somitic (dorsal) and lateral plate mesoderm. The carapacial ridge is the precursor of the carapace margin, in which the ribs will eventually terminate.

Lateral growth of the carapacial margin during early development entraps the distal extremities of the developing ribs, which extend laterally, rather than ventrally, and assume a superficial position dorsal to the scapulocoracoid. The ribs eventually terminate in the lateral margin of the dermal carapace. As they elongate, the endochondral ribs influence the surrounding tissue to form the more superficial bones of the carapace. It is especially important to recognize that the carapace itself develops from endochondral tissue, in contrast with the dermal tissue in which forms the isolated osteoderms that contribute to the dorsal armor of pareiasaurs, placodonts and some archosaurs (Burke 1989;

Table 3.3 Summary of expression patterns of genes in the carpacial ridge and limb buds of turtles versus those of other amniotes (from Kuraku et al. 2005)

Carapacial ridge			Limb bud			
Gene	Ps	Domain	Ps	Gg	Mn	Domain
Sp5	+	Mesenchyme	+	–	+	Muscle progenitor cells
CRABP-1	+	Ectoderm	–	+	+	Mesenchyme
APCDD1	+	Mesenchyme, ectoderm	+	+	+	Mesenchyme
LEF-1	+	Mesenchyme, ectoderm	+	+	+	AER, mesenchyme
Msx1	–		+	+	+	Mesenchyme
Mgf 10	–		+	+	+	Mesenchyme
MGF8	–		+	+	+	AER
Shh	–		+	+	+	ZPA
Bmp2	–		+	+	+	AER
Bmp4	–		+	+	+	Mesenchyme, AER
Msx2	–		+	+	+	Mesenchyme

Abbreviations: *Gg* chicken, *Mn* mouse, *Ps* soft-shelled turtle

Vickaryous and Hall 2008). Removal and labeling of tissue shows that somitic mesoderm contributes mesenchyme to the carapacial ridge, in contrast with the mesenchyme of the adjacent limb buds which is derived from the lateral plate mesoderm.

Underlying the unique presence of the carapacial ridge and its influence on the formation of the carapace are a host of genes and specific tissues that have been co-opted from those associated with the adjacent limb buds, and whose position and specific function have been altered from those of all other tetrapods to influence the orientation of the ribs and to form the dorsal armor. These factors have been discussed in a number of papers that examined various aspects of development.

Kuraku et al. (2005) screened for genes specifically expressed in the carapacial ridge of the Chinese soft-shelled turtle (Fig. 3.9, Table 3.3). Four are present in other vertebrate lineages but have acquired unique, de novo expression in turtles: *Sp5*; cellular retinoic acid binding protein-l (*CRABP-1*); adenomatous polyposis coli down-regulated 1 (*APCDD1*); and lymphoid enhancer-binding factor-l (*LEF-1*). The presence of β-catenin in the ectoderm of the carpacial ridge suggests that the canonical Wnt pathway is activated in the CR ectoderm and controls the initial phase of carapace development by the abrupt alteration in the expression patterns of multiple downstream genes. The unique expression of these genes in the carapacial ridge and the ridge itself can each be considered as apomorphic characters uniting all turtles. These gene expressions may be absolutely unique to turtles (autapomorphies) or potential synapomorphies with some other, as yet unrecognized, early amniote clade.

Cebra-Thomas et al. (2005) pointed out additional unique features of carapace development that may explain the rapid appearance of turtles in the fossil record without obvious intermediates. Following initiation of the carapacial ridge in Yntema stage 14, its maintenance and ability to attract rib precursor cells is dependent on fibroblast growth factor (FGF) signaling. If secretion of FGF is lost, ribs grow either anteriorly or posteriorly to attach to other ribs in which FGF signaling is maintained. Inhibitors of FGF result in migration of ribs toward the ventral body wall. FGF10 also plays a role in attracting rib rudiments. Co-ordinated growth of the carapacial plate and the ribs apparently results from a positive feedback loop caused by the induction of FGF8 in the distal tips of the ribs by FGF10-secretion from the mesenchyme of the carpacial ridge. Once in the dermis, the ribs undergo endochondral ossification. The ribs also act as a signaling center for the surrounding dermal ossification that forms the carapace.

Further aspects of the development of the carapacial ridge were discussed by Moustakas (2008). These involve support for a dermomyotomal identity of the mesenchyme of the carapacial ridge and its early expression of *Pax3, Twist1, Dermol, En1, Sim1,* and *Gremlin* genes before overt ossification expresses *Pax1.* This supports the hypothesis of mesenchymal formation of dermal bone in the carapace. It also emphases the importance of the exaptation of key genetic networks common to the limbs and vertebral column of other tetrapods in the development of the turtle shell. The innovation of the carapace does not involve the evolution of new tissue types, but rather how the developmental instructions for making certain tissues become used in new ways and places.

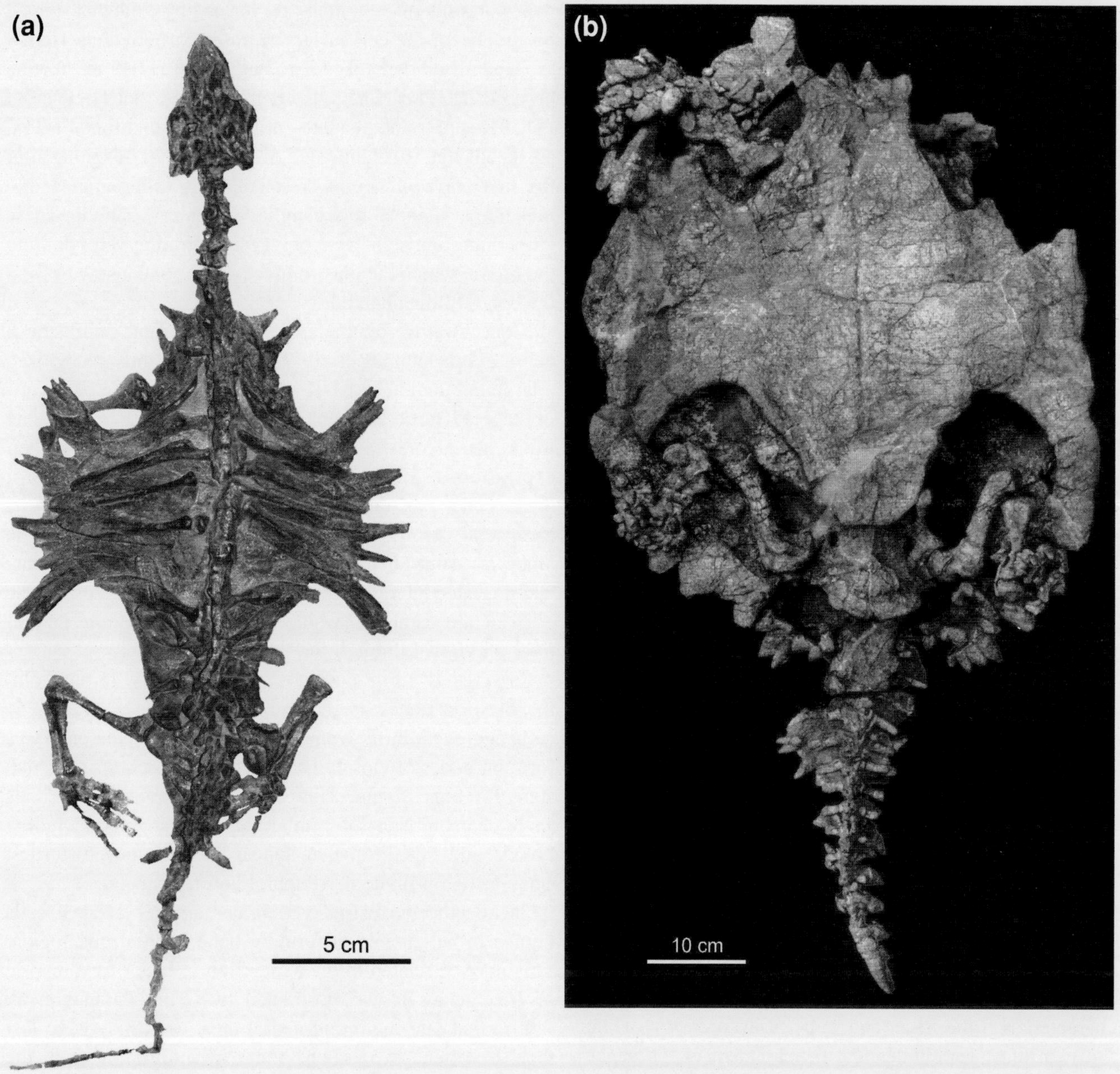

Fig. 3.10 Skeletons of two basal turtles. **a** Dorsal view of *Odontochelys*, from the Late Triassic of China, showing the lateral extension of the ribs and the rudimentary carapace (limited to small areas of ossification just above the neural spines of the vertebrae) and the more fully developed plastron with lateral spikes that presumably represent its incorporation of the gastralia (from Li et al. 2008); see interpretive drawing (Fig. 3.7a) of same specimen. **b** Ventral view of *Proganochelys*, from the Late Triassic of Germany, showing a fully developed shell with bridge on either side firmly uniting the plastron and the margin of the carapace (from Gaffney 1990)

Plastron

The turtle plastron is also a unique structure among vertebrates, which had reached an essentially modern configuration at the base of the known fossil record of chelonians (Figs. 3.10, 3.11). Developmentally, it forms before the carapace, matching its earlier temporal appearance as seen in *Odontochelys,* in which it clearly does not show the bridging structure that could link it with the carapace, as it does in *Proganochelys.* Rather, *Odontochelys* exhibits a series of laterally extending spikes, inherited from the gastralia from which it develops. Like the carapace, the plastron has maintained many aspects of its structure since its initial appearance in the fossil record. In *Proganochelys* the plastron consists of 13 bones, the median entoplastron and the paired epiplastra, hyoplastra, hypoplastra, mesoplastron I, mesoplastron II, and xiphiplastra. These have been homologized (from front to back) with the median

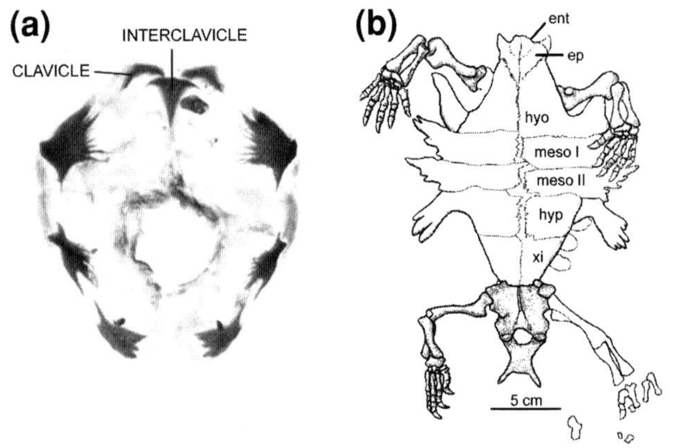

Fig. 3.11 Comparison of plastra in an embryonic extant turtle versus in a basal turtle. **a** An early stage in the development of the plastron in a modern turtle (*Trachemys scripta*) showing dermal elements of the shoulder girdle anteriorly and an early stage of the coalescence of the more posterior elements from the gastralia under the influence of neural crest cells (from Cebra-Thomas et al. 2007). **b** The fully ossified plastron of *Odontochelys*, from the Late Triassic of China, shows homologous elements (from Li et al. 2008). Note that the mesoplastra seen in *Odontochelys* are lost in most modern turtles. Images at different scales

interclavicle, the paired clavicular blades, and, more posteriorly, numerous gastralia, common to early tetrapods. All ossify intramembranously, without any cartilaginous precursors (in contrast with the endochondral ribs that are key to the formation of the carapace). There is no support for the earlier suggestion that the plastron incorporates a median, endochondral sternum.

Until recently, the early developmental processes leading to the formation of the plastron had not been determined. The bone cells of amniotes are either of somitic origin (characterizing the appendicular and axial endoskeleton and internal elements of the skull) or arise from the neural crest (characterizing the face and superficial portions of the skull). In most vertebrates, the development of the superficial bones of the skull and the shoulder girdle are influenced by neural crest cells that migrate from the dorsolateral margins of the anterior portion of the neural tube, but are not active in bone formation in the more posterior part of the body. Turtles are unique in the presence of a distinct population of neural crest cells that migrates into the area of the plastron and has re-gained the capacity to initiate ossification.

Clark et al. (2001) demonstrated that the bones of the plastron are exoskeletal and are formed by the intramembranous ossification of neural crest cells. This was unexpected because trunk neural crest cells had not been observed to form skeletal elements in other vertebrates, and cranial neural crest cells (which are also skeletogenic) are not observed to migrate more posteriorly than the collarbone and shoulder. Clark et al. (2001) established that the nine plastral bones of modern chelonians are formed by cells that stained positively for the cell surface sulfated carbohydrate HNK-l (a standard marker for avian neural crest cells) and for the protein PDGFRα. They also demonstrated that HNK-l and PDGFRα are found in a late-emigrating population of neural crest cells that is delaminated from the neural tube long after the first wave of neural crest cells has differentiated into melanocytes, dorsal root ganglia, and enteric ganglia. These late-emigrating cells were observed migrating ventrally into the region where the plastron forms and condensing to form the bones of the plastron.

Concentrating on the late-emerging trunk neural crest cells, Cebra-Thomas et al. (2007) subsequently recognized a large population of HNK-1^+ cells in the carapacial dermis of stage 17 turtle embryos. Their identity as late-emerging trunk neural crest cells was further established by their expression of FoxD3^+ and p75^+. The migratory nature of these cells was demonstrated by injecting Dil into the carapacial dermis above the neural tube of stage 17 turtle embryos. After 36–48 h, streams of cells were observed that had migrated laterally from the injection site, out toward the edge of the carapace, ventrally around the somite derivatives and down through the lateral bridge region.

In stage 18 turtle embryos (approximately 25 days old) the plastron bones are beginning to ossify. HNK-1^+/p75^+ cells aggregate in the ventral mesenchyme of the embryos, forming nodes of closely packed cells, the centers of which stain for bone tissue. This late stage population of cells share characteristics of cranial neural crest cells, including FoxD3 and PDGFRα, that appear to be unique to turtles. They also contribute to vertebral and rib cartilages.

In no other terrestrial vertebrates are neural crest cells known to be capable of forming bone in the trunk region. This may be related to the expression or lack of expression of *Hox* genes. It was postulated by Cebra-Thomas et al. (2007) that late migrating neural crest cells may have lost their *Hox* gene expression, either by emigrating from the neural tube at a later time, or by remaining in the staging area for a long time. This, in some way, enables those cells to function in the same manner as cranial neural crest cells.

Search for Chelonian Ancestors

Based on extrapolation of data from the anatomical and molecular factors of development of the carapace and plastron in modern turtles to fossils that are known from the Later Triassic, it appears extremely unlikely that either pareiasaurs or genera such as *Henodus* and *Placodus,* from among aquatic sauropterygian diapsids, exhibit plausible structural or developmental precursors to the turtle carapace. Not only do these parareptiles and diapsids form the

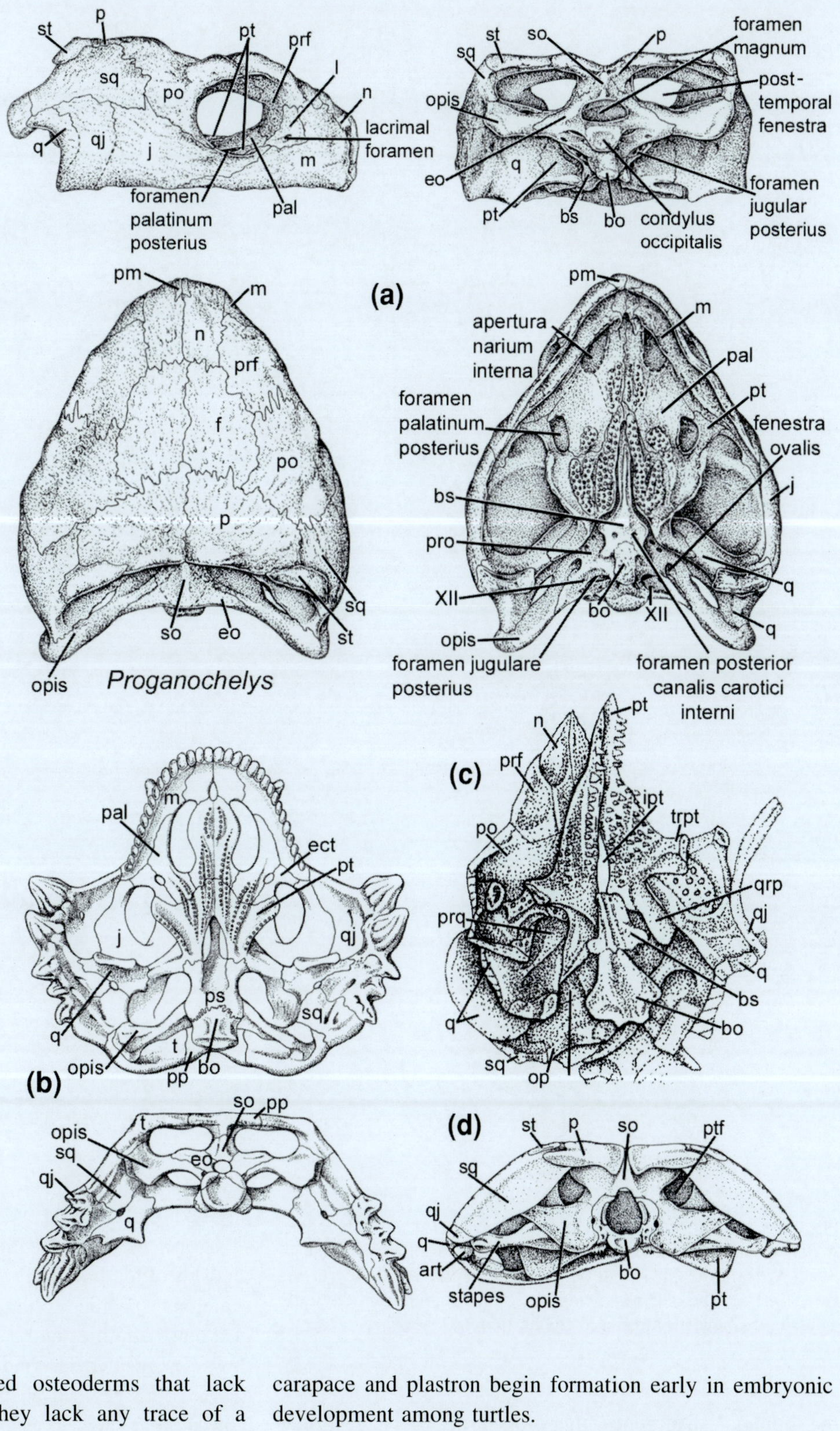

Fig. 3.12 Skulls of basal turtles and putative turtle sister taxa. **a** The turtle *Proganochelys* (12 cm in length), from the Late Triassic of Germany, in four views (from Gaffney and Meeker 1983). **b** The pareiasaur *Scutosaurus* (skull about 12 cm in length), from the Late Permian of Eastern Europe, in palatal and occipital views (from Carroll 1988). **c** The oldest known turtle, *Odontochelys*, from the Early Triassic of China, in palatal view (from Li et al. 2008). **d** Occiput of *Captorhinus* (about 5 cm in width), from the Early Permian of the southwestern United States (from Carroll 1988)

carapace from initially separated osteoderms that lack association with the ribs, but they lack any trace of a plastron, which developed and evolved prior to the carapace within the Chelonia. As pointed out by Vickaryous and Hall (2008, p. 398) "Osteoderms demonstrate a delayed onset of development compared with the rest of the skeleton, not appearing until well after hatching". In contrast, both the carapace and plastron begin formation early in embryonic development among turtles.

In any case, the ultimate solution to the question of chelonian ancestry must reside with the fossil record. *Odontochelys* provides not only knowledge of a chelonian that was at the very beginning stages of developing a carapace (ontogenetically or phylogenetically), but also pushes

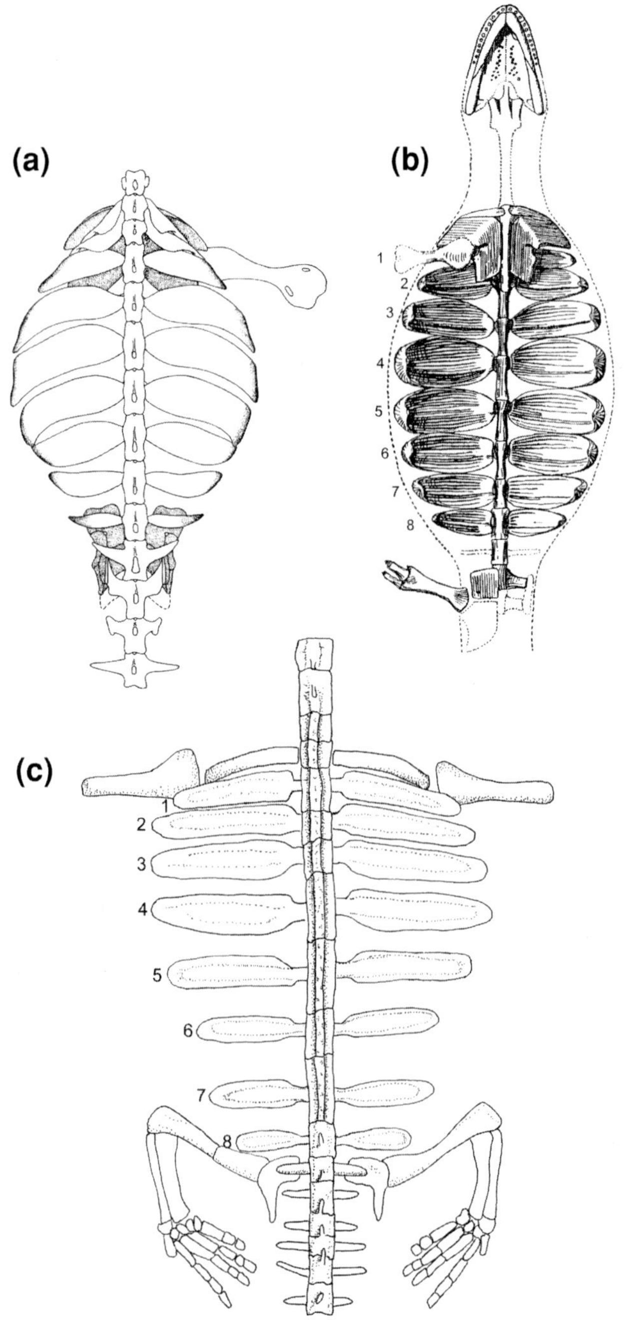

Fig. 3.13 Trunk regions of *Eunotosaurus* and *Odontochelys*. **a**, **b** *Eunotosaurus*, about 10 cm in length, from the Late Permian of South Africa: **a** dorsal view; **b** ventral view (modified from Keyser and Gow 1981). **c** The oldest known turtle, *Odontochelys*, from the Early Triassic of China (from Li et al. 2008), in dorsal view

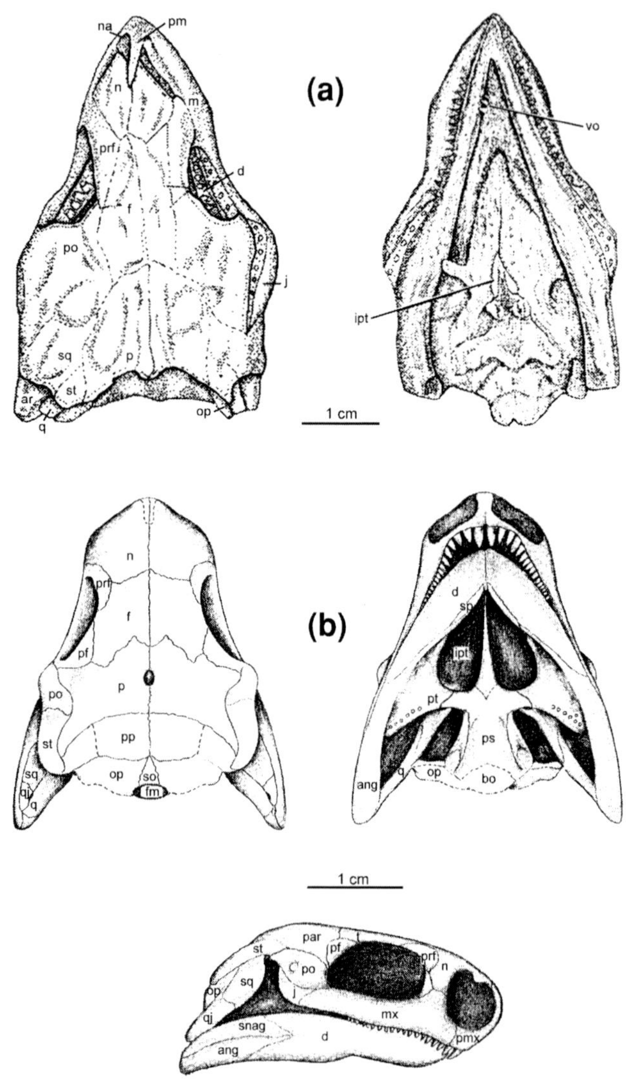

Fig. 3.14 Skulls of *Odontochelys* and *Eunotosaurus*. **a** The oldest known turtle, *Odontochelys*, from the Late Triassic of China, in dorsal and palatal views (from Li et al. 2008). **b** *Eunotosaurus*, from the Late Permian of South Africa, in dorsal, palatal, and right lateral views (modified from Keyser and Gow 1981)

the minimal divergence time for turtles back a further 20 million years.

The many, clearly significant chelonian characters that had already been achieved by the time of appearance of *Odontochelys*, 220 million years ago, suggest a considerable time interval since turtles diverged from other early amniotes. This, in itself, precludes their divergence from any taxa that had not evolved prior to the Late Triassic.

If one goes back to the basal anapsids, synapsids, and diapsids of the Late Carboniferous and Early Permian, their skeletons, specifically the skulls, are well known, and suggest close affinities and a fairly short interval of time since they diverged from one another (Figs. 3.1, 3.2, 3.12). Several lineages can be traced into the Late Permian and Triassic that show more advanced character states, several resembling those of early chelonians, including the evolution of posttemporal fenestrae, loss of teeth on the transverse flange of the pterygoid, loss of the ectopterygoid and tabular, development of a palatine fenestra and formation of a jugular process. But these changes occur convergently in

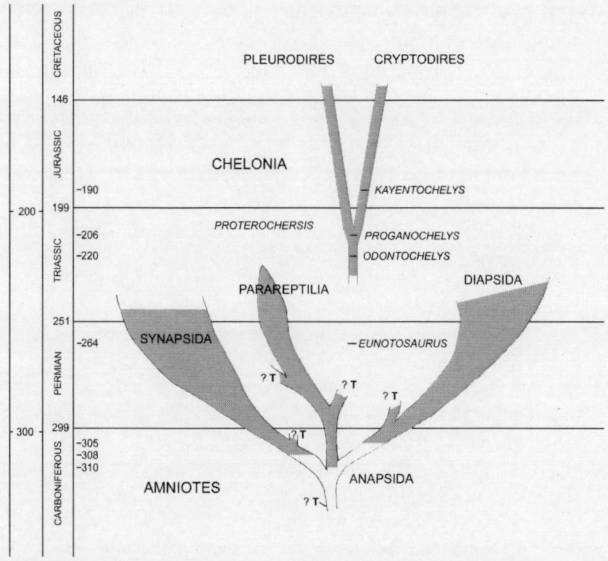

Fig. 3.15 Temporal distributions of primitive amniotes. Time scale derived from Gradstein et al. (2004). "?T" indicates uncertain divergence time

several lineages indicating that they are not homologous with one another. Surprisingly, both ancestral turtles and some, but not all of the earlier amniotes retain an unfused basicranial articulation. Despite their acquisition of some derived characters in common with turtles, the remainder of their anatomy provides no convincing evidence of sister-group affinities. It is probable that one or another taxon might be shown by cladistic analysis to have a more parsimonious pattern of character change leading toward early turtles, but none exhibit any definitive characters that are clearly homologous with the many dramatic autapomorphies that distinguish turtles.

There is, however, one genus, *Eunotosaurus* (Figs. 3.13, 3.14, 3.15), from the Late Permian of South Africa, that does possesses a number of chelonian apomorphies, notably a relatively long neck, a short trunk region with approximately 10 vertebrae, and widened, laterally extending ribs like those of *Odontochelys. Eunotosaurus* is not shown to have posttemporal fenestrae, but neither are these illustrated in *Odontochelys.* Many descriptions of *Eunotosaurus* have been published (Seeley 1892; Watson 1914; Cox 1969; Keyser and Gow 1981). Some, but not all, support chelonian affinities, but the discovery of *Odontochelys* provides a much closer basis of comparison than has previously been available, and should be exploited. *Eunotosaurus* is some 45 million years older than *Odontochelys,* but appears more turtle-like in general features than any of the other current contenders.

The most important factor that paleontologists should bear in mind when searching for the ancestry of turtles or other taxa is the necessity for recognizing one or more unique, emergent characters that may be derived from individual, contemporary taxa or antecedent clades. In contrast, the multiplicity of putative relationships that have been proposed on the basis of increasingly broad surveys of a vast number of traits and taxa demonstrate the difficulty or impossibility of using such methods for establishing specific evolutionary relationships.

Acknowledgments I would like to thank Don Brinkman for bringing together the many colleagues of Gene Gaffney for this symposium in his honor, and for arranging the publication of the many resulting lectures and discussions. I would also like to recognize the assistance of Mary-Ann Lacey for assembling and labeling the many drawing that help in understanding the nature of the putative sister taxa of turtles, and Trond Sigurdsen for their final integration in this paper. It should also be noted that understanding of the origin and evolution of turtles could not have reached the level that has been achieved to date if it were not for the continuing research of many scientists studying molecular and genetic aspects of the development of modern turtles. The financial support for assembling the data required for this paper and travel to scientific meetings where it was discussed was provided by the National Science Foundation of Canada.

References

Ax, P. (1984). *Das phylogenetische system.* Stuttgart: Fischer.

Burke, A. C. (1989). Development of the turtle carapace: Implications for the evolution of a novel bauplan. *Journal of Morphology, 1999*, 363–378.

Cao, Y., Sorenson, M. D., Kumazawa, Y., Mindell, D. P., & Hasegawa, M. (2000). Phylogenetic position of turtles among amniotes: Evidence from mitochondrial and nuclear genes. *Gene, 259*, 139–148.

Carroll, R. L. (1969). A Middle Pennsylvanian captorhinomorph, and the interrelationships of primitive reptiles. *Journal of Paleontology, 43*, 151–170.

Carroll, R. L. (1988). *Vertebrate paleontology and evolution.* New York: W. H. Freeman & Co.

Cebra-Thomas, J., Tan, F., Sistla, S., Estes, E., Bender, G., Kim, C., et al. (2005). How the turtle forms its shell: A paracrine hypothesis of carapace formation. *Journal of Experimental Zoology B, 304*, 558–569.

Cebra-Thomas, J. A., Betters, E., Yin, M., Plafkin, C., McDow, K., & Gilbert, S. F. (2007). Evidence that a late-emerging population of trunk neural crest cells forms the plastron bones in the turtle *Trachemys scripta. Evolution & Development, 9*, 267–277.

Clark, K., Bender, G., Murray, B. P., Panfilio, K., Cook, S., Davis, R., et al. (2001). Evidence for the neural crest origin of turtle plastron bones. *Genesis, 31*, 111–117.

Cox, C. B. (1969). The problematic Permian reptile *Eunotosaurus. Bulletin of the British Museum (Geology), 18*, 165–196.

Currie, P. J. (1977). A new haptodontine sphenacodont (Reptilia; Pelycosauria) from the Upper Pennsylvanian of North America. *Journal of Paleontology, 51*, 927–942.

De Beer, G. S. (1937). *The development of the vertebrate skull.* Oxford: Clarendon Press.

deBraga, M., & Rieppel, O. (1997). Reptile phylogeny and the interrelationships of turtles. *Zoological Journal of the Linnean Society, 120*, 281–354.

Gaffney, E. S. (1980). Phylogenetic relationships of the major groups of amniotes other than turtles. In A. L. Panchen (Ed.), *The terrestrial environment and the origin of land vertebrates* (pp. 593–610). London: Academic Press.

Gaffney, E. S. (1990). The comparative osteology of the Triassic turtle *Proganochelys*. *Bulletin of the American Museum of Natural History, 194*, 1–263.

Gaffney, E. S., & Jenkins, F. (2010). The cranial morphology of *Kayentachelys*, an Early Jurassic cryptodire, and the early history of turtles. *Acta Zoologica (Stockholm), 91*, 335–368.

Gaffney, E. S., & McKenna, M. C. (1979). A Late Permian captorhinid from Rhodesia. *American Museum Novitates, 2688*, 1–15.

Gaffney, E. S., & Meeker, L. J. (1983). Skull morphology of the oldest turtles: A preliminary description of *Proganochelys quenstedti*. *Journal of Vertebrate Paleontology, 3*, 25–28.

Gaffney, E. S., & Meylan, P. A. (1988). A phylogeny of turtles. In M. J. Benton (Ed.), *The phylogeny and classification of tetrapods* (pp. 147–219). Oxford: Clarendon Press.

Gardiner, B. G. (1982). Tetrapod classification. *Zoological Journal of the Linnean Society, 74*, 207–232.

Gardiner, B. G. (1993). Haematothermia: Warm-blooded amniotes. *Cladistics, 9*, 369–395.

Gilbert, S., Loredo, G., Brukman, A., & Burke, A. (2001). Morphogenesis of the turtle shell: The development of a novel structure in tetrapod evolution. *Evolution & Development, 3*, 47–58.

Gilbert, S. F., Cebra-Thomas, J. A., & Tan, F. (2004). Working hypothesis for the origin of the turtle shell. *Journal of Morphology, 260*, 294.

Goette, A. (1899). Über die Entwicklung des knöchernen Rückenschildes (Carapax) der Schildkröten. *Zeitschrift für wissenschaftliche Zoologie, 66*, 407–434.

Gradstein, F., Ogg, J., & Smith, A. (2004). *A geologic time scale 2004*. Cambridge: Cambridge University Press.

Hedges, S. B., & Poling, L. (1999). A molecular phylogeny of reptiles. *Science, 283*, 998–9001.

Hofsten, N. (1941). On the phylogeny of the Reptilia. *Zoologisca Bidrag Fran Uppsala, 20*, 501–521.

Hugall, A. F., Foster, R., & Lee, M. S. Y. (2007). Calibration choice, rate smoothing and the pattern of tetrapod diversification according to the long nuclear gene RAG-1. *Systematic Biology, 56*, 543–563.

Iwabe, N., Hara, Y., Kumazawa, Y., Shibamoto, K., Saito, Y., Miyata, T., & Katoh, K. (2005). Sister group relationship of turtles to the bird-crocodilian clade revealed by nuclear DNA-coded proteins. *Molecular Biology and Evolution, 22*, 810–813.

Keyser, A. W., & Gow, C. E. (1981). First complete skull of the Permian reptile *Eunotosaurus africanus* Seeley. *South African Journal of Science, 77*, 417–420.

Krenz, J. G., Naylor, J. P., Shaffer, H. B., & Jazen, F. J. (2005). Molecular phylogenetics and evolution of turtles. *Molecular Phylogenetics and Evolution, 37*, 178–191.

Kumazawa, Y. (2007). Mitochondrial genomes from the major lizard families suggest their phylogenetic relationships and ancient radiations. *Gene, 388*, 19–26.

Kuraku, S., Usuda, R., & Kuratani, S. (2005). Comprehensive survey of carapacial ridge-specific genes in turtles implies co-option of some regulatory genes in carapace evolution. *Evolution & Development, 7*, 3–17.

Laurin, M., & Reisz, R. R. (1995). A reevaluation of early amniote phylogeny. *Biological Journal of the Linnean Society, 101*, 59–95.

Lee, M. S. Y. (1993). The origin of the turtle body plan: Bridging a famous morphological gap. *Science, 261*, 1716–1720.

Lee, M. S. Y. (1994). The turtle's long-lost relatives. *Natural History, 103*, 63–65.

Lee, M. S. Y. (1995). Historical Burden in systematics and the interrelationships of 'parareptiles'. *Biological Reviews, 70*, 459–547.

Lee, M. S. Y. (1996). Correlated progression and the origin of turtles. *Nature, 379*, 812–815.

Lee, M. S. Y. (1997). Pareiasaur phylogeny and the origin of turtles. *Zoological Journal of the Linnean Society, 120*, 197–280.

Lee, M. S. Y. (2001). Molecules, morphology, and the monophyly of diapsid reptiles. *Contributions to Zoology, 70*, 1–19.

Lee, M. S. Y., Gow, C. E., & Kitching, J. W. (1997). Anatomy and relationships of the pareiasaur *Pareiasuchus nasicornis* from the Upper Permian of Zambia. *Palaeontology, 40*, 307–335.

Li, C., Wu, X.-C., Rieppel, O., & Wang, L.-T. (2008). An ancestral turtle from the Late Triassic of southwestern China. *Science, 456*, 497–501.

Løvtrup, S. (1977). *The Phylogeny of Vertebrates*. London: John Wiley.

Løvtrup, S. (1985). On the classification of the taxon Tetrapoda. *Systematic Zoology, 34*, 463–470.

Mannen, H., & Steven, S. L. (1999). Molecular evidence for a clade of turtles. *Molecular Phylognetics and Evolution, 13*, 144–148.

Moustakas, J. E. (2008). Development of the carapacial ridge: Implications for the evolution of genetic networks in turtle shell development. *Evolution & Development, 10*, 29–36.

Olson, E. C. (1947). The family Diadectidae and its bearing on the classification of reptiles. *Fieldiana Geology, 11*, 2–53.

Reisz, R. (1972). Pelycosaurian reptiles from the Middle Pennsylvanian of North America. *Bulletin of the Harvard University Press, 144*, 27–61.

Reisz, R. (1977). *Petrolacosaurus*, the oldest known diapsid reptile. *Science, 196*, 1091–1093.

Reisz, R. (1981). A diapsid reptile from the Pennsylvanian of Kansas. Special Publication of the Museum of Natural History, *University of Kansas, 7*, 1–74.

Reisz, R., & Laurin, M. (1991). *Owenetta* and the origin of turtles. *Nature, 349*, 324–326.

Remane, A. (1959). Die Geschichte der Tiere. In G. Heberer (Ed.), *Die evolution der organismen* (2nd ed., Vol. 1, pp. 340–422). Stuttgart: Gustav Fischer.

Rieppel, O. (2001). Turtles as hopeful monsters. *BioEssays, 23*, 987–991.

Rieppel, O. (2008). The relationships of turtles within amniotes. In J. Wyneken, M. H. Godfrey, & V. Bels (Eds.), *Biology of turtles* (pp. 345–353). Boco Raton: CRC Press.

Rieppel, O., & Reisz, R. R. (1999). The origin and early evolution of turtles. *Annual Reviews of Ecology and Systematics, 30*, 1–22.

Romer, A. S., & Price, L. I. (1940). Review of the Pelycosauria. *Geological Society of America Special Paper, 28*, 1–538.

Scheyer, T., Brüllmann, T., & Sanchez-Villagra, M. (2008). The ontogeny of the shell in side-necked turtles, with emphasis on the homologies of costal and neural bones. *Journal of Morphology, 269*, 1008–1021.

Seeley, H. (1892). On a new reptile from Welte Vreden (Beaufort West) *Eunotosaurus africanus* (Seeley). *Quarterly Journal of the Geological Society London, 48*, 583–585.

Vickaryous, M., & Hall, B. (2008). Development of the dermal skeleton in *Alligator mississippiensis* (Archosauria, Crocodylia) with comments on the homology of osteoderms. *Journal of Morphology, 269*, 398–422.

Watson, D. M. S. (1914). *Eunotosaurus africanus* Seeley, and the ancestry of the Chelonia. *Proceedings Zoological Society of London, 1914*, 1011–1020.

Werneburg, I., & Sanchez-Villagra, M. R. (2009). Timing of organogenesis support basal position of turtles in the amniote tree of life. *BMC Evolutionary Biology, 9*, 82. doi:10.1186/1471-2148-9-82.

Yntema, C. (1970). Extirpation experiments on the embryonic rudiments of the carapace of *Chelydra serpentina*. *Journal of Morphology, 132*, 235–244.

Zardoya, R., & Meyer, A. (1998). Complete mitochondrial genome suggests diapsid affinities of turtles. *Proceedings of the National Academy of Science, 95*, 14226–14231.

Chapter 4
Origin of the Turtle Body Plan: The Folding Theory to Illustrate Turtle-Specific Developmental Repatterning

Hiroshi Nagashima, Shigehiro Kuraku, Katsuhisa Uchida, Yoshie Kawashima-Ohya, Yuichi Narita, and Shigeru Kuratani

Abstract The turtle shell is comprised of a dorsal carapace and a ventral plastron, and is an autapomorphy of this group. The carapace consists of the vertebral column and ribs as well as a specialized dermis. The formation of the shell is accompanied by a change in the spatial relationship of the ribs and the pectoral girdle. Because of this rearrangement, the turtle shell has been regarded as an example of an evolutionary novelty. Understanding the changes behind this developmental repatterning will help us elucidate the evolutionary history of turtles. The change has been attributed to a deflected pattern of development of the ribs, which in normal tetrapods grow ventrally into the lateral body wall. In turtles, they grow laterally toward the primordium of the carapacial margin, called the carapacial ridge (CR), while remaining in the axial part of the embryonic body. Based on a similarity in histological configuration, the CR has been thought to possess inductive activity for rib growth, as seen in the apical ectodermal ridge of the amniote limb bud. The CR does not function as a guidance cue for rib progenitor cells but rather functions in the marginal growth of the carapacial primordium, resulting in fanned-out growth of the ribs. This peripheral and concentric expansion of the axial domain makes the lateral body wall fold inward, while the ribs cover the pectoral girdle. The turtle ribs develop along the muscle plate as in other amniotes, and do not take a different trajectory from that in other amniotes, unlike the scenario hypothesized previously. This folding enables turtles to change the apparent spatial relationships between the ribs and the pectoral girdle without altering their topological alignment and body plan as amniotes. This developmental sequence of the modern turtles aligns with a stepwise evolutionary process in the group, which is supported by the anatomy of a recently discovered fossil species, *Odontochelys*.

Keywords Carapacial ridge • Development • *Odontochelys* • *Proganochelys* • Ribs • Turtle • Shell

H. Nagashima · S. Kuraku · K. Uchida · Y. Kawashima-Ohya · Y. Narita · S. Kuratani (✉)
Laboratory for Evolutionary Morphology, RIKEN Center for Developmental Biology, Kobe 650-0047, Japan
e-mail: saizo@cdb.riken.jp

Present address:
H. Nagashima
Division of Gross Anatomy and Morphogenesis, Department of Regenerative and Transplant Medicine, Niigata University, Niigata 951-8510, Japan
e-mail: azusagawakappa@gmail.com

Present address:
S. Kuraku
Genome Resource and Analysis Unit, RIKEN Center for Developmental Biology, Kobe, 650-0047, Japan
e-mail: shigehiro-kuraku@cdb.riken.jp

Present address:
K. Uchida
Department of Marine Biology and Environmental Sciences, Faculty of Agriculture, University of Miyazaki, 1-1 Gakuen Kibanadai-nishi, Miyazaki 889-2192, Japan
e-mail: k-uchida@cc.miyazaki-u.ac.jp

Present address:
Y. Kawashima-Ohya
Laboratory of Immunology, Department of Health Pharmacy, Yokohama College of Pharmacy, 601 Matano-cho, Totsuka-ku, Yokohama 245-0066, Japan
e-mail: y.kawashima@hamayaku.ac.jp

Present address:
Y. Narita
Department of Food Business, Faculty of Health and Human Life, Nagoya Bunri University, 365 Maeda, Inazawa-cho, Inazawa 492-8520, Japan
e-mail: narita.yuuichi@nagoya-bunri.ac.jp

D. B. Brinkman et al. (eds.), *Morphology and Evolution of Turtles*, Vertebrate Paleobiology and Paleoanthropology,
DOI: 10.1007/978-94-007-4309-0_4, © Springer Science+Business Media Dordrecht 2013

Introduction

The turtle shell is the major autapomorphy for the Order Testudines and is composed of dorsal and ventral moieties called the carapace and plastron, respectively. The carapace is basically made up of thoracic vertebrae and laterally expanded ribs, whereas the plastron is a composite of nine dermal bones containing elements that are thought to be homologous to the clavicle, interclavicle and gastralia (Gegenbaur 1898; Goodrich 1930; Romer 1956). The scapula of turtles has been thought to be situated inside the ribcage, in striking contrast to the typical body plan of tetrapods, in which the scapula is normally found outside (Ruckes 1929; Burke 1989, 1991; reviewed by Hall 1998; Rieppel 2001; Gilbert et al. 2001, 2008). Based on this unique topology, the turtle shell has been regarded as an evolutionary novelty (Hall 1998; Rieppel 2001; Gilbert et al. 2001, 2008). Saltatorial evolution has been assumed to apply to turtles because the scapula can only be situated outside or inside the ribcage; there are apparently no topological intermediates (Rieppel 2001; Theißen 2006, 2009). In fact, no fossils possessing an intermediate morphology have been found so far. Even *Proganochelys*, until the recent discovery of *Odontochelys* the oldest well known fossil turtle, had already developed a fully formed shell, thus providing few clues on the evolutionary background for this topological change (Gaffney 1990).

Despite this remarkable body plan, few studies have addressed the question of scapula position. From the anatomical point of view, Ogushi (1911) assumed that the scapula entered the ribcage from its rostral end during the evolution of turtles. Watson (1914) proposed that *Eunotosaurus africanus* from the middle Permian was the ancestor of turtles and that the scapula shifted backward into the ribcage through the course of turtle evolution. Ruckes (1929) compared the development of several species of turtles and concluded that the scapula does not migrate backward during development but remains in situ where it appears; Ruckes also concluded that turtles make a specialized dorsal dermis that ensnares the developing ribs and brings them dorsal to the scapula. Based on a histological observation of shoulder muscle development in the central painted terrapin *Chrysemys picta marginata*, Walker (1947) reported that shoulder development in turtles is similar to that of *Lacerta* and that the position of the pectoral girdle does not change during development. Expanding Ruckes' observation, Burke (1989, 1991) was the first to propose a hypothesis based on modern experimental embryology, in which the carapace was assumed to have a developmental mechanism similar to that seen in limb outgrowth (see below). This view is based on a developmental repatterning of the ribs, which in other animals grow ventrally to enclose the body cavity and internal organs—in contrast with the pattern in other tetrapods turtle ribs grow toward a more lateral and more superficial direction over the scapula (reviewed by Burke 2009).

In this review we first describe how ribs develop in model animals such as the chicken and the mouse. Based on these and our original data, we then evaluate the previous hypothesis. Finally, we propose a new interpretation, the 'folding theory', to explain this unique, turtle-specific body plan.

Rib Development

Ribs develop from somites, the segments of paraxial mesoderm that form in a rostral-caudal pattern on either side of the neural tube (Seno 1961; Pinot 1969; Sweeney and Watterson 1969; Christ et al. 1974; Christ and Wilting 1992; Huang et al. 1994, 1996, 2000b, c; Kato and Aoyama 1998; Evans 2003). The newly formed epithelial somite is composed of an epithelial outer layer that lines the somitocoele, in which central mesenchymal cells reside. The ventromedial part of the somite is later de-epithelialized and differentiates into a mesenchymal component called the sclerotome, from which the axial skeletal elements, including the vertebrae and ribs, develop (Christ and Wilting 1992; Huang et al. 1994, 1996, 2000b; Evans 2003; reviewed by Christ et al. 2004). The dorsolateral part of the somite retains an epithelial nature and is named the dermomyotome, the source of skeletal muscle, connective tissue, endothelium and cartilage (Huang et al. 2000a; reviewed by Scaal and Christ 2004; Christ and Scaal 2008). Cells in the margins of a dermomyotome de-epithelialize and translocate under it to differentiate into the myotome, consisting of primitive muscle fibers spanning the rostrocaudal extent of one somitic segment (Gros et al. 2004; reviewed by Christ and Ordahl 1995; Scaal and Christ 2004; Christ and Scaal 2008). The dorsomedial and ventrolateral lips of the dermomyotome contribute to the dorsal and ventral myotome, respectively, and the rostral and caudal borders intercalate into the intermediate domain of the myotome (Gros et al. 2004). Most of the dorsal myotome contributes to the epaxial muscles, whereas the ventral component forms the hypaxial muscles (Huang and Christ 2000; Olivera-Martinez et al. 2000).

Many studies of rib development have been performed using chicken embryogenesis. Avian ribs are composed of three parts, proximal, vertebrodistal and sternodistal ribs (Aoyama et al. 2005). The proximal rib of a chicken consists of the costal head, which articulates with two adjacent thoracic centra, and the costal neck and tubercle, which articulate with the transverse processes of the corresponding vertebra (Huang et al. 2000b). The proximal rib is likely to be derived from the medial part of an epithelial somite (Olivera-Martinez et al. 2000) and, in later development, from the caudal ventromedial part of a sclerotome (Huang et al. 1994, 1996, 2000b; Aoyama and Asamoto 2000; Evans 2003). The later domain is marked by the expression of *Pax1*, the gene for a paired box-containing transcriptional factor and *Uncx4.1* (Dietrich et al. 1993, 1997; Koseki et al. 1993; Wallin et al. 1994; Dietrich and Gruss 1995; Peters et al. 1995). Consistent with cell lineage analyses, in mice bearing mutations either in *Pax1* (Koseki et al. 1993; Wallin et al. 1994), *Uncx4.1* (Leitges et al. 2000; Mansouri et al. 2000) or *Sonic hedgehog* (*Shh*) emanating from the notochord and required for differentiation of the sclerotome (Chiang et al. 1996), proximal ribs are either reduced or totally missing, whereas distal ribs exhibit rather normal morphology. Chicken embryos with a barrier placed between notochord and somites showed elimination of the proximal ribs; this phenotype was rescued by implanting Shh-producing cells medial to the somites (Aoyama et al. 2005).

Distal ribs are derived from the lateral parts of epithelial somites (Olivera-Martinez et al. 2000). These elements appear to be induced by myoblasts in the syndetome, which comprises intermyotomally positioned parts of the lateral sclerotome and serves as the common progenitor of ribs and tendons (Sensenig 1949; Christ and Wilting 1992; Huang et al. 2000b; Kato and Aoyama 1998; Aoyama and Asamoto 2000; Evans 2003; Brent et al. 2003, 2005; also see below). Somite-derived cells initially occupy the dorsomedial region of the embryonic body, called the axial domain. In the later development, ventrolateral parts of the dermomyotome, myotome and syndetome secondarily migrate ventrolaterally into the somatopleur, which is composed of lateral plate-derived somatic mesoderm and its overlying ectoderm. The ventrolateral domain of the embryonic body is termed the lateral body wall [Fig. 4.1, Gilbert 2010; also see Nagashima et al. (2007)]. The extending myotome and syndetome differentiate into the intercostal muscles and sternodistal ribs, respectively (Verbout 1985; Christ and Wilting 1992; Huang et al. 2000b, c; Brent et al. 2003; Evans 2003). Although this penetration process is crucial for sternodistal rib formation (Sweeney and Watterson 1969; Christ et al. 1983; Sudo et al. 2001; Wang et al. 2005), its mechanism has not been well understood. It has been suggested that bone morphogenetic proteins (BMPs) emanating from the somatopleure during penetration are required for the outgrowth (Sudo et al. 2001). The distribution pattern of progenitor cells of the distal ribs and myotome completely prefigures the spatial relationships between the vertebrodistal ribs, sternodistal ribs and intercostal muscles in the adult form, suggesting that the development of these elements is tightly coordinated (Fig. 4.1, Kato and Aoyama 1998; Huang et al. 2000b, c; Evans 2003).

An involvement of developing intercostal muscles in rib formation has been suggested from genetic studies using mice. In particular, knockout of *Myf5* leads to a delay in myotomal formation as well as severe defects in the distal ribs (Braun et al. 1992). *Myf5* is a member of the group of myogenic regulatory factors (MRFs), which also includes *MyoD* (also called *Myod1*), *Mrf4* (also called *Myf6* or *herculin*) and *myogenin* (also called *MyoG*). *Myf5* and *MyoD* are upstream factors functioning in the specification and maintenance of myoblasts, whereas *Mrf4* and *myogenin* are downstream factors conducting the differentiation of myoblasts into myocytes and myofibers (Braun et al. 1992; Rudnicki et al. 1992, 1993; Hasty et al. 1993; Nabeshima et al. 1993; Braun and Arnold 1995). Although *Myf5* and *MyoD* are redundant functionally, *Myf5* appears to control the development of intercostal muscles preferentially, whereas *MyoD* regulates limb muscle formation (Grass et al. 1996; Kablar et al. 1997). Consistently, the *MyoD*$^{-/-}$ mouse seems to have few defects in the ribs (Rudnicki et al. 1992). A deficiency of *Mrf4* also results in rib defects (Braun and Arnold 1995; Patapoutian et al. 1995; Zhang et al. 1995; reviewed by Olson et al. 1996), but these phenotypes are attributed to an accompanying decrease in the expression level of *Myf5*, probably caused by cis-regulatory interaction between *Myf5* and *Mrf4,* which are adjacent in their chromosome (Floß et al. 1996; Yoon et al. 1997). However, Kaul et al. (2000) proposed that it is not *Myf5* itself but unidentified genes linked to *Myf5* that cause the rib phenotype, because their *Myf5* mutants showed no malformation in rib morphology. Instead, Vinagre et al. (2010) have proposed a redundant function of *Myf5* and *Mrf4* in rib formation. As the last mutant mice bearing deficiencies in MRFs, *myogenin* mutants display moderate truncation of the ribs (Hasty et al. 1993).

Besides MRFs themselves, mutations in loci encoding upstream factors of MRFs also lead to rib malformation (for *Pax3* mutants, see Tremblay et al. 1998; Dickman et al. 1999; Henderson et al. 1999; for *Six1*$^{-/-}$ and *Six4*$^{-/-}$ double-knockout mice, see Grifone et al. 2005), suggesting that the myotome functions as the source of inductive, trophic and/or patterning interactions required for proper formation of the ribs, for which *Myf5* and *Mrf4* serve as key factors (Braun et al. 1994; Grass et al. 1996; Olson et al. 1996; Yoon et al. 1997). The nonmyogenic role of *Myf5* and *Mrf4* in rib formation appears to be directly regulated by *Hox* genes (Vinagre et al. 2010) and *Myf5/Mrf4*-expressing cells are suggested to convey their rib-forming information to the sclerotome through platelet-derived growth factors (Soriano 1997; Tallquist et al. 2000) and the fibroblast

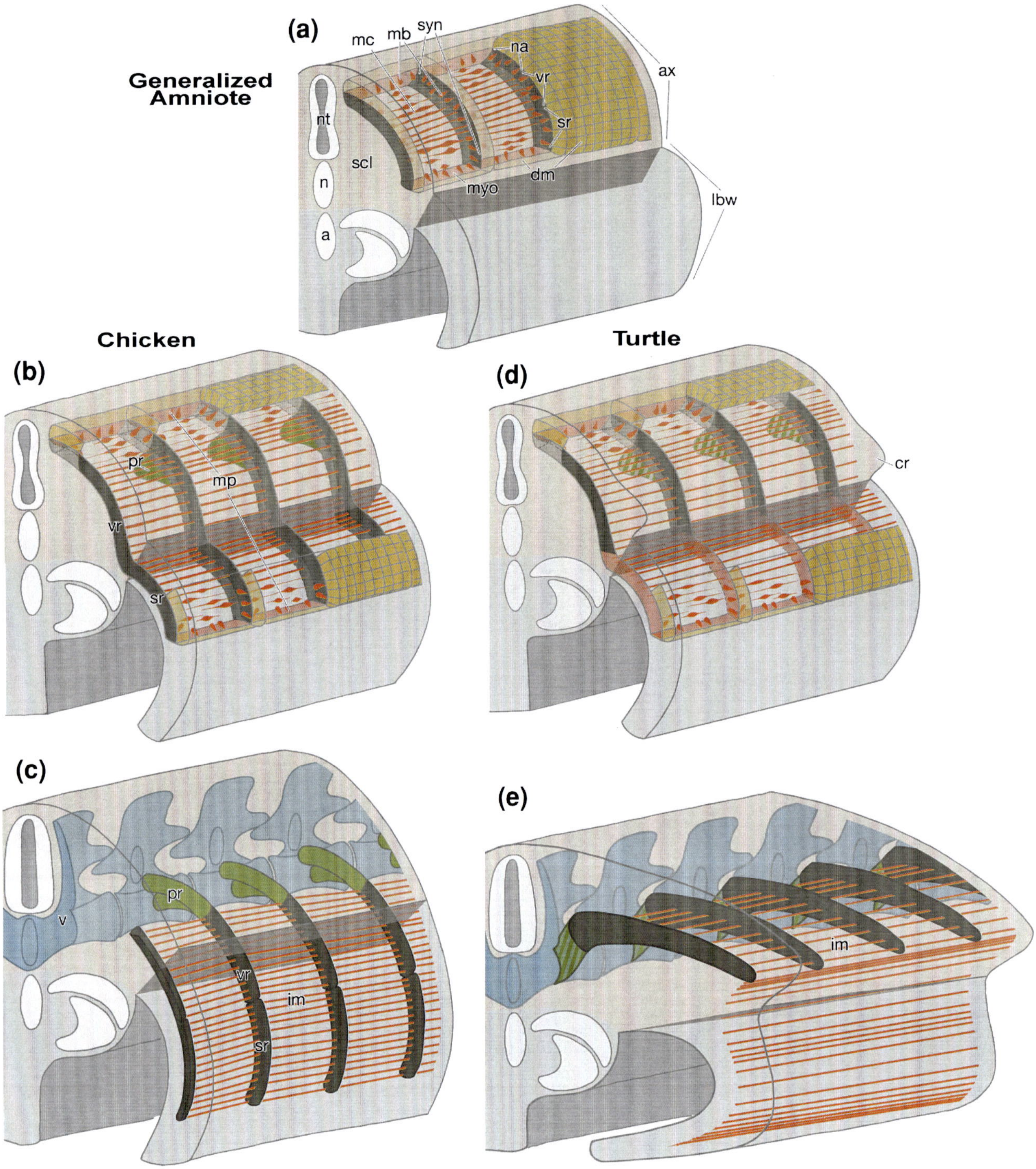

growth factor (FGF) signaling pathway (Patapoutian et al. 1995; Vinagre et al. 2010; also see Grass et al. 1996; Huang et al. 2003).

Under the positive influence of Shh, FGFs are expressed in the central myotome by direct regulation of *Myf5* (Fraidenraich et al. 1998, 2000). FGF signals are received by FGF receptor-expressing myoblasts in the rostral and caudal edges of myotomes. The myoblasts indirectly repress the expression of *Pax1* in the lateral sclerotome abutting the myoblasts and instead induce the expression of *Scleraxis* (*Scx*) via an unknown factor (Brent et al. 2003). The *Scx*-positive domain is defined as the syndetome and is thought

◀ **Fig. 4.1** Diagrams showing the developmental patterning of the trunk in chicken and turtle embryos. During early development (**a**), the embryonic body of amniotes can be divided into an axial domain (ax) and a lateral body wall (lbw), delineated by a boundary that corresponds to the longitudinal indentation appearing on the surface of the embryo. In the former domain, myoblasts (mb) appearing in the rostral and caudal edges of myotomes (myo) induce a syndetome (syn) in the intersomitic region (*green*). The medial and ventrolateral syndetomes comprise anlagen of the vertebrodistal rib (vr) and sternodistal rib (sr), respectively. As a result of these developmental mechanisms, two adjacent somites contribute to a single distal rib (reviewed by Evans 2003). At the middle stage of chicken development (**b**), the lateral body wall undergoes an invasion of somitic cells from which the sternodistal ribs develop. Myocytes form a muscle plate (mp) between two adjacent rib primordia. By the late developmental stage of the chicken embryo (**c**), ribs lie in both the axial and the lateral body wall. Here the morphological border is no longer obvious, but can be traced as a distinct cell lineage of the dermis (reviewed by Kuratani et al. 2011). At the middle stage of turtle development (**d**), rib anlagen appear in the intersomitic region, as in other amniotes, but they are arrested in the axial domain. The carapacial ridge (cr) develops in the ventrolateral edge of the axial domain. During late development of turtles (**e**), the ribs expand laterally in accordance with the lateral growth of the embryonic body. The boundary between the axial domain and the lateral body wall remains as the ventral edge of the carapacial ridge. In the top and middle panels, dermomyotomes (dm) are removed from the anterior two segments to show internal structures. Other abbreviations: *a* dorsal aorta; *im* intercostal muscles; *mc* myocyte: *n* notochord; *na* neural arch anlage; *nt* neural tube; *pr* proximal rib; *scl* sclerotome; *v* vertebra

to comprise the progenitor cells of distal ribs, as noted above (Brent et al. 2005).

Taken together, muscle-dependent induction of the distal rib and precommitment of the adult musculoskeletal pattern in early somites imply that patterning of the distal ribs in the embryonic body also largely depends on the myotomes. This view is consistent with a classical comparative anatomical idea in which ribs were assumed to have evolved in the myoseptum secondarily as attachments of trunk muscles (Goodrich 1930; Romer and Parsons 1977).

Analogy Between Limb Formation and Carapace Development

Burke (1989) examined the similarity of histological configuration between the distal tip of the limb bud and peripheral edge of the developing carapace, or the carapacial ridge (CR). The distal margin of the limb bud consists of a thickened epidermis called the apical ectodermal ridge (AER; Saunders 1948) and the underlying condensed mesenchyme termed the progress zone (Summerbell et al. 1973). This part of the limb bud plays a pivotal role in the outgrowth and proximodistal patterning of the limb (reviewed by Tabin and Wolpert 2007 and references therein). In fact, elimination of the AER results in truncation of limb skeletal elements (Saunders 1948), and the ectopic implantation of prospective limb mesenchyme into the flank induces an extra limb through formation of an ectopic AER (Saunders and Reuss 1974). These results indicate that the distal tip of the limb bud is the site of epithelio-mesenchymal interactions. In the initial induction of a limb bud, FGF10 secreted from the mesenchyme stimulates FGF8 production in the epidermis, which in turn induces expression of *Fgf10* in the mesenchyme (Crossley et al. 1996; Ohuchi et al. 1997). In these processes, transcriptional factor *Msx*, BMP growth factors, Shh signaling and Wnt signaling are also involved (Saunders and Gasseling 1968; Yokouchi et al. 1991; Ros et al. 1992; Pizette et al. 2001; Khokha et al. 2003; reviewed by Fernandez-Teran and Ros 2008). These intricate reciprocal interactions between the epidermis and mesenchyme maintain the underlying mesenchymal cells in an undifferentiated proliferative state, leading to an outgrowth of the limb (reviewed by Tabin and Wolpert 2007).

The CR appears as a longitudinal ridge on the flank of turtle embryos at the late pharyngula stage (Burke 1989). As an analogy with limb development, the CR has been assumed to induce specific patterning of rib precursor cells to place them in a dorsolateral and superficial position, which normally should migrate more ventrally into the lateral body wall (Burke 1989, 1991). As a result, the position of turtle ribs were explained as growing over the scapula, which has been suggested to develop essentially in the same way as those of other tetrapods (Ruckes 1929; Walker 1947; Burke 1989, 1991). This hypothesis was supported by the distribution of some molecules commonly found in the limb bud (fibronectin and neural cell adhesion molecule), detection of active cell proliferation in the mesenchyme and extirpation of the CR that resulted in redirection of the rib growth trajectory (Burke 1989, 1991; reviewed by Burke 2009).

This hypothesis triggered investigations of molecules commonly expressed in the CR and the limb bud. For example, expressions of *Fgf10* in the CR mesenchyme (Loredo et al. 2001) and *Fgf8* in the distal tip of the rib primordium (Cebra-Thomas et al. 2005) were observed in the red-eared slider turtle *Trachemys scripta*. Expression of *Msx* was found in the CR mesenchyme and epidermis in the European pond turtle *Emys orbicularis*, although these expressions were restricted to the CR near the limb bud (Vincent et al. 2003). Expressions of *Msx2* and *Shh* in the epidermis and of *Bmp4* and BMP antagonist *Gremlin* in the mesenchyme were reported in the late stage of CR formation in *T. scripta* (Moustakas 2008). However, these genes are not expressed in the CR of the Chinese soft-shelled turtle *Pelodiscus sinensis*, at least at Tokita-Kuratani (TK)

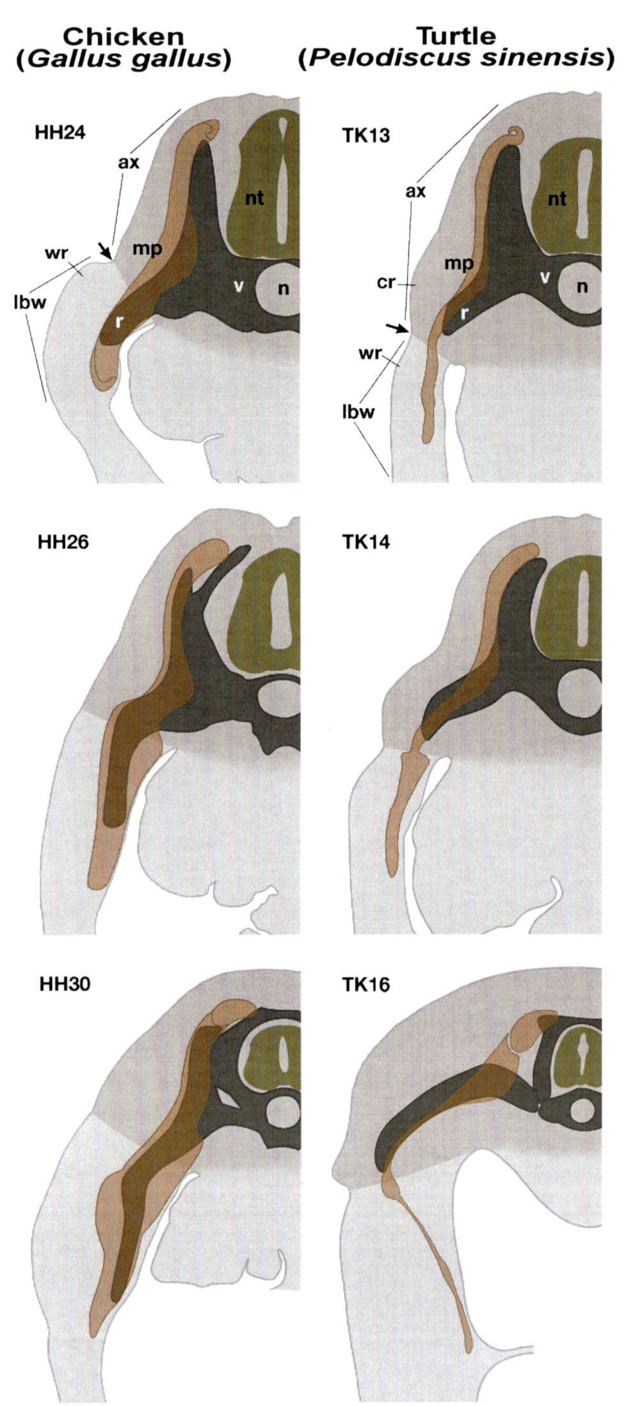

Fig. 4.2 Comparison of rib development at three stages of embryonic development in the chicken *Gallus gallus* (*left*) and the turtle *Pelodiscus sinensis* (*right*). Serial transverse sections stained with hematoxylin/eosin/alcian blue were reconstructed two-dimensionally to compare rib development in each animal. *Arrows* indicate the boundary between the lateral body wall (lbw) and the axial part (ax) of the embryonic body. A cartilaginous mesenchymal aggregation appears from HH stage 24 (Hamburger and Hamilton 1951) for the chicken and TK stage 13 (Tokita and Kuratani 2001) for *P. sinensis*. Note that in both animals, ribs (r) grow along the muscle plate (mp) and that chicken ribs are already penetrating the lateral body wall by HH stage 24. Other abbreviations: *cr* carapacial ridge; *n* notochord; *nt* neural tube; *v* vertebra; *wr* Wolffian ridge

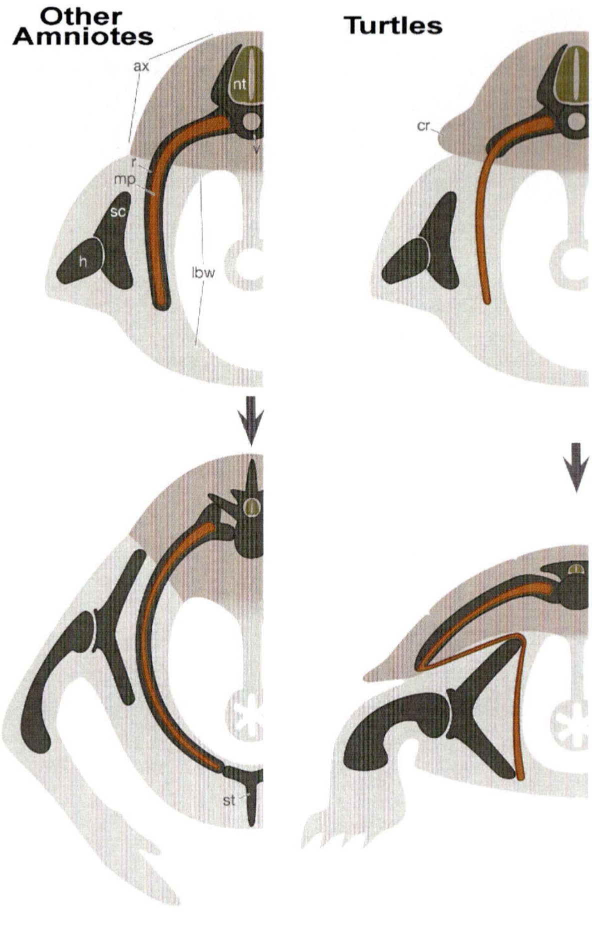

Fig. 4.3 Schematic, transverse diagrams of generalized amniotes (*left*) and turtles (*right*) showing developmental changes in spatial relationships among ribs, pectoral girdles, and associated structures. Both groups develop from nearly identical embryonic morphologies (upper pair of diagrams), except the ribs (r) of turtle embryos are arrested in the axial domain (ax) and never invade the lateral body wall (lbw). Later in the turtle development (*bottom right*), marginal growth of axial domain brings the lateral extension of ribs, inward folding of the dorsal part of the lateral body wall, and covering of the scapula (sc) by the ribs. This growth does not alter topological relationships between the ribs, the scapula and the muscle plate (mp). Other abbreviations: *cr* carapacial ridge; *h* humerus; *nt* neural tube; *st* sternum; *v* vertebra

stage 13–15 (Tokita and Kuratani 2001; Kuraku et al. 2005), when the CR and cell aggregations of the ribs appear (Fig. 4.2). Rather, based on comprehensive cDNA screening, Kuraku et al. (2005) identified four other genes specifically expressed in the CR, including *cellular retinoic acid-binding protein (Crabp)-I*, *Sp-5*, *lymphocyte enhancer factor (Lef)-1* and *Apcdd-1*, which, with the exception of *Crabp-I*, were also expressed in the turtle limb bud. Most of the orthologs of these four genes are also expressed in the limb buds of chicken and mouse embryos (Kuraku et al. 2005 and references therein). *Sp-5, Apcdd-1* and *Lef-1* are components of the canonical Wnt pathway (Takahashi et al. 2002, 2005; Weidinger et al. 2005; Shimomura et al. 2010).

And β-catenin is localized in nuclei of the CR epidermis (Kuraku et al. 2005), implying that the Lef-1/β-catenin complex functions as a transcriptional activator in CR development (reviewed by Novak and Dedhar 1999). In fact, inhibition of Lef-1 activity by overexpression of *Lef-1* lacking the β-catenin-binding site leads to arrested development of the carapacial margin (Nagashima et al. 2007).

However, no Wnt molecules have been identified upstream of the above signaling; no *Wnt* gene family expressions have been confirmed in the CR or its adjacent domain (unpublished observations by S. Kuraku and S. Kuratani), unlike the situation in the limb bud in which *Wnt3/Wnt3a and Wnt7a* are expressed (reviewed by Fernandez-Teran and Ros 2008). Instead, a role for hepatocyte growth factor (HGF) is assumed. In *P. sinensis*, *HGF* was uniquely expressed in the lateral sclerotome and somatic mesoderm during initiation of the CR but no similar expression has been confirmed in chicken or mouse embryos (Kawashima-Ohya et al. 2011). Application of an HGF antagonist under the ventrolateral dermomyotome inhibited CR formation, including the CR-specific gene expressions. In this connection, some studies have unveiled interactions between canonical Wnt signaling and an HGF cascade in a carcinogenic context; independently of Wnts, tyrosine phosphorylation of receptor Met by its ligand HGF induces accumulation of cytoplasmic β-catenin, leading to TCF/LEF-mediated gene transcription (Danilkovitch-Miagkova et al. 2001; Monga et al. 2002; Rasola et al. 2007; reviewed by Nelson and Nusse 2004). Although *Crabp-I* does not seem to be a component of the Wnt signaling cascade, expression of *Crabps* is reported to be modulated indirectly by the Wnt/β-catenin pathway (Collins and Watt 2008). Such secondary effects appear to play roles in gene induction in the CR as well. In the limb bud, HGF is also expressed in the mesenchyme but it is unlikely that HGF is involved in outgrowth of the limb bud, because a *Met*-deficient mouse developed limbs with skeletons: only limb muscle formation was impaired (Bladt et al. 1995; also see below). These gene expression patterns and their functions in the CR do not completely overlap with those in he limb bud (Kuraku et al. 2005; Kawashima-Ohya et al. 2011).

The CR develops in the most ventrolateral margin of the axial domain, as shown by histology (Burke 1989) and by DiI labeling of mesodermal cells (Figs. 4.1, 4.2, Nagashima et al. 2007), which suggests that turtle ribs are relatively shorter than those of other amniotes from morphological viewpoints. Whereas the ribs of other amniotes are patterned in both the axial domain and the lateral body wall, turtle ribs grow toward the CR, so they never penetrate the lateral body wall. This indicates that turtle ribs lack the ventral component, even if they appear to elongate laterally (reviewed by Kuratani et al. 2011).

Progenitor cells of the distalmost part of the ribs—the sternodistal regions—are thought to be contained in the somitic cell stream invading the lateral body wall (Christ and Ordahl 1995; Müller et al. 1996; Sudo et al. 2001). As an analogy to limb development, Cebra-Thomas et al. (2005) proposed a 'paracrine hypothesis', in which interaction between FGF10 in the CR mesenchyme and FGF8 in the tip of the rib anlage is assumed to bring about directed growth of turtle ribs to the CR (reviewed by Gilbert et al. 2008, Gilbert 2010). This hypothesis is based on experiments in which inhibiting the activity of FGFs in turtle embryos led not only to the degeneration of the CR but also to relocation of the ribs, which then grew into the lateral body wall as they do in other amniotes. By contrast, implantation of FGF10-coated beads between somites and the somatopleure in Hamburger-Hamilton (HH) stage 26 chick embryos (Hamburger and Hamilton 1951) inhibited growth of ribs into the somatopleure; instead, ribs were directed to the beads and stopped there (Cebra-Thomas et al. 2005). However, somite-derived cells start to migrate into the lateral body wall from HH stage 22 in the chicken embryo (Brent et al. 2003; Nagashima et al. 2005) and the distal rib anlagen can be observed in the lateral body wall of HH stages 24–26 (Fig. 4.2, Brent et al. 2003). From this observation, it seems to be too late to prohibit rib precursor cells from penetrating the somatopleure at HH stage 26 in chicken embryos. Moreover, when the CR was eliminated surgically in turtle embryos, only the turtle-specific flabellate and rostrocaudal pattern was impaired, and the ribs were arrested in the axial domain as in normal development, never invading the lateral body wall (Burke 1991; Nagashima et al. 2007). Likewise, the ectopically implanted CR could self-differentiate into an ectopic process on the turtle's back with a CR-specific histological configuration and gene expressions, but the ribs did not grow toward this structure (Nagashima et al. 2007). These results show that the CR does not have the ability to attract rib precursor cells; rather, it functions to form a fanned-out pattern of ribs through active growth at the margin of the carapacial primordium (Nagashima et al. 2007).

These results throw doubt on the analogy between limb formation and carapacial development. In fact, the developmental modes of the ribs and the limb skeleton are quite different from each other in terms of their dependence on muscle development. As stated previously, although rib formation depends entirely on muscle development, the limb skeleton can develop without muscles. Limbs develop as outgrowths of the lateral body wall and their skeletal elements, tendons and connective tissues develop from the somatic mesoderm (Shellswell and Wolpert 1977; Chevallier 1979; Christ et al. 1979, 1983; Kieny and Chevallier 1979; reviewed by Burke 2000; Winslow et al. 2007). By contrast, the limb muscles originate from somites. Muscle-free

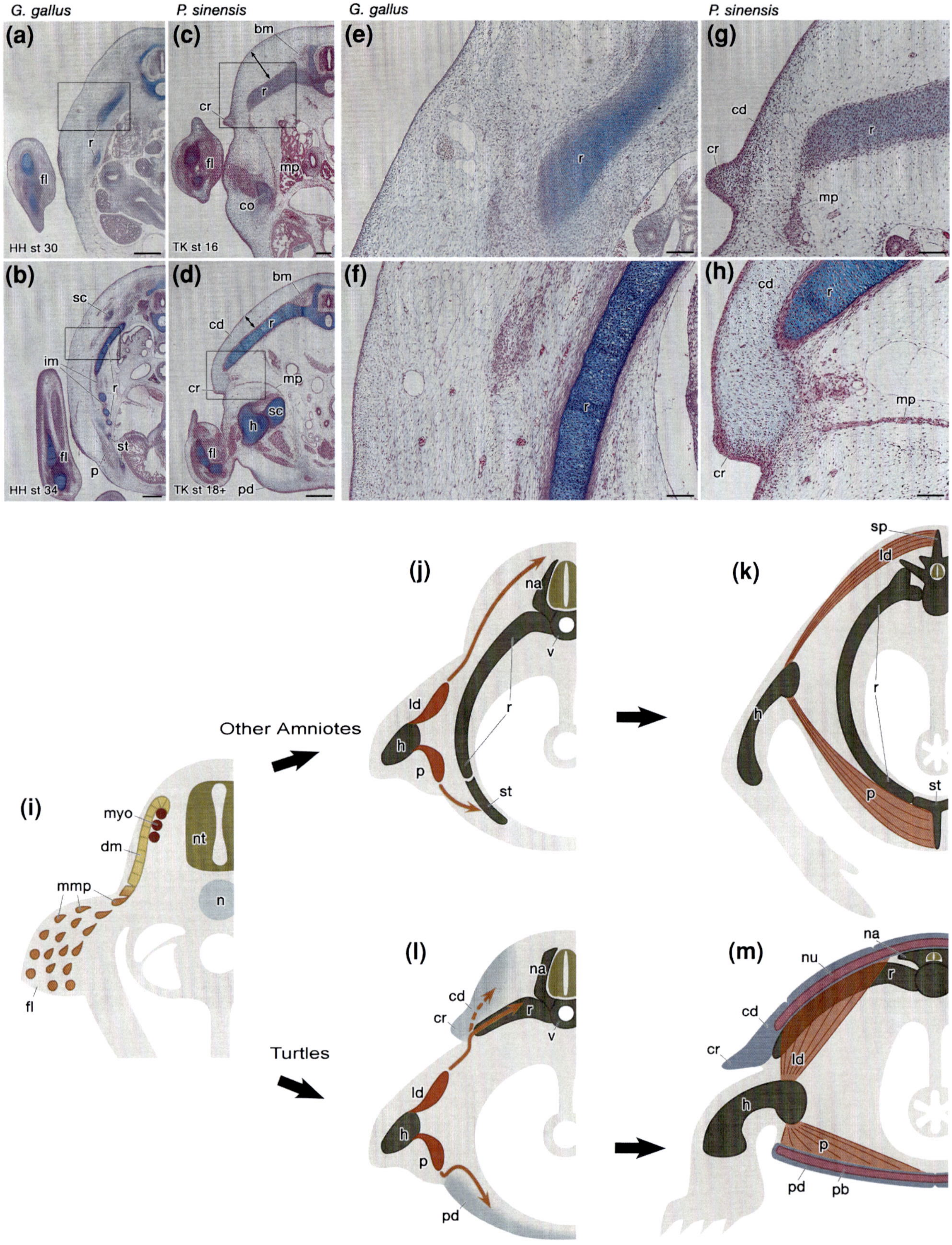
G. gallus
P. sinensis
G. gallus
P. sinensis
(a)
r
fl
HH st 30
(b)
sc
im
r
fl
st
p
HH st 34
(c)
bm
cr
r
fl
mp
co
TK st 16
(d)
bm
cd
r
cr
mp
sc
h
fl
TK st 18+
pd
(e)
r
(f)
r
(g)
cd
cr
r
mp
(h)
cd
r
mp
cr
(i)
myo
dm
mmp
fl
nt
n
Other Amniotes
Turtles
(j)
na
v
ld
r
h
p
st
(k)
sp
ld
r
h
p
st
(l)
na
cd
cr
r
v
ld
h
p
pd
(m)
na
nu
r
cd
cr
ld
h
p
pd
pb

◀ **Fig. 4.4** Development of carapacial and plastral dermis and limb muscles. **a–d** Comparison of dermal development in embryos of the chicken *Gallus gallus* (**a** and **b**) and the turtle *Pelodiscus sinensis* (**c** and **d**) Transverse sections of the rostral flank are shown for both animals. In the chicken embryo, the dermis shows little change in its morphology during development (**a** and **b**), whereas in turtles the dermis between the ribs (r) and epidermis shrinks secondarily and makes a carapacial dermis (cd) during late development (**c** and **d**). At the same time, the plastral dermis (pd) appears in the ventralmost lateral body wall (**d**). Note that the turtle ribs come to lie secondarily close to the epidermis. This is caused by shrinkage of the dermis (*arrows* in **c** and **d**), not by any change in the direction of growth of the ribs. **e–h**, Higher magnifications of boxes in **a–d**, respectively, showing the change in morphology of the dermis during late development. Alcian-blue-positive fibrous carapacial dermis begins to form from TK stage 16 onward and makes a clear boundary with the other dermis (**g** and **h**). This type of dermal arrangement cannot be observed in the chicken embryo (**e** and **f**). **i–m** Schematic drawings of development of the latissimus dorsi and pectoralis muscles. In the early embryological development of amniotes (**i**), migratory muscle precursors (mmp) migrate from the ventrolateral aspect of the dermomyotome (dm) into the forelimb bud (fl) to differentiate into mature myoblasts. In amniotes other than turtles (**j** and **k**), primordia of the latissimus dorsi (ld) and pectoralis (p) muscles grow out of the limb bud to establish their attachment to the trunk. These anlagen grow outside the ribs and sternum (*arrows* in **j**). In turtles (**l** and **m**), the anlage of the latissimus dorsi is presumed to be prevented from growing between the ribs and the epidermis because of the formation of a carapacial dermis (broken arrow in **L**). As a result, the primordium attaches to the inside of the nuchal bone (nu) developing in the carapacial dermis. The pectoralis muscle anlage attaches inside the plastral dermis, because of the formation of plastral dermis and absence of sternum in turtles. Scale bars: 500 μm for **a–d**; 100 μm for **e** and **h**. Other abbreviations: *bm* back muscles; *co* coracoid; *cr* carapacial ridge; *h* humerus; *im* intercostal muscles; *mp* muscle plate; *myo* myotome; *n* notochord; *na* neural arch; *nt* neural tube; *pb* plastral bone; *sc* scapula; *sp* spinous process; *st* sternum; *v* vertebral body

chicken limbs, created by implanting initial limb buds into the coelomic cavity, can still form limb skeletons (Kardon 1998; Kardon et al. 2003). Moreover, *Pax3* mutant, $Met^{-/-}$, $Six1^{-/-}$ and $Six4^{-/-}$ double-knockout mice fail to form limb muscles, but they do develop limb skeletons (Bladt et al. 1995; Tremblay et al. 1998; Grifone et al. 2005).

Turtle Ribs and Body Plan

Using somite extirpation experiments, Yntema (1970) showed that the turtle rib progenitors originate from the lateral part of somites, like the distal ribs of chicken (Olivera-Martinez et al. 2000). In addition, as in other amniotes, the turtle ribs develop between adjacent muscle plates or intercostal muscles (Nagashima et al. 2009), which disappear secondarily in later development.

Unique features of the turtle muscle plate appear to be one of causal factors leading to relatively short ribs. The turtle muscle plate shows a meager, thin, thread-like morphology, which is especially conspicuous in the lateral body wall (Nagashima et al. 2005). This implies that loss of rib-inducing capability in the muscle plate would result in the axial arrest of turtle rib growth. In *P. sinensis,* although the expression patterns of genes related to muscle plate development, such as *MRFs* and their upstream factors *Pax3* and *Pax7,* are similar to those in chicken (Kawashima-Ohya et al. 2011), one gene—*Myf5*—has some unusual characters. Deletion of 12 sequential nucleotides is found in *Myf5* both in the Cryptodira and in the Pleurodira (Ohya et al. 2006). This deletion seems to have occurred in evolution before the emergence of the Casichelydia (Cryptodira and Pleurodira), because it is not found in other vertebrates. The deletion site corresponds to the transactivation domain of *Myf5* (Winter et al. 1992), implying a change in the myogenic activity of turtle *Myf5*. Moreover, two kinds of splicing variants of *Myf5* are found in *P. sinensis* and *T. scripta*. The short form lacks 76 base pairs in a contiguous sequence, leading to a frameshift in the C-terminal. Therefore, this region encodes a completely different amino acid sequence from the long isoform. The short form appears to have less myogenic activity possibly functioning as a dominant negative form against the long one. These peculiar characteristics in turtle *Myf5* could explain the poor development of the muscle plate in turtle embryos (Ohya et al. 2006; also see Nagashima et al. 2009). Furthermore, it is probable that the unusual features of *Myf5* would cause truncation of the ribs, because *Myf5* has been suggested to be a key factor in rib development (Braun et al. 1992; Braun and Arnold 1995; Patapoutian et al. 1995; Zhang et al. 1995; Grass et al. 1996; Yoon et al. 1997).

Given the strong dependence of rib formation on muscle development, it is quite natural that in turtles the ribs develop along the muscle plates as do those in other amniotes, even though turtle ribs appear to erupt secondarily on the lateral aspect of the muscle plate in a late phase of development (Figs. 4.1, 4.2, Nagashima et al. 2009). In this interpretation, turtle ribs do *not* take a different route from that seen in other amniotes, unlike former assumptions (Ruckes 1929; Burke 1989, 1991; Cebra-Thomas et al. 2007; Gilbert et al. 2008). In other words, the muscle plate indicates indirectly where the ribs should be patterned, implying that the muscle plate can be regarded as a 'prospective ribcage'. Turtles do not have ribs in the lateral body wall but they do have muscle plates there. Importantly, the muscle structure is situated inside the pectoral girdle in turtles as in other amniotes, suggesting that, if turtles were to develop longer ribs in the lateral body wall, those ribs would grow along the muscle plate and could be found inside the girdle. From this perspective, the body plan and topological relationship between ribs and the girdle are completely conserved even in turtles. The positional change between the skeletal elements is achieved by folding at a hinge between the axial part and the lateral body wall (Fig. 4.3, Nagashima et al. 2009; reviewed by Kuratani et al. 2011). One cause of the folding is the marginal growth of the embryonic body caused by the CR. This growth not only results in the fanned-out arrangement of ribs along the rostrocaudal axis (Nagashima et al. 2007) but also makes the ribs overhang the scapula dorsocaudally (Nagashima et al. 2009). Therefore, one of the true evolutionary novelties in turtles is not topological changes between bony elements but the folding of the body wall mainly caused by the CR (reviewed by Kuratani et al. 2011).

There is another kind of evolutionary novelty in turtles. Muscles in the shoulder/forelimb region can be classified into two categories. The shoulder muscles that connect the shoulder girdle to the axial skeleton include the serratus anterior (AS), levator scapulae-rhomboid muscle complex (LSR). This group arises as a direct derivative of the muscle plate (Romer and Parsons 1977; Nagashima et al. 2009; Valasek et al. 2010). The limb muscles in the second group bind the arm to the trunk; the latissimus dorsi (LD) and pectoralis muscle are included in this category. This muscle group develops through migratory muscle precursors (MMPs), which delaminate from the ventrolateral part of the dermomyotome and migrate for a long distance toward the forelimb bud, where they differentiate into myocytes (reviewed by Birchmeier and Brohmann 2000; Vasyutina and Birchmeier 2006; also see Kusakabe and Kuratani 2005, 2007). In fact, in loss-of-function mutant mice for *Met*—a gene that is indispensable for delamination of MMPs from the dermomyotome—limb muscles were lost completely, whereas the shoulder muscles were preserved (Valasek et al. 2010). In turtles, limb muscles establish turtle-specific attachments to the skeleton, whereas the shoulder muscles tend to conserve the same musculoskeletal connections as those in other amniotes (Nagashima et al. 2009; reviewed by Kuratani et al. 2011).

This difference could be attributed to the timing of the connection and establishment of the developmental modules (see a review by Kuratani 2009). For instance, although the serratus anterior muscle in adult turtles runs between the scapula and inner face of the second rib, the muscle anlage connects the skeletal elements rostrocaudally at the beginning of its development as in other amniotes. However, in later development, the anlage is folded secondarily inside the ribs because the girdle is covered by the ribs. During this process, the connectivity and the morphological homology is completely conserved. The LSR shows the shared musculoskeletal connectivity with other amniotes as well. On the other hand, the LD and the pectoralis muscle never attach to the outside of the axial skeleton, unlike those in other amniotes, but to the inside of the dermal bones of carapace and plastron (Nagashima et al. 2009; reviewed by Kuratani et al. 2011). In turtles, these muscles begin to develop in the limb bud in the same way as those in other amniotes, allowing homologization to their counterparts in other amniotes (Fig. 4.4i; Walker 1947; Nagashima et al. 2009). Later in development, the muscle primordia have to grow from the forelimb bud into the trunk to make their attachments to the trunk. The muscle anlagen in other amniotes spread over the axial bones and anchor on the outside of them (Fig. 4.4i–k). However, in turtles, when these primordia emerge from the forelimb bud, highly dense dermal regions called the 'carapacial dermis' and 'plastral dermis' appear specifically in the trunk. These specialized dermal regions are characterized by their alcian blue-positive fibrous extracellular matrix (Fig. 4.4g, h). The carapacial dermis occupies the region between the ribs and epidermis and appears to prevent the LD from spreading over the ribs (Fig. 4.4l, m). Likewise, the pectoralis muscle appears to be unable to penetrate the plastral dermis. As the result, these muscle anlagen appear to be constrained to occupy its position inside the dermal bones, which develop in the carapacial and plastral dermis regions (Fig. 4.4l, m). Therefore, it appears that turtle specific muscle attachments are established by utilizing the turtle specific developmental module of the trunk.

As to the developmental origin of the trunk dermal bones in turtles, a contribution of trunk neural crest cells has been proposed, based on the expression of 'marker' genes and their proteins (Clark et al. 2001; Cebra-Thomas et al. 2007; Gilbert et al. 2007; reviewed by Gilbert et al. 2008). However, Kuratani et al. (2011) draw cautionary attention to this simplistic conclusion. Gene and protein expression patterns can only tell us the physiological characteristics of the cells and can never prove cell lineage. For example, Cebra-Thomas et al. (2007) reported that neural crest cells contributed to a part of the vertebrae and ribs of turtles (reviewed by Gilbert et al. 2008). This was based on immunohistochemistry for HNK-1, which has been shown to colocalize with early migrating crest cells in chicken and rat embryos (Rickmann and Fawcett 1985 and references therein). However, these structures are derived solely from the somites in chicken embryos (reviewed by Christ et al. 2004). In our experiments using *P. sinensis*, neither vertebrae, ribs nor plastral bones were immunostained for HNK-1 but only peripheral nerves, which is the same situation as in older chicken embryos (Kuratani et al. 2011). A part of the anterior element of plastral bones—epiplastron—could be derived from cephalic neural crest cells, because it is homologous to the clavicle in other amniotes (Gegenbaur 1898; Goodrich 1930; Romer 1956) and the anterior part of clavicle is comprised of the cephalic neural crest cells in mice (Matsuoka et al. 2005). However, only cell lineage analysis using proper labeling methods can demonstrate neural crest contribution in turtles.

Conclusions

From the above discussion, the turtle body plan is not as radical as has been assumed to date, implying that turtle evolution does not necessarily have to be postulated as saltatory. Indeed, a recently discovered fossil species, *Odontochelys*, shows an intermediate morphology in that it did not have a complete carapace whereas a plastron had already been formed (Li et al. 2008). In particular, the scapula was in a position rostral to the ribs, which were already arrested axially but did not show a fanned-out pattern. This morphology is quite similar to that of young turtle embryos that have not yet undergone body wall folding (Sánchez-Villagra et al. 2009), suggesting that *Odontochelys* had not yet acquired this developmental process (Nagashima et al. 2009). On the other hand, *Proganochelys* had a complete shell with a carapace (Gaffney 1990). Therefore, it is most likely that in the common ancestor of *Odontochelys* and *Proganochelys*, the ribs ceased penetrating the lateral body wall by unknown causes. A function for *Myf5* might be implicated in this process. Alternatively, such causes might relate to formation of the CR because both of these phenomena occur at the boundary between the axial domain and the lateral body wall. At least, we can state that after divergence of the ancestor of *Odontochelys* from the ancestor of *Proganochelys*, formation of the CR was completed with acquisition of an HGF expression domain in the later lineage. This would have resulted in marginal growth of the embryonic body leading to a flabellate pattern of the ribs and encapsulation of the scapula, which enabled carapace formation.

Acknowledgments We thank Dr. Don Brinkman for arranging the Gaffney Turtle Symposium and Festschrift publication in honor of Dr. Eugene S. Gaffney. We acknowledge Dr. Robert Carroll, Dr. Eugene S. Gaffney, and Dr. Olivier Rieppel for critical reading of the manuscript.

References

Aoyama, H., & Asamoto, K. (2000). The developmental fate of the rostral/caudal half of a somite for vertebra and rib formation: Experimental confirmation of the resegmentation theory using chick-quail chimeras. *Mechanisms of Development, 99,* 71–82.

Aoyama, H., Mizutani-Koseki, S., & Koseki, H. (2005). Three developmental compartments involved in rib formation. *International Journal of Developmental Biology, 49,* 325–333.

Birchmeier, C., & Brohmann, H. (2000). Genes that control the development of migrating muscle precursor cells. *Current Opinion in Cell Biology, 12,* 725–730.

Bladt, F., Riethmacher, D., Isenmann, S., Aguzzi, A., & Birchmeier, C. (1995). Essential role for the c-met receptor in the migration of myogenic precursor cells into the limb bud. *Nature, 376,* 768–771.

Braun, T., & Arnold, H. H. (1995). Inactivation of Myf-6 and Myf-5 genes in mice leads to alterations in skeletal muscle development. *EMBO Journal, 14,* 1176–1186.

Braun, T., Rudnicki, M. A., Arnold, H. H., & Jaenisch, R. (1992). Targeted inactivation of the muscle regulatory gene Myf-5 results in abnormal rib development and perinatal death. *Cell, 71,* 369–382.

Braun, T., Bober, E., Rudnicki, M. A., Jaenisch, R., & Arnold, H. H. (1994). MyoD expression marks the onset of skeletal myogenesis in *Myf-5* mutant mice. *Development, 120,* 3083–3092.

Brent, A. E., Schweitzer, R., & Tabin, C. J. (2003). A somitic compartment of tendon progenitors. *Cell, 18,* 235–248.

Brent, A. E., Braun, T., & Tabin, C. J. (2005). Genetic analysis of interactions between the somitic muscle, cartilage and tendon cell lineages during mouse development. *Development, 132,* 515–528.

Burke, A. C. (1989). Development of the turtle carapace: Implications for the evolution of a novel bauplan. *Journal of Morphology, 199,* 363–378.

Burke, A. C. (1991). The development and evolution of the turtle body plan. Inferring intrinsic aspects of the evolutionary process from experimental embryology. *American Zoologist, 31,* 616–627.

Burke, A. C. (2000). *Hox* genes and the global patterning of the somitic mesoderm. *Current Topics in Developmental Biology, 47,* 155–181.

Burke, A. C. (2009). Turtles…again. *Evolution & Development, 11,* 622–624.

Cebra-Thomas, J., Tan, F., Sistla, S., Estes, E., Bender, G., Kim, C., et al. (2005). How the turtle forms its shell: A paracrine hypothesis of carapace formation. *Journal of Experimental Zoology, 304B,* 558–569.

Cebra-Thomas, J. A., Betters, E., Yin, M., Plafkin, C., McDow, K., & Gilbert, S. F. (2007). Evidence that a late-emerging population of trunk neural crest cells forms the plastron bones in the turtle *Trachemys scripta. Evolution & Development, 9,* 267–277.

Chevallier, A. (1979). Role of the somitic mesoderm in the development of the thorax in bird embryos. II. Origin of thoracic and appendicular musculature. *Journal of Embryology & Experimental Morphology, 49,* 73–88.

Chiang, C., Litingtung, Y., Lee, E., Young, K. E., Corden, J. L., Westphal, H., et al. (1996). Cyclopia and defective axial patterning in mice lacking Sonic hedgehog gene function. *Nature, 383,* 407–413.

Christ, B., & Ordahl, C. P. (1995). Early stages of chick somite development. *Anatomy and Embryology, 191,* 381–396.

Christ, B., & Scaal, M. (2008). Formation and differentiation of avian somite derivatives. *Advances in Experimental Medicine and Biology, 638,* 1–41.

Christ, B., & Wilting, J. (1992). From somites to vertebral column. *Annals of Anatomy, 174,* 23–32.

Christ, B., Jacob, H. J., & Jacob, M. (1974). Origin of wing musculature. Experimental studies on quail and chick embryos. *Experientia, 30,* 1446–1449.

Christ, B., Jacob, H. J., & Jacob, M. (1979). Differentiating abilities of avian somatopleural mesoderm. *Experientia, 35,* 1376–1378.

Christ, B., Jacob, M., & Jacob, H. J. (1983). On the origin and development of the ventrolateral abdominal muscles in the avian embryo: An experimental and ultrastructural study. *Anatomy and Embryology, 166,* 87–101.

Christ, B., Huang, R., & Scaal, M. (2004). Formation and differentiation of the avian sclerotome. *Anatomy and Embryology, 208,* 333–350.

Clark, K., Bender, G., Murray, B. P., Panfilio, K., Cook, S., Davis, R., et al. (2001). Evidence for the neural crest origin of turtle plastron bones. *Genesis, 31,* 111–117.

Collins, C. A., & Watt, F. M. (2008). Dynamic regulation of retinoic acid-binding proteins in developing, adult and neoplastic skin reveals roles for β-catenin and Notch signalling. *Developmental Biology, 324,* 55–67.

Crossley, P. H., Minowada, G., MacArthur, C. A., & Martin, G. R. (1996). Roles for FGF8 in the induction, initiation, and maintenance of chick limb development. *Cell, 84*, 127–136.

Danilkovitch-Miagkova, A., Miagkov, A., Skeel, A., Nakaigawa, N., Zbar, B., & Leonard, E. J. (2001). Oncogenic mutants of RON and MET receptor tyrosine kinases cause activation of the β-catenin pathway. *Molecular and Cellular Biology, 21*, 5857–5868.

Dickman, E. D., Rogers, R., & Conway, S. J. (1999). Abnormal skeletogenesis occurs coincident with increased apoptosis in the Splotch (Sp2H) mutant: Putative roles for Pax3 and PDGFRalpha in rib patterning. *Anatomical Record, 255*, 353–361.

Dietrich, S., & Gruss, P. (1995). *Undulated* phenotypes suggest a role of *Pax-1* for the development of vertebral and extravertebral structures. *Developmental Biology, 167*, 529–548.

Dietrich, S., Schubert, F. R., & Gruss, P. (1993) Altered Pax gene expression in mouse notochord mutants: The notochord is required to initiate and maintain ventral identity in the somite. *Mechanisms of Development, 44*, 189–207.

Dietrich, S., Schubert, F. R., & Lumsden, A. (1997). Control of dorsoventral pattern in the chick paraxial mesoderm. *Development, 124*, 3895–908.

Evans, D. J. R. (2003). Contribution of somitic cells to the avian ribs. *Developmental Biology, 256*, 114–126.

Fernandez-Teran, M., & Ros, M. A. (2008). The Apical Ectodermal Ridge: Morphological aspects and signaling pathways. *International Journal of Developmental Biology, 52*, 857–871.

Floß, T., Arnold, H. H., & Braun, T. (1996). Myf-5^{m1}/Myf-6^{m1} compound heterozygous mouse mutants down-regulate Myf-5 expression and exert rib defects: Evidence for long-range *cis* effects on Myf-5 transcription. *Developmental Biology, 174*, 140–147.

Fraidenraich, D., Lang, R., & Basilico, C. (1998). Distinct regulatory elements govern Fgf4 gene expression in the mouse blastocyst, myotomes, and developing limb. *Developmental Biology, 204*, 197–209.

Fraidenraich, D., Iwahori, A., Rudnicki, M., & Basilico, C. (2000). Activation of fgf4 gene expression in the myotomes is regulated by myogenic bHLH factors and by sonic hedgehog. *Developmental Biology, 225*, 392–406.

Gaffney, E. S. (1990). The comparative osteology of the Triassic turtle *Proganochelys*. *Bulletin of the American Museum of Natural History, 194*, 1–263.

Gegenbaur, C. (1898) *Vergleichende Anatomie der Wirbelthiere*, Leipzig: Verlag von Wilhelm Engelmann.

Gilbert, S. C. (2010). *Developmental biology* (9th ed.). Sunderland: Sinauer Associates Inc.

Gilbert, S. F., Loredo, G. A., Brukman, A., & Burke, A. C. (2001). Morphogenesis of the turtle shell: The development of a novel structure in tetrapod evolution. *Evolution & Development, 3*, 47–58.

Gilbert, S. F., Bender, G., Betters, E., Yin, M., & Cebra-Thomas, J. A. (2007). The contribution of neural crest cells to the nuchal bone and plastron of the turtle shell. *Integrative and Comparative Biology, 47*, 401–408.

Gilbert, S. F., Cebra-Thomas, J. A., & Burke, A. C. (2008). How the turtle gets its shell. In J. Wyneken, M. H. Godfrey & V. Bels (Eds.), *Biology of turtles*. (pp. 1–16). Boca Raton: CRC Press.

Goodrich, E. S. (1930). *Studies on the structure and development of vertebrates*. London: Macmillan.

Grass, S., Arnold, H. H., & Braun, T. (1996). Alterations in somite patterning of *Myf-5*-deficient mice: A possible role for FGF-4 and FGF-6. *Development, 122*, 141–150.

Grifone, R., Demignon, J., Houbron, C., Souil, E., Niro, C., Seller, M. J., et al. (2005). Six1 and Six4 homeoproteins are required for Pax3 and Mrf expression during myogenesis in the mouse embryo. *Development, 132*, 2235–2249.

Gros, J., Scaal, M., & Marcelle, C. (2004). A two-step mechanism for myotome formation in chick. *Developmental Cell, 6*, 875–882.

Hall, B. K. (1998). *Evolutionary developmental biology* (2nd ed.). London: Chapman & Hall.

Hamburger, V., & Hamilton, H. L. (1951). A series of normal stages in the development of the chick embryo. *Journal of Morphology, 88*, 49–92.

Hasty, P., Bradley, A., Morris, J. H., Edmondson, D. G., Venuti, J. M., Olson, E. N. et al. (1993). Muscle deficiency and neonatal death in mice with a targeted mutation in the myogenin gene. *Nature, 364*, 501–506.

Henderson, D. J., Conway, S. J., & Copp, A. J. (1999). Rib truncations and fusions in the Sp^{2H} mouse reveal a role for Pax3 in specification of the ventrolateral and posterior parts of the somite. *Developmental Biology, 209*, 143–158.

Huang, R., & Christ, B. (2000). Origin of the epaxial and hypaxial myotome in avian embryos. *Anatomy and Embryology, 202*, 369–374.

Huang, R., Zhi, Q., Wilting, J., & Christ, B. (1994). The fate of the somitocoele cells in avian embryos. *Anatomy and Embryology, 190*, 243–250.

Huang, R., Zhi, Q., Neubüser, A., Müller, T. S., Brand-Saberi, B., & Christ, B., et al. (1996). Function of somite and somitocoele cells in the formation of the vertebral motion segment in avian embryos. *Acta Anatomica, 155*, 231–241.

Huang, R., Zhi, Q., Patel, K., Wilting, J., & Christ, B. (2000a). Dual origin and segmental organisation of the avian scapula. *Development, 127*, 3789–3794.

Huang, R., Zhi, Q., Schmidt, C., Wilting, J., Brand-Saberi, B., & Christ, B. (2000b). Sclerotomal origin of the ribs. *Development, 127*, 527–532.

Huang, R., Zhi, Q., Scmhidt, C., Brand-Saberi, B., & Christ, B. (2000c). New experimental evidence for somite resegmentation. *Anatomy and. Embryology, 202*, 195–200.

Huang, R., Stolte, D., Kurz, H., Ehehalt, F., Cann, G. M., Stockdale, F. E., et al. (2003). Ventral axial organs regulate expression of myotomal Fgf-8 that influences rib development. *Developmental Biology, 255*, 30–47.

Kablar, B., Krastel, K., Ying, C., Asakura, A., Tapscott, S. J., & Rudnicki, M.A. (1997). MyoD and Myf-5 differentially regulate the development of limb versus trunk skeletal muscle. *Development, 124*, 4729–4738.

Kardon, G. (1998). Muscle and tendon morphogenesis in the avian hind limb. *Development, 125*, 4019–4032.

Kardon, G., Harfe, B. D., & Tabin, C. J. (2003). A *Tcf4*-positive mesodermal population provides a prepattern for vertebrate limb muscle patterning. *Developmental Cell, 5*, 937–944.

Kato, N., & Aoyama, H. (1998). Dermomyotomal origin of the ribs as revealed by extirpation and transplantation experiments in chick and quail embryos. *Development, 125*, 3437–3443.

Kaul, A., Köster, M., Neuhaus, H., & Braun, T. (2000). Myf-5 revisited: Loss of early myotome formation does not lead to a rib phenotype in homozygous *Myf-5* mutant mice. *Cell, 102*, 17–19.

Kawashima-Ohya, Y., Narita, Y., Nagashima, H., Usuda, U., & Kuratani, S. (2011). Hepatocyte growth factor is crucial for development of the carapace in turtles. *Evolution & Development, 13*, 260–268.

Khokha, M. K., Hsu, D., Brunet, L. J., Dionne, M. S., & Harland, R. M. (2003). Gremlin is the BMP antagonist required for maintenance of Shh and Fgf signals during limb patterning. *Nature Genetics, 34*, 303–307.

Kieny, M., & Chevallier, A. (1979). Autonomy of tendon development in the embryonic chick wing. *Journal of Embryology & Experimental Morphology, 49*, 153–165.

Koseki, H., Wallin, J., Wilting, J., Mizutani, Y., Kispert, A., Ebensperger, C., et al. (1993). A role for Pax-1 as a mediator of notochordal signals during the dorsoventral specification of vertebrae. *Development, 119*, 649–660.

Kuraku, S., Usuda, R., and Kuratani, S. (2005). Comprehensive survey of carapacial ridge-specific genes in turtle implies co-option of some regulatory genes in carapace evolution. *Evolution & Development, 7*, 3–17.

Kuratani, S. (2009). Modularity, comparative embryology and evo-devo: Developmental dissection of evolving body plans. *Developmental Biology, 332*, 61–69.

Kuratani, S., Kuraku, S., & Nagashima, H. (2011). Evolutionary developmental perspective for the origin of the turtles: The folding theory for the shell based on the developmental nature of the carapacial ridge. *Evolution & Development, 13*, 1–14.

Kusakabe, R., & Kuratani, S. (2005). Evolution and developmental patterning of the vertebrate skeletal muscles: Perspectives from the lamprey. *Developmental Dynamics, 234*, 824–834.

Kusakabe, R., & Kuratani, S. (2007). Evolutionary perspectives from development of mesodermal components in the lamprey. *Developmental Dynamics, 236*, 410–420.

Leitges, M., Neidhardt, L., Haenig, B., Herrmann, B. G., & Kispert, A. (2000). The paired homeobox gene Uncx4.1 specifies pedicles transverse processes and proximal ribs of the vertebral column. *Development, 127*, 2259–2267.

Li, C., Wu, X., Rieppel, O., Wang, L., & Zhao, L. (2008). An ancestral turtle from the Late Triassic of southwestern China. *Nature, 45*, 497–501.

Loredo, G. A., Brukman, A., Harris, M. P., Kagle, D., Leclair, E. E., Gutman, R., et al. (2001). Development of an evolutionarily novel structure: Fibroblast growth factor expression in the carapacial ridge of turtle embryos. *Journal of Experimental Zoology, 291B*, 274–281.

Mansouri, A., Voss, A. K., Thomas, T., Yokota, Y., & Gruss, P. (2000). Uncx4.1 is required for the formation of the pedicles and proximal ribs and acts upstream of Pax-9. *Development, 127*, 2251–2258.

Matsuoka, T., Ahlberg, P. E., Kessaris, N., Iannarelli, P., Dennehy, U., Richardson, W., et al. (2005). Neural crest origins of the neck and shoulder. *Nature, 436*, 347–355.

Monga, S. P., Mars, W. M., Pediaditakis, P., Bell, A., Mulé, K., Bowen, W. C., et al. (2002). Hepatocyte growth factor induces Wnt-independent nuclear translocation of β-catenin after Met-β-catenin dissociation in hepatocytes. *Cancer Research, 62*, 2064–2071.

Moustakas, J. E. (2008). Development of the carapacial ridge: Implications for the evolution of genetic networks in turtle shell development. *Evolution & Development, 10*, 29–36.

Müller, T. S., Ebensperger, C., Neubuser, A., Koseki, H., Balling, R., Christ, B., et al. (1996). Expression of avian Pax1 and Pax9 is intrinsically regulated in the pharyngeal endoderm, but depends on environmental influences in the paraxial mesoderm. *Developmental Biology, 178*, 403–417.

Nabeshima, Y., Hanaoka, K., Hayasaka, M., Esumi, E., Li, S., Nonaka, I., & Nabeshima, Y. (1993). *Myogenin* gene disruption results in perinatal lethality because of severe muscle defect. *Nature, 364*, 532–535.

Nagashima, H., Uchida, K., Yamamoto, K., Kuraku, S., Usuda, R., & Kuratani, S. (2005). Turtle–chicken chimera: An experimental approach to understanding evolutionary innovation in the turtle. *Developmental Dynamics, 232*, 149–161.

Nagashima, H., Kuraku, S., Uchida, K., Ohya, Y. K., Narita, Y., & Kuratani, S. (2007). On the carapacial ridge in turtle embryos: Its developmental origin, function, and the chelonian body plan. *Development, 134*, 2219–2226.

Nagashima, H., Sugahara, F., Takechi, M., Ericsson, R., Kawashima-Ohya, Y., Narita, Y., et al. (2009). Evolution of the turtle body plan by the folding and creation of new muscle connections. *Science, 325*, 193–196.

Nelson, W. J., and Nusse, R. 2004. Convergence of Wnt, β-catenin, and cadherin pathways. *Science, 303*, 1483–1487.

Novak, A., and Dedhar, S. (1999). Signaling through β-catenin and Lef/Tcf. *Cellular and Molecular Life Sciences, 56*, 523–537.

Ogushi, K. (1911). Anatomische Studien an der japanischen dreikralligen Lippenschildkröte (*Trionyx japonicus*). *Morphologisches Jahrbuch, 43*, 1–106.

Ohuchi, H., Nakagawa, T., Yamamoto, A., Araga, A., Ohata, T., Ishimaru, Y., et al. (1997). The mesenchymal factor, FGF10, initiates and maintains the outgrowth of the chick limb bud through interactions with FGF8, an apical ectodermal factor. *Development, 124*, 2235–2244.

Ohya, Y. K., Usuda, R., Kuraku, S., Nagashima, H., & Kuratani, S. (2006). Unique features of Myf-5 in turtles: Nucleotide deletion, alternative splicing and unusual expression pattern. *Evolution & Development, 8*, 415–423.

Olivera-Martinez, I., Coltey, M., Dhouailly, D., & Pourqui, O. (2000). Mediolateral somitic origin of ribs and dermis determined by quail-chick chimeras. *Development, 127*, 4611–4617.

Olson, E. N., Arnold, H.-H., Rigby, P. W. J., & Wold, B. J. (1996). Know your neighbors: Three phenotypes in null mutants of the myogenic bHLH gene *MRF4*. Cell, 85, 1–4.

Patapoutian, A., Yoon, J. K., Miner, J. H., Wang, S., Stark, K., & Wold, B. (1995). Disruption of the mouse MRF4 gene identifies multiple waves of myogenesis in the myotome. *Development, 121*, 3347–3358.

Peters, H., Doll, U., & Niessing, J. (1995). Differential expression of the chicken Pax-1 and Pax-9 gene. In situ hybridization and immunohistochemical analysis. *Developmental Dynamics, 203*, 1–16.

Pinot, M. (1969). Etude expérimentale de la morphogenése de la cage thoracique chez l'embryon de poulet: mécanismes et origine du matériel. *Journal of Embryology & Experimental Morphology, 21*, 149–164.

Pizette, S., Abate-Shen, C., & Niswander, L. (2001). BMP controls proximodistal outgrowth, via induction of the apical ectodermal ridge, and dorsoventral patterning in the vertebrate limb. *Development, 128*, 4463–4474.

Rasola, A., Fassetta, M., De Bacco, F., D'Alessandro, L., Gramaglia, D., Di Renzo, M. F., et al. (2007). A positive feedback loop between hepatocyte growth factor receptor and β-catenin sustains colorectal cancer cell invasive growth. *Oncogene, 26*, 1078–1087.

Rickmann, M., & Fawcett, J. W. (1985). The migration of neural crest cells and the growth of motor axons through the rostral half of the chick somite. *Journal of Embryology & Experimental Morphology, 90*, 437–455.

Rieppel, O. (2001). Turtles as hopeful monsters. *BioEssays, 23*, 987–991.

Romer, A. S. (1956). *Osteology of the reptiles*. Chicago: University of Chicago Press.

Romer, A. S., & Parsons, T. S. (1977). *The vertebrate body*. Philadelphia: Saunders.

Ros, M. A., Lyons, G., Kosher, R. A., Upholt, W. B., Coelho, C. N., & Fallon, J. F. (1992). Apical ridge dependent and independent mesodermal domains of *GHox-7* and *GHox-8* expression in chick limb buds. *Development, 116*, 811–818.

Ruckes, H. (1929). Studies in chelonian osteology. Part II. The morphological relationships between the girdles, ribs and carapace. *Annals of the New York Academy of Sciences, 31*, 81–120.

Rudnicki, M. A., Braun, T., Hinuma, S., & Jaenisch, R. (1992). Inactivation of *MyoD* in mice leads to up-regulation of the myogenic HLH gene *Myf-5* and results in apparently normal muscle development. *Cell, 71*, 383–390.

Rudnicki, M. A., Schnegelsberg, P. N., Stead, R. H., Braun, T., Arnold, H. H., & Jaenisch, R. (1993). MyoD or Myf-5 is required for the formation of skeletal muscle. *Cell, 75*, 1351–1359.

Sánchez-Villagra, M. R., Müller, H., Sheil, C. A., Scheyer, T. M., Nagashima, H., & Kuratani, S. (2009). Skeletal development in the

Chinese soft-shelled turtle *Pelodiscus sinensis* (Testudines: Trionychidae). *Journal of Morphology, 270,* 1381–1399.

Saunders, J. W., Jr., (1948). The proximo-distal sequence of origin of the parts of the chick wing and the role of the ectoderm. *Journal of Experimental Zoology, 108*, 363–403.

Saunders, J. W., Jr., & Gasseling, M. (1968). Ectodermal-mesenchymal interactions in the origin of limb symmetry. In R. Fleischmayer & R. E. Billingham (Eds.), *Epithelial-Mesenchymal Interaction* (pp. 78–97). Baltimore: Williams and Wilkins.

Saunders, J. W. Jr., & Reuss, C. (1974). Inductive and axial properties of prospective wing-bud mesoderm in the chick embryo. *Developmental Biology, 38*, 41–50.

Scaal, M., & Christ B. (2004). Formation and differentiation of the avian dermomyotome. *Anatomy and Embryology, 208*, 411–424.

Seno, T. (1961). An experimental study on the formation of the body wall in the chick. *Acta Anatomica, 45*, 60–82.

Sensenig, E. C. (1949). The early development of the human vertebral column. *Contributions to Embryology, 33*, 23–40.

Shellswell, G. B., & Wolpert, L. (1977). The pattern of muscle and tendon development in the chick wing. In D. A. Ede, J. R. Hincliffe, & M. Balls (Eds.), *Vertebrate limb and somite morphogenesis* (pp. 71–86). Cambridge: Cambridge University Press.

Shimomura, Y., Agalliu, D., Vonica, A., Luria, V., Wajid, M., Baumer, A., et al. (2010). APCDD1 is a novel Wnt inhibitor mutated in hereditary hypotrichosis simplex. *Nature, 464,* 1043–1047.

Soriano, P. (1997). The PDGF alpha receptor is required for neural crest cell development and for normal patterning of the somites. *Development, 124*, 2691–2700.

Sudo H, Takahashi Y, Tonegawa A, Arase Y, Aoyama H, Mizutani-Koseki Y, et al. (2001). Inductive signals from the somatopleure mediated by bone morphogenetic proteins are essential for the formation of the sternal component of avian ribs. *Developmental Biology, 232,* 284–300.

Summerbell, D., Lewis, J. H., & Wolpert, L. (1973). Positional information in chick limb morphogenesis. *Nature, 224*, 492–496.

Sweeney, R. M., & Watterson, R. L. (1969). Rib development in chick embryos analyzed by means of tantalum foil blocks. *American Journal of Anatomy, 126,* 127–150.

Tabin, C., & Wolpert, L. (2007). Rethinking the proximodistal axis of the vertebrate limb in the molecular era. *Genes & Development, 21*, 1433–1442.

Takahashi, M., Fujita, M., Furukawa, Y., Hamamoto, R., Shimokawa, T., Miwa, N., et al. (2002). Isolation of a novel human gene, *APCDD1*, as a direct target of the β-catenin/T-cell factor 4 complex with probable involvement in colorectal carcinogenesis. *Cancer Research, 62,* 5651–5656.

Takahashi, M., Nakamura, Y., Obama, K., & Furukawa, Y. (2005). Identification of *SP5* as a downstream gene of the β-catenin/Tcf pathway and its enhanced expression in human colon cancer. *International Journal of Oncology, 27*, 1483–1487.

Tallquist, M.D., Weismann, K.E., Hellström, M., & Soriano, P. (2000). Early myotome specification regulates PDGFA expression and axial skeleton development. *Development, 127*, 5059–5070.

Theißen, G. (2006). The proper place of hopeful monsters in evolutionary biology. *Theory in Biosciences, 124,* 349–369.

Theißen, G. (2009). Saltational evolution: Hopeful monsters are here to stay. *Theory in Biosciences, 128*, 43–51.

Tokita, M., & Kuratani, S. (2001). Normal embryonic stages of the Chinese softshelled turtle *Pelodiscus sinensis*. *Zoological Sciences, 18*, 705–715.

Tremblay, P., Dietrich, S., Mericskay, M., Scubert, F.R., Li, Z., & Paulin, D. (1998). A crucial role for Pax3 in the development of the hypaxial musculature and the long-range migration of muscle precursors. *Developmental Biology, 203*, 49–61.

Valasek, P., Theis, S., Krejci, E., Grim, M., Maina, F., Shwartz, Y., et al. (2010). Somitic origin of the medial border of the mammalian scapula and its homology to the avian scapula blade. *Journal of Anatomy, 216*, 482–488.

Vasyutina, E. & Birchmeier, C. (2006). The development of migrating muscle precursor cells. *Anatomy and Embryology, 211,* S37–S41.

Verbout, A. J. (1985). The development of the vertebral column. *Advances in Anatomy, Embryology and Cell Biology, 90,* 1–122.

Vinagre, T., Moncaut, N., Carapuço, M., Nóvoa, A., Bom, J., & Mallo, M. (2010). Evidence for a myotomal Hox/Myf cascade governing nonautonomous control of rib specification within global vertebral domains. *Developmental Cell, 18,* 655–661.

Vincent, C., Bontoux, M., Le Douarin, N. M., Pieau, C., & Monsoro-Burq, A.H. (2003). *Msx* genes are expressed in the carapacial ridge of turtle shell: A study of the European pond turtle, *Emys orbicularis*. *Development Genes and Evolution, 213,* 464–469.

Walker, W. F, Jr. (1947). The development of the shoulder region of the turtle, *Chrysemys picta marginata*, with special reference to the primary musculature. *Journal of Morphology, 80*, 195–249.

Wallin, J., Wilting, J., Koseki, H., Fitsch, R., Christ, B., & Balling, R., (1994). The role of *Pax-1* in axial skeleton development. *Development, 120*, 1109–1121.

Wang, B., He, L., Ehehalt, F., Geetha-Loganathan, P., Nimmagadda, S., Christ, B., et al. (2005). The formation of the avian scapula blade takes place in the hypaxial domain of the somites and requires somatopleure-derived BMP signals. *Developmental Biology, 287,* 11–18.

Watson, D. S. M. (1914). *Eunotosaurus africanus* Seeley and the ancestors of the Chelonia. *Proceedings of the Zoological Society of London, 11,* 1011–1020.

Weidinger, G., Thorpe, C. J., Wuennenberg-Stapleton, K., Ngai, J., & Moon, R. T. (2005). The Sp1-related transcription factors *sp5* and *sp5-like* act downstream of Wnt/β-catenin signaling in mesoderm and neuroectoderm patterning. *Current Biology, 15,* 489–500.

Winslow, B. B., Takimoto-Kimura, R., & Burke, A. C. (2007). Global patterning of the vertebrate mesoderm. *Developmental Dynamics, 236*, 2371–2381.

Winter, B., Braun, T., & Arnold, H.-H. (1992). Co-operativity of functional domains in the muscle-specific transcription factor Myf-5. *EMBO Journal, 11,* 1843–1855.

Yntema, C. L. (1970). Extirpation experiments on the embryonic rudiments of the carapace of *Chelydra serpentina*. *Journal of Morphology, 132,* 235–244.

Yokouchi, Y., Ohsugi, K., Sasaki, H., & Kuroiwa, A. (1991). Chicken homeobox gene *Msx-1*: Structure, expression in limb buds and effect of retinoic acid. *Development, 113,* 431–444.

Yoon, J. K., Olson, E. N., Arnold, H. H., & Wold, B. J. (1997). Different *MRF4* knockout alleles differentially disrupt Myf-5 expression: *cis*-regulatory interactions at the *MRF4/Myf-5* locus. *Developmental Biology, 188,* 349–362.

Zhang, W., Behringer, R. R., & Olson, E. N. (1995). Inactivation of the myogenic bHLH gene *MRF4* results in up-regulation of myogenin and rib anomalies. *Genes & Development, 9,* 1388–1399.

Chapter 5
The Evolution of the Turtle Shell

Olivier Rieppel

Abstract This chapter traces the history of the debate on the evolution of the turtle shell, and carries the analysis of the origin of the turtle carapace forward from two complementary perspectives, viz. paleontology and developmental biology. Two alternative approaches to morphological analysis—the transformationist and the emergentist—are identified. The transformationist approach seeks to understand morphological evolution as a consequence of the gradual, step-wise transformation of the adult phenotype. The emergentist approach allows for ontogenetic deviation to result in the development of evolutionary novelties. The discovery of the so far oldest and most primitive turtle known, from the early Late Triassic of southwestern China, provides the basis for a synthesis of paleontological and developmental data in the understanding of the evolutionary origin of the turtle shell.

Keywords Carapace • Dermal bone • *Dermochelys* • Membrane bone • *Odontochelys*

Introduction

"However, this developmental specialization does not provide any hint as to the way in which [the turtle shell] evolved phylogenetically" (Carroll 1988, p. 210).

On the opening page of his '*Origin*' Darwin (1859, p. 1) used John Herschel's phrase (expressed in a letter to Charles Lyell, Feb. 20, 1836) when he called the origin of species the "mystery of mysteries." In vertebrate paleontology, it might well be the origin of turtles, and the correlated evolution of the turtle shell, that constitute (one of) the mystery of mysteries. The reason seems to be that the turtle carapace shows an intimate association of dermal ossifications with elements of the underlying axial skeleton (dorsal neural arches and dorsal ribs). Cuvier (1799, 1812) was the first to propose that the central parts of the turtle carapace, i.e., the neural and costal plates, develop through a broadening of the dorsal neural arches and of the dorsal ribs respectively, to an extent that sutural relations are established between them. While supported by Geoffroy Saint-Hilaire (1818) and others, Cuvier's views were opposed by Wiedeman (1802), Carus (1834), Rathke (1848), and Owen (1849), who thought the neural and costal plates to be derived from dermal ossifications, which secondarily fuse with the underlying axial skeleton. During the ensuing debate, which involved numerous illustrious authors such as Johann Friedrich Meckel the Younger, Thomas H. Huxley, Ernst Haeckel and Karl Gegenbaur to name just a few, a consensus on the nature of the turtle carapace remained elusive (the history of the debate was sketched in more detail by Owen 1849, Goette 1899, and Vallén 1942, amongst others).

Whereas the early debate on the nature of the carapace was based on embryological investigations, Cope (1871, p. 235) introduced a phylogenetic dimension when he separated the extant sea turtle *Dermochelys coriacea* and its fossil relatives as Athecae from all remaining turtles, which he referred to as Cryptodira. Later this division became one between Athecae and Thecophora (Dollo 1886; Atheca in Strauch 1890, p. 38), with the implication that *Dermochelys* would represent the primitive condition of the turtle dermal armor. Cope (1871, p. 232) recognized the peculiar position of the proximal articulation of the dorsal ribs "at the point of contact of two centra", and accepted the view that much of the carapace is derived from "the expansion of the ribs into an osseous upper shield", from where he went on to conclude: "The oldest Tortoises have generally the most incomplete carapace and plastron; among them the Psephoderma allied to *Sphargis* [a junior synonym of *Dermochelys* (Merrem 1820)]. Without carapace, and thus the most lizard-like of the order" (Cope 1871, p. 244).

O. Rieppel (✉)
The Field Museum, 1400 South Lake Shore Drive,
Chicago, IL 60605-2496, USA
e-mail: orieppel@fieldmuseum.org

D. B. Brinkman et al. (eds.), *Morphology and Evolution of Turtles*, Vertebrate Paleobiology and Paleoanthropology,
DOI: 10.1007/978-94-007-4309-0_5, © Springer Science+Business Media Dordrecht 2013

Psephoderma from the Rhaetian (Late Triassic) of the Bavarian Alps was described by Meyer (1858) on the basis of a dorsal shield composed of polygonal osteoderms meeting in sutures. Meyer (1858, p. 250) identified *Psephoderma* as a reptile, but rejected a close comparability of his new taxon with *Sphargis*; he based his rejection on differences in details of composition and sculpture of the dermal armor between *Psephoderma* and *Psephophorus polygonus*, which he had described earlier (Meyer 1847, p. 579). *Psephophorus* is a genus of Tertiary sea turtles close to *Dermochelys*, whereas *Psephoderma* later turned out to be a cyamodontoid placodont (Nosotti and Pinna 1989).

Dermochelys and the Derivation of the Turtle Carapace from Osteoderms

Dermochelys coriacea is characterized by the absence of a fully formed carapace, and a remarkable reduction of the plastron. In the carapace, only the nuchal plate and posterior peripherals (if correctly homologized) are ossified (Völker 1913). The dorsal ribs articulate in an intervertebral (intrasegmental) position, i.e., at the point of contact between two dorsal vertebral centra. Their shaft is somewhat broadened in the horizontal plane, but they remain widely separated from one another. The dorsal dermal armor of *Dermochelys* is composed of a mosaic of polygonal bony plates (osteoderms) that meet each other in slightly interdigitating sutures, but which do not fuse with underlying endochondral axial elements (vertebral neural larches and ribs).

Baur (1886) criticized Cope's (1871) separation of the Athecae from all other turtles as an artificial one, since in his view *Dermochelys* does not exemplify a primitive stage in the evolution of the turtle shell, but a reduced state instead. This he inferred from the strongly reduced plastron, the reduced claws, and the presence of a nuchal plate. The dorsal armor of *Dermochelys* would thus have formed by the "delamination of membrane bone from the ribs, and the dissolution of this into polygonal pieces" (Hay 1898, p. 932). Baur (1886) nevertheless accepted that ancestral turtles did have a dorsal armor composed of a mosaic of numerous osteoderms, and that the costal plates would have formed by coalescence of such osteoderms and their subsequent fusion with the underlying ribs. Such an ancestral turtle Baur (1887; see also Baur 1888) believed to be represented by *Psephoderma alpinum* from the Triassic of the Bavarian Alps [he later acknowledged the sauropterygian nature of *Pspehoderma* (Baur 1889)]. *Dermochelys* was thus recognized not as a primitive turtle, but as a model for the ancestral turtle, covered with a mosaic of polygonal osteoderms that would coalesce and fuse with the underlying axial skeleton in the evolution of the carapace (Baur 1889).

Decoupled from appeals to *Dermochelys* and *Psephoderma*, of course, the derivation of the turtle carapace from an ancestral osteoderm covering remains the most popular account of the origin of the turtle shell today. Lee (1993, p. 1719) claimed to have bridged a famous morphological gap with his theory that "the precursors of the chelonian carapace and plastron can now be seen in the osteoderms of pareiasaurs". In support of his thesis, Lee (1996, p. 813) later drew an analogy (not homology) with archosaurs and placodonts: "In these groups also, dermal armor first appears in primitive forms as a series of isolated osteoderms over the dorsal midline, and later becomes elaborated to form a rigid covering over the entire back." Scheyer et al. (2008, p. 1018) likewise presumed "that a potential [turtle] ancestor ... carried numerous osteoderms ...". And again supporting the "composite model", according to which "the turtle shell is thought to have derived from a series of intermediate forms that possessed ever increasing amounts of dermal armor that eventually fused with the underlying internal skeleton to form the carapace and plastron", Joyce et al. (2009, p. 511) concluded: "neontological data alone do not conclusively identify the process by which the turtle shell originated."

Classically, the plastron of turtles is thought to derive phylogenetically not from ventral osteoderms, but from dermal elements of the shoulder girdle (clavicles and interclavicle) and the gastral rib cage (Owen 1849; Baur 1888; Hay 1898; Versluys 1914; Zangerl 1939, 1969). Recent developmental data indicate that plastral elements form by intramembranous ossification (Gilbert et al. 2001), possibly initiated by contributing trunk neural crest cells (Clark et al. 2001; Gilbert et al. 2007, 2008); the contribution of neural crest cells to the plastron remains somewhat controversial and requires further testing (A. Burke and H. Nagashima, personal communication). Except for the nuchal plate (Cebra-Thomas et al. 2007; Gilbert et al. 2007), there is no neural crest cell contribution to the carapace (Moustakas 2008). The following discussion will concentrate on the evolution of the carapace.

The 'Duplicity Theory' of the Turtle Carapace

In his discussion of the shell morphology of *Dermochelys*, Baur picked up an observation that was first related by Gervais (1872): "According to Gervais, the mosaic-like dorsal shield of *Dermochelys* lies above the nuchal plate. The question is: what is the significance of this condition?" (Baur 1886, p. 688). This problem was later taken up by

Hay (1898), who was the first to recognize the duplex nature of the turtle shell. Hay (1898, p. 934) distinguished dermal bones from 'fascia bones' citing crocodile osteoderms as examples of the first, crocodile abdominal ribs as paradigm for the latter [note that crocodile gastralia are embedded not in subdermal tissue, but in the *erectus abdominis* muscle instead (Vickaryous and Hall 2008)]. Given the topological relationship of the nuchal plate to the overlying osteoderm shield, Hay (1898, p. 935) considered the osteoderm shield to be of dermal origin, the nuchal shield to be composed of 'fascia bone' (compare 'primary' and 'secondary' dermal bone in Gilbert et al. 2001). Similarly, the neural and costal plates he considered 'fascia bones' that enter into an intimate relation with the underlying axial skeleton: "... we may suppose that the earliest ancestors of turtles had a scaly skin, which contained osteodermal plates. Beneath these there were developed first, perhaps in the fascia of the shoulders, a nuchal bone, later other plates which in time became transformed into neuralia and costalia. As these deeper-seated fascia bones increased in importance, the osteodermal plates underwent gradual reduction. Only in *Dermochelys* have they retained anything like their early importance." Although Völker (1913) accepted Hay's (1898) distinction of two layers of ossification in the turtle carapace, he rejected the idea that the inner layer forms deep to the dermis. Völker (1913) instead distinguished the *theca*, composed primarily of elements of the inner dermal layer of ossification, from *epithecal* ossifications that develop more superficially in the dermis. The turtle shell composed of neural, costal, nuchal [of dual origin (Burke 1989a; Gilbert et al. 2001)], pygal, and marginal plates accordingly forms the theca. Epithecal ossifications are osteoderms that are superimposed on these thecal elements. The latter were known to occur not only in *Dermochelys* and its fossil relatives, but also in other fossil marine turtles with a reduced carapace (the 'epi-neural', 'supra-neural', or 'epithecal neural' ossifications capping neural plates in *Archelon* and *Toxochelys*: Case 1898; Wieland 1896, 1905, 1909; Hay 1905, 1908; see also Versluys 1914), and (as flake-like ossifications superimposed on thecal elements) in the pleurodire genus *Chelys* (= *Chelus*) (Hay 1928).

The distinction of deeper thecal from more superficial epithecal ossifications was supported by Zangerl (1939), and later was dubbed the 'duplicity theory' of the turtle carapace (Kälin 1945). Recognizing the dual nature of the carapace, the evolutionary development of the latter could no longer be seen as a mere consequence of coalescence of superficial osteoderms into larger plates, some of which secondarily fused with the underlying endoskeleton. Although some continued to maintain that scenario (e.g., Oguschi 1911), others, like Hay (1898, p. 941) envisaged an early dichotomy within turtles, one lineage ("the athecate tribe") developing an epitheca, the "second lineage of terrestrial turtles" developing a theca instead. Recognizing the untenability of the Atheca-Thecophora dichotomy within turtles, and the cryptodire status of *Dermochelys*, Versluys (1914; see also Hay 1922) hypothesized the ancestral turtle to have had an incomplete theca covered by osteoderms. In the thecophore turtles the osteoderm covering was reduced while the theca was perfected, whereas in forms like *Dermochelys*, the epitheca would be perfected whereas the theca became reduced.

The Dual Nature of the Theca

Goette (1899) investigated the ontogeny of *Eretmochelys imbricata*, finding that the neural arches of the dorsal region shift anteriorly by half a segment during development, carrying the ribs with them. The result is the intervertebral (intrasegmental) placement of the proximal rib articulations on the boundary between two vertebral centra, today recognized as a synapomorphy of turtles (Joyce 2007; Joyce et al. 2009). Goette further observed that the neural and costal plates develop from the neural arches and ribs respectively: "The ossification of the ribs starts from the perichondral bone surrounding the ribs, and proceeds within a distinctly demarcated, but uncharacteristically expanded periost" (Goette 1899, p. 415). He characterized earlier accounts by Rathke (1848), Hoffmann (1878) and Haycraft (1891), according to whom the costal plates ossify entirely within the dermis and extraneous to any periost, as "entirely mistaken" (Goette 1899, p. 416). Referring to the ossification of the dermal elements in the skull roof of the lizard *Lacerta*, where the dermal parietal skull table is covered by encrusting osteoderms, Goette (1899, p. 432) distinguished in turtles (as Hay 1898 had before him) between deeper ("thecal" sensu Völker 1913) and more superficial ("epithecal" sensu Völker 1913) ossifications. Within the theca, Goette (1899, p. 419) distinguished the nuchal, pygal and peripheral plates as true cutaneous ossifications ('*Hautknochen*') from the neural and costal plates, which are derived from the endoskeleton. According to Goette (1899), there are thus to be distinguished in the turtle shell (using Völker's 1913 terminology), firstly, the thecal and epithecal ossifications, as exemplified by the nuchal plate and overlying osteoderms in *Dermochelys*. Secondly, within the theca, there are to be distinguished the true dermal ossifications ('*Hautknochen*', *ossa investiva* that do not form through periosteal osteogenesis) from those ossifications that derive from the endoskeleton through periosteal osteogenesis.

Suspecting flaws in Goette's (1899) investigations, Newman (1905–1906) repeated them using embryos of *Chelydra serpentina* and *Graptemys geographica*. In the course of his studies, he "satisfied [himself] that the neural and costal plates actually do originate as outgrowths of a differentiated tissue that surrounds the neural and rib cartilages. Whether this differentiated tissue be true periosteum, as Goette affirms, or simply a somewhat denser portion of the connective tissue that fills the space between the epidermis and the cartilaginous skeleton, is not certain" (Newman 1905–1906, p. 733). He nevertheless sided with the paradigm first propagated by Cuvier, according to which "it would seem that the plates of the carapace have a dual origin—the neural and costals being periosteal ossifications while the nuchal, procaudals and marginals [peripherals] are dermal ossifications" (Newman 1905–1906, p. 734). The alternative, first advocated by Wiedeman (1802), Carus (1834), and Rathke (1848), was soon to be revitalized by Oguschi (1911), who in his investigations on the embryology of *Trionyx japanicus* (= *Pelodiscus sinensis*) found that the neural and costal plates have a separate origin as dermal ossifications which later secondarily fuse with the underlying axial skeleton. Additional evidence adduced by Oguschi (1911) against Goette's (1899) interpretation of the costal plates as broadened ribs is the articulation of the scapular blade on the inside of the first costal plate in front of the second dorsal rib. A position of the scapula inside the rib cage seemed unacceptable to him.

Völker (1913) again emphasized the dermal nature of the neural and costal plates, as did Versluys (1914), citing an unpublished Ph.D. thesis by Menger (1922) who corroborated Oguschi's (1911) findings that neural and costal plates have an independent ontogenetic origin in trionychids, and only secondarily fuse with the underlying endoskeleton. Studying developmental stages of *Testudo loveridgeii* (*Malacochersus tornieri*), Procter (1922) concluded that the costal plates are of dermal origin, as they continue to grow while the rib cartilage already starts to degenerate within the periosteal collar. Zangerl (1939), in contrast, confirmed Goette's (1899) observation that there is no boundary between the perichondral bone of vertebrae and ribs, and the very first ossifications of the neural and costal plates. "This seems to indicate that the neurals and costals are not true dermal ossifications. But there is one important fact which seems to have been neglected: the ribs and proc. spinosi of the vertebrae lie, in *Chelydra*, actually within the dermis. The osteoblasts which form the perichondral bone around the cartilage ribs are clearly of dermal origin" (Zangerl 1939, p. 384). Zangerl thus highlighted the fact that in turtles, endoskeletal elements (the neural arches and ribs of the dorsal vertebrae) enter the dermis as they develop and grow (see below for a more detailed discussion). This is the root cause of the debate whether the turtle carapace is, or is not, of a dual structure, combining endoskeletal and exoskeletal elements.

The issue now became one of the proper definition of dermal bone as opposed to endoskeletal ossification. Two contrasting viewpoints were articulated by Vallén (1942) and Kälin (1945). Both authors agreed that initial ossification of the neural and costal plates proceeds from the, and in continuity with the perichondral bone surrounding the dorsal neural arches and ribs, i.e., that there is no secondary fusion of dermal ossifications with the underlying axial skeleton during ontogeny. However, Vallén's (1942, p. 15) conclusion was that "the fact dermal tissue sclerifies during [the ossification of the neural and costal plates] does not imply that the respective parts of the carapace have to be considered as dermal bones (membrane bone, '*Deckknochen*'). Their morphological nature has to be judged primarily on the basis that they grow out from a perichondral bony collar, which in addition develops subdermally." This contrasts with Kälin (1945, p. 149), who took issue with Goette's (1899) observation that the neural and costal plates ossify within an uncharacteristically expanded periost. Instead, Kälin insisted that ossifications can only be truly labeled dermal bones (*ossa investiva*) if they develop directly in the dermal connective tissue ["without the activity of a periosteum, in the strict sense" (Moss 1969, p. 514)]. Kälin (1945, p. 154f) acknowledged that the ossification of costal plates starts with the formation of trabecular bone growing out of the periosteal collar of the rib, but denied that subsequent stages of expansion relate to a well-demarcated periosteum, as had been claimed by Goette (1899). Instead, he found the continuing ossification of trabecular bone to follow along structures preformed by the arrangement of collagen fibers and cells within the dermis (Kälin 1945, p. 156). These are characteristics of dermal sclerification (Moss 1969) or metaplastic ossification (Haines and Mohuiddin 1968; Vickaryous and Hall 2008), typical of true dermal bone (*ossa investiva*) according to Kälin (1945).

Kälin's (1945) observations on the development of the neural and costal plates correspond closely to the more recent report offered by Scheyer et al. (2008; see also Sánchez-Villagra et al. 2009), who again noted the primary continuity of their ossification with the perichondral bone surrounding the neural arches and dorsal ribs. The formation of neural and costal plates starts by the addition of '*Zuwachsknochen*' [membrane bone outgrowths from chondral bone (Starck 1955; Patterson 1977)] to the perichondral bone collar of the ribs and neural arches (Scheyer et al. 2008, p. 1015), but then is completed through metaplastic ossification that involves some "scattered osteocyte activity" (Scheyer et al. 2008, p. 1018): "the costals and neurals thus appear as a mixture of endoskeletal and dermal ossification." This is a compromise solution, appealing to a dual mode of origin of the neural and costal plates, that reflects the difficulty of clearly demarcating metaplastic ossification from osteoblast induced osteogenesis in the reptile dermis: "Within a given species [of reptile], or even within a single dermal sclerification, it is frequently possible to observe a number of different skeletal tissue types that grade inseparably into each other" (Moss 1969, p. 510).

Dermal Bone, Membrane Bone, and the Phylogenetic Fusion of Exoskeleton and Endoskeleton

From a modern developmental point of view (more fully discussed below), the issue whether the turtle carapace is derived from ancestral osteoderms, or forms through intramembranous ossification, either independently, or expanding from the perichondral bone collar of dorsal ribs and neural arches "is something of a red herring when it comes to the origin of turtles" (Burke 2009, p. 623). From a paleontological perspective, however, the issue remains relevant in the reconstruction of turtle ancestry (see comments, discussed above, in Lee 1993; Joyce 2007; Scheyer et al. 2008).

The claim that the turtle shell evolved from an ancestral osteoderm covering that sunk into deeper layers of the skin, eventually fusing with the underlying axial skeleton (Lee 1993, 1996; Scheyer et al. 2008; Joyce et al. 2009) implies a phylogenetic fusion of exoskeletal components with endoskeletal components. Discussions involving the concepts of exoskeleton and endoskeleton, as well as the correlated terms 'dermal bone', 'membrane bone', and 'cartilage bone' (chondral, endochondral, and perichondral bone) are traditionally marred by confusion as various authors tend to define different skeletal components on the basis of different criteria, e.g., histological, histochemical, topological, ontogenetic, or phylogenetic ones (Patterson 1977). The situation in turtles is further complicated by the fact that a thickening of the dermis in the formation of the embryonic carapacial disc, together with a deflection of rib growth to a more superficial level [under the inductive influence of the carapacial ridge (Burke 1989b; and see below)], results in endoskeletal elements (dorsal neural arches and dorsal ribs) to become embedded within the dermis. As the dorsal ribs of a turtle embryo undergo perichondral ossification, it is endoskeletal elements that initiate ossification, yet this ossification occurs in the dermis. Zangerl (1939, p. 384) explained: "The osteoblasts which form the periosteal bone around the cartilage ribs are clearly of dermal origin. ... there is, in *Chelydra*, no difference between the osteoblasts which from the periosteal bone around the ribs and those which develop the nuchal and peripheral plates. Considering this fact we cannot expect a line of demarcation between the elements of the chondro-skeleton and the shell plates."

To avoid confusion, Patterson (1977), and Starck (1979) advocated a strictly phylogenetic (rather than developmental) definition of exo- versus endoskeleton: "The morphological and phylogenetic evaluation of skeletal elements does not depend on their histogenesis, but exclusively on phylogenetic criteria" (Starck 1979, p. 13). According to Patterson (1977), an exoskeletal element (primitively comprising bone, dentine and enamel) arises at the ectoderm-mesoderm interface in the skin, or it is homologous to an element that in the ancestral condition developed in that location. Fish scales and reptile osteoderms are thus exoskeletal in origin, even if they are no longer covered with a dentine and/or enamel layer. An endoskeletal element is primitively preformed in cartilage, or it is homologous to an element which in the ancestral condition is preformed in cartilage, while in the descendant condition it ossifies directly as membrane bone. Patterson (1977) thus rejected the synonymy of 'dermal bone' and 'membrane bone', reserving the latter term for endoskeletal elements in which the ancestral ontogenetic stage of cartilage preformation has been deleted (suppressed).

The reasons why Patterson (1977) rejected the traditional synonymy of 'dermal' and 'membrane' bone become clear with his quote (Patterson 1977, p. 82) from Jollie (1962, p. 55) according to whom "A chondral bone may also have dermal extensions." On Patterson's (1977, p. 82) reading, this could mean that "any part of a cartilage bone that is not preformed in cartilage is held to be dermal. This ... has absurd results: it would lead one to regard the greater part of the vertebral column of higher teleosts as dermal" [the relation of dermal bone formation to neural crest cells was, at the time, not yet recognized (Smith and Hall 1990, 1993)]. Patterson (1977) designated extensions of cartilage bones that are not themselves preformed in cartilage as consisting of membrane bone. While still accepting the traditional synonymy of 'dermal' ('desmal') and 'membrane' bone, Starck (1979, p. 13; see also Starck 1955) called such expansions of cartilage bones '*Zuwachsknochen*' instead. Due to the peculiarities of the development of the turtle carapace (Burke 1989b), endoskeletal elements (dorsal neural arches, ribs) become embedded in the thickened dermis of the carapacial disk. Ossification results in a bony collar that covers the surface of these cartilage elements. While topologically located in the dermis, a phylogenetic perspective leads to the identification of the corresponding bone as perichondral bone. The bony trabeculae which subsequently expand from the perichondral bone through the adjacent dermis are consequently identified as 'membrane bone' sensu Patterson (1977), or '*Zuwachsknochen*' sensu Starck (1979). Hence Scheyer et al.'s (2008, p. 1015) conclusion that during their initial stage of formation, neural and costal plates are best described as '*Zuwachsknochen*'. Following Patterson (1977) and Starck (1979) in adopting a phylogenetically grounded definition of endo- and exoskeleton, the turtle carapace (theca) can thus be understood as a combination of both (Rieppel and Reisz 1999). This was, in essence, the conclusion of Vallén (1942), who considered the neural and costal plates as endoskeletal derivatives, all other thecal (and epithecal) ossifications as exoskeletal derivatives.

Advocating a "composite model" of carapace evolution, Scheyer et al. (2008, p. 1018) found ancestral osteoderms to contribute to, and complete the formation of neural and costal plates, because the initial formation of '*Zuwachsknochen*' changes to "metaplastic ossification of preformed dermal structures." Scheyer et al. (2008) grounded their conclusion in the observation that fully formed costal and neural plates share the same histology with the nuchal, pygal, and peripherals plates, which would seem to imply "that the same developmental pathways apply to all turtle shell bones, a feature presumably inherited from an osteoderm bearing ancestor" (Scheyer et al. 2008, p. 1018). According to the 'composite model', the evolution of the turtle carapace involved a phylogenetic fusion of endo- and exoskeletal elements subsequent to an "ontogenetic shift of rib and vertebral anlagen in the osteoderm-bearing ancestor into the deep dermis to overlap with those of the osteoderms. Successive fusion of dermal and endoskeletal anlagen or substitution of the former anlagen by the latter could then have been the crucial step in turtle shell development, purportedly being an example of the 'phylogenetic fusion between a dermal bone and a cartilage bone' (Patterson 1977, p. 93)."

The phylogenetic fusion of exo- and endoskeletal elements requires that the superficial dermal and deeper endoskeletal ossifications centers become directly superimposed, subsequent ossification then proceeding from a "single ossification or growth centre" (Patterson 1977, p. 93). "In my opinion, three sorts of evidence might be required before phylogenetic fusion between a dermal bone and a cartilage bone is postulated" (Patterson 1977, p. 95): "First, evidence that the primitive condition in the group is to have a distinct dermal bone and a cartilage bone at that location. Second, evidence that those bones fuse in late ontogeny in some, presumably primitive, members of the group. Third, evidence that the bones fuse in early ontogeny in some members of the group." Recognizing the distinctiveness of epithecal ossifications from the theca proper, and the derived nature of epithecal ossifications in those turtles where they occur (Zangerl 1939, 1969), no such evidence obtains for the neural and costal plates in any turtle so far investigated, fossil or extant. More important, perhaps, is the fact that none of the evidence required by Patterson (1977) in support of a putative fusion of exo- and endoskeletal elements is provided by the oldest fossil turtle currently known (Li et al. 2008).

Odontochelys, the Oldest Known Fossil Turtle: Primitive or Paedomorphic?

The major inspiration for a derivation of the turtle carapace from an ancestral osteoderm covering was historically based on the misconception of *Dermochelys* representing a primitive (rather than reduced) stage in the evolution of the turtle shell, and on the misidentification of the cyamodontoid placodont *Psephoderma* as an ancestral turtle. In more modern times, said inspiration derives primarily from *Proganochelys*, a primitive Late Triassic (Norian) turtle from Germany that documents the occurrence of osteoderms on the skull, the neck, the limbs and the tail in addition to a fully formed carapace and plastron, the two linked by a lateral bridge (Gaffney 1990). In addition, some phylogenetic analyses (e.g., Gauthier 1994; Laurin and Reisz 1995; Lee 1993, 1996; but see Hill 2005) have found turtles to be related to osteoderm-bearing reptiles. The deep entrenchment of the 'composite model' in the paleontological literature is reflected in the remarkably similar depiction of an osteoderm-covered turtle ancestor through time, as can be gleaned from Versluys (1914, Fig. 10), Hay (1922, text Figure on p. 440), Lee (1996, Fig. 1) and Joyce et al. (2009, Fig. 4).

As was noted by Joyce et al. (2009, p. 511), "testudinate affinities with armored groups of reptiles … provide phylogenetic support for the composite evolution of the turtle shell, whereas an affiliation with non-armoured amniotes … favours a de novo mode." But whereas the addition of new, integumentary characters (including osteoderm characteristics) to existing phylogenetic analyses did not provide additional support for turtle affinities with Paleozoic osteoderm-bearing reptiles such as pareiasaurs (Hill 2005), analyses of turtle relationships within amniotes overall remain controversial (see the review of the debate and references in Rieppel 2008). This renders the optimization of ancestral character states on the outgroup node of turtles, and the corresponding derivation of scenarios for the evolution of the turtle shell through outgroup comparison, equivocal. Less equivocally, however, *Proganochelys* from the Norian (Late Triassic) has lost its status as the earliest and most primitive known turtle (Gaffney 1990; Joyce 2007) to *Odontochelys* (Li et al. 2008) from the Carnian (Late Triassic) of Guizhou Province, southwestern China. *Odontochelys* is more primitive than *Proganochelys* and all other known turtles in a number of features, such as: the presence of teeth in the upper (premaxilla, maxilla) and lower (dentary) jaws; a relatively long preorbital skull with a pointed, rather than truncated snout; a distinct transverse process on the pterygoid; the absence of an acromial process on the scapula; the proximal articulation of the dorsal ribs located intersegmentally (at the midline of the centrum); free sacral ribs that are not fused with the sacral vertebrae; free caudal transverse processes that are not fused with the caudal vertebrae; a tail which with minimally 20 caudal vertebrae is longer than in any other turtle; and the presence of four (rather than only three) phalanges in digits III and IV of manus and pes. In addition, *Odontochelys* shares with *Proganochelys* some primitive features that are absent in all other known turtles, such as: the presence of denticles on the dermal palate (vomer and pterygoid); an open basicranial articulation; the development of a dorsal epiplastral process [also present in *Kayentachelys* (Joyce 2007)]; a broad and plate-like coracoid; an ilium with a short dorsal shaft; the presence of an hypoischium; and distinct gular projections on epiplastra. Whatever the relationships of turtles within amniotes may be (Rieppel 2008), there can be no doubt that amongst all the known fossil and extant turtles, *Odontochelys* is the geologically oldest and most basal one (Li et al. 2008).

However, it is not this conclusion that renders *Odontochelys* a controversial fossil, but rather the degree of development of its shell (Reisz and Head 2008). The plastron of *Odontochelys* is fully developed, comprising the anterior epiplastra and entoplastron, followed by one pair of hyoplastra, two pairs of mesoplastra, one pair of hypoplastra and one pair of xiphiplastra. Hyoplastra, mesoplastra and hypoplastra each bear laterally projecting frilled spines, a proper lateral bridge to connect with a carapace therefore being absent. Along the ventral midline, the mesoplastra are separated by a narrow, eloangate and irregularly shaped fontanelle, which exposes the serrated medial margins of the plastral elements.

The complete formation of a plastron stands in stark contrast with the incomplete development of the carapace in *Odontochelys*. Peripheral plates as well as a pygal are absent; whether a nuchal plate is present or absent remains presently unknown. This is unfortunate, for in extant turtles the nuchal plate is not only the first carapacial element to initiate ossification in ontogeny (e.g., Scheil 2003; Scheyer et al. 2008; Gilbert et al. 2008; Sánchez-Villagra et al. 2009). It is also characterized by a peculiar two-phase ossification (Menger 1922; Zangerl 1939; Burke 1989a; 'primary' and 'secondary' ossification sensu Gilbert et al. 2001, and Gilbert et al. 2008) that is correlated with the contribution of neural crest cells to its development, both features that the nuchal plate shares with the elements in the plastron (Cebra-Thomas et al. 2007; Gilbert et al. 2007; Sánchez-Villagra et al. 2009). Costal plates are likewise absent, but the dorsal ribs are expanded in the horizontal plane, corresponding to a degree of rib ossification characteristic for embryonic stage 23 of *Chelydra serpentina* (Yntema 1968). Neural plates are ossified, but at least partially ablated from the neural arches and shifted laterally during fossilization. Osteoderms are absent on the trunk, neck, tail and on the limbs. The question obviously arises whether this incomplete degree of carapace formation in *Odontochelys* is primitive (Li et al. 2008), or derived through reduction as an adaptation to aquatic habits (Reisz and Head 2008).

Odontochelys comes from marine sediments that were deposited in the Nanpanjian Trough Basin during the early to middle Carnian marine transgression; the basin remained surrounded by the Sichuan-Yunnan-Guizhou Old Land on three sides, opening into the Paleotethys towards the southwest (Wang et al. 2008). Fossilized remains of driftwood as well as other plant remains indicate the relative proximity of coastal waters (Wang et al. 2008). Limb proportions of *Odontochelys* indicate aquatic habits, most closely comparable to those extant turtles that inhabit stagnant or small bodies of water (Joyce and Gauthier 2003). *Odontochelys* certainly was not a pelagic animal as are the sea turtles and their fossil relatives, since it remained of relatively small body size and shows no sign of even an initial transformation of limbs into flippers. *Odontochelys* may accordingly have inhabited lagoonal settings along the coastline, or river delta systems. It is particularly in the sea turtles and their fossil relatives that the carapace and plastron undergo severe reduction, but the shell can also be reduced in freshwater turtles (e.g., trionychids), as well as in terrestrial turtles [e.g., *Manouria*, *Malacochersus* (Procter 1922; for a review see Pritchard 2008)]. The question thus arises whether the incomplete development of the carapace in *Odontochelys* represents an early stage in carapace evolution (Li et al. 2008), or whether the "carapace was present, but some of its dermal component were not ossified" (Reisz and Head 2008, p. 451). The latter hypothesis is based on the observation that the arrangement of the ribs in *Odontochelys* suggests the presence of a carapacial disc and ridge in the embryonic condition, and that "sea turtles and snapping turtles have greatly reduced ossification of the dermal components of the carapace, a condition similar to that seen in *Odontochelys*" (Reisz and Head 2008, p. 451).

Heterochrony (paedomorphosis) and associated shell reduction in living and fossil turtles was extensively studied by Kordikova (2000, 2002), with results that render *Odontochelys* an unlikely candidate for carapace reduction. All turtles—fossil or extant—that are characterized by a reduction of carapacial ossifications also show a reduction of plastral ossifications. Whereas the detailed patterns of plastral reductions may vary (Kordikova 2000, 2002; Pritchard 2008), *Odontochelys* shows a complete plastron except for a narrow midline fontanelle. It is possible, however, that the narrow separation of some of the plastral elements along the ventral midline was caused by postmortem dissociation. This is indicated by the serrated nature of the medial margins of the plastral plates that may have met in an interdigitating suture in the living animal. Further, a medial plastral fontanelle would be expected to be of rounded contours, and its occurrence is frequently associated with the formation of anterolateral and posterolateral fontanelles in the plastron, which are absent in *Odontochelys*. Although skeletal reductions due to paedomorphosis can be localized, it is unusual for such paedomorphotic reduction to affect only one skeletal component and none other. Those turtle lineages with the highest degree of carapace reduction show additional signs of paedomorphosis elsewhere in their skeleton. One of them is the failure of the astragalus and calcaneum to fuse in the adult, another is the reduction of claws in the digits of the forelimb (Kordikova 2000). In *Odontochelys*, claws are well developed on both manus and pes, and the astragalus and calcaneum are fused (in the paratype; sutured in the holotype), indicating an adult status for the specimen. And finally, all turtles—fossil or extant—irrespective of the degree of carapace reduction, show the intervertebral (intrasegmental) position of the proximal rib articulation. This is not the case in *Odontochelys*, where the ribs retain their primitive intersegmental position, articulating on the middle of the centrum.

Developmental Biology Illuminates the Interpretation of Fossils

Turtles are unique amongst tetrapods not only in the formation of a carapace and plastron, but also in the position of the scapula inside the rib cage (Rieppel 2001). Developmental studies indicate that the position of the scapula inside the rib cage is not the result of a backward displacement of the shoulder girdle (Ruckes 1929; Burke 1991), but of a deflection of rib growth into the carapacial ridge, i.e., the marginal zone of the carapacial disc that marks out the carapace in the turtle embryo through a thickening of the dermis (Burke 1989a, b).

Based on their developmental studies, Burke (1989a, b, 1991), Gilbert et al. (2001, 2008), Cebra-Thomas et al. (2005), and Nagashima et al. (2007, 2009) have developed a model for the de novo evolution of the turtle carapace. One of the earliest and most important developmental steps in the differentiation of the turtle body plan is the appearance of the carapacial ridge between and slightly above the anterior and posterior limb buds (Burke 1989a, b; Nagashima et al. 2007). Outgrowth of the carapacial ridge (at the ventral edge of the somatic mesoderm) causes the upper part of the lateral plate (body wall) to fold inwards, thus restricting rib growth to the axial domain (Nagashima et al. 2009). Fibroblast growth factor signaling maintains the carapacial ridge and allows it to capture the migrating rib precursor cells (Gilbert et al. 2001, 2008; Cebra-Thomas et al. 2005), which are primaxial in origin (Nagashima et al. 2007; Shearman and Burke 2009; Burke 2009). As a result, the ribs do not grow ventrally into the lateral body wall as in all other amniotes, but develop in a more superficial position within the thickened dermis of the carapacial disc. Embedded in the dermis, the cartilaginous ribs act as signaling centers for intramembranous ossification through bone morphogenetic proteins (BMPs) that are released from hypertrophic chondrocytes during endochondral ossification of the ribs (Cebra-Thomas et al. 2005; Gilbert et al. 2008). In that way, turtle ribs seem to function similarly to Meckel's cartilage, which is known to induce the ossification of the dentary around itself (Gilbert et al. 2001, p. 56). On that account, the evolution of the turtle shell does not involve the evolution of new tissue types, nor even of novel genes; instead, old developmental instructions for making certain tissue types are "used in new places" (Cebra-Thomas et al. 2005, p. 558) under the control of co-opted pre-existing regulatory genes (Kuraku et al. 2005). From a developmental and histological (rather than phylogenetic) perspective, the turtle carapace is interpreted as "a composite of endochondral axial skeleton [from the ribs (and neural arches)] plus intramembranous dermal bone" (Gilbert et al. 2008, p. 6).

The carapacial ridge, however, does not merely capture the rib precursor cells but also directs rib-growth in a fan-shaped pattern (Nagashima et al. 2007). Unlike in other amniotes, the fan-shaped arrangement of the dorsal ribs in turtles results in anterolaterally trending anterior, and posterolaterally trending posterior dorsal ribs. It is this anterolateral direction of the anteriormost dorsal rib that results in its position above the scapula. In the fossil turtle *Odontochelys*, the dorsal ribs are not arranged in such a fan-shaped pattern, the scapula consequently located in front of the anteriormost dorsal rib. Nagashima et al. (2009) hypothesize that the carapacial ridge was only incompletely developed in this taxon, extending along the flank of the embryo only and thus directing rib growth straight laterally. The completion of the carapacial ridge, encircling a carapacial disk not only laterally, but also anteriorly and posteriorly, would have occurred during a later stage of turtle evolution, resulting in the fan-shaped arrangement of the dorsal ribs, the anteriormost one capping the scapula. Barring post-mortem displacement of the ribs in *Odontochelys*, this is indeed a "fantastic example of the cross-illumination possible between paleontology and developmental biology" (Burke 2009, p. 623).

Conclusions

Rieppel and Kearney (2007) distinguished the 'generative' (here 'emergentist') from the 'transformational' approach in morphological analysis. The early appearance of a carapacial ridge, which triggers inductive epithelial-mesenchymal interactions at a new time and in a new place in the turtle embryo that affect the growth trajectory of the ribs provides an example for the emergence of an evolutionary novelty (Burke 1989a, b; Gilbert et al. 2001, 2008; Cebra-Thomas et al. 2005; Nagashima et al. 2007, 2009). This contrasts with the 'composite model' for turtle shell evolution, which views the carapace as a result of a stepwise transformation of ancestral structures, i.e., a superficial osteoderm covering of the hypothetical turtle ancestor.

The discussion above develops a number of arguments that can be made in support of the interpretation of *Odontochelys* as representing an early stage in carapace evolution—even in the absence of a well-corroborated phylogenetic hypothesis placing turtles within amniotes (Rieppel 2008). On this account, the absence of any osteoderms in *Odontochelys* conflicts with the 'composite model' of the turtle shell evolution. This model has most recently been supported with reference to ontogenetic studies and turtle shell histology (Scheyer et al. 2008), and with reference to an early turtle from the Upper Triassic Chinle deposits of New Mexico. The latter taxon, *Chinlechelys tenertesta* (Joyce et al. 2009) is known from very fragmentary and dissociated material only, identified as turtle remains predominantly on the basis of a fragment consisting of two partial dorsal vertebrae, overlying neural plate ossifications, and proximal rib articulations in an intervertebral (intrasegmental) position. This latter feature indicates that *Chinlechelys* had already achieved a degree of carapace differentiation that was not yet characteristic of *Odontochelys*. Taking *Odontochelys* to represent an early and primitive stage of carapace formation thus supports the emergentist model of the evolution of the turtle shell as a genuine evolutionary novelty.

The question remains why *Odontochelys* should have a complete ventral dermal covering while the carapace remained incomplete. For an aquatic organism, a ventral body armor offers protection against predatory attacks from below (Rieppel and Reisz 1999). Ventral body armor in an aquatic (marine) and lung-breathing organism would also perform an important hydrostatic function as 'bone ballast' (Taylor 2000, 2002). Three major strategies for buoyancy control have been identified in aquatic tetrapods: gastroliths, 'bone ballast' (in terms of relative size of skeletal elements, bone density, or pachyostosis), and blubber (Taylor 1993, 1994, 2000, 2002). Blubber is a buoyancy control mechanism restricted to marine mammals (Taylor 2002). Using gastroliths for buoyancy control would seem more versatile and would carry a lesser metabolic cost, but the effectiveness of gastroliths as a buoyancy control mechanism has been questioned (Henderson 2003, 2006; Wings 2007). Gastroliths furthermore are "rare or unknown from marine turtles" (Taylor 1993, p. 169). Developing a dermal armor for 'bone ballast' can thus be seen as an effective buoyancy control mechanism, while the hydrodynamically most advantageous position of such 'bone ballast' would certainly be a ventral one.

Acknowledgments I thank Anne Burke, Robert L. Carroll, Hiroshi Nagashima, Torsten M. Scheyer, Michael A. Taylor, and Matthew K. Vickaryous, who all offered helpful comments on an earlier draft of this paper.

References

Baur, G. (1886). Osteologische Notizen über Reptilien. *Zoologischer Anzeiger, 9*, 685–690.

Baur, G. (1887). On the morphogeny of the carapace of the Testudinata. *The American Naturalist, 21*, 89.

Baur, G. (1888). Unusual dermal ossifications. *Science, 11*, 144–145.

Baur, G. (1889). Die systematische Stellung von *Dermochelys* Blainv. *Biologisches Zentralblatt, 9*, 149–153.

Burke, A. C. (1989a). Critical feature in chelonian development: The ontogeny and phylogeny of a unique tetrapod bauplan. Ph.D Dissertation, Harvard University, Cambridge, MA.

Burke, A. C. (1989b). Development of the turtle carapace: Implications for the evolution of a novel bauplan. *Journal of Morphology, 199*, 363–378.

Burke, A. C. (1991). The development and evolution of the turtle body plan: Inferring intrinsic aspects of the evolutionary process from experimental embryology. *American Zoologist, 31*, 616–627.

Burke, A. C. (2009). Turtles ...... again. *Evolution & Development, 11*, 622–624.

Carroll, R. L. (1988). *Vertebrate Paleontology and Evolution*. New York: W. H. Freeman.

Carus, K. G. (1834). *Lehrbuch der Vergleichenden Zootomie. 2. Aufl., Bd. I.* Leipzig: Ernst Fleischer.

Case, E. C. (1898). Toxochelys. *University of Kansas Geological Survey, 4*, 370–385.

Cebra-Thomas, J. A., Betters, E., Yin, M., Plafkin, C., McDow, K., & Gilbert, S. F. (2007). Evidence that a late-emerging population of trunk neural crest cells forms the plastron in the turtle *Trachemys scriptascripta. Evolution & Development, 9*, 267–277.

Cebra-Thomas, J. A., Tan, F., Sistla, S., Estes, E., Bender, G., Kim, C., Riccio, P., & Gilbert, S. F. (2005). How the turtle forms its shell: A paracrine hypothesis of carapace formation. *Journal of Experimental Zoology (Mol Dev Evol), 304B*, 558–569.

Clark, K., Bender, G., Murray, B. P., Panfilio, K., Cook, S., Davis, R., Murnen, K., Tuan, R. S., & Gilbert, S. F. (2001). Evidence for the neural crest origin of turtle plastron bones. *Genesis, 3*, 111–117.

Cope, E. D. (1871). On the homologies of some of the cranial bones of the Reptilia, and the systematic arrangement of the class. In *Proceedings of the American Association for the Advancement of Science*, (pp. 194–247).

Cuvier, G. (1799). *Leçons d'Anatomie Comparée,* Vol. I. Paris: Boudouin.

Cuvier, G. (1812). *Recherches sur les Ossements Fossiles de Quadrupèdes,* Vol. IV. Paris: Deterville.

Darwin, Ch. (1859). *On the Origin of Species*. London: John Murray.

Dollo, L. (1886). Première note sur les chéloniens du Bruxellien (Eocène moyen) de la Belgique. *Bulletin du Musée Royale d'Histoire Naturelle de la Belgique, 4*, 75–96.

Gaffney, E. S. (1990). The comparative osteology of the Triassic turtle *Proganochelys. Bulletin of the American Museum of Natural History, 194*, 1–263.

Gauthier, J. A. (1994). The diversification of amniotes. In D. Prothero, & R. M. Schoch (Eds.), *Major Features of Vertebrate Evolution* (pp. 129–159). Knoxville: Paleontological Society.

Geoffroy, S-H. E. (1818). *Philosophie Anatomique. Des Organes Respiratoires sous le Rapport de la Détermination et de l'Identité de leurs Pièces Osseuses.* I. J.B. Baillière: Paris.

Gervais, P. (1872). Ostéologie du Sphargis Luth (Sph. coriacea). *Nouveau Archives du Muséum d'Histoire Naturelle, Paris, 8*, 199–228.

Gilbert, S. F., Loredo, G. A., Brukman, A., & Burke, A. C. (2001). Morphogenesis of the turtle shell: The development of a novel structure in tetrapod evolution. *Evolution & Development, 3*, 47–58.

Gilbert, S. F., Bender, G., Betters, E., Yin, M., & Cebra-Thomas, J. A. (2007). The contribution of neural crest cells to the nuchal bone and plastron of the turtle shell. *Integrative and Comparative Biology, 47*, 401–408.

Gilbert, S. F., Cebra-Thomas, J. A., & Burke, A. C. (2008). How the turtle gets its shell. In J. Wyneken, M. H. Godfrey & V. Bels (Eds.), *Biology of Turtles* (pp. 1–16). Boca Raton: CRC Press.

Goette, A. (1899). Über die Entwicklung des knöchernen Rückenschildes (Carapax) der Schildkröten. *Zeitschrift für wissenschaftliche Zoologie, 66*, 407–434.

Haines, R. W., & Mohuiddin, A. (1968). Metaplastic bone. *Journal of Anatomy, 103*, 527–538.

Hay, O. P. (1898). On *Protostega*, the systematic position of *Dermochelys*, and the morphogeny of the chelonian carapace and plastron. *American Naturalist, 32*, 929–948.

Hay, O.P. (1905). On the group of fossil turtles known as Amphichelydia; with remarks on the origin and relationships of the suborders, superfamilies, and families of Testudines. *Bulletin of the American Museum of Natural History, 21*, 137–175.

Hay, O. P. (1908). *The Fossil Turtles of North America.* Washington DC: Carnegie Institution.

Hay, O. P. (1922). On the phylogeny of the shell of the Testudinata and the relationships of *Dermochelys. Journal of Morphology, 36*, 421–445.

Hay, O. P. (1928). Further consideration of the shell of *Chelys* and of the constitution of the armor of turtles in general. In *Proceedings of the U.S. National Museum, 73*, pp.1–12.

Haycraft, J. B. (1891). The development of the carapace of the Chelonia. *Transactions of the Royal Society of Edinburgh, 36*, 335–342.

Henderson, D. M. (2003). Effects of stomach stones on the buoyancy and equilibrium of a floating crocodilian: A computational analysis. *Canadian Journal of Zoology, 81*, 1346–1357.

Henderson, D. M. (2006). Floating point: A computational study of buoyancy, equilibrium, and gastroliths in plesiosaurs. *Lethaia, 39*, 227–244.

Hill, R. V. (2005). Integration of morphological data sets for phylogenetic analysis: The importance of integumentary characters and increased taxonomic sampling. *Systematic Biology, 54*, 530–547.

Hoffmann, C. K. (1878). Beiträge zur vergleichenden Anatomie der Wirbelthiere. Tafel IX –XIII. *Niederländisches Archiv für Zoologie, 4*, 112–248.

Jollie, M. (1962). *Chordate Morphology.* New York: Reinhold.

Joyce, W. J. (2007). Phylogenetic relationships of Mesozoic turtles. *Bulletin of the Peabody Museum of Natural History, 48*, 3–102.

Joyce, W. G., & Gauthier, J. A. (2003). Paleoecology of Triassic stem turtles sheds new light on turtle origins. In *Proceedings of the Royal Society of London B, 271*, pp. 1–5.

Joyce, W. G., Lucas, S. G., Scheyer, T. M., Heckert, A. B., & Hunt, A. P. (2009). A thin-shelled reptile from the Late Triassic of North America and the origin of the turtle shell. In *Proceedings of the Royal Society of London B, 276*, 507–513.

Kälin, J. (1945). Zur Morphogenese des Panzers bei den Schildkröten. *Acta Anatomica, 1*, 144–176.

Kordikova, E. G. (2000). Paedomorphosis in the shell of fossil and living turtles. *Neues Jahrbuch für Geologie und Paläontologie, Abhandlungen, 218*, 399–446.

Kordikova, E. G. (2002). Heterochrony in the evolution of the shell of Chelonia. Part 1. Terminology, Cheloniidae, Dermochelyidae, Trionychidae, Cyclanorbidae and Carettochelyidae. *Neues Jahrbuch für Geologie und Paläontologie Abhandlungen, 226*, 343–417.

Kuraku, S., Usuda, R., & Kuratani, S. (2005). Comprehensive survey of carapacial ridge-specific genes in turtle implies co-option of some regulatory genes in carapace evolution. *Evolution & Development, 7*, 3–17.

Laurin, M., & Reisz, R. R. (1995). A reevaluation of early amniote phylogeny. *Biological Journal of the Linnean Society, 113*, 165–223.

Lee, M. S. Y. (1993). The origin of the turtle body plan: bridging a famous morphological gap. *Science, 261*, 1716–1720.

Lee, M. S. Y. (1996). Correlated progression and the origin of turtles. *Nature, 379*, 811–815.

Li, C., Wu, X.-C., Rieppel, O., Wang, L.-T., & Zhao, L.-J. (2008). An ancestral turtle from the Late Triassic of southwestern China. *Nature, 456*, 497–501.

Menger, W. (1922). *Ontogenie und Phylogenie des Schildkrötenpanzers.* Ph.D. Dissertation, Hessische Ludwigs-Universität, Giessen.

Merrem, B. (1820). *Versuch eines Systems der Amphibien. Tentamen sistematis amphibiorum.* Marburg.

Meyer, H.v. (1847). Mittheilungen an Professor Bronn gerichtet. *Neues Jahrbuch für Mineralogie, Geognosie, Geologie, und Petrefakten-Kunde, 1847*, 572–580.

Meyer, H. v. (1858). *Psephoderma alpinum* aus dem Dachsteinkalke der Alpen. *Palaeontographica, 6*, 246–252.

Moss, M. L. (1969). Comparative histology of dermal sclerifications in reptiles. *Acta Anatomica, 73*, 510–533.

Moustakas, J. E. (2008). Development of the carapacial ridge: implications for the evolution of genetic networks in turtle shell development. *Evolution & Development, 10*, 29–36.

Newman, H. H. (1905–1906). The significance of scute and plate "Abnormalities" in Chelonia. *Biological Bulletin, 10*, 68–114.

Nagashima, H., Kuraku, S., Uchida, K., Kawashima-Ohya, Y., Narita, Y., & Kuratani, S. (2007). On the carapacial ridge in turtle embryos: Its developmental origin, function and the chelonian body plan. *Development, 134*, 2219–2226.

Nagashima, H., Sugahara, F., Takechi, M., Ericsson, R., Kawashima-Ohya, Y., Narita, Y., & Kuratani, S. (2009). Evolution of the turtle body plan by folding and creation of new muscle connections. *Science, 325*, 193–196.

Nosotti, S., & Pinna, G. (1989). Storia delle ricerche e degli studi sui rettili placodonti. *Memorie della Società Italiana di Scienze Naturali e del Museo Civico di Storia Naturale di Milano, 24*, 29–86.

Oguschi, K. (1911). Anatomische Studien an der japanischen dreikralligen Lippenschildkröte (*Trionyx japanicus*). *Gegenbaurs Morphologisches Jahrbuch, 43*, 1–106.

Owen, R. (1849). On the development of the carapace and plastron of the chelonian reptiles. *Philosophical Transactions of the Royal Society of London, 139*, 151–171.

Patterson, C. (1977). Cartilage bones, dermal bones and membrane bones, or the exoskeleton versus the endoskeleton. In S. M. Andrews, R. S. Miles & A. D. Walker (Eds.), *Problems in Vertebrate Evolution* (pp. 77–121). London: Academic Press.

Pritchard, P. C. H. (2008). Evolution and structure of the turtle shell. In J. Wyneken, M. H. Godfrey, & V. Bels (Eds.), *Biology of the Turtles* (pp. 45–83). Boca Raton: CRC Press.

Procter, J. B. (1922). A study of the remarkable tortoise, *Testudo loveridgii* Blgr., and the morphogeny of the chelonian carapace. In *Proceedings of the Zoological Society of London, 92*, 483–526.

Rathke, H. (1848). *Uber die Entwicklung der Schildkröten*. Braunschweig: Friedrich Vieweg und Sohn.

Reisz, R. R., & Head, J. J. (2008). Turtle origins out to sea. *Nature, 456*, 450–451.

Rieppel, O. (2001). Turtles as hopeful monsters. *BioEssays, 23*, 987–991.

Rieppel, O. (2008). The relationships of turtles within amniotes. In J. Wyneken, M. H. Godfrey, & V. Bels (Eds.), *Biology of the Turtles* (pp. 345–353). Boca Raton: CRC Press.

Rieppel, O., & Kearney, M. (2007). The poverty of taxonomic characters. *Biology & Philosophy, 22*, 95–113.

Rieppel, O., Reisz, R. R. (1999). The origin and early evolution of turtles. *Annual Review of Ecology and Systematics, 30*, 1–22.

Ruckes, H. (1929). Studies in chelonian osteology. Part II. The morphological relationships between girdles, ribs and carapace. *Annals of the New York Academy of Sciences, 31*, 81–120.

Sánchez-Villagra, M. R., Müller, H., Scheil, C. A., Scheyer, T. M., Nagashima, H., & Kuratani, S. (2009). Skeletal development in the Chinese soft-shelled turtle *Pelodiscus sinensis* (Testudines: Trionychidae). *Journal of Morphology, 270*, 1381–1399.

Shearman, R. M., & Burke, A. C. (2009). The lateral somatic frontier in ontogeny and phylogeny. *Journal of Experimental Biology (Mol Dev Evol), 312B*, 603–612.

Scheil, A. A. (2003). Osteology and skeletal development of *Apalone spinifera* (Reptilia: Testudines: Trionychidae). *Journal of Morphology, 256*, 42–78.

Scheyer, T. M., Brüllmann, B., & Sánchez-Villagra, M. R. (2008). The ontogeny of the shell in side-necked turtles, with emphasis on the homologies of costal and neural bones. *Journal of Morphology, 269*, 1008–1021.

Smith, M. M., & Hall, B. K. (1990). Development and evolutionary origins of vertebrate skeletogenic and odontogenic tissues. *Biological Reviews, 65*, 277–373.

Smith, M. M., & Hall, B. K. (1993). A developmental model for the evolution of vertebrate exoskeleton and teeth: the role of cranial and trunk neural crest. *Evolutionary Biology, 27*, 387–448.

Starck, D. (1955). *Embryologie*. Stuttgart: Georg Thieme.

Starck, D. (1979). *Vergleichende Anatomie der Wirbeltiere, Bd. 2*. Berlin: Springer.

Strauch, A. (1890). Bemerkungen über die Schildkrröten-sammlung im Zoologischen Museum der kaiserlichen Akademie der Wissenschaften zu St. Petersburg. *Mémoires de l'Académie Impériale des Sciences, St. Petersburg, (7)* 38, 1–127.

Taylor, A. M. (1993). Stomach stones for feeding or buoyancy? The occurrence and function of gastroliths in marine tetrapods. *Philosophical Transactions of the Royal Society of London B, 341*, 163–175.

Taylor, A. M. (1994). Stone, bone or blubber? Buoyancy control strategies in aquatic tetrapod. In L. Maddock, Q. Bone & J. M. V. Rayner (Eds.), *Mechanics and Physiology of Animal Swimming* (pp. 151–161). Cambridge UK: Cambridge University Press.

Taylor, A. M. (2000). Functional significance of bone ballast in the evolution of buoyancy control strategies by aquatic tetrapods. *Historical Biology, 14*: 15–31.

Taylor, A. M. (2002). Origin of marine mammals. In W. F. Perrin, B. Würsig, & J. G. M. Thewissen (Eds.), *Encyclopedia of Marine Mammals* (pp. 833–837). San Diego: Academic Press.

Vallén, E. (1942). Beiträge zur Kenntnis der Ontogenie und der vergleichenden Anatomie des Schildkrötenpanzers. *Acta Zoologica, Stockholm, 23*, 1–127.

Versluys, J. (1914). Über die Phylogenie des Panzers der Schildkröten und über die Verwandtschaft der Lederschildkröte (*Dermochelys coriacea*). *Paläontologische Zeitschrift, 1*, 321–347.

Vickaryous, M. K., & Hall, B. K. (2008). Development of the dermal skeleton in *Alligator mississippiensis* (Archosauria, Crocodylia) with comments on the homology of osteoderms. *Journal of Morphology, 260*, 398–422.

Völker, H. (1913). Über das Stamm, Gliedmassen-, und Hautskelett von *Dermochelys coriacea* L. *Zoologische Jahrbücher, Abteilung für Anatomie und Ontogenie der Tiere, 33*, 431–552.

Wang, X., Bachmann, G. H., Hagdorn, H., Sander, P. M., Cuny, G., Chen, X., et al. (2008). The Late Triassic black shales from the Guanling area, Guizhou Province, south-west China: A unique marine reptile and pelagic crinoid fossil Lagerstätte. *Palaeontology, 51*, 27–61.

Wiedeman, C. R. W. (1802). Anatomische Beschreibung der Schildkröten überhaupt und der getäfelten Schildkröte (*T. tesselata* Schneid., *T. tabulata* Walbaum) insbesondere. *Archiv für Zoologie und Zootomie, 2*, 177–210.

Wieland, G. R. (1896). *Archelon ischyrios*: A new gigantic Cryptodire Testudinate from Fort Pierre Cretaceous of South Dakota. *American Journal of Science, 2*(4), 399–415.

Wieland, G. R. (1905). On marine turtles. *American Journal of Science, 20*, 325–343.

Wieland, G. R. (1909). Revision of the Protostegidae. *American Journal of Science, 27*(4), 237–251.

Wings, O. (2007). A review of gastrolith function with implications for fossil vertebrates and a revised classification. *Acta Palaeontologica Polonica, 52*: 1–16.

Yntema, C. L. A. (1968). A series of stages in the embryonic development of *Chelydra serpentina*. *Journal of Morphology, 125*, 219–252.

Zangerl, R. (1939). The homology of the shell elements in turtles. *Journal of Morphology, 65*, 383–406.

Zangerl, R. (1969). The turtle shell. In C. Gans, A. d'A. Bellairs, & T. S. Parsons (Eds.), *Biology of the Reptilia,* (Vol. 1). *Morphology A* (pp. 311–339). London: Academic Press.

Chapter 6
Three Ways to Tackle the Turtle: Integrating Fossils, Comparative Embryology, and Microanatomy

Torsten M. Scheyer, Ingmar Werneburg, Christian Mitgutsch, Massimo Delfino, and Marcelo R. Sánchez-Villagra

Abstract Herein we review a series of case studies covering the evolution and phylogenesis of turtles, and the ontogenetic development of one of the most peculiar body plans within the Craniota. Comparative analyses of skeletal development, ontogenetic timing, and bone microstructure in both extant and extinct taxa are used to document patterns and make inferences about the origin of turtles, turtle ingroup relationships, and the evolution of turtle ontogenetic development. The need for a balanced sampling of both cryptodiran and pleurodiran turtle species for future comparative studies is highlighted.

Keywords Bone histology • Comparative anatomy • *Odontochelys* • Sequence heterochrony • Testudinata • Testudines • Turtle origins

T. M. Scheyer (✉) · I. Werneburg · C. Mitgutsch · M. R. Sánchez-Villagra
Paläntologisches Institut und Museum, Universität Zürich, Karl Schmid-Strasse 4, 8006 Zürich, Switzerland
e-mail: tscheyer@pim.uzh.ch; ingmar_werneburg@yahoo.de; christian.mitgutsch@gmail.com; m.sanchez@pim.uzh.ch

M. Delfino
Dipartimento di Scienze della Terra, Università di Torino, Via Valperga Caluso 35, 10125 Turin, Italy
and
Institut Català de Paleontologia Miquel Crusafont, Universitat Autònoma de Barcelona. Edifici ICP, Campus de la UAB s/n, 08193 Cerdanyola del Vallès, Barcelona, Spain
e-mail: massimo.delfino@unito.it

Introduction

Hypotheses on the position of turtles within the amniote tree of life are contentious (Carroll 2012; Lyson et al. 2010), despite many efforts over the past century to solve this major problem in vertebrate evolution (Rieppel 2008). This situation is largely due to the peculiar morphology of the turtle body plan, especially the unique shell (Nagashima et al. 2009, 2012; Kuratani et al. 2011; Rieppel 2012) and the anapsid condition of the skull (Gaffney and Meylan 1988; Lee 1997a), which renders comparisons to other vertebrates difficult at best. In recent years, turtles have been treated either as a group of parareptiles or as diapsids; for the latter, relationships with Lepidosauromorpha, Archosauromorpha, or Sauria have been hypothesized.

The past two decades have brought refined techniques in molecular biology (Shaffer 2009) and evolutionary developmental biology, as well as in comparative anatomy, all of which have helped clarify interrelationships among living turtles (e.g., see summary by Werneburg and Sánchez-Villagra 2009). For ontogenetic approaches, several methods have been developed to analyze developmental timing data within a phylogenetic framework. One of those methods, Parsimov, has been extensively used to assess information from organogenesis or ossification patterns, for which an expanding body of data is being assembled (e.g., Sánchez-Villagra et al. 2009 and references therein). However, atomizing heterochronic data as performed in event-pair based algorithms, as well as the robustness and the value of its information content, have recently been reassessed and alternative approaches have been suggested (Ziermann 2008; Germain and Laurin 2009; Werneburg 2010; Werneburg and Sánchez-Villagra 2011).

In addition, new fossil discoveries continue to be important for documenting the evolution of turtles. Among the most important of these are fossils and taxa from Upper Triassic sediments on several continents, such as skeletons of

D. B. Brinkman et al. (eds.), *Morphology and Evolution of Turtles*, Vertebrate Paleobiology and Paleoanthropology, DOI: 10.1007/978-94-007-4309-0_6,

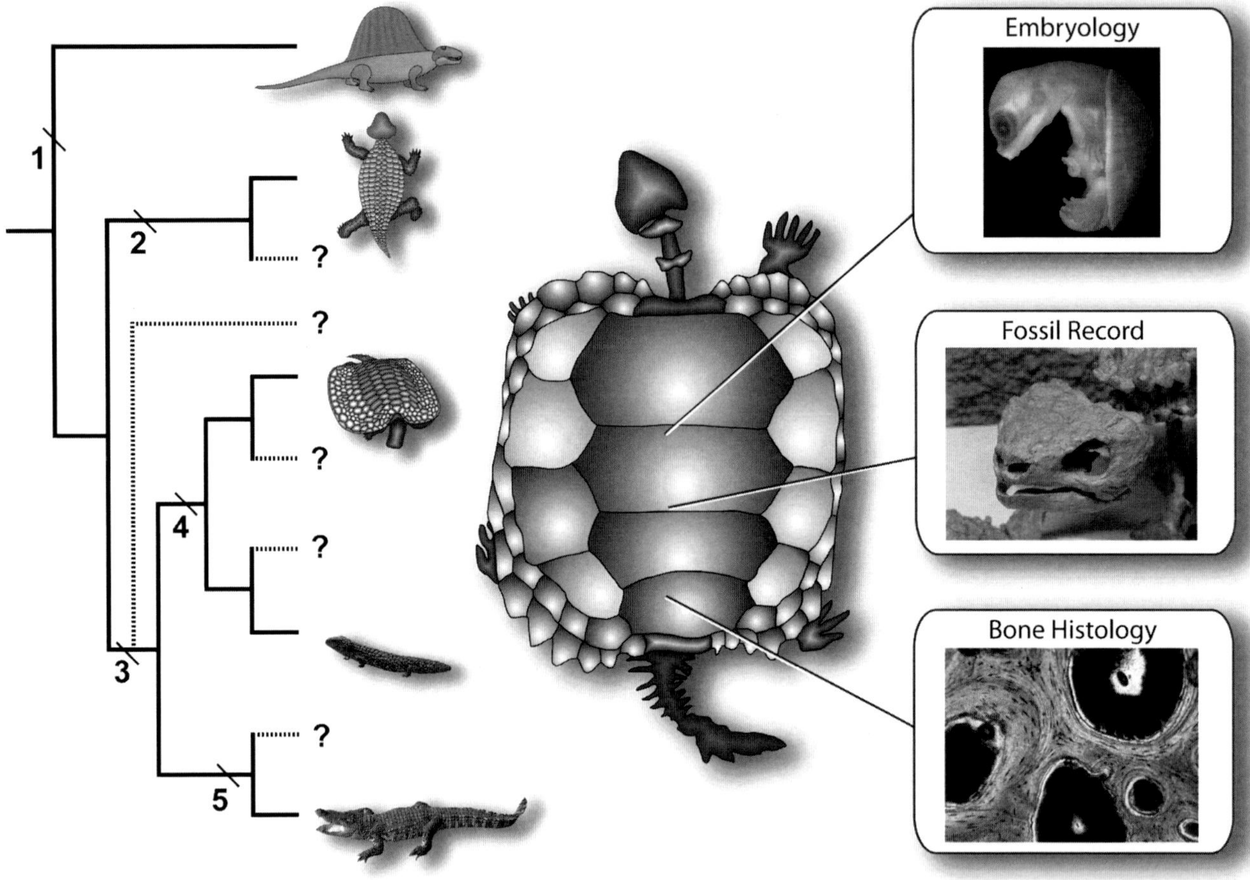

Fig. 6.1 Various hypotheses of the position of turtles within the amniote tree of life and applied methods of investigation. Numbers indicate proposed relationships between turtles and the following groups: *1* Synapsida; *2* Parareptilia; *3* Diapsida; *4* Lepidosauromorpha; *5* Archosauromorpha

Odontochelys semitestacea from China (Li et al. 2008), which are providing new insights into the early evolution of turtles.

Lately, there has been an increasing trend to integrate ontogenetic aspects into paleontology-based studies and to interpret molecular data in the light of insights yielded from fossil taxa (Raff 2007; Shubin et al. 2009; Sánchez-Villagra 2010). Fossil developmental data, as is true for neontological data, can potentially reveal systematic relationships that are not apparent from adult morphology (e.g., Hall 2005; Schoch 2009; Sánchez-Villagra 2010). In a controversial clade like turtles, it is critical to consider as many developmental and evolutionary aspects as possible, in order to have a better understanding of the group's evolutionary history. Here we review a series of case studies that integrate evolutionary, developmental, and bone histological data from fossil and extant turtles (Fig. 6.1).

As discussed below, the existence and mode of reduction of shell features is a subject of particular relevance in current discussions of turtle origins (Lee 1997a, b; Li et al. 2008; Lyson and Gilbert 2009; Reisz and Head 2008; Burke 2009; Nagashima et al. 2009, 2012; Kuratani et al. 2011; Rieppel 2012). Because the turtle shell is such an enigmatic and key autapomorphy of turtles, many of the case studies summarized below focus on that part of the skeleton.

Heterochrony Analyses and Skeletal Development in Turtles

Several methods are now available to put sequence ontogenetic data into a phylogenetic context (Hall 2005; see also review by Maxwell and Harrison 2009; Germain and Laurin 2009), with the most widely used method currently being Parsimov (Jeffery et al. 2005). Patterns of chondrification and the onset of ossification were analyzed in several turtle taxa and a variety of outgroups in two recent studies by Sánchez-Villagra et al. (2009) and Werneburg et al. (2009). In the former study, a comprehensive developmental series of the Chinese soft-shelled turtle *Pelodiscus sinensis* demonstrated that in both the cranium and postcranium, dermal

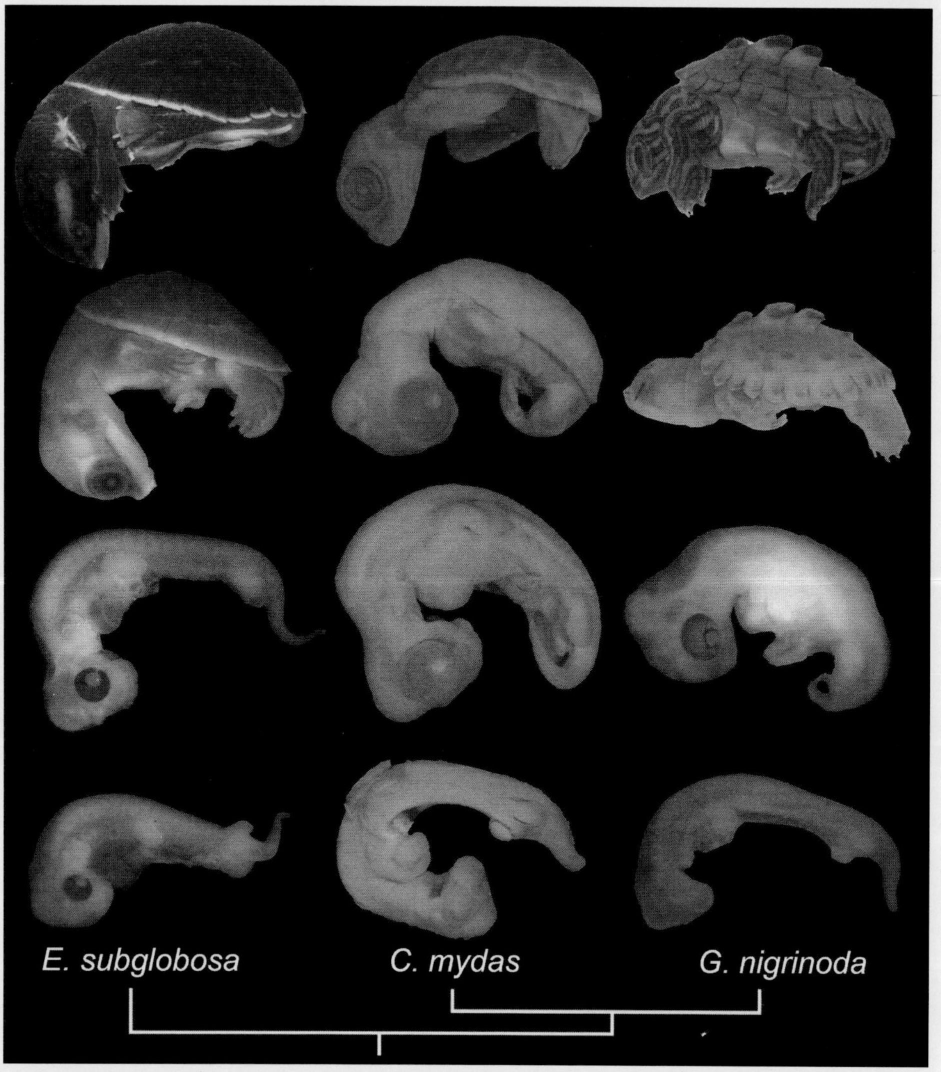

Fig. 6.2 Developmental series of three extant turtle species, demonstrating the systematic value of embryological data. Taxa are: the pleurodire *Emydura subglobosa* (*left*) and the cryptodires *Chelonia mydas* (*middle*) and *Graptemys nigrinoda* (*right*). The selected specimens [all stored at the Paläntologisches Institut und Museum, Universität Zürich, Switzerland (PIMUZ)] illustrate major steps in development such as somitogenesis, limb bud formation, and development of the carapace. Specimens not to scale

elements ossify earlier than endochondral elements. Analysis of the ossification sequence of autopodial elements in *P. sinensis* also revealed high intraspecific variation, whereas the pattern of chondrification was consistent with Shubin and Alberch's (1986) concept of primary axis and digital arch. Finally, heterochronic shifts in chondrification patterns of fore- and hind limbs were minimal and there was no evidence for the anlage of a radiale or a tibiale in this taxon (Sánchez-Villagra et al. 2009), which confirmed the lack of those elements in turtles as had been independently suggested by Fabrezi et al. (2009).

The study by Werneburg et al. (2009) examined the timing of organogenesis in several turtle species, including for the first time a pleurodire species (Fig. 6.2). By analyzing a total of 15 turtle and seven other amniote species, and using the axolotl *Ambystoma mexicanum* as an outgroup, the position of turtles as the sister group to the Archosauria/Lepidosauria clade was supported. That result was consistent with some previous, morphology-based analyses of tetrapod relationships (e.g., Gauthier et al. 1988; Laurin and Reisz 1995), but stands in contrast to most, but not all, recent molecular studies that favored a turtle-archosaur relationship (e.g., Rest et al. 2003; Iwabe et al. 2004; Lee et al. 2004) and to some morphology-based studies that supported a turtle-lepidosaur relationship (e.g., Rieppel and deBraga 1996; deBraga and Rieppel 1997; Müller 2003; Hill 2005). A preliminary analysis of turtle shell bone microstructure also hinted at a potential turtle-archosaur relationship (Scheyer 2007).

One of us developed a new referencing system to study the timing of events in development with the goal of optimizing the acquisition of embryological data for vertebrates (Werneburg 2009). This new referencing system was applied by Werneburg et al. (2009) in a study on the embryogenesis and ossification patterns in the chelid *Emydura subglobosa*, which broadened the data base on the previously neglected pleurodiran branch of turtles. By comparing data from *E. subglobosa* with five other turtle

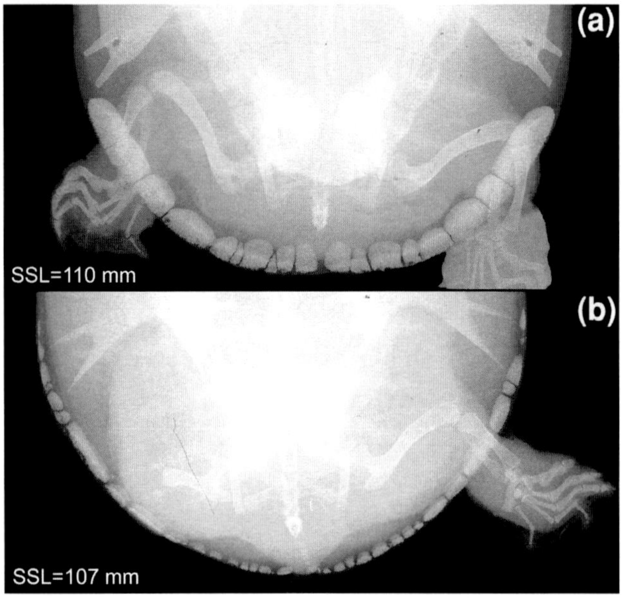

Fig. 6.3 Radiographs of the posterior portion of the body of extant *Lissemys* (Trionychidae), showing the different arrangements of peripheral ossicles around the posterior margin of the carapace that are unique to this genus within the family. **a** *Lissemys punctata* (MTD 32146) and **b** *Lissemys scutata* (MTD 40367). Pictures taken at the Museum of Zoology (Museum für Tierkunde), Senckenberg Dresden, Germany (MTD), in collaboration with Uwe Fritz. Abbreviation: SSL, straight shell length

species and a number of other tetrapod taxa from the literature, Werneburg et al. (2009) showed that late development of the neck is autapomorphic for Testudines. Within turtles, the same study further showed that (1) timing in the development of the mandibular process differed between pleurodires and cryptodires and (2) the epiplastron developed earlier in pleurodires than in cryptodires.

Homology of Peripheral Ossicles in Soft-Shelled Turtles

Most soft-shelled turtles are characterized by a complete lack of peripherals. The species of *Lissemys*, however, are an exception in having a series of peripheral ossicles posteriorly and a prenuchal anteriorly. Delfino et al. (2010) reassessed the nature of these ossicles using data generated from X-ray scans and macerated skeletons of *L. punctata* and *L. scutata* (Fig. 6.3), and by studying the microstructure of those bones histologically. All of Remane's (1952) classic criteria for homology assessment—the criteria of topology, structural equivalence, and presence of intermediate forms—were considered in testing the identity of the peripheral ossicles in *Lissemys*. Posterior peripheral ossicles were found to fulfill all of Remane's (1952) criteria. Although *Lissemys* might not be the sister group of all remaining trionychids (Joyce and Lyson 2010; but see Head et al. 2009), their posterior peripheral ossicles can still be interpreted as homologues to the peripherals of non-trionychid turtles based on 'deeply homologous' (sensu Shubin et al. 2009) underlying developmental processes, despite a lack of one-to-one correlation between the ossicles and the dorsal ribs or costal plates. The prenuchal on the other hand, in not fulfilling any of Remane's (1952) criteria, is interpreted as a neomorphic dermal bone present only in species of *Lissemys* and *Cyclanorbis*.

Ontogeny of the Shell in Side-Necked Turtles

Scheyer et al. (2008) combined histological data from shell bones of fossil and extant post-hatching turtles with ontogenetic data of extant turtles to examine development of the shell in pleurodires. Their data set included a developmental series for the extant pleurodire *Emydura subglobosa* and revealed a bimodal development for the costals and neurals (see below). No distinct ossification centers were identified for the costals or neurals. Instead, these elements start as initial outgrowths of the periosteum of the ribs and vertebral arches, respectively, before the mode of ossification switches to metaplastic ossification (i.e., ossification and incorporation of surrounding preformed and differentiated dermal tissue into the bone [Scheyer et al. 2008]). Metaplastic ossification had already been identified as the main mode of ossification in trionychid turtle shells, where the incorporation and transformation of preformed dermal soft tissue structures into the mineralized bone tissue is most obvious (Scheyer et al. 2007). In contrast, in a recent review of integumentary structures by Vickaryous and Sire (2009), which was based on several other developmental studies (Gilbert et al. 2001, 2007; Cebra-Thomas et al. 2007; summarized by Rieppel 2012), intramembraneous ossification (i.e., growth of bone spiculae into the surrounding soft tissue) was proposed to be the main mode of costal and neural development. Given these conflicting data and interpretations, here we propose that both metaplastic and intramembraneous ossification, to various degrees and at different stages in ontogeny, play a role in costal and neural formation.

The complete reduction (i.e., loss) of neurals frequently occurs among extant pleurodires (e.g., in Chelidae) and also in some fossil pleurodires, such as the pelomedusoid *Bairdemys* from the Caribbean Neogene (Wood and Díaz de Gamero 1971; Sánchez-Villagra and Scheyer 2010). The ontogenetic data assembled by Scheyer et al. (2008) revealed a heterochronic shift (i.e., postdisplacement), in neural arch and neural development in pleurodire species that lack neurals. This heterochronic shift, together with a reduction or lack of interaction between integumentary and musculoskeletal structures, was proposed to be responsible for complete neural reduction in *Emydura subglobosa*.

Implications for the Origin of Turtles

As indicated above, the systematic position of turtles within amniotes is far from being resolved. Several extinct reptilian clades have been hypothesized to be close turtle relatives (Fig. 6.1; Carroll 2012), one of them being the anapsid pareiasaurs (e.g., Lee 1997a, b), which are large, herbivorous, parareptiles from the Permian (Tsuji and Müller 2009), whose bodies were, to varying amounts, usually covered with dermal armor plates (i.e., postcranial osteoderms). A recent histological study of osteoderms of three pareiasaur taxa from South Africa showed that intramembraneous ossification is likely to be the general mode of skeletogenesis for pareiasaur postcranial armor (Scheyer and Sander 2009). Those elements share only a few histological characters with turtle shell bones and osteoderms (or with other extant eureptilian osteoderms). That finding indicates that the armor structures of turtles and pareiasaurs are not homologous, which argues against a turtle-pareiasaur relationship.

Recently, another classical morphology-based sister group relationship has been revived. In Lyson et al.'s (2010) study, turtles were recovered as the sister group to *Eunotosaurus africanus*, an un-armored, terrestrial parareptile from the Middle Permian of South Africa (Rubidge et al. 1999), which has extremely broadened ribs.

Prior to the description of the now oldest and basal-most turtle, *Odontochelys semitestacea* from Upper Triassic, marine black shales in China (Li et al. 2008), all stem turtles of comparable age (i.e., *Paleochersis talampayensis*, *Proganochelys quenstedti*, and *Proterochersis robusta*) in recent years had been interpreted as terrestrial forms (Joyce and Gauthier 2004; Scheyer and Sander 2007; Sterli et al. 2007). At present, *Chinlechelys tenertesta* from the Upper Triassic Chinle Group of New Mexico, USA, remains problematic in this regard, due to its highly fragmentary nature (Joyce et al. 2009). *Odontochelys* combines a rudimentary carapace with a fully developed and well-ossified plastron; in that sense, it can be described as being 'half-shelled'. The shell structure and depositional setting of *Odontochelys* raised the question whether turtles had an aquatic or a terrestrial origin (Rieppel and Reisz 1999; Li et al. 2008; Reisz and Head 2008; Rieppel 2012). As summarized by Lyson and Gilbert (2009, p. 133), it is still unclear whether *Odontochelys* argues for a primary aquatic origin for turtles, as Li et al. (2008) proposed, or if *Odontochelys* belongs to an early radiation of turtles into the water, as favored by Reisz and Head (2008).

Several years previously, the oldest purported fossil turtle shell fragment, belonging to *Priscochelys hegnabrunnensis* and recovered from deposits of the Germanic Muschelkalk (Middle Triassic) in southwestern Germany, had prompted Joyce and Karl (2006) to propose that turtles might have evolved in an aquatic environment. Using bone histology observation, Scheyer (2008) subsequently showed that material of *Priscochelys* represents a placodont armor fragment, instead of a turtle shell fragment and, therefore, is irrelevant to discussions on the origin of turtles.

Recent histological and embryonic studies (e.g., Scheyer et al. 2008; Sánchez-Villagra et al. 2009 and references therein) of extant turtles are potentially relevant for assessing the nature of the neurals and costals in *Odontochelys*. The broadened or expanded ribs (Li et al. 2008) of *Odontochelys* look like paedomorphic ribs of extant turtles in being slightly reduced in length, to the extent that their distal ends do not reach the lateral extensions of the hyo- and hypoplastral lateral processes of the plastron. Although the ribs of *Odontochelys* resemble turtle costals in shape and topology, they differ in lacking anterior and posterior sutural margins. A similar lack of sutures, at least in the distal part of the costals, occurs in some extant turtles; e.g., in 'geodesic' tortoises (sensu Pritchard 2008) *Manouria emys* and *Malacochersus tornieri* (Fig. 6.4). In the former species, the fontanelles in the carapace (which separate the distal ends of the costals) and plastron eventually close in adults, whereas these openings are permanent in the latter (Procter 1922). In all species of *Manouria* and *Malacochersus*, sutural contact of the carapace and the plastron remains via the bridge peripherals, to maintain the structural integrity of the shell (Procter 1922; Pritchard 2008).

Li et al. (2008, p. 499) stated that, based on the broad shape of its dorsal ribs, *Odontochelys* resembled an embryonic, extant turtle at the stage "… before marginal, nuchal and pygal elements start ossification." However, whereas peripherals and pygal elements appear relatively late in comparison to the neurals and costals (e.g., Sheil and Greenbaum 2005), the nuchal is amongst the first bones to ossify during postcranial development in extant turtles, shortly after the onset of ossification of plastral elements (contra Li et al. 2008). Even in taxa in which the peripheral shell elements are lost by developmental truncation, such as in most trionychids (e.g., Sánchez-Villagra et al. 2009) and in the leatherback *Dermochelys coriacea*, the nuchal is always consistently present and a prominent element of the thecal shell (e.g., Scheyer 2007). Furthermore, the statement that the "neural plates are not fused with broadened neural spines of dorsal vertebrae but were displaced laterally during fossilization" (Li et al. 2008, p. 498) raises the

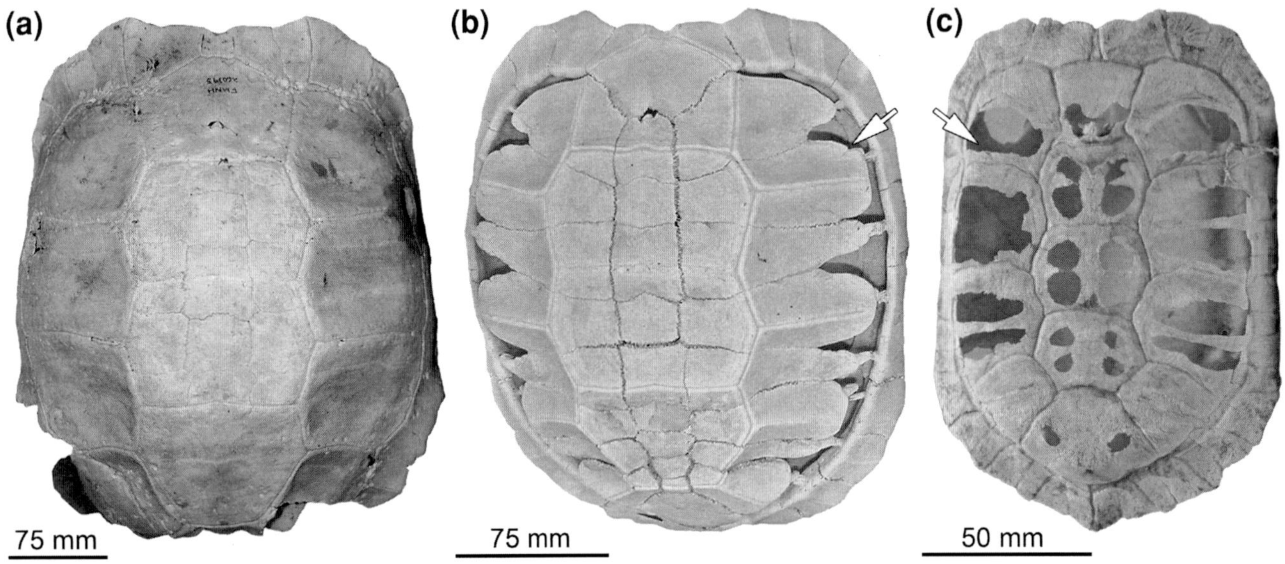

Fig. 6.4 Carapaces of the tortoises *Manouria emys* and *Malacochersus tornieri*, as examples of extant turtles in which the carapace bones may not be sutured together. **a–b** Sub-adult specimens of *Manouria Emys*: **a** example of carapace already completely ossified (FMNH 260395; note that shell is damaged posteriorly); **b** 'geodesic' shell condition (MTD D 19981) in which the costals are still paper-thin bones that largely lack sutures distally. Fontanelles (*arrow* in **b**) may close later during ontogeny. **c** Adult specimen of *Malacochersus tornieri* (MTD D 010) showing the extremely reduced condition of shell bone formation and the persisting fontanelles (*arrow*) between the costals and neurals typical for this species. Institutional abbreviations: FMNH: The Field Museum, Chicago, Illinois, USA; MTD: Museum of Zoology (Museum für Tierkunde), Senckenberg Dresden, Germany

possibility that separation of the neural plates from the corresponding neural arches could be a taphonomic effect.

The question remains whether the lack of a nuchal in the holotype skeleton of *Odontochelys* is because that bone was absent in life (i.e., not developed) or if it was present but lost postmortem, perhaps due to the same taphonomic processes that may have been responsible for shifting the neurals (see also discussion in Rieppel 2012). Based on the pictures and drawings provided by Li et al. (2008), the presence or absence of a nuchal bone also remains ambiguous in the paratype and only other published, relatively complete specimen of *Odontochelys*. Considering that in many living turtles the nuchal is an integral part of the neck muscular system (e.g., Herrel et al. 2008), if the nuchal was indeed absent in *Odontochelys* that could have significant implications for the structure and function of its neck.

Conclusions

In summary, it is essential that in a group with a peculiar body plan like turtles all sources of data are exploited by integrating paleontological and neontological data. Detailed anatomical studies can expose hidden homologies that were previously not recognized.

The timing of organogenesis as revealed by heterochrony studies using both cryptodires and pleurodires indicates a possible turtle-Sauria sister-group relationship. Turtle autopodial development, as studied for example in the trionychid *Pelodiscus sinensis*, is in accordance with the concept of a primary axis and digital arch. In addition, the posterior peripheral ossicles in the trionychid *Lissemys* were found to fulfill all of Remane's (1952) criteria of homology and, thus, were interpreted to be homologous with the peripherals in non-trionychid turtles based on 'deeply homologous' underlying developmental processes. The nature of the neural and costal bones in the turtle shell remains poorly understood; however, metaplastic and intramembraneous modes of ossification appear to contribute to their formation at different stages during ontogeny. A heterochronic shift involving post-displacement in neural arch and neural development, together with reduction or lack of interaction between the integument and underlying musculoskeleton was presented as a cause for the complete reduction (i.e., loss) of neurals in fossil and extant pleurodires. Furthermore, new fossil finds, especially of Mesozoic stem-turtles such as *Odontochelys semitestacea*, have revealed new and exciting aspects of anatomy that shed light on the early evolution of the group and the development of one of its key features, the shell.

Acknowledgments We thank Shigeru Kuratani (Kobe), Johannes Müller (Berlin), Uwe Fritz (Dresden), Alan Resetar (Chicago), and Janine M. Ziermann (Jena/Leiden), as well as all of their colleagues, for various help in accessing and studying specimens. Nadia Fröbisch (Chicago), James Gardner (Drumheller), Olivier Rieppel (Chicago), and Matt Vickaryous (Guelph) are thanked for their constructive comments on the manuscript. We also thank all of our collaborators and the SNSF, grant 31003A_127053/1 (to TMS), for supporting this research.

References

Burke, A. C. (2009). Turtles.... again. *Evolution & Development, 11*, 622–624.

Carroll, R. L. (2012). Problems of the ancestry of turtles. In D. B. Brinkman, P. A. Holroyd & J. D. Gardner (Eds.), *Morphology and evolution of turtles*. Dordrecht: Springer.

Cebra-Thomas, J. A., Betters, E., Yin, M., Plafkin, C., McDow, K., & Gilbert, S. F. (2007). Evidence that a late-emerging population of trunk neural crest cells forms the plastron bones in the turtle *Trachemys scripta*. *Evolution & Development, 9*, 267–277.

deBraga, M., & Rieppel, O. (1997). Reptile phylogeny and the interrelationships of turtles. *Zoological Journal of the Linnean Society, 120*, 281–354.

Delfino, M., Scheyer, T. M., Fritz, U., & Sánchez-Villagra, M. R. (2010). An integrative approach to examining a homology question: Shell structures in soft-shelled turtles. *Biological Journal of the Linnean Society, 99*, 462–476.

Fabrezi, M., Manzano, A., Abdala, V., & Zaher, H. (2009). Developmental basis of limb homology in pleurodiran turtles, and the identity of the hooked element in the chelonian tarsus. *Zoological Journal of the Linnean Society, 155*, 845–866.

Gaffney, E. S., & Meylan, P. A. (1988). A phylogeny of turtles. In M. J. Benton (Ed.), *The Phylogeny and Classification of the Tetrapods* (Vol. 1, pp. 157–219). Amphibians, Reptiles, Birds. Oxford: Clarendon Press.

Gauthier, J., Kluge, A. G., & Rowe, T. (1988). Amniote phylogeny and the importance of fossils. *Cladistics, 4*, 105–209.

Germain, D., & Laurin, M. (2009). Evolution of ossification sequences in salamanders and urodele origins assessed through event-pairing and new methods. *Evolution & Development, 11*, 170–190.

Gilbert, S. F., Loredo, G. A., Brukman, A., & Burke, A. C. (2001). Morphogenesis of the turtle shell: The development of a novel structure in tetrapod evolution. *Evolution & Development, 3*, 47–58.

Gilbert, S. F., Bender, G., Betters, E., Yin, M., & Cebra-Thomas, J. A. (2007). The contribution of neural crest cells to the nuchal bone and plastron of the turtle shell. *Integrative and Comparative Biology, 47*, 401–408.

Hall, B. K. (2005). *Bones and cartilage. Developmental and Evolutionary Skeletal Biology*. Amsterdam: Elsevier Academic Press.

Head, J. J., Hutchison, J. H., & Evans, D. C. (2009). The phylogenetic interrelationships and antiquity of plastomenid turtles. *Gaffney Turtle Symposium (October 17–18, 2009, Royal Tyrrell Museum, Drumheller, Canada). Abstract Volume,* 71.

Herrel, A., Van Damme, J., & Aerts, P. (2008). Cervical anatomy and function in turtles. In J. Wyneken, M. H. Godfrey & V. Bels (Eds.), *Biology of Turtles* (pp. 163–185). Boca Raton: CRC Press.

Hill, R. V. (2005). Integration of morphological data sets for phylogenetic analysis of Amniota: The importance of integumentary characters and increased taxonomic sampling. *Systematic Biology, 54*, 530–547.

Iwabe, N., Hara, Y., Kumazawa, Y., Shibamoto, K., Saito, Y., Miyata, T., et al. (2004). Sister group relationship of turtles to the bird-crocodilian clade revealed by nuclear DNA–coded proteins. *Molecular Biology and Evolution, 22*, 810–813.

Jeffery, J. E., Bininda-Emonds, O. R. P., Coates, M. I., & Richardson, M. K. (2005). A new technique for identifying sequence heterochrony. *Systematic Biology, 54*, 230–240.

Joyce, W. G., & Gauthier, J. A. (2004). Palaeoecology of Triassic stem turtles sheds new light on turtle origins. *Proceedings of the Royal Society of London, B, 271*, 1–5.

Joyce, W.G., & Karl, H.-V. (2006). The world's oldest fossil turtle: Fact versus fiction. *Fossil Turtle Research, Vol. 1, Russian Journal of Herpetology, 13*(Suppl. 1), 104–111.

Joyce, W. G., & Lyson, T. R. (2010). A neglected lineage of North America turtles fills a major gap in the fossil record. *Palaeontology, 53*, 241–248.

Joyce, W. G., Lucas, S. G., Scheyer, T. M., Heckert, A. B., & Hunt, A. P. (2009). A thin-shelled reptile from the Late Triassic of North America and the origin of the turtle shell. *Proceedings of the Royal Society of London, B, 276*, 507–513.

Kuratani, S., Kuraku, S., & Nagashima, H. (2011). Evolutionary developmental perspective for the origin of turtles: The folding theory for the shell based on the developmental nature of the carapacial ridge. *Evolution & Development, 13*, 1–14.

Laurin, M., & Reisz, R. R. (1995). A reevaluation of early amniote phylogeny. *Zoological Journal of the Linnean Society, 113*, 165–223.

Lee, M. S. Y. (1997a). Pareiasaur phylogeny and the origin of turtles. *Zoological Journal of the Linnean Society, 120*, 197–280.

Lee, M. S. Y. (1997b). Reptile relationships turn turtle. *Nature, 389*, 245–246.

Lee, M. S. Y., Reeder, T. W., Slowinski, J. B., & Lawson, R. (2004). Resolving reptile relationships. Molecular and morphological markers. In J. Cracraft & M. J. Donoghue (Eds.), *Assembling the Tree of Life* (pp. 451–467). Oxford: Oxford University Press.

Li, C., Wu, X.-C., Rieppel, O., Wang, L.-T., & Zhao, L.-J. (2008). An ancestral turtle from the Late Triassic of southwestern China. *Nature, 456*, 497–501.

Lyson, T., & Gilbert, S. F. (2009). Turtles all the way down: Loggerheads at the root of the chelonian tree. *Evolution & Development, 11*, 133–135.

Lyson, T. R., Bever, G. S., Bhullar, B.-A. S., Joyce, W. G., & Gauthier, J. A. (2010). Transitional fossils and the origin of turtles. *Biology Letters, 6*, 830–833.

Maxwell, E. E., & Harrison, L. B. (2009). Methods for the analysis of developmental sequence data. *Evolution & Development, 11*, 109–119.

Müller, J. (2003). Early loss and multiple return of the lower temporal arcade in diapsid reptiles. *Naturwissenschaften, 90*, 473–476.

Nagashima, H., Sugahara, F., Takechi, M., Ericsson, R., Kawashima-Ohya, Y., Narita, Y., et al. (2009). Evolution of the turtle body plan by the folding and creation of new muscle connections. *Science, 325*, 193–196.

Nagashima, H., Kuraku, S., Uchida, K., Kawashima-Ohya, Y., Narita, Y., & Kuratani, S. (2012). Origin of the turtle body plan—the folding theory to illustrate turtle-specific developmental repatterning. In D. B. Brinkman, P. A. Holroyd, & J. D. Gardner (Eds.) *Morphology and evolution of turtles*. Dordrecht: Springer.

Pritchard, P. C. H. (2008). Evolution and structure of the turtle shell. In J. Wyneken, M. H. Godfrey & V. Bels (Eds.), *Biology of Turtles* (pp. 45–83). Boca Raton: CRC Press.

Procter, J. B. (1922). A study of the remarkable tortoise, *Testudo loveridgii* Blgr., and the morphogeny of the chelonian carapace. *Proceedings of the Zoological Society of London, 3*, 483–526.

Raff, R. A. (2007). Written in stone: Fossils, genes and evo–devo. *Nature Reviews Genetics, 8*, 911–920.

Reisz, R. R., & Head, J. J. (2008). Turtle origins out to sea. *Nature, 456*, 450–451.

Remane, A. (1952). *Die Grundlagen des natürlichen Systems, der vergleichenden Anatomie und der Phylogenetik*. Leipzig: Akademische Verlagsgesellschaft Geest & Portig K.-G.

Rest, J. S. R., Ast, J. C., Austin, C. C., Waddell, P. J., Tibbetts, E. A., Hay, J. M., et al. (2003). Molecular systematics of primary reptilian lineages and the tuatara mitochondrial genome. *Molecular Phylogenetics and Evolution, 29*, 289–297.

Rieppel, O. (2008). The relationships of turtles within amniotes. In J. Wyneken, M. H. Godfrey, & V. Bels (Eds.), *Biology of Turtles* (pp. 345–353). Boca Raton: CRC Press.

Rieppel, O. (2012). The evolution of the turtle shell. In D. B. Brinkman, P. A. Holroyd, & J. D. Gardner (Eds.), *Morphology and Evolution of Turtles: Origin and Early Diversification*. Dordrecht: Springer.

Rieppel, O., & deBraga, M. (1996). Turtles as diapsid reptiles. *Nature, 384*, 453–455.

Rieppel, O., & Reisz, R. R. (1999). The origin and early evolution of turtles. *Annual Review of Ecology and Systematics, 30*, 1–22.

Rubidge, B. S., Modesto, S., Sidor, C., & Welman, J. (1999). *Eunotosaurus africanus* from the Ecca–Beaufort contact in Northern Cape Province, South Africa—implications for Karoo Basin development. *South African Journal of Science, 95*, 553–555.

Sánchez-Villagra, M. R. (Ed.) (2010). *Developmental Vertebrate Palaeontology. Seminars in Cell and Developmental Biology, 21.*

Sánchez-Villagra, M. R., & Scheyer, T. M. (2010). Fossil turtles from the northern Neotropics: The Urumaco sequence fauna and finds from other localities in Venezuela and Colombia. In M. R. Sánchez-Villagra, O. A. Aguilera & A. A. Carlini (Eds.), *Urumaco and Venezuelan Palaeontology—The Fossil Record of the Northern Neotropics* (pp. 173–191). Bloomington: Indiana University Press.

Sánchez-Villagra, M. R., Müller, H., Sheil, C. A., Scheyer, T. M., Nagashima, H., & Kuratani, S. (2009). Skeletal development in the Chinese soft-shelled turtle *Pelodiscus sinensis* (Testudines: Trionychidae). *Journal of Morphology, 270*, 1381–1399.

Scheyer, T. M. (2007). Comparative bone histology of the turtle shell (carapace and plastron): Implications for turtle systematics, functional morphology, and turtle origins. Ph.D. Dissertation, Mathematisch-Naturwissenschaftliche Fakultät, University of Bonn, Germany. Available from URN: http://nbn-resolving.de/urn:nbn:de:hbz:5N-12299; URL: http://hss.ulb.uni-bonn.de/diss_online/math_nat_fak/2007/scheyer_torsten.

Scheyer, T. M. (2008). Aging the oldest turtles: The placodont affinities of *Priscochelys hegnabrunnensis*. *Naturwissenschaften, 95*, 803–810.

Scheyer, T. M., & Sander, P. M. (2007). Terrestrial palaeoecology for basal turtles indicated by shell bone histology. *Proceedings of the Royal Society of London, B 274*, 1885–1893.

Scheyer, T. M., & Sander, P. M. (2009). Bone microstructures and mode of skeletogenesis in osteoderms of three pareiasaur taxa from the Permian of South Africa. *Journal of Evolutionary Biology, 22*, 1153–1162.

Scheyer, T. M., Sander, P. M., Joyce, W. G., Böhme, W., & Witzel, U. (2007). A plywood structure in the shell of fossil and living soft-shelled turtles (Trionychidae) and its evolutionary implications. *Organisms, Diversity & Evolution, 7*, 136–144.

Scheyer, T. M., Brüllmann, B., & Sánchez-Villagra, M. R. (2008). The ontogeny of the shell in side-necked turtles, with emphasis on the homologies of costal and neural bones. *Journal of Morphology, 269*, 1008–1021.

Schoch, R. R. (2009). Evolution of life cycles in early amphibians. *Annual Review of Earth and Planetary Sciences, 37*, 135–162.

Shaffer, B. S. (2009). Turtles (Testudines). In S. B. Hedges & S. Kumar (Eds.), *The Timetree of Life* (pp. 398–401). New York: Oxford University Press.

Sheil, C. A., & Greenbaum, E. (2005). Reconsideration of skeletal development of *Chelydra serpentina* (Reptilia: Testudinata: Chelydridae): Evidence for intraspecific variation. *Journal of Zoology, London, 265*, 235–267.

Shubin, N. H., & Alberch, P. (1986). A morphogenetic approach to the origin and basic organization of the tetrapod limb. *Evolutionary Biology, 20*, 319–387.

Shubin, N., Tabin, C., & Carroll, S. (2009). Deep homology and the origins of evolutionary novelty. *Nature, 457*, 818–823.

Sterli, J., de La Fuente, M. S., & Rougier, G. W. (2007). Anatomy and relationships of *Palaeochersis talampayensis*, a Late Triassic turtle from Argentina. *Palaeontographica Abt. A 281*, 1–61.

Tsuji, L. A., & Müller, J. (2009). Assembling the history of the Parareptilia: Phylogeny, diversification, and a new definition of the clade. *Fossil Record, 12*, 71–81.

Vickaryous, M., & Sire, J. Y. (2009). The integumentary skeleton of tetrapods: Origin, evolution, and development. *Journal of Anatomy, 214*, 441–464.

Werneburg, I. (2009). A standard system to study vertebrate embryos. *PLoS ONE, 4*(6): e5887. doi:10.1371/journal.pone.0005887.

Werneburg, I. (2010). Evolution and Development of Turtles: Organogenesis and Cranial Musculature. Ph.D. Dissertation, Paläontologisches Institut und Museum, Universität Zürich, Schweiz. Scidinge Hall Verlag, Zürich [ISBN: 978-3-905923-03-2].

Werneburg, I., & Sánchez-Villagra, M. R. (2009). Timing of organogenesis support basal position of turtles in the amniote tree of life. *BMC Evolutionary Biology* 9: 82, doi:10.1186/1471-2148-9-82.

Werneburg, I., & Sánchez-Villagra, M. R. (2011). The early development of the echidna, *Tachyglossus aculeatus* (Mammalia: Monotremata) and patterns of mammalian development. *Acta Zoologica (Stockholm), 92*, 75–88.

Werneburg, I., Hugi, J., Müller, J., & Sánchez-Villagra, M. R. (2009). Embryogenesis and ossification of *Emydura subglobosa* (Testudines, Pleurodira, Chelidae) and patterns of turtle development. *Developmental Dynamics, 238*, 2770–2786.

Wood, R. C., & Díaz de Gamero, M. L. (1971). *Podocnemis venezuelensis*, a new fossil pelomedusid (Testudines, Pleurodira) from the Pliocene of Venezuela and a review of the history of *Podocnemis* in South America. *Breviora, 376*, 1–23.

Ziermann, J. M. (2008). *Evolutionäre Entwicklung larvaler Cranialmuskulatur der Anura und der Einfluss von Sequenzheterochronien*. Ph.D. Dissertation, Institut für Spezielle Zoologie und Evolutionsbiologie mit Phyletischem Museum, Friedrich-Schiller-Universität, Jena.

Chapter 7
Geometric and Developmental Perspectives on the Evolution of the Skull and Internal Carotid Circulation in Turtles

Tetsuto Miyashita

Abstract Internal carotid circulation is arguably one of the most intensively analyzed morphological characters in turtle systematics and, thus, it is critical for understanding turtle phylogeny. I used landmark-based geometric morphometrics to facilitate a quantitative analysis of variation in turtle skull shape and osteological correlates of the internal carotid circulation. The analysis indicates that the position of the foramen posterior canalis carotici interni differs among eucryptodires, paracryptodires, and pleurodires, but remains relatively stable within these lineages. This supports the hypothesis that the position of the foramen posterior canalis carotici interni is a reliable character for identifying higher turtle taxa. Results from the geometric morphometric analysis are consistent with two of the three traits recently proposed to affect patterns of internal carotid circulation. However, these three traits do not fully explain six different patterns of internal carotid circulation in turtles. I identify the association between the internal carotid artery and the palatine branch of the facial nerve (CN VII) as a developmental constraint in turtles. Using this association as a reference point, embryological observations reported in the literature suggest morphogenetic processes that may govern different patterns of internal carotid circulation. When mapped on a phylogenetic tree, some characters transform more than once independently. The evolution of internal carotid circulation in turtles may have been even more complex than reconstructed in this paper, because of uncertain relationships among basal eucryptodires and because of a mosaic of character states in that critical part of the tree.

Keywords Cranial arteries • Cranial foramina • Cranial nerves • Cryptodira • Eucryptodira • Pleurodira

T. Miyashita (✉)
Department of Biological Sciences, University of Alberta,
CW 405 Biological Sciences Centre, Edmonton,
AB T6G 2E9, Canada
e-mail: tetsuto@ualberta.ca

Introduction

Internal carotid circulation has been an important source of phylogenetic characters in turtles, partly because the vessels irrigating the skull are associated with osteological correlates, such as foramina and bony canals, and partly because the pattern of the circulation is relatively conservative within each purported clade. This is a long-standing view (McDowell 1961; Albrecht 1967, 1976; Gaffney 1975a) that has been supported by subsequent morphological-based phylogenetic studies. In a pair of recent studies, Jamniczky (2008) and Jamniczky et al. (2006) provided an assessment of homologies and a comprehensive review of the phylogenetic implications of turtle internal carotid circulation. Given the additional literature entirely devoted to this character (e.g., Jamniczky and Russell 2004, 2007; Sterli and de la Fuente 2010; Sterli et al. 2010), internal carotid circulation arguably is one of the most intensively analyzed morphological traits in turtle systematics.

The advent of molecular phylogenetics has added a new facet to discussions regarding the phylogenetic utility of patterns of internal carotid circulation in turtle systematics. In a classic example of a conflict between molecular and morphological data, the phylogenetic signal of these morphological characters is inconsistent with phylogenetic trees generated from molecular data. The Trionychoidea are a clade of cryptodires supported by morphological characters, encompassing carettochelyids (pig-nosed turtles), dermatemydids (Mesoamerican river turtles), kinosternids (mud turtles), and trionychids (soft-shelled turtles) (Gaffney 1975a; Meylan and Gaffney 1989). Support for this clade partly comes from earlier suggestions (McDowell 1961; Albrecht 1967) that trionychids and kinosternoids share similar relative diameters of their cranial arterial foramina and canals. The relative dimensions of the foramina indicate a decrease in stapedial circulation and an increase in internal carotid circulation. In concert with other cranial and postcranial characters, the Trionychoidea are repeatedly

DOI: 10.1007/978-94-007-4309-0_7,

recovered as a clade in morphological phylogenetic analyses (Gaffney et al. 1991; Gaffney 1996; Shaffer et al. 1997; Brinkman and Wu 1999). However, molecular phylogenetic analyses using mitochondrial, ribosomal, and/or nuclear DNA have consistently rejected a monophyletic Trionychoidea (Shaffer et al. 1997; Fujita et al. 2004; Krenz et al. 2005; Near et al. 2005). This molecular versus morphological conflict does not entirely hinge on the reliability of the phylogenetic signal furnished by the internal carotid circulatory characters—Joyce's (2007) morphological phylogenetic analysis still supported, albeit weakly, a monophyletic Trionychoidea, even in the absence of any synapomorphies based on internal carotid circulatory morphology. Instead, this conflict in tree topologies highlights the need to reassess evidence supporting either of the hypotheses. On the morphological side of the conflict, a reassessment presents a methodological challenge of how to represent phylogenetic signals expressed in morphological characters.

Out of concern with the non-independence of internal carotid circulatory characters, Jamniczky (2008) relied on overall internal carotid circulatory pattern as the locus of homology, in contrast to Gaffney (1975a) who compared sizes of the foramina for the cranial arteries one to one. This led to an unexpected interpretation that the internal carotid circulatory pattern of kinosternoids is more similar to that of testudinoids than to that of trionychians (Jamniczky 2008; "Trionychia" sensu Gaffney 1975a). Jamniczky's (2008) analysis concluded that the conditions in kinosternoids and trionychians were autapomorphic for each clade, not synapomorphic between the two clades. Therefore, the phylogenetic signal expressed by the internal carotid circulatory morphology does not contribute to resolution of the molecular versus morphological conflict, because the character is uninformative for defining cryptodire suprafamilial relationships. The lesson from Jamniczky's (2008) reappraisal, aside from the molecular versus morphological conflict, is that rigorously tested definition and delimitation are crucial to the ability of a morphological character to reflect phylogenetic signal.

Here I look at another aspect of phylogenetic signal in turtle internal carotid circulation, namely the position of entry for the internal carotid artery into the skull. Using geometric morphometrics, I quantified variation in position of the entry of the internal carotid artery with respect to variation in positions of other cranial landmarks. Simultaneously, the morphometric analysis identified cranial landmarks that correlate with patterns of internal carotid circulation. Some of the correlated landmarks were consistent with factors recently proposed to affect patterns of internal carotid circulation. This information helps support and revises certain characters that have been widely used in turtle systematics. I also used embryological evidence to propose potential factors responsible for the anteroposterior variation in position of entry of the internal carotid artery. In this analysis, an association between the internal carotid artery and the palatine branch of the facial nerve (CN VII) is posited as a developmental constraint. Lastly, I mapped characters of the internal carotid circulation onto a turtle phylogeny.

Review of Turtle Internal Carotid Circulation and Previous Studies

General Morphology

The basic eucryptodire condition for the internal carotid artery and its main banches are shown in Fig. 7.1. Names for foramina and canals in the turtle skull follow Gaffney (1972a, 1979b). Before the cranial entry of the internal carotid artery, the stapedial artery (sa in Fig. 7.1) branches off through the foramen stapediotemporale; this foramen is vestigial in kinosternoids and absent in *Baptemys* and *Dermatemys* (McDowell 1961). After entering the skull through the foramen posterior canalis carotici interni (fpcci in Fig. 7.1) and extending along the canalis caroticus internus, the internal carotid artery (ica in Fig. 7.1) anteriorly bifurcates into the cerebral carotid artery (cca in Fig. 7.1), which enters into the foramen anterior canalis carotici interni and the palatine artery (pa in Fig. 7.1), which passes through the canalis caroticus lateralis entering into the foramen caroticus lateralis. The mandibular and orbital arteries (ma and oa, respectively, in Fig. 7.1) represent secondary major branches that originate either from the stapedial artery (general pattern; Fig. 7.1) or the palatine artery (kinosternoids; not shown). In trionychians, the terms mandibular artery and pseudopalatine artery apply to the vessels that have the same topographical origins as the cerebral carotid and palatine arteries, respectively, in other turtles (Albrecht 1967; Jamniczky 2008). In the trionychian scheme, the mandibular artery (entering into the canalis caroticus lateralis) and pseudopalatine artery (entering into the foramen anterior canalis carotici interni) branch off the internal carotid artery instead, and the orbital arteries irrigate downstream of the pseudopalatine artery (Jamniczky and Russell 2007). The trionychian scheme is independently repeated in chelonioids (Gaffney 1984; Jamniczky 2008).

Variation in Internal Carotid Circulation

Branching patterns and relative sizes of vessels constitute the primary phylogenetic signal in the internal carotid circulation of turtles. Both branching patterns and arterial

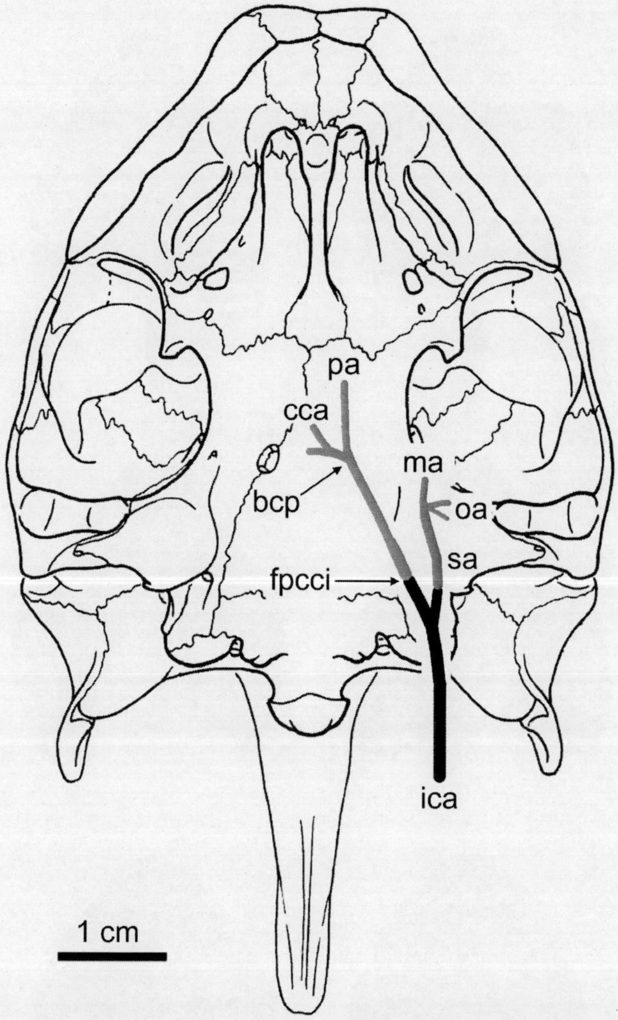

Fig. 7.1 Schematic drawing showing generalized path of the internal carotid artery and its major branches in a representative eucryptodire turtle skull (*Adocus* sp.; modified after Meylan and Gaffney 1989), in ventral view. The internal carotid artery enters the skull through the foramen posterior canalis carotici interni and gives rise to major cranial arteries. The parabasisphenoid complex encloses the bifurcation point between the cerebral and palatine arteries. Homology and terminology of vessels are based on Jamniczky (2008); terminology for foramina and canals follows Gaffney (1972a, 1979b). Shading of arteries in black and in grey indicates, respectively, portions of arteries outside of and within skull. Abbreviations: *bcp* bifurcation between the cerebral carotid and palatine arteries; *cca* cerebral carotid artery; *fpcci* foramen posterior canalis carotici interni; *ica* internal carotid artery; *ma* mandibular artery; *oa* orbital arteries; *pa* palatine artery; *sa* stapedial artery

diameters set different groups of turtles apart from each other (McDowell 1961; Albrecht 1967, 1976). Gaffney (1975a) was the first to translate this information for systematic use in turtles. In addition to differences in the relative position of the foramen posterior canalis carotici interni (discussed later in this section), he identified reduction of the stapedial artery as diagnostic for the Trionychoidea and enlargement of the stapedial artery as diagnostic for the Testudinoidea. That condition is reliably measured by the sizes of the foramina and canals through which the vessels pass (Jamniczky and Russell 2004). Thus, the foramen stapediotemporale (the dorsal opening of the stapedial canal) is smaller relative to the foramen posterior canalis carotici interni in trionychoids than in other turtles, and the reverse is true in testudinoids. Furthermore, the canalis caroticus lateralis increases in size in dermatemydids and kinosternids, which has been interpreted as a synapomorphy linking the two families within the more inclusive Kinosternoidea. However, Jamniczky (2008) showed that sizes of the foramina are correlated with each other because the cranial arteries diverge from the same source artery. Therefore, those characters are not independent but, instead, represent two conditions in the interconnected internal carotid circulatory pattern. Jamniczky (2008) argued that both conditions should be treated as states under a single character. An extensive review by Jamniczky and Russell (2007) brought the entire internal carotid circulation within the scope of analysis, and the internal carotid circulatory character in support of the Trionychoidea was revised into separate states, one autapomorphic for trionychians and the other autapomorphic for kinosternoids. Their revised character assessment neither supported nor rejected monophyly of the Trionychoidea.

Evolution of the Foramen Posterior Canalis Carotici Interni

Internal carotid circulation yields another heavily discussed, clade-specific character in turtle systematics—the location of the entrance of the internal carotid artery into the skull via the foramen posterior canalis carotici interni (Fig. 7.2). This is unexplored territory for quantitative character analysis. So far, the following four interpretations have been offered for evolution of this foramen.

(1) Gaffney (1975a) distinguished paracryptodires from eucryptodires, partly based on the more anterior location of the foramen posterior canalis carotici interni along the basisphenoid-pterygoid suture in paracryptodires. In eucryptodires, the foramen opens at or near the posterior end of the pterygoid (Gaffney 1975a; Gaffney and Meylan 1988).

(2) Revising Gaffney's (1975a) hypothesis, Evans and Kemp (1976) and Rieppel (1980) offered an alternative interpretation for the paracryptodire condition by suggesting that the more anterior position of the foramen represents a gradient, possibly a plesiomorphy with respect to the eucryptodire condition. These authors viewed paracryptodire condition as representing a gradual transition toward

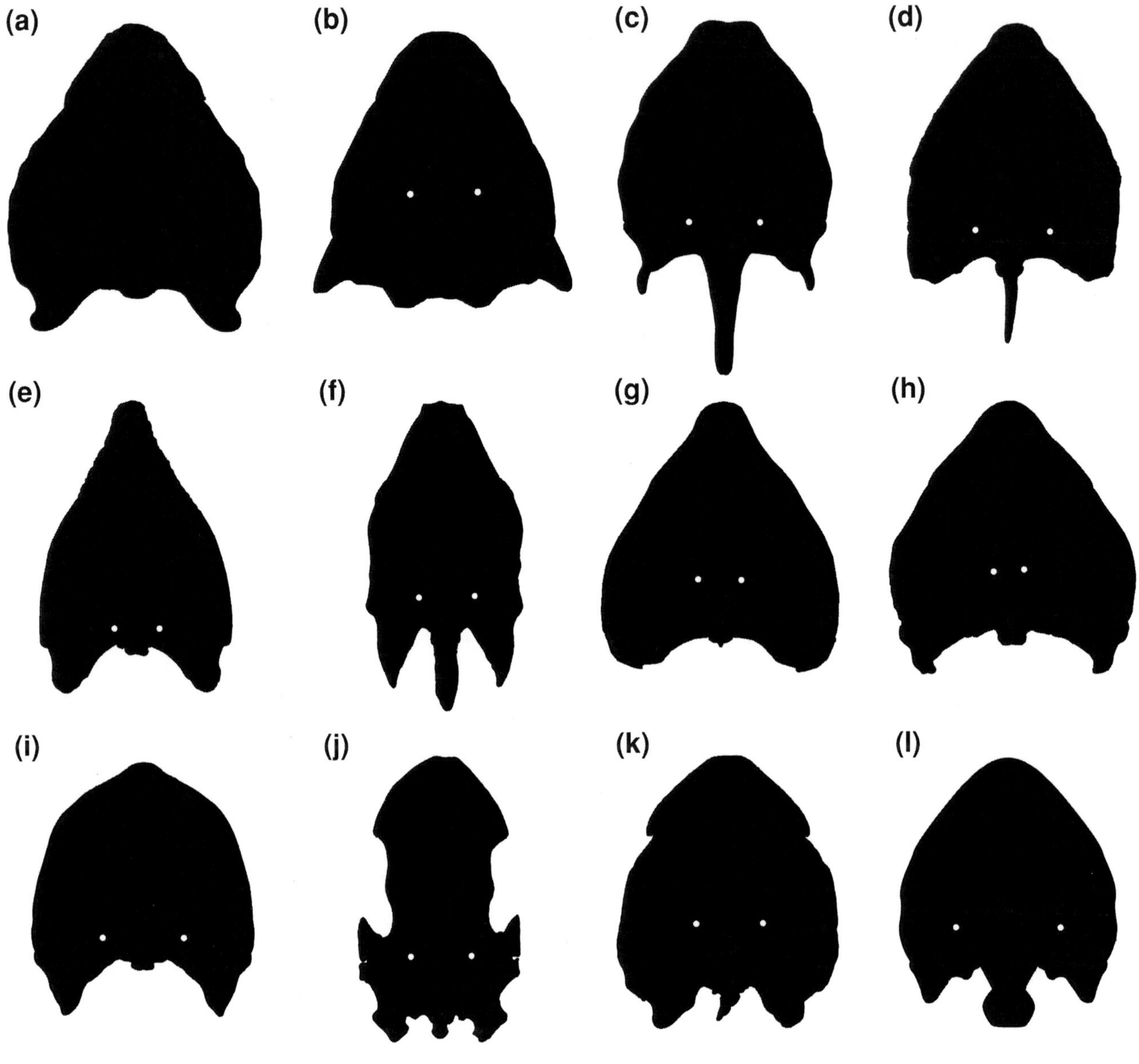

Fig. 7.2 Variations in skull morphology and in position of the foramen posterior canalis carotici interni across representative turtles. Skulls are illustrated in ventral view and are shaded *black*; *white dots* indicate the foramina posterior canalis carotici interni. Skulls and foramina are not to scale. **a**, **b** Basal turtles: **a** *Proganochelys quenstedti* (after Gaffney and Meeker 1983), illustrating the primitive condition in which the foramen posterior canalis carotici interni is absent (= pattern VI of Sterli and de la Fuente 2010); **b** *Meiolania platiceps* (Meiolaniidae; after Gaffney 1983). **c–f** Eucryptodires: **c** *Adocus* sp. (Adocidae; after Meylan and Gaffney 1989); **d** *Chelydra serpentina* (Chelydridae; after Gaffney 1979b); **e** *Solnhofia parsonsi* (Eurysternidae; after Gaffney 1979b); **f** *Lissemys punctata* (Trionychidae; after Gaffney 1979b). **g**, **h** Paracryptodires: **g** *Boremys pulchra* (Baenidae; after Brinkman and Nicholls 1993); **h** *Plesiobaena antiqua* (Baenidae; after Gaffney 1979b). **i–l** Pleurodires: **i** *Bothremys maghrebiana* (Bothremydidae; after Gaffney et al. 2006); **j** *Chelodina expansa* (Chelidae; after Gaffney 1979a); **k** *Pelomedusa subrufa* (Pelomedusidae; after Gaffney 1979b); **l** *Bairdemys venezuelensis* (Podocnemididae; after Gaffney and Wood 2002)

the eucryptodire condition and, hence, not a synapomorphy that sets paracryptodires apart as a monophyletic group.

(3) Brinkman and Nicholls (1993) postulated independent derivations for the foramen posterior canalis carotici interni among clades. In their model, the internal carotid artery initially was exposed ventrally, then later was covered by extensions of palatal elements beneath the artery. These extensions progressed from anterior to posterior in paracryptodires (thus the foramen formed anteriorly), but in the opposite direction, from posterior to anterior, in eucryptodires (thus the foramen formed posteriorly). This model predicts three transitional stages in cryptodire evolution, all

Table 7.1 Six patterns of the internal carotid circulation in turtles, based on Sterli and de la Fuente (2010)

Pattern	Systematic distribution	Split between verebral and palatine arteries	Location of foramen posterior canalis caroticus internus	Path of palatine artery
I	Eucryptodires	Covered	Pterygoid (posterior end)	Canalis caroticus lateralis
II	Paracryptodires	Covered	Basisphenoid-pterygoid suture (midway)	Canalis caroticus lateralis
III	Pleurodires	Covered	Prootic or elements underlying prootic	Canalis caroticus lateralis
IV	Macrobaenids; Synemyids; Meiolaniids	Exposed	Pterygoid (posterior end; posteriorly expanded)	Canalis caroticus lateralis
V	*Kallokibotion*; *Mongolochelys*; *Pleurosternon*; *Glyptodus*	Exposed	Absent	Canalis caroticus lateralis
VI	*Proganochelys*; *Kayentachelys*; *Condorchelys*; and others	Exposed	Absent	Interpterygoid vacuity

corroborated by fossil examples: (1) the initial stage in which the internal carotid artery was exposed ventrally (e.g., *Kallokibotion bajazidi*; Gaffney and Meylan 1992); (2) the paracryptodire condition in which the artery is underlain by palatal elements anterior to the foramen posterior canalis carotici interni, and the foramen shifts posteriorward during the course of paracryptodire evolution (Evans and Kemp 1976; Rieppel 1980; Brinkman and Nicholls 1993); and (3) the transitional state into the eucryptodire condition in which the artery is underlain by bone posteriorly, but the vessels downstream of the internal carotid artery remain exposed ventrally. For the last state, basal eucryptodires such as *Sinemys lens* (Brinkman and Nicholls 1993) and *Xinjiangchelys latimarginalis* (Brinkman and Wu 1999) have a large area in which the origin of the palatine artery from the carotid artery is exposed. The foramen posterior canalis carotici interni decreases in size in more derived eucryptodires (Meylan and Gaffney 1989) and it has disappeared in many extant forms (Brinkman and Nicholls 1993). This transition implies that enclosure of the carotid artery in eucryptodires progressed anteriorly, thus supporting independent originations for the different positions of the foramen posterior canalis carotici interni in eucryptodires and paracryptodires. The implication of the multiple transitional states in stem eucryptodires is that the position of this foramen was acquired independently in pleurodires.

(4) Most recently, Sterli and de la Fuente (2010) recognized six distinct patterns of the internal carotid circulation based on osteological correlates (patterns I–VI; see Table 7.1). In a companion paper that provided detailed studies of the basicrania of *Plesiochelys* and *Pleurosternon*, Sterli et al. (2010) suggested three main factors influenced variation in internal carotid circulation: (1) closure of the interpterygoid vacuity, leading to formation of the canalis caroticus lateralis (patterns I–V); (2) ventral expansion of the parasphenoid, leading to development of a bony flooring below the the split between the cerebral carotid and palatine arteries (patterns I–III); and (3) posterior extension of the pterygoid, leading to the foramen posterior canalis carotici interni lying within the pterygoid (patterns I and IV). Sterli and de la Fuente's (2010) patterns I–VI largely adopted the stages previously identified by Brinkman and Nicholls (1993). Therefore, Sterli and de la Fuente's (2010) scheme can be viewed as a phylogenetic reappraisal of Brinkman and Nicholls's (1993) model. Sterli and colleagues' major conclusions were: (1) basal turtles exhibiting patterns V and VI lack a foramen posterior canalis carotici interni; (2) ventral expansion of the parasphenoid trapped the internal carotid artery within the parabasisphenoid complex and resulted in the ventral covering of the cerebral carotid-palatine bifurcation (patterns I–III); and (3) posterior expansion of the pterygoid characterizes pattern I and differentiates pattern IV from pattern V, a reverse scenario from Brinkman and Nicholls's (1993) model.

Sterli et al.'s (2010) second conclusion raises an interesting possibility that the postulated expansion of the parasphenoid and ventral exposure of the parabasisphenoid complex may be positively correlated. If such a correlation can be demonstrated, it would provide a morphometric surrogate to correlate patterns I-III of the internal carotid circulation with relative degrees of the ventral parasphenoid expansion.

Sterli et al. (2010) reinforced the hypothesis by Brinkman and Nicholls (1993) that the foramen posterior canalis carotici interni might have been acquired multiple times in the evolution of turtles, because the presence of the foramen is variable among basal paracryptodires, basal eucryptodires, and stem taxa of the crown-group Testudines (Sterli and de la Fuente 2010; Sterli et al. 2010).

The Foramen Posterior Canalis Carotici Interni in Recent Cladistic Analyses of Turtles

Hirayama et al.'s (2000) cladistic analysis included the following character (their character 30): foramen posterior canalis carotici interni formed by: 0, basisphenoid only; 1, both basisphenoid and pterygoid halfway along the basisphenoid-pterygoid suture; 2, prootic only; 3, formed mostly or fully by pterygoid, foramen positioned near the posterior end of the basisphenoid). Joyce (2007) used the same character in his comprehensive morphological phylogenetics of turtles (his character 56). This character takes five independent steps from the basal condition to the three derived conditions (twice for states 2 and 3; Joyce 2007).

Although the evolution hypothesized for the posterior internal carotid foramen foramen posterior canalis carotici interni in eucryptodires and paracryptodires is complex, its position is even more variable within pleurodires. The carotid artery enters the prootic in chelids, pelomedusids, and *Araripemys*, whereas it enters the basisphenoid and pterygoid in bothremydines, cearachelyinians, euraxemydids, podocnemidids, *Arenila*, and *Sankuchemys* (Gaffney et al. 2006). Several bothremydids provide exceptions (Gaffney et al. 2006)—the foramen opens in the medial wall of the basisphenoid in *Kurmademys*, in the quadrate in *Labrostochelys*, and at the junction of the basisphenoid, pterygoid, and quadrate in *Taphrosphys* spp., *Bothremys kellyi*, and *Zolhafa*. As a result, Gaffney et al. (Gaffney et al. 2006, character 74) recognized seven states for this character. Gaffney et al. (2006) used another character to distinguish araripemydids, chelids, and pelomedusids from bothremydids, euraxemydids, and podocnemydids (character 75: posterior margin of pterygoid: 0, does not form anterior margin of foramen posterior canalis carotici interni; 1, forms anterior margin of the foramen). Importantly, the internal carotid artery passes through the prootic in all pleurodires, with the possible exception of *Kurmademys*, and the variation comes from extension of the underlying elements that cover this passage (Gaffney et al. 2006).

Phylogenetic Hypotheses on the Position of the Foramen Posterior Canalis Carotici Interni

In summary, two levels of phylogenetic hypotheses exist regarding the position of the foramen posterior canalis carotici interni. At the higher level, differences in the position of the foramen have been used to define major clades of turtles. Gaffney (1975a) differentiated paracryptodires and eucryptodires based on position of the foramen. Brinkman and Nicholls (1993), Sterli and de la Fuente (2010), and Sterli et al. (2010) implied independent origins of the foramen among pleurodires, paracryptodires, and eucryptodires. At the lower level, Gaffney et al. (2006) posited that differences in the position of the foramen also comprise a phylogenetic signal among pleurodires.

To define these phylogenetic signals at higher and lower levels, variation in the position of the foramen should be quantitatively described and the source of that variation should be identified. This information is crucial to character definition and coding strategy in a phylogenetic analysis. A single character with numerous states could confuse independent signals from multiple sources of variation. On the other hand, multiple characters based on a single source of the variation could bias the analysis toward one character state that is prerequisite to code for other characters.

Here I explore the issue of character definition and coding strategy based on the foramen posterior canalis carotici interni in turtles. The following four hypotheses are tested: (1) the foramen posterior canalis carotici interni varies in position between eucryptodires, paracryptodires and pleurodires, and therefore provides potential synapomorphies that define each clade (Gaffney 1975a; Brinkman and Nicholls 1993; Sterli and de la Fuente 2010); (2) ventral exposure of the parabasisphenoid complex is greater in patterns I-III than in patterns IV-VI (Sterli et al. 2010); (3) posterior expansion of the pterygoid is associated with the relatively posterolateral position of the foramen; and (4) variation in the position of the foramen with respect to other cranial bones within pleurodires (Gaffney et al. 2006) is due to shifts in the positions of cranial bones relative to the foramen.

Materials and Methods

Taxon Sampling

Turtle skulls often are illustrated in ventral view in descriptions, and the ventral portion of the braincase provides suitable landmarks that are relatively resistant to distortion and crushing. Significantly distorted or crushed skulls of fossil turtles were excluded from the study. Photographs were chosen over drawings. Reconstructions of fossil skulls were included if they satisfied any of the following criteria: (1) the skull only lacked fragments along the edges of bones or sutures, meaning that the original outlines of elements could be extrapolated with confidence; (2) missing or distorted parts were reconstructed from corresponding elements on the other side of the skull, where those parts were present and undamaged; or (3) the skull

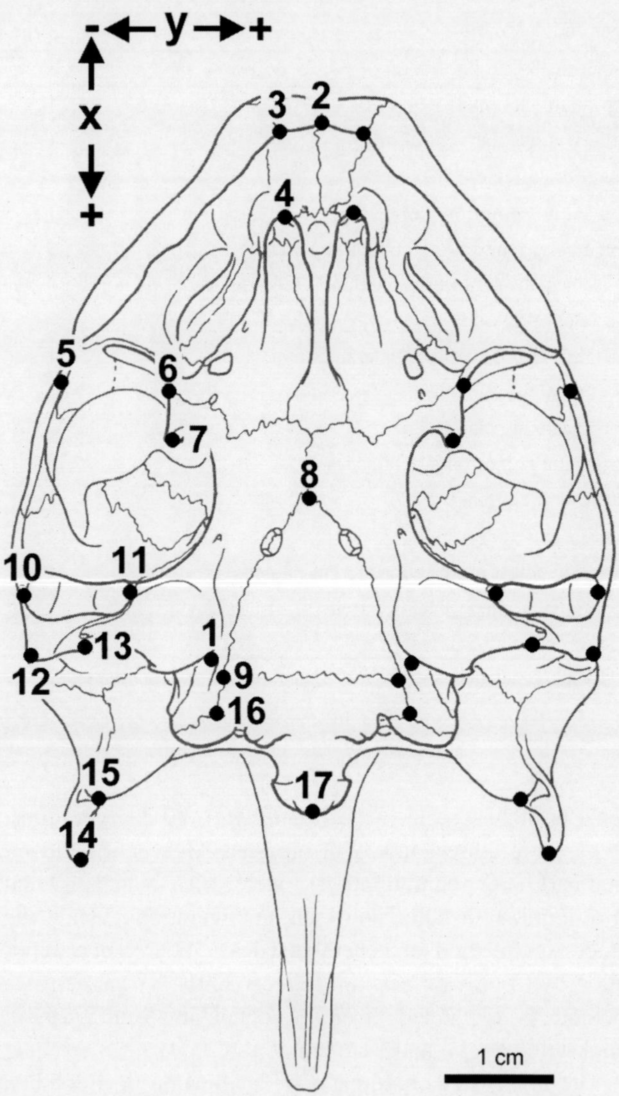

Fig. 7.3 A skull of *Adocus* sp. in ventral view (modified after Meylan and Gaffney 1989), showing configurations of the 31 landmarks used in this study. Numbers along the midline and on the anatomical right side of the skull correspond to landmarks listed in Table 7.2. Three landmarks (2, 8, and 17) are along the sagittal plane, 14 landmarks (1, 3–7, 9–16) are on the *right side*, and the other 14 landmarks (18–31; not numbered in this figure) are on the *left side* and are counterparts of those on the right. Positive and negative directions of the anteroposterior (x) and lateromedial (y) axes provide interpretation of the coordinates generated by Procrustes superimposition (see Figs. 7.6, 7.7)

could be retro-deformed along the direction of crushing. To cover the diversity of turtles adequately, I included reconstructed skulls of fossil taxa in the data set, instead of building a data set solely on osteological specimens of extant turtles. After a thorough literature search, published figures of skulls ($n = 145$) from 132 turtle taxa were collected and scanned (Appendix), the majority of which come from E. S. Gaffney's consistent and precise descriptions. Among the taxa included, intraspecific variation in skull morphology has been documented for the emydid *Pseudemys texana*, the kinosternid *Sternotherus odoratus*, and the baenid *Palatobaena cohen* (Bever 2008, 2009a, b; Lyson and Joyce 2009a). For those taxa, all illustrated specimens that document the variants were included. More than one reconstruction exists for some fossil turtles (e.g., *Mongolochelys*). If those reconstructions were based on different specimens and done by different authors, all of them were included.

The sample of turtle taxa represents a wide spectrum of morphological variation and taxonomic diversity. This paper follows the turtle phylogeny presented by Joyce (2007) and includes representatives from the three major radiations of turtles: pleurodires ($n = 39$), paracryptodires ($n = 21$), and eucryptodires ($n = 79$). Four basal turtle taxa (*Kayentachelys*, *Meiolania*, *Mongolochelys*, and *Proganochelys*) also were included in the data set ($n = 5$); these do not form a clade. *Kayentachelys* and *Meiolania* initially were placed amongst basal eucryptodires (Gaffney 1983; Gaffney et al. 1987, 2007b; Gaffney and Jenkins 2010) or as basal cryptodires (Hirayama et al. 2000), but Joyce's analysis (2007) suggests that they represent basal lineages outside crown clades of turtles. I follow this position for the purpose of this paper, because Joyce (2007) presented the most inclusive and up-to-date phylogenetic analysis of turtles.

Landmark Descriptions

In total, 31 landmarks were assigned to skulls taken from published figures (Fig. 7.3, Table 7.2) using TpsDig ver. 2.16 (Rohlf 2010). Three landmarks (2, 8, and 17) were along the sagittal plane, 14 landmarks (1, 3–7, 9–16) were on the right side, and the other 14 landmarks (18–31) were on the left side and are counterparts of those on the right. The landmarks on the left side (18–31) minimized artificial variation in position of the midline landmarks created by rotation of skull configurations during Procrustes superimposition (Zelditch et al. 2004). Beyond Procrustes superimposition and the following multivariate analysis of variance (MANOVA) and canonical variate analysis (CVA), however, the counterpart landmarks from the left side were not considered. According to Bookstein's (1991) classification, seven landmarks (6, 7, 10–14) from the right side and two landmarks from the sagittal plane (8 and 17) are Type 2 (local minima and maxima of curvature), whereas the rest are Type 1 (a juxtaposition of tissues). These landmarks delineate the right half of an entire skull from ventral view and also capture relative positions of the ventral cranial elements. When assigned to a foramen, the

Table 7.2 List of landmarks used in this study

	Landmark number and description	Type
1	Foramen posterior canalis carotici interni	1
2	Interpremaxillary suture at ventral margin of skull	1
3	Premaxilla-maxillary contact at ventral margin of skull	1
4	Internal naris (anterior extreme)	1
5	Maxilla-jugal contact at ventral margin of skull	1
6	Pterygoid at subtemporal fenestra (anterolateral extreme)	2
7	Lateral extreme of medial margin of subtemporal fenestra	2
8	Parabasisphenoid (anterior extreme)	2
9	Parabasisphenoid-basioccipital contact (lateral extreme)	1
10	Lateral extreme of quadrate condyle	2
11	Medial extreme of quadrate condyle	2
12	Quadrate (posterolateral corner on ventral side)	2
13	Pterygoid (posterolateral extreme)	2
14	Squamosal (posterior extreme)	2
15	Opisthotic-squamosal contact at posterior margin of skull	1
16	Basioccipital-exoccipital contact (lateral extreme)	1
17	Posterior extreme of occipital condyle	2

Numbers in left column correspond to landmarks labeled on Fig. 7.3. "Type" refers to the classification by Bookstein (1991)

landmark was placed at the intersection of the long axis and the axis of greatest width. Most of the landmarks were taken from the palatal or braincase floor of the skulls and, thus, satisfied coplanarity (Zelditch et al. 2004). Only landmarks 14 and 15 (and their counterpart landmarks from the left side) and landmark 17 violated the coplanarity assumption, because they were taken from horizontal planes intersecting the skull more dorsally. However, these landmarks were included because they were necessary to capture the overall shape of the skull from a ventral view. Because this study was designed to measure horizontal displacement of landmarks with respect to the foramen posterior canalis carotici interni, other sources of potential information along the dorsoventral axis intentionally were excluded.

Analysis 1: Position of the Foramen Posterior Canalis Carotici Interni

First, I tested for clade-specific positions of the foramen posterior canalis carotici interni and variation in the position of the foramen within clades (hypotheses 1 and 4) using geometric morphometrics. Therefore, taxa that lack the foramen posterior canalis carotici interni (i.e., most basal turtles, *Glyptops*, and *Pleurosternon*) were excluded from this part of the study. The reduced data set ($n = 139$) consists of one basal turtle (*Meiolania*), 80 eucryptodires, 19 paracryptodires, and 39 pleurodires. All sets of digitized landmarks, or configurations, were scaled and rotated to minimize distance between configurations using the Procrustes method of generalized least square superimposition by CoordGen ver. 6 h (Sheets 2001). The Procrustes method is explained in more detail in Rohlf (1990), Bookstein (1991), and Zelditch et al. (2004).

The Procrustes coordinates of landmarks are Euclidian projections from Kendall's shape space to the tangent space and, therefore, are not strictly interchangeable with the Procrustes distance. However, the correlation between these two was nearly perfect for this data set ($a = 0.994$; $r > 0.999$) (tpsSmall ver. 1.20; Rohlf 2003), which indicated that distance in the Kendall tangent space closely approximated the Procrustes distance. Therefore, the strong correlation allows the topology of Procrustes coordinates projected onto the Kendall tangent space to be treated as variables of the superimposed landmark data.

Using Procrustes coordinates, I performed the following four analyses to test whether position of the foramen posterior canalis carotici interni varies among eucryptodires, paracryptodires, and pleurodires: (1) Goodall's F test for significant difference in mean skull shapes between the three major clades (i.e., eucryptodires, paracryptodires, and pleurodires); (2) MANOVA (multivariate analysis of variance) of partial warp scores calculated from the Procrustes coordinates, which was followed by CVA (canonical variate analysis); (3) CVA of the Procrustes coordinates and analysis of the loadings on canonical variate axes; and (4) comparison of sample variances of the Procrustes coordinates. *Meiolania* was used to calculate Procrustes distance so as not to bias the superimposition, but it was excluded from all subsequent analyses because it does not belong to any of the subsets (i.e., eucryptodires, paracryptodires, and pleurodires) compared here.

Goodall's F was used to test for significant difference between mean skull shapes of the eucryptodire, paracryptodire, and pleurodire clades in a pair-wise comparison performed by TwoGroup ver. 6 h (Sheets 2005). F score was bootstrapped ($n = 1000$, $p = 0.001$). Similarly, a bootstrap analysis ($n = 4900$) provided 95% confidence intervals for Procrustes distance between mean skull shapes of the clades. Thin-plate spline deformation also was performed to visualize divergence of a mean skull shape for each clade from a mean skull shape of the data set. The area of deformation suggests landmarks displaced from the global mean of the data set. This test was followed by MANOVA of partial warp scores using CVAGen ver. 6 l (Sheets 2004), the purpose of which was to evaluate whether or not difference in mean skull shapes of the clades represents morphological divergence. CVA is a graphical representation of MANOVA of partial warp scores. Turtles show a wide spectrum of morphology from an anteroposteriorly long and lateromedially narrow skull to an anteroposteriorly short and lateromedially wide skull. This variation occurs within each major clade and results in superficial resemblance in skull shapes among distantly related turtles. CVA is the ordination method that maximizes distance between group means and, therefore, is capable of detecting underlying morphological differences that separate a clade containing a diverse array of skull shapes from other clades. Well-separated clusters in CVA would indicate that at least some landmarks have distinct topologies that characterize skull morphology of the clade.

Next, I used CVA of Procrustes coordinates to explore variances of the landmarks that resulted in distribution of the specimens in CVA (PAST ver. 3.2; Hammer et al. 2010). Partial warp scores provide an accurate representation of shapes defined by the landmarks, but CVA based on partial warp scores cannot answer the question as to which landmark contributed differences between groups. This is because partial warp scores describe shape space and, therefore, only have $2N{-}4$ variables ($N =$ number of landmarks) (Zelditch et al. 2004). To circumvent this problem, I compared CVA of Procrustes coordinates (which are translated into Euclidian space) with CVA of partial warp scores for the three major clades. The use of Procrustes coordinates is also justified by the nearly perfect correlation between the coordinates and Procrustes distance ($a = 0.994$; $r > 0.999$) (tpsSmall ver. 1.20; Rohlf 2003). The expectation was that the CVAs would produce plots with nearly identical data distributions. With the two results deemed to be similar enough based on the ratios of eigenvalues and the distributions of specimens, loadings of each landmark coordinate on canonical variate axes in CVA of Procrustes coordinates were interpreted as components of the difference explained by the canonical variate axis. In other words, the loadings are the contribution of landmark coordinates to the difference between mean skull shapes of the clades.

By this rationale, CVA loadings would show which landmark coordinates loaded heavily on the axes that best discriminate skull shapes of the major clades. Instead of using raw magnitudes of the loadings, contributions of the landmark coordinates to CVA of partial warp scores were inferred from relative magnitudes of the loadings. For example, the hypothesis that position of the foramen posterior canalis carotici interni can be used as a synapomorphy for each of the three major clades predicts relatively higher loading of the coordinates of the foramen on the first canonical variate axis than coordinates of other landmarks.

Finally, I divided the data set into eucryptodires, paracryptodires, and pleurodires and performed Procrustes superimposition separately for each clade. The resulting Procrustes coordinates were generated without the influence of skull configurations from other clades and, therefore, were desirable for examining variation in the skull morphology within each clade. Sample variance was calculated for each Procrustes coordinate. This information is critical for two reasons. First, landmarks that differ in position significantly between the mean skull configurations (thin-plate spline deformation and CVA) do not necessarily lead to morphological characters that distinguish one clade from another. This is because positional variation of the landmark within clades may swamp the difference between the means. An overlap between the clade-specific variations would present difficulty when evaluating a character state. Second, landmarks that are not helpful in distinguishing clades may either be highly variable or highly conserved in position within and across the subsets. Therefore, if a landmark loaded heavily on any of the canonical variate axes and varied greatly within the clades, it would suggest that the variation within the clades may be too great to detect the difference between the clades even though the mean positions differ significantly. If a landmark that loaded heavily in CVA varied little within the clades, it would suggest that the position of the landmark differs between the clades and remains stable within the clades. In this case, the difference between the clades may be translated into discreet character states. A high variation of a landmark within the clades may also be used to infer potentially useful characters below the taxonomic level examined in this study.

Analysis 2: Cranial Landmarks Correlated with Internal Carotid Circulation

After testing for the clade-specific positions of the foramen posterior canalis carotici interni [thereby supporting Sterli and de la Fuente's (2010) patterns I–IV], I compared the overall skull shapes of turtles. I evaluated if any of the other landmarks vary in correlation with Sterli and de la Fuente's patterns I-VI and tested for the morphological characters correlated with each pattern. This second set of analyses included all of the taxa from the data set ($n = 144$), but excluded the landmark representing the foramen posterior canalis carotici interni (1 and 18) for two reasons: first, to include the taxa representing patterns V and VI, which lack the foramen, and, second, to eliminate false variance transferred from the foramen to the neighboring landmarks (see discussion, "Potential Sources of Error").

I next divided the data set into two comparative groups in the following four combinations: (1) turtles with the foramen posterior canalis carotici interni (patterns I–IV) and turtles without the foramen (patterns V and VI); (2) turtles with the bifurcation of the internal carotid artery enclosed within the parabasisphenoid complex (patterns I-III) and turtles with the ventrally exposed bifurcation (patterns IV–VI); (3) turtles in which the canalis caroticus internus is restricted the pterygoid and parabasisphenoid (patterns I, II, and IV) and turtles in which the canalis caroticus internus is absent or associated with the prootic (patterns III, V, and VI); and (4) turtles with the foramen posterior canalis carotici interni associated with the basicranium (patterns II and III) and turtles with the foramen not associated with the basicranium (patterns I and IV). In each of the combinations, the morphology characterized by the alternative patterns represents a plesiomorphic condition with respect to the morphology that characterizes the initial patterns. Therefore, I performed thin-plate spline deformation of the latter patterns to the former patterns, not between the mean skull shape of each of the patterns and the global mean skull shape of the entire data set as in Analysis 1. For the last combination, turtles with patterns V and VI were excluded because the foramen is absent in theoe taxa. The analytical procedure for each combination followed Analysis 1 (steps 1–4).

Patterns V and VI were treated together throughout this section, because those patterns are correlated with a discrete morphological character, namely the presence or absence of the interpterygoid vacuity (Sterli and de la Fuente 2010).

Potential Sources of Error

The data set used in this paper includes only 21 paracryptodires, a number substantially lower than either eucryptodires or pleurodires. The Paracryptodira (pattern II) consist entirely of fossil taxa, many of which are known from incomplete skulls unsuitable for digitization. Patterns IV–VI also are greatly underrepresented in the data set, because suitable fossil skulls of macrobaenids, synemyids, and basal turtles are rare. It is unlikely that the sample size of paracryptodires, macrobaenids, synemyids, and basal turtles will dramatically increase in the near future. The only strategy to deal with unequal sample sizes is to exercise caution when interpreting variances within these groups. The most extreme example of this situation in this paper is that the sample size of turtles with the foramen posterior canalis carotici interni ($n = 139$) overwhelms that of the turtles without the foramen ($n = 6$). Goodall's F test is sensitive to an overlap of the ranges between two groups, if not too stringent a method for some geometic morphometric studies (Zelditch et al. 2004). Therefore, a P-value substantially smaller than 0.001 in this test implies that the observed difference is unlikely to be an artifact of sample size.

The large sample size of eucryptodires with pattern I ($n = 74$) is problematic, because a global mean skull shape of the data set was biased toward pattern I. Therefore, the mean skull configuration of pattern I would appear less divergent from the global mean than the means of patterns II and III, and vice versa. As a result, a thin-plate spline deformation of the true global mean to the mean of pattern I would be underrepresented. In Analysis 1, the results of the thin-plate spline deformation were interpreted with this bias in mind, and I did not invoke an argument as to which pattern diverged more from the global mean than others. In Analysis 2, the large sample size of eucryptodires precluded comparisons between pattern II and all others and between pattern III and all others, because the sample sizes of pattern IV–VI were overwhelmed by that of pattern I. As such, these comparisons largely would repeat the initial comparison among patterns I + IV, II, and III in Analysis 1.

To test for error by the observer, twenty specimens were randomly selected from the data set. I digitized the landmarks in these specimens twice with a four-month interval intervening. Using Procrustes coordinates, I compared the two sets of the digitized specimens in Goodall's F test and MANOVA of partial warp scores followed by CVA. Goodall's F test cannot reject the null hypothesis that these two samples were identical with $P > 0.9999$ ($F = 0.27$; df $= 58.00$, 2204.00; distance between means $= 0.0258$; SE $= 0.0132$). CVA of partial warp scores showed that these two samples overlapped (data not shown). Consequently, observer error in this study is considered negligible.

Another potential source of error in this study is that I compared three major clades, but did not take into account phylogenetic distance among the taxa within each clade. Phylogenetic distance cannot be calculated consistently, because no single cladistic data set includes all the turtle taxa used in this study. Eucryptodires, paracryptodires, and pleurodires each have been consistently recovered as a

Table 7.3 Results from Goodall's *F* test. Eucryptodires, paracryptodires, and pleurodires were compared pair-wise

Pair-wise Comparisons		F	df	P	D	Low D	High D	SE
Eucryptodires	Paracryptodires	19.37	58.00, 5568.00	<0.001	0.1572	0.1441	0.179	<0.01
Eucryptodires	Pleurodires	21.64	58.00, 6728.00	<0.001	0.132	0.1219	0.1485	<0.01
Paracryptodires	Pleurodires	10.4	58.00, 3248.00	<0.001	0.1374	0.1289	0.16	<0.01

Procrustes distance between means of the clades was bootstrapped to determine a 95% confidence interval (Low D and High D). Other abbreviations for column heaadings: *D* Procrustes distance between means; *df* Degree of freedom; *F* F score; *P* Probability; *SE* Standard error

monophyletic clade (e.g., Gaffney 1975a; Gaffney et al. 2006; Joyce 2007). The purposes of this study are to evaluate the previously proposed, qualitatively accepted hypothesis that the position of the foramen posterior canalis carotici interni differs markedly among these major clades (Gaffney 1975a; Brinkman and Nicholls 1993) and to find osteological correlates with the major arterial patterns (Sterli and de la Fuente 2010; Sterli et al. 2010). Both these clades and the arterial patterns have been dealt with as discrete categories in the literature and were incorporated in the design of this study as such. Instead of estimating phylogenetic distance, I used comparisons of sample variances within the clades or the arterial patterns to rule out variation within the clades or patterns (due to phylogenetic distance) as a possible source of significant difference between the means.

Finally, superimposition of landmark data may introduce a "Pinocchio effect" in which one highly variable landmark transfers its variance to other landmarks when fitting the configuration to the reference (Walker 2000). The least squares principle of the Procrustes method in scaling and rotating is vulnerable to such distortion when incorporating a landmark with exceedingly large relative displacement (Chapman 1990). Resistant-fit superimposition of repeated medians is robust to this problem, but could not be used in this study because the method departs from the Procrustes distance metric. Although it is difficult to quantify landmark variance allocated by the Pinocchio effect in a given data set, often the effect is assumed to be negligible as long as no extreme local change is observed in the sample (Zelditch et al. 2004); that is the position followed in this paper. None of the landmarks used herein marks a minimum or maximum of an extremely variable structure that could extend longer than the skull itself, which would have warranted invoking the Pinocchio effect. Nevertheless, deformation was carefully evaluated if any neighbor landmark markedly shifted.

Results

Analysis 1: Position of the Foramen Posterior Canalis Carotici Interni

A pair-wise comparison using Goodall's *F* test finds support for significant differences in mean skull shapes among eucryptodires, paracryptodires, and pleurodires (Table 7.3). Mean skull configurations and their transformation by the thin-plate spline method (Fig. 7.4) indicate the following three major differences in position of the foramen posterior canalis carotici interni: (1) the foramen posterior canalis carotici interni is more posterior in eucryptodires than in paracryptodires and pleurodires; (2) the foramen is more anterior and more medial in paracryptodires than in eucryptodires and pleurodires; and (3) the foramen is more lateral in pleurodires than in eucryptodires and paracryptodires. In association with these differences, the landmarks for the parabasisphenoid complex show anterior displacement in paracryptodires (landmarks 8 and 9) and lateral expansion in pleurodires (9). However, the anteromedial displacement of landmark 9 in paracryptodires may be due to variances allocated by the foramen posterior canalis carotici interni, which also is displaced anteromedially. The pterygoid (6, 7, 13) shows anteroposterior expansion in eucryptodires, posterolateral expansion in paracryptodires, and anteroposterior shortening in pleurodires. In pleurodires, the posterolateral extreme (13) is displaced markedly anteriorly, and the lateral extreme of the subtemporal margin of the pterygoid (7) is laterally expanded and more posterior in position. Overall skull shape (2, 3, 17) is anteroposteriorly shorter in paracryptodires and pleurodires. The portion of the skull behind the palate is relatively short anteroposteriorly and wide lateromedially in paracryptodires compared to the other two clades. These topological differences between the mean skull configurations are candidates for morphological characters to distinguish these clades. However, these differences may be swamped by variation within the clades (tested in CVA and plots of variances).

CVA of partial warp scores results in two significant canonical variate axes (Fig. 7.5), as follows: Axis 1: eigenvalue = 11.792; Wilk's $\lambda = 0.0073$; df = 116; $P < 0.001$; 55% of total variance; and Axis 2: eigenvalue = 9.7225; Wilk's $\lambda = 0.0932$; df = 57; $P < 0.001$; 45% of total variance. All of the taxa included in CVA segregate into well-defined clusters comprised of eucryptodires, paracryptodires, and pleurodires. The first canonical variate axis separates all three clusters. Amongst these, the most distant are eucryptodires and pleurodires. Paracryptodires are closer to eucryptodires than to pleurodires along the first axis. The second canonical variate axis separates

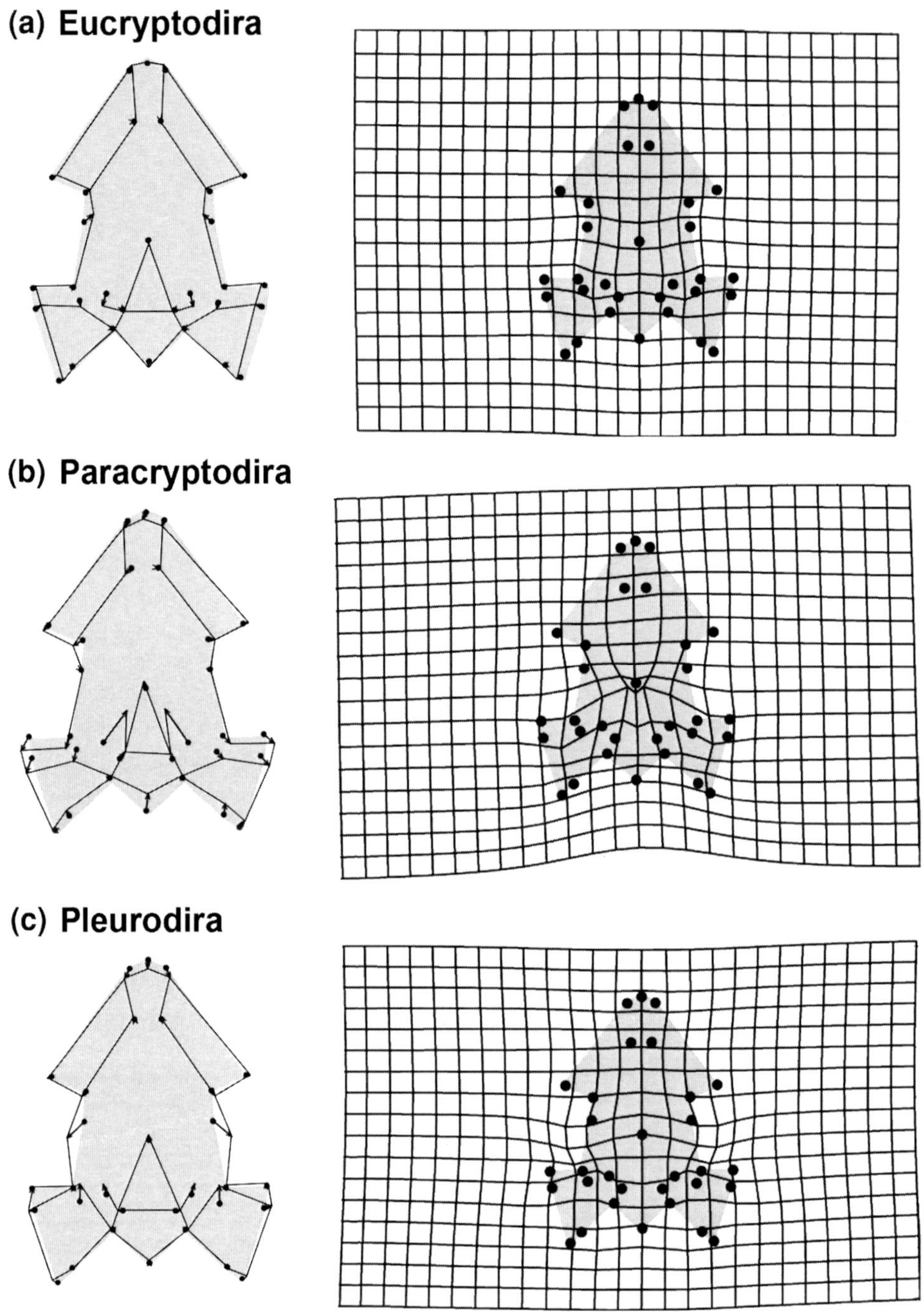

Fig. 7.4 Landmark-vector plots of mean skull configurations, from Analysis 1, based on Procrustes coordinates and their thin-plate spline deformation from the global mean skull configuration: **a** eucryptodires; **b** paracryptodires; **c** pleurodires. Grey silhouettes represent the global mean skull configuration. Small arrows in the landmark-vector plots of mean skull shapes (*left plot*) indicate displacement of landmarks as inferred by the corresponding thin-plate spline deformation (*right plot*)

paracryptodires from the rest of the data set. The test of grouping based on Mahalanobis distance between the specimens and the group means indicates that all of the specimens are grouped correctly within the clades to which they belong at the highly significant level $P < 0.01$ (data not shown). Despite superficial resemblance in overall skull shape between turtles from distantly related clades, Goodall's F and CVA do suggest divergence in skull morphology between the major clades. In other words, one or more landmarks consistently vary between the clades.

To correlate the divergence in skull morphology (CVA of partial warp scores) and the landmarks that differ in positions between mean skull configurations (Goodall's F), CVA of Procrustes coordinates was undertaken. This results in two significant canonical variate axes (Axis 1: eigenvalue = 13.01; Wilk's $\lambda = 0.0061$; df = 124; $P < 0.001$; Axis 2: eigenvalue = 10.8; Wilk's $\lambda = 0.0052$; df = 146; $P < 0.001$) and well-separated clusters in similar topographical relationships. Furthermore, the ratios of eigenvalues of the first two significant canonical variate axes

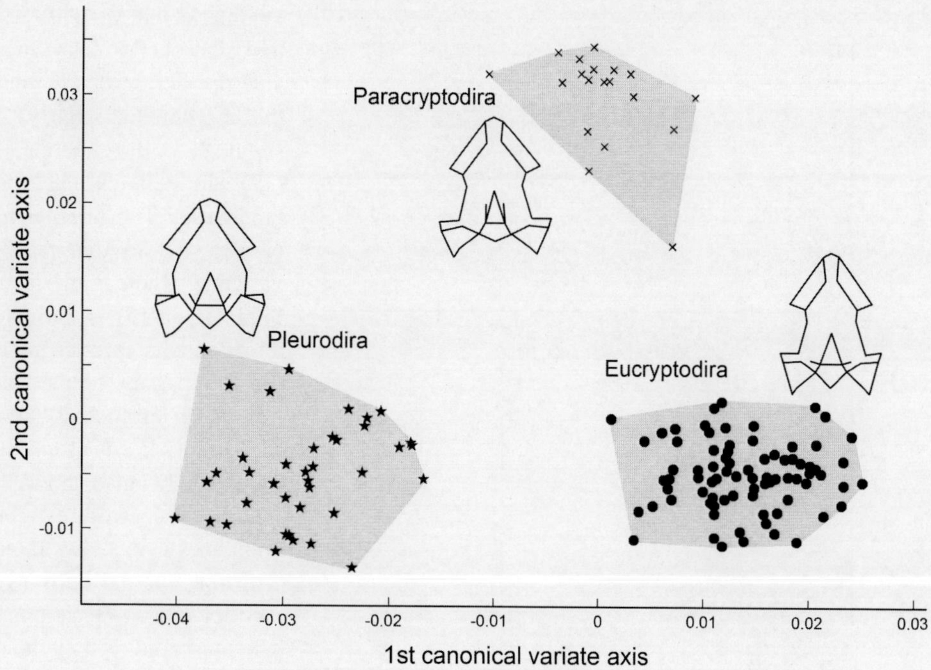

Fig. 7.5 Bivariate plot of CVA of partial warp scores, from Analysis 1, showing the two significant canonical variate axes. Eucryptodires, paracryptodires, and pleurodires segregate into their respective, well-separated clusters. Mean skull configurations (from Fig. 7.4) are depicted beside each cluster. Symbols: *solid circle* eucryptodires; × paracryptodires; *star* pleurodires

almost coincide with each other between the two CVAs (the ratio for partial warp scores = 1.21; the ratio for Procrustes coordinates = 1.2). Thus, in both CVAs, the first canonical variate axis explains approximately 20% more variance than explained by the second axis. The almost identical feature of the two CVAs allows treating loadings on CVA of Procrustes coordinates as approximated contributions of these landmark coordinates to CVA of partial warp scores.

A bivariate plot of loadings on the first two canonical variate axes (Fig. 7.6) indicates outliers that load on one or both of the axes more heavily than other landmark coordinates. Variances of these outlier coordinates contribute to the separation of the major clades in CVA (Fig. 7.5). Along the first canonical variate axis that largely discriminates pleurodires from eucryptodires and paracryptodires, six landmark coordinates score an absolute value of loading greater than 0.002: anteroposterior coordinates of the foramen posterior canalis carotici interni (1x), the snout (2x, 3x), and the posterolateral extreme of the pterygoid (13x), and both anteroposterior and lateromedial coordinates of the lateral extreme of the subtemporal margin of the pterygoid (7x, y). Along the second canonical variate axis that discriminates paracryptodires from eucryptodires and pleurodires, eight landmark coordinates score an absolute loading greater than 0.002: both anteroposterior and lateromedial coordinates of the foramen posterior canalis carotici interni (1xy), anteroposterior coordinates of the snout tip (2x, 3x), the quadrate (10x, 11x, 12x), and the occipital condyle (17x). Amongst these coordinates, three (1x, 7x, 13x) are outside a 95% ellipse. The anteroposterior coordinates of the foramen posterior canalis carotici

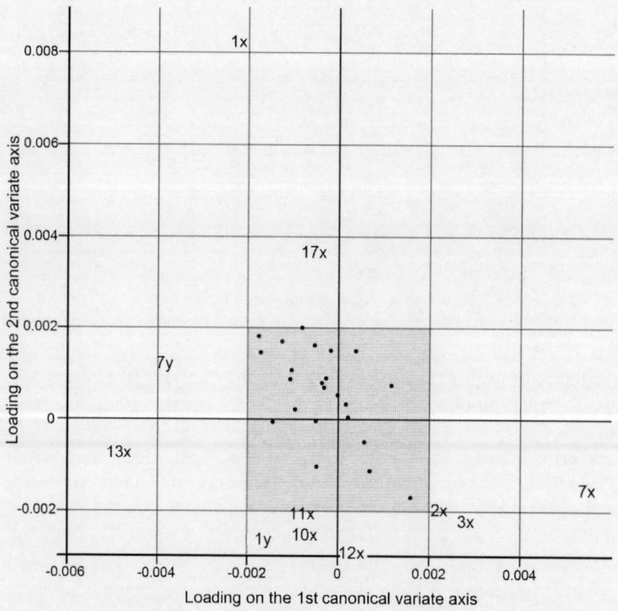

Fig. 7.6 Bivariate plot of loadings of Procrustes coordinates on the two significant canonical variate axes, from Analysis 1. The farther away from the origin (0, 0) along either or both of the axes, the higher the loading on the axis and the more important the landmark coordinate is for setting apart eucryptodires, paracryptodires, and pleurodires from each other. Shaded area represents the zone of relatively small variance (i.e., date points in that zone are less useful for differentiating clades). Data points outside the shaded area represent landmarks with relatively high loading on either of the canonical variate axes (i.e., data points that are more useful for differentiating clades). Numbers correspond to landmarks (see Fig. 7.3, Table 7.2); "x" is an anteroposterior coordinate; and "y" is a lateromedial coordinate. Only outlier landmarks are labeled

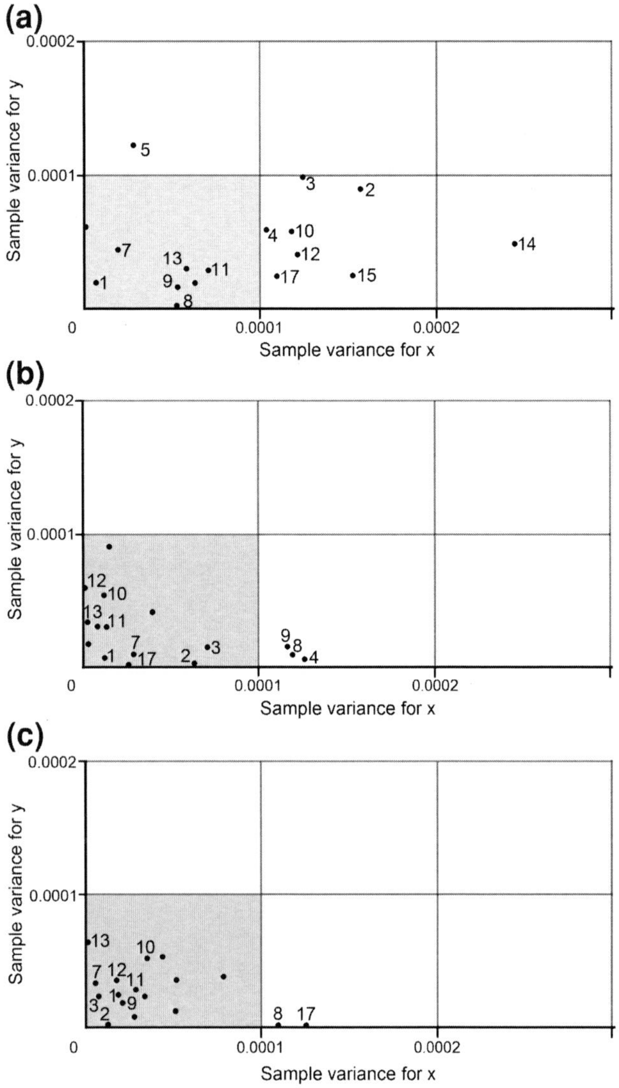

Fig. 7.7 Bivariate plots of sample variances of the landmarks calculated from Procrustes coordinates, from Analysis 1: **a** eucryptodires; **b** paracryptodires; **c** pleurodires. Only landmarks either discussed in text or associated with unusually high variances are labeled. Each number represents a landmark from the right side of a skull (Table 7.2). As in Fig. 7.3, "x" represents the anteroposterior component (sample variance of the anteroposterior coordinates); "y" represents the lateromedial axis (sample variance of the lateromedial coordinates); and shaded area represents the zone of relatively small variances to exclude outliers. Small variance means that the landmarks vary little within a clade, whereas large variances suggest significant variation of the landmarks within the clade

interni (1x) and the occipital condyle (17x) load more heavily on the second canonical variate axis than on the first axis, whereas the subtemporal extreme (7xy) and the posteromedial extreme (13x) of the pterygoid load more on the first axis than on the second axis.

Bivariate plots of sample variances for anteroposterior (x) and lateromedial (y) Procrustes coordinates identify landmarks that vary little and others that vary greatly within the clades (Fig. 7.7). In all clades, variances along the anteroposterior axis (x) tend to be greater than those along the lateromedial axis (y). In eucryptodires (Fig. 7.7a), the snout (2, 3), the quadrate (10, 12), and the occipital condyle (17) show large anteroposterior variances among other landmarks. The internal naris (4) and the occipital region (14, 15, 17) are the areas of relatively high anteroposterior variance, whereas the lateromedial width across the antorbital region (5) is variable. In contrast, the foramen posterior canalis carotici interni (1) is the least variable of all the landmarks. The parabasisphenoid complex (8, 9) shows minimal lateromedial variances, which is expected because the landmark 8 is along the midline of the skull. The pterygoid (7, 13) also shows small variances.

In paracryptodires (Fig. 7.7b), the parabasisphenoid complex (8, 9) shows large anteroposterior variances along with the internal naris (4). The foramen posterior canalis carotici interni (1) is one of the least variable landmarks. The pterygoid (7, 13), the quadrate (10–12), and the occiput (17) have relatively small variances. Anteroposterior variances of the snout (2, 3) are intermediate between the less variable and the more variable.

In pleurodires (Fig. 7.7c), almost all the landmarks shows both anteroposterior and lateromedial variances smaller than 0.001, except for anteroposterior coordinates of the anteror extreme of the parabasisphenoid complex (8) and the occipital condyle (17). The foramen posterior canalis carotici interni (1) has relatively small variances, although it is not amongst the least variable landmarks. The snout (2, 3) is the area of the least variances.

Analysis 2: Cranial Landmarks Correlated with Internal Carotid Circulation

In the second set of analyses, I identified landmarks contributing to difference in mean skull shapes between the modes of the internal carotid circulation (Sterli and de la Fuente 2010) and tested if the landmarks corroborate morphological transitions proposed to correlate with the modes of the internal carotid circulation (Sterli et al. 2010).

Goodall's F test supports significant difference in skull shapes between turtles with the foramen posterior canalis carotici interni (patterns I–IV) and those without the foramen (patterns V and VI), as follows (abbreviations as for Table 7.3): $F = 2.81$; df $= 54.00, 7614.00$; $P < 0.001$; D (95% confidence) $= 0.1181$ (Low D $= 0.1038$, High D $= 0.1660$); SE $= 0.0196$. Deformation of the mean skull shape of patterns V and VI (Fig. 7.8a) shows that the mean

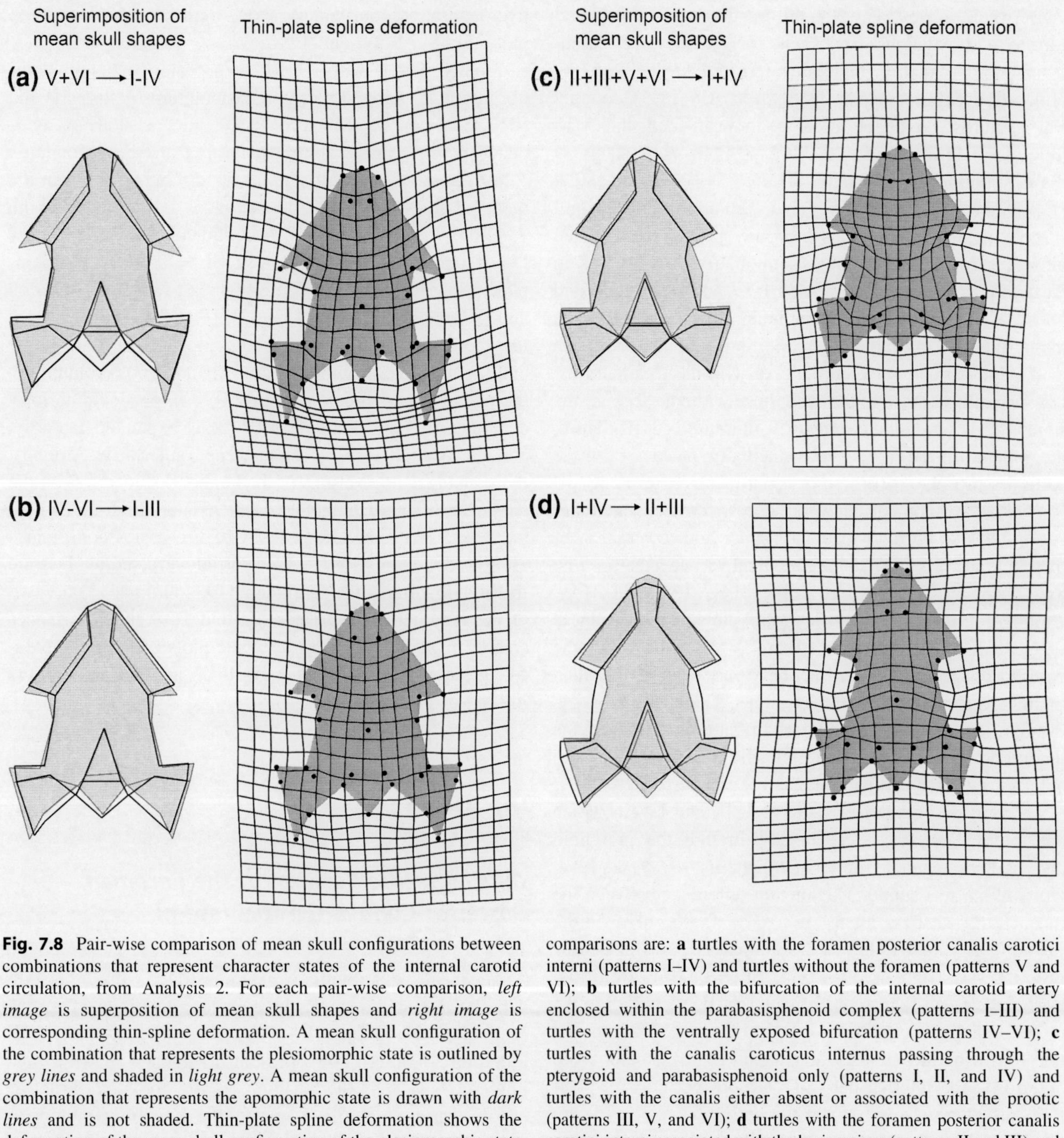

Fig. 7.8 Pair-wise comparison of mean skull configurations between combinations that represent character states of the internal carotid circulation, from Analysis 2. For each pair-wise comparison, *left image* is superposition of mean skull shapes and *right image* is corresponding thin-spline deformation. A mean skull configuration of the combination that represents the plesiomorphic state is outlined by *grey lines* and shaded in *light grey*. A mean skull configuration of the combination that represents the apomorphic state is drawn with *dark lines* and is not shaded. Thin-plate spline deformation shows the deformation of the mean skull configuration of the plesiomorphic state into the mean of the apomorphic state. Roman numerals represent patterns of the internal carotid circulation (see Table 7.1). Pair-wise comparisons are: **a** turtles with the foramen posterior canalis carotici interni (patterns I–IV) and turtles without the foramen (patterns V and VI); **b** turtles with the bifurcation of the internal carotid artery enclosed within the parabasisphenoid complex (patterns I–III) and turtles with the ventrally exposed bifurcation (patterns IV–VI); **c** turtles with the canalis caroticus internus passing through the pterygoid and parabasisphenoid only (patterns I, II, and IV) and turtles with the canalis either absent or associated with the prootic (patterns III, V, and VI); **d** turtles with the foramen posterior canalis carotici interni associated with the basicranium (patterns II and III) and turtles with the foramen not associated with the basicranium (patterns I and IV). Not to scale among a, b, c, and d

skull shape of patterns I–VI has the following characteristics: the basicranium is expanded strongly posterolaterally (landmarks 9 and 16–18) and slightly anteriorly (8); the internal naris (4) is more posterior in position; and the antorbital region (5) and the occiput (15) are shortened anteromedially. In CVA of the Procrustes coordinates that followed after comparison with CVA of partial warp scores, eleven landmark coordinates score high loadings (upper and lower 25% of the range of variation) on the single significant canonical variate axis: 4x, 5xy, 9y, 14xy, 15xy, 16xy, and 17x. Amongst these coordinates, bivariate plots of the variances for patterns I–IV and for patterns V and VI show that 4x, 5y, 15xy, 16x, and 17x have relatively small variances (smaller than the mean) in both of the comparative groups. The coordinates 5x and 14xy are highly variable and are outliers in both of the plots.

Between turtles with the bifurcation of the cerebral carotid and palatine arteries enclosed within the parabasisphenoid complex (patterns I–III) and turtles with the bifurcation exposed ventrally (patterns IV–VI), Goodall's *F* test returns significant difference in mean skull shapes, as follows: $F = 2.95$; df = 54.00, 7614.00; $P < 0.001$; D (95% confidence) = 0.0805 (Low D = 0.0638, High D = 0.1218); SE = 0.0148. The deformation of mean skull shape representing the plesiomorphic state of the exposed bifurcation (Fig. 7.8b) reveals similar trends recovered in the comparison between patterns I–IV and patterns V and VI (4, 5, 8, 9, 15–17) with three main differences. First, the magnitude of deformation is smaller in this comparison than in the comparison between the taxa with the foramen posterior canalis carotici interni and those without. Second, the snout (2, 3) is elongate anteriorly in patterns I–III. Third, the palate (6, 10–13) is lateromedially narrow in these patterns. In CVA of Procrustes coordinates after comparison with CVA of partial warp scores, fourteen landmark coordinates score high loadings (upper and lower 25% of the range of variation) on the canonical variate axis ($n = 1$): 3xy, 4x, 5xy, 6y, 9xy, 10y, 11y, 14y, 15x, and 16xy. Amongst these coordinates, bivariate plots of the variances for patterns I–III and for patterns IV–VI shows that the coordinates 2x, 3x, 9y and 16xy have variances smaller than the means in both comparative groups. The coordinates 5y, 7y, 8x, 14x, and 17x are outliers in patterns I–III, and the landmarks 5, 15, and 17 are similarly highly variable both anteroposteriorly and lateromedially in patterns IV–VI.

In comparison between patterns I, II, and IV (in which the internal carotid artery passes through the pterygoid and parabasisphenoid only) and patterns III, V, and VI, Goodall's *F* test supports significant difference between the mean skull shapes, as follows: $F = 15.06$; df = 54.00, 7614.00; $P < 0.001$; D (95% confidence) = 0.1053 (Low D = 0.0941, High D = 0.1211); SE = 0.0071. Deformation of the mean skull shape of patterns III, V, and VI to the mean skull shape of patterns I, II, and IV (Fig. 7.8c) reveals the following differences: the snout (2, 3) is elongated anteriorly; the internal naris (4) is more posterior in position; the pterygoid (6, 7) and the antorbital region (5) are displaced anteromedially; the parabasisphenoid complex (8, 9) is smaller; the pterygoid (13) is expanded posteriorly; and the occiput (14–16) is narrower lateromedially with the occipital condyle (17) displaced more posteriorly. However, in CVA of Procrustes coordinates following comparison with CVA of partial warp scores, only two landmark coordinates load heavily (upper or lower 25% of the range of variation) on the single significant canonical variate axis: 7y and 13xy. In bivariate plots of the variances, a single landmark coordinates (13x) have variances smaller than the means in both of the comparative groups. In contrast, five landmark coordinates show high variances in both plots: 5y, 8x, 14xy, and 17x.

Finally, turtles with the foramen posterior canalis carotici interni associated with the basicranium (patterns II and III) are compared with turtles with the foramen not associated with the basicranium (patterns I and IV). Goodall's *F* test returns significant support for difference between the mean skull shapes, as follows: $F = 16.01$; df = 54.00, 7344.00; $P < 0.001$; D (95% confidence) = 0.1020 (Low D = 0.0865, High D = 0.1194); SE = 0.0081. The thin-plate spline deformation from the mean shape of patterns I and IV to that of patterns II and III (Fig. 7.8d) suggests that the snout is shorter posteriorly (2, 3), the pterygoid (6, 7) and the antorbital region (5) are expanded posterolaterally, the parabasisphenoid complex (8, 9) is expanded anteriorly and laterally, the quadrate (10–12) is expanded laterally, and the occiput (14–17) is shortened anteriorly. Amongst these changes in positions of the landmarks, loadings on the single significant canonical variate axis are high for 7y, 8x, 9x, 13xy, 14x, and 17x in CVA of Procrustes coordinates after comparison with CVA of partial warp scores. Despite the high loadings of several landmark coordinates, only two of the landmark coordinates (7x and 16y) show variances smaller than the means in bivariate plots of both comparative groups. The coordinates 5y, 14xy, and 17x are outliers in both of the plots.

Discussion

Clade-Specific Position of the Foramen Posterior Canalis Carotici Interni

Analysis 1 quantitatively demonstrated that, despite superficial resemblances, three major radiations of turtles (eucryptodires: patterns I + IV; paracryptodires: pattern II; pleurodires: pattern III) represent divergence in skull morphology, and that the position of the foramen posterior canalis carotici interni is one of the important factors in each shape divergence. The divergence was supported by two observations. First, mean skull configurations for eucryptodires, paracryptodires, and pleurodires significantly differed from each other (Goodall's *F*; Table 7.3). Second, the three clades formed well-separated clusters on the CVA plot (Fig. 7.5). The position of the foramen posterior canalis carotici interni as a factor in each of the divergences was supported by three lines of evidence (Table 7.4): (1) the thin-plate spline deformations (Fig. 7.4) showed marked differences in the position of the foramen between the mean skull configurations; (2) anteroposterior and lateromedial

Table 7.4 Summary of the landmarks correlated with the patterns of the internal carotid circulation

Characters and landmark numbers	Patterns	Analysis 1			Analysis 2		
		Deformation	CVA	Test of Variance	Deformation	CVA	Test of variance
Foramen posterior canalis carotici interni (1)	**I + IV**	**Strong**	**High**	**Small**	–	–	–
	II	**Strong**	**High**	**Small**	–	–	–
	III	**Strong**	**High**	**Small**	–	–	–
Snout (2, 3)	I + IV	Strong	High	Equivoval	Strong	Equivocal	Equivocal
Internal naris (4)	**I + II + III + IV**	–	–	–	**Strong**	**High**	**Small**
Maxilla-jugal contact (5)	I + II + III + IV	–	–	–	Strong	High	Equivocal
Anterolateral extreme of pterygoid (6)	I	Weak	Low	Equivocal	–	–	–
	II	Strong	Equivocal	Equivocal	–	–	–
	III	Weak	Low	Equivocal	–	–	–
	II + III	–	–	–	Strong	Equivocal	Equivocal
Lateral extreme of pterygoid (7)	III	Strong	High	Equivoval	Strong	High	Equivocal
Parabasisphenoid complex (8, 9)	**I + II + III**	–	–	–	**Strong**	**High**	**Small**
	II	Strong	Low	Large	–	–	–
	III	Weak	Low	Small	–	–	–
Quadrate (10–12)	I + IV	Weak	High	Equivoval	Strong	Equivocal	Large
Posterolateral extreme of pterygoid (13)	**I + II + IV**	**Strong**	**High**	**Small**	**Strong**	**High**	**Small**
Occiput (14–17)	I + II + III + IV	–	–	–	Strong	High	Equivocal

Groups of the patterns represent the least inclusive combinations for the landmark based on Analysis 1 and multiple analyses from Analysis 2. Landmarks were all evaluated relative to each other. In the first column, numbers in parentheses correspond to landmarks in Fig. 7.3 and Table 7.2. Pattern numbers in the second column follow (Sterli and de la Fuente 2010; see Table 7.1). In the deformation columns, "strong" means deformation greater than the ones exhibited by the lower two-thirds of the landmarks and "weak" means deformation within lower two-thirds of landmarks. In the CVA columns, "high" refers to loading values that fall into upper and lower 25% of the range of variation and "low" refers to loading values that fall within the middle 50% of the range of variation. In the test of variance column, "large" refers to the variance larger than the mean of the variances and "small" refers to the variance smaller than the mean of the variances. Bolded entries highlight characters positively supported by all tests

coordinates of the foramen loaded heavily on the significant canonical variate axes (Fig. 7.6); and (3) despite its large variation between the clades, the foramen was amongst the least variable of the landmarks within each clade (Fig. 7.7). Taken together, the position of the foramen posterior canalis carotici interni is conserved within each clade, but differs greatly between clades when compared within the set of landmarks examined in this study. Its high loadings on the canonical variate axes and small variances within each of the three clades reject the possibility that difference in its position between the mean skull configurations were due to large variation within each clade. Topographically, the foramen posterior canalis carotici interni is more posterolateral in position in eucryptodires than in other clades. In pleurodires, the foramen is anteromedial relative to that in eucryptodires and posterior relative to that in paracryptodires. The foramen is anteromedial in position in paracryptodiers with respect to eucryptodires and pleurodires. Osteologically, these different positions can be distinguished by the elements that the foramen perforates: the pterygoid in eucryptodires, the pterygoid and basisphenoid in paracryptodires, and the prootic in most pleurodires (Gaffney 1975a; Brinkman and Nicholls 1993; Hirayama et al. 2000). This topographical variation is consistent with the results from Analysis 1.

These interpretations, together with the new classification of arterial patterns by Sterli and de la Fuente (2010), warrant reassessment of characters of the foramen posterior canalis carotici interni used in previous cladistic analyses of turtle interrelationships. The plesiomorphic state of the internal carotid circulation in turtles unambiguously is the absence of the foramen posterior canalis carotici interni in basal turtles (Sterli and de la Fuente 2010). The clade-specific positions of the foramen supported by the present analysis each represents an apomorphy. The identification of these apomorphies is consistent with Gaffney's (1975a) initial observation on difference in the position of the foramen between eucryptodires and paracryptodires, but not with Evans and Kemp's (1976) and Rieppel's (1980) hypothesized evolutionary transition from the paracryptodire condition to the eucryptodire condition. It is also consistent with independent evolutionary origins of the foramen in eucryptodires (pattern I + IV) and paracryptodires (pattern II) proposed by Brinkman and Nicholls (1993). Origin of the foramen in paracryptodires independent of eucryptodires and pleurodires is supported by the absence of the foramen in the stem paracryptodires *Glyptops* and *Pleurosternon* (Sterli et al. 2010). Because pleurodires do not form a sister clade to either eucryptodires or paracryptodires in any contemporary turtle phylogeny (Gaffney 1975a; Gaffney et al. 2006; Joyce 2007), the pleurodire condition (pattern III) also may have arisen independently. The basal position (Hirayama et al. 2000; Joyce 2007) of *Meiolania* (pattern IV), posits either an independent derivation of the foramen for this taxon or a reversal in its putative sister-taxon *Mongolochelys* (pattern V).

Independent origins of the foramen posterior canalis carotici interni challenge the homology of the foramen and may eventually complicate its terminology. However, the more important implication is that the position of cranial entrance of the internal carotid artery provides a quantitatively supported morphological character diagnostic for each major clade. Based on new information provided by Sterli and de la Fuente (2010) and Sterli et al. (2010), combined with quantitative support for each state from the present analysis, I propose the following amendment to Joyce's (2007) character 56:

The internal carotid artery passes: 0, ventral to the basicranium and palate before bifurcating into the cerebral carotid and palatine arteries (no foramen posterior canalis carotici interni); 1, between the basisphenoid and the pterygoid halfway along the basisphenoid-pterygoid suture; 2, through the prootic; 3, through the pterygoid near the posterolateral end of the parabasisphenoid. Unordered. (Remarks: In state 2, passage of the internal carotid artery may be obscured ventrally by the underlying bones. Modified from Joyce 2007, character 56.)

Variation of the Foramen Posterior Canalis Carotici Interni Within Clades

In paracryptodires, anteroposterior variance of the foramen is small relative to that of other landmarks (Fig. 7.7b). This small anteroposterior variance rejects the hypothesis (Evans and Kemp 1976; Rieppel 1980) of an anteroposterior gradient in the position of the foramen within the clade. The foramen identified as the foramen posterior canalis carotici interni by Evans and Kemp (1976) actually represents the foramen for the cerebral carotid artery (Sterli et al. 2010). *Thalassemys,* used by Rieppel (1980) in support of the gradient, is now considered a basal eucryptodire (Joyce 2007), not a paracryptodire. The foramen previously identified as the foramen posterior canalis carotici interni in paracryptodires needs further examination.

Pleurodires show a variety of conditions in the position of the foramen posterior canalis carotici interni (Gaffney et al. 2006). The anteroposterior and lateromedial variances of the foramen being smaller than those of almost all other landmarks within pleurodires (Fig. 7.7c) suggest that the position of the foramen is stable within the clade, at least relative to the other landmarks used in this study. The conservative position of the foramen within the skull is inconsistent with the osteological observation that the foramen opens into different elements (Gaffney et al. 2006). Therefore, it is hypothesized here that this variation depends on the relative dimensions and position of the elements within the skull, not a shift in the position of the internal carotid artery. The present analysis provides no evidence that the internal carotid artery significantly shifts its position of the cranial entrance relative to the skull configuration within pleurodires. For this reason, variation of the elements surrounding the foramen should not be treated at the same level as differences in the position of the foramen between eucryptodires, paracryptodires, and pleurodires.

By setting up two characters describing the position of the foramen posterior canalis carotici interni, Gaffney et al. (2006) identified both anteroposterior and lateromedial changes in the relative dimensions of the elements that surround the foramen posterior canalis carotici interni in pleurodires. Their character 74 essentially specifies that the prootic is covered ventrally by either the lateral expansion of the parabasisphenoid or the medial expansion of the quadrate in derived pleurodires. The internal carotid artery pierces whichever element expanded and underlays the prootic. This sometimes is coupled with posterior expansion of the pterygoid, which complicates Gaffney et al.'s (2006) character definition. Gaffney et al. (2006) defined another character (75) that describes participation of the pterygoid in the margin of the foramen posterior canalis carotici interni due to the posterior expansion of the element. Perhaps this complex interaction of several elements conspired against the geometric morphometric methodology used here from detecting evidence of the interactive changes in relative dimensions of the elements within pleurodires.

One strategy to code for this complex variation in a phylogenetic analysis is to formulate two characters that code for the pleurodire variation, in addition to the character (modified version of character 56 of Joyce 2007) that distinguishes position of the foramen between the major clades. The first two characters can be modified from two characters originally proposed by Gaffney et al. (2006)—one of those codes for lateromedial change in the dimensions of the parabasisphenoid complex and the quadrate, whereas the other codes for anteroposterior change in the dimensions of the parabasisphenoid complex and the pterygoid. These two characters separate the states among pleurodires and are scored as "inapplicable" for all other taxa. This approach avoids weighting the analysis in favour of the plesiomorphy and the eucryptodire and paracryptodire conditions. Seven states in Gaffney et al.'s (2006) character 74 present difficulties in recovering a strong phylogenetic signal. That character treats variation within pleurodires at the same level as it does for the independently derived states at the level of eucryptodires, paracryptodires, and pleurodires. For this reason also, multiple characters are preferred over formulating a single phylogenetic character that encompasses all the morphological variation in the position of the foramen posterior canalis carotici interni. Gaffney et al.'s (2006) characters 74 and 75 may be modified as in the following:

Foramen posterior canalis carotici interni in pleurodires: 0, surrounded by the prootic; 1, by the parabasisphenoid complex and/or the pterygoid; 2, entirely or partly by the quadrate. Unordered. (Remarks: Scored as "inapplicable" in non-pleurodires. Modified from Gaffney et al. 2006, character 74.)

Posterior margin of the pterygoid in pleurodires: 0, does not form the anterior margin of the foramen posterior canalis carotici interni; 1, forms the anterior margin of the foramen posterior canalis carotici interni. (Remarks: Scored as "inapplicable" in non-pleurodires. Modified from Gaffney et al. 2006, character 75.)

Correlations with Other Cranial Landmarks

The complex distributions of character states associated with the internal carotid artery (Table 7.1) strongly suggest multiple factors independently or interactively affect the morphology of the internal carotid circulation in turtles. The results from analyses 1 and 2 (summarized in Table 7.4) help tease apart correlations between patterns of the internal carotid circulation and other cranial landmarks and test two of the three factors proposed by Sterli et al. (2010) that facilitate patterns.

Sterli et al. (2010) proposed that the expansion of the parasphenoid ventral to the basisphenoid captured the internal carotid artery within the parabasisphenoid complex and the ventral covering of the bifurcation between the cerebral carotid and palatine arteries (patterns I–III). Evidence for the expansion of the parasphenoid comes from the basicrania of the basal eucryptodire *Plesiochelys* (pattern I) and the basal paracryptodire *Pleurosternon* (pattern V) in which the parasphenoid can be distinguished from the basisphenoid (Sterli et al. 2010). Expansion of the parasphenoid would lead to increased ventral exposure of the parabasisphenoid complex. That hypothesis is supported by Analysis 2. Larger ventral exposure of the parabasisphenoid complex was identified in the deformation from patterns IV–VI to patterns I–III. That expansion was an important factor in distinguishing patterns I–III from patterns IV–VI in CVA, and the variances were small relative to other landmarks. A similar trend of the expanded parabasisphenoid complex was detected in the comparison between patterns V and VI and patterns I–IV. However, deformation was smaller than in the comparison between patterns IV–VI and patterns I–III (Fig. 7.8). Posterolateral expansion of the parabasisphenoid complex was not explicitly supported by both CVA and the plots of variances in the comparison between patterns V and VI and patterns I–IV. All of these observations suggest that increased ventral exposure of the parabasisphenoid complex correlates with patterns I–III, which are characterized by the ventral flooring of the cerebral carotid-palatine bifurcation by the parabasisphenoid complex. The present analysis upholds a modified version of Hirayama et al.'s (2000) character 31, which was not adopted by either Gaffney et al. (2006) or Joyce (2007). The modified character can be described as follows:

Bifurcation of the internal carotid artery into the palatine artery and the cerebral carotid artery: 0, not covered ventrally; 1, covered ventrally by the parabasisphenoid complex. (Remarks: modified from Hirayama et al. 2000, character 31.)

Additionally, Sterli et al. (2010) identified posterior expansion of the pterygoid as a factor facilitating patterns I, II, and IV. That hypothesis is consistent with the results from my analyses 1 and 2. However, posterolateral expansion of the pterygoid is not necessarily correlated with the foramen forming within the pterygoid (patterns I and IV), because the foramen forms more anteriorly in paracryptodires than in any other clades (pattern II). Perhaps the posterolaterally expanded pterygoid explains the posterolateral position of the foramen posterior canalis carotici interni in eucryptodires, but this is not the case in paracryptodires. At any rate, the present analysis is consistent with the posterior expansion of the pterygoid in eucryptodires and paracryptodires, thus warranting establishment of a character based on this morphology. Joyce's (2007) character 41 codes for the presence or absence of the pterygoid-basioccipital contact. That character may be modified as in the following to include the new phylogenetic information:

Pterygoid, posterior end: 0, anterior to or lateral to the quadrate condyle; 1, as posterior as the posterior margin of the parabasisphenoid complex; 2, contacting the basioccipital posteriorly. (Remarks: modified from Joyce 2007, character 41).

Interestingly, Analysis 1 recovered weak support both for the anteriorly displaced and expanded basicranium in

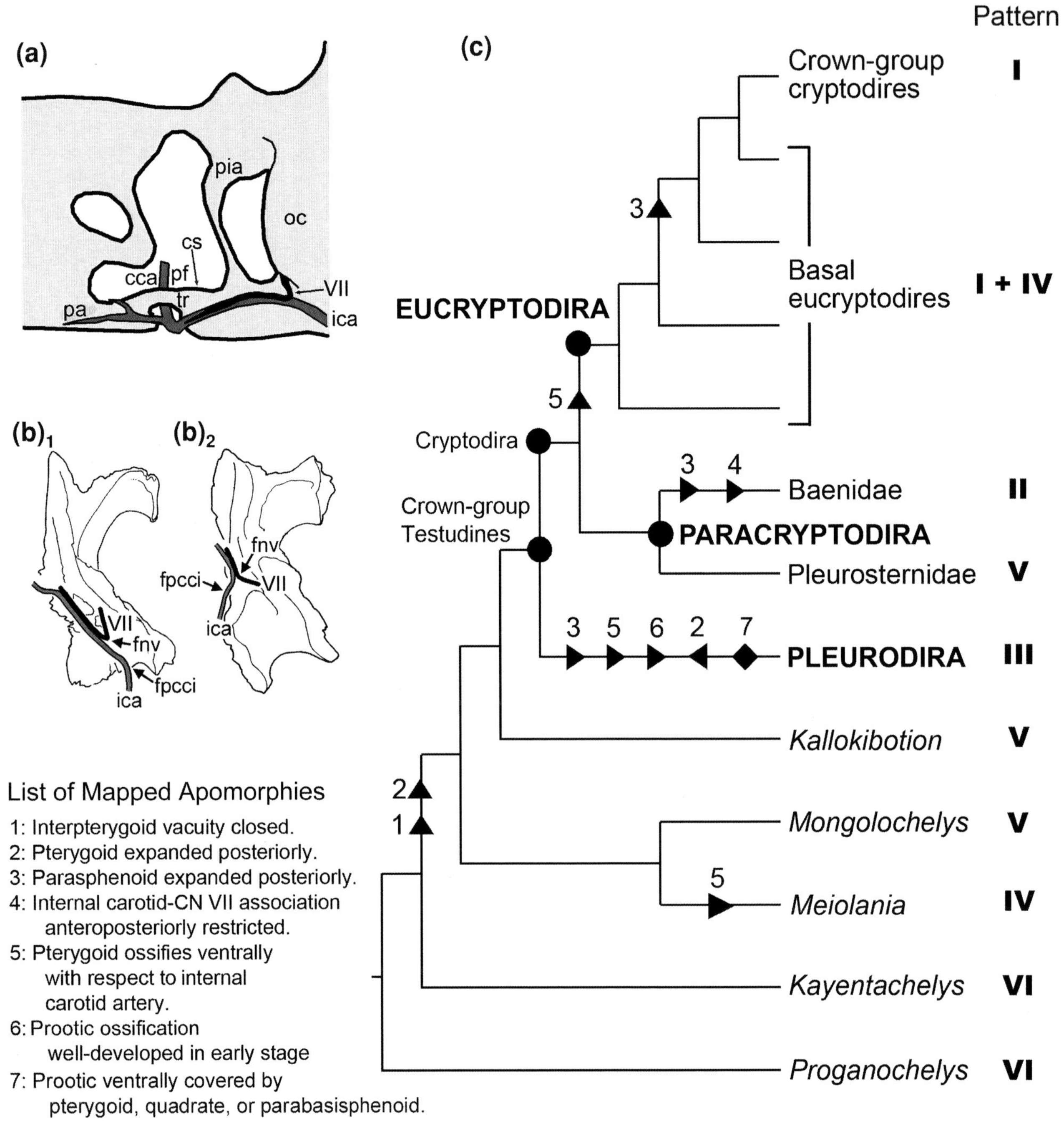

paracryptodires (pattern II) and for the laterally expanded parabasisphenoid complex in pleurodires (pattern III). These shifts of the parabasisphenoid complex may be correlated with the respective clade-specific positions of the foramen posterior canalis carotici interni. However, in pattern II, the anteroposterior component of variances was larger in the parabasisphenoid complex than other landmarks (Fig. 7.7b). In pattern III, the lateromedial component of deformation in the parabasisphenoid complex was small and did not contribute to the separation of pleurodires in CVA (Figs. 7.5c, 7.6). These possible trends within the clades would be worth testing in future.

The relatively posterior position of the internal naris correlated well with the presence of the foramen posterior canalis carotici interni (patterns I–IV). It is unlikely that the position of the internal naris directly affected the capture of the

◀ **Fig. 7.9** The evolution of morphogenetic traits that potentially facilitated the patterns of the internal carotid circulation in turtles. **a** Diagram of the orbitotemporal region of the chondrocranium of *Caretta caretta* (stage III embryo of Kuratani 1987), in *left lateral view*, showing association between the internal carotid artery and CN VII. The internal carotid artery bifurcates into the palatine artery and the cerebral carotid artery, the latter extending dorsally through the pituitary fenestra. Modified after Kuratani (1987); tissues not associated with the internal carotid circulation have been omitted for clarity. **b** Line drawings of isolated right pterygoids, both in dorsal view and with palatine artery omitted for clarity, comparing association of the internal carotid artery with the palatine branch of CN VII between a representative eucryptodire and paracryptodire: $\mathbf{b_1}$ extant *Chelydra serpentina* [Eucryptodira, Chelydridae; AMNH (American Museum of Natural History) 107388], exhibiting pattern I; $\mathbf{b_2}$ Cretaceous Baenidae indet. [Paracryptodira; MCZ (Museum of Comparative Zoology, Harvard University) 3563], exhibiting pattern II. Note how association between internal carotid artery and palatine branch of CN VII is relatively shorter in the paracryptodire than in the eucryptodire ($\mathbf{b_2}$ vs. $\mathbf{b_1}$). Both images modified after Gaffney (1975a); drawings not to same scale. **c** Morphological and developmental traits that distinguish the patterns of the internal carotid circulation, mapped under DELTRAN optimization onto a simplified version of Joyce's (2007) phylogenetic tree. Many branches and taxa from Joyce's (2007) tree have been omitted for clarity. *Solid circles* at nodes denote major clades (except for Pleurodira, which are collapsed into a single branch). Names of the three major clades considered in this study are in capital letters. Roman numerals to *right* of tree represent patterns of internal carotid circulation proposed by Sterli and de la Fuente (2010), which are defined by characters discussed in text [modified versions of character 56 from Joyce (2007) and of character 31 from (Hirayama et al. 2000)]. Arabic numbers beside branches correspond to apomorphies listed in lower left of figure. Apomorphies 1–6 are characters correlated with patterns of internal carotid circulation [modified versions of character 95 from Gaffney et al. (2006) and of character 41 from Joyce (2007); two characters originally mapped by Sterli et al. (2010); and two new characters]. Apomorphy 7 describes variation within pleurodires [modified version of characters 74 and 75 from Gaffney et al. (2006)]. See text for further details. Directions of character state transformations are indicated by arrows on stems: a crown-ward arrow indicates an acquisition, whereas a stem-ward arrow indicates a reversal. A diamond indicates a variable character. Anatomical abbreviations (A and B): *cca* cerebral carotid artery; *cs* crista sellaris (medial to the trabecula; not visible in lateral view); *fpcci* foramen posterior canalis carotici interni; *fnv* foramen pro ramo nervi vidiani for the palatine branch of CN VII; *ica* internal carotid artery; *oc* otic chamber; *pa* palatine artery; *pia* pila antotica; *pf* pituitary fenestra; *tr* trabecula; VII, CN VII (facial nerve)

internal carotid artery by the base of the skull, partly because the internal carotid artery does not directly irrigate the snout region, and partly because it is topologically far from the area entered by the artery. However, these traits may have been linked in the timing of their emergence in turtle evolution.

Other cranial landmarks found in correlation with arterial patterns failed one or more tests (Table 7.4). Typically, the variances within the clades were large. In other cases, the landmarks cannot be used to distinguish between the patterns based on CVA. These results indicate that even though the deformation of the means indicate the area of significant differences, the variability of the landmark within each group may swamp the differences. Therefore, multiple tests of loadings and variances of the landmarks are warranted in a geometric morphometric analysis. The posterolateral position of the lateral extreme of the subtemporal margin of the pterygoid may be characteristic of pleurodires. This is consistent with the acquisition of the trochlear process (= "process trochlearis pterygoidei" sensu Gaffney 1972a) in this clade. However, the variability of this landmark was high amongst other patterns. The same is also the case for the anterior elongation of the snout in patterns I and IV.

A New Hypothesis: The Internal Carotid-CN VII Constraint Hypothesis

So far, morphological correlates of the internal carotid circulation in turtles have been identified for patterns I, IV, V, and VI (Sterli and de la Fuente 2010; Sterli et al. 2010; Table 7.1). On the other hand, it is unclear what morphological novelty(ies) resulted in patterns II and III. Joyce's (2007) phylogeny (see simplified version in Fig. 7.9c) suggests that patterns II and III were derived independently from pattern V. Expansion of the parasphenoid separates patterns II and III from pattern V, but that character also distinguishes pattern I from pattern IV. The posteriorly expanded pterygoid may explain patterns I and IV, but it does not separate patterns II from patterns III, V and VI. The geometric morphometric analysis recovered weak support both for the laterally expanded parabasisphenoid complex in pattern III and for the anteriorly displaced parabasisphenoid complex in pattern II. However, lateral expansion of the parabasisphenoid complex in pattern III seems to be variable (Gaffney et al. 2006). The extent of the anterior displacement of the parabasisphenoid complex is variable in pattern II (Fig. 7.7b) and unlikely to explain the anteromedially displaced foramen posterior canalis carotici interni. In pattern I, the internal carotid artery enters the lateral region of the parabasisphenoid complex. In pattern III, the artery enters the prootic. Therefore, pattern II deviates from these patterns in that the basicranial entrance of the internal carotid artery is more anterior in position. What causes the anterior shift of the basicranial entrance of the artery in pattern II? In addition, what facilitates the parabasisphenoidal entrance in pattern I and the prootic entrance in pattern III?

The pterygoids and parabasisphenoid complexes of turtles showing any of patterns I–III tend to ossify around the internal carotid artery and its derivatives both proximal and distal to the cerebral carotid-palatine bifurcation (Sterli et al. 2010). In this region, the palatine branch of the facial nerve (CN VII) is associated with the internal carotid artery proximal (posterior) to the cerebral carotid-palatine bifurcation. In eucryptodires, paracryptodires, and pleurodires, CN VII exits the prootic laterally; the palatine branch of CN VII branches off anteroventrally and extends anteriorly via the canalis cavernosus (= "cavernous canal" of some authors) and the foramen pro ramo nervi vidian (= "vidian foramen" of some authors) to parallel the internal carotid artery; and the branch eventually passes through the pterygoid and dorsal to the palate (Gaffney 1975a, 1979b). Although there is no osteological evidence of this association in turtles of patterns V and VI, the foramen pro ramo nervi vidian opening ventrally in the pterygoids in pattern V (Evans and Kemp 1975; Gaffney 1979b; Sukhanov 2000) and in the ventrally exposed prootic in pattern VI (Gaffney et al. 1987; Gaffney 1990) indicate that this association was probably present below the palate in those turtles as well.

The association between the internal carotid artery and the palatine branch of CN VII persists from early cranial development. During the chondrocranial development of *Caretta caretta*, the palatine branch of CN VII extends parallel to the internal carotid artery anterior to the chondrocranial otic chamber all the way to the cerebral carotid-palatine bifurcation below the pituitary fenestra (Kuratani 1987; Fig. 7.9a). A similar association has been illustrated in the embryonic development of *Chelydra serpentina* (Rieppel 1990). In fact, this association is universal among osteichthyans and has been documented in numerous osteichythian embryos (e.g., Goodrich 1930), including reptiles (Bellairs and Kamal 1981; Rieppel 1987, 1988). Therefore, the association between the palatine branch of CN VII and the internal carotid artery can be considered as a constraint in the development of the internal carotid artery and CN VII, perhaps as a factor that constrains the topography of the artery throughout ontogeny. Here I call this the internal carotid-CN VII constraint hypothesis.

During chondrocranial development in turtles, the association of the palatine branch of CN VII with the internal carotid artery occurs lateroventral to the crista sellaris, posterior to the posterior end of the trabecula, and anteromedial to the otic chamber (Kuratani 1987; Rieppel 1990). At the anterior end of the association is the cerebral carotid-palatine bifurcation, ventral to the pituitary fenestra (Fig. 7.9a). The basisphenoid ossifies from paired centers in the crista sellaris and the posterior end of the trabecula, and the prootic ossifies from a center in the otic chamber (Bellairs and Kamal 1981). In addition, the parasphenoid ossifies anteroventrally or ventrally with respect to the basisphenoid developing from the crista sellaris and the cerebral carotid-palatine bifurcation of the internal carotid artery (Bellairs and Kamal 1981). Although Sterli et al. (2010) proposed that a ventral expansion of the parasphenoid trapped the bifurcation point within the parabasisphenoid complex, the topographical relationships of the chondrocranium with the parasphenoid and the internal carotid artery suggests that a posterior expansion of the parasphenoid, not a ventral expansion, better explains the trapping of the bifurcation point. Posterior expansion of the parasphenoid therefore sets patterns I–III apart from patterns IV–VI.

In eucryptodire embryos (Bellairs and Kamal 1981; Rieppel 1993; Sheil 2003, 2005; Sheil and Greenbaum 2005; Sánchez-Villagra et al. 2009), the basisphenoid continues to ossify posterolaterally before the centre of the prootic is well developed. In *Phrynops*, the one pleurodire for which an adequate cranial developmental sequence has been documented (Bona and Alcade 2009), the timing of the prootic ossification [Yntema's (1968) stage 22] is delayed with respect to that of the basisphenoid (stage 21) as in eucryptodires. However, at stage 22, the ossifying prootic of *Phrynops* is already larger than the basisphenoid. These embryological observations may explain the key difference between patterns I and III. The eucryptodire tendency to enclose the internal carotid canal within the parabasisphenoid complex is consistent both with posteromedial growth of the basisphenoid ossification and with delayed growth of the prootic ossification (Bellairs and Kamal 1981; Rieppel 1993; Sheil 2003, 2005; Sheil and Greenbaum 2005; Sánchez-Villagra et al. 2009). In contrast, the well-developed prootic reported by Bona and Alcade (2009) in *Phrynops* at stage 22 and onward may facilitate the pleurodire condition of the internal carotid artery passing through the prootic. Although not documented by Bona and Alcade (2009), a continued ossification of the prootic in the otic chamber of *Phrynops* could be expected to capture the internal carotid artery in the vicinity, along with the palatine branch of CN VII. Considering that chondochranial development has yet to be adequately studied for other pleurodires, it is uncertain both whether the internal carotid artery is captured in the above-predicted manner in *Phrynops* and other pleurodires and whether the accelerated development of the prootic documented by Bona and Alcade (2009) in *Phrynops* is universal among pleurodires. Both of these ideas could be tested by future studies of pleurodire embryological development.

There are no embryonic developmental series available for patterns II and IV–VI. However, the available osteological evidence suggests that the spatial association between the palatine branch of CN VII and the internal carotid artery was relatively anteroposteriorly short in paracryptodires. In the isolated pterygoid of a baenid shown in Fig. 7.9b, the foramen pro ramo nervi vidiani for the palatine branch of CN VII is adjacent anterolaterally to the foramen posterior canalis carotici interni (Gaffney 1975a). The morphology of the pterygoid-parabasisphenoid suture

Table 7.5 Summary of developmental traits that potentially facilitate the patterns of the internal carotid circulation in turtles

Pattern	Degree of association with CN VII	Parasphenoid	Position of pterygoid relative to internal carotid artery	Embryonic growth of prootic	Element(s) underlying prootic
I	Extensive	Posteriorly expanded	Ventral	[b]	Pterygoid
II	Restricted	Posteriorly expanded	Dorsal	[c]	Pterygoid
III	Extensive	Posteriorly expanded	Ventral	Accelerated	Pterygoid; parabasisphenoid; quadrate
IV	[c]	Not expanded	Ventral[a]	[c]	Pterygoid
V	[c]	Not expanded	Dorsal	[c]	Pterygoid
VI	[c]	Not expanded	Dorsal	[c]	Prootic exposed

See text for details and references. Pattern numbers in the second column follow (Sterli and de la Fuente 2010; see also Table 7.1).
[a] indicates pterygoid is dorsal with respect to the internal carotid artery in the fossil eurcryptodire *Dracochelys* (Gaffney and Ye 1992). Abbreviations: [b] used as a reference with which the growth of the prootic in pleurodires is compared; [c] condition unknown

indicates that the foramen posterior canalis carotici interni is at least in the anterior half of the parabasisphenoid complex. The internal carotid artery turned anteromedially past the foramen posterior canalis carotici interni, and the canalis caroticus lateralis is at the point where the internal carotid artery left the pterygoid-parabasisphenoid suture. These facts suggest that the internal carotid-CN VII association occurred at most for a distance that equals the maxmum diameter of the foramen for the trigeminal nerve in this baenid. If this is a general condition for paracryptodires having pattern II, the internal carotid-CN VII association was short and restricted anteriorly relative to those in eucryptodires and pleurodires. Given the assumption that the association with CN VII constrained the position of the internal carotid artery close to the chondrocranium, the anteriorly restricted association may be hypothesized to result in pattern II.

In patterns V and VI, CN VII exits the braincase ventrally, either through the prootic (pattern VI; see also Gaffney et al. 1987; Gaffney 1990) or through the pterygoid that underlies the prootic (pattern V; see also Evans and Kemp 1975; Gaffney 1979b; Sukhanov 2000). The association between the palatine branch of CN VII and the internal carotid artery probably was present below the palate. The internal carotid-CN VII constraint hypothesis predicts that this region was never enclosed within bones in turtles exhibiting either pattern, because the pterygoid ossified dorsal to the internal carotid artery and ventral to the prootic. The pterygoid begins to ossify along the lateral margin and continues to grow medially (Bellairs and Kamal 1981). This medial growth of the pterygoid may be guided above or below the internal carotid artery. The pterygoid invariably ossifies early and remains larger than the basisphenoid in all known developmental series of turtles (Bellairs and Kamal 1981; Rieppel 1993; Sheil 2003, 2005; Bona and Alcade 2009; Sánchez-Villagra et al. 2009), and in no turtle does the parabasisphenoid complex extend ventrally to the pterygoid. Therefore, once the pterygoid ossified above the internal carotid artery associated with the palatine branch of CN VII, this region would have been ventrally exposed as in patterns V and VI (Sterli and de la Fuente 2010). The palatine branch of CN VII presumably passed dorsal to the palate to its target through the interpterygoid vacuity in pattern VI and through the slit within the pterygoid along with the palatine artery in pattern V.

The position of the pterygoid relative to the internal carotid artery also is crucial to distinguish patterns I, III, and IV from other patterns. In eucryptodire embryos (Bellairs and Kamal 1981 and references therein; Rieppel 1993; Sheil 2003, 2005; Sheil and Greenbaum 2005), the pterygoid begins to ossify lateral to the internal carotid artery and develops medially. During the medial expansion, the anterior two-thirds to three-quarters of the pterygoid extend below the canalis caroticus internus and the palatine branch of CN VII, thereby providing an osseous floor. The osseous floor to the internal carotid artery lateral to the basisphenoid complex characterizes patterns I, III, and IV. Interestingly, the synemydid/macrobaenids *Chubutemys* and *Dracochelys* (Gaffney and Ye 1992; Gaffney et al. 2007b) represent an intermediate stage between pattern V and patterns I and IV. In these genera, the internal carotid artery is at the same level with the pterygoid and extends within a sulcus on the ventral surface of the pterygoid, which may be a consequence of the postnatal growth of the pterygoid. The sulcus for the internal carotid artery (sulcus caroticus internus) is partially exposed near the pterygoid-parabasisphenoid suture in turtles exhibiting pattern IV, including the synemydid/macrobaenids *Hangaiemys*, *Judithemys*, *Ordosemys,* and *Synemys* and the xinjiangchelyids *Annemys* and *Xinjiangchelys* (Brinkman et al. 2012). Their basal eucryptodire affinities make a plausible case for these turtles filling a gap in the transition from pattern V to pattern I, via pattern IV.

These embryological and morphological insights aided by the internal carotid-CN VII constraint hypothesis (Table 7.5) suggest two new characters and one revised character useful in turtle systematics. These characters, however, should be used with caution, because they correlate with states in the modified version (see above) of Joyce's (2007) character 56. The three characters are as follows:

Association between the internal carotid artery and the palatine branch of CN VII: 0, extensive; 1, restricted to distance shorter than one-quarter of the anteroposterior parabasisphenoidal length. (Remarks: new character.)

Position of pterygoid with respect to the internal carotid artery: 0, dorsal, the artery exposed; 1, level, the artery within the sulcus caroticus internus; 2, ventral, the artery within the canalis caroticus internus. Ordered. (Remarks: new character. Ordered because state 1 is intermediate.)

Prootic: 0, ventrally exposed; 1, exposure small, underlain ventrally by the pterygoid, the basisphenoid, and/or quadrate; 2, underlain completely by the pterygoid. Unordered. (Remarks: Correlated with posterior expansion of the pterygoid. Modified from Gaffney et al. 2006, character 95.)

To conclude this discussion, the characters were analyzed in a phylogenetic context. In addition to the characters that define patterns of the internal carotid circulation (Table 7.1), characters correlated with transition between patterns (Table 7.5) were each treated as an apomorphic condition. Together, these characters were mapped onto a simplified version of Joyce's (2007) phylogenetic tree to show the transformations discussed in the text (Fig. 7.9c). Transition from pattern VI to pattern V accompanied closure of the interpterygoid vacuity (apomorphy 1 in figure legend) and posterior expansion of the pterygoid (apomorphy 2) (Sterli et al. 2010). Pattern V is the plesiomorphic condition for the crown-group Testudines from which patterns II–IV derived. Pleurodires (pattern III) accumulated three apomorphies (3, 5, 6): posterior expansion of the parasphenoid; the pterygoid ossifying ventral with respect to the internal carotid artery; and accelerated ossification of the prootic. Only the last of these apomorphies is unique to the clade. In the same lineage, apomorphy 2 (posterior expansion of the pterygoid) reversed (Sterli et al. 2010), and the prootic through which the internal carotid artery pases may be ventrally covered by surrounding bones (apomorphy 7) (Gaffney et al. 2006). The bones underlying the prootic vary among the pterygoid, the quadrate, and the parabasisphenoid. This variation is described in two characters in text (modified versions of characters 74 and 75 from Gaffney et al. 2006). Pattern II is an apomorphic condition in paracryptodires. Two characters (apomorphies 3 and 4) are associated with the transition from pattern V to pattern II in this clade: posterior expansion of parasphenoid and anteroposteriorly restricted association between the internal carotid artery and the palatine branch of the facial nerve. Leading from the stem of the crown-group Testudines to eucryptodires, the transition from pattern V to pattern IV is accompanied one apomorphy (5): pterygoid ossifies ventrally with respect to the internal carotid artery. Proceeding crown-wards towards crown-group cryptodires, the bifurcation between the cerebral carotid and palatine arteries becomes covered ventrally in pattern I. That transition from pattern IV is associated with posterior expansion of the parasphenoid (apomorphy 3) (Sterli et al. 2010).

Under the above scenario, posterior expansion of the parasphenoid (apomorphy 3) occurred at least three times, once each in pleurodires, paracryptodires, and eucryptodires. DELTRAN optimization of this character seems reasonable, because the presence of the primitive condition in both basal eucryptodires and basal paracryptodires (Sterli et al. 2010) suggest that the ossification around the cerebral carotid-palatine bifurcation was acquired independently in both clades. Pterygoid ossification ventral to the internal carotid artery (apomorphy 5) also arose three times, once each in *Meiolania*, pleurodires, and eucryptodires. The factor(s) responsible for ventral ossification of the pterygoid having appeared multiple times in turtle evolution is (are) unknown, and will be difficult to identify because all extant turtles ossify their pterygoids ventral to the internal carotid artery. Posterior expansion of the pterygoid (apomorphy 2) developed on the stem leading to crown-group Testudines, then reversed in pleurodires. Basal eucryptodires show a mosaic of the characteristics of patterns I and IV. Phylogenetic consensus has not been reached in this part of the tree. For this reason, the character evolution of the internal carotid circulation in basal eucryptodires likely was far more complex than depicted here (Sterli et al. 2010).

Conclusions

The foramen posterior canalis carotici interni has been a feature of long-standing interest in morphological studies of the turtle skull. The results from geometric morphometric analysis reported here are consistent with previously noted difference in positions of the foramen between three major clades of turtles—eucryptodires, paracryptodires, and pleurodires. I re-defined a character describing differences in the position of the foramen that sets apart each major clade of turtles and formulated two additional characters that describe variation in the position of the foramen within Pleurodira. The trend in the position of the foramen posterior canalis carotici interni is that its position significantly differs between the three major clades, but remains relatively stable within each clade; that is in an ideal combination of properties for a synapomorphy.

The results from the geometric morphometric analysis indicate that some cranial landmarks vary in correlation with carotid arterial patterns I–VI originally identified by Sterli and de la Fuente (2010). The skull configurations distinguished by the relative shift of these landmarks are also consistent with two of the three factors proposed by Sterli et al. (2010) to affect patterns of the internal carotid circulation: (1) ventral covering of the cerebral carotid-palatine bifurcation by an expanded parasphenoid (patterns I–III); and (2) a posteriorly expanded pterygoid (patterns I, II, and IV). I revised two characters that describe each of those morphological factors.

Even then, it remains unexplained what distinguishes patterns II and III from each other and from patterns I and IV–VI. I propose the association of the internal carotid artery with the palatine branch of CN VII as a developmental constraint on morphogenetic mechanisms to explain differences among patterns I–VI. In keeping with the internal carotid-CN VII constraint hypothesis, I identified the following four developmental characters: (1) posterior expansion, not ventral expansion, of the parasphenoid as a factor in trapping the cerebral carotid-palatine bifurcation within the parabasisphenoid complex; (2) relative growth of the prootic; (3) spatial extent of the association between the internal carotid artery and the palatine branch of CN VII; and (4) the dorsoventral position of the pterygoid with respect to the internal carotid artery. Mapping these characters on to a simplified version of Joyce's (2007) phylogenetic tree, both the posterior expansion of the parasphenoid and the pterygoid ossification ventral to the internal carotid artery were inferred to have occurred at least three times. Given that basal eucryptodire interrelationship are unstable, character evolution of the internal carotid circulation complex may have been more complex than depicted here.

The internal carotid circulation is a complex system. Variations within the system may not be easily attributed to a specific source. A quantitative analysis, such as geometric morphometrics, is useful for visualizing the components of variation. However, multiple tests (e.g., CVA, plot of variances) are required to support differences that thin-plate spline deformations indicate. On the other hand, the complexity of the system implies the presence of constraint(s) and interaction of tissues during development. A comparative analysis of developmental processes can offer candidates that potentially explain the complex variations.

Acknowledgments A project like this would not have been possible without the precise, consistent, and well-illustrated descriptive works by Eugene Gaffney (American Museum of Natural History) throughout his career. Ren Hirayama (Waseda University) encouraged me in the after-hours party at an annual meeting of the Society of Vertebrate Paleontology in 2008 to do a turtle project as a reprieve from my immersion in dinosaurs, which eventually led to this paper. Heather Jamniczky (University of Calgary) sent her publications and Lara Shychoski (University of Alberta) lent her copy of Zelditch et al. (2004). Don Brinkman and Dennis Braman (Royal Tyrrell Museum of Palaeontology) provided logistical support when the preliminary version of this project was presented in the Gaffney Turtle Symposium. I thank the editors for the additional time they provided to put my thoughts together for this project. Jim Gardner (Royal Tyrrell Museum of Palaeontology) handled the manuscript. Reviews by Larry Powell (University of Calgary), Juliana Sterli (Museo Paleontológico Egidio Feruglio), and an anonymous person substantially improved the paper. Rich Palmer and Richard Fox (both University of Alberta) carefully critiqued and edited earlier versions of the manuscript. I thank Kesia Andressen for her medical support. Financial assistance came from the Alberta Ingenuity Fund and from Junichi and Kanae Miyashita.

Appendix

List of the 132 taxa and sources for the 145 skull images used for this study. The classification of turtles largely follows Gaffney et al. (2006) and Joyce (2007). All clades higher than family level, except for the crown groups, are stem-based. Within each family, genera and species are listed alphabetically. For each species, the skull image used for this study can be found in the publication cited in the corresponding source section. In some cases (e.g., Gaffney 1979b), those images were reprinted from an earlier publication. Abbreviations in column headings: *n*, number of skull images; p, patterns based on Sterli and de la Fuente (2010). An asterisk (*) denotes taxa that Sterli et al. (2010) stated likely exhibited pattern V; however, until a detailed description is presented that verifies the lack of the foramen posterior canalis carotici interni in these taxa, for the purposes of this study they are tentatively assigned pattern II. The latter is a traditional intepretation for paracryptodires and is consistent with the placement of those genera within that major clade.

Taxon	*n*	p	Sources
Testudines			
Proganochelys quenstedti	1	VI	Gaffney and Meeker (1983)
Kayentachelys aprix	1	VI	Gaffney et al. (1987)
Meiolania platyceps	1	IV	Gaffney (1983)
Mongolochelys efremovi	2	V	Sukhanov (2000); Suzuki and Tsogtbaatar (2010)
Crown-group Testudines			
Pleurodira			
Chelidae			
Batrachemys dahlia	1	III	Gaffney (1977)
Chelodina expansa	1	III	Gaffney (1977)
Chelus fimbriata	1	III	Gaffney (1977)
Emydura macquarrii	1	III	Gaffney (1977)
Hydromedusa tecifera	1	III	Gaffney (1977)
Mesoclemys gibba	1	III	Gaffney (1977)
Platemys platycephala	1	III	Gaffney (1977)
Pseudemydura umbrina	1	III	Gaffney (1977)
Pelomedusoidea			
Pelomedusidae			
Erymnochelys madagascariensis	1	III	Gaffney (1979b)
Pelomedusa subrufa	1	III	Gaffney (1979b)
Peltocephalus dumeriliana	1	III	Gaffney (1979b)
Pelusios sp.	1	III	Gaffney (1975a)
Pelusios niger	1	III	Gaffney (1979b)
Shweboemys antiqua	1	III	Wood (1970)
Araripemydidae			
Araripemys barretoi	2	III	Meylan (1996); Gaffney et al. (2006)
Euraxemydidae			
Dirqadim schaefferi	1	III	Gaffney et al. (2006)
Euraxemys essweini	1	III	Gaffney et al. (2006)
Podocnemidoidea			
Podocnemididae			
Bairdemys venezuelensis	1	III	Gaffney and Wood (2002)
Caninemys tridentata	1	III	Meylan et al. (2009)
Dacquemys paleomorpha	2	III	Gaffney (1979b); Gaffney et al. (2001b)
Podocnemis expansa	1	III	Gaffney (1979b)
Bothremydidae			
Acleistochelys mallensis	1	III	Gaffney et al. (2007a)
Araiochelys hirayamai	1	III	Gaffney et al. (2006)
Arenila krebsi	1	III	Gaffney et al. (2006)
Azabbaremys moragjonesi	1	III	Gaffney et al. (2001b)
Bothremys kellyi	1	III	Gaffney et al. (2006)
Bothremys maghrebiana	1	III	Gaffney et al. (2006)
Cearachelys placidoi	1	III	Gaffney et al. (2001a)
Foxemys mechinorum	1	III	Gaffney et al. (2006)
Kurmademys kallamedensis	1	III	Gaffney et al. (2001c)
Labrostochelys galkini	1	III	Gaffney et al. (2006)
Phosphatochelys tedfordi	1	III	Gaffney and Tong (2003)
Polysternon provinciale	1	III	Gaffney et al. (2006)
Sankuchemys sethnai	1	III	Gaffney et al. (2003)
Taphrosphys congolensis	1	III	Gaffney et al. (2006)
Taphrosphys ippolitoi	1	III	Gaffney et al. (2006)
Zalhafah bella	1	III	Gaffney et al. (2006)
Paracryptodira			
Dorsetochelys delairi	1	II*	Evans and Kemp (1975)
Pleurosternidae			
Pleurosternon bullocki	1	V	Sterli et al. (2010)
Glyptopsidae			

(continued)

(continued)

Taxon	*n*	p	Sources
Glyptops plicatulus	1	V	Gaffney (1979a)
Mesochelys durlstonensis	1	II*	Gaffney (1979a)
Baenidae			
Baena arenosa	1	II	Gaffney (1972b)
Boremys pulchra	1	II	Brinkman and Nicholls (1991)
Cedorbaena putorius	1	II	Lyson and Joyce (2009b)
Chisternon undatum	1	II	Gaffney (1972b)
Eubaena cephalica	1	II	Gaffney (1972b)
Palatobaena sp.	1	II	Gaffney (1972b)
Palatobaena cohen	4	II	Lyson and Joyce (2009a)
Peckemys brinkman	2	II	Lyson and Joyce (2009b)
Plesiobaena antiqua	2	II	Gaffney (1972b); Brinkman (2003)
Plesiobaena putorius	1	II	Gaffney (1972b)
Stygiochelys estesi	1	II	Gaffney (1972b)
Trinitichelys hiatti	1	II	Gaffney (1972b)
Eucryptodira			
Plesiochelyidae			
Plesiochelys etalloni	1	I	Gaffney (1975c)
Portlandemys mcdowelli	1	I	Gaffney (1975c)
Eurysternidae			
Angolachelys mbaxi	1	I	Mateus et al. (2009)
Solnhofia parsonsi	1	I	Gaffney (1975d)
Thalassemys moseri	1	IV	Rieppel (1980)
Sinemyidae/ Macrobaenidae			
Chubutemys coppelloi	1	IV	Gaffney et al. (2007b)
Dracochelys bicuspis	1	IV	Gaffney and Ye (1992)
Hangaiemys hoburensis	1	IV	Sukhanov (2000)
Judithemys sukhanovi	1	IV	Parham and Hutchison (2003)
Sinemys gamera	1	IV	Sukhanov (2000)

(continued)

(continued)

Taxon	*n*	p	Sources
Crown-group Cryptodira			
Toxochelyidae			
Toxochelys latiremis	1	I	Gaffney (1979b)
Cheloniidae			
Caretta caretta	1	I	Carr (1952)
Chelonia mydas	1	I	Carr (1952)
Eochelone brabantica	1	I	Gaffney (1979b)
Eretmochelys imbricata	1	I	Carr (1952)
Euclastes acutirostris	1	I	Jaril et al. (2009)
Lepidochelys kempi	2	I	Hay (1908); Carr (1952)
Dermochelyidae			
Dermochelys coriacea	1	I	Gaffney (1979b)
Chelydridae			
Chelydra serpentina	1	I	Gaffney (1972a)
Macroclemys temminckii	1	I	Gaffney (1975b)
Platysternon megacephalum	1	I	Gaffney (1975b)
Protochelydra zangerli	1	I	Erickson (1973)
Lindholmemydidae			
Mongolemys sp.	1	IV	Sukhanov (2000)
Testudinoidea			
Emydidae			
Chrysemys alabamensis	1	I	Gaffney (1979b)
Chrysemys concinna	1	I	Gaffney (1979b)
Chrysemys idahoensis	1	I	McDowell (1964)
Clemmys insculpta	1	I	Gaffney (1979b)
Deirochelys reticularia	1	I	Gaffney (1979b)
Emys orbicularis	1	I	Gaffney (1979b)
Malaclemys terrapin	1	I	Gaffney (1979b)
Malayemys subtrijuga	1	I	Gaffney (1979b)
Pseudemys texana	3	I	Bever (2008, 2009a)
Terrapene ornata	1	I	Gaffney (1979b)

(continued)

(continued)

Taxon	*n*	p	Sources
Geoemydidae			
Batagur baska	1	I	Gaffney (1979b)
Callagur borneoensis	1	I	Gaffney (1979b)
Chinemys reevesi	1	I	Gaffney (1979b)
Cuora trifasciata	1	I	Gaffney (1979b)
Geoclemys hamiltoni	1	I	McDowell (1964)
Geoemyda sp.	1	I	Gaffney (1979b)
Hardella thurgii	1	I	McDowell (1964)
Hieremys annandalii	1	I	Gaffney (1979b)
Kachuga smithii	1	I	Gaffney (1979b)
Mauremys mutica	1	I	Gaffney (1979b)
Melanochelys trijuga	1	I	Gaffney (1979b)
Morenia ocellata	1	I	Gaffney (1979b)
Ocadia sinensis	1	I	Gaffney (1979b)
Orlitia borneensis	1	I	Baur (1896)
Rhinoclemys annulata	1	I	Gaffney (1979b)
Siebenrockiella crassicollis	1	I	McDowell (1964)
Testudinidae			
Geochelone elephantopus	1	I	Gaffney (1979b)
Gopherus polyphemus	1	I	Gaffney (1979b)
Homopus areolatus	1	I	Gaffney (1979b)
Kinixys horneana	1	I	Gaffney (1979b)
Malacochersus tornieri	1	I	Gaffney (1979b)
Pyxis arachnoides	1	I	Gaffney (1979b)
Testudo graeca	1	I	Gaffney (1979b)
Trionychoidea			
Adocidae			
Adocus sp.	1	I	Gaffney (1979b)
Adocus sp.	1	I	Meylan and Gaffney (1989)
Kinosternoidea			
Dermatemydidae			
Dermatemys mawii	1	I	Gaffney (1979a)
Dermatemys sp.	1	I	Meylan and Gaffney (1989)

(continued)

(continued)

Taxon	*n*	p	Sources
Kinosternidae			
Claudius augustatus	1	I	Gaffney (1979a)
Kinosternon subrubrum	1	I	Gaffney (1979a)
Staurotypus salvinii	1	I	Williams (1952)
Sternotherus odoratus	3	I	Gaffney (1979a); Bever (2009b)
Trionychia			
Carettochelyidae			
Carettochelys insculpta	1	I	Gaffney (1979b)
Pseudanosteira sp.	1	I	Gaffney (1979b)
Puppigerus camperi	1	I	Gaffney (1979b)
Trionychidae			
Aspideretoides splendidus	1	I	Gardner et al. (1995)
Chitra indica	1	I	Gray (1855)
Cyclanorbis senegalensis	1	I	Loveridge and Williams (1957)
Cycloderma frenatum	1	I	Loveridge and Williams (1957)
Lissemys punctata	1	I	Gaffney (1979b)
Pelochelys bibroni	1	I	Gaffney (1979b)
Plastomenus thomasi	1	I	Gaffney (1979b)
Trionyx triunguis	1	I	Gaffney (1979b)

References

Albrecht, P. W. (1967). The cranial arteries and cranial arterial foramina of the turtle genera *Chrysemys*, *Sternotherus*, and *Trionyx*: A comparative study with analysis of possible evolutionary implications. *Tulane Studies of Zoology, 14*, 81–99.

Albrecht, P. W. (1976). The cranial arteries of turtles and their evolutionary significance. *Journal of Morphology, 149*, 159–182.

Baur, G. (1896). Der Schädel einer neuen grossen Schildkröte (Adelochelys) aus dem. *Zoologischen Museum in Miinchen, 12*, 314–319.

Bellairs, A.d' A., & Kamal, A. M. (1981). The chondrocranium and the development of the skull in recent reptiles. In C. Gans & T. S. Parsons (Eds.), *Biology of the Reptilia* (Vol. 11, pp. 1–263). New York: Academic Press.

Bever, G. S. (2008). Comparative growth in the postnatal skull of the extant North American turtle *Pseudemys texana* (Testudinoidea: Emydidae). *Acta Zoologica, 89*, 107–131.

Bever, G. S. (2009a). The postnatal skull of the extant North American turtle *Pseudemys texana* (Cryptodira: Emydidae), with comments on the study of discrete intraspecific variation. *Journal of Morphology, 270*, 97–128.

Bever, G. S. (2009b). Postnatal ontogeny of the skull in the extant North American turtle *Sternotherus odoratus* (Cryptodira: Kinosternidae). *Bulletin of the American Museum of Natural History, 330*, 1–97.

Bona, P., & Alcade, L. (2009). Chondrocranium and skeletal development of *Phrynops hilarii* (Pleurodira: Chelidae). *Acta Zoologica, 90*, 301–325.

Bookstein, F. L. (1991). *Morphometric tools for landmark data: Geometry and biology*. Cambridge: Cambridge University Press.

Brinkman, D. B. (2003). Anatomy and systematics of *Plesiobaena antique* (Testudines: Baenidae) from the mid-Campanian Judith River Group of Alberta, Canada. *Journal of Vertebrate Paleontology, 23*, 146–155.

Brinkman, D. B., & Nicholls, E. L. (1991). Anatomy and relationships of the turtle *Boremys pulchra* (Testudines: Baenidae). *Journal of Vertebrate Paleontology, 11*, 302–315.

Brinkman, D. B., & Nicholls, E. L. (1993). The skull of *Neurankylus eximius* (Testudines: Baenidae) and a reinterpretation of the relationships of this taxon. *Journal of Vertebrate Paleontology, 13*, 273–281.

Brinkman, D. B., & Wu, X.-C. (1999). The skull of *Ordosemys*, an Early Cretaceous turtle from Inner Mongolia, People's Republic of China, and the interrelationships of Eucryptodira (Chelonia: Cryptodira). *Paludicola, 2*, 134–147.

Brinkman, D. B., Eberth, D. A., Xu, X., Clark, J. M., & Wu, X.-C. (2012). Turtles from the Jurassic Shishugou Formation of the Junggar Basin, People's Republic of China, with comments on the basicranial region of basal eucryptodires. In D. B. Brinkman, P. A. Holroyd & J. D. Gardner (Eds.), *Morphology and evolution of turtles: Origin and early diversification* (pp. XXX–XXX). Dordrecht: Springer.

Carr, A.F. (1952). *Handbook of turtles. The Turtles of the United States, Canada, and Baja California.* Ithaca: Cornell University Press.

Chapman, R. E. (1990). Conventional Procrustes methods. In F. J. Rohlf & F. L. Bookstein (Eds.), *Proceedings of the Michigan Morphometrics Workshop* (pp. 251–267). Ann Arbor: University of Michigan Museum of Zoology.

Erickson, B. R. (1973). A new chelydrid turtle *Protochelydra zangerli* from the late Paleocene of North America. *Scientific Publication of the Science Museum of Minnesota, 2*, 1–16.

Evans, J., & Kemp, T. S. (1975). The cranial morphology of a new Lower Cretaceous turtle from southern England. *Palaeontology, 18*, 25–40.

Evans, J., & Kemp, T. S. (1976). A new turtle skull from the Purbeckian of England and a note on the early dichotomies of cryptodire turtles. *Palaeontology, 19*, 317–324.

Fujita, M. K., Engstrom, T. N., Starkey, D. E., & Shaffer, H. B. (2004). Turtle phylogeny: Insights from a novel nuclear intron. *Molecular Phylogenetics and Evolution, 31*, 1031–1040.

Gaffney, E. S. (1972a). An illustrated glossary of turtle skull nomenclature. *American Museum Novitates, 2486*, 1–33.

Gaffney, E. S. (1972b). The systematics of the North American family Baenidae (Reptilia, Cryptodira). *Bulletin of the American Museum of Natural History, 147*, 241–320.

Gaffney, E. S. (1975a). A phylogeny and classification of the higher categories of turtles. *Bulletin of the American Museum of Natural History, 155*, 391–436.

Gaffney, E. S. (1975b). Phylogeny of the chelydrid turtles: A study of shared derived characters in the skull. *Fieldiana Geology, 33*, 157–178.

Gaffney, E. S. (1975c). A taxonomic revision of the Jurassic turtles *Portlandemys* and *Plesiochelys*. *American Museum Novitates, 2574*, 1–19.

Gaffney, E. S. (1975d). *Solnhofia parsonsi*, a new cryptodire turtle from the Late Jurassic of Europe. *American Museum Novitates, 2576*, 1–25.

Gaffney, E. S. (1977). The side-necked turtle family Chelidae: A theory of relationships using shared derived characters. *American Museum Novitates, 2620*, 1–28.

Gaffney, E. S. (1979a). The Jurassic turtles of North America. *Bulletin of the American Museum of Natural History, 162*, 91–136.

Gaffney, E. S. (1979b). Comparative cranial morphology of recent and fossil turtles. *Bulletin of the American Museum of Natural History, 164*, 1–376.

Gaffney, E. S. (1983). The cranial morphology of the extinct horned turtle, *Meiolania platyceps*, from the Pleistocene of Lord Howe Island, Australia. *Bulletin of the American Museum of Natural History, 175*, 361–480.

Gaffney, E. S. (1984). Historical analyses of theories of chelonian relationship. *Systematic Zoology, 33*, 283–301.

Gaffney, E. S. (1990). The comparative osteology of the Triassic turtle *Proganochelys*. *Bulletin of the American Museum of Natural History, 194*, 1–263.

Gaffney, E. S. (1996). The postcranial morphology of *Meiolania platyceps* and a review of the Meiolanidae. *Bulletin of the American Museum of Natural History, 229*, 1–166.

Gaffney, E. S., & Jenkins, F. A. (2010). The cranial morphology of *Kayentachelys*, an Early Jurassic cryptodire, and the early history of turtles. *Acta Zoologica, 91*, 335–368.

Gaffney, E. S., & Meeker, L. J. (1983). Skull morphology of the oldest turtles: A preliminary description of *Proganochelys quenstedti*. *Journal of Vertebrate Paleontology, 3*, 25–28.

Gaffney, E. S., & Meylan, P. A. (1988). A phylogeny of turtles. In M. J. Benton (Ed.), *The Phylogeny and Classification of the Tetrapods* (Vol. 1, pp. 157–219)., *Amphibians, Reptiles, Birds* Oxford: Clarendon Press.

Gaffney, E. S., & Meylan, P. A. (1992). The Transylvanian turtle, *Kallokibotion*, a primitive cryptodire of Cretaceous age. *American Museum Novitates, 3040*, 1–37.

Gaffney, E. S., & Tong, H. (2003). *Phosphatochelys*, a new side-necked turtle (Pelomedusoides: Bothremydidae) from the Paleocene of Morrocco. *Bulletin of the American Museum of Natural History, 279*, 644–659.

Gaffney, E. S., & Wood, R. C. (2002). *Bairdemys*, a new side-necked turtle (Pelomedusoides: Podocnemididae) from the Miocene of the Caribbean. *American Museum Novitates, 3359*, 1–28.

Gaffney, E. S., & Ye, X. (1992). *Dracochelys*, a new cryptodire turtle from the Early Cretaceous of China. *American Museum Novitates, 3048*, 1–13.

Gaffney, E. S., Meylan, P. A., & Wyss, A. R. (1991). A computer assisted analysis of the relationships of the higher categories of turtles. *Cladistics, 7*, 313–335.

Gaffney, E. S., Campos, D. D. A., & Hirayama, R. (2001a). *Cearachelys*, a new side-necked turtle (Pelomedusoides: Bothremydidae) from the Early Cretaceous of Brazil. *American Museum Novitates, 3319*, 1–20.

Gaffney, E. S., Moody, R. T. J., & Walker, C. A. (2001b). *Azabbaremys*, a new side-necked turtle (Pelomedusoides: Bothremydidae) from the Paleocene of Mali. *American Museum Novitates, 3320*, 1–16.

Gaffney, E. S., Chatterjee, S., & Dudra, D. J. (2001c). *Kurmademys*, a new side-necked turtle (Pelomedusoides: Bothremydidae) from the Late Cretaceous of India. *American Museum Novitates, 3321*, 1–16.

Gaffney, E. S., Tong, H., & Meylan, P. A. (2006). Evolution of the side-necked turtles: The families Bothremydidae, Euraxemydidae, and Araripemydidae. *Bulletin of the American Museum of Natural History, 300*, 1–698.

Gaffney, E. S., Hutchison, J. H., Jenkins, F. A., & Meeker, L. J. (1987). Modern turtle origins: The oldest known cryptodire. *Science, 237*, 289–291.

Gaffney, E. S., Sahni, A., Schleich, H., Singh, S. D., & Srivastava, R. (2003). *Sankuchemys*, a new side-necked turtle (Pelomedusoides: Bothremydidae) from the Late Cretaceous of India. *American Museum Novitates, 3405*, 1–10.

Gaffney, E. S., Roberts, E., Sissoko, F., Bouare, M. L., Tapanila, L., & O'Leary, M. A. (2007a). *Acleistochelys*, a new side-necked turtle (Pelomedusoides: Bothremydidae) from the Paleocene of Mali. *American Museum Novitates, 3549*, 1–24.

Gaffney, E. S., Rich, T. H., Vickers-Rich, P., Constantine, A., Vacca, P., & Kool, L. (2007b). *Chubutemys*, a new eucryptodiran turtle from the Early Cretaceous of Argentina, and the relationship of Meiolaniidae. *American Museum Novitates, 3599*, 1–35.

Gardner, J. D., Russell, A. P., & Brinkman, D. B. (1995). Systematics and taxonomy of soft-shelled turtles (Family Trionychidae) from the Judith River Group (mid-Campanian) of North America. *Canadian Journal of Earth Sciences, 32*, 631–643.

Goodrich, E. S. (1930). *Studies on the Structure and Development of Vertebrates*. London: Constable and Company.

Gray, J. E. (1855). *Catalogue of shield reptiles in the collections of the British Museum. Part I. Testudinata (Tortoises)*. London: British Museum.

Hammer, Ø., Harper, D. A. T., & Ryan, P. D. (2010). PAST version 3.2. Retrieved May 1, 2010 from http://www.nhm.uio.no/norges/past/download.html.

Hay, O. P. (1908). The fossil turtles of North America. *Carnegie Institute Washington, Publication, 75*, 1–568.

Hirayama, R., Brinkman, D. B., & Danilov, I. G. (2000). Distribution and biogeography of non-marine Cretaceous turtles. *Russian Journal of Herpetology, 7*, 181–198.

Jamniczky, H. A. (2008). Turtle carotid circulation: A character analysis case study. *Biological Journal of the Linnean Society, 93*, 239–256.

Jamniczky, H. A., & Russell, A. P. (2004). Cranial arterial foramen diameter in turtles: Quantitative assessment of size-independent phylogenetic signal. *Animal Biology, 54*, 417–436.

Jamniczky, H. A., & Russell, A. P. (2007). Re-appraisal of patterns of turtle carotid circulation: Evidence from osteological correlates and soft tissues. *Journal of Morphology, 268*, 571–587.

Jamniczky, H. A., Brinkman, D. B., & Russell, A. P. (2006). Phylogenetic implications of turtle cranial circulation: A review. *Russian Journal of Herpetology, 13 suppl.* (*Fossil Turtle Research, 1*), 84–92.

Jaril, N.-E., de Broin, F.d. L., Bardet, N., Vacant, R., Bouya, B., Amaghzaz, M., et al. (2009). *Euclastes acutirostris*, a new species of littoral turtle (Cryptodira, Cheloniidae) from the Paleocene phosphates of Morocco (Oulad Abdoun Basin, Danian-Thanetian). *Comptus Rendus Palevol, 8*, 447–459.

Joyce, W. G. (2007). Phylogenetic relationships of Mesozoic turtles. *Bulletin of the Peabody Museum of Natural History, 48*, 3–102.

Krenz, J. G., Naylor, G. J. P., Shaffer, H. B., & Janzen, F. J. (2005). Molecular phylogenetics and evolution of turtles. *Molecular Phylogenetics and Evolution, 37*, 178–191.

Kuratani, S. (1987). The development of the orbital region of *Caretta caretta* (Chelonia: Reptilia). *Journal of Anatomy, 154*, 187–200.

Loveridge, A., & Williams, E. E. (1957). Revision of the African tortoises and turtles of the suborder *Cryptodira*. *Bulletin of the Museum of Comparative Zoology, 115*, 163–557.

Lyson, T. R., & Joyce, W. G. (2009a). A new species of *Palatobaena* (Testudines: Baenidae) and a maximum parsimony and Bayesian phylogenetic analysis of Baenidae. *Journal of Paleontology, 83*, 457–470.

Lyson, T. R., & Joyce, W. G. (2009b). A revision of *Plesiobaena* (Testudines: Baenidae) and an assessment of baenid ecology across the K/T boundary. *Journal of Paleontology, 83*, 833–853.

Mateus, O., Jacobs, L., Polcyn, M., Schulp, A. S., Vineyard, D., & Buta Neto, A. (2009). The oldest African eucryptodiran turtle from the Cretaceous of Angola. *Acta Palaeontologica Polonica, 54*, 581–588.

McDowell, S. B. (1961). On the major arterial canals in the ear-region of testudinoid turtles and the classification of the Testudinoidea. *Bulletin of the Museum of Comparative Zoology, 125*, 23–39.

McDowell, S. B. (1964). Partition of the genus *Clemmys* and related problems in the taxonomy of the aquatic Testudinidae. *Proceedings of the Zoological Society of London, 143*, 239–279.

Meylan, P. A. (1996). Skeletal morphology and relationships of the Early Cretaceous side-necked turtle, *Araripemys barretoi* (Testudines: Pelomedusoides: Araripemydidae) from the Santana Formation of Brazil. *Journal of Vertebrate Paleontology, 16*, 20–33.

Meylan, P. A., & Gaffney, E. S. (1989). The skeletal morphology of the Cretaceous cryptodire turtle, *Adocus*, and the relationships of the Tryonychoidea. *American Museum Novitates, 2941*, 1–60.

Meylan, P. A., Gaffney, E. S., & Campos, D.d. A. (2009). *Caninemys*, a new side-necked turtle (Pelomedusoides: Podocnemididae) from the Miocene of Brazil. *American Museum Novitates, 3639*, 1–26.

Near, T. J., Meylan, P. A., & Shaffer, H. B. (2005). Assessing concordance of fossil calibration points in molecular clock studies: An example using turtles. *The American Naturalist, 165*, 137–146.

Parham, J. F., & Hutchison, J. H. (2003). A new eucryptodiran turtle from the Late Cretaceous of North America (Dinosaur Provincial Park, Alberta, Canada). *Journal of Vertebrate Paleontology, 23*, 783–798.

Rieppel, O. (1980). The skull of the Upper Jurassic cryptodire turtle *Thalassemys*, with a reconsideration of the chelonian braincase. *Palaeontographica A, 171*, 105–140.

Rieppel, O. (1987). The development of the trigeminal jaw adductor musculature and associated skull elements in the lizard *Podarcis sicula*. *Journal of Zoology, 212*, 131–150.

Rieppel, O. (1988). The development of the trigeminal jaw adductor musculature in the grass snake *Natrix natrix*. *Journal of Zoology, 216*, 743–770.

Rieppel, O. (1990). The structure and development of the jaw adductor musculature in the turtle *Chelydra serpentina*. *Zoological Journal of the Linnean Society, 98*, 27–62.

Rieppel, O. (1993). Studies on skeleton formation in reptiles: Patterns of ossification in the skeleton of *Chelydra serpentina* (Reptilia: Testudines). *Journal of Zoology, 231*, 487–509.

Rohlf, F. J. (1990). Rotational fit (Procrustes) methods. In F. J. Rohlf & F. L. Bookstein (Eds.), *Proceedings of the Michigan Morphometrics Workshop* (pp. 227–236). Ann Arbor: University of Michigan Museum of Zoology.

Rohlf, F. J. (2003). TpsSmall version 1.20. Retrieved July 16, 2010 from http://life.bio.sunysb.edu/morph.

Rohlf, F.J. (2010). TpsDig version 2.16. Retrieved July 16, 2010 from http://life.bio.sunysb.edu/morph.

Sánchez-Villagra, M. R., Müller, H., Sheil, C. A., Scheyer, T. M., Nagashima, H., & Kuratani, S. (2009). Skeletal development in the Chinese soft-shelled turtle *Pelodiscus sinensis* (Testudines: Trionychidae). *Journal of Morphology, 270*, 1381–1399.

Shaffer, H. B., Meylan, P., & McKnight, M. L. (1997). Tests of turtle phylogeny: Molecular, morphological, and paleontological approaches. *Systematic Biology, 46*, 235–268.

Sheets, H. D. (2001). CoordGen version 6 h. Retrieved May 1, 2010 from http://www.canisius.edu/~sheets/morphsoft.html.

Sheets, H. D. (2004). CVAGen version 6 l. Retrieved May 1, 2010 from http://www.canisius.edu/~sheets/morphsoft.html.

Sheets, H. D. (2005). TwoGroup version 6 h. Retrieved May 1, 2010 from http://www.canisius.edu/~sheets/morphsoft.html.

Sheil, C. A. (2003). Osteology and skeletal development of *Apalone spinifera* (Reptilia: Testudines: Trionychidae). *Journal of Morphology, 256*, 42–78.

Sheil, C. A. (2005). Skeletal development of *Macrochelys temminckii* (Reptilia: Testudines: Chelydidae). *Journal of Morphology, 263*, 71–106.

Sheil, C. A., & Greenbaum, E. (2005). Reconsideration of skeletal development of *Chelydra serpentina* Reptilia: Testudinata: Chelydridae): Evidence for intraspecific variation. *Journal of Zoology, 265*, 235–267.

Sterli, J., & de la Fuente, M. S. (2010). Anatomy of *Condorchelys antique* Sterli, 2008, and the origin of the modern jaw closure mechanism in turtles. *Journal of Vertebrate Paleontology, 30*, 351–366.

Sterli, J., Müller, J., Anquetin, J., & Hilger, A. (2010). The parabasisphenoid complex in Mesozoic turtles and the evolution of the testudinate basicranium. *Canadian Journal of Earth Sciences, 47*, 1337–1346.

Sukhanov, V. B. (2000). Mesozoic turtles of Middle and Central Asia. In M. J. Benton, M. A. Shishkin, D. M. Unwin & E. N. Kurochkin (Eds.), *The Age of Dinosaurs in Russia and Mongolia* (pp. 309–367). Cambridge: Cambrdige University Press.

Suzuki, S., & Tsogtbaatar, C. (2010). A catalogue of *Mongolochelys* collected by the HMNS-MPC Joint Paleontological Expedition. *Hayashibara Museum of Natural History Research Bulletin, 3*, 119–131.

Walker, J. A. (2000). Ability of geometric morphometric methods to estimate a known covariance matrix. *Systematic Biology, 49*, 115–150.

Williams, E. E. (1952). A staurotypine skull from the Oligocene of South Dakota (Testudinata: Chelydridae). *Breviora, 2*, 1–16.

Wood, R. C. (1970). A review of the fossil Pelomedusidae (Testudines: Pleurodira) of Asia. *Breviora, 357*, 1–24.

Yntema, C. L. (1968). A series of stages in the embryonic development of *Chelydra serpentina*. *Journal of Morphology, 125*, 219–252.

Zelditch, M. L., Swiderski, D. L., Sheets, H. D., & Fink, W. L. (2004). *Geometric Morphometrics for biologists: A primer*. London: Elsevier Academic Press.

Part III
The Early Diversification of Turtles

Chapter 8
New Material of the Platychelyid Turtle *Notoemys zapatocaensis* from the Early Cretaceous of Colombia; Implications for Understanding Pleurodira Evolution

Edwin A. Cadena, Carlos A. Jaramillo, and Jonathan I. Bloch

Abstract *Notoemys zapatocaensis* is the youngest representative of the Platychelyidae, a group of Late Jurassic-Early Cretaceous pleurodires. Here we describe two new specimens of this species represented by a partial carapace and a nearly complete articulated shell. *Notoemys zapatocaensis* is different from other platychelyid turtles in having: (1) two fairly reduced lateral tuberosities on the margin of the anterior plastral lobe, (2) a shallow notch on the posterolateral margin of the epiplastra, giving a convex posterolateral edge to this bone, (3) gular scales that are rectangular in shape and much wider than long, (4) a long intergular scale that has a slight medial contact with the pectorals, resulting in a complete separation of the humeral scales, (5) a central plastral fontanelle that projects posteriorly into the xiphiplastral region, (6) a very small marginal 3, (7) a slightly shorter neural 1 than neural 2, with an exclusive lateral contact with costal 1, resulting in a complete separation of neural 2 and costal 1, (8) narrower vertebral scales, and (9) peripheral 3 lacking a posteromedial contact with costal 2. Phylogenetic analysis indicates that *N. zapatocaensis* is a sister taxon of *N. oxfordiensis*, and that *Proterochersis robusta* can be resolved in two different positions in the testudines tree: (1) with *Odontochelys semitestacea* based on the fact that both taxa share two mesoplastra meeting at midline, or (2) as the most basal pleurodire, based on a suture articulation of pelvis to shell. Anal notch shape and potentially fontanelle size are indicators of sexual dimorphism in platychelyids.

Keywords Rosablanca Formation • South America • Valanginian • Zapatoca

E. A. Cadena (✉)
Smithsonian Tropical Research Institute, Balboa, Ancon AA 0843-03092, Panama
and
Division of Vertebrate Paleontology, Florida Museum of Natural History, Dickinson Hall, University of Florida, Gainesville, FL 32611, USA
and
North Carolina State University, Raleigh, NC 27695, USA
e-mail: eacadena@ncsu.edu

C. A. Jaramillo
Smithsonian Tropical Research Institute, Balboa, Ancon AA 0843-03092, Panama
e-mail: jaramilloc@si.edu

J. I. Bloch
Division of Vertebrate Paleontology, Florida Museum of Natural History, Dickinson Hall, University of Florida, Gainesville, FL 32611, USA
e-mail: jbloch@flmnh.ufl.edu

Introduction

Turtles diverged in two infraorders (Pleurodira and Cryptodira) during the Late Triassic or earlier (Gaffney and Jenkins 2010), around 221 Ma, maximum estimated based on molecular studies (Shaffer 2009). The earliest pleurodire so far known is *Proterochersis robusta* Frass (1913) from the Late Triassic of Germany, however is important to mention here that new material is being studied from the Early to Middle Jurassic that would change the current state of *P. robusta* (Joyce and Sterli, in press). *P. robusta,* as with most other Late Triassic-Early Cretaceous pleurodires, is known only by shells. The only exception to this is *Notoemys laticentralis* Cattoi and Freiberg (1961), which is represented by a partial skull recently re-described by de Lapparent de Broin et al. (2007). Following Gaffney et al. (2006) below *Proterochersis,* all pleurodires are included in the Parvorder Megapleurodira, which is divided into Nanorders Platychelira and Eupleurodira (Cheloides and Pelomedusoides). Platychelira is represented by the single Family Platychelyidae, which includes *Platychelys oberndorferi* from the Late Jurassic (Kimmeridgian) of Germany (Wagner 1853), and the three species of *Notoemys*: *Notoemys laticentralis* from the Late Jurassic (Tithonian) of Argentina (Cattoi and Freiberg 1961; Fernandez and de la

D. B. Brinkman et al. (eds.), *Morphology and Evolution of Turtles*, Vertebrate Paleobiology and Paleoanthropology,
DOI: 10.1007/978-94-007-4309-0_8,

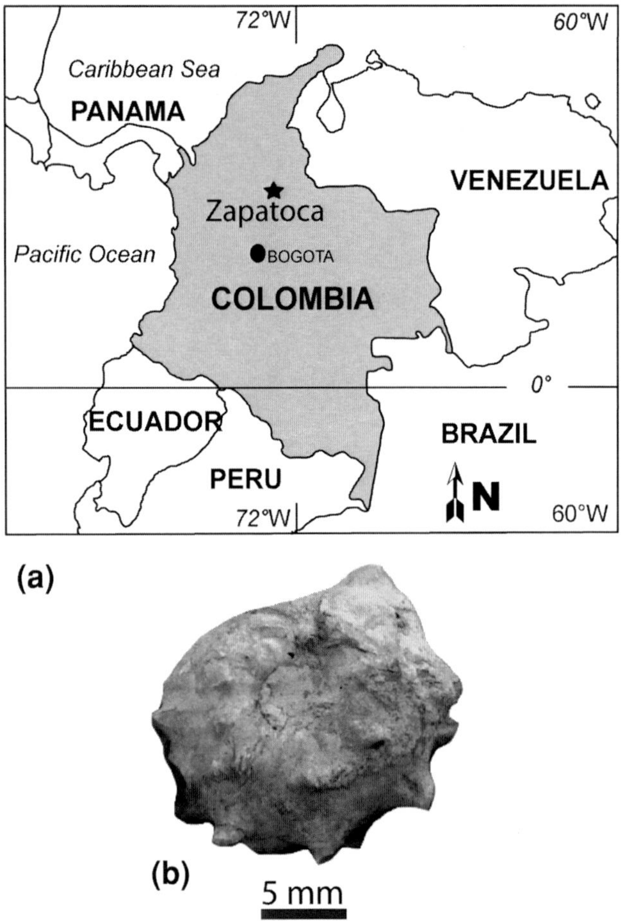

Fig. 8.1 **a** Location of the town of Zapatoca, Department of Santander, Colombia: 6° 50′ 35″N, 73° 13′ 50″W. **b** *Saynoceras verrucosum*, ammonite indicator of the base of the Late Valanginian, collected at the same layer as *Notoemys zapatocaensis* holotype and paratype (MG61)

Fuente 1988, 1994; de Lapparent de Broin et al. 2007; de la Fuente 2007), *N. zapatocaensis* from the Early Cretaceous (Valanginian) of Colombia (Cadena and Gaffney 2005), and *N. oxfordiensis* from the Late Jurassic (Oxfordian) of Cuba, known by a single poorly preserved shell (de la Fuente and Iturralde-Vinent 2001; Cadena and Gaffney 2005). None of the three species of *Notoemys* has a completely preserved anterior plastral lobe, which is unfortunate because that region has morphological features that are key to understanding the evolution of the turtle shell.

Two new specimens of *Notoemys zapatocaensis* are described here. The first is an almost complete and articulated shell (here designated as the paratype), and the second other is a partial carapace (here designated as a referred specimen). Both specimens were collected by the senior author in 2006 in Zapatoca, Colombia, from the same locality and stratum as the holotype (Fig. 8.1). The excellent preservation of the anterior plastral and carapace elements allows us to amend the diagnosis for this species, and the specimens allow a revised phylogenetic analysis of pleurodires and the most basal testudines.

Institutional Abbreviations used in this paper are: *IPN-EAC* Museo Geológico José Royo y Goméz–Instituto Colombiano de Geología y Minería-Ingeominas, Bogotá, Colombia; *MACN* Museo Argentino de Ciencias Naturales, Buenos Aires, Argentina; *MNHN* Muséum national d'histoire naturelle, Paris, France; *MNHN AC* Anatomie Comparée collections of MNHN; *MNHNCu* Museo Nacional de Historia Natural, La Habana, Cuba; and *MOZP* Museo "Prof. Dr. Olsacher" Zapala, Argentina.

Systematic Paleontology

Testudines Batsch 1788
Pleurodira Cope 1864
Platychelyidae Bräm 1965
Notoemys Cattoi and Freiberg 1961
Notoemys zapatocaensis Cadena and Gaffney 2005
(Figs. 8.2, 8.3)

Holotype: MGJRG IPN 15-EAC 140120031, nearly complete shell, missing the anteromedial region of the carapace and the anteromedial portion of the plastron; previously described and figured by Cadena and Gaffney (2005).

Newly designated paratype: MGJRG IPN 15-EAC 150620061, abbreviated as MG61 (Fig. 8.2a–d), articulated carapace and plastron, missing the right posterolateral portion of the carapace.

Newly referred specimen: MGJRG IPN 15-EAC 150620062, abbreviated as MG62, a partial central portion of a carapace including neurals 2–8, the most medial portion of costals 2–7, and an isolated medial portion of the left costal 8 that preserves an iliac scar (Fig. 8.3a, b).

Locality, horizon, and age: All three specimens are from the same locality and unit. The El Caucho Farm locality (6° 50′ 35″N, 73° 13′ 50″W) is northeast of Zapatoca town, Department of Santander, Colombia. The locality is in a limestone layer belonging to the upper segment of the shallow marine Rosablanca Formation (Guzman 1985). The occurrence of the ammonite *Saynoceras verrucosum* (F. Etayo 2008, personal communication) indicates that this part of the Rosablanca Formation corresponds to the base of the late Valanginian stage (Early Cretaceous), approximately 138 Ma according to the biochronostratigraphic framework of Ogg et al. (2008).

Revised diagnosis: *Notoemys zapatocaensis* is recognized as a pleurodire turtle on the basis of the following characteristics: (1) sutural articulation of the pelvis with the shell, (2) well-developed anal notch that is U- or V-shaped. It is a megapleurodire, based on (1) one pair of mesoplastra,

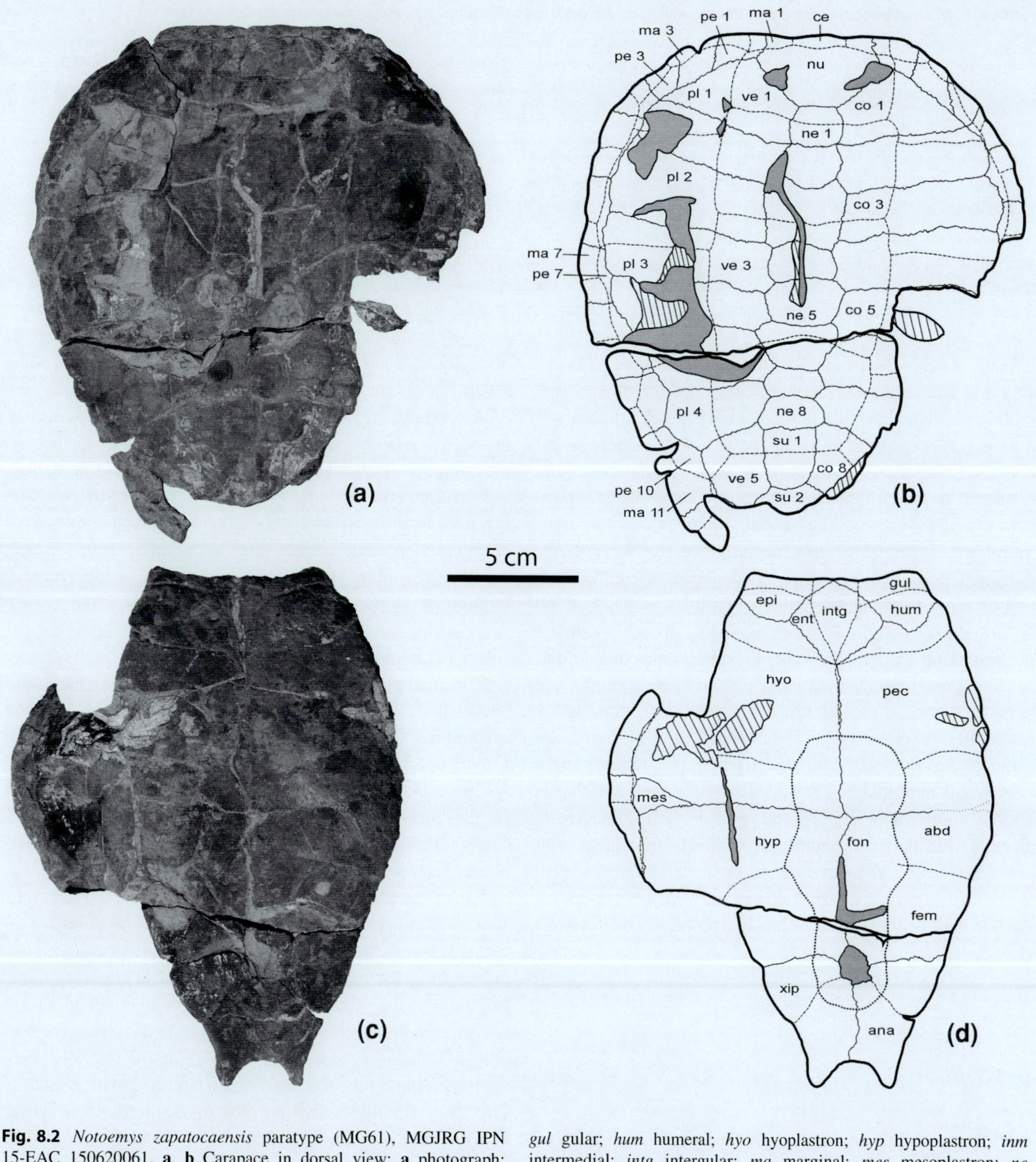

Fig. 8.2 *Notoemys zapatocaensis* paratype (MG61), MGJRG IPN 15-EAC 150620061. **a**, **b** Carapace in dorsal view: **a** photograph; **b** interpretive drawing. **c**, **d** Plastron in ventral view: **c** photograph; **d** interpretive drawing. Abbreviations: *abd* abdominal; *ce* cervical; *co* costal; *ent* entoplastron; *epi* epiplastron; *fem* femoral; *fon* fontanelle; *gul* gular; *hum* humeral; *hyo* hyoplastron; *hyp* hypoplastron; *inm* intermedial; *intg* intergular; *ma* marginal; *mes* mesoplastron; *ne* neural; *pec* pectoral scale; *pe* peripheral; *pl* pleural; *su* suprapygal; *ve* vertebral; *xip* xiphiplastron

laterally restricted that lack a medial contact, (2) a single intergular scale. *N. zapatocaensis* is a platychelyid because it has (1) a costovertebral tunnel that is very wide through its entire length, (2) an articulation tubercule on the anterior face of the first thoracic rib, (3) a carapace with posterior sides tapering medially, (4) the second neural smaller than the remainder of the neural series, (5) the thoracic vertebrae smooth and flat ventrally, hexagonal in shape with a centrolateral notch, and (6) a carapace with dorsal protuberances, located on the posterior region of the pleural and vertebral scales. It is recognized as a member of the genus *Notoemys* and differs from *Platychelys oberndorferi* in

Table 8.1 Measurements for the platychelyids, including the paratype (MG61) of *Notoemys zapatocaensis*

Taxon	CL	CW	PL	PW	CLe	CWe	PLe	PWe
Notoemys zapatocaensis MGRG IPN 15 EAC 150620061. This study	20	18	8	15	21	18	18	16
Notoemys zapatocaensis MGRG IPN 15 EAC 140120031. Figured in Cadena and Gaffney (2005)	22	20	19	14	24	20	19	16
Notoemys laticentralis MOZP 2487. Figured in Fernandez and de la Fuente (1994)	27	25	24	22	27	25	25	22
Notoemys oxfordiensis MNHNCu-P 3209. Figured in de la Fuente and Iturralde-Vinent (2001)	25	23	20	20	25	23	22	20
Platychelys oberndorferi. Figured in de Lapparent de Broin (2001)	20	17	17	13	20	17	17	13

Measurements in centimeters. Abbreviations: *CL* carapace length; *CW* carapace width; *PL* plastron length; *PW* plastron width; *CLe* total carapace length, estimated; *CWe* total carapace width, estimated; *PLe* total plastron length, estimated; *PWe* total plastron width, estimated

having (1) a wider and shorter cervical scale, (2) no supramarginal scales, (3) a smooth and relatively flatter shell with lower dorsal protuberances lacking radial striation, (4) a larger suprapygal 1, (5) the neural 3 in posterolateral contact with costal 4, (6) an iliac scar oval in shape and restricted to costal 8, and (8) a very reduced medial space between the first and the second thoracic ribs. Autopomorphies of *Notoemys zapatocaensis* are (1) an anterior plastral lobe margin with two rather reduced lateral tuberosities, almost straight in outline, (2) a shallow notch on the posterolateral margin of the epiplastra, giving a convex posterolateral edge to this bone, (3) gular scales rectangular in shape, much wider than long, (4) a long intergular scale slightly touching the pectorals medially, and completely separating the humerals, (5) a central plastral fontanelle projecting posteriorly into the xiphiplastral region, (6) a very small marginal 3, (7) a quadrangular neural 1 that is slightly shorter than neural 2 and exclusively in contact with costal 1 laterally, and neural 2 exclusively in contact with costal 2 laterally, (8) vertebral scales that are narrower than in *N. laticentralis*, *N. oxfordiensis* and *Platychelys oberndorferi,* and (9) peripheral 3 lacking posteromedial contact with costal 2.

Description

Carapace: MG61 has a shell that is cordiform in shape, with an anterior edge that is straight, its widest point at peripheral 7, and posterior lateral sides tapering medially as in the holotype and the other platychelyids. Measurements are given in Table 8.1. Posterior edges are dentate at the contact between marginal scales, as in the holotype, although this is much less pronounced than in *Platychelys oberndorferi* and slightly more pronounced than in *Notoemys laticentralis* MACN 18403. This morphological feature is unknown for *N. oxfordiensis* due to the poor preservation of its edges. Low protuberances in MG61 are located at the posterior medial region of each vertebral and pleural scales, as in the holotype and MG62. In *N. laticentralis* the protuberances are slightly lower, and they are not preserved in *N. oxfordiensis* due to the highly eroded surface of the carapace. High and very well developed protuberances are characteristics of *Platychelys oberndorferi*. The carapace surface of MG61 is smooth, with a light microvermiculation rather than the granulation seen in the holotype, similar to the condition in *N. latincentralis*. In contrast, *Platychelys oberndorferi* has a carapace surface very sculptured with radial striations originating at the center of the protuberances.

The nuchal bone of MG61 is hexagonal in shape and wider than long as in all other platychelyids and also the cryptodires *Kayentachelys aprix* Gaffney et al. (1987), *Eileanchelys waldami* Anquetin et al. (2009), *Heckerochelys romani* Sukhanov (2006), *Indochelys spatulata* Datta et al. (2000), and *Chengyuchelys baenoides* Young and Chow (1953). In contrast, all pleurodires have a nuchal bone relatively equidimensional or longer than wide.

Eight neurals are present in MG61. Neural 1 is slightly shorter than neural 2, and is the only neural in contact with costal 1, a condition also present in the holotype, although it differs from the holotype in that neural 1 is slightly larger. In contrast, the other two species of *Notoemys* and *Platychelys oberndorferi* have a neural 1 that is longer than neural 2 and in contact with costal 1 and 2 laterally, preventing an anterolateral contact of neural 2 with costal 1. This is a condition also present in *Kayentachelys aprix*, *Eileanchelys romani*, *Heckerochelys romani*, *Indochelys spatulata*, *Chengyuchelys baenoides,* and retained in the extant chelid *Chelus fimbriata* Schneider (1783). In contrast, *Brasilemys josai* de Lapparent de Broin (2000), *Araripemys barretoi* Price (1973), and one of the specimens of *Cearachelys placidoi* Gaffney et al. (2001) have a slightly more derived condition of neural 1 contacting costal 1 and 2, and neural 2 contacting costal 2 and 3. The most frequent condition for eupleurodires is to have neural 1 only contacting costal 1, and neural 2 contacting costal 1 anterolaterally. Exceptions to this include many of the chelid

Fig. 8.3 *Notoemys zapatocaensis* MGJRG IPN 15-EAC 150620062 (MG62). *Central portion* of the carapace, including neurals 2–8, the most medial portion of costals 2–7, and an isolated most? *medial portion* of the left costal 8 preserving the iliac scar. See areas shadowed in *light grey* in the turtle sketch. **a** Ventral view. **b** Dorsal view

genera, the nomen dubium *Eusarkia rotundiformis* Bergounioux (1952), and the podocnemidid *Bairdemys venezuelensis* Wood and de Gamero (1971), which completely lack the neural series. Neural 3 in the paratype (MG61) of *N. zapatocaensis,* as in *N. laticentralis,* is large, almost octagonal in shape, and in contact with costal 4 posterolaterally. In the holotype of *N. zapatocaensis*, neural 3 lacks the right posterolateral contact with costal 4; the same asymmetrical pattern is present in the *Platychelys oberndorferi* specimen figured in de Lapparent de Broin (2001, Fig. 1). However the holotype of *Platychelys oberndorferi* lacks a posterolateral contact with costal 4 on both sides, thus exhibiting a more rectangular shape than in the other platychelyids. Neurals 4–8 exhibit the same shape and sutural contacts as the holotype, specimen MG62 (Fig. 8.3a, b), *N. laticentralis,* and *Platychelys oberndorferi*, although this latter taxon exhibits neurals slightly more rectangular in shape. The neural series is unrecognizable in *N. oxfordiensis* due to its poor preservation.

Suprapygal 1 is rectangular in shape, slightly longer than wide as in the holotype, similar to *Platychelys oberndorferi, Condorchelys antiqua* Sterli (2008), *Indochelys spatulata,* and *Kayentachelys aprix*. In contrast, the suprapygal 1 in *N. laticentralis* is trapezoidal in shape, wider posteriorly than anteriorly. Suprapygal 1 is absent in most of the eupleurodires. Suprapygal 2 is only preserved in the paratype (MG61) of *N. zapatocaensis,* where it exhibits the same pentagonal shape as the holotype and the other platychelyids. The pygal of *N. zapatocaensis,* missing in the paratype (MG61) but preserved and previously described in the holotype, has a medial notch on its posterior edge. Based on a reexamination of the holotype of *N. laticentralis*

MACN 18403 by the senior author of this paper, such a notch is also present, although shallower, in that specimen. This new interpretation differs from previous studies (de la Fuente and Iturralde-Vinent 2001), which considered the posterior pygal notch to be absent in *N. laticentralis*.

Eight sets of costals are complete in both sides of the carapace of the paratype (MG61) of *Notoemys zapatocaensis*, with the sets on the right side slightly broken laterally. The shape of the costals is similar to the holotype and the other platychelyids. MG62 preserves the left costal 8 with the iliac scar slightly oval, rounded and restricted to this costal, as in *N. laticentralis*. This seems to be also the condition in *N. oxfordiensis*, although this region is badly preserved in specimens of that taxon. In contrast, *Platychelys oberndorferi* has an elongated iliac scar extending onto costal 8, suprapygal, and the medial margin of the peripherals. Eleven peripheral bones are recognized on the left side of the paratype (MG61). Peripherals 1–3 are in medial contact with costal 1. Peripheral 3 lacks a posteromedial contact with costal 2, differing in this feature from the other platychelyids and other testudines in which peripheral 3 contacts costal 2. The presence of a small peripheral 3 restricted between peripherals 2 and 4 in the holotype was defined as a potential diagnostic characteristic of *N. zapatocaensis* (Cadena and Gaffney 2005), but this feature is not present in the paratype (MG61) described here, which shows a well-developed peripheral 3 on both sides of the carapace. Thus the condition in the holotype is reinterpreted as a pathology of that specimen, as was initially done by Cadena and Gaffney (2005). Peripherals 5–7 are longer than wide, whereas peripheral 8 and 10 are slightly larger than peripheral 9 and 11, as in the holotype and *N. laticentralis*.

The cervical scale in the paratype (MG61) of *Notoemys zapatocaensis*, as in the holotype and *N. laticentralis*, is rectangular in shape, much wider than long. The cervical is slightly shorter in many of the specimens of *Platychelys oberndorferi*. This is the primitive condition also present in *Proterochersis robusta, Kayentachelys aprix, Indochelys spatulata, Eileanchelys waldami, Heckerochelys romani*, and *Chengyuchelys baenoides* (at least for its middle cervical). In *Dortoka vasconica* de Lapparent de Broin and Murelaga (1996), the cervical is almost equidimensional and in chelids it is slightly longer than wide, except in the extant species *Hydromedusa tectifera* Cope (1870), which has a large cervical enclosed between marginals 1, pleurals 1, and vertebral 1. All pelomedusoids turtles lack a cervical scale. The condition is unknown for euraxemydids.

Five vertebral scales are clearly visible on the dorsal aspect of the carapace in the paratype (MG61) of *Notoemys zapatocaensis*; vertebrals 1–3 are almost rectangular in shape as in the holotype and *Platychelys oberndorferi*, and much narrower than in *N. laticentralis, Proganochelys quenstedti, Proterochersis robusta, Kayentachelys aprix, Heckerochelys romani, Indochelys spatulata*, and *Eileanchelys waldami*. The condition is unknown in *N. oxfordiensis*. Vertebral 4 is nearly hexagonal in shape as in *Platychelys oberndorferi*, and much narrower than in *N. laticentralis, Proganochelys quenstedti, Proterochersis robusta, Kayentachelys aprix, Heckerochelys romani, Indochelys spatulata*, and *Eileanchelys waldami*. The sulcus between vertebrals 3 and 4 is on neural 6 and costal 6, as in *Platychelys oberndorferi* and the other two species of *Notoemys*, as well as in most of the primitive testudines for which five neurals are recognized (Character 74, Joyce 2007; erroneously defined for vertebrals 2 and 3). In eucryptodires and eupleurodires, the sulcus between vertebrals 3 and 4 is on costal 5. This characteristic is related to the narrowing of the vertebral scales. Vertebral 5, although preserved only anterolaterally in the paratype (MG61) of *N. zapatocaensis*, seems to be heptagonal, as in the holotype and *N. laticentralis*. In contrast it is octagonal in *Platychelys oberndorferi*. The pattern of reduction in the width of vertebral scales described for *N. zapatocaensis* and *Platychelys oberndorferi* is also shared by eupleurodires and eucryptodires.

Four pleural scales are visible on the left portion of the carapace of MG61, and are the same shape as in the holotype, *Notoemys laticentralis*, and *Platychelys oberndorferi*. Pleural 4 is more nearly rectangular than pentagonal as is the condition in *P. oberndorferi*. *P. oberndorferi* also has straighter medial edges for all pleurals. Twelve marginal scales are visible on the left side of the carapace of MG61. Marginal 1 lacks contact with pleural 1, as is also the condition for the holotype. In contrast, *N. laticentralis* has a marginal 1 contacting pleural 1 posteriorly. Marginal 2 has the same shape and size as the holotype, slightly longer than in *N. laticentralis* and *P. oberndorferi*. In both the paratype (MG61) and the holotype of *N. zapatocaensis*, marginal 3 is smaller compared to other platychelyids and testudines. Marginals 4–8, marginal 10, and marginal 12 are longer than wide, rectangular in shape. Based on a reinterpretation of the elements that Cadena and Gaffney (2005) erroneously identified as marginals 9 and 11 as marginals 10 and 12, this is also the case in the holotype. Marginals 10–12 of *N. laticentralis, Kayentachelys aprix*, and *Condorchelys antiqua* also share the pattern described for *N. zapatocaensis*. In contrast, in *P. oberndorferi* these marginal scales are slightly more pentagonal in shape. Marginals 9 and 11 are pentagonal in shape for the paratype (MG61) of *N. zapatocaensis*, the holotype (erroneously identified as marginals 8 and 10 by Cadena and Gaffney 2005), *N. laticentralis, Kayentachelys aprix, Condorchelys antiqua, Heckerochelys romani*, and eupleurodires for which the posterior series of marginals are more equidimensional due to an increase in the size of the peripherals.

Plastron: The anterior plastral lobe of the paratype (MG61) of *Notoemys zapatocaensis* is shorter than the posterior lobe, and has a straight anterior edge with very reduced tuberosities in both lateral corners and a slight concavity at the medial margin. In *Platychelys oberndorferi*, the anterior edge exhibits a very short tuberosity at the midline of the plastron, whereas *N. oxfordiensis* lacks tuberosities on the anterior edge. Both *P. oberndorferi* and *N. oxfordiensis* have a slightly more convex anterior plastral edge than *N. zapatocaensis*. In the case of *N. laticentralis* the arrangement of bones and scales at the margin of the anterior plastral lobe remains unknown because neither the holotype MACN 18403 nor MOZP 2487, figured in de la Fuente (2007), completely preserve this aspect. The primitive condition exhibited by *Odontochelys semitestacea*, *Proganochelys quenstedti*, and *Proterochesis robusta* is the presence of an anterior plastral lobe edge highly decorated with large tuberosities, defining a very dentate anterior margin. Tuberosities persist, although they are much more reduced in number and size, in *Kayentachelys aprix* and *Chengyuchelys baenoides*. They have disappeared completely in *Indochelys spatulata*, which has a very straight anterior edge. *Dortoka vasconica* and most other eupleurodires have a very convex anterior plastral lobe, although some exceptions are present, for example the bothremydid *Taphrosphys* spp., which has a nearly straight anterior plastral edge, or the short projections at the intergular–gular scales region of some pelomedusidids.

The entoplastron of *Notoemys zapatocaensis* is diamond-shaped, slightly touching the edge of the anterior plastral lobe, and completely separating the epiplastra. The most primitive condition, seen ventrally in *Odontochelys semitestacea,* is an entoplastron with an extensive participation in the edge of the anterior plastral lobe, and both epiplastra meeting at midline posteriorly to the entoplastron. *Proganochelys quenstedti, Proterochersis robusta, Paleochersis talampayensis*, Rougier et al. (1995), *Kayentachelys aprix,* and *Indochelys spatulata* show a more progressive condition in which the entoplastron completely separates the epiplastra. A more advanced condition is present in *N. zapatocaensis* where the entoplastron does not participate in the edge of the anterior plastral lobe. *Eileanchelys waldami, Heckerochelys romani, Chengyuchelys baenoides, Platychelys oberndorferi*, *N. oxfordiensis*, and *N. laticentralis* show an entoplastron more withdrawn from the anterior edge of the plastral lobe, with both epiplastra having a short midline contact anterior to the entoplastron. The length of the contact of the epiplastra anterior to the entoplastron is greater in *Dortoka vasconica* and eupleurodires. A graphical reconstruction of the entoplastron and the epiplastra relationships for some testudines is shown in Fig. 8.4.

The epiplastron in the paratype (MG61) of *Notoemys zapatocaensis* is trapezoidal in shape with a convex posterior edge as in *Chengyuchelys baenoides* and *Heckerochelys romani.* In *N. oxfordiensis*, *N. laticentralis*, *Platychelys oberndorferi, Dortoka vasconica,* and eupleurodires, the posterior edge of the epiplastron is straight to slightly concave, and it is highly concave in *Proganochelys quenstedti* and *Proterochersis robusta.* The hyoplastron and hypoplastron are similar to other platychelyids in shape, but are distinctive in that the central fontanelle extends from the central portion of the hyoplastra to the anteromedial part of the xiphiplastra, and is completely filled by bone that is thinner than the rest of the shell. The outline of the central fontanelle is marked by a sulcus as is also the case in the *N. laticentralis* specimen MOZP 2487, which was figured in de Lapparent de Broin et al. (2007, Fig. 1d). The presence of the central fontanelle in the holotype of *N. zapatocaensis* remains uncertain, because the margins of the hyoplastra and the xiphiplastra are broken at the midline. However if the central fontanelle existed in the holotype of *N. zapatocaensis*, it would have been restricted to the central portion of the plastron and would not have extended posteriorly into the xiphiplastral region, since a sulcus is absent and no differences in bone thickness are present that would indicate that the fontanelle has been secondarily filled in. *N. oxfordiensis*, *P. oberndorferi*, the stem testudine *Sichuanchelys* sp. indet., and *Indochelys spatulata* share with the holotype of *N. zapatocaensis* the presence of a central fontanelle restricted to the area between hyoplastra and hypoplastra and, in the case of *P. oberndorferi* and the stem testudine *Sichuanchelys* sp. indet., a posterior fontanelle restricted to the area between the hypoplastra and xiphiplastra bones. This latter fenestra is unknown for *N. oxfordiensis* and absent in *I. spatulata* and *Heckerochelys romani.* In contrast, *N. laticentralis* and the paratype (MG61) of *N. zapatocaensis* share the presence of a large central fontanelle posteriorly projected toward the xiphiplastral region. Stem testudines such as *Odontochelys semitestacea, Proganochelys quenstedti*, *Proterochersis robusta*, *Paleochersis talampayensis*, and *Kayentachelys aprix,* lack plastral fontanelles. This is also the case in *Dortoka vasconica* and eupleurodires, with the exception of *Araripemys barretoi* which has central and posterior fontanelles. The fontanelles are characteristic of early ontogenetic stages in almost all modern turtles, and filled by bone in later stages; examples of this process are seen in *Podocnemis lewyana* MNHN 1994-286, and *Chelus fimbriata* MNHN AC 5176.

The mesoplastra in the paratype (MG61) of *Notoemys zapatocaensis* are triangular in shape and wider than long, lack a midline contact, and are smaller than in *N. latincentralis*, *N. oxfordiensis* and *Platychelys oberndorferi.* The primitive condition seen in *Odontochelys semitestacea* is two pairs of mesoplastra meeting at the midline of the

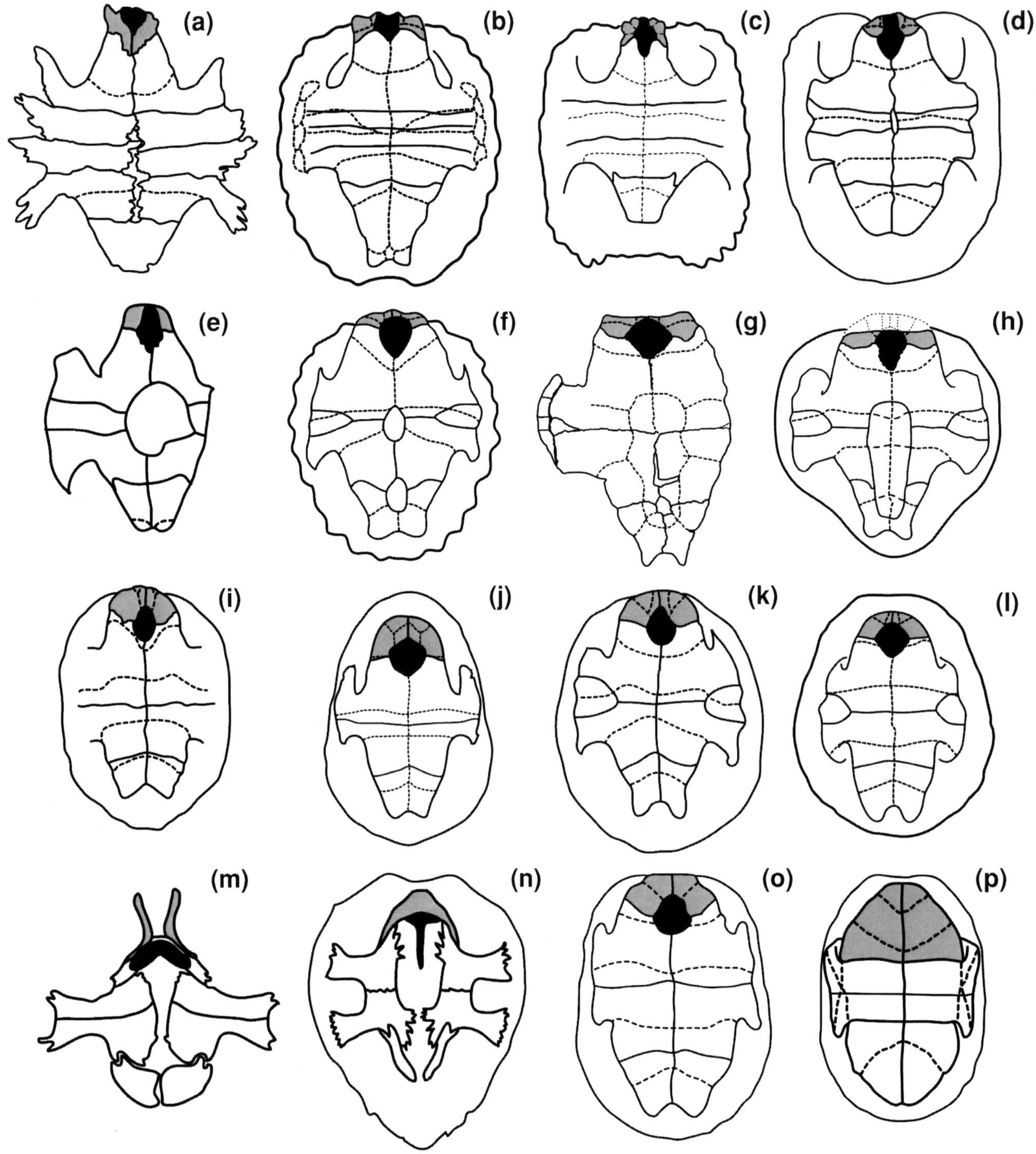

Fig. 8.4 Entoplastron and epiplastra relationship in testudines. Sketches of the plastron were redrawn from previous publications, indicated after the species name. Entoplastron shadowed in *black*, and epiplastra in *gray*. **a** *Odontochelys semitestacea* Li et al. (2008). **b** *Proterochersis robusta* Joyce (2007). **c** *Proganochelys quenstedti* Joyce (2007). **d** *Kayentachelys aprix* Gaffney (1990). **e** *Indochelys spatulata* Datta et al. (2000). **f** *Platychelys oberndorferi* de Lapparent de Broin (2000). **g** *Notoemys zapatocaensis* this study. **h** *Notoemys laticentralis* de Lapparent de Broin et al. (2007). **i** *Dortoka vasconica* de Lapparent de Broin et al. (2004). **j** *Chelodina oblonga* Joyce (2007). **k** *Bonapartemys bajobarrealis* de Lapparent de Broin and De la Fuente (2001). **l** *Podocnemis sextuberculata* Joyce (2007). **m** *Apalone ferox* Joyce (2007). **n** *Eretmochelys imbricata* Joyce (2007). **o** *Mauremys leprosa* Claude et al. (2003). **p** *Kinosternon leucostomun* Joyce (2007)

plastron, a condition considered by Li et al. (2008) as a diagnostic characteristic for this genus. However, this characteristic is not exclusive to *Odontochelys semitestacea* since it is also present in *Proterochersis robusta*. In contrast, *Proganochelys quenstedti*, *Kayentachelys aprix*, *Eileanchelys waldami*, *Heckerochelys romani*, and *Chengyuchelys baenoides* have only one mesoplastral pair, meeting at midline of the plastron; in the case of *Sichuanchelys* sp. indet. and *Indochelys spatulata*, the mesoplastra reach the lateral border of the central fontanelle. *Dortoka vasconica*, chelids, and *Araripemys barretoi* lack mesoplastra; all other eupleurodires have one pair of mesoplastra that are laterally restricted, almost equidimensional, and lack a midline contact.

The posterior plastral lobe in the paratype (MG61) of *Notoemys zapatocaensis* is markedly concave, in contrast to the flat surface of the holotype. The lateral edges of the lobe are slightly rounded with two shallow embayments; the anterior embayment is at the lateral aspect of the sutural contact between the hypoplastron and xiphiplastron, and the posterior one at the lateral end of the sulcus between the femoral and the anal scale. This is also the condition for the holotype. *P. oberndorferi* and *N. laticentralis* have a less marked embayment on the lateral edges of the posterior plastral lobe. In *N. oxfordiensis* the condition remains unknown since the most of the posterior plastral lobe is missing.

The xiphiplastra in the paratype (MG61) of *Notoemys zapatocaensis* have a deep U-shaped anal notch with posterior tips, similar to specimen MOZP-2487 of *N. laticentralis*. In contrast the holotype of *N. zapatocaensis* and the specimen of *Platychelys oberndorferi* figured in de Lapparent de Broin (2001, Fig. 1b) have a shallow, wide V-shaped anal notch, lacking well-developed posterior tips. *Odontochelys semitestacea* lacks a xiphiplastral anal notch, exhibiting a narrowly rounded to straight posterior edge. *Proterochersis robusta* has an interanal? scale at the most posterior margin of the plastron, creating a very narrow anal notch. All eupleurodires have a well-developed anal notch but it is variable in size, shape and depth in each family or genus.

The intergular scale in the paratype (MG61) of *Notoemys zapatocaensis* is pentagonal, elongated in shape, longer than wide, and reaches the posteromedial corner of the entoplastron, as in the bothremydid *Ummulisani rutgersensis* figured in Gaffney et al. (2006, Fig. 269). In contrast, *N. laticentralis*, *N. oxfordiensis*, and *Platychelys oberndorferi* have an intergular scale that extends to the posteromedial margin of the entoplastron, a condition much less advanced than in *Dortoka vasconica* and most of the eupleurodires, where the intergular only covers the most anteromedial corner of the entoplastron; in the case of the podocnemidid *Erymnochelys madagascariensis* the very small intergular is restricted between the gulars. The intergular scale remains unknown for *Odontochelys semitestacea,* and for other primitive testudines such as *Progranochelys quenstedti*, *Proterochersis robusta,* and *Heckerochelys romani*. *Chengyuchelys baenoides* differs in having two small intergulars. The gulars in the paratype (MG61) of *N. zapatocaensis* are almost rectangular in shape, much wider than long, a condition not seen elsewhere within the testudines. This condition is intermediate between the short, square, and more laterally positioned gulars of *Proganochelys quenstedti* and *Proterochersis robusta* and the triangular, more medially positioned gulars of *N. oxfordiensis*, *P. oberndorferi*, *Dortoka vasconica*, and most of the eupleurodires. The humeral scales of MG61 are completely separated medially by the intergular, such as in the bothremydid *Ummulisani rutgersensis*; thus they are smaller than in other platychelyids, *Proganochelys quenstedti*, *Proterochersis robusta*, *Chengyuchelys baenoides*, *Heckerochelys romani, D. vasconica*, and the eupleurodires. The humeropectoral sulcus in *N. zapatocaensis* is concave, slightly in contact with the posterior corner of the entoplastron as in *N. laticentralis*, while it is more posteriorly positioned in *N. oxfordiensis*, *P. oberndorferi*, *Odontochelys semitestacea*, *Proterochersis robusta*, *Proganochelys quenstedti*, *Ch. baenoides, H. romani*, and *D. vasconica* figured in de Lapparent de Broin and Murelaga (1999, Fig. 4). In eupleurodires and *D. vasconica* figured in de Lapparent de Broin et al. (2004, pl. III, 4), the humeropectoral sulcus extends further anteriorly over the posterior region of the entoplastron. The pectoroabdominal, the abdominofemoral, and the femoroanal sulci in the paratype (MG61) of *N. zapatocaensis*, as well as in *N. laticentralis,* are interrupted at the midline of the plastron by the large central fontanelle.

Phylogenetic Analysis

In order to perform a cladistic analysis, we included *Notoemys zapatocaensis* and 19 other taxa (principally pleurodires) in a matrix of 84 characters in total; 61 of them are skull/lower jaw characters (60 taken from Joyce 2007, and one taken from Cadena et al. 2010), 23 are shell characters. There are 19 in-group taxa and 1 out-group taxon (*Odontochelys semitestacea*). See Appendix 1 for the list of characters and Appendix 2 for the character-taxon matrix. The shell characters were taken and in some cases modified from previously published character matrices and detailed systematic studies, including de Lapparent de Broin and de la Fuente (2001), de la Fuente and Iturralde-Vinent (2001), de la Fuente (2003), Cadena and Gaffney (2005), Joyce (2007), and Li et al. (2008). A few of these characters are new to this study and were defined based on direct examination of extant or fossil specimens. We constructed the character-taxon matrix using Mesquite Version 2.72

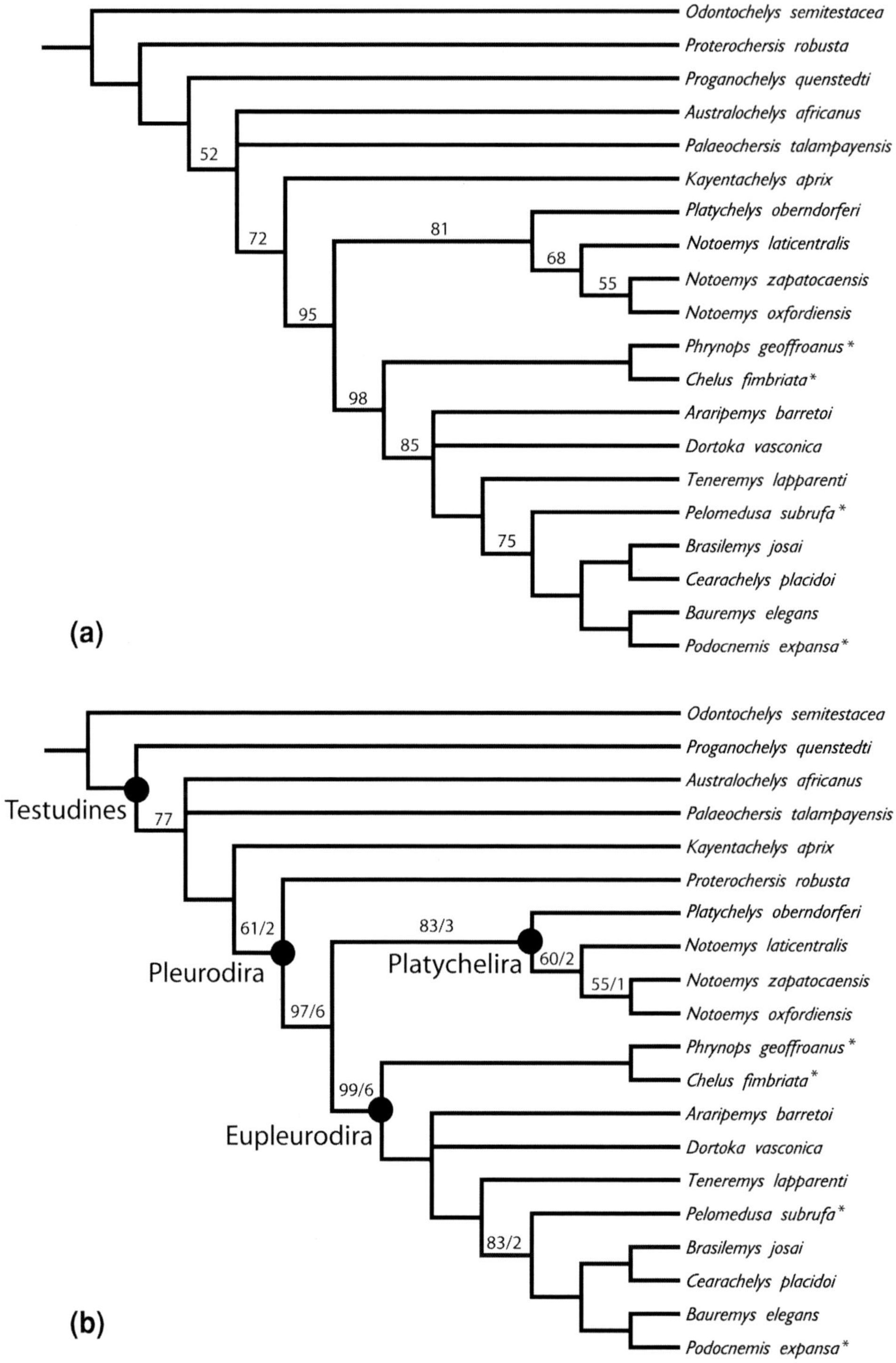

Fig. 8.5 Two alternative phylogenetic relationships among pleurodires and basal testudines recovered by this study. **a** Single most parsimonious cladogram, based on analysis in which all characters were unordered and equally weighted, using heuristics search (10,000 replicates) parsimony algorithm. **b** Strict consensus (our preferred phylogenetic hypothesis) of two most parsimonious cladograms, based on analysis in which all characters were unordered, character 81 was weighted four, and rest of characters had a weight of one. Extant taxa are indicated with an *asterisk*. Bootstrap values percentages (upper numbers) were obtained using 1000 replicates. Bremer decay indices (lower numbers) were obtained using TreeRot version 3 (Sorenson and Franzosa 2007)

(Maddison and Maddison 2009). For the phylogenetic analysis we used the parsimony algorithm of PAUP 4.0b10 (Swofford 2002). All characters were equally weighted and unordered in a first run. A second run was performed with all characters remaining unordered, but with weighting character 81 with a value of 4 using the set weight character option in PAUP; the remaining characters retained an equal weight of 1. The reason to weight character 81 is that it is an undisputably consistent character in the evolution of the pleurodires, with no homoplasy known; see discussion in Gaffney et al. (2006, Character 133, p. 620). Multistate characters were treated as polymorphic. Finally, we

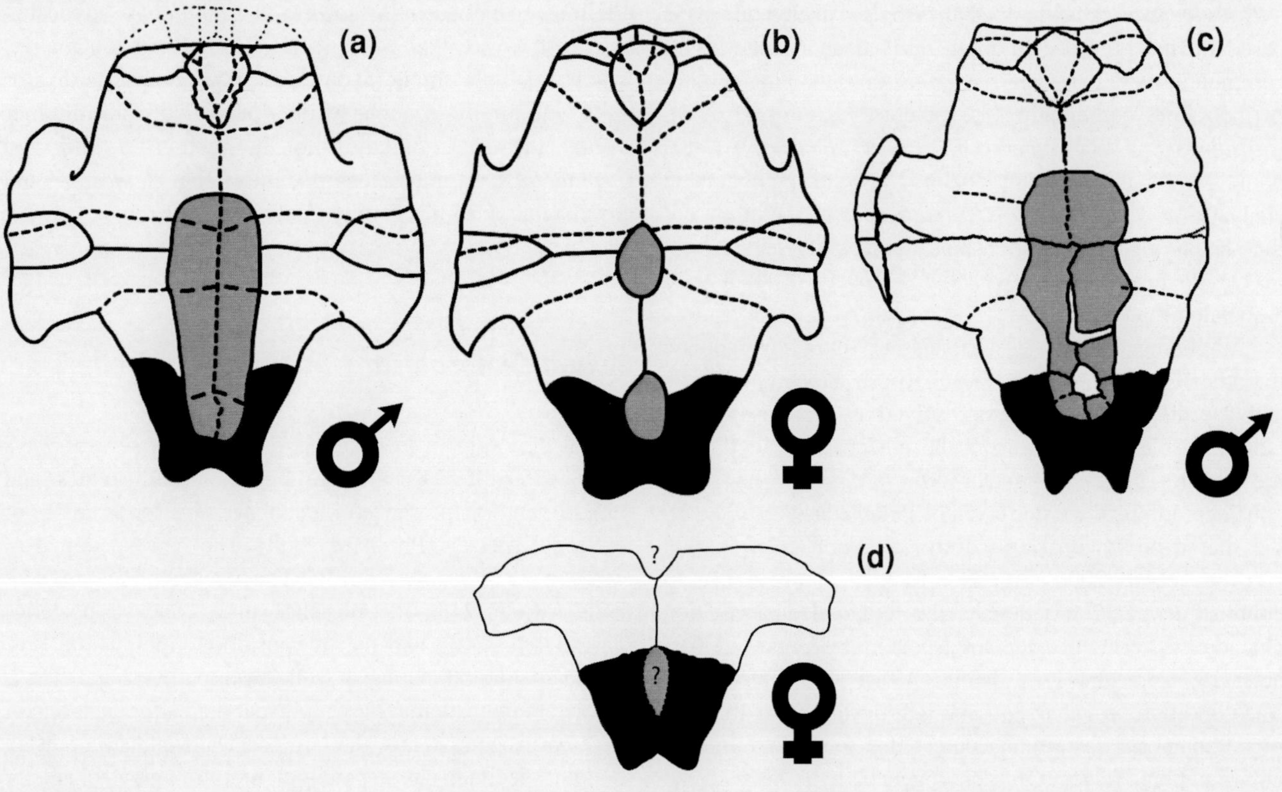

Fig. 8.6 Differences in the xiphiplastra and fontanelles within platychelyids, potentially related to sexual dimorphism. Males (*circle* with *arrow* symbol) are characterized by a concave posterior plastral lobe, long and narrow posterior xiphiplastral tips, a well-developed anal notch in a U shape, and a large central fontanelle. Females (*circle* with *cross down* symbol) are characterized by a flat posterior plastral lobe, short and wide posterior xiphiplastral tips, a V-shaped anal notch, and two interrupted fontanelles. **a** *Notoemys laticentralis* figured in de Lapparent de Broin et al. (2007). **b** *Platychelys oberndorferi* redrawn from de Lapparent de Broin (2000). **c** *N. zapatocaensis* this study. **d** *N. zapatocaensis* figured in Cadena and Gaffney (2005)

performed a heuristics search (10,000 replicates), obtaining boostrap percentages for 1000 replicates, and Bremer decay support indices using TreeRot Version 3 (Sorenson and Franzosa 2007).

Discussion

Phylogenetic Results

The phylogenetic analysis resulted in a single most parsimonious tree (Fig. 8.5a), with a tree length of 104, consistency index (CI) = 0.85, retention index (RI) = 0.90, and homoplasy index (HC) = 0.15. *Proterochersis robusta* is resolved at the base of the cladogram below *Odontochelys semitestacea*, a position mainly influenced by the presence of two pairs of mesoplastra meeting at the midline of the plastron (Character 78), which is a condition lost independently in derived pleurodires and cryptodires. A second run weighting the most unique pleurodire character, the sutural articulation of pelvis to shell (Character 81), with a minimum value of four resulted in two most parsimonious trees. The consensus tree (Fig. 8.5b) has a consistency index (CI) = 0.84, a retention index (RI) = 0.89, a homoplasy index (HC) = 0.16, and tree length of 109. *P. robusta* is placed in this tree as the most basal pleurodire, the same result as obtained by Gaffney et al. (2006, Fig. 292) it is the hypothesis that we favor here.

The Platychelyidae (*Platychelys* and *Notoemys* spp.) are a monophyletic clade, the same result obtained by Gaffney et al. (2006, Fig. 292). In contrast, the phylogenetic hypothesis presented by Joyce (2007, Fig. 18) shows that *N. latincetralis* is a sister taxon to the Pleurodira, and not just to the Platychelyidae. *Notoemys zapatocaensis* and *N. oxfordiensis* are sister taxa based on the fact that both share an intergular scale covering most of the entoplastron posteriorly (Character 76). However this relationship is not strongly supported by Bremer indices values (see Fig. 8.5b), with only one extra step required to collapse the *N. zapatocaensis–N. oxfordiensis* branch, creating a polytomy? between the three species of *Notoemys*. Cadena and Gaffney (2005) hypothesized that *N. zapatocaensis* is more closely related to *N. laticentralis* than to *N. oxfordiensis*. This hypothesis is based on the presence of more derived characteristics in common in these two taxa (see description) than in the other two possible combinations of taxa (*N. zapatocaensis–N. oxfordiensis* or *N. laticentralis–N. oxfordiensis*). At this point we favor the result obtained by our phylogenetic analysis (*N. oxfordiensis–N. zapatocaensis*), pointing out that the phylogenetic relationships among platychelyids will be clearly resolved only with the discovery of skull-shell associated material.

Continuing with the analysis of the phylogenetic hypothesis presented here (Fig. 8.5b), *Araripemys barretoi* and *Dortoka vasconica* are resolved as the most basal pelomedusoids, differing from the others in that both lack mesoplastra bones (Character 78). *A. barretoi* has been considered to be the sister taxon of Pelomedusidae based on the shared presence of an extreme temporal emargination (Gaffney et al. 2006, Fig. 292; Meylan et al. 2009). In contrast, Joyce (2007) defined the temporal emargination character in terms of squamosal-postorbital contact (Character 18, Appendix 1), a definition that we adopted here. The condition in *A. barretoi* and Pelomedusidae is the lack of a squamosal-postorbital contact due to upper temporal emargination. However, our phylogenetic hypothesis shows that even though both *A. barretoi* and Pelomedusidae share the same squamosal-postorbital contact condition, this single character is not enough to support a closer relationship. In the case of *D. vasconica,* our phylogenetic hypothesis places this taxon within the Pelomedusoides clade, as was also suggested by de Lapparent de Broin and Murelaga (1999). In contrast, the hypothesis put forward by Gaffney et al. (2006, Fig. 292) placed this taxon below the clade including the Chelidae and Pelomedusoides. The placement of *D. vasconica* and *Teneremys lapparenti* obtained here must wait for the discovery of more complete shells and skulls in order to be strongly supported.

Finally, in our preferred phylogenetic hypothesis *Pelomedusa subrufa* (representing Pelomedusidae) is the sister taxon of the most diverse group of middle Cretaceous-Cenozoic eupleurodires-denominated Panpodocnemidae (following França and Langer 2006). Within Panpodocnemidae, *Brasilemys josai* and *Cearachelys placidoi* (the most basal and better preserved bothremydid) are more closely related, sharing a neural 1 that contacts costals 1 and 2, and a neural 2 that only contacts costal 2. This is in contrast to *Bauremys elegans*, *Podocnemis expansa* and the rest of podocnemidids which have the neural 1 only contacting costal 1 and neural 2 contacting costal 1 anterolaterally (except for *Bardemys venezuelensis* which lacks the complete neural series). Previous phylogenetic hypotheses have placed *Br. josai* with *Hamadachelys escuilliei* and *Portezueloemys patagonica* as the closer relatives to Podocnemidae (Cadena et al. 2010). We attribute the closer relationship between *Br. josai* and *C. placidoi* advocated here to the way that we built our character-taxon matrix, which uses only single taxon rather than composite taxa or the whole family or genus as previous studies did. In other words, the closer relationship between *Br. josai* and *C. placidoi* does not necessarily imply that *Br. josai* is the closer relative to the Family Bothremydidae.

Sexual Dimorphism

Sexual dimorphism in turtles is expressed in several ways such as a difference in size between adult males and females, and the presence of a concave plastron in males of terrestrial species (Pritchard 2008). The new specimen of *Notoemys zapatocaensis* (MG61) shares with *N. laticentralis* MOZP 2487 the presence of a concave posterior plastral lobe, long and narrow posterior xiphiplastral tips, and an anal notch that has a well-developed U shape. These characters indicate that they represent males for each one of these two species. Thus, the large central fontanelle present in both specimens is a potential morphological character associated with sexual dimorphism, in this case representing males. In contrast, the holotype of *N. zapatocaensis* Cadena and Gaffney (2005) and the specimen of *Platychelys oberndorferi* figured in de Lapparent de Broin (2001, Fig. 1b) share a posterior plastral lobe that is flat and has a narrow V-shaped anal notch, as well as smaller central and posterior fontanelles, indicating that they represent females for each one of these two species. This could be also the case for the holotype of *N. oxfordiensis* figured in de la Fuente and Iturralde-Vinent (2001, Fig. 3). A graphic reconstruction of the xiphiplastron for platychelyids, as well as their differences potentially associated to sexual dimorphism, is shown in Fig. 8.6.

The identification of morphological variations associated with sexual dimorphism in fossil turtles has important implications in phylogenetic analysis. For example, de Lapparent de Broin et al. (2007) noted that *Notoemys laticentralis* differs from the rest of platychelyids in the wider and longer central fontanelle, a condition that was interpreted above as representing a potential male morphological condition for *N. laticentralis* and *N. zapatocaensis,* and possibly for all males of this clade, making this characteristic useless for phylogenetic or systematic purposes at least at the species or genus taxomomic level.

Acknowledgments Fieldwork and this paper were supported by the Smithsonian Paleobiology Endowment Fund, and the Florida Museum of Natural History Miss Lucy Dickinson Fellowship. Specials thanks go to V. Lamus for his help during the fieldwork. Thanks to T. Gaona and J. Arenas (Muséo Geológico José Royo y Gómez, Bogotá,

Colombia) and Lapparent de Broin (Muséum national d'Histoire naturelle, Paris, France) for access to collections, and to M. Iturralde from the Museo Nacional de Historia Natural, La Habana, Cuba, for the excellent pictures of the holotype of *Notoemys oxfordiensis*. Thanks to D. Brinkman and C. Marion for editing this manuscript, to the Smithsonian Tropical Research Institute (Center for Tropical Paleoecology and Archeology, Panama), to the Vertebrate Paleontology Lab of the Florida Museum of Natural History for access to the preparation of the fossil, and to W. Joyce for access to the character-taxon matrix used in the phylogenetic analysis. Thanks to R. Rueda and M. Gonzalez for their continued support and inspiration.

Appendices

Appendix 1

Description of the characters used in the phylogenetic analysis. Characters 60–83 represent carapace and plastron characters and are described here, whereas characters 1–59 were taken directly from Joyce (2007). Characters were polarized with respect to *Odontochelys semitestacea*, the source of the character, and if the character was modified or new is also indicated at the end of the description. Character 84 represents the only skull character, added from Cadena et al. (2010).

(A) Carapace

60. Ossification in the dermal component of the carapace: absent (0); present (1). Modified from Burke (2009).
61. Cervical scale(s): middle cervical wider than long (0); middle cervical as long as wide (1), cervical absent (2). Character modified from de Lapparent de Broin and Murelaga (1999) and Joyce (2007, Character 70).
62. Lateral arrangement between neural 1 and 2, and costals 1 and 2: neural 1 contacts costals 1 and 2, neural 2 only contacts costal 2 (0); neural 1 and costal 1 exclusively in contact with each other, neural 2 only contacts costal 2 (1); neural 1 contacts costals 1 and 2, neural 2 contacts costals 2 and 3 (2); neural 1 only contacts costal 1, neural 2 contacts costal 1 anterolaterally (3); neural series absent (4). New character.
63. Carapace posteriorly notched: present (0); absent (1). Pygal notch character from Cadena and Gaffney (2005).
64. Supramarginal scales: full series of twelve, on both sides of the carapace (0); incomplete series, restricted to the anterior margin on both sides of the carapace (1); absent (2). Character modified from Cadena and Gaffney (2005).
65. Posterior lobe of the carapace: same width as the anterior lobe or slightly wider (0); tapering medially (1). Character modified from Cadena and Gaffney (2005).
66. Articulation tubercule on the anterior face of the first thoracic rib: absent, smooth anterior face (0); present (1). Character modified from Cadena and Gaffney (2005).
67. Thoracic vertebrae: cylindrical, longer than wide, keeled ventrally (0); smooth and flat ventrally, hexagonal in shape with central lateral notch (1). Character modified from Cadena and Gaffney (2005).
68. Axillary process: contacts peripherals only (0); contacts costal 1 or the sutural contact between costals 1 and 2 (1). Character modified from Cadena and Gaffney (2005).
69. Suprapygal 1: parallel-sided (0); tapers anteriorly (1); absent (2). Character modified from Cadena and Gaffney (2005).
70. Vertebral scales 2 and 3: hexagonal in shape, much wider than long (0); rectangular in shape, slightly wider than long (1); hexagonal or rectangular, as long as wide or longer than wide (2). Character modified from Cadena and Gaffney (2005).
71. Medial contact of the posterior costals: absent (0); present (1); present due to complete absence of neural series (2). Character modified from Joyce (2007, Character 68).
72. Lateral position of the sulcus between vertebrals 3 and 4 in taxa with five vertebrals: sulcus positioned on costal 6 (0); sulcus positioned on costal 5 (1). Character modified from Joyce (2007, Character 74).

(B) Plastron

73. Posterior epiplastral process: present (0); absent (1). Character from Cadena and Gaffney (2005).
74. Posterior entoplastral process: present (0); absent (1). Character from Cadena and Gaffney (2005).
75. Entoplastron participation in the anterior margin of the plastron in ventral view: wide participation (0); short participation (1); lacking participation (2). Character modified from Cadena and Gaffney (2005).
76. Intergular scale(s): covering slightly the anterior portion of the entoplastron (0); covering most of the entoplastron posteriorly (1); not covering the entoplastron (2). Character modified from Cadena and Gaffney (2005).
77. Anterior plastral lobe margin: defined by tuberosities, dentate margin (0); very reduced tuberosities, straight to slightly dentate margin (1); lacking tuberosities, smooth, highly convex margin (2). New character.
78. Mesoplastra: two pairs meeting at the midline of the plastron (0); one pair of mesoplastra, with midline contact or reaching the central fontanelle margin (1); one pair of mesoplastra, wider than long, without midline contact, (2); one pair of mesoplastra, as long as

wide, without midline contact, (3); mesoplastra absent (4). Character modified from Cadena and Gaffney (2005).

79. Central plastral fontanelle: absent (0); present (1). Character modified from Cadena and Gaffney (2005).
80. Anal notch: absent,with straight to slightly concave posterior edge of the xiphiplastra (0); present, well-developed in open U- or V-shape (1). Character modified from Joyce (2007, Character 87).
81. Sutural articulation of pelvis to shell: absent (0); present (1). Character modified from Joyce (2007, Character 125).
82. Iliac scar: absent (0); extends from costals onto the peripherals (1); restricted to costal 8 (2); positioned on costal 8 and pygal, sometimes reaching costal 7 (3). Character modified from Joyce (2007, Character 127).
83. Shape of ilium articular site: narrow and pointed posteriorly (0); oval (1). Character from de la Fuente and Iturralde-Vinent (2001).

(C) Skull

84. Pterygoid, cavum pterygoidei = fossa podocnemidoid of de Lapparent de Broin (2000): absent (0); shallow and slightly hidden anteromedially by the underlapping basisphenoid medially and laterally by the pterygoid (1); deep and partially to totally covered by the pterygoid flange (posterolateral wings of the pterygoid) (2). Character taken from Cadena et al. (2010).

Appendix 2

Character matrix (20 taxa and 84 characters) used for phylogenetic analysis (Nexus file as Supplementary Data 1). Multistate 0 or 1 is represented by "a", and 'not applicable' by "–".

Odontochelys semitestacea

000???????0???0?????????????????0?0??00????0?00??
??????????0??????????????0000 000–??0

Proganochelys quenstedti

000000000000–000000000000000000000000?00000000
00000000000000010?000000?0??00100100000?0

Proterochersis robusta

???
????????10?010000?0??00100001110?

Australochelys africanus

???????10?0???00?0???0??00?0??001001??10000000??
1?0?100?0????????????????????????????

Palaeochersis talampayensis

00000?01000???00000000??00001?0?1001?010000000
0?1?0?100????1??0?0???????001?1??00??0

Kayentachelys aprix

0000101111 0??1000000110000101001101 1?01000000
000100?1010000100120 0?0000010101 10000?0

Platychelys oberndorferi

???
?????????1001111110100112012111110?

Notoemys laticentralis

?????????????????????????????2110???11?0??????01?
0?1010???10012111110001 1???211121?

Notoemys zapatocaensis

???
????????10112111101001121121 1121?

Notoemys oxfordiensis

???
????????1001211?1????1?212211121?

Araripemys barretoi

1—101011111110003011100002110211123111101000
1 001001101?0011?2120?0?22011022241a11310

Dortoka vasconica

???
????????112120001221111202401131?

Brasilemys josai

1—10101111111000101 11???????0211123111 1010001
001001 101????12012000122 11????????????1

Cearachelys placidoi

1—101011111110101011100002110211123111101000
100100110100?11201200?122011120230 11??0

Bauremys elegans

1—10101111111000 10?11000021 1021?1231111010001
00?001?0?0001123?2000122111120230 11?12

Podocnemis expansa

1—10101111111000101110000211021112311111 0100a
100100 11010001123120001221111202301 1312

Pelomedusa subrufa

1—101011111110003011100002110211123111101000
10010011012001123120001221111202301 1310

Phrynops geoffroanus

01100111110111 01–20111000010102110231111 01000
00110011012010114120 00122211120240 11310

Chelus fimbriata

1—0011110011101020 1110 0001010211023111 10100
00001001101?0101101200012211112024011310

Teneremys lapparenti

???
????????11212 0??1221111 2023??1???

References

Anquetin, J., Barrett, P. M., Jones, M. E., Moore-Fay, S., & Evans, S. E. (2009). A new stem turtle from the Middle Jurassic of Scotland: New insights into the evolution and palaeoecology of basal turtles. *Proceedings of the Royal Society B, 276*, 879–886.

Batsch, A. C. (1788). *Versuch einer Anleitung, zur Kenntniß und Geschichte der Thiere und Mineralien.* Jena: Akademische Buchhandlung.

Bergounious, F. M. (1952). Les chéloniens fossils de Gafsa, Appendix in Arambourg, C., Les vertebrés fossils des gisements de phosphates (Maroc-Algérie-Tunisie). *Service Géologique, Protectoral de la République Française au Maroc, Notes et Mémoires, 92*, 377–396.

Bräm, H. (1965). Die Schildkröten aus dem oberen Jura (Malm) der Gegend von Solothurn. *Schweizerische Palaöntologische Abhandlungen, 83*, 1–190.

Burke, A. (2009). Turtles…again. *Evolution & Development, 11*, 622–624.

Cadena, E. A., Bloch, J., & Jaramillo, C. (2010). New podocnemid turtle (Testudines: Pleurodira) from the Middle–Upper Paleocene of South America. *Journal of Vertebrate Paleontology, 30*, 367–382.

Cadena, E. A., & Gaffney, E. S. (2005). *Notoemys zapatocaensis*, a new side-necked turtle (Pleurodira: Platychelyidae) from the Early Cretaceous of Colombia. *American Museum Novitates, 3470*, 1–19.

Cattoi, N., & Freiberg, M. (1961). Nuevo Hallazgo de chelonia extinguidos en la Republica Argentina. *Physis, 22*, 202.

Claude, J., Paradis, E., Tong, H., & Auffray, J. (2003). A geometric morphometric assessment of the effects of environment and cladogenesis on the evolution of the turtle shell. *Biological Journal of the Linnean Society, 79*, 485–501.

Cope, E. D. (1864). On the limits and relations of the Raniformes. *Proceedings of the Academy of Natural Sciences of Philadelphia, 16*, 181–183.

Cope, E. D. (1870). Seventh contribution to the herpetology of tropical America. *Proceedings of the American Philosophical Society, 11*, 147–169.

Datta, P. M., Manna, P., Ghosh, S., & Das, D. (2000). The first Jurassic turtle from India. *Paleontology, 43*, 9–109.

de la Fuente, M. S. (2003). Two new pleurodiran turtles from the Portezuelo Formation (Upper Cretaceous) of Northern Patagonia, Argentina. *Journal of Paleontology, 77*, 559–575.

de la Fuente, M. S. (2007). Testudines. In Z. Gasparini, L. Salgado & R. Coria (Eds.), *Patagonian Mesozoic Reptiles* (pp. 50–86). Bloomington: Indiana University Press.

de Lapparent de Broin, F. (2000). The oldest pre-Podocnemidid turtle (Chelonii, Pleurodira), from the Early Cretaceous, Ceará state, Brasil, and its environment. *Threeballs del Museu de Geología de Barcelona, 9*, 43–95.

de Lapparent de Broin, F. (2001). The European turtle fauna from the Triassic to the Present. *Dumerilia, 4*, 155–216.

de Lapparent de Broin, F., & de la Fuente, M. (2001). Oldest world Chelidae (Chelonii, Pleurodira), from the Cretaceous of Patagonia. *Comptes Rendus de l'Académie des Sciences de Paris, 333*, 463–470.

de Lapparent de Broin, F., & Murelaga, X. (1996). Une nouvelle faune de chéloniens dans le Crétacé Supérieur européen. *Comptes Rendus de l'Académie des Sciences de Paris, série IIa, 323*, 729–735.

de Lapparent de Broin, F., & Murelaga, X. (1999). Turtles from the Upper Cretaceous of Laño (Iberian Peninsula). *Estudios del Museo de Ciencias Naturales de Alava, 14*(Número Especial 1), 135–211.

de Lapparent de Broin, F., Murelaga, X., & Codrea, V. (2004). Presence of Dortokidae (Chelonii, Pleurodira) in the earliest Tertiary of the Jibou Formation, Romania: Paleobiogeographical implications. *Acta Paleontologica Romaniae, 4*, 203–215.

de Lapparent de Broin, F., de La Fuente, M., & Fernandez, M. (2007). *Notoemys laticentralis* (Chelonii, Pleurodira), Late Jurassic of Argentina: New examination of the anatomical structures and comparisons. *Revue de Paléobiologie, 26*, 99–136.

de la Fuente, M. S., & Iturralde-Vinent, M. (2001). A new pleurodiran turtle from the Jagua Formation (Oxfordian) of Western Cuba. *Journal of Paleontology, 75*, 860–869.

Fernandez, M. S., & de la Fuente, M. S. (1988). Nueva tortuga (Cryptodira: Thalassemydidae) de la Formación Vaca Muerta (Jurasico, Tithoniano) de la Provincia del Neuquen, Argentina. *Ameghiniana, 25*, 129–138.

Fernandez, M. S., & de la Fuente, M. S. (1994). Redescription and phylogenetic position of *Notoemys*: The oldest Gondwanian pleurodiran turtle. *Neues Jahrbuch fur Geologie und Palaontology Abhandlungen, 193*, 81–105.

Fraas, E. (1913). *Proterochersis,* eine pleurodire Schildkroete aus dem Keuper. *Jahreshefte der Gesellschaft für Naturkunde in Württemberg, 69*, 13–90.

França, M., & Langer, M. (2006). Phylogenetic relationships of the Bauru Group turtles (Late Cretaceous of South Central Brazil). *Revista Brasileira de Paleontología, 9*, 365–373.

Gaffney, E. S. (1990). The comparative osteology of the Triassic turtle *Proganochelys*. *Bulletin of the American Museum of Natural History, 155*, 389–436.

Gaffney, E. S., & Jenkins, F. A., Jr. (2010). The cranial morphology of *Kayentachelys*, an Early Jurassic cryptodire, and the early history of turtles. *Acta Zoologica, 91*, 335–368.

Gaffney, E. S., Hutchison, J. H., Jenkins, F. A, Jr, & Meeker, L. J. (1987). Modern turtle origins: The oldest known cryptodire. *Science, 237*, 289–291.

Gaffney, E. S., de Almeida Campos, D., & Hirayama, R. (2001). *Cearachelys*, a new side-necked turtle (Pelomedusoides: Bothremyididae) from the Early Cretaceous of Brazil. *American Museum Novitates, 3319*, 1–20.

Gaffney, E. S., Tong, H., & Meylan, P. (2006). Evolution of the side-necked turtles: The families Bothremydidae, Euraxemydidae, and Araripemydidae. *Bulletin of the American Museum of Natural History, 300*, 1–698.

Guzman, G. (1985). Los Grifeidos infracretacicos Aetostreon couloni y *Ceratrostreon boussingaulti*, de la Formación Rosablanca, como indicadores de oscilaciones marinas. In S. Etayo (Ed.), *Proyecto Cretácico* (pp. 1–16). Bogota: Publicacion especial del Ingeominas.

Joyce, W. (2007). Phylogenetic relationships of Mesozoic turtles. *Bulletin of Peabody Museum of Natural History, 48*, 1–100.

Joyce, W., & Sterli, J. (2012). Congruence, non-homology, and the phylogeny of basal turtles. *Acta Zoologica, 93*, 149–159.

Li, C., Wu, X.-C., Rieppel, O., Wang, L., & Zhao, L. (2008). An ancestral turtle from the Late Triassic of Southwestern China. *Nature, 456*, 97–501.

Maddison, W., & Maddison, D. (2009). Mesquite: A modular system for evolutionary analysis. Version 2.5(build j77).

Meylan, P. A., Gaffney, E. S., & De Almeida Campos, D. (2009). *Caninemys*, a new side-necked turtle (Pelomedusoides: Podocnemididae) from the Miocene of Brazil. *American Museum Novitates, 2639*, 1–26.

Ogg, J. G., Ogg, G., & Gradsteing, F. (2008). *The concise Geologic Time Scale*. Cambridge: Cambridge University Press.

Price, L. I. (1973). Quelonio amphichelydia no Cretaceo inferioe do nordeste do Brazil. *Revista Brasileira de Geociencias, 3*, 84–96.

Pritchard, P. (2008). Evolution and structure of the turtle shell. In J. Wyneken, M. Godfrey & V. Bels (Eds.), *Biology of turtles* (pp. 45–83). Boca Raton: CRC press.

Rougier, G. W., de la Fuente, M., & Arcucci, A. B. (1995). Late Triassic turtles from South America. *Science, 268*, 855–858.

Schneider, J. G. (1783). *Allgemeine Naturgeschichte der Schildkröten, nebst einem System* (364pp). Leipzig: Verseichnisee der einzelnen Arten.

Shaffer, H. B. (2009). Turtles (Testudines). In S. B. Hedges & S. Kumar (Eds.), *The timetree of life* (pp. 398–401). Oxford: Oxford University Press.

Sorenson, M., & Franzosa, E. (2007). TreeRot Version 3. Accessed May 10, 2009, available at http://people.bu.edu/msoren/TreeRot.html.

Sterli, J. (2008). A new, nearly complete stem turtle from the Jurassic of South America with implications for turtle evolution. *Biology Letters, 22*, 1–4.

Sukhanov, V. B. (2006). An archaic turtle, *Heckerochelys romani* gen. et sp. nov., from the Middle Jurassic of Moscow Region, Russia. In I. G. Danilov & J. F. Parham (Eds.), *Fossil Turtle Research* (pp. 112–118). St. Petersburg : Zoological Institute of Russian Academy of Sciences [*Russian Jounal of Herpetology, 13*(Suppl. 1)].

Swofford, D. L. (2002). PAUP*. Phylogenetic analysis using Parsimony (*and other methods). Version 4.0b10. Sinauer Associates, Sunderland, Massachusetts.

Wagner, A. (1853). Beschreibung einer fossilen Schildkröte und etlicher anderer Reptilien- Überreste aus den lithographischen Schiefern und dem grünen Sandsteine von Kehlheim. *Abhandlungen der Bayerischen Akademie der Wissenschaften, Mathematisch-physikalische Klasse, 7*(1), 25–37.

Wood, R. C., & de Gamero, M. D. (1971). *Podocnemis venezuelensis*, a new fossil pelomedusoid (Testudines, Pleurodira) from the Pliocene of Venezuela and a review of the history of *Podocnemis* in South America. *Breviora, 376*, 1–23.

Young, C. C., & Chow, M. C. (1953). New fossil reptiles from Szechuan, China. *Acta Scientia Sinica, 2*, 216–243.

Chapter 9
Spoochelys ormondea gen. et sp. nov., an Archaic Meiolaniid-Like Turtle from the Early Cretaceous of Lightning Ridge, Australia

Elizabeth T. Smith and Benjamin P. Kear

Abstract The Lower Cretaceous (lower to middle Albian) Griman Creek Formation deposits of Lightning Ridge in central-eastern Australia are famous for producing opalised fossils. Much of this material is poorly documented but recent assessments suggest a diverse assemblage of mainly non-marine vertebrates, invertebrates and plants. This biota is associated with a Gondwanan high-latitude zone that would have been subject to cool-temperate conditions. Turtle remains are particularly common at Lightning Ridge, comprising several distinct lineages including aquatic chelids and peculiar meiolaniid-like taxa—meiolaniids were spectacular horned turtles known from the Australian region and South America. *Spoochelys ormondea* gen. et sp. nov. shares some distinctive skeletal traits with this group (e.g., cranial scute pattern, incisura columellae auris confluent with the Eustachian tube) but also retains remarkably plesiomorphic cranial structures (e.g., an interpterygoid vacuity and short inferior parietal process) that are otherwise characteristic of Triassic and Jurassic stem turtles. The placement of meiolaniids and their sister lineages within Testudines is controversial in recent phylogenies. To test the relationships of *Spoochelys*, we used the two most comprehensive published data sets of fossil Testudinata and rescored a number of characters. Separate Maximum Parsimony and Bayesian analyses of both matrices supported recognition of *Spoochelys* as a primitive testudinatan but could not confirm its relationships with the meiolaniid clade. Nevertheless, the persistence of surprisingly archaic taxa such as *Spoochelys* into the Early Cretaceous of Australia implies survival of an ancient Pangean lineage, and brings into question long-held assumptions of Laurasian affinities for the meiolaniid-like turtles of Gondwana.

Keywords Albian • Gondwana • High-paleolatitude • Meiolaniidae • Terrestrial • Relict taxon

Introduction

Until very recently, the Australian Mesozoic (Early Cretaceous, Albian) turtle record consisted of only five named taxa—three marine protostegids (Molnar 1991; Kear and Lee 2006; Kear 2006a), and two non-marine turtles: *Otwayemys cunicularis* Gaffney et al. 1998, based on shell remains, cranial fragments and vertebrae from the Eumeralla Formation of Cape Otway, Victoria; and *Chelycarapookus arcuatus* Warren 1969, from the Merino Group of Casterton, Victoria, which is represented by a solitary steinkern. Smith (2009, 2010) also provided accounts of new non-marine turtle fossils from the Lower Cretaceous (middle Albian) opal fields of Lightning Ridge in New South Wales, Australia (Fig. 9.1), including indeterminate chelid pleurodires and several undescribed taxa that shared features in common with meiolaniids—extinct terrestrial horned turtles known from the Cretaceous-Paleogene of Argentina and Neogene-Holocene of Australia/Southwest Pacific islands (White et al. 2010; see also Gaffney 1996 for a species summary).

The Lightning Ridge vertebrate assemblage is dominated by turtle bones, which are usually recovered as a biproduct of opal mining and sufficiently robust to survive mechanical excavation and processing. The individual elements are occasionally articulated but mostly occur as associated or isolated fragments. Among the most spectacular turtle fossils found at Lightning Ridge are the articulated skull and a

E. T. Smith (✉)
Vertebrate Palaeontology Laboratory, University of New South Wales, PO Box 1, Kensington, NSW 2052, Australia
and
Australian Opal Centre, PO Box 229, Lightning Ridge, NSW 2834, Australia
e-mail: elizabethtsmith@exemail.com.au

B. P. Kear
Palaeobiology Programme, Department of Earth Sciences, Uppsala University, Villavägen 16, SE-752 36 Uppsala, Sweden
e-mail: benjamin.kear@geo.uu.se

D. B. Brinkman et al. (eds.), *Morphology and Evolution of Turtles*, Vertebrate Paleobiology and Paleoanthropology,
DOI: 10.1007/978-94-007-4309-0_9,

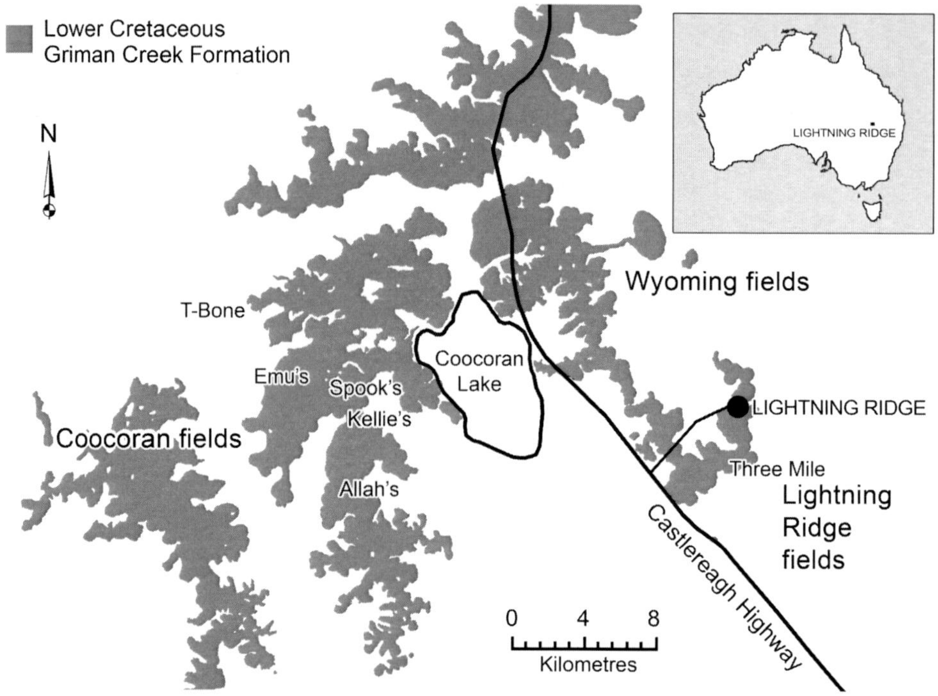

Fig. 9.1 Locality map showing Griman Creek Formation and Lightning Ridge opal fields, indicating some of the many sites that have produced opalised turtle specimens

second partial braincase of a new taxon described herein, *Spoochelys ormondea* gen. et sp. nov. These were discovered by opal miner Ormie Molyneux in the mid-1990s on the Coocoran Opal Fields, about 35 km due west of the Lightning Ridge township. Such specimens, together with a third fragmentary skeleton comprising associated cranial/ postcranial bones, are intriguing because they display morphological similarities to meiolaniids and prompt a re-evaluation of this enigmatic lineage.

The phylogenetic relationships of meiolaniids are contentious, and two conflicting hypotheses have been advocated: meiolaniids are stem testudinatans related to basal taxa such as *Mongolochelys efremovi* Khosatzky 1997 (Joyce 2007; Sterli 2008); or meiolaniids, together with *Otwayemys* and *Chubutemys copelloi* Gaffney et al. 2007 from the Lower Cretaceous of Patagonia are sister taxa of Laurasian Jurassic—Paleocene "sinemydids-macrobaenids" (see Brinkman and Wu 1999; Parham and Hutchinson 2003) and thus represent primitive eucryptodires (Gaffney et al. 1998; Hirayama et al. 2000; de Lapparent de Broin and Molnar 2001; Brinkman et al. 2012).

Eugene Gaffney's seminal works on Meiolaniidae (Gaffney 1983, 1985, 1992, 1996; Gaffney and McNamara 1990; Gaffney et al. 1984, 1992) are a legacy of outstanding observational, descriptive and analytical skill, yet he was first to acknowledge that his diagnosis of relationships relied primarily on the highly derived *Meiolania platyceps* Owen 1886, from Pleistocene deposits of Lord Howe Island, Southwest Pacific (Gaffney 1996, p. 72). It must be emphasized that parts of the skeleton of *Meiolania* are still poorly understood and that stratigraphically older meiolaniid remains are scrappy and often difficult to interpret.

The discovery of abundant meiolaniid-like turtle fossils in the Lower Cretaceous deposits at Lightning Ridge is therefore significant because it provides not only the best preserved Mesozoic potential exemplars of this lineage but also facilitates a critical evaluation of the phylogenetic inconsistencies surrounding Meiolaniidae.

Institutional abbreviations: AM, Australian Museum, Sydney; LRF, Australian Opal Centre, Lightning Ridge; MPEF, Museo Paleontológico Egidio Feruglio, Trelew; NHMUK, The Natural History Museum, London; NMV, National Museum of Victoria, Melbourne.

Systematic Paleontology

Testudinata sensu Joyce 2007
Spoochelys gen. nov.

Type species: *Spoochelys ormondea* sp. nov.

Etomology: 'Spook' referring to the locality at Spook's Field that yielded the most complete remains; and '-chelys' for turtle.

Diagnosis: Medium-sized (snout-tail length approximately 400–500 mm) primitive testudinatan differentiated from basal taxa such as *Proganochelys* Baur 1887, *Palaeochersis* Rougier, de la Fuente, and Arcucci 1995, and *Australochelys* Gaffney and Kitching 1994 by the presence of an incipient antrum postoticum, development of a thin columella auris with a distinct footplate, loss of the wide occipital plate,

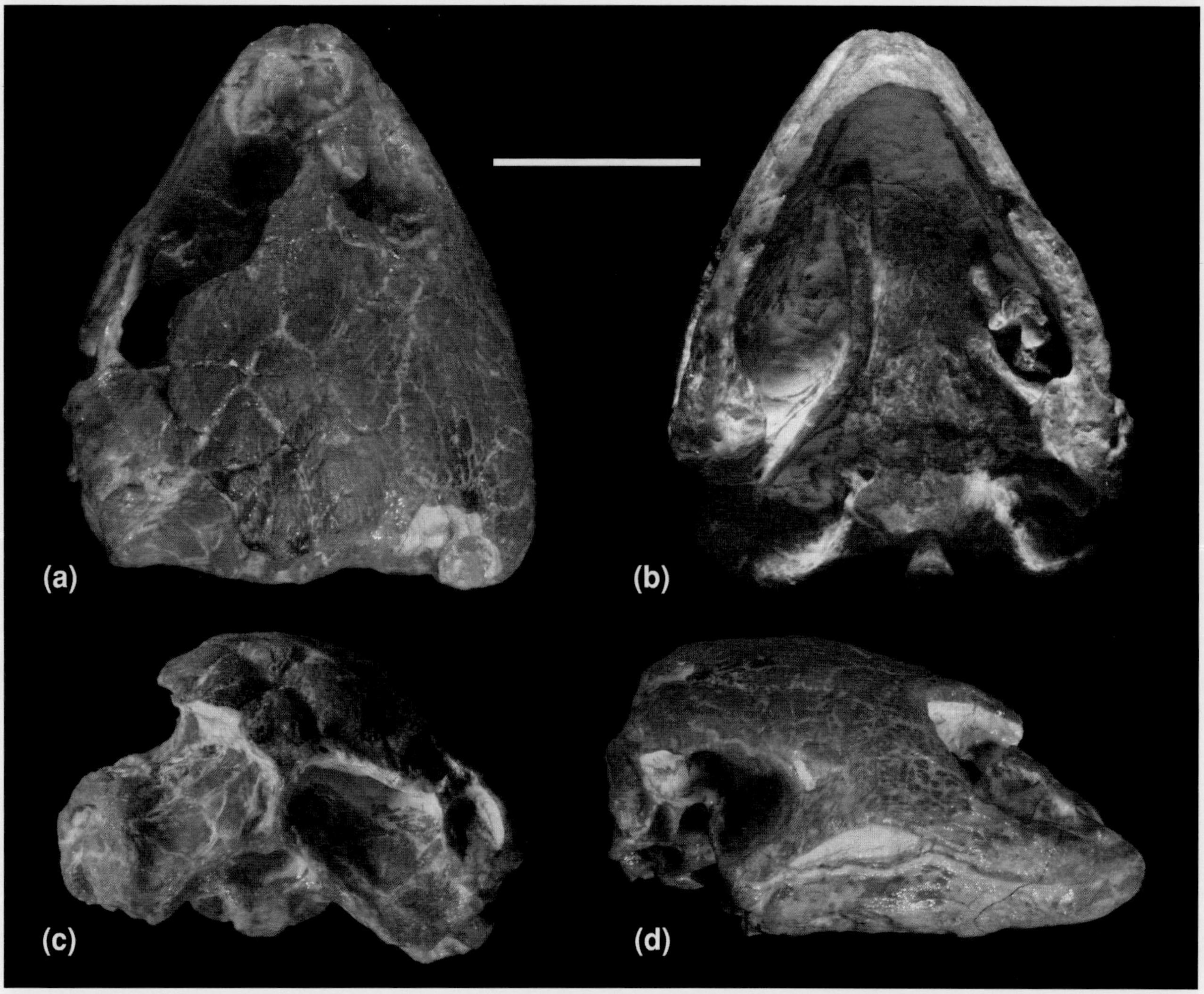

Fig. 9.2 Referred skull of *Spoochelys ormondea* gen. et sp. nov., original specimen LRF-TH450 from Spook's Field (Coocoran), Lightning Ridge, NSW, Australia. Photographs in **a** dorsal, **b** ventral, **c** posterior, and **d** lateral views. Scale bar = 20 mm

fusion of the basipterygoid articulation, and formed cervical articulations. *Spoochelys* can be excluded from crown-Testudinata (Testudines sensu Joyce 2007) by maintenance of a both prefrontal-postorbital and parietal-squamosal contact, poorly developed processus inferior parietalis, incipient antrum postoticum, absence of a distinct processus trochlearis oticum, retention of an interpterygoid vacuity, poorly developed crista supraoccipitalis, flat rostrum basisphenoidale, paired ventral basisphenoid-basioccipital tubercles, transverse processes of the cervical vertebrae positioned towards the midline of the centrum, and the presence of an acromial ridge on the scapula. Distinction from potential testudine sister taxa including *Mongolochelys*, *Chubutemys*, *Otwayemys*, meiolaniids, and *Kallokibotion bajazidi* Nopsca 1923 can be made by the absence of a distinct processus trochlearis oticum, retention of an interpterygoid vacuity, and poorly developed crista supraoccipitalis; *Spoochelys* differs from *Patagoniaemys gasparinae* Sterli and de la Fuente 2011 in transverse processes of the cervical vertebrae being positioned towards the midline of the centrum, and presence of supramarginal scutes on the anterior carapace.

Spoochelys ormondea sp. nov.
(Figs. 9.2, 9.3, 9.4, 9.5, 9.6, 9.7, 9.8)

Holotype: Associated cranial-postcranial elements: quadrate, AM F121643; supraoccipital, AM F121646; anterior peripherals, AM F121579, AM F121580, AM F121581; anterior costal fragments, AM F121686, AM F121687; caudal vertebra, AM F121641; scapula AM F121587; ulna AM F121621; pedal phalanges, AM F121613.

Holotype unit, locality, and age: The holotype specimens were discovered in a mineral claim at 'T-Bone Extension', a designated mining locality within the Coocoran Opal Fields, about 35 kms west of Lightning Ridge,

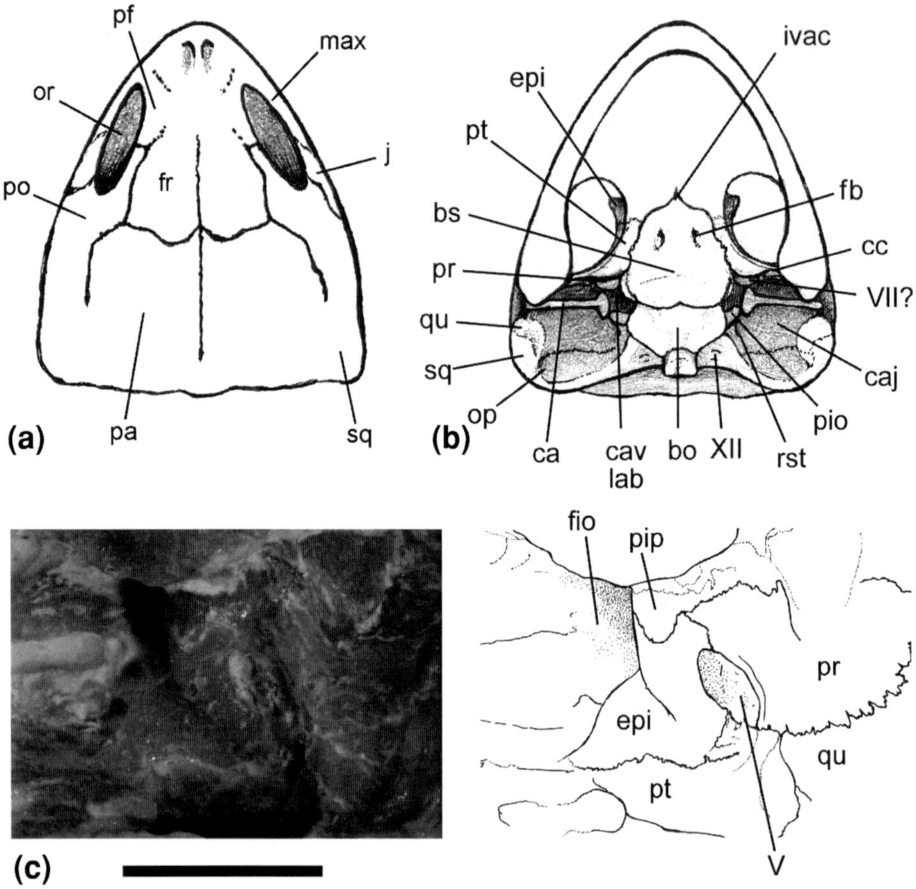

Fig. 9.3 Skull material of *Spoochelys ormondea* gen. et sp. nov. Skull reconstruction in **a** dorsal and **b** ventral views. **c** Photograph (*left*) and interpretive drawing (*right*) of the lower temporal fossa anterolateral oblique view showing epipterygoid and associated structures (original specimen LRF-TH450). Scale bar = 10 mm. Abbreviations defined in Appendix 4

northwestern New South Wales, Australia (Fig. 9.1). Remains referable to *Spoochelys* have also been found in other mine sites throughout the Coocoran and adjacent areas. The Lightning Ridge opalized fossils derive from the "Finch clay facies" within the Wallangulla Sandstone Member of the Griman Creek Formation (Holmes and Senior 1976; Watkins 1984). The "Finch clay facies" or "opal dirt" is considered to be either early to middle Albian (*Coptospora paradoxa* spore-pollen Zone; Burger 1980), or middle Albian in age (Dettmann et al. 1992), and at Lightning Ridge, is a subsurface unit (<30 m deep) unconformably overlain by Neogene silcritic sediments of the Cumborah Gravel (Senior and Chadderton 2007). The "Finch clay facies" comprises fine-grained, deeply weathered volcanogenic clays (60–65% silica-rich kaolinite, ~20% smectite with low alumina content, and 5% illite: Watkins 1984) that occur in discontinuous lenses up to 5 meters thick. Precise stratigraphic correlation and dating of these layers has not been undertaken and thus the chronological relationships of their constituent fossil assemblages are unclear. Nevertheless, the "Finch clay facies" sequences have yielded a rich record of plants (Smith and Smith 1999), non-marine mollusks (Hocknull 2000; Hamilton-Bruce and Kear 2010), fish (Smith and Smith 1999), aquatic reptiles (Molnar and Willis 2001; Kear 2006b; Smith 2010), dinosaurs (see recent summary in Agnolin et al. 2010), pterosaurs (Smith and Smith 1999; Kear et al. 2010), birds (Molnar 1999), and mammals (Archer et al. 1985, Flannery et al. 1995; Clemens et al. 2003; Musser 2005) that represent elements of an Early Cretaceous Gondwanan high latitude (60–70°S) biota (Dettmann et al. 1992). Invertebrate fossils, vertebrate taphonomy and sedimentary evidence has been used to infer a freshwater deltaic to estuarine depositional setting, and plant/isotopic data indicates a seasonal, cool-temperate climate (Dettmann et al. 1992).

Etomology: The species epithet 'ormondea' is in honor of Ormie Molyneux who discovered the most complete cranial specimens.

Referred specimens: Skull LRF-TH 450/LRF 450 (cast); braincase LRF-TH 451/LRF 451 (cast); carapace and plastron sections, LRF 341, LRF 343, LRF 344, LRF 462; LRF 738, LRF 738a, LRF 738b, LRF 1650, AM F121592; third cervical vertebrae, AM F127984, AM F127990; fourth cervical vertebrae, AM F68254, AM F112750; fifth cervical vertebrae, LRF 464; sixth cervical vertebra, LRF 944; seventh cervical vertebrae, AM F127972, AM F127996; eighth cervical vertebrae, AM F72276; sacral and caudal vertebrae, LRF 016, LRF 019, LRF 407, LRF 735, AM

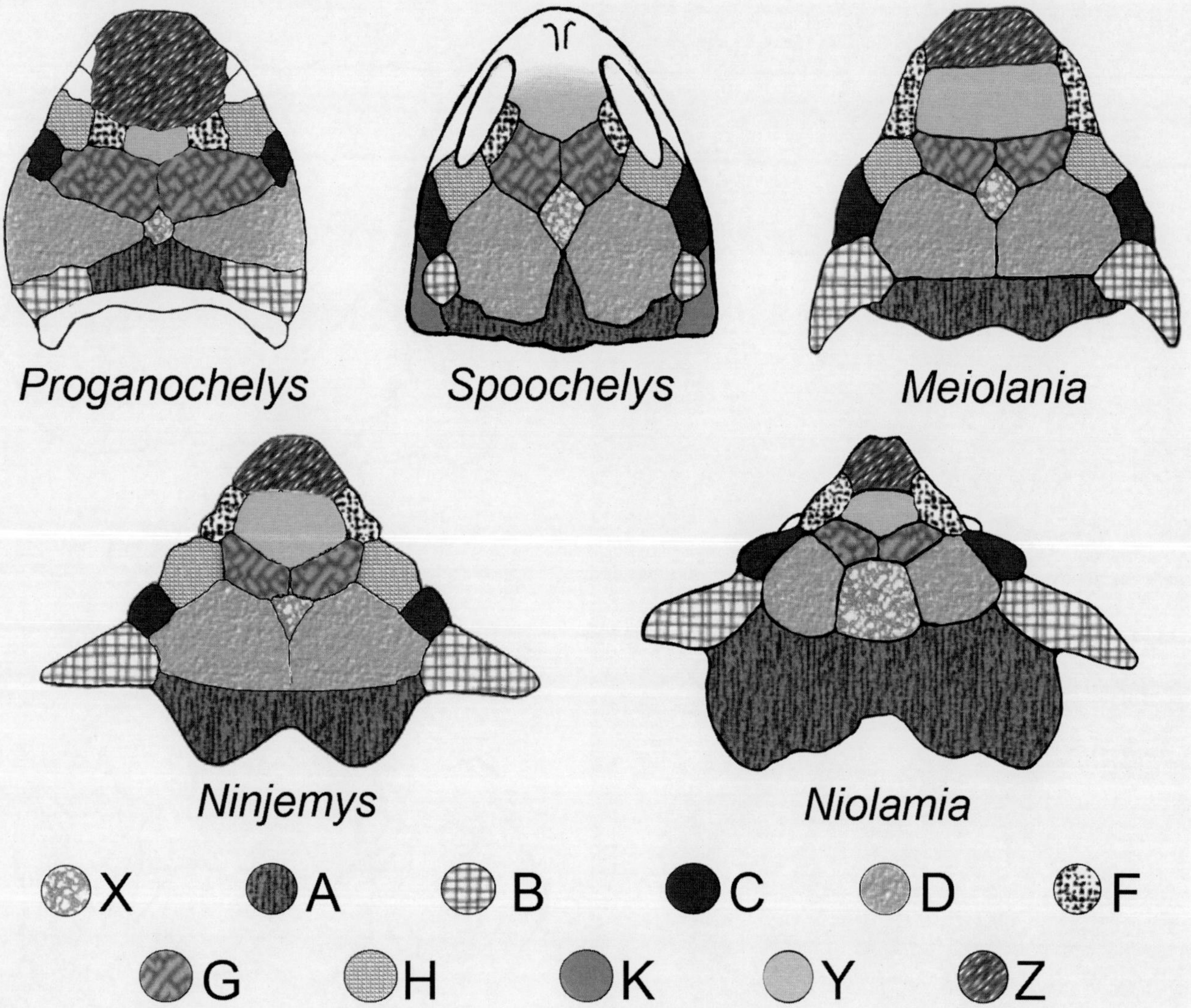

Fig. 9.4 Comparison of scute patterns in *Proganochelys*, *Spoochelys* and selected meiolaniid taxa. Positional terminology follows Gaffney (1983, 1996) but is not used to imply homology

F66772, AM F121603, AM F121638, AM F127978; pectoral girdle, LRF 081, AM F66776, AM F66777, AM F127942, AM F127954; humeri, LRF 078, AM F68310, AM F68321, AM F127940, AM F127956; ulnae, LRF 079, AM F121621, AM F127944; pelvic girdle elements, AM F66779, AM F121614, AM F121615, AM F121622, AM F127948, AM F127949; femur, LRF 420, LRF 1216, AM F121617, AM F127943; tibia, AM F127958; tarsus and pes, AM F121964.

Remarks: The holotype specimens were excavated from a single 'paleochannel' by opal miners Rob and Debbie Brogan and constitute an associated series based on proximity (the quadrate, supraoccipital, anterior peripherals, and scapula-acromion were within immediate contact), taphonomy (minimal evidence of breakage or edge-rounding indicative of transport), and ontogeny (all elements including the additional costal fragments, caudal vertebra, ulna, and pedal phalanges are from an osteologically mature animal). The lack of multiple identical elements also implies derivation from a single individual. Opalised fossils recovered from within the bounds of registered opal mining claims are classed as Schedule 1 Minerals under the Mining Act NSW of 1992 and are thus estranged from Crown property rights. We have elected to designate AM F121643, AM F121646, AM F121579, AM F121580, AM F121581, AM F121686, AM F121687, AM F121641, AM F121587, AM F121621, and AM F121613 as the holotype because these specimens have been formally donated to a public collection at the Australian Museum, Sydney. Title of the other articulated cranial remains described herein has been retained by the private claim holder. However, high definition hydrostone casts (LRF R450, LRF R451) have been accessioned by the Australian Opal Centre, Lightning Ridge, Australia, and the original opalized specimens are presently held on loan under the temporary numbers LRF-TH 450 and

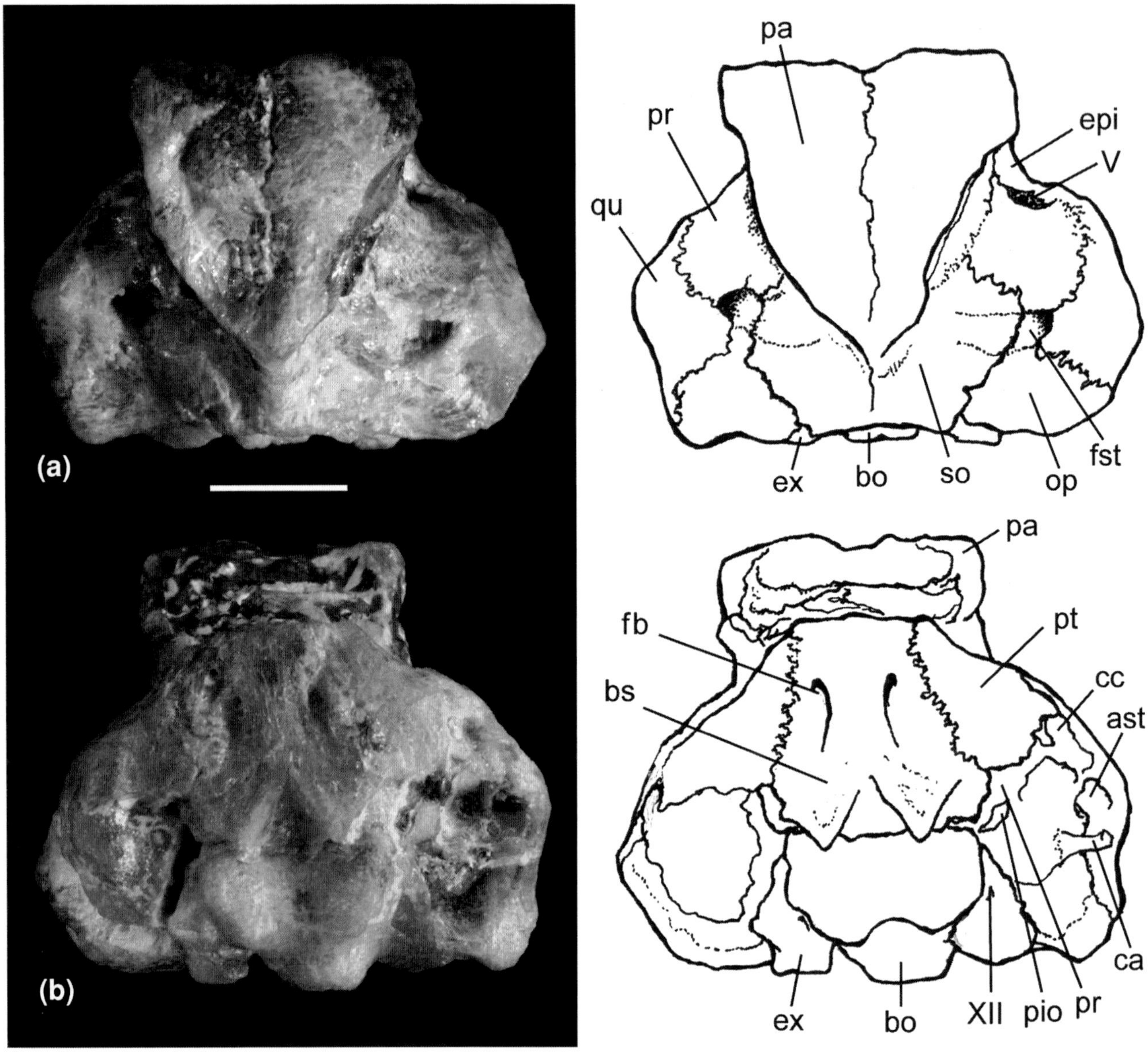

Fig. 9.5 Referred braincase of *Spoochelys ormondea* gen. et sp. nov., original specimen LRF-TH451 from Emu's Field (Coocoran), Lightning Ridge, NSW, Australia. **a** Dorsal view: photograph (*left*) and interpretive drawing (*right*). **b** Ventral view: photograph (*left*) and interpretive drawing (*right*). Scale bar = 10 mm. Abbreviations defined in Appendix 4

LRF-TH 451 respectively. The Australian Opal Centre is an accredited Australian Federal Government repository for significant natural history and cultural artifacts.

Description and Comparisons

Skull and scute pattern: The skull of *Spoochelys ormondea* is 48 mm in maximum anteroposterior length (based on LRF R450/LRF-TH 450: Figs 9.2, 9.3). The general cranial structure is very robust and lacks both the dorsoventral flattening and deep emargination of the temporal region evident in many more derived turtles. The dermal roof of LRF-TH 450 is coarsely ornamented and bears deeply incised scute sulci that recall the patterning of *Proganochelys quenstedti* Baur 1887 (see Gaffney 1990, p. 29, Fig. 17) but more closely resemble *Meiolania* (Gaffney 1983); we therefore employ the interpretive nomenclature developed for this taxon by Gaffney (1983; 1996: see Fig. 9.4). The central, diamond-shaped scute X would have covered the parietal and bordered the G scutes on the frontals. The D scutes apparently met on the midline behind scute X but were separated posteriorly by a triangular projection of scute A which covered the posterior midline of

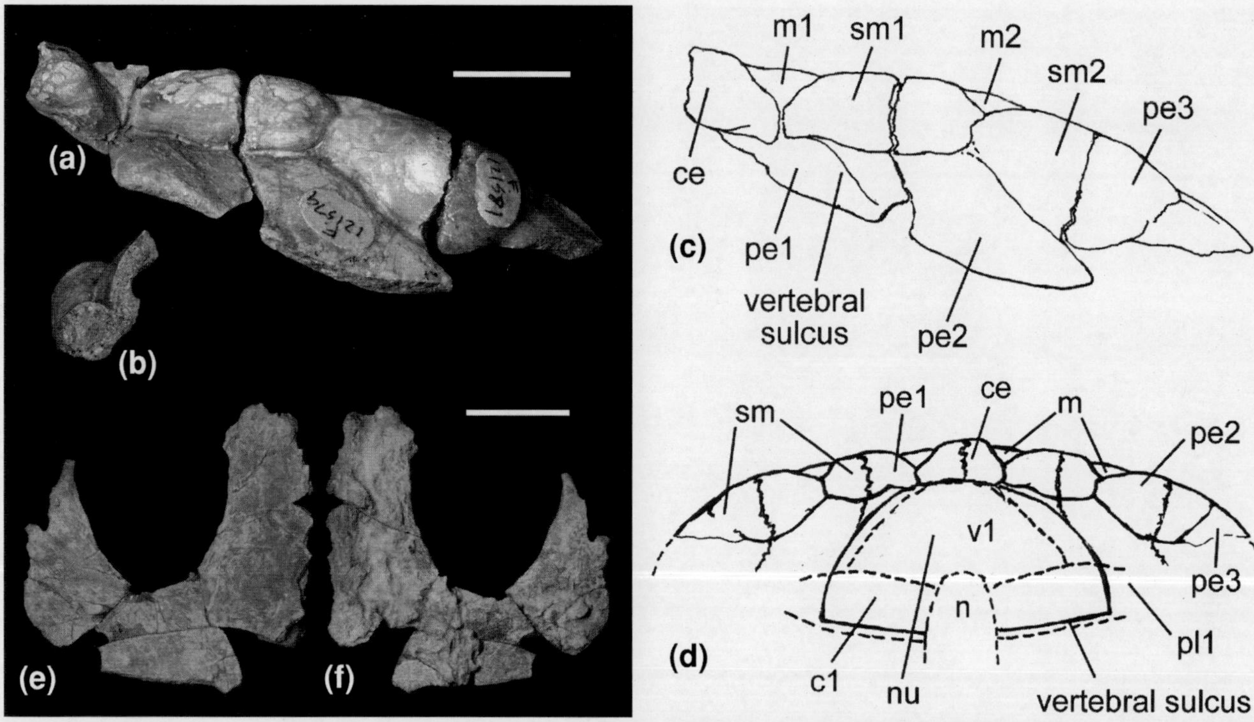

Fig. 9.6 Shell material of *Spoochelys ormondea* gen. et sp. nov. **a** Photograph of anterior right section of the carapace (AMF121579, AMF121580, AMF121581). **b** Photograph of lateral view of peripheral (AMF121579). **c** Interpretive diagram of AMF121579, AMF121580, AMF121581. **d** Reconstruction of anterior part of carapace. **e, f** anterior plastral lobe (LRF1650) in **e** ventral and **f** dorsal views. Scale bars = 20 mm. Abbreviations defined in Appendix 4

the parietals. In *Mongolochelys*, additional scutes have been reconstructed along the posterior margin of the skull but these are more uniform in size (Sukhanov 2000, p. 347, Fig. 17.28). Scute F adjoined scute G anterolaterally along the orbital margin, and scute Y overlapped the posterior part of the frontal. Scutes H and C were oriented over the postorbital with B and K in the squamosal region. There is no evidence of a prominent cranial 'horn' formed by scute B (Gaffney 1983, p. 398). However, this area of the skull is strongly delineated and could have supported a low 'boss-like' structure. The J2 scute on the jugal would have been large, contacting J1 over the quadratojugal and I over the maxilla respectively.

The anterior nasal region of *Spoochelys* is poorly preserved in LRF-TH 450 and it is impossible to determine if nasal bones were present (Fig. 9.2a). The prefrontal forms the anterodorsal margin of the orbit, contacting the postorbital; it is unclear if prefrontals met on the midline or extended ventrally to contact the palatal elements.

The maxilla in LRF R450/LRF-TH 450 contributes to the ventral orbital margin and extends into the floor of the orbital cavity (Fig. 9.2a, d). Its contact with the jugal is horizontally oriented such that external exposure of the maxilla is acuminate as in *Meiolania* (Gaffney 1983). The maxilla separates the anterior section of the jugal from the ventral border of the cheek, comparable to the condition in the basal testudinatan *Palaeochersis talampayensis* Rougier, de la Fuente, and Arcucci 1995 (Sterli et al. 2007).

Although damaged anteriorly, the frontals are visible in LRF R450/LRF-TH 450 where they are excluded from the orbital rim by the prefrontal-postorbital suture. The parietals of LRF R450/LRF-TH 450 contact the frontals and postorbitals, and extend to the squamosal posterolaterally (although precise bone edges are unclear: Figs. 9.2a, 9.3a). The processus inferior parietalis is short, reaching the epipterygoid and prootic. *Spoochelys* closely resembles both *Meiolania* and *Kayentachelys aprix* Gaffney, Hutchinson, Jenkins, and Meeker 1987 (Sterli and Joyce 2007) in lacking the bony cavum epiptericum, weak development of the processus inferior parietalis, and exclusion of the parietal from the trigeminal foramen (Fig. 9.3c).

The postorbital of LRF R450/LRF-TH 450 is narrow along the posterodorsal margin of the orbital fossa but extends far back on the skull roof, articulating with the quadratojugal and jugal along its posteroventral edge; its posterior suture with the squamosal is indeterminate but can be inferred on the basis of positioning. A postorbital-quadratojugal contact is widely manifest in primitive turtles, including *Proganochelys* (Kordikova 2002), *Palaeochersis* (Sterli et al. 2007), *Kayentachelys* (Gaffney et al. 1987),

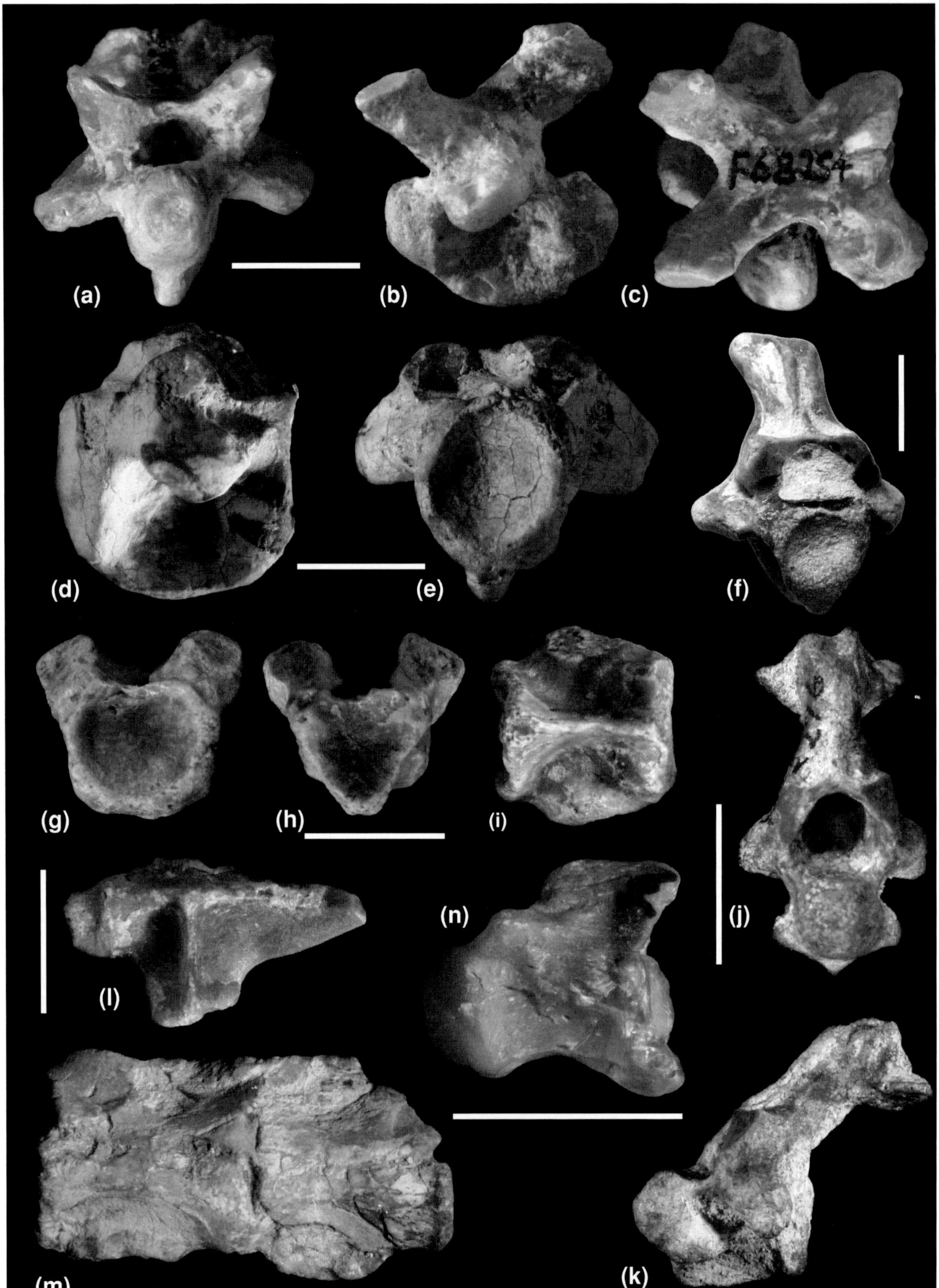
(a)
(b)
(c)
(d)
(e)
(f)
(g)
(h)
(i)
(j)
(l)
(n)
(m)
(k)

◀ **Fig. 9.7** Vertebrae of *Spoochelys ormondea* gen. et sp. nov. **a–k**, Cervical vertebrae: **a–c** fourth cervical (AMF68254) in **a** anterior, **b** lateral, and **c** dorsal views; **d**, **e** fifth cervical (LRF464) in **d** lateral and **e** anterior views; **f** sixth cervical (LRF944) in anterolateral view; **g–i**, seventh cervical (AMF127996) in **g** anterior, **h** posterior, and **i** ventral views; **j**, **k** eighth cervical (AMF72276) in (**j**) anterior and (**k**) lateral views. **l** Sacral vertebra (AMF121603) in ventral view. **m** Fused sacral complex (LRF019) in ventral view. **n** Posterior caudal (LRF735) in lateral view. Scale bars = 10 mm

Meiolania (Gaffney 1983) and Asian Jurassic-Cretaceous eucryptodires such as *Annemys levensis* Sukhanov and Narmandakh 2006, and *Mongolochelys* (Sukhanov 2000).

The jugal constitutes about one-third of the orbital rim and extends to the postorbital together with the quadratojugal posteriorly in the cheek region (visible in LRF R450/LRF-TH 450: Fig. 9.2d). The jugal's ventral margin is shallowly emarginated. The quadratojugal of LRF R450/LRF-TH 450 encircles most of the cavum tympani and projects below the incisura columellae auris (Fig. 9.2d); it contacts the squamosal dorsally (reminiscent of the arrangement in *Proganochelys*: see Gaffney 1990). A small opening occurs between the quadratojugal, jugal and postorbital as in *Proganochelys* (Gaffney 1990, p. 50); Kordikova (2002, p. 238) suggested that this might be ontogenetic.

The squamosal sutures to the parietal medially in LRF450/LRF-TH450, and also the quadratojugal, participating in the rim of the cavum tympani and forming most of the incipient antrum postoticum (Fig. 9.2a, c, d). Posteriorly the squamosal descends as a bony buttress behind the incisura columellae auris; its relationship to the opisthotic on the dorsal surface of the otic chamber is unknown.

Palatal complex: The anterior component of the otic chamber in *Spoochelys* (see LRF-TH 451) is narrow and there is no evidence of a typical cryptodiran processus trochlearis oticum (Fig. 9.5). The mandibular condyle is positioned in front of the basisphenoid-basioccipital suture similar to primitive pleurodires (Gaffney et al. 2006: see Fig. 9.2b). The quadrate-quadratojugal contact forms the periphery of the well-developed cavum tympani. The incisura columellae auris is primitively positioned dorsal to the mandibular condyle, and opens posteroventrally, incorporating both the stapes and Eustachian tube (Figs. 9.2d, 9.3b, 9.5b). Medially, the sutural surface for the prootic (well preserved in the holotype AM F121643) is extensive, separating the canalis stapedio-temporale (stapedial artery) and canalis cavernosus (lateral head vein).

The epipterygoid contacts the prootic above the foramen nervi trigemini (see LRF-TH 451 in Fig. 9.3). The vertical epipterygoid-prootic suture excludes the parietal from the trigeminal foramen similar to the condition in *Meiolania* (Gaffney 1983, p. 419, Fig. 47; Gaffney and Jenkins 2010, p. 18; Fig. 9.6). Ventrally, the epipterygoid meets the pterygoid and dips below the level of the basisphenoid rostrum. It also contributes to the posteromedial margin of the infraorbital fossa and the anterior and ventral edges of the foramen nervi trigemini. A pronounced semicircular flange extends anterolaterally and invades the subtemporal fossa beyond the processus inferior parietalis (see also *Kayentachelys*: Sterli and Joyce 2007; Gaffney and Jenkins 2010).

The pterygoids in LRF R450/LRF-TH 450 (Figs. 9.2c, 9.3b, 9.5b) meet along the midline but are separated posteriorly by a V-shaped space consistent in shape and position with the interpterygoid vacuity of *Proganochelys* (Gaffney 1990, p. 36, Fig. 25). They abut the basisphenoid close to the foramen carotici cerebralis posterior (sensu Sterli et al. 2010; and see Brinkman et al. 2012) and extend posterolaterally across the medial surfaces of the quadrate to just above the mandibular articulation (there is no quadrate ramus concavity sensu Gaffney et al. 2007). The pterygoids are widely separated from the rear section of the basisphenoid-basioccipital complex, leaving the cavum labyrinthicum, posterior section of the canalis cavernosus (which opens at the level of the fenestra ovalis), and prootic exposed ventrally (consistent with the plesiomorphic condition in *Proganochelys*: see Gaffney et al. 2006, pp. 610–611).

Basicranium: The holotype supraoccipital (AM F121646) is morphologically identical to those preserved in the referred skulls (LRF R450/LRF-TH 450, LRF R451/LRF-TH 451). Posterior continuation of the parietal suture in LRF-TH 451 suggests that the supraoccipital did not have extensive dorsal exposure (Fig. 9.5a), although definitive bone contacts are not visible in LRF-TH 450 and this part of the skull might have been fused (Fig. 9.2a). Indeed, the dorsal crest in AM F121646 is gracile and finishes well short of the dermal roof. The crista supraoccipitalis is very short and the supraoccipital's ventral component expands broadly across the roof of the otic chamber. There is a clear contact with the posterior edge of the foramen stapedio temporalis—a distinctive feature of paracryptodires (Gaffney 1979a, b).

The exoccipitals represent the lateral walls of the foramen magnum and are separated from the occipital condyle and basioccipital tubercles. Their posterior edges are deeply indented and the foramina nervi hypoglossi pierce the exoccipital-basioccipital suture and enter the exoccipital in LRF-TH 451 (see Fig. 9.5a).

The basioccipital contributes to the posteromedial margin of the cavum labyrinthicum. The widely divergent basioccipital tubercles delineate a median groove and are joined posteriorly by a bony flange in both LRF R450/

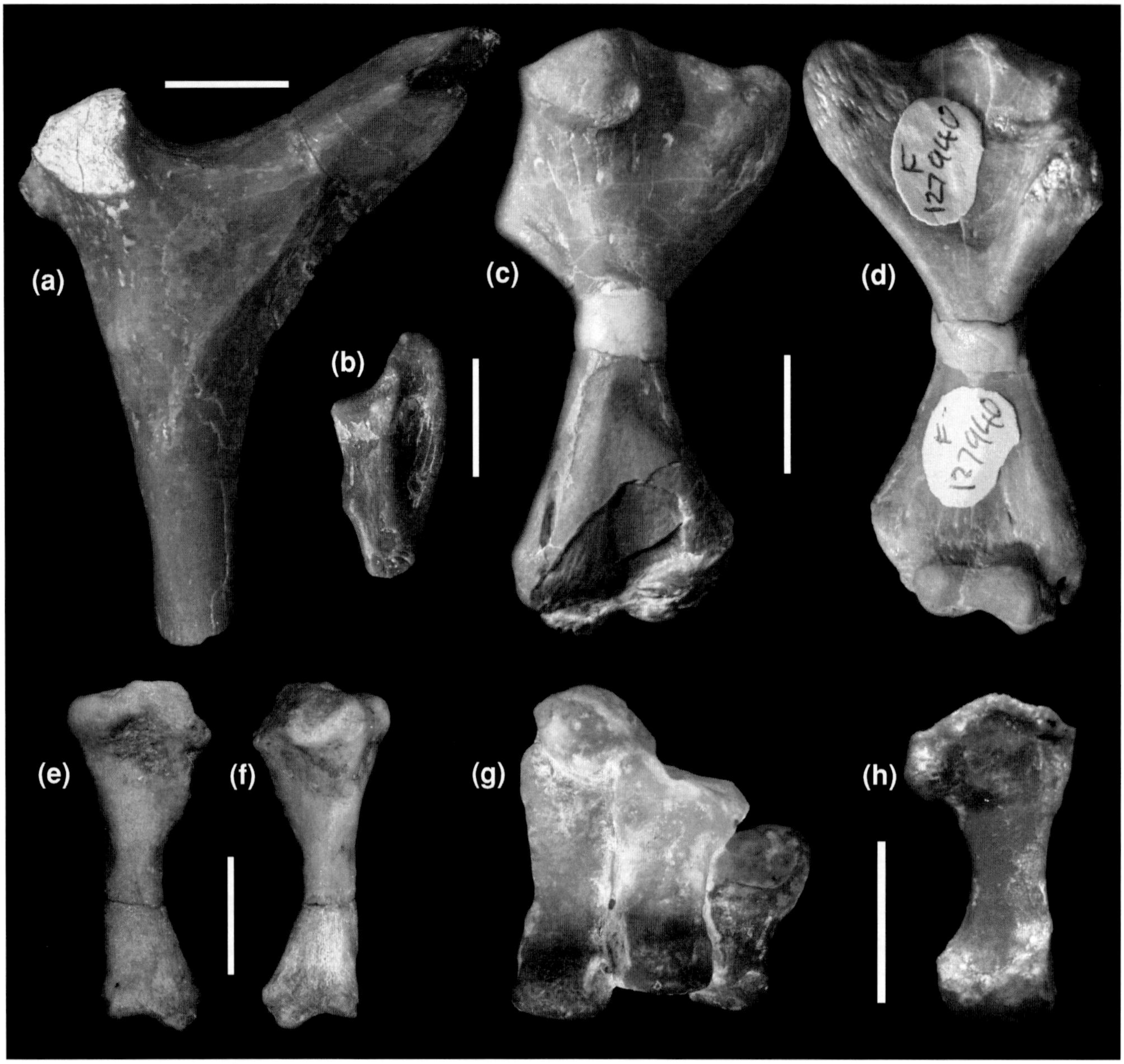

Fig. 9.8 Limb material of *Spoochelys ormondea* gen. et sp. nov. **a** Left scapula and acromion (AMF121587) in lateral view. **b** Proximal section of left ulna (AMF127944) in dorsal view. **c**, **d** left humerus (AMF127940) in **c** anterodorsal and **d** ventral views. **e**, **f** Right tibia (AMF127958) in **e** dorsal and **f** ventral views. **g** Articulated metatarsals I–III (AMF127961). **h** Isolated metatarsal II (AMF127964) showing heavy overlap tubercle. Scale bars = 10 mm in **a–f**, 5 mm in **g**, **h**

LRF-TH 450 and LRF R451/LRF-TH 451 (Figs. 9.2c, 9.3b, 9.5b). The condylar neck is short and supports a subtriangular condyle composed entirely of the basioccipital [*Kayentachelys* and more derived turtles alternatively incorporate a substantial contribution from the exoccipital (Gaffney and Jenkins 2010, p. 19)].

The prootic of LRF R450/LRF-TH 450 and LRF R451/LRF-TH 451 forms the entire posterior border of the foramen nervi trigemini (similar to *Meiolania*: see Gaffney 1983, p. 431, Fig. 47). It contacts the epipterygoid and parietal dorsally and participates in the thin anterior wall of the otic chamber, foramen stapedio temporale, cavum labyrinthicum, and cavum acustico-jugulare (Figs. 9.2c, 9.5a). Articulation with the processus interfenestralis of the opisthotic is weak and the fenestra ovalis is not floored.

The opisthotic (visible in LRF R450/LRF-TH 450 and LRF R451/LRF-TH 451: Figs. 9.2c, 9.3b, 9.5a, b) forms the posterolateral section of the otic chamber and contacts the exoccipital towards the dorsal edge of the foramen magnum. The opisthotic-supraoccipital suture runs anterolaterally across the roof of the otic chamber and reaches the foramen stapedio-temporale. *Spoochelys* lacks a bony floor to the cavum labyrinthicum. The opisthotic-quadrate suture of *Spoochelys* crosses the roof of the cavum acustico-jugulare anteromedially, potentially permitting contribution of the opisthotic to the aditus canalis stapedio temporalis (see also *Meiolania*: Gaffney 1983, p. 443).

The basisphenoid of LRF R451/LRF-TH 451 (Fig. 9.5b) is massive with the paired foramen caroticum basispenoidale positioned level with the front of the otic chamber (shallow grooves mark the path of the external carotid arteries). The basisphenoid-basioccipital contact is underlapped by twin triangular tubercles (see Fig. 9.3). Adjoining processes contribute to the medial margin of the cavum labyrinthicum. The dorsum sellae is subdivided by a midline crest extending ventrally to the rostrum basisphenoidale.

Hyolaryngeal skeleton: A component of the right stapes is preserved in LRF-TH 451 (Fig. 9.5b). The footplate is expanded but there is no bone below the columella auris, suggesting a cartilaginous termination.

The complete right cornu branchiale I of LRF R450/LRF-TH 450 is slender and mediolaterally curved (Fig. 9.2b) similar to the cornu branchiale I of *Testudo graeca* Thomson 1932 (Schumacher 1973, p.159, Fig. 38) and cornu branchiale II of *Meiolania* (Gaffney 1983, p. 424, Fig. 42).

Mandible: The mandible of LRF R450/LRF-TH 450 is exposed in lateral and ventromedial views (Fig. 9.2b). The dentary symphysis appears to be well fused and there is no trace of a suture. The splenial extends along the lower rim of the mandibular ramus as in *Proganochelys* (Gaffney 1990, p. 99, Fig. 61B) and *Kayentachelys* (Sterli and Joyce 2007, p. 690, Fig. 9B$_2$), and there is an oblique splenial-prearticular suture. The short dentary-splenial contact bridges the Meckelian sulcus and forms a medial wall to the foramen intermandibularis medius [a state also evident in *Meiolania* (see Gaffney 1983, p. 450, Fig. 61C)].

Carapace and plastron: The carapace of *Spoochelys* (Fig. 9.6a–c) is represented by anterior peripherals (AM F121579, AM F121580, AM F121581), costals (AM F121686, AM F121687) and other fragments (LRF 407, LRF 738, LRF 738a, LRF 738b, AM F121592) that are referred to this taxon based on morphological compatability with the holotype. As a composite these suggest a vaulted carapace roof with projected anterior margin. The first pleural sulcus is shallowly incised on the second peripheral, and the first peripheral seems to have projected lateral to the cervical scute as in *Notoemys zapatocaensis* Rueda and Gaffney 2005. The posterior sulcus of the first vertebral scute follows the rear margin of the first costals and is oriented perpendicular to the first pleural sulcus (based on LRF 343, LRF 344, and AM F121686). The position of the cervical scute adjoins a row of at least two elliptical supramarginal scutes (AM F121579, AM F121580, AM F121581), which are intersected by small marginals anteriorly. The medial supramarginal scute sulci are heavily incised and the supramarginals straddle the peripheral sutures.

A partial articulated plastron (LRF 1650, which is missing the anterior extremity: Fig. 9.8d) and other isolated elements (LRF 341, LRF 343, LRF 344, LRF 462) have been assigned to *Spoochelys* based on association with some peripherals and vertebrae. The external surfaces of all of these lack obvious scute sulci [a defining feature of *Meiolania* (see Gaffney 1996)] and are ornamented only by fine incisions and small fossae. The hyoplastron has weakly inflected axillary buttresses. The plastron-carapace connections appear to have been ligamentous but were supplemented by plastral "pegs" anteriorly. Part of the rear section of the entoplastron is preserved in LRF 1650 but it is not clear whether lateral mesoplastra were present.

Vertebrae: Examples of cervical vertebrae three through eight (LRF 464, LRF 944, AM F68254, AM F72276, AM F112750, AM F127972, AM F127984, AM F127990, AM F127996) are associated with other postcranial elements (see Fig. 9.7a–l). The centra are ventrally pinched and the transverse processes are positioned along the midline. The cervical articulation pattern is 2((3((4))5))6))7((8), with the fourth centrum being biconvex and keeled ventrally. The accompanying neural arch manifests a near-continuous postzygapophyseal surface and a prominent cavity incises the posterior face of the neural spine. The fifth and sixth cervical centra are also keeled but bear procoelous articular surfaces with small parapophyses on the lower edge of the cotyle. The seventh cervical is biconcave and the anteriorly projecting parapophyses are buttressed by robust laminae. The eighth cervical is biconvex and of subequal length to the preceding vertebrae. It has large parapophyses and a prominent neural pedicle for articulation with the nuchal [also present in *Otwayemys* (Gaffney et al. 1998; Hirayama et al. 2000)].

Some sacral vertebrae (LRF 019, AM F121603) are ascribed to *Spoochelys* because they articulate with proximal caudals. The sacral series comprises two primary vertebra fused to the ?tenth thoracic and first caudal centrum (although sacro-caudal ossification is not clear in AM F121603; Fig. 9.7m). The presence of an amalgamated sacro-caudal complex is shared with *Meiolania* (Gaffney 1996, p. 31) and some primitive testudinatans (Hoffstetter and Gasc 1969) including *Proterochersis robusta* Fraas 1913, p. 24, Fig. 5) and *Platychelys orbendorferi* Wagner 1853. The transverse processes of the first caudal in *Spoochelys* (LRF 735) are positioned high on the front of the centrum, and its posterior articular face forms a deep circular cotyle [see also *Meiolania* (Gaffney 1985, p. 18, Fig. 14)].

A number of opisthocoelous testudinatan caudals (LRF 016, LRF 407, AM F66772, AM F121638, AM F127978) have been found at Lightning Ridge (procoelous and biconvex examples are comparatively very rare) and are clearly compatable with the holotype (AM F121641). They manifest prominent haemal facets for the chevron bones, and in more distal vertebrae, the keels are variably fused to the centrum and extend posteroventrally behind the post-zygapophyses (Fig. 9.7n).

Pectoral girdle and forelimb: The scapulocoracoid complex of *Spoochelys* is triradiate; the holotype scapula (AMF121587: Fig. 9.8a) has a distinct spur on the elongate acromial process, which is also observable in *Meiolania* [AM F6110 (Gaffney 1996, p. 22, Fig. 17)]. The prominent acromial ridge imparts a triangular acromial cross-section similar to that reported in *Proganochelys*, *Kayentachelys*, *Mongolochelys*, and *Meiolania* (Joyce 2007). The scapular process is rod-like and medioventrally flexed as in *Proganochelys* and *Palaeochersis* (Lee 1996; Sterli et al. 2007). The glenoid neck is short and the scapular spine is separated from the acromial process by an angle of $\sim 120^{\circ}$ (consistent with a domed carapace [Walker 1973]). The coracoid shaft is very short and the coracoid flares broadly (LRF 081).

Some extremely robust humeri (LRF 078, AM F68310, AM F68321, AM F127940, AM F127956) are compatible with the holotype ulna (AM F121587). These exhibit a shallow intertubercular fossa and rounded head bordered by a proximally extensive lateral process (Fig. 9.8c, d). The distal extremity bears subspherical articular condyles that are ventrally directed. The ectepicondylar canal is closed and the ectepicondylar foramen opens just lateral to the capitellum.

The holotype (AM F121587) and referred (LRF 079, AM F121621, AM F127944) ulnae have a deep notch in the olecranon process (See Fig. 9.8b) and an accompanying dorsomedial trench on the shaft, implying tight cohesion with the radius and possible adaptation towards terrestrial locomotion (see Gaffney 1996, pp. 44–45).

Pelvic girdle and hindlimb: Fragmentary pelvic elements include ventral sections of several ilia (AM F66779, AM F121614, AM F121615, AM F121622, AM F127948, AM F127949), some pieces of femora (LRF 420, LRF 1216, AM F121617, AM F127943) and other distal hindlimb elements (AM F127958, AM F127961) have been attributed to *Spoochelys* because of association with shell material. Little is discernable about the pelvic structure, although a section of the femoral shaft indicates robustness.

The stout tibia (AM F127958: Fig. 9.8e, f) has a pronounced proximal expansion and adjacent dorsomedial protuberance on the shaft similar to *Notoemys laticentralis* and "pleurodires" (sensu Fernandez and Fuente 1994, p. 100). A medial prominence on the distal articulation suggests tight cohesion with the astragalus. The tarsals comprise a fused, block-like astragalocalcaneum with conjoined tibial and fibular articulation surfaces (AM F127961). The astragalar component is ventromedially expanded and hemispherical inferring a plantigrade stance (Gaffney 1996, pp. 253–254). The associated squat metatarsals (II–V) have heavy overlapping tubercles (as in *Meiolania*; Gaffney 1996, p. 63, Fig. 52) and cylindrical, non-divided distal articular ends (Fig. 9.8g, h). The distal phalanges (AM F121613) are very short with deeply concave proximal articulations.

Phylogenetic Relationships of *Spoochelys ormondea*

To investigate the relationships of *Spoochelys*, a series of phylogenetic analyses were conducted using revised versions of the two most comprehensive published Testudinatan data sets that incorporate a substantial number of meiolaniid taxa—Sterli and de la Fuente (2011) and Gaffney et al. (2007). Because these phylogenies use contrasting character interpretations and advocate markedly different topologies, each was tested separately using both Maximum Parsimony and Bayesian multivariate methods. Maximum Parsimony trees (MPTs) and bootstrap frequencies (1000 replicates with maxtrees set at 10×10^5 and >70% considered the cut-off measure for strong clade support [see Felsenstein 2004]) were tested using heuristic searches in PAUP*ver. 4.0b10 (Swofford 2002) with TBR (tree-bisection-reconnection) branch swapping and 100 random-addition replicates (maximum number of saved trees = 10×10^5). DELTRAN character state optimization was preferentially employed because it assumes character state asymmetry; this is pertinent when dealing with morphological data where a constant rate of change between states is unlikely (Angnarsson and Miller 2008). However, unequivocal synapomorphies were deemed to be those shared by both DELTRAN and ACCTRAN outputs. To assess the improvement in resolution if highly fragmentary taxa are not considered, "pruned" analyses excluding taxa with >50% missing data (see taxon lists in Appendix 3) were also undertaken using each data matrix. Branch (Bremer) support values were calculated with the aid of TreeRot.ver3 (Sorenson and Franzosa 2007), although, decay indices for the large data set of Sterli and de la Fuente (2011) were also sought by retaining trees with additional steps under enforced backbone conditions (see discussion in Worthy and Lee 2008).

Bayesian analyses were performed using MrBayes ver. 3.1 (Ronquist and Huelsenbeck 2003) with the obligate standard discrete model [$M_{K\text{-}V}$ (Lewis 2001)] for morphology, which assumes equal change between all character states. Characters were treated as unordered (default) following the parameters of the original analyses. In all cases the rates variation parameter was set to "gamma" (permitting hetrogeneity between characters) and coding to "variable" (Clarke and Middleton 2008). Two simultaneous runs and four Markov Chains (one cold and three heated using default heating values) were applied for 5×10^6 generations, with sampling every 1000th generation. The first 500 sampled trees were discarded from each run as burn-in. The remainder were used to construct a majority-rule consensus with Posterior Probability (PP) >0.95 deemed as strong, and PP = 0.90–0.95 as moderate clade support respectively (see Kolaczkowski and Thornton 2006). In order to test improvements of tree resolution, subsequent runs were also conducted with taxa incorporating >50% missing data excluded.

Parsimony Analysis of the Sterli and de la Fuente (2011) Data Matrix

The data set of Sterli and de la Fuente (2011) was modified from Joyce (2007) and incorporates taxon/character additions from Sterli (2008). Gaffney and Jenkins (2010, p. 359) also suggested that a number of undisclosed state changes could affect results from the original Joyce (2007) matrix and accordingly alter the relationships of meiolaniids vis-à-vis *Meiolania* in Joyce (2007), Sterli (2008), and other derivative studies such as Anquetin et al. (2009). However, only four characters (three scored by Joyce 2007) were discussed in detail and we found that none of these required recoding in Sterli and de la Fuente (2011). Nevertheless, we added *Spoochelys* and rescored a number of states for *Meiolania*, *Otwayemys*, and *Chubutemys* (characters 8, 13, 38, 41, 81, 82, 99, 104, 108, 110, 114, 119, 125, 128, 148, 151; see supplementary NEXUS file in Appendix 1) because the code lines of Sterli and de la Fuente (2011) conflicted with our observations from the original specimens and the literature (see Appendix 2 for a discussion of the individual characters).

Analysis of the revised 152 character matrix incorporating all taxa (rooted using "*Sphenodon_punctatus*" as the user-specified outgroup) and with all characters designated as unordered following Sterli and de la Fuente (2011), yielded trees (>10×10^5 MPTs of Length [L] = 493; Consistency Index [CI] = 0.3996, Rescaled Consistency Index [RCI] = 0.3135; see strict consensus tree in Fig. 9.9a) with weak bootstrap/Bremer support (<50/1) for all named high-level clades other than Testudinata (65/4), Pleurodira + *Platychelys* Wagner 1853 + *Caribemys* de la Fuente and Iturralde-Vinent 2001 + *Notoemys* Rueda and Gaffney 2005 (83/7), Meiolaniidae (83/1), Baenidae (91/4), Chelonioidea excluding *Santanachelys* Hirayama 1998 (97/4), Testudinoidea (67/4), and Trionychia (100/5). Deletion of taxa with >50% missing data improved some confidence intervals and resulted in a paraphyletic arrangement of basal taxa around *Spoochelys* (72826 MPTs; L = 452; CI = 0.4314, RCI = 0.3399; Fig. 9.9b): bootstrap/Bremer analayses returned Testudinata excluding *Proganochelys* (74/3), Pleurodira (100/26), Paracryptodira (57/3), Baenidae (52/3), Chelonioidea excluding *Santanachelys* (100/5), Kinosternoidea (86/3), and Trionychia (100/25).

Our topologies consistently placed *Spoochelys* as a primitive testudinatan, excluded from crown Testudinata (Testudines sensu Joyce 2007; the following subscript numbers equate to characters listed in Appendix 1) plus all other higher taxa on the basis of its incipient antrum postoticum$_{38}$, absence of a distinct processus trochlearis oticum$_{40}$, presence of an interpterygoid vacuity$_{44}$, poorly developed crista supraoccipitalis$_{52}$, and flat rostrum basisphenoidale$_{71}$. Close affinity between *Spoochelys* and *Meiolania* was not advocated by our phylogenies. Rather, meiolaniids formed a sister lineage to Testudines (sensu Joyce 2007) united by a procoelous articulation between cervicals VII and VIII$_{125}$ ("?" in *Niolamia* Ameghino 1899); this state is secondarily modified in many pleurodires and eucryptodires (Joyce 2007; Sterli and de la Fuente 2011). Monophyly of *Meiolania* + *Niolamia* was resolved on the symplesiomorphic presence of a heavily sculptured prefrontal region with prominences and bosses$_8$ and the presence of a tail club$_{125}$, together with the derived development of an interpterygoid slit. Notably, these taxa failed to nest with *Mongolochelys*, *Otwayemys*, and *Patagoniaemys* (sensu Sterli and de la Fuente 2011). Nevertheless, *Otwayemys* + *Patagoniaemys* did form as sister taxa in 80% of the returned MPTs based on their derived contact of the first dorsal rib with the well developed anterior bridge buttresses$_{127}$ (although this state is unknown in *Chubutemys*, *Niolamia*, and *Spoochelys*). Further sampling from the first 33583 MPTs collapsed during initial runs of the full Sterli and de la Fuente (2011) data set showed that *Mongolochelys* + *Otwayemys* + *Patagoniaemys* + Meiolaniidae, or *Chubutemys* + *Mongolochelys* + *Otwayemys* + *Spoochelys* + *Patagoniaemys* + Meiolaniidae could alternatively be monophyletic with only three additional steps (L = 496; CI = 0.3972, RCI = 0.3107 under constraint conditions). The character states supporting these clades were somewhat ambiguous: *Mongolochelys* + *Otwayemys* + *Patagoniaemys* + Meiolaniidae united by the presence of a squamosal-supraoccipital contact$_{22}$, exposure of the supraoccipital on the dorsal skull roof$_{53}$, elongate epiplastra extensively contacting the hyoplastra posteromedially$_{97}$ (all known only in *Mongolochelys* and *Meiolania*), and retention of the central plastral fontanelle into

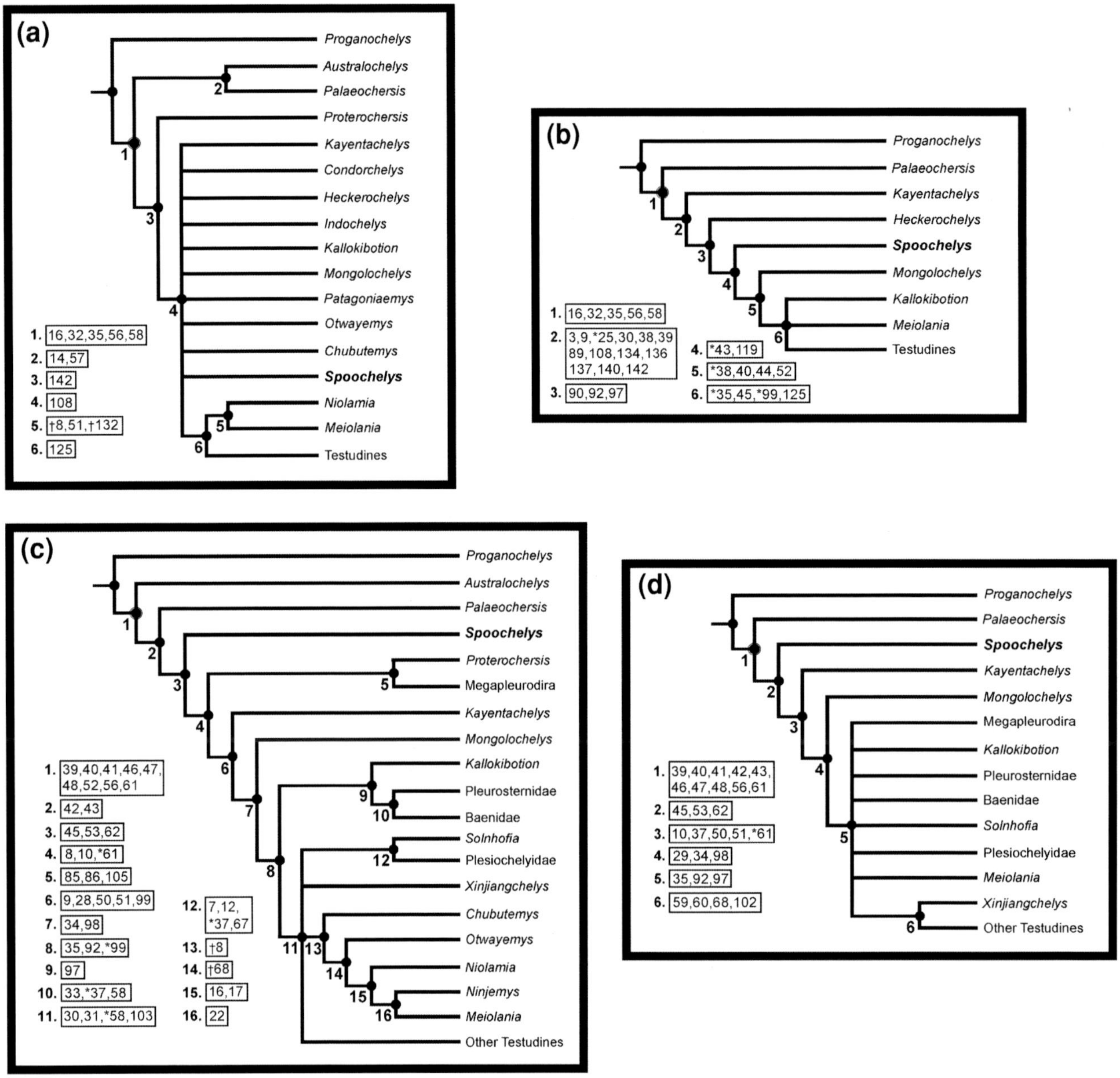

Fig. 9.9 Maximum Parsimony phylogenies of turtles incorporating *Spoochelys ormondea* gen. et sp. nov. and rescored data for *Meiolania* and some other taxa (see Appendices 1 and 2). Strict consensus trees were derived using the Joyce (2007) matrix/parameters as modified by Sterli and de la Fuente (2011: **a**, **b**), or alternatively, that of Gaffney et al. (2007: **c**, **d**). Results from analyses that excluded taxa with >50% missing data are depicted in **b** and **d**. *Boxed numbers* represent character states at selected nodes (•); *state code 2; † plesiomorphic. Bootstrap/Bremer support values for major clades are given in the text

adulthood$_{90}$ ("?" in *Niolamia* but widely manifest in Eucryptodires); incorporation of *Spoochelys* and *Chubutemys* was resolvable on the basis of a primitive frontal contribution to the orbit$_{10}$ ("?" in *Otwayemys* + *Patagoniaemys* + *Niolamia*), and an apomorphic procoelous articulation between cervicals II and III$_{119}$ ("?" in *Niolamia* but present throughout Testudines) respectively.

Forcing these taxa within Eucryptodira (recreating the higher-level topology of Gaffney et al. 2007) required trees to be 40 steps longer (non-parametric test in PAUP $p < 0.0001$) and relied solely upon derived traits (elongate epiplastra extensively contacting the hyoplastra posteromedially$_{97}$, procoelous articulation between cervicals II and III$_{119}$), which have been independently acquired numerous times throughout Testudinata (Joyce 2007; Sterli and de la Fuente 2011). Close homology between *Spoochelys* + meiolaniids, or affinity with eucryptodires was therefore not supported by our version of the Sterli and de la Fuente (2011) data set, even when constraint conditions were implemented.

Parsimony Analysis of the Gaffney et al. (2007) Data Matrix

Spoochelys was included in this 111 character matrix to further examine its potential relationships with meiolaniids and Eucryptodira. Code lines for *Otwayemys*, *Chubutemys*, *Meiolania*, *Niolamia* and *Ninjemys* Gaffney 1992 had to be rescored for this analysis (specifically characters 3,10, 20, 26–31, 33, 34, 36–38, 45–51, 53, 56, 74, 92, 106–109; see supplementary NEXUS file in Appendix 1) because of conflict with our interpretations of anatomy and state distributions (see specific comments in Appendix 2). Successive runs of the data with all characters unordered and the combined higher-taxon "Synapsida/Diapsida" designated as the user-defined out-group (sensu Gaffney et al. [2007]; 4 MPTs of L = 241; CI = 0.5212, RCI = 0.3829) resolved *Spoochelys* outside of Testudines (sensu Joyce 2007) as a stem turtle (see strict consensus tree in Fig. 9.9c) but otherwise yielded a comparable ingroup topology to that of Gaffney et al. (2007). "Pruning"of taxa with >50% missing data (3 MPTs; L = 233; CI = 0.5494, RCI = 0.3826) destabilised the ingroup structure but retained *Spoochelys* interopolated between basal Testudinata (*Proganochelys* and *Palaeochersis*, sensu Joyce et al. 2004) and *Kayentachelys* + all more derived turtles (Fig. 9.9d). Conversely, constraining *Spoochelys* + *Mongolochelys* + *Otwayemys* + *Chubutemys* + Meiolaniidae to form a monophyletic group outside of Eucryptodira (sensu Sterli and de la Fuente 2011) yielded trees that were only three steps longer and had no significant difference (4 MPTs; L = 244; CI = 0.5246, RCI = 0.3743, non-parametric test in PAUP p > 0.5127). Bootstrap/Bremer support values were accordingly low (<50/1) at nodes other than *Australochelys* + other Testudinata (95/4); *Palaeochersis* + other Testudinata (82/2); *Palaeochersis* + *Spoochelys* + *Kayentachelys* + *Mongolochelys* + *Kallokibotion* + Testudines (85/2); *Proterochersis* + Megapleurodira (63/1); Pleurosternidae + Baenidae (54/2); Meiolaniidae (96/2); *Meiolania* + *Ninjemys* (74/1); Cryptodira (55/1); and Trionychoidea + Testudinoidea (88/4). These results showed some refinement once highly fragmentary taxa (missing >50% data) were removed: *Palaeochersis* + other Testudinata (100/10); *Spoochelys* + other Testudinata (87/2); *Kayentachelys* + *Mongolochelys* + *Kallokibotion* + Testudines (74/2); *Xinjiangchelys* Yeh 1986 + *Sinemys* Wiman 1930 + *Hangaiemys* Sukhanov and Narmandankh 1974 + *Dracochelys* Gaffney and Ye 1992 + *Ordosemys* Brinkman and Peng 1993 + *Judithemys* Parham and Hutchinson 2003 + Cryptodira (51/1); *Sinemys* + *Hangaiemys* + *Dracochelys* + *Ordosemys* + *Judithemys* + Cryptodira (71/3); Cryptodira (56/2); and Trionychoidea + Testudinoidea (89/3).

Our analyses differentiated *Spoochelys* from basal testudinatans including *Proganochelys* and *Palaeochersis* (the following bold subscript numbers refer to characters used in Gaffney et al. 2007) by the presence of an antrum postoticum$_{\mathbf{45}}$, loss of the wide occipital plate$_{\mathbf{53}}$, and development of a thin columella auris with a distinct footplate$_{\mathbf{62}}$. Exclusion of *Spoochelys* from Testudines (sensu Joyce 2007) was advocated by its plesiomorphic prefrontal-postorbital contact$_{\mathbf{8}}$, a poorly developed processus inferior parietalis$_{\mathbf{10}}$, and the presence of paired ventral basisphenoid-basioccipital tubercles$_{\mathbf{61}}$. Interestingly, this topology implies independent acquisition of traits that represented synapomorphies for *Spoochelys* + *Otwayemys* + *Chubutemys* + Meiolaniidae (incisura columellae auris confluent with the Eustachian tube$_{\mathbf{44}}$, central articulations fully formed$_{\mathbf{65}}$, biconvex fourth cervical centrum$_{\mathbf{70}}$, and the possession of a ligamentous attachment between the carapace and plastron$_{\mathbf{96}}$) or *Spoochelys* + *Otwayemys* (ventral process present on the posterior cervical centra$_{\mathbf{74}}$) when constraint conditions were enforced. *Chubutemys* + *Otwayemys* + Meiolaniidae were alternatively united by a single symplesiomorphy (presence of a prefrontal-postorbital contact$_{\mathbf{8}}$) in the tree structure envisaged by Gaffney (1983, 1996), Gaffney et al. (1998), (2007). Placement of these taxa within Eucryptodira was resolved on the basis of partial coverage of the canalis caroticus internus by the basisphenoid-pterygoid contact (unknown in *Otwayemys* and *Ninjemys*)$_{\mathbf{30}}$, formation of the canalis caroticus internus by the pterygoid distal to the foramen basisphenoidale (unknown in *Otwayemys* and *Ninjemys*)$_{\mathbf{31}}$, foramen basisphenoidale formed by both the basisphenoid and pterygoid with its positioning anterior to the foramen posterius canalis carotici interni (unknown in *Otwayemys* and *Ninjemys*)$_{\mathbf{89}}$, and the absence of mesoplastra (recorded only in *Otwayemys* and *Meiolania*)$_{\mathbf{103}}$. In light of these results, we cannot unequivocally establish a direct relationship between *Spoochelys* and Meiolaniidae using parsimony analyses of the Gaffney et al. (2007) data matrix, nor can we advocate definitive assignment of *Chubutemys* + *Otwayemys* + Meiolaniidae to Eucryptodira given our character state rescorings.

Bayesian Analyses of Both Data Matrices

Topologies produced by the Bayesian analyses were broadly comparable to those of the Maximum Parsimony trees for both revised data sets (see Fig. 9.10). However, resolution at most nodes was weak and only a few high-level clades returned significant support. Sterli and de la Fuente (2011) robustly derived Testudinata (PP = 1.00), *Palaeochersis* + all higher Testudinata (PP = 0.99),

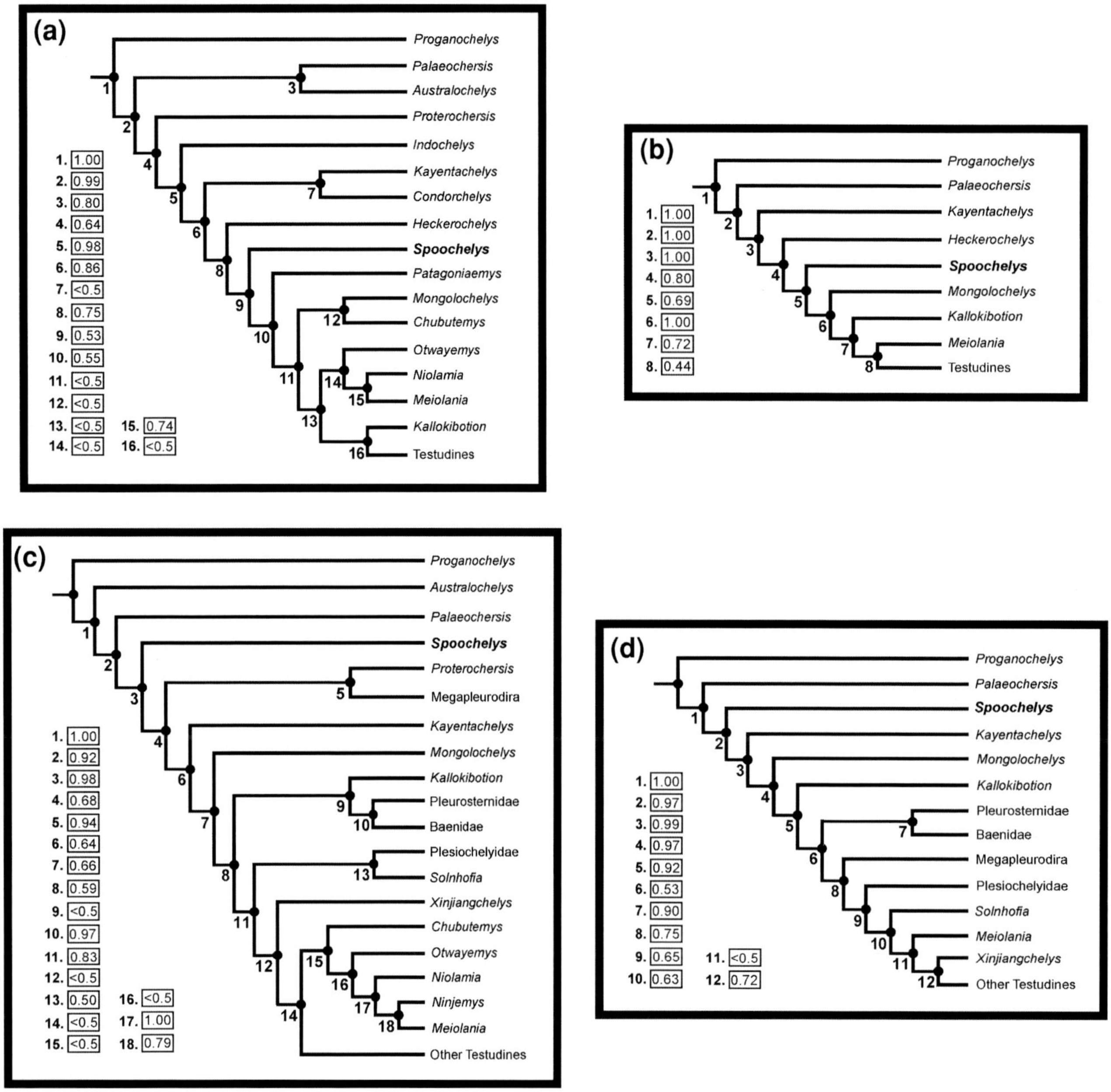

Fig. 9.10 Bayesian phylogenies of turtles incorporating *Spoochelys ormondea* gen. et sp. nov. and rescored data for *Meiolania* and some other taxa (see Appendices 1 and 2). Trees were derived using the Joyce (2007) matrix/parameters as modified by Sterli and de la Fuente (2011: **a**, **b**), or alternatively, that of Gaffney et al. (2007: **c**, **d**). Results from analyses that excluded taxa with >50% missing data are depicted in **b** and **d**. *Boxed numbers* represent posterior probabilities at selected nodes (•)

Indochelys Datta et al. 2000 + all higher Testudinata (PP = 0.98), *Platychelys* + *Notoemys* + *Caribemys* + Pleurodira (PP = 1.00), Baenidae (PP = 0.90), Chelonioidea excluding *Santanachelys* (PP = 1.00), and Trionychia (PP = 0.99); whereas Gaffney et al. (2007) yielded *Palaeochersis* + all higher Testudinata (PP = 1.00), *Spoochelys* + all higher Testudinata (PP = 0.92), *Kayentachelys* + all higher Testudinata (PP = 0.98), *Proterochersis* + Megapleurodira (PP = 0.94), Paracryptodira (PP = 0.97), Meiolaniidae (PP = 1.00), Eucryptodira excluding Plesiochelyidae + *Solnhofia* + *Xinjiangchelys* (PP = 0.91), *Judithemys* + Chelonioidea + Chelydridae + Chelomacryptodira (PP = 0.94), Chelonioidea + Chelydridae + Chelomacryptodira (PP = 0.99), Chelydridae + Chelomacryptodira (PP = 0.95), and Chelomacryptodira (PP = 1.00). Pruning of taxa with >50% missing data did not substantially improve resolution at these nodes: Sterli and de la Fuente (2011) supporting Testudinata (PP = 1.00), *Palaeochersis* + all higher Testudinata (PP = 1.00), *Kayentachelys* + all higher Testudinata (PP = 1.00),

Mongolochelys + all higher Testudinata (PP = 1.00), Testudines (PP = 1.00), Pleurodira (PP = 1.00), Paracryptodira (PP = 0.92), Baenidae (PP = 0.90), Chelonioidea excluding *Santanachelys* (PP = 1.00), and Trionychia (PP = 1.00); and Gaffney et al. (2007) providing *Palaeochersis* + all higher Testudinata (PP = 1.00), *Spoochelys* + all higher Testudinata (PP = 0.97), *Kayentachelys* + all higher Testudinata (PP = 0.99), *Mongolochelys* + all higher Testudinata (PP = 0.92), Paracryptodira (PP = 0.90), Eucryptodira excluding Plesiochelyidae + *Solnhofia* + *Meiolania* + *Xinjiangchelys* (PP = 0.99), Chelonioidea + Chelydridae + Chelomacryptodira (PP = 0.98), Chelydridae + Chelomacryptodira (PP = 0.94), and Chelomacryptodira (PP = 1.00). However, slight alterations in topology did occur (e.g., deletion of *Proterochersis* prompted Megapleurodira to shift within *Kallokibotion* + other Testudines when using the Gaffney et al. (2007) data set; PP = 0.75), indicating that uncertainty within the tree structure could be influenced by the exclusion of fragmentary taxa.

Our Bayesian analyses universally placed *Spoochelys* as a basal testudinatan (more derived than *Proganochelys* + *Austrochelys* + *Palaeochersis*: PP = 0.92–1.00), but excluded it from *Mongolochelys* + *Meiolania* + Testudines (PP = 0.92, taxa missing 50% data deleted) or inconclusively from higher turtles (PP = 0.75, all taxa included) with Sterli and de la Fuente (2011). Gaffney et al. (2007) alternatively returned *Spoochelys* outside of Testudines (PP = 0.98, all taxa included) or *Kayentachelys* + *Mongolochelys* + *Kallokibotion* + Testudines (PP = 0.99, taxa missing 50% data deleted). Monophyly of *Chubutemys* + *Otwayemys* + Meiolaniidae was not substantiated, the lineage either being paraphyletic in runs of Sterli and de la Fuente (2011) or garnering no support (PP ≥ 0.50) from the original taxon suite of Gaffney et al. (2007). Conversely, Meiolaniidae showed strong support (PP = 1.00). Interpolation of Meiolaniidae within Eucryptodira was not advocated by our analyses; although, resolution amongst basal testudines/eucryptodires was weak (PP ≤ 0.50–0.83) and thus we cannot unanimously reject this topology when using the Gaffney et al. (2007) data set.

Conclusions

Spoochelys ormondea gen. et sp. nov, was a small-bodied (ca. 400–500 mm total length), surprisingly archaic turtle that is represented by skull remains and attributed postcranial elements from the Lower Cretaceous (Albian) Griman Creek Formation of Lightning Ridge in southeastern Australia. Phylogenetic analyses of these specimens using rescored versions of the most comprehensive published data sets of fossil Testudinata infer that *Spoochelys* forms part of the primitive stem turtle radiation, but otherwise the taxon cannot be unequivocally placed within a discrete clade. Nonetheless, *Spoochelys* shares some key derived features with meiolaniids (cranial scute patterning and confluence of the incisura columellae auris with the Eustachian tube), a controversial group of Gondwanan non-marine turtles that have been classified as either basal testudinatans (Joyce 2007) or eucryptodires (Gaffney 1996; Gaffney et al. 1998, 2007; Hirayama et al. 2000; Lapparent de Broin and Molnar 2001). Paradoxically, our results demonstrate that neither hypothesis is robustly supported by revisions of the competing data sets. In accordance we defer definitive placement of *Spoochelys* with a recommendation that the taxon be regarded as Testudinata *incertae sedis*, pending discovery of more complete material and/or the re-evaluation of existing specimens using revised character matrices.

Spoochelys was probably a terrestrial turtle as suggested by its apparently vaulted carapace, closely articulating radius and ulna, plantar expansion of the astragalocalcaneum [inferring a plantigrade stance (Gaffney 1990)], and short metacarpals/metatarsals (see definition for the "terrestrial field" in Joyce and Gauthier 2004). The attributed opisthocoelous caudal vertebrae with their angled centrum articulations and massive transverse processes also imply substantial dorsolateralis muscle mass and a propensity for powerful tail movements. Whether this imparted some swimming ability is unknown, although it is tempting to speculate (based on the progressive enlargement of the haemal keels) that a 'club-like' caudal ossification might have been present [vis-à-vis *Meiolania* (see Gaffney 1985)]. Lateral retraction of the neck also appears to have been feasible as indicated by the dorsoventrally tall, concavo-convex cervical articulations and chelid-like neural arch specializations. The large size, shape and position of the transverse processes may be associated with enlargement of the longissimus musculature, particularly the *m. testocervicalis lateralis*. This is comparable to modern pleurodires (Hoffstetter and Gasc 1969), and might reflect a predatory lateral "strike" capacity (see Hoffstetter and Gasc 1969; Pritchard 1984), adaptation for cropping vegetation (if *Spoochelys* was herbivorous), or defensive mechanism for lateral neck retraction (perhaps implied by the rounded anterior extension of the carapace).

The high-latitude, cool temperate (mean sea surface temperature ~16°C [Stevens and Clayton 1971; Dettmann et al. 1992]) palaeoenvironment envisaged for southern Australia during the middle-late Albian is atypical for turtles, whose distribution is climatically restricted particularly in relation to egg incubation and hatchling survival (Mrosovsky 1980). Indeed, the occurrence of fossil chelids at Lightning Ridge, a group that today occupies seasonally low temperature habitats (see Smith 2010 and references therein), could infer an analogous specialization in *Spoochelys*. Irrespectively, ectothermic reptiles are generally excluded from polar regions (Tarduno et al. 1998), and accordingly the Australian Early Cretaceous biota is thought to have been thermally isolated with limited faunal/floral interchange from Antarctica/South America prior to climatic warming in the Cenomanian-Turonian (Archer et al. 1985; Dettmann et al. 1992; Flannery et al. 1995; Molnar 1999; Henderson et al. 2000; Clemens et al. 2003; Thulborn and Turner 2003; Rich and Vickers-Rich 2003; Musser 2005; Kear et al. 2006a, 2006b; Rowe et al. 2008; Rich et al. 2009). Given its remarkably primitive skeletal anatomy, *Spoochelys* might therefore represent a relic taxon whose Pangean ancestry probably extended well back into the Jurassic or even earlier.

Acknowledgments Our sincere thanks to Don Brinkman for his endless patience and editorial help. Walter Joyce, Juliana Sterli and Igor Danilov provided comprehensive reviews. Henk Godthelp, Michael Archer, Suzanne Hand, Gene Gaffney, Tom Rich, Patricia Vickers-Rich, Lesley Kool, Karen Black, Anna Gillespie, Robin Beck, P. M. Datta, Asish Kumar Ray, Subhash Sen, Mike Lee, Arthur White, Robert Jones, Alex Ritchie, Robert Smith, Jenni Brammall, Dave Barclay, Brett and Peter Barton, Stefan Bedenikovic, Vicki Bokros, Rob and Debbie Brogan, Paul Burza, Peter Drackett, Jack Fahey, Dave Galman, Matthew Goodwin, Bill Kotru, Ormie and Donella Molyneux, Marcel and Sam Miltenburg, George and Bill Mulder, Dave Roussell and Lalja Pedersson, Graeme and Christine Thomson, Steve Turner, Joe Walker and Larry White provided discussions, preparation, access to specimens, photographs and graphical work. Financial support for this research was provided by The Royal Tyrrell Museum of Paleontology for ETS, and The Australian Research Council and Uppsala University for BPK.

Appendices

Appendix 1

Revised character scores (Nexus format) used for phylogenetic analyses. State codes for *Spoochelys ormondea* gen. et sp. nov. were based on specimens described herein. Modifications made for other taxa were drawn from published descriptions and direct examination of the following specimens: *Chubutemys copelloi*, MPEF-PV1236; Gaffney et al. (2007); *Otwayemys cunicularius*, NMV P186116; Gaffney et al. (1998); *Meiolania platyceps*, AM F208a, AM F208b, AM F401, AM F43183, AM F49141, AM F57984, AM F61110, AM F18314, AM F18364, AM F18668, AM F18775, AM F81940, AM F81949, AM F81965, AM F82180; Gaffney (1983, 1996); Joyce (2007); *Ninjemys oweni*, NHMUK R391 (cast); Gaffney (1992, 1996); *Niolamia argentina*, AMF L1418 (cast); Gaffney (1996); Sterli and de la Fuente (2011).

(A) Revised Matrix from Sterli and de la Fuente (2011)

Spoochelys_ormondea
???????1?0010001000000?1?????????01201301120 0??
0??00001101?00?10?0??0011(01)010???????01???1??????
0?0??????????????????0100001110 1????0?111?100???????
?????0

Chubutemys_copelloi
???100011001100?000?00?1?????1110?120211?121000
00001??1101200?????????110?1?????0????????????????
??????????????????1?????01???0???11?1?11??????????
??

Otwayemys_cunicularius
??????????????????????????????????2?2?1??????????
????????????????0???001?0???0?0?2?0?1101???0??2000
0000010001?1?00?1????? 011??00?1?1????????????????

Meiolania_platyceps
000000000100100010001 0101(02)0000101002200311 12
11000(01)01110210220011 0(01)2?0001100100?10??00210
?11010100102000 0?00000????00001001111 10(01)?0001
111100101–01?10010

(B) Revised matrix from Gaffney et al. (2007)

Spoochelys_ormondea
??????0?00010000010??0?????000?0000001111111 11
1?00010?1?00011001 2?001000102???00??0??????1????10
??????????????

Chubutemys_copelloi
?0??11?0??0010000???0001110?111101?01011111?11
11???11011?21021?012?1????0??2????????????0?????2???
??????????????

Otwayemys_cunicularius
???????????????00????0?????????????????????1???1?????
??????????????012?00?0?0102???00???0?00?????21?01200
010001??00

Meiolania_platyceps
0101110010001101110001011100111?0110011111110
111011111110220 2?0011000100000111000110000011 0??
2110110001000?????

Ninjemys_oweni
??0111?????01?0111??01011??????????????111111?111
???1?1?1????????????????????11??????????????????????
???????????

Niolamia_argentina
010?1100??001?0111??000111??111001?0011111 11??
???1111111?2102????????????????1?????????????????????
?????????????

Appendix 2

Notes on descriptions and scores for select characters used in phylogenetic analyses.

(A) Characters from Sterli and de la Fuente (2011)

Character 8. [Prefrontal E] Prefrontals heavily sculptured with prominences and bosses (Rougier et al. 1995): (0) = present; (1) = absent. Some specimens of *Meiolania* (AM F18668) display a pronounced boss in the prefrontal region (Gaffney 1983, p. 405) and a similar structure is observable in *Ninjemys* (see Gaffney 1996, p. 79, Fig. 66). The scoring for *Meiolania* is therefore changed to "0".

Character 12. [Parietal B] Parietal contact with the pterygoid, epipterygoid, or palatine: (0) absent, trigeminal foramen only developed as a notch; (1) present, trigeminal foramen clearly formed by the descending process of the parietal. Note that the parietal contacts the epipterygoid in *Spoochelys* (see Fig. 9.6b) and *Meiolania* (Gaffney 1983, p. 419) but is excluded form the trigeminal foramen by an epipterygoid-prootic contact.

Character 13. [Parietal C] Length of anterior extension of the lateral braincase wall: (0) short, inferior parietal process only produces a narrow strut anterior to trigeminal foramen, palatine contact absent; (1) elongate, inferior parietal process produces an extended process anterior to trigeminal foramen, palatine contact commonly present. Recoded from "−" to "0" for *Meiolania* following Joyce (2007).

Character 36. [Quadrate B + C of Joyce (2007)]. Development of the cavum tympani: (0) shallow but not developed anteroposteriorly; (1) shallow but anteroposteriorly developed; (2) deep and anteroposteriorly developed. Gaffney et al. (1998, p. 16) describe the cavum tympani of *Otwayemys* as "elongate, deep and smoothly continuous"; score changed to "2".

Character 38. [Quadrate E] Antrum postoticum: (0) absent; (1) incipient; (2) fully developed. The antrum postoticum is commonly scored as present in *Meiolania* (Joyce 2007; Gaffney et al. 2007; Anquetin et al. 2009; Sterli 2008; Sterli and de la Fuente 2011); however (Gaffney 1983, pp. 416–417) reported that there was no discrete chamber corresponding to the antrum postoticum in this taxon (score changed to "0"), and indeed the squamosal and opisthotic are exposed posteriorly within the cavum tympani (see Gaffney 1983, pp. 408 and 443 respectively).

Character 41. [Epipterygoid A] Epipterygoid: (0) present, rod-like; (1) present, laminar; (2) absent. The epipterygoid of *Meiolania* lacks the free rod-like dorsal portion evident in *Kayentachelys* (see Sterli and Joyce 2007) and alternatively forms a broad sutural contact with the processus parietalis inferior (Gaffney 1983, Sterli and Joyce 2007). Because of this difference we advocate recoding of *Meiolania* from "0" to "1".

Character 44. [Pterygoid C] Interpterygoid vacuity: (0) present; (1) absent. See discussion in Appendix 2, part B, character 29.

Character 51. [Pterygoid J] Intrapterygoid slit: (0) absent; (1) present. Gaffney (1983, 1996) and Gaffney et al. (2007) hypothesised that the 'intrapterygoid slit' is secondarily developed from the closed basicranial condition of eucryptodires. Sterli et al. (2010, 1344) alternatively suggested that this opening could be a modified interpterygoid vacuity. This interpretation requires confirmation via comprehensive revision of all meiolaniid taxa but would create interdependence with character 44 (presence of the interpterygoid vacuity).

Character 66. [Canalis caroticum B] Foramen caroticum laterale: (0) the arteria palatina enters the skull through the interpterygoid vacuity; (1) foramen caroticum laterale located between the basisphenoid and the pterygoid with the foramen posterius canalis carotici interni formed by the basisphenoid; (2) the canalis carotici interni is not completely floored allowing to see the foramen caroticum laterale and the foramen basisphenoidale in ventral view, foramen posterius canalis carotici interni is formed by the pterygoid or the pterygoid and the basisphenoid; (3) the arteria palatina branches from the arteria carotici interna inside the skull, so the foramen caroticum laterale is formed by the pterygoid as the foramen posterius canalis carotici interni; (4) the arteria palatina branches from the arteria carotici interna inside the skull, foramen posterius canalis carotici interni is formed by the pterygoid and the prootic; (5) foramen caroticum laterale absent; (6) foramen caroticum laterale formed within the pterygoid, posterior canalis carotici interni formed midway between pterygoid and basisphenoid. Gaffney and Jenkins (2010, pp. 335–336) specified terminology for the various basipterygoid foramina, indicating that the foramen caroticum basisphenoidale are those openings that pierce the body of the basisphenoid. In contrast, Joyce (2007) and the later character sets of Sterli (2008) and Sterli and de la Fuente (2011) use the term foramen posterius canalis carotici interni (which should be in part formed by the pterygoid: sensu Gaffney and Jenkins 2010, p. 336) to define the entry point of the internal carotid artery. We follow Gaffney and Jenkins (2010) in describing the paired foramina in the basisphenoid of *Spoochelys* as the caroticum basisphenoidale, and note that the presence of an interpterygoid vacuity together with the lack of any accessory foramina along the basisphenoid-pterygoid suture implies state "0"—entry of the arteria palatina into the skull via the interpterygoid vacuity.

Character 73. [Carapace A] Carapacial scutes: (0) present; (1) partially present; (2) absent. The carapace of *Spoochelys* is fragmentary but scute sulci are present on all of the preserved sections. Because the full extent of the scute coverage is unknown we have scored this taxon as "0/1".

Character 81. [Costal A] Medial contact of costal I: (0) absent; (1) present. Disposition of the intercostal sutures is unknown in *Meiolania* (Gaffney 1996, p. 24) thus scoring is changed from "0" to "?".

Character 82. [Costal B] Medial contact of posterior costals: (0) absent; (1) medial contact of up to three posterior costals present; (2) medial contact of all costals present. Disposition of the intercostal sutures is unknown in *Meiolania* (Gaffney 1996, p. 24) thus scoring is changed from "0" to "?".

Character 99. [Mesoplastron A] Number and size of mesoplastra: (0) one or two pairs of mesoplastra present that fully hinder any contact between hyoplastra and hypoplastra; (1) one reduced pair of mesoplastra present that allows partial contact between the hyoplastra and hypoplastra; (2) mesoplastra absent. Mesoplastra are generally thought to be absent in *Meiolania* (recoded "2"), which has a wide transverse contact between the hypo- and hyoplastra (Megirian 1992, p. 102; Gaffney 1996, p. 38).

Character 104. [Plastral scutes B] Midline sulcus: (0) straight; (1) distinctly sinuous. *Meiolania* is recoded from "0" to "?" because scute margins are unclear (Gaffney et al. 1996, p. 33).

Character 108. [Extragular C] Anterior plastral tuberosities: (0) present; (1) absent. Gaffney (1996, p. 33) described a pair of gular projections on the anterior lobe of the plastron in *Meiolania* and noted their smaller size but overall similarity to those of *Proganochelys* (score changed to "0"). Coding is altered from "0" to "1" for *Otwayemys*, which possesses extragular scales but apparently not plastral tuberosities (Gaffney et al. 1998, p. 14, Fig. 10).

Character 110. [Humeral A] Plastral scale set 3—humerals: (0) one pair present; (1) two pairs present, subdivided by plastral hinge. *Meiolania* is recoded from "0" to "?" because scute margins are unclear (Gaffney et al. 1996, p. 33).

Character 114. [Inframarginal A] Inframarginal scutes: (0) more than two pairs present, plastral scutes do not contact marginals; (1) two pairs present (axillaries and inguinals), limited contact between plastral scutes and marginals present; (2) absent, unrestricted contact between plastral scutes and marginals present. Gaffney et al. (1998, p. 15, Fig. 11) reconstructed a series of axillary, inguinal, and one set of inframarginals scutes in *Otwayemys* (score changed from "?" to "1").

Character 119. [Cervical articulation A] Cervical articulation: (0) articulations not formed; (1) articulations formed. Gaffney et al. (2007, p. 21) reported that at least cervical eight in *Chubutemys* has fully formed articulations (score changed to "1").

Character 125. [Cervical articulation G] Cervical articulation between VII and VIII: (0) 7(8; (1) 7)8. Gaffney et al. (2007, p. 21) stated that cervical eight in *Chubutemys* has biconvex articulations (score changed to "0").

Character 128. [Dorsal rib B] Contact of dorsal ribs IX and X with costals: (0) present; (1) absent. Gaffney (1996, pp. 25–26) considered it likely that the tenth dorsal rib did not articulate with a costal plate in *Meiolania*. We have therefore recoded this taxon as "0/1" to acknowledge manifestation of an intermediary condition.

Character 148. [Manus A] Phalangeal formula of manus and pes: (0) most digits with three elongate phalanges; (1) most digits with two shortened phalanges. *Meiolania* has only two short phalanges in the manus and pes (see Gaffney 1996, p. 50, Figs. 39 and 60); character recoded "1".

Character 151. [Pes A] Claw of fifth digit: (0) present; (1) absent. Gaffney (1996, p. 60) suggested that the fifth pedal digit was absent in *Meiolania* because metatarsal V lacks a distal articular surface (recoded "1").

(B) Characters from Gaffney et al. (2007)

Character 3. [NA] Nares divided: (0) yes; (1) no. Nares are divided by the premaxilla in *Niolamia* following Sterli and de la Fuente (2011) thus the score is changed from "1" to "0". Structure of the narial aperture is indeterminate in *Chubutemys*, which is recoded "?". The nares are clearly divided in *Meiolania*, although not completely in all specimens (Gaffney 1983, p. 401), scores are altered from "0/1" to "0".

Character 4. [NA] Nasomaxillary sinus: (0) absent; (1) present. *Proganochelys* and meiolaniids share a trough-like cavity in the anterodorsal surface of the maxilla. In *Proganochelys* this extends through the lacrimal bone as the lacrimal duct, which is floored by the maxilla (Gaffney 1990, p. 49, Fig. 40). Conversely in *Meiolania*, the maxillary trough is called the "nasomaxillary sinus" and leads into a blind canal roofed by the nasal (Gaffney 1983, p. 429, Fig. 45). The anterior section of the "nasomaxillary sinus" in *Meiolania* and *Ninjemys* could represent a component of the primitive lacrimal duct. However, we have not changed the state coding because homology is unclear.

Character 10. [PA] Processus inferior parietalis: (0) small to absent; (1) large. The processus inferior parietalis is weakly developed and the epipterygoid extends further into the lower temporal fossa than the relatively small descending process of the parietal in *Spoochelys* and *Meiolania* (see Gaffney 1983, pp. 419 and 431, Fig. 47). The processus inferior parietalis is also described as "small" in *Kayentachelys* (Gaffney and Jenkins 2010, p. 339). Score changed from "1" to "0" for *Meiolania*.

Character 18. [SQ] Broad sq-qj contact: (0) absent, narrow; (1) broad contact below cavum tympani. Gaffney et al. (2007) assigned *Niolamia* state "1" despite much of this part of the skull being morphologically inconclusive (Woodward 1901; Gaffney 1996). The sq-qj contact below the cavum tympani is only definitively known in *Meiolania* and is inferred for *Ninjemys* and *Warkalania* on the basis of a 'bony sheet' below the cavum tympani (Gaffney 1996).

Character 20. [PO] PO-SQ contact: (0) present; (1) absent. Scoring for *Chubutemys* is modified from "0" to "?" because this contact is not clearly preserved (Gaffney et al. 2007, p. 15).

Character 26. [PAL] Teeth: (0) present; (1) absent. *Ninjemys* is recoded from "1" to "?" because the palate is incomplete (Gaffney 1992, p. 3, Fig. 1).

Character 27. [PAL] Foramen palatinum posterius large: (0) small to moderate; (1) very large. *Ninjemys* is recoded from "1" to "?" because the palate is incomplete (Gaffney 1992, p. 3, Fig. 1).

Character 28. [PT] Vertical flange on processus pterygoideus externus: (0) absent; (1) present. Joyce (2007) and Sterli and Joyce (2007) interpreted this feature as highly variable and of limited phylogenetic significance. In contrast, Gaffney (1979, 1996), Gaffney and Meylan (1988), Gaffney et al. (1987, 1991), and Gaffney and Jenkins (2010) advocate its use as a cryptodiran synapomorphy. In *Meiolania* "the vertical flange on the lateral margin of the processus … is absent" but a "swelling" of the bone is hypothesised to mark its original position (Gaffney 1983, p. 427). Because this does not strictly constitute a "vertical flange" we have changed the score from "1" to "0" for *Meiolania* and "1" to "?" for both *Ninjemys* and *Niolamia* since this part of the skull is not adequately preserved in either taxon (see Gaffney 1983, p. 431; Gaffney 1996, p. 77; Gaffney 1992, p. 3, Fig. 1).

Character 29. [PT] Interpterygoid vacuity: (0) widely, slightly open; (1) completely closed by medial extension of PT. Gaffney (1983, 1996) and Gaffney et al. (2007) argued that the "intrapterygoid slit" of meiolaniids developed as a neomorph from the primitive eucryptodiran condition in which the interpterygoid vacuity had completely closed. *Dracochelys bicuspis* Gaffney and Ye 1992 was forwarded as evidence of a intermediary state (Gaffney 1996, p. 119) where the basipterygoid-pterygoid suture is separated laterally to create paired slit-like openings for the palatine arteries (Gaffney and Ye 1992, p. 9). However, maintenance of at least a transverse midline contact between the pterygoid lamellae and rostrum basisphenoidale is a fundamental feature of the eucryptodiran basicranium (Gaffney 1996; Gaffney et al. 2007). Modification of the eucryptodiran state into the "intrapterygoid slit" arrangement of meiolaniids requires retraction of the pterygoid lamellae from the basisphenoid rostrum, and ventral displacement of the posterior pterygoids and canalis cavernosus, creating a condition similar to basal turtles (e.g., *Proganochelys*, *Palaeochersis* and *Kayentachelys*: Sterli et al. 2010, p. 1341). A different interpretation based on Sterli et al. (2010, p. 1344) and the palatal morphology of *Spoochelys* could involve formation of the "intrapterygoid slit" via progressive posterior closure of the pterygoid midline suture. Gaffney et al. (2007) described the pterygoid-basisphenoid contact in *Chubutemys* as being "damaged by crushing" but apparently manifesting lateral expansions of the suture to form paired foramina. *Ninjemys* is altered from "1" to "?" because the palate is incomplete (Gaffney 1992, p. 3, Fig. 1).

Character 30. [PT] Canalis caroticus internus: posterior to foramen basisphenoidale: (0) absent; (1) at least partially covered by BS-PT; (2) covered by prootic. *Ninjemys* is recoded from "1" to "?" because the palate is incomplete (Gaffney 1992, p. 3, Fig. 1).

Character 31. [PT] Canalis caroticus internus at least partially formed by PT: (0) no; (1) yes, distal to foramen basisphenoidale (= foramen carotici cerebralis posterius sensu Sterli et al. 2010). *Ninjemys* is recoded from "1" to "?" because the palate is incomplete (Gaffney 1992, p. 3, Fig. 1).

Character 32. [PT] Canalis caroticus lateralis vs canalis caroticus internus: (0) lateralis equal to or larger than internus; (1) lateralis smaller than internus. Coding for *Meiolania* should be "1" because the "intrapterygoid slit" (= foramen caroticus laterale with terminology sensu Sterli et al. 2010) is considerably larger than the canalis caroticus internus (see discussion in Gaffney 1983, pp. 447–449, and 436, Fig. 53). However, Gaffney et al. (2007) recorded *Meiolania* as "?", presumably because of uncertainty surrounding the identification and ancestral dimensions of the canalis caroticus lateralis. Indeed, homology between these structures is unclear. For example, the slit-like foramen caroticus laterale of primitive eucryptodires pierce the transverse pterygoid-rostrum basisphenoidale suture and the paired arteria palatina are recessed (Brinkman and Nicholls 1993; Sukhanov 2000; Sterli et al. 2010; Brinkman et al. 2012). In contrast, the rostrum basisphenoidale does not suture to the pterygoids in *Meiolania* and the passages for the palatine arteries were presumably walled medially by the sagittal ridge from the basisphenoid (see Gaffney 1983, p. 430).

Character 33. [PT] Foramen posterius canalis carotici interni: (0) not formed by basisphenoid and pterygoid; (1) formed by basisphenoid and pterygoid, located midway along basisphenoid. *Ninjemys* is recoded from "1" to "?" because the basicranium is incomplete (Gaffney 1992, p. 3, Fig. 1).

Character 34. [PT] Middle ear with ossified floor formed by posteromedial pterygoid process: (0) process absent; (1)

present. *Ninjemys* is recoded from "1" to "?" because the basicranium is incomplete (Gaffney 1992, p. 3, Fig. 1).

Character 36. [PT] Deep cavity on quadrate ramus: (0) absent; (1) present. *Ninjemys* is recoded from "1" to "?" because the palate is incomplete (Gaffney 1992, p. 3, Fig. 1).

Character 37. [PT] BS-PT space: (0) interpterygoid vacuity widely open; (1) distinct space, perhaps a foramen caroticus laterale (fcl), present; (2) closed, pt-bs suture, foramen posterius canalis carotici internus (fpcci) possibly present. *Ninjemys* is recoded from "1" to "?" because the palate is incomplete (Gaffney 1992, p. 3, Fig. 1). *Niolamia* and *Meiolania* changed from from "1" to "0" (see discussion for characters 29 and 32).

Character 38. [PT] Intrapterygoid slit: (0) absent, small; (1) extensive, covering foramen carotici cerebralis posterior (fccp). See discussion for characters 29 and 32. *Ninjemys* is recoded from "1" to "?" because the palate is incomplete (Gaffney 1992, p. 3, Fig. 1).

Character 45. [QU] Antrum postoticum: (0) absent; (1) present. *Ninjemys* is recoded from "1" to "?" because the cavum tympani is incomplete (Gaffney 1992, p. 3, Fig. 1). *Meiolania* is changed "1" to "0" because the antrum postoticum is not developed (Gaffney 1983, pp. 416–417); *Niolamia* "1" to "?" following Sterli and de la Fuente (2011).

Character 46. [QU] Covers opisthotic laterally: (0) no; (1) yes. *Ninjemys* is recoded from "1" to "?" because the otic chamber is incomplete (Gaffney 1992, p. 3, Fig. 1).

Character 47. [QU] Cranioquadrate space: (0) relatively open; (1) a well-defined canal. *Ninjemys* is recoded from "1" to "?" because the basicranium is incomplete (Gaffney 1992, p. 3, Fig. 1).

Character 48 [OP] Processus paroccipitalis: (0) loosely articulated; (1) tightly fused. *Ninjemys* is recoded from "1" to "?" because the otic chamber is incomplete (Gaffney 1992, p. 3, Fig. 1).

Character 49. [OP] Fenestra perilymphatica: (0) relatively large; (1) relatively small. *Ninjemys* is recoded from "0" to "?" because the otic chamber is incomplete (Gaffney 1992, p. 3, Fig. 1).

Character 50. [PR] Anterior surface otic chamber: (0) faces more anterodorsally; (1) faces more anteriorly, anteroventrally (processus trochlearis oticum). *Ninjemys* is recoded from "1" to "?" because the otic chamber is incomplete (Gaffney 1992, p. 3, Fig. 1).

Character 51. [PR] Otic chamber thickness: (0) anterior wall thin; (1) anterior wall thick. *Ninjemys* is recoded from "1" to "?" because the otic chamber is incomplete (Gaffney 1992, p. 3, Fig. 1).

Character 53. [SO] Wide occipital plate: (0) wide occipital plate with depressions; (1) absent. *Ninjemys* is recoded from "1" to "?" because the otic chamber is incomplete (Gaffney 1992, p. 3, Fig. 1).

Character 56. [BS] Basipterygoid articulation: (0) open articulation; (1) sutured joint. *Ninjemys* is recoded from "1" to "?" because the braincase is incomplete (Gaffney 1992, p. 3, Fig. 1).

Character 70. [CE] 4th cervical articulation: (0) amphicoelous; (1) biconvex; (2) opisthocoelous. Gaffney et al. (1998) suggested that the fourth cervical articulation of *Otwayemys* was probably procoelous based on tentative identification of a biconcave fifth (?) cervical centrum. This has been altered to "?" because of ambiguity.

Character 74. [CE] Strong ventral process: (0) absent on all centra; (1) present on posterior centra. *Otwayemys* is recoded from "0" to "1" because the eighth cervical vertebra (NMVP187261) is keeled (Gaffney et al. 1998, p. 20).

Character 92. [CAR] 9th costal bone: (0) present; (1) absent. The exact number of costals bones is unknown in *Meiolania* (Gaffney 1996, p. 24) thus scoring is changed to "?".

Character 106. [PLA] Gular scales: (0) 2 pairs; (1) 1 pair. *Otwayemys* is recoded from "?" to "0" based on Gaffney et al. (1998, p. 12).

Character 107. [PL] Inframarginals: (0) 3–4 pairs; (1) 0–2 pairs (axillary, inguinal). *Otwayemys* is recoded from "?" to "1" based on Gaffney et al. (1998, p. 13).

Character 108. [PLA] Abdominal scute relative to midline: (0) meets on midline; (1) withdrawn from midline. Medial contact between the abdominal scutes cannot be reconstructed in *Meiolania* (Gaffney 1996, p. 33) or *Otwayemys* Gaffney et al. (1998, p. 13) thus states for both taxa are scored "?".

Character 109. [PLA] Pectoral scute: (0) not on epiplastron; (1) on epiplastron. *Meiolania* is recoded from "0" to "?" because scute margins are unclear (Gaffney et al. 1996, p. 33).

Character 111. [PLA] Midline sulcus sinuous: (0) no; (1) yes. *Meiolania* is recoded from "0" to "?" because scute margins are unclear (Gaffney et al. 1996, p. 33).

Appendix 3

Inventory of "Pruned" taxa and their missing data (percentage).

(A) Taxa from Sterli and de la Fuente (2011)

Proterochersis robusta 73%
Australochelys africanus 75%
Condorchelys antiqua 65.8%
Indochelys spatulata 86.2%

Siamochelys peninsularis 73.7%
Platychelys oberndorferi 63.2%
Caribemys oxfordiensis 74.3%
Notoemys laticentralis 53.3
Dorsetochelys delairi 60.5%
Portlandemys mcdowelli 59.9%
Mesodermochelys undulatus 62.5%
Hoplochelys crassa 65.8%
Peltochelys durlstonensis 71.7%
Sandownia harrisi 53.9%
Chubutemys copelloi 60.5%
Otwayemys cunicularis 72%
Patagoniaemys gasparinae 73%
Niolamia argentina 79.6%

(B) Taxa from Gaffney et al. (2007)

Australochelys africanus 59.5%
Proterochersis robusta 79.3%
Otwayemys cunicularis 68.5%
Chubutemys copelloi 51.4%
Ninjemys oweni 72.9%
Niolamia argentina 57.7%

Appendix 4

Anatomical abbreviations

ast Aditus canalis stapedio-temporalis
bo Basioccipital
bs Basisphenoid
c Costal
ca Columella auris
caj Cavum acustico jugulare
cav Lab cavum labyrinthicum
cc Canalis cavernosus
ce Cervical scute
epi Epipterygoid
ex Exoccipital
fb Foramen basisphenoidale
fio Foramen interorbitale
fr Frontal
fst Foramen stapedio-temporale
ivac Interpterygoid vacuity
j Jugal
m Marginal scute
max Maxilla
n Neural
nu Nuchal
op Opisthotic
or Orbit
pa Parietal
pe Peripheral
pf Prefrontal
pio Processus interfenestralis
pip Processus inferior parietalis
pl Pleural scute
po Postorbital
pr Prootic
pt Pterygoid
qu Quadrate
rst Recessus scalae tympani
sm Supramarginal scute
so Supraoccipital
sq Squamosal
v Vertebral scute
V Foramen nervi trigemini
VII? Foramen nervi facialis?
XII Foramen nervi hypoglossi

References

Agnolin, F. L., Ezcurra, M. D., Pais, D. F., & Salisbury, S. W. (2010). A reappraisal of the Cretaceous non-avian dinosaur faunas from Australia and New Zealand: Evidence for their Gondwanan affinities. *Journal of Systematic Palaeontology, 8*, 257–300.

Ameghino, F. (1899). Sinopsis geologico-paleontologica. Suplemento (adiciones y correcciones). La Plata, Julio de, pp. 1–13.

Angnarsson, I., & Miller, J. A. (2008). Is ACCTRAN better than DELTRAN? *Cladistics, 24*, 1032–1038.

Anquetin, J., Barrett, P. M., Jones, M. E., Moore-Fay, S., & Evans, S. E. (2009). New stem turtle from the Middle Jurassic of Scotland: New insights into the evolution and palaeoecology of basal turtles. *Proceedings of the Royal Society of London. Series B, 276*, 879–886.

Archer, M., Flannery, T. F., Ritchie, A., & Molnar, R. E. (1985). First Mesozoic mammal from Australia—an Early Cretaceous monotreme. *Nature, 318*(6044), 363–366.

Baur, G. (1887). Ueber den Ursprung der Extremitaeten der Ichthyopterygia. *Bericht des Oberrheinischen Geologischen Vereins, 20*, 17–20.

Brinkman, D. B., & Peng, J. H. (1993). *Ordosemys leios*, n. gen., n. sp., a new turtle from the Early Cretaceous of the Ordos Basin, Inner Mongolia. *Canadian Journal of Earth Sciences, 30*, 2128–2138.

Brinkman, D. B., & Wu, X.-C. (1999). The skull of *Ordosemys*, an Early Cretaceous turtle from Inner Mongolia, People's Republic of China, and the interrelationships of Eucryptodira (Chelonia, Cryptodira). *Paludicola, 2*, 134–137.

Brinkman, D., Eberth, D., Xu, X., Clark, J., & Wu, X.-C. (2012). Turtles from the Jurassic Shishugou Formation of the Junggar Basin, People's Republic of China, with comments on the basicranial region of basal eucryptodires. In D. B. Brinkman, P. A. Holroyd & J. D. Gardner (Eds.) *Morphology and evolution of turtles*. Dordrecht: Springer.

Burger, D. (1980). Palynology of the lower Cretaceous Surat Basin. *Bureau of Mineral Resources, Geology and Geophysics Bulletin, 189*, 1–106.

Clarke, J. M., & Middleton, K. M. (2008). Mosaicism, modules, and the evolution of birds: Results from a Bayesian approach to the study of morphological evolution using discrete character data. *Systematic Biology, 57*, 185–201.

Clemens, W. A., Wilson, G. P., & Molnar, R. E. (2003). An enigmatic (synapsid?) tooth from the Early Cretaceous of New South Wales, Australia. *Journal of Vertebrate Palaeontology, 23*, 232–237.

Datta, P. M., Manna, P., Ghosh, S. C., & Das, P. (2000). The first Jurassic turtle from India. *Palaeontology, 43*, 99–109.

de la Fuente, M. S., & Iturralde-Vinent, M. (2001). A new pleurodiran turtle from the Jagua Formation (Oxfordian) of Western Cuba. *Journal of Paleontology, 75*, 860–869.

de Lapparent de Broin, F., & Molnar, R. E. (2001). Eocene chelid turtles from Redbank Plains, southeast Queensland, Australia. *Geodiversitas, 23*, 41–79.

Dettmann, M. E., Molnar, R. E., Douglas, J. G., Burger, D., Fielding, C., Clifford, H. T., et al. (1992). Australian Cretaceous terrestrial faunas and floras: Biostratigraphic and biogeographic implications. *Cretaceous Research, 13*, 207–262.

Felsenstein, J. (2004). *Inferring phylogenies*. Sunderland: Sinauer Associates, Inc.

Fernandez, M. S., & de la Fuente, M. S. (1994). Redescription and phylogenetic position of *Notoemys*: The oldest Gondwanian pleurodiran turtle. *Neues Jahrbuch Geologie und Palaeontologie Abhandlungen Stuttgart, 193*, 81–105.

Flannery, T. F., Archer, M., Rich, T. H., & Jones, R. (1995). A new family of monotremes from the Cretaceous of Australia. *Nature, 377*, 418–420.

Fraas, E. (1913). *Proterochersis*, eine pleurodire Schildkroete aus dem Keuper. *Jahreshefte der Gesellschaft fur Naturkunde in Wurttemberg, 69*, 13–90.

Gaffney, E. S. (1979a). Comparative cranial morphology of recent and fossil turtles. *Bulletin of the American Museum of Natural History, 164*, 65–376.

Gaffney, E. S. (1979b). The Jurassic turtles of North America. *Bulletin of the American Museum of Natural History, 162*, 91–136.

Gaffney, E. S. (1983). The cranial morphology of the extinct horned turtle *Meiolania platyceps*, from the Pleistocene of Lord Howe Island, Australia. *Bulletin of the American Museum of Natural History, 175*, 361–480.

Gaffney, E. S. (1985). The cervical and caudal vertebrae of the cryptodiran turtle, *Meiolania platyceps*, from the Pleistocene of Lord Howe Island, Australia. *American Museum Novitates, 2805*, 1–29.

Gaffney, E. S. (1990). The comparative osteology of the Triassic turtle *Proganochelys*. *Bulletin of the American Museum of Natural History, 194*, 1–263.

Gaffney, E. S. (1992). *Ninjemys*, a new name for *"Meiolania" owenii* (Woodward), a horned turtle from the Pleistocene of Queensland. *American Museum Novitates, 3049*, 1–10.

Gaffney, E. S. (1996). The postcranial morphology of *Meiolania platyceps* and a review of the Meiolaniidae. *Bulletin of the American Museum of Natural History, 229*, 1–165.

Gaffney, E. S., & Jenkins, F. A. (2010). The cranial morphology of *Kayentachelys*, an early Jurassic cryptodire, and the early history of turtles. *Zoologica Scripta, 91*, 335–368.

Gaffney, E. S., & Kitching, J. W. (1994). The most ancient African turtle. *Nature, 369*, 55–58.

Gaffney, E. S., & McNamara, G. (1990). A meiolaniid turtle from the Pleistocene of northern Queensland. *Memoirs of the Queensland Museum, 28*, 107–113.

Gaffney, E. S., & Meylan, P. A. (1988). A phylogeny of turtles. In M. J. Benton (Ed.), *The phylogeny and classification of the tetrapods.Volume 1. Amphibians, reptiles, birds* (pp. 157–219). Oxford: Clarendon Press.

Gaffney, E. S., & Ye, X. (1992). *Dracochelys*, a new cryptodiran turtle from the Early Cretaceous age. *American Museum Novitates, 3048*, 1–13.

Gaffney, E. S., Balouet, J. C., & de Broin, F. (1984). New occurrences of extinct meiolaniid turtles from New Caledonia. *American Museum Novitates, 2800*, 1–6.

Gaffney, E. S., Hutchinson, J. H., Jenkins, F. A., & Meeker, L. (1987). Modern turtle origins: the oldest known cryptodire. *Science, 237*, 289–291.

Gaffney, E. S., Meylan, P. A., & Wyss, A. R. (1991). A computer assisted analysis of the relationships of the higher categories of turtles. *Cladistics, 7*, 313–335.

Gaffney, E. S., Archer, M., & White, A. (1992). *Warkalania*, a new meiolaniid turtle from the Tertiary Riversleigh deposits of Queensland, Australia. *The Beagle, Records of the Northern Territory Museum of Arts and Sciences, 9*, 35–48.

Gaffney, E. S., Kool, L., Brinkman, D. B., Rich, T. H., & Vickers-Rich, P. (1998). *Otwayemys*, a new cryptodiran turtle from the Early Cretaceous of Australia. *American Museum Novitates, 3233*, 1–28.

Gaffney, E. S., Tong, H., & Meylan, P. A. (2006). Evolution of the side-necked turtles: The families Bothremydidae, Euraxemydidae and Araripemydidae. *Bulletin of the American Museum of Natural History, 300*, 1–698.

Gaffney, E. S., Rich, T. H., Vickers-Rich, P., Constantine, A., Vacca, R., & Kool, L. (2007). *Chubutemys*, a new eucryptodiran turtle from the Early Cretaceous of Argentina, and the relationships of the Meiolaniidae. *American Museum Novitates, 3599*, 1–35.

Hamilton-Bruce, R. J., & Kear, B. P. (2010). A possible succineid land snail from Lower Cretaceous non-marine deposits of the Griman Creek Formation at Lightning Ridge, New South Wales. *Alcheringa, 34*, 325–331.

Henderson, R. A., Crampton, J. S., Dettman, M. E., Douglas, J. G., Haig, D., Shafik, S., et al. (2000). Biogeographical observations on the Cretaceous biota of Australia. *Memoirs of the Association of Australian Palaeontologists, 23*, 355–404.

Hirayama, R. (1998). Oldest known sea turtle. *Nature, 392*, 705–708.

Hirayama, R., Brinkman, D. B., & Danilov, I. G. (2000). Distribution and biogeography of non-marine Cretaceous turtles. *Russian Journal of Herpetology, 7*, 181–198.

Hocknull, S. A. (2000). Mesozoic freshwater and estuarine bivalves from Australia. *Memoirs of the Queensland Museum, 45*, 405–426.

Hoffstetter, R., & Gasc, J. P. (1969). Vertebrae and ribs of modern reptiles. In C. Gans, A. D. Bellairs, & T. Parsons (Eds.), *Biology of the Reptilia Vol 1, Morphology* (pp. 201–310). New York: Academic Press.

Holmes, G. G., & Senior, B. R. (1976). Lightning ridge opal field. 25th International Geological Congress Sydney Excursion Guide 7B, pp. 13

Joyce, W. G. (2007). Phylogenetic relationships of Mesozoic turtles. *Bulletin of the Peabody Museum of Natural History, 48*, 3–102.

Joyce, W. G., & Gauthier, J. A. (2004). Palaeoecology of Triassic stem turtles sheds new light on turtle origins. *Proceedings of the Royal Society London, 271*, 1–5.

Joyce, W. G., Parham, J. F., & Gauthier, J. A. (2004). Developing a protocol for the conversion of rank-based taxon names to phylogenetically defined clade names, as exemplified by turtles. *Journal of Paleontology, 78*, 989–1013.

Kear, B. P. (2006a). Reassessment of *Cratochelone berneyi* Longman, 1915, a giant Early Cretaceous sea turtle from Australia. *Journal of Vertebrate Paleontology, 26*, 779–783.

Kear, B. P. (2006b). Plesiosaur remains from Cretaceous high-latitude non-marine deposits in southeastern Australia. *Journal of Vertebrate Paleontology, 26*, 196–199.

Kear, B. P., & Lee, M. S. (2006). A primitive protostegid from Australia and early sea turtle evolution. *Biology Letters, 2*, 116–119.

Kear, B. P., Schroeder, N. I., Vickers-Rich, P., & Rich, T. H. (2006a). Early Cretaceous high latitude marine reptile assemblages from southern Australia. *Paludicola, 5*, 200–205.

Kear, B. P., Schroeder, N. I., & Lee, M. S. (2006b). An archaic crested plesiosaur in opal from the lower Cretaceous high-latitude deposits of Australia. *Biology Letters, 2*, 615–619.

Kear, B. P., Deacon, G. L., & Siverson, M. (2010). Remains of a Late Cretaceous pterosaur from the Molecap Greensand of Western Australia. *Alcheringa, 34*, 273–279.

Khosatzky, L. I. (1997). Big turtle of the Late Cretaceous of Mongolia. *Russian Journal of Herpetology, 4*, 148–154.

Kolaczkowski, B., & Thornton, J. W. (2006). Is there a star tree paradox? *Molecular Biology and Evolution, 23*, 1819–1823.

Kordikova, E. G. (2002). Comparative morphology of the palate dentition in *Proganochelys quenstedti* Baur 1887 from the Upper Triassic of Germany and chelonian ancestry. *Jahreshber Palaeontologie Abhandlung Stuttgart, 225*, 195–249.

Lee, M. S. (1996). The homologies and early evolution of the shoulder girdle in turtles. *Proceedings of the Royal Society of London. Series B, 2663*, 111–117.

Lewis, P. O. (2001). A likelihood approach to estimating phylogeny from discrete morphological character data. *Systematic Biology, 50*, 913–925.

Megirian, D. (1992). *Meiolania brevicollis* sp. nov. (Testudines: Meiolaniidae): A new horned turtle from the Australian Miocene. *Alcheringa, 16*, 93–106.

Molnar, R. E. (1991). Fossil reptiles in Australia. In P. Vickers-Rich, J. M. Monaghan, R. F. Baird & T. H. Rich (Eds.), *Vertebrate palaeontology of Australasia* (pp. 605–702). Melbourne: Monash University Publications Committee.

Molnar, R. E. (1999). Avian tibiotarsi from the Early Cretaceous of Lightning Ridge, New South Wales. In Y. Tomida, T. H. Rich, & P. Vickers-Rich (Eds.), Proceedings of the 2nd Gondwanan Dinosaur Symposium. *Natural Science Museum Monographs Tokyo, 15*, 197–209.

Molnar, R. E., & Willis, P. M. (2001). New crocodyliform material from the Early Cretaceous Griman Creek Formation, at Lightning Ridge, New South Wales. In G. C. Grigg, F. Seebacher & C. E. Franklin (Eds.), *Crocodilian biology and evolution* (pp. 75–82). Chipping Norton: Surrey Beatty and Sons.

Mrosovsky, N. (1980). Thermal biology of sea turtles. *American Zoology, 20*, 531–547.

Musser, A. M. (2005). Investigations into the evolution of Australian mammals with a focus on Monotremata. PhD Dissertation, University of New South Wales.

Nopcsa, F. (1923). On the geological importance of the primitive reptilian fauna of the upper-most Cretaceous of Hungary; with a description of a new tortoise (*Kallokibotion*). *Quarterly Journal Geological Society, 79*, 100–116.

Owen, R. (1886). Description of fossil remains of two species of of a megalanian genus (*Meiolania*) from "Lord Howe's Island". *Philosophical Transactions of the Royal Society of London, 177*, 471–480.

Parham, J. F., & Hutchinson, J. H. (2003). A new eucryptodiran turtle from the Late Cretaceous of North America (Dinosaur Provincial Park, Alberta, Canada). *Journal of Vertebrate Palaeontology, 23*, 783–798.

Pritchard, P. C. (1984). Piscivory in turtles, and evolution of the long-necked Chelidae. *Symposium of the Zoological Society of London, 52*, 87–110.

Rich, T. H., & Vickers-Rich, P. (2003). Diversity of Early Cretaceous mammals from Victoria, Australia. *Bulletin of the American Museum of Natural History, 285*, 36–53.

Rich, T. H., Vickers-Rich, P., Flannery, T. F., Kear, B. P., Cantrill, D., Komarower, P., et al. (2009). An Australian multituberculate and its palaeobiogeographical implications. *Acta Palaeontologica Polonica, 54*, 1–6.

Ronquist, F., & Huelsenbeck, J. P. (2003). MRBAYES 3: Bayesian phylogenetic inference under mixed models. *Bioinformatics, 19*, 1572–1574.

Rougier, G. W., de la Fuente, M. S., & Arcucci, A. B. (1995). Late Triassic turtles from South America. *Science, 268*, 855–858.

Rowe, T., Rich, T. H., Vickers-Rich, P., Springer, M., & Woodburne, M. O. (2008). The oldest platypus and its bearing on divergence timing of the platypus and echidna clades. *PNAS, 1054*, 1238–1242.

Rueda, E. A., & Gaffney, E. S. (2005). *Notoemys zapatocaensis*, a new side-necked turtle (Pleurodira: Platychelyidae) from the Early Cretaceous of Colombia. *American Museum Novitates, 3470*, 1–19.

Schumacher, G. H. (1973). The head muscles and hyolaryngeal skeleton of turtles and crocodilians. In C. Gans & T. S. Parsons (Eds.), *Biology of the Reptilia* (pp. 101–199). New York: Academic Press.

Senior, B. R., & Chadderton, L. T. (2007). Natural gamma radiation and exploration for precious opal in Australia. *Australian Gemmologist, 23*, 160–176.

Smith, E. T. (2009). Terrestrial and freshwater turtles of Early Cretaceous Australia. PhD Dissertation, University of New South Wales.

Smith, E. T. (2010). Early Cretaceous chelids from Lightning Ridge, New South Wales. *Alcheringa, 34*, 375–384.

Smith, E. T., & Smith, R. (1999). *Black opal fossils of Lightning Ridge*. Sydney: Kangaroo Press.

Sorenson, M. D., & Franzosa, E. A. (2007). *TreeRot, version 3*. Boston: Boston University.

Sterli, J. (2008). A new, nearly complete stem turtle from the Jurassic of South America with implications for turtle evolution. *Biology Letters, 4*, 286–289.

Sterli, J., & de la Fuente, M. S. (2011). A new turtle from the La Colonia Formation (Campanian-Maastrichtian), Patagonia, Argentina, with remarks on the evolution of the vertebral column in turtles. *Palaeontology, 54*, 63–68.

Sterli, J., & Joyce, W. G. (2007). The cranial anatomy of the Early Jurassic turtle *Kayentachelys aprix*. *Acta Palaeontologica Polonica, 52*, 675–694.

Sterli, J., de la Fuente, M. S., & Rougier, W. G. (2007). Anatomy and relationships of *Palaeochersis talampayensis*, a Late Triassic turtle from Argentina. *Palaeontographica Abteilung A. Stuttgart, 281*, 1–61.

Sterli, J., Muller, J., Anquetin, J., & Hilger, A. (2010). The parabasisphenoid complex in Mesozoic turtles and the evolution of the testudinate basicranium. *Canadian Journal of Earth Sciences, 47*, 1337–1346.

Stevens, G. R., & Clayton, R. N. (1971). Oxygen isotope studies on Jurassic and Cretaceous belemnites from New Zealand and their biogeographic significance. *New Zealand Journal of Geology and Geophysics, 14*, 829–897.

Sukhanov, V. B. (2000). Mesozoic turtles of Middle and Central Asia. In M. J. Benton, M. A. Shishkin, D. M. Unwin & E. N. Kurochkin (Eds.), *The age of dinosaurs in Russia and Mongolia* (pp. 309–367). Cambridge: Cambridge University Press.

Sukhanov, V. B., & Narmandankh, P. (1974). New Early Cretaceous turtle from continental deposits of the northern Gobi. Mesozoic and Cenozoic Biostratigraphy of Mongolia. *The Joint Soviet-Mongolian Palaeontological Expedition, Transactions*, 1, 192–200 (in Russian).

Sukhanov, V. B., & Narmandankh, P. (2006). New taxa of Mesozoic turtles from Mongolia. *Fossil Turtle Research, 1*, 119–127.

Swofford, D. L. (2002). PAUP*. Phylogenetic analysis using parsimony (*and other methods). *Version* 4. Sunderland: Sinauer Associates.

Tarduno, J. A., Brinkman, D. B., Renne, P. R., Cottrell, R. D., Scher, H., & Castillo, P. (1998). Evidence for extreme climatic warmth from Late Cretaceous Arctic vertebrates. *Science, 282*, 2241–2244.

Thomson, J. S. (1932). The anatomy of the tortoise. *Scientific Proceedings of the Royal Society of Dublin, 20*, 359–461.

Thulborn, A., & Turner, S. (2003). The last dicynodont. *Proceedings of the Royal Society of London. Series B, 270*, 985–993.

Wagner, A. (1853). Beschreibung einer fossilen Schildkröte und etlicher anderer Reptilien- Überreste aus den lithographischen Schiefern und dem grünen Sandsteine von Kehlheim. *Abhandlungen der Bayerischen Akademie der Wissenschaften, Mathematisch-physikalische Klasse*, 7(1).

Walker, W. F. (1973). The locomotor apparatus of Testudines. In C. Gans & T. S. Parsons (Eds.), *Biology of the Reptilia Vol 4,* Morphology (pp. 1–100). New York: Academic Press.

Warren, J. W. (1969). A fossil chelonian of probable Lower Cretaceous age from Victoria, Australia. *Memoirs of the National Museum of Victoria, 29*, 23–29.

Watkins, J. (1984). Future prospects for opal mining in the Lightning Ridge region. *New South Wales Geological Survey Report, 119*, 1–50.

White, A. W., Worthy, T. H., Hawkins, S., Bedford, S., & Spriggs, M. (2010). Megafaunal meiolaniid horned turtles survived until early human settlement in Vanuatu, Southwest Pacific. *PNAS, 107*, 15512–15516.

Wiman, C. (1930). Fossile Schildkröten aus China. *Paleontologica Sinica, Series C, 6*, 1–56.

Woodward, A. S. (1901). On some extinct reptiles from Patagonia, of the genera *Miolania*, *Dinilysia* and *Genyodectes*. *Proceedings of the Zoological Society London, 1901*, 169–184.

Worthy, T. H., & Lee, M. S. Y. (2008). Affinities of Miocene waterfowl (Anatidae: *Manuherikia*, *Dunstanetta* and *Mioadorna*) from the St Bathans fauna, New Zealand. *Palaeontology, 51*, 667–708.

Yeh, H.-K. (1986). A Jurassic turtle discovered for the first time in the Xinjiang Uygur Autonomous Region. *Vertebrata PalAsiatica, 24*, 171–181.

Chapter 10
Turtles from the Jurassic Shishugou Formation of the Junggar Basin, People's Republic of China, with Comments on the Basicranial Region of Basal Eucryptodires

Donald B. Brinkman, David A. Eberth, Xing Xu, James M. Clark, and Xiao-Chun Wu

Abstract Five fossil turtle species (?*Sichuanchelys* sp., three species of *Xinjiangchelys*, and an indeterminate species of *Annemys*) are present in the Shishugou Formation (late Middle to early Late Jurassic) of the Junggar Basin, northwestern China. Two of these, *X. radiplicatoides* sp. nov. and *Annemys* sp., are each represented by an associated skull and shell. These demonstrate that the Xinjiangchelyidae, as currently defined, encompasses different grades of evolution. *X. radiplicatoides* is primitive in features of the basicranial region, lack of emargination of the skull roof, and presence of an inflated postorbital region. *Annemys* sp., which has a low skull with deeply emarginated temporal and cheek regions and large foramina palatinum posterius, is similar to basal eucryptodires from the Early Cretaceous of Asia, particularly *Hangaiemys*. Early stages in the evolution of the basicranial region in eucryptodires are documented by the well-preserved basicranial region of *X. radiplicatoides* and *Annemys* sp. The slit-like structure of the foramen palatinum posterius in *X. radiplicatoides* is consistent with the interpretation that this opening developed by closure of the interpterygoid vacuity around the palatine artery. Processes of the basisphenoid that extend laterally into the pterygoid, identified here as basipterygoid processes, are well developed in *Xinjiangchelys* and most Early Cretaceous sinemydids/macrobaenids. Although a high taxonomic diversity of turtles is present in the Shishugou Formation, diversity at individual localities is low, often with a single taxon being present or overwhelmingly dominant, and most localities differ in the kinds of turtles that are dominant at that locality. This pattern of high alpha diversity (total diversity within a unit), low diversity within individual localities, and high beta diversity (between-locality diversity within a unit) is unusual in turtle assemblages, and suggests that the paleoecology of the Shishugou Formation has unusual aspects compared to similarly diverse turtle assemblages, where diversity at a locality typically reflects total diversity within the unit.

Keywords *Annemys* • China • Junggar Basin • Jurassic • *Sichuanchelys* • Shishugou Formation • Xinjiangchelyidae • *Xinjiangchelys*

D. B. Brinkman (✉) · D. A. Eberth
Royal Tyrrell Museum of Palaeontology,
Box 7500, Drumheller, AB T0J 0Y0, Canada
e-mail: don.brinkman@gov.ab.ca

D. A. Eberth
e-mail: david.eberth@gov.ab.ca

X. Xu
Key Laboratory of Evolutionary Systematics of Vertebrates,
Institute of Vertebrate Paleontology and Paleoanthropology
Chinese Academy of Sciences, 42 Xi-Zhi-Men-Wai,
Beijing 100044, People's Republic of China
e-mail: xu.xing@ivpp.ac.cn

J. M. Clark
George Washington University, Lisner Hall RM 340,
2023 G Street NW, Washington, DC 20052, USA
e-mail: jclark@gwu.edu

X.-C. Wu
Earth Sciences Section, Canadian Museum of Nature,
PO Box 3443 Station D, Ottawa, ON K1P 6P4, Canada
e-mail: xcwu@mus-nature.ca

Introduction

Fossil turtles from Asia are important to our understanding of the rise of modern turtle clades and assemblages, because many extant groups of eucryptodires are thought to have originated there (Hutchison 2000; Hirayama et al. 2000; Sukhanov 2000). These include testudinoids, which first occurred in Asia in the Early Cretaceous (Brinkman et al. 2008), and trionychians, which were present in Asia in the Jurassic (Gaffney and Meylan 1992a; Danilov and Parham 2006). In addition, fossil turtles from Asia have shown that a diverse assemblage of eucryptodires existed there during the Middle-Late Jurassic (Danilov and Parham 2008; Rabi et al. 2010). Thus, the Jurassic of Asia is key for understanding the initial stages of eucryptodire evolution.

D. B. Brinkman et al. (eds.), *Morphology and Evolution of Turtles*, Vertebrate Paleobiology and Paleoanthropology,
DOI: 10.1007/978-94-007-4309-0_10,

Jurassic turtles were first reported from Asia by Young and Chow (1953), who described six taxa on the basis of seven specimens recovered in 1951 during construction of the Chengdu–Chungking Railway in the Province of Sichuan, southern China. Three of these species were placed in monospecific genera: *Chengyuchelys baenoides* Young and Chow 1953; *Tienfuchelys tzuyangensis* Young and Chow 1953; and *Sinaspideretes wimani* Young and Chow 1953. The remaining three were included in the genus *Plesiochelys* Rütimeyer 1873, a basal eucryptodire largely based on plesiomorphic features of the shell and otherwise known from the Jurassic of Europe. The species referred to *Plesiochelys* were *P. latimarginalis* Young and Chow 1953, *P. radiplicatus* Young and Chow 1953, and *P. chungkingensis* Young and Chow 1953. The ages of the specimens described by Young and Chow (1953) are uncertain, because the Chengdu–Chungking Railway cuts through both Middle and Upper Jurassic sediments and detailed locality information was not recorded for any of the specimens. Subsequent papers further documented the diversity and distribution of turtles in the Jurassic of Sichuan (Ye 1963, 1973; Ye and Fang 1982; Fang 1987; Ye and Pi 1997). Four additional taxa were recognized from the Late Jurassic: *Plesiochelys tatsuensis* Ye 1963; *P. kwanganensis* Ye 1963; *P. oshanensis* Ye 1973; and *P. jingyanensis* Ye and Fang 1982. Documentation of Middle Jurassic turtles in China was greatly increased by the discovery of a rich vertebrate assemblage in Zigong, Sichuan, from which three new turtle taxa were described on the basis of shells: *Chengyuchelys zigongensis* Ye 1982; *C. dashanpuensis* Fang 1987; and *Sichuanchelys chowi* Ye and Pi 1997. The Zigong taxa are distinctly more primitive compared with the Late Jurassic forms in their retention of a mesoplastron, and this feature was used to place them in a separate family, the Chengyuchelyidae Ye 1990.

The first Chinese Jurassic turtle recognized from outside of Sichuan, *Xinjiangchelys junggarensis* Ye 1986a, was described on the basis of a complete carapace from a locality near Jiangjunmiao, in the northern part of the Junggar Basin, Xinjiang Uygur Autonomous Region, in the northwestern part of the country. This specimen was regarded as Middle Jurassic in age because of the associated fauna, particularly the presence of tritylodonts (Zhao 1980); that age estimate has been corroborated by later geological studies (Eberth et al. 2001).

The description of *Xinjiangchelys junggarensis* in northwestern China and subsequent discoveries of Middle Jurassic turtle material from elsewhere in central Asia (Kaznyshkin 1988), led to the recognition that the Jurassic eucryptodires from Asia were part of a distinct radiation of primitive eucryptodires not directly related to the European *Plesiochelys*. Kaznyshkin (1988) accordingly transferred all Asian material previously included in *Plesiochelys* to the genus *Xinjiangchelys*. In a later paper, *Xinjiangchelys* was also placed in its own family, the Xinjiangchelyidae Nessov 1990 (in Kaznyshkin et al. 1990).

Explorations by the Canada–China Dinosaur Project, an international expedition that worked in the Junggar Basin of Xinjiang from 1987 to 1990, led to discoveries in the northern part of the basin, north of Qitai, of additional Jurassic turtle-bearing localities clustered in two areas about 90 km apart. Here we refer these as the "Jiangjunmiao field area" and the "Wucaiwan field area", the former near the abandoned settlement of Jiangjunmiao and the latter near Huoshaoshan (see Fig. 10.1a–c). The Jiangjunmiao field area includes the holotype locality of *Xinjiangchelys junggarensis*. No additional material was collected from that quarry but two additional localities of comparable age were identified. The first yielded a non-diagnostic carapace found during explosive excavation of a large sauropod. The second comprised a rich bone bed (herein called the ''Jiangjunmiao turtle bonebed'') dominated by disarticulated and partially articulated turtle elements in a small channel setting. Variation in surface texture of shell elements indicates that at least two kinds of turtles are present, although most belong to a turtle with distinctive plications on the carapace. In addition to shell elements, significant cranial remains were recovered. In the Wucaiwan field area, remains of at least 14 individuals were collected from a single stratigraphic interval near Pingfengshan (see locality map and detailed description in Peng and Brinkman 1993, Fig. 1, p. 2016). Exposures in this area originally were referred to the Qigu Formation by Peng and Brinkman (1993); subsequently it was recognized that the Qigu Formation is restricted to the southern margin of the Junggar Basin and the temporally equivalent sediments from the northern margin of the basin instead belong to the Shishugou Formation (Fig. 10.1d). Two turtle taxa were recognized by Peng and Brinkman (1993) from Pingfengshan: *X. latimarginalis*, represented by 12 specimens, including a nearly complete carapace and some associated postcrania, and an indeterminate eucryptodire, provisionally identified as "*Xinjiangchelys* sp.", represented by a single plastron.

Peng and Brinkman (1993) followed Kaznyshkin (1988) in placing *Xinjiangchelys junggarensis*, the type specimen of "*Plesiochelys*" *latimarginalis*, and an incomplete shell and series of isolated elements from a Middle Jurassic locality in Kirgizstan within the species *X. latimarginalis*. Subsequent studies concluded that the range of variation in *X. latimarginalis*, as defined by Peng and Brinkman (1993), exceeded that expected for a single turtle species (Matzke et al. 2004, 2005; Nessov 1995). Consequently, Nessov (1995) erected the species *X. tianshanensis* for material from Kirgizstan and Matzke et al. (2004) partitioned the Junggar material into two species: *X. junggarensis*, which they restricted to the holotype specimen, and an unnamed congener for the Pingfengshan specimens originally assigned to *X. latimarginalis* by Peng and Brinkman (1993).

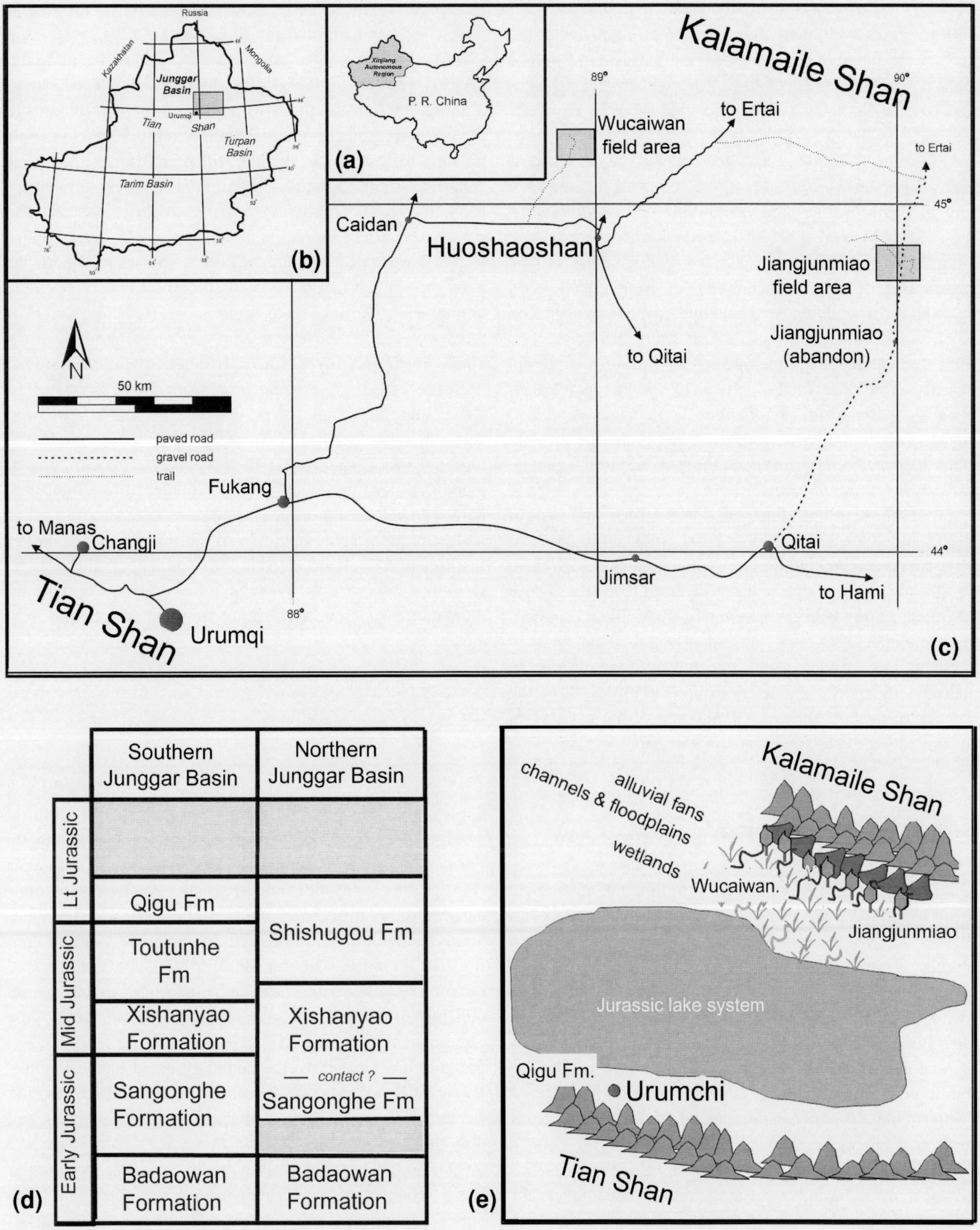

Fig. 10.1 Locality, stratigraphic, and depositional information for Jurassic turtle-bearing localities in Junggar Basin. **a–c** locality maps: **a** location of Xinjiang Uygur Autonomous Region in northwestern People's Republic of China; **b** location of Junggar Basin within Xinjiang; **c** locations of Wucaiwan and Jiangjunmiao field areas in northern part of Junggar Basin. **d** Stratigraphic chart showing Jurassic formations in northern and southern parts of the Junggar Basin; note that the Middle-Late Jurassic boundary lies within the Shishugou Formation. **e** Paleoenvironmental reconstruction of the Junggar Basin during the time of deposition of the upper part of the Shishugou Formation, showing approximate positions of present-day Wucaiwan and Jiangjunmiao field areas

Brinkman et al. (2008) partially supported Matzke et al. (2004) approach by recognizing *X. latimarginalis* as a distinct taxon, but felt that that the Pingfengshan specimens could not be distinguished from *X. junggarensis* and, contrary to Matzke et al. (2004), included them in the latter species.

Matzke et al. (2004, 2005) documented the presence of three additional *Xinjiangchelys* species from the southern Junggar Basin, in the Qigu Formation: *X.* cf. *radiplicatus*; *X. qiguensis* Matzke et al. 2004; and *X. chowi* Matzke et al. 2005. *X.* cf. *radiplicatus* shared with X. *radiplicatus* from the Late Jurassic of Sichuan the presence of strong plications on the carapace. *X. qiguensis* and *X. chowi* were distinguished on the basis of features of the shell. As a result, Brinkman et al. (2008) concluded that five species of *Xinjiangchelys* are present in the Jurassic formations of the Junggar Basin: *X. junggarensis*; *X.* cf. *radiplicatus*; *X. qiguensis*; *X. chowi*; and *X.* sp. A slightly greater diversity was reported by Rabi et al. (2010), who recognized another two additional genera, *Sichuanchelys* and *Annemys*.

Expeditions to the Junggar Basin co-led by two of us (JC and XX) from 2000 to 2007 collected a series of turtle specimens, including two incomplete skeletons from the Shishugou Formation in the Wucaiwan field area. Those specimens, along with the previously undescribed fragmentary remains from the turtle-dominated bone bed at Jiangjunmiao, provide significant new information on the anatomy, diversity, and relationships of turtles from the Junggar Basin and allow a reevaluation of the patterns of diversification of early eucryptodires in Asia.

Osteological Terminology

Aside from two exceptions, here we follow the anatomical terminology for turtle skulls developed by Gaffney (1972, 1979b). The first exception is the term "foramen caroticum laterale". That term has been used to refer both to a foramen opening into the braincase and to a more posterior foramen opening through the roof of the palate, although in recent papers (e.g., Gaffney et al. 2007) the former definition has been adopted. Because this term has been used to apply to non-homologous structures, Sterli et al. (2010) proposed a new term, the "foramen caroticum cerebralis posterior", for the posterior opening. We follow Sterli et al. (2010), although for consistency we adhere to Gaffney (1972, 1979b) in using Latin terms for anatomical structures in the turtle skull and, thus, prefer the modified term "foramen posterius canalis carotici cerebralis (fpccc)".

The second exception concerns the "basipterygoid process". A basipterygoid process was not recognized in turtles by Gaffney (1972, 1979b), but a laterally projecting triangular process of the basisphenoid in the Early Cretaceous paracryptodire *Pleurosternon bullockii* was identified as a basipterygoid process by Evans and Kemp (1975). Gaffney (1979a) questioned that identification, arguing that the structure was unlikely to be homologous to the basipterygoid process of primitive reptiles because of its position relative to the dorsum sellae and foramen anterius canalis carotici interni—in primitive reptiles, the basipterygoid process is anterior to these structures, whereas the process identified by Evans and Kemp (1975) as a "basipterygoid process" is more posterior. Subsequently, it has been recognized that a basipterygoid process is widespread among early turtles, being present in, for example, *Proganochelys* and *Kayentachelys* among others (Gaffney 1990; Sterli and Joyce 2007). Because of the widespread presence of a basipterygoid process in those basal turtles, we feel that the triangular process present in *Pleurosternon* and other basal turtles is correctly identified as the basipterygoid process and refer to it by that term, although to recognize that this identification remains to be confirmed, here we enclose that term in quotes.

Institutional abbreviations used in this paper are: IVPP, Institute of Vertebrate Paleontology and Paleoanthropology, Beijing, China; WCW, Wucaiwan (used as part of the field number for specimens collected from the Wucaiwan field area).

Geology

The Shishugou Formation consists of a 450 m thick sequence dominated by fluvial-lacustrine beds. Fossil turtle remains described here are from exposures of the Shishugou Formation in the Jiangjunmiao and Wucaiwan field areas (Fig. 10.1a–c). In both field areas, the formation is dominated by alluvial facies (Eberth et al. 2001). The Jiangjunmiao Formation is considered to span the Middle-Late Jurassic boundary (Eberth et al. 2001), and unpublished radiometric dates from the middle and top of the formation indicate the upper Jurassic section is restricted to the Oxfordian (Clark et al. 2006) Although turtle material was found below and above the boundary, most of the identifiable specimens that could be placed in stratigraphic position are from the upper portion of the formation and, thus, are Oxfordian in age (Fig. 10.1d).

During the Jurassic, the highlands bordering the Junggar Basin (Kalamaili Shan Mountains in the north and Tian Shan Mountains in the south) drained into the central part of the basin. The Shishugou Formation was deposited in wetlands between the Kalimeili Mountains in the north and a central lake, whereas the Qigu Formation was deposited farther to the south along the margins of the Tian Shan

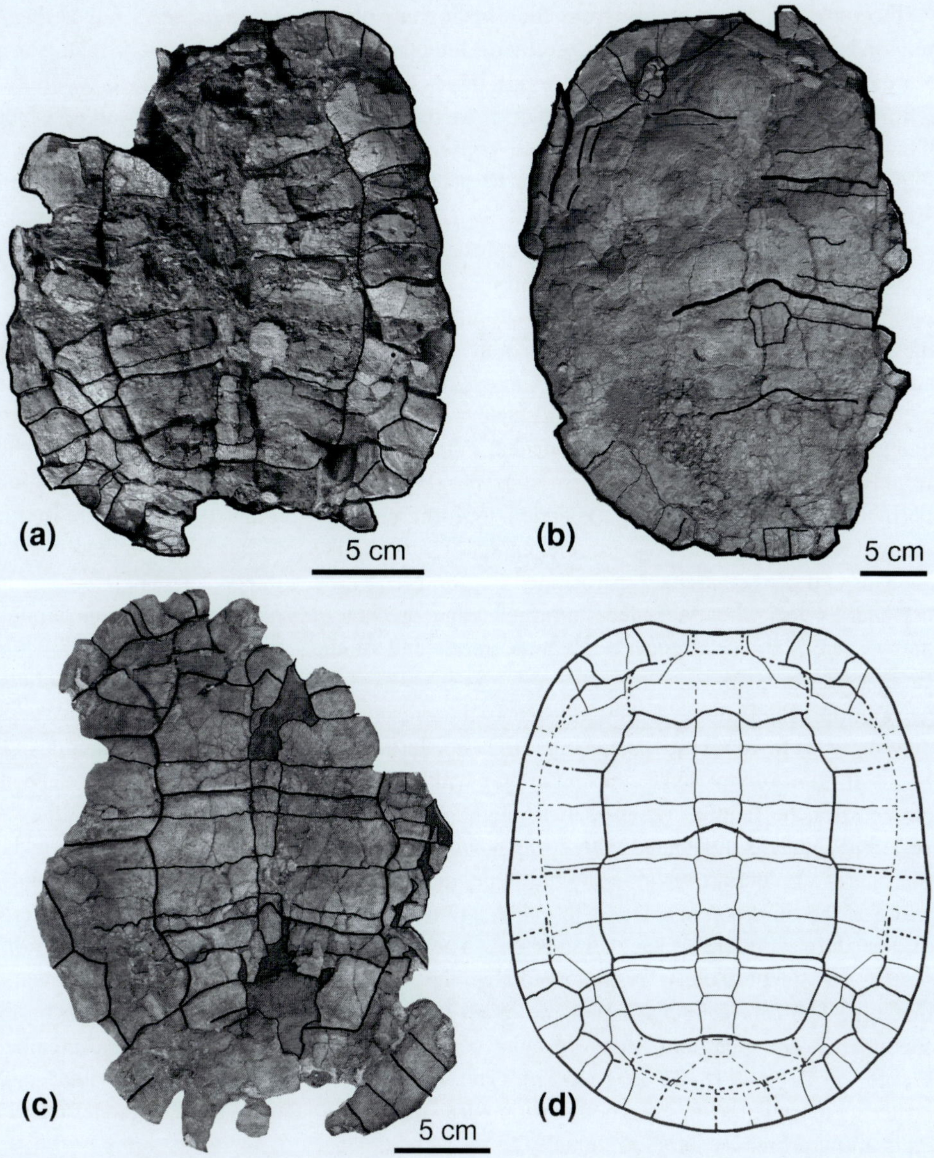

Fig. 10.2 Carapaces of ?*Sichuanchelys* sp., Wucaiwan field area, northern Junggar Basin, Xinjiang, China; upper part of Shishugou Formation; Middle (Callovian) and/or Late (Oxfordian) Jurassic. **a–c** Photographs of original specimens, all in dorsal view: **a** holotype (IVPP V18101); **b** referred specimen (IVPP V18102); **c** referred specimen (IVPP V18103). **d** Reconstruction of carapace in dorsal view, based on all three of the figured specimens. Images at different scales

Mountains (Fig. 10.1e). The climate was seasonally dry (Eberth et al. 2001).

Systematic Paleontology

Order Testudines Batsch 1788
Megaorder Cryptodira Cope 1868
?*Sichuanchelys* sp.
(Fig. 10.2)

Unit, locality and age: Shishugou Formation, Wucaiwan field area, northern Junggar Basin, Xinjiang Uygur Autonomous Region. The stratigraphic position of the holotype relative to the Middle-Late Jurassic boundary is not known, so it may be either late Middle (Callovian) or early Late (Oxfordian) Jurassic in age.

Referred specimens: Three incomplete carapaces, both with their dorsal surfaces exposed: IVPP V18101 (field number WCW-04-57), incomplete carapace missing portions of margins and with dorsal surface exposed (Fig. 10.2a); IVPP V18102 (field number WCW-01-5), missing about the lateral half of its right side (Fig. 10.2b); and IVPP V18103 (field number WCW-03-33), missing much of its margins (Fig. 10.2c). Both specimens are from exposures of the Shishugou Formation in the Wucaiwan field area.

Occurrence: Known only from the Shishugou Formation in the Wucaiwan field area, northern Junggar Basin, Xinjiang. Stratigraphic positions for all three specimens within the Shishugou Formation relative to the Middle-Upper Jurassic boundary are uncertain; consequently, the temporal range of this species can only be listed as Callovian and/or Oxfordian.

Description: ?*Sichuanchelys* sp. is represented by three carapaces of moderate size, ranging from 22.5 to 36 cm long (Fig. 10.2a–c). Although none is complete, together they allow for the major carapacial features to be reconstructed (Fig. 10.2d).

The carapace is sub-rectangular in shape, with the width being about 80% of its length. A nuchal emargination is weakly developed (Fig. 10.2a) or absent (Fig. 10.2b). Peripherals one and two are expanded to form a flange along the anterolateral edge of the carapace, giving the anterior end of the carapace a truncated appearance. These peripherals lack the thickened, upturned edge present in many species of *Xinjiangchelys*. The posterior end of the carapace is rounded.

A prominent feature of the carapace is the presence of extremely wide vertebral scutes (Fig. 10.2a–c). In the case of the third vertebral scute, the maximum width is nearly three times the length. The presence of greatly expanded vertebral scutes is associated with correspondingly reduced (i.e., narrow) pleural scutes. In the holotype, the maximum width of the third pleural scute is less than a third the width of the third vertebral scute (Fig. 10.2a). The remaining scutes are incompletely documented by the available material. The cervical scute is not preserved in any of the specimens. The marginal series is most completely preserved in IVPP V18102 (Fig. 10.2b). The anterior and posterior marginal scutes are both restricted to the peripherals and are broadly exposed dorsally.

Sutures are largely obscured but those on the anterior portion of the shell are distinct on IVPP V18103 (Fig. 10.2c). As well, sutures between most of the costals and the anterolateral peripherals can be identified on either the holotype or IVPP V18103, allowing the reconstruction shown in Fig. 10.2d.

Remarks: ?*Sichuanchelys* sp. is distinctive in the presence of unusually wide vertebral scutes. In the proportions of the vertebral and pleural scutes, ?*Sichuanchelys* sp. is similar to *Sichuanchelys chowi* from the Middle Jurassic of Sichuan: both taxa have exceptionally broad vertebral scutes and narrow pleural scutes. They differ in that *Sichuanchelys chowi* has a deep nuchal emargination, whereas that feature is reduced or absent in ?*Sichuanchelys* sp. As well, the shape of the vertebral scutes differs and the pleural scutes are relatively larger in ?*Sichuanchelys* sp. The morphology of the plastron, which currently is unknown in ?*Sichuanchelys* sp., will provide a test of the inclusion of this species in *Sichuanchelys* rather than *Xinjiangchelys*, because the former primitively retains a mesoplastron, indicating that it is basal relative to *Xinjiangchelys* and other Jurassic turtles from Xinjiang, all of which have lost the mesoplastron.

Capaxorder Eucryptodira Gaffney 1975

Family Xinjiangchelyidae Nessov in Kaznyshkin et al. 1990

Remarks: Xinjiangchelyidae are basal eucryptodires from the Middle Jurassic-Early Cretaceous of Asia that have a plastron that articulates with the peripherals via a series of peg- and socket-joints. They are derived relative to Chengyuchelyidae in the absence of mesoplastra. They can be differentiated from members of the Plesiochelyidae, a group of basal eucryptodires from the Jurassic of Europe, on the basis of a series of features of the plastron and carapace (Peng and Brinkman 1993). These include the presence of a plastron in which the inguinal and axillary buttresses are in the same plane as the bridge and do not, or only slightly, contact the costals, the presence of a weak connection between the carapace and plastron via a series of peg and socket joints, the extension of the pleural-marginal sulcus onto the antero-lateral peripheral, the presence of a significant dorsal thickening of the lateral edge of peripherals two to seven, and highly expanded seventh to eleventh peripherals. In these features xinjiangchelyids are similar to Early Cretaceous sinemydids/macrobaenids such as *Dracochelys* Gaffney and Ye 1992 and *Hangaiemys* Sukhanov and Narmadakh 1974. Xinjiangchelyids are primitive relative to sinemydids/macrobaenids, in the presence of amphicoelous cervical vertebrae, a more robust plastron with large anterior and posterior lobes, and epiplastra that are large and bear distinct dorsal epiplastral processes. Thus the Xinjiangchelyidae are best interpreted as a paraphyletic group that includes basal members of a radiation of eucryptodires, with the Sinemydidae and Macrobaenidae as successively more derived grades (Rabi et al. 2010).

Genus *Xinjiangchelys* Ye 1986a

Type species: *Xinjiangchelys junggarensis* Ye 1986a.

Referred species: *Xinjiangchelys chowi* Matzke et al. 2005; *X. chungkingensis* (Young and Chow 1953); *X. jinyanensis* (Ye and Fang 1982); *X. latimarginalis* (Young and Chow 1953); *X. oshanensis* (Ye 1973); *X. qiguensis* Matzke et al. 2004; *X. radiplicatus* (Young and Chow 1953); *X. radiplicatoides* sp. nov.; and *X. tianshanensis* Nessov 1995.

Occurrence: Middle Jurassic-Early Cretaceous of Asia: Xinjiang (Middle Jurassic-Early Cretaceous) and Sichuan (Middle and Late Jurassic), People's Republic of China; Fergana Basin (Middle Jurassic), Kirgizstan; and Krasnoyarski Kria, Siberia (Middle Jurassic), Russia (see Rabi et al. 2010, Tables 1, 2 and references therein).

Revised diagnosis: Genus of Xinjiangchelyidae characterized by the following unique combination of shell features: carapace thick; vertebral scutes subequal in width and

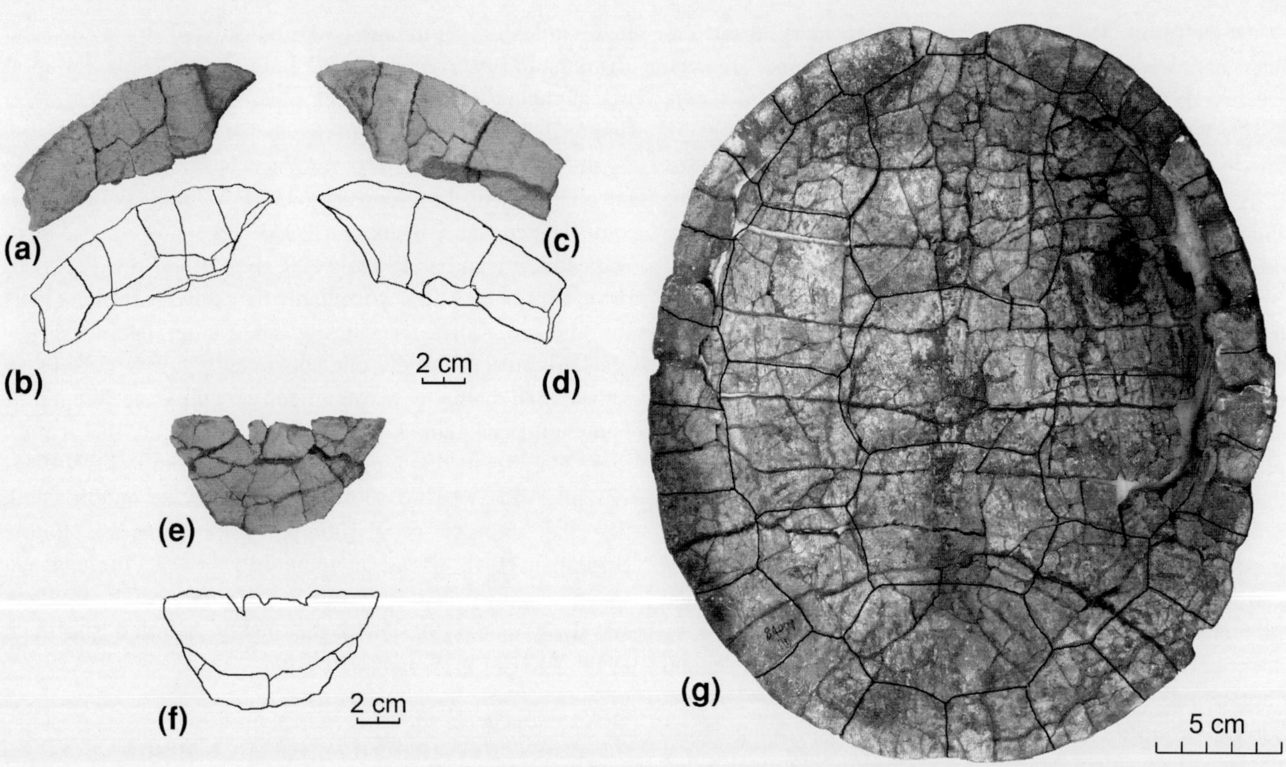

Fig. 10.3 Carapaces of *Xinjiangchelys junggarensis* Ye 1973 (Xinjiangchelyidae); northern Junggar Basin, Xinjiang, China; upper part of Shishugou Formation; Late Jurassic (Oxfordian). **a–f** Referred and fragmentary carapace (IVPP V18100), from the Pingfengshan locality, Wucaiwan field area: **a–d** first to third left peripherals: **a** photograph and **b** interpretive drawing, both in dorsal view; **c** photograph and **d** interpretive drawing, both in ventral view; **e**, **f** suprapygal 2: **e** photograph and **f** interpretive drawing, both in dorsal view. **g** Holotype and nearly complete carapace, (IVPP V7648), from the Jiangjunmiao field area, in dorsal view. Images at different scales

length or wider than long; and epiplastra form a transversely-oriented bar across anterior lobe of plastron.

Remarks: *Annemys* was included in *Xinjiangchelys* by Matzke et al. (2004), but is excluded here because it has a relatively thin shell and narrow vertebral scutes. *Xinjiangchelys* is similar to *Shartagemys* Sukhanov and Narmandakh 2006 in the pattern of scutes on the carapace, but we also recognize the latter as a distinct genus because the orientation of its epiplastra differs.

Xinjiangchelys junggarensis Ye 1973
(Fig. 10.3)

Synonymies: *Xinjiangchelys latimarginalis* (in part): Kaznyshkin 1988, p. 28; Peng and Brinkman 1993, p. 2016, Figs. 3–12; Sukhanov 2000, pp. 313–314, Fig. 17.1; Rabi et al. 2010, p. 262, Fig. 1, Table 1. *Xinjiangchelys junggarensis*: Matzke et al. 2004, p. 1295, Figs. 12, 14; Brinkman et al. 2008, pp. 42–43, Fig. 42.

Referred specimens: Thirteen specimens: IVPP V18100 (field number WCW-04-36): fragmentary carapace consisting of first three left peripherals in articulation, an incomplete costal, and second suprapygal (Fig. 10.3a–f); and a dozen IVPP specimens (IVPP V9537-1 to -9 and -11 to -13), shell and postcranial material, originally described and assigned to *Xinjiangchelys latimarginalis* (Peng and Brinkman 2003, pp. 2016–2022, Figs. 3–11). All referred specimens are from the Pingfengshan locality, Wucaiwan field area, northern Junggar Basin, Xinjiang Uygur Autonomous Region; upper part of Shishugou Formation; Late Jurassic (Oxfordian).

Occurrence: Known from two localities in the upper part (Oxfordian) of the Shishugou Formation, in the northern Junggar Basin, Xinjiang: holotype locality in the Jiangjunmiao field area (Ye 1973); and Pingfengshan locality (for the above-listed, referred specimens) in the Wucaiwan field area.

Revised diagnosis: Species of *Xinjiangchelys* different from *X. latimarginalis*, *X. tianshanensis*, *X. qiguensis*, *X. chowi*, and *X. jinyanensis* in having a carapace that is an elongate oval, rather than round, in outline. Different from *X. tianshanensis* and *X. qiguensis* in having strongly expanded anterolateral peripherals and the first peripheral having a contact with the first costal. Different from *X. chungkingensis*, *X. radiplicatus*, and *X. radiplicatoides* in having carapace with unornamented, smooth external surface; also differs from last two species in having vertebral scutes that are wider than long. Different from *X. oshanensis* in having a larger entoplastron that is located largely posterior to the epiplastra.

Description: IVPP V18100 is a fragmentary, disarticulated carapace (Fig. 10.3a–f). Individual elements are well preserved. The first three left peripherals (Fig. 10.3a–d) are preserved in articulation. Their proportions match those in the holotype carapace of *Xinjiangchelys junggarensis* (Fig. 10.3g) and the other referred specimens from the Pingfengshan locality originally described as *X. latimarginalis* by Peng and Brinkman (1993, Figs. 3a, 4a, 5a, 6). The first peripheral is mediolaterally expanded and, therefore, is longer than wide. The suture for the nuchal is curved and its articular surface for the first costal is short. The second peripheral is more nearly equi-dimensional and its articular surface for the first costal is longer than that on the first peripheral. Ventrally, the articular surface for the axillary buttress of the hyoplastron is represented by a distinct groove that extends well onto the second peripheral. This groove just meets the medial edge of the peripheral, indicating that the axillary buttress would have contacted the first costal. Costal fragments are present, but cannot be accurately positioned and are not figured here. The second suprapygal (Fig. 10.3e–f) is nearly complete. Distinct sutural surfaces for the pygal and eleventh peripherals are present. The sulci show that the twelfth marginal scute extended onto the posterior edge of the element and the eleventh marginal scute overlapped the lateral corner.

Remarks: IVPP V18100 is interpreted as an additional specimen of the taxon that Peng and Brinkman (1993) included in the species *Xinjiangchelys latimarginalis*. Matzke et al. (2004, 2005) argued that the Pingfengshan population was incorrectly identified as *X. latimarginalis* and, instead, provisionally referred those specimens to *X.* '*latimarginalis*' sensu Peng and Brinkman (1993). The Pingfengshan taxon differs from unequivocal specimens of *X. latimarginalis* most distinctly in the proportions and outline of the carapace, which is elongate and oval in the Pingfengshan taxon versus more rounded in the holotype of *X. latimarginalis* (cf., Peng and Brinkman 1993, Figs. 3a, 4a vs. Young and Chow 1953, plates 2–3). The carapace of the Pingfengshan taxon is most similar to the holotype carapace of *X. junggarensis* (cf., Peng and Brinkman 1993, Figs. 3a, 4a vs. Fig. 10.3g), which prompted Brinkman et al. (2008) to transfer the Pingfengshan specimens to that species. In addition, the Pingfengshan carapaces and holotype of *X. junggarensis* differ from *X. latimarginalis* and *X. tianshanensis* in having a subrectangular first peripheral with a distinct contact with the first costal. In *X. latimarginalis* and *X. tianshanensis*, the first peripheral is more triangular in shape and has, at most, a point contact with the first costal. The most nearly carapace from Pingfengshan (IVPP V9537-1; Peng and Brinkman 1993, Fig. 3a, b, 4a, b) differs from the holotype of *X. junggarensis* in having greatly expanded anterolateral peripherals giving a truncated anterior edge; the possibility that this difference is taxonomically significant was considered by Peng and Brinkman (1993). As noted by Rabi et al. (2010), variation of characters within a species must be considered in evaluating their taxonomic significance. Variation in the shape of the anterior end of the carapace is documented by six specimens from Pingfengshan: IVPP V9537-1, -13, -9, -5, and -3 (Peng and Brinkman 1993, Figs. 4a, 5a, 6a, b, d, respectively) and IVPP V18100 (Fig. 10.3a–d). These all have an elongate first peripheral that only barely contacts the first costal. The greatly expanded anterolateral peripherals are present in only one specimen (IVPP V9537-1), so rather than being a taxonomically significant feature as Peng and Brinkman (1993) initially considered, that feature more likely is just an individual variant. Both IVPP V9537-13 and IVPP V18100 show a more rounded anterior end, like the holotype of *X. junggarensis*. Thus, we follow Brinkman et al. (2008) in including the 12 Pingfengshan specimens previously referred to *X. latimarginalis* by Peng and Brinkman (1993) and the newly reported specimen IVPP V18100 in *X. junggarensis*.

Xinjiangchelys radiplicatoides sp. nov.
(Figs. 10.4, 10.5, 10.6, 10.7, 10.8, 10.9, 10.10)

Holotype: IVPP V18104 (field number WCW-01-19): incomplete skeleton consisting of the shell and skull in association and incomplete left shoulder girdle, femur, and tibia preserved in place within the body cavity (Figs. 10.4, 10.5). Surfaces of the shell have been cleaned of matrix, so the dorsal surface of the carapace and the ventral surface of the plastron are visible. The carapace and plastron are largely complete, although the carapace has been distorted post-mortem. The skull is relatively complete, but distorted, and many of its sutures are obscured by poor preservation of the bone surfaces.

Holotype unit, locality, and age: Upper part of Shishugou Formation; Wucaiwan field area, northern Junggar Basin, Xinjiang Uygur Autonomous Region; Late Jurassic (Oxfordian).

Etymology: Species name refers to the morphological similarity to *Xinjiangchelys radiplicatus*.

Referred specimens: Less nearly complete specimens (isolated shell plates, girdle and skull bones, and fragmentary skulls) of *Xinjiangchelys radiplicatoides* sp. nov. are abundant in the Jiangjunmiao turtle bonebed. Costal plates with the diagnostic plicate ornament dominate the locality. Other elements from the locality are identified through comparisons with the holotype skeleton. Referred specimens from the Jiangjunmiao turtle bonebed are included in the catalogue number IVPP V9539, with individual specimens being designated by a unique suffix, as follows: IVPP V9539-1, braincase (Fig. 10.6); IVPP V9539-2, isolated parietal (Fig. 10.7a, b); IVPP V9539-3, isolated postorbital (Fig. 10.7e, f); IVPP V9539-4, isolated quadrate (Fig. 10.7c, d); IVPP V9539-5, isolated dentary

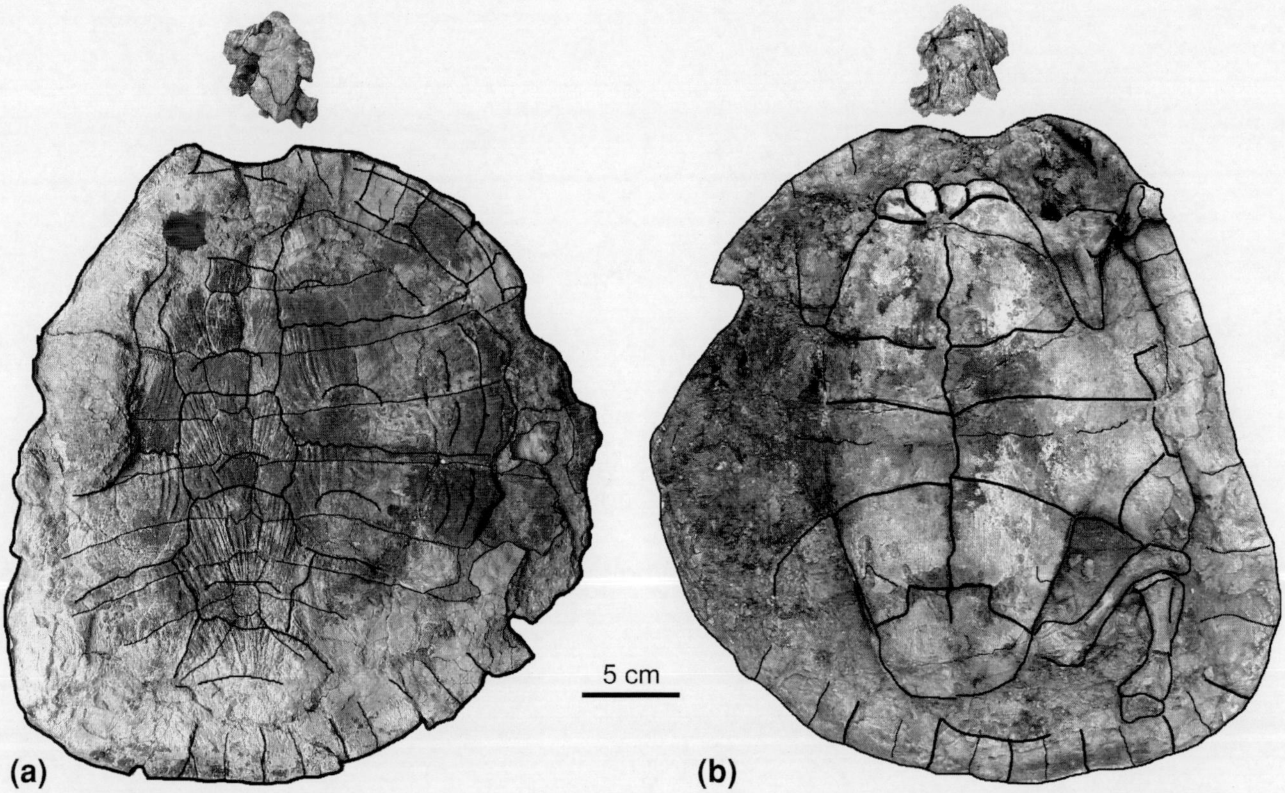

Fig. 10.4 Holotype skeleton (IVPP V18104) of *Xinjiangchelys radiplicatoides* sp. nov. (Xinjiangchelyidae); Wucaiwan field area, northern Junggar Basin, Xinjiang, China; upper part of Shishugou Formation; Late Jurassic (Oxfordian). **a** Dorsal view; and **b** ventral view. Images at same scale

Fig. 10.5 Skull and lower jaws of holotype (IVPP V18104) of *Xinjiangchelys radiplicatoides* sp. nov. (Xinjiangchelyidae); Wucaiwan field area, northern Junggar Basin, Xinjiang, China; upper part of Shishugou Formation; Late Jurassic (Oxfordian). **a** Dorsal view; **b** ventral view; **c** occipital view; and **d** left lateral view. Abbreviations: *bo* basioccipital; *bs* basisphenoid; *fr* frontal; *na* nasal; *pa* parietal; *po* postorbital; *prf* prefrontal; *pt* pterygoid; *qu* quadrate; *so* supraoccipital. Images at same scale

(Fig. 10.7g, i); IVPP V9539-6, nuchal (Fig. 10.8a); IVPP V9539-7, first peripheral (Fig. 10.8b); IVPP V9539-8, second peripheral (Fig. 10.8c); IVPP V9539-9, first costal (Fig. 10.8d, e); IVPP V9539-10, first and second costals in articulation (Fig. 10.8f, g); IVPP V9539-11, isolated third costal (Fig. 10.8h); IVPP V9539-12, isolated third costal (Fig. 10.8i); IVPP V9539-13, pygal, second suprapygal, and eleventh peripheral in articulation (Fig. 10.8j, k); IVPP V9539-14, right and left epiplastra (Fig. 10.9a, b and c, d, respectively); IVPP V9539-15, hyoplastron (Fig. 10.9e); IVPP V9539-16, humerus (Fig. 10.10a–c); IVPP V9539-17, scapula (Fig. 10.10d, e). CT scans of the referred braincase, IVPP V9539-1, are available on the Digimorph Web site at: http://digimorph.org/specimens/Xinjiangchelys_radiplicatoides/.

Occurrence: Known from two localities in the Shishugou Formation, northern Junggar Basin, Xinjiang: holotype locality (Oxfordian) in Wucaiwan field area; and Jiangjunmiao turtle bonebed (Callovian or Oxfordian) in Jiangjunmiao field area.

Diagnosis: Species of *Xinjiangchelys* similar to *X. radiplicatus* and different from other congeners in having carapace ornamented with plications extending forward

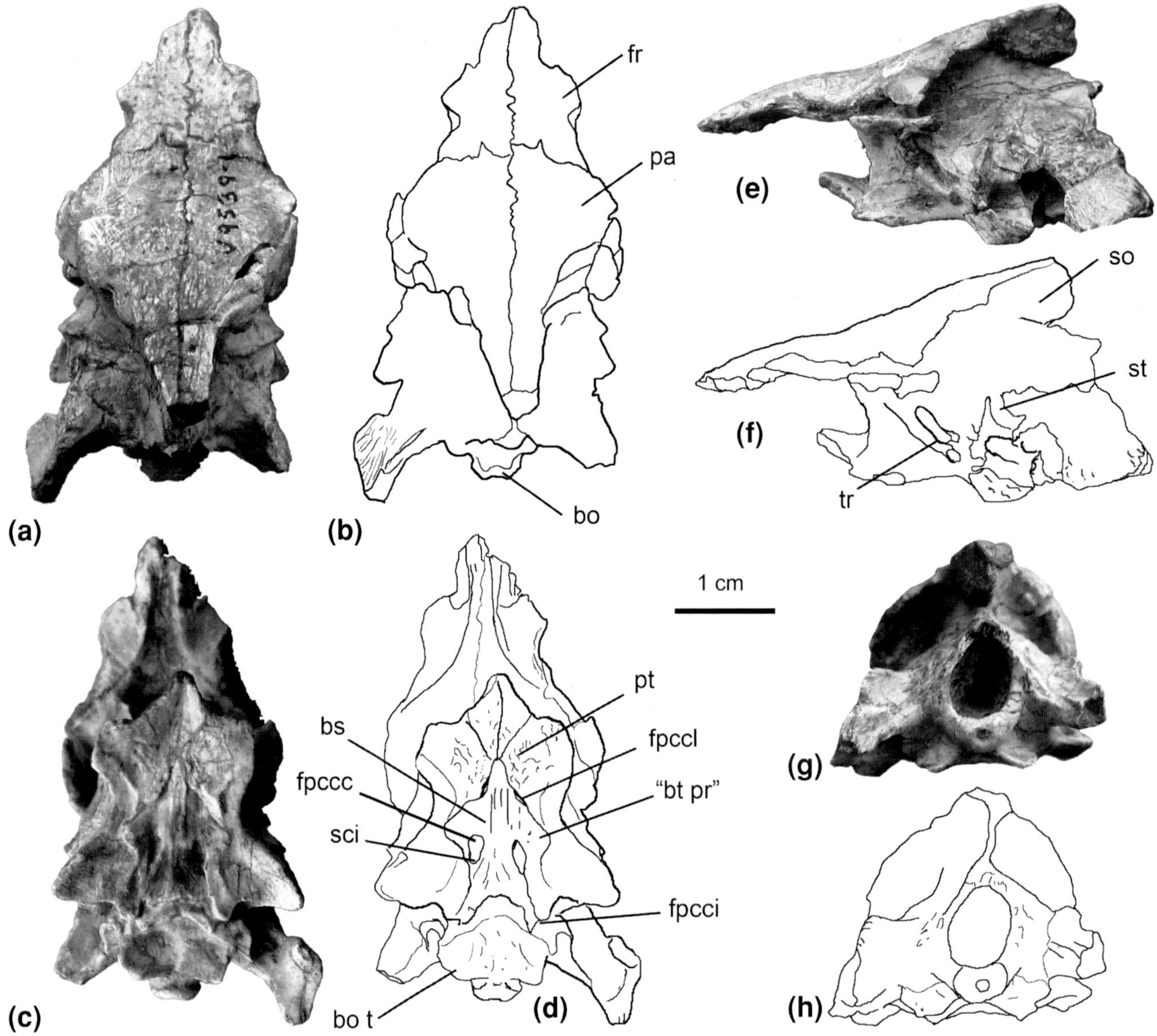

Fig. 10.6 Incomplete, referred skull (IVPP V9539-1) of *Xinjiangchelys radiplicatoides* sp. nov. (Xinjiangchelyidae); Jiangjunmiao turtle bonebed, Jiangjunmiao field area, northern Junggar Basin, Xinjiang, China; upper part of Shishugou Formation; Middle (Callovian) or Late (Oxfordian) Jurassic. **a**, **b** Dorsal view: **a** photograph and **b** interpretive drawing; **c**, **d** ventral view: **c** photograph and **d** interpretive drawing; **e**, **f** left lateral view: **e** photograph and **f** interpretive drawing; **g**, **h** posterior view: **g** photograph and **h** interpretive drawing. Note that the size of the foramen palatinum posterius is inferred from the portion of the border for the foramen preserved on the pterygoid. Abbreviations: *bo* basioccipital; *bo t* basioccipital tubercula; *bs* basisphenoid; *bt pr* basipterygoid process; *fpccc* foramen posterior canalis caroticus cerebralis; *fr* frontal; *fpccl* foramen posterius canalis caroticus lateralis; *fpcci* foramen posterius canalis caroticus internus; *fpp* foramen palatinum posterius; *pa* parietal; *pt* pterygoid; *sci* sulcus caroticus internus; *so* supraoccipital; *st* medial edge of canalis stapedio-temporalis; *tr* trigeminal foramen. Images at same scale

from growth centers on the posterior of the vertebral and pleural scutes and the presence of vertebral scutes in adults that are subequal in width and length. Differs from *X. radiplicatus* in that plications forming the carapace ornament are more numerous and less distinct, radiate outwards from a smooth area on the posterior portions of the vertebral and pleural scutes, and are crossed by ridges that parallel the margins of the scutes.

Description

Skull: The skull is triangular in dorsal view and nearly as wide as it is long (Figs. 10.5, 10.6). The orbits are located far forward and are well exposed dorsally. The skull roof slopes upwards from the orbit, so that in lateral profile the skull appears wedge-shaped. As well, in dorsal view, the lateral edge

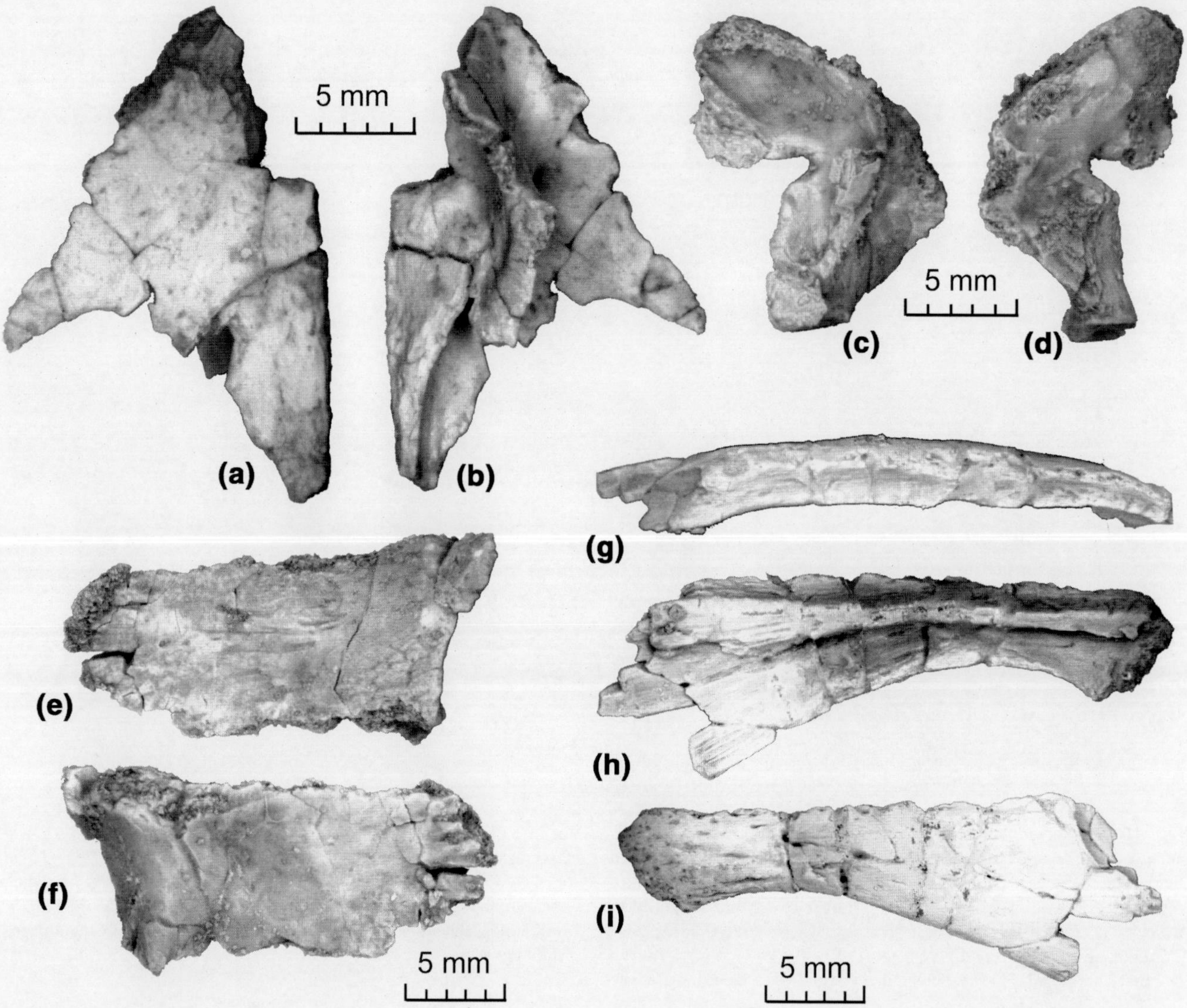

Fig. 10.7 Isolated, referred skull elements of *Xinjiangchelys radiplicatoides* sp. nov. (Xinjiangchelyidae); Jiangjunmiao turtle bonebed, Jiangjunmiao field area, northern Junggar Basin, Xinjiang, China; upper part of Shishugou Formation; Middle (Callovian) or Late (Oxfordian) Jurassic. **a**, **b** Right parietal (IVPP V9539-2) in **a** dorsal view and **b** ventral view; **c**, **d** left quadrate (IVPP V9539-4) in **c** lateral view and **d** medial view; **e**, **f** right postorbital (IVPP V9539-3) in **e** dorsal view and **f** ventral view; **g–i** left dentary (IVPP V9539-5) in **g** occlusal view, **h** medial view, and **i** external view. Images at different scales

of the skull slopes outwards so that the posterior end of the skull is wider than the orbital region. As a result, the height and width of the posterior end of the skull are both approximately double the equivalent dimensions in the region of the orbits.

The skull roof is incompletely preserved, meaning the degree of emargination of the temporal and cheek regions is uncertain. However, at their greatest extent, the emargination of these regions is less than in *Annemys* (see below, *Annemys* sp. account).

Nasals, visible in IVPP V18104, are triangular bones wedged between the frontals and prefrontals. A small contact is present between opposite nasals anterior to the frontals.

Frontals are large elements that contact the nasals anteriorly, the maxilla and prefrontal anterolaterally, the parietal posteriorly, and the postorbital posterolaterally. A lateral lappet extends between the prefrontal and postorbital to form a portion of the orbital border. Narrow anterior processes of the frontals separate the prefrontals, as is typical for basal eucryptodires.

The parietal is the largest element in the skull roof. The frontal–parietal suture extends transversely across the skull roof well posterior to the orbits. Laterally the parietal has a long contact with the postorbital. An isolated parietal, IVPP V9539-2 (Fig. 10.7a, b), preserves a long lateral end that extended past the deepest point of the temporal emargination. Thus, a contact with the squamosal is likely, although the presence of a contact between these bones cannot be directly confirmed. The descending branch of the parietal, fully preserved in IVPP V9539-2 and visible in lateral

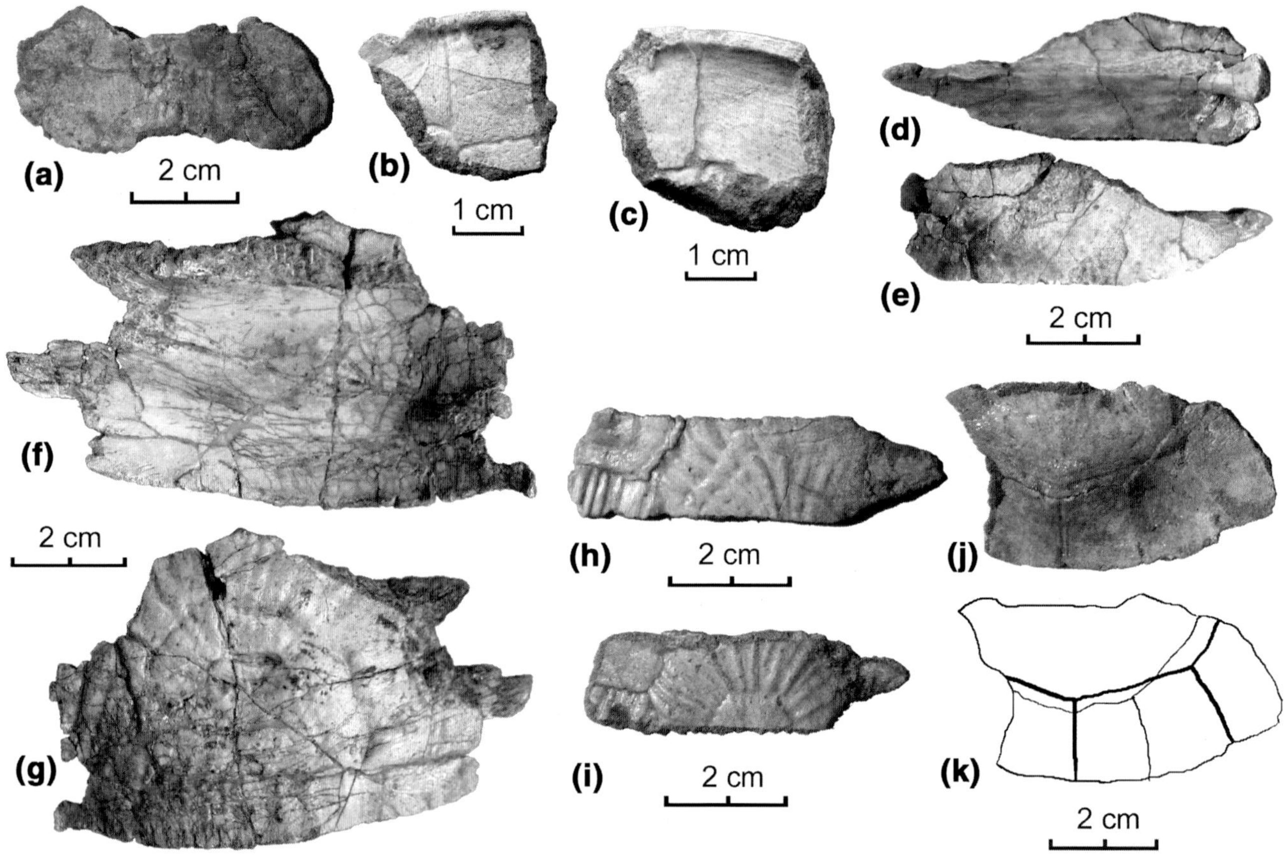

Fig. 10.8 Referred carapace elements of *Xinjiangchelys radiplicatoides* sp. nov. (Xinjiangchelyidae); Jiangjunmiao turtle bonebed, Jiangjunmiao field area, northern Junggar Basin, Xinjiang, China; *upper part* of Shishugou Formation; Middle (Callovian) or Late (Oxfordian) Jurassic. **a** Nuchal (IVPP V9539-6) in dorsal view; **b** first peripheral (IVPP V9539-7) in dorsal view; **c** second peripheral (IVPP V9539-8) in dorsal view; **d, e** first right costal (IVPP V9539-9) in **d** visceral view and **e** dorsal view; **f, g** first and second right costals in articulation (IVPP V9539-10) in **f** visceral view and **g** dorsal view; **h** third right costal (IVPP V9539-11) in dorsal view; **i** third right costal (IVPP V9539-12) in dorsal view; **j, k** pygal, second suprapygal, and eleventh peripheral in articulation (IVPP V9539-13): **j** photograph and **k** interpretive drawing, both in dorsal view. Images at different scales

view in IVPP 9539-1, is well developed (Fig. 10.6e, f). This contacted the pterygoid and epipterygoid, contributing to a fully enclosed trigeminal foramen.

The postorbital is poorly preserved in IVPP V18104 and missing in IVPP V9539-1, but IVPP V9539-3, an isolated postorbital from the Jiangjunmiao turtle bonebed, shows that this is an elongate, rectangular element with a thickened area bordering the orbit (Fig. 10.7e, f).

The portions of the skull formed by the jugal, maxilla, premaxilla, and prefrontal are present in IVPP V18104, but because the sutures delineating these bones could not be identified the exact contacts of these bones are unknown. The squamosal and quadratojugal are not preserved.

The anterior end of the palate is not preserved in any of the available specimens. The posterior portion is preserved in both IVPP V18104 and IVPP V9539-1 (Figs. 10.5, 10.6). In IVPP V9539-1 the palatines form a triangular area anterior to the basisphenoid and pterygoid (Fig. 10.6c, d). The palatines barely contact the basisphenoid, preventing a midline contact between the pterygoids. The posterior edge of the foramen palatinum posterius is present on the anterior edge of the pterygoid and that foramen is of moderate size.

The pterygoids are nearly completely preserved in IVPP 9539-1, missing only the processus pterygoideus externus and their posterolateral tips (Fig. 10.6c, d). The pterygoids extend posteriorly along the basisphenoid, just contacting the anterolateral corner of the basioccipital.

The basisphenoid is an elongate element that separates the pterygoids and has a broad contact with the basioccipital (Fig. 10.6c, d). The V-shaped anterior end of the basisphenoid separates the pterygoids at the midline. Ventrally an elongate, rounded ridge extends posteriorly from this process to the posterior edge of the basisphenoid. A prominent feature of the basisphenoid is the "basipterygoid process," a triangular lateral processes that underlies the pterygoid, extending laterally to a ridge that runs posteriorly from the lateral tip of the processus pterygoideus externus. The ventral surface of the basisphenoid lacks the distinct

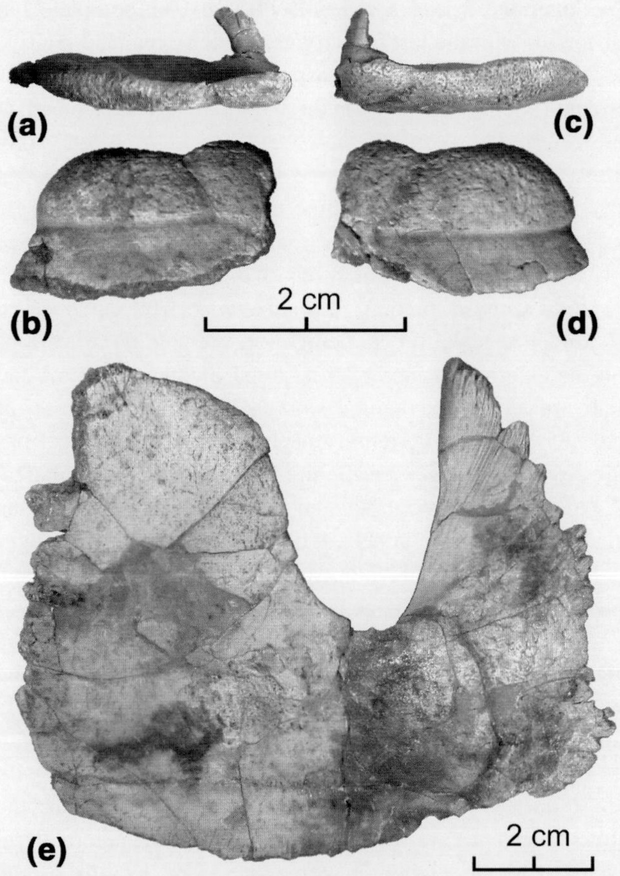

Fig. 10.9 Isolated, referred plastral elements of *Xinjiangchelys radiplicatoides* sp. nov. (Xinjiangchelyidae); Jiangjunmiao turtle bonebed, Jiangjunmiao field area, northern Junggar Basin, Xinjiang, China; upper part of Shishugou Formation; Middle (Callovian) or Late (Oxfordian) Jurassic. **a–d** Epiplastra (IVPP V9539-14): **a**, **b** right epiplastron in **a** anterior view and **b** ventral view; **c**, **d** left epiplastra in **c** anterior and **d** ventral view; **e** left hyoplastron (IVPP V9539-15) in ventral view. Images at different scales

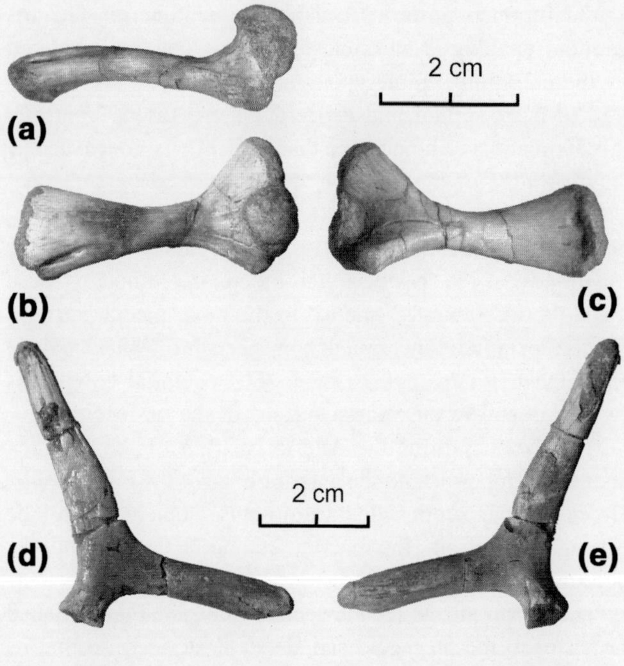

Fig. 10.10 Isolated, referred humerus and scapula of *Xinjiangchelys radiplicatoides* sp. nov. (Xinjiangchelyidae); Jiangjunmiao turtle bonebed, Jiangjunmiao field area, northern Junggar Basin, Xinjiang, China; upper part of Shishugou Formation; Middle (Callovian) or Late (Oxfordian) Jurassic. **a–c** right humerus (IVPP V9539-16) in **a** posterior view, **b** dorsal view, and **c** ventral view; **d**, **e** left scapula (IVPP V9539-17) in **d** anterior view and **e** posterior view. Images at different scales

pits like those present in *Ordosemys leios* Brinkman and Peng 1993b (Brinkman and Wu 1999).

The foramina and canals of the basicranial region are distinct in IVPP V9539-1. CT scans of this specimen allow the relative size and position of the common carotid artery and its major branches to be identified (Fig. 10.6d). The foramen posterius canalis caroticus internus is located at the posterior end of the contact between the pterygoid and basisphenoid. The canalis caroticus internus runs forward between the basisphenoid and pterygoid to open into groove in the basisphenoid, referred to as the sulcus caroticus internus by Sukhanov (2000). This groove extends forward to the foramen posterius canalis carotici cerebralis (fpccc in Fig. 10.6d). Because the foramen posterius canalis carotici cerebralis is the opening in the basisphenoid through which the cerebral branch of the internal carotid artery would have entered the sella turcica, this groove would have been occupied by the internal carotid artery. Slit-like openings between the basisphenoid and the pterygoid anterior to the "basipterygoid processes" are identified as the foramen posterius canalis caroticus lateralis, the opening through which the palatine artery enters the skull. Thus the palatine artery would have branched off from the internal carotid artery in the area of the sulcus caroticus internus and passed ventral to the basisphenoid to reach the foramen palatinum posterius. A shallow groove leads from the anterior end of the sulcus caroticus internus to the foramen posterius canalis caroticus lateralis. This likely marks the path of the palatine artery in this region.

The braincase and otic capsule are visible in dorsal view in IVPP V18104 (Fig. 10.5a). The external surface of the braincase is fully exposed in IVPP V9539-1 (Fig. 10.6), although sutures are difficult to identify and the elements lateral to the stapedial foramina are missing. Reflecting the inflated posterior region of the skull, the supraoccipital crest is tall (Figs. 10.5c, 10.6g, h). Although the posterior end is not preserved, it would have extended at most only a short distance past the occipital condyle. The presence of a short supraoccipital crest is a widespread feature of basal cryptodires, also being present for example in *Kayentachelys* Gaffney et al. 1987, *Kallokibotion* Nopcsa 1923, and *Chubutemys* Gaffney et al. 2007. In IVPP V9539-1 the medial edge of the left stapedial foramen is present, showing that this foramen was large in diameter compared to the foramen posterius canalis carotici interni. The trigeminal opening is an elongate, narrow opening bordered by the pterygoid, epipterygoid, and parietal.

The basioccipital and exoccipitals are fused. Ventrally this element has a broad, area that is shallowly concave. The anterior edge of the concave area is marked by a distinct ridge that lies just posterior to the basioccipital-basisphenoid suture. The lateral edges of the concave area are identified as the basioccipital tubercula, although these are little differentiated relative to the more medial surface. The basioccipital tubercula floor a space into which a pair of foramina nervi hypoglossi open. The occipital condyle is located dorsal to the ventral surface of the basioccipital.

The prootic forms the anterior margin of the foramen stapedio-temporale and the anterior edge of the otic capsule. However, the sutures delineating this element cannot be distinguished from fractures so its full extent cannot be determined.

The opisthotic, as seen in ventral view, is largely situated posterior to the pterygoid and slopes posteriorly so that its lateral tip is posterior to the occipital condyle. The fenestra postotica, which is bordered by the opisthotic, exoccipital, pterygoid, and quadrate, is a circular opening that faces primarily ventrally.

The quadrate is best represented by an isolated element from the Jiangjunmiao turtle bonebed (IVPP V9539-4; Fig. 10.7c, d). This is rather tall, with a shallow incisura columellae auris. The cavum tympani has a gently curved anterior edge. The anterior surface of the quadrate bears a pitted area indicating that it contributed to the processus trochlearis oticum. Features on the internal surface of the quadrate can be identified through comparison with *Chelydra* as illustrated by Gaffney (Gaffney 1972, Fig. 18). A dorsal groove leading from the insisura columellae auris is identified as the canalis stapedio-temporalis, and a ventrally oriented groove is identified as the canalis cavernosus. Below the canalis cavernosus a sutural surface for the pterygoid is present.

An isolated dentary (IVPP V9539-5; Fig. 10.7g, i) is long and slender, and has a narrow triturating surface. Sutural surfaces show that the splenial extended about halfway along the medial surface of the dentary. The coronoid did not contribute to the formation of the triturating surface. The anterior end of the dentary is expanded, so the ventral surface of the element is concave in lateral view.

Vertebrae: The vertebral column is poorly represented in the available material. Although the skull was preserved in association with the shell in IVPP V18104, no cervical or caudal vertebrae are apparent in that specimen. Isolated cervical centra of two distinct morphologies are present in the Jiangjunmiao turtle bonebed, but there is no way of associating either of those cervical morphs with *Xinjiangchelys radiplicatoides*.

Carapace: The carapace of IVPP V18104 is complete, but distorted, with the left posterolateral edge expanded relative to the right (Fig. 10.4a, b). Based on this specimen numerous isolated shell elements from the Jiangjunmiao turtle bonebed are referred to *Xinjiangchelys radiplicatoides* (Figs. 10.8, 10.9). The undistorted right side of the carapace indicates that the shell is subcircular in outline and is approximately as wide as it is long. In its strongly rounded profile, the carapace of *X. radiplicatoides* is similar to *X. chowi*, *X. qiguensis*, *X. tianshanensis*, and *X. latimarginalis*. The carapace of *X. junggarensis* differs in being more rectangular. Plications on the external surface of the carapace extend forward and outward from a hexagonal smooth area near the posterior edge of each of the vertebral and pleural scutes. The plications are crossed by ridges paralleling the smooth hexagonal area. As demonstrated by isolated costal elements from the Jiangjunmiao turtle bonebed, both the plications and ridges are most strongly developed in juvenile individuals and become subdued with age (Fig. 10.8).

The nuchal is best documented by IVPP V9539-6, from the Jiangjunmiao turtle bonebed (Fig. 10.8a). This short and wide element bears a distinct nuchal emargination anteriorly and a concave notch on its posterior edge for the first neural.

Eight neurals are present. The first is hexagonal with short posterolateral edges, the second is rectangular, the third to seventh are hexagonal and have short anterolateral edges, and the eighth is rectangular. The lengths of the neurals decrease posteriorly, with the sixth neural being only slightly longer than wide.

The first costal is subrectangular in shape and bears a sharp ridge ventrally that extends along the length of the element (Fig. 10.8d–g). The sutural surface on the anterior face of this ridge shows that the first thoracic rib was elongate and reached the lateral edge of the bone. No evidence for a contact of the costal with the hyoplastron is present. Costals two to six extend relatively straight laterally and show little expansion of their lateral ends (Fig. 10.8h, i). Costals eight differ in that they curve posteriorly and have an expanded lateral edge. The shape of costals seven is intermediate between costals six and eight.

Peripherals one to three and nine to eleven are preserved in IVPP V18104; additionally, an isolated first peripheral (IVPP V9539-7; Fig. 10.8b) and an isolated second peripheral (IVPP V9539-8; Fig. 10.8c) are available from the Jiangjunmiao turtle bonebed. The anterior peripherals have thickened lateral edges, like those of *Xinjiangchelys junggarensis*. This extends at least to the fourth peripheral and likely farther. The first peripheral has a broad area of contact with the first costal. The axillary buttress extends onto the second peripheral. Posterolateral peripherals are thin elements that are nearly square in dorsal view. The anterior extend of the bridge is unclear, however, posteriorly it extends to the eighth peripheral.

The posterior end of the carapace is best documented by an articulated second suprapygal, pygal, and eleventh peripheral (Fig. 10.8j, k). Two suprapygals are present. The first is narrow relative to the second and does not contact the peripherals. The second suprapygal extends to the tenth peripheral and has a short area of contact with the eighth costal. The pygal is nearly square.

The pattern of scutes on the carapace is documented by IVPP V18104 (Fig. 10.4a). The cervical scute is wide. Vertebral scutes are approximately square and are much narrower than the pleural scutes. The lateral edges of the second and third vertebral scutes are nearly straight. The fourth vertebral scute has a more distinctly angled lateral edge. The fifth vertebral scute is only slightly wider than the fourth. For the fifth vertebral scute, its posterior edge crosses the second suprapygal, whereas its anterior edge crosses the first suprapygal. The first pleural extends onto peripherals one and two. The lateral margin of the pleural scutes posterior to this could not be followed on IVPP V18104, but isolated posterolateral peripherals from the Jiangjunmiao turtle bonebed show that the lateral tip of the pleural scutes extended onto the peripherals in the posterior region of the shell. Both the eleventh and twelfth marginal scutes extend onto the second suprapygal. The pygal is subdivided by the sulcus separating the twelfth marginal scutes.

Plastron: The plastron of *Xinjiangchelys radiplicatoides* is documented by the holotype skeleton (Fig. 10.4b) and by the isolated elements from the Jiangjunmiao turtle bonebed (Fig. 10.9). The anterior lobe, posterior lobe, and bridge are subequal in length. Both the anterior and posterior lobes have truncated ends. Plastral fenestrae are absent in adults, although these were well developed in juveniles.

As in other species of *Xinjiangchelys*, the epiplastra are short and wide, so together they form a transversely-oriented bar that extends across the anterior lobe of the plastron. Isolated epiplastra (Fig. 10.9a–d) from the Jiangjunmiao turtle bonebed show that distinct dorsal epiplastral processes are present. The entoplastron is a large, shield-shaped element. Its anterior end is continuous with the anterior edge of the hyoplastron, so together the entoplastron and hyoplastron form a nearly straight sutural surface for the epiplastra.

The hyoplastron has a rectangular ventral surface and a bridge area that is weakly angled relative to this surface. The inguinal buttress and the bridge are differentiated by the texture of the bone: the bridge area, like the ventral surface, has weakly developed texture but the inguinal buttress is smooth, likely because this area is not covered by scutes. These two areas are in approximately the same plane. The lateral edge of the hyoplastron bears a series of pegs that fit in sockets in the bridge peripherals. These are best developed anteriorly, but extend along the full length of the lateral edge of the hyoplastron. The buttress does not extend onto the costals.

The hypoplastron largely mirrors the hyoplastron in general shape, although the bridge portion of the hypoplastron is narrower than the bridge portion of the hyoplastron. The axillary buttress, which is differentiated from the bridge area by the smooth surface, is in about the same plane as the bridge region. The buttress extends to the eighth peripheral, but does not extend onto the costals.

Xiphiplastra are present in IVPP V18104, and these form the posterior half of the posterior lobe of the plastron.

Sulci are generally distinct. The gular and intergular scutes are located entirely on the epiplastra and are transversely oriented. The mid-line sulcus is weakly sinusoidal anteriorly and nearly straight posteriorly. The humeral-pectoral sulcus extends transversely across the base of the anterior lobe. The pectoral-abdominal sulcus is located near the posterior edge of the hyoplastron. The abdominal-femoral sulcus crosses the base of the posterior lobe; this sulcus is gently convex and curves forward. The femoral-anal sulcus has a blunt anterior end located just anterior to the hypoplastron–xiphiplastron suture. Four inframarginals are present, each extending about halfway up the bridge region.

Pectoral girdle: The acromion process of the scapula is preserved in specimen IVPP V18104; based on that specimen, an isolated scapula from the Jiangjunmiao turtle bonebed is referred to *Xinjiangchelys radiplicatoides* (Fig. 10.10d, e). The acromion and scapular processes are rod-like and between them form an obtuse angle. The acromion process is slightly shorter than the humerus.

Fore limb: The humerus is visible in ventral view in IVPP V18104; based on that skeleton, isolated humeri from the Jiangjunmiao turtle bonebed are referred to *Xinjiangchelys radiplicatoides* (Fig. 10.10a–c). The humerus is a stout bone, with a slightly upturned head. In end view, the two trochanters form nearly a 90° angle with one another. The ectepicondylar foramen is fully enclosed.

Hind limb: The hind limb is represented by a left femur and tibia of IVPP V18104, preserved in articulation (Fig. 10.4b). The femur is longer than the humerus and has a less distinctly upturned head. The tibia is shorter than the femur.

Remarks: *Xinjiangchelys radiplicatoides* sp. nov. is one of the most osteologically fully documented basal eucryptodires and it provides the best evidence for a close affiliation between *Xinjiangchelys* species from Xinjiang and Sichuan. *X. radiplicatus* was erected by Young and Chow (1953) on the basis of the central part of a carapace from the Late Jurassic of Sichuan. Later, a complete shell was described by Ye (1986b). The latter specimen is here interpreted as a juvenile, because of its relatively small size and well developed costal-peripheral fenestrae. *X. radiplicatus* was diagnosed largely on the basis of a distinctive ornamentation on the carapace. This consists of ridges extending forwards from the middle of the posterior edges of the scutes. The presence of a closely related taxon in the southern part of the Junggar Basin was reported by Maisch et al. (2003), who referred fragmentary carapace elements from the Toutunhe and Qigu formations with a similar distinctive plicated surface texture to *X. cf. radiplicatus*. Specimen IVPP V18104 and the isolated elements from the Jiangjunmiao turtle bonebed confirm that the Xinjiang species is similar to *X. radiplicatus*, but specifically distinct.

Xinjiangchelys radiplicatoides differs from *X. radiplicatus* in the pattern of ornament on the carapace. In *X. radiplicatoides* the plications are more numerous and less distinct and extend outward from a smooth rectangular area near the posterior end of the carapace. In *X. radiplicatus* the plications are coarser and converge on a point near the posterior edge of the carapace. As well, these species differ in that *X. radiplicatoides* has ridges paralleling the margin of the scutes crossing the plications. These differences are interpreted here as specifically significant, rather than a result of variation within a species, because the range of variation present in specimens from the Jiangjunmiao turtle bonebed does not include any examples of the ornament pattern seen in *X. radiplicatus*.

As well as being similar in their development of plications on the surface of the carapace, adults of *Xinjiangchelys radiplicatus* and *X. radiplicatoides* are similar to one another and different from other species of *Xinjiangchelys* in having vertebral scutes that are subequal in width and length and smaller than the pleural scutes. In other species of *Xinjiangchelys*, the vertebral scutes generally are twice as wide as they are long and are larger than the pleural scutes.

As indicated by the undistorted left half of the holotype, the carapace in *Xinjiangchelys radiplicatoides* was distinctly round in outline. The presence of a round carapace is widespread in *Xinjiangchelys*, also being present in *X. chowi*, *X. tianshanensis*, *X. qiguensis*, and *X. latimarginalis*. This shape differs from that of *X. junggarensis* which has a more elongate, oval carapace. However, the carapace in the juvenile specimen of *X. radiplicatus* described by Ye (1986b) is oval and longer than wide, indicating that the proportions of the shell in that species changes during growth.

Two specimens, IVPP V18104 and IVPP V9636-1, provide the first evidence of a relatively complete skull for a species of *Xinjiangchelys*. The skull differs in its proportions from that of *Annemys levensis*, described by Sukhanov (2000) and *Annemys* sp. described below, which are the only other Asian Jurassic cryptodires represented by relatively complete cranial material. In *Annemys* the height and width of the postorbital portion of the skull does not expand posteriorly, so that the post-orbital region is more parallel sided. The presence of an inflated post-temporal region can be interpreted as a primitive feature, because that is widespread in basal cryptodires, being present for example in *Kallokibotion* (Gaffney and Meylan 1992b) and *Chubutemys* (Gaffney et al. 2007). Expansion of the temporal region likely was a strategy for increasing the mass of the adductor musculature in the absence of an elongate supraoccipital crest.

The skull of *Xinjiangchelys radiplicatoides* is notably primitive in its basicranial region. The pointed anterior end of the basisphenoid is reminiscent of the cultriform process of primitive reptiles and the slit-like shape of the foramen caroticum laterale is suggestive of a reduced interpterygoid vacuity. The latter identification is consistent with the interpretation presented by Sterli et al. (2010) that in basal turtles, such as *Proganochelys*, the palatine artery would have entered the skull through the interpterygoid vacuity and that the canalis caroticus lateralis formed as a result of closure of the interpterygoid vacuity around that artery.

The basicranial region of *Xinjiangchelys radiplicatoides* is similar to that of *Chubutemys* in the shape and proportions of the basisphenoid, presence of a slit-like opening between the basisphenoid and the pterygoid, and presence of a distinct sulcus carotici leading to the canalis caroticus cerebralis. They differ in that the "basipterygoid process" of *Chubutemys* is relatively smaller and the posterior portion of the canalis caroticus cerebralis is less fully covered. The similarities between these two taxa are consistent with the interpretation that *Chubutemys* is a eucryptodire at about the same grade of evolution as *X. radiplicatoides*, but may also be because they are both plesiomorphic (Smith and Kear 2012).

The quadrate is different from that of *Annemys* and similar to *Kayentachelys* in having a shallow incisura columellae (Sterli and Joyce 2007; Gaffney and Jenkins 2010). Thus, in this feature as well, *Xinjiangchelys radiplicatoides* is primitive relative to *Annemys*. Although the anterior edge of the otic capsule of *X. radiplicatoides* is not completely visible on either specimen, the rugose anterior surface on the quadrate demonstrates that a processus trochlearis oticum was present, as would be expected in a basal eucryptodire.

Xinjiangchelys chowi Matzke, Maisch, Sun, Pfretzschner, and Stöhr 2005
(Fig. 10.11)

Referred specimen: IVPP V18105 (field number WCW-02-61): incomplete carapace, preserving the left half and much of the median area, exposed in dorsal view (Fig. 10.11a, b), plus associated postcranial elements, including an incomplete cervical vertebra (Fig. 10.11c, d). From "Pingfen Shan J23" locality, Wucaiwan field area, northern Junggar Basin, Xinjiang Uygur Autonomous Region; upper part of Shishugou Formation; Late Jurassic (Oxfordian). This is the only referred specimen known for the species.

Occurrence: Known from two localities in the Junggar Basin, Xinjiang: holotype locality near Urumqui, southern part of basin, in the Middle Jurassic (Callovian) Toutunhe

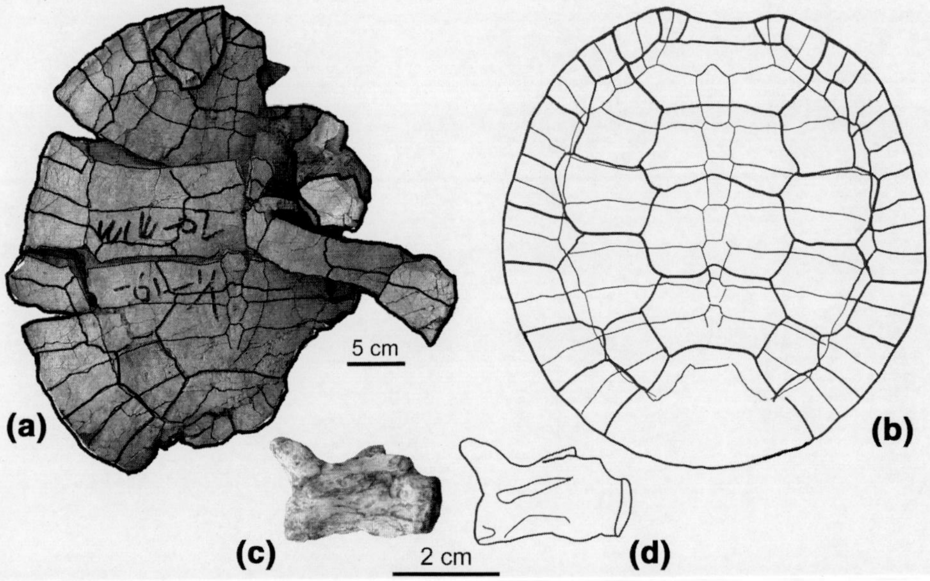

Fig. 10.11 Associated carapace and cervical vertebra (IVPP V18105) of *Xinjiangchelys chowi* Matzke et al. 2005 (Xinjiangchelyidae); "Pingfen Shan J23" locality, Wucaiwan field area, northern Junggar Basin, Xinjiang, China; upper part of Shishugou Formation; Late Jurassic (Oxfordian). **a** Carapace in dorsal view; **b** reconstruction of carapace, based on figured specimen, in dorsal view; **c**, **d** cervical vertebra in left lateral view: **c** photograph and **d** interpretive drawing. Images at different scales

Formation (Matzke et al. 2005); and "Pingfen Shan J23" locality (for above-listed referred specimen), Wucaiwan field area, northern part of basin, in the Upper Jurassic (Oxfordian) upper part of the Shishugou Formation.

Diagnosis: See Matzke et al. (2005)

Description: The carapace of the referred specimen, IVPP V18105, is of moderate size, low domed, and round in outline, with a weakly developed nuchal emargination (Fig. 10.11a, b). The external surface of the carapace is smooth and sulci on the carapace are generally distinct.

The carapace is partially disarticulated, allowing many of the sutures to be identified with confidence. The nuchal is broad and relatively short. Seven neurals can be identified, of which the posterior six are strongly hexagonal. The sutures delineating the eighth costal and suprapygals cannot be differentiated, so it is uncertain if an eighth neural was present and if the neural series contacted the suprapygals. The first nine peripherals can be differentiated. The first peripheral has only a very narrow area of contact with the first costal. A striking feature of the carapace is the presence of wide bridge peripherals that are approximately as wide as long and are broadly exposed in dorsal view. This is in contrast to the typical xinjiangchelyid condition, in which the bridge peripherals are narrower than they are long and are poorly exposed in dorsal view.

Sulci generally are distinct. The cervical scute is short and wide. The vertebral scutes are wider than long and have strongly angled lateral edges. Unlike ?*Sichuanchelys* sp., which also has wide vertebral scutes, the pleural scutes are also wide, their maximum width being about a half the maximum width of the vertebral scutes. The pleural scutes extend onto the peripheral series and the fourth, seventh, and ninth marginal scutes extend onto the costals.

A single cervical vertebra (Fig. 10.11c, d) was found in association with the carapace. The centrum is amphicoelous and elongate. The transverse process is a short, flange-like structure that extends from about the middle of the centrum to its anterior end. A sharp mid-ventral ridge is present. Both in its proportions and development of the transverse process, the cervical vertebrae are similar to those of *Xinjiangchelys junggarensis* as described by Peng and Brinkman (1993, Figs. 3e–g, 8).

Remarks: *Xinjiangchelys chowi* was described by Matzke et al. (2005) on the basis of a shell from the Middle Jurassic (Callovian) Toutunhe Formation on the southern margin of the Junggar Basin. It was diagnosed largely on the basis of the presence of large plastral fenestrae. Matzke et al. (2005) considered this to be a diagnostic feature of the species, rather than a developmental feature, because the carapace was of large size. However subsequent studies have questioned the diagnostic reliability of that feature (Rabi et al. 2010), because in turtles the size of plastral fenestrae typically changes during development. Despite uncertainty about the taxonomic significance of the presence of large plastral fenestrae in the holotype, *X. chowi* is here considered a valid species because of the presence of mediolaterally expanded bridge peripherals that are broadly exposed in dorsal view. This is most strongly expressed by the fifth and sixth peripherals, which are wider than they are long. These proportions contrast with the typical condition

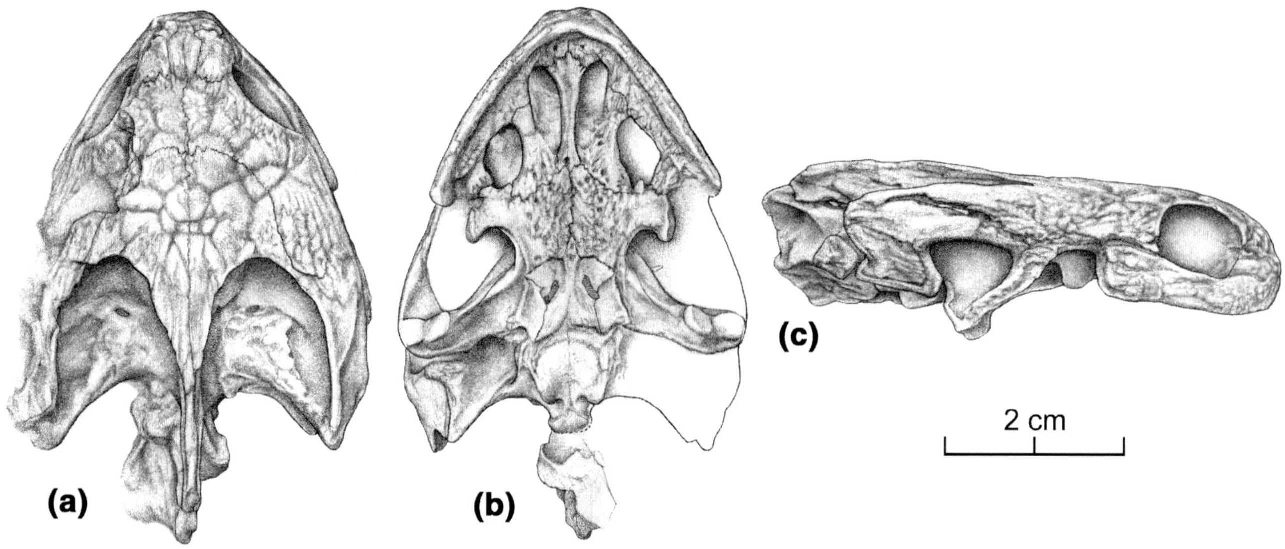

Fig. 10.12 Illustration of skull (part of skeleton IVPP V18106) of *Annemys* sp. (Xinjiangchelyidae); Wucaiwan field area, northern Junggar Basin, Xinjiang, China; upper part of Shishugou Formation; Middle (Callovian) or Late (Oxfordian) Jurassic. **a** Dorsal view; **b** ventral view; **c** right lateral view. Matrix adhering to skull is not fully illustrated. Images at same scale

in basal cryptodires, in which the bridge peripherals are narrower than long and poorly exposed in dorsal view. In having wide peripherals, *X. chowi* is most similar to *X. qiguensis*. However *X. qiguensis* has distinct plications extending posteriorly from the vertebral and pleural sulci. Specimen IVPP V18105 is included in *X. chowi*, because no such plications are present.

The occurrence of *Xinjiangchelys chowi* in both the upper beds of the Shishugou Formation of the Wucaiwan area, which are Late Jurassic (Oxfordian) in age, and the Middle Jurassic (Callovian) Toutunhe Formation suggests there was no major change in turtle assemblages across the Middle-Late Jurassic boundary in the present-day Junggar Basin region.

Genus *Annemys* Sukhanov and Narmandakh 2006
Annemys sp.
(Figs. 10.12, 10.13, 10.14, 10.15)

Synonymy: *Annemys* sp.: Rabi et al. 2010, p. 263, Fig. 1g, h, Table 1.

Referred specimens: IVPP V18106 (field number WCW-04-10), incomplete skeleton consisting of a moderately complete skull (Figs. 10.12, 10.13, 10.14), four cervical vertebrae in articulation (Fig. 10.15), and an incomplete carapace (not figured) in two parts, the first part including four disarticulated anterior costals and a series of anterolateral peripherals in articulation and the second part including a series of posterolateral peripherals in articulation. CT scans of the skull from this specimen are available on the Digimorph Web site at: http://digimorph.org/specimens/Annemys_sp/. IVPP V18107 (field number WCW05A-30), fragmentary carapace consisting of the nuchal and first peripheral in articulation and two disarticulated costals (not figured). Both specimens are from the Wucaiwan field area, northern Junggar Basin, Xinjiang Uygur Autonomous Region, and their positions within the Shishugou Formation relative to the Middle-Upper Jurassic boundary are uncertain.

Occurrence: *Annemys* sp. (sensu this study) is known only from the Middle (Callovian) and/or Upper (Oxfordian) Jurassic Shishugou Formation, in the Wucaiwan field area, northern Junggar Basin. Outside of the Jungarr Basin, the genus is known from the Late Jurassic both in the nearby Turpan Basin, Xinjiang, and in Mongolia (Rabi et al. 2010, Table 1 and references therein).

Description

Skull: The skull of IVPP V18106 is nearly complete and little crushed (Figs. 10.12, 10.13, 10.14). It has an elongate, subtriangular shape in dorsal view (Figs. 10.12a, 10.13a). Its width across the quadrates is about two-thirds of its length to the end of the supraoccipital crest. In lateral view (Figs. 10.12c, 10.14), the skull is low, its height being about a quarter of its width across the quadrates. The orbits are large, located far forward, and face primarily anterolaterally. Both the temporal and cheek regions are deeply emarginated: the temporal emargination extends just anterior to the otic capsule and the cheek emargination extends to about midway along the height of the orbit. The supraoccipital crest is an elongate, thin, vertical flange of bone that extends posteriorly beyond the squamosals. The dorsal surface of the skull bears well-defined sulci that show that a complex arrangement of scutes was present on the skull.

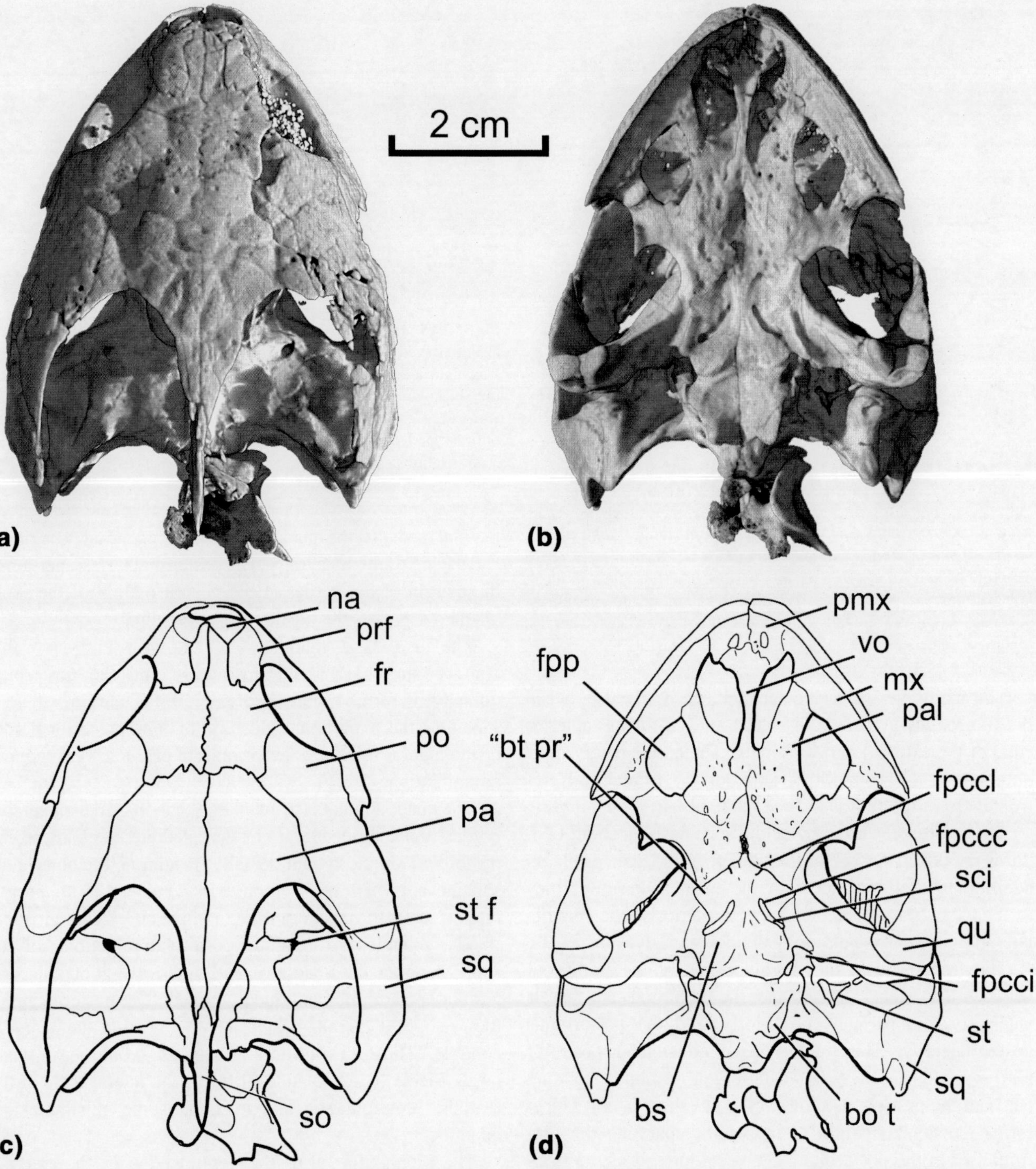

Fig. 10.13 CT volume renderings and interpretive drawings, in dorsal and ventral views, of skull (part of skeleton IVPP V18106) of *Annemys* sp. (Xinjiangchelyidae); Wucaiwan field area, northern Junggar Basin, Xinjiang, China; upper part of Shishugou Formation; Middle (Callovian) or Late (Oxfordian) Jurassic. **a**, **b** Surface CT scans in **a** dorsal view and **b** ventral view; **c**, **d** interpretive drawings in **c** dorsal view and **d** ventral view. Abbreviations: *bo t* basioccipital tubercula; *bs* basisphenoid; *bt pr* basipterygoid process; *fpccc* foramen posterius canalis caroticus cerebralis; *fpcci* foramen posterius canalis caroticus internus; *fpccl* foramen posterius canalis caroticus lateralis; *fpp* foramen palatinum posterius; *fr* frontal; *mx* maxilla; *na* nasal; *pa* parietal; *pal* palatine; *pmx* premaxilla; *po* postorbital; *prf* prefrontal; *pt* pterygoid; *qu* quadrate; *sci* sulcus caroticus internus; *so* supraoccipital; *sq* squamosal; *st* stapes; *st f* stapedial foramen; *vo* vomer. Images at same scale

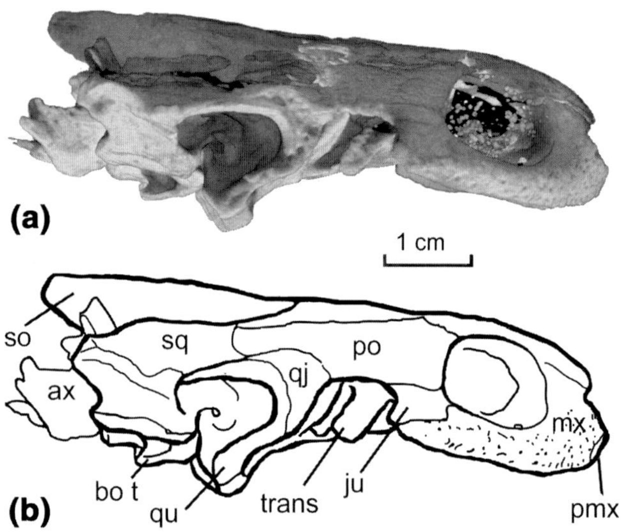

Fig. 10.14 CT volume renderings and interpretive drawings, in right lateral view, of skull (part of skeleton IVPP V18106) of *Annemys* sp. (Xinjiangchelyidae); Wucaiwan field area, northern Junggar Basin, Xinjiang, China; *upper part* of Shishugou Formation; *Middle* (Callovian) or *Late* (Oxfordian) Jurassic. **a** Surface CT scan; **b** interpretive drawing. Abbreviations: *ax* axis; *bo t* basioccipital tubercula; *ju* jugal; *mx* maxilla; *pmx* premaxilla; *po* postorbital; *qj* quadratojugal; *qu* quadrate; *so* supraoccipital; *pr pt* processus pterygoideus externus. Images at same scale

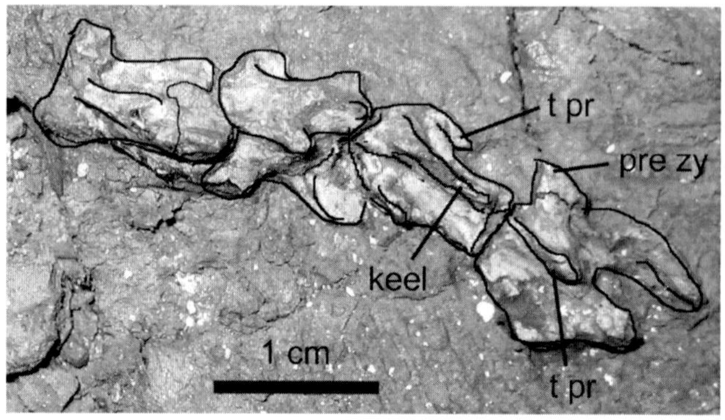

Fig. 10.15 Photograph of four articulated cervical vertebrae (part of skeleton IVPP V18106) of *Annemys* sp. (Xinjiangchelyidae); Wucaiwan field area, northern Junggar Basin, Xinjiang, China; *upper part* of Shishugou Formation; *Middle* (Callovian) or *Late* (Oxfordian) Jurassic. Vertebrae are partly embedded in *rock slab*; for clarity, outlines and major structures for each vertebra are traced in *black*. Abbreviations: *keel* mid-ventral keel; *pre zy* prezygapophysis; *t pr* transverse process

The nasals are subtriangular elements wedged between the prefrontals and maxilla, where they form the anterior edge of the external narial opening. The nasals barely meet one another at the midline, excluding the frontal and prefrontal from the margins of the external narial openings.

The prefrontals are of moderate size, forming about two-thirds the dorsal surface of the skull anterior to the orbits. In addition to their contact with the nasals anteriorly, they contact the frontals medially and posteriorly, the maxilla laterally, and the vomer and palatine ventrally.

The frontals form the dorsal portion of the skull roof between the orbits. They contact the prefrontals and nasals anteriorly, the parietals posteriorly, and the postorbitals posterolaterally. The frontal has a rectangular posterior portion, and an elongate anterior projection extending between the prefrontals to reach the nasals. A lateral lappet forms the dorsal margin of the orbit, separating the prefrontal from the postorbital. The posterior end of the frontals is located posterior to the orbits.

The parietals are the largest bones in the skull roof. They contact the frontal anteriorly, postorbital laterally, squamosal posterolaterally, and the supraoccipital posteriorly. Although the temporal region is deeply emarginated, the parietal has a lateral process that barely touches the squamosal, excluding the postorbital from the upper temporal emargination.

The postorbital is a large rectangular element that forms a portion of the border of the orbit anteriorly and contacts the parietal medially, the quadratojugal and squamosal posteriorly, and the jugal anteroventrally. The postorbital forms a portion of the margin of the border of the ventral margin of the skull, where it separates the jugal and quadratojugal.

The jugal is a triangular element in lateral view that contacts the maxilla anteroventrally and the postorbital dorsally. It forms a portion of the orbital margin, where it separates the maxilla and postorbital. The posterior border of the jugal forms the anterior margin of the lower temporal emargination.

The quadratojugal is a small, crescent-shaped element that borders the anterior margin of the quadrate and contacts the postorbital anteriorly and dorsally. It minimally contacts the squamosal posteriorly. Ventral to its contact with the postorbital, it forms the posterior margin of the cheek emargination.

The squamosal forms the posterolateral corner of the skull, capping the quadrates and anteriorly contacting the parietal, postorbital, and quadratojugal. The posterior tip of the squamosal extends posterior to the level of the occipital condyle. Thus the antrum postoticum would have been large. Medially the squamosal forms the lateral edge of the temporal emargination and it contacts the opisthotic and quadrate.

The premaxilla forms the ventral edge of the external narial opening and the anterior edge of the labial ridge. Ventrally, the premaxilla forms the anterior edge of the palate, contacting the vomer posteriorly and the maxilla laterally. The premaxilla forms a small portion of the margin of the internal narial opening between the vomer and the maxilla. Prepalatine foramina are located within the ventral surface of the premaxilla.

The maxilla forms most of the anterolateral surface of the skull. In external view, the maxilla contacts the jugal posteriorly, the prefrontal dorsally, the nasal anterodorsally, and the premaxilla anteroventrally. Between the jugal and prefrontal, the maxilla forms the ventral margin of the orbit. The suborbital portion of the maxilla is narrow and has a nearly straight ventral edge. In palatal view, the maxilla forms most of the triturating surface. This is a narrow, flat surface without a distinct lingual ridge. The contacts with the palatine and pterygoids are short and widely separated by the large foramen palatinum posterius.

The palate is completely exposed ventrally and the sutures between individual bones are clear (Figs. 10.12b, 10.13b). The internal narial openings are oval in shape, with their long axis oriented anteroposteriorly. The vomer forms a narrow, flat-topped bar of bone that separates the internal narial openings. The foramen palatinum posterius is large, exceeding the internal narial opening in size. The foramen palatinum posterius is bordered by the pterygoid, palatine, and maxilla. The proportion of the border of the foramen formed by the pterygoid is about half that formed by the palatine. A rugose area is present on the ventral surface of the palate between the foramen palatinum posterius and basisphenoid. This rugose area is largely located on the pterygoid, but also extends onto the palatine and the posterior tip of the vomer.

In ventral view the palatine is a triangular element, wide posteromedially and tapering to a narrow bar that separates the internal narial opening from the foramen palatinum posterius. The palatines contact one another at the midline between the pterygoid and vomer.

The pterygoids form the ventral surface of the palate posterior to the palatines. The pterygoids are very thin at the midline and a small unossified area between the pterygoids is present anterior to the basisphenoid. A midline contact between the pterygoids is present between the basisphenoid and the palatines. The pterygoid extends to the posterior edge of the basisphenoid, but does not contact the basioccipital. The processus pterygoideus externus is well developed and bears a large, flat lateral surface. A ridge extends from the lateral tip of the process to the posterior edge of the pterygoid, presumably marking the medial extent of insertion of the pterygoideus muscle. The posterolateral wing of the pterygoid slopes strongly laterally, so the posterior edge of the pterygoid is located at the level of the posterior edge of the basisphenoid and the jaw joint is located at about the same level as the posterior end of the basisphenoid. A deep fossa is present posterior to the quadrate ramus of the pterygoid.

The basisphenoid is a rectangular element, having a broad contact with the basioccipital. In ventral view the two bones are barely distinguished, but the contact between them can be recognized as following a groove that extends across these elements. The short, V-shaped, anterior end of the basisphenoid separates the pterygoids at the midline. A prominent feature of the basisphenoid is the "basipterygoid processes." These are triangular processes that extend laterally to barely contact the C-shaped ridge that runs posteriorly from the lateral tip of the processus pterygoideus externus. The ventral surface of the basisphenoid lacks a distinct pair of pits like those present in *Ordosemys leios*, although a small, rugose area is present at about the same position.

The canals and foramina in the basicranial region are distinct. As described above for *Xinjiangchelys radiplicatoides*, the foramen posterius canalis carotici interni is located at the posterior end of the contact between the pterygoid and basisphenoid. The canalis caroticus internus runs forward between the basisphenoid and pterygoid to open at the posterior edge of the "basipterygoid process." A deep groove, the sulcus caroticus internus, extends forward from this canal to the foramen posterius canalis caroticus cerebralis. A separate opening for the palatine artery, the foramen posterius canalis caroticus lateralis, is located farther anteriorly between the basisphenoid and the pterygoid. Rather than being a slit-like opening like that of *X. radiplicatoides*, this is a distinct foramen that leads into a well-defined canal. The canalis caroticus cerebralis and the canalis caroticus lateralis are subequal in diameter.

The basioccipital is subrectangular in ventral view, with the basioccipital tubercula oriented anteroposteriorly along the lateral edges of the bone. The basisphenoid–basioccipital suture is located at about the level of the posterior end of the pterygoid, so the basioccipital is located almost entirely posterior to the pterygoid. Also extending well posterior to the pterygoid is the opisthotic, which means the latter bone is largely visible in ventral view.

The quadrate has a complex structure. In lateral view, the quadrate forms the anteroposteriorly elongate and sub-triangular cavum tympani (Fig. 10.14). The incisura columellae auris is open posteriorly and is very narrow. Thus, the Eustachian tube must have been posterior to the quadrate. In ventral view, the quadrate forms the condylus mandibularis, which has the saddle-shaped joint surface typical of turtles.

The otic capsule is fully exposed in dorsal view (Figs. 10.12a, 10.13a). This is elongate and has a flat dorsal surface. A large stapedial foramen is present. The anterior edge of the otic capsule is not exposed, which means the size of the processus trochlearis oticum is uncertain. Sutures between the prootic, opisthotic, and quadrate cannot be identified.

Cervical vertebrae: IVPP V18106 includes a cervical series containing four vertebrae preserved in articulation (Fig. 10.15). These vertebrae are amphicoelous and elongate. The transverse processes are short, flange-like structures located near the anterior end of the centrum. A sharp midventral ridge is present. In proportions, the cervical vertebrae are similar to the one described above for

Xinjiangchelys radiplicatoides in that the flange-like transverse processes are located near the anterior end of the centrum.

Carapace: Carapace elements (not figured) preserved in IVPP V18106 include four disarticulated anterior costals, five anterolateral peripherals in articulation, and the posterior edge of the carapace. A second specimen, IVPP V18107 (also not figured) includes four disarticulated costals, three in exposed in dorsal view and one exposed in ventral view. As in the holotype specimen of *Annemys latiens* Sukhanov and Narmandakh 2006 described by Sukhanov (2000), the carapace is thin and smooth. In dorsal view the carapace is longer than wide and has a rounded posterior end. The anterolateral peripherals are narrow and have an upturned lateral edge.

Also, the pattern of scutes is like that of *A. latiens* in that vertebral scutes two and three are narrow and show little angulation of their lateral edges. In IVPP V18107 low plications extend posteriorly from the anterior edge of the vertebral scutes.

Remarks: *Annemys* was erected by Sukhanov and Narmandakh (2006) on the basis of specimens from the Late Jurassic of Mongolia. It was differentiated from *Xinjiangchelys* on the basis of features of the carapace, particularly the presence of a relative thin shell with narrow vertebral scutes two and three. In these features it is similar to sinemydids/macrobaenids, but differs from them in having a broader anterior lobe of the plastron. IVPP V18106 is included in the genus *Annemys*, because it is has a relatively thin carapace with narrow vertebral scutes and is Jurassic in age.

Inclusion of IVPP V18106 in *Annemys* also is supported by similarities in the general proportions of the skull with that of *Annemys levensis* as described by Sukhanov (2000). Both are relatively low, have deep temporal and cheek emarginations, and a long supraoccipital crest. In these features, the skull of *Annemys* is different from that of *Xinjiangchelys radiplicatoides*, which has an expanded temporal region, weakly emarginated temporal and cheek region, and a short supraoccipital crest.

Two species of *Annemys* have been named, *A. latiens* and *A. levensis*. These differ primarily in features of the carapace, particularly the presence of a relatively thin shell in *A. latiens* compared to *A. levensis*. The shell of IVPP V18106 is not sufficiently well preserved to be assigned to either species. The skull of IVPP V18106 differs from that of *A. levensis* in that the prefrontals do not contact one another at the midline anterior to the frontals. In *A. levensis* the prefrontals meet at the midline between the frontals and nasals (Sukhanov 2000, Fig. 17.2). Also, in *A. levensis* the postorbital forms a portion of the posterior margin of the skull, thereby separating the parietal and squamosal, whereas in IVPP V18106 the parietal has a long lateral process that contacts the squamosal. As reconstructed, *A. levensis* differs from IVPP V18106 in the presence of a relatively small foramen palatinum posterius. However, as noted by Rabi et al. (2010) the skull of *A. levensis* appears to be rather poorly preserved, so a larger opening is possible. In the basicranial region, IVPP V18106 differs from *A. levensis* in that the carotid canals are closed posteriorly instead of being open as in *A. levensis*. Because the skull of *A. latiens* is not known, it is not possible to resolve whether IVPP V18106 is attributable to that species. Thus, we consider IVPP V18106 to be specifically indeterminate within *Annemys*.

In features of its skull and shell, *Annemys* is at a more derived grade of evolution than *Xinjiangchelys radiplicatoides*. Derived skull features of *Annemys* include the presence of a low temporal region, the deeply emarginate skull, the long and blade-like supraoccipital crest, the long incisura columella, circular shape of the foramen caroticum laterale, and a relatively short basisphenoid that is truncated anteriorly. The carapace of *Annemys* is derived relative to *X. radiplicatoides* in being thin and in having narrow vertebral scutes. In all these features, *Annemys* both is derived relative to *X. radiplicatoides* and is similar to Early Cretaceous eucryptodires from Asia, such as *Ordosemys* Brinkman and Peng 1993a, *Sinemys* Brinkman and Peng 1993b, *Hangaiemys* Sukhanov and Narmandakh 1974, and *Dracochelys* Gaffney and Ye 1992. In addition, as noted by Rabi et al. (2010) *Annemys* shows particularly strong similarities to the Early Cretaceous sinemydid *Hangaiemys* in the proportions of the skull, development of the supraoccipital crest, and presence of a greatly enlarged foramen palatinum posterius; this combination of features suggests a close relationship between the two genera. Whether similarities in the skulls of *Annemys* and *Hangaiemys* compared to *X. radiplicatoides* reflect true relationships must await the discovery of additional cranial material from Jurassic eucryptodires from Asia. At present, however, the substantial similarities between *Annemys* and *Hangaiemys* appear to support a recent suggestion that the Xinjiangchelyidae are a paraphyletic assemblage of basal cryptodires (Rabi et al. 2010).

Discussion

The Shishugou Formation Turtle Assemblage

Although taxonomic uncertainty remains, a high level of diversity is present among Jurassic turtles in the northern part of the Junggar Basin. At least five species are now known from the Shishugou Formation: ?*Sichuanchelys* sp.; *X. junggarensis*; *X. radiplicatoides*; *X. chowi*; and *Annemys* sp. The plastron described by Peng and Brinkman (1993, pp. 2023–2025, Figs. 13, 14) as *Xinjiangchelys* sp. may be an additional taxon. One additional taxon has been reported from the Qigu Formation, *X. qiguensis*, although the taxonomic significance of the differences between this and *X. chowi* has been questioned (Rabi et al. 2010). This level of diversity is similar to that of other well-documented Jurassic turtle assemblages. For example, in the Late Jurassic of Sichuan eight named turtle taxa are present. Significantly, in both the Junggar and Sichuan assemblages many of the taxa are represented by single specimens, suggesting that the diversity of these assemblages is not yet fully documented by available samples.

Despite a high level of diversity of turtles in the Shishugou Formation, diversity at individual localities is low. Typically, when multiple individuals are found together, no more than two taxa are present and one of those is overwhelmingly dominant. For example, in the Jiangjunmiao turtle bonebed two taxa are present, but most of the identifiable elements are from *Xinjiangchelys radiplicatoides*. At the Pingfengshan locality Peng and Brinkman (1993) recognized the presence of two taxa based on 14 specimens, with all but one of these being *X. junggarensis*. This pattern appears to be typical for the Late Jurassic of Central Asia. Similarly, in the adjacent Turpan Basin, a locality containing multiple thousand individuals of *Annemys* has been reported (Wings and Joyce 2009). This pattern of high alpha diversity (total diversity within a unit), low diversity within individual localities, and high beta diversity (between-locality diversity within a unit) is unusual when compared with most turtle assemblages of the Cretaceous. The paleoecological significance of this pattern is uncertain. Given the paleoenvironmental and paleoclimatic setting, it is likely that short-term environmental conditions, such as wet intervals following periods of drought, leading to explosive expansion of low diversity communities played a factor. Additional taphonomic and paleoecological studies are necessary to resolve this.

Implications for Evolution of the Basicranial Region in Basal Eucryptodires

The basicranial region has long been recognized as an important source of information for interpreting the interrelationships of turtles (Gaffney 1979b). Most recently, the evolution of the braincase in turtles was reviewed by Sterli et al. (2010), who placed each of six patterns of carotid circulation described by Sterli and de la Fuente (2010) into a phylogenetic framework. In the most basal pattern, represented by *Proganochelys*, the carotid artery splits into cerebral and palatine branches ventral to the skull, the cerebral branch enters the braincase through a foramen (the foramen posterius canalis caroticus cerebralis) in the basisphenoid, and the palatine branch extends forward through the interpterygoid vacuity. In a slightly more derived pattern, represented by *Kallokibotion*, the interpterygoid vacuity has closed around the palatine artery so that the artery enters the skull through a distinct canal, called the canalis caroticus lateralis. In pleurodires, the internal carotid and its cerebral and palatine branches are fully encased in bone and a single foramen is present posteriorly. Sterli et al. (2010) interpreted this as a result of expansion of the parasphenoid portion of the basisphenoid, trapping the internal carotid between the basisphenoid and the pterygoid.

In the most basal paracryptodires, such as *Pleurosternon* and *Glyptops*, the carotid artery branches into the cerebral and palatine arteries in a shallow concavity along the side of the basisphenoid and, slightly more anteriorly, the two branches enter canals though separate openings. In baenids, a single opening is present along the side of the basisphenoid. Sterli et al. (2010) hypothesize that the paracryptodire condition also formed by expansion of the parasphenoid portion of the basisphenoid to trap the internal carotid and its bifurcation between the basisphenoid and the pterygoid.

Sterli et al. (2010) did not incorporate the eucryptodire pattern into this evolutionary sequence because the early stages of the evolution of the eucryptodire basicranial region were too poorly documented. The skulls of *Xinjiangchelys radiplicatoides* and *Annemys* sp. described here provide additional data that document the early stages of evolution of the basicranial region in eucryptodires (Fig. 10.16). Specifically, these new skulls show that in the most primitive eucryptodires there is still a recognizable interpterygoid vacuity and the foramen posterius canalis caroticus lateralis is located a long distance from the foramen posterior canalis caroticus cerebralis. Also, as discussed in the next section, the "basipterygoid process" is still well developed.

In eucryptodires, the carotid artery enters the skull through a single foramen, called the foramen posterior canalis caroticus internus, located either between the pterygoid and basisphenoid or within the pterygoid. In basal eucryptodires, such as *Chubutemys* (Fig. 10.16a), xinjiangchelyids (Fig. 10.16b, d), and Early Cretaceous sinemydids/macrobaenids from Asia (Fig. 10.16c, e, f), the canalis caroticus internus does not completely enclose the carotid artery and its branches. Instead, the carotid artery exits ventral to the basisphenoid prior to the point at which the carotid artery branches into the cerebral and palatine arteries. Thus, these arteries are not fully encased in bone. In these turtles, the cerebral artery would have continued forward in the sulcus caroticus internus, a well-defined groove that leads to the foramen posterius canalis caroticus cerebralis. There is no distinct groove for the palatine artery, so the exact course of this artery is unknown. However, it would have branched off from the carotid artery somewhere between the anterior opening of the carotid canal

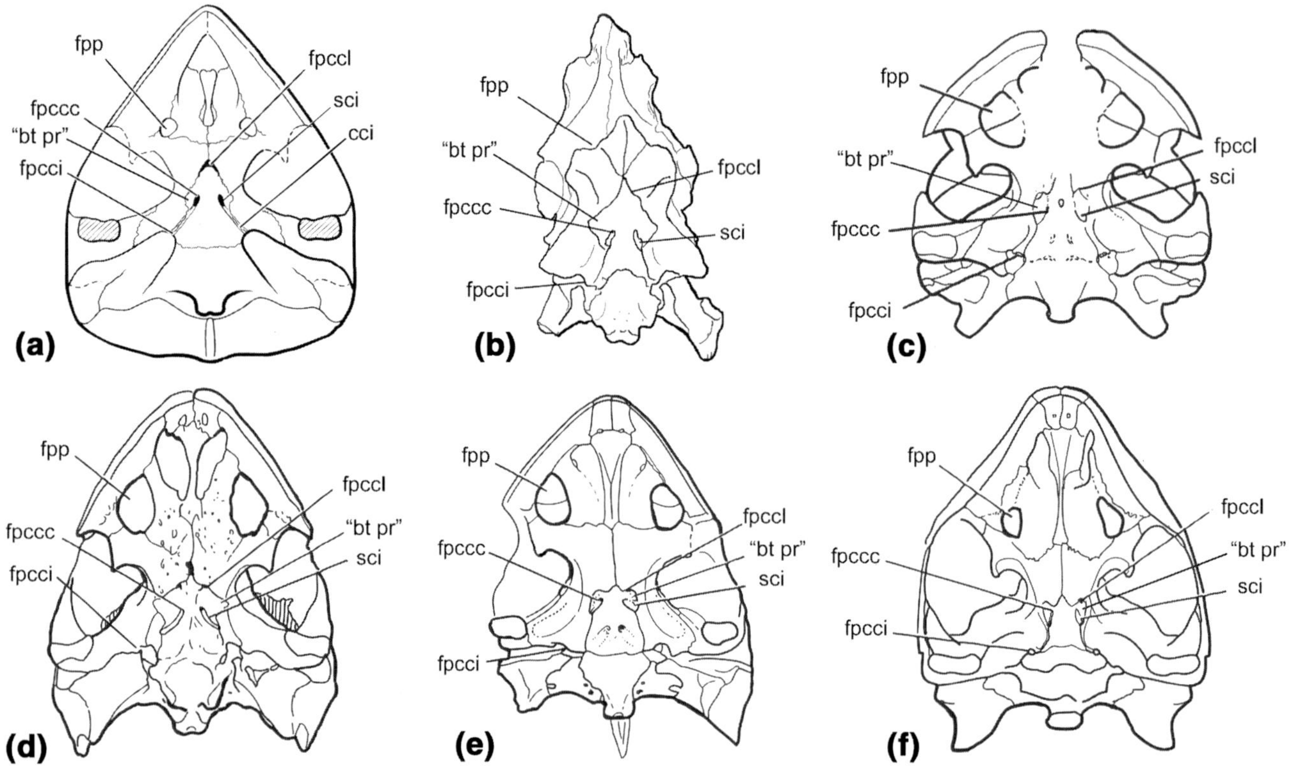

Fig. 10.16 Palatal views of skulls of reprersentative basal cryptodires to show differences in structure of the basicranial region. **a** *Chubutemys*, redrawn from Gaffney et al. (2007); **b** *Xinjiangchelys radiplicatoides* sp. nov., drawn from IVPP V9539-1 (see Fig. 10.6c, d); **c**, reconstruction of *Sinemys gamera*, based on IVPP V9532-11; **d** *Annemys* sp, drawn from IVPP V18106 (see Fig. 10.13d); **e** *Hangaiemys*, redrawn from Sukhanov (2000); **f** reconstruction of *Ordosemys* sp., based on IVPP V12092. Abbreviations: *bt pr* basipterygoid process; *cci* canalis caroticus internus, *fpccc* foramen posterior canalis carotis cerebralis; *fpcci* foramen posterior canalis caroticus internus; *fpccl* foramen posterior canalis caroticus lateralis; *fpp* foramen palatinum posterior; *sci* sulcus caroticus internus. Images at different scales

and the foramen posterior canalis carotici cerebralis, and passed below the basisphenoid to enter the cranial cavity through the foramen posterior canalis caroticus lateralis, which is located between the pterygoid and basisphenoid. In *Cubutemys* (Fig. 10.16a) and *Xinjiangchelys radiplicatoides* (Fig. 10.16b) the foramen posterior canalis caroticus lateralis is a slit-like space that can be interpreted as a reduced interpterygoid vacuity. This also appears to be the case in *Dracochelys* (Gaffney and Ye 1992, Fig. 5) and *Annemys* (Fig. 10.16d). In contrast, *Ordosemys* (Fig. 10.16f), and *Sinemys* (Fig. 10.16c) are more derived in that the interpterygoid vacuity is more fully constricted to form the canalis caroticus lateralis. Also, these taxa are more derived in that the foramen posterius canalis caroticus lateralis is located closer to the foramen posterius canalis caroticus cerebralis.

A more derived eucryptodire condition is seen in *Adocus* Cope1868 and *Judithemys* Parham and Hutchison 2003, in which the area of exposure of the internal carotid and its branches is restricted to a small foramen-like opening in the area just above the point at which the artery branches (Meylan and Gaffney 1989; Parham and Hutchison 2003). The foramen posterius canalis caroticus lateralis and the foramen posterius canalis cerebralis are visible in this opening. In the fully derived condition, the carotid artery enters the skull through a foramen on the posterior edge of the skull and the more anterior portion of the artery, and its branches are fully encased in bone. Because the exposed portion of the carotid artery and its branches lie ventral to the basisphenoid, it is likely that restriction of the space between the foramen anterior canalis caroticus internus and the foramen canalis caroticus lateralis was a result of expansion of the pterygoid, rather than the parasphenoid portion of the basisphenoid (Sterli et al. 2010).

The "Basipterygoid Process" in Basal Eucryptodires

A separate question of significance for interpreting the early evolution of the basicranial region in eucryptodires is the development of the "basipterygoid process." This structure was first documented in *Pleurosternon* by Evans and Kemp (1975) and subsequently recognized in *Glyptops* by Gaffney (1979a). It can also be recognized in *Chubutemys* (Gaffney et al. 2007, Fig. 4). In eucryptodires from Asia, this structure is present in *Sinemys* (Fig. 10.16c) and *Ordosemys* (Fig. 10.16f), as well as *Xinjiangchelys radiplicatoides* (Fig. 10.16b) and *Annemys* sp. (Fig. 10.16d) as described above. In these xinjiangchelyids and sinemydids/macrobaenids from Asia, the "basipterygoid process" is a large and obvious feature of the basisphenoid. The position of this process relative to the reduced interpterygoid vacuities in *X. radiplicatoides* and *Chubutemys* is consistent with this structure being correctly identified as the basipterygoid process. Gaffney (1979a) noted that its position relative to the dorsum sellae and foramen anterius canalis carotici interni differs from that in primitive reptiles. In primitive reptiles, the basipterygoid process is anterior to these structures, whereas the process that Evans and Kemp (1975) identified as the basipterygoid process is posterior to the same structures. However, it was subsequently recognized that an undoubted basipterygoid process is present in the primitive turtle *Proganochelys* (Gaffney 1990). In *Proganochelys* the basipterygoid process is located posterior to the dorsum sellae (see Gaffney 1990, Fig. 45). Thus, the unusual position of the basipterygoid process in *Proganochelys* can be interpreted as a feature of turtles that is retained in basal cryptodires and the "basipterygoid process" is correctly interpreted as homologous with the basipterygoid process of more basal reptiles.

Acknowledgments We thank Wang Hai-Jun for his preparation of many of the specimens described here. The Jiangjunmiao turtle bonebed was discovered by Tang Zhi-Lu, and assistance in collecting from this locality was provided by Dennis Braman. The braincase of *Xinjiangchelys radiplicatoides* from the Jiangjunmiao turtle bonebed was prepared by Clive Coy. The authors would like to acknowledge the staff of the Institute of Vertebrate Paleontology and Paleoanthropology for support in the field while collecting material described here from the Wucaiwan Locality and for assistance while we studied material in Beijing. The holotype of *X. radiplicatoides* was collected by Catherine Forster. We also acknowledge the help and support of Zhao Xi-Jin in sharing his expertise and understanding of the geology and fossil localities of Xinjiang. Andreas Matzke provided useful insights on the anatomy and evolution of eucryptodires in Asia. Jim Parham, Walter Joyce, Ben Kear, and Igor Danilov all reviewed the paper and made comments that led to its improvement; their time and efforts are greatly appreciated. CT scans were done at the University of Texas High-Resolution X-ray CT Facility, Austin, Texas, USA. Drawings of the *Annemys* skull in Fig. 10.12 were made by Donna Sloan of the Royal Tyrrell Museum of Palaeontology. This project was supported by the National Natural Science Foundation of China, the National Science Foundation Division of Earth Sciences of the USA grant EAR 0310217, George Washington University, the National Geographic Society, and the Chinese Academy of Sciences (to XX and JC).

References

Batsch, A. J. G. C. (1788). *Versuch einer Anleitung, zur Kenntniß und Geschichte der Thiere und Mineralien*. Jena: Akademische Buchhandlung.

Brinkman, D. B., & Peng, J.-H. (1993a). *Ordosemys leios*, n. gen., n.sp., a new turtle from the Early Cretaceous of the Ordos Basin, Inner Mongolia. *Canadian Journal of Earth Sciences, 30*, 2128–2138.

Brinkman, D. B., & Peng, J.-H. (1993b). New material of *Sinemys* (Testudines, Sinemydidae) from the Early Cretaceous of China. *Canadian Journal of Earth Sciences, 30*, 2139–2152.

Brinkman, D. B., & Wu, X.-C. (1999). The skull of *Ordosemys*, an Early Cretaceous turtle from Inner Mongolia, P. R. of China, and the interrelationships of Eucryptodira (Chelonia, Cryptodira). *Paludicola, 2*, 134–147.

Brinkman, D. B., Li, J.-L., & Ye, X.-K. (2008). Order Testudines. In J.-L. Li, X.-C. Wu, & F.-C. Zhang (Eds.), *The Chinese Fossil Reptiles and their Kin* (pp. 35–102). Beijing: Science Press.

Clark, J. M., Xu, X., Eberth, D. E., Forster, C. A., Malkus, M., Hemming, S., et al. (2006). The Middle–Late Jurassic terrestrial transition: New discoveries from the Shishugou Formation, Xinjiang, China. In P. M. Barrett & S. E. Evans (Eds.), *9th international symposium Mesozoic terrestrial ecosystems and biota* (p. 153). Manchester: Cambridge Publications, Natural History Museum.

Cope, E. D. (1868). On some Cretaceous Reptilia. *Proceedings of the Academy of Natural Sciences, Philadelphia, 1868*, 233–242.

Danilov, I. G., & Parham, J. F. (2006). Redescription of '*Plesiochelys*' *tatsuensis* from the Late Jurassic of China, with comments on the antiquity of the crown clade Cryptodira. *Journal of Vertebrate Paleontology, 26*, 573–580.

Danilov, I. G., & Parham, J. F. (2008). A reassessment of some poorly known turtles from the Middle Jurassic of China, with comments on the antiquity of extant turtles. *Journal of Vertebrate Paleontology, 28*, 306–318.

Eberth, D. A., Brinkman, D. B., Chen, P. J., Yuan, F. T., Wu, S. Z., Li, G., et al. (2001). Sequence stratigraphy, paleoclimate patterns, and vertebrate fossil preservation in Jurassic-Cretaceous strata of the Junggar Basin, Xinjiang Uygur Autonomous Region, People's Republic of China. *Canadian Journal of Earth Science, 38*, 1627–1644.

Evans, J., & Kemp, T. S. (1975). The cranial morphology of a new Lower Cretaceous turtle from southern England. *Palaeontology, 18*, 25–40.

Fang, Q.-R. (1987). A new species of Middle Jurassic turtles from Sichuan. *Acta Herpetologica Sinica, 6*, 65–69.

Gaffney, E. S. (1972). An illustrated glossary of turtle skull nomenclature. *American Museum Novitates, 2486*, 1–33.

Gaffney, E. S. (1975). A phylogeny and classification of the higher categories of turtles. *Bulletin of the American Museum of Natural History, 155*, 389–436.

Gaffney, E. S. (1979a). The Jurassic turtles of North America. *Bulletin of the American Museum of Natural History, 162*, 91–136.

Gaffney, E. S. (1979b). Comparative cranial morphology of recent and fossil turtles. *Bulletin of the American Museum of Natural History, 164*, 65–375.

Gaffney, E. S. (1990). The comparative osteology of the Triassic turtle *Proganochelys*. *Bulletin of the American Museum of Natural History, 194*, 1–263.

Gaffney, E. S., & Jenkins, F. (2010). The cranial morphology of *Kayentachelys*, an early cryptodire, and the early history of turtles. *Acta Zoologica (Stockholm), 91*, 335–368.

Gaffney, E. S., & Meylan, P. A. (1992a). *Sinaspideretes* is not the oldest trionychid turtle. *Journal of Vertebrate Paleontology, 12*, 257–259.

Gaffney, E. S., & Meylan, P. A. (1992b). The Transylvanian turtle, *Kallokibotion*, a primitive cryptodire of Cretaceous age. *American Museum Novitates, 3040*, 1–37.

Gaffney, E. S., & Ye, X.-K. (1992). *Dracochelys*, a new cryptodiran turtle from the Early Cretaceous of China. *American Museum Novitates, 3048*, 1–13.

Gaffney, E. S., Hutchison, J. H., Jenkins, F. A., & Meeker, L. J. (1987). Modern turtle origins: The oldest known cryptodire. *Science, 237*, 289–291.

Gaffney, E. S., Rich, T. H., Vickers-Rich, P., Constantine, A., Vacca, R., & Kool, L. (2007). *Chubutemys*, a new Eucryptodiran turtle from the Early Cretaceous of Argentina, and the relationships of the Meiolaniidae. *American Museum Novitates, 3599*, 1–35.

Hirayama, R., Brinkman, D. B., & Danilov, I. G. (2000). Distribution and biogeography of non-marine Cretaceous turtles. *Russian Journal of Herpetology, 7*, 181–198.

Hutchison, J. H. (2000). *Diversity of Cretaceous turtle faunas of eastern Asia and their contribution to the turtle faunas of North America* (pp. 27–38). Paleontological Society of Korea Special Publication No, 4.

Kaznyshkin, M. N. (1988). [Late Jurassic turtles of northern Fergana (Kirghiz SSR)]. *Vestnik Zoologii, 5*, 26–32. [In Russian.]

Kaznyshkin, M. N., Nalbandyan, L. A., & Nessov, L. A. (1990). *Yezhegodnik Vsesoyuznogo Paleontologicheskogo Obshchestva* [Middle and Late Jurassic turtles of Fergana (Kirghiz SSR)], *33*, 185–204. [In Russian.]

Maisch, M. W., Matzke, A. T., Pfretzschner, H.-U., Sun, G., Stöhr, H., & Grossmann, F. (2003). Fossil vertebrates from the Middle and Upper Jurassic of the southern Junggar Basin (NW China)—Results of the Sino-German Expeditions 1999–2000. *Neues Jahrbuch für Geologie und Paläontologie onatshefte, 2003*, 297–313.

Matzke, A. T., Maisch, M. W., Sun, G., Pfretzschner, H.-U., & Stöhr, H. (2004). A new xinjiangchelyid turtle (Testudines, Eucryptodira) from the Jurassic Qigu Formation of the southern Junggar Basin, Xinjiang, North-West China. *Palaeontology, 47*, 1267–1299.

Matzke, A. T., Maisch, M. W., Sun, G., Pfretzschner, H.-U., & Stöhr, H. (2005). A new Middle Jurassic xinjiangchelyid turtle (Testudines; Eucryptodira) from China (Xinjiang, Junggar Basin). *Journal of Vertebrate Paleontology, 25*, 63–70.

Meylan, P. A., & Gaffney, E. S. (1989). The skeletal morphology of the Cretaceous cryptodiran turtle, *Adocus*, and the relationships of the Trionychoidea. *American Museum Novitates, 2941*, 1–60.

Nessov, L. A. (1995). On some Mesozoic turtles of the Fergana Depression (Kyrgyzstan) and Dzhungar Alatau Ridge (Kazakhstan). *Russian Journal of Herpetology, 2*, 134–141.

Nopcsa, F. (1923). On the geological importance of the primitive reptilian fauna of the uppermost Cretaceous of Hungary; with a description of a new tortoise (*Kallokibotion*). *Quarterly Journal of the Geological Society, 79*, 100–116.

Parham, J. F., & Hutchison, J. H. (2003). A new eucryptodiran turtle from the Late Cretaceous of North America (Dinosaur Provincial Park, Alberta, Canada). *Journal of Vertebrate Paleontology, 23*, 783–798.

Peng, J.-H., & Brinkman, D. B. (1993). New material of *Xinjiangchelys* (Reptilia: Testudines) from the Late Jurassic Qigu Formation (Shishugou Group) of the Pingfengshan locality, Junggar Basin, Xinjiang. *Canadian Journal of Earth Sciences, 30*, 2013–2026.

Rabi, M., Joyce, W. G., & Wings, O. (2010). A review of the Mesozoic turtles of the Junggar Basin (Xinjiang, Northwest China) and the paleobiogeography of Jurassic to Early Cretaceous Asian Testudinates. *Palaeobiodiversity and Palaeoenvironments, 90*, 259–273.

Rütimeyer, L. (1873). Die fossilen Schildkröten von Solothurn. *Neue Denkschrift der allgemeinen schweizerischen naturforschenden Gesellschaft, 25*, 1–185.

Smith, E. T., & Kear, B. P. (2012). *Spoochelys ormondea* gen. et sp. nov, an archaic meiolaniid-like turtle from the Early Cretaceous of Lightning Ridge, Australia. In D. B. Brinkman, P. A. Holroyd, & J. D. Gardner (Eds.), *Morphology and evolution of turtles* (pp. XXX–XXX). Dordrecht: Springer.

Sterli, J., & de la Fuente, M. S. (2010). Anatomy of *Condorchelys antiqua* Sterli, 2008 and the origin of the modern jaw closure mechanism in turtles. *Journal of Vertebrate Paleontology, 30*, 351–366.

Sterli, J., & Joyce, W. G. (2007). The cranial anatomy of the Lower Jurassic turtle *Kayentachelys aprix*. *Acta Palaeontologica Polonica, 52*, 675–694.

Sterli, J., Müller, J., Anquetin, J., & Hilger, A. (2010). The parabasisphenoid complex in Mesozoic turtles and the evolution of the testudinate basicranium. *Canadian Journal of Earth Sciences, 47*, 1337–1346.

Sukhanov, V. B. (2000). Mesozoic turtles of Middle and Central Asia. In M. J. Benton, M. A. Shishkin, D. M. Unwin, & E. N. Kurochkin (Eds.), *The Age of Dinosaurs in Russia and Mongolia* (pp. 309–367). Cambridge: Cambridge University Press.

Sukhanov, V. B., & Narmandakh, P. (1974). Mesozoic and Cenozoic Faunas and biostratigraphy of Mongolia [New Early Cretaceous turtle from continental deposits of the northern Gobi]. *The Joint Soviet-Mongolian Paleontological Expedition. Transactions, 1*, 192–200. [In Russian.]

Sukhanov, V. B., & Narmandakh, P. (2006). New taxa of Mesozoic turtles from Mongolia. *Fossil Turtle Research, 1*, 119–127.

Wings, O., & Joyce, W. G. (2009). An exceptionally large Jurassic turtle taphocoenosis from Xinjiang Autonomous province, China. *Journal of Vertebrate Paleontology, 29*(Supplement to No. 3), 202A.

Ye, X.-K. (1963). Fossil turtles of China. *Palaeontologica Sinica, 150*, 1–112.

Ye, X.-K. (1973). [Fossil turtles from Wu'erho (Urho). Reports of the paleontological expedition to the Xinjiang. 2. The Pterosaurian fauna from Wu'erho (Urho).] *Academia Sinica, Institute of Vertebrate Palaeontology and Palaeoanthropology, Memoir 11*, 811. [In Chinese.]

Ye, H.-K. (1982). Middle Jurassic turtles from Sichuan, SW China. *Vertebrata PalAsiatica, 20*, 282–290.

Ye, X.-K. (1986a). A Jurassic turtle from Junggar, Xinjiang. *Vertebrata PalAsiatica, 24*, 171–181.

Ye, X.-K. (1986b). New material of *Plesiochelys radiplicatus* with preliminary discussion of related problems. *Vertebrata Pal Asiatica, 24*, 269–273.

Ye, X.-K. (1990). Chengyuchelyidae, n. fam., Middle Jurassic turtles of China. *Studia Palaeocheloniologica, 3*, 33–40.

Ye, X.-K., & Fang, Q.-R. (1982). A new species of fossil *Plesiochelys* from Jingyan, Sichuan. *Acta Herpetologica Sinica, 1*, 63–68.

Ye, X.-K., & Pi, X. (1997). A new genus of Chengyuchelyidae from Dashanpu, Zigong Sichuan. *Vertebrata PalAsiatica, 35*, 182–188.

Young, C. C., & Chow, M. C. (1953). New fossil reptiles from Szechuan, China. *Acta Scientia Sinica, 2*, 216–229.

Zhao, X.-J. (1980). Mesozoic vertebrate-bearing beds and stratigraphy of northern Xinjiang. *Institute of Vertebrate Paleontology and Paleoanthropology, Memoir, 15*, 1–20.

Chapter 11
Rediscovery of the Carapace of the Lost Holotype of the Purbeck Turtle *Chelone obovata* Owen 1842

Andrew R. Milner

Abstract *Chelone obovata* was named and described but not figured by Owen (1842) on the basis of a specimen (or specimens) from the Purbeck Limestone Group in the J. Chaning Pearce Collection. This material was subsequently lost to the scientific community and *C. obovata* has been treated as a nomen dubium in the few later works that mention it. One part—the carapace—has been relocated in the collections of the National Museum of Wales in Cardiff. It is preserved in ventral aspect and is a carapace of the Purbeck Limestone Group turtle taxon now widely referred to as *Hylaeochelys latiscutata* (Owen) Lydekker 1889a. Although the species-name *obovata* is senior to *latiscutata*, it has the status of nomen oblitum and should not replace the widely used term, but should be listed in future synonymies.

Keywords Cretaceous • Morphology • Purbecklimestone • Systematics • Testudines

Introduction

Although fossil turtles had been recognised in the Lower Cretaceous Purbeck Limestone (now the Purbeck Limestone Group) early in the Nineteenth Century, the first specimens were not described and named until 1842. In his report to the British Association meeting of 1841, Sir Richard Owen described four turtle specimens from the Purbeck Limestone as representing three new taxa (Owen 1842). One of these, *Platemys bullockii*, was based on a single plastron from the Bullock Collection then recently purchased by the British Museum. It was incorrectly attributed to the Eocene London Clay from Sheppey, Kent (Owen 1850), and was later renamed *Pleurosternon bullockii*. It was correctly assigned to the Purbeck Limestone by Lydekker and Boulenger (1887) and has subsequently remained the holotype of the commonest Purbeck turtle.

The history of Owen's (1842) other three original specimens, which were held in private collections, has been far less clear or satisfactory. These comprised two specimens, one from Sir Philip Egerton's collection and one from the J. S. Bowerbank Collection that formed the types of *Tretosternon punctatum*, together with a single specimen from the J. Chaning Pearce Collection that formed the holotype of *Chelone obovata*. All three were described but not figured by Owen (1842, pp. 165–172). When Owen (1853) later named, described, and figured the Purbeck turtles in detail, he used a completely new suite of specimens and taxa, and no mention was made of the 1842 material or taxa. All three 1842 specimens then effectively vanished for 150 years.

Lydekker and Boulenger (1887) and Lydekker (1889b) noted the non-availability of the Egerton and Bowerbank specimens, but the binomen *Tretosternon punctatum* became widely used on the basis of substitute specimens suggested but not formalised by Lydekker (e.g., Delair 1958). Milner (2004, pp. 1444–1445) located the Bowerbank plastron in the collections of The Natural History Museum, London, and determined it as attributable to *Hylaeochelys latiscutata* (Owen) Lydekker 1889a. The Egerton specimen is still lost and Milner (2004, p. 1445) proposed it as the lectotype of *Tretosternon punctatum*, now a nomen dubium unless the specimen is relocated.

Chelone obovata was briefly mentioned by Woodward (1909), Delair (1958) and Milner (2004), but the original material remained unlocated. Delair (1958, p. 49) suggested that the taxon be provisionally reassigned to *Pleurosternum* as a distinct species, *P. obovata*, but it was treated as a nomen dubium by Milner (2004) pending rediscovery of the only specimen.

A. R. Milner (✉)
Department of Palaeontology, The Natural History Museum, Cromwell Road, London SW7 5BD, UK
e-mail: andrew.milner@nhm.ac.uk

D. B. Brinkman et al. (eds.), *Morphology and Evolution of Turtles*, Vertebrate Paleobiology and Paleoanthropology,
DOI: 10.1007/978-94-007-4309-0_11, © Springer Science+Business Media Dordrecht 2013

Institutional abbreviations used in the paper are as follows: NHMUK, The Natural History Museum, London, England; NMW, National Museum of Wales, Cardiff, Wales.

The Holotype of *Chelone obovata*

The original specimen of *Chelone obovata* was described by Owen (1842, p. 170) as "the most complete and beautiful specimen of a fossil turtle that I have yet seen … from … Purbeck, in the collection of Channing [sic] Pearce esq. of Bradford, Wiltshire." It comprised a complete carapace in ventral aspect about 27 cm long, a separate and largely complete plastron in ventral aspect with some damage in its central region, and hind limb elements including at least one recognisable femur and a tibia. Owen did not explicitly state that all the material was on a single slab, and it may have been on more than one slab although believed to be a single specimen. Although the relative positions of these elements were not described, a single slab bearing such a specimen would be unusual and should be readily identifiable from Owen's description. Clearly, no subsequent worker up to and including myself (Milner 2004) had seen any specimen, even one without a history, corresponding to this.

The J. Chaning Pearce Collection

The early history of *Chelone obovata* is part of the history of the J. Chaning Pearce Collection, which merits a short summary. Joseph Chaning Pearce (born July 1811) pursued a career as a doctor in the south-west of England in the 1830s and 1840s. He also built up one of the largest private collections of British fossils outside London. In 1845, he moved to Montague House, Langridge, north of Bath, Somerset, and set aside part of the house as a private museum for his collection (De la Beche 1848, pp. xxi–xxii). A large, detailed written catalogue was also produced (Pearce 1847). Unfortunately, shortly after setting up his museum, he died in May 1847 of kidney stones at the age of 35 (De la Beche 1848, pp. xxi–xxii) leaving a widow, Amelia, and three children, Mary Ann, Samuel, and Joseph (junior). Amelia later remarried a Mr Bush and the family remained near Bath until at least 1858 (later census data records at least two daughters of the second marriage born at Bath: Agnes in 1852 and Rosella in 1858 [1871 and 1901 census data in Graham 2010]). Census data for 1871 also records that Joseph's younger son Joseph Chaning Pearce (junior) qualified as a General Practitioner and moved to the London area where he married in 1867 and that his first child was born at Dulwich (1871 census data in Graham 2010). However later children are recorded as born at The Manor House, Brixton, Surrey, and it thus appears that the family moved to Brixton (then a village outside London) in about 1870, possibly when Joseph junior set up a medical practice (1881 census data in Graham 2010). Throughout this period of 23 years following Joseph senior's death, the fossil collection had remained in the family and they set up a new private museum at the Manor House. This museum was visited by the Geological Association in 1885, a visit subsequently described by Woodward (1885) and it appears that the collection was still intact then. The family moved in 1886 to Ramsgate in Kent, where their residence was also named Montague House (1891 census data in Graham 2010). Joseph junior died in 1925 (Times obituary in Graham 2010). I have found no record of the fate of the collection for the 29 years between 1886 and 1915, so have been unable to determine whether it was retained by the family in exhibition or storage or whether it passed to a third party when they moved to Ramsgate.

The collection resurfaces in 1915 when much of it was purchased by the Bristol Museum and Art Gallery (Eberle and Bolton 1916). There is no record of the vendor, but in the same year, the Geological Survey purchased a J. Chaning Pearce ichthyosaur from the collection of the Royal Geological Society of Cornwall (M. Howe, personal communication). It is unclear whether the ichthyosaur was still part of the original collection or had parted company with it some years before in one of the house moves, it being a large specimen. When Bristol acquired the main collection, they also acquired Pearce's original handwritten catalogue which is still in their archives (Pearce 1847). Perusal of the relevant Bristol catalogues covering 1915–1935 suggests that no systematic attempt was made to reregister the collection in the next 20 years and much of it has only gradually been reregistered since 1945.

The Fate of the Holotype of *Chelone obovata*

No specimen corresponding to Owen's original description of *Chelone obovata* is present in the collections at the Bristol City Museum and Art Gallery today. In my review of Purbeck turtles (Milner 2004), I had assumed that the specimen was destroyed on 24 November 1940 when the exhibition area of the museum was demolished by incendiary and high-explosive bombs. The stored fossil material and archives were undamaged in the raid, but there is no surviving turtle specimen with a carapace and plastron on one slab. However on a visit to Bristol in 2008, I was able to examine Chaning Pearce's (1847) original catalogue. This catalogue is a book with hand-ruled pages, each made up of three columns with the entries spaced down each page. The pages and collection are ordered stratigraphically. In the Purbeck section of the catalogue, on page 32, is the following relevant entry in Pearce's handwriting (personal observation): "Fossil No.12" (in column 1), "Chelone obovata" (in column 2), "Perfect carapace 10¾ inches long and 9½ inches wide, also the plastron with femur and tibiae of both hinder [sic] legs. An exceedingly fine and perfect specimen." (in column 3).

Next to this record in the catalogue is a pencil annotation: "Sent to Cardiff Museum 3.3.33 see Ex.Reg. p. 125". It thus appeared that the specimen, unrecognised as a holotype, had been exchanged with Cardiff Museum (now the National Museum of Wales) in March 1933. "Ex.Reg" presumably represents an Exchange Register for the 1930s, but this could not be located. However, it became clear that the specimen was not in the Bristol City Museum and Art Gallery when it was bombed and, thus, it might still survive.

Correspondence and discussion with Mr Tom Sharpe at the National Museum of Wales, Cardiff, established that a batch of specimens from Bristol, including one listed as *Chelone obovata*, was accessioned in October 1933. It appears that little information about the specimen had been forwarded and it transpired that the Museum now has in its possession a single turtle specimen of unknown provenance in an exhibit representing Mesozoic faunas. Mr Sharpe kindly provided a photograph of this specimen and it was immediately clear that this was the carapace and tibia of the lost holotype of *Chelone obovata*. However, no other fossil turtle material is present in the collections at the National Museum of Wales.

The Cardiff Specimen (NMW 33.401.G1)

The rediscovered carapace-bearing slab (Figs. 11.1, 11.2) is now catalogued as NMW 33.401.G1. The specimen comprises a carapace in internal (i.e., ventral or visceral) aspect with several of the posterior elements missing, but those are represented by natural moulds of their dorsal surfaces. A single, small, elongate bone at the posterior margin of the carapace may be a distal long bone, perhaps the possible tibia reported by Owen (1842). The horizon of collection is unknown, but the appearance of the limestone is typical of many specimens collected from the middle part of the Stair Hole Member of the Durlston Formation (Westhead and Mather 1996), as found in the quarries west of Swanage. In older terminology, these beds were referred to as the Intermarine Member ("DB112-145" of Clements 1993). The slab appears to have been chiselled with a small narrow chisel to fit the shape of the carapace as a single exercise, rather than to be a slab cut from a larger block. The significance of this will be discussed later.

Much of Owen's (1842) description is comparative in relation to turtles from the Late Cretaceous and Early Palaeogene, and only parts of it are relevant to the problem of recognising the individual specimen. Owen (1842, p. 171) described the carapace as 10¾ inches long and 9½ inches wide; the dimensions of NMW 33.401.G1 correspond to those values. Three sections from Owen's (1842) description are reproduced here with annotations for comparison with the specimen as illustrated in Figs. 11.1 and 11.2. Owen (1842, p. 170) wrote that the carapace "… although obtusely pointed behind, begins to contract to that extremity only at its posterior third part; it gradually widens though the two anterior thirds of its extent and is broadest at the junction of the fifth and sixth ribs; the contour being obversely ovate, or with the broader end turned downwards. This modification of form arises, not from the superior length of the fifth and sixth pairs of ribs, but from the breadth of their sternal appendages, called marginal plates." The widest point two-thirds of the way from the front at the level of the fifth and sixth ribs, together with the wide peripheral bones in this region can be seen in Figs. 11.1 and 11.2.

Of the suprapygals and pygal, Owen (1842, pp. 170–171) wrote "The eleventh six-sided plate [suprapygal 1], which resembles a triangle, with truncated angles, and is wedged between the last pair of ribs, is here divided by a transverse suture into two nearly equal parts. The twelfth plate [suprapygal 2] is nearly twice as broad as long, and has convex lateral margins: the thirteenth vertebral plate [pygal], or the last of the marginal plates, is relatively broader than in existing turtles, and has its posterior margin more feebly emarginate." In NMW 33.401.G1 suprapygal 1 is preserved largely as bone, whereas suprapygal 2 and the pygal are preserved as imprints of the dorsal surface (Figs. 11.1, 11.2), but the shapes of all three match Owen's (1842) description.

The most specific feature of Owen's (1842) description that is matched in NMW 33.401.G1 concerns the apparent asymmetry of the costal outgrowths of the first pair of ribs. Owen (1842, p. 170) wrote "The first rib …. very gradually contracts into its dentiform extremity, which on the [morphological] left side appears to have been separated by a narrow membranous space from the anterior marginal plates, but not on the right side." This asymmetry is in fact due to local crushing on the morphological right side, but the differential in the size of the spaces in the dermal armour is clear (Figs. 11.1, 11.2) and unusual.

In conclusion, not only does the provenance and determinable history suggest that NMW 33.401.G1 is the lost holotype carapace of *Chelone obovata*, but it matches Owen's (1842) description in all respects including slight asymmetry due to crushing.

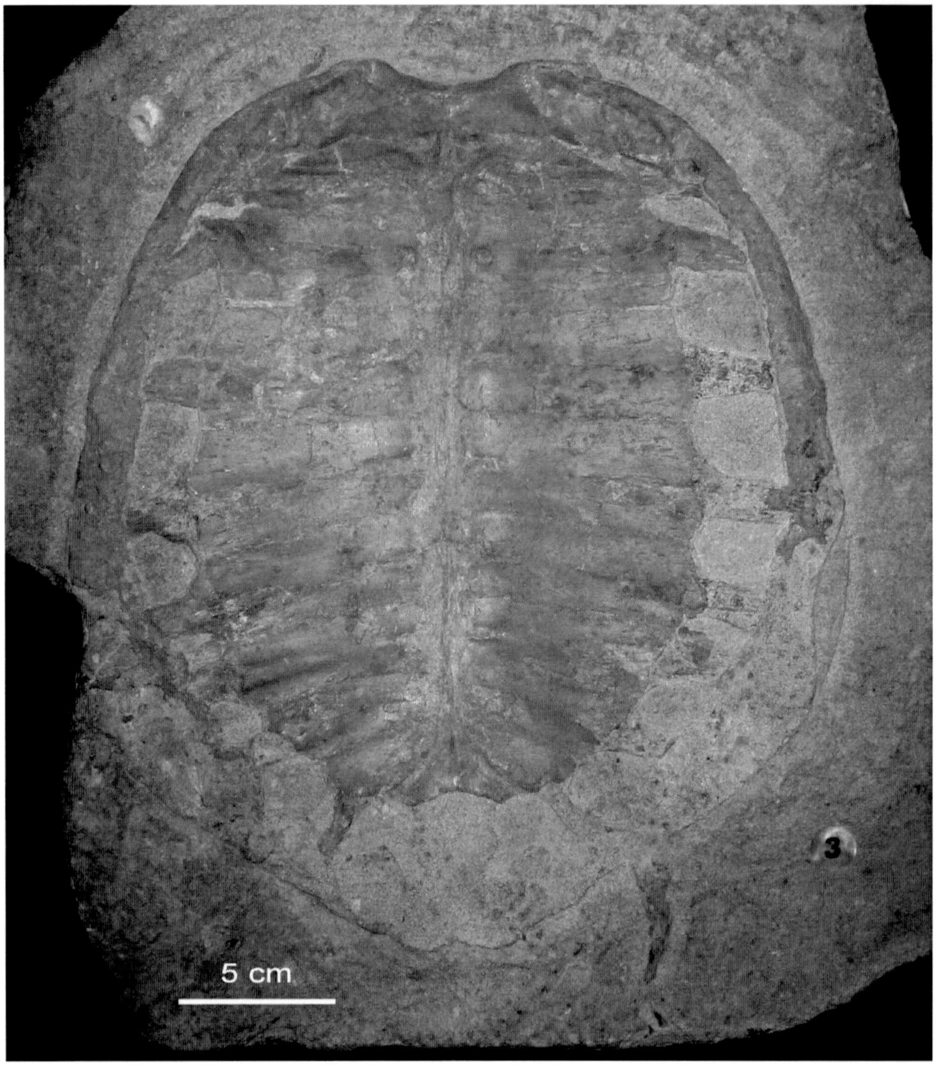

Fig. 11.1 *Hylaeochelys latiscutata* (Owen). Photograph of carapace in internal (visceral) aspect and possible tibia. Specimen NMW 33.401.G1, rediscovered part of the holotype of *Chelone obovata* Owen 1842 nomen oblitum. See Fig. 11.2 for interpretive drawing

Taxonomic Identity of NMW 33.401.G1

Identification of NMW 33.401.G1with one of the Purbeck shell-taxa recognised by Milner's (2004) revision is not problematic, although the external (i.e., dorsal) surface of the carapace is not directly visible. Milner (2004) recognised four shell-taxa, namely the solemydid *Helochelydra anglica* (widely referred to as *Tretosternum punctatum* in the literature), the pleurosternids *Pleurosternum bullockii* and '*Glyptops*' *typocardium*, and the plesiochelyid *Hylaeochelys latiscutata*. Of these, *Pleurosternum bullockii* is the most common form, comprising over 80% of specimens seen by Milner (2004, p. 1465); the other taxa make up the other 16%. The general shape of NMW 33.401.G1 is most similar to that of *Hylaeochelys latiscutata*, being relatively rounded with a shallow anterior emargination. The incomplete growth of the costal plates over the ribs at such a large size is also characteristic of the Purbeck *Hylaeochelys* and can be seen in specimens such as NHMUK R1640 (Milner 2004, text-fig. 12). That feature does not occur in large specimens of the other shell-taxa from Purbeck.

Of the other Purbeck taxa, the carapace of *Pleurosternon* generally is slightly more elongate and oval in outline, with no anterior emargination at all, and it does not have the wide posterolateral marginals found in NMW 33.401.G1 and *Hylaeochelys*. The carapace of '*Glyptops*' *typocardium* is more similar to NMW 33.401.G1, but the anterior emargination is deeper with more rounded shell margins on either side of it. The carapace shape of *Helochelydra* ('*Tretosternon*') from Purbeck is unknown, but dorsal surface impressions of NMW 33.401.G1 show that its dermal surface was not pustulate as in *Helochelydra*.

In conclusion the rediscovered holotype carapace of *Chelone obovata* is entirely consistent with attribution to the Purbeck shell-taxon that Milner (2004, p. 1462) referred to as *Hylaeochelys latiscutata*.

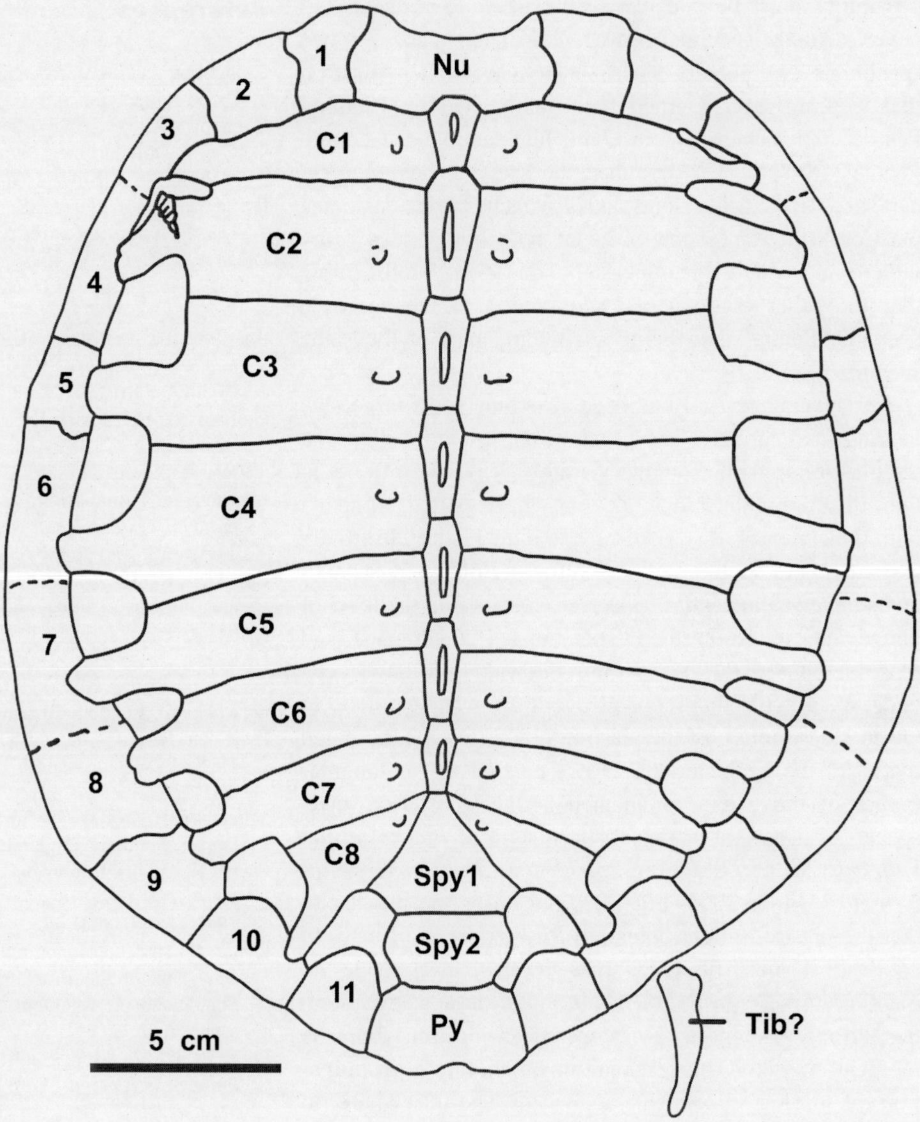

Fig. 11.2 *Hylaeochelys latiscutata* (Owen). Interpretive drawing of carapace in internal (visceral) aspect and possible tibia. Specimen NMW 33.401.G1, rediscovered part of the holotype of *Chelone obovata* Owen 1842 nomen oblitum. See Fig. 11.1 for photograph. Abbreviations: *C*1–*C*8 costals 1–8, *Nu* nuchal, *Py* pygal, *Spy*1 suprapygal 1, *Spy*2 suprapygal 2, *Tib*? possible tibia, 1–11 peripherals 1–11

The Nomenclature Problem

The rediscovery of part of the holotype of *Chelone obovata* Owen 1842 requires resolution of the nomenclatoral implications, this being one of the most senior names applied to Purbeck turtle material. The use of *Chelone* is unproblematic—this name was first used by Brongniart (1805) for the living Green Turtle and was used by Owen (1842) at a time when it was a 'catch-all' taxon for assorted fossil and living turtles. However, that generic name is no longer applicable to any Mesozoic turtle. In contrast, *obovata* Owen 1842 is clearly senior to *latiscutata* Owen 1853 (original spelling *latiscutatum*). However, its subsequent history permits it to be defined as a nomen oblitum. Since Owen's description in 1842, the only references to *obovata* are in a catalogue by Woodward (1909), in the review of Purbeck reptiles by Delair (1958), and as a nomen dubium in the review by Milner (2004). It has not been used as a defined valid taxon since Owen's first paper. In accordance with the International Code of Zoological Nomenclature (ICZN 1999) the name should therefore be treated as a nomen oblitum and *latiscutata* should continue to be the senior valid name (the nomen protectum) for the Purbeck *Hylaeochelys* material. In future of course, *Chelone obovata* should be listed in synonymies of *Hylaeochelys latiscutata* as a senior nomen oblitum preceding the valid nomenclature.

The Missing Plastron and Femur

The uniform chiselling around the carapace of NMW 33.401.G1 suggests a single coherent exercise in preparation rather than a break-up of a larger slab and the possibility must be considered that *Chelone obovata* was never a single specimen on a single large slab, but two specimens on separate slabs—a carapace and a plastron that were merely assumed to be associated by Chaning Pearce and Owen. In fact, in his catalogue Chaning Pearce (1847, p. 32) described the specimen as "Perfect carapace 10¾ inches long and 9½ inches wide, <u>also</u> the plastron with femur and tibiae of both hinder legs" (my underlining added for emphasis). The "also" might imply that the other components were added to the carapace specimen, rather than being an integral part of the same specimen.

For several years, I had been searching for a large slab bearing both carapace and plastron, but it is now clear that the plastron must be sought separately. It is certainly not in the collections at Cardiff and must have parted company with the carapace at an earlier stage in its history. I initially assumed that I should now search for an isolated plastron of *Hylaeochelys latiscutata*. However, considering the level of knowledge in 1842, the association of a carapace and plastron on separate slabs from Purbeck may have been undertaken with more confidence than was justified. In Owen's description of the plastron (1842, p. 171), he noted that "… the xiphisternals…are separated by an angular fissure of the extent of an inch at their posterior interspace…." This is a characteristic feature of *Pleurosternon bullockii* (Milner 2004, Figs. 3, 7A), not *Hylaeochelys latiscutata* (Milner 2004, Fig. 7B), and raises the possibility that the plastron might belong to *Pleurosternon* instead. In contrast, it must be noted that Owen (1842) made no mention of mesoplastral elements, and if these were absent, that strongly suggests a *Hylaeochelys* plastron. For the moment, this issue must remain open pending re-examination of further specimens at Bristol and elsewhere to determine if any can be interpreted as matching Owen's (1842) description.

Acknowledgments I would like to thank Tim Ewin (Bristol Museum and Art Gallery; now at The Natural History Museum, London) for directing me to the Chaning Pearce catalogue, Tom Sharpe and Cindy Howells (National Museum of Wales) for locating the specimen and giving me access to it, and Mike Howe (Geological Survey) for information about the Chaning Pearce ichthyosaur in the collections at the Geological Survey, Keyworth. The paper was improved by the refereeing of Steven Sweetman, David Brockhurst, and Richard Moody.

References

Brongniart, A. (1805). *Essai d'une classification naturelle des reptiles*. Paris: Baudoouin.

Clements, R. G. (1993). Type-section of the Purbeck Limestone Group, Durlston Bay, Swanage, Dorset. *Proceedings of the Dorset Natural History and Archaeological Society, 114*, 181–206.

De la Beche, H. T. (1848). Anniversary address of the president. *Quarterly Journal of the Geological Society of London, 2*, xxi–cxx.

Delair, J. B. (1958). The Mesozoic reptiles of Dorset. Part 1. *Proceedings of the Dorset Natural History and Archaeological Society, 79*, 47–72.

Eberle, J. F., & Bolton, H. (1916). *Report of the Museum and Art Gallery Committee*. Bristol Museum and Art Gallery: City and County of Bristol.

Graham, G. H. (2010). Dr. Joseph Chaning Pearce 1844. Accessed September 8, 2010, from http://members.cox.net/ggthomp01/josephcpearce1844.html.

International Commission for Zoological Nomenclature (1999). *International Code of Zoological Nomenclature*. London: International Trust for Zoological Nomenclature.

Lydekker, R. (1889a). On certain chelonian remains from the Wealden and Purbeck. *Quarterly Journal of the Geological Society, 45*, 511–518.

Lydekker, R. (1889b). *Catalogue of the fossil Reptilia and Amphibia in the British Museum (Natural History)*. Part III. *The Order Chelonia*. London: British Museum (Natural History).

Lydekker, R., & Boulenger, G. A. (1887). Notes on Chelonia from the Purbeck, Wealden and London Clay. *Geological Magazine, 4*(3), 270–275.

Milner, A. R. (2004). The turtles of the Purbeck Limestone Group of Dorset, Southern England. *Palaeontology, 47*, 1441–1467.

Owen, R. (1842). Report on British fossil reptiles, Part II. *Report for the British Association for the Advancement of Science, Plymouth, 1841*(11), 60–204.

Owen, R. (1850). Monograph on the fossil Reptilia of the London Clay, and of the Bracklesham and other Tertiary Beds. Part II. Supplement 1, Chelonia (*Platemys*). *Palaeontographical Society Monograph, 3*, 1–4.

Owen, R. (1853). Monograph on the fossil Reptilia of the Wealden and Purbeck Formations. Part 1. Chelonia. *Palaeontographical Society Monograph, 7*, 1–12.

Pearce, J. C. (1847). Catalogue of the J. Chaning Pearce Collection of British Fossils. *Department of Geology Bristol Museum, Geology Manuscript No. 49. Accession No. 3324.*

Westhead, R. K., & Mather, A. E. (1996). An updated lithostratigraphy for the Purbeck Limestone Group in the Dorset type-area. *Proceedings of the Geologists' Association, 107*, 117–128.

Woodward, H. B. (1885). Visit to the Geological Museum of Dr J. Chaning Pearce, at the Manor House, Brixton. *Proceedings of the Geological Association, 9*(4), 165–168.

Woodward, A. S. (1909). Note on a chelonian skull from the Purbeck beds at Swanage. *Proceedings of the Dorset Natural History Society and Antiquaries Field Club, 30*, 143–144.

Chapter 12
Kappachelys okurai gen. et sp. nov., a New Stem Soft-Shelled Turtle from the Early Cretaceous of Japan

Ren Hirayama, Shinji Isaji, and Tsuyoshi Hibino

Abstract *Kappachelys okurai* gen. et sp. nov. is named and described based on two isolated carapacial elements (right seventh costal and left seventh peripheral) from the Lower Cretaceous (?upper Neocomian) Akaiwa Formation of west-central Honshu, Japan. *Kappachelys* is a small turtle (shell length ~10 cm) that exhibits a unique combination of three features: coarse and deep vermiculate sculpture on carpace; no scute sulci; and well-developed peripherals. The form of the sculpture and lack of scute sulci both suggest affinities with the Trionychidae (soft-shelled turtles), whereas the plesiomorphic retention of well-developed peripherals indicates *Kappachelys* lies outside the Trionychidae. Given this combination of primitive and derived features, we interpret *Kappechelys* as a stem trionychid. In the same region of Japan, the overlying Lower Cretaceous (Aptian) Kitadani Formation contains some of the oldest known, unequivocal trionychid fossils. Based on its slightly older age, similar geographical distribution, and more primitive shell morphology, *Kappachelys* could be ancestral to the trionychids of the Kitadani Formation.

Keywords Early Cretaceous • Japan • *Kappachelys* • Trionychidae • Trionychoidea

R. Hirayama (✉)
School of International Liberal Studies,
Waseda University, Tokyo, Japan
e-mail: renhirayama@waseda.jp

S. Isaji
Natural History Museum and Institute, Chiba, Japan
e-mail: isaji@chiba-muse.or.jp

T. Hibino
Shiramine Institute of Paleontology, Ishikawa, Japan
e-mail: s-kyoiku@city.hakusan.lg.jp

Introduction

Soft-shelled turtles (Trionychidae) have perhaps the most bizarre morphology among all turtles. Their earliest records can be traced back to the Early Cretaceous (Albian or Aptian) of Asia (e.g., Nessov 1995; Hirayama et al. 2000; Hirayama 2002; Danilov and Vitek 2009, 2012; Vitek and Danilov 2010, Fig. 1). The earliest known trionychids are characterized by a unique shell morphology (e.g., distinct external sculpture of ridges, pits, and grooves; scutes and scute sucli absent; peripheral bones absent; plastron reduced; no bony bridge between carapace and plastron) that is maintained in later members of the clade. No obvious transitional form between trionychids and other turtles has been ever described.

Lower Cretaceous, non-marine sediments of the Tetori Group distributed around Mt. Hakusan on west-central Honshu, Japan, have yielded abundant specimens of non-marine turtles, including trionychids (Hirayama et al. 2000; Hirayama 2000, 2005, 2006, 2008; Isaji et al. 2005, 2006). Here we report a presumed stem trionychid of probable late Neocomian age from the Akaiwa Formation (Tetori Group), based on two diagnostic carapace bones collected from the Oarashidani locality in the Shiramine area, Hakusan City, southeastern Ishikawa Prefecture (Fig. 12.1).

Geological Setting

In the Shiramine area of Ishikawa Prefecture, the Tetori Group is divided into four formations: the Gomijima, Kuwajima, Akaiwa, and Kitadani formations in ascending order (Maeda 1961). These non-marine formations are conformable, but they have proven difficult to date because none of their rocks are interbedded with marine strata containing index fossils.

D. B. Brinkman et al. (eds.), *Morphology and Evolution of Turtles*, Vertebrate Paleobiology and Paleoanthropology,
DOI: 10.1007/978-94-007-4309-0_12,

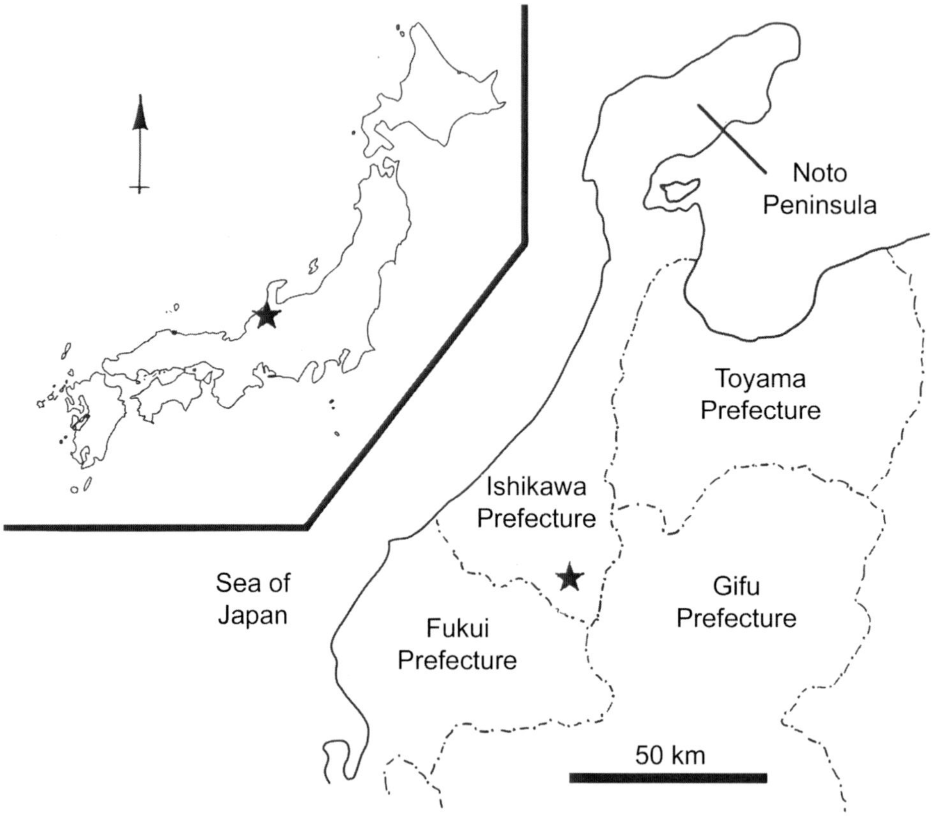

Fig. 12.1 Map depicting location of the Oarashidani fossil locality (*star*), holotype locality for *Kappachelys okurai* gen. et sp. nov. in exposures of the Lower Cretaceous (?late Neocomian in age) Akaiwa Formation, located in the Shiramine area, Hakusan City, southeastern Ishikawa Prefecture, west-central Honshu, Japan

About 30 km southeast of the Oarashidani locality, ammonites of earliest Cretaceous age (Sato et al. 2008) have been recovered from the Mitarai Formation in the Shokawa area, Gifu Prefecture. Although the Mitarai Formation does not crop out in the Shiramine area, regional stratigraphic correlations place it below exposures of the Gomijima Formation (Maeda 1961). Based on this correlation, the age of the base of the Tetori Group as exposed in the Shiramine area is constrained to the earliest Cretaceous. The Gomijima Formation is a basal conglomerate containing brackish molluscs. The overlying Kuwajima Formation records a transition from brackish to fluvial depositional environments.

The Akaiwa Formation, which is the source for the turtle fossils reported here, is a fluvial unit that has been divided into two members (Ishikawa Prefecture Board of Education 1978, Fig. 12.2). The lower or "alternation member" consists of alternating beds of mudstones, fine sandstones, and coarse sandstones interbedded with conglomerates; this member was deposited in a meandering river system. The upper or "sandstone member" consists of massive sandstones deposited in a braided river system.

The overlying Kitadani Formation is important for constraining the minimum age of the Tetori Group in the Shiramine area. Non-marine mollusks from the formation suggest an age of late Hauterivian to late Aptian (Isaji 1993), whereas charophytes suggest a slightly older Barremian age (Kubota 2005). Based on the maximum and minimum age estimates reported above, we estimate the Akaiwa Formation in the Shiramine area as being late Neocomian in age.

Stratigraphically, the Oarashidani locality occurs in the lower or "alternation member" of the Akaiwa Formation (Fig. 12.2, section 10), in fine-grained sandstones that lie about 10 m below the first conglomerate bed, which is a distinct marker bed in that region (Ishikawa Prefecture Board of Education 1978; M. Okura, pers. comm.). Blocks of sandstone excavated from the locality have yielded turtle bones, fish scales, and isolated teeth of small ornithopods and crocodiles (M. Okura, pers. comm.). Geographically, the Oarashidani locality occurs in an outcrop along a woodland path in the Shiramine area, Hakusan City. It is only about 2 km east of the more richly fossiliferous Kuwajima Kaseki-kabe or "Fossil Cliff" locality, which occurs stratigraphically lower in the uppermost part of the Kuwajima Formation (Fig. 12.2, section 7); for details about the latter locality and its vertebrate fossils, see Kusuhashi (2008 and references therein).

Specimens included in this paper are housed in the following collections: FPDM, Fukui Prefectural Dinosaur Museum, Katsuyama City, Fukui Prefecture, Japan; RH, Ren Hirayama private collection, Tokyo, Japan; SBEI, Shiramine Board of Education, Hakusan City, Ishikawa Prefecture, Japan.

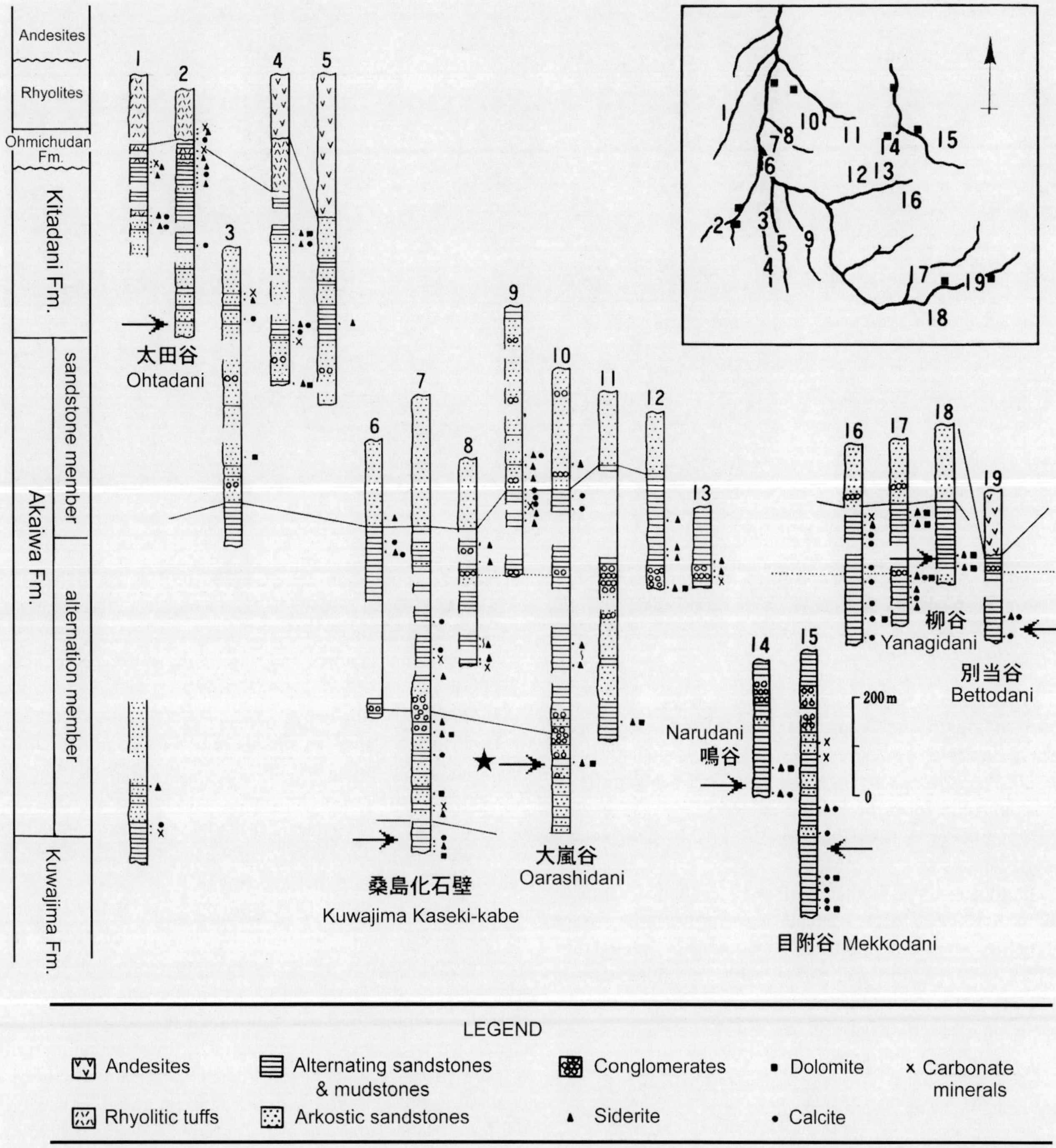

Fig. 12.2 Stratigraphic sections and occurrences of turtle fossils through the Tetori Group, in the Shiramine area, Hakusan City, southeastern Ishikawa Prefecture, west-central Honshu, Japan. *Arrows* indicate turtle fossil-bearing layers; *star* denotes holotype locality for *Kappachelys okurai* gen. et sp. nov. Geological data after the Ishikawa Prefecture Board of Education (1978). Inset map shows geographic locations of sections in the Shiramine area

Systematic Paleontology

Testudines Batsch 1788
Cryptodira Cope 1868
Trionychia Hummel 1929

Diagnosis: Anterior medial process of frontals absent; processus trochlearis oticum includes a substantial contribution from the parietal; dentary pocket present; high coronoid process located near middle of mandible; large processus retroarticularis on mandible; second to seventh cervicals

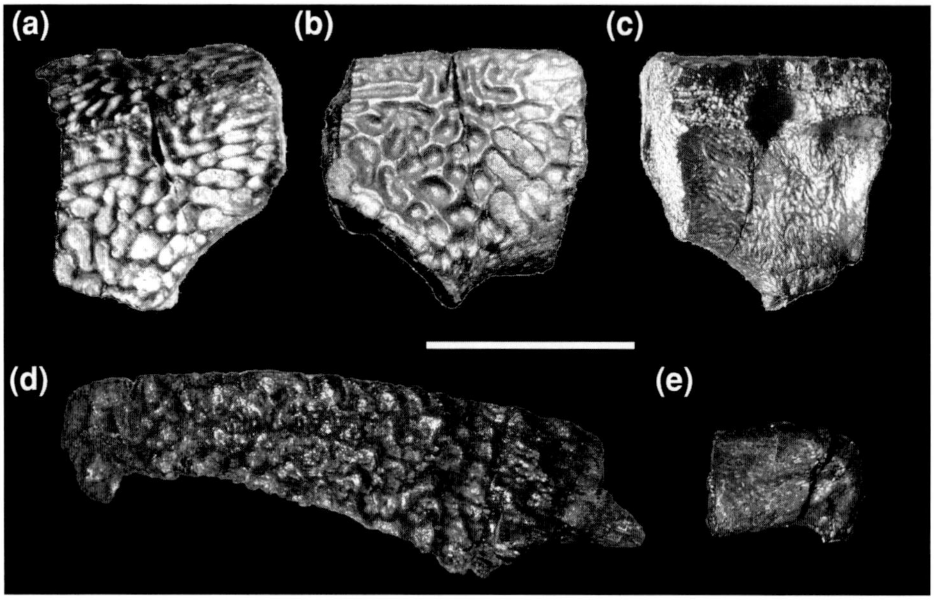

Fig. 12.3 *Kappachelys okurai* gen. et sp. nov., from the Lower Cretaceous (?upper Neocomian) Akaiwa Formation, Oarashidani fossil locality, Ishikawa Prefecture, Japan. **a–c** SBEI 1728 (holotype), left seventh peripheral: **a** lateral view, with anterior to left; **b** dorsal view, with anterior to left; and **c** visceral view, with anterior to right. **d**, **e** SBEI 1729, right seventh costal: **d** dorsal view of entire specimen; **e** ventral view of proximal portion. Images at same magnification; scale bar = 1 cm

opisthocoelous; last cervical opisthocoelous; prominent pitted or vermiculate shell sculpture (Hirayama 2000, 2005).

Kappachelys gen. nov.

Type species: *Kappachelys okurai* sp. nov.

Etymology: From "kappa", a ghost in Japanese folklore, largely inspired by the native, extant soft-shelled turtle (*Pelodiscus sinensis*), plus "chelys", Greek for turtle, a common suffix for turtle generic names.

Distribution: Early Cretaceous (probably late Neocomian), Japan.

Diagnosis: Small-bodied turtle with estimated carapace length of 10 cm. Differs from all other known turtles in the following unique combination of carapace characters: retains well-developed peripherals with a prominent pit in visceral surface for receipt of distal end of rib; scute sulci absent; and external surface bears extremely coarse and deep vermiculate sculpture.

Kappachelys Okurai gen. et sp. nov.
(Figs. 12.3, 12.4b, 12.5b, 12.6b)

Holotype: SBEI 1728, an isolated, seventh left peripheral (Figs. 12.3a–c, 12.4b).

Holotype unit, locality, and age: Lower or alternation member, Akaiwa Formation, Tetori Group; Oarashidani locality, Shiramine area, Hakusan City, southeastern Ishikawa Prefecture, Honshu, Japan; Early Cretaceous, probably late Neocomian.

Etymology: Specific epithet honors Mr. Masatoshi Okura who collected the only known specimens on 02 December 1994.

Referred specimen: SBEI 1729, an isolated, right seventh costal (Figs. 12.3d, e, 12.5b, 12.6b), from the holotype locality and horizon.

Distribution: Known only from the holotype locality and horizon.

Diagnosis: As for genus.

Description: Both specimens are isolated, nearly complete, and three-dimensionally preserved carapacial bones. The exterior surfaces of both specimens (e.g., Fig. 12.3a, b, d) are covered with coarse, vermiculate sculpture consisting of irregular ridges and deep depressions. No traces of scute sulci are present on either specimen.

The holotype, SBEI 1728 (Fig. 12.3a–c), is left seventh peripheral that measures about 12 mm long and wide. Its visceral surface (Fig. 12.3c) bears a pronounced pit that, in life, received the distal end of a thoracic rib. The form of this pit is typical of turtles in which the peripherals and costals are in ligamentous contact. The dorsomedial margin of the peripheral is damaged, so it is uncertain whether it could have been sutured with the adjacent costal. However, the form of the distal end of the referred costal (see below) argues against a sutured peripheral-costal contact. The ventral margin of the peripheral anteriorly bears a facet for ligamentous articulation with the hypoplastron. No trace of a plastral buttress is developed.

The referred specimen, SBEI 1729 (Fig. 12.3d, e), is a right seventh costal that measures 28 mm wide (i.e., along the long axis), 5 mm along its medial border, and 10 mm along its lateral margin. Its distal portion bears a pronounced free, thoracic rib end that projects beyond the lateral (distal) margin of the overlying costal plate. In transverse profile, the distal margin of the costal ends in a thin blade as in

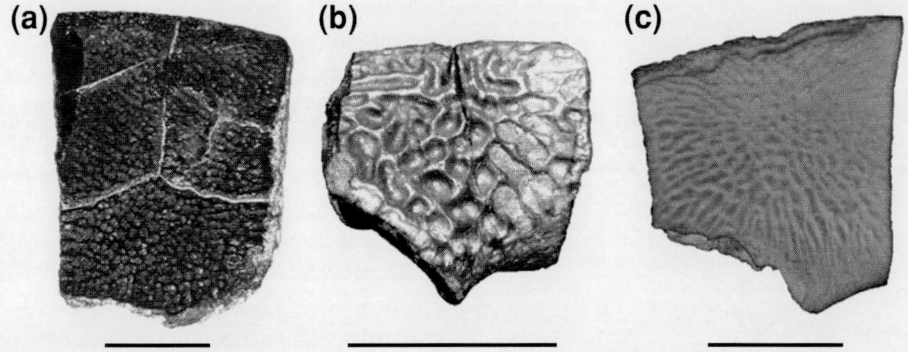

Fig. 12.4 Peripherals of representative trionychian turtles (dorsal views). **a** Trionychia fam., gen., et. sp. indet. (SBEI 529), right third peripheral, from the Lower Cretaceous Kuwajima Formation, Ishikawa Prefecture, Japan. **b** *Kappachelys okurai* get. et sp. nov. (SBEI 1728, holotype), left seventh peripheral, from the Lower Cretaceous Akaiwa Formation, Ishikawa Prefecture, Japan. **c** Extant *Carettochelys insculpta* (RH 917), left seventh peripheral. Images at different magnifications; scale bars = 1 cm

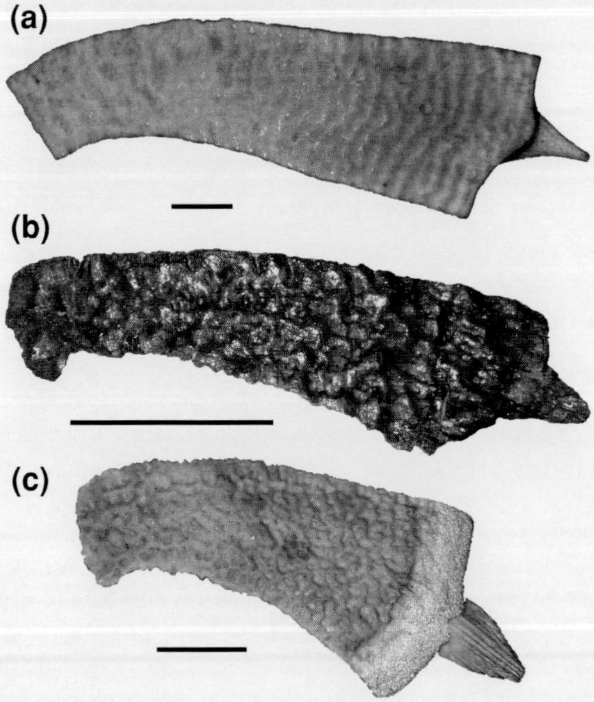

Fig. 12.5 Right seventh costals of representative trionychian turtles (dorsal views). **a** Extant *Carettochelys insculpta* (RH 917). **b** Early Cretaceous *Kappachelys okurai* gen. et sp. nov. (SBEI 1729). **c** Extant *Pelodiscus sinensis* (RH 976). Images at different magnifications; scale bars = 1 cm

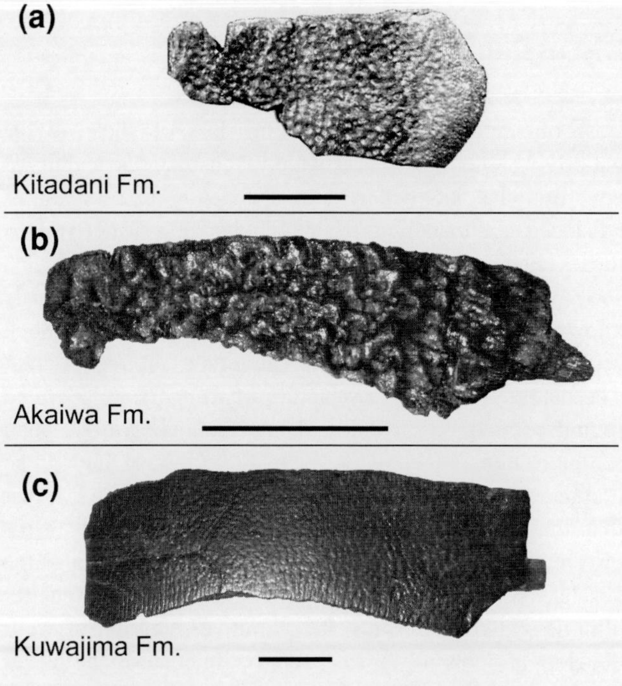

Fig. 12.6 Comparison of costals and stratigraphic succession of trionychian turtles during the Early Cretaceous, through the upper part of the Tetori Group in west-central Japan. **a** Trionychidae gen. et sp. indet. (FPDM-V 127), from the Kitadani Formation, Katsuyama, Fukui Prefecture. **b** *Kappachelys okurai* gen. et sp. nov. (SBEI 1729), from the Akaiwa Formation, Hakusan, Ishikawa Prefecture. **c** Trionychia fam., gen., et sp. indet. (SBEI 1708), from the Kuwajima Formation, Kaseki-kabe locality, Hakusan, Ishikawa Prefecture. Images at different magnifications; scale bars = 1 cm

carettochelyids, which suggests a ligamentous contact with the peripheral, rather than a cartilaginous connection with the rim of soft tissue that extends beyond the bony carapacial disc as in trionychids. Ventrally, the thoracic rib head is pronounced (Fig. 12.3e) as in most other trionychoids, except for adocids. The ventral surface of the costal lacks a distinct rib thickening as in trionychids.

Differences in the relative sizes of the two specimens indicate that the peripheral came from a larger individual than the costal. Nevertheless, in absolute terms both specimens are small and we estimate that the carapace of *Kappachelys* was only about 10 cm long.

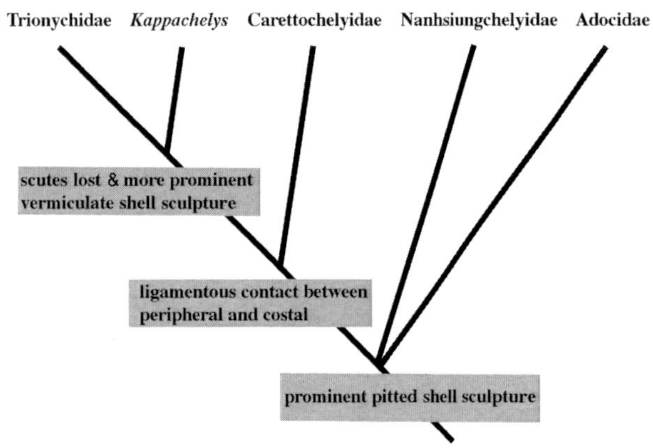

Fig. 12.7 Hand-drawn tree depicting proposed relationships of *Kappachelys okurai* gen. et sp. nov. within Trionychia, based on shell characters

Discussion

Loss of scutes (and corresponding loss of sucli on the underlying shell plates) is widely regarded as a derived condition that has occurred independently in trionychids, advanced carettochelyids (Carettochelyidae), and derived sea turtles such as dermochelyids and protostegids (e.g., Meylan 1988; Hirayama and Chitoku 1996). Among previously known turtles, the combination of scute loss and distinct shell sculpture is limited to trionychids and some carettochelyines. The coarse, vermiculate sculpture of *Kappachelys* is more reminiscent of trionychids than carettochelyines. Shell sculpture in carettochelyids (Figs. 12.4c, 12.5a) tends to be dominated by small tubercles often arranged in rows, rather than the ridge-and-groove style of ornament typical for trionychids (Fig. 12.5c) and seen in *Kappachelys*. Although the absence of scute sulci and the form of the sculpture argue for affinities with trionychids, the primitive retention of well-develop peripherals in *Kappachelys* indicates that it lies outside the Trionychidae. Based on that trio of primitive and derived features, we provisionally interpret *Kappachelys* as a stem trionychid (Fig. 12.7). It is worth acknowledging that although no known fossil and most extant trionychids lack peripherals, extant species of *Lissemys* have a series of small, irregular-shaped bony ossifications called "peripheral ossifications" around the posterior margin of their carapace (e.g., Scheyer et al. 2012, Fig. 6.3). Although these peripheral ossifications have been interpreted as being homologous with peripheral bones of non-trionychids (e.g., Meylan 1987; Scheyer et al. 2012), they could be secondarily acquired neomorphs. Regardless of their homologies, the peripheral ossifications of *Lissemys* are much simpler than the peripheral described here for *Kappachelys* and do not preclude the latter genus from being a stem trionychid.

Elsewhere in west-central Japan, fossils of even more primitive trionychian turtles with scale sulci are quite abundant in the Kuwajima Formation, which is dated as Neocomian and underlies the Akaiwa Formation (Hirayama 2005). All known trionychian specimens from the Kuwajima Formation retain scute sulci (Figs. 12.4a, 12.6c) and, thus, are more primitive than *Kappachelys* and trionychids. The first occurrence of unequivocal trionychid fossils is in the Kitadani Formation (Fig. 12.6a), which is dated as Aptian and overlies the Akaiwa Formation (Hirayama 2002). In other words, fossils currently known from the upper part of the Tetori Group in west-central Japan appear to document a transition from non-trionychid trionychians in the Kuwajima Formation, through stem trionychids (*Kappachelys*) in the Akaiwa Formation, and culminating with trionychids in the Kitadani Formation (Fig. 12.6). This sequence suggests that trionychids might have evolved in Asia from a small-bodied, *Kappachelys*-like ancestral form during a rather short geologic time span around the Barremian stage.

Trionychids are also unique among turtles in having a plywood-like internal structure to their shell (Scheyer et al. 2007). Microscopic observations of *Kappachelys* and other early trionychid shells will be published elsewhere by Y. Nakajima of Tokyo University.

Conclusions

A new genus and species, *Kappachelys okurai*, is described based on two isolated carapacial bones (a left seventh peripheral and a right seventh costal), from the Lower Cretaceous (probably late Neocomian in age) Akaiwa Formation in Ishikawa Prefecture, west-central Japan. Even though *Kappachelys* is known by limited shell material, those specimens exhibit a trio of characters (derived pattern of shell ornament and derived absence of scute sulci; primitive retention of well-developed peripherals) that allow *Kappachelys* to be interpreted as a stem trionychid. The oldest known unequivocal trionychids are slightly younger than *Kappachelys* and some of those occur in the same region of Japan. Consequently, *Kappachelys* could be close to the ancestry of trionychids. The fossils and interpretations presented herein demonstrate that the Tetori Group is important for tracing the early evolutionary history of trionychian turtles.

Acknowledgments We are especially grateful to Masatoshi Okura for the collection and preparation of turtle fossils, including the two specimens of *Kappachelys*. We also thank: T. Sonoda (Ibaraki University, Mito), M. Onodera (Kagoshima University, Kagoshima), and R. Obata (Wakasa-cho) for their help with excavations at the Oarashidani fossil locality; I. G. Danilov (Zoological Institute of the Russian Academy of Sciences, St. Petersburg) for access to comparative specimens; Y. Nakajima (Tokyo University, Tokyo) for sharing preliminary results from his histological studies on early trionychid shells; and I. G. Danilov, P. A. Meylan (Eckerd College), J. F. Parham

(Alabama Museum of Natural History), and an anonymous reviewer for their comments and corrections on the submitted version of our manuscript. Field work at the Oarashidani fossil locality was supported by a Grant-in-Aid for Specially Promoted Research (awarded to M. Manabe of the National Science Museum, Tokyo: grant no. 12800018) from the Japanese Ministry of Education and Science.

References

Batsch, A. J. G. C. (1788). *Versuch einer Anleitung, zur Kenntniß und Geschichte der Thiere und Mineralien.* Jena: Akademische Buchhandlung.

Cope, E. D. (1868). On the origin of genera. *Proceedings of the Academy of Natural Sciences of Philadelphia, 1868*, 242–300.

Danilov, I. G., & Vitek, N. S. (2009, October 17–18). Cretaceous trionychids of Asia: A review of record and biogeography. *Gaffney Turtle Symposium* (*Royal Tyrrell Museum, Drumheller, Canada*). *Abstract Volume*, 52–58.

Danilov, I. G., & Vitek, N. S. (2012). Cretaceous trionychids of Asia: An expanded review of their record and biogeography. In D. B. Brinkman, P. A. Holroyd, & J. D. Gardner (Eds.), *Morphology and evolution of turtles.* Dordrecht: Springer.

Hirayama, R. (2000). Fossil turtles. In H. Matsuoka (Ed.), *Fossils of Kuwajima "Kaseki-kabe" (Fossil-bluff), Scientific report on a Neocomian (Early Cretaceous) fossil assemblage from the Kuwajima Formation, Tetori Group, Shiramine, Ishikawa, Japan* (pp. 75–92). Shiramine Village Board of Education, Ishikawa Prefecture, Japan (in Japanese with English abstract).

Hirayama, R. (2002). [Preliminary report of the fossil turtles from the Kitadani Formation (Early Cretaceous) of the Tetori Group Group of Katsuyama, Fukui Prefecture, Central Japan.] *Bulletin of the Fukui Prefectural Dinosaur Museum, 1*, 29–40 (In Japanese with English abstract).

Hirayama, R. (2005). New materials of non-marine turtles from the Early Cretaceous Tetori Group of former Shiramine-mura of Hakusan City, Ishikawa Prefecturee, Central Japan. In *Scientific report on fossil animals of Kuwajima "Kaseki-kabe" (Fossil-bluff) from the Kuwajima Formation, Tetori Group, Hakusan, Ishikawa, Japan* (pp. 12–20). Shiramine Village Board of Education, Ishikawa Prefecture, Japan (In Japanese with English abstract).

Hirayama, R. (2006). Review of fossil turtles of Japan. *Fossils (Palaeontological Society of Japan), 80*, 47–59 (In Japanese with English abstract).

Hirayama, R. (2008). Trionychid turtles (Trionychoidea; Testudines) from the Early Cretaceous of Central Japan. *Journal of Vertebrate Paleontology, 28* (Suppl. to No.3), 91A.

Hirayama, R., & Chitoku, T. (1996). Family Dermochelyidae (Superfamily Chelonioidea) from the Upper Cretaceous of North Japan. *Transactions and Proceedings of the Palaeontological Society of Japan, New Series, 184*, 597–622.

Hirayama, R., Brinkman, D. B., & Danilov, I. G. (2000). Distribution and biogeography of non-marine Cretaceous turtles. *Russian Journal of Herpetology, 7*, 181–198.

Hummel, K. (1929). Die fossilen Weichschildkröten (Trionychia). Eine morphologisch-systematische und stammesgeschichtliche studie. *Geologische und Palaeontologische Abhandlungen, 16,* 359–487.

Isaji, S. (1993). *Nippononaia ryosekiana* (Bivalvia, Mollusca) from the Tetori Group in central Japan. *Bulletin of the National Science Museum, Series C (Geology & Paleontology), 19*, 65–71.

Isaji, S., Okazaki, H., Hirayama, R., Matsuoka, H., Barrett, P. M., Tsubamoto, T., et al. (2005). Depositional environments and taphonomy of the bone-bearing beds of the Lower Cretaceous Kuwajima Formation, Tetori Group, Japan. *Bulletin of the Kitakyushu Museum of Natural History, Series A (Natural History), 3*, 123–133.

Isaji, S., Matsushita, A., & Hirayama, R. (2006). Chelonian eggshell fossils from the Lower Cretaceous Kuwajima Formation of the Tetori Group, Central Japan. *Paleontological Research, 10*, 29–36.

Ishikawa Prefecture Board of Education (1978). Geology of the Shiramine Area. In *Research report of the silicified wood sites from the Tetori Group of the Tetori Valley* (pp. 5–76). Ishikawa Prefecture Board of Education, Kanazawa, Japan.

Kubota, K. (2005). Charophyte gyrogonites from the Lower Cretaceous Kitadani Formation of the Tetori Group in the Takinamigawa area, Katsuyama City, Fukui Prefecture, central Japan. *Paleontological Research, 9*, 203–213.

Kusuhashi, N. (2008). Early Cretaceous multituberculate mammals from the Kuwajima Formation (Tetori Group), central Japan. *Acta Palaeontologica Polonica, 58,* 379–390.

Maeda, S. (1961). On the geological history of the Mesozoic Tetori Group in Japan. *Japanese Journal of Geology and Geography, 32*, 375–396.

Meylan, P. A. (1987). The phylogenetic relationships of soft-shelled turtles (Family Trionychidae). *Bulletin of the American Museum of Natural History, 186*, 1–101.

Meylan, P. A. (1988). *Peltochelys* Dollo and the relationships among the genera of the Carettochelyidae (Testudines: Reptilia). *Herpetologica, 44*, 440–450.

Nessov, L. A. (1995). On some Mesozoic turtles of the Fergana Depression (Kyrgistan) and Dzhungar Alatau Ridge (Kazakhstan). *Russian Journal of Herpetology, 2,* 134–141.

Sato, T., Asami, T., Hachiya, K., & Mizuno Y. (2008). Discovery of *Neocosmoceras*, a Berriasian (early Cretaceous) ammonite, from Mitarai in the upper reaches of the Shokawa River in Gifu Prefecture, Japan. *Bulletin of the Mizunami Fossil Museum, 34*, 77–80.

Scheyer, T. M., Sander, P. M., Joyce, W. G., Bohme, W., & Witzel, U. (2007). A plywood structure in the shell of fossil and living soft-shelled turtles (Trionychidae) and its evolutionary implications. *Organisms, Diversity & Evolution, 7*, 136–144.

Scheyer, T. M., Werneburg, I., Mitgutsch, C., Delfino, M., & Sánchez-Villagra, M. R. (2012). Three ways to tackle the turtle: Integrating fossils, comparative embryology, and microanatomy. In D. B. Brinkman, P. A. Holroyd & J. D. Gardner (Eds.), *Morphology and evolution of turtles.* Dordrecht: Springer.

Vitek, N. S., & Danilov, I. G. (2010). New material and a reassessment of soft-shelled turtles (Trionychidae) from the Late Cretaceous of Middle Asia and Kazakhstan. *Journal of Vertebrate Paleontology, 20*, 383–393.

Chapter 13
Morphology and Relationships of *Brachyopsemys tingitana* gen. et sp. nov. from the Early Paleocene of Morocco and Recognition of the New Eucryptodiran Turtle Family: Sandownidae

Haiyan Tong and Peter Meylan

Abstract The first cryptodiran turtle with a pitted palate, *Brachyopsemys tingitana* gen. et sp. nov., is described on the basis of skulls and a lower jaw from the early Paleocene (Danian) of the Ouled Abdoun Basin, Morocco. The new taxon shares a number of apomorphic features with *Sandownia harrisi* from the Aptian of the Isle of White, England, *Angolachelys mbaxi* from the Turonian of Angola, and an unnamed turtle from the Albian Glen Rose Formation in Texas. A phylogenetic analysis provides evidence that they form a monophyletic clade which we name the Sandownidae fam. nov. The phylogenetic relationships of Sandownidae are problematic. Their affinities among the Eucryptodira remain uncertain; a well-preserved shell of a member of the family may be useful in determining relationships. The family appears to be a bottom-dwelling group that occurred in near-shore marine environments along the coast of the Atlantic across the Cretaceous-Tertiary boundary. Its evolutionary history is connected to the development of the Atlantic Ocean.

Keywords *Brachyopsemys* • Cryptodira • Early Paleocene • Morocco • Sandownidae • Phylogeny

H. Tong (✉)
30 Rue Carnot, 94270 Le Kremlin-Bicêtre, France
e-mail: htong09@yahoo.fr

H. Tong · P. Meylan
Division of Paleontology, American Museum of Natural History, Central Park West at 79th Street, New York, NY 10024, USA
e-mail: meylanpa@eckerd.edu

P. Meylan
Natural Sciences, Eckerd College, 4200 54th Avenue S, St. Petersburg, FL 33711, USA

Introduction

Among recently described or discovered Cretaceous and early Tertiary turtles are four cryptodire skulls that all exhibit a remarkable suite of characters including an extensive skull roof, an extensive secondary palate with a large contribution by the jugal, and a reduced or absent ventral exposure of the basisphenoid. This paper is an attempt to determine if these highly functional characters (broad palate and complete skull roof) result from shared common ancestry or if they might be the result of convergent evolution. The four taxa that show these characters are *Sandownia harrisi* from the Aptian of the Isle of Wight, England (Meylan et al. 2000), *Angolachelys mbaxi* from the Turonian of Angola (Mateus et al. 2009), an unnamed turtle ("the Glen Rose turtle") from the Albian Glen Rose Formation of Texas (Barck 1992; Vineyard and Jacobs 2009), and the turtle described as a new genus and species in this paper.

Sandownia harrisi is a remarkably well preserved skull and partial lower jaw that has been described in detail (Meylan et al. 2000). Its phylogenetic position was proposed on the basis of a set of analyses that were relatively limited in scope and it was considered to be a member of the Trionychoidea.

Angolachelys mbaxi is known from a single, poorly preserved skull that is badly broken dorsally. The type description includes a brief treatment of the morphology of the skull, a dentary, two procoelous cervicals, and an ungual phalanx (Mateus et al. 2009). There is no data set in the type description (a data set was made available to us by the authors). A single tree selected by the authors from among 60 MPT was used to argue for recognition of the "Angolachelonia" including *A. mbexi*, *S. harrisi*, *Solnhofia parsonsi*, and the Glen Rose turtle.

The Glen Rose turtle is yet to be named and fully described but it has been presented at several conferences (Vineyard 1999; Vineyard and Jacobs 2009) and is highly relevant to the problem at hand. Derived characters were listed by Vineyard and Jacobs (2009) and several of these are shared with the other three taxa of interest here.

DOI: 10.1007/978-94-007-4309-0_13,

The fourth taxon is named below. It is known from two complete skulls, and a third skull that is missing the skull roof but has associated lower jaws. These specimens were collected by local people in the Ouled Abdoun Phosphate Basin, Morocco. During the last decade, abundant fossil turtles have been discovered in the Maastrichtian (Late Cretaceous) to Ypresian (Early Eocene) phosphate deposits of the Ouled Abdoun Basin. They include pleurodiran bothremydids and cryptodiran chelonioid sea turtles (Karl et al. 1998; Tong and Hirayama 2002, 2004, 2008; Gaffney and Tong 2003; Hirayama and Tong 2003; Gaffney et al. 2006; Nour-Eddine et al. 2009; Tong et al. in press). According to the local collectors, one of the skulls described here, American Musuem of Natural History (AMNH) 30001, was collected in the 'dalle couche 2' (Slab of layer 2) of the 'Recette 4' (Quarry 4), in the Ouled Abdoun Basin. Shark teeth found in the matrix surrounding the specimens during preparation included *Palaeogaleus brivesi*, Lamniformes indet., *Ixobatis mucronata*, *Plicatosyllium* sp., *Dasiatis hexagonalis*, *Danogaleus gueriri*, Rhinobatoid indet., *Youssoubatis ganntourensis*, *Ctenopristis nougareti* and *Squalus* sp. This composition of Danian selachians mixed with some Maastrichtian elements indicates a Danian age (Cappetta, pers. com. 2001). These turtles were mentioned as a 'pitted cryptodire' in Gaffney et al. (2006). Two of three skulls and the lower jaw have been fully prepared and are described herein. The specimens are housed in the American Museum of Natural History (AMNH). It is certain that this material represents an undescribed form and the occasion of its description provides an opportunity to reconsider this bizarre group of turtles. Thus, this paper has three purposes: (1) to describe a remarkable new cryptodire from the Paleocene of Morocco; (2) to examine the question of monophyly of a set of possibly related Cretaceous and early Tertiary cryptodires with extensive secondary palates and extensive skull roofs; and (3) to determine the best hypothesis of relationship among these turtles and other cryptodires.

Systematic Palaeontology

Order Testudines Linnaeus 1758
Suborder Cryptodira Cope 1868
Parvorder Eucryptodira Gaffney 1975
Superfamily Incertae Sedis
Family Sandownidae nov.

Type genus: *Sandownia* Meylan, Moody, Walker and Chapman 2000

Included genera: *Sandownia* Meylan, Moody, Walker and Chapman 2000; *Angolachelys* Mateus, Jacobs, Polcyn, Schulp, Vineyard, Neto and Antunes 2009; *Brachyopsemys* gen. nov.

Diagnosis: Centrocryptodiran turtles (otic trochlea present and internal carotid artery located deep within pterygoid) with the jugal making a significant contribution to the triturating surface; pterygoids partly or completely cover basisphenoid; premaxillae partly or completely fused and slightly protuberant anteriorly; prefrontal L-shaped, with a posteriolateral process along the ventral margin of the orbit; external process of the pterygoid absent and occipital condyle with anterior location in between the articular surfaces of the quadrate. Long symphysis of dentary; broad triangular lower triturating surface that slopes ventrally from the coronoid process and lacks labial and lingual ridges (known in *Sandownia harrisi* and the taxon described below).

Remarks: Mateus et al. (2009) named the unranked taxon "Angolachelonia" and defined it as the clade originating from the most recent common ancestor of *Angolachelys mbaxi* and *Solnhofia parsonsi*. Our phylogenetic studies (see below) suggest that, as defined, Angolachelonia is a junior synonym of Parvorder Eucryptodira (sensu Gaffney and Meylan 1988). We have found strong support for monophyly of the clade consisting of *Angolachelys mbexi*, *Sandownia harrisi* and the taxon described below and chose to recognize this relationship through the use of the family group name, Sandownidae, with the well-preserved type genus *Sandownia*. Given that the name "Angolachelonia" is non-Linnaean, it cannot be construed as a family-group name and as defined may be a synonym of an ordinal group name. We find it more useful to construct this new family-group name based on character diagnoses rather than on a name based on a phylogeny that was selected from among 60 MPT (Mateus et al. 2009). It is clear that the Glen Rose turtle is a member of the family Sandownidae.

Brachyopsemys gen. nov.

Type species: *Brachyopsemys tingitana* sp. nov.

Etymology: Brachyops: Greek, short face; emys: Greek, turtle.

Diagnosis: Skull broad and low, wider than long, rounded anteriorly in dorsal view, with a very extensive skull roof (more extensive than other sandownids), very shallow temporal emargination with parietal-squamosal contact; orbits facing anterolaterally, placed forward and closely spaced; large triturating surface with a pair of deep pits in the jugal (unique among cryptodires, seen also in the pleurodire family Bothremydidae); unique among turtles in having prefrontal-jugal contact present below orbit excluding maxilla from orbit margin and a medial process of the jugal that forms a large, flat, horizontal, plate-like flange within the fossa temporalis inferior; unique among described members of the family Sandownidae in having the incisura columellae auris open.

Brachyopsemys tingitana sp. nov.
(Figs. 13.1, 13.2, 13.3, 13.4).

Holotype: A complete skull (AMNH 30001).

Holotype locality: Recette 4, Ouled Abdoun Basin, Morocco.

Holotype unit and age: Dalle Couche 2, early Paleocene (Danian).

Etymology: From Moretania Tingitana, an ancient Roman province corresponding to northern Morocco.

Referred specimens: An incomplete skull, lacking the skull roof, with associated lower jaw (AMNH 30558); a complete skull (AMNH 30554).

Occurrence: Early Paleocene (Danian) of Morocco.

Diagnosis: Same as for genus.

Measurements: see Table 13.1.

Description

Preservation: AMNH 30001 is a well preserved complete skull without deformation. Some areas within the fossa temporalis remain filled with matrix. AMNH 30558 is a nearly complete skull, lacking only the skull roof, with an associated nearly complete lower jaw. This specimen has been prepared to reveal much of the braincase, orbit floor and otic region. AMNH 30554 is a complete skull that remains to be prepared.

General aspect: The skull is low and broad, with a rough external surface and scute sulci on the skull roof but without obvious ornamentation. It has a nearly rounded shape as seen from above but the lateral margins are parallel. The skull roof is complete, with very shallow temporal emargination, which is even shallower than that of *Sandownia harrisi* (Meylan et al. 2000), but comparable to that of *Erquelinnesia gosseleti* (Zangerl 1971). The crista supraoccipitalis does not extend beyond the posterior margin of the skull roof so is not visible dorsally, but it does extend significantly beyond the condylus occipitalis. The apertura narium externa is oval-shaped and wider than high; facing forward. The orbits are oval and quite small when compared to those of extant chelonioids. They are placed very anteriorly, facing mainly forward and are very close to each other. The interorbital bar is very narrow as in *S. harrisi*. In lateral view, the skull roof slopes gently from the interorbital bar posteriorly. The cheek region, below the orbit, is convex laterally to form a rounded swelling, giving the skull a 'chubby-cheeked' aspect. The cheek emargination is very shallow, shallower than in *Sandownia*, and is much lower than the lower rim of the orbit. In ventral view, the triturating surface is very large and wide, with a pair of deep pits located laterally. The condylus occipitalis lies anteriorly, in line between the condylus mandibularis of the two quadrates.

Skull roof scales: Some weak sulci are visible in AMNH 30001. On the midline, there is a pair of frontal scales between the orbits, followed by a pair of frontoparietals and a single parietal scale. The supraorbital scale lies posterior to the orbit and lateral to the frontoparietal scale. Posterior to it and lateral to the parietal scale is a temporal scale. One suborbital scale can be distinguished on each side under the orbit. Below it lies the maxillary scale (Figs. 13.1, 13.2). Skull scale terminology follows Tong and Hirayama (2002).

Prefrontal: Both prefrontals are complete in AMNH 30001. The prefrontal is L-shaped with the vertical part contributing to the narrow interorbital bar and a horizontal part that forms a significant portion of the anterior and ventral orbit margin. This condition is as in *Sandownia harrisi*, in contrast to cheloniids, trionychids and most other turtles in which the prefrontal has no laterally directed horizontal branch. The vertical ramus contacts the frontal dorsally between the orbits; ventrally, it forms the upper rim of the apertura narium externa. The surface between the orbits and the apertura narium externa is flat and nearly vertical. The horizontal process extends from the vertical portion laterally to form the anteroventral margin of the orbit and the anterior margin of the foramen orbito-nasale. It is divided into two branches by the foramen orbito-nasale: a smaller medial branch turns posteriorly to contact the vomer medially and the palatine posteriorly, and a larger lateral branch that forms the ventral orbit margin and contacts the jugal laterally, palatine posteriorly, and maxilla anteroventrally on the skull surface. In addition, there is a very short prefrontal-maxilla contact posterolaterally on the anterior edge of the orbital floor between the palatine and jugal. The prefrontal of *S. harrisi* differs in the longer dorsal branch, the smaller contribution to the external nares and the absence of prefrontal-jugal contact. The foramen orbito-nasale in *Brachyopsemys* is a large oval-shaped opening (6.5 × 4 mm) lying on the anteromedial corner of the orbit floor as in *Chelonia mydas*. In trionychids such as *Amyda cartilaginea* and *Lissemys punctata*, this foramen is much larger and more posteriorly placed.

Nasal: The nasal is clearly absent.

Frontal: Both frontals are complete in AMNH 30001. It is a rather small and trapezoidal-shaped bone forming the dorsal rim of the orbit. On the skull roof, the frontal contacts the prefrontal anteriorly, the parietal posteriorly with a long and transversal suture, and the postorbital posteriolaterally by a short suture. The ventral surface of the frontal bears a thick, blunt parasagittal ridge which separates the fossa orbitalis laterally from the sulcus olfactorius medially. The sulcus olfactorius is narrow anteriorly and widens posteriorly. The morphology of the frontal in AMNH 30001 is similar to *Sandownia harrisi*. Trionychids also have a very small frontal that contributes to the dorsal rim of the orbit, while in cheloniids the frontal is larger.

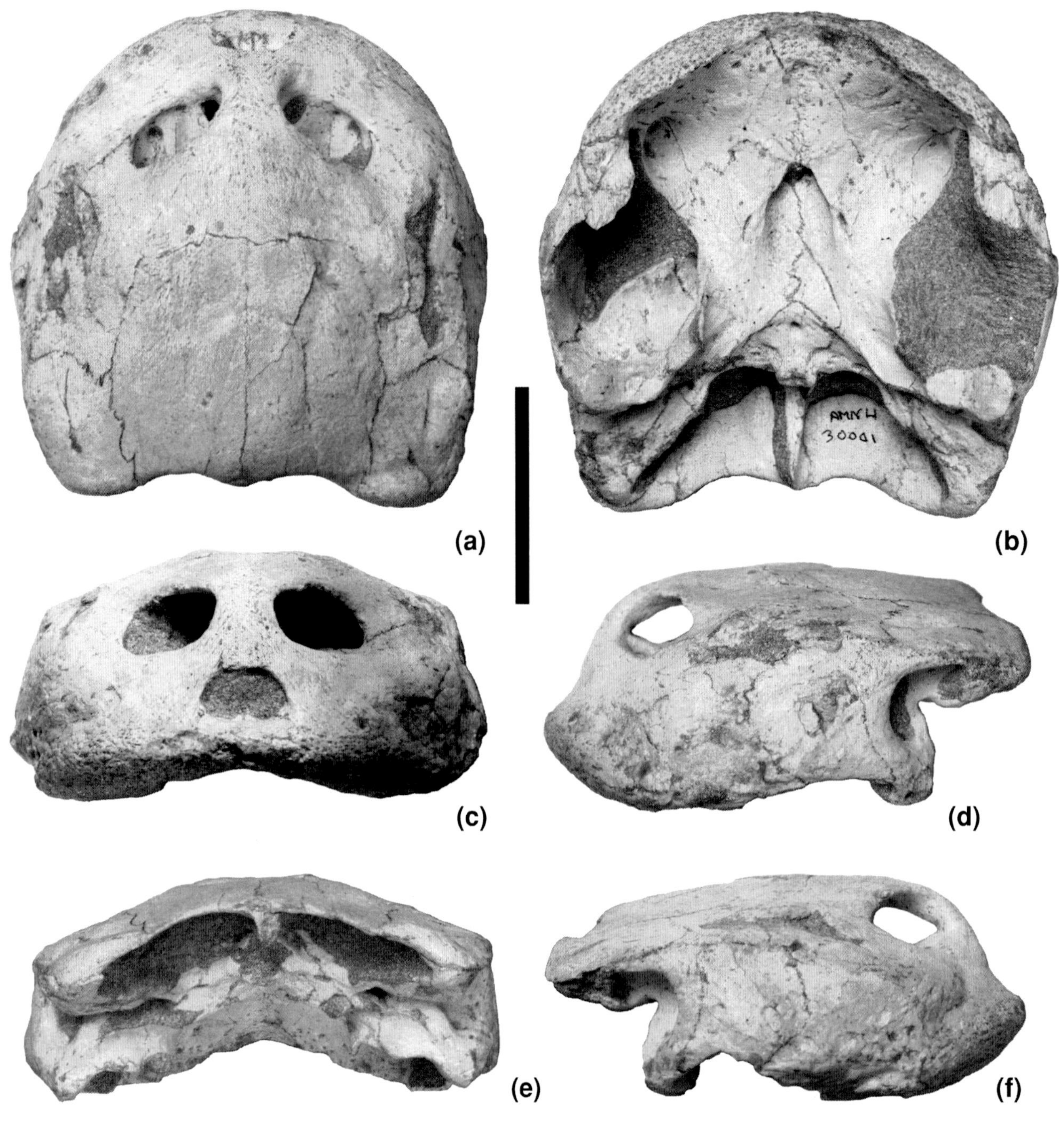

Fig. 13.1 *Brachyopsemys tingitana* gen. et sp. nov., from the Danian (early Palaeocene) of Morocco. AMNH 30001, holotype skull, photographs in: **a** dorsal, **b** ventral, **c** anterior, **d** left lateral, **e** posterior, and **f** right lateral views. Scale bar = 50 mm

Parietal: The parietals are complete in AMNH 30001 but are not fully prepared ventrally. Only small ventral portions of the processus inferior parietalis are preserved in AMNH 30558. The parietal is a large and anteroposteriorly elongated bone as seen from above, producing a large part of the skull roof. On the skull roof, it forms the medial portion of the very shallow temporal emargination and contacts the frontal anteriorly, the postorbital anterolaterally, the squamosal posterolaterally and the supraoccipital posteromedially. Contact with the squamosal is extensive. The very complete skull roof with parietal-squamosal contact is similar to *Sandownia harrisi*, but the temporal emargination in the latter is deeper. A parietal-squamosal contact is also present in cheloniids and other turtles with an extensive skull roof (see Discussion).

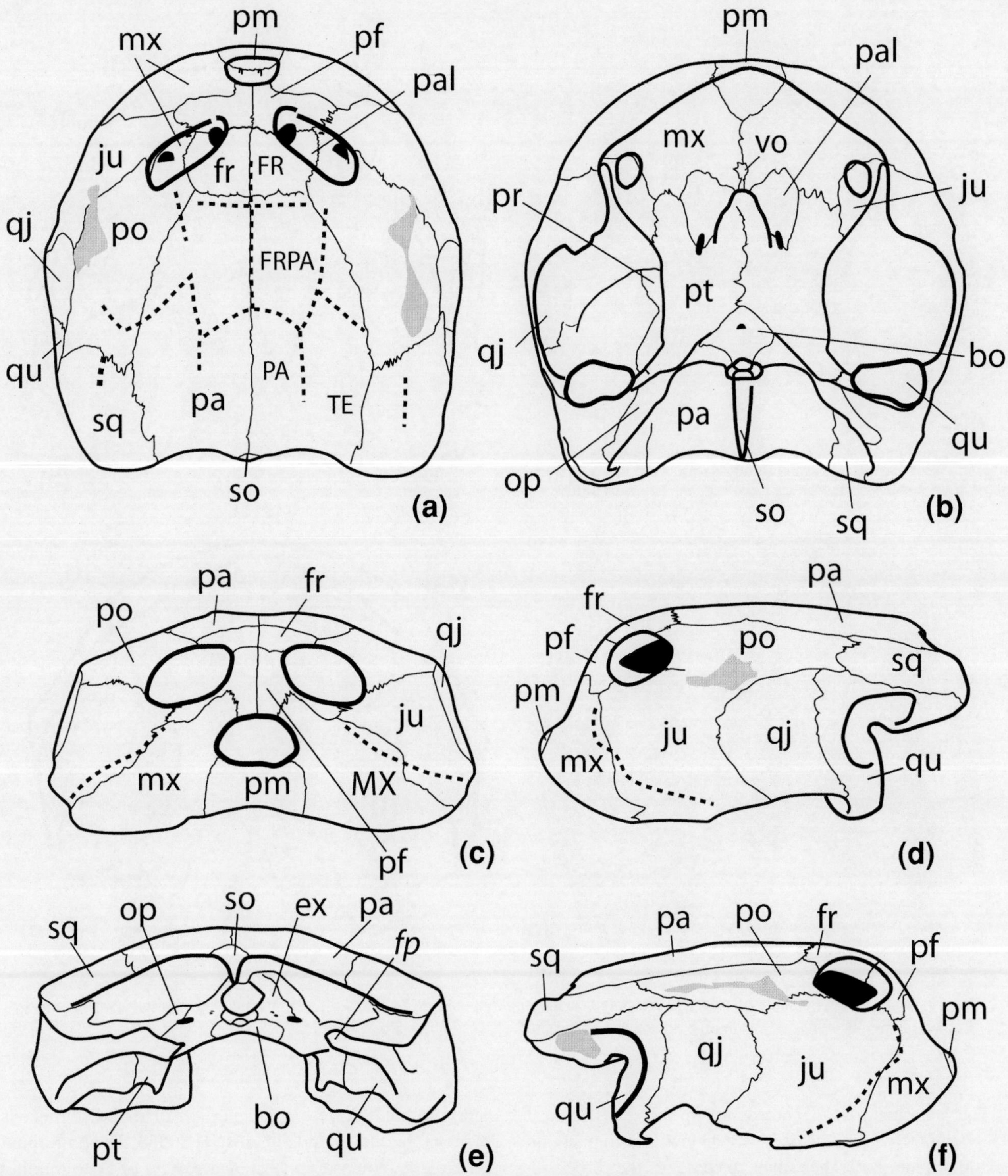

Fig. 13.2 *Brachyopsemys tingitana* gen. et sp. nov., from the Danian (early Palaeocene) of Morocco. AMNH 30001, holotype skull, interpretive drawings in: **a** dorsal, **b** ventral, **c** anterior, **d** left lateral, **e** posterior, and **f** right lateral views. Scale bar = 50 mm. Abbreviations: *bo* basioccipital; *ex* exoccipital; *FRPA* frontoparietal scale; *fr* frontal; *FR* frontal scale; *ju* jugal; *mx* maxilla; *MX* maxillary scale; *op* opisthotic; *pa* parietal; *PA* parietal scale; *pal* palatine; *pf* prefrontal; *pm* premaxilla; *po* postorbital; *pr* prootic; *pt* pterygoid; *qj* quadratojugal; *qu* quadrate; *so* supraoccipital; *sq* squamosal; *TE* temporal scale; *vo* vomer

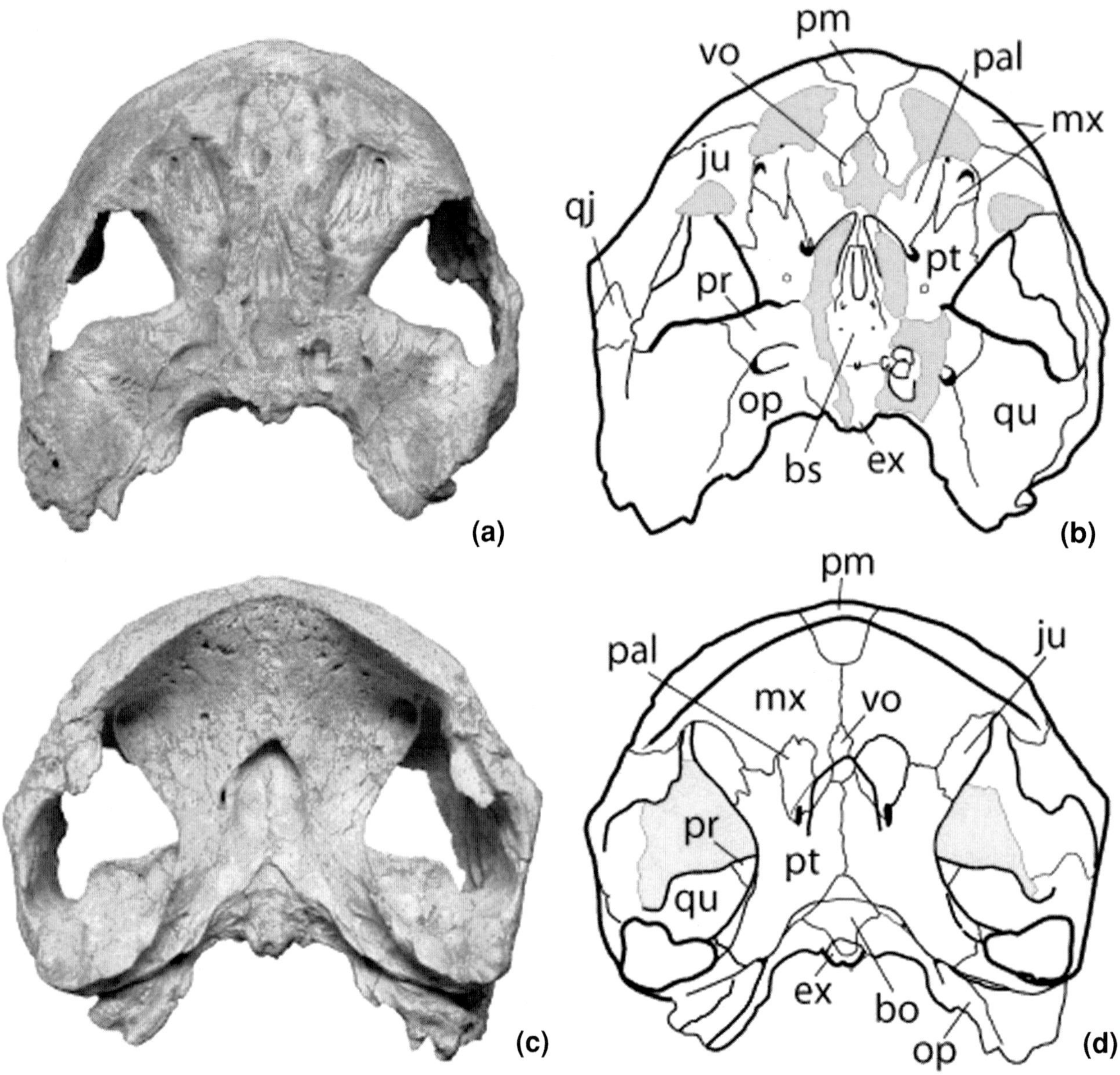

Fig. 13.3 *Brachyopsemys tingitana* gen. et sp. nov., from the Danian (early Palaeocene) of Morocco. AMNH 30554, referred skull, photographs (*left*) and interpretive drawings (*right*) in: **a–b** dorsal, and **c–d** ventral views. Abbreviations see Fig. 13.2

The anterior part of processus inferior parietalis in AMNH 30001 is free of matrix; it is low and contacts the palatine ventrally. The small pieces of the processus inferior parietalis preserved in AMNH 30558 show the contact with the palatine anteroventrally by a rather long suture, the pterygoid ventrally, and the prootic posteroventrally. Ventrally, between the pterygoid and the prootic, the parietal forms the upper margin of the foramen nervi trigemini.

Postorbital: Both postorbitals are slightly damaged on the skull roof of AMNH 30001, but the outline can be easily made out. Only a very small portion of the postorbital is preserved on the left side of AMNH 30558. The postorbital is a large anteroposteriorly elongate sheet of bone forming the posterior margin of the orbit. The postorbital has typical contacts with the frontal anteromedially, the parietal posteromedially, the jugal anterolaterally, the quadratojugal laterally and the squamosal posteriorly. In addition, there is a short postorbital-quadrate contact between the quadratojugal and the squamosal only on the left side of AMNH 30001. This is considered as an anomaly, since there is usually a quadratojugal-squamosal contact preventing

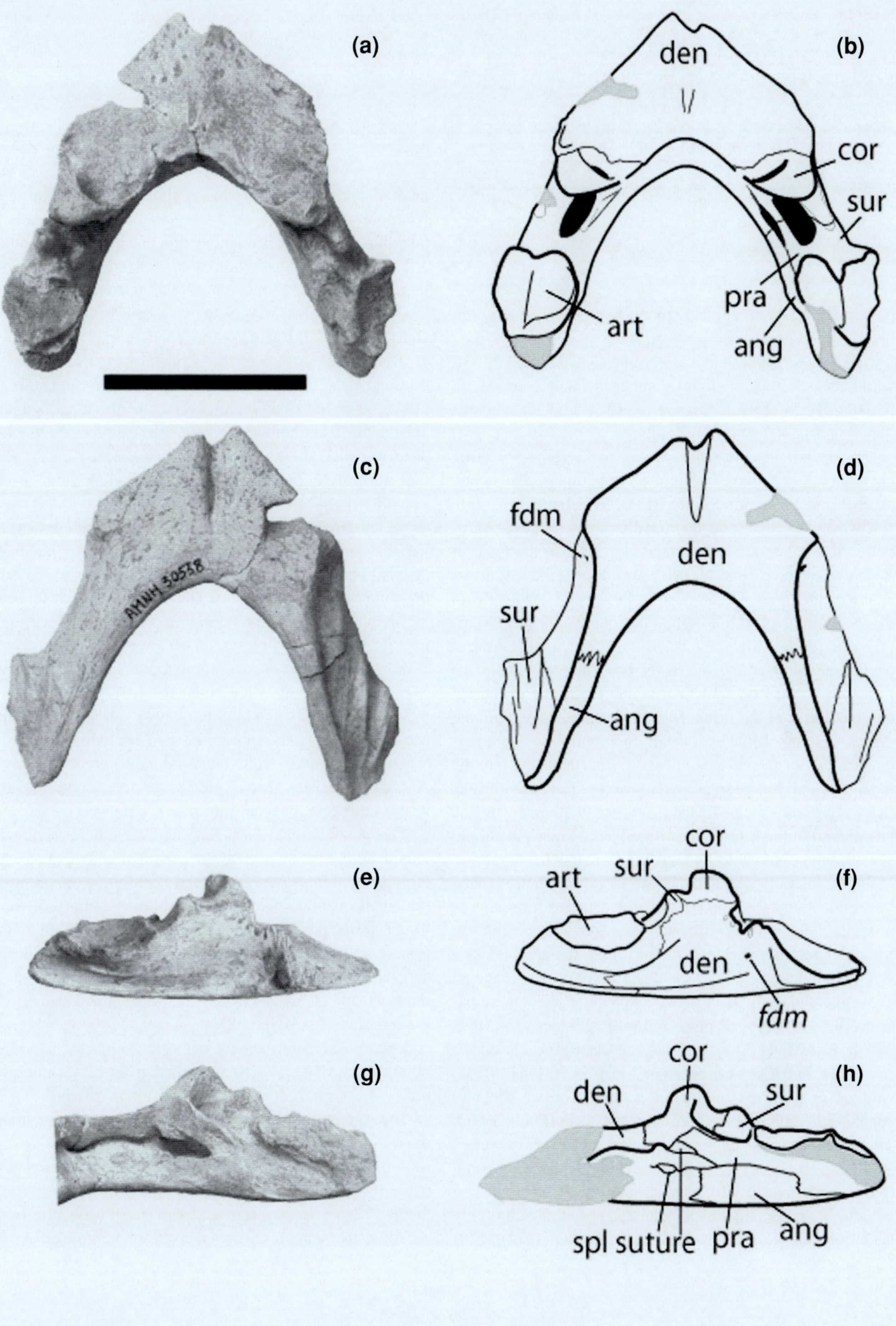
(a)
(b)
den
cor
sur
pra
art
ang
(c)
AMNH 30538
(d)
fdm
den
sur
ang
(e)
(f)
cor
art
sur
den
fdm
(g)
(h)
cor
den
sur
spl suture pra ang

◀**Fig. 13.4** *Brachyopsemys tingitana* gen. et sp. nov., from the Danian (early Palaeocene) of Morocco. AMNH 30558, lower jaw, photographs (*left*) and interpretive drawings (*right*) in: **a–b** dorsal, **c–d** ventral, **e–f** right lateral, and **g–h** lingual views. Scale bar = 50 mm. Abbreviations: *ang* angular; *art* articular; *cor* coronoid; *den* dentary; *fdm* foramen dentofaciale majus; *pra* prearticular; *spl* splenial; *sur* surangular

Table 13.1 Measurements of *Brachyopsemys tingitana* gen. et sp. nov., from the Danian (early Palaeocene) of Morocco (in mm.)

		Length	Width	Height
Skull	AMNH 30001	115	116	45
	AMNH 30558	122	123	–
Lower jaw	AMNH 30558	90	98	30

postorbital-quadrate contact among turtles. The fossa orbitalis is completely open posteriorly.

Jugal: The right jugal is complete in AMNH 30001, while the left one is slightly damaged on its surface. In AMNH 30558, both jugals are nearly complete, with the dorsal part damaged on both sides.

The jugal is unusually large and strongly convex dorsolaterally; it makes a large contribution to the lateral skull surface as in *Sandownia harrisi* (Meylan et al. 2000). But where the jugal and the maxilla have an equal contribution to the lateral skull surface in *S. harrisi*, the jugal of *Brachyopsemys* is much larger than the maxilla due to its anterior enlargement. The jugal has the usual contacts with the postorbital posterodorsally, the quadratojugal posteriorly, and the maxilla anteroventrally. A remarkable contact occurs medially, along the lower orbit margin, where the jugal meets the lateral process of the prefrontal excluding the maxilla from the orbit. This condition is unique among turtles. The jugal makes a small contribution to the anterolateral part of the orbit floor, where it contacts the maxilla medially and the pterygoid posteriorly. Laterally in the fossa temporalis inferior, there is a small foramen on the lateral wall of the triturating surface lateral to the palatal pit. This foramen is visible on both sides in AMNH 30558.

Ventrolaterally the jugal makes a large contribution to the lower cheek margin where it turns medially to form a large process, which is a flat, horizontal plate-like flange within the fossa temporalis inferior. Posterior to the flange, there is a well-developed lateral knob on the lower margin of the jugal above the anterior part of the very shallow cheek emargination. This knob is well preserved on both sides of AMNH 30558. In AMNH 30001, the knob is clearly visible on the right side, but less well-developed than in AMNH 30558; while it is damaged on the left side. A similar structure appears to be present in the Glen Rose turtle (Barck 1992). This structure may be unique among turtles, however, a similar knob is developed in some species of *Palatobaena* (Lyson and Joyce 2009). This structure may be associated with an M. zygomaticomandibularis, a muscle that has only been identified in trionychids (Schumacher 1973). This muscle extends from the medial surface of the jugal, quadratojugal and quadrate to the lateral surface of the dentary and surangular. The lower jaw of *Brachyopsemys* has a very large area for muscle insertion ventral to and posterior to the coronoid process that seems a very likely candidate for insertion of this muscle. The medial process of the jugal decribed above would provide additional area for origin of a large M. zygomaticomandibularis. Small knobs in a smilar location on the jugal of *Apalone ferox* have been illustrated by Dalrymple (1977, Fig. 5).

The ventral surface of the jugal forms the posterolateral portion of the triturating surface lateral to the maxilla and pterygoid. Within the ventral surface of the jugal is a deep palatal pit. This conical and dorsolaterally directed pit is deep; it makes the jugal convex dorsolaterally on the skull roof surface to give the skull a 'chubby-cheeked' aspect. Laterally the pit is bordered by a vertical, thin ridge with a sharp margin. On the ventral surface, the jugal contacts the maxilla anteromedially and the pterygoid posteromedially. A jugal-palatine contact is absent as in *S. harrisi* and trionychids, in contrast to most cheloniids in which this contact is present. The morphology and contacts of the jugal, maxilla, palatine and pterygoid on the triturating surface in *Brachyopsemys* are closely comparable to *S. harrisi*, however, the palatal pit is absent in the latter. Similar palatal pits are known in some bothremydids (Gaffney et al. 2006), but have never been reported in the Cryptodira.

Quadratojugal: Both quadratojugals are complete in AMNH 30001. In AMNH 30558, the left quadratojugal is nearly complete, but the right one is damaged.

The quadratojugal is a vertical, flat bone covering the lateral side of the skull between the jugal and quadrate. It is large as in *Sandownia harrisi* and cheloniids, in contrast to the small quadratojugal of trionychids. Ventrally, it forms the posterior portion of the very shallow cheek emargination. It contacts the jugal anteriorly, the postorbital dorsally and the quadrate posteriorly. It contacts the squamosal on the right side of AMNH 30001, but this contact is absent on the left side of AMNH 30001.

Squamosal: Both squamosals are complete in AMNH 30001. In AMNH 30558, a small portion of squamosal is preserved on the left side, the right one is missing.

The squamosal is a roughly rectangular flat bone, longer than wide, covering the posterolateral corner of the skull roof and contributing to the lateral portion of the temporal margin. It contacts the postorbital anteriorly, the parietal medially by a long suture and the quadrate ventrally. In addition, the squamosal contacts the quadratojugal on the right side of AMNH 30001. The squamosal forms the roof of the antrum postoticum, but the sutures on the ventral surface, posterior to the cavum tympani are somewhat confusing, the antrum postoticum seems to be formed ventrally by the quadrate. If this interpretation is correct, there is no squamosal-opisthotic contact. The large size and the contact of the squamosal on the skull roof are comparable to *Sandownia* and cheloniids, and other turtles with an extensive skull roof; but different from the cone-shaped squamosal of trionychids.

Premaxilla: The premaxilla is complete in AMNH 30001. AMNH 30558 has nearly complete premaxillae, with the dorsal surface only slightly damaged.

In both AMNH 30001 and AMNH 30558, the right and left premaxillae are completely fused as in Trionychia (Meylan and Gaffney 1989) and *Erquelinnesia* (Zangerl 1971); there is no midline suture visible on dorsal nor on ventral surface. The premaxillae are partially fused in *Sandownia* (Meylan et al. 2000). Dorsally the premaxilla forms the lower rim of the aperture narium externa and contributes to the floor of the fossa nasalis where it contacts the maxilla posterolaterally. The dorsal exposure of the premaxilla is wide anteriorly and sends a median posterior process between the maxillae. The dorsal surface of the premaxilla bears a low and blunt parasagittal ridge. The ridges of the right and left premaxillae together produce a midline grove. This structure is clearly visible in AMNH 30001, in the prepared part of the fossa nasalis, but its extent cannot be determined at the present state of preparation. This surface is damaged in AMNH 30558.

In ventral view, the premaxilla contributes to the anteromedial portion of the triturating surface and its labial ridge. It contacts the maxilla posterolaterally. There is no premaxilla-vomer contact on either the dorsal or ventral surface. A premaxilla-vomer contact is present in most cryptodiran turtles including *Sandownia harrisi*, but trionychids usually have no vomer-premaxilla contact due to the reduction of the ventral portion of the vomer and/or the development of the foramen intermaxillaris. Cheloniids may not have a vomer-premaxilla contact in ventral view due to medial expansion of the palatine process of the maxillae, but it may be retained dorsally in the floor of the fossa nasalis (Gaffney 1979, p. 87). Thus the condition in our specimens is more similar to that of trionychids.

Maxilla: Both maxillae are complete in AMNH 30001. In AMNH 30558, they are nearly complete, but the dorsal process is broken off on both sides.

Dorsally, the maxilla makes up the lateral margin of the apertura narium externa and most of the fossa nasalis floor, but it does not reach the orbital margin in contrast to all other turtles. The maxilla has a reduced exposure on the skull surface, where it contacts the premaxilla medially, the prefrontal dorsomedially and the jugal posterolaterally. The lateral and ventral surfaces of the maxilla and the premaxilla bear numerous nutritive foramina, suggesting that they were covered by a large, thick rhamphotheca. However, the nutritive foramina are larger, but sparser on the triturating surface than on the lateral surface, which suggests that the rhamphotheca was thicker on the feeding surface than on the lateral surface. The extension of this epidermal jaw covering on the skull surface is marked by a wide and shallow sulcus on the lateral surface of the maxilla, along the maxilla-jugal suture. Such a sulcus is also present in *Sandownia harrisi* (Meylan et al. 2000).

In the fossa nasalis, the maxilla contacts the premaxilla anteromedially, the vomer posteromedially and the palatine posterolaterally. The maxillae share a common midline contact between the premaxilla and vomer on the fossa nasalis floor. Although the maxilla does not contribute to the orbit margin, it is exposed on the orbit floor as in other turtles, but posterior to the jugal-prefrontal contact. The exposure is triangular, narrowed posteriorly. There are two foramina lying on its anterior end: the large lateral one (about 2 mm in diameter) is the foramen supramaxillare, whereas the very small medial one is interpreted here as the foramen supraorbitale. The maxilla does not reach the foramen orbito-nasale, in contrast to *Chelonia mydas*. Maxillary contacts on the orbit floor include the jugal anterolaterally and laterally, the prefrontal anteromedially, the palatine medially and the pterygoid posteriorly.

In ventral view, the maxilla, premaxilla, vomer, palatine, jugal and pterygoid, all contribute to the very large, concave triturating surface. This surface is bordered by a very thick labial ridge. The labial ridge made by the premaxilla in the middle is very low; it increases in height posterolaterally in the maxilla. The triturating surface has no medial ridge and no obvious posterior limit. Laterally, the maxilla slopes into a very deep pit on the triturating surface. The pit lies near the lateral border of the triturating surface and is formed mainly by the jugal. The maxilla contributes to its medial portion. On the triturating surface, the maxilla contacts the premaxilla anteromedially, the vomer posteromedially, the palatine and the pterygoid posteriorly, and the jugal laterally. In addition, the maxillae share a long common midline contact between the premaxilla and the vomer. It has a very small contribution to the apertura narium interna margin between the vomer and the palatine. A maxillary midline contact is commonly seen in trionychids because of the reduction of the vomer. It may also occur in some living cheloniids (*Caretta caretta*) or as an individual variation (Gaffney 1979). This contact is absent in *Sandownia harrisi*.

Vomer: The vomer is complete in AMNH 30001. In AMNH 30558, the dorsal surface of the vomer is damaged; the remaining part is totally free of matrix.

The vomer is a small hourglass-shaped element. Ventrally, it has a very small exposure on the triturating surface behind the long maxilla midline suture. It forms the anteromedial margin of the apertura narium interna. It contacts the maxilla anteriorly and laterally on the triturating surface and on the floor of the fossa nasalis, but apparently a vomer-palatine contact is prevented by the maxilla. A reduced vomer with a long midline suture between the maxillae is found in trionychids and carettochelyids, whereas ventral exposure of the vomer is large in cheloniids, and also in *Sandownia harrisi*. Dorsally, the vomer sends a sagital process (vomerine pillar) to meet the dorsal portion of the vomer between the palatines. The vomerine pillar is short, its posterior end being anterior to the anterior edge of the apertura narium interna. This structure is visible in AMNH 30558, as well as in AMNH 30001 by CT scan.

Dorsally, on the orbit floor, the vomer contacts the prefrontal anterolaterally and the palatine laterally. Posteriorly, there is a short vomer-pterygoid contact preventing palatine midline contact.

Palatine: Both palatines are complete in both AMNH 30001 and AMNH 30558. In AMNH 30001, the ventral surface is fully visible; but the posterior end of the dorsal side is still covered by matrix. The palatine is free of matrix on both dorsal and ventral sides in AMNH 30558.

The palatine is a C-shaped bone in cross section. The ventral exposure of the palatine, on the triturating surface is roughly diamond-shaped, lying posteromedial to the maxilla. It contributes to the posteromedial portion of the triturating surface and most of the apertura narium interna margin. The apertura narium interna is located at about the middle of the palatal surface of the skull, the position being comparable to that of *Sandownia harrisi*. On the palatal surface, the palatine contacts the maxilla anteriorly and the pterygoid posteriorly. There is no palatine-jugal contact on the palatal surface as in *S. harrisi* and trionychids. The palatines lie very closely together on the triturating surface; however, the vomer and maxilla prevent a palatine midline contact. On the floor of the fossa nasalis, the palatines are separated from one another by the vomer anteriorly and by the pterygoid posteriorly.

The dorsal shelf of the palatine contributes to the medial portion of the orbit floor, where it contacts the prefrontal anteriorly, both medial and lateral to the foramen orbito-nasale. It meets the vomer medially, the maxilla anterolaterally, and the pterygoid posteriorly. It seems that there is no palatine midline contact on the dorsal surface. Anteriorly, the palatine makes up the posterior margin of the foramen orbito-nasale; and posteriorly, the medial rim of the foramen palatinum posterius. The foramen palatinum posterius is a large and anteroposteriorly elongate opening lying on the posterior end of the palatine, on the palatine-pterygoid suture as in *Sandownia harrisi*. This foramen is usually located on the palatine-maxilla or palatine-pterygoid suture in trionychids, but lateral to the palatine; it is absent in cheloniids.

The dorsal process of the palatine appears to be a thin sheet that covers the anterior part of the lateral wall of the braincase. However, the pterygoid contacts the vomer dorsally and ventrally and appears to make up most of the thickness of the side wall in AMNH 30558. The palatine appears only to contribute to the thickness of the anterior part of the side wall. A pterygoid-vomer contact excludes palatine contact of the descending process of the parietal.

Epipterygoid: In AMNH 30558, there is a fragment of bone on the left side wall of the braincase between the foramen palatinum posterius and the foramen nervi trigemini that is located in the approximate position of the epipterygoid. However, this could be a fragment of the descending process of the parietal. Thus, at this time it is not possible to say if the epipterygoid was present or absent.

Pterygoid: Both pterygoids are complete in AMNH 30001, but only the ventral side is free of matrix. They are very slightly damaged in AMNH 30558 and prepared on both ventral and dorsal sides.

In ventral view, the pterygoid is anteroposteriorly elongate, forming the posterior part of the secondary palate between the jugal and palatine where it has a short contact with the maxilla anteriorly, as in *Sandownia harrisi*. Medially the pterygoid forms the roof of the apertura narium interna where they share a long, common, midline suture between the vomer and the basioccipital. The pterygoid completely covers the basisphenoid ventrally, so the latter is not visible in palatal view. Anteriorly, in the apertura narium interna, the pterygoid contacts the vomer anteromedially and the palatine anterolaterally in AMNH 30558. These contacts are covered by matrix in AMNH 30001. The processus pterygoideus externus seen in most cryptodires is absent. The lateral margin of the pterygoid is a high and thin vertical ridge extending from the jugal anteriorly to the quadrate posteriorly. Just anterior to the processus articularis of the quadrate along this ridge is a small protuberance. This structure is well preserved in AMNH 30001, but slightly damaged in AMNH 30558. Medial to this protuberance is a deep anteromedially-posterolaterally directed groove, probably for pterygoideus musculature. It is bordered laterally by a ridge that forms the lateral margin of the pterygoid, which is continuous with the processus articularis of the quadrate. Posteromedially the groove is bordered by a low and thin ridge formed by the quadrate and pterygoid. This groove extends posterolaterally to the posterior surface of the processus articularis. The pterygoid contacts the basioccipital posteromedially, the exoccipital posteriorly and the quadrate posterolaterally. A small foramen posterius canalis carotici interni lies under the posteromedial ridge of the pterygoid-quadrate groove, on the pterygoid-exoccipital suture. The foramen is well preserved and visible in AMNH 30558, but remains covered by matrix in AMNH 30001. It is formed mainly by the pterygoid; the exoccipital forms the dorsomedial margin. A foramen carotico-pharyngeale as seen in *Sandownia harrisi* is absent.

On the dorsal surface, the pterygoid contributes to the posteriolateral portion of the orbital floor where it contacts the jugal anterolaterally, the maxilla anteriorly, and the palatine anteromedially. The foramen palatinum posterius is visible dorsally in AMNH 30558, lying on the posteriomedial corner of the orbit floor, on the pterygoid-palatine suture. On both sides a clear, round depression is visible posteriolateral to the foramen palatinum posterius. A midline contact of the pterygoids is visible on the dorsal surface, between the vomer and the basisphenoid. Dorsally, the pterygoid contacts the parietal, and posterodorsally it contacts the prootic and quadrate. The crista pterygoidea is very thick and seems to form the entire lower portion of the sidewall of the braincase. The foramen cavernosum is a small slit-like opening. The sulcus cavernosus is narrow.

Quadrate: The quadrates are complete in AMNH 30001; they are damaged on both sides in AMNH 30558.

Laterally, the quadrate forms the C-shaped cavum tympani. The cavum is small and deep, with deep dorsal and anterior walls. The incisura columellae auris is widely open posteroventrally as in cheloniids such as *Chelonia mydas*, in contrast to the closed incisura of *Sandownia harrisi* and trionychids. A small tubercle is present on the lower margin of the incisura columellae auris, which is more developed in AMNH 30558 than in AMNH 30001. A rounded and shallow depression is present anterior to the incisura columellae auris. The antrum postoticum is preserved in both sides of AMNH 30001, but it is full of matrix; it is also partly preserved on the left side of AMNH 30558. It is moderate in size. In lateral view, in addition to the usual contacts with the quadratojugal anteriorly and the squamosal dorsally, the quadrate contacts the postorbital anterodorsally by a short suture on the left side of AMNH 30001.

Dorsally, the quadrate forms the lateral portion of the very large processus trochlearis oticum and contributes to the lateral margin of the foramen stapedio-temporale (see prootic). Ventrally, the quadrate forms the low processus articularis. The posteromedial surface of the processus contains a concavity that is continuous with the anteromedial-posterolaterally directed groove of the pterygoid. A small foramen chorda tympani inferius lies within the posterior end of this groove, posterodorsal to the condylus mandibularis. The condylus mandibularis is roughly triangular; being wider than long; it is divided by a parasagittal furrow into two facets: a larger lateral one and a smaller medial one. In AMNH 30558, both facets are slightly convex. In AMNH 30001, the lateral facet is convex while the medial facet is flat. The quadrate contacts the pterygoid anteromedially, the prootic anterodorsomedially, and the opisthotic posteromedially. The fenestra postotica is a relatively small and oval opening formed by the pterygoid ventromedially, the quadrate laterally, the exoccipital medially and the opisthotic dorsally. It is closed medially and separated from the foramen jugulare posterius.

Basisphenoid: The basisphenoid is not visible in AMNH 30001; it is nearly complete in AMNH 30558 and exposed only on the dorsal surface. In ventral view, in both specimens it is completely covered by the pterygoid and basioccipital, apparently due to the shortening of the posterior part of the skull and extreme anterior location of the occipital condyle. *Sandownia* also has reduced ventral exposure of the basisphenoid, but a small basisphenoid exposure is visible in ventral view, surrounded by the pterygoid. The basisphenoid completely covered by pterygoid ventrally is also present in a solemydid referred to *Helochelydra nopcsai* by Joyce et al. (2011).

In dorsal view, the basisphenoid is anteroposteriorly elongate, forming the floor of the cavum cranii. The trabeculae meet in the midline to form the rostrum basisphenoidale. The sella turcica is a rather deep and anteroposteriorly elongate depression which is open anteriorly between converging lateral margins. Some bone is missing at the anterior end of the rostrum. The foramen anterius canalis carotici internae lie at the posterior end of the sella turcica. They are small and more widely spaced than in cheloniids [*C. mydas*, *Euclastes wielandi* (AMNH 30022) from the Danian of Morocco phosphates, Hirayama and Tong 2003]. The rostrum basisphenoidale is slightly damaged at its anterior end; it is relatively long, although not as rod-like as seen in cheloniids. The general morphology of the rostrum and sella turcica is comparable to that of *Platysternon megacephalum* (Gaffney 1979, Fig. 63), but different from the short and widely separated trabeculae of *Sandownia harrisi*, *Ferganemys verzilini* (Nessov 1977), and trionychids. The rostrum is slightly raised from the pterygoids allowing the passage of a rather large canal between the basisphenoid and pterygoids. These canals converge on the midline from each side. The anterior opening of the canals, which lies under the rostrum, is interpreted as the foramen anterius canalis caroticus laterale. The anterior location of this foramen is similar to the condition in *S. harrisi*, and to some extent to *Kallokibotion bajazidi* (Gaffney and Meylan 1992); but different from *Ferganemys*, *Amyda cartilagineus* and *Dermatemys mawii* in which this foramen is located more posteriorly and lateral to the dorsum sellae (Gaffney 1979, Figs. 58, 59). The foramen anterius canalis caroticus laterale is larger than the foramen anterius canalis carotici interni, as in *S. harrisi*.

The dorsum sellae is high and does not conceal the sella turcica, similar to Moroccan *Euclastes wielandi* (AMNH 30022). The processus clinoideus is broken off on both sides, but the remaining base indicates their position at the dorsolateral edge of the dorsum sellae. The posterior foramen nervi abducentis lies posterior to the processus clinoideus; the anterior foramen nervi abducentis is visible on the lateral surface of the basisphenoid, under the processsus clinoideus. The basis tuberculi basalis is a small tubercule lying on the basisphenoid/basioccipital suture. The basisphenoid contacts the pterygoid ventrally, the prootic dorsolaterally and the basioccipital posteriorly.

Basioccipital: The basioccipital is complete in both specimens, but only in AMNH 30558 are both the dorsal and ventral surfaces prepared. The basioccipital is an arrow-shaped bone in ventral view. There are some differences between AMNH 30001 and AMNH 30558. In 30001, the ventral surface of the basioccipital is shorter and concave with a rounded depression in the middle. While in AMNH 30558, the basioccipital is longer anteroposteriorly; a slight transverse crest divides the basioccipital into an anterior triangular and flat part, and a slightly concave posterior part; there is no rounded depression on it. Posteriorly, the basioccipital forms the ventral portion of condylus occipitalis. The condylus is well ossified, differing from the cartilaginous condyle seen in the holotype of *Sandownia harrisi*. The tuberculum basioccipitale is absent. The condylus occipitalis lies in line with the condylus mandibularis of the quadrate, as in *S. harrisi*. On the ventral surface, the basioccipital contacts the pterygoid anteriorly by an inverse V-shaped suture and the exoccipital laterally.

The dorsal exposure of the basioccipital on the floor of the cavum cranii is a small triangle. It bears most of the basis tuberculi basalis which is an anteroposteriorly elongated tubercle on the basioccipital-basisphenoid suture. In dorsal view, the basioccipital contacts the basisphenoid anteriorly, the opisthotic anterolaterally and the exoccipital posterolaterally.

Exoccipital: Both exoccipitals are complete in AMNH 30001, but damaged in AMNH 30558. The exoccipital forms the dorsal portion of the condylus occipitalis, the lateral margin of the foramen magnum and extends laterally to the medial margin of the fenestra postotica. It contacts the pterygoid ventrolaterally, the basioccipital ventromedially, the opisthotic dorsolaterally and the supraoccipital dorsally. The foramen jugulare posterius is a small oval opening on the exoccipital/opisthotic suture but is mostly formed by exoccipital. It is completely enclosed and separated from the fenestra postotica by exoccipital and opisthotic as in *Chelydra serpentina* (Gaffney 1979, Fig. 95), but differs from the condition in *Sandownia harrisi*. Two foramina nervi hypoglossi of equal size, lie on the posterior surface of the exoccipital. They are also visible on the dorsal surface of exoccipital in the floor of the cavum cranii of AMNH 30558.

Supraoccipital: The supraoccipital is complete but not fully prepared in AMNH 30001; the anteromedial part is still covered by matrix. In AMNH 30558, only a small portion of the supraoccipital is preserved on the left side. As in *Sandownia harrisi*, the dorsal exposure of the supraoccipital on the skull roof is very small. It is diamond-shaped and posterior to the parietals. The crista supraoccipitalis is complete on AMNH 30001. It is long but completely covered by parietals dorsally, as in *Erquelinnesia gosseleti*. The supraoccipital forms the upper rim of the foramen magnum and contacts the parietal anterodorsally, the prootic and opisthotic laterally and the exoccipital ventrally.

Prootic: Both prootics are complete in AMNH 30001, but due to the very complete skull roof, its dorsal surface is not prepared. Only the ventral side of the right prootic is visible. In AMNH 30558, both prootics are prepared on ventral and dorsal sides; the left prootic is nearly complete; the right one is damaged.

The prootic forms the medial part of an obvious processus trochlearis oticum. The processus is very large and curved. It is convex dorsally and concave ventrally with a sharp anteroventral edge. The processus is formed by the prootic medially and the quadrate laterally, the prootic contribution to the process being less than that of the quadrate. The size of the process is comparable to that of *Euclastes wielandi* from Morocco, but smaller than that of *Tasbacka ouledabdounensis* (Tong and Hirayama 2002) and *Erquelinnesia gosseleti* (Zangerl 1971). In *E. wielandi* the processus has a nearly flat ventral surface and it is formed equally by prootic and quadrate. In this new taxon, the ventral surface of the processus is more concave and mostly formed by the quadrate. In the Trionychoidae, there is "large portion of processus trochlearis oticum made up by parietal" (Meylan and Gaffney 1989, p. 46). *Adocus* sp. also has an unusually large and developed processus trochlearis oticum (Meylan and Gaffney 1989) that is formed by the parietal, prootic and quadrate, as in trionychids, and it is dorsally concave, unlike the case in our specimens in which the process is dorsally convex.

The foramen stapedio-temporale is complete on the left side of AMNH 30558, and damaged on the right side. This area is not prepared in AMNH 30001. The foramen lies on the prootic-opisthotic-quadrate suture. It is a large opening (3 mm in diameter anteroposteriorly) and extending medially from it is a large and deep groove. It is much larger than the foramen anterius canalis carotici interni. The difference in size between these two foramina seems to be comparable to *Plesiochelys* spp. and *Portlandemys mcdowelli* (Gaffney 1976). The foramen and the groove are formed mainly in the prootic. The quadrate contribution is limited to the anterolateral margin of the foramen and the opisthotic contributes to the posterolateral margin. On the dorsal surface of the otic chamber, the prootic contacts the supraoccipital posteromedially, the opisthotic posteriorly and the quadrate laterally. A prootic-parietal contact is not preserved in AMNH 30558, and not visible in AMNH 30001. In AMNH 30558, the supraoccipital and the medial part of the prootic are missing on the right side, exposing the cavum labyrinthicum. The fenestra ovalis and the recessus labyrinthicus prooticus are visible within.

Anteroventrally, the prootic forms the dorsal margin of the foramen nervi trigemini, which is well preserved on the left side of AMNH 30558. This foramen is a small rounded opening. The prootic contacts the pterygoid ventromedially.

Opisthotic: Both opisthotics are complete in AMNH 30001 and on the left side of AMNH 30558. The right opisthotic is damaged in AMNH 30558. Only the posterior surface is exposed between the exoccipital and quadrate in AMNH 30001, it is fully prepared in AMNH 30558.

The opisthotic is a blade-like bone, forming the dorsal margin of the fenestra postotica, part of the dorsal roof of the cavum acustico-jugulare, and dorsolateral margin of the foramen jugulare posterius. On the dorsal surface, it contributes also to the posterolateral margin of the foramen stapedio-temporale. The opisthotic contacts the prootic anteriorly, the quadrate laterally, the supraoccipital dorsomedially and the exoccipital ventromedially.

In the right cavum labyrinthicum of AMNH 30558, the processus interfenestralis, the recessus labyrinthicus opisthoticus, and the canalis semicircularis posterius are exposed. The hiatus acousticus and the foramen jugulare anterius are visible on both sides.

Lower jaw (Fig. 13.4): The lower jaw of AMNH 30558 is nearly complete; with the left ramus slightly damaged.

Dentary: The right dentary is complete; the left one is slightly damaged. The left and right dentaries are fused as in most turtles. The dentary forms a large triturating surface and most of lateral surface of the ramus. It contacts the coronoid dorsally, the surangular posteriodorsolaterally and the angular posteroventrally. The triturating surface is large, wide and almost flat. The labial margin forms an obtuse angle; the lingual margin is parallel to the anterior portion of the labial margin. The symphysis is long and forms the highest point of the triturating surface. The surface slopes downward from the back to the front and from the symphysis laterally. There is neither lingual nor labial ridge on the lower triturating surface, but a median notch is present at the anterior margin. However, the posterolateral border of the triturating surface bears a small ridge which delimits a small depression posteromedial to it, anterolateral to the processus coronoideus. This small depression faces the deep pit in the jugal on the upper triturating surface when the lower jaw is attached to the skull. Very large nutrient foramina on the anterior part of the lower triturating surface suggest that the rhamphotheca was thicker anteriorly than posteriorly. The morphology of the triturating surface is very different from cheloniids which have an extensive secondary palate such as *Erquelinnesia gosseleti*. (Zangerl 1971), *Euclastes* spp. (Zangerl 1953; Hirayama and Tong 2003), *Tasbacka* spp. (Nessov 1987) and pitted pleurodires such as *Bothremys* spp. (Gaffney and Zangerl 1968; Gaffney et al. 2006). The lower jaw is quite comparable to that of *Sandownia harrisi* in that (1) the symphysis of the dentary is long, (2) the lower triturating surface is broad and triangular and slopes ventrally from the coronoid process, (3) labial and lingual ridges are absent, and (4) there is a clear constriction anterior to the coronoid. However, *Sandownia* differs in that is has a longer symphysis and stronger constriction.

The ventral surface of the lower jaw is perfectly flat and horizontal with a midline groove under the triturating surface. The groove is narrow posteriorly and widens anteriorly resulting in a median notch that is seen in dorsal view. The lingual surface of the dentary bears a deep sulcus cartilaginis Meckelii. The foramen alveolare inferius is a small oval opening lying on the bottom of the sulcus. At the posterior end of the sulcus, there is a small sutural area along the lower margin just opposite of another larger sutural area on the coronoid. These suture surfaces are presumably for contact with the splenial which appears to have been lost on both sides.

The lateral surface of the lower jaw posterior to the triturating surface bears a deep longitudinal groove along its lower margin, which is bordered by a sharp horizontal ridge inferiorly. The lateral surface of the dentary above the groove is generally convex laterally, with a shallow depression just below the small lateral ridge of the lower triturating surface mentioned above. In this depression is a small, rounded foramen dentofaciale majus. A blunt and very low anterodorsally–posteroventrally directed swelling lies under the coronoid process. A more developed ridge is present along the dentary-surangular suture (see surangular).

The lateral depression of the dentary extends onto the surangular and all the way to the posterior end of the retroarticular process. It would appear to form an enlarged area for insertion of a very well developed zygomaticomandibularis muscle as in trionychids. In particular, the sharp horizontal ridge along the ventral margin of the dentary is in the same location as a ridge that marks the ventral limit of insertion of this muscle in a jaw of *Apalone ferox*.

Coronoid: The left coronoid is complete, while the right one is slightly damaged. The coronoid alone forms the rather tall processus coronoideus which lies at about midlength of the ramus. The anterior surface of the process bears a deep pit in the posterior part of the lower triturating surface. There is a small foramen on the bottom of the pit on the left side, while three foramina are present in the right pit. The tall coronoid process lying at the midpoint of the jaw ramus length is present in trionychids, *Adocus* sp. (Meylan and Gaffney 1989) and carettochelyids [*Carettochelys insculpta* (Gaffney 1979), *Anosteira maomingensis* (Tong et al. 2010)]. The process of AMNH 30558 is more developed than that of cheloniids, such as *Chelonia mydas* and *Lepidochelys kempii*, but not as tall as in members of the Trionychoidae, such as *Adocus* sp., *Amyda cartilaginea*, and carettochelyids. The coronoid process is formed by the coronoid only in AMNH 30558, like that in carettochelyids, but unlike in *Adocus* sp. and *A. cartilaginea* (Meylan and Gaffney 1989, Fig. 9) and *Lissemys punctata*, in which the dentary makes up part of the process. *Sandownia harrisi* also has a tall coronoid process formed only by the coronoid, which is similar to AMNH 30558.

Posteriorly, the coronoid contributes to the margin of the fossa Meckelii; ventromedially it forms part of the upper margin of the sulcus cartilaginis Meckelii. On the lateral margin of the fossa Meckelii, the coronoid forms, together with the surangular posteriorly, a dorsal process. The coronoid contacts the dentary anteriorly and ventrolaterally, the surangular posterolaterally and the prearticular posteromedially. A sutural surface can be seen on the lower margin of the medial surface of the coronoid, which is presumably for the splenial contact.

Surangular: Both surangulars are complete. The surangular forms the posterolateral part of the jaw ramus. A well developed ridge along the dentary-surangular suture extends posteriorly onto the surangular and roughly parallel to the lateral margin of the area articularis mandibularis. There is a small, apparently unnamed foramen under the ridge, and above the dentary-surangular suture. Anterodorsally, the surangular contributes to the posterolateral rim of the fossa Meckelii and the small process on the lateral margin of the fossa (see coronoid). The surangular contacts the coronoid anterodorsally, the dentary anteroventrally, the articular posterodorsally and the angular posteroventrally. The foramen nervi articulotemporalis is a small, anterior-facing opening that lies under the anterolateral part of the area articularis mandibularis.

Prearticular: The right prearticular is complete; the left is a little damaged. The prearticular is a sheet of bone contributing to the medial rim of the fossa Meckelii. It contacts the coronoid anteriorly, the articular posteriodorsally and the angular ventrally. There is a suture surface on the anteroventral margin of the prearticular, which is a continuation of a similar surface on the coronoid and is opposite to a sutured surface on the dentary. These suture surfaces are presumably for the splenial contact, although the splenial is not preserved.

Articular: Both articulars are slightly damaged at their posteromedial surface. The articular forms the large area articularis mandibularis which is made up by two facets: the larger lateral facet and smaller median one. Both medial and lateral facets are slightly concave and separated by an anteroposteriorly elongated swelling. Posterior to the articular area, the processus retroarticularis is damaged, so the whole shape and length cannot be determined. The articular contacts the surangular ventrolaterally, the prearticular ventromedially. The foramen posterius chorda tympani is not preserved on either side.

Angular: Both angulars are almost complete. The angular makes up the posteroventral part of the lower jaw. It contacts the dentary anteriorly, the prearticular dorsomedially and the surangular dorsolaterally. Ventrolaterally, the angular makes up, with the dentary, a sharp and horizontal ridge at the lower margin of the ramus, above and along this ridge lies a long and deep groove.

Splenial: The splenial itself is not preserved on either side, but sutured surfaces on the coronoid, prearticular and dentary posterior to the sulcus cartilaginis Meckelii indicate its presence. These sutured surfaces can be seen on both sides of the lower jaw, but are better preserved on the right side. The presence of the splenial is primitive since it is absent in most recent turtles but it occurs commonly in Mesozoic forms (Gaffney 1979).

Discussion

Morphology of Brachyopsemys

This is the first report of palatal pits in a member of the Cryptodira. Palatal pits were previously known only in one group of pleurodires; they characterize the subtribe Bothremydina in the family Bothremydidae (Gaffney et al. 2006). This subtribe includes five genera and nine species, and has a wide geographic distribution (North America, North Africa, Southern Europe and Middle East) and a long geological range (Late Cretaceous to Eocene). Within the Bothremydina is one genus (two species) that lack palatal pits, and they are considered to be secondarily lost. Both *Brachyopsemys* and members of the Bothremydini have a low, broad skull and extensive secondary palate. The general morphology and position of the palatal pits in *Brachyopsemys* is comparable to those of Bothremydina, such as *Bothremys maghrebiana* (Fig. 13.5b). However, when compared in detail, some differences can be noted in the morphology and composition of the secondary palate. *Brachyopsemys* has a complete secondary palate without midline separation. A similar condition is also known in cheloniids (*Erquelinnesia*, *Tasbacka*, *Euclastes*, *Puppigerus*) and the eurysternid, *Solnhofia parsonsi*. The secondary palate of Bothremydina is divided by a midline groove. In addition, the palate of *Brachyopsemys* has a significant contribution by the pterygoid, in contrast to Bothremydini. The exact composition of the pits also differs. In Bothremydina the palatal pits are formed by the maxilla, jugal and

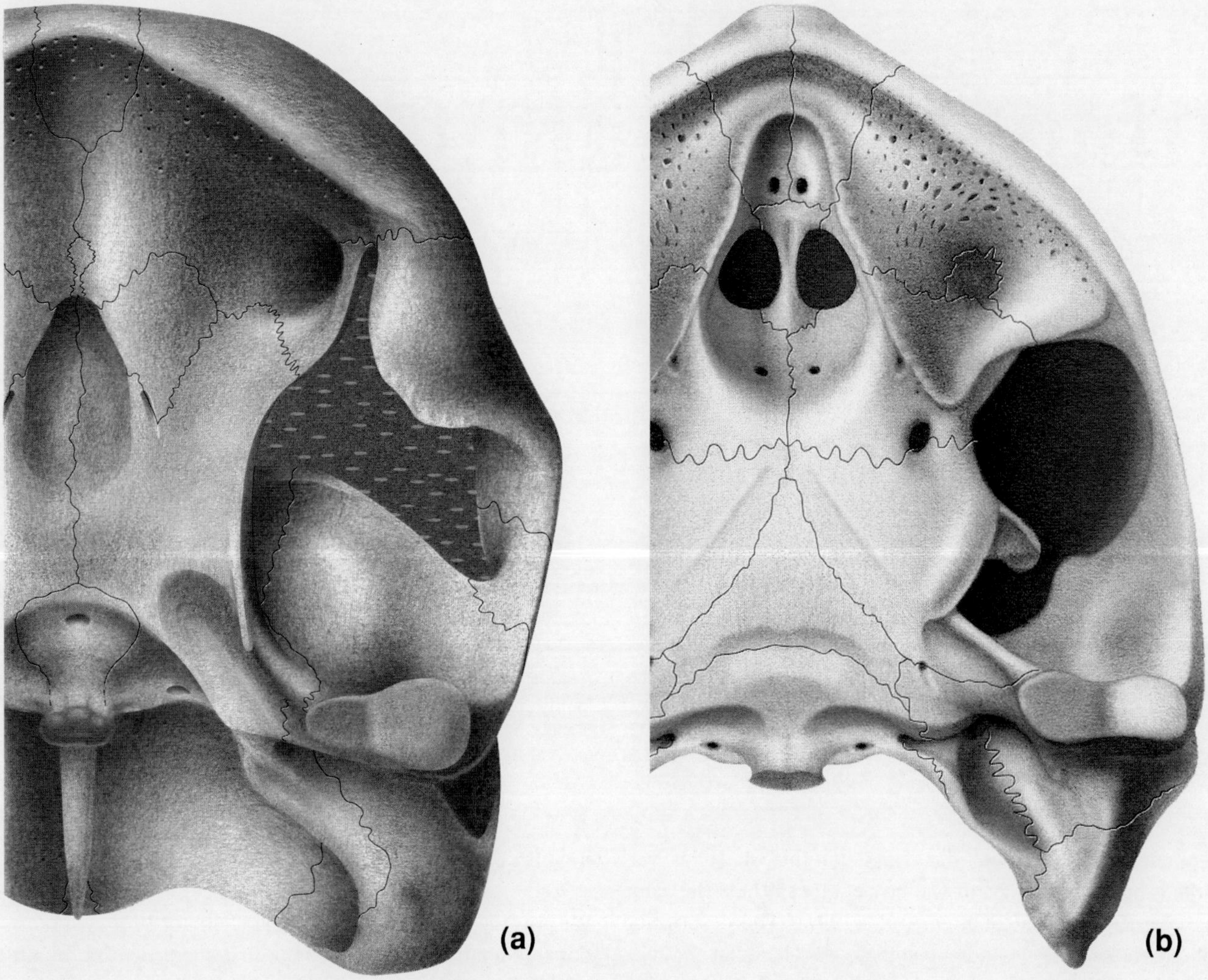

Fig. 13.5 Comparison of a pitted palate cryptodire (*Brachyopsemys*) and pleurodire (*Bothremys*). **a** Reconstruction of palate of *Brachyopsemys*. **b** Reconstruction of palate of *Bothremys maghrebiana* (after Gaffney et al. 2006). Not to scale

palatine; while in *Brachyopsemys*, the pits are formed by the maxilla and jugal only, without a contribution from the palatine. The jugal contributes to the palatal pit laterally in *Brachyopsemys* rather than posteriorly or only on the bottom of the pit, as seen in Bothremydina.

Among turtles with pitted palates, the shape and the depth of the pit varies, from a wide and relatively shallow depression to a well-defined and deep rounded pit. The appearance of a secondary palate in turtles appears to be a trophic modification with the development of the triturating area into a broad crushing surface (Meylan et al. 2000). Turtles with enlarged triturating surface usually have durophagous diet. The function of such a pit is likely to help to hold slippery shells (brachiopods, mollusks) to facilitate crushing. Variation in the shape and size of the pits probably limits the size range of the prey.

Phylogenetic Analysis

In an attempt to understand the phylogenetic position of this new taxon, we scored it for the data set developed by Gaffney et al. (1991) and expanded by Shaffer et al. (1997) and Near et al. (2005). This data set is further expanded here to include additional taxa and new characters. Characters 1–41 are those of Gaffney et al. (1991) and are described in that paper. Characters 42–115 are described in Shaffer et al. (1997). Characters 116–142 were used in analyses to place fossils in Near et al. (2005) but were not described in that paper. They are included here (Appendices 13.1 and 13.2) along with descriptions and scores of seven additional characters that were added to help resolve the position of *Brachyopsemys*. Thus, scores for a total of 149 characters are given in

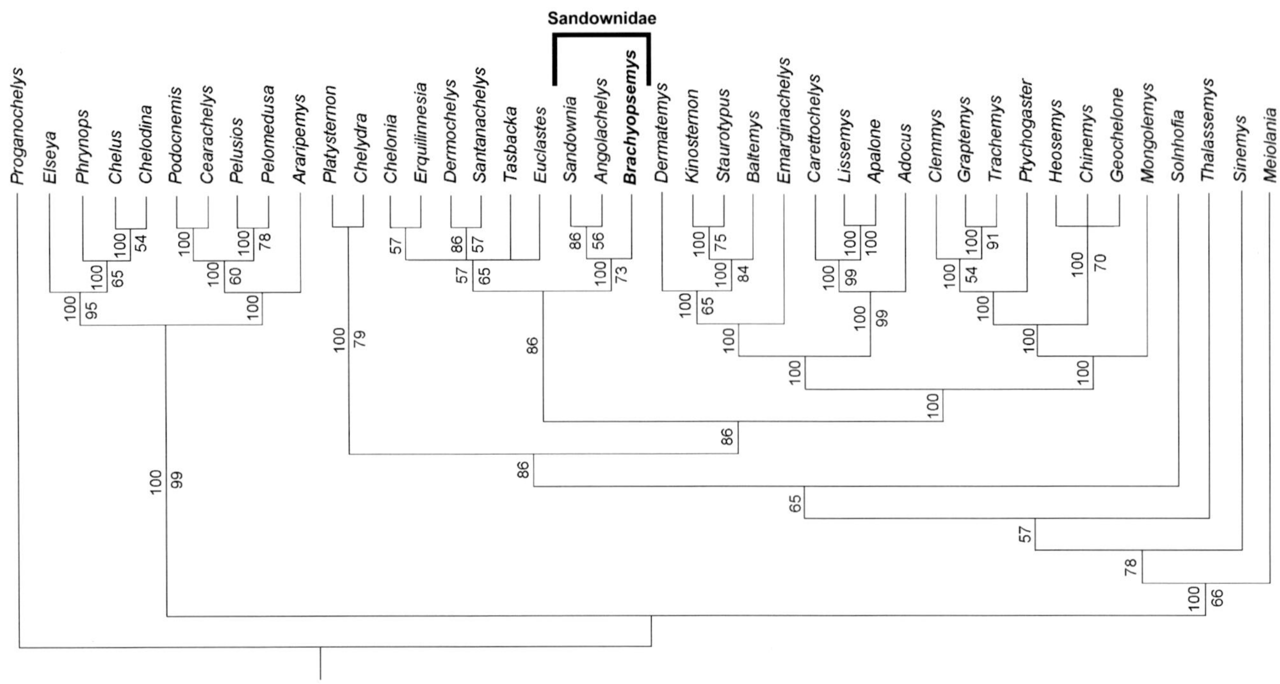

Fig. 13.6 Majority-rule consensus tree of 65 trees of 366 steps each, resulting from a heuristic search of the data set given in Appendix 2 (excluding the shell taxa *Peltochelys*, *Aspideretes maortuensis*, *Hadrianus*, *Hoplochelys*, and *Pelusios rusingae*), showing inferred relationships of the new genus *Brachyopsemys* (*bolded*) and the new family Sandownidae. Majority-rule values are given to *left* of each branch; bootstrap values greater than 50 (for 100 replicates) are given to *right* of each branch

Appendix 13.2. During the course of this study, it was noted that character 23 of Gaffney et al. (1991) is the same as character 109 of Shaffer et al. (1997). The description of character 109 is more complete so character 23 was effectively deleted from the data set by scoring all taxa as "?". This method for eliminating character 23 means that characters were not renumbered which we expect will facilitate comparisons among studies.

In order to examine the possiblity that *Brachyopsemys* might be related to *Sandownia* and *Angolachelys* and to test the hypothesis that these taxa may be more closely related to *Solnhofia* than to other cryptodires [i.e., members of the "Angolachelonia" of Mateus et al. (2009)], we have included *Solnhofia* on the basis of Gaffney (1975) and Joyce (2000), *Thalassemys moseri* on the basis of Rieppel (1980), *Erquelinnesia* on the basis of Zangerl (1971), *Angolachelys* on the basis of Mateus et al. (2009), *Euclastes* on the basis of Hirayama and Tong (2003); *Tasbacka* on the basis of Tong and Hirayama (2002) and Nessov (1987), and *Brachyopsemys* based on the current study. The 'Glen Rose turtle' is not included in our cladistic analysis.

Danilov and Parham (2005) have noted that a skull previously interpreted as *Lindholmemys* (Nessov 1986; Shaffer et al. 1997) is an adocid, necessitating removal of this genus from the data set.

Results of Phylogenetic Analysis

We are able to code 58 of 149 characters for *Brachyopsemys*. For this analysis five shell taxa that seemed to be of little relevance to this study (*Peltochelys*, *Aspideretes maortuensis*, *Hadrianus*, *Hoplochelys* and *Pelusios rusingae*) were excluded. A heuristic search using PAUP* 4.0b of the data set given in Appendix 13.2 produced 65 equally parsimonious trees of 366 steps each. A majority-rule consensus tree is given in Fig. 13.6. The percent of 366 trees supporting each node is given above the preceding stem, values for 100 bootstrap replicates is indicated below each stem.

Monophyly of Sandownidae

The monophyly of Sandownidae is supported by a series of characters including the significant contribution made by the jugal to the upper triturating surface, the posterior extension of the pterygoids to partly or completely cover the basisphenoid, the partial or complete fusion of the premaxillae that are slightly protuberant anteriorly, the absence of the external process of the pterygoid, the absence of the foramen praepalatinum, and the L-shaped prefrontal that sends a posteriolateral process along the ventral margin of the orbit (this reaches the jugal in *Brachyopsemys*).

Table 13.2 Characters shared by *Brachyopsemys*, *Sandownia*, *Angolachelys*, and the unnamed Glen Rose turtle

Characters	*Brachyopsemys*	*Sandownia*	*Angolachelys*	Glen Rose turtle
Very short face with orbits facing forward	Yes	Yes	Yes	Yes
Extensive skull roof with shallow temporal emargination	Yes	Yes	Yes	Yes
Skull roof extensively scaled	Yes	Yes	?	?
Squamosal-parietal contact	Present	Present	Present	Present
Long postorbital contacting squamosal	Yes	Yes	Yes	Yes
Lateral process of prefrontal extends to middle of ventral margin of orbit	Yes	Yes	Yes	?
Prefrontal-jugal contact excluding maxilla from orbit margin	Present	Absent	Absent	?
Extensive secondary palate with internal nares located at midlength of skull	Yes	Yes	Yes	Yes
Thick, rounded labial ridge of upper triturating extending across premaxilla and maxilla	Yes	Yes	Yes	Yes
Fused premaxillae	Completely	Partially	Completely ?	?
Vomer on palate	Very small	Large	Large ?	?
Vomer-premaxilla contact	Absent	Present	?	Present
Palatines excluded from midline contact on palate	Yes	Yes	?	Yes
Jugal contribution of lateral part of palate	Yes	Yes	Yes	Yes
Maxilla-pterygoid contact excludes palatine-jugal contact	Yes	Yes	Yes	Yes
Palatal pit	Present	Absent	Absent	Absent
Processus externus pterygoidei	Absent	Absent	Absent	Absent
Foramen palatinum posterius	Present	Present	Absent ?	Present
Pterygoids cover basisphenoid ventrally	Completely	Partially	Completely	Completely
Occipital condyle in line with condylus madibularis of quadrate	Yes	Yes	Yes	Yes
Incisura columellae auris	Open	Closed	Closed	?
Foramen jugulare posterius and fenestra postotica	Separated by bone	Confluent	?	?
Long canalis caroticus lateralis with foramen anterior canalis caroticus lateralis located anteriorly	Yes	Yes	?	?
Long symphysis of dentary	Yes	Yes	Yes	?
Broad triangular lower triturating surface that slopes ventrally from the coronoid process and lacks labial and lingual ridges	Yes	Yes	?	?
Tall, posteriorly hooked coronoid process	Yes	Yes	?	?

In addition, all three taxa show a complete skull roof in which parietal-squamosal contact excludes the postorbital from temporal exposure (Table 13.2).

The Phylogenetic Position of Sandownidae

Brachyopsemys and the other members of the Sandownidae can be referred to the Cryptodira on the basis of a very well-developed processus trochlearis oticum for the main adductor tendon of the jaw adductor musculature (Gaffney and Meylan 1988). The morphology of the pterygoid and in particular the condition of the palatine artery and internal carotid artery suggest referral to the Polycryptodira (Gaffney 1996). Although monophyly of the family seems firmly supported, the position of the Sandownidae among the Polycryptodira is problematic. Meylan et al. (2000) focused on the trionychoid-like features of *Sandownia harrisi* and employed an analysis of limited scope to suggest that this form was a trionychoid related to *Carettochelys* and the Trionychidae. However, those authors acknowledged that *Sandownia* has features of the Chelonioidea. Joyce (2007) included *Sandownia* in his cladistic analysis of Mesozoic turtles, and suggested that *Sandownia* has "no firm affinities with any group of Cryptodires". He also stated that "placement of *Sandownia harrisi* as sister to '*Talassemys marina*' within Clade 20 only requires one step more than its placement as a basal representative of Trionychoidea" (Joyce 2007, p. 66). His choice among alternative options for placement of Sandownia was based in part on stratigraphic considerations. In spite of the fact that three species of sandownid are now known from skulls, we have to agree with the conclusion of Joyce (2007) stated above. Although the phylogenetic utility of the skull of turtles has been firmly established over the last four decades, the observation that no firm conclusion can yet be made about the phylogenetic position of the Sandownidae suggests that there may be occasions in which it will be availability of well preserved shells that will result in resolution of turtle phylogeny. The trionychoid and chelonioid features of the Sandownidae are discussed below.

Two important characters bearing on the phylogenetic position of the Sandownidae merit further discussion. Our assessment of the material described by Mateus et al. (2009) suggests that nasals are probably absent from *Angolachelys mbexi*. However, the authors interpreted the material to include nasals. Nasals are also reported as present in the Glen Rose turtle and "not expanded laterally and displaced anteriorly by the prefrontals" (Barck 1992, p. 18). If the Sandownidae were derived from a primitive chelonioid stem, one might expect nasals to be present basally. Similarly, Mateus et al. (2009) have interpreted the foramen palatinum posterius to be absent in *Angolachelys* as in chelonioids (they are present in *Sandownia* and *Brachyopsemys*). We consider the preservation in this area of the palate of *Angolachelys* to be questionable. Absence of the foramen palatinum posterius would be a chelonioid feature but presence in *Sandownia* and *Brachyopsemys* would require a reversal of this character.

Trionychian Features of Sandownidae

Although *Brachyopsemys* has some contribution by the palatine to the sidewall of the braincase, an important character of Trionychoidae and Kinosternidae, the contribution cannot be considered to be as significant as in *Adocus* sp. (Meylan and Gaffney 1989) or other taxa that have been referred to these two groups. Rather the dorsal plate of the palatine overlaps the lateral surface of the anterior part of a greatly thickened crista pterygoidea forming a thin veneer on the lateral surface. There is no contact to the descending process of the parietal as is seen in Trionychia and kinosternids. This condition differs from both chelonioids and trionychoids, but is somewhat closer to trionychoids. In fact, the contribution of the palatine to the sidewall of the braincase is related to the length of the sidewall itself and the position of the palatine. In trionychoids, the sidewall of the braincase is long, the ventral portion being made by the palatine anteriorly and pterygoid posteriorly. While the chelonioids have a very short and tall sidewall of the braincase, in which the ventral portion is made up entirely by pterygoid. The relative position of the palatine to the braincase in *Brachyopsemys* specimens thus more closely resembles the condition in trionychoids. The fact that the palatine does not contribute to the whole thickness of the sidewall of the braincase is due to the anterior position of the pterygoids. The palatine midline contact can be thus considered as secondarily lost, probably due to the forward shift of the basicranium. The type description of *Angolachelys* indicates that the palatine does not reach the sidewall of the braincase.

Members of Sandownidae also resemble the Trionychia in the absence of the external process of the pterygoid (character 77), the reduced vomer (not reduced in *Sandownia harrisi*), the maxillae meeting on the midline between the premaxilla and the vomer, and partial or complete fusion of the premaxillae (character 81). Fused premaxillae are also present in the cheloniid *Erquelinnesia gosseleti* (Zangerl 1971). The completely fused premaxillae in *Brachyopsemys* and *Angolachelys* are more derived than partially fused premaxilla in *Sandownia* (Meylan et al. 2000). The tall coronoid process located near the middle of the mandibular ramus is also present in Trionychoidae, such as the adocid *Adocus* (Meylan and Gaffney 1989), the nanhsiungchelyid *Zangerlia neimongolensis* (Brinkman and Peng 1996) and the carettochelyid *Anosteira maomingensis* (Tong et al. 2010). There is vomer to pterygoid contact (character 75) in *Sandownia*, *Angolachelys* and *Brachyopsemys* as in Trionychia.

The zygomaticomandibularis muscle is only reported from the Trionychidae (Schumacher 1973). However, surveys of the jaw musculature of living turtles are still incomplete and careful study of the homology of this muscle and the identification of criteria for its detection in fossil skulls remain to be established. However, the possibility that further study of this muscle might help to elucidate relationships of the Sandownidae should not be overlooked. *Sandownia harrisi* would appear to have the same depressed area below the coronoid on the lateral surface of the jaw (Meylan et al. 2000, Fig. 4c).

Other important characteristics of the Trionychia are clearly absent from members of Sandownidae including parietal contribution to the processus trochlearis oticum (character 74) and the enlarged basisphenoid (character 79). In fact the basisphenoid is reduced (at least in ventral exposure) in sandownids. Furthermore, *Brachyopsemys* has a large foramen stapedio-temporale which is much larger than the foramen anterius canalis carotici interni. This suggests that it lacks 'decrease of stapedial circulation and increase of carotid circulation', a feature previously considered a synapomorphy of Trionychoidea (Meylan and Gaffney 1989). Recent studies on the cranial circulation of turtles do not support the monophyly of Trionychoidea

(Trionychoidae plus Kinosternoidae) (Jamniczky and Russell 2007; Jamniczky 2008), as is also the case for recent molecular phylogenies (Shaffer et al. 1997; Fujita et al. 2004). The enlarged foramen stapedio-temporale in *Brachyopsemys* may represent a primitive condition as seen in Testudinoidea, primitive cheloniids and also plesiochelyids.

Chelonioid Features of Sandownidae

The most apparent suite of chelonioid characters of Sandownidae is a set of correlated characters of the skull roof. Because of strong parietal-squamosal contact (character 15), temporal exposure of the postorbital (character 111) is absent as is the presence of a short postorbital exposed by temporal emargination (character 142). Although contact of the parietal and squamosal is present in *Proganochelys quenstedti* (Gaffney 1990) and primitive cryptodires such as *Dorsetochelys delairi* (Evans and Kemp 1976), *Mesochelys durlstonensis* (Evans and Kemp 1975), and some baenids (Gaffney 1979), posterior exposure of a long postorbital in chelydrids and some potentially basal chelonioids (i.e., *Toxochelys latiremis*) suggests that the extensive skull roof, that includes parietal-squamosal contact in some members of the Chelonioidea, may be a reversal to the primitive condition.

The antrum postoticum (character 140) is small in all members of the Sandownidae for which it is known as is the case in chelonioids. Similarly the foramen praepalatinum is absent from the palate (character 145) in both groups. This foramen is also absent in Trionychia but in this case absence is due to the presence of a large intermaxillary foramen.

The fortuitous preservation of *Brachyopsemys* specimen AMNH 30558 allows study of the floor of the braincase and reveals what may be the most important chelonioid-like characters. The trabeculae of the basisphenoid clearly converge and join anterior to a small sella turcica. Although not as close as in *Chelonia mydas*, the foramen anterius canalis carotici interni are relatively close together (character 64) and furthermore the dorsum sellae does not overhang or conceal the sella turcica (character 65) either in *Sandownia harrisi* or in *Brachyopsemys*. However, it is also clear from this specimen that the foramen caroticum laterale in *Brachyopsemys* is not enlarged as it is in chelonioids (character 91); see also Hirayama (1994, character 29).

Brachyopsemys and *Sandownia harrisi* are also chelonioid-like in have clearly defined scale areas on the skull roof. The presence-absence of skull roof scales was not included in the data matrix, but it should be noted that within the Trionychia scales are absent from the skull roof and some clades have a distinctive sculpture pattern (carettochelyids) that is not seen in members of the Sandownidae. Some nanhsiungchelyids [*Nanhsiungchelys wuchingensis* (Ye 1966)] have both skull roof scales and a sculptured skull roofing surface.

The chelonioid affinity of sandownids is further supported by the lack of shell ornamentation. Although no shell material has been associated with *Brachyopsemys*, and *Sandownia* (Meylan et al. 2000), shell fragments collected along with *Angolachelys* (Mateus et al. 2009) and the Glen Rose turtle (Barck 1992) have no ornamentation on the surface. All Trionychoidae have a sculptured shell surface.

Relationships Within Sandownidae

Within the Sandownidae, *Sandownia* and *Angolachelys* appear more closely related to each other than to *Brachyopsemys*. *Sandownia* and *Angolachelys* share the derived feature, closure of the incisura columellae auris. Meylan et al. (2000) suggested that this feature in *Sandownia harrisi* supported its referal to the Trionychia. The presence of an open incisura in *Brachyopsemys* suggests that closure has occurred within the Sandownidae.

Barck (1992) reported an undetermined eucryptodiran turtle (the Glen Rose turtle) from the Early Cretaceous of Texas. The wide and low skull, extensive skull roof, forward position and closely spaced orbits, large triturating surface and condylus occipitalis located between the condylus mandibularis of the quadrate resemble *Brachyopsemys* and other members of the Sandownidae. The skull from Texas is not described in detail, however, it was included in the Mateus et al. (2009) data set and was resolved as the sister group of *Sandownia* plus *Angolachelys*. In a separate study, it is resolved as the sister taxon of (*Euclastes* [*Sandownia* (*Angolachelys*+*Erquelinnesia*)]) (Vineyard and Jacobs 2009). When this species is fully described, inclusion in a phylogenetic analysis will almost certainly place it within Sandownidae.

Distribution of Sandownidae

The four members of the family (*Sandownia*, the Glen Rose turtle, *Angolachelys* and *Brachyopsemys*) have a known geological range from the Aptian to Paleocene. The oldest representative, *Sandownia* is from the early Aptian Lower Lobster beds within the Lower Greensand Formation, Isle of Wight, England. The rich microflora dominated by nonmarine taxa and low diversity of marine components

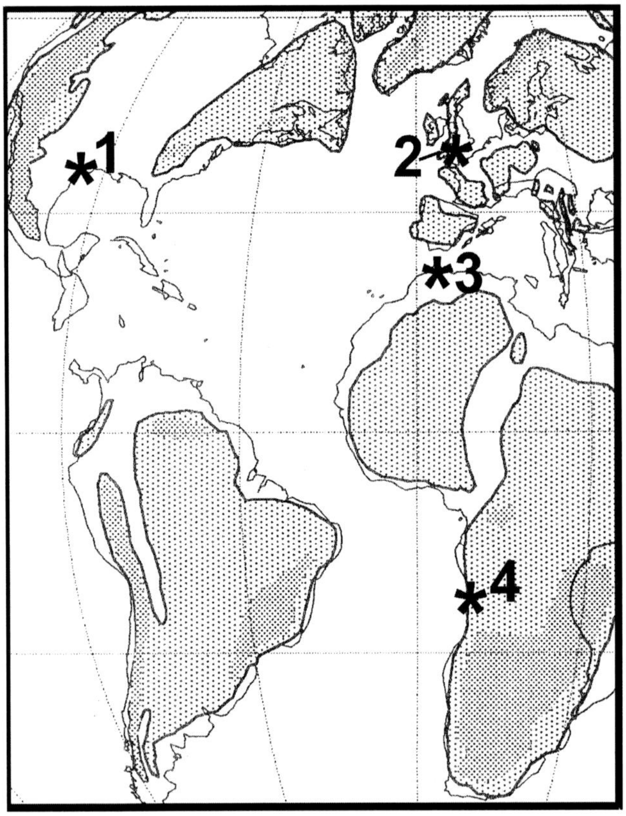

Fig. 13.7 Paleogeographical reconstruction of Maastrichtian stage (Smith et al. 1994), showing the position of the members of the family Sandownidae. *1* Glen Rose turtle; *2 Sandownia*; *3 Brachyopsemys*; *4 Angolachelys*

indicates a restricted, near-shore marine environment, with reduced salinity (Meylan et al. 2000). The second taxon, the Glen Rose turtle, comes from the Glen Rose Formation of Hood County, in central Texas. The Glen Rose Formation consists of limestones and marls deposited in marginal marine environments. It is of early Albian age and has yielded abundant invertebrates, vertebrates and plant remains (Barck 1992; Vineyard and Jacobs 2009). *Angolachelys* is the only member of the group from the south Atlantic. It is from the Turonian of Iembe, in the Kawanza Basin, Angola. The shallow marine sediments of the Tadi Beds of the Itombe Formation at the Iembe site and nearby localities have yielded ammonites, sharks, bony fishes, mosasaurs, plesiosaurs and sauropod dinosaurs (Jacobs et al. 2006; Mateus et al. 2009). *Brachyopsemys* is the youngest member of the group. The phosphate sediments of the Ouled Abdoun Basin were deposited in the long, narrow gulf opening to the Atlantic margin during the latest Cretaceous and early Tertiary. Abundant vertebrate remains, consisting mostly of marine taxa, have been collected from that basin (Gaffney et al. 2006). From a paleogeographical and paleoecological point of view, this family appears to be a bottom-dwelling group that occurred in near-shore, shallow marine environments along the coast of the Atlantic (Fig. 13.7). Its evolutionary history is likely to be connected to the development of the Atlantic Ocean. The addition of the Paleocene *Brachopsemys* to the group indicates that this is another family of turtles that survived the Cretaceous-Tertiary boundary.

Acknowledgments We thank F. Escuillié (Gannat) and B. Segaoui (Erfoud) for providing the specimens,; O. Mateus (Lisbon) for the data set of *Angolachelys*; E. S. Gaffney (New York), W. Joyce (Tübingen) and R. Hirayama (Tokyo) for reviewing the manuscript; J. Pfaller (Gainesville) for discussions of the zygomaticomandibularis, and the Singer-Polygnac Fondation and Eckerd College for financial support. AMNH 30001 was prepared by J. Klausen, J. Shumsky and J. Kelly (New York). The reconstruction of the palate was produced by F. Ippolito after A. Montalvo (New York). Special thanks to Gene Gaffney for providing opportunities to forward our careers.

Appendices

Appendix 13.1

Description of Characters: the matrix used in the study is a further amended matrix based on those of Gaffney et al. (1991) (characters 1–39) and Shaffer et al. (1997) (characters 40–115). Characters 116–142 were used to place fossils in Near et al. (2005), but were not described in that paper. The remainder have been added for the purposes of this study.

116. *Number of bony dermal callosities in the plastron reduced.* Absent = 0, present = 1. Primitively the turtle plastron is composed of one midline element (entoplastron) and at least five pairs of bones (epi-, hyo-, meso-, hypo- and xiphiplastra) although some forms may have more than one pair of mesoplastra (*Odontochelys semitestacea* and *Proterochersis robusta*). In some clades, only a deep dermal ossification represents these elements and an overlying superficial dermal callosity is absent (Meylan 1987, character 9). Three or fewer pairs of dermal callosities in the plastron, as seen in members of the Trionychinae, is considered derived.

117. *Suprascapular fontanelles.* Absent = 0, present = 1. Primitively the adult turtle carapace is closed at the junction of the nuchal, first costal and first neural bones. An opening in this region, as seen in certain members of the Trionychinae is considered derived.

118. *Shape of epiplastron.* Triangular = 0, I-shaped = 1, J-shaped = 2. Primitively the epiplastron of turtles is a more or less triangular element with a variable exterior margin. In certain trionychids it is reduced to an I-shaped (1) or J-shaped (2) deep dermal ossification.

119. *Bridge length.* Long bridge = 0, short bridge = 1. In most turtles, including most primitive forms, the bridge is long, about one-third of the total plastron length. In those

forms in which it is shorter, one-quarter or less of plastron length as in *Chelydra serpentina*, *Staurotypus* or *Apalone*, this is considered a derived condition.

120. *Medial process of jugal.* Present = 0, absent = 1. Most turtles have a medially directed process of the jugal that articulates with the pterygoid or pterygoid and palatine (Gaffney 1979, p. 79). Where this medial process is absent, as in *Dermochelys coriacea*, this is considered a derived condition.

121. *Coracoid length.* Dorsal process of scapula longer than coracoids = 0, coracoids longer than dorsal process of scapula = 1. Turtles have a triradiate pectoral girdle in which the dorsal process of the scapula is normally the longest of the three processes. Those forms in which the coracoid is longer than the dorsal process of the scapula, as in *Dermochelys*, *Lepidochelys kempi* or *Podocnemis*, are considered to have a derived condition.

122. *Wide angle between dorsal process and acromion process of scapula.* About right angle = 0, angle > 105 = 1. The dorsal process of the scapula forms an approximate right angle in *Proganochelys* and most turtles. Where this wider (>105° as in most testudinids) this can be considered a derived condition.

123. *Pectineal process of pubis expanded.* No = 0, yes = 1. In most turtles, including *Proganochelys*, the pectineal process (lateral pubic process of Gaffney 1990, Fig. 144) is narrower than the inter-pubic contact. Those forms in which this process is as wide or wider than the interpubic contact, as in *Lissemys*, *Apalone*, and *Dermochelys*, can be considered to have a derived condition.

124. *Humerus length.* Femur longer than humerus = 0, femur shorter than humerus = 1. In most turtles including *Proganochelys* (Gaffney 1990, Fig. 151) the femur is longer than the humerus. In certain marine species (i.e., *Dermochelys*, *Lepidochelys*) and land tortoises (*Geochelone*, *Gopherus*) the femur is shorter than the humerus. The latter is considered as a derived condition.

125. *Simple, flat carpals and tarsals.* Multifaceted carpals and tarsals = 0, simple flat carpals and tarsals = 1. In most turtles including *Proganochelys* (Gaffney 1990, Figs. 161, 174) the carpals and tarsals are multifaceted, complex elements. In forms with paddle-like limbs, these elements are simple, thin discs of bone.

126. *Epiplastral lip.* Absent = 0, present = 1. Only the most primitive turtles possess a dorsal process of the epiplastron (see character 112). Where this process is absent, the epiplastra are normally only slightly thickened at the shell margin. Where the epiplastron is greatly thickened along its anterior margin, as in many members of the Testudinidae, this is considered as a derived condition.

127. *Pleuro-marginal scute sulcus coincident with costoperipheral suture.* No = 0, yes or on costals = 1. In most turtles including *Proganochelys* (Gaffney 1990, Fig. 86) there are scute sulci on the peripheral bones marking the medial limits of the marginal scutes. Where supramarginal scutes are absent, these are pleuro-marginal scute sulci. When these sulci lie on the costoperipheral suture (as in some members of the Testudinidae) or on the costals (posteriorly in *Adocus*), this can be considered a derived condition.

128. *Bridge peripheral height.* Bridge peripherals the same size as remaining peripherals = 0, bridge peripherals twice as tall as the remaining peripherals = 1. Although the bridge of *Proganochelys* is not well known (Gaffney 1990, p. 127), in most turtles, including primitive forms, the bridge peripherals are approximately the same size as the more anterior and posterior peripherals. Those forms in which the bridge peripherals are approximately twice as tall as the remaining peripherals (as in *Gopherus* and *Geochelone*), can be considered to have a derived condition.

129. *Coracoid fan-shaped.* No = 0, yes = 1. The coracoid of most turtles is longer than wide and in the Casichelyida is somewhat wider distally than proximally. Those forms that have a very short but distally very wide (i.e., fan-shaped) coracoid (as in *Gopherus* or *Kinixys*), can be considered to have a derived condition.

130. *Palatine contributes to triturating surface of upper jaw.* Absent = 0, present = 1. In most turtles including *Proganochelys* the palatine does not appear on the triturating surface. Those forms in which it makes a significant contribution to the triturating surface (as in *Graptemys* or *Podocnemis*) can be considered to have a derived condition.

131. *Foramen caroticopharyngeale.* Small = 0, large = 1. The foramen caroticopharyngeale is a small opening in the pterygoid through which the arteria caroticopharyngeale exits the skull. In most turtles in is very small [i.e., *Chelydra* (Gaffney 1979, Fig. 9)] or not identifiable (*Proganochelys*). Those forms that have an enlarged foramen caroticopharyngeale [as in *Adocus* (Meylan and Gaffney 1989, Fig. 5, labeled as "foramen basisphenoidale") and *Clemmys* (Gaffney 1979, Fig. 47)] can be considered to have a derived condition.

132. *Carapacial contacts of inguinal buttress.* Not reaching costals = 0, reaching costals = 1. In primitive turtles including *Proganochelys* the plastral buttresses do not reach the overlying costal bones of the carapace. When they do, there are two sets of contacts visible for the inguinal buttress. It may contact only the fifth costal (as in *Podocnemis* or *Deirochelys*) or it may contact both the fifth and sixth costals (as in *Hardella*). These are derived conditions.

133. *Musk duct absent from axillary region.* Absent = 0, present = 1. Musk ducts absent in *Progaonchelys* but are present in a variety of turtles and in many forms are visible in the shell as foramina (*Podocnemis*) or grooves (kinosternids) in the vicinity of the bridge. Since there are forms with only axillary or only inguinal musk ducts present, the occurrence of musk ducts is treated as two characters.

This character will be problematic for fossils of those clades in which the living members have musk ducts that do not leave any trace in the skeleton (i.e., Trionychidae). Presence of axillary musk ducts can be considered derived.

134. *Musk duct absent from inguinal region*. See character 133. Absent = 0, present = 1. Presence of inguinal musk ducts can be considered derived.

135. *Hinge at hyo-hypoplastral suture*. Absent = 0, present = 1. Primitively the hyo- and hypoplastra are sutured along their contact. Forms in which this contact is kinetic (as in *Terrapene*) can be considered to have a derived condition.

136. *Pectoral scales (set 4) reach entoplastron*. No = 0, yes = 1. Primitively the turtle plastron is covered by seven sets of scales (Hutchison and Bramble 1981) as is the case in *Proganochelys* (Gaffney 1990, Fig. 92). The fourth pair, the pectorals, in *Proganochelys* and most turtles is located well posterior to the entoplastron. Those forms in which it reaches the entoplastron (as in *Emys* and *Podocnemis*) can be considered to have a derived condition.

137. *Last pair of marginal scutes contacts suprapygal*. No = 0, yes = 1. In most turtles the last pair of marginal scales remain unfused and in contact with only the peripheral elements. Those forms that have the last marginals (either paired or fused) that contact the suprapygal (as in *Hardella*) can be considered to have a derived condition.

138. *Quadrate-basioccipital contact*. No = 0, yes = 1. In *Proganochelys* and other primitive turtles, the quadrate is located well lateral to the basicranium, lacks a medial process, and is not in contact with the basioccipital. Those forms with a large medial process of the quadrate that contacts the basioccipital (as in *Podocnemis*) can be considered to have a derived condition (See also Gaffney et al. 2006, p. 600).

139. *Jugal contributes to triturating surface of upper jaw*. No = 0, yes = 1. In most turtles including *Proganochelys* the jugal does not appear on the triturating surface. Those forms in which it makes a significant contribution to the triturating surface (as in *Bothremys* or *Sandownia*) can be considered to have a derived condition.

140. *Antrum postoticum*. Absent = 0, small = 1, large = 2. The antrum postoticum is absent from *Proganochelys* because the incisura columellae auris is widely open. The presence of a small antrum postoticum (as in *Podocnemis expansa* or *Chelonia*) is considered derived and a moderate to large antrum (as in *Pelusios* or *Gopherus*) is considered further derived. (See also Gaffney et al. 2006, p. 592).

141. *Quadrate contacts basisphenoid but not basioccipital*. No = 0, yes = 1. In *Proganochelys* and other primitive turtles, the quadrate is located well lateral to the basicranium, lacks a medial process, and is not in contact with the basisphenoid or basioccipital. Those forms with a large medial process of the quadrate that contacts the basisphenoid but not the basioccipital (as in *Pelusios*) can be considered to have a derived condition that differs from that described in character 138 above.

142. *Short postorbital exposed by temporal emargination*. No = 0, yes = 1. Primitively the skull roof of turtles is not emarginate and the postorbital is not exposed by emargination. Those forms in which a small postorbital is exposed by temporal emargination (as in *Pelusios* or *Trachemys*) can be considered to have a derived condition.

143. *Prefrontal-palatine contact*. Present = 0, absent = 1. In *Proganochelys* the prefrontal apparently contacts the palatine medial to the foramen orbito-nasale (Gaffney 1990, Fig. 42b) as is the case in most cryptodires. Those forms in which this contact is absent can be considered to have a derived condition.

144. *Jugal exposed on temporal margin*. No = 0, yes = 1. Primitively the jugal is not exposed by temporal emargination, those forms in which it is exposed can be considered to have a derived condition.

145. *Foramen praepalatinum*. Present = 0, absent = 1, foramen intermaxillaris = 2. The presence of the F. praepalatinum in *Proganochelys* remains uncertain, however, as they are common to most Casichelydia their presence can be considered primitive. Two derived conditions are recognized. One in which the premaxilla is well-ossified and there is no foramen intermaxillaris and one in which a foramen intermaxillaris is well developed.

146. *Crista supraoccipitalis projecting well posterior to occipital condyle*. No = 0, yes = 1. In *Proganochelys*, the supraoccipital is very short and there is no posterior projection. In casichelyidians (*Chelus*, *Pelusios*) the supraoccipital is very short and does not extend well posterior to the occipital condyle. In others (*Apalone*, *Chelydra*, *Chelonia*), it extends well posterior to the occipital condyle.

147. *Maxillae meet on the midline*. No = 0, yes = 1. In *Proganochelys* and most turtles the premaxillae and vomers separate the maxilla on the midline of the palate. Where the maxillae do meet on them midline as in *Caretta* or *Phrynops*, this can be considered a derived condition.

148. *Basisphenoid with reduced ventral exposure*. No = 0, yes = 1. Primitively turtles have a broad and long basisphenoid. Those forms in which this exposure is much reduced or absent can be considered to have a derived condition.

149. *L-shaped prefrontal as seen from anterior*. No = 0, yes = 1. Absent primitive, present derived. The prefrontal normally forms the anteriomedial margin of the orbit. Those forms in which it has an L-shape and contributes to the ventromedial margin of the orbit are considered to have a derived condition.

Appendix 13.2

Data set used to estimate phylogenetic positions of *Brachyopsemys* and Sandownidae.

Proganochelys: 00000 00000 00000 00000 00?00 00000 00000 00000 00?00 00000 00000 0000? 00000 00000 ?0000 00000 00000 00?00 00000 00000 00000 ?0000 00000 00000 00000 0??00 00000 0?000 0000? 0000

Elseya: 11111 00001 10010 01001 11?10 11111 11101 11110 00400 11110 00001 0000? 00000 00000 ?0000 01000 00000 00000 00000 01000 11000 00010 011?1 00000 ?00?0 00011 01110 1000? 01001 0100

Phrynops: 11111 00001 10010 01001 11?10 11111 11101 11110 00400 11111 11001 0001? 00000 00000 ?0000 01000 00000 00000 00000 01000 11000 00010 011?1 00000 000?0 00010 01??0 ??00? 01101 0100

Chelus: 11111 00001 10010 01101 11?10 11111 11101 11110 00400 11111 11111 0001? 00000 00000 ?0000 01000 00000 00000 00000 01000 11000 00010 011?1 00000 ?00?0 00011 03??0 1000? 01000 0000

Chelodina: 11111 00001 10011 01001 11?10 11111 11101 11110 00600 11111 11111 0000? 00000 00000 ?0001 10000 ?0000 01000 00000 00000 00000 01000 11000 00010 111?1 00000 ?00?0 00010 01000 1000? 11101 0000

Podocnemis: 11?11 00001 10011 01111 11?10 11111 01101 11111 11100 00000 00001 0001? 00000 00000 ?0000 01000 00000 00000 00000 01000 10000 00010 01101 00000 100?0 00011 01110 10101 00101 1001

Pelusios: 11?11 00001 10011 01111 11?10 11110 01101 11111 11211 00000 00001 0001? 00000 00000 ?0000 01000 00000 00000 00000 01000 10000 00010 11101 00000 ?00?0 00011 01010 00002 11100 0000

Pelomedusa: 11?11 00001 10011 01111 11?10 11111 01101 11111 11311 00000 00001 0001? 00000 00000 ?0000 01000 00000 00000 00000 01000 10000 00010 11101 00000 ?00?0 00011 02110 00000 11100 0000

Platysternon: 11111 11111 11101 10110 00?11 00111 10111 00101 00600 00000 00000 11111 00000 00000 00000 01000 00000 00000 00000 00100 01100 01100 11101 00000 00000 00011 00??0 ??00? 00000 1010

Chelydra: 11111 11111 11101 10110 00?11 00111 10111 00101 00500 00000 00000 11111 00000 00010 00000 01000 00000 00000 00000 00000 00100 01100 11101 00010 00000 00010 00??0 ??002 00000 1000

Chelonia: 11111 11111 11100 10110 00?11 00111 10111 11101 00700 00000 00000 0010? 11110 00000 00000 00000 10000 01000 00100 01101 01100 00000 11011 01011 00??0 ??001 00001 1020

Dermochleys: 11111 11111 11100 10110 00?11 00111 10111 11101 00?00 00000 00000 0010? 11111 10000 ?0000 00000 0?000 000?? 100?? 0000? 00100 01101 01100 ?0?01 11011 0?010 00??0 ??000 00101 0000

Dermatemys: 11111 11111 11101 10110 00?11 01111 10111 11101 00700 00000 00000 00000 00000 01111 10100 01000 00000 00011 11000 01000 00110 01101 11101 00000 00000 00010 00??0 ?0000 01000 1000

Kinosternon: 11111 11111 11101 10110 00?11 01111 10111 11101 00700 00000 00000 00000 00000 01111 10000 01000 00000 00011 11111 10111 00110 01101 11101 00010 00000 00011 001?1 ?0000 01101 1000

Staurotypus: 11111 11111 11101 10110 00?11 01111 10111 11101 00600 00000 00000 10000 00000 01111 10000 01000 00000 00011 11111 10111 00110 01101 11101 00010 00000 00011 001?1 ?0000 01011 1000

Carettochelys: 11111 11111 11101 10110 00?11 01111 10111 11101 10900 00000 00000 0011? 02000 01111 ?1011 10111 1?000 000?? 000?1 01100 ?11?0 01101 11101 00000 100?0 00010 00??1 ??001 01002 1000

Lissemys: 11111 11111 11101 10110 00?11 01111 10111 11101 10800 00000 00000 00110 00000 01110 11011 10111 1?111 111?? 000?0 01000 ?01?0 01101 11101 00100 101?0 0?010 00??1 ??000 01112 1100

Apalone 11111 11111 11101 10110 00?11 01111 10111 11101 10800 00000 00100 00110 00000 01110 11011 10111 1?111 111?? 000?0 01000 ?01?0 01101 11101 11210 101?0 0?010 00??1 ??000 01112 1000

Clemmys 11111 11111 11101 10110 00?11 01111 10111 11101 00400 00000 00002 00000 00000 00000 00000 01000 00000 00010 00000 01000 11000 01101 11111 00000 00000 00010 12101 100?? 01000 1000

Graptemys 11111 11111 11101 10110 00?11 01111 10111 11101 00400 00000 00002 0000? 00000 00000 ?0000 01000 00000 00010 00000 01000 11101 11101 11111 00000 00000 00011 02000 000?? 01001 1000

Trachemys 11111 11111 11101 10110 00?11 01111 10111 11101 00400 00000 00002 00000 00000 00000 00000 01000 00000 00010 00000 01000 11001 11101 11111 00000 00000 00011 02000 000?? 01000 1000

Heosemys 11111 11111 11101 10110 00?11 01111 10111 11102 00500 00000 00001 0000? 00000 00000 ?0000 01000 00000 00010 00000 01000 11100 01101 11111 00000 00000 00010 ?1110 110?? 01??? ????

Chinemys 11111 11111 11101 10110 00?11 01111 10111 11102 00500 00000 00001 0000? 00000 00000 ?0000 01000 00000 00010 00000 01000 11100 01101 11111 00000 00000 00011 01110 1100? 0100? 1000

Geochelone 11111 11111 11101 10110 00?11 01111 10111 11102 00500 00000 00001 00010 00000 00000 00000 00000 00000 00010 00000 01000 11100 01101 11111 00001 01010 11120 01000 110?? 01000 1000

Adocus 11111 11111 11101 10110 00?11 01111 1011? ?1101 00?00 00000 00000 0000? 00000 0?101 ?1110 0100?

01000 00000 00000 00000 00100 ?1101 11111 00000 100?0 00011 10??0 11000 01000 1000

Emarginachelys 11111 11?11 11101 1?110 0??1? ?1111 1011? ??1?? ?0?00 0000? 00000 ??00? 00000 0?111 ?0100 0100? 00000 00?10 ?0000 0100? 00110 ?110? 11111 00000 00000 01000 00??0 ?0000 00000 1000

Meiolania 11111 11111 11100 10000 01?10 ?0111 10100 00000 00?00 00000 00000 ?000? 0?000 000?0 ?0000 0100? 00000 00?0? 00000 00000 ?0100 00000 0111? 00000 00000 00020 00??0 ??0?? 00000 0010

Sinemys 11?11 11111 11101 1?000 0??1? ?1111 10110 000?? 1??01 0??01 00000 ?111? ????0 00??? ?100? ?10?? ?0000 0???? 00000 ?10?? 0110? ?1000 111?? ????0 ?0?00 000?0 ?0110 000?? 0110? 0?00

Peltochelys ????? ????? ????? ????? ????? ??111 1???? ????? ????? ????? ????? ??0?? ?2??? ????? ????? ??0?? ?1000 ???00 ??001 ????? 00?0? ????? ??1?? 0000? ????? 0?0?? ???0? ????? ????? ????

Mongolemys 11111 11?11 11101 11110 0??1? ??111 101?? ????0 0???? 00?00 ?0001 0?00? 00??? 000?? ??000 0100? 00000 ?0?10 ?0000 ??0?0 0?10? 0??0? 111?? ????? ????? 0001? 11110 00001 01000 ?000\

Sandownia 11111 1?111 11100 11110 0??11 00??? ????? ????0 ??00 00?00 00?0? ???1? ??001 001?? ??001 00?0? 0???? ?0??? 10??? ?10?? ??1?? 0??0? 0???? ????0 ????? ????1 1???? ??01? 00001 0011

A. maortuensis ????? ????? ????? ????? ????? ??111 10111 ??1?? 1???? ????? ????0 0?1?? 0???? ????? ?1??? ??1?? ??111 ???1? ??010 ????? ????? ????? ??1?? 1121? 1?1?? 0??0? ?0??0 ????? ????? ????

Santanachelys 11111 1?111 11100 11110 0??11 00??? ????? ????1 ???00 00?00 00?0? ???0? 1?111 1?0?? ??001 01?0? 0???? ?0??? 1???? ?10?? ??1?? 0??0? 0???? 00001 11111 0?000 001?? ??0?? 00?0? 1?00

Hadrianus ????? ????? ????? ????? ????? ??111 10??? ????? 0???? ????? ????0 ??0?? ????? ????? ????? ??0?? ?0000 ??10 ??00? ???0? 2??0? ????? ?11?1 0000? ?10?? 1102? ?0??0 00??? ????? ????

Ptychogaster ?1111 11?11 1110? ????0 0???? ??111 10??? ????? 0???? ????? ????1 ??0?? ????? ????? ??0?? ?100? ?0000 ???10 ??00? ????0 2?00? ????? ?11?? 0000? ????? 000?? ??11 100?? 0???? ????

Baltemys ????? ????? ????? ????? ????? ??111 10??? ????? ????? ????? ????0 0?0?? ?0??? ????? ????? ??0?? ?0000 ???11 ??121 ????1 2??1? ????? ?11?? 0001? ????? 000?? ?01?0 00??? ????? ????

Hoplochelys ????? ????? ????? ????? ????? ??111 10??? ????? ????? ????? ????0 0?0?? ?0??? ????? ????? ??0?? ?0000 ???10 ??110 ????0 0??1? ????? ?11?? 0001? ????? 000?? ?01?0 10??? ????? ????

Araripemys 11111 00?01 10011 01111 0??10 ?0111 11101 10111 11?00 00000 00100 0010? 00000 00000 ?0000 01010 00001 00?10 00000 0100? 10000 00010 111?0 00000 ?0010 0?0?0 00??0 000?0 0011? ?000

P. rusingae ????? ????? ????? ????? ????? ??110 0???? ????? 11??? ????? ????? ??0?? ????? ????? ????? ??0?? ?0000 ????? ??0?0 ????? 1???? ????? ????? ?0??? ????? ??0?? ????0 ????? ????? ????

Cearachelys 11111 00001 1001? ?1111 11?10 ?1111 01101 ??1?? 1??10 00000 00000 0?01? 00110 000?? ??000 0100? 00000 ???00 00000 ?01?? 10000 00?10 111?? 00000 ????? 000?1 0?110 001?1 0010? 0000

Brachyopsemys 11111 11111 11100 10110 0??1? 00??? ????? ????0 ???00 00?00 00?0? ???0? ??000 001?? ??000 00?0? 1???? ?0??? 00??? ?10?? ??1?? 0??0? 0???? ????0 ????? ????1 0???? ??011 00101 1111

Erquilinnesia 11111 11111 11100 1?110 0??1? 00111 10??? ????? 0??00 00?00 0??00 0?10? ??1?1 1?0?0 ??00? 0000? 10000 ?0??? 10000 ?1000 ??10? ????? 0110? 0?000 ??0?1 01011 00??0 ?0001 00?01 1020

Thalassemys 11111 11111 11100 00000 00?1? 00?11 1?000 ????? ???00 00?00 00?00 0??0? ?000? 0?0?? ?00?1 0100? 000?0 00??0 0000? ?10?? ??10? 0110? 0??01 00?0? ????? ?0??? 0???0 0?001 00100 1000

Solnhofia 11111 11111 11101 10010 00?1? 0?111 1?0?? ?01?0 10?00 00000 00?00 0??0? ??01? 000?? ?0000 0100? 00000 00??0 00000 ?10?? ?0100 01100 0??0? 00000 00??0 ?00?1 00110 0?001 00100 1000

Tasbacka 11111 11111 11100 10110 00?1? 00111 101?? ??1?? ???00 00?00 00?00 ??10? ??11? 100?? ??000 0100? 0?000 ?0??? 00??? ?10?? ??1?0 0??0? 0?11? 00000 ??01? ?10?0 00??0 ??001 00101 1020

Euclastes 11111 11111 11100 10110 00?1? 00??? ????? ????? ????? 00?00 00?0? ???0? ??11? 1?0?? ??000 01?0? 0???? ?0??? 0???? ?10?? ??1?? 0??0? 0???? 0???0 ????? ????0 0???? ??001 00101 1020

Angolachelys 11111 11111 11100 10?10 0??1? ?0??? ??1?? ????? ????? 00?00 00?0? ???1? ????? ?00?? ??0?1 00?0? 1???? ?0??? ?0??? ?10?? ??1?? ???0? 0???? ????0 ????? ????1 0???? ??010 00100 ??11

References

Barck, A. (1992). Paleontology of the Glen Rose Formation (Lower Cretaceous), Hood Country, Texas. *The Texas Journal of Science, 44*, 3–24.

Brinkman, D. B., & Peng, J.-H. (1996). A new species of *Zangerlia* (Testudines: Nanhsiungchelyidae) from the Upper Cretaceous redbeds at Bayan Mandahu, Inner Mongolia, and the relationships of the genus. *Canadian Journal of Earth Sciences, 33*, 526–540.

Dalrymple, G. H. (1977). Intraspecific variation in the cranial feeding mechanism of turtles of the genus *Trionyx* (Reptilia, Testudines, Trionychidae). *Journal of Herpetology, 11*, 255–285.

Danilov, I. G., & Parham, J. F. (2005). A reassessment of the referral of an isolated skull from the Late Cretaceous of Uzbekistan to the stem-testudinoid turtle genus *Lindholmemys*. *Journal of Vertebrate Paleontology, 25*, 784–791.

Evans, J., & Kemp, T. S. (1975). The cranial morphology of a new Lower Cretaceous turtle from southern England. *Palaeontology, 18*, 25–40.

Evans, J., & Kemp, T. S. (1976). A new turtle skull from the Purbeckian of England and a note on the early dichotomies of cryptodiran turtles. *Palaeontology, 19*, 317–324.

Fujita, M. K., Engstrom, T. N., Starkey, D. E., & Shaffer, H. B. (2004). Turtle phylogeny: Insights from a novel nuclear intron. *Molecular Phylogenetics and Evolution, 31*, 1031–1040.

Gaffney, E. S. (1975). *Solnhofia parsonsi*, a new Cryptodiran turtle from the Late Jurassic of Europe. *American Museum Novitates, 2576*, 1–25.

Gaffney, E. S. (1976). Cranial morphology of the European Jurassic turtles *Portlandemys* and *Plesiochelys*. *Bulletin of the American Museum of Natural History, 157*, 489–543.

Gaffney, E. S. (1979). Comparative cranial morphology of recent and fossil turtles. *Bulletin of the American Museum of Natural History, 164*, 67–376.

Gaffney, E. S. (1990). The comparative osteology of the Triassic turtle *Proganochelys*. *Bulletin of the American Museum of Natural History, 194*, 1–263.

Gaffney, E. S. (1996). The postcranial morphology of *Meiolania platyceps* and a review of the Meiolaniidae. *Bulletin of the American Museum of Natural History, 229*, 1–166.

Gaffney, E. S., & Meylan, P. A. (1988). A phylogeny of turtles. In M. J. Benton (Ed.), *The phylogeny and classification of the tetrapods* (pp. 157–219). Oxford: Clarendon Press.

Gaffney, E. S., & Meylan, P. A. (1992). The Transylvanian turtle, *Kallokibotion*, a primitive Cryptodire of Cretaceous age. *American Museum Novitates, 3040*, 1–37.

Gaffney, E. S., Meylan, P. A., & Wyss, A. R. (1991). A computer assisted analysis of the relationships of the higher categories of turtles. *Cladistics, 7*, 313–335.

Gaffney, E. S., & Tong, H. (2003). *Phosphatochelys*, a new side-necked turtle (Pelomedusoides: Bothremydidae) from the Paleocene of Morocco. *Bulletin of the American Museum of Natural History, 13*, 644–659.

Gaffney, E. S., & Zangerl, R. (1968). A revision of the chelonian genus *Bothremys* (Pleurodira: Pelomedusidae). *Fieldiana Geology, 16*, 193–239.

Gaffney, E. S., Tong, H., & Meylan, P. A. (2006). Evolution of the side-necked turtles: The families Bothremydidae, Euraxemydidae, and Araripemydidae. *Bulletin of the American Museum of Natural History, 300*, 1–700.

Hirayama, R. (1994). Phylogenetic systematics of chelonioid sea turtles. *The Island Arc, 3*, 270–284.

Hirayama, R., & Tong, H. (2003). *Osteopygis* (Testudines: Cheloniidae) from the Lower Tertiary of the Ouled Abdoun phosphate basin, Morocco. *Palaeontology, 46*, 845–856.

Hutchison, J. H., & Bramble, D. M. (1981). Homology of the plastral scales of the Kinosternidae and related turtles. *Herpetologica, 37*, 73–85.

Jacobs, L. L., Mateus, O., Polcyn, M. J., Schulp, A. S., Antunes, M. T., Morais, M. L., et al. (2006). The occurence and geological setting of Cretaceous dinosaurs, mosasaurs, plesiosaurs, and turtles from Angola. *Journal of Paleontological Society of Korea, 22*, 91–110.

Jamniczky, H. A. (2008). Turtle carotid circulation: A character analysis case study. *Biological Journal of the Linnean Society, 93*, 239–256.

Jamniczky, H. A., & Russell, A. P. (2007). Reappraisal of patterns of nonmarine Cryptodiran turtle carotid circulation: Evidence from osteological correlates and soft tissues. *Journal of Morphology, 268*, 571–587.

Joyce, W. G. (2000). The first complete skeleton of *Solnhofia parsonsi* (Cryptodira, Eurysternidae) from the Upper Jurassic of Germany and its taxonomic implications. *Journal of Paleontology, 74*, 684–700.

Joyce, W. G. (2007). Phylogenetic relationships of Mesozoic turtles. *Bulletin of the Peabody Museum of Natural History, 48*, 3–102.

Joyce, W. G., Chapman, S. D., Moody, R. T. J., & Walker, C. A. (2011). The skull of the solemydid turtle *Helochelydra nopcsai* from the Early Cretaceous of the Isle of Wight (UK) and a review of Solemydidae. *Special Papers in Palaeontology, 86*, 75–97

Karl, H.-V., Tichy, G., & Ruschark, H. (1998). *Osteopygoides priscus* n. gen. n. sp. und die Taxonomie und Evolution der Osteopygidae (Testudines: Chelonioidea). *Mitteilungen der Geologischen und Palaontologischen Landesmuseum Joanneum, 56*, 329–350.

Lyson, T. R., & Joyce, W. G. (2009). A revision of *Plesiobaena* (Testudines: Baenidae) and an assessment of baenid ecology across the K/T boundary. *Journal of Paleontology, 83*, 833–853.

Mateus, O., Jacobs, L., Polcyn, M., Schulp, A. S., Vineyard, D., Neto, A. B., et al. (2009). The oldest African Eucryptodiran turtle from the Cretaceous of Angola. *Acta Palaeontologica Polonica, 54*, 581–588.

Meylan, P. A. (1987). The phylogenetic relationships of soft-shelled turtles (Family Trionychidae). *Bulletin of the American Museum of Natural History, 186*, 1–101.

Meylan, P. A., & Gaffney, E. S. (1989). The skeletal morphology of the Cretaceous Cryptodiran turtle, *Adocus*, and the relationships of the Trionychoidea. *American Museum Novitates, 2941*, 1–60.

Meylan, P. A., Moody, R. T. J., Walker, C. A., & Chapman, S. D. (2000). *Sandownia harrisi*, a highly derived trionychoid turtle (Testudines: Cryptodira) from the Early Cretaceous of the Isle of Wight, England. *Journal of Vertebrate Paleontology, 20*, 522–532.

Near, T. J., Meylan, P. A., & Shaffer, H. B. (2005). Assessing concordance of fossil calibration points in molecular clock studies: An example using turtles. *The American Naturalist, 165*, 137–146.

Nessov, L. A. (1977). Skull morphology of the Early Cretaceous turtle belonging to the family Adocidae. *Trudy Zoologicheskogo Instituta AN SSSR, 74*, 75–79.

Nessov, L. A. (1986). Some Late Mesozoic and Paleocene turtles of Soviet Middle Asia. *Studia Geologica Salamanticensia, volumen Especial 2. Studia Palaeocheloniologica, 2*, 7–22.

Nessov, L. A. (1987). The Paleogene sea turtles of southern Kazakhstan and the phylogenetic relationships between the Toxochelyidae and the Cheloniidae. *Palaeontological Journal, 4*, 76–87.

Nour-Eddine, J., de Lapparent de Broin, F., Bardet, N., Vacant, R., Bouya, B., Amaghzazc, M., & Meslouh, S. (2009). Euclastes acutirostris, a new species of littoral turtle (Cryptodira, Cheloniidae) from the Palaeocene phosphates of Morocco (Oulad Abdoun Basin, Danian-Thanetian). *Compte Rendu Palevol, 8*, 447–459.

Rieppel, O. (1980). The skull of the Upper Jurassic cryptodire turtle *Thalassemys*, with a reconsideration of the chelonian braincase. *Palaeontographica, 171*, 105–140.

Schumacher, G.-H. (1973). The head muscles and hyolaryngeal skeleton of turtles and crocodilians. In: C. Gans & T. S. Parsons (Eds.), *Biology of the Reptilia* (pp. 101–199). London, New York: Academic Press.

Shaffer, H. B., Meylan, P. A., & McKnight, M. L. (1997). Tests of turtle phylogeny: Molecular, morphological, and paleontological approches. *Systematic Biology, 46*, 235–268.

Smith, A. G., Smith, D. G., & Funnell, B. M. (1994). *Atlas of Mesozoic and Cenozoic coastlines*. Cambridge: Cambridge University Press.

Tong, H., & Hirayama, R. (2002). A new species of *Tasbacka* (Testudines: Cryptodira: Cheloniidae) from the Paleocene of the Ouled Abdoun phosphate basin, Morocco. *Neues Jarrbuch fur Geologie und Palaeontologie, Monatsheft, 2002*, 277–294.

Tong, H., & Hirayama, R. (2004). First Cretaceous dermochelyid turtle from Africa. *Revue de Paléobiologie, spécial, 9*, 55–59.

Tong, H., & Hirayama, R. (2008). A new species of *Argillochelys* (Testudines: Cryptodira: Cheloniidae) from the Ouled Abdoun phosphate basin, Morocco. *Bulletin de la Société Géologique de France, 179*, 623–630.

Tong, H., Hirayama, R., & Tabouelle, J. (in press). *Puppigerus camperi* (Testudines: Cryptodira: Cheloniidae) from the Ypresian (Early Eocene) of Ouled Abdoun Basin, Morocco. *Bulletin de la Société Géologique de France*.

Tong, H., Zhang, J.-Y., & Li, J.-J. (2010). *Anosteira maomingensis* (Testudines: Carettochelyidae) from the Late Eocene of Maoming, Guangdong, southern China: New material and redescription. *Neues Jahrbuch für Geologie und Paläontologie Abhandlungen, 256*, 279–290.

Vineyard, D. (1999). Turtle skulls from the Early Cretaceous of Texas. *Journal of Vertebrate Paleontology, 19*(Suppl. 3), 82A.

Vineyard, D. P., & Jacobs, L. L. (2009). A new eucryptodiran turtle from the Mid-Cretaceous (Albian) Glen Rose Formation of Texas. *Gaffney Turtle Symposium* (*October* 17–18, 2009, *Royal Tyrrell Museum, Drumheller, Canada*), *Abstract Volume*, pp. 181–186.

Ye, X.-K. (1966). A new Cretaceous turtle of Nanhsiung, northern Kwangtung. *Vertebrata Palasiatica, 10*, 191–200.

Zangerl, R. (1953). The vertebrate fauna of the Selma Formation of Alabama. Part VI: The turtles of the Family Toxochelyidae. *Fieldiana: Geology Memoirs, 3*, 137–277.

Zangerl, R. (1971). Two toxochelyid sea turtles from the Landenian sands of Erquelinnes (Hainaut) of Belgium. *Institut Royal des Sciences Naturelles de Belgique, Mémoires, 169*, 1–32.

Part IV
Pleurodire Diversity and Biogeography

Chapter 14
A New Long-Necked Turtle, *Laganemys tenerensis* (Pleurodira: Araripemydidae), from the Elrhaz Formation (Aptian–Albian) of Niger

Paul C. Sereno and Sara J. ElShafie

Abstract An articulated skull and postcranial skeleton of a pelomedusoid turtle, *Laganemys tenerensis* gen. et sp. nov., is described from the Lower Cretaceous (Aptian–Albian) Elrhaz Formation in Niger. *Laganemys* has a proportionately long skull, which increases in depth anteriorly, from the occiput to the snout. The thin flat carapace and plastron are covered with fine sulcus-and-ridge texture. The carapace has a deep nuchal embayment anteriorly, a small mesoplastron laterally, and three median fenestrae. The cervical series is nearly as long as the carapace with specialized joints to enhance lateral flexion between cervicals 2 and 3 and cervicals 6 and 7. The relatively long tail is composed of at least 26 vertebrae. Forelimbs and hind limbs have long and relatively straight unguals. Discovered in a fluvial setting, *Laganemys* would have been an adept long-necked aquatic predator in still waters. A suite of derived features unites *Laganemys tenerensis* with *Araripemys barretoi*, a pelomedusoid from northeastern Brazil of similar form, habits and geologic age. These genera provide additional evidence of faunal exchange between South America and Africa in the mid Cretaceous (ca. 110 Mya) prior to the advent of deep waters in the central Atlantic Ocean.

Keywords *Araripemys* • Aquatic predation • Pelomedusidae • Pelomedusoides • Pleurodira

P. C. Sereno (✉) · S. J. ElShafie
Department of Organismal Biology and Anatomy, University of Chicago, 1027 East 57th Street, Chicago, IL 60637, USA
e-mail: dinosaur@uchicago.edu

S. J. ElShafie
e-mail: selshafie@uchicago.edu

Introduction

Pleurodires are less speciose than cryptodires among living turtles and are restricted in geographic range to freshwater habitats in the southern hemisphere. Pleurodiran diversity, habitats, and geographic range, however, were considerably greater in the past and appear to have peaked from the Late Cretaceous (ca. 100 Mya) through the Paleocene (ca. 55 Mya) (Gaffney et al. 2006). Preceding this rich record of pleurodiran diversity, however, is the first half of pleurodiran history. About 200 million years ago, pleurodires and cryptodires diverged from a common casichelydian ancestor, and for the ensuing 100 Mya the pleurodire fossil record is comparatively thin (Gaffney et al. 2006). The most complete pleurodire predating the Late Cretaceous is *Araripemys barretoi*, a small, flat-bodied, thin-shelled freshwater turtle (Fig. 14.1) known from several skeletons from the Araripe Basin of northeastern Brazil (Price 1973; Maisey 1991; Meylan 1996). Mid Cretaceous in age (Aptian–Albian), *Araripemys barretoi* is a pivotal species within Pleurodira and slightly older (ca. 110 Mya) than most other pleurodires. Extant pleurodires are divided into Chelidae and Pelomedusoides, and phylogenetic analysis has placed *Araripemys* as the outgroup to other pelomedusoids (Meylan 1996; Gaffney et al. 2006). Initially described on the basis of a partial shell (Price 1973), *Araripemys* is now known from several acid-prepared specimens that have allowed for a fairly complete cranial and postcranial osteological description (Meylan and Gaffney 1991; Meylan 1996; Gaffney et al. 2006).

Thin shell pieces with low, textured ornamentation and lightly impressed scute grooves similar to those in *Araripemys* were discovered in the mid Cretaceous (Aptian–Albian) Elrhaz Formation in Niger. On these fragmentary remains, Broin (1980) erected a new genus and species, *Taquetochelys decorata*, and mentioned the possible existence of a second species that might be referable to *Araripemys*. More recently, Fuente and Broin (1997) tentatively referred to *Araripemys* the anterior portion of a carapace and a fragmentary costal

D. B. Brinkman et al. (eds.), *Morphology and Evolution of Turtles,* Vertebrate Paleobiology and Paleoanthropology,
DOI: 10.1007/978-94-007-4309-0_14, © Springer Science+Business Media Dordrecht 2013

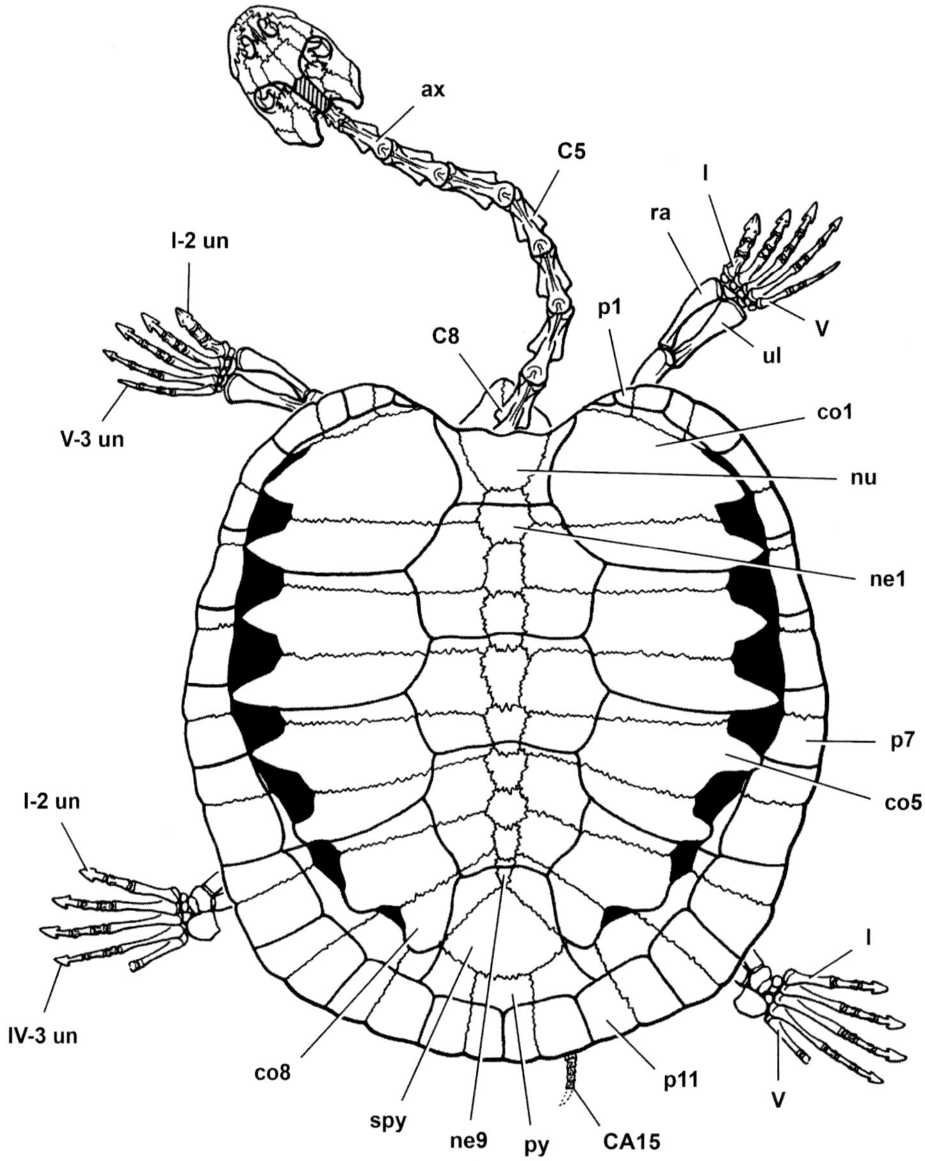

Fig. 14.1 Skeletal reconstruction of *Araripemys barretoi* Price 1973 in dorsal view (modified from Meylan and Gaffney 1991). Abbreviations: *I, IV, V* digits I, IV, V; *ax* axis; *C5, 8* cervical vertebra *5, 8*; *CA15* caudal vertebra *15*; *co1, 5, 8* costal *1, 5, 8*; *ne1, 9* neural *1, 9*; *nu* nuchal; *p1, 7, 11* peripheral *1, 7, 11*; *py* pygal; *ra* radius; *spy* suprapygal; *ul* ulna; *un* ungual

with similar pitted ornamentation from Paleocene deposits in northwestern Argentina. To date, these are the only reports on *Araripemys*-like turtles from outside the Araripe Basin since *A. barretoi* was described.

In 2000 a nearly complete turtle skeleton was recovered in the Elrhaz Formation (Aptian–Albian) of Niger (Sereno and ElShafie 2009). It was found 15–20 cm below the mid section of a skeleton of the large spinosaurid dinosaur, *Suchomimus tenerensis* (Sereno et al. 1998). At the time of its discovery, a portion of the left side of the shell and possibly the left hind limb broke away and disintegrated. The cross-section of the shell was exposed on the trench wall under the dinosaur skeleton, and the turtle and central portion of the dinosaur skeleton were collected in the same field jacket.

Preparation of the specimen revealed a remarkably complete articulated skeleton, including a skull and bones rarely preserved intact, such as the hyoids, carpus, manus and pes (Figs. 14.2, 14.3b, 14.4, 14.5, 14.6, 14.7, 14.8, 14.9, 14.10, 14.11, 14.12, 14.13, 14.14a, 14.15, 14.16, 14.17, 14.18, 14.20, 14.21, 14.22, 14.23). We describe this specimen below as the holotype of *Laganemys tenerensis* gen. et sp. nov, the first reasonably complete remains of a basal pelomedusoid discovered since description of *Araripemys barretoi* by Price in 1973.

The Elrhaz Formation is composed almost exclusively of medium-grained fluvial sandstone and is known for exquisite preservation of vertebrate material. Even by this preservational standard, the new thin-shelled turtle is exceptional. The much larger dinosaur immediately above the turtle is well preserved but only partially articulated; some of its bones

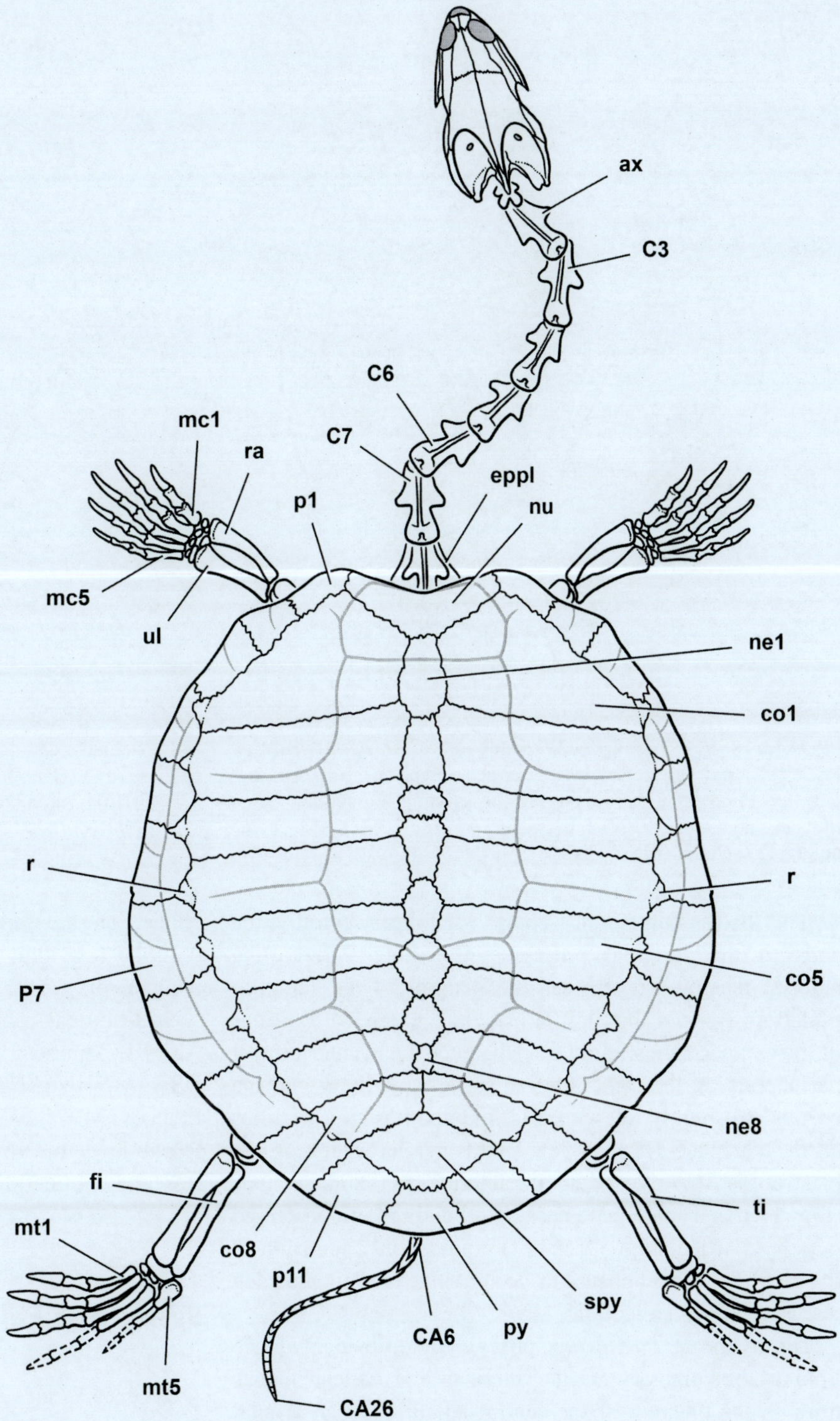

Fig. 14.2 Skeletal reconstruction of *Laganemys tenerensis* gen. et sp. nov. in dorsal view based on the holotypic skeleton (MNN GAD28). Visible scute margins indicated. Abbreviations: *ax* axis; *C3, 6, 7* cervical vertebra 3, 6, 7; *CA6, 26* caudal vertebra 6, 26; *co1, 5, 8* costal 1, 5, 8; *eppl* epiplastron; *fi* fibula; *mc1, 5* metacarpal 1, 5; *mt1, 5* metatarsal 1, 5; *ne1, 8* neural 1, 8; *nu* nuchal; *p1, 7, 11* peripheral 1, 7, 11; *py* pygal; *r* rib; *ra* radius; *spy* suprapygal; *ti* tibia; *ul* ulna

were transported postmortem. The turtle, on the other hand, is preserved with a high degree of natural articulation, including the skull, hyoids, cervical series, fore- and hind-limbs. It must have been buried quickly with only minor displacement, perhaps during transport and final burial of the overlying dinosaur skeleton.

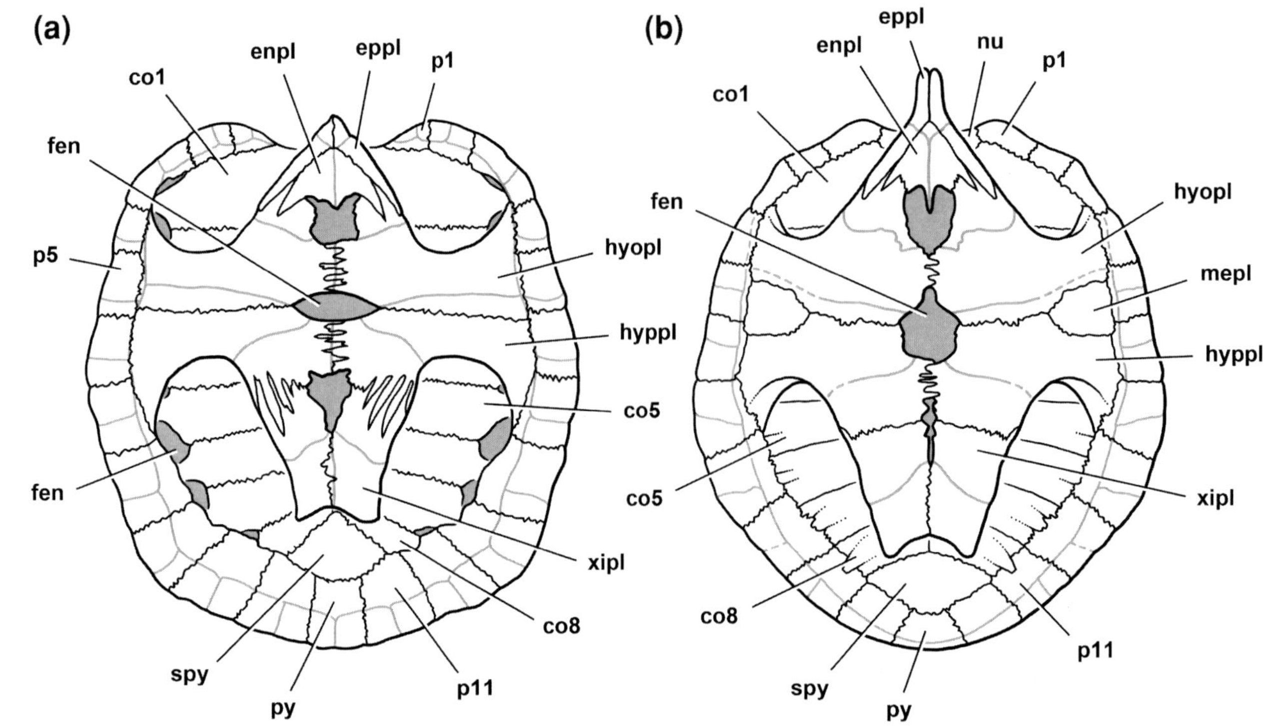

Fig. 14.3 Shell reconstruction in ventral view of *Araripemys barretoi* Price 1973 and *Laganemys tenerensis* gen. et sp. nov. **a** *Araripemys barretoi* Price 1973 based on several specimens (modified from Meylan and Gaffney 1991). **b** *Laganemys tenerensis* based on the holotypic skeleton (MNN GAD28). Visible scute margins indicated. Abbreviations: *co1, 5, 8* costal 1, 5, 8; *enpl* entoplastron; *eppl* epiplastron; *fen* fenestra; *hyopl* hyoplastron; *hyppl* hypoplastron; *mepl* mesoplastron; *nu* nuchal; *p1, 5, 11* peripheral 1, 5, 11; *py* pygal; *spy* suprapygal; *xipl* xiphiplastron

The shell is only slightly compressed dorsoventrally, the ends of the acromial processes projecting through the anterior plastral fenestra and the margins of the carapace pulled away from the plastral bridge (Fig. 14.13). Portions of the carapace are slightly ajar; the nuchal and anteriormost peripherals on the right side are shifted anteriorly and the costals are slightly telescoped in mid and posterior sections of the carapace (Fig. 14.11). Right and left sides of the plastron is slightly spread in anterior and mid sections (Fig. 14.13). The delicate neck, skull and tail are preserved largely in articulation. Of the extremities, only a portion of the right forelimb appears to be missing; the left hind leg was lost during collection.

Discovery of *Laganemys* provides additional evidence linking contemporary South American and African faunas prior to the opening of the central Atlantic Ocean around 100 Mya (Maisey 1993; Sereno et al. 2004). The vertebrate fauna from the Santana Formation in the Araripe Basin of Brazil and from the Elrhaz Formation in the Illumeden Basin of Niger are regarded as roughly comparable in age (Aptian–Albian, ca. 110 Mya). Their depositional settings, however, differ, the former predominantly lacustrine and the latter strictly fluvial. The remains of fish and thin-shelled turtles like *Araripemys*, as a result, are much more common in Araripe sediments than in the fluvial sandstones from Niger. The extraordinary preservation of the turtle we describe here thus adds an important taxon to this faunal comparison.

Institutional abbreviations used in this paper are: AMNH, American Museum of Natural History, New York, USA; MNHN, Musée National d'Histoire Naturelle, Paris, France; MNN, Museum National du Niger, Niamey, République du Niger; and UCRC, University of Chicago Research Collection, Chicago, Illinois, USA.

Systematic Paleontology

Testudines Linnaeus 1758
Pleurodira Cope 1865
Pelomedusoides de Broin 1988
Family Araripemydidae Price 1973 (= Araripemyidae Broin 1980)

Type genus and species: *Araripemys barretoi* Price 1973 (Figs. 14.1, 14.3a).

Distribution: Mid Cretaceous (Aptian–Albian) rocks in the Araripe Basin (Santana Formation) of Brazil and Illumeden Basin (Elrhaz Formation) of Niger.

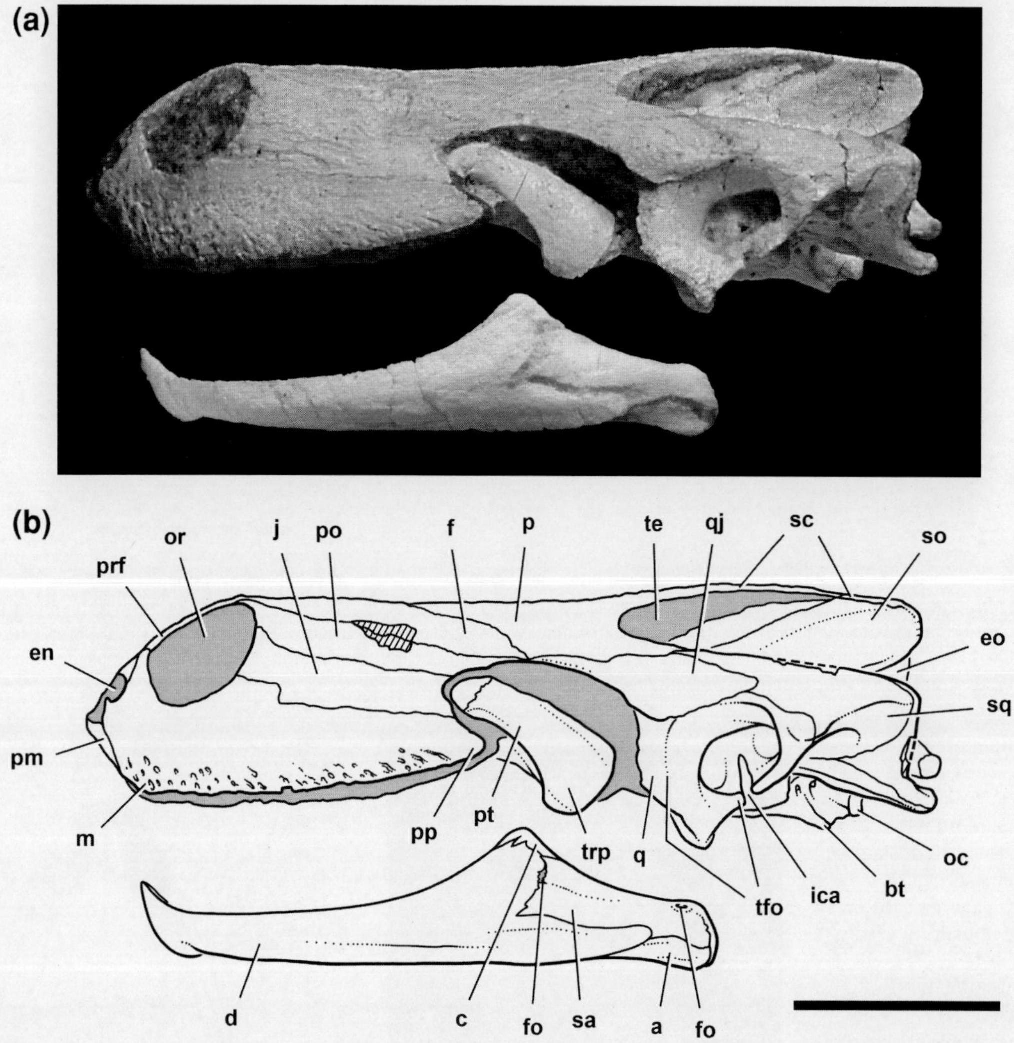

Fig. 14.4 Skull of *Laganemys tenerensis* gen. et sp. nov. (MNN GAD28) in left lateral view (reversed). **a** Photograph. **b** Line drawing. Scale bar = 1 cm. Abbreviations: *a* angular; *bt* basal tuber; *c* coronoid; *d* dentary; *en* external naris; *eo* exoccipital; *f* frontal; *fo* foramen; *ica* incisure of the columella auris; *j* jugal; *m* maxilla; *oc* occipital condyle; *or* orbit; *p* parietal; *pm* premaxilla; *po* postorbital; *pp* posterior process; *prf* prefrontal; *pt* pterygoid; *q* quadrate; *qj* quadratojugal; *sa* surangular; *sc* supraoccipital crest; *so* supraoccipital; *sq* squamosal; *te* temporal emargination; *tfo* tympanic fossa; *trp* trochlear process

Diagnosis: Basal pleurodires of modest body size (adult carapace length 10–30 cm) characterized by an elongate basisphenoid (50% of cranial length in the midline); carapace very thin (approximately 1 mm) with very low profile; fine-grained pit and ridge-and-sulcus texture on the external surface of the carapace, ventral aspect of the peripherals, and the external surface of the plastron; broad nuchal embayment; neural 3 with only four sutural contacts (neurals 2, 4; right and left costal 3); rib tips exposed on carapace with their distal tips projecting between adjacent peripherals (costals 1–4) or toward the central body of a single peripheral (costals 5–8); epiplastron strap-shaped; three median plastral fenestrae; scute margins lightly incised or absent on the carapace and plastron; cervical series (C1–8) elongate (90% of the length of the carapace); mid cervical transverse processes broad-based and subtriangular; postatlantal postzygapophyses partially or completely joined in the midline; cervical epipophyses wedge-shaped with a transverse distal margin and positioned adjacent to one another near the midline; and metacarpal 2 with a flange-like lateral buttress for metacarpal 3 along the proximal two-thirds of the shaft.

Remarks: Unable to refer *Araripemys* to any existing family, Price (1973) erected a monotypic Family Araripemydidae, which at that time was redundant with the genus and therefore carried no particular phylogenetic information. Subsequently, de Broin (1980) referred a second genus *Taquetochelys* to the Araripemydidae, although neither she nor later authors who described similar shell fragments from Argentina (Fuente and de Broin 1997) provided a familial diagnosis. Meylan (1996) also listed the family without diagnosis, referring to it an unnamed taxon from the Santana fauna (Gaffney and Meylan 1991). Fielding et al. (2005) erected a second species, *Araripemys arturi*, based on fragmentary and

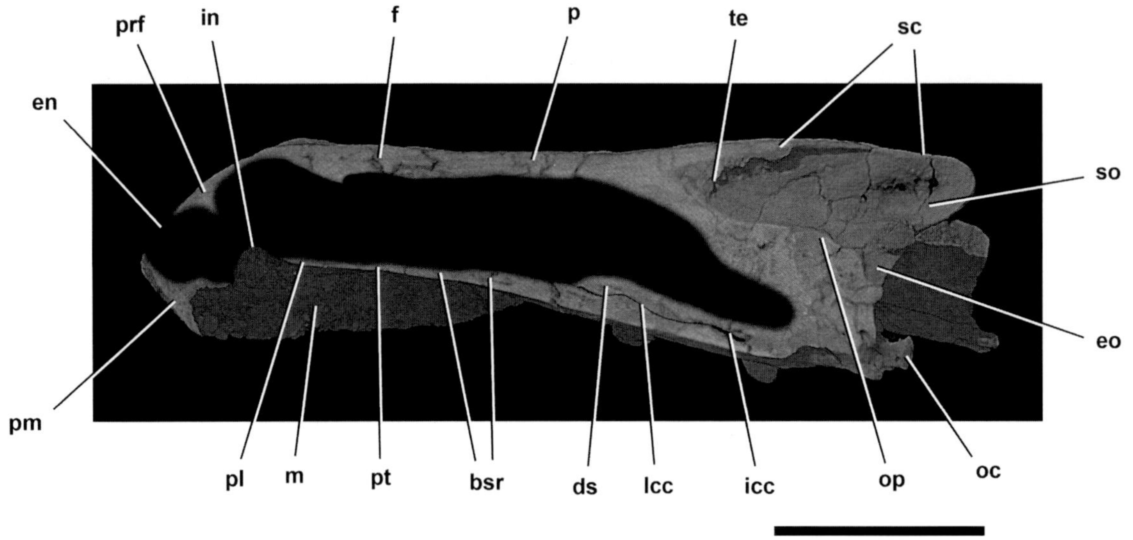

Fig. 14.5 Computed-tomographic reconstruction of the cranium of *Laganemys tenerensis* gen. et sp. nov. (MNN GAD28) in parasagittal cutaway view. Cross section is to the *left* of the midline in left lateral view. Scale bar = 1 cm. Abbreviations: *bsr* basisphenoid rostrum; *ds* dorsum sellae; *en* external naris; *eo* exoccipital; *f* frontal; *icc* internal carotid canal; *in* internal naris; *lcc* lateral carotid canal; *m* maxilla; *oc* occipital condyle; *op* opisthotic; *p* parietal; *pl* palatine; *pm* premaxilla; *prf* prefrontal; *pt* pterygoid; *sc* supraoccipital crest; *so* supraoccipital; *te* temporal emargination

Fig. 14.6 Cranium of *Laganemys tenerensis* gen. et sp. nov. (MNN GAD28) in dorsal view. **a** Photograph. **b** Line drawing. Scale bar = 1 cm. Abbreviations: *ce* cheek emargination; *en* external naris; *eo* exoccipital; *f* frontal; *j* jugal; *m* maxilla; *op* opisthotic; *or* orbit; *p* parietal; *po* postorbital; *prf* prefrontal; *q* quadrate; *qj* quadratojugal; *so* supraoccipital; *sq* squamosal; *te* temporal emargination

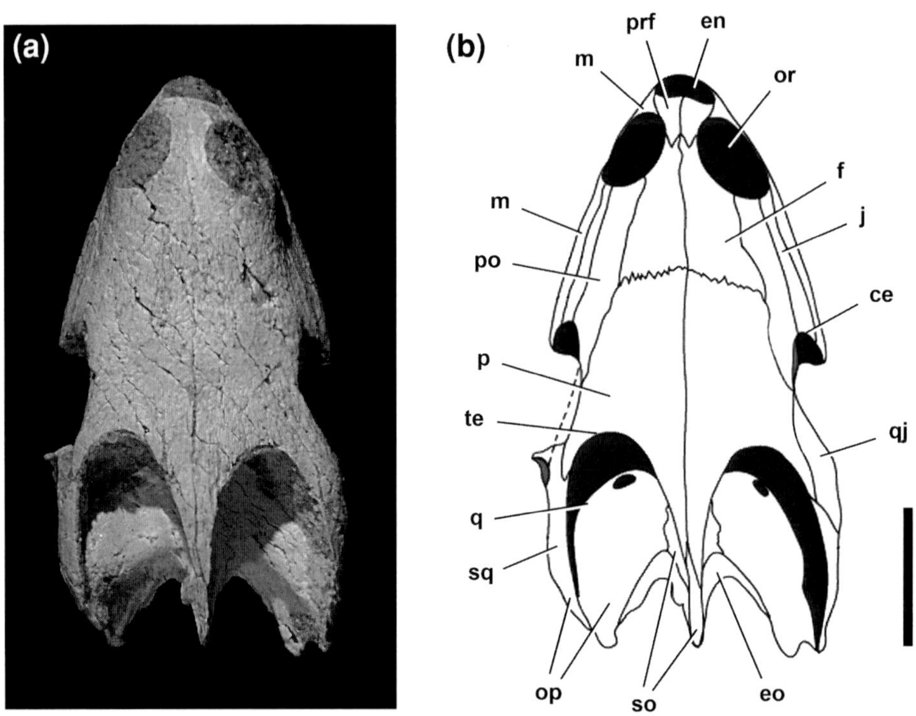

immature material. Subsequent review regarded this poorly established taxon as a *nomen dubium* (Gaffney et al. 2006). As far as we are aware, we are the first to diagnose Araripemydidae as a nonredundant taxon. At present it contains the type genus *Araripemys* from Brazil and a second genus from Niger described below. We include only derived characters shared by both genera in the diagnosis given above.

Broin (1980) introduced "Araripemyidae," a variant spelling of the familial name Araripemydidae (Gaffney et al. 2006). Although Broin's variant is one of two spelling options recognized by the Code of Zoological Nomenclature for generating a familial name based on a genus, it was not the option chosen by the original author (Price 1973).

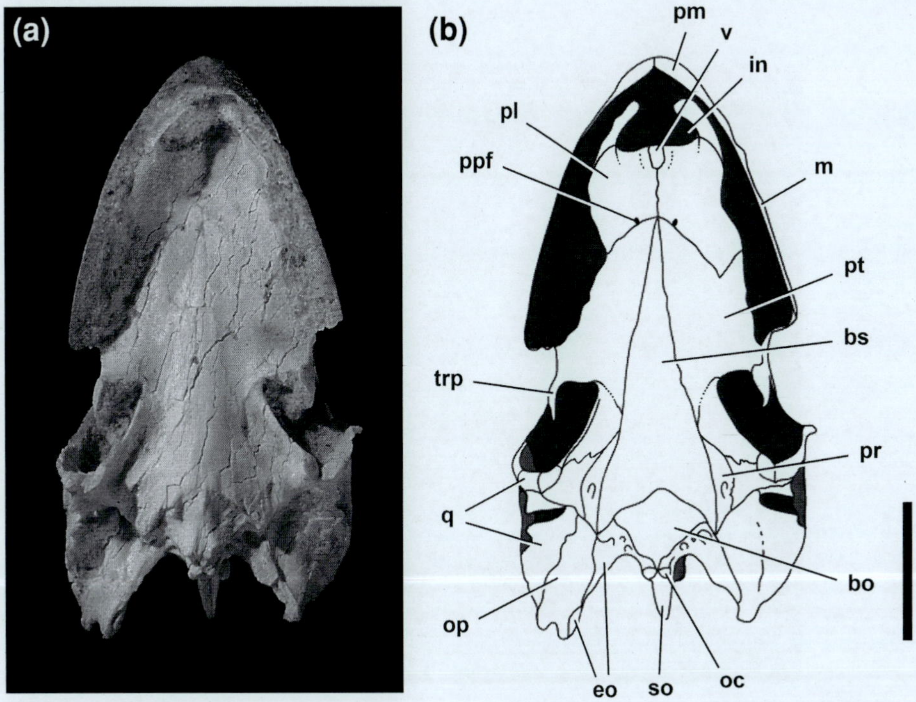

Fig. 14.7 Cranium of *Laganemys tenerensis* gen. et sp. nov. (MNN GAD28) in ventral view. **a** Photograph. **b** Line drawing. Scale bar = 1 cm. Abbreviations: *bo* basioccipital; *bs* basisphenoid; *eo* exoccipital; *in* internal naris; *m* maxilla; *oc* occipital condyle; *op* opisthotic; *pl* palatine; *pm* premaxilla; *ppf* posterior palatine foramen; *pr* prootic; *pt* pterygoid; *q* quadrate; *so* supraoccipital; *trp* trochlear process; *v* vomer

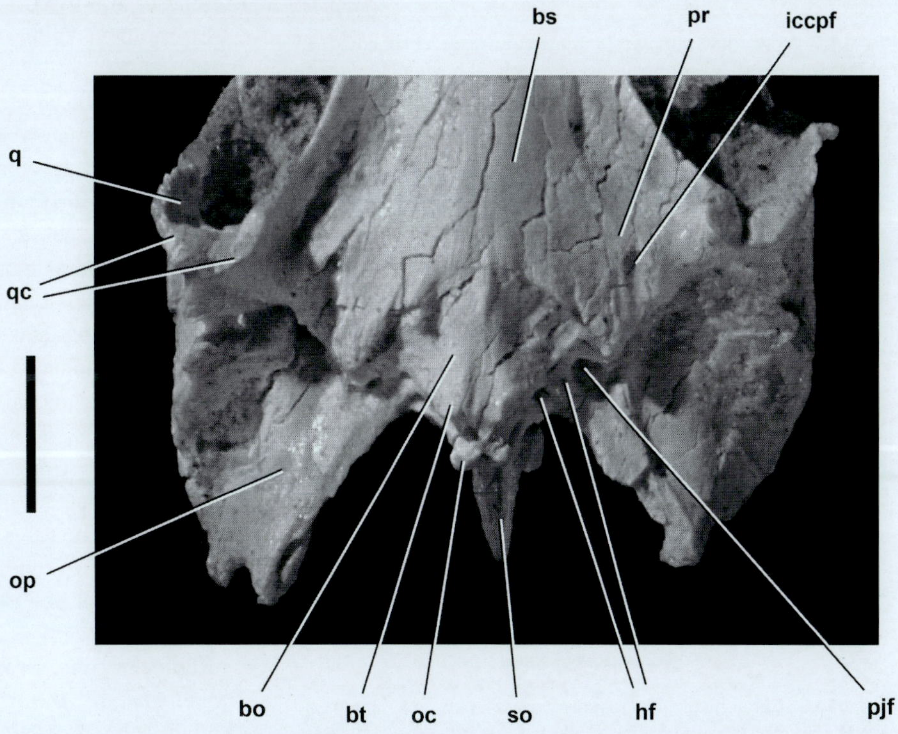

Fig. 14.8 Basicranium of *Laganemys tenerensis* gen. et sp. nov. (MNN GAD28) in ventral view. Scale bar = 5 mm. Abbreviations: *bo* basioccipital; *bs* basisphenoid; *bt* basal tuber; *hf* hypoglossal foramina; *iccpf* internal carotid canal, posterior foramen; *oc* occipital condyle; *op* opisthotic; *pjf* posterior jugular foramen; *pr* prootic; *q* quadrate; *qc* quadrate condyle; *so* supraoccipital

Araripemys Price 1973
Araripemys barretoi Price 1973
(Figs. 14.1, 14.3a).

Holotype locality, **unit**, **and age**: 2 kms northeast of Santana do Cariri, Ceará State, Brazil; Romualdo Member, Santana Formation, Araripe Basin; Aptian–Albian (Price 1973).

Revised diagnosis: Basal pleurodire of small size (adult carapace length 20–30 cm) with semicircular nuchal; costal 1 contributing to the anterior margin of the carapace (separating the nuchal and peripheral 1); peripheral 1 small, subtriangular; posterior margin of the carapace extended, with rectangular peripherals (long axis radial) that cover all but the distal shank, ankle and pes of an extended hind limb; mesoplastron

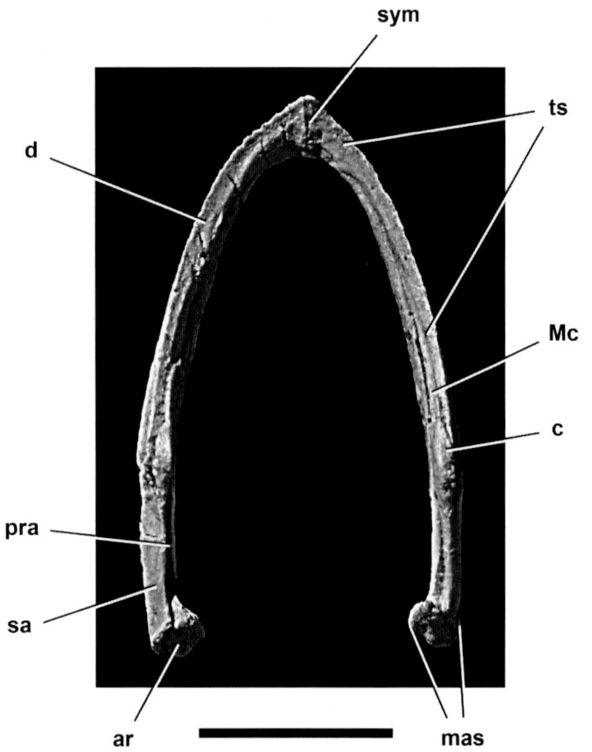

Fig. 14.9 Lower jaw of *Laganemys tenerensis* gen. et sp. nov. (MNN GAD28) in dorsal view. Scale bar = 1 cm. Abbreviations: *ar* articular; *c* coronoid; *d* dentary; *mas* mandibular articular surface; *Mc* Meckel's canal; *pra* prearticular; *sa* surangular; *sym* symphysis; *ts* triturating surface

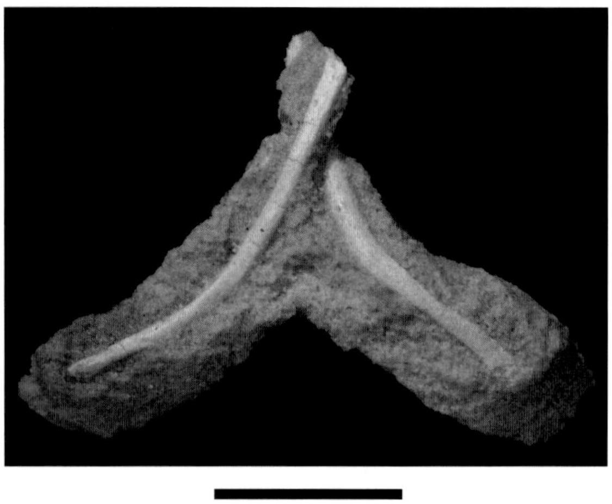

Fig. 14.10 Ceratohyals of *Laganemys tenerensis* gen. et sp. nov. (MNN GAD28) in ventral view. Scale bar = 1 cm

absent; deeply interdigitating suture between the hypo- and xiphi-plastron; dorsal 1 firmly sutured to nuchal; arrowhead-shaped ungual phalanges on manus and pes.

Remarks: Meylan (1996, p. 20) and Gaffney et al. (2006, p. 35) provided diagnoses for this genus and species. The revised diagnosis restricts cited features to those that are, or could potentially be, autapomorphic.

Laganemys gen. nov.

Type species: *Laganemys tenerensis* sp. nov.

Etymology: *Lagano* (Greek), pancake; *emys* (Greek), turtle.

Diagnosis: Same as for type species.

Laganemys tenerensis sp. nov.
(Figs. 14.2, 14.3a, 14.4, 14.5, 14.6, 14.7, 14.8, 14.9, 14.10, 14.11, 14.12, 14.13, 14.14a, 14.15, 14.16, 14.17, 14.18, 14.20, 14.21, 14.22, 14.23).

Holotype: MNN GAD28, a nearly complete articulated skull and postcranial skeleton lacking the left posterolateral corner of the carapace and plastron, a few anterior caudal vertebrae, the right forelimb distal to the humerus, and the left hind limb.

Holotype locality, unit, and age: 16° 26′ 16.3 N, 9° 7′ 3.6 E (field locality 94, 2000 Expedition to Niger), Gadoufaoua, approximately 125 kms east of Agadez, Niger Republic; GAD 5 level in the Elrhaz Formation, Illumeden Basin; Aptian–Albian (Taquet 1976). Discovered in close association with the spinosaurid *Suchomimus tenerensis* (Sereno et al. 1998).

Etymology: *Tenere*, from Ténéré Desert; *ensis* (Latin), from.

Diagnosis: Basal pleurodire of small size (adult carapace length approximately 15 cm), skull very elongate (length more than five times maximum width) with snout increasing in depth anteriorly; maxilla with long posterior process ventral to the cheek emargination; postorbital excluded from the temporal emargination by quadratojugal-parietal contact; parietal-squamosal contact along the temporal emargination; parietal with posterolateral process; dentary ramus gently arched ventrally with squared distal end (chin) in lateral view; nuchal V-shaped; neural 3 small and rectangular with long axis oriented transversely, neural 8 small and rectangular (long axis sagittal), and neural 9 absent; epiplastron J-shaped; mesoplastron present and pentagonal; main forelimb bones (radius, metacarpal 2–4) considerably shorter (45–60%) than comparable hind limb bones (tibia, metatarsal 2–4); and metacarpal 2 with medially divergent distal condyles.

Remarks: de Broin (1980) erected a new taxon, *Taquetochelys decorata*, from the Elrhaz Formation (Aptian–Albian) of Niger. The material consists of 10 shell fragments (MNHN GDF838-848). The right hypoplastron (MNHN GDF847; Fig. 14.14b) was selected as the holotype (de Broin 1980, pl. III, Fig. 10); the other shell fragments were designated as paratypes (de Broin 1980, pl. III, Figs. 2–9, 11a, 11b).

The hypoplastron and other shell fragments were collected in isolation in the late 1960s and early 1970s during French expeditions to an area in the Ténéré Desert known as

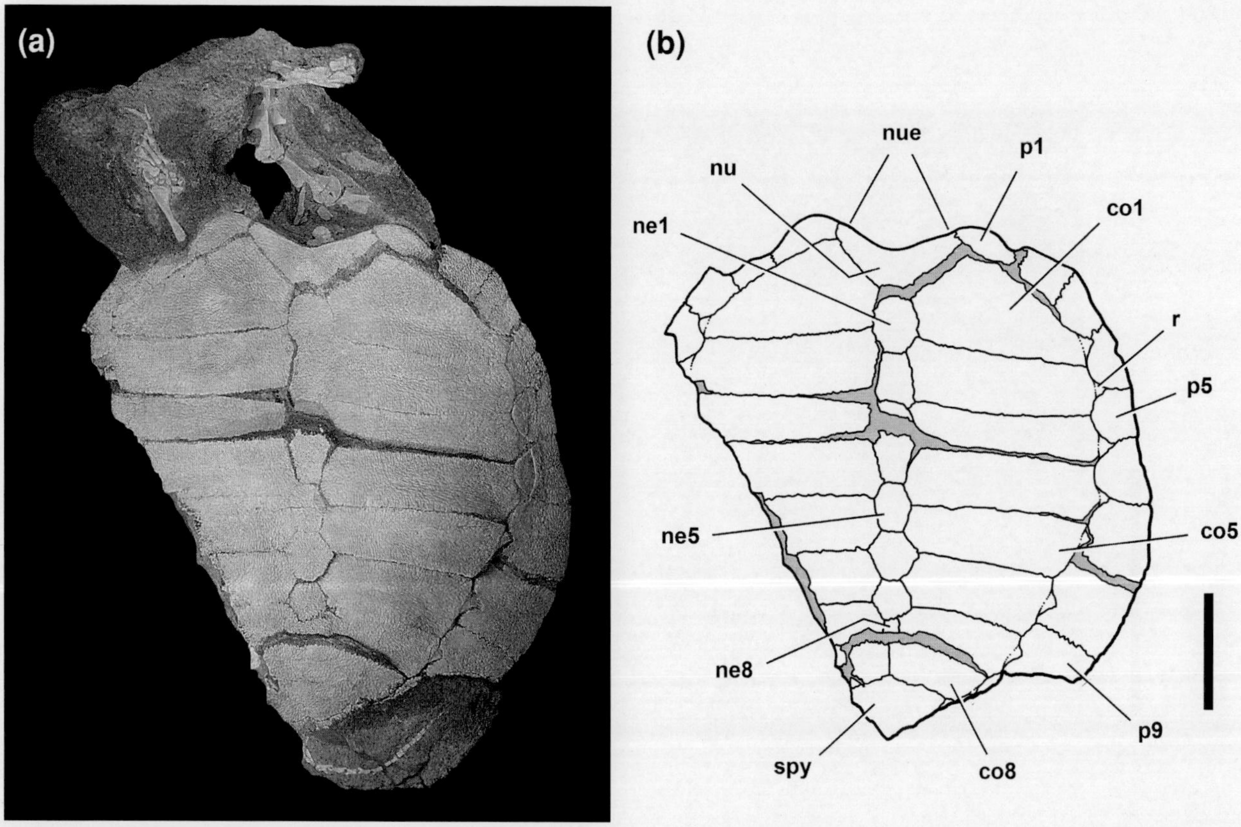

Fig. 14.11 Carapace of *Laganemys tenerensis* gen. et sp. nov. (MNN GAD28) in dorsal view. **a** Photograph. **b** Line drawing. Scale bar = 3 cm. Abbreviations: *co1, 5, 8* costal 1, 5, 8; *ne1, 5, 8* neural 1, 5, 8; *nu* nuchal; *nue* nuchal emargination; *p1, 5, 9* peripheral 1, 5, 9; *r* rib; *spy* suprapygal

Gadoufaoua (Taquet 1976). Unlike the other turtle remains described from Gadoufaoua by de Broin (1980), no locality information is given for the shell fragments referred to *T. decorata*, which may not pertain to a single individual given the abundance of disarticulated vertebrate remains in most exposures of the Elrhaz Formation. The new taxon, *Laganemys tenerensis*, was discovered at a new locality a number of kilometers away from previous localities (Taquet 1976, Figs. 7, 8). Nonetheless, as the holotype and only known specimen of was discovered in the same formation and region as the hypodigm of *T. decorata*, their taxonomic status as valid genera and species must be carefully considered.

Shape and textural differences are apparent between the holotypic hypoplastron of *T. decorata* and that in *L. tenerensis*, although both are thin, covered with a fine-grained ornamentation, and lie adjacent to a small mesoplastron (Fig. 14.14). Laterally the sutural margin for peripherals 5 and 6 in *L. tenerensis* is divided discretely into two parts, the suture for peripheral 6 angling posteromedially at approximately 30°; in *T. decorata* the margin is gently convex (Fig. 14.14). Anteriorly the sutural margin for the mesoplastron is different. In *L. tenerensis* the suture has a discrete angle of approximately 50°, giving the mesoplastron a distinctive pentagonal shape; in *T. decorata* this margin is more irregular (Fig. 14.14). The ramus of the hypoplastron between the mesoplastron and posterior embayment, as a result, is proportionately narrower in *L. tenerensis* than in *T. decorata*.

Posteriorly the margin of the embayment for the hind limb also shows differences. The contour of the embayment is more deeply arched in *L. tenerensis*, with a smooth margin approximately twice the width of that in *T. decorata* (Fig. 14.14). In *Araripemys barretoi*, the smooth margin is weakly developed or absent and the embayment has a broad contour as in *T. decorata*. The smooth margin, in addition, tapers to a point farther medially in *L. tenerensis*, medial to the apex of the embayment (Fig. 14.14).

A low texture is present across the surface of the plastron in both taxa, but it differs in the size and prominence of the pattern of dimples and pits and their organization into radiating ridges and sulci. We confine our comments to the hypoplastron, which de Broin (1980) designated as the holotype of *T. decorata*. The texture on the hypoplastron of *L. tenerensis* is dominated by ridge-and-sulcus texture, which covers most of the plate except for a transverse band in the posteromedial portion of the plate, where it merges into small subspherical pits less than 0.5 mm in diameter

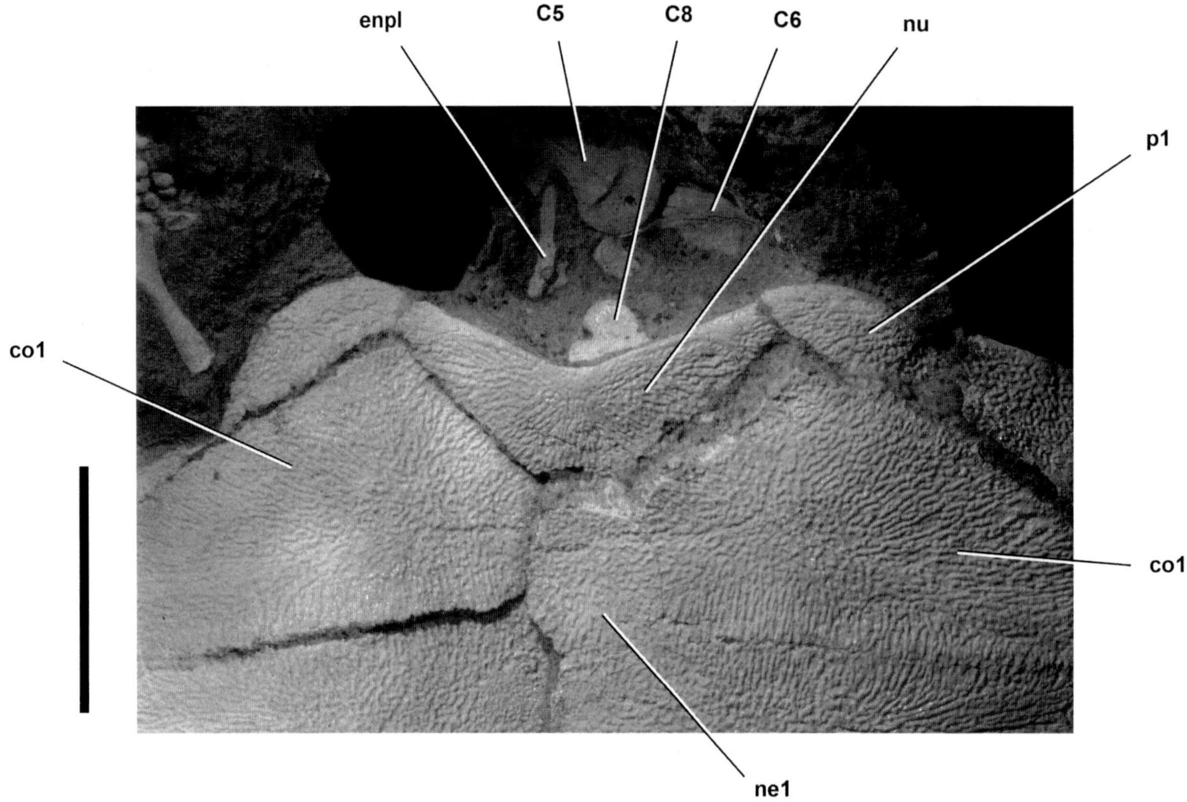

Fig. 14.12 Anterior portion of the carapace of *Laganemys tenerensis* gen. et sp. nov. (MNN GAD28) in dorsal view. Scale bar = 2 cm. Abbreviations: *C5, 6, 8* cervical vertebra 5, 6, 8; *co1* costal 1; *enpl* entoplastron; *ne1* neural 1; *nu* nuchal; *p1* peripheral 1

(Fig. 14.14a). The ridge-and-sulcus texture is organized into parallel ridges and grooves in the dorsal and posteromedial portions of the plate, and the ridges in general seem to emanate from what may constitute an ossification center near the apex of the embayment. In *T. decorata*, in contrast, the texture is dominated by subspherical dimples that merge into a ridge-and-sulcus texture only near the anterior and lateral margins of the hypoplastron (Fig. 14.14b). This dimpled texture, similar to the surface texture of a basketball, extends without diminution to the edge of the smooth inset margin of the posterior embayment. In *L. tenerensis*, in contrast, the ornamentation is noticeably reduced near the edge of the inset margin (Fig. 14.14a). Orientation of ridge-and-sulcus texture around a center is much less pronounced in *T. decorata*.

The shape and textural differences outlined above are noticeable and seem more substantial than variation attributable to age differences, individual variation, or sexual dimorphism. Although the material referred to *T. decorata* looks different than the holotype of *L. tenerensis*, we agree with Gaffney et al. (2006, p. 111) that the dimpled surface texture and other features of the holotypic hypoplastron of *T. decorata* are difficult to establish as autapomorphies justifying taxonomic distinction. Were the holotype closer in form to that of *L. tenerensis*, we may have been able to refer the new material to *T. decorata*, despite the very limited range of potentially diagnostic features in the holotypic hypoplastron. The range of differences does not allow that option. Furthermore, as additional araripemydids on Africa and elsewhere come to light, poorly established taxa such as *T. decorata* will only invite future taxonomic problems. We therefore regard the genus *Taquetochelys* and species *T. decorata* as *nomina dubia*.

Description

The skull and postcranial skeleton of *Laganemys* are well exposed and form the basis for this description. Computed-tomographic (CT) imaging of the cranium has revealed internal structure (Fig. 14.5). Future work on the cranium will include details of its neurovascular passages and endocranial volume. Future imaging of the postcrania will reveal structures currently obscured by matrix including the internal form of the shell, girdles and proximal limb bones.

We use "Romerian" rather than veterinarian terminology for orientation (e.g., "anterior" vs. "cranial") and refer to a trunk vertebra as a "dorsal" rather than a "thoracic"

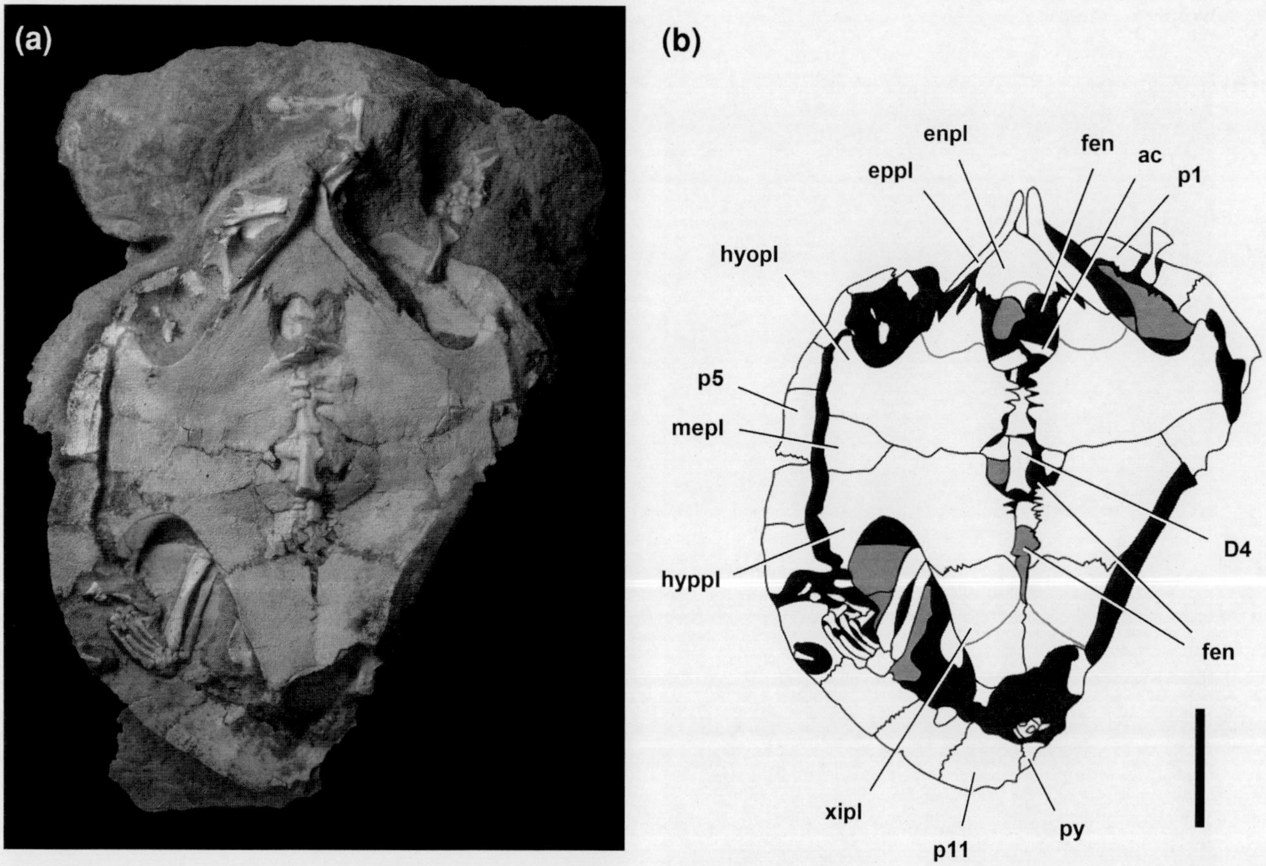

Fig. 14.13 Plastron of *Laganemys tenerensis* gen. et sp. nov. (MNN GAD28) in ventral view. **a** Photograph. **b** Line drawing. Scale bar = 3 cm. Abbreviations: *ac* acromion of the scapula; *D4* dorsal vertebra 4; *enpl* entoplastron; *eppl* epiplastron; *fen* fenestra; *hyopl* hyoplastron; *hyppl* hypoplastron; *mepl* mesoplastron; *p1, 5, 11* peripheral 1, 5, 11; *py* pygal; *xipl* xiphiplastron

(Romer 1956; Wilson 2006). In the skull, we follow the anatomical terminology summarized by Gaffney (1972) but express all terms in English. For the subset of specialized cranial structures sometimes expressed in Latin (Gaffney 1972), the Latin equivalents are given in parentheses on first usage. Neither standardized veterinarian terminology nor Latin have been shown to enhance accuracy in anatomical communication over Romerian equivalents (Wilson 2006).

Skull: The skull of *Laganemys* is proportionately longer than in *Araripemys*, due largely to the extension in the middle portion of the skull. In lateral view of the skull of *Laganemys* (Fig. 14.3), the orbit and cheek emargination are separated by a long sheet of bone. In *Araripemys*, in contrast, the posterior margin of the orbit and anterior margin of the cheek emargination are near one another (Meylan 1996). Likewise, in lateral view in *Laganemys* the cheek emargination is situated entirely anterior to the temporal emargination (Fig. 14.4), whereas in *Araripemys* they broadly overlap. As a result the trochlear process of the pterygoid (processus trochlearis pterygoidei) is situated anterior to the temporal fossa (fossa temporalis superior) in *Laganemys* (Fig. 14.4), whereas in *Araripemys* it is exposed in dorsal view of the skull through the temporal fossa (Meylan 1996).

The depth of the anterior end of the cranium in *Laganemys* is very distinctive as well. The cranium is deepest at the orbits (Fig. 14.2; Table 14.1) and increases approximately by 20% at its anterior end (Fig. 14.4); this is not the case in *Araripemys* (Meylan 1996). The cheek emargination is dorsoventrally deep in both *Laganemys* and *Araripemys*, but in the former the embayment extends farther anteriorly, resulting in a longer pointed posterior process on the maxilla (Fig. 14.4). The cheek emargination is also visible as an embayment along the lateral side of the skull in dorsal view (Fig. 14.6). The temporal emargination is proportionately narrower in *Laganemys* (Fig. 14.6), whereas in *Araripemys* the fossa is roughly as long as wide in dorsal view of the skull (Meylan 1996).

The subcircular orbits are slightly longer anteroposteriorly than deep and are directed laterally as much as dorsally as in *Araripemys* (Figs. 14.4, 14.6). The aspect of the orbits that sets *Laganemys* apart is the gentle telescoping, or eversion, of the posterior half of the orbital margin.

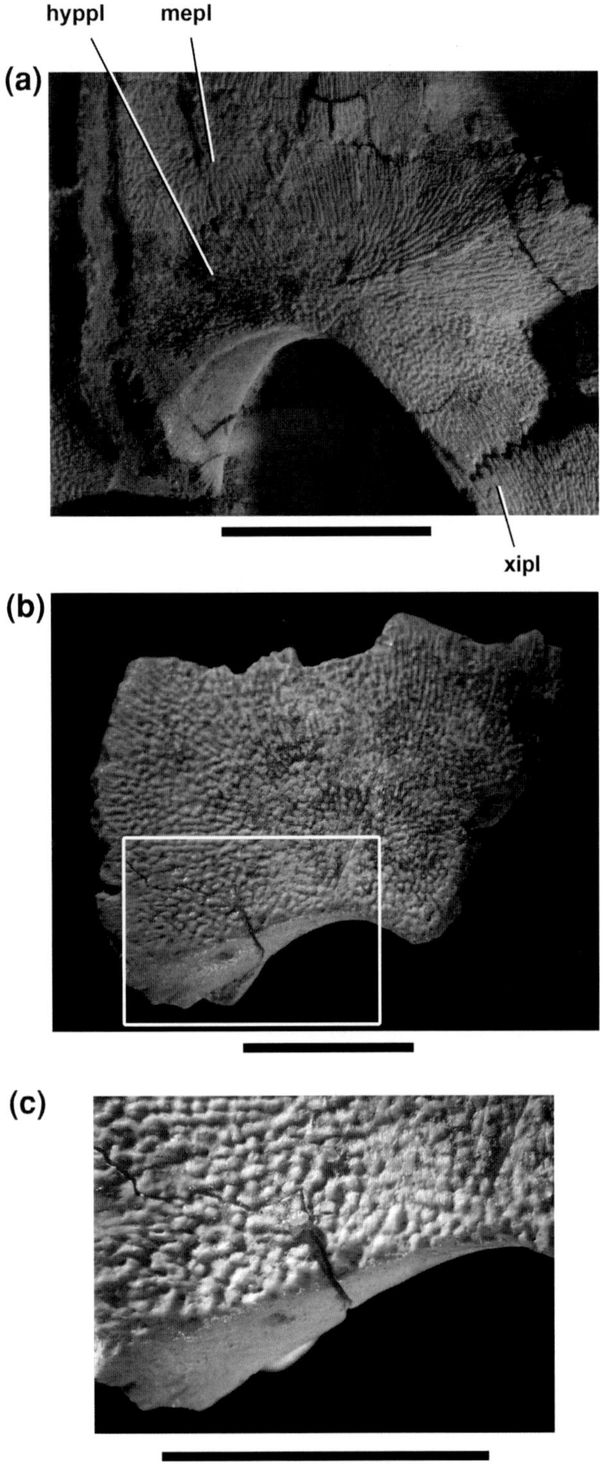

Fig. 14.14 Hypoplastron of *Laganemys tenerensis* gen. et sp. nov. (MNN GAD28) and *Taquetochelys decorata* (MNHN GDF847) in ventral view. **a** Right hypoplastron of *Laganemys tenerensis*. **b** Right hypoplastron of *Taquetochelys decorata*. **c** Magnified view of posterior margin of the right hypoplastron of *Taquetochelys decorata*. Scale bar = 2 cm in **a** and 1 cm in **b** and **c**. Abbreviations: *hyppl* hypoplastron; *mepl* mesoplastron; *xipl* xiphiplastron

The skull roof is composed of the usual set of paired roofing elements; nasals are absent as in other pelomedusoids (Meylan 1996; Gaffney et al. 2006). The supraoccipital crest (crista supraoccipitalis) extends as far posteriorly as the posterolateral process, the latter formed by the squamosal and opisthotic (Fig. 14.6). In both *Laganemys* and *Araripemys*, the supraoccipital crest and posterolateral processes extend far posterior to the occipital condyle.

In ventral view the labial ridge formed by the premaxilla and maxillae is narrower than in *Araripemys* (Fig. 14.7). The labial ridges of upper and lower jaws (Fig. 14.9), thus, are more V-shaped than U-shaped. The basisphenoid contribution to the palate equals or exceeds that of other palatal bones, none of which bear teeth. The ramus of the lower jaw (Figs. 14.4, 14.9) is considerably longer and more slender in *Laganemys* as compared to *Araripemys* (Meylan 1996; Gaffney et al. 2006).

Dorsal skull roof: The *premaxilla* is fused with its opposite medially and to the maxilla posteriorly, obliterating most of its external sutures. A portion of the premaxilla–maxilla suture appears to be preserved on the right side, marking the edge of a transversely narrow, deep bone in external view. Its palatal sutures and surface are recessed above and anterior to the labial ridge and are obscured by matrix (Figs. 14.5, 14.7). Exposed sutural contacts, thus, are limited to the maxilla. In anterior view the premaxillae join along the ventral margin of the external naris (apertura narium externa) and form a subtriangular dorsomedian process that partially divides the opening. The external surface of the bone is pitted. The labial ridge curves dorsally near the midline, forming with its opposite a V-shaped notch to receive the pointed anterior end of the mandible (Fig. 14.9). *Araripemys* does not appear to have a similar premaxillary notch (Meylan 1996).

The *maxilla* contacts the premaxilla and prefrontal anteriorly and the jugal posteriorly, forming the lateral margin of the external naris and the ventral margin of the orbit. The anterior end of the maxilla extends dorsally along the anterior margin of the orbit as a slightly raised, tapering prong. In *Araripemys*, the maxilla extends dorsally adjacent to the external naris, not the orbit (Meylan 1996). On the posterior side of the orbit, the maxilla-jugal suture steps ventrally before passing posteriorly (Fig. 14.4). The posterior position of the maxilla-jugal suture on the orbital margin in *Laganemys* differs from the ventral position of the suture in *Araripemys* (Meylan 1996). The maxilla forms most of the posterior process under the cheek emargination. The external surface is pitted above the labial ridge, with some of the pits forming shallow concavities under the orbit (Fig. 14.4). The posterior process is marked by low posteroventrally inclined ridges and some small foramina.

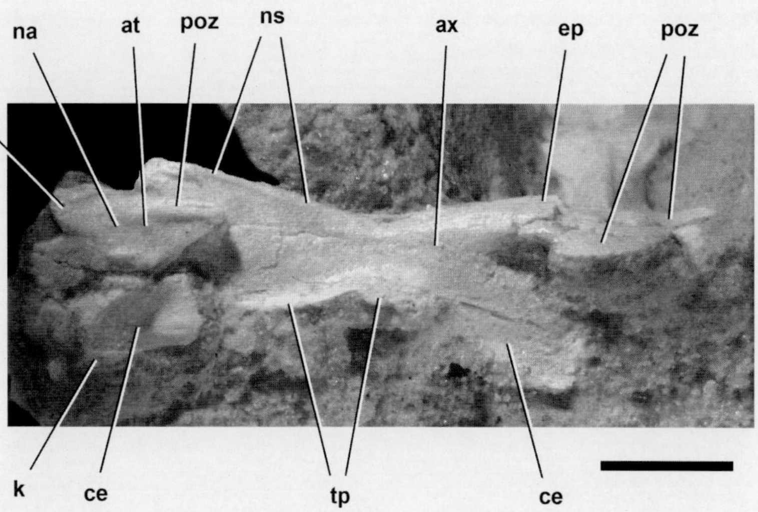

Fig. 14.15 Atlas and axis of *Laganemys tenerensis* gen. et sp. nov. (MNN GAD28) in left lateral view. Scale bar = 5 mm. Abbreviations: *at* atlas; *ax* axis; *ce* centrum; *ep* epipophysis; *k* keel; *na* neural arch; *ns* neural spine; *poz* postzygapophyses; *prz* prezygapophysis; *tp* transverse process

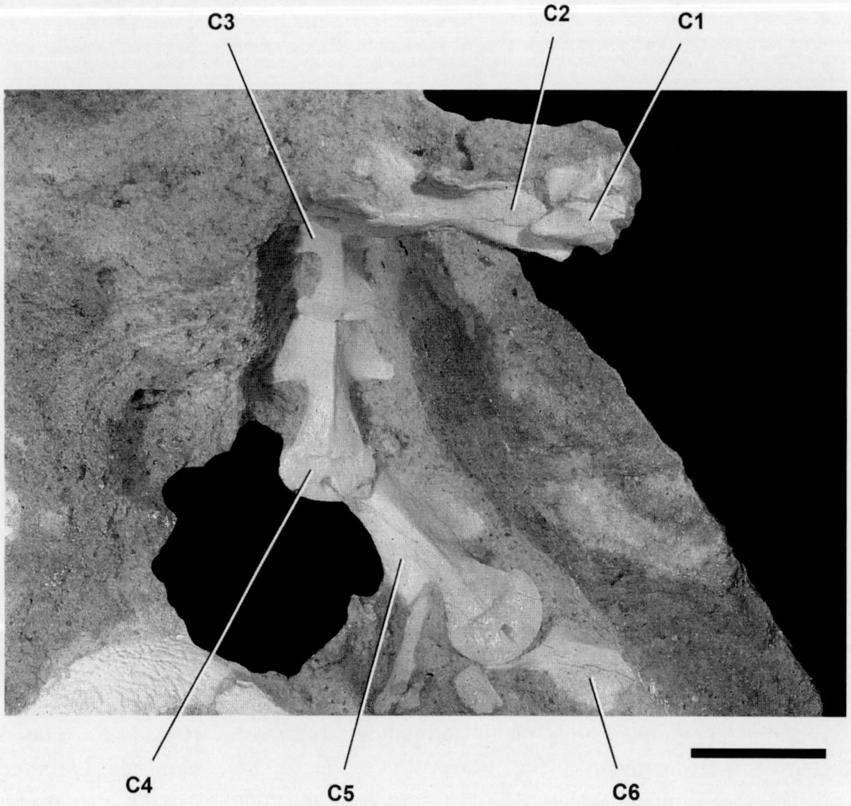

Fig. 14.16 Anterior and mid cervical vertebrae of *Laganemys tenerensis* gen. et sp. nov. (MNN GAD28) in lateral and dorsal view. Scale bar = 1 cm. Abbreviations: *C1–6* cervical vertebrae 1–6

In ventral view, a portion of the maxilla-palatine suture is exposed on the left side of the palate. The maxilla shares the triturating surface along its length with the palatine. In *Araripemys*, in contrast, the narrower triturating surface is limited to the maxilla (Meylan 1996). The internal nares together form a subtriangular opening with a broad, transverse posterior margin. The maxilla forms the lateral side of each internal naris and is separated from the footplate of the vomer by the palatine.

The *jugal* is a strap-shaped bone that extends from the posterior margin of the orbit to the deepest notch of the cheek emargination (Fig. 14.4). Its contacts include the postorbital dorsally, the maxilla ventrally and the quadratojugal posteriorly. Unlike *Araripemys*, the jugal margin of the orbit is distinctly everted. The jugal contacts the base of the trochlear process of the pterygoid but does extend onto the palate as in *Araripemys* (Meylan 1996; Gaffney et al. 2006).

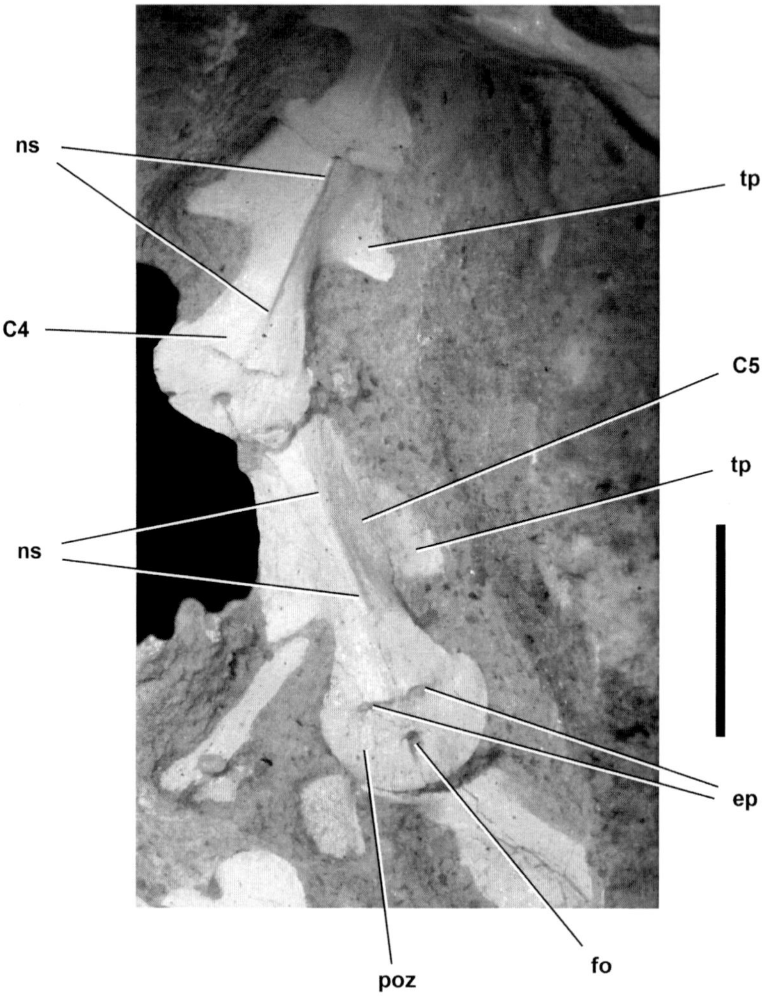

Fig. 14.17 Cervical vertebrae 4 and 5 of *Laganemys tenerensis* gen. et sp. nov. (MNN GAD28) in dorsal view. Scale bar = 1 cm. Abbreviations: *C4, 5* cervical vertebra 4, 5; *ep* epipophysis; *fo* foramen; *ns* neural spine; *poz* postzygapophyses; *tp* transverse process

The *postorbital*, like the jugal, is strap-shaped, extending from the orbit to the cheek emargination. The postorbital contacts the frontal and parietal dorsomedially and the jugal ventrally. The posterior tip of the postorbital contacts the anterior tip on the quadratojugal along the cheek emargination. The orbital margin formed by the postorbital is telescoped like that of the adjacent jugal. Unlike *Araripemys* and most pelomedusoids, the postorbital does not reach the superior temporal fossa (Fig. 14.6).

The *quadratojugal*, which is best preserved on the right side, is a diamond-shaped bone contacting the parietal dorsally and the squamosal and quadrate posteriorly. Ventrally the quadratojugal borders the posterior half of the cheek emargination. Dorsally it fails to separate the parietal and squamosal, which meet along the temporal emargination, an unusual sutural configuration among pleurodires.

The *squamosal* contacts the quadratojugal and a slender process of the parietal anteriorly and the quadrate ventrally and the exoccipital (Fig. 14.4). It forms the majority of the lateral rim of the temporal emargination and extends posteriorly along the posterolateral side of the skull. In this region it joins the opisthotic to form a prominent posteriorly projecting process, as in *Araripemys* and other pelomedusoids such as *Euraxemys* (Meylan 1996; Gaffney et al. 2006).

The *prefrontal*, a subtriangular plate that is thin in cross-section (Fig. 14.5), forms the posterior margin of the external naris and the anteromedial margin of the orbit (Fig. 14.6). The prefrontal is gently transversely arched along the narial opening, where it contacts the maxilla. It extends posterodorsally as a tapering plate, its distal tip overlapping the frontal and terminating in a V-shaped posterior suture. Comparatively, *Araripemys* has been shown with both a V-shaped (Meylan 1996) and interdigitating (Gaffney et al. 2006) prefrontal–frontal suture. The orbital margin of the prefrontal, which is not telescoped, is restricted to the anteromedial margin by an ascending process of the maxilla. This differs from the condition in *Araripemys*, where the prefrontal borders the ventral orbital margin (Meylan 1996; Gaffney et al. 2006). The contacts of the prefrontal within the orbit are visible only in the CT scan.

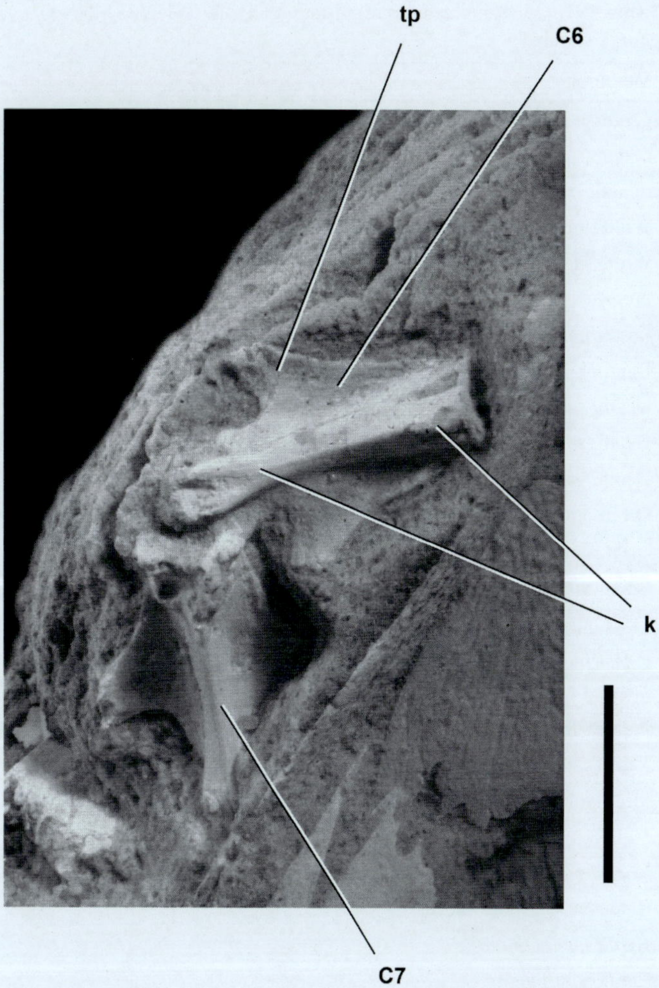

Fig. 14.18 Cervical vertebrae 6 and 7 of *Laganemys tenerensis* gen. et sp. nov. (MNN GAD28) in ventral view. Scale bar = 1 cm. Abbreviations: *C6, 7* cervical vertebra 6, 7; *k* keel; *tp* transverse process

The *frontal* is overlapped anteriorly by the prefrontal, abutts laterally against the postorbital, and meets the parietal posteriorly along an interdigitating suture (Fig. 14.6). The frontoparietal suture is near the posterior margin of the orbit in *Araripemys*, whereas in *Laganemys* it is located approximately one-third of the distance along the dorsal skull roof between the orbit and temporal emargination. The orbital rim formed by the frontal is gently everted.

The *parietal* extends across the posterior two-thirds of the skull roof from its interdigitating contact with the frontal anteriorly to the sides of the supraoccipital on the supraoccipital crest posteriorly (Fig. 14.6). In parasagittal section, both the frontal and parietal are relatively thick (Fig. 14.5). The parietal meets the postorbital and quadratojugal laterally along a nearly straight suture, with its distal tip contacting the squamosal. The posterior margin of the parietal is deeply embayed by the temporal emargination, forming a distinctive posterolateral process. In *Araripemys* there is no parietal-quadratojugal or parietal-squamosal contact and no development of a posterolateral process (Meylan 1996; Gaffney et al. 2006).

Palate: The outer margin of the palate, including nearly the entire triturating surface for the mandible, is formed by the premaxilla and maxilla. The remainder of the palate is composed of the quadrate, pterygoid, palatine and vomer (Fig. 14.7) and slopes posteroventrally at approximately 15° from the internal naris (apertura narium interna) to the ventral surface of the occiput (Fig. 14.5).

The internal narial opening is at least partially divided posteriorly by the *vomer*, a bone that was not preserved in the acid-prepared crania of *Araripemys* (Meylan 1996; Gaffney et al. 2006). The vomer is absent in many extant pelomedusids (Gaffney 1979; Gaffney et al. 2006), but this does not appear to be the case among araripemydids. In *Laganemys* the posterior portion of the bone is preserved as a tongue-shaped median element, presumably the posterior ends of fused right and left vomers (Fig. 14.7). The footplate appears to overlap right and left palatines, although their contact is developed as an interdigitating suture.

The *palatine* forms the posterior margin of the internal naris, which is depressed centrally where the palatine and footplate of the vomer meet (Fig. 14.7). The palatine

Table 14.1 Measurements (mm) of the holotypic skeleton (MNN GAD28) of the mid Cretaceous pelomedusoid turtle *Laganemys tenerensis* gen. et sp. nov.

Measurement		Code
Cranium		
Length, premaxilla to distal tip of supraoccipital crest	40	I
Length, premaxilla to occipital condyle	37	A
Width, maxillary flange to opposite	20	B
Width, across quadrate condyles	20	–
Preorbital length	3	–
Postorbital length, orbital margin to tip of supraoccipital crest	31	–
Depth, orbital roof to maxillary labial ridge	10	–
Depth, parietal roof to maxillary labial ridge	8	–
Depth, supraoccipital crest to quadrate condyle	13	G
Depth, supraoccipital crest to occipital condyle	10	K
Orbit, anteroposterior diameter	6	D
Orbit, dorsoventral diameter	5	J
Interorbital width	3	C
External nares, width	4	E
Posterior palate, width between inferior temporal fossae	9	N
Internal nares, width	6	F
Premaxilla–maxilla labial ridge, length	21	–
Lower Jaws		
Midline length	28	A
Symphysis to coronoid process, length	20	B
Mid length, depth	3	–
Ramus at coronoid process, width	2	C
Hyoid Ossification		
Ceratohyal length	(26)	
Axial Skeleton		
Carapace, maximum length	144	
Carapace, midline length	137	
Carapace, maximum width	140	
Plastron, maximum length	130	
Plastron, maximum width	101	
Plastral bridge, minimum anteroposterior length	38	
Cervical 1, intercentrum length	5	
Cervical 2, centrum length (without odontoid)	16	
Cervical 4, centrum length	17	
Cervical 6, centrum length	17	
Cervical 7, centrum length	17	
Dorsal 2, centrum length	15	
Dorsal 3, centrum length	13	
Dorsal 4, centrum length	13	
Caudal 4, centrum length	5	
Caudal 6, centrum length	4	
Caudal 12, centrum length	2	
Caudal 19, centrum length	2	
Caudal 21, centrum length	1	

(continued)

Table 14.1 (continued)

Measurement	Code
Appendicular Skeleton	
Radius length	16
Ulna length	16
Metacarpal 1 length	(5)
Metacarpal 2 length	6
Metacarpal 4 length	8
Metacarpal 5 length	7
Manual digit I, phalanx 1 length	4
Manual digit II, phalanx 2 length	3
Manual digit II, ungual length	6
Manual digit III, phalanx 2 length	4
Manual digit III, ungual length	7
Manual digit IV, phalanx 1 length	3
Manual digit IV, phalanx 2 length	4
Manual digit IV, ungual length	6
Manual digit V, ungual length	4
Tibia length	27
Metatarsal 1 length	13
Metatarsal 2 length	15
Metatarsal 3 length	17
Metatarsal 4 length	15
Pedal digit I, phalanx 1 length	5
Pedal digit I, ungual length	7
Pedal digit II, phalanx 1 length	4
Pedal digit II, ungual length	6
Pedal digit III, phalanx 1 length	4
Pedal digit III, ungual length	5

Measurements of paired structures are from the right side, except those from the forelimb. In right column, letter code after cranial measurements corresponds to measurement diagrams for the cranium and lower jaws in Gaffney et al. (2006, Appendices 4, 8); in same column, symbol "–" indicates no corresponding letter code

descends laterally to meet the maxilla along the margin of the palate. At their junction, the palatine sends a prong-shaped process anteriorly over the maxilla along the lateral side of the internal naris. The palatine is overlapped by a very short triangular process of the maxilla, before the suture turns posteriorly along the triturating surface. In *Araripemys* the palatine does not contribute to the triturating surface, which is composed solely of the maxilla (Meylan 1996; Gaffney et al. 2006). In *Laganemys* the medial one-half of the triturating surface is formed by the palatine, a subtle lingual ridge apparent on the palatine indicating the medial margin of the keratinous sheath.

The palatines meet along an interdigitating median suture, which extends posteriorly to meet the apex of the basisphenoid and the anteromedial corners of the pterygoids at a single point (Fig. 14.7). A posterior palatine foramen (foramen palatinum posterius) is located to each side of the midline within the palatine near the palatine-pterygoid suture. The left foramen is positioned slightly more anteriorly than the right foramen, which lies on the anterior side of the interdigitating palatine-pterygoid suture. This foramen is larger and situated farther laterally on the palate in *Araripemys* (Meylan 1996; Gaffney et al. 2006). The palatine extends posteriorly in the mid palate, forming a V-shaped suture with the pterygoid before reaching the lateral side of the palate. Contact in this region with the jugal is obscured by matrix.

The *pterygoid* contacts the palatine anteriorly along a suture, most of which is interdigitated. The long medial suture with the basisphenoid appears to be fused (Fig. 14.7). Other contacts near the basisphenoid include the prootic and quadrate. Lateral contact with the jugal is partially exposed at the base of the trochlear process within the cheek emargination (Fig. 14.4). The trochlear process angles posteroventrally from the jugal-pterygoid suture, extending posteroventrally as a hatchet-shaped, pendant flange below the cheek emargination. The lateral surface is lightly textured and separated from the palatal surface by a distinct edge. The expanded distal end of the trochlear process is dorsoventrally convex.

A crescentic, flange of the pterygoid is elevated slightly from the palatal surface, lying between the basisphenoid and inferior temporal fossa (Fig. 14.7). The arcuate medial margin of this flange extends from the edge of the trochlear process and curves along the lateral margin of the basisphenoid and prootic, and then forms an interdigitating suture with the quadrate near the mandibular condyles (condylus mandibularis). This crescentic palatal surface is more clearly demarcated than in other pleurodires, with the medial edge invaginated into the basisphenoid.

The *quadrate* in lateral view forms the smooth curved wall of the tympanic fossa (cavum tympani) and contacts in this region the quadratojugal and squamosal (Fig. 14.2). In ventral view, quadrate forms the mandibular condyles and extends medially to contacts the prootic and opisthotic. A fissure, the incisure columella auris, is present on the ventral wall of the tympanic fossa. The fossa extends posteromedially into the braincase as the postotic antrum, an internal space hidden in lateral view by the posterior rim of the tympanic fossa.

The relatively flat mandibular condyles are deeply cleft, with the medial condyle facing ventrolaterally at approximately 45° to the sagittal plane (Figs. 14.7, 14.8). The lateral condyle is flat and faces anteroventrally at approximately 45° to a transverse plane. The condition is similar to that in *Araripemys* (Meylan 1996; Gaffney et al. 2006). Contact with the pterygoid is initiated just above the condyles and extends medially until reaching the prootic.

Braincase: The *prootic* contacts the quadrate laterally, the opisthotic posteriorly, and a small margin of the supraoccipital posteromedially and the basisphenoid medially (Figs. 14.6, 14.7, 14.8). The prootic, exposed as a crescent between the quadrate and basisphenoid in ventral view, is marked by the posterior opening of the internal carotid artery (foramen posterior canalis carotici interni), the principal arterial supply to the brain. Although the prootic is fused to adjacent elements, the fused sutures are still discernible.

In dorsal view, the *opisthotic* contacts the supraoccipital medially, the prootic anteriorly, and the quadrate laterally. The opisthotic forms the core of a prominent paroccipital process (processus paroccipitalis) that extends posteriorly as far as the supraoccipital crest (Figs. 14.5, 14.6). In ventral view, the opisthotic is bordered laterally by the quadrate and medially by the exoccipital.

The *supraoccipital* forms the narrow apex of the relatively large, teardrop-shaped foramen magnum. The supraoccipital extends from this point on the foramen magnum posteriorly, forming the base and posterior tip of the supraoccipital crest. In sagittal view, the supraoccipital crest does not extend posteriorly as far as the paroccipital process and posterior extremity of the squamosal (Fig. 14.5). The very thin central portion of the crest thickens toward its ventral margin, which is concave and confluent with the dorsal border of the foramen magnum.

The *exoccipitals* form the prominent, thin sidewalls of the foramen magnum, joining the supraoccipital dorsally and the basioccipital ventrally. Between these contacts, the exoccipital extends posterolaterally along the proximal one-half of the paroccipital process. Gaffney et al. (2006) describe a small foramen opening on the exoccipital-opisthotic suture on the paroccipital process. This foramen is not present in *Laganemys*. In the body of the bone, however, are three foramina, a large posterior jugular foramen (foramen jugulare posterius) and two smaller foramina for the hypoglossal nerve (foramina nervi hypoglossi; Fig. 14.8). The exoccipital contacts the basisphenoid just anterior to these foramina. The exoccipital extends posteriorly to form a rounded occipital condyle (condylus occipitalis) that is crescentic in posterior view. Between opposing crescentic condyles, lies a central fossa. As in Pelomedusidae, the occipital condyles are composed exclusively of the exoccipitals. In *Araripemys*, in contrast, the basisphenoid forms a small median ventral condyle. The differing structure of the occipital condyles between *Laganemys* and *Araripemys* is one of the few features which would link one of these genera with another pelomedusoid and thus is indicative of homoplasy.

The *basioccipital* is exposed ventrally as a transversely broad and convex, diamond-shaped bone. Anteriorly near its contact with the basisphenoid, there is a shallow median keel. The keel dissipates posteriorly, with a pair of low wedge-shaped basal tubera (tuberculum basioccipitale) developed in conjunction with the exoccipitals. The basioccipital has a small triangular median process extending between the tubera, but it stops short of the occipital condyles (Fig. 14.8). In *Araripemys*, in contrast, the basioccipital has been shown to contribute the ventral one-third of the occipital condyle (Gaffney et al. 2006). This portion of the skull is well preserved and highlights a significant structural difference.

The *basisphenoid* forms a long, narrow-shaped ramus that extends along the midline for over half the skull length in ventral view, dividing the pterygoids to each side and establishing a point contact with the palatines (Fig. 14.7). The basisphenoid is relatively long in *Laganemys* and *Araripemys* compared to other pleurodires.

Lower jaw: A complete mandible was present and consists of six articulated bones, the dentary, coronoid, surangular, angular, prearticular, and articular (Figs. 14.4b,

14.9). The splenial is not present, a condition common to all pelomedusoids. The conjoined lower jaws have the form of a slightly bowed, pointed arch in dorsal or ventral view (Fig. 14.9). The articular surface for the quadrate condyles (area articularis mandibularis) is set slightly medial to the lateralmost arc of the dentary ramus. The dentary ramus is proportionately long to match the proportionately long cranium. The coronoid region, for example, is positioned in the posterior 25% of the lower jaw, and the depth of the dentary ramus at mid length is only 3 mm (Table 14.1). The lower jaw differs from that of *Araripemys* by its much longer proportions and ventral curvature, pointed anterior end, and absence or reduction of internal and external foramina.

The *dentary* ramus is slender with a very thin triturating edge forming its dorsal margin. In lateral view the axis of the ramus curves ventrally toward the symphysis, its anterior end squared and slightly expanded to form a "chin" ventrally and a pointed apex dorsally. The dorsal apex fits into a notch in the upper jaws between the premaxillae. A suture separates the dentaries at the symphysis as in *Araripemys* and some pelomedusids (Meylan 1996). As in *Araripemys*, the dentary extends far posteriorly, terminating in a pair of tongue-shaped processes, one at the root of the coronoid rise and a second longer one lateral to the adductor fossa (Fig. 14.4b). Sutural contacts of the dentary include the coronoid, surangular and angular. In medial view, the triturating surface is well developed along the entire upper margin of the dentary and is broadest at mid length, angling approximately 45° ventromedially. Anteriorly and posteriorly, the triturating surface narrows in width and is more strongly inclined. Meckel's groove (sulcus cartilaginis meckelii) first appears near the symphysis and deepens posteriorly.

A subtriangular *coronoid* is exposed laterally and, to a greater extent, medially. It forms the top of the coronoid rise and encloses Meckel's canal medially. The posterior margin of the coronoid joins the prearticular on the anterior rim of the adductor fossa. Sutural contacts of the coronoid include the dentary, surangular and prearticular.

The *surangular* is best exposed in lateral view, where it contacts the dentary, coronoid, angular and articular. The auriculotemporal foramen (foramen nervi auriculotemporalis), if correctly identified as such, is very small and located near the dorsal margin of the surangular (Fig. 14.4b, fo). In most basal pelomedusoids and *Araripemys* (Meylan 1996), in contrast, the auriculotemporal foramen is considerably larger and located on the lateral aspect of the surangular, serving as an exit from the adductor fossa for a branch of the chorda tympani nerve (Gaffney 1972, Fig. 17). *Foxemys* and all later pelomedusoids, however, lack this foramen (Gaffney et al. 2006), and so there remains some doubt regarding the identity of the small, dorsally positioned foramen in *Laganemys*. At the base of the coronoid process, a horizontal groove leads anteriorly to the sutural triple-junction between the surangular, dentary and coronoid (Fig. 14.4b). A gap at this sutural junction may have served as another opening in the sidewall of the mandible in *Laganemys*.

The *angular* is a strap-shaped bone best exposed in medial view along the ventral margin of the jaw. The angular tapers to a slender process anteriorly at mid length along the dentary ramus, enclosing Meckel's canal (fossa meckelii) posteriorly and tapering to a point along the ventral margin of Meckel's groove anteriorly. Posteriorly the angular maintains its width as it extends under the articular at the jaw articulation. Sutural contacts include the dentary, surangular, prearticular and articular. Along the angular–prearticular suture on both sides, a small internal mandibular foramen (foramen intermandibularis caudalis) is visible.

The *prearticular* is another strap-shaped postdentary bone exposed only in medial view of the lower jaw. Its contacts include the dentary, angular, coronoid and articular. The prearticular forms the medial wall of the adductor fossa and extends to the end of the lower jaw under the articular.

The *articular* is a small wedge-shaped bone that contacts the angular, surangular and prearticular (Fig. 14.9). The angular and surangular overlap the articular laterally and remain as a distinctive flange at the end of the lower jaw, separated from the articular by a suture filled with matrix. This angular–surangular flange appears to contact the lateral quadrate condyle. The articular, which is positioned medial to this flange, has an articular surface that is angled posteroventrally at approximately 45° from the horizontal, as seen in medial or lateral view of the lower jaw. The articular surface is ovate, a little deeper than broad and divided into two transversely concave facets by a rounded median keel. The keel articulates with the groove between the flat quadrate condyles. Below the articular surface, the posterior foramen for the chorda tympani nerve (foramen posterius chorda tympani) enters the articular to access the adductor chamber (Gaffney 1972).

Hyoid: A pair of curved, rod-shaped bones is preserved in association ventral to the cervical series (Fig. 14.10). We identify these as ceratohyals (= cornua), which in turtles and other reptiles are often curved, rod-shaped bones that broaden and flatten slightly toward their proximal articular ends (Romer 1956). The distal end of the ceratohyal is rod-shaped and more slender than the flattened proximal end, the articular head of which appears to have broken away.

The ceratohyal measures 23 mm in length and may have been 26 mm long as a complete element, or approximately 65% of skull length (Table 14.1). The ceratohyal in *Araripemys* has a more robust shaft and is proportionately shorter, measuring approximately 50% of skull length (Gaffney et al. 2006, Fig. 31B).

Shell ornamentation: A radiating ridge-and-sulcus texture is present on most of the external surfaces of the carapace and plastron. Each ridge or sulcus is spaced approximately 0.5 mm apart and typically radiates from a central locus on each plate (Figs. 14.11, 14.12, 14.13a, 14.21). The ridge-and-sulcus texture also occurs on the ventral surface of the peripherals. On some of the peripherals, the texture is limited to the outer margin; adjacent to the bridge of the plastron, the texture covers the entire ventral surface (Figs. 14.14a, 14.23). The ventral surface of costal and neural plates on the interior aspect of the shell is smooth. The ridge-and-sulcus texture, thus, is present on surfaces with external exposure, and a similar texture is present on the external surfaces of the skull. In *Araripemys* the external texture on the shell is pitted to a greater degree, although we regard fine-grained pit and ridge-and-sulcus texture as a synapomorphy of the Araripemydidae. The more subdued shell textures among trionychids are easily distinguished from that in araripemydids. Nonetheless, araripemydids and trionychids share similarities in shell ornamentation, reduction or loss of scutes, and a relatively flat shell profile, all of which may constitute adaptations to an aquatic lifestyle.

Carapace: The *carapace* is complete except for a left portion of the posterior half that was broken away and lost upon discovery of the specimen. The carapace and plastron are in natural articulation with only minor disarticulation and dorsoventral crushing. Minor anteroposterior shortening has occurred with some imbrication of costals 3 and 4 and 7 and 8 (Fig. 14.11). Minor transverse flattening has occurred, creating gaps laterally at the plastral bridge and in the midline between the right and left hypo- and hyo-plastron (Fig. 14.13). When these postmortem distortions are removed, the carapace has a distinctly oval shape with two low convexities along the margin of peripherals 1–3 and greatest breadth at peripheral 7 (Fig. 14.2). The carapace in *Araripemys* differs from *Laganemys* with its more circular profile, deeper nuchal embayment, greater posterior extension relative to the hind limbs, presence of a series of fenestrae, and retention of a small ninth neural (Fig. 14.1).

In *Laganemys* the length of the carapace (144 mm) slightly exceeds its width (140 mm) as preserved (Table 14.1), and the vertical height within the shell appears to have been approximately 20 mm. The carapace, which has a thickness of approximately 1 mm, is composed of 1 nuchal, 11 paired peripherals, 8 neurals, 8 paired costals, 1 suprapygal, and 1 pygal (Figs. 14.2, 14.11).

The broad *nuchal* plate has a deep anterior embayment (Fig. 14.12). This feature is also present in *Araripemys* (Fig. 14.1) and is included among synapomorphies of Araripemydidae. The condition of *Laganemys* differs from *Araripemys* by retention of the primitive nuchal-peripheral 1 contact on the periphery of the carapace, which excludes costal 1 from the margin of the carapace. In *Araripemys* costal 1 forms a short section of the anterior margin of the carapace, separating the nuchal and peripheral 1. We regard the condition in *Araripemys* as an autapomorphy for the genus. The nuchal in *Laganemys* has a unique boomerang shape, with a transverse span equal to one-fourth the width of the carapace (Figs. 14.2, 14.11, 14.12). The broader nuchal embayment in *Araripemys* incorporates all of the margin of peripheral 1, the nuchal retaining a subtrapezoidal shape (Fig. 14.1). Gaffney et al. (2006) regarded nuchal shape in *Araripemys* as distinctive, along with its well developed sutural contact with the first dorsal vertebra. That contact is not currently exposed in *Laganemys*.

The *neural series* is limited to eight plates with a neural formula of $6 > 6 > 4 < 6 < 6 < 6 < 6 > 5$. Neural 7 is hexagonal with 6 sutural contacts. Neural 8 is very small and probably has 5 contacts (neural 7, costals 7, 8) (Fig. 14.2). *Araripemys* has a neural formula of $6 > 6 > 4 < 6 < 6 < 6 > 6 > 3$ and shows less reduction of the posterior neural series than *Laganemys*. All specimens of *Araripemys* have nine neurals, with neural 8 hexagonal and neural 9 small and triangular (Fig. 14.1). A comparison of neural formulae among basal pleurodires suggests that *Laganemys* has lost neural 9 and reduced the size of neural 8. Neural 3 in both *Laganemys* and *Araripemys* only has one costal contact (costal 3), although it retains a hexagonal shape (Figs. 14.1, 14.2). The hexagonal neural 3 in pelomedusoids, in contrast, maintains two costal contacts that subdivide its lateral margin ($4 < 6 < 6 < 6 < 6 < 6 < 6 < 6$). Neural 3 in *Laganemys* is noteworthy as well for its shape; its long axis is oriented transversely (Fig. 14.2). The neural and pygal series in *Laganemys* are separated by a substantial median contact between opposing costal 8 plates (Figs. 14.2, 14.11). In *Araripemys* intercostal contact at the posterior end of the neural series is variable. Some specimens show substantial median contact between opposing costal 8 plates equivalent to that in *Laganemys*, whereas in other specimens such contact is limited to the convergence of sutures to a point junction (Fig. 14.1; Meylan 1996). Intercostal contact separating the posteriormost neural and suprapygal is absent in *Euraxemys* but present in some pelomedusids, podocnemidids and bothremydids (Gaffney et al. 2006).

The *pygal series* consists of a single suprapygal and pygal. The single suprapygal plate in *Laganemys* and *Araripemys* (Figs. 14.1, 14.2, 14.11, 14.20) may represent a derived condition, as two suprapygals are present in the basal pleurodire *Notoemys* (Meylan 1996; Rueda and Gaffney 2005).

The *costal series* consists of eight plates, the first of which is much broader than the others as in *Araripemys*

(Fig. 14.1). The added breadth gives costal 1 a subtriangular shape, whereas costals 2–8 have subparallel anterior and posterior margins (Figs. 14.11, 14.12). Costal 8 meets its opposite in the midline as described above. The distal end of a dorsal rib projects from the distal margin of each costal plate, as seen in dorsal and ventral views of the carapace (Figs. 14.2, 14.3, 14.11). The distal ends of the ribs extending from costals 1–4 insert between peripherals, whereas the rib ends on costals 5–8 insert into a notch in the middle of the adjacent peripheral. Unlike the carapace of *Araripemys*, which has a series of small fenestrae between costal and peripheral plates (Fig. 14.1), the rib ends in *Laganemys* insert into the peripheral series closing all such gaps. The presence of fenestrae in the carapace in *Araripemys* cannot be attributed to immaturity, given their uniform presence in many specimens. Likewise, we regard the holotypic and only known specimen of *Laganemys* as a mature individual, given the fusion of all sutures in the vertebral column and the coossification or tight articulation of cranial sutures. The presence of fenestrae in the carapace in *Araripemys* and the broad extension of the posterior margin of the carapace may represent derived conditions and constitute additional specialization related to an aquatic lifestyle.

The first plate in the *peripheral series* is crescentic (Fig. 14.12). The other peripherals vary in shape: they are rectangular along the anterior margin (peripherals 2, 3), triangular where they are separated by a rib end (peripherals 3–5), subquadrate with a medial notch to accommodate a rib end (peripherals 7–10), and subquadrate (peripheral 11). The anterior margin of the carapace is scalloped with a convex edge formed by peripheral 1 and a second convexity at the junction of peripherals 2 and 3 (Fig. 14.2).

A lightly incised *scute pattern* is visible on the carapace, indicating the presence of 5 vertebral, 4 pleural and probably 12 marginal scutes, a common condition among turtles. The scutes, like the underlying peripherals, are reduced along the anterior margin of the carapace; there is no evidence for nuchal or supramarginal scutes anterior to the vertebral series in either *Araripemys* or *Laganemys* (Figs. 14.1, 14.2).

Dorsal vertebra 3 and 4 are exposed through the second plastral fenestra in ventral view of the shell (Fig. 14.13). These centra, approximately 14 mm in length, have a rounded ventral keel and tab-shaped processes extending from each side of the posterior end of the centrum.

Plastron: The plastron preserves all elements on one side or the other except for the distal tip of the xiphiplastron. Measuring 130 mm in length, the plastron nearly equals the length of the carapace (Table 14.1). The 11 bones of the plastron include an entoplastron and paired epi-, hyo-, hypo-, meso-, and xiphi-plastra. There are three median plastral fenestrae, the anterior two much larger than the posteriormost. The fenestrae are separated by plastral plates, which join along deeply interdigitating sutures (Figs. 14.3b, 14.13). The plastral bridge to the carapace spans nearly one-third the length of the plastron. Postmortem compression of the shell has disarticulated both axillary and inguinal buttresses (Fig. 14.13).

The *epiplastron* is a narrow strap-shaped plate, the posterior two-thirds of which lies at a 45° angle from the midline forming the lateral plastral margin alongside the ento- and hyo-plastron (Figs. 14.3b, 14.13). The posterior tip inserts into a notch in the hyoplastron. The anterior one-third of the element curves anteriorly to meets its opposite at the midline. The conjoined epiplastra project approximately 1 cm anterior to the entoplastron and approximately 2 cm anterior to the nuchal embayment on the carapace. In width (6 mm) and length (2 cm), the conjoined epiplastral projection roughly corresponds to that of a cervical centrum (Table 14.1). The broad nuchal embayment and opposing epiplastral process were probably involved in enhancing support or mobility of the long cervical series. A strap-shaped epiplastron is a synapomorphy of Araripemydidae. A J-shaped epiplastron is an autapomorphy of *Laganemys*.

The *entoplastron* is pointed anteriorly and posteriorly in the midline and has gently convex lateral margins, and a pointed posteromedian process extending into the anterior plastral fenestra. The bone thus is shaped like an arrowhead (Figs. 14.3b, 14.13). The entoplastron contacts the hypoplastron laterally along an interdigitating suture, the last notch of which accommodates a slender entoplastral prong. In *Araripemys* there is only a single broader prong inserting into a notch in the hyoplastron (Meylan 1996).

The *hyoplastron* forms most of the anterior one-half of the plastron, expanding anteriorly to meet the ento- and epiplastron and contributing to the lateral border of the anterior and middle plastral fenestrae (Figs. 14.3b, 14.13). The anterior edge of the hyoplastron has a concave embayment to accommodate the forelimb that extends posteriorly nearly as far as the posterior rim of the anterior plastral fenestra. The thickened lateral (axillary) buttress extends anteriorly along the lateral margin of the carapace as far as costal 1 and attaches to the middle of peripheral 3. The anterior margin of the buttress is smooth. The hyoplastra meet in the midline along a deeply interdigitating suture, separating anterior and middle plastral fenestrae. This suture is short (approximately 1 cm) with four triangular prongs of similar length (5 mm) on each side. Posteriorly the hyoplastron meets the hypo- and meso-plastron along an interdigitating suture.

The *mesoplastron* is a pentagonal plate located on the lateral margin of the plastral bridge between the hyo- and hypo-plastron (Fig. 14.3b). Its lateral margin contacts peripherals 5 and 6, and its medial apex is located between the hyo- and hypo-plastron. In *Laganemys* the mesoplastron

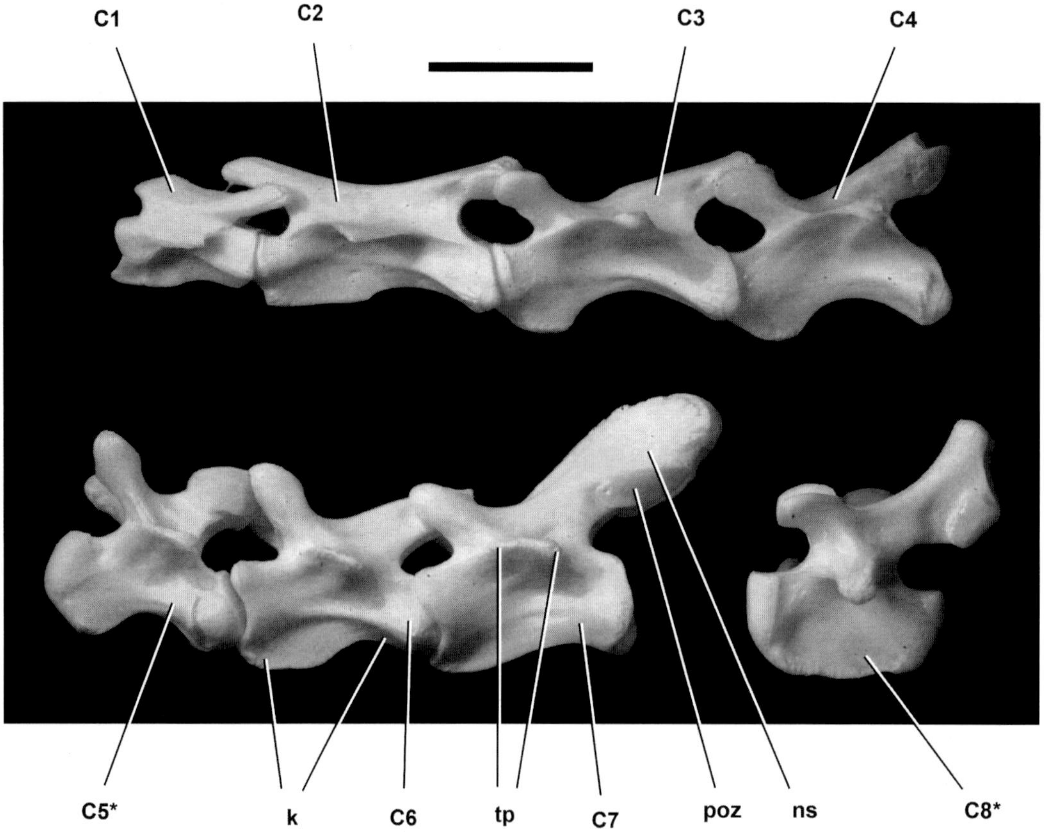

Fig. 14.19 Cervical vertebrae of the extant South American fringed turtle, *Chelus fimbriatus* (UCRC RV4), in lateral view. Asterisk indicates vertebrae with biconvex centra (C5, C8). Scale bar = 2 cm. Abbreviations: *C1–8* cervical vertebrae 1–8; *k* keel; *ns* neural spine; *poz* postzygapophyses; *tp* transverse process

is retained as a significant element of the plastron, forming approximately 30% of the plastral bridge (Fig. 14.3b). In *Araripemys*, in contrast, the mesoplastron is absent (Fig. 14.3a), although it is retained in many pelomedusoids.

The *hypoplastron* contacts the hyo- and meso-plastron anteriorly and xiphiplastron posteriorly along interdigitating sutures (Figs. 14.3b, 14.13, 14.14a, 14.23). The suture between the hypo- and xiphi-plastron runs from the posterior embayment to the posterior plastral fenestra. Except for an anterior marginal prong on the xiphiplastron, the suture between the hypo-, and xiphi-plastron is finely interdigitating, in contrast to the much longer interdigitating processes present in *Araripemys* (Fig. 14.3a). Like the hyo-, the hypo-plastron joins its opposite with six or seven triangular prongs of similar length (5 mm), separating middle and posterior plastral fenestrae. The posterior edge of the hypoplastron has a deeply concave embayment to accommodate the hind limb that extends anteriorly as far as the anterior edge of the posterior plastral fenestra. The posterior edge of the hypoplastron has a smooth margin that broadens as it approaches the lateral (inguinal) buttress, which extends as far posteriorly as peripheral 7 (Figs. 14.3b, 14.14a, 14.23).

The *xiphiplastron* forms the posterior one-fourth of the plastron, tapering in width posteriorly (Figs. 14.3b, 14.13, 14.23). The posterior palatal fenestra forms a narrow median fissure, separating the anterior one-third of the xiphiplastron from its opposite. More posteriorly, the xiphiplastron joins its opposite along a finely interdigitating median suture (Figs. 14.3b, 14.23). The posterior margin of the plastron follows a broad concave arc between prominent posterolateral corners of the xiphiplastra.

A lightly incised *scute pattern* is visible on the plastron that reveals the presence of scutes that once covered the plastron. In *Araripemys* a pair of anterior (humeral) scutes were reported on the anterior ends of the epiplastra (Fig. 14.3a; Meylan 1996), although later it was scored as a median intergular scute (Gaffney et al. 2006) as in extant turtles (Pritchard 1979). The most anterior scute on the plastron in *Laganemys* appears to be a median intergular scute (Fig. 14.3b.

Five pairs of scutes (humeral, pectoral, abdominal, femoral, anal; Pritchard 1979) cover the remainder of the plastron in both *Laganemys* and *Araripemys* (Fig. 14.3), although in some areas in *Laganemys* the scute margin is so faintly incised that it cannot be followed.

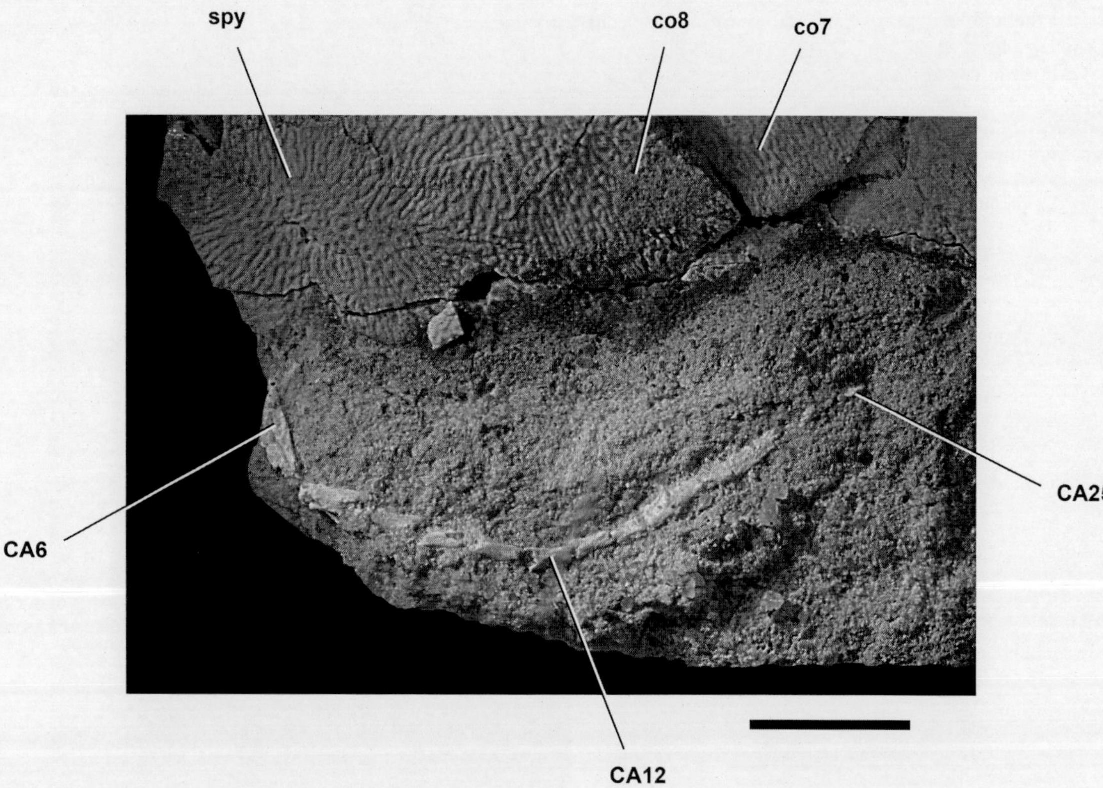

Fig. 14.20 Caudal vertebrae of *Laganemys tenerensis* gen. et sp. nov. (MNN GAD28) in dorsal view. Scale bar = 1 cm. Abbreviations: *CA6, 12, 25* caudal vertebra 6, 12, 25; *co7 8* costal 7, 8; *spy* suprapygal

Cervical vertebrae: The cervical vertebrae are preserved largely in articulation between the skull (removed during preparation) and the anterior hiatus of the shell. Cervical vertebrae 1, 2 and 4–7 are exposed in ventral view, whereas cervical vertebrae 3–6 and 8 are exposed in dorsal view.

The cervical vertebrae are joined along an S-shaped curve with greatest intervertebral angles between vertebrae with articular ends adapted for bending (Figs. 14.11, 14.13). The partially retracted position of the cervical series, thus, is a reflection of relative intervertebral mobility in life during retraction and extension of the head. The two vertebral joints with a high angle of excursion (approximately 90° cervicals 2 and 3; approximately 80°, cervicals 6 and 7) lie in transverse planes parallel to that of the shell as expected in a "side-necked" pleurodire. As in *Araripemys*, postatlantal cervical centra are elongate, roughly the same length, and approximately 25% longer than mid dorsal centra (Table 14.1). The cervical column in both *Araripemys* and *Laganemys* is approximately 90% of the midline length of the carapace, as measured from the nuchal embayment to the distal edge of the pygal (Figs. 14.1, 14.2).

The *atlas* has a short centrum approximately one-third the length of more posterior cervical centra. It has tab-shaped prezygapophyses, broad-based subtriangular postzygapophyses, and blunt transverse processes that rise from the base of the neural arch (Fig. 14.15). A low ridge is present on the dorsal edge of the prezygapophyseal process. The dorsal edge of the postzygapophysis is rugose for ligament attachment. The ventral aspect of the atlantal centrum is pinched anteriorly to form a low ventral keel. The form of the posterior centrum face is not well exposed but appears to be concave. The morphology of the atlas does not depart markedly from that in *Chelus* (Fig. 14.19).

The centrum of the *axis* is V-shaped in cross-section at mid length. A strong ventral keel arises at mid length along the centrum and gains in depth anteriorly, extending as a median prong under the atlantal centrum. The posterior portion of the axial centrum arches ventrally. The anterior face of the axial centrum is not well exposed but is probably convex and fitted to the concave posterior face of the atlantal centrum. The posterior face of the axial centrum is strongly convex. The axial centrum in *Laganemys*, thus, is biconvex, which also appears to be the case in *Araripemys* (Meylan 1996) and other pelomedusoids.

Broad-based, thin transverse processes projects from the anterior two-thirds of the axial centrum (Fig. 14.15). In dorsal view, the transverse process is subtriangular and broadest posteriorly. In lateral view, the transverse process is canted posteroventrally at its attachment to the neural arch and is deflected just below the horizontal as it extends laterally. The axial neural spine, like the ventral keel, arises

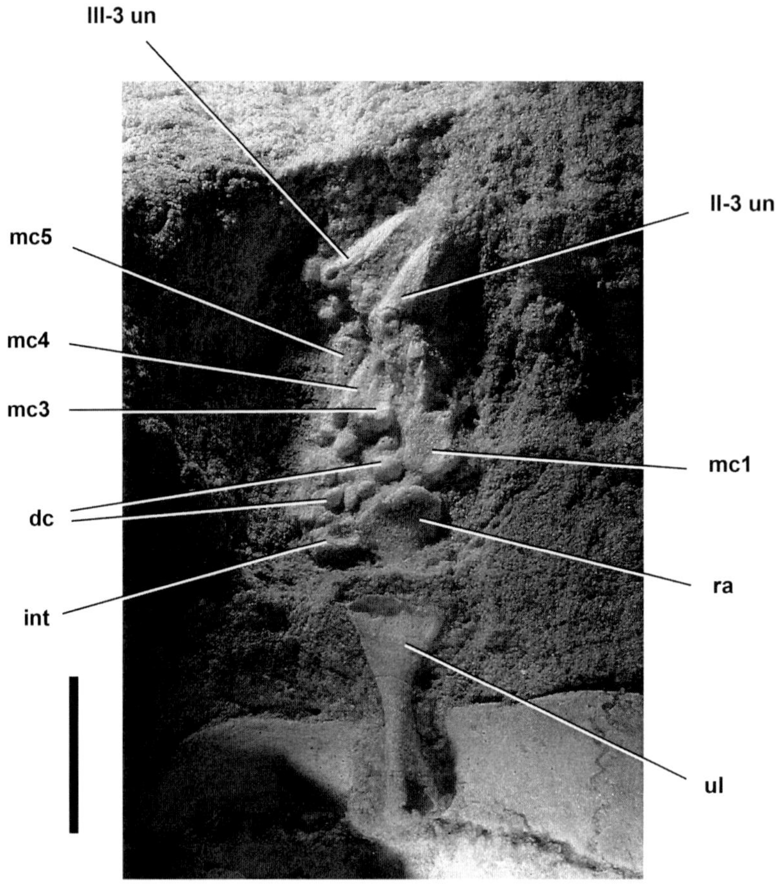

Fig. 14.21 Left forelimb of *Laganemys tenerensis* gen. et sp. nov. (MNN GAD28) in dorsal view. Scale bar = 1 cm. Abbreviations: *II*, *III* digit II, III; *dc* distal carpal; *int* intermedium; *mc1*, *3–5* metacarpals 1, 3–5; *ra* radius; *ul* ulna; *un* ungual

at mid length along the centrum and gains in depth anteriorly, extending as a median prong over the atlantal neural arch (Fig. 14.15). The prezygapophyses are flattened, laterally facing facets to each side of the prominent spine, the dorsal margin of which is slightly swollen.

The axial postzygapophyses are fused in the midline, forming a broad, horizontal fan-shaped process that encloses a subquadrate opening. This opening is largest in the axis. In postaxial cervicals 3–7, the opening is present as a smaller rounded foramen. Despite fusion of the postzygapophyses in cervical vertebrae 2–8, the articular facets themselves on the underside of the united process may remain separate; most of the articular facets are not fully exposed. The articular facet on the underside of the left axial postzygapophysis is exposed; it is flat, oval and separated from its opposite in the midline. The epipophyseal process originates as a low ridge on the dorsolateral aspect of the neural arch at mid length. The ridge gains in depth and width posteriorly, terminating as a posteriorly facing, wedge-shaped process, separated from its opposite by a median trough.

The centra of *postaxial cervical vertebrae* are procoelous, with the transverse concavity being stronger than the dorsoventral concavity (Figs. 14.15, 14.17). The anterior centrum face of cervical 7 is broader and more deeply concave from side to side than in other postaxial cervical vertebrae, enhancing the potential for transverse rotation of cervical 6. Cervical 6 as preserved is flexed at an angle of approximately 80° relative to cervical 7 (Fig. 14.18). The transverse diameter of the anterior face of cervical 7 (6 mm) is one-third greater than that of cervical 6 (4 mm), and its deeply concave articular surface is bounded to each side by a swollen articular rim. The posterior centrum face is convex but does not form a smooth ball in finished bone. In this regard, ossification seems less complete that in many other turtles. A ventral keel rises at mid length along the centra of the postaxial cervicals and increases in depth toward the anterior centrum face, which as a result has a subtriangular rather than circular shape. The ventral keel splits posteriorly into two rounded ridges that pass to the posteroventral corners of the posterior centrum face, resulting in an inverted subtriangular shape as in *Araripemys* (Meylan 1996). Cervical 7 has a well developed ventral keel that extends along the entire length of the centrum and is pendant anteriorly. A circular pit is present on the left side of the keel (Fig. 14.18).

In mid cervical vertebrae, a subtriangular transverse process extends horizontally from the base of the neural arch, projecting more prominently and extending farther posteriorly along the side of the centrum. In cervical 6, the

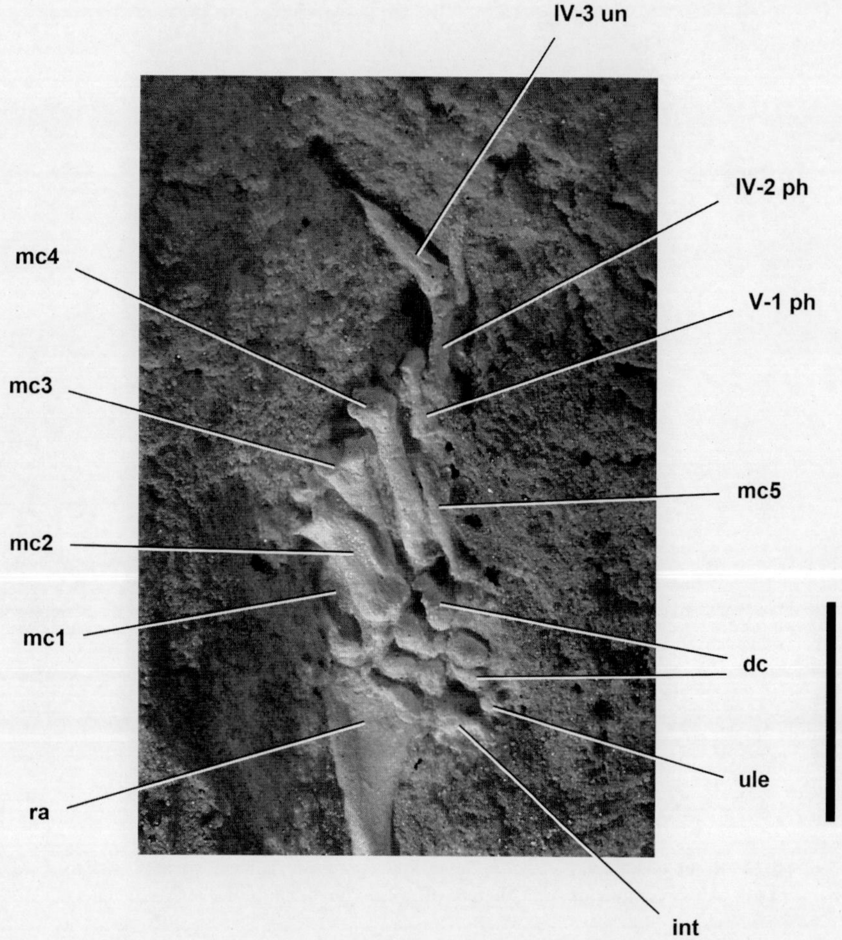

Fig. 14.22 Left carpus and manus of *Laganemys tenerensis* gen. et sp. nov. (MNN GAD28) in ventral view. Scale bar = 1 cm. Abbreviations: *IV*, *V* digit IV, V; *dc* distal carpal; *int* intermedium; *mc1–5* metacarpals 1–5; *ph* phalanx; *ra* radius; *ule* ulnare; *un* ungual

transverse process extends as a long subtriangular flange. In cervical 7 the anterior portion of the flange is reduced whereas the tip projects farther laterally (Fig. 14.18).

In mid cervical vertebrae, the neural spine is developed along the neural arch as a low crest that is strongest at mid length. The spine splits to each side of the midline posteriorly, forming the medial edge of wedge-shaped epipophyseal processes that are separated in the midline by a narrow groove. In postaxial cervicals, the epipophyseal process begins as a low ridge on the lateral aspect of the neural arch at mid length. The ridge forms the lateral margin of a broad, low epipophyseal process and joins the edge of the postzygapophysis. The flat, horizontal prezygapophyses of cervical 5 are exposed in articulation against the flat, conjoined postzygapophyses of cervical 4. Unlike the subdivided axial postzygapophyses, the articular surface of postaxial postzygapophyses extends without break across the underside of the subcircular, conjoined process.

The postzygapophyses on cervical 8 are separated distally by a median notch (Fig. 14.12). The epipophyses on cervical 8 are narrower and spaced farther from the midline than in other postaxial cervical vertebrae. In *Araripemys*, in contrast, the postzygapophyses of cervical 8 have been reconstructed as fully fused with broad, adjacent epipophyses and a small median foramen (Meylan 1996). In *Laganemys* postzygapophyses of this form are limited to cervical vertebrae 3–7. Fusion of the postzygapophyses characterizes *Euraxemys*, whereas the condition of the postzygapophyses in cervical 8 is variable in chelids and podocnemidids (Gaffney et al. 2006). The postzygapophyses are separate in pelomedusids and are widely spaced in the basal pleurodire *Notoemys* (Meylan 1996; Gaffney et al. 2006).

Dorsal vertebrae: The mid dorsal vertebrae are partially exposed through the plastral fenestrae (Fig. 14.13). Dorsal 4, which has been displaced ventrally through the middle fenestra, is the most visible. Centrum length decreases in the anterior portion of the dorsal series. The centrum of dorsal 2 is nearly as long as the postatlantal cervical centra, whereas the length of dorsal 4 is only approximately 75% the length of the cervical 7. The centrum of dorsal 4 is V-shaped in cross-section with a prominent, but rounded, ventral keel. It appears to be amphicoelous, although neither end is fully exposed. A dorsoventrally compressed parapophyseal process is present near the anterior end of the centrum and attaches to the head of the dorsal rib (Fig. 14.13). The anterior margin of the rib

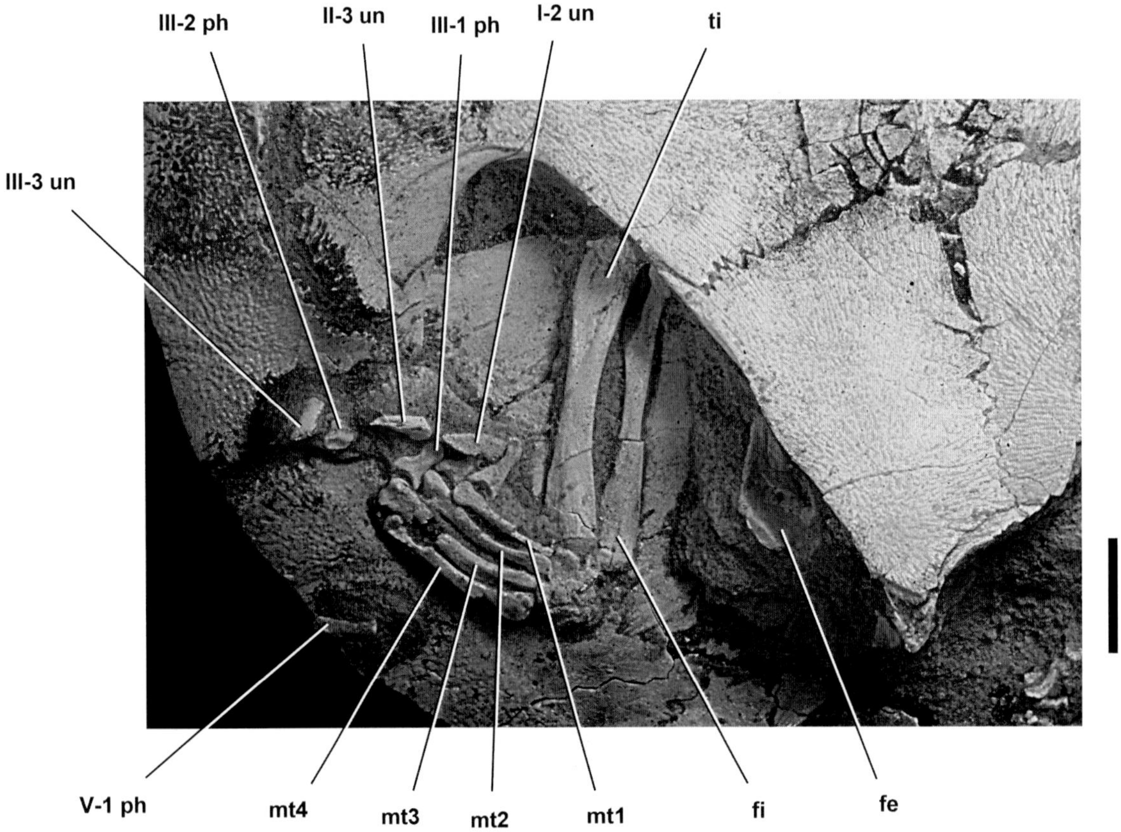

Fig. 14.23 Right distal hind limb of *Laganemys tenerensis* gen. et sp. nov. (MNN GAD28) with pes in ventral and lateral view. Scale bar = 1 cm. Abbreviations: *I–III, V* digits I–III, V; *fe* femur; *fi* fibula; *mt1–4* metatarsals 1–4; *ph* phalanx; *ti* tibia; *un* ungual

extends across the intervertebral joint to attach to a lateral flange projecting from the side of dorsal 3. The centra and keel of more anterior dorsal vertebrae are approximately one-half the depth of that in dorsal 4. The intervertebral attachment of the dorsal rib can also be seen in these vertebrae.

The dorsal vertebrae in *Laganemys* are proportionately longer than in *Araripemys* (Meylan 1996) and have projecting processes for attachment to the ribs. In *Araripemys* dorsal vertebrae 2–9 and their associated ribs are more fully incorporated into the carapace.

Caudal vertebrae: Most of the caudal series is preserved and exposed. The anterior caudal vertebrae are close to their natural position posterodorsal to the posterior embayment of the xiphiplastra (Fig. 14.13). Two anterior caudal vertebrae are exposed in ventral view; there may be a few additional caudal vertebrae embedded in the matrix closer to the sacrum. One of the exposed pair of anterior caudal vertebrae, tentatively identified as caudal 4, has a centrum length of 5 mm and a posterior diameter of 2 mm. This centrum is more elongate than comparable centra at the base of the tail in *Araripemys*, and the subtriangular transverse process is better developed (Meylan 1996). The articular faces of the centra are flat or slightly convex. The anterior caudal vertebrae, like those in mid and distal regions of the tail, are not procoelous as in pelomedusids. The transversely compressed centrum has a shallow median groove that diverges toward the posterior sides of the distal end, terminating in a pair of parasagittal prominences. Although these prominences are well positioned as articulations for haemal arches, no ossified chevrons are present as is also the case in *Araripemys* (Meylan 1996).

The remainder of the tail comprises an articulated series of 21 vertebrae that measure approximately 45 mm in length (Fig. 14.20). This portion of the tail has been displaced anterodorsally, so its proximal end lies dorsal to the pygal plate. The first caudal in this section of the tail is estimate to be caudal 6; the series tapers in diameter, terminating in a tiny nubbin of bone representing caudal 26. Caudal 20 is missing, a gap of appropriate length in its place in the otherwise articulated series.

The tail is longer in *Laganemys* than *Araripemys* (Figs. 14.1, 14.2). It extends well beyond the posterior margin of the carapace. Tail length has been augmented in *Laganemys* by increasing the number and length of caudal vertebrae (Table 14.1). In *Araripemys* there are less than 20 caudal vertebrae, whereas in *Laganemys* there are at least 26 caudal vertebrae. In mid and distal caudal vertebrae in *Laganemys*, the centra have lengths between two and three times

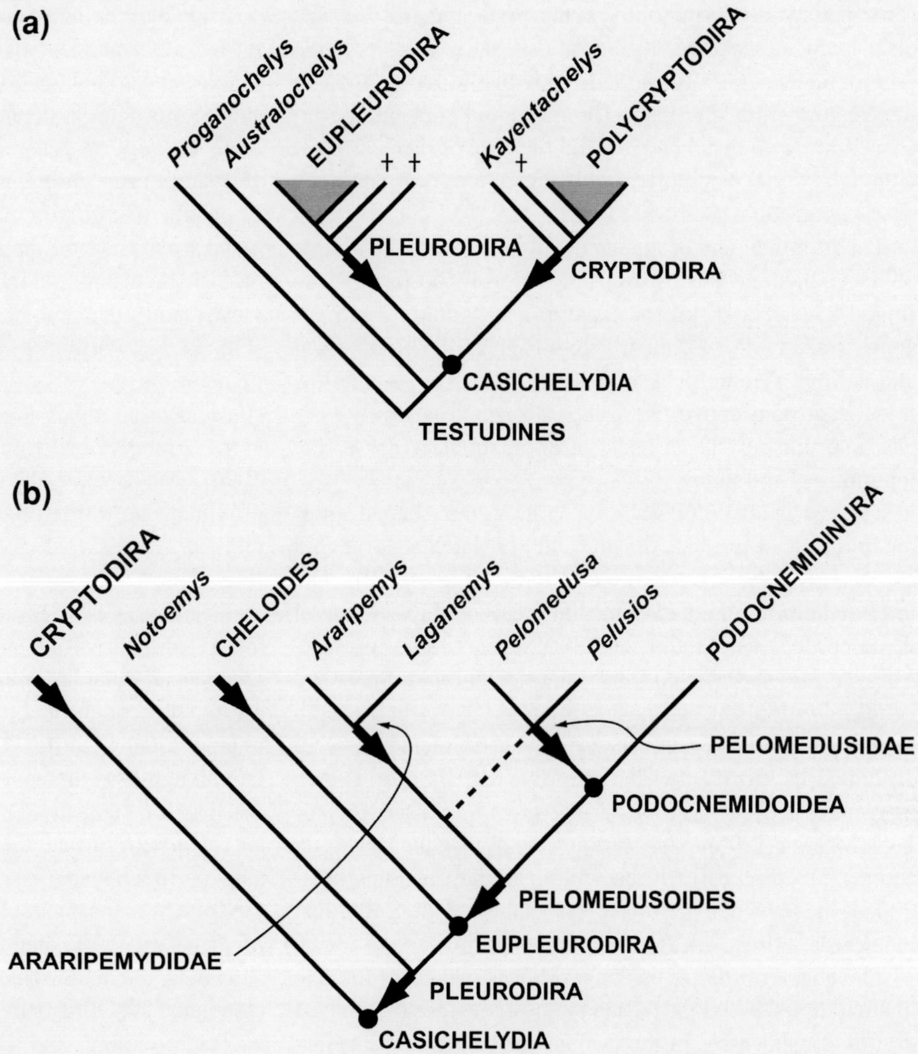

Fig. 14.24 **a** Suggested node-stem triplet of phylogenetic definitions to stabilize the meaning of higher taxa as had long been effected in classifications by Gaffney (from Sereno 2005). **b** Cladogram showing the phylogenetic relationships of araripemydid turtles within Pleurodira, based on reanalysis of data in Gaffney et al. (2006). *Dashed line* in **b** indicates a slightly less parsimonious position for Araripemydidae. *Dots* indicate node-based definitions; *arrows* indicate stem-based definitions; *daggers* indicate extinct taxa; *tone* indicates crown pleurodires and crown cryptodires (see Table 14.2 for phylogenetic definitions)

centrum diameter, whereas they appear to have subequal dimensions in *Araripemys* (Meylan 1996). Zygapophyseal articulations angle ventromedially at approximately 45–60° in *Laganemys* and are present in posterior caudal vertebrae back to approximately caudal 20, after which the centra are little more than cylinders. There are no neural spines in mid and distal caudal vertebrae.

Forelimb: The left radius and ulna, carpus and manus are preserved extending from the plastral embayment for the forelimb (Figs. 14.21, 14.22). The humerus is not present, and the edge of peripheral 2 is inserted between the proximal ends of the radius and ulna. This portion of the forelimb, thus, is partially disarticulated with some distal displacement of the radius, carpus and manus relative to the ulna. The preserved orientation of the forearm and manus, nevertheless, is typical of extant aquatic turtles. The forearm and manus in most turtles is capable of a palm-down orientation with the pollex on the medial side of the manus (Figs. 14.1, 14.2). In many aquatic turtles, however, supination at the elbow reverses the position of the forearm and manus to a posterior and palm-up position, respectively. This is the case in the holotypic specimen of *Laganemys*. The left manus is in ventral view (palm-up) with the pollex on its lateral side when looking down on the carapace (Figs. 14.11a, 14.20).

The *radius* is fully exposed in ventral (anterior) view (Figs. 14.13, 14.22). Like the ulna, its proximal end is considerably narrower than the paddle-shaped distal end, which is relatively more expanded than in *Araripemys* (Meylan 1996). The distal end expands to a width of 6 mm, or approximately 40% of radial length (Table 14.1). The proximal end appears to have a shallow saddle-shaped articular surface for the medial humeral condyle, and the shaft narrows slightly in mid section. In *Araripemys*, in contrast, the radial shaft does not narrow appreciably in mid section, and a prominent ridge is developed along the medial side of the shaft and distal end (Meylan 1996). In *Laganemys* the ventral (posterior) aspect of the distal end is marked by a shallow subtriangular fossa (Fig. 14.22). The distal end is beveled mediodistally at a greater angle (approximately 60°) to the long axis of the radius than in *Araripemys* (Meylan 1996). The distal end contacts the intermedium, two distal carpals, and the base of metacarpal 1, although some postmortem shifting among carpals may have occurred (Figs. 14.21, 14.22).

The proximal end of the *ulna* is more expanded than that of the radius and has a fossa on its dorsal (posterior) side (Fig. 14.21). The distal end expands to a width of 7 mm, or approximately 45% of ulnar length (Table 14.1). The distal end is more symmetrical than that of the radius. A shallow fossa is present above the distal end, which is gently beveled laterodistally at an angle of approximately 85° to the long axis of the ulna.

The *carpus* is composed of a large wedge-shaped intermedium, an ellipsoidal ulnare, and six flattened or ovoid carpals, as best seen in ventral view (Fig. 14.22). The intermedium has a concave medial facet fitted to the lateral distal corner of the radius and thus appears to be in place. The ventral aspect of both the intermedium and ulnare are concave with a finished nonarticular surface of periosteum facing ventrally. A similar set of carpal elements were described in *Araripemys*, but only five carpals were shown in reconstruction, these being interpreted as distal carpals 1–5 (Meylan 1996). The specific identity of the distal six carpals in *Laganemys* is uncertain, as they are not positioned in sequence in direct association with individual metacarpals (Figs. 14.21, 14.22).

The *metacarpals* are the longest bones in each of the five manual digits, and digit IV and metacarpal 4 are the longest digit and metacarpal in the manus, respectively, as in *Araripemys* (Figs. 14.1, 14.2, 14.21, 14.22, Table 14.1). In *Laganemys* as in *Araripemys*, the forearm and manus have similar proportions relative to one another; the radius is 76 versus 73% of the length of digit IV in *Laganemys* and *Araripemys*, respectively. The phalanges, however, comprise a greater proportion of manual length in *Laganemys* than in *Araripemys*; metacarpal 4 is 50 versus 66% of radial length, respectively, whereas metacarpal 4 is 62 versus 91% of the length of the phalanges in digit IV, respectively (Table 14.1). Individual phalanges, likewise, have more elongate proportions in *Laganemys* (Fig. 14.22). The first phalanx in manual digits I–IV is considerably longer than broad in *Laganemys*, whereas these proximal phalanges have subquadrate proportions in *Araripemys* (Meylan 1996). The unguals in *Laganemys*, likewise, taper from a subcylindrical proximal articular surface to flattened distal apices, whereas in *Araripemys* the unguals are shaped like arrowheads with distinctive prongs on each side for the ungual sheath. Thus, the nonterminal and ungual phalanges in *Laganemys* comprise a greater proportion of manual length and are relatively more slender than in *Araripemys*.

Mid-shaft diameter decreases from metacarpal 1 to 5, the former stout and dorsoventrally compressed and the latter slender and rod-shaped (Fig. 14.22). Although *Araripemys* shows this lateral decrease in metacarpal diameter, it is more striking in *Laganemys*. Metacarpal 1, for example, has very stout proportions, its mid-shaft width (4 mm) 60% of its length (6 mm). In *Araripemys* the mid-shaft width of metacarpal 1 is only approximately 40% of its length. The base of metacarpal 1 in *Laganemys* is V-shaped in dorsal view, and the broad shaft is transversely concave (Fig. 14.22). Portions of the distal condyles were lost during preparation.

Metacarpal 2 has a characteristic shape, similar to, but more strongly expressed, than in *Araripemys* (Meylan 1996). There is a distinctive lateral buttress along the proximal two-thirds of the metacarpal, distal to which the divided metacarpal condyles are deflected medially (Fig. 14.22). We regard the flange-like lateral buttress on metacarpal 2 as a synapomorphy uniting araripemydids. We regard the strong medial deflection of the distal condyles as an autapomorphy of *Laganemys*. As far as we are aware, the condition of metacarpal 2 in araripemydids is unique among turtles (Gaffney 1990). Proximal to the lateral buttress, there is a subtriangular overlap facet for the base of metacarpal 3. When metacarpal bases overlap *en echelon* in turtles, this occurs from medial to the lateral side of the metacarpus as seen in dorsal view (Gaffney 1990). Thus, the base of metacarpal 2 should overlap that of metacarpal 3 in natural articulation, not the reverse as shown in the reconstruction of the manus in *Araripemys* (Meylan 1996, Fig. 10A).

The proximal end of metacarpal 3 is dorsoventrally compressed. Its distal end is broader than deep and has divided condyles with a saddle-shaped articular surface for the proximal phalanx (Fig. 14.22). Metacarpal 4 also has a dorsoventrally compressed base, but its distal end is subquadrate and only weakly divided into condyles for articulation with the proximal phalanx. Metacarpal 5 appears to have a subquadrate base but flattens dorsoventrally toward its single distal articular condyle.

The *phalanges* in the manus compose a formula of 2-3-3-3-3 as is common among subaquatic turtles. In manual digit I, the medial one-half of the first phalanx and most of the ungual was lost during preparation. In each manual digit, the proximal phalanx is shorter than the second phalanx. All nonterminal manual phalanges have articular ends that are broader than deep, divided condyles, and well formed collateral ligament pits. All manual digits terminate with slender, gently arched unguals with dorsoventrally flattened tips that would have borne horny ungual sheaths (Figs. 14.21, 14.22). Although Meylan (1996) suggested the terminal phalanx of manual digit V was not covered with an ungual sheath in *Araripemys*

(Fig. 14.1), in *Laganemys* its form is similar to that of the other terminal phalanges (Fig. 14.2). We suspect that in both genera it was clawed in life in a fashion similar to manual digits I–IV.

Hind limb: The right hind limb is preserved largely in articulation extending from the plastral embayment. The proximal end of the femur is exposed under the lateral edge of the xiphiplastron and extends anteriorly within the shell. The bones of the crus are exposed within the plastral embayment in ventral (posterior) view, with the fibula medial to the tibia. The pes is pressed against peripherals 8 and 9 and exposed in ventral and medial views (Fig. 14.23). The tarsus and metatarsal 5 are poorly preserved, and the phalanges of digit IV and all but part of the proximal phalanx of digit V are missing. Pedal digits I and II are flexed and preserved in articulation. The distal two phalanges of pedal digit III are displaced approximately 1 cm from the proximal phalanx, and a portion of the proximal phalanx of digit V is displaced approximately 2 cm to the edge of the carapace.

The *tibia* is not bowed as in *Araripemys* (Meylan 1996) but rather tapers in width from lateral and medial sides toward the mid-shaft (Fig. 14.23). The medial margin of the shaft of tibia in *Laganemys* is slightly concave rather than convex as in *Araripemys*. An elongate subtriangular fossa is present on the ventral (posterior) aspect of the proximal end of the tibia. The distal end is rounded with a deep articular surface for the astragalus. The *fibula* is more slender relative to the tibia and more constricted at mid-shaft than in *Araripemys* (Meylan 1996). Its maximum mid-shaft diameter (2 mm) is one-half that of the maximum width of either the proximal or distal ends.

The *metatarsals* are the longest bones in pedal digits I–IV, and digit III and metatarsal 3 are the longest digit and metatarsal in the pes as in *Araripemys* (Figs. 14.1, 14.2, 14.23, Table 14.1). The metatarsals are flattened dorsoventrally with *en echelon* overlap of their proximal ends. Metatarsal 1 has rotated so that the narrow lateral side of its shaft is facing dorsally (Fig. 14.21). Metatarsals 1–3 have transversely broad, divided distal condyles, whereas metatarsal 4 has a narrower single distal condyle. Metatarsals 2 and 4 are subequal in length.

The *phalanges* in the pes compose a formula of 2-3-3-?-?, a count for the inner digits that is widespread among turtles (Gaffney 1990). The pedal unguals are broader and more dorsoventrally compressed than those in the manus but similar in form, lacking well formed lateral processes for attachment of the ungual sheath. In *Araripemys* the pedal unguals are shown as increasing in length from digits I to IV (Meylan 1996). In *Laganemys*, in contrast, the longest ungual is on pedal digit I, with progressively shorter unguals on pedal digits II and III (Figs. 14.2, 14.23).

Phylogenetic Position

Higher Level Taxonomy Within Pleurodira

The meaning of suprageneric taxon names at the base of Pleurodira is important, given an increasing number of extinct genera and hypotheses of relationship. Classifications of pleurodiran turtles using Linnaean categorical ranks have been around for more than a century and have survived recent revolution in systematic methods (Gaffney 1984; Broin 1988; Gaffney et al. 2006). Phylogenetic systematists (e.g., Hennig 1966), nonetheless, long have questioned the utility of ranks and redundant taxa, and a school of "phylogenetic taxonomy" has emerged that defines taxa on the basis of phylogenetic relationships (de Queiroz and Gauthier 1990, 1992). Additional conventions have been proposed, such as allying "commonly used" higher taxa with crown clades, and these conventions have been applied to turtle taxonomy (Joyce et al. 2004).

Both traditional and phylogenetic approaches, ironically, have generated turtle taxonomies that seem to dwell as much on excess as utility. Gaffney et al. (2006) erected a taxonomy for pleurodires rife with new ranked suprageneric taxa of questionable use, such as "Parvorder Minipleurodira" and "Subtribe Nigeremydina," many of which label poorly supported nodes. In their best-case phylogenetic scenario (with eight "shell-based" genera eliminated), many of the new or resurrected suprageneric taxa have no home on a consensus tree summarizing 590 equally parsimonious trees—just two steps beyond their preferred 382-step minimum-length cladogram. "Magnafamily Pelomedusera", in addition, contains taxa (Araripemydidae, Pelomedusidae) that are not sister taxa on their preferred cladogram.

The rank-free phylogenetic scheme erected by Joyce et al. (2004; Joyce 2007), on the other hand, restricts Pleurodira to crown members (the largest clade bounded by extant taxa), with "Panpleurodira" erected as an unwelcome replacement for Pleurodira in classifications such as Gaffney et al. (2006). Pelomedusoides is also recognized as a crown group (for Pelomedusidae + Podocnemidae) by excluding stem taxa. These stem taxa, however, were included when this taxon was coined (Broin 1988) and in subsequent usage (Meylan 1996; Gaffney et al. 2006). As before, a new taxon "Panpelomedusoides" was coined as a replacement to accommodate the ousted stem taxa. In this way, the traditional use of higher taxa that include extinct species is sacrificed to define taxa on present-day survivors.

Many turtle taxonomists have taken a "wait-and-see" approach in response to these conflicting taxonomic

Table 14.2 Phylogenetic definitions for Casichelydia (crown turtles) and principal higher level taxa within Pleurodira utilizing node-stem triplets for nomenclatorial stability (Fig. 14.24b)

Taxon	Definitional type	Phylogenetic definition
Casichelydia Gaffney 1975	Node (crown clade)	The least inclusive clade containing *Pelomedusa subrufa* (Bonnaterre 1789) and *Testudo graeca* Linnaeus 1758
Cryptodira Cope 1868	Stem	The most inclusive clade containing *Testudo graeca* Linnaeus 1758 but not *Pelomedusa subrufa* (Bonnaterre 1789)
Pleurodira Cope 1865	Stem	The most inclusive clade containing *Pelomedusa subrufa* (Bonnaterre 1789) but not *Testudo graeca* Linnaeus 1758
Eupleurodira Gaffney and Meylan 1988	Node (crown clade)	The least inclusive clade containing *Chelus fimbriatus* (Schneider 1783), *Pelomedusa subrufa* (Bonnaterre 1789), *Podocnemis expansa* (Schweigger 1812)
Cheloides Gray 1825	Stem	The most inclusive clade containing *Chelus fimbriatus* (Schneider 1783) but not *Pelomedusa subrufa* (Bonnaterre 1789), *Podocnemis expansa* (Schweigger 1812)
Pelomedusoides Broin 1988	Stem	The most inclusive clade containing *Pelomedusa subrufa* (Bonnaterre 1789), *Podocnemis expansa* (Schweigger 1812) but not *Chelus fimbriatus* (Schneider 1783)
Araripemydidae Price 1973	Stem	The most inclusive clade containing *Araripemys barretoi* Price 1973 but not *Chelus fimbriatus* (Schneider 1783), *Pelomedusa subrufa* (Bonnaterre 1789), *Podocnemis expansa* (Schweigger 1812)
Podocnemidoidea Cope 1868	Node (crown clade)	The least inclusive clade containing *Pelomedusa subrufa* (Bonnaterre 1789), *Podocnemis expansa* (Schweigger 1812)
Pelomedusidae Cope 1868	Stem (crown clade)	The most inclusive clade containing *Pelomedusa subrufa* (Bonnaterre 1789) but not *Podocnemis expansa* (Schweigger 1812)
Podocnemidinura Cope 1868	Stem	The most inclusive clade containing *Podocnemis expansa* (Schweigger 1812) but not *Pelomedusa subrufa* (Bonnaterre 1789)

Dagger marks higher taxa without extant representatives. Commas between author and year are omitted in phylogenetic definitions to avoid confusion (Sereno 2005)

schemes (Bickham et al. 2007). An alternative way forward has advocated the use of phylogenetic definitions to reflect and stabilize, rather than reorganize, the phylogenetic content in traditional classificatory schemes (Sereno 2005). If, for example, Pleurodira and Cryptodira long have been recognized as subclades of Casichelydia (Gaffney et al. 1987), complementary phylogenetic definitions can stabilize that relationship. A heuristic phylogenetic taxonomy can reduce present and future ambiguity regarding the taxonomic content of traditional suprageneric taxa, such as Araripemydidae.

For our subsequent discussion of relationships, we provide phylogenetic definitions for a higher taxonomy of Pleurodira (Fig. 14.24b, Table 14.2), building on the definition of Pleurodira initially proposed in Sereno (2005) (Fig. 14.24a). This taxonomy stabilizes three widely recognized dichotomies with phylogenetic definitions formulated as node-stem triplets: Casichelydia = Pleurodira + Cryptodira; Eupleurodira = Cheloides + Pelomedusoides; Podocnemidoidea = Pelomedusidae + Podocnemidinura. We do not recommend the use of Pelomedusoidea or Podocnemoidae as in Meylan (1996); the former taxon may engender confusion with Pelomedusoides in the vernacular ("pelomedusoids"), and the latter is redundant with Podocnemidinura as used in Gaffney et al. (2006). Contrary to Joyce et al. (2004), we recognize a stem-based Pelomedusoides to reflect its consistent use in the literature. Likewise, we recommend use of the vernacular "pelomedusoids" for Pelomedusoides, as is common in the literature and as used in the present paper. The proposed definitions also respect the hierarchy of suffixes recommended in traditional taxonomy (ICZN 1999).

Relationships Among Basal Pleurodires

Once the morphology of *Araripemys barretoi* became better known, its status as a basal pleurodire has not been questioned (de Broin 1980, 1988; Meylan and Gaffney 1991; Hirayama 1991). Among extant pleurodires, *Araripemys* has always been considered to be closer to Podocnemidoidea than Chelidae. The basic question is where *Araripemys*—or now the Araripemydidae—is positioned relative to podocnemidoid pleurodires (Fig. 14.24b). Does Araripemydidae lie outside all podocnemidoids, or is it more closely related to pelomedusids?

Meylan (1996) and Gaffney et al. (2006) considered the position of *Araripemys barretoi* among basal pleurodires. Both analyses placed *Araripemys* outside Podocnemidoidea as the basal sister taxon within Pelomedusoides (Fig. 14.24b). Meylan's (1996) analysis was based on 35 characters in *Araripemys* and 14 other pleurodire taxa (6 extant taxa, 8 extinct), for which he reported 5 minimum-length trees of 56 steps. He cited three synapomorphies that unite Podocnemidoidea to the exclusion of *Araripemys*: (1) neural-suprapygal contact eliminated by median intercostal contact; (2) closure of the incisure of the columella auris; and (3) frontal interorbital suture transverse rather than anteriorly pointing.

Rerunning the data matrix shows that only 32 of the 35 characters are parsimony-informative (27, 32 and 34 are uninformative) and that there are 8 rather than 5 minimum-length trees of 56 steps. The topology near *Araripemys*, nonetheless, is as reported by Meylan (1996). However, only the first two of the three characters he cited as synapomorphies supporting this topology are in the character list and matrix. Indeed, the basal position of *Araripemys* within Pelomedusoides is based entirely on these two synapomorphies (characters 22, 35); two additional steps are required to collapse this node.

In the data matrix of Meylan (1996), several characters are scored incorrectly for *Araripemys*. These include jugal–quadratojugal contact (character 2; present rather than absent), the presence of a vomer (character 3; unknown rather than present), and presence/size of the mesoplastron (character 27; absent rather than unknown). In addition, we added *Laganemys* to this matrix to determine its affect, as it differs from *Araripemys* in character state scores in four characters (characters 2, 14, 27, 33). Rerunning the matrix with adjusted data for *Araripemys* and/or including *Laganemys* does not alter the key results of the analysis by Meylan (1996); two synapomorphies that are absent in *Araripemys* (and now also *Laganemys*) support the basal position of Araripemydidae within Pelomedusoides.

Gaffney et al. (2006, Fig. 288) obtained a similar topology with *Araripemys* as the basal taxon within Pelomedusoides based on a larger matrix of 175 characters in *Araripemys* and 40 other taxa, for which they reported a single minimum-length tree of 382 steps. One additional step collapses the basal position of *Araripemys* within Pelomedusoides. There was no discussion of the character evidence, however weak, for the basal position of *Araripemys* within Pelomedusoides. Rather Gaffney et al. (2006, p. 653) highlighted a synapomorphy of Podocnemidinura absent in both *Araripemys* and Pelomedusidae—the partial or complete covering of the prootic by adjacent cranial bones (character 94). This character, however, is not a synapomorphy supporting Podocnemidinura in their analysis as they noted elsewhere. This overlapping three-state character was left unordered, and as a result the two derived states optimize at nodes other than Podocnemidinura on either of their preferred cladograms (Gaffney et al. 2006, Figs. 288, 292). A second shortest tree was obtained after addition of eight "shell-based" taxa with character data largely limited to the shell; *Araripemys* now joins Pelomedusidae as the sister clade to Podocnemidinura within Podocnemidoidea (Fig. 14.24b, dashed lines). The character evidence supporting this relationship also was not discussed.

Araripemydidae as Basal Pelomedusoids

From the above it is clear that the precise position of *Araripemys* and the Araripemydidae among basal pleurodires is poorly established. The following questions remain:

(1) Which characters unite Podocnemidoidea to the exclusion of Araripemydidae in the analysis of Gaffney et al. (2006)?
(2) What happened to the pair of synapomorphies that functioned in this manner in the earlier analysis of Meylan (1996)?
(3) What is the character evidence that links *Araripemys* and Pelomedusidae, when "shell-based" taxa are added to the analysis?
(4) What effect, if any, does a second well preserved araripemydid, *Laganemys*, have on phylogenetic resolution?

Question 1. In the analysis of Gaffney et al. (2006), three homoplastic characters weakly unite Podocnemidoidea to the exclusion of *Araripemys*: procoelous caudal vertebrae (character 129); carapace with nuchal embayment (character 154); and a small, laterally positioned mesoplastron (character 158). *Laganemys* confirms the absence of procoelous caudal vertebrae in araripemydids. The evolution of procoelous caudal vertebrae, however, has an ambiguous distribution on the shortest tree (Chelidae, Pelomedusidae + Podocnemoidae). Procoelous caudal vertebrae thus might have evolved earlier within Pleurodira only to have been lost in araripemydids. Indeed, this is the optimization of this character when "shell-based" taxa are added to the analysis. A third poorly defined character state ("formed centra, but articulations vary") was also listed for character 129; perhaps fortunately, no taxa were scored with this condition.

The nuchal embayment (character 154) evolved or has been lost half a dozen times in the analysis (Gaffney et al. 2006); in the shortest tree, it is optimized as an unambiguous reversal uniting Podocnemidoidea. When additional "shell-based" taxa are added to the analysis, the supporting reversal no longer exists. This character does not provide convincing support.

The coding and optimization of states regarding the mesoplastron (character 158) are problematic. This four-state character is a coding chimera (Sereno 2007). "Absent" is mixed with two shape states (rectangular, equidimensional) and one based on topology (median contact). The supporting transformation for Podocnemidoidea in the shortest tree is a partial reversal, the reappearance of a small mesoplastron. To make matters more homoplastic, *Laganemys* has a small mesoplastron (absent in *Araripemys*). When "shell-based" taxa are added to the analysis, mesoplastron transformations do not lend any support for Podocnemidoidea, so this character does not maintain a basal position for Araripemydidae within Pelomedusoides.

Question 2. The pair of podocnemidoid synapomorphies cited by Meylan (1996) is listed as characters 52 and 141 in Gaffney et al. (2006). Closure of the incisure of the columella auris now has three states (character 52). The condition is open in Araripemydidae, as confirmed in *Laganemys* (Fig. 14.4). The main difference in this connection is that Meylan (1996) scored chelids as having an open incisure, whereas Gaffney et al. (2006) described a closed incisure for chelids, eliminating the possibility this character could function as a podocnemidoid synapomorphy.

Meylan (1996) highlighted the loss of neural–suprapygal contact as a podocnemidoid synapomorphy absent in *Araripemys*, and Gaffney et al. (2006) scored *Araripemys* as primitive in this regard (character 141). Meylan (1996), nonetheless, described the condition in *Araripemys* as variable, some specimens showing broad costal contact between the last neural and suprapygal as in *Laganemys* (Figs. 14.2, 14.11). Gaffney et al. (2006) scored *Araripemys* as primitive (neural–suprapygal contact present) and parsed the character into four overlapping states of increasing costal contact. Despite the possibly erroneous character state score for *Araripemys* (Fig. 14.1), no transformation of this character, whether ordered or not, unites podocnemidoids over *Araripemys* on their preferred tree.

Question 3. With "shell-based" taxa added to the matrix, two unambiguous synapomorphies support *Araripemys* + Pelomedusidae (Gaffney et al. (2006, Fig. 292), namely, extreme temporal emargination (character 14) and hypoplastron–costal 1 contact (character 148). Derived temporal emargination is homoplasious, appearing four times independently in pleurodires as noted by Gaffney et al. (2006). *Laganemys* (Fig. 14.4), like *Araripemys*, has deep temporal emargination, but it would not be scored as such by their criterion (narrow orbit-to-temporal distance). Contact between the hyoplastron and costal 1 is a three-state coding chimera (Sereno 2007). The absence of contact is mixed with the two states describing where contact occurs in some taxa. The supporting synapomorphy is a reversal from one of these states of contact to absence of contact in *Araripemys*, although the condition in pelomedusids is scored as variable. This character cannot be observed in *Laganemys* and is missing in more than half of the taxa in the analysis. In sum, it is not surprising that the link between *Araripemys* and Pelomedusidae collapses with one additional step in tree length.

Question 4. Adding *Laganemys* to the analysis of Gaffney et al. (2006), either with or without "shell-based" taxa, results in a slight decrease in resolution regarding the position of Araripemydidae (see Appendix for character state scores). When reanalyzed without "shell-based" taxa, two polytomies (six minimum-length trees) are present, one of which involves Araripemydidae, Pelomedusidae and Podocnemidoidea. When "shell-based" taxa are added, there is similarly no resolution between these same three taxa. While *Laganemys tenerensis* has increased our knowledge of Araripemydidae, it has not helped to resolve the relationship of the family in the context of available phylogenetic studies.

Phylogenetic Resolution

The description of *Laganemys tenerensis* has brought to light considerable new evidence regarding the unique morphology of an unusual transAtlantic clade of aquatic pleurodires. The relationship between *Laganemys* and *Araripemys* is strong, even before the many new synapomorphies listed in the above diagnoses are added to the analysis. While the family rests comfortably within Pelomedusoides among basal pleurodires (Fig. 14.24b), resolution of its relationship with two other major clades within Pelomedusoides (Pelomedusidae, Podocnemidoidea) is not possible based on available data. This circumstance may persist, given the completeness of araripemydid material, until other basal pleurodires come to light.

Biogeographic Significance

Intercontinental vicariant events during the Cretaceous have been proposed to account for the transAtlantic distribution of well known extant pelomedusoid pleurodires, namely the pairs of closely related genera *Pelusios* and *Pelomedusa* on Africa and the genera *Peltocephalus* and *Podocnemis* on South American (Baur 1993). The fifth and last extant pelomedusoid genus, *Erymnochelys*, resides on Madagascar and is most closely related to the South American pelomedusoids (Gaffney and Meylan 1988; Noonan 2000). Although this Madagascar-South America connection has been cited as evidence against large-scale transAtlantic vicariance (Noonan 2000), one potential paleogeographic scenario for the break-up sequence of Gondwana envisions the early geographic isolation of Africa from other southern land areas, including South America and Madagascar

("Africa-first" hypothesis; Sereno et al. 2004; Sereno and Brusatte 2008).

Much more problematic for a vicariant explanation for extant pelomedusoid distribution is the widely distributed clade of extinct bothremydids, which lie outside Podocnemidinura, the clade including the South American genera *Peltocephalus* and *Podocnemis* (Fig. 14.24b; Noonan 2000; Gaffney et al. 2006). Bothremydids are known not only from Africa, South America, and Madagascar, but also from North America and Eurasia (Gaffney et al. 2006). The Pangaeic distribution of bothremydids in the Late Cretaceous strongly suggests that the pair of surviving genera on African constitute a relict distribution (Maisey 1993). At least four pelomedusoid lineages are already recorded before the close of the Early Cretaceous (Gaffney et al. 2006).

Laganemys provides one of the closest links to date between the similar age mid Cretaceous faunas (ca. 110 Mya; Aptian–Albian) recovered in the Araripe Basin in Brazil and the Illumeden Basin in Niger, just prior to the advent of deep waters in the central Atlantic Ocean (ca. 100 Mya). In the Araripe Basin, the best known vertebrate fauna comes from concretions in the Romualdo Member of the Santana Formation, which consists of an alternating sequence of lacustrine and fluvial sediments that show occasional marine incursions (Maisey 1993). Numerous specimens of *Araripemys barretoi* have been recovered from the Romualdo Member (Meylan 1996). The diverse Lagertätten from this member includes many soft-bodied invertebrates, although dinosaurs are rare and fragmentary (Naisch et al. 2004).

The rarity of dinosaurs in the Araripe Basin inhibits comparison to the dinosaur and crocodylomorph-rich fauna from the Illumeden Basin of Niger (Taquet 1976; Sereno et al. 1998, 2007; Sereno and Larsson 2009). The comparable strata in the Illumeden Basin are freshwater fluvial sediments of the Elrhaz Formation. Recent finds in Niger include well preserved specimens of the notosuchian crocodylomorph *Araripesuchus wegeneri* (Sereno and Larsson 2009), which is very close in morphology to *Araripesuchus gomesii* (Price 1959) from the Romualdo Member of the Santana Formation. Although differing in only minor ways, these two species of *Araripesuchus* may not be sister taxa; the genus *Araripesuchus* is speciose and broadly distributed across South America, Africa and Madagascar (Sereno and Larson 2009).

Laganemys tenerensis and *Araripemys barretoi*, in contrast, are clearly closest relatives among known pleurodires. Their distribution on each side of the encroaching waters of the mid Atlantic suggests that there was active faunal exchange between these landmasses immediately prior to the Late Cretaceous.

Function

Cervical Reach and Aquatic Feeding

The cervical column in both *Araripemys* and *Laganemys* is approximately 90% of the midline length of the carapace and composed of elongate vertebrae of similar length (Figs. 14.1, 14.2). The marked nuchal embayment at the anterior end of the carapace suggests that this long neck was capable of near vertical excursion.

Considerable lateral mobility with the series is suggested by the form and orientation of the zygapophyseal joints. The fused horizontal postzygapophyseal articular surface is very broad, allowing considerable excursion by the opposing narrower prezygapophyses (Fig. 14.17). The cervical series is preserved along an S-shaped curve, with hyperflexion occurring largely in a horizontal plane between cervical vertebrae 2 and 3 and 6 and 7 (Figs. 14.2, 14.16, 14.18).

Among pleurodires, chelids may provide an extant analog for understanding the function of the proportionately long and flexible cervical series for prey capture in araripemydids. *Chelus fimbriatus*, the matamata or fringed turtle, is a specialized suction feeder that uses fast neck extension and marked bucco-pharyngo-esophageal expansion ("gape and suck") for capture of elusive aquatic prey (Wise et al. 1989; Lemell et al. 2002). *Chelodina longicollis*, the common snake-necked turtle, is less specialized but also uses fast neck extension and bucco-pharyngeal expansion for elusive aquatic prey capture (Van Damme and Aerts 1997). Expansion of the oropharyngeal spaces creates inertial suction that draws prey toward and into the mouth. In both of these chelids, however, the hypoid apparatus is hypertrophied to handle rapid oropharyngeal expansion; there are two sets of ossified ceratobranchials that are enlarged and joined into a rigid basket under the pharynx (Aerts et al. 2001; Lemell et al. 2002). Araripemydids clearly do not have a hypertrophied hyoid apparatus and thus are unlikely to be such specialized aquatic feeders.

Rapid strike neck extension and bucco-pharyngeal expansion, however, also characterizes less specialized, shorter-necked aquatic feeders such as *Chelydra serpentina*, the American snapping turtle (Bramble 1978; Lauder and Prendergast 1992), and *Terrapene nelsoni*, the spotted box turtle (Summers et al. 1998). These cryptodires use less exaggerated bucco-pharyngeal expansion to offset the motion of the head toward the prey, so as not to induce any water flow in the vicinity of the prey. These species, in effect, are underwater "ram-feeders" (Lauder and Prendergast 1992), a feeding function they perform with a pair of ossified rod-shaped ceratobranchials in the floor of the buccal cavity.

These extant analogs, in sum, suggest that araripemydids were long-necked aquatic feeders with rapid strike capability for capture of fish and other elusive prey. Like their extant snake- and shorter-necked analogs, however, they were probably opportunistic feeders with a diet that also included carrion and a range of planktonic and benthic invertebrates (Chessman 1984; Ernst and Barbour 1989). The delicate structure and low profile of their shell and its reduced scutation suggests that araripemydids were fully aquatic and occupied still or slow-moving freshwater habitats similar to those occupied by trionychids, or soft-shelled turtles (Ernst and Barbour 1989).

Limb Proportions

The relative size of the forelimb and hind limb differ markedly between *Laganemys* and *Araripemys*. *Araripemys* exhibits what is likely the primitive and common condition among turtles, in which forelimb bones are slightly shorter than comparable hind limb bones. This comparison is necessarily excludes the humerus and femur, because the length of these bones is not yet known in *Laganemys*.

As shown in reconstruction (Fig. 14.1; Meylan 1996), the radius is approximately 81% of tibial length, and the longest metacarpal (metacarpal 4) is approximately 82% of the longest metatarsal (metatarsal 3) in *Araripemys*. In *Laganemys* the forelimb is considerably shorter than the hind limb. The radius is approximately 59% of tibial length, and the longest metacarpal (metacarpal 4) is approximately 47% of the longest metatarsal (metatarsal 3). In sum, principal forelimb bones in *Laganemys* are 20–30% shorter than in *Araripemys* when measured against comparable hind limb bones (Fig. 14.2).

To determine if these differing limb proportions are due to forelimb reduction or hind limb enlargement, we compare available limb bone lengths to maximum carapace length in these two closely related genera. In *Araripemys* the radius and tibia are approximately 13 and 16% of carapace length, respectively. In *Laganemys* the radius and tibia are approximately 11 and 19% of carapace length, respectively. By this comparison, the radius seems slightly shorter and the tibia slightly longer relative to carapace length in *Laganemys*. Thus it appears from these measurements that both factors, forelimb reduction and hind limb enlargement, may have generated the limb disparity observed in *Laganemys* as compared to *Araripemys*. Enhanced limb disparity, the functional meaning of which is unknown, is regarded here as an autapomorphy for *Laganemys*.

Conclusions

We describe a new long-necked turtle from mid Cretaceous rocks in Niger, *Laganemys tenerensis*, which is closely related to a turtle of similar age from the Araripe Basin in Brazil, *Araripemys barretoi*. These genera provide additional evidence for faunal exchange between South America and Africa in the mid Cretaceous (ca. 110 Mya) prior to the advent of deep waters in the central Atlantic Ocean (Maisey 1993; Sereno and Brusatte 2008).

Their mutual affinity is apparent in the many features they share, not least an extremely long neck and a flat, uniquely textured shell. The position of Araripemydidae among pleurodires remains unresolved. Character data clearly establishes the family as a member of Pelomedusoides. The broad exposure of the prootic on the ventral surface of the braincase in both *Laganemys* and *Araripemys* is one the more convincing plesiomorphies shared with pelomedusids. Podocnemidoid pleurodires, in contrast, cover this bone with others.

Outstanding features of *Laganemys* compared to *Araripemys* involve the elongate skull and the relatively short forelimb. Araripemydids likely lived in slow-moving fluvial and lacustrine habitats as opportunistic feeders capable of fast strike pursuit of elusive prey.

Acknowledgments We would like to thank C. Abraczinskas for executing Figs. 14.2, 14.3, 14.4, 14.11 and 14.24 and for her assistance in layout and formatting of the other figures, R. Masek of the Fossil Lab at the University of Chicago for his extraordinary preparation of the specimen, E. S. Gaffney and F. de Broin de Lapparent for access to specimens in their care and discussions on fossil turtles, the personnel of the High-Resolution X-ray Computed Tomography Facility at The University of Texas at Austin for CT scans of the skull, the field crew of the 2000 Expedition to Niger for discovery of the specimen, and A. Maga and O. Ide of the Institut de Recherche en Sciences Humaines, University of Niamey, Niger, for support and permission for field work. This research was supported by the Biological Sciences Collegiate Division at the University of Chicago and the Association for Women in Science (to SJE) and by the Packard Foundation, the Whitten-Newman Foundation, and the Island Fund of the New York Community Trust (to PCS). The manuscript was reviewed by M. P. J. Ryan, E. S. Gaffney, and T. Konishi.

Appendix

Character scores used in this analysis for *Laganemys tenerensis* gen. et sp. nov. and *Araripemys barretoi*.

(A) Character scores for *Laganemys tenerensis* gen. et sp. nov. for 175 characters in the analysis of Gaffney et al. (2006), all scored from the holotypic skeleton (MNN GAD28).

11110000100012101010000000001110100000033000??
1?10130000201011111111001100300110111111?00?0110
0000110000?13?110211000100000111210???1???1142
1?022111212111?2?201111000 1?10120?5

(B) Character states for *Araripemys barretoi* altered from those given in Gaffney et al. (2006).
Character 30: state 0 changed to state 1.
Character 40: state 0 to state 3.
Character 147: state 3 to ?.

References

Aerts, P., Van Damme, J., & Herrel, A. (2001). Intrinsic mechanics and control of fast cranio-cervical movements in aquatic feeding turtles. *American Zoologist, 41*, 1299–1310.

Baur, A. M. (1993). Africa–South American relationships: A perspective from the Reptilia. In P. Goldblatt (Ed.), *Biological relationships between Africa and South America* (pp. 244–288). New Haven: Yale University Press.

Bickham, J. W., Parham, J. F., Philippen, H. D., & Rhodin, A. G. J. (2007). Turtle taxonomy: Methodology, recommendations, and guidelines. *Chelonian Research Monographs, 4*, 73–84.

Bonnaterre, J. P. (1789). *Tableau Encyclopédique et Méthodique des Trois Règnes de la Nature*. Paris: Panckoucke.

Bramble, D. M. (1978). Functional analysis of underwater feeding in the snapping turtle. *American Zoologist, 18*, 230–261.

Chessman, B. C. (1984). Food of the snake-necked turtle, *Chelodina longicollis* (Shaw) (Testudines: Chelidae) in the Murray Valley. *Australian Wildlife Research, 11*, 573–578.

Cope, E. D. (1865). Third contribution to the herpetology of tropical America. *Proceedings of the Academy of Natural Sciences of Philadelphia, 17*, 185–198.

Cope, E. D. (1868). On the origin of genera. *Proceedings of the Academy of Natural Sciences of Philadelphia, 20*, 242–300.

de Broin, F. (1980). Les tortues de Gadoufaoua (Aptien du Niger); aperçu sur la Paléobiogéographie des Pelomedusidae (Pleurodira). *Mémoires de la Société Géologique de France, 139*, 39–46.

de Broin, F. (1988). Les tortues et le Gondwana. Examen des rapports entre le fractionnement du Gondwana au Crétacé et la dispersion géographique des tortues pleurodires á partir du Crétacé. *Studia Geologica Salmanticensia. Studia Palaeocheloniologica, 2*, 103–142.

de la Fuente, M., & de Lapparent de Broin, F. (1997). An Araripemys-like decorated pleurodire turtle in the Paleocene of northwestern Argentine. *Geobios, 30*, 235–242.

de Queiroz, K., & Gauthier, J. (1990). Phylogeny as a central principle in taxonomy: Phylogenetic definitions of taxon names. *Systematic Zoology, 39*, 307–322.

de Queiroz, K., & Gauthier, J. (1992). Phylogenetic taxonomy. *Annual Review of Ecology and Systematics, 23*, 449–480.

Ernst, C. H., & Barbour, R. W. (1989). *Turtles of the world*. Washington, DC: Smithsonian Institution Press.

Fielding, S., Martill, D. M., & Naish, D. (2005). Solnhofen style soft-tissue preservation in a new species of turtle from the Crato Formation (Early Cretaceous, Aptian) of north-east Brazil. *Palaeontology, 48*, 1301–1310.

Gaffney, E. S. (1972). An illustrated glossary of turtle skull nomenclature. *American Museum Novitates, 2486*, 1–33.

Gaffney, E. S. (1975). A phylogeny and classification of the higher categories of turtles. *Bulletin of the American Museum of Natural History, 155*, 389–436.

Gaffney, E. S. (1979). Comparative cranial morphology of recent and fossil turtles. *Bulletin of the American Museum of Natural History, 164*, 1–376.

Gaffney, E. S. (1984). Historical analysis of theories of chelonian relationship. *Systematic Biology, 33*, 283–301.

Gaffney, E. S. (1990). The comparative osteology of the Triassic turtle *Proganochelys*. *Bulletin of the American Museum of Natural History, 194*, 1–263.

Gaffney, E. S., Hutchinson, J. H., Jenkins, F. A., Jr., & Meeker, L. J. (1987). Modern turtle origins: The oldest known cryptodire. *Science, 237*, 289–291.

Gaffney, E. S., Tong, H., & Meylan, P. A. (2006). Evolution of the side-necked turtles: The families Bothremydidae, Euraxemydidae, and Araripemydidae. *Bulletin of the American Museum of Natural History, 300*, 1–698.

Gaffney E.S., & Meylan, P.A. (1988). A phylogeny of turtles. In M. J. Benton (Ed.) *The Phylogeny and Classification of the Tetrapods: Amphibians, Reptiles, Birds* (pp. 157–219). Oxford: Clarendon Press.

Gray, J. E. (1825). A synopsis of the genera of reptiles and amphibia, with a description of some new species. *Annals of Philosophy, 10*, 193–217.

Hennig W. (1966). *Phylogenetic Systematics*. Urbana: University of Illinois Press, pp. 263.

Hirayama, R. (1991). *Phylogenetic relationship of Araripemys (family Araripemydidae; Pleuroria; Testudinata)*. Abstracts of the 140th regular meeting of the Palaeontological Society of Japan, June 22, 23, 1991, Chiba.

ICZN. (Ed.) (1999). *International Code of Zoological Nomenclature*. London: The International Trust for Zoological Nomenclature.

Joyce, W. G. (2007). Phylogenetic relationships of Mesozoic turtles. *Bulletin of the Peabody Museum of Natural History, 48*, 3–102.

Joyce, W. G., Parham, J. F., & Gauthier, J. A. (2004). Developing a protocol for the conversion of rank-based taxon named to phylogenetically defined clade names, as exemplified by turtles. *Journal of Paleontology, 78*, 989–1013.

Lemell, P., Lemell, C., Snelderwaard, P., Gumpenberger, M., Wocheslander, R., & Weisgram, J. (2002). Feeding patterns of *Chelus fimbriatus* (Pleurodira: Chelidae). *Journal of Experimental Biology, 205*, 1495–1506.

Lauder, G. V., & Prendergast, T. (1992). Kinematics of aquatic prey capture in the snapping turtle *Chelydra serpentina*. *Journal of Experimental Biology, 164*, 55–78.

Linnaeus, C. (1758). *Systema Naturae per Raegna Tria Naturae. Volume 1. Regnum Animale* (10th ed.) Photographic facsimile. London: Trustees, British Museum (Natural History).

Maisey, J. G. (1991). *Santana fossils: An illustrated Atlas*. Neptune, NJ: Tropical Fish Hobbyist Publications.

Maisey, J. G. (1993). Tectonics, the Santana Lagerstätten, and the implications for late Gondwanan biogeography. In P. Goldblatt (Ed.), *Biological relationships between Africa and South America* (pp. 435–454). New Haven: Yale University Press.

Meylan, P. A. (1996). Skeletal morphology and relationships of the Early Cretaceous side-necked turtle, *Araripemys barretoi* (Testudines: Pelomedusoides: Araripemydidae), from the Santana Formation of Brazil. *Journal of Vertebrate Paleontology, 16*, 20–33.

Meylan, P. A., & Gaffney, E. S. (1991). Araripemys Price, 1973. In J. G. Maisey (Ed.), *Santana fossils: An illustrated Atlas* (pp. 326–334). Neptune, NJ: Tropical Fish Hobbyist Publications.

Naish, D., Martill, D. M., & Frey, E. (2004). Ecology, systematics and biogeographical relationships of dinosaurs, including a new theropod, from the Santana Formation (?Albian, Early Cretaceous) of Brazil. *Historical Biology, 16*, 57–70.

Noonan, B. P. (2000). Does the phylogeny of pelomedusoid turtles reflect vicariance due to continental drift? *Journal of Biogeography, 27*, 1245–1249.

Price, L. I. (1959). Sobre um cocodrilideo Notosuquio do Cretácico Brasileiro. *Departamento Nacional da Produção Mineral, Divisão de Geología e Mineralogía, Río de Janeiro, 188*, 7–55.

Price, L. I. (1973). Quelônio amphichelydia do Cretáceo inferior do nordeste do Brasil. *Brazileira de Geociências, 3*, 84–96.

Pritchard, P. C. H. (1979). *Encyclopedia of turtles*. Neptune, NJ: Tropical Fish Hobbyist Publications.

Romer, A. S. (1956). *Osteology of the Reptilia*. Chicago: University of Chicago Press.

Rueda, E. A. C., & Gaffney, E. S. (2005). *Notoemys zapatocaensis*, a new side-necked turtle (Pleurodira: Platychelyidae) from the Early Cretaceous of Colombia. *American Museum Novitates, 3470*, 1–19.

Schneider, J. G. (1783). *Allgemeine Naturgeschichte der Schildkröten, nebst einem Systematischen Verseichnisse der einzelnen Arten und zwey Kupfern*. Leipzig: Johan Gotfried Müllersche Buchhandlung.

Schweigger, A. F. (1812). Prodromus monographiae Cheloniorum, Pt. 1. *Königsberger Achiv für Naturwissenschaft und Mathematik, 1812*, 271–458.

Sereno, P. C. (2005). The logical basis of phylogenetic taxonomy. *Systematic Biology, 54*, 595–619.

Sereno, P. C. (2007). Logical basis for morphological characters in phylogenetics. *Cladistics, 23*, 565–587.

Sereno, P. C., & Brusatte, S. L. (2008). Basal abelisaurid and carcharodontosaurid theropods from the Lower Cretaceous Elrhaz Formation of Niger. *Acta Palaeontologica Polonica, 53*, 15–46.

Sereno, P. C., & ElShafie, S. (2009). The unusual South American pelomedusoid turtle, *Araripemys*, discovered in Africa. *Journal of Vertebrate Paleontology, 11*(Suppl), 180A.

Sereno, P. C., & Larsson, H. C. E. (2009). Cretaceous Crocodyliforms from the Sahara. *ZooKeys, 28*, 1–143.

Sereno, P. C., Beck, A. L., Dutheil, D. B., Gado, B., Larsson, H. C. E., & Lyon, G. H. (1998). A long-snouted predatory dinosaur from Africa and the evolution of spinosaurids. *Science, 282*, 1298–1302.

Sereno, P. C., Wilson, J. A., & Conrad, J. L. (2004). New dinosaurs link southern landmasses in the Mid-Cretaceous. *Proceedings of the Royal Society of London B, 271*, 1325–1330.

Summers, A. P., Darouian, K. F., Richmond, A. M., & Brainerd, E. L. (1998). Kinematics of aquatic and terrestrial prey capture in *Terrapene carolina*, with implications for the evolution of feeding in cryptodire turtles. *Journal of Experimental Zoology Part A, Comparative Experimental Biology, 281*, 280–287.

Taquet, P. (1976). Géologie et paléontologie du gisement de Gadoufaoua (Aptian du Niger). *Cahiers de Paléontologie, 1976*, 1–191.

Van Damme, J., & Aerts, P. (1997). Kinematics and functional morphology of aquatic feeding in Australian snake-necked turtles (Pleurodira; *Chelodina*). *Journal of Morphology, 233*, 113–125.

Wilson, J. A. (2006). Anatomical nomenclature of fossil vertebrates: Standardized terms or 'lingua franca'? *Journal of Vertebrate Paleontology, 26*, 511–518.

Wise, S. C., Formanowicz, D. R., Jr., & Brodie, E. D., Jr. (1989). Matamata turtles ambush but do not herd prey. *Journal of Herpetology, 23*, 297–299.

Chapter 15
Two Synchronic and Sympatric Bothremydidae Taxa (Chelonii, Pleurodira) in the Late Cretaceous Site of "Lo Hueco" (Cuenca, Spain)

Adán Pérez-García, Francisco Ortega, and Xabier Murelaga

Abstract Although the Bothremydidae are very abundant in the Cretaceous fossil sites of Spain, they are mainly represented by disarticulated and fragmentary remains that are difficult to assign to genus or species. A complete plastron from the fossil site of "Lo Hueco" (Fuentes, Cuenca) is analyzed here. This specimen is assigned to Foxemydina, but its characters indicate that it does not belong to *Elochelys*, the genus previously identified in this site. This suggests the synchronic and sympatric presence of two Bothremydidae genera in this European Late Cretaceous fossil site.

Keywords Bothremydidae • Foxemydina • Late Cretaceous • Lo Hueco • Turtle

Introduction

Several Campanian–Maastrichtian fossil sites have been identified in the Late Cretaceous of Spain. Representatives of four groups of turtles have been recognized, two of them assigned to Pleurodira (Dortokidae and Bothremydidae) and two to Cryptodira (Solemydidae and an indeterminated taxon from the fossil site of "Lo Hueco", in Fuentes, Cuenca) (de Lapparent de Broin 2001; Pérez García 2009; Pérez García et al. 2009a). Bothremydidae have been cited in numerous Spanish localities: Armuña in Segovia (Jiménez et al. 1990; Jiménez Fuentes 1992), Laño and "Sierra de la Tesla" in Burgos (de Lapparent de Broin and Murelaga 1996, 1999; Murelaga et al. 2005; Berreteaga Escudero 2008), Fontllonga and Biscarri in Lleida (Murelaga et al. 1998), Korres in Alava (Pereda-Suberbiola et al. 1999), Chera in Valencia (Company 2004), Arén and Serraduy in Huesca (Murelaga and Canudo 2005) and "Lo Hueco" in Cuenca (Pérez García et al. 2008, 2009b, 2010). However, the fossil site of "Lo Hueco", discovered in 2007, is the only one in which articulated remains of individuals of this group of turtles have been found. In this site, shells of the hitherto poorly recorded European taxon *Elochelys* have been identified. Specifically these are *Elochelys convenarum* (Pérez García et al. 2010), of which only one specimen, from Haute-Garonne (southern France), was previously known (Laurent et al. 2002).

Bothremydidae are one of the most abundant groups of turtles in the Campanian–Maastrichtian fossil sites of Western Europe (de Lapparent de Broin and Murelaga 1999). One of their best-known representatives is *Polysternon*, not only because it was the first to be identified, but also because it has been recognized in Spain and in France by relatively abundant material. Two species of this genus has been described, *Polysternon provinciale* (Matheron 1869), identified in France (Nopcsa 1931; de Broin 1977; Buffetaut et al. 1996; de Lapparent de Broin and Werner 1998; Tong and Gaffney 2000; Gaffney et al. 2006) and *Polysternon atlanticum* (de Lapparent de Broin and Murelaga 1996), until now exclusively identified in Spain (de Lapparent de Broin and Murelaga 1996, 1999). This genus has been assigned to Foxemydina (Gaffney et al. 2006), taxon that also includes the European bothremydids *Foxemys mechinorum* (Tong et al. 1998), *Elochelys perfecta* (Nopcsa 1931), and *Elochelys convenarum*, (Laurent et al. 2002), being the known record of the first two species exclusive of France. The Bothremydina *Rosasia soutoi* (Carrington da Costa 1940), from Portugal, is the other bothremydid genera present in the Late Cretaceous of Western Europe (Carrington da Costa 1940; Antunes and

A. Pérez-García (✉)
Departamento de Paleontología, Facultad de Ciencias Geológicas, Universidad Complutense de Madrid, C/José Antonio Novais 2, 28040 Ciudad Universitaria Madrid, Spain
e-mail: paleontologo@gmail.com

F. Ortega
Grupo de Biología, Facultad de Ciencias, UNED. C/Senda del Rey 9, 28040 Madrid, Spain
e-mail: fortega@ccia.uned.es

X. Murelaga
Departamento de Estratigrafía y Paleontología, Facultad de Ciencia y Tecnología, Universidad del País Vasco/Euskal Herriko Unibertsitatea, Apartado 644, 48080 Bilbao, Spain
e-mail: xabier.murelaga@ehu.es

D. B. Brinkman et al. (eds.), *Morphology and Evolution of Turtles*, Vertebrate Paleobiology and Paleoanthropology,
DOI: 10.1007/978-94-007-4309-0_15, © Springer Science+Business Media Dordrecht 2013

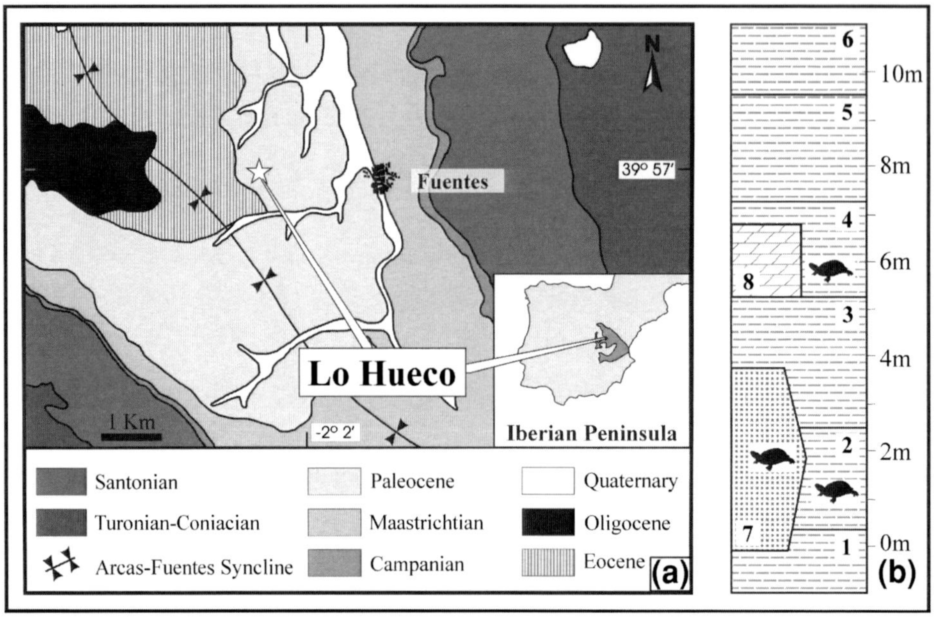

Fig. 15.1 Geographical and geological context of the Campanian–Maastrichtian fossil site of "Lo Hueco" (Cuenca, Spain). **a** Simplified geological map. **b** Simplified stratigraphical column of the fossil site area showing the three levels in which turtle fossils have been found. *1* green marly mudstones; *2* and *4*, grey marly mudstones; *3* and *5* red marly mudstones; *6*, brown marly mudstones; *7*, channel-shaped structure; *8*, dolomitized area. Modified from Ortega et al. (2008a, b, c)

de Broin 1988; de Lapparent de Broin 2001; Gaffney et al. 2006).

Due to their fragmentary state, little of the Spanish remains can be identified at a generic or specific level. However, *Polysternon atlanticum* has been described on the basis of plates and isolated bones from the late Campanian site of Laño (Condado de Treviño, Burgos) (de Lapparent de Broin and Murelaga 1996, 1999) and this taxon was later cited, with doubts, in the Maastrichtian site of Chera (Valencia) (Company 2004). *Polysternon* sp. has been cited in the Late Cretaceous of Korres (Alava) (Pereda-Suberbiola et al. 1999) and in the Maastrichtian of Fontllonga and Biscarri (Lleida) (Murelaga et al. 1998). Except some vague allusions to the possible presence of *Elochelys* or *Rosasia* in the Campanian–Maastrichtian fossil site of Armuña (Segovia) and Laño (Burgos) (Jiménez et al. 1990; Jiménez Fuentes 1992), the only confirmed record of another taxon in Spain is that of *Elochelys convenarum* in the Campanian–Maastrichtian fossil site of "Lo Hueco" (Cuenca) (Pérez García et al. 2008, 2010).

Little is known about the limits of the morphological variability of some of these taxa, because they are, so far, mainly known by disarticulated and fragmentary remains. The identification of *Elochelys convenarum* in "Lo Hueco" (Pérez García et al. 2010) has revealed that some of the characters previously considered diagnostic for this taxon (Laurent et al. 2002), and also used in the diagnosis of other representatives of Bothremydini, are highly dependent on intraspecific variation. The turtle remains from this site (see Pérez García et al. 2009b, Fig. 2, 2010, Fig. 2) provide relevant information on variability, systematics and paleobiogeographical relationships between the European bothremydids. Previously only *Elochelys* has been recognized in "Lo Hueco", we present here a new specimen from this site which could be a non-*Elochelys* member of Foxemydina. For its study, we perform a comparative analysis with the other taxa of Bothremydidae from Western Europe, and this specimen is incorporated in the phylogenetic hypothesis proposed by Gaffney et al. (2006).The unequivocal presence of two taxa of Bothremydidae in the same fossil site had not been previously documented in the Late Cretaceous of Europe. This possibility complicates the generic identification of incomplete Bothremydidae remains in European fossil sites.

Institutional abbreviations used in this text are: HUE, Collection "Lo Hueco", deposited at the Museo de las Ciencias de Castilla-La Mancha, Cuenca, Spain; MDE, Musée des Dinosaures, Espéraza, Aude, France; MHNM, Muséum d'Histoire Naturelle de Marseille, Marseille, France.

Geographical and Geological Context

"Lo Hueco" site is located in the eastern flank of the Arcas-Fuentes Syncline, in the south-western branch of the Iberian Ranges (Ramírez del Pozo et al. 1975), in Fuentes (Province of Cuenca, Spain). This fossil site is located in the upper part of the "Margas, Arcillas y Yesos de Villalba de la Sierra" Formation, in the "Garumn" facies (Ortega et al. 2008a, b, c; Barroso-Barcenilla et al. 2009; Fig. 15.1a). This stratigraphic position and its palaeontological content indicate that it was deposited in the late Campanian-lower Maastrichtian interval (Ortega et al. 2008a, b, c; Barroso-Barcenilla et al. 2009). A succession of versicolor marly mudstone levels, with an extremely rich and diverse autochthonous fossil concentration, partially modified by a sandy channel structure, also with abundance of fossil remains, can be observed in this outcrop (Fig. 15.1b) (see more data related to geology and taphonomy of the fossil site in Barroso-Barcenilla et al. 2009). Besides the discovery of many remains attributed to “fishes”, amphibians, squamate lizards, crocodiles, and dinosaurs (Ortega et al. 2008a, b, c; Barroso-Barcenilla et al. 2009; Escaso et al. 2009), turtles are one of the groups of vertebrates best represented, and are both abundant and well preserved. Members of two different groups of turtles have been identified. The presence of a large cryptodiran turtle is represented by few fragmentary elements, differing from the representatives of Solemydidae (Pérez García et al. 2009a), hitherto the only representative of this major group known in the Iberian Campanian–Maastrichtian record. The record of Bothremydidae is very abundant (Pérez García et al. 2008, 2009b, 2010). It is represented by hundreds of specimens from three levels: a channel-shaped structure and two levels of grey marly mudstones (Fig. 15.1b). The turtle elements commonly appear disarticulated; with very different patterns of preservation, ranging from plates in which its “pelomedusoid” ornamentation is perfectly preserved to others in with the abrasion has eliminated the surface texture. Some turtle remains are partially or totally covered by a ferruginous scab, while others are free of it. Some specimens preserve partial skeletons or complete shells, disarticulated and scattered in some individuals but articulated in others, as in a carapace attributed to *Elochelys convenarum* found in the upper level of grey marly mudstones, which is the most complete shell of *Elochelys* known (Pérez García et al. 2010). The presence of *Elochelys* is also recognized in the lower level of grey marly mudstones, but another articulated specimen of Foxemydina which is not assignable to this taxon has been also found in this level. This specimen is studied in this paper.

The palaeoenvironmental interpretation of the site suggests a near coast muddy flood plain crossed by distributary sandy channels environment, exposed to brackish to fresh water aquatic influences (Barroso-Barcenilla et al. 2009). This interpretation is consistent with the environment in which these turtle taxa are thought to have lived because it has been suggested that some genera of Bothremydidae were inhabitants of freshwater, often near the coast (de Broin 1988).

Systematic Paleontology

Chelonii Brongniart (1800) (Latreille 1800)
Pleurodira (sensu Gaffney et al. 2006)
Pelomedusoides (Cope 1868)
Bothremydidae Baur (1891)
Bothremydodda Baur (1891) (sensu Gaffney et al. 2006)
Bothremydini (Gaffney et al. 2006)
Foxemydina (Gaffney et al. 2006)
Genus Indet.
(Fig. 15.2)

Referred specimen: HUE-1220 is an almost complete plastron, little deformed, that lacks the right mesoplastron (Fig. 15.2).

Description: The plastron is elongated, with a length of 395 mm. Although the ornamentation on the whole ventral surface is not available, this has been preserved in some regions. The ornamentation is the typical pelomedusoid ornamentation (de Broin 1977; Gaffney et al. 2006): smooth, with small dichotomous discontinuous grooves, which can anastomise, forming polygons millimeters in size. The specimen is completely ossified, lacking fontanelles (Fig. 15.2a, b).

The width of the anterior lobe is slightly larger than that of the posterior lobe, while the length of the anterior lobe is smaller than that of the posterior lobe. Both lobes are wider than long, the width/length ratio in the anterior one being 2.21 and in the posterior lobe being 1.28. The morphology of the anterior lobe is rounded, its anterior margin being nearly straight. In the posterior region of the plastron an anal notch is present, 3.82 times wider than long. The length of the bridge is greater than that of each of the two lobes.

A pair of epiplastra, the entoplastron, a pair of hyoplastra, the left mesoplastra, a pair of hypoplastra and a pair of xiphiplastra can be recognized. The length of the epiplastral symphysis is shorter than the length of the entoplastron. In ventral view (Fig. 15.2b, d), the entoplastron is rhombic, with convex edges, wider than long. Despite the deformation, it appears that the rear edge of the entoplastron reaches approximately at the same level that the anterior part of the bridge. The craniocaudal length of the hyoplastra is similar to that of the hypoplastra. Between these plates, a pair of mesoplastra would be laterally placed, but only the left is preserved. The mesoplastra are semicircular, longer than wide. The xiphiplastra are shorter than the hyoplastra or the hypoplastra.

In ventral view, the sulci allow the limits of the scutes to be recognized. The sagittal sulcus is situated on the sagittal suture along almost its entire length, not overlapping with this suture when it has deviations. This sulcus follows a straight line between the posterior margin of the intergular scute and the anal notch. The only odd scute is the intergular, which is longer than wide, having its maximum width in the anterior region. This scute enters in the anterior part of the entoplastron. All other scutes are pairs. The gulars, small and triangular in shape, cover only a part of the epiplastra. They are separated from each other by the intergular scute. The humerals cover much of the anterior half of the entoplastron, contacting each other in the sagittal plane. The humero-pectoral sulcus crosses the entoplastron, the epiplastra, and enters in the latero-anterior edge of the hyoplastra. The pectoral scutes cover the rear region of the entoplastron and are posteriorly extended to the contact with the mesoplastra, but do not overlap these plates. The

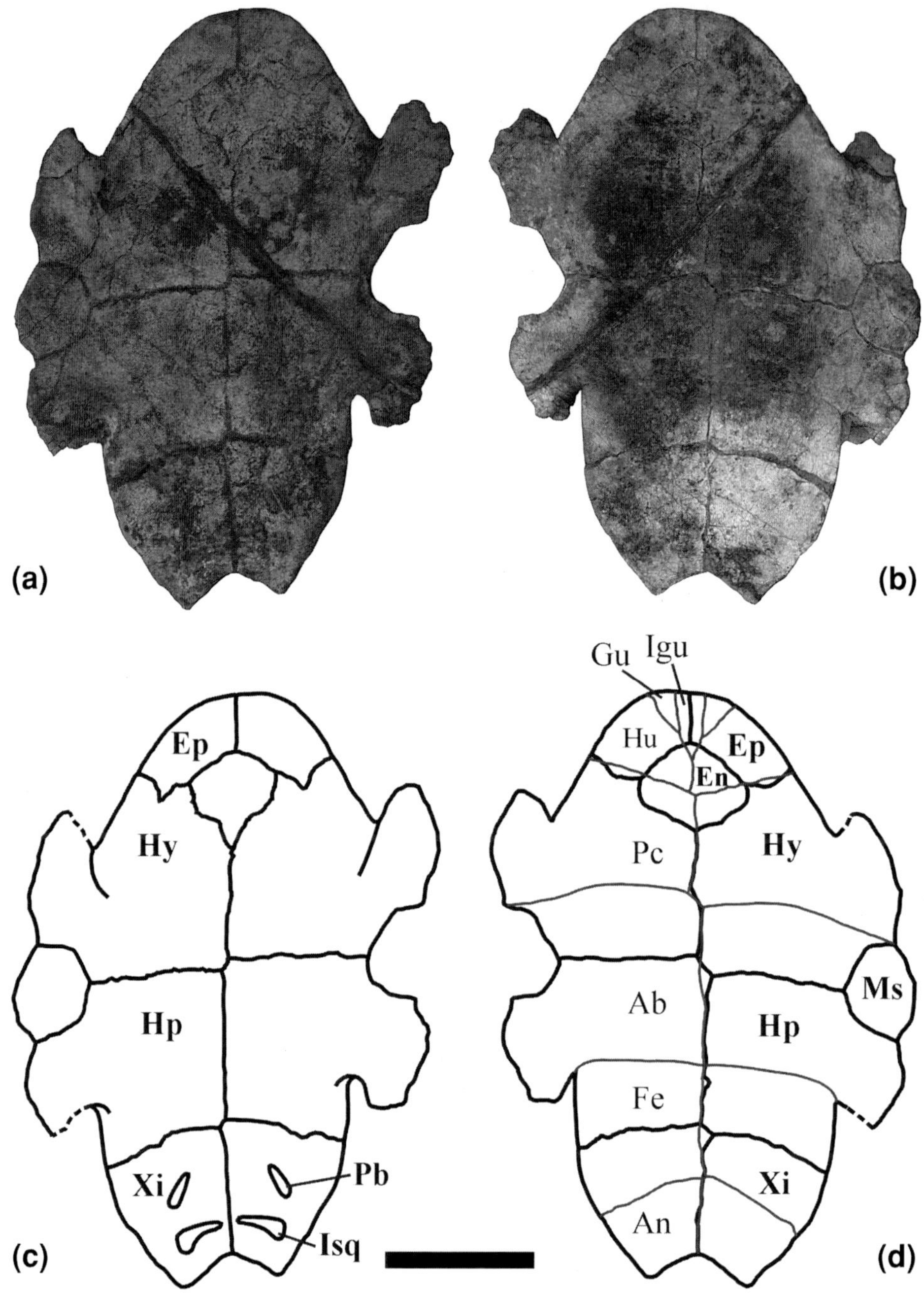

Fig. 15.2 Specimen HUE-1220. Plastron assigned to Foxemydina indet. from the Campanian–Maastrichtian fossil site of "Lo Hueco" (Cuenca, Spain). **a**, **c** Dorsal view: **a** photograph and **c** interpretive drawing. **b**, **d**, Ventral view: **b** photograph and **d** interpretive drawing. In interpretive drawings, thick lines represent sutures between plates and narrow lines represent borders of scutes. Abbreviations: *Ab* abdominal scute, *An* anal scute, *En* entoplastron, *Ep* epiplastron, *Fe* femoral scute, *Gu* gular scute, *Hp* hypoplastron, *Hu* humeral scute, *Hy* hyoplastron, *Igu* intergular scute, *Isq* ischium scar, *Ms* mesoplastron, *Pb* pubis scar, *Pc* pectoral scute, *Xi* xiphiplastron. Scale bar equals 100 mm

abdominal scutes cover the posterior third of the hyoplastra, mesoplastra and the anterior half of the hypoplastra. The rear half of these plates and the anterior of the xiphiplastra are covered by the femoral scutes. The anal scutes are placed over the rear half of the xiphiplastra.

In dorsal view (Fig. 15.2a, c), the epiplastra have a slight epiplastral thickening, restricted to the posterior region of the symphysis. In the central part of the posterior border of the epiplastra a small extension that is directed toward the posterior region is developed. In the posterior edge of the entoplastron a narrow extension that generates a slightly elongated morphology is also developed. This is a little longer than wide, unlike the shape of the entoplastron as observed in ventral view. The lateral buttresses of the hyoplastra and of the hypoplastra are well developed. The scar of the pubis has an elongated morphology and is anteromedially oriented, with an acute anterior tip. The scar of the ischia, which has a subtriangular morphology, is tangential to the axial plane. The posterior tip of this scar is anterior to the anal notch.

Discussion

HUE-1220 shares with Megapleurodira the union of the epiplastra in the sagittal plane, the presence of a single intergular scute, the presence of a pair of mesoplastra laterally located, and the absence of axillary and inguinal scutes

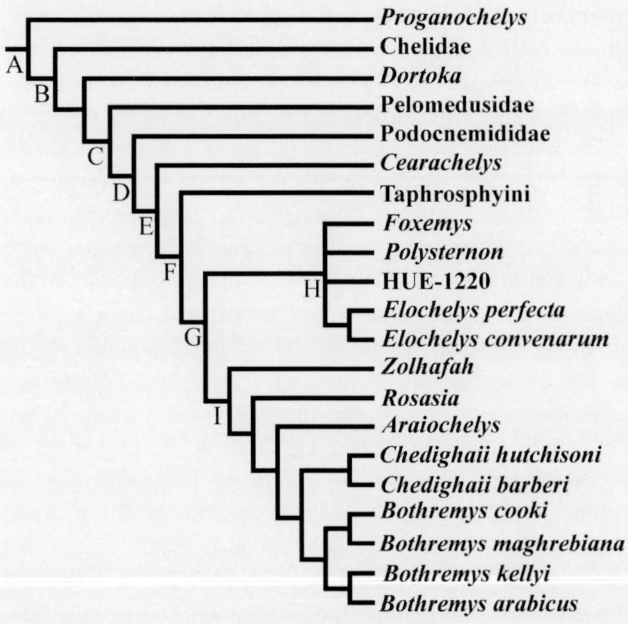

Fig. 15.3 Strict consensus of the 86 most parsimonious trees of 299 steps performed using TNT (Goloboff et al. 2008). Nodes: *A*, Chelonii; *B*, Pleurodira; *C*, Pelomedusoides; *D*, Podocnemidera; *E*, Bothremydidae; *F*, Bothremydodda; *G*, Bothremydini; *H*, Foxemydina; *I*, Bothremydina. Retention index (RI) = 0.699 and Consistency index (CI) = 0.764

(Gaffney et al. 2006). This specimen shares with Podocnemidoidea the presence of pectoral scutes that overlap the entoplastron (Gaffney et al. 2006) and with Bothremydidae the wide and short morphology of the anterior lobe of the plastron (Antunes and de Broin 1988; de Lapparent de Broin and Murelaga 1996, 1999). Thus a phylogenetic analysis focused on the resolution of the members of Bothremydidae and incorporating the plastron from "Lo Hueco" was performed. For this analysis, the matrix of characters proposed by Gaffney et al. (2006) was used. All the characters present in the original analysis and all the representatives of this tribe that are, a priori, the ingroup of the analysis are retained (see Appendix). The resolution of conflicts outside Bothremydidae is beyond the objectives of this work. Therefore, we have simplified the data matrix by eliminating some taxa external to the Bothremydini that introduce a high instability in the final outcome of the analysis (see Gaffney et al. 2006). It should be noted that one of the taxa closely related to the specimen analyzed here, the representative of Foxemydina *Polysternon atlanticum*, was not considered in the original matrix (Gaffney et al. 2006) because it is represented by isolated plates. It is also not included here because it can not be confirmed that all the isolated plates previously assigned to this species (de Lapparent de Broin and Murelaga 1996, 1999) belong unequivocally to a single taxon. The matrix has been processed using the software of cladistic analysis TNT (Goloboff et al. 2008). For the phylogenetic analysis, all characters were considered unordered and equally weighted.

The parsimony analysis produces a set of 86 maximum parsimony trees in which HUE-1220 always belong to Foxemydina (Fig. 15.3) and the distribution of the taxa included in Bothremydini is completely stable. Also, in all the trees obtained, the two subtribes proposed by Gaffney et al. (2006), Foxemydina and Bothremydina, are recognized.

HUE-1220 is grouped with the members of the tribe Bothremydini (Foxemydina + Bothremydina sensu Gaffney et al. 2006). The members of this clade share many synapomorphies, although none of them are features of the plastron so cannot be observed on the specimen from "Lo Hueco".

The Foxemydina (sensu Gaffney et al. 2006) is represented in Europe by *Foxemys*, *Polysternon*, and *Elochelys*, and the Bothremydina (sensu Gaffney et al. 2006) is represented in Europe by *Rosasia*. In the cladistic analysis, HUE-1220 is placed in the Foxemydina. The placement of HUE-1220 within the Foxemydina is supported by three features of the plastron. One of these is the degree of overlap of the entoplastron by the pectoral scutes. In *Rosasia* the sulcus is located in a posterior position so the pectoral scute does not overlap the entoplastron (Antunes and de Broin 1988; Gaffney et al. 2006). The representatives of the Foxemydina and HUE-1220 differ from *Rosasia* in that the pectoral scutes overlap the entoplastron. However, the position of this sulcus in the representatives of Foxemydina varies. In *Polysternon atlanticum*, although the humero-pectoral sulcus crosses the entoplastron it is posterior to the epi-hyoplastral suture (de Lapparent de Broin and Murelaga 1999). In *Polysternon provinciale* the humero-pectoral sulcus not only can be situated on the epiplastra, or on the epiplastron-hyoplastron suture, it may even be slightly down on this suture (de Broin 1977). In HUE-1220 and *Elochelys* the humeral-pectoral sulcus lies anterior to the epi-hyoplastral suture, except the lateral end, where it enters in the hyoplastra (de Lapparent de Broin and Murelaga 1999; Gaffney et al. 2006). This feature has been considered diagnostic of *Elochelys*, suggesting that HUE-1220 should be included in that genus. However, this condition can also be observed in other representatives of Foxemydina. *Foxemys* was originally described as having a humero-pectoral sulcus that crosses the entoplastron posteriorly and then cuts the epi-hyoplastral suture at about half the length of that suture (Tong et al. 1998). This condition differs little from that generally observed in *Polysternon provinciale*. However in some specimens of *Foxemys*, including the holotype, the humero-pectoral sulcus is entirely anterior to the epi-hyoplastral suture (Gaffney et al. 2006) except the lateral end, where it extends onto the hyoplastra (see Tong et al. 1998, Fig. 2; Gaffney et al. 2006, Fig. 259). This position is similar to that observed in the plastron from "Lo Hueco". A second feature that supports placement of HUE-1220 in the Foxemydina rather than with *Rosasia* in the Bothremydina is the position of the pectoral-abdominal sulcus relative to the mesoplastra. As was described above, the pectoro-abdominal sulcus of HUE-1220 differs from that of *Rosasia*: in *Rosasia* this sulcus overlaps the mesoplastra while in HUE-1220 the sulcus barely reaches the mesoplastra. In this feature HUE-1220 is similar to *Polysternon provinciale*, and *Foxemys* (Gaffney et al. 2006). A third feature is the posterior extent of the entoplastron. HUE-1220 is similar to *Polysternon* and *Foxemys* in that the posterior edge of the entoplastron reaches approximately the same level as the anterior part of the bridge. This condition differs from that present in *Rosasia*, in which the entoplastron is more cranially placed (Gaffney et al. 2006).

Within the Foxemydina HUE-1220 forms an unresolved polycotomy with *Foxemys*, *Polysternon* and *Elochelys*. All members of Foxemydina share two synapomorphies: the presence of pectoral scutes on epiplastra or on the epiplastron-hyoplastron suture (condition of character that differs from that observed in *Polysternon atlanticum*), and the location of the pectoro-abdominal sulcus anterior to mesoplastra.

Elochelys is represented by two species, *Elochelys convenarum* and *Elochelys perfecta*. These are grouped together because of the presence of a large intergular scute separating the gulars and the humerals, a character-state shared with no other members of Foxemydina. Since the size of the intergular scute in HUE-1220 is similar to that of *Foxemys* and *Polysternon* rather than *Elochelys*, It is excluded from *Elochelys*.

In this analysis, the comparison with *Foxemys* is limited because that genus is only diagnosed by two cranial autapomorphies: the ventral rim of the orbit is absent; and the exposure of the prootic is small and is surrounded by pterygoid, basisphenoid, and quadrate, with the foramen nervi facialis exposed.

Polysternon is defined by five autapomorphies: the triturating surfaces of the maxilla are triangular, wider posteriorly than anteriorly; a nuchal embayment of the carapace is present; the condylus mandibularis is distinctly anterior to the plane of the basioccipital-basisphenoid suture; the surface texture of the shell has fine, striated ridges; and the anterior lobe of the plastron is short, wide at base, and semicircular in outline, with a width/length ratio of 2.0 or less. Only one of the features diagnostic for *Polysternon* in this cladistic analysis, the width/length ratio of the anterior lobe of the plastron, can be compared with those of HUE-1220. The character state of HUE-1220 differs from that of *Polysternon*. Thus HUE-1220 it is excluded from *Polysternon*.

A plastral character that has traditionally been considered diagnostic of *Foxemys* is a relatively large intergular scute, greater than that observed in *Polysternon* (Gaffney et al. 2006; Tong et al. 1998). In most of the specimens of *Foxemys* this scute extends for half the length of the entoplastron, although in the holotype, MDEt 10, and in the MHNM uncataloged specimen cited and attributed to *Foxemys* by Gaffney et al. (2006) the intergular scute is smaller and reaches only the anterior third of the entoplastron (Gaffney et al. 2006). In HUE-1220 the intergular covers a small portion of the surface of the entoplastron. In this feature HUE-1220 is different from *Foxemys* and similar to *Polysternon provinciale* (de Broin 1977).

In other features, HUE-1220 shows a combination of features that are similar to either *Foxemys* or *Polysternon*. HUE-1220 is similar to *Foxemys* and *Polysternon atlanticum* and different from *Polysternon provinciale* in the shape of the lateral borders of the posterior plastral lobe and in the shape of the anal notch. *Polysternon provinciale* is diagnosed by the presence of rounded lateral borders of the posterior plastral lobe while in *Foxemys* and *Polysternon atlanticum* these borders are straight (Gaffney et al. 2006). The condition observed in HUE-1220 is similar to that of *Foxemys* and *Polysternon atlanticum* (de Lapparent de Broin and Murelaga 1999; de Lapparent de Broin 2001; Gaffney et al. 2006). The anal notch of HUE-1220 is like that of *Foxemys* (Tong et al. 1998; Gaffney et al. 2006) and *Polysternon atlanticum* (de Lapparent de Broin and Murelaga 1999) in being large, short, and wide. *Polysternon provinciale* differs in that the anal notice is relativley narrow and deep (de Broin 1977; Gaffney et al. 2006).

Comparison with *Polysternon atlanticum* is limited because of inadequate understanding of the morphology of that species. In fact, a comprehensive diagnosis of this species has not been formulated because it is only known through disarticulated material. The only diagnostic character of *Polysternon atlanticum* that can be compared with HUE-1220 is size. The maximum length estimated for the shells of *Polysternon atlanticum* is 32 cm (de Lapparent de Broin and Murelaga 1999), similar to that observed in the representatives of the genus *Elochelys* (20–25 cm for *Elochelys perfecta* and 34–35 cm for *Elochelys convenarum*) (Laurent et al. 2002; Pérez García et al. 2010) but significantly lower than that of HUE-1220, *Polysternon provinciale*, and *Foxemys* (de Broin 1977; Tong et al. 1998; Gaffney et al. 2006). Thus in this feature, HUE-1220 differs from *Polysternon atlanticum* and is similar to *Foxemys* and *Polysternon provinciale*.

A series of features shared by HUE-1220 and *Polysternon provinciale* and *Foxemys* are also present. The entoplastron of HUE-1220 is as large as in *Polysternon provinciale* and *Foxemys*. In *Polysternon provinciale* and in HUE-1220 the entoplastron is slightly wider than long. In most specimens of *Foxemys* it is wider than long but in others it is as wide as long (Gaffney et al. 2006). The pubic and ischiatic scars are similar to those observed in *Foxemys* and in *Polysternon*, both in morphology as in disposition. The length of the epiplastral symphysis is more than half of that of the entoplastron, as in *Polysternon provinciale*, differing slightly from the known condition to *Foxemys*, in which is about one-third to one-half (Tong et al. 1998; Gaffney et al. 2006). In the plastron from "Lo Hueco" the gular scutes do not reach the entoplastron. In most of the specimens of *Polysternon provinciale* this condition is similar, but some specimens in which the gulars reach this plate are known. In *Foxemys* the gulars are in contact with the entoplastron but in none of them do the gulars enter onto this plate (de Broin 1977; Gaffney et al. 2006). In HUE-1220 the entoplastron is slightly larger than the mesoplastra. In *Polysternon* the mesoplastra are also slightly smaller than the entoplastron (Gaffney et al. 2006). In some specimens of *Foxemys* the mesoplastra are smaller than the entoplastron but in others the size of these plates is the same (Tong et al. 1998; Gaffney et al. 2006).

Although HUE-1220 shares some features with members of the Foxemydina, particularly *Polysternon provinciale* and *Foxemys*, the limited information available on a plastron does not allow an assignment to generic or specific level. As well, taxonomic resolution of this specimen is prevented by a poor understanding of the range of variation within species of Foxemydina. The available information on the specific variability of other taxa, such as of *Elochelys perfecta*, is very limited. The problem affects even questions about the validity of *Foxemys*. Although it is thought that *Polysternon provinciale* and *Foxemys mechinorum* belong to two different genera (Tong et al. 1998; Gaffney et al. 2006), it has been proposed that these could actually be two species of the genus *Polysternon* (de Lapparent de Broin 2000, 2001; Murelaga and Canudo 2005). There is no doubt that *Foxemys mechinorum* and *Polysternon provinciale* are sister taxa (see Gaffney et al. 2006), regardless of the attribution to the same genus or to two different genera. It has been argued that the separation into two genera is more consistent with the present separation between other pleurodiran genera (Gaffney et al. 2006). It has even been argued that *Polysternon atlanticum* seems to be more similar to *Foxemys* than to *Polysternon* (Tong et al. 1998; Gaffney et al. 2006). Therefore, the discovery of more bothremydid material is necessary to formulate a more comprehensive diagnosis of some of these taxa and provide a greater knowledge of their interspecific and intraspecific variability and their relationships. In this sense, the material attributed to *Elochelys* in "Lo Hueco", shows that the intraspecific variability of *Elochelys convenarum* is higher than previously thought (Pérez García et al. 2010), affecting even characters considered diagnostic not only for *Elochelys convenarum*, but also characters previously employed to diagnose other representatives of Foxemydina. It has also been observed in specimens of *Elochelys convenarum* that the variability in the morphology and dimensions of the intergular scute varies considerably. This region of the plastron is where the autapomorphy of this taxon in the phylogenetic analysis performed here have been recognized. In the holotype, this scute is very wide in the rear part (Laurent et al. 2002). In others specimens, as in HUE-4913, the width of its intergular scute is constant along its entire length (Pérez García et al. 2010). In others, as in HUE-994, this scute is much narrow in the posterior region than in the anterior region, separating the humeral scutes on the entoplastron. However, the observed variability is, so far, different from that described for some pleurodiran taxa such as the podocnemidid *Neochelys arenarum* (de Broin 1977), in which in some specimens the intergular scute separates the humerals and in others these are in contact. Therefore, with the current knowledge, it is concluded the plastron from "Lo Hueco" is an undetermined member of Foxemydina, that differs from *Elochelys*.

This is the first robust demonstration of the synchronic and sympatric coexistence of two different genera of Bothremydidae in the same European Late Cretaceous ecosystem. The recognition of the synchronic and sympatric coexistence of two bothremydid taxa indicates that more caution should be taken in the taxonomic assignment of fragmentary bothremydid samples. Usually, the recognition of a bothremydid taxon in a locality has involved the attribution of isolated material to this taxon. For example, the identification of the only known skull of *Polysternon provinciale*, from Villeveyrac (France), was only based on the prior recognition of this taxon, by postcranial material, in the same Formation and geographic region (Buffetaut et al. 1996). de Lapparent de Broin and Werner (1998) and Tong and Gaffney (2000) supported this attribution, but Gaffney et al. (2006) have argued that it is possible to modify this identification with further discoveries. In the case of the only known skull of *Rosasia soutoi*, their identification was based on an association with a bothremydid peripheral plate (Antunes and de Broin 1988), that is not diagnostic on specific or generic level, and in the fact that this bothremydid skull was found in a locality where only *Rosasia* had been recognized (Antunes and de Broin 1988; Gaffney et al. 2006). *Polysternon atlanticum* has been described by abundant disarticulated material found in the same locality (de Lapparent de Broin and Murelaga 1996, 1999). However, many of the elements assigned to this species are indistinguishable from those of *Rosasia*, *Foxemys* or *Elochelys*.

Conclusions

Bothremydidae are one of the most abundant groups of turtles in the Campanian–Maastrichtian record from Western Europe. Although several members have been recognized, so far, the unequivocal presence of more than one bothremydid taxa in the same level of a European fossil site has not been demonstrated.

In Spain, although abundant bothremydid material has been recognized in several localities, the fossil site of "Lo Hueco" (Fuentes, Cuenca) is the only site where articulated specimens have been found. The presence of representatives of the genus *Elochelys* was recognized there previously.

An articulated plastron, found in a level that includes *Elochelys*, is here described. This specimen does not share the autapomorphic character that diagnoses the plastron of *Elochelys*: the presence of a large intergular scute that separates the humeral scutes. It does show a feature of the humero-pectoral scute that has been described in the Foxemydina only for the genus *Elochelys*, the location of this scute mainly anterior to the epi-hyoplastral suture except the lateral end, but this feature is considered variable in some representatives of this group of turtles. Therefore this plastron is assigned to an undetermined Foxemydina different to *Elochelys*.

This taxonomic conclusion demonstrates that two representatives of Bothremydidae are present in the "Lo Hueco" locality. The possible presence of more than one taxon of Bothremydidae in European Late Cretaceous fossil sites should be taken into account in the taxonomic assignment of fragmentary material belonging to this group of turtles.

Acknowledgments Senior author (APG) is funded by the FPU Subprogramme of the "Ministerio de Ciencia e Innovación" (ref. AP2007-00873). Authors thank Emiliano Jiménez Fuentes, Pedro Seyferth R. Romano and Marcelo S. de la Fuente for comments and suggestions This investigation is part of the activity of the research project "Geología y Paleontología del yacimiento de "Lo Hueco" (Cretácico Superior. Fuentes, Cuenca)" (CGL2009-10766) and of that of the research group "Registro geológico de periodos críticos: factores paleoclimáticos y paleoambientales" (GI 910161). The Willi Hennig Society sponsors TNT cladistics software.

Appendix

Characters coded for HUE-1220, added to the data matrix of Gaffney et al. (2006).

Mesoplastra, 2; Trapezoidal entoplastron, 1; Epiplastra meet on midline, 1; Dorsal epiplastral process, 1; Intergular scales, 1; Gular projections, 1; Anterior lobe of plastron short, wide at base, and semicircular in outline, 1; Pectoral scales on entoplastron, 1; Pectoral scales on epiplastron, 1; Pectoral-abdominal scale sulcus crosses mesoplastron, 1; Size and shape of ischial suture, 1; Posterior lobe wider than pelvis, 1; Intergular scale, 0; Axillary/inguinal scales, 1; Abdominal scale narrow, 0; Anal notch, 1.

References

Antunes, M. T., & de Broin, F. (1988). Le Crétacé terminal de Beira Litoral, Portugal: Remarques stratigraphiques et écologiques; étude complémentaire de *Rosasia soutoi* (Chelonii, Bothremydidae). *Ciências da Terra (UNL), 9*, 153–200.

Barroso-Barcenilla, F., Cambra-Moo, O., Escaso, F., Ortega, F., Pascual, A., Pérez-García, A., et al. (2009). New and exceptional discovery in the upper cretaceous of the Iberian Peninsula: The palaeontological site of "Lo Hueco," Cuenca, Spain. *Cretaceous Research, 30*, 1268–1278.

Baur, G. (1891). Notes on some little known American fossil tortoises. *Proceedings of the Academy of Natural Sciences of Philadelphia, 43*, 411–430.

Berreteaga Escudero, A. (2008). *Estudio estratigráfico, sedimentológico y paleontológico de los yacimientos con fósiles de vertebrados del Cretácico final de la Región Vasco-Cantábrica.* Ph.D. Dissertation, Universidad del País Vasco/EHU.

Brongniart, A. (1800). Essai d'une classification naturelle des reptiles. *Bulletin des Science par la Société Philomathique de Paris, 2*(81–82), 89–91.

Buffetaut, E., Costa, G., Le Loeuff, J., Martin, M., Rage, J. C., Valentin, X., et al. (1996). An early campanian vertebrate fauna from the Villeveyrac basin (Hérault, southern France). *Neues Jahrbuch für Geologie und Paläontologie Monatshefte, 1*, 1–16.

Carrington da Costa, J. (1940). Un novo quelónio fóssil. *Comunicações dos Serviços Geológicos de Portugal, 21*, 105–125.

Company, J. (2004).*Vertebrados continentales del Cretácico superior (Campaniense-Maastrichtiense) de Valencia.* Ph.D. Dissertation, Universidad de Valencia.

Cope, E. D. (1868). On the limits and relations of the raniformes. *Proceedings of the Academy of Natural Sciences of Philadelphia, 16*, 181–183.

de Broin, F. (1977). Contribution à l'étude des Chéloniens. Chéloniens continentaux du Crétacé et du Tertiaire de France. *Mémoires du Muséum National d'Histoire Naturelle, C38*, 1–366.

de Broin, F. (1988). Les tortues et le Gondwana. Examen des rapports entre le fractionnement du Gondwana et la dispersion géographique des tortues pleurodires à partir du Crétacé. *Studia Geologica Salmanticensia. Studia Palaeocheloniologica, 2*, 103–1042.

de Lapparent de Broin, F. (2000). The oldest pre-Podocnemidid turtle (Chelonii, Pleurodira), from the early Cretaceous, Ceará state, Brasil, and its environment. *Treballs del Museu de Geología de Barcelona, 9*, 43–95.

de Lapparent de Broin, F. (2001). The European turtle fauna from the Triassic to the present. *Dumerilia, 4*, 155–216.

de Lapparent de Broin, F., & Murelaga, X. (1996). Une nouvelle faune de chéloniens dans le Crétacé supérieur européen. *Comptes Rendus de l'Académie des Sciences, 323*(*Série 2a*), 729–735.

de Lapparent de Broin, F., & Murelaga, X. (1999). Turtles from the Upper Cretaceous of Laño (Iberian Peninsula). *Estudios del Museo de Ciencias Naturales de Álava, 14*(N.E. 1), 135–211.

de Lapparent de Broin, F., & Werner, C. (1998). New Late Cretaceous turtles from the western desert. *Egypt. Annales de Paléontologie, 84*(2), 131–214.

Escaso, F., Ortega, F., Sanz, J. L., Pérez García, A., & Gasulla, J. M. (2009). New material of Rhabdodon from the upper Campanian-lower Maastrichtian of "Lo Hueco" (Cuenca, Spain). *Abstract Volume 7th Annual Meeting of the European Association of Vertebrate Palaeontologists*, 26.

Gaffney, E. S., Tong, H., & Meylan, P. A. (2006). Evolution of the side-necked turtles: The families Bothremydidae, Euraxemydidae, and Araripemydidae. *Bulletin of the American Museum of Natural History, 300*, 1–700.

Goloboff P., Farris J. S., & Nixon K. C. (2008). TNT a free program for phylogenetic analysis. *Cladistics, 24*, 1–13.

Jiménez, E., Martin, S., Mulas, E., & Pérez, E. (1990). Yacimientos con quelonios del Cretácico español. *Acta Geológica Salmanticensia, 68*, 173–176.

Jiménez Fuentes, E. (1992). Quelonios fósiles de Castilla y León. In E. Jiménez Fuentes & J. Civis Llovera (Eds.), *Vertebrados fósiles de Castilla y León* (pp. 71–100). Aquilafuente: Universidad de Salamanca

Latreille, P. A. (1800). *Histoire naturelle des Salamandres de France, précédé d'un tableau méthodique des autres Reptiles indigènes.* Paris: Villier.

Laurent, Y., Tong, H., & Claude, J. (2002). New side-necked turtle (Pleurodira: Bothremydidae) from the Upper Maastrichtian of the Petites-Pyrénées (Haute-Garonne, France). *Cretaceous Research, 23*, 465–471.

Matheron, P. (1869). Notice sur les reptiles fossiles des dépôts fluvio-lacustres crétacés du bassin à lignite de Fuveau. *Bulletin de la Société géologique de France, 2*(26), 781–795.

Murelaga, X., Pereda Suberbiola, X., Astibia, H., & de Lapparent de Broin, F. (1998). Primeros datos sobre las tortugas del Cretácico superior de Lleida. *Geogaceta, 24*, 239–242.

Murelaga, X., García Garmilla, F., & Pereda-Suberbiola, X. (2005). Primeros restos de vertebrados del Cretácico superior de Quecedo de Valdivielso (Burgos). *Geogaceta, 37*, 195–198.

Murelaga, X., & Canudo, J. I. (2005). Descripción de los restos de quelonios del Maastrichtiense superior de Aren y Serraduy (Huesca). *Geogaceta, 38*, 51–54.

Nopcsa, F. (1931). Sur des nouveaux restes de Tortues du Danien du Midi de la France. *Bulletin de la Société Géologique de France, 1*, 223–236.

Ortega, F., Sanz, J. L., Barroso-Barcenilla, F., Cambra-Moo, O., Escaso, F., García-Oliva, M., et al. (2008a). El yacimiento de macrovertebrados fósiles del Cretácico Superior de "Lo Hueco" (Fuentes, Cuenca). *Publicaciones del Seminario de Paleontología de Zaragoza, 8*, 331–344.

Ortega, F., Sanz, J.L., Barroso-Barcenilla, F., Cambra-Moo, O., Escaso, F., García-Oliva, M., et al. (2008b). El yacimiento de Vertebrados del Cretácico Superior de "Lo Hueco" (Fuentes, Cuenca, España). *Actas del III Congreso Latinoamericano de Paleontología de Vertebrados*, 187.

Ortega, F., Sanz, J.L., Barroso-Barcenilla, F., Cambra-Moo, O., Escaso, F., García-Oliva, M., et al. (2008c). Macrovertebrados del Cretácico Superior de "Lo Hueco" (Fuentes, Cuenca). *Libro de resúmenes de las XXIV Jornadas de la Sociedad Española de Paleontología*, 165–166.

Pereda-Suberbiola, X., Murelaga, X., Baceta, J. I., Corral, J. C., Badiola, A., & Astibia, H. (1999). Nuevos restos fósiles de vertebrados continentales en el Cretácico Superior de Álava (Región Vasco-Cantábrica): sistemática y posición estratigráfica. *Geogaceta, 26*, 79–82.

Pérez García, A. (2009). Historia evolutiva de las tortugas del Mesozoico de la Península Ibérica. In H. Dopazo & A. Navarro (Eds.), *Adaptación y Evolución: 150 años después del Origen de las Especies* (pp. 430–436). Valencia: Obrapropia.

Pérez García, A., Ortega, F., & Murelaga, X. (2008). Sobre la presencia del género *Elochelys* (Chelonii, Bothremydidae) en la Península Ibérica. *Libro de resúmenes de las XXIV Jornadas de la Sociedad Española de Paleontología*, 177–178.

Pérez García, A., Ortega, F., & Murelaga, X. (2009a). Una probable tortuga Pancryptodira del Cretácico Superior de Lo Hueco (Cuenca, España). *Paleolusitana, 1*, 365–371.

Pérez García, A., Ortega, F., & Murelaga, X. (2009b). Preliminary analysis of the bothremydid turtles from "Lo Hueco" (Upper Cretaceous. Cuenca, Spain). *Gaffney Turtle Symposium (October 17–18, 2009, Royal Tyrrell Museum, Drumheller, Canada), Abstract Volume*, 64–70.

Pérez García, A., Ortega, F., & Murelaga, X. (2010). Ampliación de la distribución geográfica y temporal de *Elochelys convenarum* (Chelonii, Bothremydidae) en el Cretácico Superior de la Península Ibérica. *Ameghiniana, 47*, 307–315.

Ramírez del Pozo, J., Portero, J. M., & Olivé, A. (1975). Fuentes, 635 (24–25). Mapa Geológico de España 1:50.000. Segunda Serie. Instituto Geológico y Minero de España.

Tong, H., & Gaffney, E. S. (2000). Description of the skull of *Polysternon provinciale* (Matheron, 1869), a side-necked turtle (Pelomedusoides: Bothremydidae) from the Late Cretaceous of Villeveyrac, France. *Oryctos, 3*, 9–18.

Tong, H., Gaffney, E. S., & Buffetaut, E. (1998). Foxemys, a new side-necked turtle (Bothremydidae: Pelomedusoides) from the Late Cretaceous of France. *American Museum Novitates, 3251*, 1–19.

Chapter 16
New Information about Pelomedusoides (Testudines: Pleurodira) from the Cretaceous of Brazil

Pedro S. R. Romano, Gustavo R. Oliveira, Sergio A. K. Azevedo, Alexander W. A. Kellner, and Diogenes de Almeida Campos

Abstract Brazilian turtle remains date from the Cretaceous and have been recovered from in 11 different basins. Two of these are of particular importance because of the richness of species and specimens: Araripe (Early Cretaceous) and Bauru (Late Cretaceous). Here we present information based on new material that adds to our understanding of the diversity of turtles from Araripe Basin and provides a basis for discussion of the taxonomic status of some species from Bauru Basin. A new specimen from the Araripe Basin that is from the Crato Formation, although generically indeterminate is proposed to be the oldest representative of the clade Podocnemidera. This allocation would extend the stratigraphic range of the Podocnemidera to the Aptian/Albian, matching that of its sister group, the Pelomedusera. New specimens from the Bauru Basin allow a better understanding of the morphology of the shell in *Roxochelys* and an assessment and interpretation of diagnostic features used to distinguish Bauru Basin endemic forms. Our preliminary examination of this material leads us to conclude that the diversity described in this basin is overestimated. As consequence, we argue that Bauru Basin includes only two well diagnosed species of turtles: *Roxochelys wanderleyi* and *Bauruemys elegans*.

Keywords Alpha-taxonomy • Araripe Basin • Bauru Basin • Shell morphology • Side-necked turtles

P. S. R. Romano (✉)
Departamento de Biologia Animal, Museu de Zoologia João Moojen, Universidade Federal de Viçosa, Campus Universitário, Viçosa, MG 36570-000, Brazil
e-mail: psrromano@gmail.com

G. R. Oliveira · S. A. K. Azevedo · A. W. A. Kellner
Departamento de Geologia e Paleontologia, Museu Nacional, Universidade Federal do Rio de Janeiro, Quinta da Boa Vista, São Cristóvão, Rio de Janeiro, RJ 20940-040, Brazil
e-mail: gustavoliveira@gmail.com

S. A. K. Azevedo
e-mail: sazevedo@mn.ufrj.br

A. W. A. Kellner
e-mail: kellner@mn.ufrj.br

D. de Almeida Campos
Departamento Nacional de Produção Mineral, Museu de Ciências da Terra, Avenida Pasteur 404, Urca, Rio de Janeiro, RJ 22290-902, Brazil
e-mail: diogenes.campos@dnpm.gov.br

Introduction

The record of side-necked turtle from the Cretaceous of Brazil includes almost all families of Pelomedusoides; the single exception is the African Pelomedusidae. The oldest records of species with associated shell-skull specimens of Araripemydidae, Euraxemydidae, and Bothremydidae are from the Early Cretaceous (Aptian/Albian) of the Araripe Basin (Fielding et al. 2005; Oliveira and Kellner 2005; Oliveira and Romano 2007; Oliveira et al. 2009). The Podocnemididae from the Late Cretaceous of Bauru Basin include some of the earliest species of this family (Kischlat 1994; Kischlat et al. 1994; França and Langer 2005, 2006; Gaffney et al. 2006; Oliveira and Romano 2007; Romano et al. 2009). In addition, one of the oldest members of the crown-group Eucryptodira, *Santanachelys gaffneyi*, is from Aptian/Albian of Araripe Basin and one of the oldest Eupleurodira (sensu Gaffney et al. 2006) that was recently collected from the Sergipe-Alagoas Basin was dated as Barremian (Gallo et al. 2009). As a consequence, the Brazilian Cretaceous record constitutes an important source of knowledge for understanding the evolution of turtles as a whole. However, the lack of alpha level taxonomic studies, with some species being poorly defined, leads to a confusing scenario. Fortunately, recently described materials increases the number of fossil pelomedusoid turtle species that are well known (e.g., Gaffney et al. 2006; Meylan et al. 2009) and the

D. B. Brinkman et al. (eds.), *Morphology and Evolution of Turtles*, Vertebrate Paleobiology and Paleoanthropology, DOI: 10.1007/978-94-007-4309-0_16, © Springer Science+Business Media Dordrecht 2013

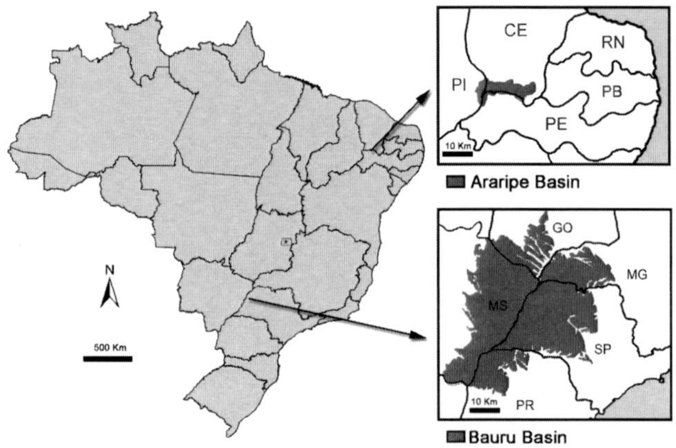

Fig. 16.1 Map of Brazil showing locations of the Araripe (*upper right*) and Bauru (*lower right*) basins. Abbreviations: *CE* Ceará State, *GO* Goiás State, *MG* Minas Gerais State, *MS* Mato Grosso do Sul State, *PB* Paraíba State, *PE* Pernambuco State, *PI* Piauí State, *PR* Paraná State, *RN* Rio Grande do Norte State, *SP* São Paulo State

phylogenetic relationships, biogeographical context, and morphological variation of some of the species have become better understood due to recent studies (e.g., Gaffney et al. 2001, 2006; de la Fuente 2003; Romano and Azevedo 2006, 2007; França and Langer 2006; Meylan et al. 2009; Oliveira et al. 2009; Romano et al. 2009; Riff et al. 2010; Cadena et al. 2010; Romano et al. 2012). As a result, the diagnostic characters are better defined and, finally, our understanding of Brazilian turtles can be more adequately established.

The Lower Cretaceous Santana Group (Crato and Romualdo formations), Araripe Basin, and the Upper Cretaceous Adamantina and Serra da Galga formations, Bauru Basin (sensu Bertini et al. 2006), have both produced important specimens for the study of turtle evolution (Fig. 16.1). Turtles have also been collected from the Cretaceous basins of Sergipe-Alagoas, Potiguar, Parnaíba, and São Luís; and the Cenozoic basins of Paraná, Pernambuco-Paraíba, São José de Itaboraí, and Acre. Twenty-three species have been recognized in Brazilian fossil record, only four of which are represented by living species. Eleven species from Cretaceous of Brazil that have been formally described include: *Apodichelys lucianoi*, *Araripemys barretoi*, *Bauruemys brasiliensis*, *Bauruemys elegans*, *Brasilemys josai*, *Cambaremys langertoni*, *Cearachelys placidoi*, *Euraxemys essweini*, *Roxochelys harrisi*, *Roxochelys wanderleyi*, and *Santanachelys gaffneyi*.

Objectives and Organization

Here we present new material collected in the Araripe and Bauru basins. The material from the Araripe Basin comes from the Crato Formation and includes the first occurrence of a Podocnemidera. New specimens from the Bauru Basin increases our knowledge of morphological variation in shells of species from this basin allowing for a discussion of the validity of diagnostic features used to distinguish shell-only materials. We also argue that some poorly represented species should be synonymized. We follow the phylogenetic taxonomy of Gaffney et al. (2006) as it is the most comprehensive nomenclature proposed for Pelomedusoides to date, and because it is consistent with the established classical taxonomic nomenclature.

The MSc Dissertation work by Campos (1977) and subsequent brief communication (Campos and de Broin 1981) provided the foundation for this study. Generally, the published literature on Brazilian fossil turtles is of little practical value (see a complete list in Oliveira and Romano 2007). Research on the diversity of turtles of Brazil includes: a list of vertebrate species in South America (Mones 1986); a catalog of specimens deposited in the Museu de Ciências da Terra of the Departamento Nacional de Produção Mineral, Rio de Janeiro, Brazil (Melo and Schwanke 2006); a historical study of the discovery of turtles in Brazil (Oliveira and Romano 2007); a preliminary discussion of the taxonomic status of Bauru Basin turtles (Romano et al. 2009); and the first announcement of Podocnemidera remains from Crato Formation (Oliveira et al. 2009).

Eugene Gaffney and collaborators are undertaking a study of the evolution of the Podocnemididae and are describing several new species (Gaffney, 2009, personal communication). To complement their study, we focus our discussion on questions of alpha-taxonomy and post-cranium morphology. As consequence, we do not provide a phylogeny because half of the species do not have enough information to document their phylogenetic relationships, as shown in the phylogeny of França and Langer (2006). However, in order to test our previous conclusions on the interpretation of diagnostic characters, some of which are based on proportions and relative positions of structures, we present a preliminary morphometric analyses of some of the species discussed here.

Institutional abbreviations used in this paper are: AMNH (Americam Museum of Natural History, New York, USA), BSP (Bayerische Staatssammlung für Paläeontologie und Historische-Geologie, Munich, Germany), CPP (Centro de Pesquisas Paleontológicas "Llewellyn Ivor Price", Uberaba, Brazil), DGM (Divisão de Geologia e Mineralogia, Departamento Nacional de Produção Mineral, Rio de Janeiro, Brazil), FR (Forschungsinstitut Senckenberg, Frankfurt, Germany), MCT (Museu de Ciências da Terra, Departamento Nacional de Produção Mineral, Rio de Janeiro, Brazil), MCZ (Museum of Comparative Zoology of Harvard University, Cambridge, USA), MN (Museu Nacional, Universidade Federal do Rio de Janeiro, Rio de Janeiro, Brazil), MPSC (Museu Paleontológico de Santana do Cariri, Santana do Cariri, Brazil), PCHP (Peter C. H. Pritchard personal

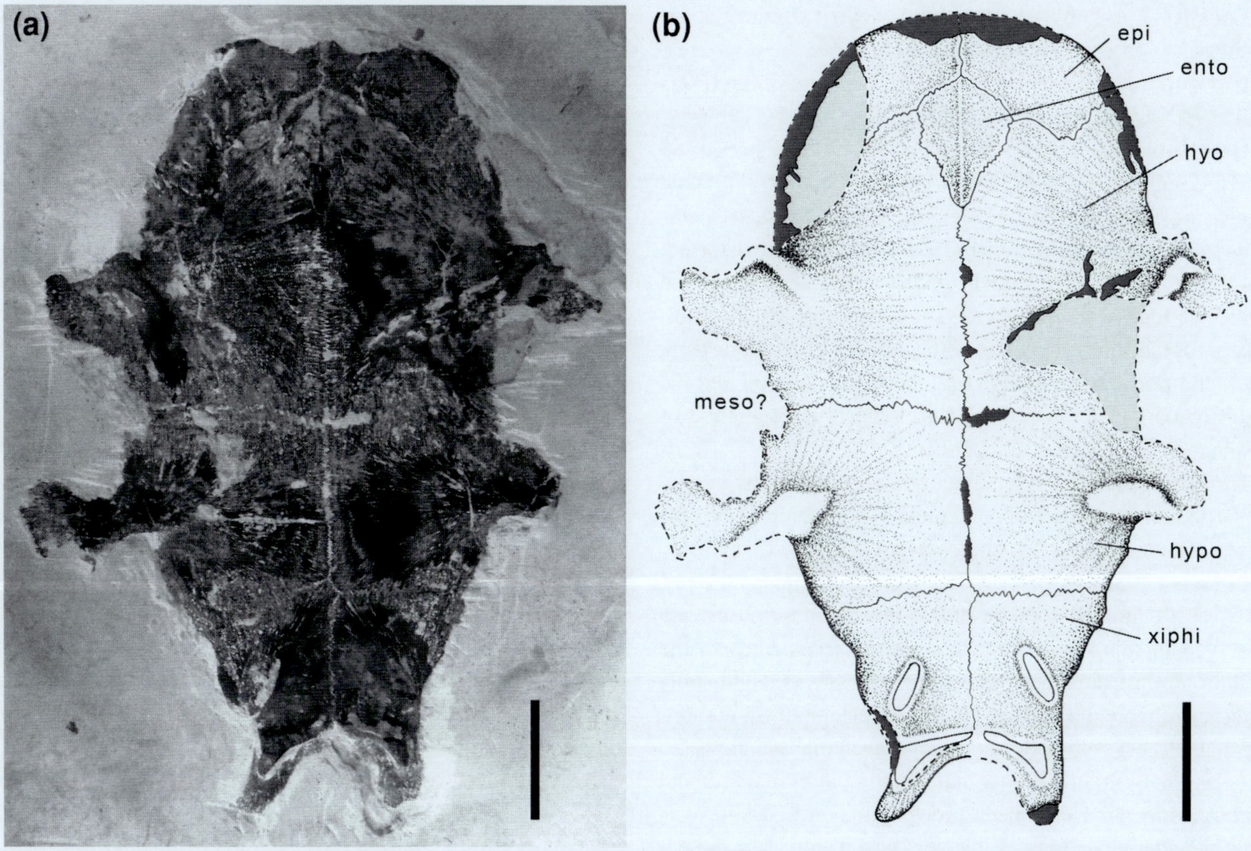

Fig. 16.2 Visceral view of AMNH 30652, plastron of an indeterminate Podocnemidera. **a** Photograph. **b** Interpretive drawing. Abbreviations: *epi* epiplastron, *ento* entoplastron, *hyo* hyoplastron, *hypo* hypoplastron, *meso* mesoplastron, *xiphi* xiphiplastron. *Light gray* represents impression of bones and *dark gray* represents sedimentary matrix. Scale bars equal 20 mm

collection), SMNK-PAL (Staatliches Museum für Naturkunde, Karlsruhe, Germany), THUg (Teikyo Heisei University, Chiba, Japan), UFRGS (Universidade Federal do Rio Grande do Sul, Porto Alegre, Brazil).

New Material From Araripe Basin

The presence of Podocnemidera in the Crato Formation is documented by specimen AMNH 30652 (Fig. 16.2). For comparison the following specimens were examined: *Araripemys barretoi* DGM 756-R (holotype), DGM 364-LE, MPSC 137 R, MN 6637-V, MN 6949-V, MN 6743-V, MN 6745-V; *Cearachelys placidoi* MN 6760-V, MPSC-R 2120 (holotype); *Euraxemys essweini* MN 6919-V (= *Caririemys violetae*; see discussion in this subchapter); *Bauruemys elegans* MCT 1492-R (holotype), MN 6674-V, MN 6762-V, MN 6772-V, MN 6796-V; *Roxochelys wanderleyi* DGM 216-R (holotype), MCT 1722-R, MCT 1787-R (see details in the next subchapter); "*Bauruemys brasiliensis*" DGM 214-R (holotype; see discussion in the next subchapter); plus some specimens described in the literature: *Araripemys barretoi* SMNK-PAL 3979; *Cearachelys placidoi* THUg 1798, BSP 1976 I 160; *Euraxemys essweini* FR 4922 (holotype); *Santanachelys gaffneyi* THUg 1386 (holotype); "*Cambaremys langertoni*" CPP 0252 (holotype; see discussion in the next subchapter); *Chelus fimbriata* PCHP 1337; *Phrynops geoffroanus* MCZ 146145; *Pelomedusa subrufa* unnumbered; *Peltocephalus dumerilianus* PCHP 1351; and *Podocnemis unifilis* PCHP 1051.

Systematic Paleontology

Order Testudines Linnaeus 1758
Suborder Pleurodira Cope 1864
Hyperfamily Pelomedusoides Cope 1868
cf. Podocnemidera Cope 1868
Genus Indet.
(Fig. 16.2)

Referred specimen: AMNH 30652; a nearly complete plastron, lacking the lateral portions, anterior portion of epiplastra and both mesoplastra, preserved in a laminated limestone, visible in visceral view (Fig. 16.2).

Locality: probably Santana do Cariri, Ceará State, northeastern Brazil.

Horizon and age: Crato Formation. Aptian/Albian (Pons et al. 1990; Neumann and Cabrera 1999).

Description: Specimen AMNH 30652 is a well preserved nearly complete plastron that is exposed in visceral view. It measures 125 mm in total length. Although most specimens found in the Crato *lagerstätte* are crushed and distorted, AMNH 30652 is three-dimensionally preserved. The bone surface is also well preserved, showing a brownish color. Except for the anterior portion of the left epi- and hyoplastron, the lateral portion of the right hyo- and hypoplastron and the posterior part of the xiphiplastron, there is no evidence of breakage at the edges of the elements.

AMNH 30652 is clearly a pleurodire turtle as indicated by the articulation scars for the pubis and ischium on the xiphiplastron, a classic synapomorphy of this clade (Gaffney and Meylan 1988). AMNH 30652 is probably made up of 11 bones as in most Pelomedusoides, but no mesoplastra are preserved in this specimen. There are no fontanelles in the midline of plastron as in *Araripemys*, *Pelomedusa*, and some chelids. Probably the midline contact of the epiplastra is short, like that of most Pelomedusoides. As seen in internal view, the entoplastron is slightly longer than wide, as in *Euraxemys* and in contrast to most Pelomedusoides and chelids, where it is slightly wider than long. However, this feature can appear differently ventral view. The hyoplastron forms the axillary notch and the hypoplastron forms the inguinal notch. The xiphiplastron forms most of the posterior lobe, which tapers posteriorly. The articulation of pubis is elongated as in *Bauruemys elegans* (Suárez 1969) and "*Bauruemys brasiliensis*" (Staesche 1937) (Kischlat 1994; França and Langer 2005). The articulation of ischium is subtriangular, resembling the condition in chelids. It is not possible to see the scales on the plastron, since this specimen is exposed in ventral view.

Remarks: Overall the taxonomy of turtles from the Araripe Basin is well resolved. The unique exception is *Caririemys violetae* Oliveira and Kellner 2007a, b, which was described on the basis of a single specimen (MN 6919-V) consisting of an incomplete carapace, right pelvis and right femur. According to Oliveira and Kellner (2007a), *Caririemys violetae* could represent the same taxon as FR 4922. Since then FR 4922 was described and named *Euraxemys essweini* Gaffney et al. (2006). The only differences between *Caririemys* and *Euraxemys* are that neural 1 contacts costal 1 and costal 2 contacts neural 3. Those features, however, could be interpreted as ontogenetical differences or even as individual variations since they are reported in living *Pelusios* (Pritchard 1988). Based on the fact that the holotype of *Caririemys violetae* represents a young individual that does not show enough features to distinguish it from FR 4922 at the species level we here consider this species as a junior synonym of *Euraxemys essweini*.

In comparison with other pleurodiran turtles (Fig. 16.3; Table 16.1), AMNH 30652 can be seen to lack plastron fontanelles as is also the case in most Pelomedusoides. This is in contrast to *Araripemys* and *Pelomedusa,* which have three and one fontanelles respectively (Kischlat and Campos 1990; Meylan 1996; Oliveira and Kellner 2005; Gaffney et al. 2006). There is a relatively short midline contact of the epiplastra, as in *Euraxemys*, but in contrast to *Cearachelys*, which does have a long midline contact (Gaffney et al. 2006). The anterior lobe in AMNH 30652 is rounded and very similar to *Euraxemys*, *Cearachelys* and most pleurodires and differing strongly from the pointed anterior notch of *Araripemys*. In visceral view, the entoplastron of AMNH 30652 is slightly longer than wide, in contrast to *Cearachelys* where it is slightly wider than long. The entoplastron in AMNH 30652 is similar to *Euraxemys*, and different from the trapezoidal shaped entoplastron of *Cearachelys*, and the V-shaped element of *Araripemys*. The hyo- and hypoplastra are rectangular bones with a shallow articulation. The hyoplastron has a shallow anteriorly contact with epi- and entoplstron, and the hypoplastron has a shallow articulation with xiphiplastron, in contrast to the deeper condition found in *Araripemys*.

Morphometric analysis: It is difficult to identify turtles based only on the bones of the plastron. The turtle shell only presents a limited number of characters and shows great deal of individual variation (e.g., Williams 1954; Pritchard 1988; Romano 2008). A recent example of the lack of accuracy in the establishment of new species based only on fragmentary shell material is *Araripemys* "*arturi*", which is a junior synonym of *A. barretoi*, as pointed out independently by Gaffney et al. (2006) and Oliveira and Kellner (2007b). Therefore in order to more objectively observe differences between AMNH 30652 and other well-known species of Pleurodira we applied the thin-plate spline method using 11 landmarks of the plastron in nine species (*Chelus fimbriata*, *Phrynops geoffroanus*, *Pelomedusa subrufa*, *Araripemys barretoi*, *Euraxemys essweini*, *Cearachelys placidoi*, *Baurumeys elegans*, *Peltocephalus dumerilianus*, and *Podocnemis unifilis*) (Fig. 16.4; Table 16.2). The thin-plate spline is a simple procedure used to visualize all displacements of all landmarks relative to all others as a deformation grid (Zelditch et al. 2004). The landmarks were digitalized using TPS Dig 2.12 (Rohlf 2008) and then superimposed using Procrustes standardization (the generalized least square method). All landmarks were collected in ventral view of the specimens and we estimated (by mirroring) the ventral view of AMNH 30652 based on the sutures observed in visceral view. This procedure could introduce error on the landmarks coordinates digitalized, especially on landmark 5 (anterior contact between hyoplastra meeting the entoplastron), but this error does not compromise the analysis because it is the same if

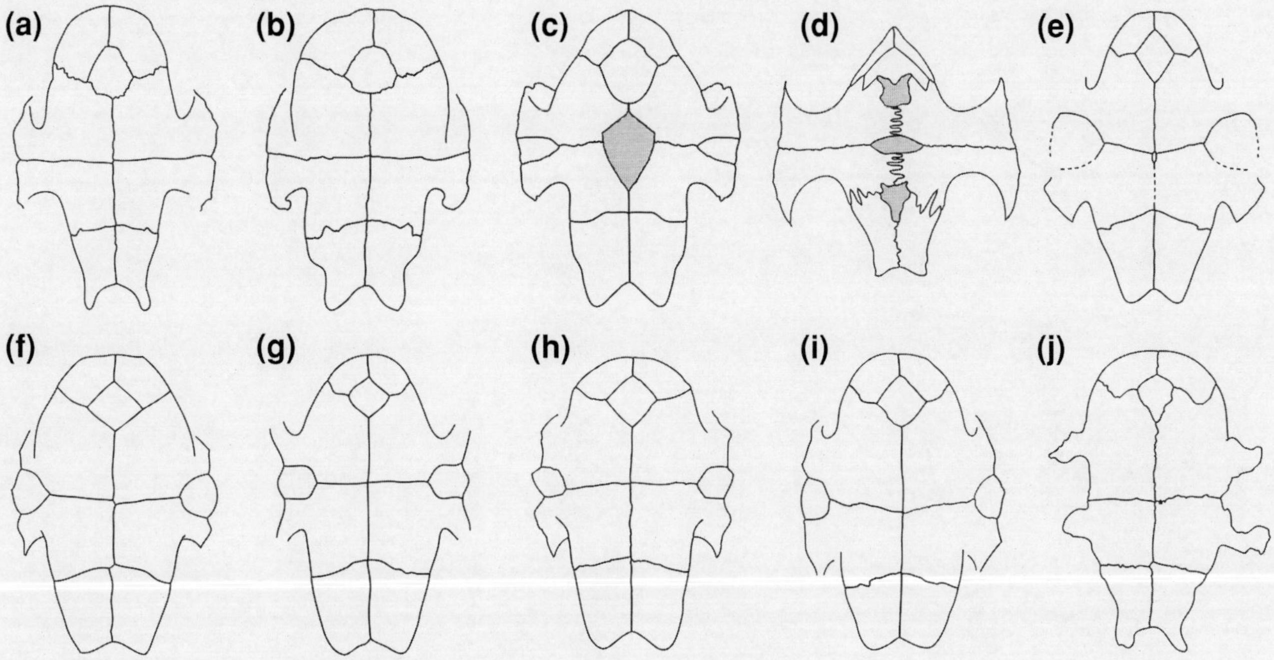

Fig. 16.3 Plastrons of some side-necked turtle species. All in ventral view and not too scale. **a** *Chelus fimbriata* PCHP 1337 (modified from Pritchard and Trebbau 1984, p. 101). **b** *Phrynops geoffroanus* MCZ 146145 (modified from Pritchard and Trebbau 1984, p. 112). **c** *Pelomedusa subrufa* unumbered (modified from Williams 1954, p. 3). **d** *Araripemys barretoi* reconstruction based on AMNH 22550, 22556, and 24453 (modified from Meylan 1996, p. 23). **e** *Euraxemys essweini* FR 4922 (modified from Gaffney et al. 2006, p. 540). **f** *Cearachelys placidoi* THUg 1798 (modified from Gaffney et al. 2006, p. 544). **g** *Baurumeys elegans* reconstruction based on MN 6674-V, 6762-V, and 6772-V (modified from Romano and Azevedo 2007). **h** *Peltocephalus dumerilianus* PCHP 1351 (modified from Pritchard and Trebbau 1984, p. 84). **i** *Podocnemis unifilis* PCHP 1051 (modified from Pritchard and Trebbau 1984, p. 66). **j** Indeterminate Podocnemidera AMNH 30652 (estimated from visceral view)

Table 16.1 Characteristics of the plastron of AMNH 30652 compared with described side-necked turtles from the Araripe Basin and Pelomedusidae and Podocnemididae

	Mesoplastron	Fontanelles	Ento-hyoplastron contact	Hypo-xiphiplastron contact	Anterior lobe	Entoplastron
Chelidae	Absent	Absent	Shallow	Moderate	Rounded	Slightly wider than long
Pelomedusidae	Present	Absent in *Pelusios,* one present in *Pelomedusa*	Shallow	Moderate	Moderately rounded	Slightly wider than long
Araripemys	Absent	Three	Deep	Deep	Pointed	V-shaped
Euraxemys	Present	Absent	Shallow	Shallow	Rounded	Slightly longer than wide
Cearachelys	Present	Absent	Shallow	Shallow	Rounded	Slightly wider than long
Podocnemididae	Present	Absent	Shallow	Moderate	Rounded	Slightly wider than long
AMNH 30652	Not preserved	Absent	Shallow	Shallow	Rounded	Slightly longer than wide

compared to all specimens. We also performed a Principal Component Analysis (PCA) over the Procrustes residuals matrix in order to determine if there is structure in the data without considering the phylogenetic signal in it. The superimposition of the landmarks used to generate the Procrustes residuals matrix, the thin-plate spline deformation grid, and the PCA were all performed using PAST (Hammer et al. 2001).

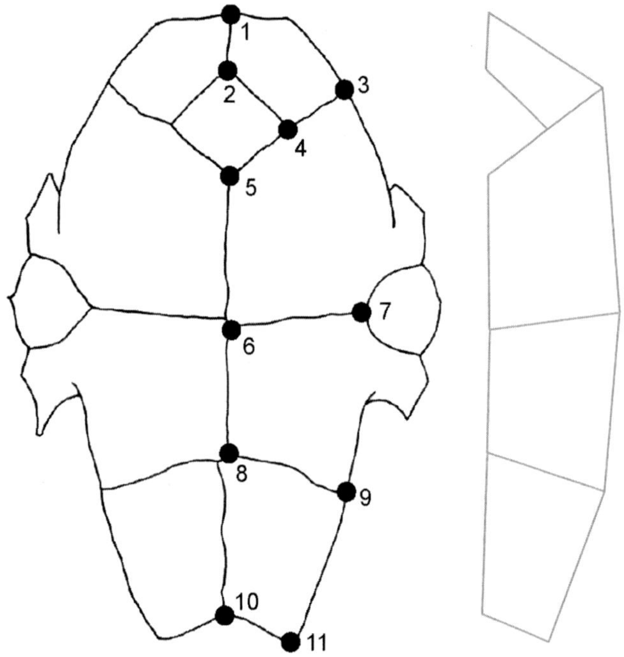

Fig. 16.4 The landmarks of the plastron used in morphometric analyses, as shown in *Cearachelys placidoi* THUg 1798 in ventral view (*left*) with its representation of the shape (*right*). See landmark descriptions in Table 16.2

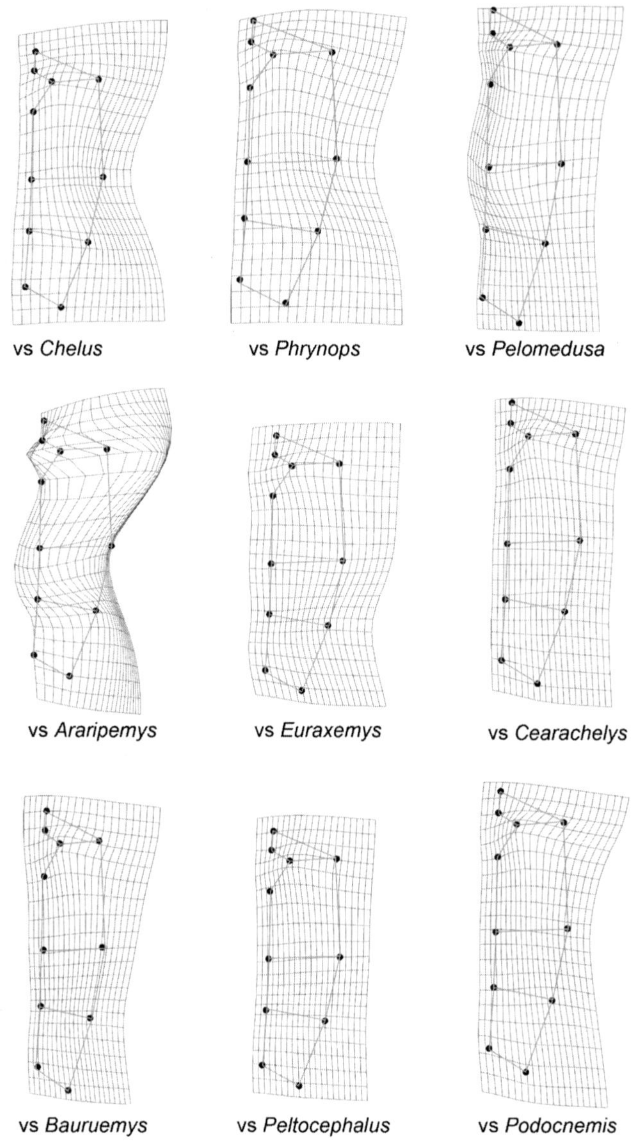

Fig. 16.5 Thin-plate spline deformation grids of AMNH 30652 in comparison to other Pleurodiran turtles

Table 16.2 Description of landmarks of the bones of the plastron used in this study

Landmark	Description	Type
1	Medial anterior contact between both epiplastra	2
2	Medial posterior contact between both epiplastra	1
3	External lateral contact between epi- and hyoplastron	2
4	Internal lateral contact between epi- and hyoplastron	1
5	Anterior contact between hyoplastra and the entoplastron	1
6	Medial contact between hyo- and hypoplastra	1
7	Lateral contact between hyo- and hypoplastra and the mesoplastron	1
8	Medial contact between hypo- and xiphiplastra	1
9	Lateral contact between hypo- and xiphiplastra	2
10	Medial posterior contact between xiphiplastra	2
11	Most posterior point of xiphiplastron	3

The results of thin-plate spline (Fig. 16.5) clearly shows four patterns of distortion in the grid of AMNH 30652 in relation to the species mentioned above: (1) expansion between landmarks 3 and 4 and 8 and 10, and contraction between 7 and 9 compared to the chelid species (more evident in *Chelus* than *Phrynops*); (2) a great expansion between landmarks 3 and 7 with little contraction on the other landmarks compared to *Pelomedusa*; (3) a great expansion of the landmark 3, 4, and 6, and moderate expansion between 10 and 11, and overall lateral contraction if compared to *Araripemys*; and (4) contraction between landmarks 2, 4, and 5 and 6, 7, 8, and 9 with little or none expansion between other landmarks when compared to the Podocnemidera turtles (less deformed in *Cearachelys* and *Peltocephalus* than the others). These results indicate that the overall shape of AMNH 30652 is similar to that of the Podocnemidera, which is showed in the PCA (Fig. 16.6). It is interesting to stress that the 2-dimension plot over the first and second PCs indicates that the overall shape patterns in the bones of the plastron can be linked to differentiation of groups inside Pleurodira, with Podocnemidera well separated from Araripemydidae, Pelomedusidae and Chelidae.

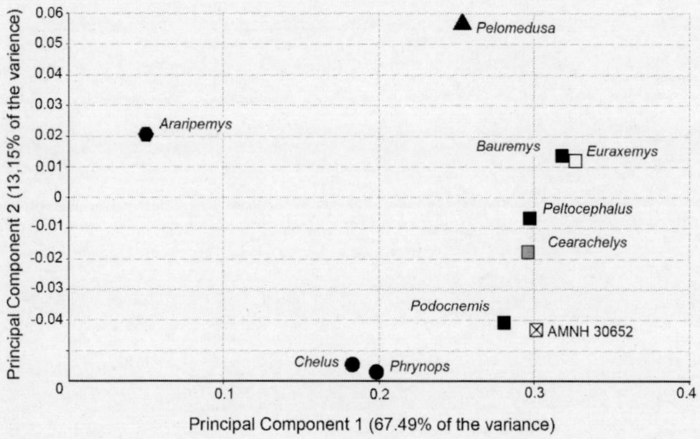

Fig. 16.6 Plot of the first 2 principal components of the covariance matrix calculated over the Procrustes residuals matrix. Symbols: AMNH 30652 (*crossed square*), Podocnemididae (*black square*), Bothremydidae (*gray square*), Euraxemydidae (*white squares*), Araripemydidae (*black hexagon*), Pelomedusidae (*black triangle*), Chelidae (*black circle*)

New Materials from Bauru Basin

The new specimens from the Bauru Basin, MCT 1722-R and MCT 1787-R (Figs. 16.7, 16.8), are examples of *Roxochelys wanderleyi*. The specimens examined for comparison are listed above. In addition comparison was made with all *Bauruemys elegans* specimens mentioned in Romano and Azevedo (2007) and also the illustrated Brazilian Cretaceous specimens available in the literature (see Oliveira and Romano 2007 and Romano et al. 2009 for a complete list of references).

Systematic Paleontology

Order Testudines Linnaeus 1758
Suborder Pleurodira Cope 1864
Hyperfamily Pelomedusoides Cope 1868
Family Podocnemididae Cope 1868
Genus *Roxochelys* Price 1953
Roxochelys wanderleyi Price 1953
(Figs. 16.7, 16.8)

Specimens: MCT 1722-R and MCT 1787-R, two nearly complete shell with carapace and plastron separated from each other with visceral and dorsal views visible (Figs. 16.7, 16.8).

Locality: Presidente Prudente, São Paulo State, southeastern Brazil (Mezzalira 1959, 1989, 2000; Suárez 1973; Romano et al. 2009).

Horizon and age: Presidente Prudente Formation, Maastrichtian (Fernandes and Coimbra 1996, 2000; Fernandes 2004).

Description: MCT 1722-R is represented by a nearly complete shell prepared in both visceral and dorsal views with separated carapace and plastron and measuring 315 mm in total length. The carapace is composed of the nuchal bone, seven neural bones, eight costal bones, pygal, suprapygal, and eleven peripheral bones. The shape of carapace is nearly circular as in *Bauruemys elegans* and *Lapparentemys vilavilensis*. The nuchal is fragmented but it is possible to observe that it is wider than long, as in *R. wanderleyi* holotype (DGM 216-R), differing from *B. elegans*, which has a slightly wider nuchal, and from *Cambaremys langertoni,* which has a longer nuchal (Fig. 16.9). The neural series is incomplete with costal bones 7 and 8 preventing a contact with suprapygal. This condition is the same in *Bauruemys*, whereas in smaller specimens of this species the pair of 6th costal bone also prevents the contact with suprapygal (Fig. 16.9). The first four neural bones show the same morphology as the *R. wanderleyi* holotype (DGM 216-R), although the fourth neural bone is relatively bigger in MCT 1722-R than in DGM 216-R. The first neural bone is rectangular, neural bones 2–6 are hexagonal. The posterior portion of the 7th neural bone is broken but this bone is probably also pentagonal. The first costal bone contacts neurals 1 and 2, as in DGM 216-R and *C. langertoni*, differing from *B. elegans* where it contacts only the first neural (Fig. 16.9). The peripheral bones and the dermal scutes do not show any distinguishing features and are similar to the morphology of DGM 216-R. The plastron is well preserved and shows the generalized organization seen in most Podocnemidera. There is no mesoplastron preserved and it is not possible to determinate if there is reduced and laterally positioned mesoplastra (as in most Pelomedusoides) in this specimen. The humeral-pectoral scute overlies the entoplastron and the hyoplastron as in DGM 216-R and differing from *Bauruemys* which overlies the entoplastron, the epiplastron, and the suture between epi and hyoplastron (Fig. 16.10). The femoral scute is shorter than the abdominal scute differing from *Bauruemys* which is exactly the opposite (Fig. 16.10).

MCT 1787-R is better preserved than MCT 1722-R and is also represented by a complete shell prepared in both visceral and dorsal views with separated carapace and plastron. It is slightly distorted at the left side. The shell measures 375 mm in total length. All distinguishing features described above for MCT 1722-R can be seen in MCT 1787-R in addition the mesoplastra is preserved and shows the general morphology observed in most Pelomedusoides (Fig. 16.10).

Remarks: The first collections of fossil turtle remains from the Bauru Basin were made during the last decade of the Nineteenth Century and the beginning of the Twentieth Century. These are fragmentary shells found in Mato Grosso do Sul State (Derby 1896) and São Paulo State

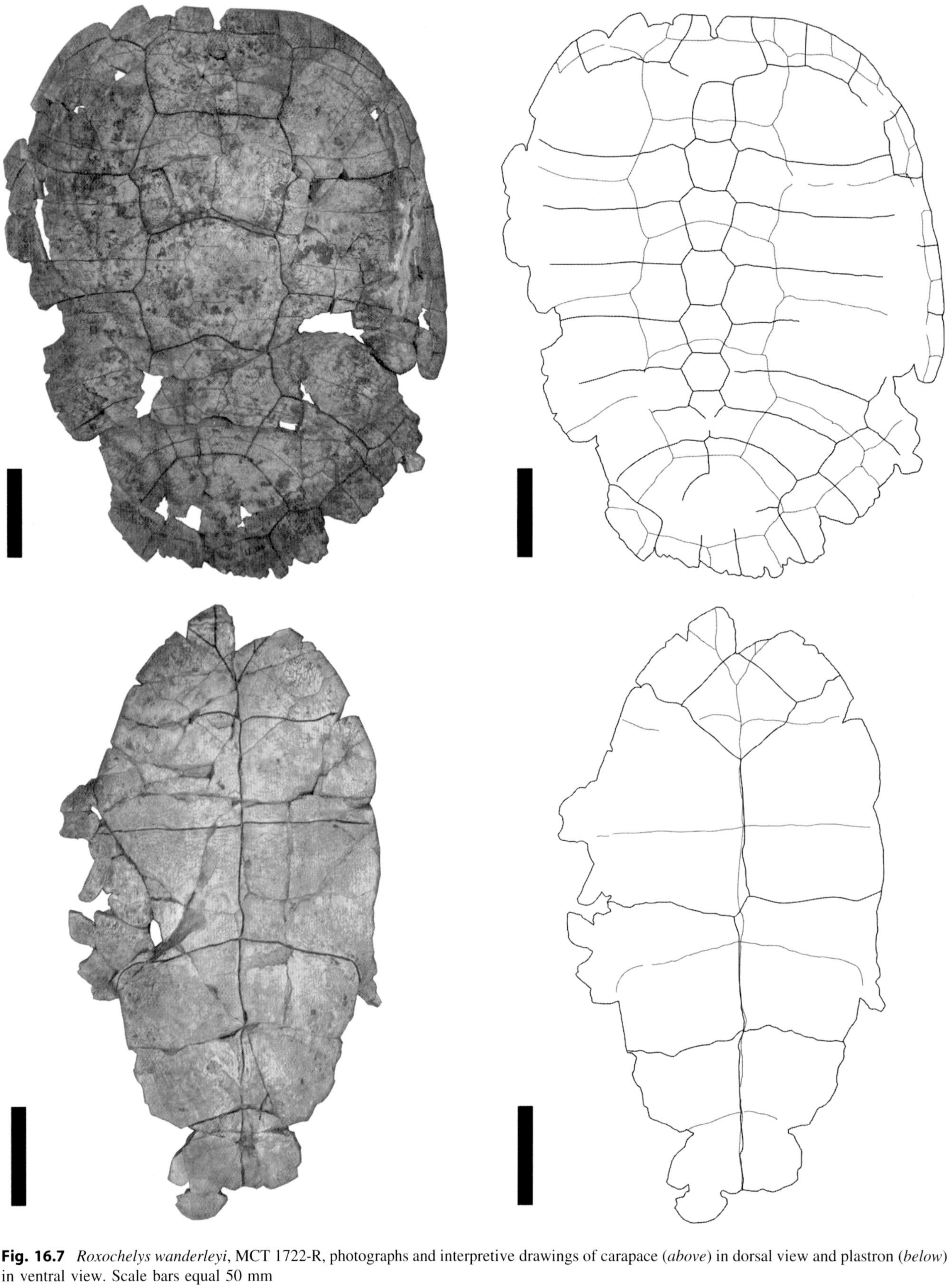

Fig. 16.7 *Roxochelys wanderleyi*, MCT 1722-R, photographs and interpretive drawings of carapace (*above*) in dorsal view and plastron (*below*) in ventral view. Scale bars equal 50 mm

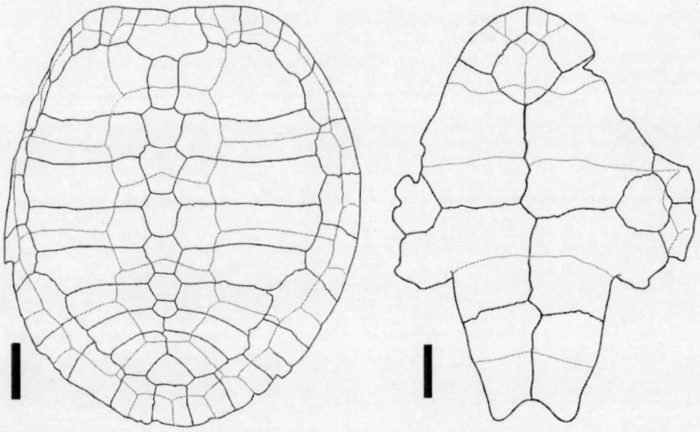

Fig. 16.8 *Roxochelys wanderleyi,* MCT 1787-R, reconstruction of carapace (*left*) in dorsal view and plastron (*right*) in ventral view. Scale bars equal 50 mm

(von Ihering 1911). Several specimens have been reported subsequently from other localities (see Mezzalira 1989, 2000; Oliveira and Romano 2007; Romano et al. 2009), but, in most cases, are fragmentary postcranial elements that cannot be accurately identified. Although a large number of articulated turtles have also been collected, most consist only of shell material (Romano et al. 2009). Five podocnemidid species have been recognized: *Roxochelys harrisi* (Pacheco 1913), *Bauruemys brasiliensis* (Staesche 1937), *R. wanderleyi* Price 1953, *B. elegans* (Suárez 1969), and *Cambaremys langertoni* França and Langer 2005.

Pacheco (1913) provided the first formal description of a fossil turtle from the Bauru Basin based on a right xiphiplastron and two fragments of the "marginal portion of the carapace", which he named *Podocnemis harrisi*. This author identified this material as a pleudodiran and attributed the species to *Podocnemis* given the lack of diagnostic features that would allow the recognition of a new genus. However, Schmidt (1931) indicated that the reference of this species to *Podocnemis* was unjustifiable. The generic assignment of this species has varied. It has been included in *Naiadochelys* (von Huene 1927; Staesche 1929), *Taphrosphrys* (von Huene and 1929), *Podocnemis* (Staesche 1937) and cf. *Roxochelys* (Price 1953). In addition, Moraes Rego (1935) reported that two species are present in Bauru Basin (*Podocnemis harrisi* and *Naiadochelys* sp.), ignoring the fact that both reports were based on the same specimen (the holotype of *P. harrisi*) (Price 1953). Moreover, Candeiro et al. (2006, p. 927) reported that the holotype of *P. harrisi* is DGM 287, but it is known that Pacheco (1913) did not indicate a collection number for the type specimen and that the type specimen is lost (Price 1953).

Wanderley (1936) reported on turtle remains from west of São Paulo State. Photographs of these were sent to Karl von Staesche to identify. On the basis of two photographs of the single carapace, two photographs of a fragment of the plastron and one photograph of three associated costals, Staesche (1937) described a new species named *Podocnemis brasiliensis*. This author assumed the specimens were from the same species and that the carapace and plastron would probably be the same specimen, given the coincidence of proportions. Simpson (1943) considered *P. brasiliensis* as a synonym junior of *P. harrisi* and after a carefully inspection on the specimens housed on DNPM collections Price (1953) concluded that the photographs used by Staesche (1937) consisted of at least two different species. Thus, Price (1953) erected a new genus and species, *Roxochelys wanderleyi*, based on the portion of the carapace (holotype: DGM 216-R) and retained the name *Podocnemis brasiliensis* for the plastron (lectotype: DGM 214-R). In the same work, Price indicated that the type-material of *P. harrisi* could not be found nor directly compared to *Roxochelys* because the xiphiplastron of the holotype of *R. wanderleyi* is not preserved. However, he emphasized that, due to the large thickness of the plastron, *P. harrisi* would be more closely related to *Roxochelys*, which disagreed with Staesche (1937) who indicated that *P. harrisi* would be a species of *Podocnemis* based on the position of the ischium scar.

The first cranium-based species was erected by Suárez (1969a, b) and named *Podocnemis elegans*. Suárez (1969b) illustrated but did not indicate the number of the holotype (a nearly complete shell and crushed skull) and the paratype (a skull). Candeiro et al. (2006) mistakenly indicated that UFRGS 148 and MN 4487-V would be the type materials of this species. The former specimen is not a turtle specimen as P. Romano found while visiting UFRGS vertebrate fossils collection (in fact, this acronym has not been used in this collection for a long time and the number 148 is reserved for a dicynodont; Schultz 2010, personal communication). The second specimen correspond to cervical vertebra (probably of *Bauruemys elegans*) referred to in Kischlat (1994). In our examination of the DNPM collection we found one carapace (specimen MCT 1492-R) that matches with holotype description. The paratype is currently on loan to Americam Museum of Natural History, New York. The type-locality of this species, informally called "Tartaruguito" (literally "turtle in the rock") site is the richest turtle locality in Brazil and several complete or nearly complete specimens with associated cranium and post cranium have been collected from this locality (Suárez 2002; Oliveira and Romano 2007; Romano and Azevedo 2007). Although this species is represented by several specimens, a complete description of the taxon has not been published. However, its phylogenetic relationships have been accessed via cladistic analysis in recent studies (Romano and Azevedo 2006; França and Langer 2006; Meylan et al. 2009; Cadena et al. 2010). Moreover, Romano and Azevedo (2007) conducted a statistical analysis of several topotype specimens of this species and concluded that they are from a single population. A redescription of this species (and two new species from Peirópolis, Minas

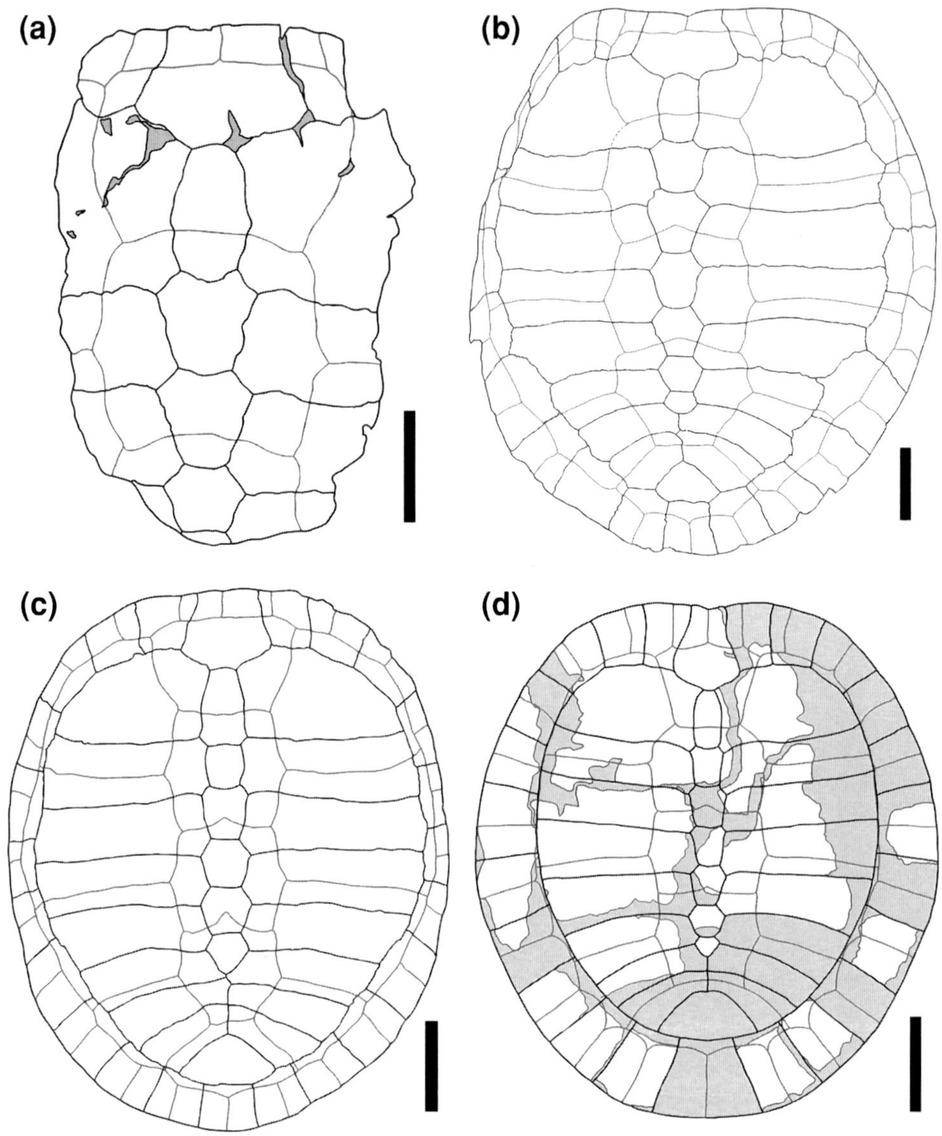

Fig. 16.9 Reconstruction of the carapace of three species of Bauru Basin side-necked turtles. **a** *Roxochelys wanderleyi* (based on DGM 216-R). **b** *Roxochelys wanderleyi* (based on MCT 1787-R). **c** *Bauruemys elegans* (based on MN 6674-V). **d** *Cambaremys langertoni* (based on CPP-0252; modified from França and Langer 2005). Missing parts depicted in *gray*. Scale bars equal 50 mm

Gerais State, Brazil) is being undertaken by Gaffney's pleurodiran working group (Gaffney 2009, personal communication).

Recently, a new species, named *Cambaremys langertoni*, has been proposed (França and Langer 2005). The authors, however, indicate that "both features that putatively distinguish *Cambaremys* from "*Podocnemis*" *brasiliensis* [anal notch shallower and pelvic sutures of plastron less transversely expanded in *Cambaremys*] could be explained if the former was considered a younger individual" (França and Langer 2005, p. 408). As emphasized by França and Langer (2005), Wood and Diaz de Gamero (1971) demonstrated (in *Bairdemys venezuelensis*) that older individuals would be expected to have a more deeply incised anal notch. Moreover, "no autapomorphic feature was recognized for *Cambaremys*" (França and Langer 2005, p. 408) and the differences observed can be explained as a result of ontogenetic change. So, the erection of a new taxon is probably unnecessary.

Although the taxonomy of turtle species from the Bauru Basin has been discussed a consensus has not been reached. Wood and Diaz de Gamero (1971) concluded that *Podocnemis harrisi* cannot be distinguished as a chelid or a pelomedusoid and conclude that this species should be treated as a *nomen vanum*. They also indicate that *P. brasiliensis* is clearly a pelomedusoid *incertae sedis* that should be treated as Pelomedusoides gen. et. sp. indet. (the same circumstance applies to "*Podocnemis argentinensis*" Cattoi and Freinberg 1958). More recently, Kischlat (1994) and Kischlat et al. (1994) undertook a major taxonomic revision of Bauru species, erecting a new genus (*Bauruemys*) and bringing to an end the discussion of the presence of *Podocnemis* in Cretaceous sediments. Kischlat et al. (1994) also indicated the probable existence of a new species of *Roxochelys*, but did

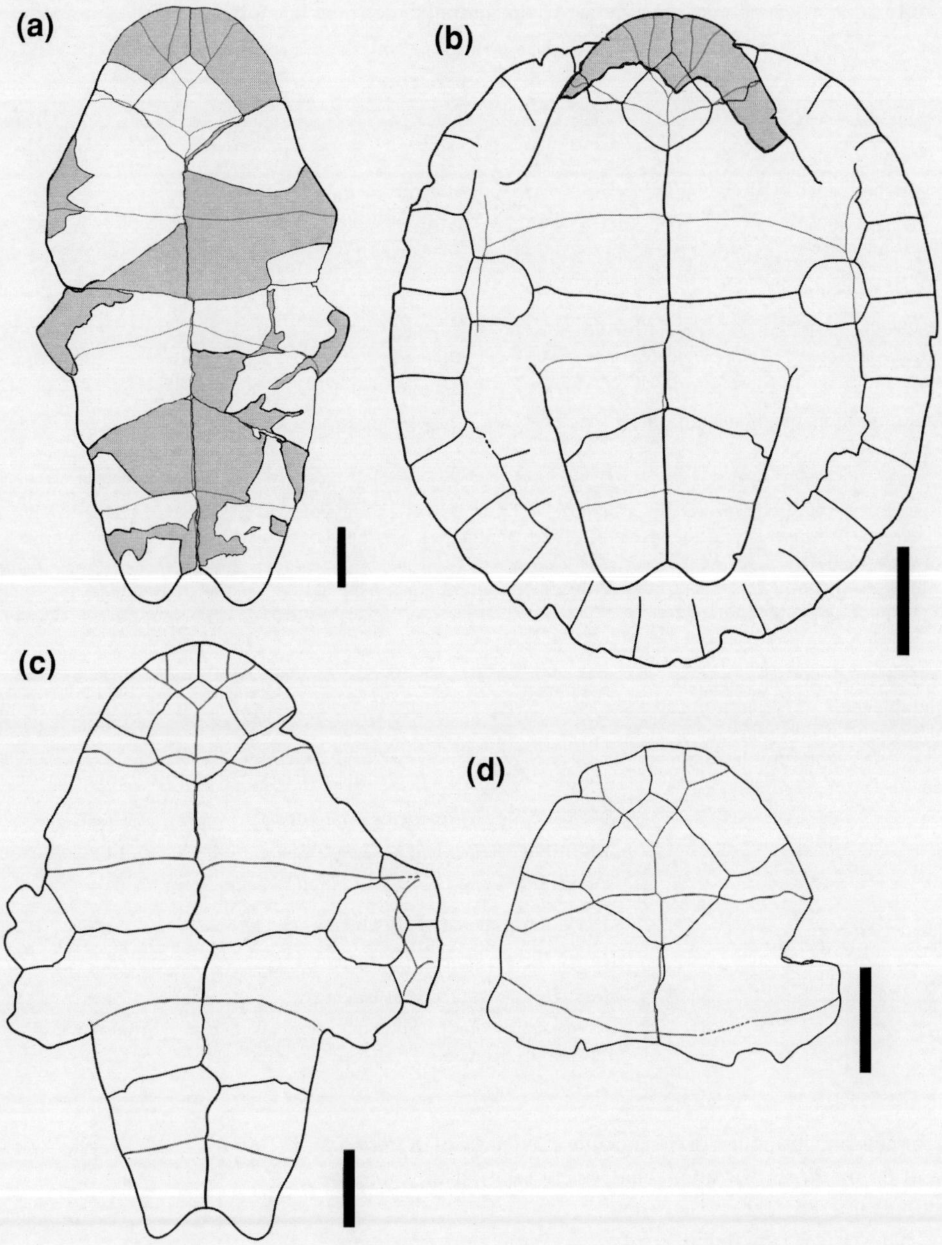

Fig. 16.10 Reconstruction of the plastron of three species of Bauru Basin side-necked turtles. **a** *Bauruemys brasiliensis* (based on DGM 214-R). **b** *Bauruemys elegans* (based on MN 6674-V, with missing parts (depicted in *gray*) based on MN 6772-V). **c** *Roxochelys wanderleyi* (based on MCT 1787-R). **d** *Roxochelys wanderleyi* (based on DGM 216-R). Missing parts depicted in *gray*. Scale bars equal 50 mm

not illustrate or mentioned the specimens that this was based on. Kischlat (1994) confirmed Price's (1953) attribution of *R. harrisi* and *R. wanderleyi* to *Roxochelys* and included the species *B. elegans* and, doubtfully, *B. brasiliensis* in his new genus *Bauruemys*.

As noted by Romano (2008) in a report of some phenotypic plasticity in *Bauruemys elegans*, the fragility of some characters used to diagnose the three genera described from the Bauru Basin should be carefully evaluated. The examination of the new specimens of *Roxochelys wanderleyi* MCT 1722-R and MCT 1787-R allow the refinement of character distributions in the five proposed species (Table 16.3).

The diagnoses of *Roxochelys harrisi*, *Bauruemys brasiliensis*, and *Cambaremys langertoni* are currently unsatisfactory. The description of these species is a result of taxonomic splitting and the lack of cladistic criteria (the case of *R. harrisi* and *B. brasiliensis*) or has been neglected (the case of *C. langertoni*). As consequence, such species lack autapomorphies and are based on a conjunct of shell characteristics that show a large amount of plasticity in living species. However, species should be regarded as a hypothesis about the evolutionary relationships and the ability to distinguish a group of individual organisms; and such species do not attend to this aim. Thus, the lack of resolution of the phylogenetic placement of *C. langertoni*

Table 16.3 Comparison of some species of side-necked turtles from Bauru Basin

	Roxochelys harrisi	*Roxochelys wanderleyi*	*Baurumeys brasiliensis*	*Bauruemys elegans*	*Cambaremys langertoni*
Bone thickness	Thick	Thick	Thick	Thin	Thin
Number of neural bones	?	7	?	6[a]	7
Vertebral sulci II-III and III-IV	?	Straight	?	Cranially convex	Straight
Nuchal	?	Wider than long	?	Slight wider than long	Longer than wide
Form of neural 2	?	Hexagonal	?	Squared	Hexagonal
Contact of costal 1	?	Neurals 1 and 2	?	Only neural 1	Neurals 1 and 2
Femoral scute	?	Short than abdominal	Longer than abdominal	Longer than abdominal	?
Position of humeral-pectoral sulci	?	Over hyoplastron	Over the suture between epi- and hyoplastron	Over the suture between epi- and hyoplastron	?

Specimens analyzed: *Roxochelys wanderleyi* (DGM 216-R [holotype], MCT 1722-R, MCT 1787-R); *Baurumeys brasiliensis* (DGM 214-R [lectotype]); *Baurumeys elegans* (MCT 1492-R [holotype], MN 6674-V, 6761-V, 6762-V, 6772-V, 6782-V, 6789-V, 6795-V, 6796-V, 6797-V, 6800-V, 6807-V, 7017-V)

[a] One specimen of *Bauruemys elegans* has a 7th neural bone: MN 7017-V (see Romano 2008)

Missing data recorded as "?"

Table 16.4 Measurements of the thickness of epiplastron and the lenght of the suture between epi- and hyoplastron in some Bauru Basin specimens

Specimen	Thickness of epiplastron	Length of the suture between epi- and hyoplastron
DGM 216-R (*Roxochelys wanderleyi* holotype)	15.3	35.6
MCT 1722-R (*Roxochelys wanderleyi*)	7.9	34.8
MCT 1787-R *Roxochelys wanderleyi*	14.2	39.5
DGM 214-R (*Bauruemys brasiliensis* lectotype)	12.2	35.5
DGM unnumbered cf. *Bauruemys elegans*	11.7	38.6
MN 6772-V (*Bauruemys elegans*)	8.2	32.2
MN 6761-V (*Bauruemys elegans*)	7.9	32.2

All measures are in mm and were taken using Mitutoyo digital micrometer (Stainness-Hordened) of 150 mm by Romano

(França and Langer 2006) and the exclusion of *R. harrisi* and *B. brasiliensis* from all analyses of relationships may be due to this neglect.

The generic distinctness of *Roxochelys* and *Bauruemys* proposed originally by Kischlat (1994) and Kischlat et al. (1994) is corroborated by later phylogenetic studies (França and Langer 2006; Meylan et al. 2009). As predicted putatively by Kischlat et al. (1994), *Roxochelys* seams to be more closely related to living podocnemidid than *Bauruemys*.

The thickness of the plastron and carapace, which is noticeably larger in *Roxochelys*, has being used as a key character to identify and separate these two genera in all studies referred to above. However, we observed that in some specimens housed on DNPM and MN the thickness is correlated to the length of the specimens and not to a species-level distinctiveness (Table 16.4; Fig. 16.11). An allometric growth relationship between the length of the suture between epi- and hyoplastron and its thickness may be present and it seems to be unreasonable to use the

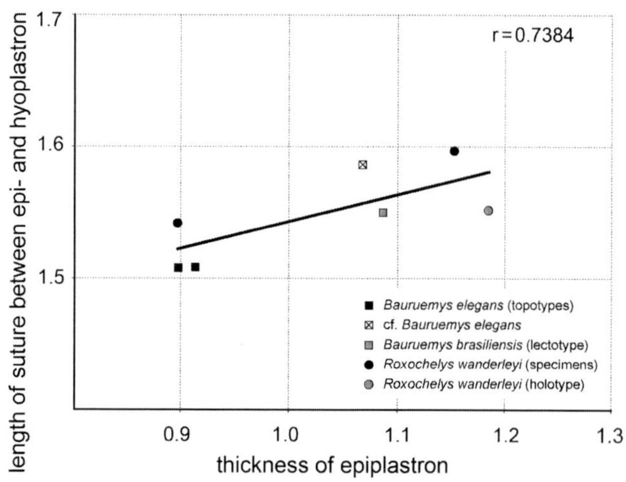

Fig. 16.11 Linear regression of the log-transformed (base 10) measurements of the thickness of epiplastron (absciss) versus the length of the suture between epi- and hyoplastron (ordinate). See measurements in Table 16.4

thickness of shell bones as a diagnostic feature to segregate *Roxochelys* and *Bauruemys*. It is more liable to assume the thickness of shell is due to distinct ontogenetic stages or even taphonomic process.

The turtle shell, in general, is characterized by a conservative morphology and shows a great deal of individual variation as well. As a consequence, it is difficult to look for susceptible homology tests through phylogenetic approach based on shell characters. Thus, it is a risky idea to establish new taxa based only on fragmentary and little known shell specimens. This is the case of four of the five proposed species from the Bauru Basin and, based on the current knowledge, we believe that the best approach is to lump the poor-known species that cannot be satisfactory differentiated. Of course, this decision will lead us to the proposition of synonymies or to continue considering some species as *nomina dubia* or *nomina vanum*.

Conclusions

AMNH 30562 shares more characteristics with *Euraxemys essweini*: the absence of fontanels in the midline of the plastron, short midline contact of the epiplastra, entoplastron slightly longer than wide, and the xiphiplastron forming most of the posterior lobe. However, no apomorphic feature could be determinate in AMNH 30562 that supports attribution of this specimen to that species or the proposition of a new species. Moreover, the morphometrical analyses indicate Podocnemidera affinities of AMNH 30652, particularly with *Cearachelys* and *Peltocephalus*. Therefore, we can conclude that AMNH 30652 is probably the most ancient Podocnemidera identified so far, extending the chronological distribution of this clade to Aptian/Albian and as old as its sister group Pelomedusera.

MCT 1722-R and MCT 1787-R are new specimens of *Roxochelys wanderleyi* and extend the chronological and biogeographic distribution of this species (to the Coniacian-Maastricthian and to more localities in the Bauru Basin). Previously (Romano et al. 2009) we argued that (1) *Bauruemys brasiliensis* and *Bauruemys elegans* might represent the same species; (2) that since *Cambaremys langertoni* was based on a young individual it is probably a junior synonym of *Roxochelys wanderleyi* (and not *B. brasiliensis* as presumed by França and Langer 2005) and the differences observed in the single known individual of this species is due to ontogeny; and (3) that it is impossible to distinguish adequately *Roxochelys harrisi* from the other Bauru Basin species and, due to the Principle of Priority, this species is potentially the senior synonym of *Roxochelys wanderleyi* and *Bauruemys brasiliensis*. So, as it is almost impossible to establish a neotype for this species, we agree with Wood and Diaz de Gamero (1971) that *R. harrisi* is a *nomen vanum* (contra de Broin's 1991 opinion that *R. harrisi* is a senior synonym of *R. wanderleyi*). However, the idea of lumping species names is not in agreement with all taxonomists working on the group. Even so, it is clear that *R. harrisi* and *B. brasiliensis* are problematic name bearing entities that should be considered *nomina dubia* due to the lack of good diagnostic characteristics. Although we conclude that *C. langertoni* does not present enough characters to support a different genus and species and that it should be considered a junior synonym of *R. wanderleyi*, the single specimen that represents this species shows some differences that allow the interpretation that this species is *incertae sedis* instead (a preferable choice for some colleagues). Nevertheless, with the better understandings of *Roxochelys* diagnostic characters it is possible to conclude that only *R. wanderleyi* and *Bauruemys elegans* are well defined species. Thus, with five species the diversity of the Bauru Basin has been probably overestimated and with an improved ability to distinguish known species the taxonomic diversity of Bauru Basin turtles is becoming better resolved.

Finally, the material reported on in this paper allows a refinement of the minimum age estimation of the Pleurodira phylogenetic tree. The minimum time of divergence of the node between the Podocnemidera and Pelomedusera is proposed as Aptian/Albian. Also, one of the oldest undisputed Podocnemididae (*Roxochelys*) is Coniacian-Maastricthian. This is an important source of information to the current enterprise of calibration of time of divergence of taxonomic groups for studies undertaken by molecular systematic researchers.

Acknowledgments We are grateful to D. Brinkman (Royal Tyrrell Museum of Palaeontology) for inviting and encouraging us to submit this chapter and also for all the assistance during Gaffney Turtle Symposium in 2009. We would like to thank E. Gaffney, J. Maisey, C. Mehling, and J. Galkin (FARB collections, Division of Paleontology, American Museum of Natural History) and R. Cassab and R. Machado (Museu de Ciências da Terra, Departamento Nacional de Produção Mineral) for access to collections and to loan the material under their care. M. Oliveira (Museu Nacional/UFRJ) is thanked for the illustration of AMNH 30652. D. Brinkman (Royal Tyrrell Museum of Palaeontology) read earlier drafts of this paper and made a heedful revision contributing to its improvement. We are also grateful to M. Ryan (Cleveland Museum of Natural History), A. Pérez-García (Universidad Complutense de Madrid), and E. Gaffney (American Museum of Natural History) for their careful reviews of this manuscript. We thank the support from CNPq (Conselho Nacional de Desenvolvimento Científico e Tecnológico) grants to G. Oliveira (#140812/2007-5), P. Romano (#142330/2006-0), and A. Kellner (#304965/2006-5); FAPERJ (Fundação Carlos Chagas Filho de Amparo a Pesquisa do Rio de Janeiro) grant to A. Kellner (#E-26/152.885/2006); and Royal Tyrrell Museum Cooperating Society travel grants to G. Oliveira and P. Romano.

References

Bertini, R. J., Santucci, R. M., Toledo, C. E. V., & Menegazzo, M. C. (2006). Taphonomy and depositional history of an Upper Cretaceous turtle-bearing outcrop from the Adamantina Formation, Southwestern São Paulo State. *Revista Brasileira de Paleontologia, 9*, 181–186.

Cadena, E., Bloch, J., & Jaramillo, C. (2010). New podocnemidid turtle (Testudines; Pleurodira) from the Middle–Late Paleocene of Tropical South America. *Journal of Vertebrate Paleontology, 30*, 367–382.

Campos, D. A. (1977). *Tartarugas fósseis do Brasil*. Master degree Dissertation, Universidade Federal do Rio de Janeiro.

Campos, D. A., & de Broin, F. (1981). Tartarugas fósseis do Brasil. *Anais da Academia Brasileira de Ciências, 53*, 210–211.

Candeiro, C. R. A., Martinelli, A. G., Avilla, L. S., & Rich, T. H. (2006). Tetrapods from the Upper Cretaceous (Turonian-Maastrichtian) Bauru Group of Brazil: A reappraisal. *Cretaceous Research, 27*, 923–946.

Pacheco, D. J. (1913). Notas sôbre a geologia do valle do rio Grande a partir da fóz do rio Pardo até a sua confluencia com o rio Paranahyba. Exploração do rio Grande e seus affluentes. In *Relatorio da Commissão Geográphica e Geológica* (pp. 33–38). São Paulo: São José dos Dourados

de Broin, F. (1991). Fóssil Turtles from Bolivia. In R. Suarez-Soruco (Ed.), Fósiles y Facies de Bolivia (Vol. 12, pp. 509–527). Volume 1, Vertebrados. *Revista Técnica Yacimentos Petrolíferos Fiscales de Bolívia*.

de la Fuente, M. S. (2003). Two new pleurodiran turtles from the Portezuelo formation (Upper Cretaceous) of Northern Patagonia, Argentina. *Journal of Paleontology, 77*, 559–575.

Derby, A. O. (1896). Nota sobre a geologia e paleontologia de Matto-Grosso. *Archivos do Museu Nacional, 9*, 59–88.

Fernandes, L. A. (2004). Mapa litoestratigrráfico da parte oriental da Bacia Bauru (PR, SP, MG), escala 1:1.000.000. *Boletim Paranaense de Geociências, 55*, 53–66.

Fernandes, L. A., & Coimbra, A. M. (1996). A Bacia Bauru (Cretáceo Superior, Brasil). *Anais da Academia Brasileira de Ciências, 68*, 195–205.

Fernandes, L. A., & Coimbra, A. M. (2000). Revisão estratigráfica da parte oriental da Bacia Bauru (Neocretáceo). *Revista Brasileira de Geociências, 30*, 717–728.

Fielding, S., Martill, D. M., & Naish, D. (2005). Solnhofen-style soft-tissue preservation in a new species of turtle from the Crato formation (Early Cretaceous, Aptian) of North-East Brazil. *Paleontology, 48*, 1301–1310.

França, M. A. G., & Langer, M. C. (2005). A new freshwater turtle (Reptilia, Pleurodira, Podocnemidae) from the Upper Cretaceous (Maastrichtian) of Minas Gerais, Brazil. *Geodiversitas, 27*, 391–411.

França, M. A. G., & Langer, M. C. (2006). Phylogenetic relationships of the Bauru Group turtles (Late Cretaceous of South-central Brazil). *Revista Brasileira de Paleontologia, 9*, 365–373.

Gaffney, E. S., & Meylan, P. A. (1988). A phylogeny of turtles. In M. J. Benton (Ed.), *The phylogeny and classification of the tetrapods, Vol. 1, Amphibians, Reptiles, Birds* (pp. 157–219). Systematics Association Special Volume 35A.

Gaffney, E. S., Campos, D. A., & Hirayama, R. (2001). *Cearachelys*, a new side-necked turtle (Pelomedusoides: Bothremydidae) from the Early Cretaceous of Brazil. *American Museum Novitates, 3319*, 1–20.

Gaffney, E. S., Tong, H., & Meylan, P. A. (2006). Evolution of the side-necked turtles: The families Bothremydidade, Euraxemydidae, and Araripemydidae. *Bulletin of the American Museum of Natural History, 300*, 1–698.

Gallo, V. G., Silva, H. M. A., Ramos, R. R. C., Petra, R., & Romano, P. S. R. (2009). A tartaruga Pleurodira mais antiga do Brasil. In *Anais do XXI Congresso Brasileiro de Paleontologia* (pp. 180–181). Belém, PA.

Hammer, O., Harper, D. A. T., & Ryan, P. D. (2001). Past: Paleontological statistics software package for education and data analysis. *Palaeontologia Electronica, 4*, 1–9.

von Huene, F. (1927). Contribución a la paleogeografía de Sud-América—II. Las relaciones paleogeograficas de Sud América durante el Cretáceo Superior. *Academina Nacional de Ciencias de Córdoba, 30*, 256–294.

von Huene, F. (1929). Los Saurisquios y Ornitisquios del Cretáceo Argentino. *Anales del Museo de La Plata*, Tomo III, 2°ser. 196 pp.

von Ihering, R. (1911). Fósseis de São José do Rio Preto (Estado de São Paulo). *Revista do Museu Paulista, 8*, 41–146.

Kischlat, E.-. E. (1994). Observações sobre *Podocnemis elegans* Suarez (Chelonii, Pleurodira, Podocnemididae) do Neocretáceo do Brasil. *Acta Geológica Leopoldensia, 39*, 345–351.

Kischlat, E.-E., & Campos, D. A. (1990). Some osteological aspects of Araripemys barretoi Price, 1973 (Chelonii, Pleurodira, Araripemydidae). In *Atas do I Simpósio sobre a Bacia do Araripe e Bacias Interiores do Nordeste* (pp. 387–395). Crato, CE.

Kischlat, E.-E., Barbarena, M. C., & Timm, L. L. (1994). Considerações sobre a queloniofauna do Grupo Bauru, Neocretáceo do Brasil. In *Boletim do III Simpósio sobre o Cretáceo do Brasil* (pp. 105–107). Rio Claro, SP.

Melo, M. S., & Schwanke, C. (2006). Curadoria do acervo de Testudines Linnaeus, 1758 do Museu de Ciências da Terra do Departamento Nacional de Produção Mineral, Rio de Janeiro. In V. Gallo, P. M. Brito, H. M. A. Silva, & F. J. Figueiredo (Eds.), *Paleontologia de Vertebrados: Grandes temas e contribuições científicas* (pp. 315–330). Rio de Janeiro: Interciência.

Meylan, P. A. (1996). Skeletal morphology and relationships of the early cretaceous side-necked Turtle, *Araripemys barretoi* (Testudines: Pelomedusoides: Araripemydidae), from the Santana Formation of Brazil. *Journal of Vertebrate Paleontology, 16*, 20–33.

Meylan, P. A., Gaffney, E. S., & Campos, D. A. (2009). *Caninemys*, a new side-necked turtle (Pelomedusoides: Podocnemididae) from the Miocene of Brazil. *American Museum Novitates, 3639*, 1–26.

Mezzalira, S. (1959). Nota preliminar sobre as recentes descobertas paleontológicas no Estado de São Paulo no período de 1958–1959. *Notas prévias, Instituto Geográfico e Geológico, 2*, 1–9.

Mezzalira, S. (1989). *Os fósseis do Estado de São Paulo*. São Paulo: Instituto Geológico, Série Pesquisa.

Mezzalira, S. (2000). Os fósseis do Estado de São Paulo—Parte II—Período 1987 (parcial) -1996. *Boletim do Instituto Geológico, 15*, 23–25.

Mones, A. (1986). Palaeovertebrata sudamericana. Catálogo sistemático de los vertebrados fósiles de America del Sur. Parte I. Lista preliminar y bibliografia. *Courier Forschungsinstitut Senckenberg, 82*, 1–625.

Moraes Rego, L. F. (1935). Camadas cretaceas do sul do Brasil. *Annuario da Escola Polytechnica, 4*, 231–274.

Neumann, V. H., & Cabrera, L. (1999). Una Nueva Propuesta Estratigráfica para la Tectonosecuencia Post-Rifte de la Cuenca de Araripe, Nordeste de Brasil. In *Simpósio sobre o Cretáceo Do Brasil, 5* (pp. 279–285), Anais, Serra Negra: UNESP.

Oliveira, G. R., & Kellner, A. W. A. (2005). Note on a plastron (Testudines, Pleurodira) from the Lower Cretaceous Crato Member, Santana Formation, Brazil. *Arquivos do Museu Nacional, 63*, 523–528.

Oliveira, G. R., & Kellner, A. W. A. (2007a). A new side-necked turtle (Pleurodira, Pelomedusoides) from the Santana Formation (Early Cretaceous), Araripe Basin, Northeastern Brazil. *Zootaxa, 1425*, 53–61.

Oliveira, G. R., & Kellner, A. W. A. (2007b). Taxonomic status of *Araripemys "arturi"* Fielding, Martill & Naish, 2005 (Testudines, Pleurodira, Araripemydidae). In I. S. Carvalho, R. C. T. Cassab,

C. Schwanke, M. A. Carvalho, A. C. S. Fernandes, M. A. C. Rodrigues, et al. (Eds.), *Paleontologia: Cenários de Vida* (pp. 393–399). Rio de Janeiro: Editora Interciência.

Oliveira, G. R., & Romano, P. S. R. (2007). Histórico dos achados de tartarugas fósseis do Brasil. *Arquivos do Museu Nacional, 65*, 113–133.

Oliveira, G. R., Romano, P. S. R. Azevedo, S. A. K., & Kellner, A. W. A. (2009). The first Podocnemidera remains (Testudines, Pleurodira) from the Crato Formation, Araripe Basin, Brazil. In *Gaffney turtle symposium (October 17–18, 2009, Royal Tyrrell Museum, Drumheller, Canada), Abstract Volume* (pp. 130–140).

Pons, D., Berthou, P. Y., & Campos, D. A. (1990). Quelques observations sur la palynologie de l'Aptien supérieur et de l'Albien do bassin d'Araripe (N-E du Brasil). In *Simposio sobre a Bacia do Araripe e das Bacias Interiores do Nordeste* (Vol. 1, pp. 142–252). Crato: Universidade Regional do Cariri.

Price, L. I. (1953). Os quelônios da Formaçâo Bauru, Cretáceo terrestre do Brasil meridional. *Boletim da Divisão de Geologia e Mineralogia, Departamento Nacional de Produção Mineral, 147*, 1–34.

Pritchard, P. C. H. (1988). A survey of neural bone variation among recent chelonian species, with functional interpretations. *Acta Zoologica Cracoviensia, 31*, 625–686.

Pritchard, P. C. H., & Trebbau, P. (1984). *The Turtles of Venezuela*. Society for the Study of Amphibians and Reptiles.

Riff, D., Romano, P. S. R., Oliveira, G. R., & Aguilera, O. (2010). Neogene crocodile and turtle fauna in northern South America. In C. Hoorn, H. Vonhof, & F. Wesselingh (Eds.), *Amazonian, landscape and species evolution: A look into the past* (pp. 259–280). Oxford: Wiley Blackwell.

Rohlf, F. J. (2008). *TPS Dig Version 2.12*. New York: Stony Brook.

Romano, P. S. R. (2008). An unusual specimen of *Bauruemys elegans* and its implications for the taxonomy of the side-necked turtles from Bauru Basin (Upper Cretaceous of Brazil). *Journal of Vertebrate Paleontology, 28* (suppl. 3), 133A–134A.

Romano, P. S. R., & Azevedo, S. A. K. (2006). Are extant podocnemidid turtles relict of a widespread Cretaceous ancestor? *South American Journal of Herpetology, 1*, 175–184.

Romano, P. S. R., & Azevedo, S. A. K. (2007). Morphometric analysis of the Upper Cretaceous Brazilian side-necked turtle *Bauruemys elegans* (Suárez, 1969) (Pleurodira, Podocnemididae). *Arquivos do Museu Nacional, 65*, 395–402.

Romano, P. S. R., Oliveira, G. R., Azevedo, S. A. K., & Campos, D. A. (2009). Lumping the podocnemidid turtles species from Bauru Basin (Upper Cretaceous of southeastern of Brazil). In *Gaffney turtle symposium (October 17–18, 2009, Royal Tyrrell Museum, Drumheller, Canada), Abstract Volume* (141–152).

Romano, P. S. R., Oliveira, G. R., & Azevedo, S. A. K. (2012). Filogenia e biogeografia de Pleurodira (Testudines). In V. Gallo, P. M. Brito, H. M. A. Silva, & F. J. Figueiredo (Eds.), *Paleontologia de Vertebrados: Relações entre América do Sul e África* (pp. 225–241). Rio de Janeiro: Interciência.

Schmidt, K. P. (1931). A fossil turtle from Peru. *Field Museum of Natural History Geological Series, 4*(8), 251–254.

Simpson, G. G. (1943). Turtles and the origin of the fauna of Latin America. *American Journal of Science, 241*, 413–429.

Staesche, K. (1929). Schildkrötenreste aus der Oberen Kreide Patagoniens. *Palaeontographica, 72*, 103–123.

Staesche, K. (1937). Podocnemis brasiliensis n. sp. aus der oberen Kreide Brasiliens. *Neues Jahrbuch der Mineralogie, Geologie und Paläeontologie (Abteilung B), 77*, 291–309.

Suárez, J. M. (1969a). Um quelônio da Formação Baurú. In *Anais do XXIII Congresso Brasileiro de Geologia* (pp. 167–176). Salvador, BA.

Suárez, J. M. (1969b). Um quelônio da Formação Bauru. *Boletim da Faculdade de Filosofia, Ciências e Letras de Presidente Prudente, 2*, 35–54.

Suárez, J. M. (1973). *Contribuição à geologia do extremo oeste do Estado de São Paulo*. Ph.D. Dissertation, Faculdade de Filosofia, Ciências e Letras de Presidente Prudente.

Suárez, J. M. (2002). Sítio fossilífero de Pirapozinho, SP: extraordinário depósito de quelônios do Cretáceo. In C. Schobbenhaus, D. A. Campos, E. T. Queiroz, M. Winge, & M. L. C. Berberet-Born (Eds.), *Sítios Geológicos e Paleontológicos do Brasil* (pp. 49–54). Rio de Janeiro: Departamento Nacional de Produção Mineral.

Wanderley, A. L. (1936). Estudos geológicos e paleontológicos na variante Araçatuba-Jupiá. In E. P. Oliveira (Ed.), *Relatorio annual do director. Anno de 1935* (pp. 103–110). Rio e Janeiro: Departamento Nacional da Producção Mineral, Serviço Geologico e Mineralogico.

Williams, E. E. (1954). A key and description of the living species of the genus *Podocnemis* (Sensu Boulenger) (Testudines, Pelomedusidae). *Bulletin of the Museum of Comparative Zoology, 3*, 279–295.

Wood, R. C., & Diaz de Gamero, M. L. (1971). *Podocnemis venezuelensis*, a new fossil pelomedusid (Testudines, Pleurodira) from the Pliocene of Venezuela and a review of the history of Podocnemis in South America. *Breviora, 376*, 1–23.

Zelditch, M. L., Swiderski, D. L., Sheets, H. D., & Fink, W. L. (2004). *Geometric morphometrics for biologists: A primer*. London: Elsevier Academic Press.

Chapter 17
Nostimochelone lampra gen. et sp. nov., an Enigmatic New Podocnemidoidean Turtle from the Early Miocene of Northern Greece

Georgios L. Georgalis, Evangelos Velitzelos, Dimitrios E. Velitzelos, and Benjamin P. Kear

Abstract A new podocnemidoidean turtle, *Nostimochelone lampra* gen. et sp. nov., was recently recovered from littoral marine-estuarine sediments of the lower Miocene Zeugostasion Formation, near the village of Nostimo in northwestern Macedonia, Greece. This new taxon is characterized by a mosaic of primitive and derived features most notably the presence of a broad embayment on the anterior carapace margin, which involves both the nuchal (whose width > length) and first pair of peripherals, a continuous series of six markedly elongate and very narrowed hexagonal neural bones, extension of the axillary buttress onto the midline of the anteroposteriorly elongate costal I (leaving a concave scar) and also laterally across the peripheral II-peripheral III suture, medial contact of the humeral scutes (implying a small intergular), and extensive overlap of the pectoral scutes on the entoplastron, probably extending to the epiplastral-hyoplastral suture. Conclusive phylogenetic placement of *Nostimochelone* is difficult to establish because the remains are incompletely preserved. Nevertheless, its discovery is significant because it represents both the first record of a pleurodiran turtle from Greece and also one of only a handful of fossil podocnemidoidean occurrences thus far documented from the Neogene of Europe.

G. L. Georgalis
School of Chemical Engineering, Aristotle University of Thessaloniki, 54124 Thessaloniki, Greece
e-mail: dimetrodon82@yahoo.gr

E. Velitzelos · D. E. Velitzelos
Department of Historical Geology and Paleontology, National and Kapodistrian University of Athens, 15784 Athens, Greece
e-mail: velitzel@geol.uoa.gr

D. E. Velitzelos
e-mail: dvelitzel@geol.uoa.gr

B. P. Kear (✉)
Palaeobiology Programme, Department of Earth Sciences, Uppsala University, Villavägen 16, SE-752 36 Uppsala, Sweden
e-mail: benjamin.kear@geo.uu.se

Keywords Burdigalian • Littoral marine • Mediterranean Europe • Neogene • Podocnemidoidea

Introduction

Podocnemidoideans (= Podocnemidoidea sensu Gaffney et al. 2006) represent one of the most diverse and geographically widespread clades of pleurodiran turtles with a fossil record extending back to the Early Cretaceous (Albian) of South America (see de Lapparent de Broin 2000a). Gaffney et al. (2006) provided a comprehensive overview of comparative morphology between the recognized lineages, and reinforced affinity of the constituent family-level clades—Bothremydidae (Albian-Eocene of Africa/Madagascar, the Middle East, India, South America, North America and Europe) and Podocnemididae (Cenomanian-Holocene of Africa/Madagascar, India, South America, and Europe). Bothremydidae + Podocnemididae monophyly was also demonstrated by the cladistic analyses of [Broin in] Antunes and de Broin (1988), Meylan (1996), de Lapparent de Broin and Murelaga (1999), de Lapparent de Broin (2000a), and França and Langer (2006). Since then, further studies have expanded our understanding of bothremydid diversity (e.g., Gaffney et al. 2007, 2009a, b) and attempted to clarify phylogenetic relationships amongst podocnemidids (e.g., Vargas-Ramirez et al. 2008; Meylan et al. 2009; Cadena et al. 2010).

In Europe, the documented record of podocnemidoideans has focused on the extensive radiation of bothremydids from the Late Cretaceous to Paleogene (see Gaffney et al. 2006 for summary); however, a relatively short-lived migration of podocnemidids also entered southern Europe during the Eocene and there are isolated reports of pelomedusoids from the Oligocene (de Lapparent de Broin 2001 and references therein). In contrast, European Neogene podocnemidoidean remains are very rare with only a partial carapace (which cannot be located at present), described as "*Podocnemis*" *lata* Ristori 1895, from the Miocene of Malta. de Lapparent de

D. B. Brinkman et al. (eds.), *Morphology and Evolution of Turtles*, Vertebrate Paleobiology and Paleoanthropology,
DOI: 10.1007/978-94-007-4309-0_17, © Springer Science+Business Media Dordrecht 2013

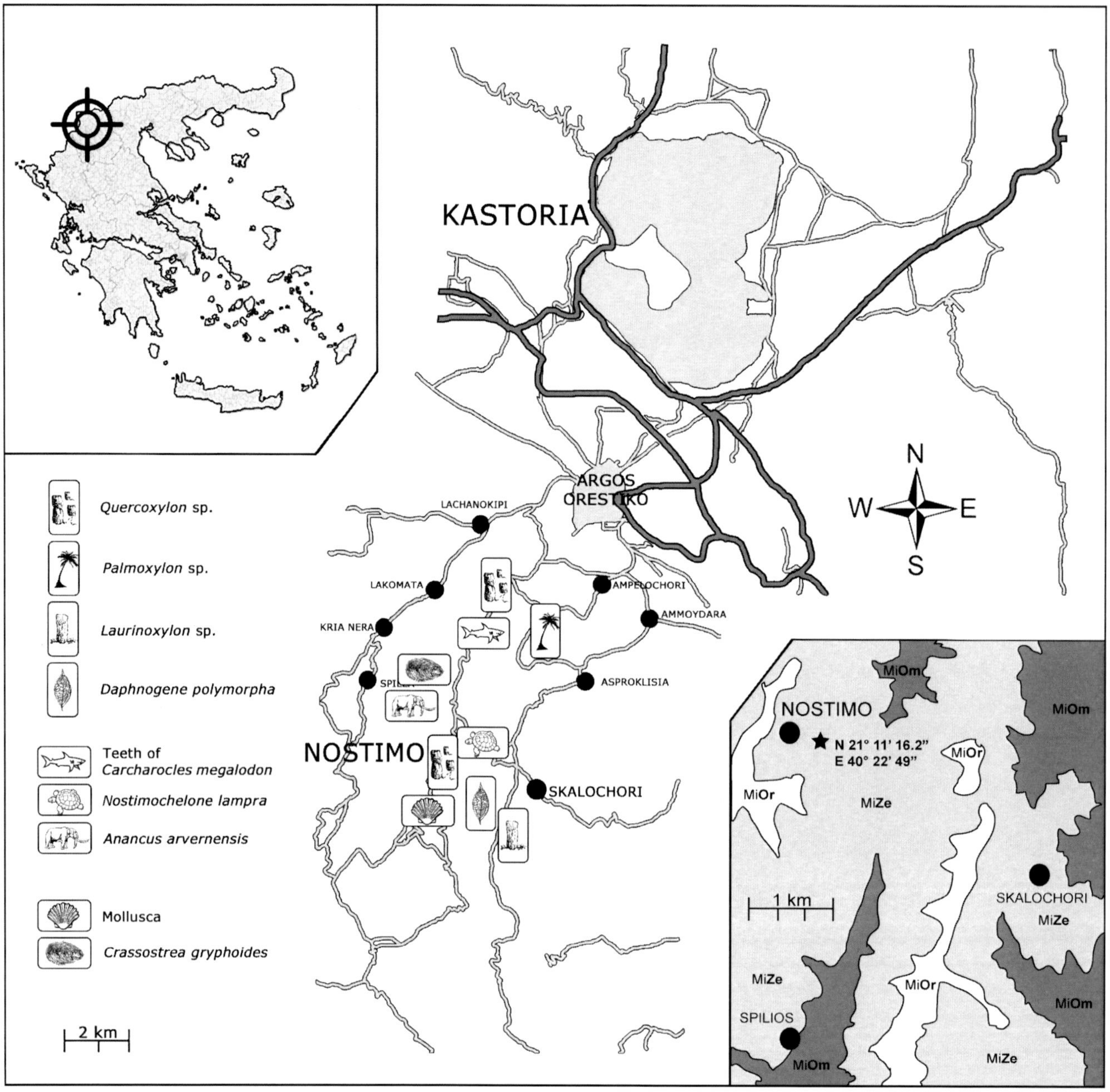

Fig. 17.1 Locality map showing the distribution of fossil sites and identifiable macrofossils (see symbols along *left side* of larger map for plant, shark, reptile, and mammals fossils), including the holotype of *Nostimochelone lampra* gen et sp. nov. (NMP V1) found near the village of Nostimo, Greece. Enlargement at bottom right indicates position of the type locality (*star*) relative to outcrops of the Zeugostasion (Mi**Ze**), Omorfokklissia (Mi**Om**), and Orlia (Mi**Or**) formations (geological mapping based on Savoyat et al. 1971)

Broin and Werner (1998) and de Lapparent de Broin (2000b, 2001) referred this specimen to Bothremydidae, an assignment that conflicts with the hypothesized lower Eocene (Ypresian) stratigraphical limit of the group (see Gaffney et al. 2006). In accordance, Gaffney et al. (2006) designated "*P.*" *lata* a *nomen dubium* because the holotype possessed no uniquely distinguishing features, and also regarded other reports of Miocene bothremydids from North America (Collins and Lynn 1936; Gaffney and Zangerl 1968; for an alternative reinterpretation see Weems and Knight 2012) and Oman (Roger et al. 1994) to be inconclusive. The recent recovery of an articulated carapace and plastron belonging to a podocnemidoidean turtle from lower Miocene (Burdigalian) littoral marine sediments of the Zeugostasion Formation (or "Zeugostasion Series" sensu Wielandt-Schuster et al. 2004) near the village of Nostimo in the Mesohellenic Basin of northwestern Macedonia (Fig. 17.1), is therefore significant. It constitutes both the first pleurodiran turtle from Greece and also one of only a handful of Neogene podocnemidoidean occurrences thus far documented from Europe.

The pre-Messinian (latest Miocene, ca. 7 million years ago) record of fossil turtles from the Aegean rim of Greece and western Turkey is very poorly known (see de Lapparent de Broin 2001). The stratigraphically oldest remains were tentatively attributed to primitive freshwater testudinoids (cf. *Palaeochelys* von Meyer 1847) and recovered from undifferentiated middle Oligocene to middle Miocene sediments near the Sea of Marmara in Turkey (Schleich 1994). Conversely, a rich chronicle of terrestrial testudinids has been found in Greece but ranges in age from the late Miocene in Macedonia, Attica and Samos through to Pliocene–Pleistocene in the Eastern Aegean islands [e.g., Lesvos; diagnostic material is often referred to the genera *Testudo* Linnaeus 1758 and *Cheirogaster* Bergounioux 1935 (see de Lapparent de Broin 2002 for summary)]. The early Miocene pleurodiran turtle described in this chapter therefore constitutes one of the most ancient testudine fossils presently known from the Aegean region.

Institutional Abbreviations: AMNH, American Museum of Natural History, New York, USA; NMP, Nostimo Museum of Palaeontology, Nostimo Kastorias, Macedonia, Greece; TUB, Technischen Universität Berlin, Berlin, Germany. The Nostimo Museum of Palaeontology is formally affiliated with and has a shared specimen catalogue with the Geological and Palaeontological Museum at the National and Kapodistrian University of Athens.

Geological Setting

The only known specimen of *Nostimochelone* (NMP V1) was uncovered by heavy machinery during excavations for a road cutting just outside of Nostimo (east towards the village of Skalochori), approximately 30 km (by road) southwest of Kastoria in northwestern Macedonia, Greece. Characteristic fine-grained silcrete-conglomerate matrix encasing the bones, together with the occurrence of coeval marine molluscs [*Crassostrea gryphoides* (Schlotheim 1813)], shark teeth [*Carcharocles megalodon* (Agassiz 1833)], and large quantities of silicified wood [including the palm fossil "*Palmoxylon*" sp. (see Velitzelos and Velitzelos 1999)] are consistent with derivation from the Zeugostasion Formation (phonetically translated as "Zevgostation" by Savoyat et al. 1971). Wielandt-Schuster et al. (2004) alternatively termed this unit the "Zeugostasion Series" and considered it part of a geographically more widespread, unnamed formation capping the marine sequences within the Mesohellenic Basin. The Zeugostasion Formation is the primary rock unit cropping out around Nostimo and comprises mainly sandy marls with intervening thin clastic sandstone and fossiliferous conglomerate layers; the latter are about 1–2 m thick and contain numerous bivalves and gastropods (see Savoyat et al. 1971; Georgiades-Dikeoulia et al. 2000). The Zeugostasion Formation forms part of an extensive lower-middle Miocene transgressive series that locally incorporates the underlying Omorfokklissia Formation (sic "Omorfokklissia Series" sensu Wielandt-Schuster et al. 2004) and terminates with the uppermost Orlia Formation (see Savoyat et al. 1971). Paleoenvironmental interpretations suggest a subtropical coastal (littoral) setting associated with one or more high-energy fluvial outflows (Georgiades-Dikeoulia et al. 2000; Wielandt-Schuster et al. 2004). Although detailed geological mapping of the Neogene deposits immediately surrounding Nostimo has yet to be undertaken, a late Burdigalian age is inferred for the Zeugostasion Formation on the basis of foraminiferal assemblage (Savoyat et al. 1971) and Sr isotope data from adjacent sections (see Wielandt-Schuster et al. 2004).

Systematic Paleontology

Testudines Batsch 1788
Pleurodira Cope 1864
Podocnemidoidea Cope 1868 (sensu Gaffney et al. 2006)
Nostimochelone gen. nov.

Type species: *Nostimochelone lampra* sp. nov., type species by monotypy.

Etymology: *Nostimo*—(Nόστιμο), refers to the type locality for the taxon; and—*chelone* (Χελώνη; Greek), "turtle".

Diagnosis: *Nostimochelone* is assigned to Podocnemidoidea because it possesses a sutural articulation of the pelvis to the shell, an osseous connection between the carapace and plastron with well-developed axillary and inguinal processes contacting the costal bones, laterally placed, rounded mesoplastra, and an extensive overlap of the pectoral scutes on the entoplastron. Among podocnemidoideans *Nostimochelone* can be diagnosed by the following combination of character states: nuchal wider than long (width = 1.25 × length) with a broadly embayed anterior edge (concavity extending laterally to the medial parts of the first pair of peripherals); a continuous series of six markedly elongate and very narrow neurals with regular, hexagonal outline and short anterolateral sides; anteroposteriorly elongate costal I (midline length = 2.3 × that of costal II) with a concave internal axillary process scar extending to the midline of the plate and laterally crossing the juncture between peripheral II and peripheral III; inguinal buttress contacts costal V but does not extend beyond its lateral extremity; humeral scutes with a midline contact over the entoplastron; pectoral-abdominal sulcus extending well anterior to the mesoplastron.

Fig. 17.2 Holotype specimen (NMP V1) of *Nostimochelone lampra* gen. et sp. nov. **a** Underside (visceral surface) of carapace in matrix block, in oblique ventral, posterior, and slightly *left* lateral view; anterior end of carapace is to *top* and slightly to *right* of figure. **b** Natural internal cast of carapace with broken portions of the neural arches embedded in the matrix; in dorsal view, with anterior end to *top* of figure. Scale bars equal 100 mm in (**a**) and 50 mm in (**b**)

Nostimochelone lampra sp. nov.
(Figs. 17.2, 17.3)

Holotype: NMP V1, partial carapace (lacking posterior margin) and plastron, the latter is adhered to a natural matrix endocast of the carapace, which incorporates embedded fragments of the vertebrae and other bones.

Horizon, unit, locality, and age: Undifferentiated sediments of the Zeugostasion Formation; Mesohellenic Basin, immediately outside of Nostimo village (N 21° 11′ 16.2″ E 40° 22′ 49″), southwest of Kastoria, northwestern Macedonia, Greece; late Miocene (late Burdigalian) in age.

Etymology: The species designation, *lampra* (Λαμπρός; Greek), English translation "bright", alludes to this discovery helping illuminate the fossil turtle record of Greece and also honours Lampros Georgalis (Thessaloniki) who facilitated initial examination of the holotype specimen.

Diagnosis: As for the genus.

Description

Nostimochelone lampra was a mid-sized podocnemidoidean turtle, with estimated carapace length exceeding 350 mm (maximum preserved length/width = 323.7/277 mm); consistent with currently extant podocnemidoideans including *Erymnochelys* Baur 1888 and *Podocnemis* Wagler 1830 (sensu Gaffney 1988), both of which range from around 300 to over 400 mm long (see Wood 1976, 2003; de Lapparent de Broin and Negri 1993; Carvalho and Bocquentin 2002). The holotype (NMP V1) comprises an undistorted carapace and plastron encased in a block of hard, fine-grained silcrete-conglomerate. The specimen was broken into two pieces during initial excavation revealing the carapace in internal aspect (Fig. 17.2a), and producing a counterpart internal mould (Fig. 17.2b) with embedded vertebrae, peripherals, and the plastron exposed on the opposing surface (Fig. 17.3e). The carapace (Fig. 17.3a) is complete anteriorly but lacks most of its posterior margin. Its dorsal aspect is totally covered in matrix making examination of the external sutures, scute sulci, and ornamentation impossible. In addition, the lateral rows of peripherals were irregularly sheared into fragments when the shell was split open making their morphology (e.g., including positioning of the musk ducts) and intervening sutures difficult to interpret. Nevertheless, remnants of at least peripherals I–VII (and up to XI as impressions) are recognizeable on the right hand side (when oriented in life position) of the carapace and counterpart whilst peripherials I and II are present on the left. Observation of the carapace block in oblique view (Fig. 17.2a) suggests that the shell was shallowly domed and probably quite flat dorsally with a maximum bone thickness of around 6 mm. The plastron is broken along its anterior margin with only the posterior part of the entoplastron, the hyoplastra, hypoplastra, right mesoplastron, and right xiphiplastron remaining as identifiable elements. The entire left posterolateral segment of the

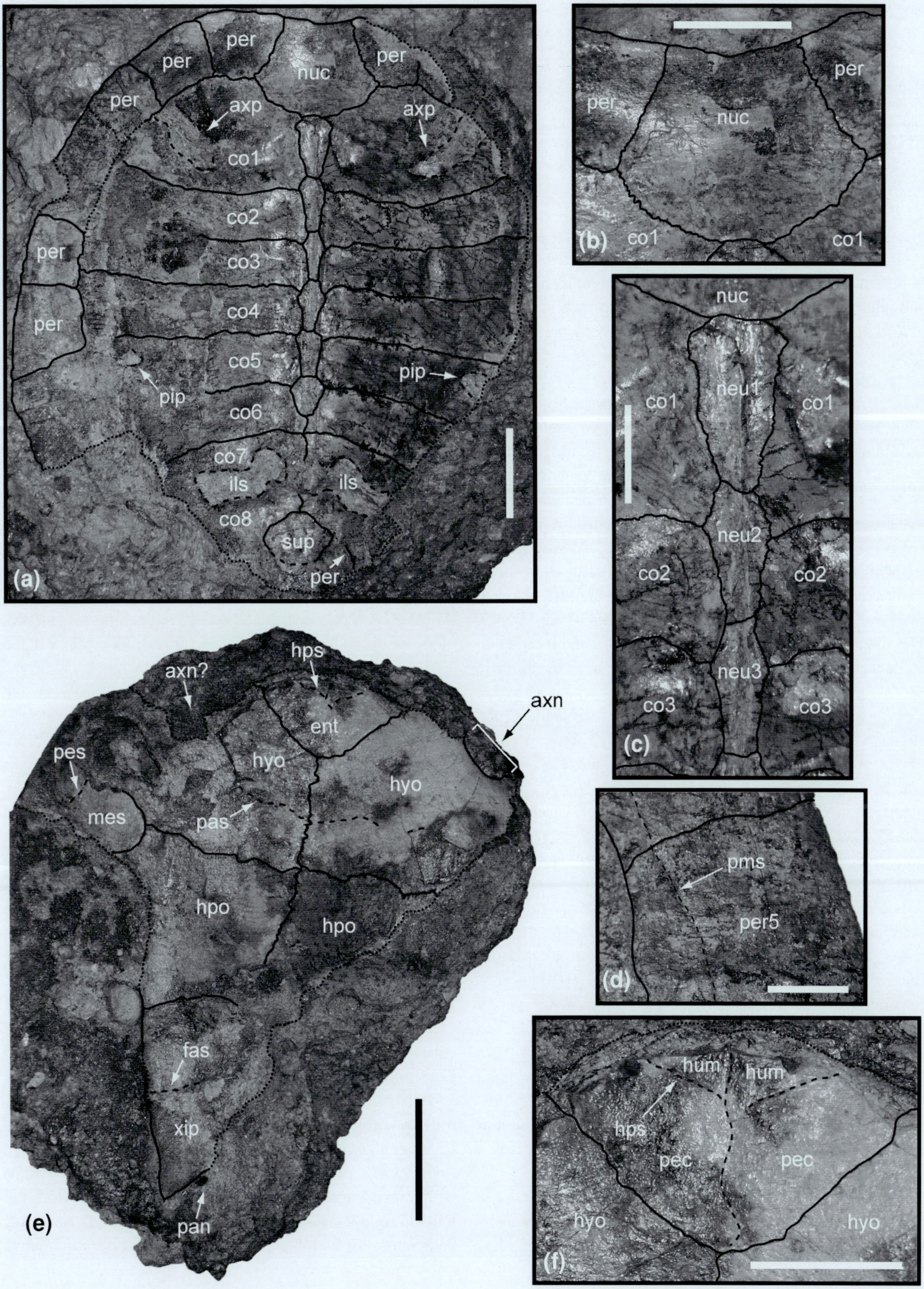
(a)
per
per
per
nuc
per
axp
axp
co1
co2
co3
co4
co5
co6
co7
co8
per
per
pip
pip
ils
ils
sup
per
(b)
per
nuc
per
co1
co1
(c)
nuc
neu1
co1
co1
neu2
co2
co2
neu3
co3
co3
(d)
pms
per5
(e)
axn?
hps
axn
ent
pes
hyo
hyo
mes
pas
hpo
hpo
fas
xip
pan
(f)
hum
hum
hps
pec
pec
hyo
hyo

◀ **Fig. 17.3** Holotype (NMP V1) carapace and plastron of *Nostimochelone lampra* gen. et sp. nov. (**a–d**) Underside (visceral surface) of carapace in ventral view: (**a**) entire preserved portion of carapace, with enlargements of the (**b**) nuchal, (**c**) neurals I–III, and (**d**) *right* peripheral V from the counterpart internal cast. (**e, f**) External surface of plastron in ventral view: (**e**) entire preserved portion of plastron, with enlargement of the (**f**) entoplastron. Scale bars equal 50 mm in (**a, e**); 30 mm in (**b, f**); 20 mm in (**c**); and 10 mm in (**d**). *Unbroken lines* = finished bone edge or suture; *dashed line* = attachment scar or scute sulcus; *stippled line* = edge of preserved bone. Abbreviations: *axp* anterior axillary process scar, *axn* axillary notch, *co1-co8* costals one through eight, *ent* entoplastron, *fas* femoral-anal sulcus, *hpo* hypoplastron, *hps* humero-pectoral sulcus, *hum* humerals, *hyo* hyoplastron, *ils* iliac sutural scar, *mes* mesoplastron, *neu1-neu3* neurals one through three, *nuc* nuchal, *pan* position of anal notch, *pas* pectoral-abdominal sulcus, *pec* pectorals, *per* peripherals, *per5* peripheral five, *pes* lateral peripheral suture, *pip* posterior inguinal process scar, *pms* pleural-marginal sulcus, *sup* suprapygal, *xip* xiphiplastron

plastron is missing and most of the external bony surface posterior to the entoplastron is heavily weathered, perhaps indicating that the shell had come to rest upon its back with subsequent exposure and erosion on the surface.

Carapace: The carapace of NMP V1 is incomplete but appears to have been rounded in outline and narrow at peripheral I, becoming wider around costals V–VII where the peripherals seem to have increased in width (Fig. 17.3a). Its bones are all strongly sutured and there are no fontanelles. The well-preserved nuchal is roughly trapezoidal in shape and slightly wider than long (66.1 mm wide, 52.7 mm midline length to suture), thus conforming to the common condition within Eupleurodira (see de Lapparent de Broin 2000a; Gaffney et al. 2006). Its posterior margin is evenly rounded and its anterior edge forms a transversely broad nuchal embayment, which is enhanced by medial inflexion of the adjacent peripherals (Fig. 17.3b). The internal surface of the nuchal is smooth and there is no evidence of an articulation with the cervical or thoracic neural spines as occurs in *Araripemys* Price 1973 (Meylan 1996), some primitive turtles including meiolaniids, and advanced marine chelonioids (Joyce 2007). There is a series of six well-defined neurals (Fig. 17.3c), the first five of which are elongate and extremely narrow (length ~50% greater than width; length/width measurements based on the clearly visible internal sutures of neurals I–V = 34.2/18.9, 28.3/16.5, 29.4/12.8, 28.7/16.8, 27.2/16.3 mm); neural VI is only slightly longer than wide (length/width = 25/21.7 mm). All of the neurals are regularly hexagonal in shape and have short anterolateral sides (neural formula = $4 < 6 < 6 < 6 < 6 < 5$). However, neural I contacts only four other bones (nuchal, left and right costal I, and neural II) thus conforming to the "quadrangular" position (sensu de Lapparent de Broin and Murelaga 1999; Gaffney et al. 2006). The posterior section of the carapace is badly damaged and there has been some infilling of both fractures and broken bone with epoxy resin for conservation and display purposes. Nevertheless, most of the anterior and left lateral sutural edges of the suprapygal can be delineated allowing for a tentative reconstruction of its shape (Fig. 17.3a).

There are eight pairs of costals (Fig. 17.3a). Costal I is lobate in outline and anteroposteriorly elongate with a midline length (65 mm on both the left and right sides) more than double (× 2.3) that of costal II (28 mm); this derived state is manifest elsewhere in bothremydids, some *Podocnemis* spp., and various chelids (see Gaffney et al. 2006). The broken sutural contact of the anterior axillary process extends medially as a curving scar (rounded at its apex) to a point approximately halfway across the internal surface of costal I, close to its midline (Fig. 17.3a). Laterally the axillary process scar crosses the border of the carapace at the contact between peripherals II and III. Costals II–VII are strap-like and have sub-parallel anterior and posterior suture lines; the shape of costal VIII cannot be accurately discerned but the presence of an intervening suture (and position of neural VI) indicates that the posterior costal bones probably met along the midline. A short contact for the inguinal process is present on the incomplete lateral extremities of the left and right costal V. All of the costal rib articulations have been broken off. An elongate iliac sutural scar extends across both costal VII and VIII but does not make any obvious contact with the suprapygal [absence of contact between the iliac scar and suprapygal has been considered derived amongst Pelomedusoides (Antunes and de Broin 1988; de Lapparent de Broin and Werner 1998; de Lapparent de Broin and Murelaga 1999)].

The first, second, third, and probably fourth peripherals are in contact with costal I (Fig. 17.3a); however, the remaining peripheral series is badly damaged making all of the subsequent sutures difficult to discriminate. Peripheral I is narrow and trapezoidal in outline. Peripherals III–VII are more rectangular with the length of each bone exceeding its breadth. The lateral, free borders of the peripherals are smoothly rounded and a portion of the anterior pleural-marginal sulcus is visible on the counterpart, situated just below the costal-peripheral suture (Fig. 17.3d). Little else can be discerned of the dorsal part of the shell and it is difficult to determine if there was any ornamentation on the carapace.

Plastron: The plastron is visible only in external (ventral) view (Fig. 17.3e). From what is preserved it appears to have been concave along the midline and longer than wide (maximum preserved length/width = 291.9/67.51 mm) with a broad anterior, and tapering posterior lobe. Both the left and right lateral extremities have been lost. The anterior components of both the entoplastron and hyoplastra are also broken off and there are no recognizable remnants of the epiplastra. The remaining portion of the entoplastron (Fig. 17.3f) is wider than long (maximum width/length = 67.5/37.8 mm as preserved in external view) and has a smoothly convex posterior margin that extends well behind the level of the axillary notches. The anteromedial surface of the entoplastron would have been covered by the humerals with the pectorals broadly underlying most of the central and posterior areas of the bone (the latter state is a potential synapomorphy for Podocnemidoidea although the amount of pectoral overlap is variable (Gaffney et al. 2006). The humeropectoral sulcus is v-shaped (angled at ~140°), and intersects the entoplastron far in front of its posterior margin and probably close to the midline—implying both that the intergular scute was small and did not extend posteriorly to separate the humerals, and also that the pectoral scutes probably extended laterally to the epiplastral-hyoplastral suture. Weathering has obliterated the lateral ends of the humeropectoral sulcus and most of the axillary notches. However, the hyoplastra were clearly wider than long (maximum width/length measured between the sutures and axillary notch on the better preserved right side = 99/86.7 mm) and firmly sutured to both the mesoplastra and hypoplastra indicating an akinetic shell [amongst pleurodires kinetic plastral hinges are found only in the extant pelomedusid *Pelusios* Wagler 1830 (see Bramble and Hutchison 1981 for discussion)]. A section of the pectoral-abdominal sulcus is visible in the caudal half of both hyoplastra and situated well anterior of the mesoplastron. It therefore probably did not cross the mesoplastral plate, although, where it reached the bridge cannot be discerned from the present specimen [intersection of the mesoplastra by the pectoral-abdominal sulcus is considered primitive within Pleurodira (Gaffney et al. 2006, p. 641)]. The mesoplastron itself is incomplete but was clearly quite small (maximum width 42.3 mm), medially rounded and laterally placed (see Gaffney et al. 2006. pp. 628–629 for distribution of mesoplastral character states); its lateral suture to the peripheral series is preserved along the right anterolateral margin on the plastron. The hypoplastra are slightly shorter than the hyoplastra in length (maximum dimension measured between the outer edge of hyoplastral-xiphiplastral sutures on the right side = 70 mm) and were presumably covered by parts of both the abdominal and femoral scutes in life. Poor preservation has rendered most of the abdominal-femoral junction indistinct and the inguinal regions are missing. The oblique hypoplastral-xiphiplastral suture is clearly exposed on the right hand side and a component of the femoral-anal sulcus is evident indicating a near perpendicular orientation relative to the lateral edge of the bone. The sharply tapered distal extremity of the xiphiplastron (angled at ~50°) is laterally deflected implying a deep, triangular anal notch; this is a derived trait amongst pleurodires (Gaffney et al. 2006).

Phylogenetic Affinities of *Nostimochelone lampra*

The lack of adequate cranial or appendicular remains, together with the obscured carapace scute sulci of the holotype, precludes examination of *Nostimochelone lampra* using existing phylogenetic data sets—de Lapparent de Broin and Murelaga (1999) [LBM] and Gaffney et al. (2006) [GTM] are employed here because they focused on the broader relationships of podocnemidoideans and incorporated a substantial number of shell characters (numbering cited below refers to the relevant character[s] in these phylogenies). Nevertheless, it is possible to infer the placement of our new taxon based on established character definitions and the holotype specimen as preserved. For example, *N. lampra* clearly pertains to Pleurodira as indicated by its possession of a sutural articulation of the pelvis to the shell (LBM, 23; GTM, 133), which is a classic defining trait of pleurodiran turtles (see Gaffney and Meylan 1988; de Lapparent de Broin 2000a; de la Fuente and Iturralde-Vinet 2001; Joyce 2007), and the presence of laterally positioned, apparently "equidimensional" mesoplastra (LBM, 17; GTM, 158), which is considered a synapomorphy for "Panpleurodira" by Joyce (2007) or the magnafamily Podocnemidera by Gaffney et al. (2006). Placement within Podocnemidoidea is supported by the extensive overlap of the pectoral scutes on the entoplastron (LBM, 20; GTM, 165). Gaffney et al. (2006) suggested that this trait might be synapomorphic for podocnemidoideans, although, Meylan et al. (2009) listed it as diagnostic feature for podocnemidids because bothremydids tend to retain a more primitive condition (see Gaffney et al. 2006, p. 630); a conclusion that we support.

Proportions of the nuchal (width/length ratio = 1.25) in *N. lampra* are consistent with attribution to Eupleurodira [LBM, 0; GTM, 139 (see Gaffney et al. 2006)], and together with the presence of a broad nuchal-peripheral I embayment [GTM, 154; note that Gaffney et al. (2006) defined the character for the nuchal only], serve to distinguish this new taxon from some Afro-Eurasian podocnemidids such *Neochelys* Bergounioux 1954 from the Eocene of Europe (de Broin 1977; de Lapparent de Broin 2003), and remains attributed to *Shweboemys* Swinton 1939 from the late Eocene of Egypt (Andrews 1903; von Reinach 1903; Dacqué 1912; de Lapparent de Broin 2000b) and Miocene of Saudi Arabia (de Broin 1982). Both *Neochelys* and *Shweboemys* lack a nuchal embayment (*sensu stricto*), but Andrews' (1903, pl. VIII, Fig. 2A) reproduction of the nuchal in

"*Shweboemys*" *antiqua* (Andrews 1903) also depicts a more primitive width/length ratio of greater than 2 (see Gaffney and Wood 2002 for comments on the questionable monophyly of *Shweboemys* spp.). Paradoxically, a shallow nuchal embayment is present in "*Podocnemis*" *stromeri* von Reinach 1903 (see von Reinach 1903, pl. I; Dacqué 1912, p. 288, text-Fig. 6), which was synonymised with "*Shweboemys*" *antiqua* by de Lapparent de Broin (2000b).

The regular, hexagonal neurals (LBM, 8; GTM, 145) of *N. lampra* are typically eupleurodiran, with limitation of neural I to only four lateral contacts against the nuchal, first costals and neural II (= "quadrangular" sensu de Lapparent de Broin and Murelaga 1999) potentially derived for Pelomedusoides (LBM, 9 [Lapparent de Broin and Murelaga 1999]; however, Gaffney et al. (2006) thought that this feature was plesiomorphic amongst pleurodires (GTM, 144). Conversely, reduction of the neural series to only six continuous bones (LBM, 10; GTM, 146)—this condition also can be defined in terms of the terminal neural contacting the sixth pair of costals (see GTM, 141)—is a derived state variably manifest within Podocnemidoidea (de la Fuente 2003; França and Langer 2005; Gaffney et al. 2006). For example, six neurals are present in the Eurasian Late Cretaceous (Campanian–Maastrichtian) bothremydids *Kurmademys* Gaffney et al. 2001 (Gaffney et al. 2006) and *Elochelys convenarum* Laurent, Tong and Claude 2002 (Laurent et al. 2002); although, seven are evident in the type species *Elochelys perfecta* Nopcsa 1931 (see Gaffney et al. 2006, p. 562, Fig. 270C), six in *Chupacabrachelys* Lehman and Wick 2010 and six or seven in shells referred to "*Podocnemis*" *barberi* Schmidt 1940/"*Podocnemis*" *alabamae* Zangerl 1948, all from the Late Cretaceous (Campanian) of North America (see Gaffney et al. 2006, pp. 553–554; taxonomic assessment in Gaffney et al. 2009a), and "*Podocnemis*" *parva* Haas 1978a/"*Podocnemis*" *judea* Haas 1978b from the Late Cretaceous (Cenomanian) of Israel (see Gaffney et al. 2006, p. 110). Gaffney et al. (2006, pp. 563, 566–567, Figs. 273A, 274A) also reported an indeterminate pelomedusoid shell (AMNH 30550), possibly attributable to the bothremydid *Galianemys* Gaffney et al. 2002, from the Late Cretaceous (Cenomanian) of Morocco that displays six neurals but with the terminal bone contacting the seventh costal pair. Similarly, de Lapparent de Broin and Werner (1998) recorded an incomplete bothremydid carapace (TUB Vb-648; ?*Arenila krebsi* de Lapparent de Broin and Werner 1998 or Pelomedusoides indet. in Gaffney et al. 2006, p. 108) with six neurals from the Late Cretaceous (Maastrichtian) of Egypt. França and Langer (2005, and references therein) noted that either six or seven neurals can occur in fossil and living podocnemidid taxa. In addition, other morphotypes exist in which the neural series is either discontinuous (e.g., loosing anterior contact with the nuchal and permitting a median suture between the first costals as in *Stereogenys* Andrews 1901; see Andrews 1903, p. 116, pl. VII, A) or absent (e.g., *Bairdemys* Gaffney and Wood 2002; see Wood and Diaz de Gamero 1971, p. 4, Figs. 1, 6; Weems 2009, p. 192 alternatively figured six neurals in *Bairdemys* remains from the U.S.A.). *Portezueloemys* de la Fuente 2003 and *Bauremys* Kischlat 1994 from the Late Cretaceous (Maastrichtian) of South America both exhibit six neurals but typically with only the 7th and 8th costals in contact (de la Fuente 2003; França and Langer 2005). In contrast, the extant *Erymnochelys* from Madagascar, together with the African Miocene-Pliocene podocnemidids *Turkanemys* Wood 2003 and *Kenyemys* Wood 1983, all manifest six neurals variably articulating with the 6th–8th costal pairs (see Zangerl 1948; Wood 2003). Andrews (1903) described comparable neural structures in "*Shweboemys*" *antiqua* (see also "*Podocnemis*" *stromeri* in von Reinach 1903, plates I, V, VIII), and Jain (1986) hypothesized that six neurals were present in the Late Cretaceous (Maastrichtian) "*S.*" *pisdurensis* (Jain 1977) from India. Some specimens of *Neochelys arenarum* de Broin 1977 from the early Eocene (Ypresian) of France are also known to have six neurals (de Broin 1977) while other species have seven (de Broin 1977; Jiménez Fuentes 1993). Ristori (1895, pl. I, Fig. 1) reconstructed the posterior-most neural in the fragmentary holotype of "*Podocnemis*" *lata* as being in contact with the sixth pair of costals (see also Gaffney et al. 2006, p. 115); however, poor preservation of this material (all but one of the proceeding neurals are missing) prevents more informative comparisons.

Nostimochelone lampra displays a uniquely elongate and transversely compressed neural outline (neurals I–V length/width ratios = 1.8, 1.7, 2.3, 1.7 and 1.7 respectively; see Fig. 17.3a, c). Similar shapes have been figured in European Cretaceous-Paleogene dortokids (in which long and short neurals alternate, e.g., *Ronella* Lapparent de Broin 1999 in Gheerbrant et al. 1999; see carapace photographs in de Lapparent de Broin et al. 2004, p. 214, pl. II, Figs. 1, 4), some bothremydids including *Kurmademys* (in which neural length/width ratios appear to range up to 2 based on the carapace reconstruction of Gaffney et al. 2006, p. 543, Fig. 257), and extant Australian chelids (e.g., *Elseya* Gray 1867) where extremely narrow, reduced neurals may be exposed along the midline of the shell (see Thomson and Georges 1996, p. 83, Fig. 1a–d).

The prominent suture scars extending from the internal midline of the first costals in *N. lampra* represent contact points for the axillary processes from the hyoplastra (LBM, 12; GTM, 148), which is a common trait amongst pleurodires (see de Lapparent de Broin and Werner 1998) and indicates a well-sutured articulation between the carapace and plastron (LBM, 2). Conversely, *N. lampra* differs from the typical eupleurodiran condition in the continuation of these sutures laterally across the juncture between peripheral II and III (LBM, 13; GTM, 149). de Lapparent de Broin and Werner (1998) reported that the axillary buttress scars usually only cross peripheral III in podocnemidoideans, but may pass close to the anterior edge of the bone in taxa such as *Podocnemis*. Gaffney et al. (2006) alternatively stated that the axillary process can reach the peripherals II-III suture in some chelids and podocnemidids, citing Tronc and Vuillemin (1974) for an example in *Erymnochelys* (see also character scorings for this taxon in the matrix of Meylan et al. 2009). de Lapparent de Broin and Werner (1998, p. 168) inferred that positioning of the axillary process scar at the "limit of peripherals 2 and 3" was primitive for pleurodires because it occurs in the Jurassic taxon *Platychelys* Wagner 1853. Gaffney et al. (2006) also stated that anterior continuation of the axillary process contact over peripheral II is found in *Proganochelys* Baur 1887 and thus appears to be primitive for turtles.

Although the entoplastron of *Nostimochelon lampra* is incomplete, the clearly defined intersection of the humeropectoral sulcus is positioned well anteriorly and indicates a broad overlap by the pectoral scutes potentially as a far as the epiplastral-hyoplastral suture (LBM, 20; GTM, 165). This accords with the derived condition indicative of Podocnemidoidea and more concisely Podocnemididae—thus excluding pelomedusids in which the humeropectoral sulcus passes along the posterior edge or well behind the entoplastron (see Zangerl 1948; de Lapparent de Broin and Werner 1998; de Lapparent de Broin 2008, p. 116, pl. III, Fig. 3). Confluence of the humeral scutes suggests that the intergular was probably small, as in most podocnemidids (GTM, 170); although overlap of the anterior entoplastron by the intergular has been shown to be variable at genus-level (see Zangerl 1948, p. 43; Wood 2003, pp. 123 and 121, Fig. 4.10), and in some cases, the contact between the humerals can be much reduced [e.g., "*Podocnemis*" *stromeri* (Zangerl 1948)] or lost altogether [e.g., *Stereogenys* and "*Shweboemys*" *pisdurensis* (Andrews 1903; Zangerl 1948; Jain 1986)].

Conclusions

Nostimochelone lampra gen. et sp. nov. is a new pleurodiran turtle from the early Miocene (Burdigalian) of Greece, a region where the pre-Messinian (prior to latest Miocene ~ 7 million years ago) fossil record of Testudines is virtually unknown (see de Lapparent de Broin 2001). The holotype and only recovered specimen displays a combination of primitive and derived traits reminiscent of podocnemidoideans, and especially African podocnemidids such as *Erymnochelys* and other closely related forms. However, *N. lampra* does not conform to any current genus-level definition, especially in its uniquely narrow neural outline, and we therefore propose generic distinction for this taxon but retain it as Podocnemidoidea *incertae sedis* pending discovery of more diagnostic material. The discovery of *N. lampra* is particularly significant because it represents the first report of a pleurodire from Greece and also one of the youngest stratigraphical occurrences of Podocnemidoidea from within Europe. Association of its remains with a littoral marine to tidal estuarine depositional setting (if autochthonous) also suggests an analogous lifestyle to that envisaged for bothremydids (see summary in Gaffney et al. 2006) and some podocnemidid taxa (e.g., *Bairdemys* from the Oligocene–Miocene of the U.S.A., Puerto Rico and Venezuela; MacPhee and Wyss 1990; Gaffney and Wood 2002; Weems 2009), which are often associated with brackish water and near-shore marine settings. Such habitats are consistent with the predominantly warm, humid coastal palaeoenvironments reconstructed for eastern Mediterranean Europe in the early-middle Miocene (Suc et al. 1999; Kovar-Eder et al. 2008), and support the possibility of a Neogene Tethyan dispersal (rather than post-Gondwanan vicariance; Romano and Azevado 2006 and references therein) for podocnemidoidean turtles from Africa into Europe and Asia (sensu Meylan et al. 2009).

Acknowledgments We are indebted to Petros Papakonstantinou (Nostimo Museum of Paleontology) for his assistance with access to NMP V1 and information on the type locality/horizon. France de Lapparent de Broin generously contributed both her knowledge and time towards earlier drafts of this work. Eugene Gaffney, Adán Pérez-García, and Márton Rabi also contributed helpful reviews. Many thanks to Lampros Georgalis (Thessaloniki) for facilitating travel to the Nostimo Museum of Palaeontology for GLG and BPK. This research was supported financially by The Palaeontological Association and La Trobe University.

References

Agassiz, L. (1833–1844). *Recherches sur les Poisons Fossiles (5 volumes with supplements)*. Paris, France: Imprimerie de Patitpierre, Neuchâtel et Soleure.

Andrews, C. W. (1901). Preliminary note on some recently discovered extinct vertebrates from Egypt (Part II). *Geological Magazine, 8*, 436–444.

Andrews, C. W. (1903). On some pleurodiran chelonians from the Eocene of the Fayum, Egypt. *Annals and Magazine of Natural History (Series 8)*, 11, 115–122.

Antunes, M. T., & de Broin, F. (1988). Le Crétacé terminal de Beira Litoral, Portugal: Remarques stratigraphiques et écologiques étude complémentaire de *Rosasia soutoi* (Chelonii, Bothremydidae). *Ciências Terra, 9*, 153–200.

Batsch A. J. G. C. (1788). *Versuch einer Anleitung, zur Kenntniß und Geschichte der Thiere und Mineralien*. Jena: Akademische Buchhandlung.

Baur, G. (1887). Ueber den Ursprung der Extremitäten der Ichthyopterygia. *Bericht des Oberrheinischen Geologischen Vereins, 20*, 17–20.

Baur, G. (1888). Osteologische Notizen über Reptilien. *Zoologischer Anzeiger, 296*, 1–5.

Bergounioux, F. -M. (1935). Contribution á l' étude paléontologiques des Chéloniens. *Mémoires de la Société Géologique de France (Nouvelle Série), mémoire 25*, 1–207.

Bergounioux, F.-M. (1954). Les Chéloniens fóssiles des terrains tertiaires del Vénétie. *Mémoire degli Istituti di Geologica e Mineralogia dell'Universitá di Padova, 18*, 1–115.

Bramble, D. M., & Hutchison, J. H. (1981). A reevaluation of plastral kinesis in African turtles of the genus *Pelusios*. *Herpetologica, 37*, 205–212.

Cadena, E. A., Bloch, J. I., & Jarmillo, C. A. (2010). New podocnemidid turtle (Testudines: Pleurodira) from the middle-upper Paleocene of South America. *Journal of Vertebrate Paleontology, 30*, 367–382.

Carvalho, P., Bocquentin, J., & de Lapparent de Broin, F. (2002). Une nouvelle espèce de *Podocnemis* (Pleurodira, Podocnemididae) provenant du Néogène de la formation Solimões, Acre, Brésil. *Geobios, 35*, 677–686.

Collins, R. L., & Lynn, W. G. (1936). Fossil turtles from Maryland. *Proceedings of the American Philosophical Society, 76*, 151–174.

Cope, E. D. (1864). On the limits and relations of the Raniformes. *Proceedings of the Academy of Natural Sciences of Philadelphia, 16*, 181–183.

Cope, E. D. (1868). On the origin of genera. *Proceedings of the Academy of Natural Sciences of Philadelphia, 20*, 242–300.

Dacqué, E. (1912). Die fossilen Schildkröten Aegyptens. *Geologische und Palaeontologische Abhandlungen, 14*, 275–337.

de Broin, F. (1977). Contribution à l'étude des Chéloniens; chéloniens continentaux du Crétacé et du Tertiaire de France. *Mémoires du Muséum National d'Histoire Naturelle, Série C Géologie, 38*, 1–366.

de Broin, F. (1982). Chelonia, Testudines. In H. Thomas, S. Sen, M. Khan, B. Battail, & G. Ligabue (Eds.), *The Lower Miocene fauna of Al-Sarrar (Eastern Province, Saudi Arabia). ATLAL, Journal of Saudi Arabian Archaeology, 5*, 117–118.

de la Fuente, M. S. (2003). Two new pleurodiran turtles from the Portezuelo formation (Upper Cretaceous) of northern Patagonia, Argentina. *Journal of Paleontology, 77*, 559–575.

de la Fuente, M. S., & Iturralde-Vinet, M. (2001). A new pleurodiran turtle from the Jagua formation (Oxfordian) of western Cuba. *Journal of Paleontology, 75*, 860–869.

de Lapparent de Broin, F. (2000a). The oldest pre-podocnemidid turtle (Chelonii: Pleurodira) from the Early Cretaceous, Ceará State, BrazilCeará State, Brazil and its environment. *Treballs del Museu de Geologia de Barcelona, 9*, 43–85.

de Lapparent de Broin, F. (2000b). African chelonians from the Jurassic to the present: Phases of development and preliminary catalogue of the fossil record. *Palaeontologia Africana, 36*, 43–95.

de Lapparent de Broin, F. (2001). The European turtle fauna from the Triassic to the present. *Dumerilia, 4*, 155–218.

de Lapparent de Broin, F. (2002). A giant tortoise from the late Pliocene of Lesvos Island (Greece) and its possible relationships. *Annales Géologiques des Pays Helléniques, 39*, 99–130.

de Lapparent de Broin, F. (2003). *Neochelys* sp. (Chelonii, Erymnochelyinae), from Silveirinha, early Eocene, Portugal. *Ciências da Terra, 15*, 117–132.

de Lapparent de Broin, F. (2008). Miocene chelonians from south-western Namibia. *Memoir of the Geological Survey of Namibia, 20*, 107–145.

de Lapparent de Broin, F., & Murelaga, X. (1999). Turtles from the Upper Cretaceous of Laño (Iberian Peninsula). *Estudios del Museo de Ciencias Naturales de Alva, 14(1)*, 135–212.

de Lapparent de Broin, F., Bocquentin J., & Negri, F. R. (1993). Gigantic Turtles (Pleurodira, Podocnemididae) from the Late Miocene-Early Pliocene of South Western Amazon. *Bulletin de l'Institut Français d'Études Andines, 23*, 657–670.

de Lapparent de Broin, F., & Werner, C. (1998). New Late Cretaceous turtles from the Western Desert, Egypt. *Annales de Paléontologie, 84*, 131–214.

de Lapparent de Broin, F., Murelaga Bereikua, X., & Codrea, V. (2004). Presence of Dortokidae (Chelonii, Pleurodira) in the earliest Tertiary of the Jibou formation, Romania: Palaeobiogeographical implications. *Acta Palaeontologica Romaniae, 4*, 203–215.

França, M. A. G., & Langer, M. C. (2005). A new freshwater turtle (Reptilia, Pleurodira, Podocnemidae) from the Upper Cretaceous (Maastrichtian) of Minas Gerais, Brazil. *Geodiversitas, 27*, 391–411.

França, M. A. G., & Langer, M. C. (2006). Phylogenetic relationships of the Bauru Group turtles (Late Cretaceous of south-central Brazil). *Revista Brasileira de Paleontologia, 9*, 365–373.

Gaffney, E. S. (1988). A cladogram of the pleurodiran turtles. *Acta Zoologica Cracoviensia, 31*, 487–492.

Gaffney, E. S., & Meylan, P. A. (1988). A phylogeny of turtles. In M. J. Benton (Ed.), *The phylogeny and classification of the tetrapods. Volume 1. Amphibians, reptiles, birds* (pp. 157–219). Oxford: Clarendon Press.

Gaffney, E. S., & Wood, R. C. (2002). *Bairdemys*, a new side-necked turtle (Pelomedusoides: Podocnemididae) from the Miocene of the Caribbean. *American Museum Novitates, 3359*, 1–28.

Gaffney, E. S., & Zangerl, E. S. (1968). A revision of the chelonian genus *Bothremys* (Pleurodira: Pelomedusidae). *Fieldiana Geology, 16*, 193–239.

Gaffney, E. S., Chatterjee, S., & Rudra, D. K. (2001). *Kurmademys*, a new side-necked turtle (Pelomedusoides, Bothremydidae) from the late Cretaceous of India. *American Museum Novitates, 3321*, 1–16.

Gaffney, E. S., Tong, H., & Meylan, P. A. (2002). *Galianemys*, a new side-necked turtle (Pelomedusoides, Bothremydidae) from the late Cretaceous of Morocco. *American Museum Novitates, 3379*, 1–20.

Gaffney, E. S., Tong, H., & Meylan, P. A. (2006). Evolution of the side-necked turtles: The families Bothremydidae, Euraxemydidae, and Araripemydidae. *Bulletin of the American Museum of Natural History, 300*, 1–698.

Gaffney, E. S., Roberts, E. M., Sissoko, F., Bouaré, M. L., Tapanila, L. M., & O'Leary, M. A. (2007). *Acleistochelys*, a new side-necked turtle (Pelomedusoides, Bothremydidae) from the Paleocene of Mali. *American Museum Novitates, 3549*, 1–24.

Gaffney, E. S., Hooks, G. E., & Schneider, V. P. (2009a). New material of North American side-necked turtles (Pleurodira, Bothremydidae). *American Museum Novitates, 3655*, 1–26.

Gaffney, E. S., Krause, D. W., & Zalmout, I. S. (2009b). *Kinkonychelys*, a new side-necked turtle (Pelomedusoides, Bothremydidae) from the late Cretaceous of Madagascar. *American Museum Novitates, 3662*, 1–25.

Georgiades-Dikeoulia, E., Velitzelos, E., & Koskeridou, E. (2000). The *Crassostrea gryphoides* Schlot. Miocene banks of Greece as palaeoenvironmental indicators. *Geological Society of Greece, Special Publications, 9*, 101–108.

Gheerbrant, E., Codrea, V., Hosu, A., Sen, S., Guernet, C., de Lapparent de Broin, F., et al. (1999). Découverte de vertébrés dans les Calcaires de Rona (Thanétien ou Sparnacien), Transylvanie, Roumanie: Les plus anciens mammiféres cénozo d'Europe Orientale. *Ecologae Geologicae Helvetiae, 92*, 517–535.

Gray, J. E. (1867). Description of a new Australian tortoise (*Elseya latisternum*). *Annals and Magazine of Natural History (Series 3) 20*, 43–45.

Haas, G. (1978a). A Cretaceous pleurodire turtle from the surroundings of Jerusalem. *Israel Journal of Zoology, 27*, 20–33.

Haas, G. (1978b). A new turtle of the genus *Podocnemis* from the lower Cenomanian of Ein Yabrud. *Israel Journal of Zoology, 27*, 169–175.

Jain, S. L. (1977). A new fossil pelomedusid turtle from the Upper Cretaceous Pisdura sediments, Central India. *Journal of the Plaeontological Society of India, 20*, 360–365.

Jain, S. L. (1986). New pelomedusid turtle (Pleurodira: Chelonia) remains from Lameta formation (Maastrichtian) at Dongargon,

Central India, and a review of the pelomedusids from India. *Journal of the Plaeontological Society of India, 31,* 63–75.

Jiménez Fuentes, E. (1993). Aclaraciones sobre el status de *Neochelys zamorensis*, Pelomedúsido (Reptilia, Chelonii) de Pequeña Talla del Eoceno de Zamora (España). *Studia Geologica Salmanticensia, 28,* 141–153.

Joyce, W. G. 2007. Phylogenetic relationships of Mesozoic turtles. *Bulletin of the Peabody Museum of Natural History, 48,* 3–102.

Kischlat, E. E. (1994). Observaçôes sobre *Podocnemis elegans* Suarez (Chelonii, Pleurodira, Podocnemididae) do Neocretáceo do brazil. *Acta Geologica Leopoldensia, 17,* 345–351.

Kovar-Eder, J. H., Jechorek, H. Kvaček, Z., & Parashiv, V. (2008). The integrated plant record: An essential tool for reconstructing Neogene zonal vegetation in Europe. *Palaios, 23,* 97–111.

Laurent, Y., Tong, H., & Claude, J. (2002). New side-necked turtle (Pleurodira: Bothremydidae) from the Upper Maastrichtian of the Petites-Pyrénées (Haute-Garonne, France). *Cretaceous Research, 23,* 465–471.

Lehman, T. M., & Wick, S. L. (2010). *Chupacabrachelys complexus,* n. gen. n. spn. sp. (Testudines: Bothremydidae), from the Aguja formation (Campanian) of West Texas. *Journal of Vertebrate Paleontology, 30,* 1709–1725.

Linnaeus, C. (1758). Systema Naturae per regna tria naturae, secundum classes, ordines, genera, species, cum characteribus, differentiis, synonymis, locis. Holmiae, Laurentii Salvii, 824 p.

MacPhee, R. D. E., & Wyss, A. R. (1990). Oligo-Miocene vertebrates from Puerto RicoPuerto Rico, with a catalog of localities. *American Museum Novitates, 2865,* 1–45.

Meylan, P. A. (1996). Skeletal morphology and relationships of the Early Cretaceous side-necked turtle, *Araripemys barretoi* (Testudines: Pelomedusoides: Araripemydidae), from the Santana formation of Brazil. *Journal of Vertebrate Paleontology, 16,* 20–33.

Meylan, P. A., Gaffney, E. S., & de Almeida Campos, D. (2009). *Caninemys*, a new side-necked turtle (Pelomedusoides: Podocnemididae) from the Miocene of Brazil. *American Museum Novitates, 3639,* 1–26.

Nopcsa, F. (1931). Sur des nouveaux restes de Tortues du Danien du Midi de la France. *Bulletin de la Société Géologique de France, 1,* 223–236.

Price, L. I. (1973). Quelônio amphichelydia no Cretáceo inferior do nordeste do Brazil. *Revista Brasileira de Geociências, 3,* 84–96.

Ristori, G. (1895). Di un nuovo Chelonio fossile del Miocene dell'Isola di Malta. *Atti della Società Toscana di Scienze Naturali, Memoire, 14,* 3–17.

Roger, J., Pickford, M., Thomas, H., de Lapparent de Broin, F., Tassy, P., Van Neer, W., et al. (1994). Découverte de vertébrés fossiles dans le Miocéne de la région du Huqf au Sultanat d'Oman. *Annales de Paléontologie, 80,* 253–273.

Romano, P. S. R., & Azevado, S. A. K. (2006). Are extant podocnemidid turtles relicts of a widespread Cretaceous ancestor? *South America Journal of Herpetology, 1,* 175–184.

Savoyat, E., Monopolis, D., Bizon, G., & Yannetakis, C. P. (1971). *Geological Map of Greece. Nestorion Sheet. Scale 1:50,000.* Athens: Institute of Geology and Mineral Exploration, Athens (formerly Institute for Geology and Subsurface Research, Athens).

Schleich, H. H. (1994). Fossil schildkröten und krokodilreste aus dem tertiar Thrakiens (W-Turkei). *Courier Forschungsinstitut Senckenberg, 173,* 137–151.

Schlotheim, E. F. (1813). Beiträge zur Naturgeschichte der Versteinerungen in geognostischer Hinsicht. *Leonard's taschenbuch für die gesammte mineralogie, 7*(1813), 3–134.

Schmidt, K. P. (1940). A new turtle of the genus *Podocnemis* from the Cretaceous of Arkansas. *Field Museum of Natural History Geology Series, 8,* 1–12.

Suc, J.-P., Fauquette, S., Bessedik, M., Bertini, A., Zheng, Z., Clauzon, G., et al. (1999). Neogene vegetation changes in West European and West circum-Mediterranean areas. In J. Augusti, L. Rook, & P. Andrews (Eds.), *The evolution of Neogene terrestrial ecosystems in Europe* (pp. 378–388). Cambridge, MA: Cambridge University Press.

Swinton, W. E. (1939). A new fossil fresh-water tortoise from Burma. *Records of the Geological Survey of India, 74,* 548–551.

Thomson, S., & Georges, A. (1996). Neural bones in Australian chelid turtles. *Chelonian Conservation and Biology, 2,* 82–86.

Tronc, E., & Vuillemin, S. (1974). Contribution a l'étude de la fauna endémique Malgasche: étude ostéologique de *Erymnochelys madagascariensis* Grandidier, 1867 (Chelonien, Pelomedusidae). *Bulletin de l'Académie Malagache, 51,* 189–206.

Vargas-Ramirez, M., Castaño-Mora, O. V., & Fritz, U. (2008). Molecular phylogeny and divergence times of ancient South American and Malagasy river turtles (Testudines: Pleurodira: Podocnemididae). *Organisms, Diversity and Evolution, 8,* 388–398.

Velitzelos, E., & Velitzelos, D. (1999). Rational and museum prominence of the paleontological findings of Western Macedonia. *Elliniko Panorama, 15,* 188–205.

von Meyer H. (1847). *Palaeochelys bussenensis* im älteren Süsswasserkalk. *Jahreshefte des Vereins für Vaterländische Naturkunde in Württemberg, 3,* 167–168.

von Reinach, A.V. (1903). Schildkrötenreste aus dem ägyptischen Tertiär. *Abhandlungen herausgegeben von der Senckenbergischen Naturforschenden Gesellschaft (Frankfurt), 29,* 1–64.

Wagler, J. (1830). *Natürliches System der Amphibien, mit Vorangehender Classification der Säugethiere und Vögel.* Tübingen: J. C. Cotta'achen, München, Stuttgart.

Wagner, A. (1853). Beschreibung einer fossilen Schildkröte und etlicher anderer Reptilien Überreste aus den lithographischen Schiefern und dem grünen Sandsteine von Kehlheim. *Abhandlungen der Königlichen Bayerischen Akademie der Wissenschaften, 7,* 241–264.

Weems, R. E., & Knight, J. L. (2012). A new species of *Bairdemys* (Pelomedusoides: Podocnemididae) from the Oligocene (early Chattian) Chandler Bridge formation of South Carolina, USA, and its paleobiogeographic implications for the genus. In D. B. Brinkman, P. A. Holroyd, & J. D. Gardner (Eds.), *Morphology and evolution of turtles* (pp. XXX–XXX). Dordrecht: Springer.

Wielandt-Schuster, U., Schuster, F., Harzhauser, M., Mandic, O., Kroh, A., Röge, F., et al. (2004). Stratigraphy and paleoecology of Oligocene and early Miocene sedimentary sequences of the Mesohellenic Basin (NW Greece). *Courier Forschungsinstitut Senckenberg, 248,* 1–55.

Wood, R. C. (1976). *Stupendemys geographicus*, the world's largest turtle. *Breviora, 436,* 1–31.

Wood, R. C. (1983). *Kenyemys williamsi*, a fossil pelomedusid turtle from the Pliocene of Kenya. In A. J. G. Rhodin & K. Miyata (Eds.), *Advances in herpetology and evolutionary biology. Essays in honor of Ernest E. Williams* (pp. 74–85). Cambridge, MA: Museum of Comparative Zoology, Harvard University.

Wood, R. C. (2003). Fossil turtles from Lothagam. In M. G. Leakey & J. M. Harris (Eds.), *Lothagam. The dawn of humanity in Eastern Africa* (pp. 115–136). New York: Columbia University Press.

Wood, R. C., & Diaz de Gamero, M. L. (1971). *Podocnemis venezuelensis*, a fossil pelomedusid (Testudines, Pleurodira) from the Pliocene of Venezuela and a review of the history of *Podocnemis* in South America. *Breviora, 376,* 1–23.

Zangerl, R. (1948). The vertebrate fauna of the Selma formation of Alabama. Part II. The pleurodiran turtles. *Fieldiana Geology Memoirs, 3,* 23–56.

Chapter 18
A New Species of *Bairdemys* (Pelomedusoides: Podocnemididae) from the Oligocene (Early Chattian) Chandler Bridge Formation of South Carolina, USA, and Its Paleobiogeographic Implications for the Genus

Robert E. Weems and James L. Knight

Abstract A new species of podocnemidid pleurodire turtle, *Bairdemys healeyorum*, is described from the upper Oligocene Chandler Bridge Formation of South Carolina, USA, on the basis of a nearly complete shell found with associated skull fragments, lower jaw, girdle elements, and limb elements. This is the first record of an Oligocene pleurodire turtle from North America. The shell and lower jaw are unique in detail but similar in overall morphology to equivalent parts of *Bairdemys venezuelensis* and *B. sanchezi*. The plastron of this new species also shares many features with the previously described Maryland Miocene species of podocnemidid, "*Taphrosphys*" *miocenica*, so the Miocene taxon is here referred to *Bairdemys*. In the Oligocene and Miocene, North and South America were separated by broad expanses of salt water, so *Bairdemys* probably was salt-water tolerant and reached North America by "island hopping" from South America across the Caribbean Sea.

Keywords Calvert Formation • Pungo River Formation • Shell • Skull • Venezuela

Introduction

A largely complete shell and other skeletal elements of a side-neck turtle were discovered in 1989 by Craig and Alice Healey, two experienced museum volunteers, during development of the Crowfield Plantation neighborhood near Ladson in Dorchester County, South Carolina, northwest of the city of Charleston (Fig. 18.1). The specimen was recovered from the upper Oligocene (lower Chattian) Chandler Bridge Formation (Fig. 18.2). The Chandler Bridge Formation, although typically only 2–4 feet thick, is widespread in the Ladson-Summerville region. It long went unrecognized for lack of outcrops, but major suburban expansion along the Interstate 26 corridor from Charleston northwest through Ladson and Summerville, South Carolina created many new excavations throughout that area, especially in the 1970s and 1980s. It was during this building boom that the Chandler Bridge Formation was recognized as a new and paleontologically important stratigraphic unit (Sanders et al. 1982). Based primarily on its cetacean fauna, the Chandler Bridge Formation can be assigned to the early part of the late Oligocene (early Chattian), which places its age at about 27–28 Ma.

The Chandler Bridge Formation was deposited in a coastal marine setting, with estuarine, lagoonal, beach, and shallow marine environments being represented (Sanders et al. 1982; Weems and Sanders 1986; Katuna et al. 1997; Cicimurri and Knight 2009). There apparently was little fluvial input into this depositional environment (*contra* Katuna et al. 1997) because the preserved vertebrate fauna overwhelmingly consists of marine sharks and rays (Cicimurri and Knight 2009), marine bony fishes (e.g., Fierstine and Weems 2009), cetaceans, dugongs, sea turtles, and sea birds (Sanders 1980). Other animal remains are encountered only rarely, such as land birds, land mammals, and salt-water tolerant crocodilians (Erickson and Sawyer 1996). Equally rare are the remains of gopher tortoises (one specimen; Franz and Franz 2004), soft-shelled turtles (one specimen; ChM PV4882, unpublished), and the side-neck turtle described here (two specimens). No remains have been found that are referable to typical freshwater fish (such as sturgeons, gars, ictalurid catfishes, etc.), amphibians, most kinds of freshwater turtles (emydids, kinosternids, chelydrids), alligators, or aquatic freshwater mammals.

R. E. Weems (✉)
Paleo Quest, 14243 Murphy Terrace, Gainesville, VA 20155, USA
e-mail: rweems@usgs.gov

J. L. Knight
South Carolina State Museum, 301 Gervais Street, Columbia, SC 29202, USA
e-mail: jim.knight@scmuseum.org

D. B. Brinkman et al. (eds.), *Morphology and Evolution of Turtles,* Vertebrate Paleobiology and Paleoanthropology,
DOI: 10.1007/978-94-007-4309-0_18, © Springer Science+Business Media Dordrecht 2013

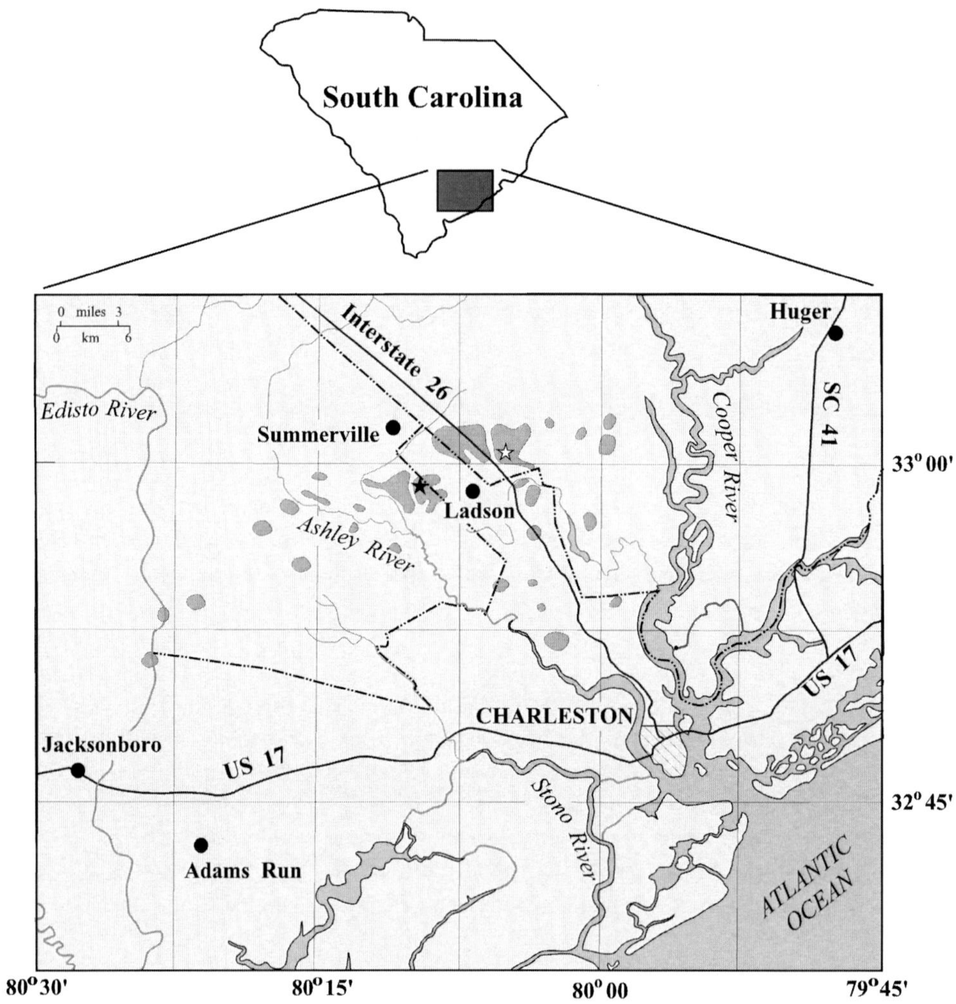

Fig. 18.1 Map of the Charleston region in the southeastern part of South Carolina, USA, showing the distribution of the Chandler Bridge Formation (shaded in *medium gray*), and the locations where the holotype (*white star*) and referred specimen (*black star*) of *Bairdemys healeyorum* were found. City of Charleston shown by *diagonal hachure*. Geology adapted from Weems and Lewis (2002)

Systematic Paleontology

Order Testudines Linnaeus 1758
Suborder Pleurodira Cope 1864
Hyperfamily Pelomedusoides Cope 1868a
Family Podocnemididae Cope 1868b
"Shweboemys Group" sensu Meylan 1996
Bairdemys Gaffney and Wood 2002

Type species: *Bairdemys hartsteini* Gaffney and Wood 2002

Referred species: *Bairdemys venezuelensis* (Wood and DÍaz de Gamero 1971), *Bairdemys hartsteini* Gaffney and Wood 2002, *Bairdemys sanchezi* Gaffney et al. 2008, *Bairdemys winklerae* Gaffney et al. 2008, *Bairdemys miocenica* (Collins and Lynn 1936; new combination), *Bairdemys healeyorum*, sp. nov. (this study).

Occurrence: Oligocene of South Carolina; Miocene of Venezuela, Puerto Rico, and the United States (Maryland and North Carolina).

Revised diagnosis (expanded from Gaffney and Wood 2002): A "*Shweboemys* Group" Pelomedusoides turtle (sensu Meylan 1996) known from skull, jaw, shell, and some girdle and limb elements; secondary palate shorter than in all "*Shweboemys* Group" except "*Shweboemys*" *gaffneyi* (per Gaffney and Wood 2002); medial edges of palatal cleft curved as in "*Shweboemys*" *gaffneyi*; ventral convexity on triturating surface larger than in all other "*Shweboemys* Group"; eustachian tube separated by bone from rest of *fenestra postotica* in contrast to all other known Podocnemididae; *antrum postoticum* extremely small and slit-like in contrast to all other "*Shweboemys* Group"; frontal and prefrontal strongly convex on dorsal surface in contrast to all other "*Shweboemys* Group"; basisphenoid separated from palatines by medially meeting pterygoids as in "*Shweboemys*" *antiqua* (per Gaffney and Wood 2002) and *S. pilgrimi*; basioccipital longer than in *Shweboemys pilgrimi*; jugal-pterygoid contact prevents palatine-parietal contact. Intergular, gular, and humeral scutes greatly reduced and restricted to the far anterior plastron, pectoral

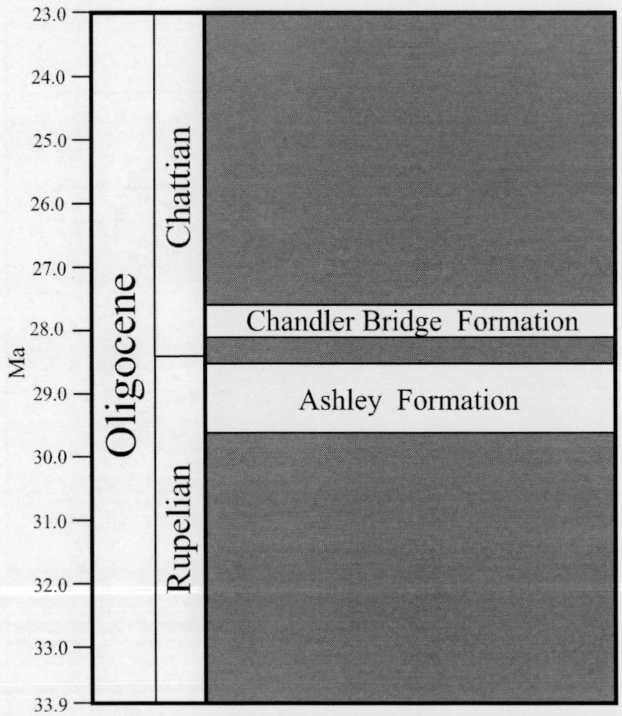

Fig. 18.2 Chart showing the position and age of the Chandler Bridge Formation within the Oligocene stratigraphic column of the Charleston region in the southeastern part of South Carolina, USA (after Weems et al. 2006). Most of the fossil vertebrate material in this area has come from the Chandler Bridge Formation (deposited in near-shore to lagoonal environments) and the Ashley Formation (deposited in a mid-shelf environment)

scutes compensatorially greatly expanded forward. Caput of humerus strongly elongated into an elliptical shape, its long axis close to parallel with the humerus shaft.

Bairdemys healeyorum sp. nov.
(Figs. 18.3, 18.4, 18.5, 18.6, 18.7, 18.8, 18.9, 18.10, 18.11, 18.12)

Holotype: South Carolina State Museum specimen SC 90.16, a largely complete skeleton, but lacking most of the skull and the distal limbs (Figs. 18.3a–d, 18.4, 18.6, 18.9, 18.10).

Holotype unit, locality, and age: Chandler Bridge Formation; northeastern shore of Crowfield Lake (now submerged), northeast of Ladson, Dorchester County, South Carolina, USA (Fig. 18.1); late Oligocene, early Chattian (Fig. 18.2).

Referred specimen: Charleston Museum specimen ChM PV4794, a partial anterior carapace from Bed 2 of the Chandler Bridge Formation. Found by Albert E. Sanders and Peter S. Coleman, 07 July 1979, north side of Ladson Road (County Road 230) about 0.12 mile (0.2 km) east of Chandler Bridge Creek, Dorchester County, South Carolina, USA.

Diagnosis: A species of *Bairdemys* differing from *Bairdemys hartsteini* and *Bairdemys sanchezi* in its relatively narrower lower jaw, from *Bairdemys venezuelensis* and *Bairdemys sanchezi* in the absence of a raised ridge along the midline of the fused dentaries, from *Bairdemys winklerae* in its relatively narrow quadrate articular surfaces and in the short length of the quadrate shaft above the articular surface which produces a very stocky appearance in lateral view, and from *Bairdemys miocenica* in its relatively wider and shorter anterior plastral lobe, relatively wider nuchal, and less robust humeral shaft.

Description

Skull: Fragments of a right maxilla and both quadrates were found, which suggest that the skull was preserved with the rest of the skeleton but was destroyed at the time the specimen was unearthed. The preserved piece of the right maxilla is very imperfect (Fig. 18.3b) but clearly lay below the posterior margin of the orbit and had a palatal shelf component that formed the edge of a secondary palate. The depth of this bone is less than found in *Bairdemys hartsteini*, *B. venezuelensis*, or in *B. sanchezi*, but is similar to the rearmost portion of the maxilla of *B. winklerae*. This suggests that the maxilla extended well up along the lower posterior edge of the orbit as in *B. winklerae*, presumably reducing the degree of contribution of the jugal to the orbital rim. Only the distal ends of the quadrates are preserved; the left is more complete than the right (Fig. 18.3c, d). The articular surface is relatively narrow side-to-side as in *B. hartsteini*. Also as in *B. hartsteini*, the length of the quadrate shaft above its articular surface is relatively short and therefore stouter in lateral appearance than in *B. venezuelensis*, *B. winklerae*, or *B. sanchezi*.

Lower jaw: The lower jaw is nearly complete (Fig. 18.4a) and shows an extensive fusion and rearward expansion of the triturating portion of the dentaries, which is typical of turtles that have a matching secondary palate on the roof of the mouth. This particular lower jaw morphology is very similar to the jaw morphology of *Bairdemys venezuelensis* and *B. sanchezi*, and all three species represent an intermediate stage in what appears to be a morphocline between the less specialized lower jaw of podocnemidids such as ?*Roxochelys* and the highly specialized jaw of shweboemines such as *Stereogenys* (Fig. 18.5). The bothremydid *Bothremys cooki* also has an expanded secondary palate and triturating surface on the lower jaws (Fig. 18.5), but its lower jaw specializations (including a very high and broad lingual ridge and rising labial ridges that form large, deep, cone-shaped pits) are quite

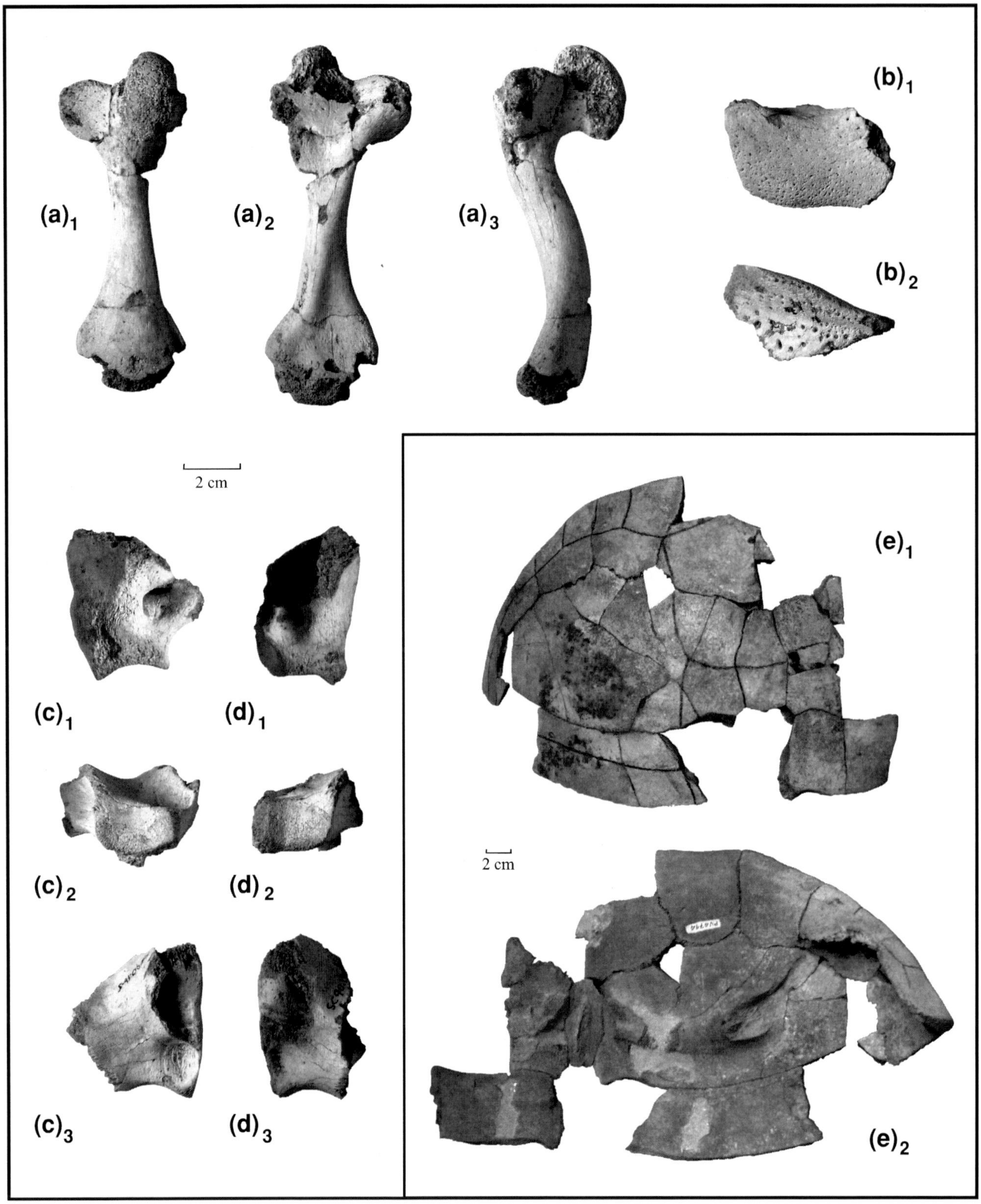

Fig. 18.3 Skeletal elements of *Bairdemys healeyorum*, from the Chandler Bridge Formation (late Oligocene), South Carolina, USA. $\mathbf{a}_1$–$\mathbf{d}_3$ Non-shell elements from holotype skeleton (SC 90.16): a humerus in $\mathbf{a}_1$ internal, $\mathbf{a}_2$ antero-lateral, and $\mathbf{a}_3$ posterolateral views; b right maxilla in $\mathbf{b}_1$ external and $\mathbf{b}_2$ ventral views, both with anterior end to right; c left quadrate in $\mathbf{c}_1$ posterior, $\mathbf{c}_2$ ventral, and $\mathbf{c}_3$ anterior views; d right quadrate in $\mathbf{d}_1$ posterior, $\mathbf{d}_2$ ventral, and $\mathbf{d}_3$ anterior views. e Anterior portion of referred carapace (ChM PV4794) in $\mathbf{e}_1$ external and $\mathbf{e}_2$ visceral views

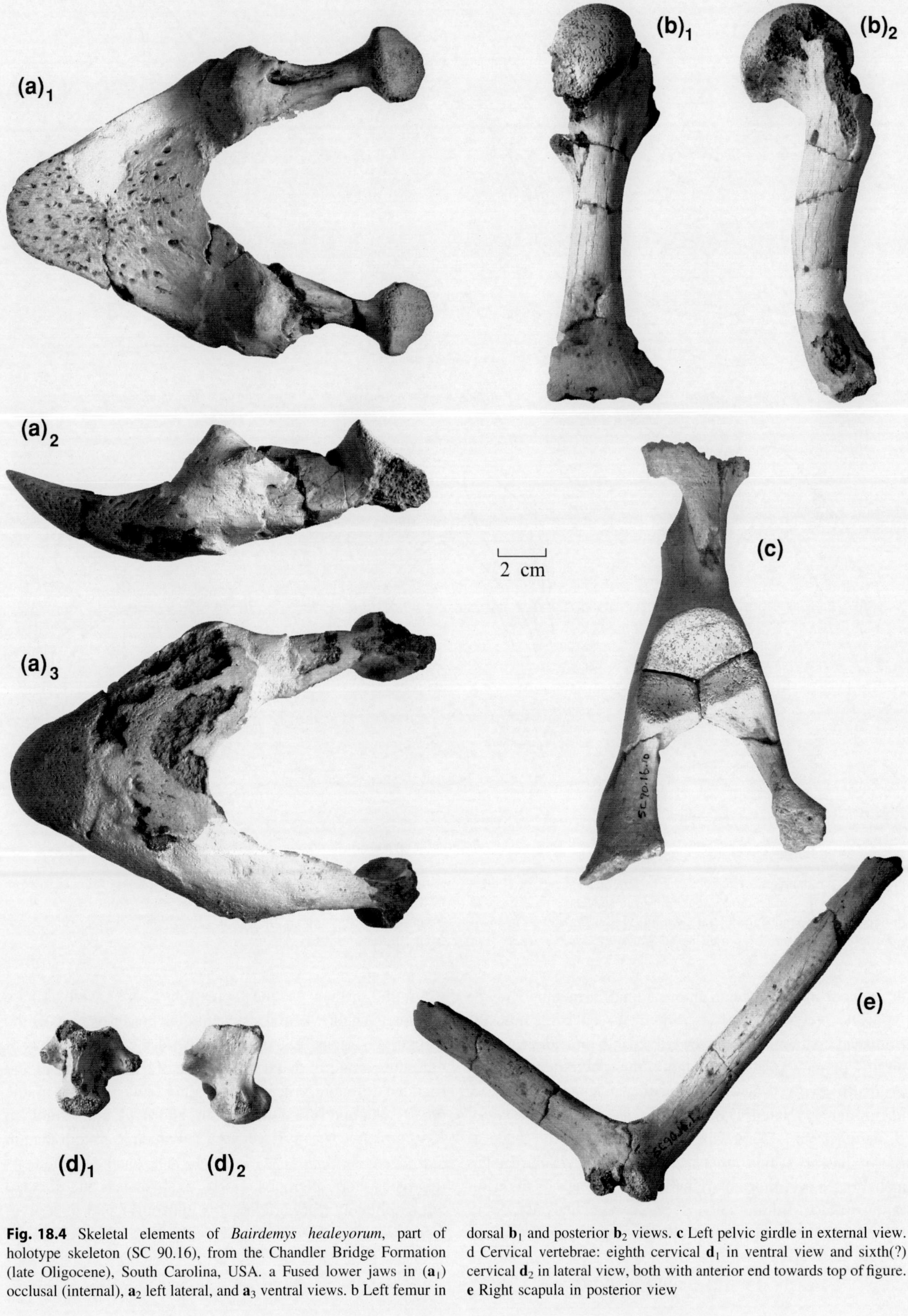

Fig. 18.4 Skeletal elements of *Bairdemys healeyorum*, part of holotype skeleton (SC 90.16), from the Chandler Bridge Formation (late Oligocene), South Carolina, USA. a Fused lower jaws in (**a**$_1$) occlusal (internal), **a**$_2$ left lateral, and **a**$_3$ ventral views. b Left femur in dorsal **b**$_1$ and posterior **b**$_2$ views. **c** Left pelvic girdle in external view. d Cervical vertebrae: eighth cervical **d**$_1$ in ventral view and sixth(?) cervical **d**$_2$ in lateral view, both with anterior end towards top of figure. **e** Right scapula in posterior view

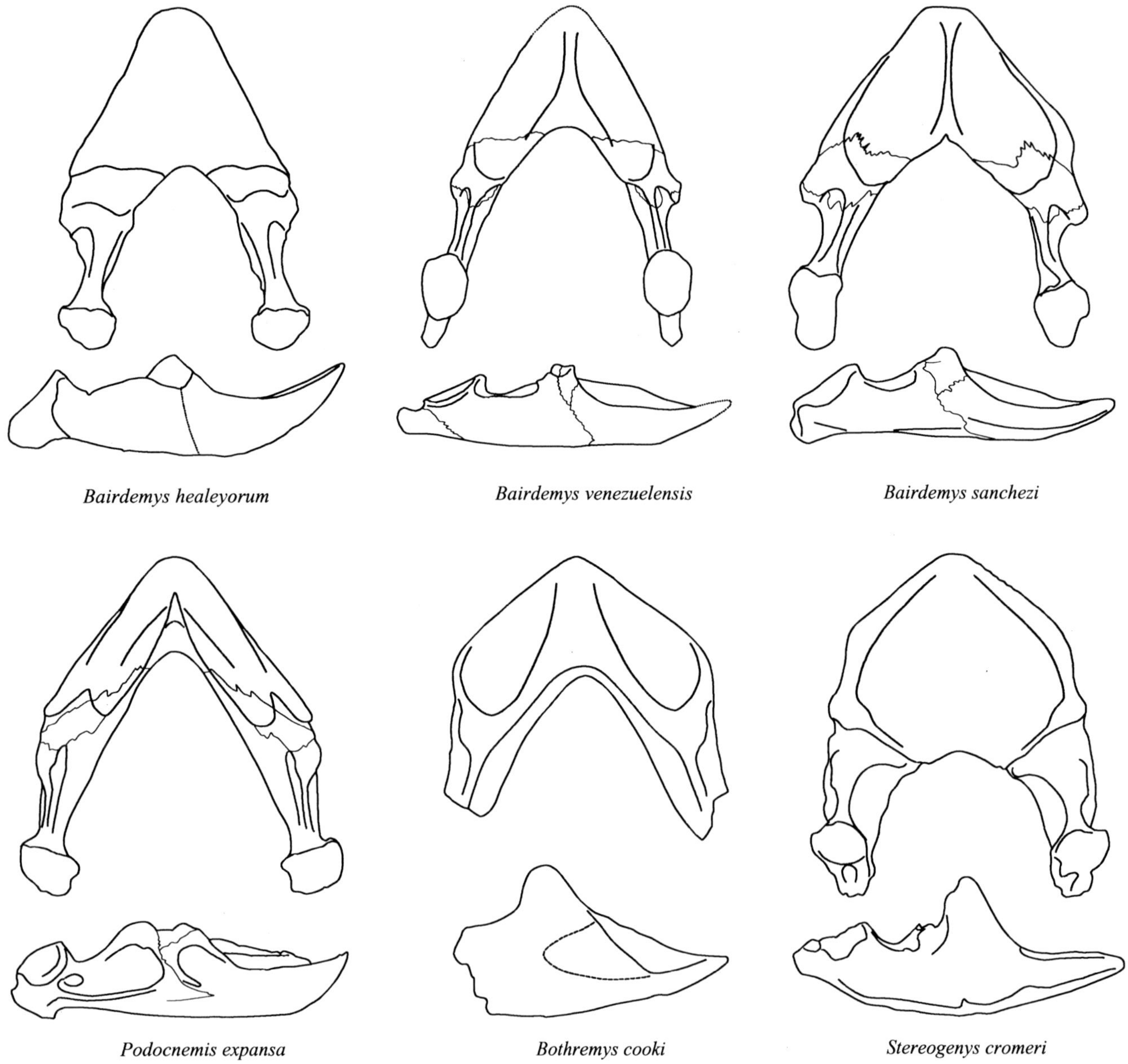

Fig. 18.5 Comparison of lower jaws in different species of *Bairdemys* and other podocnemidid turtles. For each species, jaws are shown in occlusal (*top*) and right lateral (*bottom*) views. *Bairdemys healeyorum* based on material reported here; *Bairdemys venezuelensis* from Sánchez-Villagra and Winkler (2006); *Bairdemys sanchezi* from Gaffney et al. (2008); *Podocnemis expansa* from Hay (1908); *Bothremys cooki* from Gaffney et al. (2006); and *Stereogenys cromeri* from Andrews (1906)

different from the specializations found in members of the "*Shweboemys* group." It is difficult to envision how the *Bothremys* pattern could have evolved into the "*Shweboemys* group" pattern, so almost certainly these two lineages independently evolved analogous feeding mechanisms and are not intimately related. The anterior angle formed by the jaw of *B. healeyorum* is somewhat more acute than the angle of *B. venezuelensis* and *B. sanchezi,* and *B. healeyorum* also has a shallower posterior emargination at the back of the triturating surface. Additionally, *B. healeyorum* has only a very low and very broad medial swelling along the midline of the triturating surface, and the depressions to either side of it are extremely shallow and restricted to the posterior parts of the triturating surface. The triturating surface extends backward onto the coronoids as far as the base of the pronounced and rounded coronoid processes. The jaw ramus formed by the prearticular and surangular is short and stout, comparable to *B. venezuelensis* and *B. sanchezi* but much less stout than in *Stereogenys cromeri*. The joint surface on the articular is relatively short anteroposteriorly, oval-shaped, and directed more obliquely rearward than the joint surface in *B. venezuelensis* and *B. sanchezi*.

Fig. 18.6 External (dorsal) view of carapace of *Bairdemys healeyorum*, part of holotype skeleton (SC 90.16), from the Chandler Bridge Formation (late Oligocene), South Carolina, USA

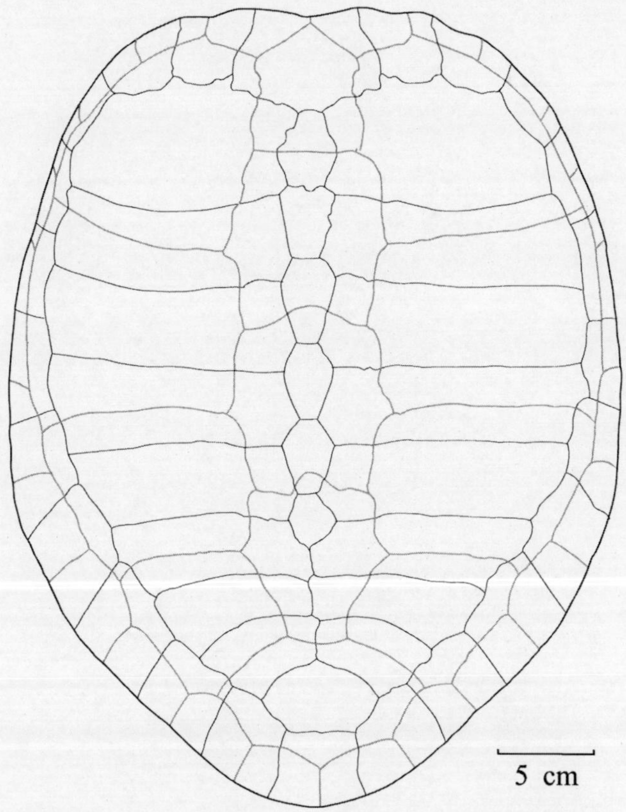

Fig. 18.7 Reconstruction of carapace of *Bairdemys healeyorum* in external (dorsal) view, based on the holotype (SC 90.16) and referred (ChM PV4794) specimens

Vertebrae: The centra and parts of the transverse processes of two procoelous cervical vertebrae are preserved. One, by comparison with the cervicals of *Podocnemis*, is probably the sixth. It is rather elongate and has a ventral keel at the anterior end of the centrum that tapers away posteriorly (Fig. 18.4d_2). In posterior view, its posterior convex articular surface is rather heart-shaped in outline. The other vertebra, with a broken but obviously pronounced and basally elongated ventral keel, is almost certainly the eighth (Fig. 18.4d_1). Its posterior convex articular surface is broadly U-shaped and much shallower dorsoventrally than the posterior articular surface on the sixth(?) cervical vertebra.

Carapace: The fourth left costal, most of the first right costal, the first, second, and third right peripherals, and about half of the nuchal were not found. Otherwise the carapace is essentially complete (Fig. 18.6) and allows a complete restoration of its appearance in life (Fig. 18.7). The carapace as preserved is slightly too wide to attach snugly to the plastron ventrally, indicating that it became somewhat flattened and spread during burial. This flattening has been taken into account in the restoration of the shell in dorsal and ventral views. The sulcal grooves are easily seen on most parts of the carapace, allowing accurate placement of the dorsal scute boundaries.

For the most part, the carapace is that of a typical generalized podocnemidid turtle like *Bauruemys* and ?*Roxochelys*, and it also shows much similarity to the shell of bothremydids like *Foxemys* (Fig. 18.8). The scute pattern, however, is unusual in that the first vertebral scute extends far forward, almost but not quite separating the anteriormost marginal scutes. This condition approaches that found in *Shweboemys pisdurensis* (Fig. 18.8), in which the anterior vertebral scute extends fully to the front of the shell and completely separates the anteriormost marginal scutes (Jain 1986). There is no indication that a cervical scute was present.

One slight tendency toward skeletal specialization in *Bairdemys healeyorum* is that only the six anteriormost neurals are present; the seventh and eighth are not developed. This represents an early stage in an evolutionary trend in this genus that culminated in *B. venezuelensis* (Fig. 18.8), in which all neurals are lost from the carapace and the costals all meet each other directly along the midline of the shell (Wood and Díaz de Gamero 1971). The nuchal is large and wider than long. On the ventral side of the carapace (Fig. 18.9), the sutural scar for attachment of the anterior plastral buttress extends far medially toward the midline along the first costal to about the middle of that element. In

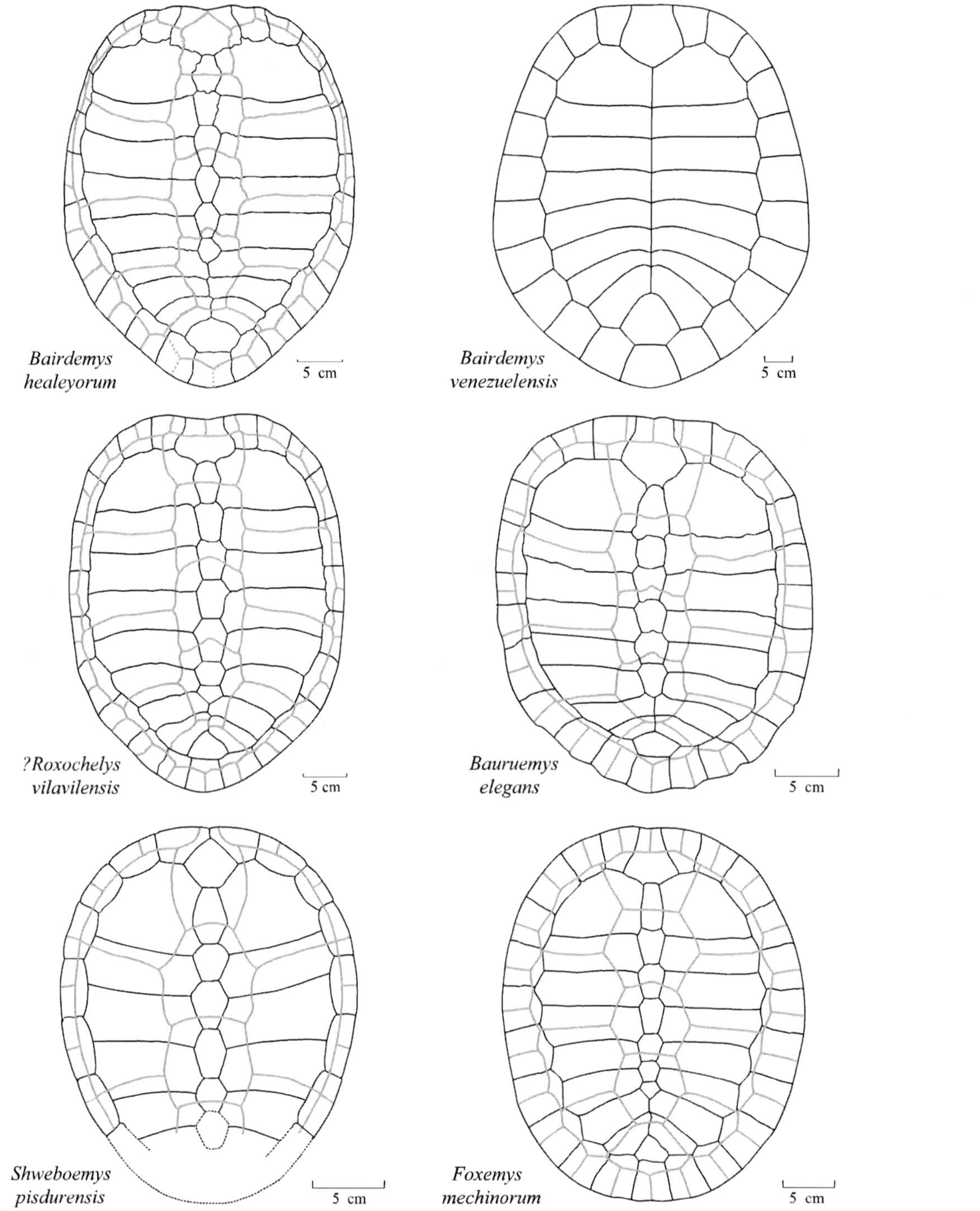

Fig. 18.8 Comparison of carapaces in *Bairdemys* and other podocnemid turtles, all in external (dorsal) view. *Bairdemys healeyorum* based on material reported here; *Bairdemys venezuelensis* from Wood and Díaz de Gamero (1971); *?Roxochelys vilavilensis* from de Broin (1971); *Bauruemys elegans* from Suarez (1969); *Shweboemys pisdurensis* from Jain (1986); and *Foxemys mechinorum* from Tong et al. (1998)

contrast, the posterior plastral buttress is less strongly developed, extending medially along the fifth costal for only about a fourth of its length. The sutural scar for attachment of the distal end of the ilium to the carapace is located on the posterior part of the seventh costal and on the anterior part of the eighth costal, about one-fourth of their lengths from the midline.

The partial anterior carapace in the Charleston Museum collection (CM-PV4794) mostly represents the part of the carapace not present in the type specimen (Fig. 18.3e).

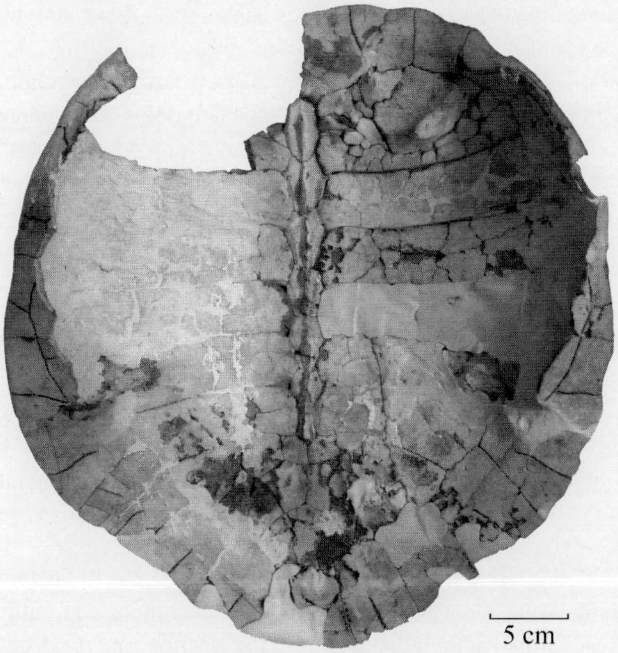

Fig. 18.9 Visceral (ventral or internal) view of carapace of *Bairdemys healeyorum*, part of holotype skeleton (SC 90.16), from the Chandler Bridge Formation (late Oligocene), South Carolina, USA

Fig. 18.10 External (ventral) view of plastron of *Bairdemys healeyorum*, part of holotype skeleton (SC 90.16), from the Chandler Bridge Formation (late Oligocene), South Carolina, USA

It represents a somewhat smaller individual, however, and was recovered from a different locality, so it clearly is not a part of the individual represented by the type specimen. Its surface is somewhat better preserved than that of the type, and shows a faintly developed nodular surface texture that is largely worn away on the type.

Plastron: The plastron is virtually complete except for some distal parts of the plastral bridges that were broken away (Fig. 18.10); its ventral surface is poorly preserved. In most areas the sulcal grooves are readily discernable, and this allows an essentially complete restoration of the plastron in ventral view (Fig. 18.11). The mesoplastra are subround in shape, relatively small, and located far from the midline. The plastron is broadly attached to the carapace by a plastral bridge that extends from the front edge of the fourth peripheral to as far back as the anterior portion of the eighth peripheral. The ratios of the anterior lobe, plastral bridge, and posterior lobe are roughly 1:2:2. The midline of the plastron is sutured along its entire length. The xiphiplastra are wide, flattened, elongated, and rounded at their posterior ends with a fairly large and rounded anal notch between them. The sutural scars where the ischium and pubis attach to the plastron are both on the xiphiplastra; the anterior ischial suture being about twice as large as the posterior pubic suture.

The abdominal, femoral, and anal scutes on the central and posterior part of the plastron are unexceptional, but the anterior scutes are much reduced and compressed toward the anterior midline region. The intergular scute is pentagonal in shape, largely confined to the anterior border region of the epiplastra, and extends backward onto the entoplastron only slightly. The gulars likewise are greatly reduced to small, triangular-shaped scales that occupy less than half the anteroposterior length of the epiplastra. The humeral scutes also are much compressed anteroposteriorly, mostly occupying the central region of the epiplastra, the anterior one-third of the entoplastron, and extending back barely onto the anteriormost edges of the hyoplastra. This forward compression of the anterior scutes is balanced by a pronounced forward expansion of the pectoral scutes, which occupy most of the anterior half of the hyoplastra, the rear two-thirds of the entoplastron, and the posterior portions of the epiplastra. This combination of traits is a distinctive feature of *Bairdemys* (Fig. 18.12). ?*Roxochelys*, *Dacquemys* and some bothremydids show a strong tendency in this direction, but no other pleurodire is known to have taken this trend to such an extreme. *Stereogenys* and *Shweboemys* do not show any pronounced forward compression of the intergular and gular scutes, but they do show a comparably strong forward expansion of the pectoral scutes that is made possible because the humeral scutes are very greatly

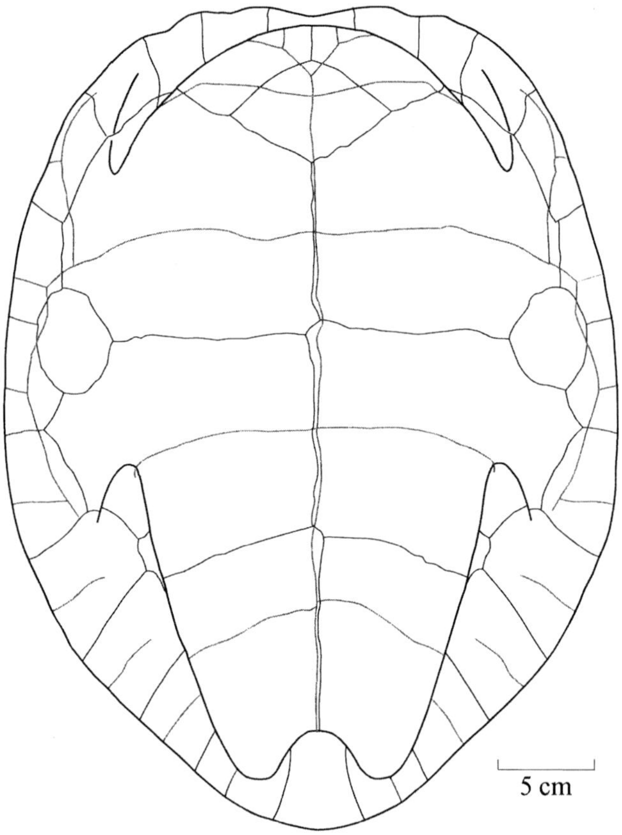

Fig. 18.11 Reconstruction of plastron of *Bairdemys healeyorum* in external (ventral) view, based on the holotype (SC 90.16)

reduced, so much so that they become separated along the midline by the intergular scutes (Fig. 18.12). This unusual condition is found elsewhere among podocnemidoid turtles only in the Taphrosphyini.

Pectoral girdle: Most of the right scapula was recovered, but the distal ends of its long and slender dorsal and acromial processes were broken away (Fig. 18.4e). The dorsal process is fairly rounded distally in cross-section, but the shorter acromial process is distinctly flattened and slightly curved into a C-shape toward its internal side. These processes form an angle of about 100°. Neither coracoid was recovered.

Front limb: The only part of the front limb recovered was the right humerus. Its distal end is somewhat eroded, but otherwise it is fairly complete and provides a good idea of its appearance. The distal end of the humerus is distinctly downturned. The entepicondyle and ectepicondyle are inconspicuously developed, but the ectepicondylar groove is deep and well developed along the external margin. At its proximal end, the lateral and medial processes are connected to the head of the humerus by relatively thin but rounded bony bridges. The head of the humerus (humeral caput) is elongated into a flattened ellipse with its long axis nearly parallel to the axis of the humeral shaft (Fig. 18.3a). Extreme elongation of the caput serves to channel most movement on the caput into the plane of the long axis, which in this case would have resulted in an up and down rowing motion. Turtles adapted to walking or to a wide range of motions in the forelimbs while swimming tend to have a rounded humeral caput (e.g., *Chelydra*, *Terrapene*, *Actinemys*, *Chrysemys*, *Gopherus*, *Testudo*, and some trionychid turtles) (Hay 1908, Fig. 595; Olsen 1968), whereas turtles with an elongated humeral caput generally show strong aquatic adaptation and a tendency to use the forelimbs as rowing organs (e.g., *Kinosternon*, *Pseudemys*, and some trionychids) (Hay 1908, Fig. 661, pl. 3; Olsen 1968). The marine cheloniid sea turtles, all of which are highly specialized for rowing, also show pronounced elongation of the caput (e.g., *Lepidochelys*; Hay 1908, pl. 2). Therefore, the extreme elongation of the humeral caput in *B. healeyorum* almost certainly was a specialization for swimming or even for rowing and not for walking.

Pelvic girdle: Both the left and right sides of the pelvic girdle were recovered (Fig. 18.4c). The ischium and pubis were sutured at their distal ends to the plastron, and the distal end of the ilium was similarly sutured along a broad attachment surface to the carapace. All three bones contribute to the acetabulum, with the ischium and pubis each contributing about one-fifth of its area and the ilium about three-fifths.

Hind limb: The only part of the hind limb recovered was the left femur (Fig. 18.4b), which is damaged and thus provides only limited information about this bone. However, enough is preserved to show that it is distinctly larger than the humerus, as is typical for freshwater turtles. The caput is partly eroded away; it is somewhat elongated, but not so elongated as the head of the humerus. Relative to the base of the trochanter major, the base of the trochanter minor diverges from the shaft much farther down and at a greater angle. The overall proportions of the proximal femur are rather reminiscent of the proportions of the femora of soft-shelled turtles, suggesting strong aquatic specialization. The femur is strongly down-turned at its distal end.

Bairdemys miocenica (Collins and Lynn 1936) new combination

Synonymy: *Taphrosphys miocenica*,, (Collins and Lynn 1936).

Holotype: United States National Museum specimen USNM 13784, anterior plastron (epiplastra, entoplastron, and hyoplastra) of a single individual.

Holotype locality, unit, and age: One-quarter mile (0.4 km) south of Camp Roosevelt along the Calvert Cliffs, Calvert County, Maryland, USA; "Zone 10" of Shattuck (1904), Calvert Formation; early middle Miocene (early Langhian, ca. 15–16 Ma; Weems and Edwards 2007).

Referred material: Nuchal and humerus described and figured in Zug (2001).

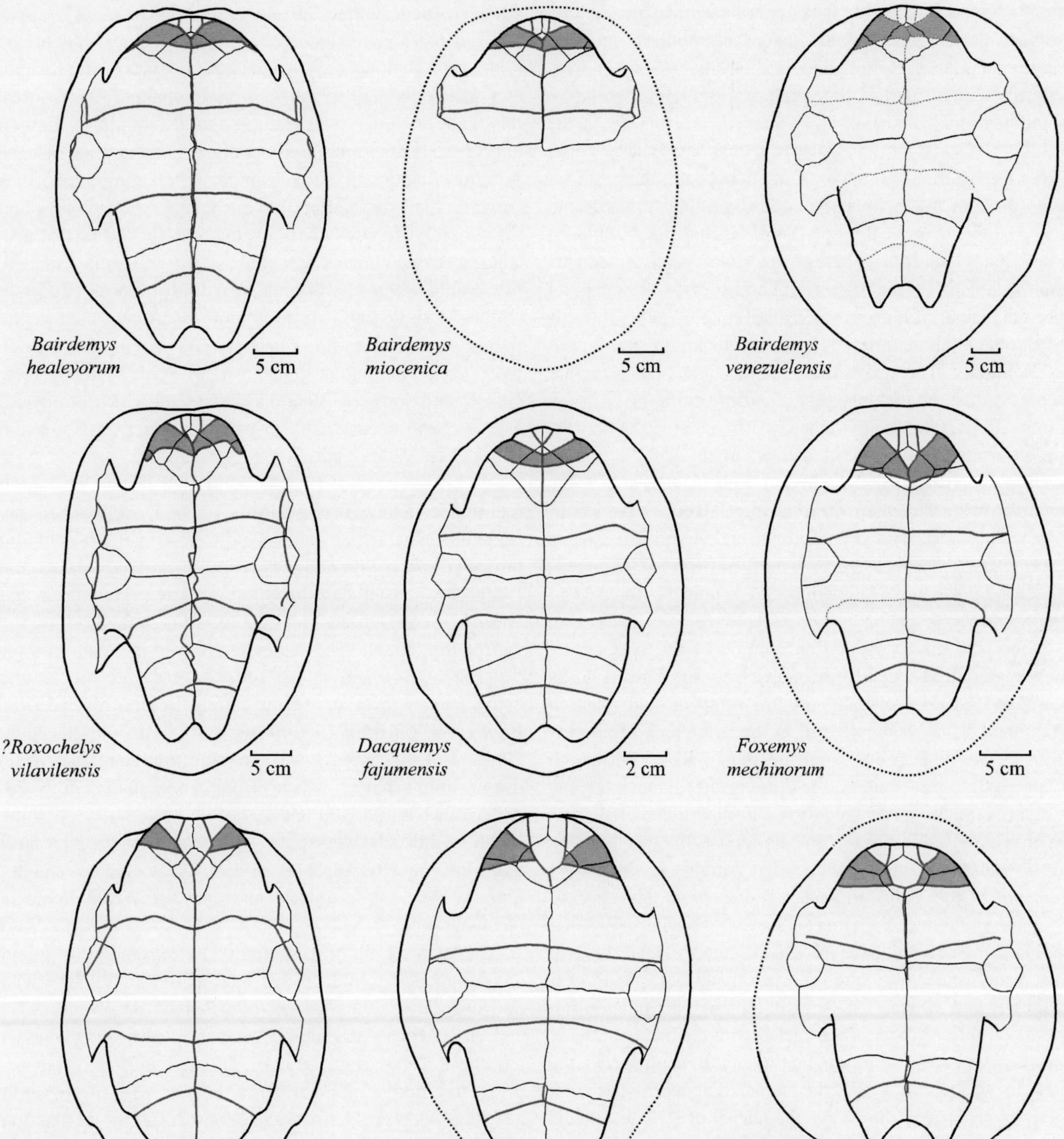

Fig. 18.12 Comparison of the plastron in different species of *Bairdemys* and other podocnemidid turtles, all in external (ventral) view. *Bairdemys healeyorum* based on material reported here; *Bairdemys miocenica* from Collins and Lynn (1936); *Bairdemys venezuelensis* from Gaffney et al. (2006); ?*Roxochelys vilavilensis* from de Broin (1971); *Dacquemys fajumensis* from Andrews (1906); *Foxemys mechinorum* from Tong et al. (1998) *Stereogenys libyca* from Andrews (1906); *Shweboemys pisdurensis* from Jain (1986); and *Taphrosphys congolensis* from Gaffney et al. (2006)

Diagnosis: Anterior lobe of plastron short, broad and rounded with well developed but rather thin axillary buttresses, maximum length of hyoplastron about equal to its breadth, mesoplastral contact with hyoplastron indicates that mesoplastra were relatively small, located far from the midline, and rather polygonal in shape. Intergular scute small, pentagonal, mostly located on the epiplastra and only slightly projecting backward onto the entoplastron; gular scutes small, triangular, and located far forward and entirely on only a small part of the epiplastra; humeral scutes also relatively small and forwardly located, meeting along the midline beneath the intergular; pectoral scutes large, covering the anterior two-thirds of the hyoplastra, the posterior two-thirds of the entoplastron, and extending forward onto the posterior part of the epiplastra. Nuchal much wider at rear than at front, being anteriorly constricted by encroachment of the first peripherals; cervical scute not present. Caput of humerus elongated roughly along the axis of the humeral shaft, which is rather thick and stout.

Remarks: The proportions of the gular and intergular scutes on the anterior plastron of *Bairdemys miocenica* are quite similar to, though not exactly identical with, the proportions of the gulars and intergular of *Bairdemys healeyorum* (Fig. 18.12). This strongly suggests that the anterior pleurodire plastron from the Calvert Formation of Maryland, described by Collins and Lynn (1936) as "*Taphrosphys*" *miocenica*, was derived from the Oligocene species *Bairdemys healeyorum* and should be referred to that genus.

Since the discovery and description of this species, no new material has been forthcoming from the Calvert Cliffs of Maryland. However, pleurodire carapace and plastron fragments have been collected in the Lee Creek Mine near Aurora, North Carolina from the Pungo River Formation (Zug 2001); this unit is age-equivalent to much of the Calvert Formation. As the North Carolina material is from at or near the same stratigraphic horizon as the type plastron from Maryland, in all likelihood it pertains to *Bairdemys miocenica* and helps to better characterize this species. Notable is the nuchal element (Zug 2001) which shows that *B. miocenica* did not have a cervical scute and therefore can be firmly placed among the pelomedusoid pleurodires (Gaffney and Meylan 1988). Also, although the relationship of the posterior border of the nuchal with the first neural and first costals is very similar to that seen in *B. healeyorum*, the anterior constriction of the nuchal between the first peripherals clearly shows that the nuchal of *B. miocenica* is not specifically referable to *B. healeyorum*. The humerus is close to that of *B. healeyorum* and like it has a very elongate caput (Zug 2001), but it differs in that the shaft is relatively thicker and stouter in its construction.

Discussion

Only the femur and humerus of *Bairdemys healeyorum* are known, so the more distal parts of the limbs cannot be evaluated. These two proximal limb elements, however, strongly suggest that *Bairdemys* had legs that were specialized for aquatic locomotion. Because the head of the humerus was elongated into an extreme ellipse with its long axis nearly parallel to the shaft as in cheloniid sea turtles, this bone obviously was designed to accommodate motion almost exclusively in a vertical plane. This implies that the forelimb functionally was a flipper. The strong similarities between the proximal end of the femur of *Bairdemys* and the proximal femora of trionychid soft-shelled turtles also suggest strong aquatic adaptation in the rear limbs. Indeed, this combination of front and rear limb traits is strikingly similar to the combination of limb functions found in the living pig-nosed turtle, *Carettochelys insculptata*, which lives in rivers, lagoons, and estuaries of southern New Guinea and northern Australia. Interestingly, *Carettochelys* also is found occasionally in brackish water (Georges and Rose 1993; Visser and Zwartepoorte 2005).

No occurrences of Eocene pleurodire turtles have been reported from anywhere in North America, and this strongly suggests that the last of the North American bothremydid turtles (*Taphrosphys* and *Bothremys*?) survived through the Paleocene in southeastern North America (Hutchison and Weems 1998; Gaffney et al. 2006) but then died out by the end of that epoch. The distinctive lower jaw and anterior plastron of *Bairdemys* bear little similarity to any of the better known members of the family Bothremydidae (i.e., *Bothremys*, *Chedighaii*, *Taphrosphys*) (Gaffney et al. 2006, 2009), and this greatly weakens any argument that *Bairdemys* might have evolved from some unknown North American bothremydid ancestor. Similarly, although some bothremydids survived in the northern African region until the Miocene (Roger et al. 1994), derivation is equally unlikely from any of the late surviving Old World members of this group.

The greatest plastral similarity to *Bairdemys* is found among podocnemidid turtles, such as the South American podocnemidine ?*Roxochelys vilavilensis* (de Broin 1971) and the African erymnochelyine *Dacquemys fajumensis* (Andrews 1906) (Fig. 18.12). Although *Bairdemys* shares distinctive derived cranial characteristics with *Stereogenys* and *Shweboemys* (Gaffney and Wood 2002), in its plastron *Bairdemys* is more similar to other podocnemid turtles in that its humeral scutes meet along the midline and are not separated by the intergular (Fig. 18.12). This trait, along with the less strongly expanded secondary palate and less robust lower jaw found in *Bairdemys*, indicates that the *Bairdemys* lineage diverged from the *Shweboemys-Stereogenys* lineage well before the mutual ancestor of the latter two genera developed the distinctive plastral scute arrangement present in both of them. The oldest well documented member of the *Shweboemys-Stereogenys* lineage is *Shweboemys pisdurensis* from the Maastrichtian of India (Jain 1986), so the *Bairdemys* lineage therefore must have

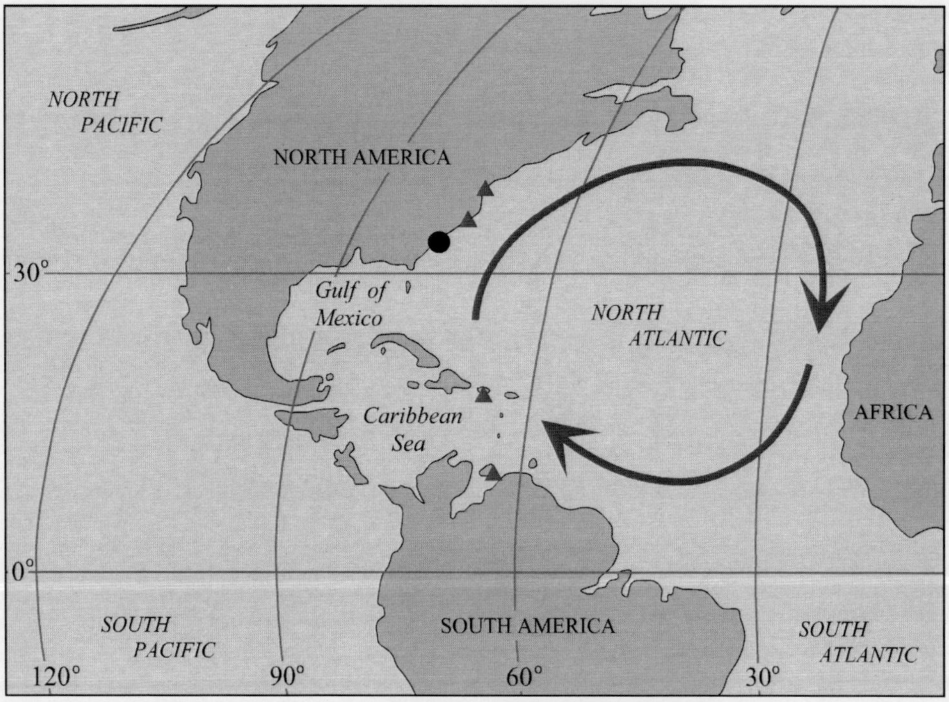

Fig. 18.13 Oligocene (*circle*) and Miocene (*triangle*) occurrences of *Bairdemys* shown on an Oligocene paleogeographic map of the Caribbean and adjacent regions. Along the Atlantic seaboard of North America, shallow seas in the middle Miocene transgressed westward across the modern landscape about 50 miles (80 km) farther inland than they did in the late Oligocene. From south to north, localities are in present day Venezuela, Puerto Rico, South Carolina, North Carolina, and Maryland. Paleogeographic land–sea relationships are adapted from Global Paleogeographic Views of Earth History—Late Precambrian to Recent http://jan.ucc.nau.edu/~rcb7/35moll.jpg. The anatomical similarities between the *Bairdemys* lineage and the Afro-Asian *Shweboemys-Stereogenys* lineage strongly suggest that all of these genera were derived from ancestors that originated in Africa and/or India. There is little chance that the ancestors of the *Bairdemys* lineage crossed the Atlantic directly from Africa to North America, because the strongly clockwise current pattern in the North Atlantic (*green arrows*) would have precluded dispersal in that direction. It is far more likely that the *Bairdemys* ancestors crossed the Atlantic Ocean from Africa to South America and then "island hopped" from South America to North America

split from the *Shweboemys-Stereogenys* lineage well before the end of the Cretaceous.

Bairdemys healeyorum from the upper Oligocene Chandler Bridge Formation is the oldest known species within the *Bairdemys* lineage. This is not what would have been anticipated, because the anatomical similarities between the *Bairdemys* lineage and the Afro-Asian *Shweboemys-Stereogenys* lineage, as well as mutual similarities with the Old World Erymnochelyinae, indicate that all of these genera were derived from a common ancestor that lived in Africa and/or southern Asia (de Lapparent de Broin 2000). Thus, given the Late Cretaceous and Paleogene paleogeography of the Atlantic basin, the ancestor of the *Bairdemys* lineage must have come to the Americas either by making a daunting "sweepstakes" crossing of the Atlantic Ocean directly from Africa to North America in Oligocene time or else by making an Oligocene or earlier crossing of the Atlantic from Africa to South America, followed by an "island-hopping" crossing of the Caribbean from South America to North America. The sweepstakes model seems very unlikely because the North Atlantic is dominated by a strong clockwise current flow, induced by the Coriolis effect, that would have been present even well before the Oligocene (Fig. 18.13). Therefore, any direct "sweepstakes" crossing of the Atlantic from Africa (or Europe) to North America in the Oligocene would have had to occur against a strong prevailing oceanic current going in the opposite direction. In contrast, a crossing from Africa to South America not only would have been much shorter, it also would have been made easier by going with the direction of the prevailing current. Thus, even though there is no fossil evidence so far to support a dispersal route first from Africa to South America, and then from South America to North America, this by far seems the more plausible possibility.

Once established in the southeastern United States, colonization by *Bairdemys* apparently was successful for about 17 million years, from the late Oligocene through the middle Miocene. Probably Neogene climatic deterioration in the southeastern United States, which began in earnest at the beginning of the late Miocene, heralded the demise of these tropically adapted turtles in North America. In tropical Venezuela, however, *Bairdemys* continued to thrive throughout most or all of the late Miocene (Gaffney et al. 2008), and the related genus *Shweboemys* even survived in tropical Burma until the Pliocene or Pleistocene (Jain 1986).

Modern side-neck turtles primarily occupy fresh water habitats, but prolonged tolerance of salt water has been documented in at least some chelids (e.g., *Hydromedusa*, *Pelusios*, *Chelodina*; Frazier 1986). Salt-water tolerance has been suggested for the extinct podocnemidid *Stupendemys* (Wood 1976) and for a number of species within three clades of the Pelomedusoides (Bothremydini, Taphrosphyrini, and the *Shweboemys* Group) (Meylan et al. 2009). The scarcity of remains of *Bothremys* in the shallow marine to estuarine strata of Maryland, North Carolina, and South Carolina, and its absence so far in deposits of the same age in Virginia, strongly suggests that the preferred habitat of this turtle was fresh water. Even so, such a preference in habitat by no means precluded a tolerance to prolonged immersion in salt-water and the potential for wide dispersal that this trait would have presented. A strong tolerance for salt water, coupled with its exceptional specialization for swimming, may well explain how *Bairdemys*, uniquely among Paleogene podocnemidids, was able to spread across the salt waters of the Atlantic to South America and from there across the Caribbean region to Puerto Rico and the southeastern United States (Fig. 18.13).

There is another possible factor that may have contributed to the successful spread of *Bairdemys* across the Caribbean. There may have existed, around the beginning of the Oligocene, a land-bridge that connected the Greater Antilles islands with South America along the course of the present-day submerged Aves Ridge (GAARlandia of Iturralde-Vinent and MacPhee 1999). If this land bridge (or a chain of closely spaced islands in the same location) did exist then, it would have greatly reduced the distances that *Bairdemys* needed to navigate across salt water to reach North America from South America during the Oligocene.

Acknowledgments Special thanks go to Craig and Alice Healey of West Columbia, South Carolina, who for so many years conscientiously collected for, donated to, and volunteered at the South Carolina State Museum. The description of this fossil turtle is but small payment for their many contributions. Vance McCollum and Curtis Bentley also are acknowledged for their many assistances at the Crowfield site. The authors gratefully acknowledge the help of Eugene S. Gaffney (American Museum of Natural History) in determining the taxonomic placement of this turtle and Albert E. Sanders (Charleston Museum) for his discussions on the stratigraphic setting of the Chandler Bridge Formation and for making the Charleston Museum specimen of *Bairdemys healeyorum* available for study. We also thank France de Lapparent de Broin, Eugene S. Gaffney, and Takuya Konishi for thorough and very helpful reviews of the manuscript of this paper.

References

Andrews, C. W. (1906). Order Chelonia. In Trustees of the British Museum (Eds.), *A descriptive catalogue of the Tertiary vertebrata of the Fayûm, Egypt* (pp. 275–306). London: British Museum of Natural History.

Cicimurri, D. J., & Knight, J. L. (2009). Late Oligocene sharks and rays from the Chandler Bridge formation, Dorchester County, South Carolina, USA. *Acta Palaeontologica Polonica, 54*, 627–647.

Collins, R. L., & Lynn, W. G. (1936). Fossil turtles from Maryland. *Proceedings of the American Philosophical Society, 76, 151–174.*

de Broin, F. (1971). Une espèce nouvelle de Tortue pleurodire (?*Roxochelys vilavilensis* n.sp.sp.) dans les Crétacé superieur de Bolivie. *Bulletin de la Societe Géologique de France* (*7ème*), *12*, 445–452.

de Lapparent de Broin, F. (2000). African chelonians from the Jurassic to the present: Phases of development and preliminary catalogue of the fossil record. *Palaeontologica Africana, 36*, 43–82.

Erickson, B. R., & Sawyer, G. T. (1996). The estuarine crocodile *Gavialosuchus carolinensis* n. sp (Crocodilia: Eusuchia) from the late Oligocene of South Carolina, North America. *Monographs of the Science Museum of Minnesota, 3*, 1–47.

Fierstine, H. L., & Weems, R. E. (2009). Paleontology of the Oligocene Ashley and Chandler Bridge formations of South Carolina, 4: Analysis and new records of billfishes (Perciformes: Xiphioidei). *Palaeo Ichthyologica, 11*, 43–88.

Franz, R., & Franz, F. E. (2004). Gopher tortoise evolution: East vs. west, a possible paradigm shift. Abstracts of the 29th Annual Meeting and Symposium of the Desert Tortoise Council (February 20–23, 2004), http://www.deserttortoise.org/abstract/abstracts2004/2004abs14.html.

Frazier, J. G. (1986). Epizoic barnacles on pleurodiran turtles: Is the relationship rare? *Proceedings of the Biological Society of Washington*, *99*, 472–477.

Gaffney, E. S., & Meylan, P. A. (1988). A phylogeny of turtles. In M. J. Benton (Ed.), *The phylogeny and classification of the tetrapods. Vol. 1. Amphibians, reptiles, birds*. Systematics association special volume N35A (pp. 157–219). Oxford: Clarendon Press.

Gaffney, E. S., & Wood, R. C. (2002). *Bairdemys*, a new side-necked turtle (Pelomedusoides: Podocnemididae) from the Miocene of the Caribbean. *American Museum Novitates*, *3359*, 1–28.

Gaffney, E. S., Tong, H., & Meylan, P. A. (2006). Evolution of the side-necked turtles: The families Bothremydidae, Euraxemydidae, and Araripemydidae. *Bulletin of the American Museum of Natural History*, *300*, 1–698.

Gaffney, E. S., Scheyer, T. M., Johnson, K. G., Bocquentin, J., & Aguilera, O. A. (2008). Two new species of the side necked turtle genus, *Bairdemys* (Pleurodira, Podocnemididae) from the Miocene of Venezuela. *Palaeontologische Zeitschrift*, *82*, 209–229.

Gaffney, E. S., Hooks, G. E. III, & Schneider, V. P. (2009). New material of North America side-necked turtles (Pleurodira: Bothremydidae). *American Museum Novitates*, *3655*, 1–26.

Georges, A., & Rose, M. (1993). Conservation biology of the pig-nosed turtle, *Carettochelys insculpta*. *Chelonian Conservation and Biology*, *1*, 3–12.

Hay, O. P. (1908). *Fossil turtles of North America*. Carnegie Institute of Washington, Publication (Vol. 75, pp. 1–568).

Hutchison, J. H., & Weems, R. E. (1998). Paleocene turtle remains from South Carolina. In A. E. Sanders (Ed.), Paleobiology of the Williamsburg Formation (Black Mingo Group; Paleocene) of South Carolina. *Transactions of the American Philosophical Society*, *88*, 165–195.

Iturralde-Vinent, M. A., & MacPhee, R. D. E. (1999). Paleogeography of the Caribbean region: Implications for Cenozoic biogeography. *Bulletin of the American Museum of Natural History, 238*, 1–95.

Jain, S. L. (1986). New pelomedusid turtle (Pleurodira: Chelonia) remains from Lameta Formation (Maastrichtian) at Dongargaon, central India, and a review of pelomedusids from India. *Journal of the Palaeontolical Society of India, 31*, 63–75.

Katuna, M. P., Geisler, J. H., & Colquhoun, D. J. (1997). Stratigraphic correlation of Oligocene marginal marine and fluvial deposits across the middle and lower coastal plain, South Carolina. *Sedimentary Geology, 108*, 181–194.

Meylan, P. A. (1996). Skeletal morphology and relationships of the Early Cretaceous side-necked turtle, *Araripemys barretoibarretoi* (Testudines: Pelomedusoides: Araripemydidae), from the Santana Formation of Brazil. *Journal of Vertebrate Paleontology, 16*, 20–33.

Meylan, P. A., Gaffney, E. S., & de Almeida Campos, D. (2009). *Caninemys*, a new side-necked turtle (Pelomedusoides: Podocnemididae) from the Miocene of Brazil. *American Museum Novitates, 3639*, 1–26.

Olsen, S. J. (1968). Fish, amphibian and reptile remains from archaeological sites. *Papers of the Peabody Museum of American Archaeology and Ethnology, 56*, 1–156.

Sánchez-Villagra, M. R., & Winkler, J. D. (2006). Cranial variation in *Bairdemys* turtles (Podocnemididae: Miocene of the Caribbean region) and description of new material from Urumaco, Venezuela. *Journal of Systematic Palaeontology, 4*, 241–253.

Sanders, A. E. (1980). Excavation of Oligocene marine fossil beds near Charleston, South Carolina. *National Geographic Research, 12*, 601–621.

Sanders, A. E., Weems, R. E., & Lemon, E. M. Jr. (1982). The Chandler Bridge Formation; a new Oligocene stratigraphic unit in the lower Coastal Plain of South Carolina. In Contributions to Stratigraphy. *U.S. Geological Survey Bulletin, 1529-H*, 105–124.

Shattuck, G. B. (1904). Geological and paleontological relations, with a review of earlier investigations. In W. B. Clark, G. B. Shattuck, & W. H. Dall (Eds.), *The Miocene deposits of Maryland*. Maryland Geological Survey, Miocene (Vol. 1, pp. xxxiii–cxxxvii).

Suarez, J. M. (1969). Um quelônio da Formação Bauru. *Departamento de Geografia da Faculdade de Filosofia, Ciências e Letras de Presidente PrudentePresidente Prudente, no. 2, 23*, 168–176.

Tong, H., Gaffney, E. S., & Buffetaut, E. (1998). *Foxemys*, a new side-necked turtle (Bothremydidae: Pelomedusoides) from the Late Cretaceous of France. *American Museum Novitates, 3251*, 1–19.

Visser, G., & Zwartepoorte, H. (2005). Reproduction of the pig-nosed turtle *Carettochelys insculpta* (Ramsay, 1886) at the Rotterdam Zoo. *Radiata, 14*, 3–12.

Weems, R. E., & Edwards, L. E. (2007). The age and provenance of "*Eschrichtius*" *cephalus* Cope (Mammalia: Cetacea). *Journal of Vertebrate Paleontology, 27*, 752–756.

Weems, R. E., & Lewis, W. C. (2002). Structural and tectonic setting of the Charleston, South Carolina region: Evidence from the Tertiary stratigraphic record. *Bulletin of the Geological Society of America, 114*, 24–42.

Weems, R. E., & Sanders, A. E. (1986). The Chandler Bridge Formation (upper Oligocene) in the Charleston region, South Carolina. *Geological Society of America, Centennial Field Guide (Southeast. Sect.), 6*, 323–326.

Weems, R. E., Harris, W. B., Sanders, A. E., & Edwards, L. E. (2006). Correlation of Oligocene sea level cycles between Western Europe and the southeastern United States. In G. Camoin, A. Droxler, C. Fulthorpe, & K. Miller (Eds.), *Sea level changes: Records, processes, and modeling—SEALAIX'06, Giens, 25-29/09/06. Abstract Book* Association Sedimentology France (Vol. 55, pp. 205–206).

Wood, R. C. (1976). *Stupendemys geographicus*, the world's largest turtle. *Breviora, 436*, 1–31.

Wood, R. C., & Díaz de Gamero, M. L. (1971). *Podocnemis venezuelensis*, a new fossil pelomedusid (Testudines, Pleurodira) from the Pliocene of Venezuela and a review of the history of *Podocnemis* in South America. *Breviora, 376*, 1–23.

Zug, G. R. (2001). Turtles of the Lee Creek Mine (Pliocene: North Carolina). In C. E. Ray & D. J. Bohaska (Eds.), *Geology and paleontology of the Lee Creek Mine, North Carolina*, III. *Smithsonian Contributions to Paleobiology* (Vol. 90, pp. 203–218).

Part V
Diversity, Biogeography, and Paleobiology of Late Cretaceous and Tertiary Turtles

Chapter 19 Preliminary Overview of Late Cretaceous Turtle Diversity in Eastern Central Europe (Austria, Hungary, and Romania)

Márton Rabi, Mátyás Vremir, and Haiyan Tong

Abstract A preliminary overview of the relatively poorly known Late Cretaceous turtle faunas from eastern Central Europe is given, including brief descriptions both of historically significant and of more recently collected taxa and specimens from the Santonian-Campanian of Hungary, the Campanian of Austria, and the Maastrichtian of Romania. Eastern Central European Late Cretaceous turtle faunas are similar to contemporaneous Western European faunas in their low taxonomic diversity and in being composed almost exclusively of continental forms that are considered endemic to Europe. The eastern Central European fauna includes two pleurodire lineages: the primitive Dortokidae, represented by a separate regional lineage that includes Dortokidae gen. et sp. nov. from the Maastrichtian of Transylvania (Romania) and other indeterminate dortokids from the Santonian of Hungary and Campanian of Austria; and the Bothremydidae, represented by *Foxemys* sp., a member of the Foxemydina that otherwise is known only from Western Europe. Cryptodires are represented by a conservative turtle, *Kallokibotion bajazidi* (Nopcsa, Quarterly Journal of the Geological Society of London 79:100–116, 1923a), from the Maastrichtian of Transylvania and by other previously unknown or unrecognised, closely related taxa from the Santonian of Hungary and the Campanian of Austria. The absence of *Kallokibotion* in Western Europe, coupled with the presence in both regions of the bothremydid *Foxemys* and of separate dortokid lineages, indicate that turtle faunas in Western and eastern Central Europe were partially separated from one another during the Late Cretaceous. "*Senonemys suemegensis*" (Bohn, Földtani Közlöny 96:111–118, 1966) is the only turtle known from Upper Cretaceous marine rocks in eastern Central Europe, but because the material was not adequately described and it now appears to be lost, the identity and relationships of this taxon remain unknown. Many of the turtle-bearing localities in eastern Central Europe show differences in the relative abundances of specimens and faunal compositions that likely are due to a combination of ecological and biogeographical factors and regional extinctions.

Keywords Austria • Bothremydidae • Cryptodira • Dortokidae • Europe • *Foxemys* • Fossil turtles • Hungary • *Kallokibotion* • Late Cretaceous • Pleurodira • Romania

Introduction

A solid understanding of turtle diversity in the Late Cretaceous is important for understanding the selectivity of the K–Pg extinction event among major continental and marine vertebrate groups. Turtles are especially relevant in this regard, because many turtle lineages survived into the Paleogene, their skeletal elements usually can be identified to at least family level, and, thanks to their robust shells and skulls, they have a considerably better fossil record compared both to other K–Pg tetrapod survivors, such as amphibians, squamates, crocodiles, birds, and mammals, and to groups that went extinct, such as marine reptiles, non-avian dinosaurs, and pterosaurs. Compared to other Late Cretaceous turtle faunas in the Northern Hemisphere, those of Europe have been relatively less intensively studied than those of North America or Asia, although they

M. Rabi (✉)
Department of Paleontology, Eötvös Loránd University, 1117 Pázmány Péter sétány 1/C, Budapest, Hungary
e-mail: iszkenderun@gmail.com

M. Vremir
Department of Natural Sciences, Transylvanian Museum Society, 2-4, Strada Napoca, 400009 Cluj-Napoca, Romania
e-mail: vremirmatyi@yahoo.co.uk

H. Tong
30 Rue Carnot, 94270 Le Kremlin-Bicêtre, France

H. Tong
Division of Paleontology, American Museum of Natural History, Central Park West at 79th Street, New York, NY 10024, USA
e-mail: htong09@yahoo.fr

D. B. Brinkman et al. (eds.), *Morphology and Evolution of Turtles,* Vertebrate Paleobiology and Paleoanthropology,
DOI: 10.1007/978-94-007-4309-0_19, © Springer Science+Business Media Dordrecht 2013

certainly warrant similar attention. Turtle faunas of this age in the eastern part of Central Europe—specifically Austria, Hungary, and Romania—have attracted even fewer researchers, in part, because after the death of the famous vertebrate paleontologist Baron Ferenc Nopcsa in 1933, no local specialist replaced him for many decades and, in part, because many localities have only been (re)discovered in the last two decades. Our contribution here helps fill this gap by providing a preliminary overview of the Santonian-Maastrichtian turtle faunas of Austria, Hungary, and Romania. Following brief summaries of previous work and the relevant geology, in this paper we (1) present brief systematic and descriptive accounts for all known taxa and then (2) discuss the paleobiogeography, relative abundances, and palaeoecology of those turtles.

Institutional abbreviations used in this paper are as follows: BMNH, British Museum of Natural History (now the Natural History Museum), London, United Kingdom; EME, Transylvanian Museum Society, Cluj-Napoca, Romania; LPB (FGGUB), Laboratory of Paleontology, Bucharest, Faculty of Geology and Geophysics, University of Bucharest, Romania; GBA, Geological Survey of Austria, Wien, Austria; IPUW, Institute of Paleontology, University of Wien, Austria; MÁFI, Hungarian Geological Survey, Budapest, Hungary; MCDRD, Daco-Roman Civilisation Museum, Deva, Romania; MTCO, Crisurilor Country Museum, Natural History Department, Oradea, Romania; MTM, Hungarian Natural History Museum, Budapest, Hungary; UBB, Biology–Geology Faculty of Babeş-Bolyai University, Cluj-Napoca, Romania.

Historical Summary of Late Cretaceous Turtle Studies in Eastern Central Europe

The first mention of Late Cretaceous turtles from the region belongs to Bunzel (1871) who, besides other reptilian remains, reported indeterminate and isolated turtle shell plates that were recovered during underground mining in the lower Campanian coal-bearing complex [now called the Grünbach Formation (Herman and Kvaček 2007)] of the Gosau Group, at Muthmannsdorf, near Wiener Neustadt, Lower Austria (Fig. 19.1). Seeley (1881) revised the Gosau vertebrate fauna, including the turtles. He erected the new species *Emys neumayri*; no holotype was designated, but in the same paper both a costal 1 (Seeley 1881, Fig. 13, pl. XXXI, Fig. 27, pl. XXVII) and a hyoplastron (Seeley 1881, Fig. 16, pl. XXX) were referred to the species. A second species, *Pleuropletus suessi*, was named on the basis of specimens that Seeley (1881) identified as a skull element, two costal plates, and a scapula (Bunzel 1871, Figs. 4 and 5, pl. VI; Seeley 1881, Fig. 15, pl. XXX). The skull element and the costals turned out to be a skull roof fragment and two ilia, respectively, of the nodosaurid dinosaur *Struthiosaurus austricus* (Pereda-Suberbiola and Galton 2001), whereas Seeley's huge "chelonian scapula" is unidentifiable. Finally, Seeley (1881) also reported other, smaller-bodied species of *Emys*, but none of those were named and no specimens were figured.

The most important work on Late Cretaceous turtles from the former area of the Austrian-Hungarian Monarchy was done by the legendary Hungarian paleontologist Baron Ferenc Nopcsa. In his first scientific paper, Nopcsa (1897) reported turtle remains from the Transylvanian Haţeg Basin (now part of Romania). Koch (1900) referred turtle material from Szentpéterfalva (now Sinpetru), Haţeg Basin, to *Emys* sp. Several turtle specimens, including incomplete shells of juveniles and carapace fragments, girdles, limb bones, and vertebrae of adults, were collected by Ottokár Kadic in 1915, from around Valiora in the Haţeg Basin; these are housed in the Geological Institute of Hungary and, although Nopcsa planned to study them, he never published on the material (Kadic 1916). Later, on the basis of skulls, jaws, and postcranial elements that he had collected, Nopcsa (1923a) erected the new genus and species *Kallokibotion bajazidi*. A detailed description of that species and the naming of a second species, *K. magnificum*, were presented later in the same year (Nopcsa 1923b). Nopcsa also studied the Muthmannsdorf reptile fauna of Lower Austria, and he briefly mentioned that *Emys neumayri* likely was a synonym of *Kallokibotion* (Nopcsa 1926). The Baron's death in 1933 was followed by a long pause in Late Cretaceous chelonian research in eastern Central Europe. Over the next six decades, the only notable publications were the report of questionable *Kallokibotion* remains from Grünbach am Schneeberg in Lower Austria (Thenius 1962), the preliminary study of specimens housed in the Geological Institute of Hungary that had been collected by Kadic in 1915 from the Haţeg Basin (Młynarski 1966), and a revision of the *Kallokibotion* specimens housed in the Natural History Museum in London (Gaffney and Meylan 1992). Beginning in the second half of the 1990s, collection of new material from old localities and discoveries of new fossiliferous localities in both Romania and Hungary has resulted in a surge of publications (Codrea and Vremir 1997; Vremir 2004; Vremir and Codrea 2009; Rabi and Botfalvai 2006; Rabi and Tong 2007; Rabi 2009). Even though this new phase of research is at a preliminary stage, it is apparent that the distribution (both temporal and geographical) and taxonomic diversity of Late Cretaceous turtles in eastern Central Europe is greater than previously recognized. As currently understood, turtle faunas from the region comprise of both pleurodires (dortokids and bothremydids) and cryptodires (*Kallokibotion* and related forms).

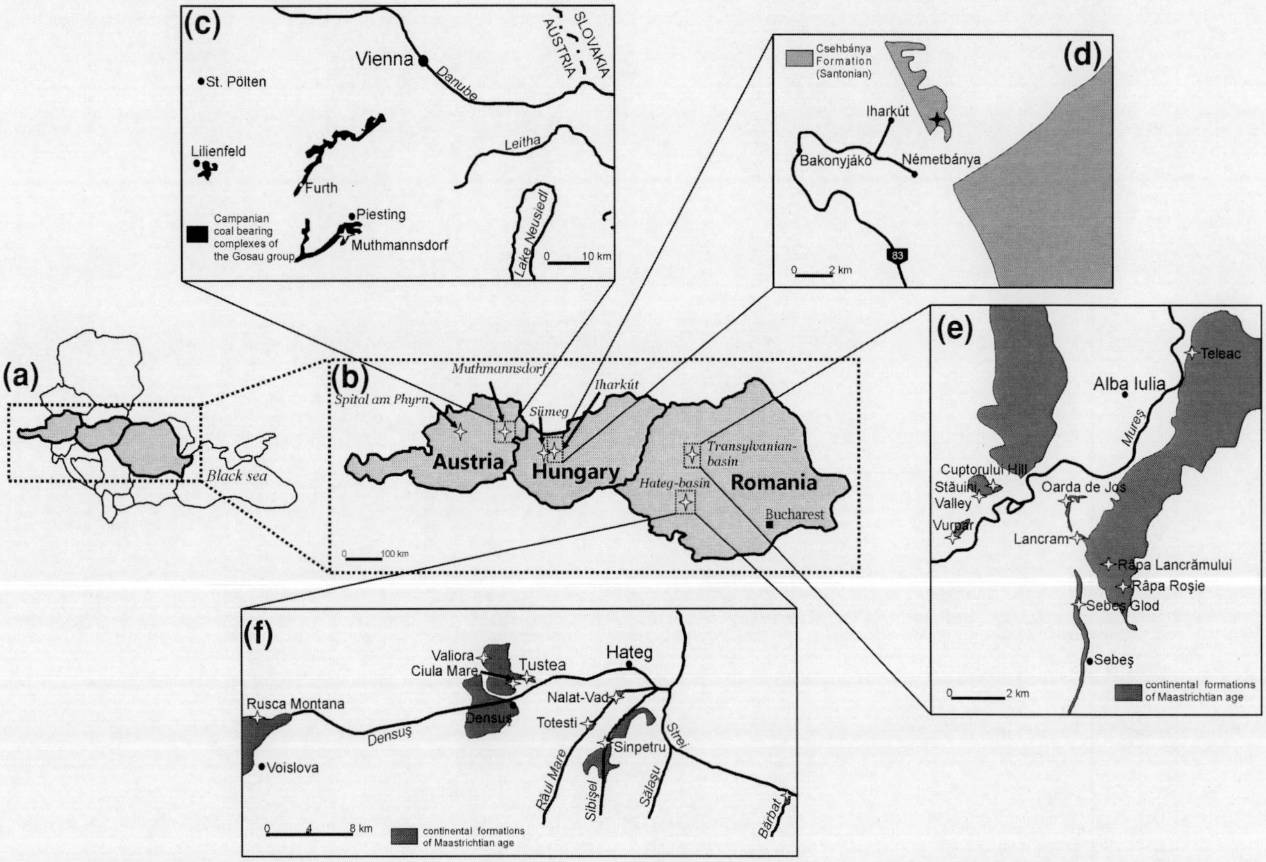

Fig. 19.1 Geographic locations of Late Cretaceous fossil turtle localities in eastern Central Europe (Austria, Hungary, and Romania) and distributions of the fossiliferous formations. *Stars* correspond to turtle localities mentioned in text and Tables 19.1, 19.2, 19.3, 19.4. **a** Map of eastern Central Europe with Austria, Hungary, and Romania shaded *grey*; **b** close-up of the study area with the main turtle localities and areas indicated; **c** close-up of the Muthmannsdorf region in Lower Austria; **d** close-up of the Iharkút region in Hungary; **e** close-up of the Transylvanian Basin in Romania; **f** close-up of the Haţeg Basin in Romania. Maps at different scales

Geological Setting

Four different areas in Austria, Hungary, and Romania contain Upper Cretaceous sedimentary formations that have yielded turtle remains (Fig. 19.1, Tables 19.1, 19.2, 19.3, 19.4). These formations range in age from Santonian to Maastrichtian and represent different paleoenvironments. The vertebrate faunas in each formation contain a number of endemic elements at the generic or specific level.

Csehbánya Formation (Santonian) in Hungary

The oldest known Late Cretaceous turtles in eastern Central Europe come from the Santonian Csehbánya Formation of Hungary. Vertebrate fossils were discovered in 2000 in outcrops of the Csehbánya Formation at the Iharkút open pit bauxite mine, in the Bakony Mountains (Transdanubian Range), north-central Hungary (Fig. 19.1b, d). Most of the vertebrate fossils, including the turtles, come from an approximately 3 m thick, lenticular body of coarse sand, sandstone, organic-rich silt, clay, and a breccia containing rip-up clay clasts and dolomite pebbles. The Csehbánya Formation in general consists of channel and alluvial plain deposits, including sandstone bodies and paleosoil horizons, and the bone beds probably were deposited in a shallow channel or pond that was episodically filled by debris flows (Ősi and Mindszenty 2009). A Santonian age for the formation is supported by palynological (Knauer and Siegl-Farkas 1992) and paleomagnetic (Szalai 2005) data. The Iharkút vertebrate fauna includes fishes, amphibians, turtles, lizards, crocodiles, dinosaurs, pterosaurs, and birds (e.g., Makádi et al. 2006; Ősi and Rabi 2006; Ősi and Mindszenty 2009).

Ugod Limestone Formation (Campanian) in Hungary

The Campanian Ugod Limestone Formation is the only other formation in Hungary that has yielded a Late

Table 19.1 Late Cretaceous turtle occurrences in Hungary and Austria

Locality	Taxa	Material	Stratigraphic unit and age	Sediments and depositional environment	Associated biota	Notes	References
Iharkút, Bakony Mts., Hungary	Dortokidae indet.	Dortokidae indet.: isolated shell plates and ilia	Csehbánya Formation (Santonian)	Lenticular beds of sand, sandstone, and silty clay; channel fills and alluvial plains	Fish, amphibians, lizards, crocodyliforms, pterosaurs, dinosaurs, birds	*Foxemys* specimens most abundant; dortokids and kallokibotiid-like specimens much less common	Knauer and Siegl-Farkas (1992); Szalai (2005); Makádi et al. (2006); Ősi and Rabi (2006); Rabi and Botfalvai (2006); Rabi and Tong (2007); Ősi and Mindszenty (2009); Rabi (2009); Rabi et al. (in review)
	Foxemys sp.	*Foxemys* sp.: isolated skulls, lower jaws, incomplete shells, girdle bones					
	cf. Kallokibotioninae indet.	cf. Kallokibotioninae indet.: isolated shell plates					
Kecskevár quarry, near Sümeg, Hungary	*Senoemys suemegensis*	Carapace steinkern, with peripherals on left and right preserved as bone	Ugod Limestone Formation (Campanian)	Gosau type limestone; shallow marine rudist reef (upper photic zone)	Rudists, corals, echinoderms, calcareous algae	Here regarded as nomen dubium. Specimen lost, but published description may be consistent with Bothremydidae	Bohn (1966)
Muthmannsdorf and Grünbach am Schneeberg, Lower Austria	Dortokidae indet., (Muthmannsdorf) cf. Kallokibotioninae indet. (Muthmannsdorf and Grünbach am Schneeberg)	Isolated carapace and plastral elements	Grünbach Formation, Gosau Group (early Campanian)	Coal seams, coaly siltstone, sandstone; near-shore swampy and lacustrine lowland environment	Lizards, crocodyliforms (*Doratodon*, Eusuchia indet.) pterosaurs, dinosaurs (*Zalmoxes*, *Struthiosaurus*)	Some material originally described as *Emys neumayri* or *Emys* sp. by Seeley (1881)	Bunzel (1871); Seely (1881); Nopcsa (1926); Buffetaut (1979); Wellnhofer (1980); Gemel and Rauscher (2000); Pereda-Suberbiola and Galton (2001); Sachs and Hornung (2006)
Wuhrberg Mt., Spital am Pyhrn, Windischgarsten, Upper Austria	cf. Kallokibotioninae indet.	Carapace steinkern, with some anterior peripherals preserved as bone	Gosau Group (early Campanian)	Coal seams, coaly siltstone, sandstone, near-shore swampy and lacustrine lowland environment	Unknown	Collected in 1905, but has not been described	This study

Table 19.2 Late Cretaceous turtle occurrences in the Transylvanian Basin (Alba county) in Romania

Locality	Taxa	Material	Stratigraphic unit and age	Depositional environment	Associated biota	Notes	References
Stăuini/Cheii Valley (Kolcspatak), Alba, Romania	Testudines indet.	Carapace fragment	Top Bozeş Formation (latest Campanian)	Marine, turbidites within flyschoid sequence	*Trochacteon* sp., *Inoceramus balticus*, other invertebrates	Small undiagnostic fragment	Vremir (2004)
Vurpăr (Borberek), Vinţu de Jos, Alba, Romania	Dortokidae gen. et sp. nov. *Kallokibotion bajazidi* *Kallokibotion* sp.	Various endo- and exoskeletal parts; incomplete plastron with associated carapace remains	Lower part of "Şard" Formation (early Maastrichtian)	Terrestrial, overbank silty claystone (calcareous paleosol); sandy channel fills	Rare eusuchians (*Allodaposuchus*); frequent nodosaurids (*Struthiosaurus transylvanicus*) and ornithopods (*Zalmoxes shqiperorum*, *Z. robustus*); terrestrial gastropods	Turtle fossils frequent in paleosols at about 50–70 m above top of "Vurpăr" Formation	Vremir (2001, 2004); Codrea and Dica (2005); Therrien (2005); Vremir and Codrea (2009); Codrea et al. (2010)
Cuptorului Hill, Stăuini, Alba, Romania	Dortokidae gen. et sp. nov. *Kallokibotion bajazidi* *Kallokibotion* sp.	Disarticulated exoskeleton; isolated carapace fragments	Lower part of "Şard" Formation (early Maastrichtian)	Terrestrial, overbank silty claystone (calcareous paleosol)	Nodosaurids (*Struthiosaurus transylvanicus*), Dinosauria indet., *Hadrosaurichnus*?	Turtle fossils frequent in paleosol unit at 50–80 m above top of "Vurpăr" Formation	Vremir (2001, 2004); Therrien et al. (2002)
Teleac (Telek), Alba, Romania	Dortokidae gen. et sp. nov. *Kallokibotion* sp.	Peripherals; epi-and entoplastral fragments	Upper part of Sebeş Formation (?late Maastrichtian)	Terrestrial, red silty claystone, sandstone	Ornithopod dinosaurs	–	Vremir (2001, 2004); Vremir and Codrea (2009)
Oarda de Jos, (Alsó Váradja) Alba, Romania	Dortokidae gen. et sp nov. *Kallokibotion* sp.	Various isolated exo- and endoskeletal elements	Lower part of Sebeş Formation (late early Maastrichtian)	Fluvio-paludal, siltstone, claystone, calcareous mudstone, sandstone of swampy-riverine habitat	Incarbonised wood, freshvwater snails, cray-fish, fish, amphibians, lacertilians, crocodyliforms, dinosaurs, birds, avian eggs, multituberculates	Isolated turtle bones from ponded floodplain and crevasse splay deposits	Vremir (2004); Vremir and Codrea (2009); Codrea et al. (2010)

Table 19.3 Late Cretaceous turtle occurrences in the Transylvanian (Alba County), Rusca Montană (Caraş-Severin County), and Haţeg (Hunedora County) basins in Romania

Locality	Taxa	Material	Stratigraphic unit and age	Depositional environment	Associated biota	Notes	References
Lancrăm (Lámkerék), Sebeş, Alba, Romania	*Kallokibotion* sp. Testudines indet. (Dortokidae?)	Small, isolated exoskeletal elements	Lower part of Sebeş Formation (late early Maastrichtian)	Channel deposits, (conglomerates, sandstone, and siltstone); channel-dominated fluvial system of streamside gallery-woodland environment	Mainly coniferous wood, stumps, logs; dinosaurs (*Telmatosaurus*, *Zalmoxes*, titanosaurs); dinosaurian footprints	Isolated and transported parautochthonous elements	Vremir (2001, 2004); Vremir and Codrea (2002, 2009); Codrea et al. (2010)
Secaş Valley- and Feţei Hill (Szölohegy), Sebeş, Alba, Romania	*Kallokibotion* sp. Testudines indet.	Small, isolated exoskeletal elements	Lower and middle parts of Sebeş Formation (early late Maastrichtian)	Terrestrial, red claystone, siltstone, and, sandstone	Dinosaurs	Isolated elements, mainly in red paleosol facies	Vremir (2004)
Sebeş Glod, Alba, Romania	*Kallokibotion bajazidi* *Kallokibotion sp.* Testudines indet. (Dortokidae?)	Associated carapace fragments; isolated peripherals and costals	Lower part of Sebeş Formation (late early Maastrichtian)	Red calcareous claystone-mudstone (floodplain) and, channel-fill, cross-bedded sandstone	Crocodyliforms, pterosaurs (Azhdarchidae), dinosaurs (*Zalmoxes*, titanosaurs, dromeosaurids), birds	Rarely articulated; long subaerial exposure in overbank environment	Vremir (2001, 2004); Vremir and Codrea (2009); Codrea et al. (2010); Csiki et al. (2010)
Râpa Lancrăm and Râpa Roşie (Lámkerékimart and Vörösmart or Roteberg), Sebeş, Alba, Romania	*Kallokibotion bajazidi* *Kallokibotion* sp. Testudines indet.	Hypoplastron; isolated and mostly fragmentary exoskeletal elements	Middle and upper parts of Sebeş Formation (early late up to ?latest Maastrichtian)	Red, silty claystone and coarse, mainly cross-bedded channel deposits	Crocodyliforms, pterosaurs (Azhdarchidae), and dinosaurs (Telmatosaurus, *Zalmoxes*, titanosaurs, ankylosaurs)	Mainly isolated and transported paraautochtonous elements	Codrea and Vremir (1997); Jianu et al. (1997); Vremir (2001, 2004); Vremir et al. (2009); Vremir and Codrea (2009); Codrea et al. (2010)
Negoiu, Lunca Cernii de Jos, Rusca Montană Basin, Romania	*Kallokibotion* sp.	One peripheral plate	Rusca Montană Formation (Maastrichtian)	Terrestrial red claystone	Crocodyliforms, dinosaurs (*Zalmoxes*, titanosaurs), multituberculates	Isolated elements	Vremir (2004); Codrea (pers. comm. 2008)
Ciula Mică and Ciula Mare (Kis es Nagy Csula) western Haţeg Basin, Romania.	*Kallokibotion bajazidi* *Kallokibotion* sp. Testudines indet.	Mostly isolated exoskeletal elements	Middle member of Densuş-Ciula Formation (Maastrichtian)	Terrestrial and fluvial clastics	Dinosaurs	Articulated exoskeletal parts in overbank environment	Kadic (1916); Vremir (2004)

Table 19.4 Late Cretaceous turtle occurrences in the Haţeg Basin (Hunedora County) in Romania

Locality	Taxa	Material	Stratigraphic unit and age	Depositional environment	Associated biota	Notes	References
Pui, Bărbat Valley, (Puj, Borbátviz) eastern Haţeg Basin, Romania	Dortokidae gen. et sp. nov. *Kallokibotion bajazidi* *Kallokibotion* sp. Testudines indet.	Carapace with articulated plastron; various exo and endoskeletal elements	?Sinpetru Formation (?early Maastrichtian)	Red claystone, mudstone, siltstone; terrestrial mainly calcareous paleosol and fluvial	Fish, amphibians lizards, crocodyliforms pterosurs, dinosaurs (*Zalmoxes*, *Telmatosaurus*, *Magyarosaurus*, velociraptorinae, troodontidae?), multituberculates	*Kallokibotion* more common than dortokids	Itterbeeck et al. (2004); Vremir (2004); Therrien (2005); Vremir and Codrea (2009)
Sinpetru (Szentpéter falva), Sibişel Valley, central Haţeg Basin, Romania.	Dortokidae gen. et sp. nov. *Kallokibotion bajazidi* (including *K.* "*magnificum*") *Kallokibotion* sp. Testudines indet.	Almost complete and partial articulated skeletons; isolated carapacial and plastral remains; numerous endoskeletal elements	Sinpetru Formation (early late Maastrichtian)	Terrestrial; various paleosol and channel fills; floodplain depression of braided river	Amphibians, lacertilians, crocodyliforms, pterosaurs (Ornithocheiridae, Pteranodontidae), dinosaurs (*Zalmoxes*, *Telmatosaurus*, *Magyarosaurus*, *Struthiosaurus*, small theropods), birds, multituberculates	Holotypes of *Kallokibotion bajazidi* and *K.* "*magnificum*"; mainly articulated carapace and plastron, rarely associated with other elements; in channel and overbank environments; Dortokids in channel facies	Nopcsa (1897, 1923a, b); Koch (1900); Szalai (1934); Młynarski (1966); Grigorescu (1983); Groza (1983); Vremir (2004); Therrien (2005); Vremir and Codrea (2009)
Unciuc, Toteşti, and Vadu (Nălaţ-Vad), Râul Mare Valley, central Haţeg Basin, Romania.	*Kallokibotion bajazidi* *Kallokibotion* sp.	More or less complete and articulated skeletons; isolated carapacial and plastral remains	Sinpetru Formation (late early Maastrichtian)	Terrestrial, fluvio-paludal environment; channel fills, crevasse splays, levee, floodplain of swampy riverine habitat	Amphibians, lizards, pterosaurs, dinosaurs (ornithopods, nodosaurids, titanosaurs, theropods), birds, dinosaurian eggs, multituberculates	Mainly articulated postcranial skeletons in certain horizons in fine channel fills	Codrea et al. (2002); Vremir (2004); Van Itterbeeck et al. (2004); Therrien (2005); Vremir and Codrea (2009)
Vălioara, Tuştea, Boiţa (Velióra, Tustya és Bojicza) and Livezile, northern Hateg Basin, Romania	Dortokidae gen. et sp. nov. *Kallokibotion bajazidi* *Kallokibotion* sp.	Mostly isolated exoskeletal elements	Middle member of Densuş-Ciula Formation (early-?late Maastrichtian)	Terrestrial (mainly moderately drained floodplain deposits) and fluvial (channel fills)	Fish, amphibians, lizards, crocodyliforms, pterosaurs dinosaurs, dinosaurian egg clutches, multituberculates	Mainly isolated and disarticulated skeletal parts in channels and crevasses	Kadic (1916); Vremir (2004); Therrien (2005); Grigorescu (2010)

Cretaceous turtle specimen. This rudistid limestone unit is exposed in the Sümeg area of the Bakony Mountains, about 70 km east of the Iharkút site (Fig. 19.1b). Shortly after fluvial deposition in the Santonian (represented by the Csehbánya Formation), a transgressive event began that initially laid down coal-bearing sequences, followed by shallow water deposits and pelagic marls in the Campanian. The gradual transgression of these marine waters over a Triassic dolomite plateau led to the creation of an extensive rudist platform, which became the Ugod Limestone Formation (Haas 1979). The formation mostly contains calcareous algae, corals, echinoderms, and rudists; however, an internal mold of a turtle carapace was found in the Kecskevár quarry, east of the town of Sümeg. Bohn (1966) erected the new genus and species *Senonemys suemegensis* on the basis of this material, but unfortunately the specimen currently is lost.

Coal-bearing Complex of the Gosau Group (Early Campanian) in Austria

The Gosau Group is a pericontinental to shallow marine succession of Late Cretaceous and Paleocene age that fills several basins along the northern margin of the Eastern Alpine zone. A rich reptile fauna is known from the coal-bearing complex (also called the Grünbach Formation) of the Gosau Group. This vertebrate-bearing unit is made up of interbedded coal seams, coaly siltstones, sandstones, and conglomerates that were deposited under freshwater to nearshore marine conditions (Kvaček and Herman 2004; Sachs and Hornung 2006). The age of the Grünbach Formation is early Campanian, based on foraminifers and palynomorphs (Summesberger et al. 2007). Nearly all vertebrate remains (including all but one of the turtle specimens) were recovered during the latter half of the 19th Century from coaly siltstones in the underground mine of Muthmannsdorf, 15 km east of Wiener Neustadt, Lower Austria (Fig. 19.1c). Additionally, Thenuis (1962) mentioned the presence of questionable *Kallokibotion* remains from Grünbach am Schneeberg, but we have not studied that material. Besides turtles, the Grünbach Formation fauna includes choristoderes, lizards, crocodiles, dinosaurs, and pterosaurs (Bunzel 1871; Seeley 1881; Nopcsa 1926; Buffetaut 1979; Wellnhofer 1980; Pereda-Suberbiola and Galton 2001; Sachs and Hornung 2006). Plant fossils indicate a subtropical climate with seasonal aridity and a paleoenvironment of coastal freshwater ponds and oxbow lakes surrounded by swampy lowlands (Kvaček and Herman 2004; Herman and Kvaček 2007). About 200 km west of Muthmannsdorf, an internal mold of a turtle carapace was recovered from probably coeval, coaly siltstone beds at Wuhrberg Mountain, near Spital am Pyhrn, south of Windischgarsten, Upper Austria (Fig. 19.1b).

Various Formations (Latest Campanian and Maastrichtian) in Romania

Fossiliferous Upper Cretaceous continental deposits have been known in Romania since Nopcsa's (1905, 1923a) initial descriptions of the classic dinosaur fauna from the Haţeg Basin. Recent explorations have determined that these fossil-bearing deposits are widely distributed spatially, across the western margin of the Transylvanian Basin and also within several small, intermontane basins (for a recent review see Codrea et al. 2010). These units were deposited following the latest Cretaceous (Laramidian or late Getian) orogenetic phase that produced the early structures of the Carpathian Mountains, and they represent the oldest molasse deposits created by erosion of sediments from those newly-emerged mountains (Săndulescu 1984; Willingshofer et al. 2001; Krézsek and Bally 2006). A single carapace fragment has been reported (Vremir 2004; Table 19.2) from the uppermost Campanian Top Bozeş Formation, but it is too fragmentary to be identified beyond Testudines indet. All other Late Cretaceous turtle-bearing localities in Romania are Maastrichtian in age and occur in several foramations exposed along the southwestern edge of the Transylvanian Basin and in the smaller Haţeg and Rusca Montană basins (e.g., Codrea et al. 2010; Grigorescu 2010; Fig. 19.1, Tables 19.2, 19.3, 19.4).

The continental "red beds" exposed in the Alba Iulia-Sebeş area (southwestern Transylvanian Basin), have been interpreted as belonging to several distinct lithostratigraphic units and their age has been reported as either Oligo-Miocene or Late Cretaceous [for an overview see Codrea and Dica (2005)]. Until recently, few studies had attempted to solve these stratigraphical problems. Two recent studies (Codrea and Dica 2005; Codrea et al. 2010), proposed a new lithostratigraphical scheme that recognized two uppermost Cretaceous (Maastrichtian) continental formations, both of which yield vertebrate remains. The lower, spatially restricted unit is the Vurpăr Formation (VPF) and is early Maastrichtian in age. The overlying Şard Formation (SDF) is Maastrichtian and, possibly, up to Priabonian in age. In addition, two other continental "red bed" units were redefined in the Alba Iulia-Sebeş area: the Bărăbanţ Formation (BTF), regarded as Oligocene in age and exposed north-northwest of Alba Iulia; and the Sebeş

Formation (SBF), regarded as late Oligocene to middle Miocene in age (Grigorescu 1992; Codrea and Dica 2005; Codrea et al. 2008) and well developed between the Teleac and Sebeş localities along the east side of the Mures Valley (Codrea et al. 2010).

In the light of the most recent field investigations and acquisition of new data, the above stratigraphic model requires revision, particularly regarding the relationship between the Maastrichtian VPF–SDF complex and the so-called "Oligo-Miocene" SBF. Investigations of new outcrops that yielded well-preserved and age-diagnostic vertebrate fossils (e.g., pertosaurs) indicates, in contrast to previous opinions, a Maastrichtian age for the Sebeş Formation as well (Vremir et al. 2009; Csiki et al. 2010). According to these data, a sole lower Maastrichtian-lower Paleogene continental unit must be considered, comprising the recently defined and basically coeval "Vurpăr", "Şard" and "Sebeş" formations, which were all developed in the same depositional setting, but from different source areas. Vertebrate fossils are distributed in certain levels within the "Vurpăr", "Şard", and "Sebeş" formations. Most of the fossil turtle specimens have been recovered from the Vurpăr, Oarda de Jos, Sebeş-Glod, and Râpa Roşie localities (Vremir 2004; Fig. 19.1e).

The Haţeg Basin represents a Late Cretaceous syn-orogenic, extensional basin formed during the Laramidian/Getian phase, located on top of the Getic basement. Nopcsa (1905) identified two different facies that are now recognized as formal lithostratigraphic units of early Maastrichtian-?Paleocene age (Grigorescu 1992, 2010). The Sinpetru Formation (south-central part of Haţeg Basin) consists predominantly of green and red argillaceous sandstones, siltstones, mudstones, and conglomerates, but it is completely devoid of volcanoclastic components. In the Densuş-Ciula Formation (western and northern parts of Haţeg Basin), which is characterized by the additional presence of volcanoclastic components, three lithostratigraphic sequences (members) are recognized. Vertebrate fossils are distributed throughout the Sinpetru Formation, whereas they are restricted to the middle member of the Densuş-Ciula Formation (Grigorescu 2010). Most of the turtle fossils originate from localities around Pui, Sinpetru, Totesti, Vadu, Valioara, and Tusţea (Fig. 19.1f).

Upper Cretaceous continental deposits in the Rusca Montană Basin (an intermontane basin west of the Haţeg Basin), were formed in a similar manner as those in the Haţeg Basin and are underlain by upper Campanian-lowermost Maastrichtian flyschoid marine deposits. The Rusca Montană Formation is a conglomeratic unit that ranges from several hundred up to 2000 m in thickness. The lower part of the formation consists of interbedded sandstones and claystones, whereas the upper part contains several pyhroclastic and detrital units interbedded with coal (Grigorescu 1992). The scarce vertebrate fossils are confined to the upper part of the formation, in reddish clay- and siltstones or, less commonly, in coal-rich fluvio-lacustrine facies that also contain a rich paleofloral assemblage. The only turtle fossil from the Rusca Montană Formation is a peripheral plate of *Kallokibotion* sp. that was collected from the Negoiu locality (Vremir 2004; Table 19.3).

Systematic Paleontology by Rock Unit

Turtles from the Csehbánya Formation (Santonian) in Hungary

Testudines Linnaeus 1758
Pleurodira Cope 1864
Megapleurodira Gaffney et al. 2006
Dortokidae de Lapparent de Broin and Murelaga 1996
Dortokidae indet.
(Fig. 19.2)

Referred specimens: MTM V 2010. 200.1., left costal 1; MTM V 2010.196.1., left costal 5; MTM V 2010.197.1., indeterminate costal; MTM V 2010.245.1., indeterminate neural; MTM V 2010.246.1., indeterminate peripheral; MTM V 2010.222.1., left hyoplastron; MTM V 2010.203.1., right xiphiplastron; MTM V 2010.160.1. and MTM V 2010.162.1., ilia. All found isolated from one another.

Occurrence: Csehbánya Formation (Santonian); Iharkút bauxite pit, Bakony Mountains, western Hungary.

Description and comparison: The most distinctive feature of the shell elements is their micro-reticulate ornament. As in other dortokids, this ornament consists of fine, longitudinal ridges oriented latero-medially or radially on the costals and peripherals, and antero-posteriorly on the plastron (Fig. 19.2j). The ridges often surround minute pits. The neurals and the medial third of the costals additionally are decorated with sharp crests that extend antero-posteriorly (Fig. 19.2a, c, e). These crests are reminiscent of those found in the Late Cretaceous Dortokidae gen. et sp. nov. from Romania (see account, below) and *Dortoka vasconica* (de Lapparent de Broin and Murelaga 1996) from Spain, but unlike those in the Paleocene *Ronella botanica* (Gheerbrant et al. 1999–2000) from Romania (de Lapparent de Broin and Murelaga 1996, 1999; de Lapparent de Broin et al. 2004; Vremir and Codrea 2009). Scars preserved on visceral surfaces of several costals indicate that the axillary process contacted costal 1 slightly lateral to the head of the second thoracic rib and that the inguinal process was attached to the carapace via costal 5 as in other dortokids (Fig. 19.2d, b, respectively). However, MTM V 2010.196.1. (Fig. 19.2b) shows that the inguinal suture was confined to

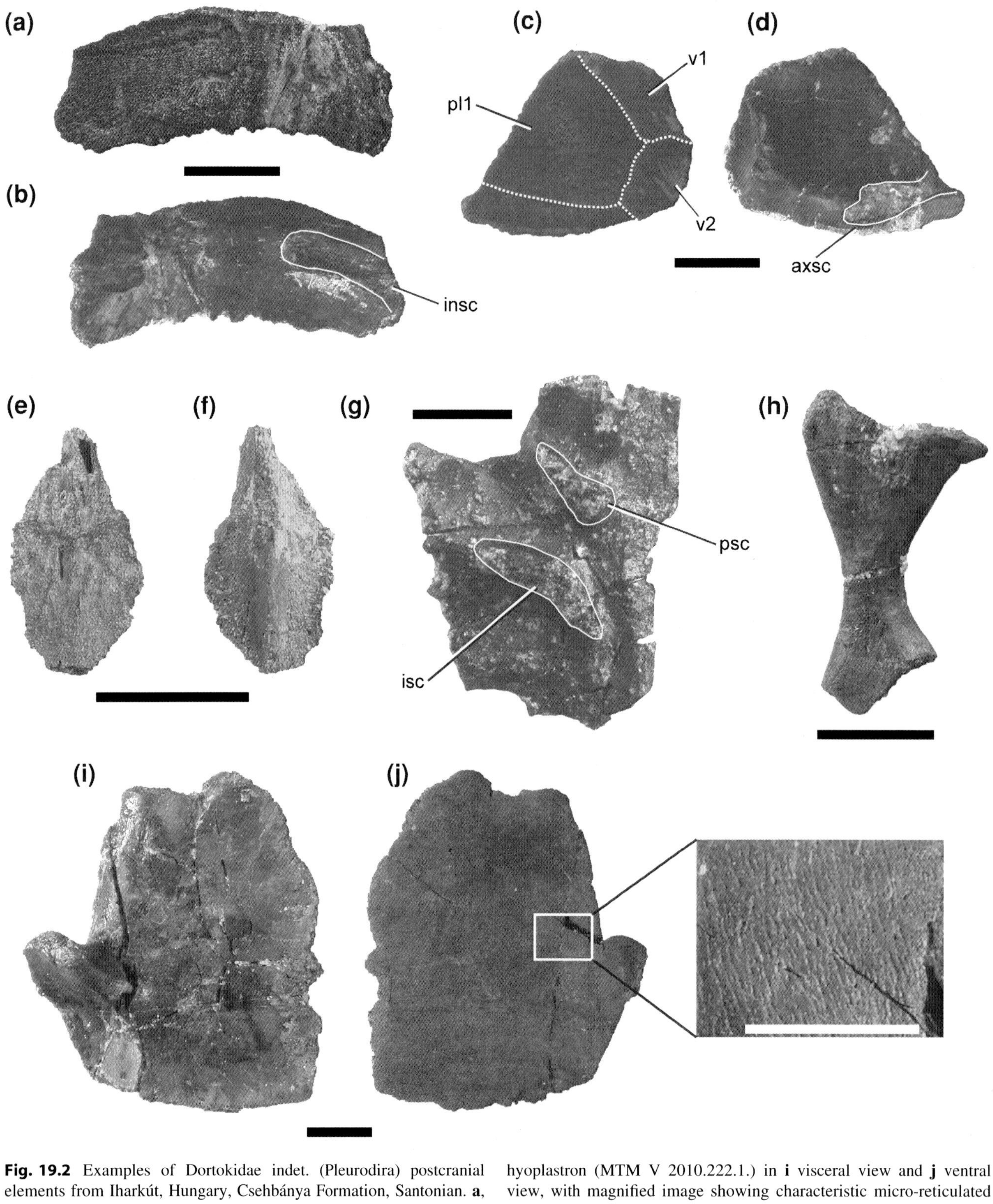

Fig. 19.2 Examples of Dortokidae indet. (Pleurodira) postcranial elements from Iharkút, Hungary, Csehbánya Formation, Santonian. **a**, **b** Left costal 5 (MTM V 2010.196.1.) in **a** dorsal view and **b** visceral view; **c**, **d** left costal 1 (MTM V 2010. 200.1.) in **c** dorsal view and **d** visceral view; **e**, **f** neural (MTM V 2010.245.1.) in **e** dorsal view and **f** visceral view; **g** right xiphiplastron (MTM V 2010.203.1.) in visceral view; **h** right ilium (MTM V 2010.162.1.) in medial view; **i**, **j** left hyoplastron (MTM V 2010.222.1.) in **i** visceral view and **j** ventral view, with magnified image showing characteristic micro-reticulated external ornament of Dortokidae. Scute sulci (**c**) indicated by *dotted, white lines*; attachment scars (**b**, **d**, and **g**) outlined by *thin, solid, white lines*. Abbreviations: *axsc* axillary scar, *insc* inguinal scar, *isc* ischiac scar, *pl*1 pleural 1, *psc* pubic scar, *v*1 *vertebral* 1, *v*2 vertebral 2. Images at different magnifications; all scale bars equal 1 cm

costal 5 and not prolonged onto costal 4, as in *Dortoka*. The sulci generally are sharp and elevated from the plane of the plates. The figured costal 1(MTM V 2010.200.1) shows that the pleural scute 1 did not overlap onto costal 2 (Fig. 19.2c) as in Dortokidae gen. et sp. nov. and *Ronella*, but unlike in *Dortoka*. The pubic scar lies entirely on the xiphiplastron (Fig. 19.2g) and, similarly to Dortokidae gen. et sp. nov. and *Ronella*, it does not overlap onto the hypoplastron. Based on our preliminary investigations, the Iharkút dortokid is more closely related to Dortokidae gen. et sp. nov. than to *Dortoka*, but more detailed comparisons are needed to clarify its relationship with *Ronella*.

Remarks: The family Dortokidae is a primitive group of freshwater pleurodires that is endemic to Europe and is known only by shell material. de Lapparent de Broin and Murelaga (1999) considered Pelomedusoides as the sister taxon of the Dortokidae. Later both de Lapparent de Broin et al. (2004) and Gaffney et al. (2006) suggested a more basal position for Dortokidae relative to Eupleurodira, but more derived than *Platychelys* and *Notoemys*. The family was first described from the late Campanian-early Maastrichtian of Spain and France, and originally contained a single genus and species, *Dortoka vasconica* (de Lapparent de Broin and Murelaga 1996, 1999). Later the family was reported from the late Barremian of Spain, the Cenomanian of France, and the latest Paleocene of Romania; the last occurrence yielded a new taxon, named as *Ronella botanica* (de Lapparent de Broin and Murelaga 1999; de Lapparent de Broin et al. 2004, Vullo et al. 2010). Recently, Vremir and Codrea (2009) identified a new dortokid, informally referred to here as Dortokidae gen. et sp. nov., based on shell material from the Maastrichtian of Romania. Although further work is needed, at present it appears that the eastern Central European dortokids represent a separate lineage from the dortokids in Western Europe.

Eupleurodira Gaffney and Meylan 1988 (sensu Gaffney et al. 2006)

Pelomedusoides Cope 1868
Bothremydidae Baur 1891
Foxemys Tong et al. 1998
Foxemys sp.
(Figs. 19.3, 19.4)

Referred specimens: MTM V 2010.86.1., MTM V 2010.87.1., MTM V 2010.215.1., MTM V 2010.216.1., fragmentary and nearly complete skulls; MTM V 2010.221.1., right otic region consisting of quadrate, squamosal, ophistotic, and prootic; MTM V 2010.219.1., lower jaw; MTM V 2010.89.1., left incomplete lower jaw; MTM V 2010.212.1., anterior part of carapace; MTM V 2010.211.1., posterior part of carapace; MTM V 2010.214.1., posterior fragment of carapace consisting of left costals 7 and 8; MTM V 2010.207.1., anterior lobe of plastron; MTM V 2010.206.1., xiphiplastron; MTM V 2010.166.1, ilium; MTM V 2010.177.1., scapula; MTM V 2010.169.1., pubis; MTM V 2010.187.1.,left humerus; MTM V 2010.190.1., femur. All found isolated from one another.

Occurrence: Csehbánya Formation (Santonian); Iharkút bauxite pit, Bakony Mountains, western Hungary.

Description and comparison: Thanks to an abundance of well-preserved specimens, this pleurodire has the best known skeletal anatomy for any of the turtles from the Csehbánya Formation. Apart from a few skull bones, distal limb elements, and caudal vertebrae, most of the skeleton of this taxon can be reconstructed; thus, it also represents one of the most completely known bothremydids described so far. The skull of this turtle (Fig. 19.3) is characterized by the following features: wide triturating surfaces on the upper and lower jaws, but without any jugal contribution to the upper triturating surfaces; wide prefrontals meeting along the midline; large and dorsolaterally placed orbits; slight temporal emargination; a pentagonal basisphenoid that contacts the quadrate; a narrow, posteriorly open, and slit-like incisura columella auris; and a long supraoccipital process extending much farther posteriorly than the posterior tip of the squamosals (only seen in MTM V 2010.87.1.; not figured here). This combination of characters is present only in members of the subtribe Foxemydina, a freshwater group of the pleurodiran family Bothremydidae that is restricted to the Late Cretaceous of Europe (Gaffney et al. 2006). The skulls from Iharkút are most similar to those of *Foxemys mechinorum* Tong et al. 1998 from the Late Cretaceous of southern France in having an anteriorly shifted condylus mandibularis, a basioccipital shorter than the basisphenoid (Fig. 19.3b), and a low ventral orbital rim; this trio of characters differs from *Polysternon provinciale* Matheron 1869, which is the only other foxemydine for which the skull has been described (Buffetaut et al. 1996; Tong et al. 1998; Tong and Gaffney 2000; Gaffney et al. 2006). The shell also shares more traits with *F. mechinorum*; for example, it lacks the typical ornament (parallel striations) of *Polysternon* and the lateral borders of the posterior plastral lobe are straight (Fig. 19.4a). Several other postcranial specimens, including shell elements, limb bones, and girdle elements, are also referred to this taxon (Fig. 19.4b–e). Some characters, including a basioccipital concavity that barely extends onto the basisphenoid (Fig. 19.3b), a dentary with a considerably wider symphysis and more rounded anterior tip (Fig. 19.3f), and the presence of a nuchal notch, indicate that the Iharkút *Foxemys* represents a new species. This will be formally named and described in detail elsewhere (Rabi et al., in review).

Remarks: The Bothremydidae are a diverse family of pleurodiran turtles that were widely distributed during the Early Cretaceous-Eocene in Gondwana (except Australia) and Euramerica. In Europe they were previously known

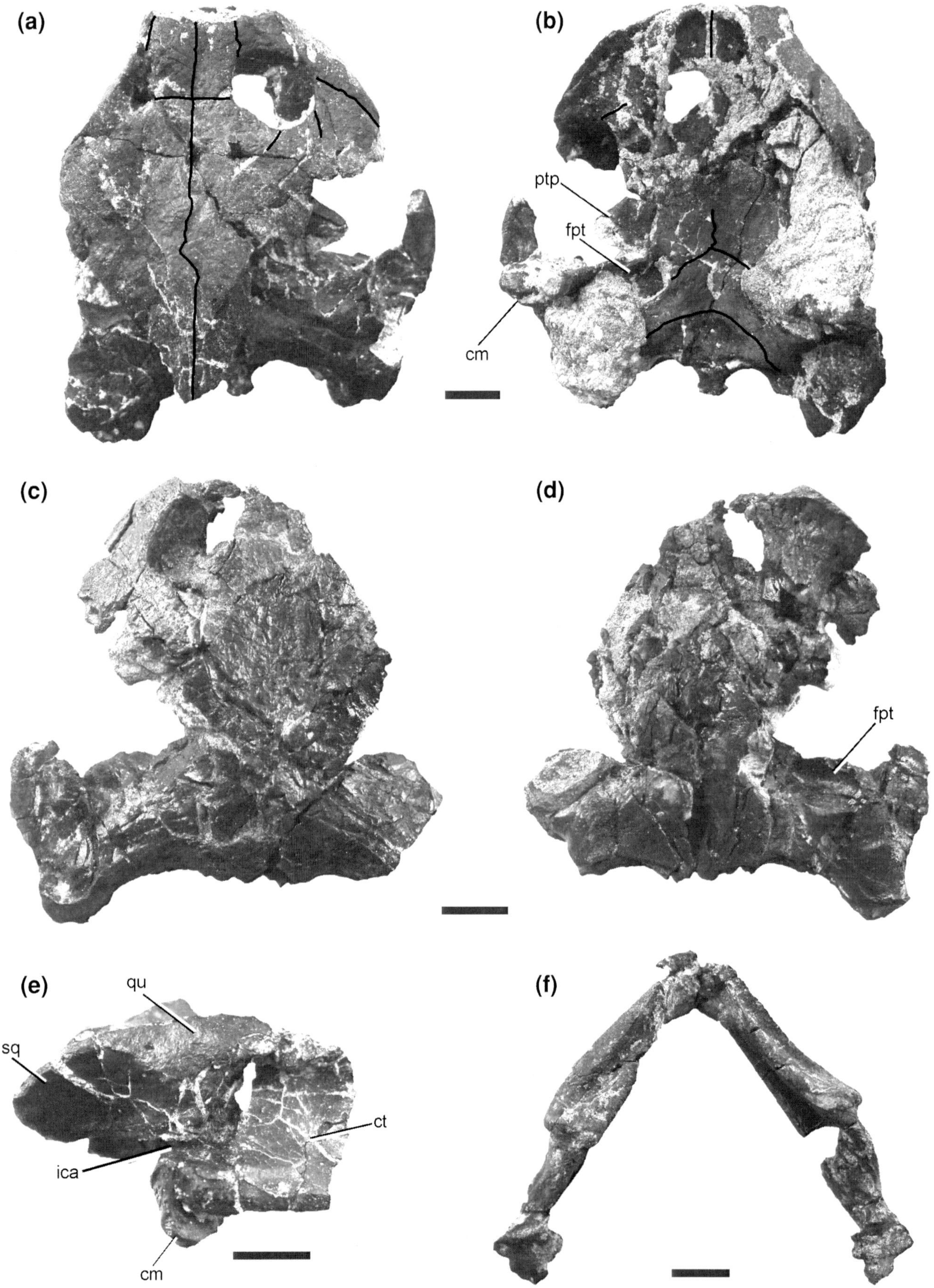
(a)
(b)
ptp
fpt
cm
(c)
(d)
fpt
(e)
qu
sq
ct
ica
cm
(f)

◀ **Fig. 19.3** Examples of *Foxemys* sp. (Pleurodira; Bothremydidae) cranial remains from Iharkút, Hungary, Csehbánya Formation, Santonian. **a**, **b** Skull (MTM V 2010.215.1.) in **a** dorsal view and **b** ventral view, both with some sutures indicated by *black lines*; **c**, **d** skull (MTM V 2010.216.1.) in **c** dorsal view and **d** ventral view; **e** right otic chamber region (MTM V 2010. 221.1) consisting of quadrate, squamosal, ophistotic, and prootic, in lateral view and showing the posteriorly open, slit-like incisura columella auris characteristic of Foxemydina; **f** nearly complete mandibles (MTM V 2010.219.1.) in dorsal view. Abbreviations: *cm* condylus mandibularis of the quadrate, *ct* cavum tympani, *fpt* fossa pterygoidea, *ica* incisura columellae auris, *ptp* processus trochlearis pterygoidei, *qu* quadrate, *sq* squamosal. Images at different magnifications; all scale bars equal 1 cm

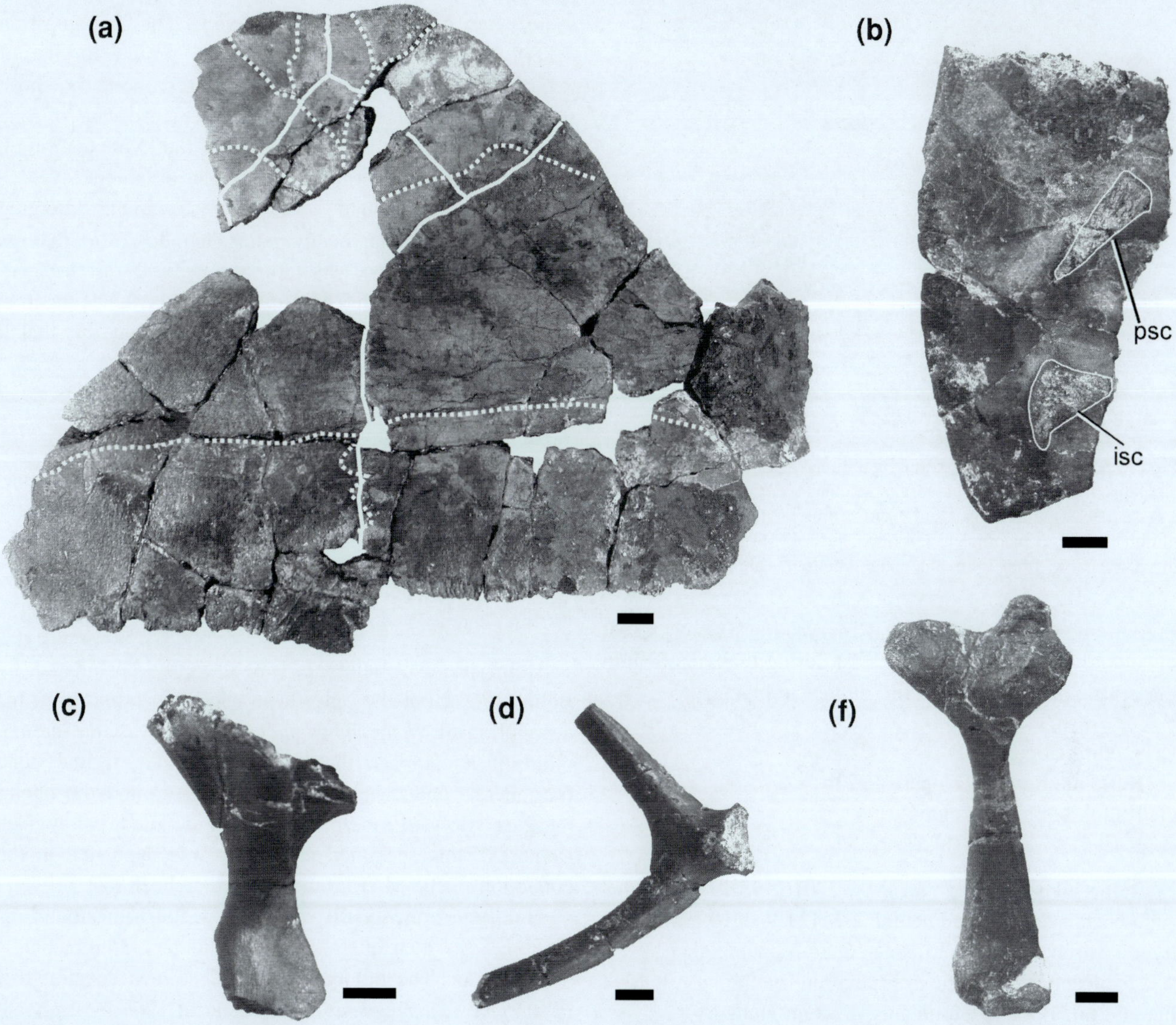

Fig. 19.4 Examples of *Foxemys* sp. (Pleurodira; Bothremydidae) postcranial remains from Iharkút, Hungary, Csehbánya Formation, Santonian. **a** Anterior lobe of plastron (MTM V 2010.207.1.) in ventral view, with scute sulci indicated by *dotted*, *white lines* and sutures indicated by *thick*, *solid*, *white lines*; **b** left xiphiplastron (MTM V 2010.206.1.) in visceral view, with attachment scars outlined by *thin*, *solid*, *white lines*; **c** right ilium (MTM V 2010.166.1.) in medial view; **d** incomplete right scapula (MTM V 2010.177.1.) in posterior view; **e** left humerus (MTM V 2010. 187.1.) in ventral view. Abbreviations: *isc* ischiac scar, *psc* pubic scar. Images at different magnifications; all scale bars equal 1 cm

only from France, Spain, and Portugal, where they are represented by two closely related groups: the subtribe Foxemydina (including the genera *Foxemys*, *Polysternon,* and *Elochelys*) and the subtribe Bothremydina (including *Rosasia*) (de Lapparent de Broin and Murelaga 1996, 1999; Gaffney et al. 2006). As shown by the posteriorly open incisura columellae auris of the quadrate, the bothremydid from Iharkút is closer to Foxemydina than to Bothremydina. The Foxemydina inhabited freshwater environments and have been considered to be endemic to Western Europe. The new

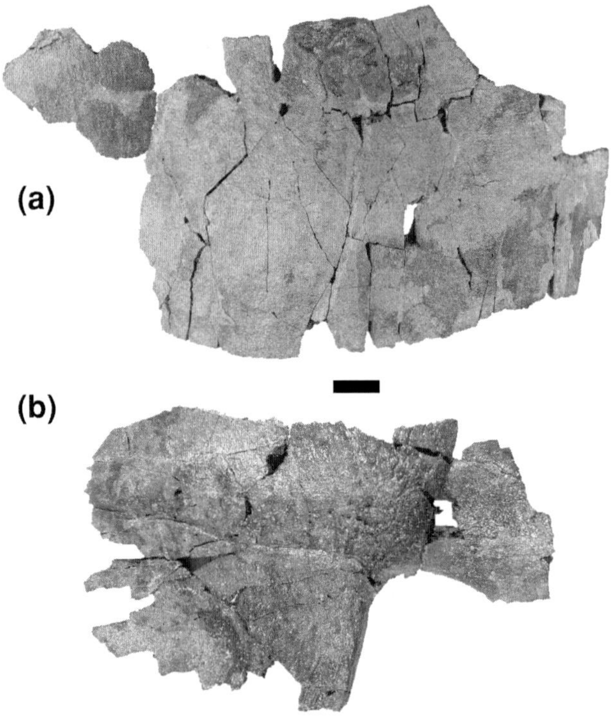

Fig. 19.5 Examples of cf. Kallokibotioninae indet. (Cryptodira) shell elements from Iharkút, Hungary, Csehbánya Formation, Santonian. **a** Associated right costals 2 and 3 and neural 2 (MTM V 2010.195.1.) in dorsal view; **b** incomplete left hypoplastron (MTM V 2010.194.1.) in ventral view. Images at same magnification; scale bar equals 1 cm

Iharkút specimens indicate that foxemydines were present outside of Western Europe, at least during the Santonian in the central part of the western Tethyan archipelago.

Cryptodira Cope 1868
cf. Kallokibotioninae Nopcsa 1923b
cf. Kallokibotioninae indet.
(Fig. 19.5)

Referred specimens: MTM V 2010.195.1., associated right costals 2 and 3 and neural 2; MTM V 2010.194.1., left hypoplastron. Both found isolated from one another.

Occurrence: Csehbánya Formation (Santonian); Iharkút bauxite pit, Bakony Mountains, western Hungary.

Description and comparison: The carapace fragment MTM V 2010. 195.1. consists of right costals 2 and 3 and neural 2 (Fig. 19.5a). Its dorsal surface bears a distinctive ornament of fine, but macroscopically visible vermiculations and tubercles. This ornament is somewhat reminiscent of that in solemydids from the Late Cretaceous of France and Spain, except that the vermiculations are much finer and smaller. In many other features, the shell bones are most similar to *Kallokibotion bajazidi* from the Maastrichtian of Romania. Although the suboval neural 2 in MTM V 2010.195.1. contrasts with the more usual coffin-shaped neural 2 in *Kallokibotion*, assessing the significance of that difference must await the findings of a study (Vremir and Rabi, in preparation) on neural variation in a large sample of undescribed *Kallokibotion* material from Romania. The costals are domed and, as reconstructed, the vertebral scutes were wide as in *Kallokibotion*. Besides preserving sucli for vertebrals 2 and 3, MTM V 2010. 195.1. also preserves sulci for the posterior third of pleural 2 and the anterior half of pleural 3. As far as can be determined, the postions of these scutes are similar to *Kallokibotion*. The hypoplastron (MTM V 2010.194.1.; Fig. 19.5b) is similar to that of *Kallokibotion* in the following features: bone is short; inframarginal scute present; anterior process of the xiphiplastral suture markedly overlaps the hypoplastron on the inner side; and vermiculate external ornament.

Remarks: Although these two specimens are too incomplete to unequivocally refer them to *Kallokibotion*, based on their preserved morphology and on the geographical and temporal proximity between the Hungarian and Romanian localities, it seems likely that the Ihakút specimens belong to a *Kallokibotion*-like turtle.

Turtles from the Ugod Limestone Formation (Campanian) in Hungary

Testudines Linnaeus 1758
"*Senonemys suemegensis*" Bohn 1966

Holotype: No holotype was formally designated by Bohn (1966), but the only turtle specimen reported in his description of "*Senonemys suemegensis*" was the internal mold of a carapace that preserved some original bone (peripherals and distal ends of costals) around the lateral margins (Bohn 1966, pl. VIII). As detailed below (see Remarks), this specimen reportedly was deposited in the collection of the MÁFI, but it has since been lost.

Holotype unit, locality and age: Ugod Limestone Formation; Kecskevár quarry, Sümeg, Hungary; Campanian.

Remarks: The holotype and only known specimen of "*Senonemys suemegensis*" was found in 1963 by a private collector in shallow marine, Gosau-type limestone deposists of the Campanian Ugod Limestone Formation, in the Kecskevár limestone quarry, east of the town of Sümeg, Hungary. It was then transported to the Geological Institute of Hungary (MÁFI) and, shortly thereafter, it was described by Bohn (1966). Unfortunately, because Bohn (1966) did not provide an inventory number for the specimen it has been difficult to trace its history within the collection. To further complicate matters, at some later point in time, the specimen appears to have been accidentally mixed with a trionychoid shell. Not surprisingly, we have not been able to re-locate the specimen and, for the time being, it must be considered lost.

The only published account of the specimen is Bohn's (1966) publication, which is unsatisfactory in that the description is inconsistent with the photos and it contains many erronous interpretations about the morphology of the specimen. Furthermore, he included "*Senonemys*" in the "Emydae" as a close relative of the Late Jurassic sea turtle *Thalassemys*. Based on the photo (pl. VIII) and the description published by Bohn (1966) the outline of the shell appears to be complete, with a total length of 34 cm and a total width of 31 cm. The peripherals on the left side, the lateral peripherals on the right side, and the lateral ends of the costals are preserved as bone. It also appears that "*Senonemys*" had an unreduced shell that was low and wide, a nuchal emargination was present, the costals were relatively short antero-posteriorly with the exception of costal 1, and the plastron contacted the carapace via costal 1 and 5. A potentially significant feature is the presence of a pair of well-developed triangular scars at the level of costals 7 and 8 that tentatively could be interpreted as sutures for the ilia. If the pelvis was indeed sutured to the shell, that argues for "*Senonemys*" being a pleurodire. Other characters, although needing detailed verification, are consistent with bothremydid affinities for "*Senonemys*", as is the temporal and geographical proximity of the locality to the Santonian Iharkút fauna. The lost specimen of "*Senonemys suemegensis*" clearly is from a turtle, but considering the many problems with Bohn's (1966) description (e.g., no holotype was explicitely designated, no diagnostic characters were listed, and inconsistencies between his description and figures) and that the holotype is both poorly preserved and lost, we propose that "*Senonemys suemegensis*" be considered a *nomen dubium* within Testudines *incertae sedis* until such time as the material can be re-located and properly studied. Even though its taxonomic identity is uncertain, "*Senonemys*" is interesting because it documents the only known occurrence of a turtle from marine limestone deposits in eastern Central Europe.

Turtles from the Gosau Group (Early Campanian) in Austria

Pleurodira Cope 1864
Megapleurodira Gaffney et al. 2006
Dortokidae de Lapparent de Broin and Murelaga 1996
Dortokidae indet.
(Fig. 19.6)

Referred specimens: IPUW 4750/64, left costal 1; IPUW 4750/65, incomplete right costals ?6 + 7.

Occurrence: Coal-bearing beds of the Grünbach Formation, Gosau Group, (early Campanian); Muthmannsdorf underground coalmine (now covered in), Lower Austria.

Fig. 19.6 Examples of Dortokidae indet. (Pleurodira) shell elements from Muthmannsdorf, Lower Austria, coal-bearing complex of Grünbach Formation, Gosau Group, early Campanian. **a**, **b** Left costal 1 (IPUW 4750/64) in **a** visceral view, showing scar for sutural contact with the axillary buttress and **b** dorsal view, showing micro-reticulated external ornament and pleural 1 not extending onto costal 2; **c** right costals ?6 + 7 (IPUW 4750/65) in dorsal view. *Thin, solid, white lines* indicate both attachment scar (**a**) and suture (**c**); scute sulci (**b** and **c**) indicated by *dotted, white lines*. Abbreviations: *axsc* axillary scar, *pl*1 pleural 1, *v*1 vertebral 1. Images at same magnification; scale bar equals 1 cm

Description and comparison: Ornament on external surfaces of the plates from Muthmannsdorf (Fig. 19.6b, c) is identical to that of the Iharkút dortokid in being micro-reticulated or gently pitted and in that these structures are arranged in a latero-medial direction. On the medial third of costal 1 (Fig. 19.6b), antero-posteriorly oriented crests also are present, as in the Iharkút dortokid and Dortokidae gen. et sp. nov. On that same specimen, pleural 1 does not reach costal 2, unlike in *Dortoka*, but similar to the condition in the Hungarian and Romanian dortokids (including *Ronella*). The axillary buttress was sutured to costal 1 (Fig. 19.6a) in the same position as in Dortokidae gen. et sp. nov. and the Iharkút dortokid. As seen on the right probable costals 6 + 7 (Fig. 19.6c), costal 6 is anteroposteriorly short as in *Dortoka*, but unlike in *Ronella*. To sum up, the dortokid from Austria seems to share more characters with the Iharkút dortokid and Dortokidae gen. et sp. nov., than with the Western European *Dortoka*.

Remarks: A few other specimens from Muthmannsdorf housed in the collections of the University of Vienna also belong to Dortokidae and will be described elswhere (Vremir and Rabi, in preparation).

Cryptodira Cope 1868
cf. Kallokibotioninae Nopcsa 1923b
cf. Kallokibotioninae indet.
(Fig. 19.7)

Fig. 19.7 Examples of cf. Kallokibotioninae indet. (Cryptodira) shell material from Muthmannsdorf, Lower Austria (**a–d**) and Spytal am Phyrn, Wuhrberg Mountain, Upper Austria (**e** and **f**), both in coal-bearing complex of the Gosau Group, early Campanian. **a–c** Left costal 1 (IPUW 76/XXI/40): **a** photograph in dorsal view, with magnified image showing characteristic tuberculate external ornament of *Kallokibotion*; **b** photograph in visceral view; **c** reproduction of original illustration (as "*Emys neumayri*") in dorsal view (from Seeley 1881, Fig. 13, pl. XXXI); **d** right hyoplastron (IPUW 4750/47) in ventral view; **e**, **f** internal mold of carapace with original anterior peripherals (GBA 2010/068/0001) in **e** dorsal view and **f** left lateral view. Abbreviations: *axsc* axillary scar, *spe*1 suture for peripheral 1, *v*1, vertebral 1. Images at different magnifications; scale bars equal 1 cm for **a–d** and for **e** and **f**

Synonym: *Emys neumayri* (Seeley 1881, p. 698, Fig. 13, pl. XXXI, Fig. 27, pl. XXVII, Fig. 16, pl. XXX).

Referred specimens: IPUW 76/XXI/40, left costal 1 (Seeley 1881 [as *Emys neumayri*], Fig. 13, pl. XXXI, Fig. 27, pl. XXVII); IPUW 4750/21, right costal 4; IPUW 4750/7, left costal 8; IPUW 2349/106, left hyoplastron (Seeley 1881 [as *E. neumayri*], Fig. 16, pl. XXX); IPUW 4750/47, right hyoplastron; GBA 2010/068/0001, internal mold of carapace with original anterior peripherals.

Occurrence: Coal-bearing beds of the Gosau Group (early Campanian): Grünbach Formation at Muthmannsdorf underground coalmine (now covered in) and, questionably, at Grünbach am Schneeberg (Thenius 1962), Lower Austria, and in unspecified beds at Wuhrberg Mountain, near Spytal am Phyrn, Upper Austria.

Description and comparison: All isolated plates from Muthmannsdorf externally bear tiny vermiculate or tuberculate ornamentation or are indented with minute pits (Fig. 19.7a), as in *Kallokibotion bajazidi* from the Maastrichtian of Romania or the *Kallokibotion*-like form reported above from the Santonian of Hungary. The left costal 1, IPUW 76/XXI/40 (Seeley 1881, Fig. 27, pl. XXVII, Fig. 13, pl. XXXI; Fig. 19.7a–c), is further similar to that of *Kallokibotion* in showing a sutured axillary process anterolaterally and wide vertebral scutes. Based on the preserved sutural surface forming the leading edge of costal 1, the nuchal was larger and more trapezoidal than the unusually small and square nuchal previously reported for *Kallokibotion bajazidi*. However, the nuchal of *K. bajazidi* has been described only from a single specimen (BMNH R4919; Gaffney and Meylan 1992, Figs. 21, 22) and new material from Romania (see below) indicates that the nuchal was more variable in this species and could have been more similar to the general trapezoidal shape found in most other turtles. The costal 1 from Muthmannsdorf also strongly resembles MAFI Ob3142 (not figured), a *Kallokibotion* costal 1 from the Maastrichtian of Romania, and differs only in being less domed and in having the nuchal suture slightly shorter. The other isolated plates from Muthmannsdorf, including the left hyoplastron figured by Seeley (1881, Fig. 16, pl. XXX) and the right hyoplastron figured here (Fig. 19.7d) are identified as cf. Kallokibotioninae indet. based on their particular ornament. An undescribed specimen (GBA 2010/068/0001; Fig. 19.7e, f) housed in the Geological Survey of Austria and collected in 1905 from the Gosau beds at Wuhrberg Mountain, near Spytal am Phyrn, south of Windischgarsten, Upper Austria, preserves the internal mold of the carapace and the original anterior peripherals. This specimen is also preliminarily referred to cf. Kallokibotioninae indet. on the basis of the high-domed shape of the shell and the vermiculate ornament on the external bony surfaces. All of these Austrian specimens will be described in detail elsewhere (Rabi, in preparation).

Remarks: The first turtle elements from the Campanian coal-bearing Gosau beds of Austria, found in an underground coal mine near Muthmannsdorf, were briefly mentioned and figured by Bunzel (1871) and referred to Chelonia indet. Later Seeley (1881) studied the Gosau reptilian fauna and erected *Emys neumayri* on the basis of a costal 1 (Seeley 1881, Fig. 13, pl. XXXI, Fig. 27, pl. XXVII) and a left hyoplastron (Seeley 1881, Fig. 16, pl. XXX), but he did not designate a holotype. Nopcsa (1926) noted that the material of *E. neumayri* may belong to *Kallokibotion*, but did not provide any explanation for that assignment. We largely confirm Nopcsa's suggestion in that most of the turtle plates from Muthmannsdorf belong to a *Kallokibotion*-like taxon, including the ones figured by Seeley (1881), and we consider *E. neumayri* an invalid name. Seeley (1881) differentiated other species of *Emys* mainly on the basis of their smaller size. Although we have not been able to locate those specimens, based on our current understanding of the Gosau Group turtle fauna it is likely that most of Seeley's (1881) small "*Emys*" specimens were small-sized, presumably subadult, *Kallokibotion*-like individuals and that the remainder belonged to Dortokidae. Thenius (1962) also reported questionable *Kallokibotion* remains from Grünbach am Schneeberg but we have not been able to study these specimens.

Turtles from Various Formations (Maastrichtian) in Romania

Pleurodira Cope 1864
Megapleurodira Gaffney et al. 2006
Dortokidae de Lapparent de Broin and Murelaga 1996
Dortokidae gen. et sp. nov.
(Fig. 19.8)

Synonym: *Muehlbachia nopcsai* nomen invalidum: Vremir and Codrea (2009, pp. 122–124); Vremir (2010, pp. 648–649), Figs. 12, 13.

Referred specimens: EME VP-102/a, incomplete plastron (left hyo-hypoplastron and incomplete right xiphiplastron) and associated carapace fragments; EME VP-102/b, left hyoplastron and incomplete right hyoplastron; EME VP-103, left costal 1; EME VP-104, right costal 8; EME OdA-51, left costal 1; EME OdA-52, right hyoplastron; EME OdA-53, neural 5; UBB ODA N-2, right costals 1 + 2; UBB ODA N-6, right hyoplastron; UBB OdA-63, incomplete left xiphiplastron of a juvenile; UBB OdA-64, left epiplastron; OdA-65, entoplastron; UBB OdA-66, right peripheral 11; UBB Oda-73, right peripheral 2; UBB OdA-75, right costal 3 or 4; EME P-7, left costal 6; EME VP-108, incomplete left costal 1; EME VP-109, plastral fragment;

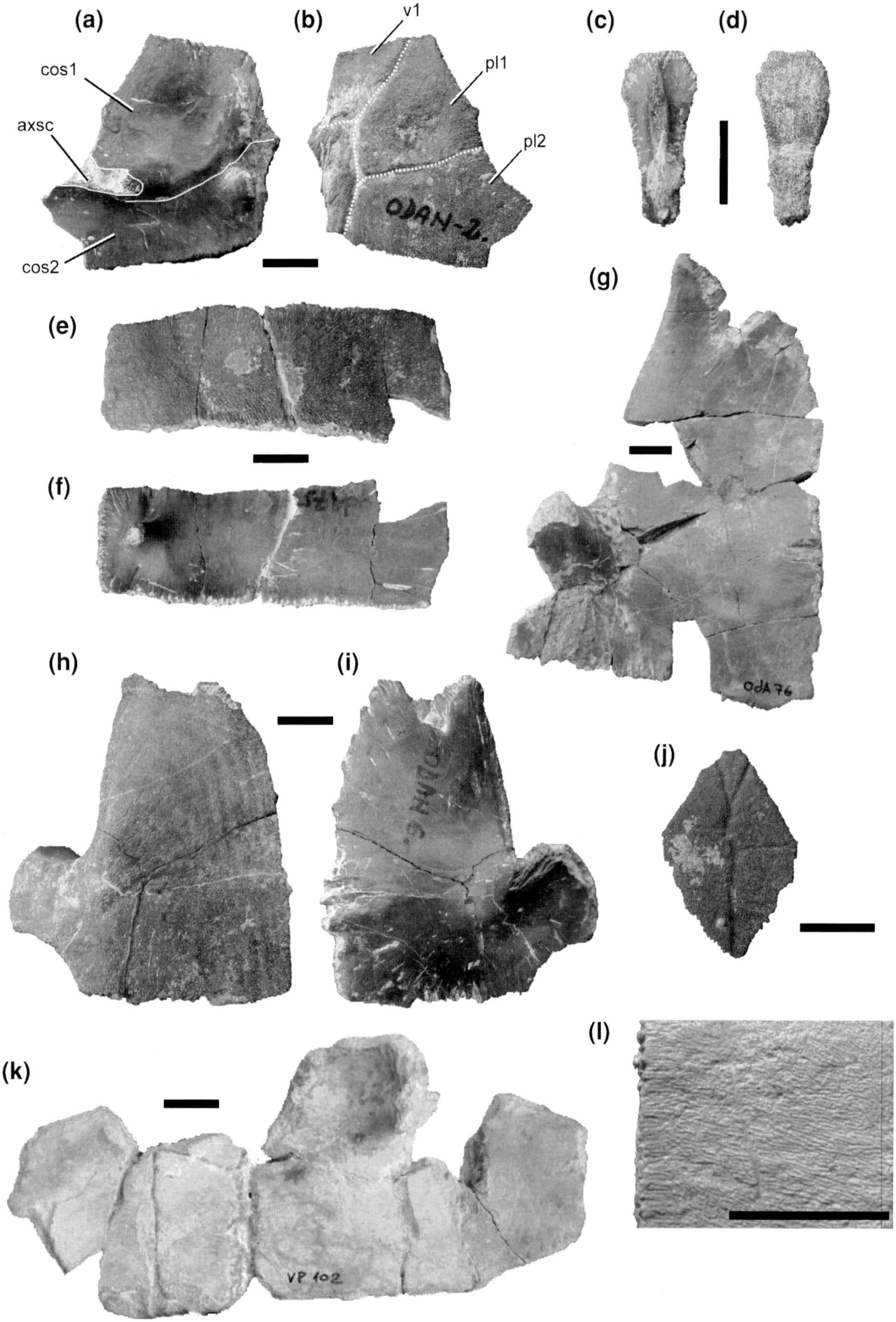
(a)
(b)
(c)
(d)
v1
cos1
pl1
axsc
pl2
ODAN-2.
cos2
(e)
(f)
(g)
(h)
(i)
(j)
(k)
(l)
OdA 76
VP 102

◀ **Fig. 19.8** Examples of Dortokidae gen. et sp. nov. (Pleurodira) shell elements from two localities and formations in the Transylvanian Basin, Romania: Oarda de Jos (**a–j**) in the "Sebeş" Formation, early? Maastrichtian, and Vurpăr (**k** and **l**) in the "Şard" Formation, late early Maastrichtian. **a**, **b** Fused right costals 1 + 2 (UBB ODA N-2) in **a** visceral view with *thin, solid, white lines* indicating both attachment scar and suture between first and second costals, and **b** dorsal view showing pleural 1 extending slightly extending onto costal 2 and with scute sulci indicated by *dotted, white lines*; **c**, **d** neural 5 (EME OdA-53) in **c** visceral view and **d** dorsal view; **e**, **f** right costal 3 or 4 (UBB OdA-75) in **e** dorsal view and **f** visceral view; **g** left hyoplastron of a large morphotype (UBB OdA-76) in visceral view; **h**, **i** right hyoplastron (UBB ODA N-6) in **h** ventral view and **i** visceral view; **j** entoplastron (OdA-65) in visceral view; **k** incomplete plastron (part of EME VP-102/a), consisting of left hyo-hypoplastron and incomplete right xiphiplastron, in visceral view; **l** magnified view showing micro-reticulate ornament on external shell surface. Abbreviations: *axsc* axillary scar, *cos*1 costal 1, *cos*2 costal 2, *pl*1 pleural 1, *pl*2 pleural 2, *v*1 vertebral 1. Images at different magnifications; all scale bars equal 1 cm

EME VP-110, posterolateral peripheral; EME VP-111 and 112, disarticulated plastral fragments; EME OdA-54, peripheral; EME Tc-5, incomplete entoplastron with sutured hyoplastral fragment; EME Tc-6, costal fragment; EME P-8, right hyoplastron of a subadult; UBB ODA N-10, 11, and 58-61, peripherals; UBB ODA N-16, pygal; UBB OdA-57, 70-72, and 74, costal fragments; UBB OdA-76, left hyoplastron of a large morphotype; UBB OdA-56, ?phalange; LPB (FGGUB) R.2230, left hyoplastron; LPB (FGGUB) R.1960/a, b, d, left hypoplastron, peripheral, and indeterminate plastral fragment; LPB (FGGUB) unnumbered, articulated carapace, plastron, scapulae, and pelvis; LPB (FGGUB) field no. 3/1998/1-3, peripheral, fused peripherals, and left xiphiplastron; LPB (FGGUB) field no. 08/2/1-2, left hyoplastron and left xiphiplastron; LPB (FGGUB) R.1440, pubis and ischium; LPB (FGGUB) R.1476, ilium; LPB (FGGUB) R.1749, ilium; LPB (FGGUB) R.1470, humerus; LPB (FGGUB) R.1477, cervical vertebrae.

Occurrence: Limited to Maastrichtian deposits in the Transylvanian and Haţeg basins of Romania (Tables 19.2, 19.4). Transylvanian Basin, Alba District: Vurpăr, bluish-spotted and red calcareous paleosol (overbank facies) in lower part of "Şard" Formation, late early Maastrichtian; Oarda de Jos, brownish grey and silty claystones (fluvial and paludal facies) in lower part of "Sebeş" Formation, early? Maastrichtian; Cuptorului Hill, red calcareous paleosol in lower part of "Şard" Formation, late early Maastrichtian; and Teleac, cross-laminated and reddish grey sandstones in middle part of "Sebeş" Formation, late? Maastrichtian. Haţeg Basin, Hunedoara District: Pui, dark red paleosoils (overbank facies) in ?Sinpetru Formation, Maastrichtian; Sinpetru, sandy and micro-conglomeratic channel fill in Sinpetru Formation, early Maastrichtian; and Tusţea, red silty claystone (paleosoil) in middle member of Densuş-Ciula Formation, Maastrichtian.

Description and comparison: The basic shell ornament in Dortokidae gen. et sp. nov. is micro-reticulate, consisting of the fine and longitudinal ridges typical for Dortokidae (de Lapparent de Broin and Murelaga 1999). However, the ornament in Dortokidae gen. et sp. nov. (Fig. 19.8l) is better developed and more closely resembles that of *Dortoka* than *Ronella*. Also, as in *Dortoka*, sharp, closely spaced, and more or less parallel crests complete the ornament on the neurals and costals, spanning the proximal third (anteriorly) or half (posteriorly) of those plates. Minute pits, possibly related to fungal infections (see Rothschild et al. 2012) are sometimes present on plastral elements and costals.

Pleural 1 (Fig. 19.8b) has the same length as or is shorter than the corresponding element in *Dortoka*, but it is longer than in *Ronella*. Pleural 1 further resembles *Dotoka* in that it does not reach or only slightly overlaps costal 2 proximally. Pleural 2 is enlarged posterolaterally, like in *Dortoka*, and reaches the anterolateral border of costal 5. Dortokidae gen. et sp. nov. has a strong axillary process, like in *Dortoka*, and it covers between one-third (Fig. 19.8a) to one-half of costal 1. This variability probably is due to sexual dimorphism; in many extant turtles, the extent of the axillary process is related to the degree of shell elevation, with males having a concave plastron and more convex and, consequently, higher carapace, versus females having a flatter plastron and a lower carapace. As in *Ronella*, the axillary process clearly did not reach the second thoracic rib (Fig. 19.8a). The nuchal plate is wider and slightly shorter than in *Dortoka*, but similar to that of *Ronella*. Vertebral 1 is *Dortoka*-like, being narrow posteriorly and not overlapping the lateroposterior corners of the nuchal. As in *Ronella*, neurals 1 and 2 are elongate and rectangular or oval in outline, whereas the mid-neurals (particularly neural 5; Fig. 19.8c, d) are elongate, hexagonal, and posteriorly tapered, with long posterolateral sides. The wide and triangular iliac scar on the underside of costal 8 is anteriorly enlarged, extends close to or contacts costal 7 (as in *Ronella*), is excluded from the peripheral (as in *Dortoka*), but is slightly prolonged onto the suprapygal. The pygal is small and ranges in outline from rectangular to square; it is similar to the pygal of *Ronella*, but is much shorter than the elongate, trapezoidal pygal typical of *Dortoka*.

Dortokidae gen. et sp. nov. has a long and narrow plastron. The anterior plastral lobe resembles that of *Dortoka* in being longer than in *Ronella* (Fig. 19.8h, i). The pectorals medially overlap the entoplastron and may extend even farther to reach the entoplastral point (Fig. 19.8j); as in *Ronella*, the pectorals also expand anterolaterally onto the hyoplastron. The outline of the entoplastron (Fig. 19.8j) is similar to *Dortoka* in being more elongate and rhomboidal than in *Ronella*. Compared to *Ronella*, however, the location of the humero-pectoral suture is slightly more derived in being situated in the mid-section of the plate. A short and oval pubic scar lies exclusively on the xiphiplastron; this differs from the condition in *Dortoka*, in which the pubic scar extends anteriorly onto the hypoplastron. The femoro-anal suture lies across the anterior half of the xiphiplastron, as in *Ronella*. The xiphiplastron is narrow, with parallel margins in its posterior half and has a wide anal notch.

Remarks: Although the name "*Muehlbachia nopcsai*" has been proposed (Vremir and Codrea 2009) for this new dortokid genus and species, that name cannot be used because it appeared in a conference abstract volume that does not satisfy the ICZN requirements. The same name also appeared in a paper by Vremir (2010). This new taxon will be properly named and described elsewhere.

The presence of a dortokid-like taxon in Transylvania (Romania) was first recorded from the lower Maastrichtian vertebrate locality of Vurpăr (Vremir 2004). Rare, but better preserved specimens subsequently were recognized from several other lower Maastrichtian localities in the Alba Iulia and Sebeş regions (e.g., Oarda de Jos and Teleac) of the Transylvanian Basin and from Pui, Sinpetru, and Tustea localities in the Haţeg Basin. This small-sized pleurodire clearely pertains to a new dortokid taxon characterized by a mixture of endo and exoskeletal features that are reminiscent of both the Late Cretaceous (Campanian–Maastrichtian) southwestern European genus *Dortoka* and the latest Paleocene (Thanetian) Romanian genus *Ronella*, but differs from these genera and from older dortokid material reported above from the Santonian of Hungary and the early Campanian of Austria.

The phylogenetic position of Dortokidae gen. et sp. nov. is unclear at the moment, because it shows a transitional morphology between *Dortoka* and *Ronella*. It seems that in the Late Cretaceous of Romania, Hungary, and Austria a separate dortokid lineage was present compared to Western Europe. Whether *Ronella* is directly descended from this eastern Central European lineage is ambiguous at the moment. According to de Lapparent de Broin et al. (2004), and based on certain more primitive characters observed in *Ronella* compared to *Dortoka,* they appear to represent two different lineages that may have separated as early as the latest Jurassic or Early Cretaceous. Dortokidae gen. et sp. nov. shows a mosaic of primitive and more derived characters found in both *Dortoka* and *Ronella*, pointing to a more complex evolutionary history for the family. A series of characters observed in the Santonian Dortokidae specimens from Iharkút also suggest a mosaic evolution and raises the possibility that certain features that have been considered diagnostic at the generic level for dortokids might instead be expressions of individual or intraspecific variation. Further study and comparison with other European dortokids is needed to help answer to these questions.

Cryptodira Cope 1868
Kallokibotioninae Nopcsa 1923b
Kallokibotion Nopcsa 1923a

Type species: *Kallokibotion bajazidi* (Nopcsa 1923a).

Occurrence: Within Romania, limited to Maastrichtian deposits in the Haţeg, Rusca Montană, and Transylvanian basins (Tables 19.2, 19.3, 19.4). Haţeg Basin, Hunedoara District: Pui, red and occasional calcareous paleosols (overbank facies) in the ?Sinpetru Formation, Maastrichtian; Unciuc, Totesti, and Vadu, all in Sinpetru Formation, late early Maastrichtian; Ciula Mare and Ciula Mică, both in middle member of Denşus-Ciula Formation, early Maastrichtian; Vălioara, alluvial silty claystone and reddish paleosol in middle member of Denşus-Ciula Formation, early-?late Maastrichtian; and Boiţa, Tuştea, and Livezile, all in middle member of Denşus-Ciula Formation, Maastrichtian. Rusca Montană Basin, Caraş-Severin County: Negoiu, Lunca Cernii de Jos, in Rusca Montană Formation, Maastrichtian. Transylvanian Basin, Alba District: Vurpăr, Stăuini, and Cuptorului Hill, all in red calcareous paleosols in lower part of "Şard" Formation, early-?late Maastrichtian; Oarda de Jos, Lancrăm, and Secaş Valley, mix of alluvial siltstones, sandstones, and micro-conglomerates in lower part of "Sebeş" Formation, early-?late Maastrichtian; Sebeş Glod, Sebeş area, red and occasional calcareous paleosoils in lower section of "Sebeş" Formation, late early Maastrichtian; Teleac, Feţei Hill, Râpa Lancrăm, and Râpa Roşie, all in Sebeş area, mostly cross-laminated, reddish grey sandstones and micro-conglomerates, occasional brownish red claystones, middle and upper parts of "Sebeş" Formation, late (probably including latest) Maastrichtian.

Diagnosis: See Gaffney and Meylan (1992, p. 3).

Kallokibotion bajazidi (Nopcsa 1923a)
(Fig. 19.9)

Synonymies: *Kallokibotion bajazidi*: Nopcsa 1923b, p. 104; Grigorescu 1983, p. 41; Groza 1983, p. 55, pl. 6, Figs. 4, 5; Gaffney and Meylan 1992, pp. 3–5, Table 1, Figs. 1–32; Codrea and Vremir 1997, pp. 233, 236, Fig. 2 a–c, pl. I, Figs. 1–3; Vremir 2004, p. 148. *Kallokibotium bajazidi*: Nopcsa 1923b, p. 12, Figs. 1, 2, pl. I, Figs. 1–30, pl. II, Figs. 1–6, pl. III, Figs. 1–3, 5–7, 9, 10, 12, pl. IV, Fig. 1; Szalai 1934, p. 103; Jianu et al. 1997, p. 222. *Kallokibotium magnificum*: Nopcsa 1923b, p. 14, Figs. 3, 4, pl. IV, Fig. 2. *Pleurosternon* sp.: Młynarski 1966, p. 242, pl. XV, Fig. 2. *Kallokibotion* aff. *bajazidi*: Młynarski 1966, p. 243, pl. XV, Fig. 3.

Lectotype: Nopcsa (1923a, b) did not designate a holotype. A lectotype was designated by Gaffney and Meylan (1992, p. 3): BMNH R4916, consisting of most of

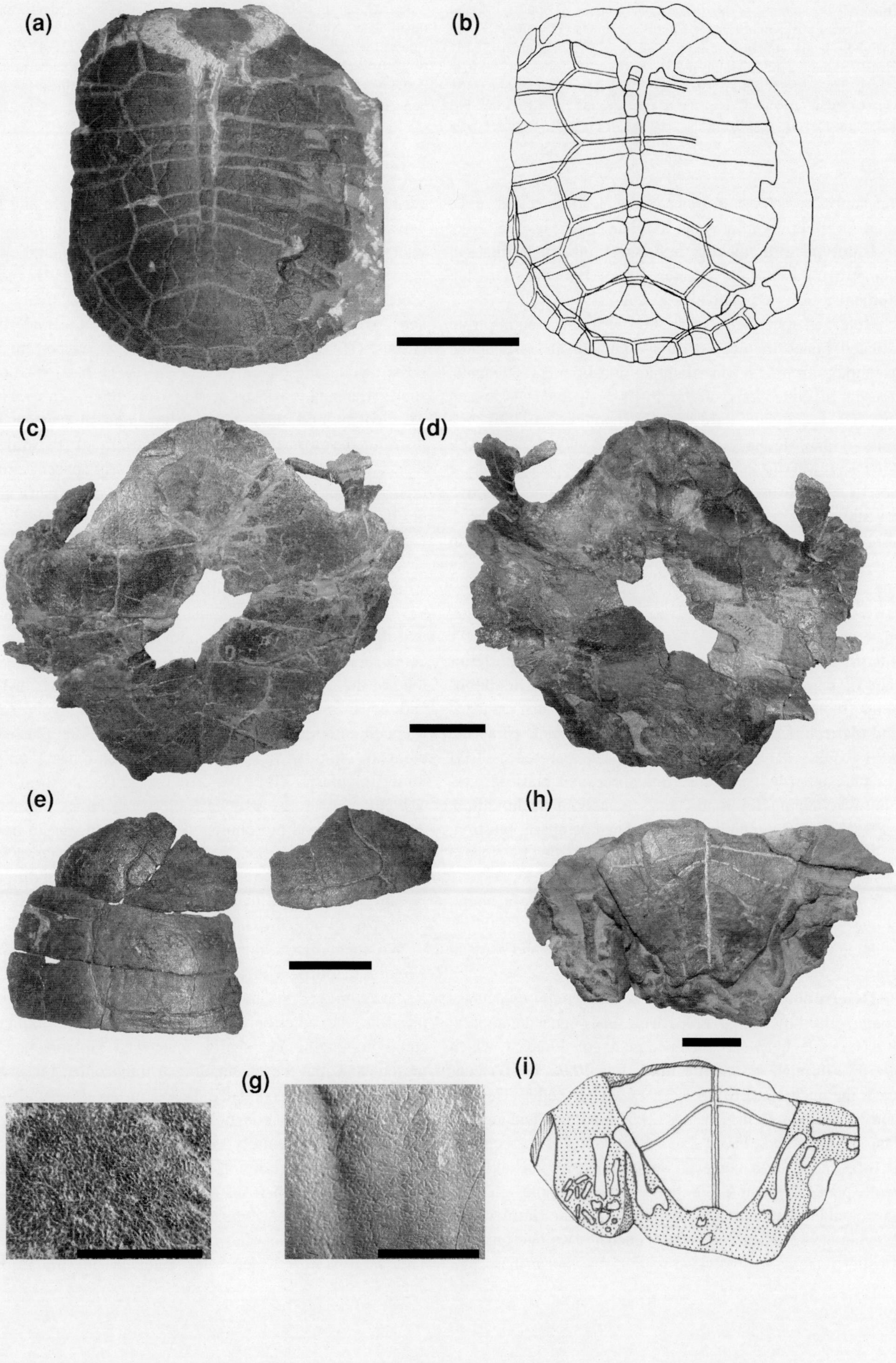
(a)
(b)
(c)
(d)
(e)
(h)
(f)
(g)
(i)

◀ **Fig. 19.9** Examples of *Kallokibotion bajazidi* (Nopcsa 1923a) (Cryptodira) postcranial remains from the Sinpetru Formation, Haţeg Basin, Romania. **a**, **b** Carapace (UBB-V437) from Sinpetru, early Maastrichitian: **a** photograph and **b** interpretative drawing, both in dorsal view; **c**, **d** plastron (UBB-TOT 1/1) from Totesti, late early Maastrichtian: **c** ventral view and **d** visceral view; **e** left costals 1–3 and right costal 1 (UBB-TOT 1/8-10) from Totesti, late early Maastrichtian, in dorsal view; **f**, **g** magnified views of shell showing tuberculate external ornament characteristic for *Kallokibotion*; **h**, **i** posterior part of shell with associated hind limbs (EME-1232) from Sinpetru, late early Maastrichtian: **h** photograph and **i** interpretative drawing, both in ventral view. Images at different magnifications; scale bars equal 5 cm for **a–e**, **h** and **i** and 1 cm for **f** and **g**

the carapace, the posterior lobe of the plastron, pelvis, and a caudal vertebra (Gaffney and Meylan 1992, Figs. 23, 24).

Holotype unit, locality and age: Sinpetru Formation; Sinpetru, Sibisel Valley, Haţeg Basin, Hunedora District, Romania; early Maastrichtian in age.

Referred specimens: Over 300 specimens, ranging from isolated bones to more or less complete and articulated skeletons, from 22 Maastrichtian localities in Romania, housed in collections of the BMNH (includes type and referred specimens published by Nopcsa (1923b) and Gaffney and Meylan (1992), EME, LPB (FGGUB), MÁFI, MCDRD, MTCO, and UBB. Most of the specimens have not been published or studied in detail; a comprehensive inventory and evaluation of those specimens is in progress (Rabi and Vremir, in preparation). Characters discussed in our preliminary account here are based on the following subset of specimens: UBB-V437, carapace (Fig. 19.9a, b); UBB-V16 (RR36), right hypoplastron; UBB-TOT 1/1, plastron (Fig. 19.9c, d); EME OdA-50, costals; UBB-TOT 1/8-10, anterior costals (Fig. 19.9e); EME-1232, posterior part of carapace and plastron with articulated hindlimbs (Fig. 19.9h, i); LPB (FGGUB) R.2086, associated carapace and plastron; LPB (FGGUB) R.1081, incomplete carapace; LPB (FGGUB) R.1079, incomplete carapace; LPB (FGGUB) R.1465, articulated carapace and plastron, cranium, vertebrae, forelimbs, and an incomplete hindlimb; LPB (FGGUB) unnumbered, articulated carapace, plastron, and limbs; and BMNH type specimens published by Nopcsa (1923b) and Gaffney and Meylan (1992).

Occurrence: Same as for genus (see generic account, above).

Diagnosis: Same as for genus (see Gaffney and Meylan 1992, p. 3).

Description and comparison: *Kallokibotion bajazidi* is characterised by a large, rather wide, and rounded carapace. It has a wide and well-developed pentagonal nuchal, which narrows through its anterior third (Fig. 19.9a, b). By contrast, the nuchal in BMNH R4919 was depicted by Gaffney and Meylan (1992) (Figs. 21, 22) as small and rectangular. Our observations based on LPB (FGGUB) specimens R.1465 and R.2086, however, indicate that the nuchal was a wide plate and that it lies immediately behind a small, rectangular plate virtually identical to what Gaffney and Meylan (1992) identified as the "nuchal"; we interpret this smaller plate as a subdivided neural. *K. bajazidi* has eight variably elongated neurals (neurals 5–7 are wide and hexagonal, whereas neural 8 is pentagonal), two suprapygals, a widened hexagonal pygal, eight pairs of costals, and eleven pairs of peripherals that are slightly overlapped by pleural scutes. The hexagonal vertebral scutes are relatively wide and cover up to the medial two-thirds of the underlying costals. The pleurals are progressively reduced in size posteriorly along the carapace (Fig. 19.9a, b, e). The basic shell ornament consists of fine (less than 1 mm across) vermiculate and tuberculate protrusions that are evenly distributed across the surface (Fig. 19.9f, g). Plastral elements are usually decorated with closely-spaced, minute pits. The scute sulci usually are wide and deep, particularly on the peripherals, and become more pronounced with increased body size.

The plastron usually is elongate and oval in outline, with strong and widened articular buttresses. The mesoplastra have a widened and triangular shape, contact the lateral two-thirds of the hyo- and hypoplastral sutures, but are not connected to one another across the midline. The anterior lobe of the plastron is half the length of the posterior lobe, and consists of short epiplastra and a relatively small, rounded entoplastron. The posterior lobe of the plastron is elongate, slightly tapered posteriorly, and has relatively long xiphiplastra (Fig. 19.9c, d, h, i). The plastron of *K. "magnificum"* (e.g., BMNH R4930) resembles the above-described plastral morphology in being elongate, in having a short anterior lobe and a relatively longer posterior lobe, and in having sharply narrowed xiphiplastra, but it differs in having mesoplastra that are rectangular in outline and contact one another axially, even in large specimens.

We recognize a third plastral morphotype for *Kallokibotion* that differs substantially from both *K. bajazidi* and *K. "magnificum"* in having a rounded and extremely wide plastron, the anterior and posterior lobes approximately equal in length, and sharply narrowed xiphiplastra. The mesoplastra may be rectangular or trapezoidal, but always contact one another axially. Based on some recently collected specimens, it is obvious that such morphological differences (particularly on the plastron) cannot be explained by sexual dimorphism or by ontogenetic change. Most of the medium-sized, elongate specimens belonging to the *K. bajazidi* and *K. "magnificum"* morphs originate from alluvial channel fills and, especially, overbank sediments. The large-sized, wider, and rounded plastral morph, however, is more common in fluvial and paludal sediments. The possibility that these may be ecomorphotypes is intriguing, but this cannot be answered until a comprehensive re-evaluation of all available fossil material is done.

Remarks: *Kallokibotion* was a medium to large sized (shell length up to 60 cm), semi-aquatic to terrestrial turtle. It is the most widespread and best documented turtle in the Maastrichtian of Romania, because it is known by abundant specimens from numerous localities and stratigraphic units in the Haţeg, Rusca Montană, and Transylvanian basins. Outside of Romania, *Kallokibotion* or *Kallokibotion*-like taxa are reliably known only from the Santonian Csehbánya Formation in Hungary and the Campanian Gosau Group in Austria (Nopcsa 1926; Thenius 1962; Gemel and Rauscher 2000; Sachs and Hornung 2006; this study). Based on this regionally-limited distribution, *Kallokibotion* can be considered an endemic component of Late Cretaceous terrestrial vertebrate faunas in eastern Central Europe.

In his second contribution on *Kallokibotion*, Nopcsa (1923b) distinguished two species: *K. bajazidi* and *K. magnificum.* According to Nopcsa (1923b), the latter differs from the type species by its smaller and narrower nuchal, narrower vertebral scutes, rather short intergulars, and mesoplastra contacting across the midline. Gaffney and Meylan (1992) argued that such differences (if they even really exist) are better considered as intraspecific variations and, for that reason, they considered *K. magnificum* to be a junior synonym of *K. bajazidi.* Their intepretations, however, were based only on specimens at the BMNH. Although those specimens included most of the ones that Nopcsa (1923a, b) used in his papers, the specimens are dissociated and most of them are poorly preserved (or not properly prepared)—particularly the exoskeletal parts—therefore important features remained hidden. In such circumstances, it is difficult to observe the supposed differences between the two species, even on the type material. For those reasons, we suggest that the status of *K.* "*magnificum*" deserves further consideration. Fortunately, articulated and well-preserved specimens have recently been collected from the Haţeg Basin (Toteşti, Nălaţ-Vad, Sinpetru, and Pui). These new specimens show additional morphologic details that will be critical for reassessing the specific status of *K.* "*magnificum*" and the possibility that several ecomorphs were present in the Maastrichtian of Romania.

Kallokibotion has long been regarded as a primitive turtle, but its precise relationships remain contentious. In the original descriptions, Nopcsa (1923a, b) classified *Kallokibotion* within the "Amphichelydia", which was a name used by earlier workers for primitive turtles. More recent studies have regarded *Kallokibotion* as the sister-taxon of all other Selmacryptodira (Gaffney and Meylan 1992; Gaffney et al. 2007 and references therein), placed it in a clade with *Tretosternon* as the sister taxon of Paracryptodira + Eucryptodira (Hirayama et al. 2000), or placed it on the stem leading to the Cryptodira-Pleurodira dichotomy [Joyce (2007); this arrangement influenced Sterli (2008) and Anquetin et al. (2009)]. *Kallokibotion* shares only plesiomorphic characters with other primitive taxa, such as *Mongolochelys*, Meiolaniidae, Pleurosternidae, and Solemydidae—consequently, its relationships with those taxa and with more derived turtles are challenging to resolve. The only real agreement among contemporary workers is that *Kallokibotion* is a Late Cretaceous relic and the last surviving member of a lineage that likely dates back to the Jurassic. Preparation and detailed study of the above-mentioned new *Kallokibotion* specimens may help shed new light on the relationships and evolutionary history of the genus.

Discussion

Paleobiogeography of Eastern Central European Turtle Faunas

Turtle faunas of eastern Central Europe and Western Europe during the Late Cretaceous are partly similar, but some degree of biogeographical separation is apparent between them (Fig. 19.10). Noteworthy is that solemydids and the "chelydroid-like" forms (sensu de Lapparent de Broin 2001) are absent from eastern Central Europe (de Lapparent de Broin and Murelaga 1999). The composition of the Late Cretaceous turtle faunas of Austria, Hungary, and Romania is a mixture of Gondwanan descendants (Bothremydidae: de Lapparent de Broin 2000; Rabi et al. in preparation) and survivors from the Early Cretaceous of Europe (Dortokidae: de Lapparent de Broin et al. 2004). The geographic origin for the other main taxon, *Kallokibotion*, is uncertain due to its disputed phylogenetical relationships (e.g., Gaffney and Meylan 1992; Hirayama et al. 2000; Joyce 2007).

Bothremydidae are considered to be the most diverse turtle family in the Late Cretaceous of Europe. They are best known from the early Campanian to late Maastrichtrian of Portugal, Spain, and France, where the family is represented by the subtribe Foxemydina (including *Polysternon provinciale*, *P. atlanticum*, *Foxemys mechinorum*, *Elochelys perfecta,* and *E. convenarum*) and the Bothremydina (including *Rosasia soutoi*) (Nopcsa 1931; Antunes and Broin 1988; de Lapparent de Broin and Murelaga 1996, 1999; Tong et al. 1998; Tong and Gaffney 2000; Laurent et al. 2002; Company 2004; Gaffney et al. 2006; Pérez-García et al. 2010). The only bothremydid currently known from eastern Central Europe is an inderterminate species of

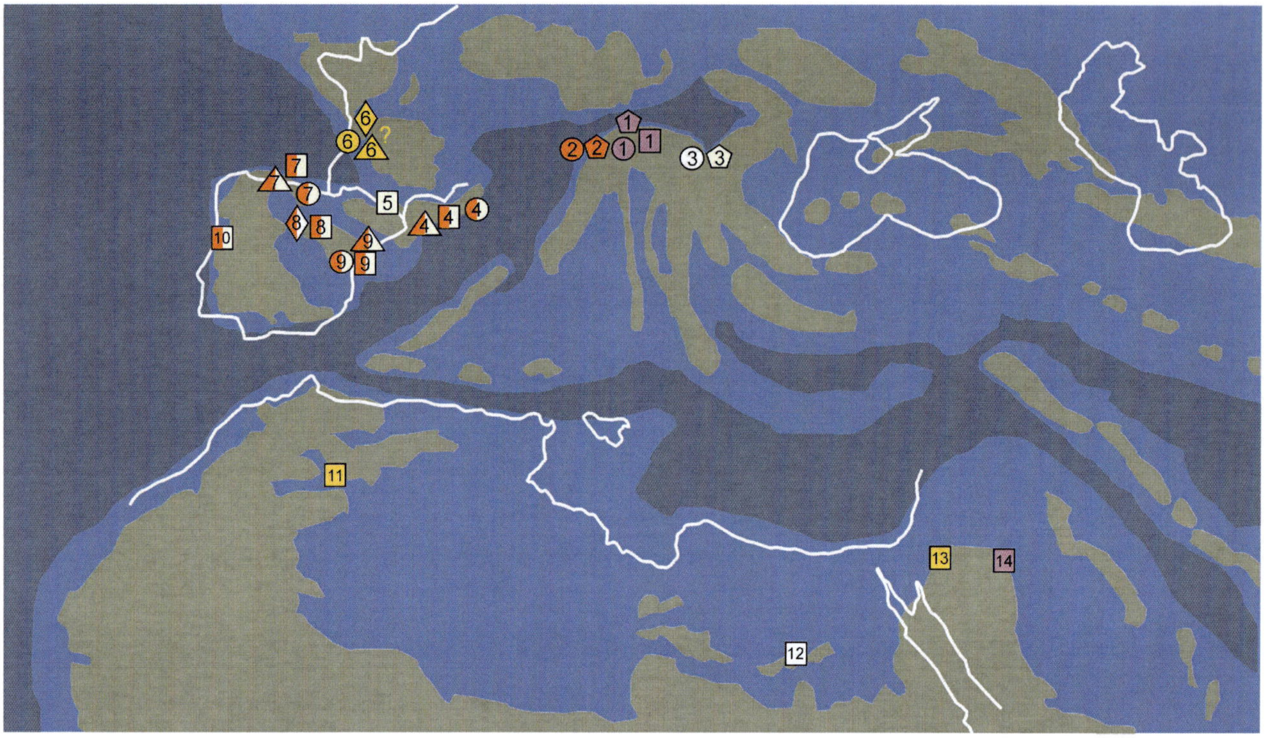

Fig. 19.10 Paleobiogeographic distribution of Late Cretaceous non-marine turtle families in Europe and North Africa. For symbols, shape denotes family and color denotes geological age (see explanatory legend below map) and Arabic numeral corresponds to locality(ies), as follows: *1* Iharkút, Hungary (this paper); *2* Muthmannsdorf, Austria (this paper); *3* multiple localities in Haţeg and Transylvanian basins, Romania (this paper); *4* Aix-en Provence, Fox-Amphoux, Fuveau, Villeveyrac, and Massecaps in southern France (Buffetaut et al. 1996; de Lapparent de Broin and Murelaga 1999; Buffetaut 2005; Gaffney et al. 2006); *5* Cassagnou, France (Laurent et al. 2002); *6* Charentes, France (Vullo et al. 2010); *7* Laño, Spain (de Lapparent de Broin and Murelaga 1999); *8* Lo Hueco, Spain (Garcia et al. 2009, 2010); *9* Chera, Spain (Company 2004); *10* Aveiro, Portugal (Gaffney et al. 2006); *11* Erfoud, Morocco (Gaffney et al. 2006); *12* Ammonite Hills, Egypt (de Lapparent de Broin and Werner 1998); *13* Ein Yarburd and Jerusalem, Israel (Gaffney et al. 2006); *14* El-Huseiniyah, Jordan (Zalmout et al. 2005)

Foxemys, which is represented by skeletal remains from the Csehbánya Formation of Hungary. The presence of this genus in the Santonian of Hungary indicates a Late Cretaceous radiation of Gondwanan tetrapods into eastern Central Europe, similar to the event demonstrated in Western Europe on the basis of hybodont sharks, simoliophid snakes, charcharodontosaurid theropod dinosaurs and, possibly, ziphodont crocodiles during the Cenomanian in France (Vullo et al. 2005, 2007). The Foxemydina are endemic to Europe (Gaffney et al. 2006) and they probably evolved from more primitive, as yet undiscovered, European ancestors that had an African origin. The time of dispersal from Africa into Europe cannot be precisely given, but it could have occurred between the Cenomanian and the Santonian based on an unpublished cladogram by two of us (MR and HT). Related to the recognition of these ghost lineages is that dispersal routes for European bothremydids also remain unclear. The presence of *Foxemys* during the Santonian in Hungary implies that migration from Africa was not only possible via Ibero-Armorica, but also via the more eastern Apulian micro-continent. An east to west dispersal of *Foxemys* within Europe, from Hungary to France and Spain, is most consistent with currently known occurrences. The Hungarian occurrence (Santonian) pre-dates the oldest Western European occurrences (late Campanian–early Maastrichtian) for *Foxemys* and the oldest Late Cretaceous (Cenomanian) turtle fauna of Ibero-Armorica lacks bothremydids (Vullo et al. 2010). However, we caution that this hypothesis cannot be confirmed because nothing is known of non-marine turtle faunas in Western Europe between the Cenomanian and Campanian.

The primitive pleurodiran family Dortokidae was first identified from the late Campanian–early Maastrichtian of Spain and France, where it is represented by *Dortoka vasconica* (de Lapparent de Broin and Murelaga 1996, 1999);

since then, the family also has been reported from the Barremian of Spain and the Maastrichtian-Paleocene of Romania (de Lapparent de Broin et al. 2004; Vremir and Codrea 2009). Indeterminate dortokid specimens reported here from the Santonian Csehbánya Formation and the Campanian Grünbach Formation apparently are closely related to the Romanian Maastrichtian Dortokidae gen. et sp. nov. and, collectively, they represent a separate lineage compared to dortokids from Western Europe. Clarification of the relationships of this lineage with the Romanian Paleocene *Ronella* and an exploration of the paleobiogeographical implications will be presented elsewhere (Vremir and Rabi, in preparation).

Kallokibotion originally was considered endemic to the Maastrichtian of Transylvania but further study and material from the Csehbánya Formation (Santonian) of Hungary and the Gosau Group (Campanian) of Austria indicate that the genus was somewhat more widely distributed in the region (see also Nopcsa 1926; Thenius 1962). However, other records of possibly related forms from the Paleocene of France (Gaffney and Meylan 1992) and the Maastrichtian of Russia (Averianov and Yarkov 2004) cannot be assigned to *Kallokibotion*. Primitive cryptodires are also known from France and Spain, where they are represented by *Solemys vermiculata* and *S. gaudryi* (de Lapparent de Broin and Murelaga 1996, 1999; Company 2004; Buffetaut 2005; Marmi et al. 2009; Tortosa et al. 2009). However, *Solemys* (based on its shell) is not closely related to *Kallokibotion*, because no derived characters unite the two genera. Their only shared features are plesimorphic at the level of Paracryptodira; e.g., presence of a mesoplastron, presence of inframarginals, paired intergulars, and loss of dorsal process of epiplastron (de Lapparent de Broin and Murelaga 1999).

Presence/Absence and Relative Abundances of Turtle Taxa

Interesting patterns can be observed in the presence/absence and relative abundances of turtle taxa in the Upper Cretaceous formations of central Eastern Europe. In the Iharkút outcrop of the Hungarian Csehbánya Formation, bothremydid (*Foxemys* sp.) fossils are among the most abundant vertebrate fossils and this taxon is by far the most abundant among turtles. The skeletal elements of *Foxemys* sp. also are well preserved, whereas dortokid specimens are rarer and are only represented by shell or girdle elements. *Kallokibotion*-like turtles are even rarer, being known by only a few shell elements. For the Muthmannsdorf material from Austria (Grünbach Formation), the *Kallokibotion*-like form is dominant, dortokids are much less common, and bothremydids are absent. In the Haţeg Basin of Romania most of the specimens belong to *Kallokibotion*, whereas dotokids are the dominant turtle in the Sebeş-Alba region; bothremydids seem to be completely absent from both areas.

The reasons for the differences in the presence/absence and relative abundances of the three turtle taxa may be at least partially due to paleoecological factors, however, paleobiogeography also needs has to be taken into account. One paleoecological factor that can be expected to have exerted a major influence on turtles in the region is the variety of depositional environments. These varied from the complex fluvial system of the Csehbánya Formation (Hungary), to oxbow lakes, coastal ponds, and swampy lowlands of the Grünbach Formation (Austria), to floodplain, fluvial, and marshy habitats of the various Romanian formations (Kvaček and Herman 2004; Therrien 2005; Herman and Kvaček 2007; Ősi and Mindszenty 2009). Facies-dependent distributions of fossil turtles are not unique to the Late Cretaceous of Europe; it has been demonstrated elsewhere, for example, in the Maastrichtian of North Dakota (Lyson and Joyce 2009). The two turtle taxa known from the Maastrichtian of Romania never cooccur in the same layer, even when they are both present at the same outcrop: *Kallokibotion* is linked to seasonally terrestrial paleoenvironments (overbank facies and paleosoil horizons), whereas Dortokidae gen. et sp. nov. is linked to lacustrine and paludal facies. Bothremydids are considered to be more aquatic (de Lapparent de Broin and Murelaga 1999; Gaffney et al. 2006); consequently, their dominance in the channel deposits of the Csehbánya Formation is not surprising.

Although the facies-dependence of turtle taxa in Late Cretaceous localities in eastern Central Europe is poorly understood and not always consistent, roughly contemporaneous localities in Western Europe also indicate the influence of paleoenvironment on the changing relative abundances of turtle families. At Laño (Spain) the relative abundance of turtle taxa varies among different beds, which, in turn, correspond to certain paleoenvironments within the same site and formation (de Lapparent de Broin and Murelaga 1999). In floodplain deposits cropping out at Massecaps (Cruzy, Hérault, France), *Foxemys* is by far the most common turtle and the most abundant vertebrate, whereas *Solemys* is less common (Buffetaut et al. 1999; de Lapparent de Broin and Murelaga 1999; Buffetaut 2005; Gaffney et al. 2006). In the dark clays formed in estuarine to lagoonal environments at Villeveyrac (Hérault, France) *Polysternon provinciale* has a much better preservation, including complete shells and skulls, compared to the sympatric *Solemys*, which is represented by shell elements only (Buffetaut et al. 1996). Company (2004) has shown that in two stratigraphically equivalent horizons at Chera (Valencia, Spain) the paludal beds contain large numbers of bothremydids, few dortokids and even fewer solemydids, whereas in the more terrestrial beds solemydids are the most common turtle (though turtles generally are rare in the latter beds). In eastern Central Europe, the dominance of *Kallokibotion*-like turtles in the swampy and lacustrine facies of the Campanian Grünbach Formation and their co-occurrence there with dortokids contrasts with the pattern in the Maastrichtian of Romania where the two taxa occur in different facies; thus, at least for those two groups, the picture appears more complex.

The potential role of biogeographical factors and time in the distribution of turtles during the Late Cretaceous in eastern Central Europe cannot be dismissed. For example, the absence of post-Santonian bothremydids in the region is puzzling, considering that they were present there during the Santonian and were present in Western Europe during the Campanian and Maastrichtian. Bothremydids may never have colonized the areas of present day Romania and Austria or they may have gone extinct in those areas before the Campanian.

Paleoecology

Differences summarized above in the relative abundances of turtle taxa found in localities representing different paleoenvironments indicate that the taxa filled three separate ecological roles. The dominance of *Foxemys* in channel deposits indicates that it was able to inhabit deep and strongly flowing rivers, in addition to other aquatic environments. This is consistent with its large size, flat shell, and limb bones, features that are similar to extant *Podocnemis* which are also good swimmers. Dortokids are more common in deposits accumulated under lower hydrodynamic conditions, such as ponds, slow-flowing creeks, swamps, and oxbow lakes on the floodplain area. Such preferred habitats are consistent with their smaller size, because among extant aquatic turtles those with smaller bodies tend to prefer quieter water bodies. *Kallokibotion* is most common in paleosoils and its articulated skeletons also occur in those rocks; those occurrences suggest it could have led a more terrestrial lifestyle, although that does not necessarily mean that it was not partly aquatic. Apparently it lived alongside rivers and on more elevated ground, but it could have also found habitats in shallow pools and ponds, probably on the bottom because it could not have been a strong swimmer. That *Kallokibotion* was better adapted for a semi-terrestrial lifestyle is indicated by its high and domed shell, dorosoventrally expanded skull, the wide angle formed by the scapular and acromial processes (Depecker et al. 2006), and the strongly bent diaphysis of the femur. Undescribed material from the Haţeg Basin reveals that *Kallokibotion* had more slender and elongate unguals than primitive terrestrial taxa such as *Meiolania* and *Proganochelys* (Gaffney 1990; Joyce and Gauthier 2004; Scheyer and Sander 2007; Rabi and Vremir, in preparation); at presence, the significance of this difference for interpreting the lifestyle of *Kallokibotion* is unclear. de Lapparent de Borin and Murelaga (1999) inferred a similar ecological separation for the Laño turtle fauna (and consequently for other western European faunas) where the role of *Kallokibotion* could have been occupied by *Solemys*. A semi-terrestrial lifestyle for *Solemys* also is supported by taphonomical data (e.g., Marmi et al. 2009; although those authors used the term 'semi-aquatic', in our opinion that means the same as semi-terrestrial).

Concerning their feeding strategies, *Foxemys* with its very wide triturating surface, could have led a durophagous lifestyle, similar to other closely related bothremyidids (Gaffney et al. 2006). Dortokids are as yet unknown in this respect, because their skulls have not been discovered. The narrow triturating surface of *Kallokibotion* (Gaffney and Meylan 1992, Rabi and Vremir, in preparation) suggests an omnivorous or herbivorous diet, which would be consistent with its inferred semi-terrestrial lifestyle.

Predation on turtles has not been unequivocally demonstrated in any of the localities, but based on overall faunal composition (i.e., presence of both turtles and their potential predators), Iharkút and some Transylvanian localities (e.g., Sinpetru, Vurpâr, and Oarda de Jos) seem the most likely sites where evidence of predation could be expected. In such assembleges *Foxemys* (Iharkút only) and/or dortokids could have been potential prey for generalist aquatic crocodiles (e.g., *Allodaposuchus*; Nopcsa 1928; Delfino et al. 2008) or even for the large-bodied, freshwater, riverine mosasaurs of Iharkút (Makádi 2005).

Summary

Since Nopcsa's time, our knowledge of the historically less intensively studied Late Cretaceous turtle faunas of eastern Central Europe has increased over the past 15 years thanks to recent discoveries and revision of older collections. As for Western European faunas, the diversity appears to be low at the family level and the faunas consist of groups endemic to Europe. Future work may be able to demonstrate higher diversity at lower taxonomic levels (i.e., genus and species) in the study areas. Compared to France and Spain, the eastern Central European localities cover a relatively longer time interval ranging from the Santonian to the late

Maastrichtian; nevertheless no significant changes can be detected in their composition. With its Santonian age, the Hungarian Csehbánya Formation provides an exceptional insight into a virtually unknown interval of turtle diversity in Europe.

In the Late Cretaceous of eastern Central Europe, pleurodires are represented by two extinct groups: (1) the more primitive Dortokidae, represented by the Maastrichtian Dortokidae gen. et sp. nov. and two indeterminate, but presumably closely related taxa (one Santonian and the second Campanian) and (2) the Bothremydidae, represented by an interterminate Santonian species of *Foxemys* sp. that is closely related to the younger type species *F. mechinorum* from France. Cryptodires are represented by the primitive relict form *Kallokibotion bajazidi* (Mastrichtian) and other *Kallokibotion*-like turtles of Santonian and Campanian age. We have shown that dortokids and, probably, kallokibotionines had a considerably greater geographical and temporal range than previously thought. Eastern Central European faunas were partly similar to Western European ones, but the endemic nature of the kallokibotionines and the presence of a different dortokid lineage indicate some degree of biogeographical separation between the two areas; these could represent distinct faunal provinces. Despite the fact that the Hungarian, Austrian, and Romanian assemblages correspond to different time intervals, as well as paleogeographical units and paleoenvironments, the taxonomic composition is generally uniform, but with significant differences in the relative abundance of taxa. Dortokids were present in the whole region and their remains are most abundant in the paludal and lacustrine facies of Romania. The single bothremydid taxon is only known from the Csehbánya Formation of Hungary, where it is very common. Kallokibotionines appear to be most frequent in the overbank facies of the Haţeg Basin, Romania, but they are also relatively well represented in the swampy facies of the Austrian Gosau Group (but the sample sizes of the two areas are highly unequal). Despite the archipelago-like paleogeography reconstructed for the area (Pereda-Suberbiola 2009 and references therein), differences in the relative abundances of Late Cretaceous turtles are not necessarily indicative exclusively of a biogeographical pattern, but also reflect differences among the turtle taxa in their preferred habitats.

Eastern Central European Late Cretaceous turtle faunas indicate at least two different influences: dortokids likely descended from European Early Cretaceous ancestors (de Lapparent de Broin et al. 2004), whereas the Foxemydina had a more recent African origin in the Late Cretaceous. *Kallokibotion* is either a relict of Early Cretaceous pleurosternids or a late survivor of a conservative Jurassic lineage. Its survival up to the end of the Late Cretaceous presumably was favored by isolation on the archipelago.

Acknowledgments We especially thank Jim Gardner for the invitation to contribute to this volume and for carefully checking and correcting several versions of the manuscript. Zoltán Csiki, Igor Danilov, and Massimo Delfino are thanked for their constructive reviews. Zoltán Csiki also assisted MR and MV in studying the turtle collection of the University of Bucharest. Vlad Codrea is thanked for providing access to the turtle material at the Babes-Bolyai University (Cluj-Napoca, Romania). Fieldwork in Iharkút was suppored by the Hungarian Research Fund (OTKA T–38045, PD 73021), the Jurassic Foundation, the Hantken Foundation, and the Pro Renovanda Cultura Hungariae Foundation. This project also was funded by the Synthesys Program, the SECyT-NKTH, and the Eötvös Loránd University.

References

Anquetin, J., Barret, P., Jones, M. E. H., Moore-Fay, S., & Evans, S. E. (2009). A new stem turtle from the Middle Jurassic of Scotland: New insights into the evolution and paleoecology of basal turtles. *Proceedings of the Royal Society B, 276*, 879–886.

Antunes, M. T., & De Broin, F. (1988). Le Crétacé terminal de Beira Litoral, Portugal: Remarques stratigraphiques et écologiques, étude complémentaire de *Rosasia soutoi* (Chelonii, Bothremydidae). *Ciências da Terra, 9*, 153–200.

Averianov, A. O., & Yarkov, A. A. (2004). New turtle remains from the Late Cretaceous and Paleogene of Volgograd region, Russia. *Russian Journal of Herpetology, 11*, 41–50.

Baur, G. (1891). Notes on some little known American fossil tortoises. *Proceedings of the Academy of Natural Sciences of Philadelphia, 43*, 411–430.

Bohn, P. (1966). *Senonemys sümegensis* nov. gen., nov. sp. – ein neuer Schildkrötenfund aus Ungarn. *Földtani Közlöny, 96*, 111–118.

Buffetaut, E. (1979). Revision der Crocodylia (Reptilia) aus den Gosau-Schichten (Ober-Kreide) von Österreich. *Beiträge zur Paläontologie von Österreich, 6*, 89–105.

Buffetaut, E. (2005). Late Cretaceous vertebrates from the Saint-Chinian area (Southern France): A review of previous research and an update of recent finds. *Acta Palaeontologica Romaniae, 5*, 39–48.

Buffetaut, E., Costa, G., Le Loeuff, J., Martin, M., Rage, J.-C., Valentin, X., et al. (1996). An early campanian vertebrate fauna from the Villeveyrac Basin (Hérault, southern France). *Neues Jahrbuch für Geologie und Paläontologie Monatshefte, 1996*, 1–16.

Buffetaut, E., Le Loeuff, J., Tong, H., Duffaud, S., Cavin, L., Garcia, G., et al. (1999). Un nouveau gisement de vertébrés du Crétacé supérieur à Cruzy (Hérault, Sud de la France). *Comptes Rendus de l'Académie des Sciences de Paris, Sciences de la Terre et des Planètes, 328*, 203–208.

Bunzel, E. (1871). Die Reptilfauna der Gosau-Formation in der Neuen Welt bei Wiener-Neustadt. *Abhandlungen der Geologische Reichsanstalt, 5*, 1–18.

Codrea, V., & Dica, P. (2005). Upper Cretaceous-lowermost Miocene lithostratigraphic units in Alba Iulia-Sebeş-Vintu de Jos area (SW Transylvanian Basin). *Studia Universitatis Babes-Bolyai, Geologia, 50*, 19–26.

Codrea, V., & Vremir, M. (1997). *Kallokibotion bajazidi* Nopcsa (Testudines, Kallokibotidae) in the Red Strata of Râpa Roşie - Sebeş (Alba county). *Sargetia, Acta Musei Devensis, seria Stiintele Naturii, 17*, 233–238.

Codrea, V., Smith, T., Dica, P., Folie, A., Garcia, G., Godefroit, P., et al. (2002). Dinosaur egg nests, mammals and other vertebrates from a new Maastrichtian site of the Hateg Basin (Romania). *Comptes Rendus Palevol, 1*, 173–180.

Codrea, V., Murzea-Jipa, C., & Venczel, M. (2008). A sauropod vertebra at Râpa Roşie (Alba district). *Acta Palaeontologica Romaniae*, *6*, 43–48.

Codrea, V., Vremir, M., Jipa, C., Godefroit, P., Csiki, Z., Smith, T., et al. (2010). More than just Nopcsa's Transylvanian dinosaurs: A look outside the Haţeg Basin. *Palaeogeography, Palaeoclimatology, Palaeoecology*, *293*, 391–405.

Company, J. (2004). Vertebrados continentales del Cretácico Superior (Campaniense-Maastrichtiense) de Valencia (Ph.D. dissertation, Universitad de València, Valencia).

Cope, E. D. (1864). *On the limits and relations of the Raniformes. Proceedings of the Academy of Natural Sciences of Philadelphia, 16*, 181–183.

Cope, E. D. (1868). On the origin of genera. *Proceedings of the Academy of Natural Sciences of Philadelphia*, *20*, 242–300.

Csiki, Z., Vremir, M., Brusatte, S., & Norell, M. (2010). *An aberrant island-dwelling theropod dinosaur from the Late Cretaceous of Romania. Proceedings of the National Academy of Sciences USA*, *107*, 15357–15361.

de Lapparent de Broin, F. (2000). African chelonians from the Jurassic to the present: Phases of development and preliminary catalogue of the fossil record. *Palaeontologia Africana*, *36*, 43–82.

de Lapparent de Broin, F. (2001). The European turtle fauna from the Triassic to the Present. *Dumerilia*, *4*(3), 155–216.

de Lapparent de Broin, F., & Murelaga, X. (1996). Une nouvelle faune de chéloniens dans le Crétacé Supérieur européen. *Comptes Rendus de l'Académie des Sciences, Sér. IIA*, *323*, 729–735.

de Lapparent de Broin, F., & Murelega, X. (1999). Turtles from the Upper Cretaceous of Laño (Iberian peninsula). *Estudios del Museo de Ciencias Naturales de Alava, 14* (*Número Especial 1*) 135–211.

de Lapparent de Broin, F., & Werner, C. (1998). New Late Cretaceous turtles from the Western Desert, Egypt. *Annales de Paleontologie*, *84*, 131–214.

de Lapparent de Broin, F., Murelega Bereikua, X., & Codrea, V. (2004). Presence of Dortokidae (Chelonii, Pleurodira) in the earliest Tertiary of the Jibou Formation, Romania: Palaeobiogeographical implications. *Acta Palaeontologica Romanie*, *4*, 203–215.

Delfino, M., Codrea, V., Folie, A., Dica, P., Godefroit, P., & Smith, T. (2008). A complete skull of *Allodaposuchus precedens* NOPCSA, 1928 (Eusuchia) and a reassessment of the morphology of the taxon based on the Romanian remains. *Journal of Vertebrate Paleontology*, *28*, 111–122.

Depecker, M., Berge, C., Penin, X., & Renous, S. (2006). Geometric morphometrics of the shoulder in extant turtles (Chelonii). *Journal of Anatomy*, *208*, 35–45.

Gaffney, E. S. (1990). The comparative osteology of the Triassic turtle *Proganochelys*. *Bulletin of the American Museum of Natural History*, *194*, 1–263.

Gaffney, E. S., & Meylan, P. A. (1988). A phylogeny of turtles. In M. J. Benton (Ed.) *The phylogeny and classification of the tetrapods. Volume* 1: *Amphibians, reptiles, birds. Systematics association special volume, 35A* (pp. 157–219). Oxford: Clarendon Press.

Gaffney, E. S., & Meylan, P. A. (1992). The Transylvanian turtle *Kallokibotion*, a primitive cryptodire of Cretaceous age. *American Museum Novitates*, *3040*, 1–37.

Gaffney, E. S., Tong, H., & Meylan, P. A. (2006). Evolution of the side-necked turtles: The families Bothremydidae, Euraxemydidae, and Araripemydidae. *Bulletin of the American Museum of Natural History*, *300*, 1–698.

Gaffney, E. S., Rich, T. H., Vickers-Rich, P., Constantine, A., Vacca, R., & Kool, L. (2007). *Chubutemys*, a new eucryptodiran turtle from the Early Cretaceous of Argentina, and the relationships of the Meiolaniidae. *American Museum Novitates*, *3599*, 1–35.

García, P. A., Ortega, F., & Murelaga, X. (2009). Una probable tortuga Pancryptodira del Cretácico Superior de Lo Hueco (Cuenca, España). *Paleolusitana*, *1*, 365–371.

García, P. A., Ortega, F., & Murelaga, X. (2010). Ampliación de la distribución geográfica y temporal de *Elochelys convenarum* (Chelonii, Bothremydidae) en el Cretácico Superior de la Península Ibérica. *Ameghiniana*, *47*, 307–315.

Gemel, R., & Rauscher, K. (2000). Fossile Schildkröten aus Österreich. In Biologiezentrum des Ober Österreich. Landesmuseums (Hrsg.): *Die Europäische Sumpfschildkröte* (pp. 63–86), Stapfia *69*, Neue Folge *149*.

Gheerbrant, E., Codrea, V., Hosu, A., Sen, S., Guernet, C., de Lapparent de Broin, F., et al. (1999–2000). Découverte en Transylvanie (Roumanie) de gisements à vertébrés dans les calcaires de Rona (Thanétien ou Sparnacien): Les plus anciens mammifères cénozo d'Europe Orientale. *Eclogae Geoogicae Helvetiae*, *1999*, *92*, 517–535.

Grigorescu, D. (1983). Cadrul stratigrafic si paleoecologic al depozitelor continentale cu dinosaurieni din bazinul Haţeg. *Sargetia, Acta Musei Devensis, seria Stiintele Naturii*, *13*, 37–47.

Grigorescu, D. (1992). Nonmarine Cretaceous formations of Romania. In N. J. Mater & P. J. Chen (Eds.), *Aspects of nonmarine cretaceous geology* (pp. 142–164). Beijing: China Ocean Press.

Grigorescu, D. (2010). The Latest Cretaceous fauna with dinosaurs and mammals from the Haţeg Basin—a historical overview. *Palaeogeography, Palaeoclimatology, Palaeoecology*, *293*, 271–282.

Groza, I. (1983). Rezultatele preliminare ale cercetărilor intreprinse de către Muzeul judeţean Hunedoara—Deva în stratele cu dinosaurieni de la Sînpetru—Haţeg. *Sargetia, Acta Musei Devensis, seria Stiintele Naturii*, *13*, 49–66.

Haas, J. (1979). The Ugod Limestone Formation in the Bakony Mountains. *Annals of the Hungarian Geological Institute*, *61*, 1–149.

Herman, A. B., & Kvaček, J. (2007). Early Campanian Grünbach flora of Austria: Systematic composition and paleoclimatic interpretations. *Acta Paleobotanica*, *47*, 37–55.

Hirayama, R., Brinkman, D. B., & Danilov, I. G. (2000). Distribution and biogeography of non-marine Cretaceous turtles. *Russian Journal of Herpetology*, *7*, 181–198.

Jianu, C.-M., Mészáros, M., & Codrea, V. (1997). A new collection of Haţeg and Râpa Roşie material (Dinosauria, Crocodilia, Chelonia) in the Cluj-Napoca University. *Sargetia, Acta Musei Devensis, seria Stiintele Naturi*, *17*, 219–232.

Joyce, W. G. (2007). Phylogenetic relationships of Mesozoic turtles. *Bulletin of the Peabody Museum of Natural History*, *48*, 3–102.

Joyce, W. G., & Gauthier, J. A. (2004). Paleoecology of Triassic stem turtles sheds new light on turtle origins. *Proceedings of the Royal Society of London B*, *271*, 1–5.

Kadic, O. (1916). Jelentés az 1915, évben végzett ásatásaimról: II A valiorai dinosaurusok gyűjtése. *A Magyar Királyi Földtani Intézet Évi Jelentések 1915-ről*, 573–576.

Knauer, J., & Siegl-Farkas, Á. (1992). Palynostatigraphic position of the Senonian beds overlying the Upper Cretaceous bauxite formations of the Bakony Mountains. *Annual Report of the Hungarian Geological Institute*, *1990*, 463–471.

Koch, A. (1900). A Magyar Korona országainak kövült gerincesállat maradványainak rendszeres átnézete. *A Magyar Orvosi és Természettudományi Vizsgálatok Munkálatai*, 1–538.

Krézsek, C. S., & Bally, A. W. (2006). The Transylvanian basin (Romania) and its relation to the Carpathian fold and thrust belt: Insights in gravitational salt tectonics. *Marine and Petroleum Geology*, *23*, 405–446.

Kvaček, J., & Herman, A. (2004). The Campanian Grünbach flora of Lower Austria: Paleoecological interpretations. *Annalen des Naturhistorischen Museums in Wien, 106A*, 91–101.

Laurent, Y., Tong, H., & Claude, J. (2002). New side-necked turtle (Pleurodira: Bothremydidae) from the Upper Maastrichtian of the Petites-Pyrénées (Haute-Garonne, France). *Cretaceous Research, 23*, 465–471.

Linnaeus, C. (1758). *Systema Naturae* (10th ed., Vol. 1). Laurentii Sylvii.

Lyson, T. R., & Joyce, W. G. (2009). Paleoecology of two stratigraphically equivalent Latest Cretaceous (Maastrichtian) turtle thanatocoenoses. *Gaffney Turtle Symposium (October 17–18, 2009, Royal Tyrrell Museum, Drumheller). Abstract Volume*, 107–114.

Makádi, L. (2005). A new aquatic varanoid lizard from the Upper Cretaceous of Hungary. *Kaupia – Darmstädter Beiträge zur Naturgeschichte* (Abstract Volume, 3rd Annual Meeting of the EAVP), *14*, 127.

Makádi, L., Botfalvai, G., & Ősi, A. (2006). Late Cretaceous continental vertebrate fauna from the Bakony Mountains. I: Fishes, amphibians, turtles, lizards. *Földtani Közlöny, 136*, 487–502.

Marmi, J., Vila, B., & Galobart, Á. (2009). *Solemys* (Chelonii, Solemydidae) remains from the Maastrichtian of Pyrenees: Evidence for a semi-aquatic lifestyle. *Cretaceous Research, 30*, 1307–1312.

Matheron, P. (1869). Notice sur les reptiles fossiles des dé pots fluvio-lacustres crétacés du bassin à lignite de Fuveau. *Mémoires de l'Académie Impériale des Sciences, Belle-Lettres et Arts de Marseille, 2*(26), 781–795

Młynarski, M. (1966). Die fossilen Schildkröten in dem Ungarischen Sammlungen. *Acta Zoologica Cracoviensia, 11*, 223–288.

Nopcsa, F. (1897). Vorlaufiger Bericht über das Auftreten oberer Kreide im Hatszeger Tale in Siebenbürgen. *Verheindlung der Kaiserlischen und Königlischen Geologischen Reichs Anstalt*, Vienna, 273–274.

Nopcsa, F. (1905). A Gyulafehérvár, Déva, Ruszkabánya és a Romániai határ közé eső vidék geológiája. *A Magyar Királyi Földtani Intézet Évkönyve, 14*, 82–254.

Nopcsa, F. (1923a). On the geological importance of the primitive reptilian fauna of the uppermost Cretaceous; with a description of a new tortoise (*Kallokibotion*). *Quarterly Journal of the Geological Society of London, 79*, 100–116.

Nopcsa, F. (1923b). *Kallokibotion*, a primitive amphychelidean tortoise from the uppermost Cretaceous of Hungary. *Paleontologia Hungarica, 1*, 1–34.

Nopcsa, F. (1926). Die Reptilien der Gosau in neuer Beleuchtung. *Centralblatt für Mineralogie, Geologie und Paläontologie, 15*, 520–523.

Nopcsa, F. (1928). Palaeontological notes on Reptilia. 7. Classification of the Crocodilia. *Geologica Hungarica, Series Palaeontologica, 1*, 75–84.

Nopcsa, F. (1931). Note préliminaire sur quelques tortues du Danien du Midi de la France. *Annales de Muséum d'Histoire Naturelle de Marseille, 22*, 109–113.

Ősi, A., & Mindszenty, A. (2009). Iharkút, Dinosaur-bearing alluvial complex of the Csehbánya Formation. In E. Babinszky (Ed.) *Cretaceous sediments of the Transdanubian range. Field guide of the geological excursion organized by the Sedimentological Subcommission of the Hungarian Academy of Sciences and the Hungarian Geological Society* (pp. 51–63). Budapest: Hungarian Geological Institute.

Ősi, A., & Rabi, M. (2006). The Late Cretaceous continental vertebrate fauna from the Bakony Mountains II: Crocodiles, dinosaurs (Theropoda, Aves, Ornithischia), pterosaurs. *Földtani Közlöny, 136*, 503–526.

Pereda-Suberbiola, X. (2009). Biogeographical affinities of Late Cretaceous continental tetrapods of Europe: A review. *Bulletin de la Société Géologique de France, 180*, 57–71.

Pereda-Suberbiola, X., & Galton, P. (2001). Reappraisal of the nodosaurid ankylosaur *Struthiosaurus austriacus* Bunzel from the Upper Cretaceous Gosau Beds of Austria. In K. Carpenter (Ed.), *The armored dinosaurs* (pp. 173–210). Bloomington: Indiana University Press.

Pérez-Garcia, A., Ortega, F., & Murelaga, X. (2010). Systematics of the European Upper Cretaceous turtle *Elochelys* (Panpleurodira, Bothremydidae). *8th Meeting of the European Association of Vertebrate Paleontologists, Aix-en Provence, Abstract Volume*, 68.

Rabi M. (2009). An update of Late Cretaceous chelonians and crocodilians from Central Europe. *Journal of Vertebrate Paleontology, 29* (Suppl 6), 168A.

Rabi, M., & Botfalvai, G. (2006). A new bothremydid (Chelonia: Pleurodira) fossil assemblage from the Late Cretaceous (Santonian) of Hungary—additional studies in historical paleobiogeography of Late Cretaceous bothremydids. *Hantkeniana, 5* (4th Annual Meeting of the European Association of Vertebrate Paleontologists, Abstract Volume), 61–65.

Rabi, M., & Tong, H. (2007). Shell reconstruction of the bothremydid turtle (Testudines: Pleurodira) from the Late Cretaceous of Iharkút, Hungary. *5th Annual Meeting of the European Association of Vertebrate Paleontologists, Abstract Volume*, 49–50.

Rabi, M., Tong, H., & Botfalvai, G. (in review). A new species of the side-necked turtle *Foxemys* (Pelomedusoides: Bothremydidae) from the Late Cretaceous of Hungary and the historical biogeography of the Bothremydini. *Geological Magazine*.

Rothschild, B., Schultze, H.-P., & Pellegrini, R. (2012). Osseous and other hard tissue pathologies in turtles and abnormalities of mineral deposition. In D. B. Brinkman, P. A. Holroyd & J. D. Gardner (Eds.), *Morphology and evolution of turtles* (pp. XXX–XXX). Dordrecht: Springer.

Sachs, S., & Hornung, J. (2006). Juvenile ornithopod (Dinosauria: Rhabdodontiidae) remains from the Upper Cretaceous (Lower Campanian Gosau Group) of Muthmannsdorf (Lower Austria). *Geobios, 39*, 415–425.

Săndulescu, M. (1984). *Geotectonica României*. Bucureşti: Editura Tehnică.

Scheyer, T. M., & Sander, P. M. (2007). Shell bone histology of turtles indicates terrestrial paleoecology of basal turtles. *Proceedings of the Royal Society B, 274*, 1885–1893.

Seeley, H. G. (1881). The reptile fauna of the Gosau Formation preserved in the Geological Museum of the University of Vienna. *Quarterly Journal of the Geological Society of London, 3*, 620–702.

Sterli, J. (2008). A new, nearly complete stem turtle from the Jurassic of South America with implications for turtle evolution. *Biology Letters, 4*, 286–289.

Summesberger, H., Machalski, M., & Wagreich, M. (2007). First record of the Late Campanian heteromorph ammonite *Nostoceras hyatti* from the Alpine Cretaceous (Grünbach, Gosau Group, Lower Austria). *Acta Paleontologica Polonica, 57*, 443–451.

Szalai, T. (1934). Die fossilen schildkröten Ungarns. *Folia Zoologica et Hydrobiologica, 6*(2), 97–192.

Szalai, E. (2005). Paleomagnetic studies in Iharkút. Manuscript, Eötvös Loránd University, Department of Environmental Geology, Budapest.

Thenius, E. (1962). Niederösterreich im Wandel der Zeiten. *Grundzüge der Erd- und Lebensgeschichte von Niederösterreich. 2. Auflage*, 1–126.

Therrien, F. (2005). Palaeoenvironments of the latest Cretaceous (Maastrichtian) dinosaurs of Romania: Insights from fluvial deposits and paleosols of the Transylvanian and Haţeg basins. *Palaeogeography, Palaeoclimatology, Palaeoecology, 218*, 15–56.

Therrien, F., Jianu, C.-M., Scarlat, B., Weishampel, D. B., & King, J. W. (2002). Palaeoenvironmental reconstruction of latest Cretaceous dinosaur-bearing formations of Romania: Preliminary results. *Sargetia, Acta Musei Devensis, seria Stiintele Naturii, 19*, 33–59.

Tong, H., & Gaffney, E. S. (2000). Description of the skull of *Polysternon provinciale* (Matheron, 1869), a side-necked turtle (Pelomedusoides: Bothremydidae) from the Late Cretaceous of Villeveyrac, France. *Oryctos, 3*, 9–18.

Tong, H., Gaffney, E. S., & Buffetaut, E. (1998). *Foxemys*, a new side-necked turtle (Bothremydidae: Pelomedusoides) from the Late Cretaceous of France. *American Museum Novitates, 3251*, 1–19.

Tortosa, T., Dutour, Y., Cheylan, G., Tong, H., & Buffetaut, E. (2009). New Late Cretaceous dinosaur localities in Provence (southeastern France). *7th Meeting of the European Association of Vertebrate Paleontologists, Berlin, Abstract Volume*, 65.

Van Itterbeeck, J., Săsăran, E., Codrea, V., Săsăran, L., & Bultynck, P. (2004). Sedimentology of the Upper Cretaceous mammal- and dinosaur-bearing sites along the Râul Mare and Barbat rivers, Haţeg Basin, Romania. *Cretaceous Research, 25*, 517–530.

Vremir, M. (2001). Palaeontology, palaeoecology and taphonomy of the Late Cretaceous macrovertebrates of Alba Iulia-Sebeş area (Transylvania). M.Sc. thesis, Babes Bolyai University, Cluj-Napoca, Romania.

Vremir, M. (2004). Fossil turtle found in Romania—overview. *Annals of the Hungarian Geological Institute, 2002*, 143–152.

Vremir, M. (2010). New faunal elements from the late Cretaceous (Maastrichtian) continental deposits of Sebeş area (Transylvania). *Terra Sebus. Acta Musei Sabesiensis, 2*, 635–684.

Vremir, M., & Codrea, V. (2002). The first Late Cretaceous (Maastrichtian) dinosaur footprints from Transylvania (Romania). *Studia Universitatis Babeş-Bolyai, Geologia, 2*, 27–36.

Vremir, M., & Codrea, V. (2009). Late Cretaceous turtle diversity in Transylvanian and Haţeg basins (Romania). *The 7th International Symposium of Paleontology, Cluj-Napoca, Romania, Abstract Volume*, 122–124.

Vremir, M., Unwin, D. M., & Codrea, V. (2009). *A giant Azhdarchid (Reptilia, Pterosauria) and other Upper Cretaceous reptiles from Râpa Roşie-Sebeş (Transylvanian basin, Romania) with a reassessment of the age of the "Sebeş Formation"*. The 7th International Symposium of Paleontology, Cluj-Napoca, Romania, Abstact Volume, 125–128.

Vullo, R., Néraudeau, D., Allain, R., & Cappetta, H. (2005). Un nouveau gisement à microrestes de vertébrés continentaux et littoraux dans le Cénomanien inférieur de Fouras (Charente-Maritime, Sud-Ouest de la France). *Comptes Rendus Palevol, 4*, 95–107.

Vullo, R., Neraudeau, D., & Lenglet, T. (2007). Dinosaur teeth from the Cenomanian of Charentes, western France: Evidence for a mixed Laurasian-Gondwanan assemblage. *Journal of Vertebrate Paleontology, 27*, 931–943.

Vullo, R., Lapparent de Broin, F., Néraudeau, D., & Durrieu, N. (2010). Turtles from the Early Cenomanian paralic deposits (Late Cretaceous) of Charentes, France. *Oryctos, 9*, 37–48.

Wellnhofer, P. (1980). Flugsaurierreste aus der Gosau-Kreide von Muthmannsdorf (Niederösterreich)—ein Beitrag zurKiefermechanik der Pterosaurier. *Mitteilungen der Bayerischen Staatssammlung für Paläontologie und Historische Geologie, 20*, 95–112.

Willingshofer, E., Andriessen, P., Cloething, S., & Neubauer, F. (2001). Detrital fission track thermochronology of Upper Cretaceous syn-orogenic sediments in the South Carpathians (Romania): Inferences on the tectonic evolution of a collisional hinterland. *Basin Research, 13*, 379–395.

Zalmout, I. S., Mustafa, H. A., & Wilson, J. A. (2005). *Karkaemys arabicus*, a new side-necked turtle (Pleurodira, Bothremydidae) from the Upper Cretaceous Wadi Umm Ghudran Formation of Karak, Jordan. *Contributions from the Museum of Paleontology of the University of Michigan, 31*, 155–177.

Chapter 20
Re-Assessment of Late Campanian (Kirtlandian) Turtles from the Upper Cretaceous Fruitland and Kirtland Formations, San Juan Basin, New Mexico, USA

Robert M. Sullivan, Steven E. Jasinski, and Spencer G. Lucas

Abstract The fossil turtles from the Upper Cretaceous Fruitland and Kirtland formations (late Campanian; Kirtlandian) have been known for more than 100 years. We re-assess and revise these Late Cretaceous testudine taxa from the San Juan Basin, New Mexico, USA, and discuss their biostratigraphic distribution. We recognize the following valid taxa as present: the bothremydid *Chedighaii hutchisoni*; the pleurosternid *Compsemys* sp.; the baenodds *Denazinemys nodosa* and *Scabremys* gen. nov., established for the distinct species *S. ornata*, previously included in *Denazinemys*; and *Boremys grandis*, though it is a rare taxon. The non-baenodd baenid *Neurankylus baueri* is recognized as a valid species. Two additional non-baenodd baenid taxa, *Thescelus hemispherica* and *T. rapiens*, are retained as distinct species and are not considered synonymous with *T. insiliens*. We also recognize a small indeterminate kinosternoid similar to that reported from the Campanian of Mexico. The two adocids *Adocus bossi* and *A. kirtlandius* are retained as distinct species. The nanhsiungchelyid *Basilemys gaffneyi* sp. nov. is established, whereas *Basilemys nobilis* is considered a nomen dubium because it lacks the diagnostic features that would allow referral to any known valid species. We recognize three trionychids: *Aspideretoides austerus* and *A. robustus* (new combination), and an unnamed plastominine. We synomize *Aspideretoides fontanus* and *A. vorax* with *A. austerus*. *Aspideretoides ovatus* is considered a subadult of *A. robustus*. The unnamed plastominine may represent a new genus and species. Turtles of the Fruitland-Kirtland formations resemble other late Campanian turtle assemblages from western North America, and are part of the characteristic vertebrate fauna of the Kirtlandian land-vertebrate age. The upper Fruitland and lower Kirtland formations (Hunter Wash local fauna) have greater turtle taxonomic diversity than the upper Kirtland Formation (Willow Wash local fauna). This apparent decrease in taxonomic diversity is interpreted as being real and reflects a shift in depositional (channel) environments to a more terrestrial one, a pattern which is seen in other North American Late Cretaceous settings.

Keywords Fruitland Formation • Kirkland Formation • Late Cretaceous • Late Campanian • San Juan Basin • New Mexico

R. M. Sullivan (✉) · S. E. Jasinski
Section of Paleontology and Geology,
The State Museum of Pennsylvania, 300 North Street,
Harrisburg, PA 17120-0024, USA
e-mail: rsullivan@state.pa.us

S. E. Jasinski
e-mail: sej139@yahoo.com

S. G. Lucas
New Mexico Museum of Natural History and Science,
1801 Mountain Road NW, Albuquerque, NM 87104, USA
e-mail: spencer.lucas@state.nm.us

Introduction

Fossil turtles have been known from the Fruitland and Kirtland formations, San Juan Basin, New Mexico, USA, for more than 100 years. Oliver Perry Hay was the first to name testudine taxa from these Upper Cretaceous strata (Hay 1908, 1910), and Charles W. Gilmore was the first to critically assess the species of fossil turtles from the Fruitland and Kirtland formations in a subsequent series of papers (Gilmore 1916, 1919, 1935). Wiman (1933) also published a paper on these and other fossil turtles from the Upper Cretaceous and Paleocene strata of the San Juan Basin, based on a collection at the University of Uppsala, Sweden, that was purchased through the private collector Charles H. Sternberg.

Few detailed studies on Fruitland-Kirtland turtles have been published since. Armstrong-Ziegler (1978) listed turtle taxa and later (Armstrong-Ziegler 1980) reported on a few fragmentary specimens in the collections of the Museum of Northern Arizona, Flagstaff. Mateer (1981) reviewed the Kirtland Formation turtles, and other "mega reptiles," from

D. B. Brinkman et al. (eds.), *Morphology and Evolution of Turtles,* Vertebrate Paleobiology and Paleoanthropology,
DOI: 10.1007/978-94-007-4309-0_20, © Springer Science+Business Media Dordrecht 2013

the Uppsala collection, originally described in a series of papers by Wiman (1930, 1931, 1932, 1933). Lucas (1981) and Hunt and Lucas (1992) briefly mentioned the turtles from the Fruitland and Kirtland formations and provided a list of taxa. This was followed by a summary of New Mexico Cretaceous vertebrates, where they again listed turtles from both the Fruitland and Kirtland formations (Hunt and Lucas 1993). Most recently, McCord (1996) provided a cursory assessment of Late Cretaceous through early Eocene turtles from the San Juan Basin based on a small collection amassed by the University of Arizona, Laboratory of Paleontology, Tucson. That collection has since been transferred to the New Mexico Museum of Natural History and Science, Albuquerque.

Since the 1970s, collecting efforts in the Fruitland and Kirtland formations in the San Juan Basin resulted in the recovery of more than 400 turtle specimens to date (early 2010); over 200 catalogued specimens are in each of the collections of the New Mexico Museum of Natural History and Science (Albuquerque) and the State Museum of Pennsylvania (Harrisburg). These turtle specimens range from fragmentary carapace and plastron material to nearly complete shells. Some limb, girdle, and vertebral material has been recovered together with some isolated mandibular remains. No shells with articulated axial and appendicular skeletons have been recovered from Cretaceous or Paleocene deposits of the San Juan Basin.

Here, we document all the known turtle taxa from the upper Fruitland and Kirtland formations and assess their biostratigraphic occurrences in light of this increased sample size. We list all known (as of 2010) previously referred, and newly referred, turtle specimens from these two formations in Appendix .

Institutional abbreviations used in this chapter are: AMNH, American Museum of Natural History, New York, New York, USA; CMN, Canadian Museum of Nature, Ottawa, Ontario, Canada; FMNH, Field Museum of Natural History, Chicago, Illinois, USA; KUVP, University of Kansas, Lawrence, Kansas, USA; LSUMG, Louisiana State University, Museum of Natural Sciences, Baton Rouge, Louisiana, USA; MNA, Museum of Northern Arizona, Flagstaff, Arizona, USA; NMMNH, New Mexico Museum of Natural History, Albuquerque, New Mexico, USA; PMU, Paleontologiska Museet, Uppsala, Sweden; ROM, Royal Ontario Museum, Toronto, Ontario, Canada; SECCP, Secretaria de Educacion y Cultura, Coleccion Paleontologica, Coahuila, Mexico; SMP, The State Museum of Pennsylvania, Harrisburg, Pennsylvania, USA; TMM, Texas Memorial Museum, Austin, Texas, USA; TTU, Texas Tech University Museum, Lubbock Texas, USA; UALP, University of Arizona, Tucson, Arizona, USA; USNM, National Museum of Natural History, Smithsonian Institution, Washington, D. C., USA.

Systematic Paleontology

Testudines Linnaeus 1758
Pleurodira Cope 1865
Bothremydidae Baur 1891
Chedighaii Gaffney et al. 2006
Chedighaii hutchisoni Gaffney et al. 2006

Holotype: KUVP 14765, skull.

Holotype locality, unit, and age: KUVP loc. #35, NW¼, SW¼, Sec. 34, T24N, R13W, San Juan County, New Mexico; Hunter Wash Member, Kirtland Formation; late Campanian (early Kirtlandian).

Remarks: Gaffney et al. (2006) named and described *Chedighaii hutchisoni* for a nearly complete skull from the lower Kirtland Formation (Hunter Wash Member). No shell (carapace and plastron) material has been identified as *Chedighaii hutchisoni*, although some fragments (e.g., of a xiphiplastron) of "*Naiadochelys*" *ingravata* Hay (1908), presumably from the Chaco Canyon region, may pertain to this species (Gaffney et al. 2006).

Cryptodira Cope 1868
Paracryptodira Gaffney 1975
Pleurosternidae Cope 1868
Compsemys Leidy 1856
Compsemys sp.
(Fig. 20.1)

Referred specimens: See Appendix for complete list of referred specimens.

Occurrence: Within the San Juan Basin, New Mexico, known from upper part of Fruitland Formation and throughout the Kirtland Formation; late Campanian (Kirtlandian). See "Remarks", below, for occurrences outside of San Juan Basin.

Remarks: Armstrong-Ziegler (1978) listed *Compsemys* sp. as present in the Fruitland Formation without comment, and later she referred a carapace fragment (MNA Pl. 1648) to this genus based on its "closely set, flat-topped pustulae" sculpturing (Armstrong-Ziegler 1980, p. 16). McCord (1996) also recognized *Compsemys* from the Fruitland and Kirtland formations based on two fragmentary specimens (UALP 14391 and UALP 14393, now NMMNH P-49819 and P-49827, respectively). Only a single specimen from the Fruitland Formation in the collections of the NMMNH and SMP has been identified as *Compsemys* sp., NMMNH P-22741. However, 15 specimens of *Compsemys* sp. have been recovered from the Kirtland Formation (see Appendix), most of which are from the upper part (De-na-zin Member).

Specimens identified here as *Compsemys* sp. all bear the purported distinctive sculpturing, consisting of fine tubercles, some of which are co-joined to form short sinuous strands. SMP VP-1892, a peripheral, clearly shows this

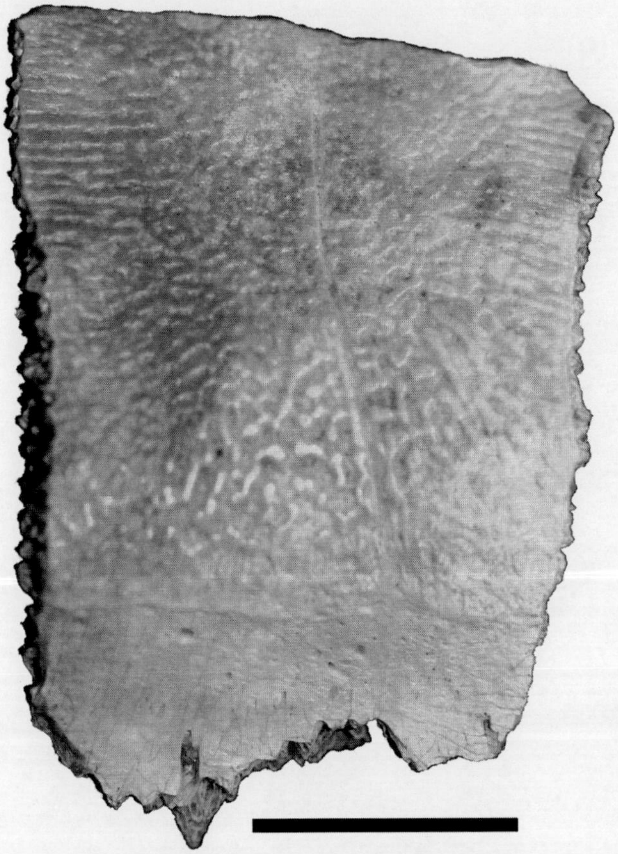

Fig. 20.1 *Compsemys* sp. SMP VP-1892, peripheral, in dorsal view. Bar scale = 1 cm

sculpturing (Fig. 20.1). We note, however, that the sculpturing of *Compsemys* is very similar, if not identical, to stem plastominines, and that the two taxa cannot be easily be separated on that feature alone (see below).

Compsemys victa Leidy 1856, known primarily from the Paleocene of North America, has also been identified in the Upper Cretaceous (late Maastrichtian; Lancian) Laramie Formation of Colorado (Hutchison and Holroyd 2003). We are unable to assign any of the Fruitland and Kirtland material to this species due to their incomplete nature. *Compsemys* sp. has also been identified from the Smoky Hollow Member of the Straight Cliffs Formation (late Turronian) of Utah (Eaton et al. 1999), and is known from various Campanian–Maastrichtian age units in the North American Western Interior.

Baenidae Cope 1882
Baenodda Gaffney and Meylan 1988
Denazinemys Lucas and Sullivan 2006
Denazinemys nodosa (Gilmore 1916)
(Fig. 20.2)

Holotype: USNM 8345 (Fig. 20.2a, b), nearly complete carapace and plastron.

Holotype locality, unit, and age: Locality 60 of Bauer (1916), two miles northwest of Ojo Alamo store (= Willow Wash), San Juan County, New Mexico; De-na-zin Member, Kirtland Formation; late Campanian (late Kirtlandian).

Referred specimens: See Appendix for complete list of referred specimens.

Occurrence: San Juan Basin, New Mexico; Hunter Wash and De-na-zin members, Kirtland Formation; late Campanian (Kirtlandian).

Revised diagnosis: Differs from *Scabremys ornata* (Gilmore 1935) (see below) by the following features: carapace sub-triangular with widest dimension posteriorly; prepleurals present and contacting the first vertebral medially; first vertebral irregular hexagonal shape with greatest width posteriorly, contacting anterior second vertebral; extracervicals lateral to primary cervical, gular sub-divided; and carapace nodes irregular and not forming distinct anteroposteriorly-directed ridges.

Remarks: Lucas and Sullivan (2006) reviewed specimens attributed to the form-genus "*Baena*", including both species "*B.*" *nodosa* and "*B.*" *ornata*. *Denazinemys* is known primarily from the Late Cretaceous of New Mexico, but has been reported as far south as Big Bend, Texas (Tomlinson 1997), suggesting that *Denazinemys* was restricted to the southern part of the Western Interior (Lucas and Sullivan 2006). It should be noted that the specimens designated as cf. *Denazinemys* (= "*Baena*") *nodosa*, from the Campanian lower and upper shale members of the Aguja Formation, are too incomplete for species recognition (contra Tomlinson 1997), and assignment to other baenids could be made in the absence of other characters (see below).

We regard these two turtles, *Denazinemys nodosa* and *Scabremys* (= *Denazinemys*, in part) *ornata*, as distinct genera and species (see below). We note that many of the incomplete specimens do not exhibit the suite of characters that would permit identification to the genus level. Typically, only the node-like sculpturing of the carapace is seen on incomplete, isolated material, so even reference to genus based on carapace fragments is not at all certain.

The holotype of *Denazinemys nodosa* (USNM 8345) is from the De-na-zin Member (upper Kirtland Formation), whereas the holotype of *Scabremys ornata* (USNM 13229) is from the Hunter Wash Member (lower Kirtland Formation). The fragmentary material (carapace fragments) referred to *D. nodosa* cannot be referred to this species because of its incomplete nature, contrary to Lucas and Sullivan (2006). See the Appendix for revised list of specimens referred to *D. nodosa*.

Scabremys gen. nov.

Synonymies: *Baena*: Leidy 1870 [in part]. "*Baena*": Gaffney 1972 [in part]. *Denazinemys*: Lucas and Sullivan 2006 [in part].

Fig. 20.2 *Denazinemys nodosa* (Gilmore 1916). **a**, **b** USNM 8345 (holotype), nearly complete carapace and plastron: **a** carapace, in dorsal view; **b** plastron, in ventral view. **c**, **d** SMP VP-1869: **c** carapace, in dorsal view; **d** plastron, in ventral view. Bar scale = 10 cm

Type species: *Scabremys ornata* (Gilmore 1935).

Etymology: From the Latin stem "scabr" meaning rough, in reference to its unusually rough, nodular sculpturing on the carapace; and from the Greek "emys," meaning turtle.

Occurrence: San Juan Basin, New Mexico; Hunter Wash Member, Kirtland Formation; late Campanian (early Kirtlandian).

Diagnosis: Same as for species.

Remarks: *Denazinemys* is considered a monotypic taxon based on a set of apomorphic characters cited above. Previous inclusion of the species *D. ornata* in *Denazinemys* was largely a result of it previously being placed with "*Baena*" based on having nodose sculpturing on the external surface of the carapace. However, nodose sculpturing is rather widespread among baenids, and is present, in various degrees, on the carapaces of the type species of *Denazinemys*, *Boremys* Lambe 1906, and *Thescelus* Hay 1908. The holotype of *D. ornata* has a number of unique characters that exclude it from the genus *Denazinemys* and other baenodd taxa.

Scabremys ornata (Gilmore 1935) new combination.
(Fig. 20.3)

Synonymies: *Baena ornata*: Gilmore 1935, p. 165, Figs. 7, 8. "*Baena*" *ornata* (Gilmore): Gaffney 1972, pp. 302–303. *Denazinemys ornata* (Gilmore): Lucas and Sullivan 2006, pp. 226–227, Fig. 3.

Holotype: USNM 13229 (Fig. 20.3), nearly complete carapace and plastron.

Holotype locality, unit, and age: Three miles northeast of Hunter's Store (Bisti Post Office), SW ¼, T 24N, R13W, San Juan County, New Mexico; Hunter Wash Member, Kirtland Formation; late Campanian (early Kirtlandian).

Referred specimens: None.

Revised diagnosis: Differs from *Denazinemys nodosa* by the following features: carapace oval, with widest dimension midway along carapace; prominent midline ridge formed by three or four raised ridges, separated by narrow grooves, that extend for nearly the entire length of the carapace; no prepleurals; first vertebral irregular hexagonal shape with greatest width anteriorly, contacting the cervical and both first marginals; extracervicals absent, and gulars not subdivided.

Remarks: Gilmore (1935) established the taxon *Baena ornata* based on USNM 13229 (Fig. 20.3), a nearly complete carapace and plastron, from the Hunter Wash Member of the Kirtland Formation. The description and observations made by Gilmore (1935) are mostly sound and need not be repeated here in their entirety. Suffice it to say there are some features that need to be emphasized and are noted here with their corresponding additional character number (characters 137–142) added to those of Joyce (2007).

The shape of the shell of *Scabremys ornata* is significantly different from that of *Denazinemys nodosa* in that it retains the more primitive rounded/oval shape and has its widest part midway, rather than at the posterior part of the carapace (character 137). The posterior scalloping of the carapace is less prominent and more restricted compared to *D. nodosa*. The external nodose surface texture in this taxon is extreme, and is far greater than that of *D. nodosa*. Unlike *D. nodosa*, the nodes are more elongated and ridge-like with a prominent compound ridge running along the midline juxtaposed with the neurals (character 138), which we score as a derived character. The anterior section of the carapace lacks prepleurals (character 139), which are prominently present in *D. nodosa* and *Boremys*. The anterior margin of vertebral 1 (character 140) is widest (primitive) in *Neurankylus* Lambe 1902, *Trinitichelys* Gaffney 1972, *Plesiobaena* Gaffney 1972, and *Scabremys*; the opposite condition (shortest) is seen in the other taxa. The presence of extracervicals (character 141) is seen only in *D. nodosa*, *Baena arenosa* Leidy 1870, and *Chisternon undatum* (Leidy 1872) and is a derived feature that is absent in all other baenids. *S. ornata* is the only baenid taxon without subdivided gulars (character 142), which also is a derived feature.

Using the data matrix of Joyce (2007) we extracted the baenids (*Neurankylus eximius* Lambe 1902, *Trinitichelys hiatti* Gaffney 1972, *Plesiobaena antiqua* (Lambe 1902), *Boremys pulchra* (Lambe 1906), *Chisternon undatum*, *Baena arenosa*, and *Denazinemys nodosa*), added a hypothectical ancestor together with *Scabremys ornata*, and scored all nine taxa for six additional characters cited above. We ran a phylogenetic analysis using PAUP 4.0b10 (Swofford 2002). Our data matrix, which did not include skull characters, was subjected to a heuristic search with 1000 replicates. All characters were unordered and unweighted. The best tree length obtained was 81, with a consistency index of 0.9506, and a retention index of 0.8095. It placed *Scabremys ornata* as the sister taxon to (*Boremys pulchra* [*Chisternon undatum* (*Baena arenosa* + *Denazinemys nodosa*)]). The strict consensus tree of three trees grouped the taxa as (*Plesiobaena antiqua* [*Scabremys ornata* (*Boremys pulchra* [*Chisternon undatum* (*Baena arenosa* + *Denazinemys nodosa*)])]) with the hypothetical ancestor, *Neurankylus eximius*, and *Trinitichelys hiatti* forming a basal polytomy (Fig. 20.4). Our interpretation that *Scabremys ornata* is clearly distinct from *Denazemys nodosa* and other baenodd turtles is supported by this analysis.

Lucas and Sullivan (2006) re-evaluated the holotype of *Scabremys* ("*Baena*") *ornata* (USNM 13229) from the lower Kirtland Formation (Hunter Wash Member), and placed it in the new genus *Denazinemys*. No other specimens have been referred to this species from New Mexico.

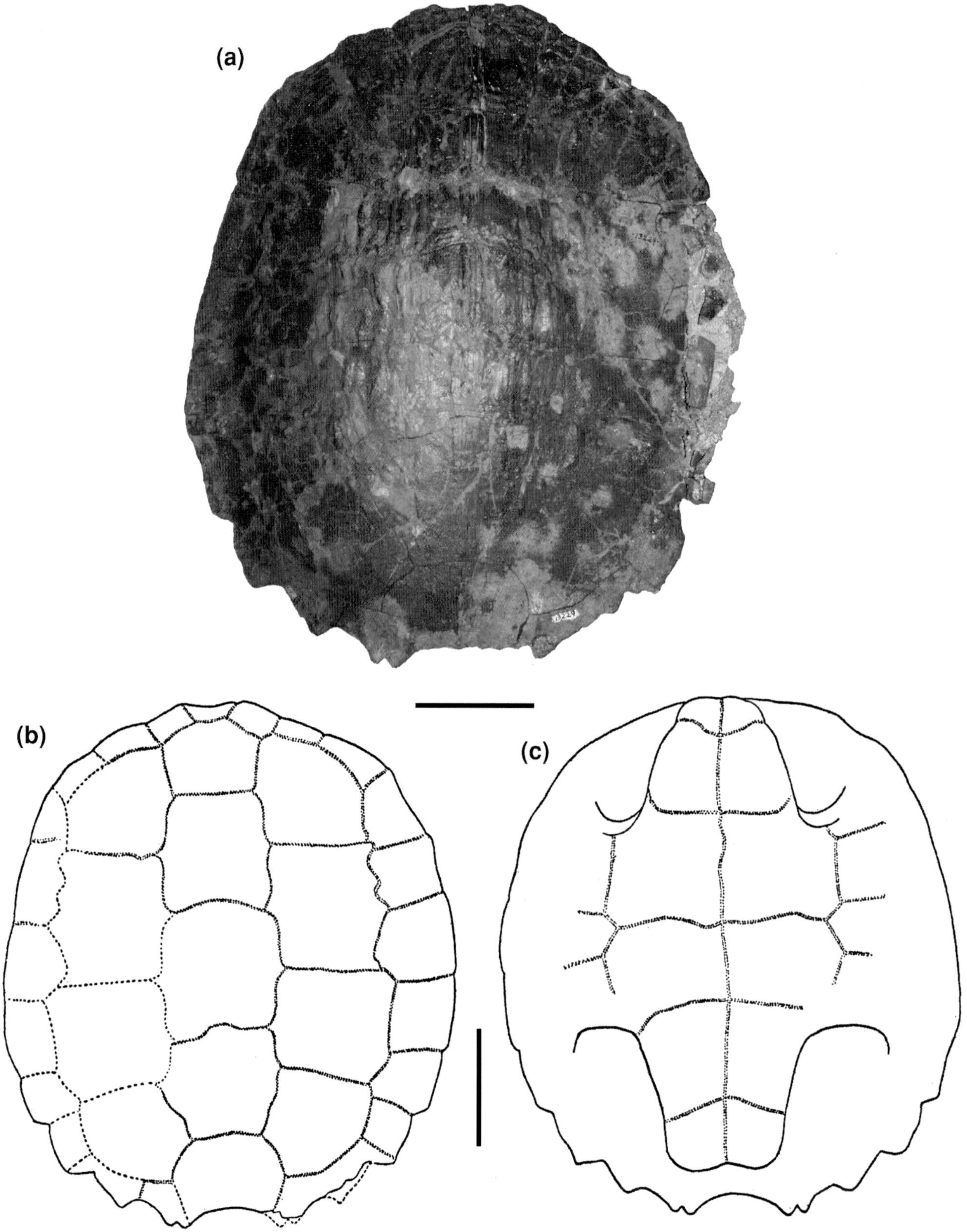

Fig. 20.3 *Scabremys* gen. nov., *S. ornata* (Gilmore 1935); USNM 13229 (holotype), nearly complete carapace and plastron. **a** Photograph of carapace, in dorsal view; **b**, **c** *line* illustrations, from Gilmore (1935): **b** carapace, in dorsal view; **c** plastron, in ventral view. Photograph and drawings at different magnifications; bar scales = 10 cm

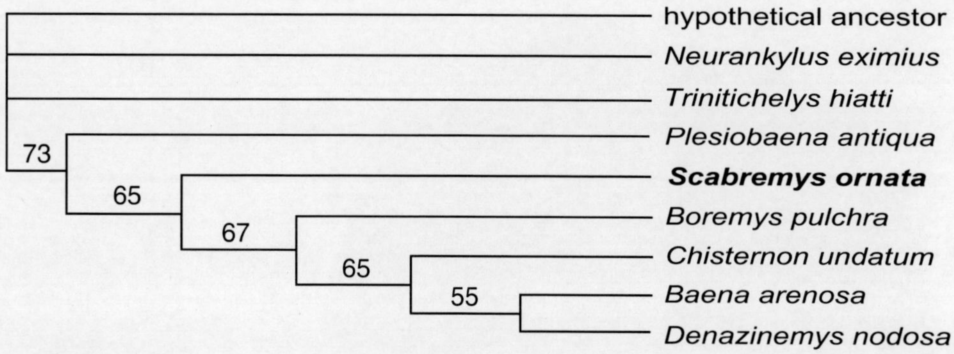

Fig. 20.4 A strict concensus tree of three trees for the Baenidae, showing the phylogenetic position of *Scabremys ornata*. Bootstrap values are indicated *above lines*. Data matrix extracted from Joyce (2007), with six additional characters (see text)

However, one specimen (LSUMG V-1136), consisting of a nearly complete carapace and reportedly from the "lower part of the Paleocene Black Peaks Formation," southwest of Sombrero Peak, was identified as "*Baena*" cf. "*B.*" *ornata* by Tomlinson (1997). We note here that the rocks of the lower part of the Black Peaks Formation are now considered Maastrichtian age and not Paleocene in age (Fowler 2009, personal communication). This age is consistent with the fact that LSUMG V-1136 co-occurs with *Hoplochelys* Hay 1908, a taxon which, in New Mexico, is present in the early Maastrichtian Naashoibito Member, Ojo Alamo Formation (Lehman 1981; Jasinski et al. 2011). However, no specimens of *Scabremys* (= "*Denazinemys*") *ornata* are presently known from either the De-na-zin Member (Kirtland Formation) or the Naashoibito Member (Ojo Alamo Formation). Based on the illustration of Tomlinson (1997, p. 37, Fig. 3.7), LSUMG V-1136 would seem to have features that are referable to both *D. nodosa* (presence of prepleurals) and *S. ornata* (oval shape of carapace). We therefore regard this specimen as an indeterminate baenid (see below). For now *Scabremys ornata* is known only from the holotype (USNM 13229, Fig. 20.3) and is restricted to the lower part of the Kirtland Formation (Hunter Wash Member).

Boremys Lambe 1906
Boremys grandis Gilmore 1935
(Fig. 20.5)

Synonymies: *Boremys grandis*: Gilmore 1935, p. 170. *Boremys pulchra*: Lambe 1906 [in part], p. 189. *Boremys pulchra* (Lambe 1906): Gaffney 1972 [in part] p. 296.

Holotype: USNM 12979, nearly complete carapace (Fig. 20.5a) and plastron.

Holotype locality, unit, and age: Three miles northeast of Hunter's Store (Bisti Post Office), SW ¼, T 24N, R13W, San Juan County, New Mexico; Hunter Wash Member, Kirtland Formation; late Campanian (early Kirtlandian).

Referred specimens: See Appendix for complete list of referred specimens.

Occurrence: San Juan Basin, New Mexico; Hunter Wash Member, Kirtland Formation; late Campanian (early Kirtlandian).

Remarks: Lambe (1906) named *Boremys pulchra* from the Dinosaur Park Formation (Brinkman 2005) based on an incomplete carapace and plastron. Gaffney (1972), in his review of the Baenidae, synonymized *Boremys grandis* Gilmore 1935 with *B. pulchra*. However, he also noted that the supermarginal scutes seen in both the holotype (USNM 12979, Fig. 20.5a) and PMU.R16 (Fig. 20.5b = Ex. 9 of Wiman 1933, p. 11) may be a reason to consider it a valid species, though he considered this doubtful based on limited evidence. In a subsequent chapter, Brinkman and Nicholls (1991) recognized *B. grandis* as a distinct species based on: (1) the proliferation of supramarginal scales and (2) its large size. They also tentatively considered the arrangement of cervical scales to be a unique feature of *B. grandis*, because that is not seen in any specimens of *B. pulchra* (Brinkman and Nicholls 1991).

SMP VP-1565 (Fig. 20.5c) is tentatively referred to *Boremys grandis* based on a small plastral fragment consisting of a section of the right gular and epiplastron. If large size is a valid distinguishing character, then on that basis assignment to *B. grandis* is all but certain as it is consistent with the size of the holotype specimen (USNM 12979). *Boremys pulchra* is half the size of *B. grandis* (Gilmore 1935; Brinkman and Nicholls 1991). Aside from the holotype, and the Uppsala specimen (PMU.R16), this is the only other specimen that is referable to *B. grandis*, making it an extremely rare species. The holotype and the referred specimens are all from the lower Kirtland Formation (Hunter Wash Member).

Incertae sedis
Neurankylus Lambe 1902
Neurankylus baueri Gilmore 1916
(Fig. 20.6)

Synonymies: *Neurankylus baueri*: Gilmore 1916, p. 290. *Neurankylus eximius*: Lambe 1902 [in part]; Gaffney 1972, p. 291.

Fig. 20.5 *Boremys* specimens. **a**, **b** *Boremys grandis* Gilmore 1935: **a** USNM 12979 (holotype), nearly complete carapace and plastron, carapace only, in dorsal view; **b** PMU.R16, incomplete carapace, in dorsal view. **c** cf. *Boremys* sp., SMP VP-1565, right gular and epiplastron fragment, in ventral view. Specimens at different magnifications; bar scales = 1 cm

Fig. 20.6 *Neurankylus baueri* Gilmore 1916. **a**, **b** USNM 8344 (holotype), nearly complete carapace and plastron: **a** carapace, in dorsal view; **b** plastron, in ventral view. **c**, **d** SMP VP-2379, nearly complete carapace and plastron: **c** carapace, in dorsal view; **d** plastron, in ventral view. Specimens at different magnifications; bar scales = 10 cm

Holotype: USNM 8344 (Fig. 20.6a, b), nearly complete carapace and plastron.

Holotype locality, unit, and age: Ah-shi-sle-pah Wash (= Meyers Creek), middle branch, San Juan County, New Mexico; Hunter Wash Member, Kirtland Formation; late Campanian (Kirtlandian).

Referred specimens: See Appendix for complete list of referred specimens.

Occurrence: San Juan Basin, New Mexico; Hunter Wash and De-na-zin members, Kirtland Formation; late Campanian (Kirtlandian).

Revised diagnosis: Differs from all other species of *Neurankylus* in having the following combination of characters: gulars well-developed and deep; intergulars U-shaped and in contact with humerals posteriorly (thereby excluding the gulars from contact across the midline).

Remarks: Gilmore (1916) established the species *Neurankylus baueri* based on USNM 8344 (Fig. 20.6a, b), which is a complete carapace and plastron, from the Hunter Wash Member of the Kirtland Formation. He stated that it differed from the holotype of *N. eximius* (CMN 1504) in the number of costals (8 vs. 9), but noted that Hay (1908) considered this feature to be the result of individual variation, and not of any taxonomic significance. The ninth costal is probably an atavism, a condition where a primitive trait is occasionally expressed (Brinkman and Joyce 2010, pers. comm.). Unfortunately, nowhere in Gilmore's description of the holotype of *N. baueri* did he identify any characters that would allow *N. baueri* to be distinguished from *N. eximius*. However, Larson et al. (2012) re-diagnosed *N. eximius* based on new material and distiguished *N. eximius*, in part, as having "intergulars only barely separating gulars with little or no shared sulcus with the humerals, and sigmoid intergular–gular sulcus oriented anterolaterally to posteromedially, creating a heart shape." This differs from the U-shaped intergulars and gulars not in contact with the mid-line in *N. baueri*. This condition is essentially the same as seen in ROM 864, as illustrated by Larson et al. (2012, Fig. 21.1d).

We note here that ROM 864, a specimen collected by Charles H. Sternberg and sold to Ward's Natural Science Establishment, and later purchased by the Royal Ontario Museum in 1933, is said to be from the Fruitland Formation, however, it is more likely to be from the lower Kirtland Formation (Hunter Wash Member).

Wiman (1933) described a number of specimens collected by C. H. Sternberg, and most, if not all of them, are from the lower Kirtland Formation (although he noted that one came from the "Ojo Alamo Formation," but based on its preservation, we consider that highly unlikely).

A newly collected specimen, SMP VP-2379 (Fig. 20.6c, d), consists of a carapace and plastron that are incomplete, slightly crushed, and distorted. The carapace is cracked, and a large portion of the right side is missing (Fig. 20.6c). The plastron (Fig. 20.4d) is nearly complete and has the same arrangement of gular and intergulars as in the holotype of *Neurankylus baueri* (Fig. 20.4b). Overall, SMP VP-2379 is larger, more robust, and appears to have a more oval shape, compared to the holotype of *N. baueri* (Fig. 20.6a, b). However, when accounting for the distortion, our calculations of length-to-width ratios for both the holotype of *N. baueri* and SMP VP-2379 are very close: 1.15–1.27, respectively.

Sullivan and Lucas (2006) considered *Neurankylus baueri* to be distinct from *N. eximius* primarily based on having the first suprapygal shorter and wider than that in *N. eximius*, but we now acknowledge this is probably due to individual variation. We accept the rediagnosis of *N. eximius* by Larson et al. (2012) and note that the San Juan Basin Fruitland-Kirtland specimens all have U-shaped intergulars and the gulars do not contact the midline.

Thescelus Hay 1908
Thescelus hemispherica Gilmore 1935
(Fig. 20.7)

Holotype: USNM 12818 (Fig. 20.7a, b), incomplete carapace and plastron.

Holotype locality, unit, and age: Three miles northeast of Hunter's Store (Bisti Post Office), SW ¼, T 24N, R13W, San Juan County, New Mexico; Hunter Wash Member, Kirtland Formation; late Campanian (early Kirtlandian).

Referred specimen: PMU.R23 (Fig. 20.7c, d), incomplete carapace and plastron.

Occurrence: San Juan Basin, New Mexico; Hunter Wash Member, Kirtland Formation; late Campanian (early Kirtlandian).

Remarks: Gilmore (1935) established the species *Thescelus hemispherica* based on USNM 12818 (Fig. 20.7a, b), an incomplete carapace and plastron from what is now known as the Hunter Wash Member of the Kirtland Formation. He distinguished *T. hemispherica* from *T. insiliens* based on: (1) "the bosslike ornamentation of the carapace;" (2) "the relatively wider vertebrals;" (3) "posterior border of carapace without constructions;" and (4) "the nuchal less deeply excavated" (Gilmore 1935, pp. 176–177). Gilmore (1935, p.177) further distinguished *T. hemispherica* from *T. rapiens* Hay 1908 in having: (1) "a median depression along the back;" (2) relatively wider vertebrals;" and (3) "rough sculpture of the carapace." The features used by Gilmore (1935) to distinguish *T. hemispherica* from *T. rapiens* may be, in part, problematic (for example, the relative widths of the vertebrals) and the two taxa may be synonymous. However, presently we retain them as distinct species, pending further evidence (see account below for *T. rapiens*).

Thescelus rapiens Hay 1908
(Fig. 20.8)

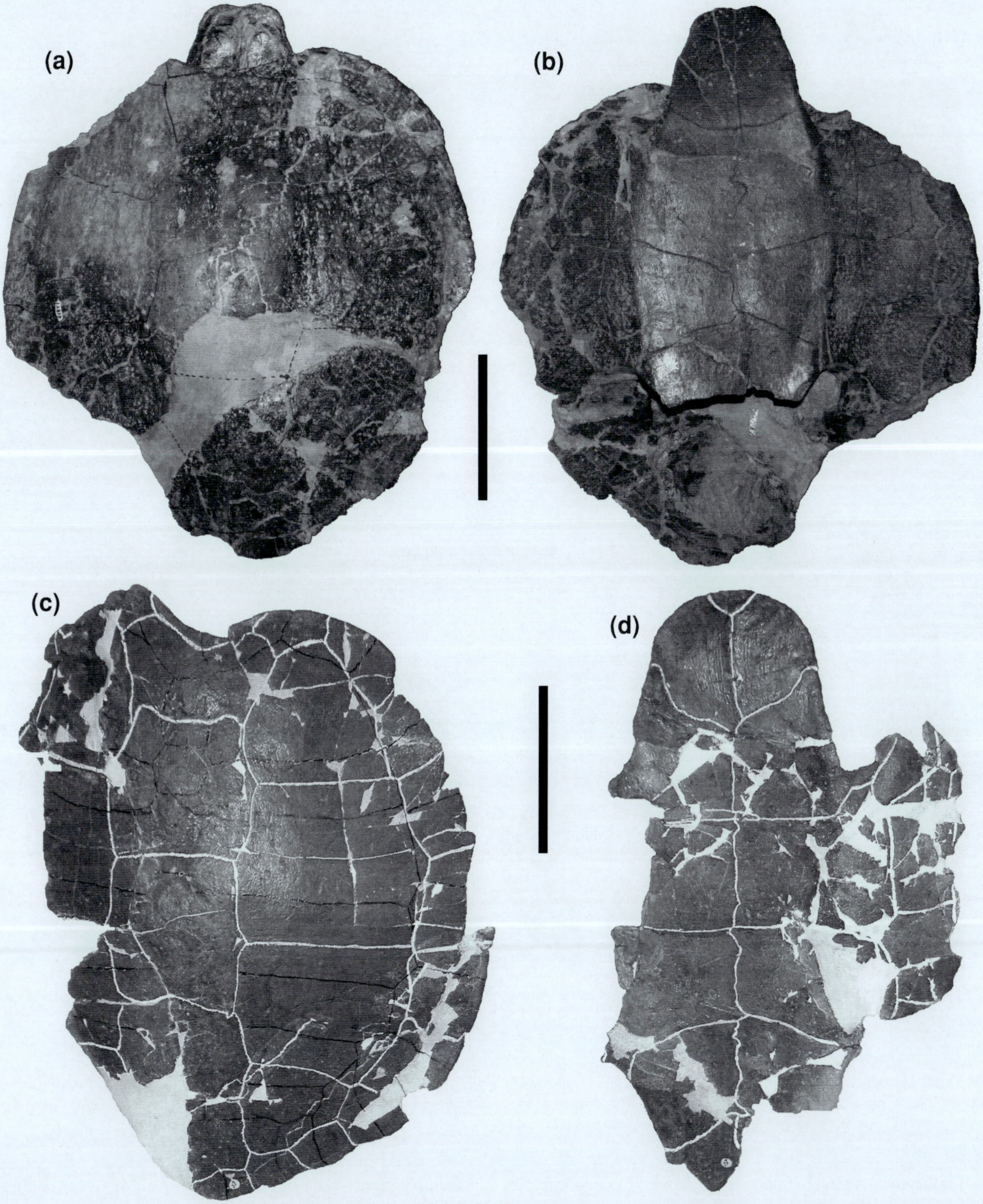

Fig. 20.7 *Thescelus hemispherica* Gilmore 1935. **a**, **b** USNM 12818 (holotype), incomplete carapace and plastron: **a** carapace, in dorsal view; **b** plastron, in ventral view. **c**, **d** PMU.R23, incomplete carapace and plastron: **c** carapace, in dorsal view; **d** plastron, in ventral view. Specimens at different magnifications; bar scales = 10 cm

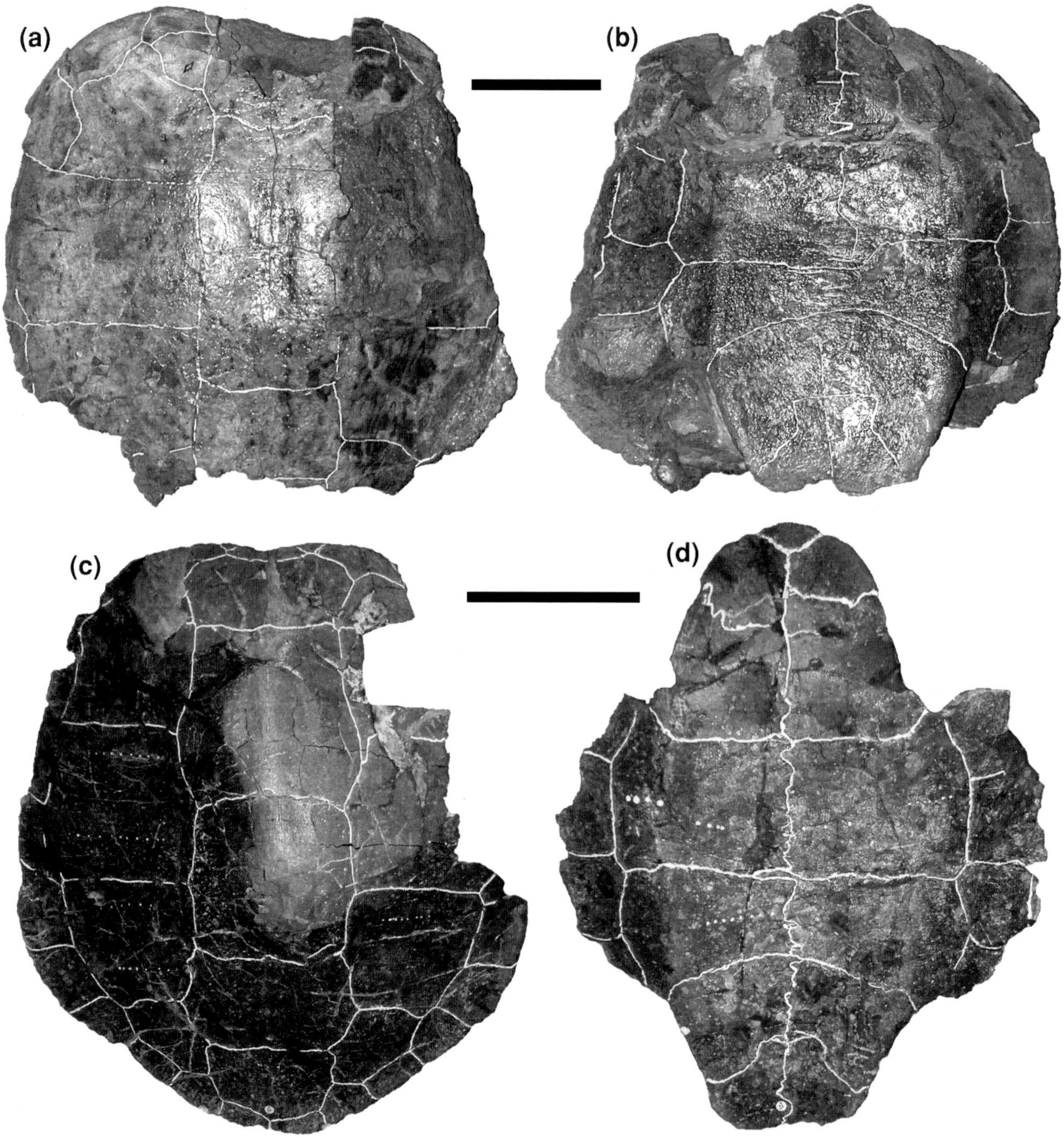

Fig. 20.8 *Thescelus rapiens* Hay 1908. **a**, **b** AMNH 6066 (holotype), incomplete carapace and plastron: **a** carapace, in dorsal view; **b** plastron, in ventral view. **c**, **d** PMU.R22, incomplete carapace and plastron: **c** carapace, in dorsal view; **d** plastron, in ventral view. Specimens at different magnifications; bar scales = 10 cm

Holotype: AMNH 6066 (Fig. 20.8a, b), incomplete carapace and plastron.

Holotype locality, unit, and age: Ojo Alamo, San Juan County, New Mexico; De-na-zin Member, Kirtland Formation; late Campanian (late Kirtlandian).

Referred specimen: PMU. R22 (Fig. 20.8c, d), incomplete carapace and plastron.

Occurrence: San Juan Basin, New Mexico; Hunter Wash and De-na-zin members, Kirtland Formation (late Campanian; Kirtlandian).

Remarks: Hay (1908) established the species *Thescelus insiliens* and *T. rapiens* for specimens from the "Laramie" beds of Wyoming (AMNH 1108) and from the "Laramie" deposits at Ojo Alamo (AMNH 6066), respectively. The holotype of

T. insiliens is from what is now considered the Lance Formation (late Maastrichtian; Lancian). The holotype of *T. rapiens* was collected at the same locality as the holotype of the dinosaur *Kritosaurus navajovius* (Brown 1910), which we know to be in the De-na-zin Member (late Campanian; late Kirtlandian) of the Kirtland Formation (Gilmore 1916; Lehman 1981).

Gaffney (1972), in reviewing this taxon, provided a restoration of *Thescelus insiliens* based on two specimens from different geologic horizons and different geographic regions. He recognized 12 characters that serve to diagnose the genus, which he considered to be monotypic. Three species were placed into synonymy: *T. rapiens*, *T. hemispherica* and *Baena longicauda* Russell 1934 (Gaffney 1972). However, Sullivan and Lucas (2006) noted differences in patterns of bones-to-scutes in PMU.R22 and PMU.R23, a pattern that differs from the stratigraphically higher *T. insiliens*. Moreover, we can further distinguish *T. insiliens* from the two Kirtland taxa based on the constricted posterior border of the carapace and on the femoral notch width of the plastron, which is greater than the humeral notch width.

The remaining characters that Gilmore (1935) used to differentiate the three species of *Thescelus* are: (1) presence of ornamentation on the carapace (*T. hemispherica*); (2) relatively wider vertebrals (*T. hemispherica*); (3) less deeply excavated nuchal (*T. hemispherica* and *T. rapiens*?); and (4) median depression along the back (*T. rapiens*). Re-assessment suggests that these features are questionably useful for determining phylogenetic relationship and taxonomic identity. The ornamentation, where preserved, is not pronounced. The significance of the "relatively wider vertebrals" is unclear and may be variable. It is our opinion that all members of *Thescelus* had excavated nuchals, and the degree to which they are excavated is variable and taxonomically insignificant. Lastly, the median depression along the midline of the carapace may be an artifact of preservation.

Although it is evident that we can clearly separate the stratigraphically younger species (*Thescelus insiliens*) from the two Kirtland Formation species (*T. rapiens* and *T. hemispherica*), it remains difficult to differentiate the latter two species from one another. It may be that *T. hemispherica* is restricted to the lower Kirtland Formation (Hunter Wash Member), and *T. rapiens* is from both the lower and the upper Kirtland Formation (Hunter Wash and De-na-zin members). Based on the material in the Uppsala collection, we tentatively place PMU R.22 in *T. rapiens*, as per Wiman (1933), and PMU R.23 in *T. hemispherica*, as per Gilmore (1935).

Thescelus sp.

Referred specimen: SMP VP-2100 (not figured), nearly complete plastron with carapace fragments. Collected from SMP locality 421, Alamo Mesa (southwest), SW ¼, Sec. 27, T24N, R12W, San Juan County, New Mexico, within the Hunter Wash Member of the Kirtland Formation.

Remarks: SMP VP-2100 is assigned to *Thescelus* sp., because it lacks specific characters that would permit assignment to either *T. rapiens* or *T. hemispherica*. Even so, we note that it is morphologically similar to PMU.R22, which we tentatively referred (see above) to *T. rapiens*.

Baenidae Indeterminate

Referred specimens: See Appendix for complete list of referred specimens.

Remarks: Based on the descriptions and illustrations provided by Tomlinson (1997), we consider specimens from the Big Bend region of Texas to be insufficient for any genus or species assignment. The fragmentary specimens from New Mexico cannot be assigned to either *Denazinemys nodosa* or *Scabremys ornata* with any level of confidence. Moreover, we note that some specimens of *Thescelus* and *Boremys* have nodose texture on sections of the carapace similar to that seen in *D. nodosa* and *S. ornata*, making isolated fragments of carapace material impossible to identify to genus level. For now we tentatively consider all this material to be Baenidae indeterminate.

Eucryptodira Gaffney 1975
Kinosternoidea Gaffney and Meylan 1988
Genus and species indeterminate
(Fig. 20.9)

Referred specimens: See Appendix for complete list of referred specimens.

Occurrence: San Juan Basin, New Mexico; De-na-zin Member, Kirtland Formation; late Campanian (late Kirtlandian).

Description: The first specimen, SMP VP-1907 (not figured), consists of an incomplete left hypoplastron bearing a prominent sulcus that separates the femoral from the abdominal. It is broken along the medial side, so the medial half of the left hypoplastron is missing. The sutural contact with the hyoplastron is preserved anteriorly. Posteriorly the distal end is also broken. The ventral (external) surface is flat and bears a smooth, fine-textured surface, consisting of short, entwined, fiber-like sculpturing with numerous minute foramina. The internal (visceral) surface is also smooth. It is relatively thick and becomes pinched at the femoral notch. It measures 0.53 cm in maximum thickness.

SMP VP-2004 (Fig. 20.9a–d) consists of a right hypoplastron and two incomplete costals. The right hypoplastron consists of the anterior portion where, in life, it would have articulated with the hypoplastron. It is slightly thicker than SMP VP-1907, measuring 0.58 cm. The suture surface is preserved along the anterior and medial parts of the element, as well as along the posterior part where it would have articulated with the xiphiplastron. It is broken

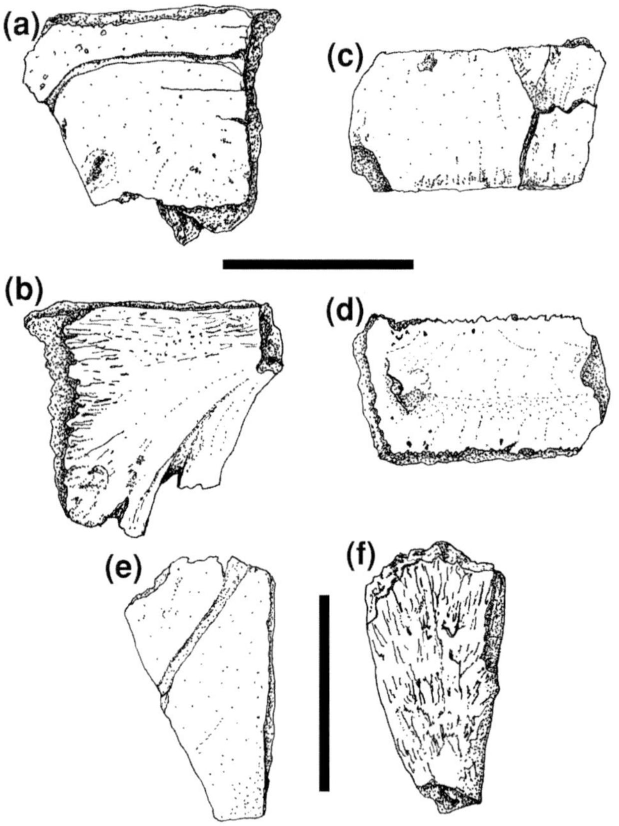

Fig. 20.9 Kinosternoidea indeterminate. **a–d** SMP VP-2004, right hypoplastron and two incomplete costals: **a**, **b** anterior part of right hypoplastron: **a** ventral view; **b** visceral view; **c**, **d** right costal 5: **c** dorsal view; **d** visceral view. **e**, **f** SMP VP-2009, nearly complete right xiphiplastron: **e** ventral view; **f** visceral view. Specimens at different magnifications; bar scales = 1 cm

laterally. The two incomplete costals are also broken laterally. The dorsal (external) surface of one is partly eroded, and there is no trace of any sulci, so it is identified as costal 1. The other (Fig. 20.9c, d) bears sulci of the vertebrals (Fig. 20.9c), but lacks that of the pleurals, so it is tentatively identified as right costal 5. The internal surfaces of both costals preserve the costal ribs, the heads of which are broken medially.

SMP VP-2009 (Fig. 20.9e, f) is a nearly complete right xiphiplastron; only the posterior (caudal) tip of the element is broken. The medial edge has a prominent sutural surface, and the anterior edge is marked by an interdigitating sutural surface where it would have articulated with the right hypoplastron. The maximum thickness is along the medial suture, and the bone tapers laterally. In medial cross-section view, the distal part of the xiphiplastron gently curves upward (dorsally). The texture is the same as seen in the other specimens. The femora-anal sulcus extends just lateral from the midline, posterolaterally to the edge, anterior to the midsection of the element.

SMP VP-2533 (not figured) is tentatively identified as a right costal 8. The element has an elongate, sub-trapezoidal shape. The medial end is narrow, with the lateral end wider. The anterior, posterior, and medial edges are strongly sutured. The lateral edge is broken anteriorly and thins laterally, lacking any sutural surface. Dorsally, there are two scallop-like sulci, which we think are the vertebral 4 (towards the midline) and pleural 4 (laterally), with vertebral 5 lying posteriorly. Another sulcus extends posteriorly from between the two scalloped impressions. A small crescent-shaped sulcus lies lateral to the posterior edge of the previous one. Neither of the sulci is identifiable, and these demarcations may not be actual sulci. Internally, there is a prominent, thin costal rib.

The last specimen, SMP VP-3274 (not figured), is an incomplete ?left costal 2. It is prominently sutured on the anterior, medial and posterior sides. It is thickest toward the midline and thins laterally. It is estimated that it represents less than half of the complete costal. Dorsally, there is a prominent sulcus. Internally, the head of the rib is close to the midline. The external surface texture is the same as in the preceding specimens.

Remarks: Five specimens are here referred to the Kinosternoidea based on features of the carapace and plastron as well as the distinctive smooth sculpturing. They are identical to specimens recently reported by Brinkman and Rodriguez de la Rosa (2006) from the Campanian Cerro del Pueblo Formation of Mexico. These specimens share the same smooth surface sculpturing, consisting of fine, short, entwined, fiber-like structures with numerous minute foramina.

Brinkman and Rodriguez de la Rosa (2006) referred the Cerro del Pueblo Formation specimens to the Kinosternoidea, genus and species indeterminate, based on the presence of two derived features cited by Hutchison (1991). These are: (1) the contact between the plastral bridge and the carapace is reduced to peripherals four to six; and (2) the vertebrals are distinctly hexagonal. Although none of the SMP specimens preserves the region between the plastral bridge and the carapace, the presence of the second feature is confirmed based on the costal material. We note, too, that the right hypoplastron (SMP VP-2004) is identical to that of SEPCP 48/485 illustrated by Brinkman and Rodriguez de la Rosa (2006, Fig. 4B).

Sankey (2006) referred material from the Aguja Formation of Texas to the Kinosternoidea as cf. *Hoplochelys*, but this material does not conform to the SMP specimens and is not considered further.

All five SMP specimens were recovered at a single collecting site, called the "John Burris Microsite," located within the general locality SMP loc. 350 (east branch of Hunter Wash [west end]), in the De-na-zin Member, Kirtland Formation. The presence of a kinosternoid in the De-na-zin Member (Kirtland Formation) that is similar to the Cerro del Pueblo Formation kinosternoid, further supports the interpretation that the upper Kirtland Formation may be correlative to the lower part of the Cerro del Pueblo Formation, which thus may be partly of Kirtlandian age (Sullivan and Lucas 2006).

Adocidae Cope 1870
Adocus Cope 1868
Adocus bossi Gilmore 1919
(Fig. 20.10a, b)

Holotype: USNM 8613 (Fig. 20.10a, b), incomplete carapace and nearly complete plastron.

Holotype locality, unit, and age: Head of Ah-shi-sle-pah Wash, Sec. 3, T22N, R10W, San Juan County, New Mexico; Hunter Wash Member, Kirtland Formation; late Campanian, early Kirtlandian.

Referred specimen: USNM 8577, nearly complete plastron and incomplete carapace.

Occurrence: San Juan Basin, New Mexico; Hunter Wash Member, Kirtland Formation; late Campanian (early Kirtlandian).

Remarks: Gilmore (1919) named two species of *Adocus* from the Hunter Wash Member of the Kirtland Formation, *A. bossi* (Fig. 20.10a, b), and *A. kirtlandius* (Fig. 20.10c, d). Gilmore (1919) distinguished *A. bossi* from *A. kirtlandius* (below) in having: (1) coarse carapace surface texture (three "pits" within 5 mm); (2) seven neurals; (3) eighth costals with wide proximal ends; (4) vertebral scutes longer than wide; (5) long anal scutes; (6) rounded posterior lobe of the plastron; and (7) larger size. A few of these characters may be related to ontogeny, specifically the coarser surface texture, larger size, length of the anal scutes and possibly the shape of the eighth costals. The remaining characters still serve to distinguish the two species.

Adocus kirtlandius Gilmore 1919
(Fig. 20.10c, d)

Holotype: USNM 8593 (Fig. 20.10c, d), incomplete carapace and nearly complete plastron.

Holotype locality, unit, and age: South side of Ah-shi-sle-pah Wash, Sec. 4, T22N, R10W, San Juan County, New Mexico; Hunter Wash Member, Kirtland Formation; late Campanian (early Kirtlandian).

Occurrence: San Juan Basin, New Mexico; Hunter Wash Member, Kirtland Formation; late Campanian (early Kirtlandian).

Remarks: The holotype of *Adocus kirtlandius* (USNM 8593) consists of an incomplete shell and it is the smaller of the two *Adocus* holotypes from the Kirtland Formation. Gilmore (1919) distinguished this species from *A. bossi* in having: (1) finer surface texture (four or five "pits" within 5 mm); (2) eight neurals; (3) eighth costals with narrow proximal ends; (4) vertebral scutes wider than long; (5) anal scutes short; (6) posterior lobe of the plastron squared-off; and (7) smaller size.

Wiman (1933), in his paper concerning the turtles collected by C. H. Sternberg housed in the Uppsala collection, did not address, or document, specimens of *Adocus*.

Mateer (1981) also did not comment on *Adocus*, other than noting it was similar to *Basilemys* Hay 1902. There is an uncatalogued incomplete plastron in the Uppsala collection (labeled "carapace" and listed as Sternberg's No. 22) that has been identified as *Adocus*. However, the anterior portion of the plastron conforms to *Denazemys nodosa* in having divided gulars and extragulars. Thus, there are no specimens of *Adocus* among the material in the Uppsala collection. Armstrong-Ziegler (1980) cited MNA Pl.1646, isolated carapace and plastron fragments, as pertaining to *Adocus* sp., but further noting the carapace texture to have "three rows of pits in a 5-mm. line," thus putting the specimen within the textural definition of *A. bossi*. However, we have not seen this specimen, so we cannot confirm this observation.

Adocus sp.
(Fig. 20.11)

Referred specimens: See Appendix for complete list of referred specimens.

Remarks: Due to difficulties in discriminating between the two species of *Adocus*, all the incomplete and fragmentary material is assigned to *Adocus* sp. We note here that a single incomplete peripheral (SMP VP-1328) is noteworthy because it preserves the color pattern on the dorsal (external) side (Fig. 20.11). This pattern consists of a patchwork of dark polygons separated by narrower, more prominent and more linear, light-colored segments of the carapace. This pattern is too regular to be due to leaching and/or weathering. Color patterns on fossil turtle shells are rare and having been reported twice before, once in *Neurankylus* sp. from the Paleocene (Puercan) of New Mexico and once in *Chrysemys picta* (Schneider 1783) from the Miocene (Barstovian) of Nebraska (Holman and Sullivan 1981; Sullivan et al. 1988).

Nanhsiungchelyidae Yeh 1966
Basilemys Hay 1902
Basilemys nobilis Hay 1910

Holotype: USNM 6555, fragments of carapace and plastron (hypoplastron and xiphiplastron).

Holotype locality, unit, and age: Ojo Alamo, San Juan County, New Mexico; "Ojo Alamo beds," 50 feet above the lower conglomerate in Naashoibito Member, Ojo Alamo Formation; early Maastrichtian (late "Edmontonian").

Remarks: Hay (1910) established *Basilemys nobilis* based on carapace and plastron material, most notably parts of the hypoplastron and xiphiplastron, from the "Ojo Alamo" beds. The material upon which this species is based is not diagnostic and cannot be distinguished from other species of *Basilemys* (*B. variolosa* [Cope 1876], *B. praeclara* Hay 1910, and

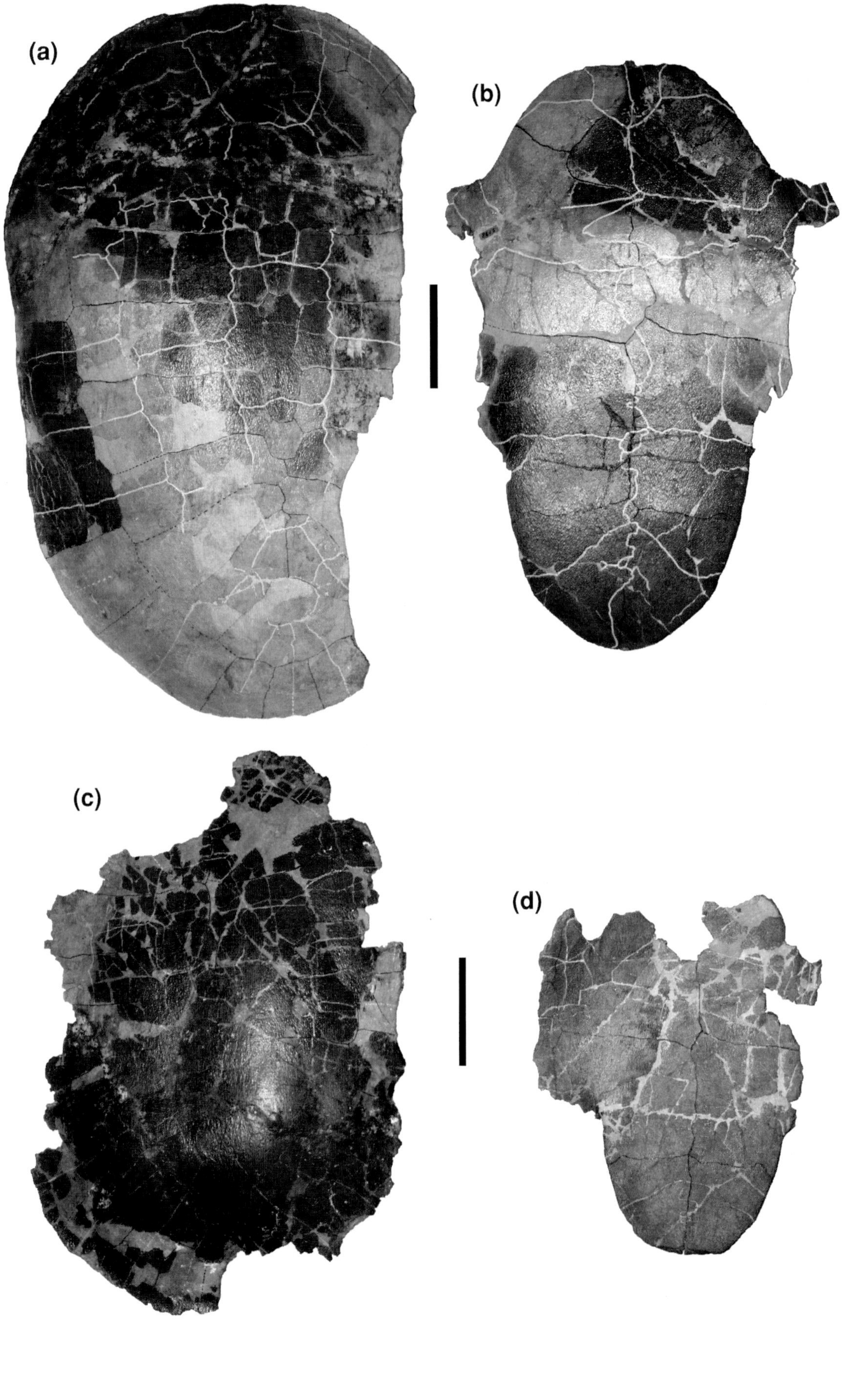
(a)
(b)
(c)
(d)

◀**Fig. 20.10** *Adocus* species from the Fruitland and Kirkland formations. **a**, **b** *Adocus bossi* Gilmore 1919; USNM 8613 (holotype), partial carapace and nearly complete plastron: **a** carapace, in dorsal view; **b** plastron, in ventral view. **c**, **d** *Adocus kirtlandicus* Gilmore 1919; USNM 8593 (holotype), partial carapace and nearly complete plastron: **c** carapace, in dorsal view; **d** plastron, in ventral view. Specimens at different magnifications; bar scales = 10 cm

Fig. 20.11 *Adocus* sp. SMP VP-1328, peripheral fragment with color pattern preserved, in dorsal view. Bar scale = 1 cm

B. sinuosa Riggs 1906). Moreover, the specimen (USNM 11084) previously, and provisionally, referred to *B. nobilis*, was noted by Gilmore (1935, p. 179) as being impossible to identify with certainty. We agree with Gilmore (1935) that the holotype is inadequate, so we regard *B. nobilis* as a nomen dubium.

Basilemys gaffneyi sp. nov.
(Fig. 20.12)

Synonymies: *Basilemys nobilis*: Hay 1910 [in part]; Wiman 1933, Fig. 7; Mateer 1981, p. 68, text-figs. 3, 4. *Basilelmys nobilis*: Hay 1910 [in part]; Gilmore 1935, p. 178, Figs. 13, 14.

Holotype: USNM 11084 (Fig. 20.12a, b), nearly complete carapace and plastron (see Remarks).

Holotype locality, **unit**, **and age**: Two miles above Hunter's Store (Bisti Post Office), San Juan County, New Mexico; Hunter Wash Member, Kirtland Formation; late Campanian (Kirtlandian).

Diagnosis: Differs from all other species of *Basilemys* in having an undivided gular.

Referred specimens: PMU.R29 (Fig. 20.12c, d), nearly complete carapace and plastron; SMP VP-3368 (not figured), incomplete plastron. See also Appendix .

Occurrence: San Juan Basin, New Mexico; upper part of Fruitland Formation and lower part of Kirtland Formation; late Campanian (Kirtlandian).

Remarks: The distinguishing features of "*Basilemys nobilis*" were exclusively derived from two referred specimens (PMU.R29 and USNM 11084), which we here consider to be *B. gaffneyi*. Hay (1910) distinguished "*B. nobilis*" as having an inner slope of the xiphiplaston that drops off rapidly compared to *B. praeclara* and *B. variolosa*, but we regard this difference as taxonomically insignificant, because this severity in slope is probably due to individual variation. Wiman (1933) described a nearly complete shell (PMU.R29a,b; Fig. 20.12c, d) collected by Charles H. Sternberg from the lower Kirtland Formation (Hunter Wash Member) at Ah-shi-sle-pah Wash (formerly Meyers [= Myers] Creek) and identified it as "*B. nobilis.*" Wiman (1933) mistakenly identified a pair of marginals as posterior inframarginals. Two years later, Gilmore (1935) described a nearly complete carapace and plastron, USNM 11084 (Fig. 20.12a, b), which is also assigned to "*Basilemys nobilis*". Later, Langston (1956) contrasted "*B. nobilis*" with *B. variolosa*, concluding it has a number of features (apomorphies) that set it apart from *B. variolosa* based on PMU.R29 (Fig. 20.12c, d). These include: (1) more oval shell; (2) large triangular inframarginal scale interposed basely between peripherals 6 and 7; and (3) "intergular" scale undivided.

Armstrong-Ziegler (1980) reported a putative ?*Basilemys* sp. from the Fruitland Formation based on shell fragments (MNA Pl.1647). Her description of the surface texture suggests that these fragments are not *Basilemys*, but instead may be referable to *Adocus*. However, we have not seen this material and are unable to confirm this.

Mateer (1981) noted that the Uppsala specimen (PMU.R29) bore little similarity to USNM 11084 (Fig. 20.12a, b), differing only in the absence of an anteriorly expanded first vertebral (= central). However, our examination of USNM 11084 challenges Mateer's contradictory statement, which belies the fact that the two specimens in question are nearly identical in all respects.

Brinkman and Nicholls (1993) recognized four species of *Basilemys* (*B. variolosa*, "*B. nobilis*," *B. praeclara* and *B. sinuosa*). Both *B. variolosa* and "*B. nobilis*" lack inframarginal scales, whereas both *B. praeclara* and *B. sinuosa* are united by v-shaped extragulars (meeting at the

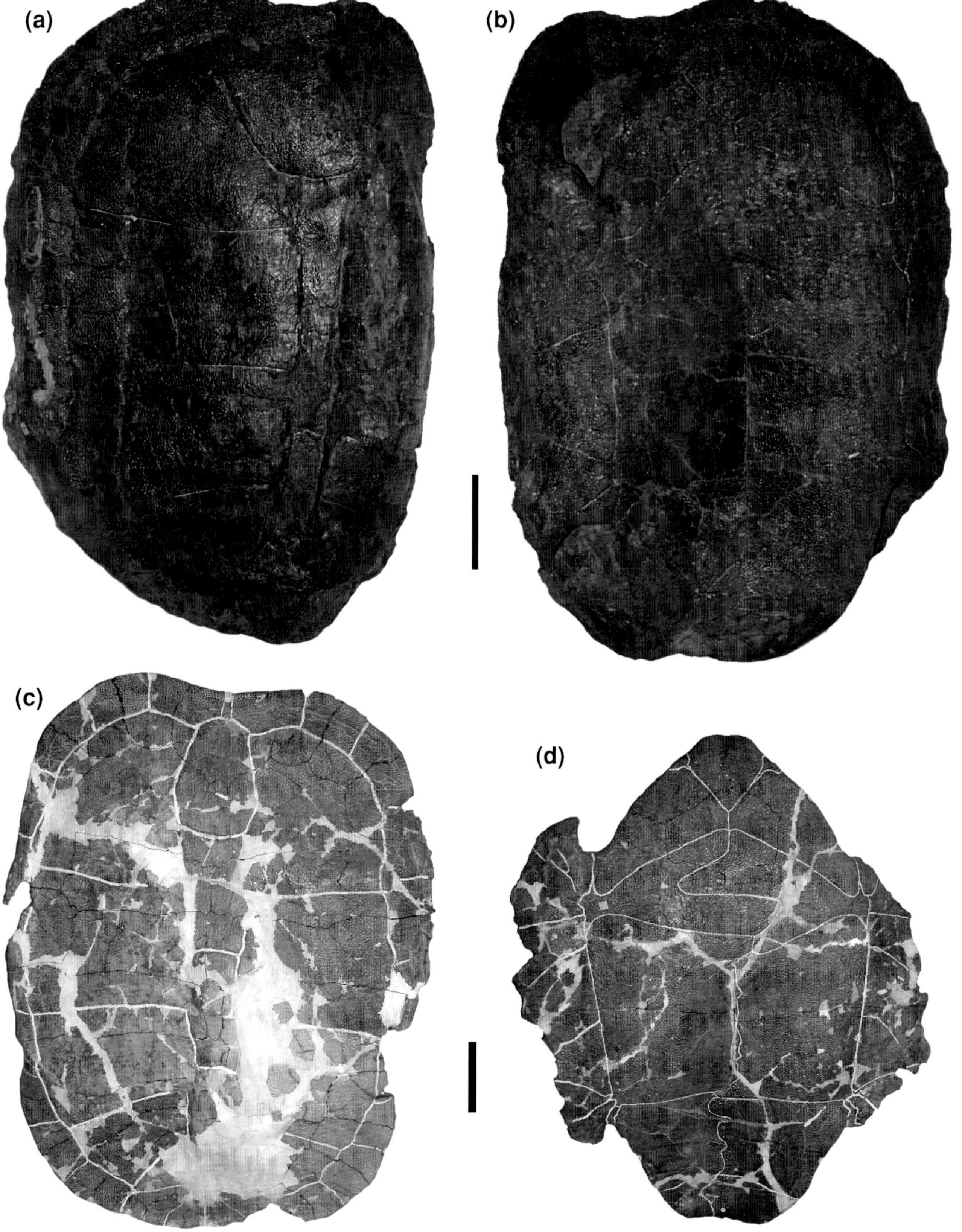

Fig. 20.12 *Basilemys gaffneyi* sp. nov. **a**, **b** USNM 11084 (holotype), nearly complete carapace and plastron: **a** carapace, in dorsal view; **b** plastron, in ventral view. **c**, **d** PMU.R29, incomplete carapace and plastron: **c** carapace, in dorsal view; **d** plastron, in ventral view. Specimens at different magnifications; bar scales = 10 cm

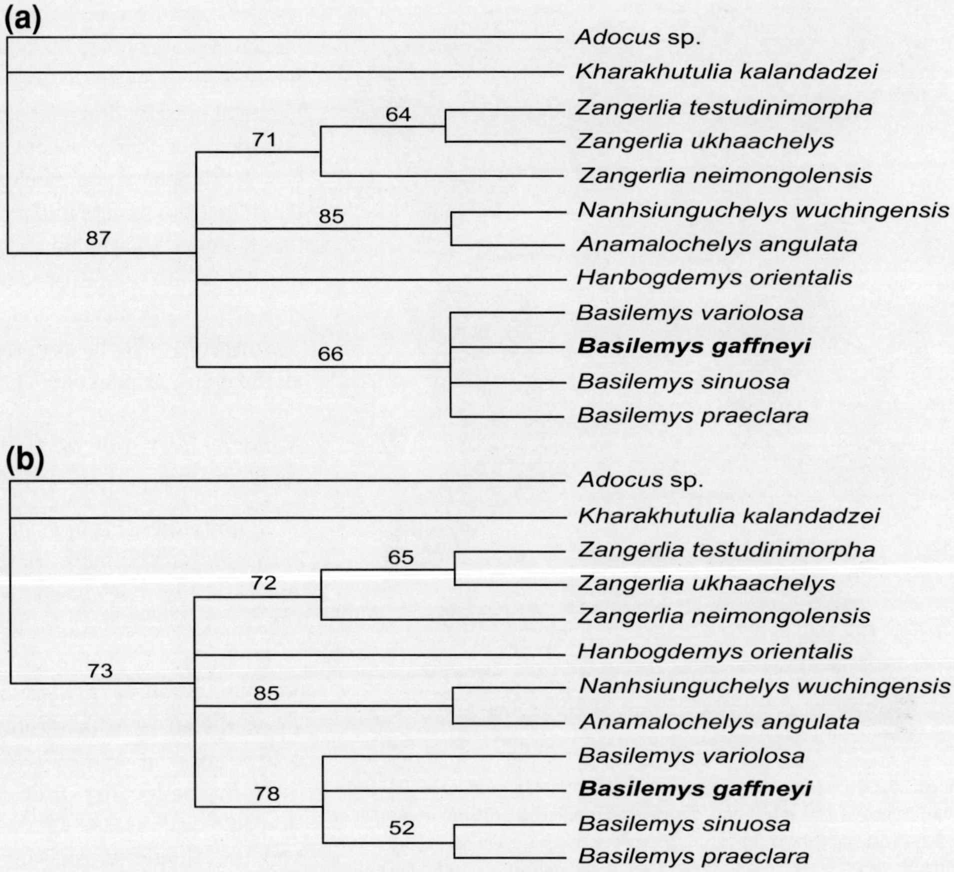

Fig. 20.13 Two cladograms for the Nanhsiungchelyidae, showing the phylogenetic relationship of *Basilemys gaffneyi* sp. nov. **a** Best tree (tree length = 53) excluding geography; **b** best tree (tree length = 55), including geography. Data matrix extracted from Joyce and Norell (2005) with two additional characters (see text)

midline); extragulars almost separating humerals; strongly sinuous course of the median sulcus; and pygal bone wider than long (Brinkman and Nicholls 1993).

The phylogenetic relationships of Mesozoic turtles have been recently reviewed by Joyce (2007). More specifically, there have been a number of studies concerning the taxa and phylogenetic relationships of taxa assigned to the Nanhsiungchelyidae. These include: Brinkman and Nicholls (1993), Brinkman and Peng (1996), Hirayama et al. (2001), Joyce and Norell (2005), Sukhanov and Narmandakh (1975), (1977), (2006), Sukhanov et al. (2008), and Yeh (1966).

We ran two phylogenetic analyses using PAUP 4.0b10 (Swofford 2002), incorporating the characters used by Joyce and Norell (2005) and the supplemental characters of Sukhanov et al. (2008). We scored the characters for *Basilemys gaffneyi*. We included characters 40 (sculpturing of the shell surface with relatively big and irregular pits and grooves) and 41 (overlapping of scales on the dorsal surface of the plastral lobes) from Sukhanov et al. (2008) and changed character 40 (geographic distribution) from Joyce and Norell (2005) to character 42. We amended character 37 (inframarginals) and character state (1) to be two or one pair. Characters were unordered and had equal weight. The two analyses produced the following: one without geography—yielding 18 most parsimonious trees, with a tree length = 53, consistency index = 0.8364, retention index = 0.7714; rescaled consistency index = 0.6434; and the other with geography—also yielding 18 most parsimonious trees, with a tree length = 55, consistency index = 0.8364, retention index = 0.7692. Bootstrap 50% majority rule consensus trees were generated for both runs (Fig. 20.13). The North American genus *Basilemys* forms a distinct clade, even in the analysis without coding for geography (Fig. 20.13a).

We are of the opinion that the differences between the holotype of *B. gaffneyi* (USNM 11084) and PMU.R29 are taxonomically insignificant. Both specimens are from the same geographic region and stratigraphic horizon (Hunter Wash Member, Kirtland Formation). We chose USNM 11084 as the holotype, because it is the more complete and better-preserved specimen.

Trionychidae (Fitzinger 1826)
Aspideretoides Gardner et al. 1995
Aspideretoides austerus (Hay 1908)
(Figs. 20.14, 20.15c)

Synonymies: *Aspideretes austerus*: Hay 1908, p. 495. *Aspideretes fontanus*: Hay 1908, p. 494. *Aspideretes vorax*: Hay 1908, p. 496.

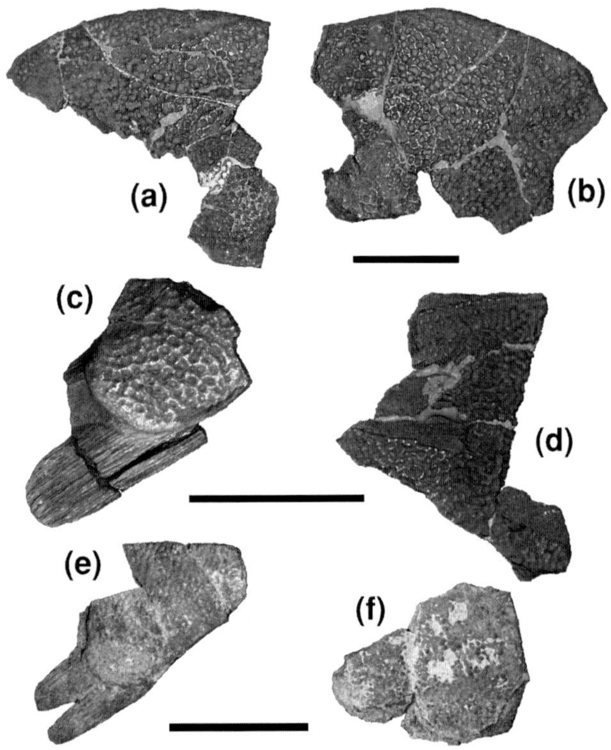

Fig. 20.14 *Aspideretoides austerus* (Hay 1908). **a–d**, AMNH 6068 (holotype): **a** and **b** left and right parts of anterior portion of carapace, each preserving lateral portion of nuchal and adjacent costals, both in dorsal view; **c** and **d** sections of right hypoplastron, both in ventral view. **e**, **f** SMP VP-1717: **e** lateral portion of right hypoplastron, in ventral view; **f** associated hypoplastron fragment, in ventral view. Specimens at different magnifications; bar scales = 5 cm

Holotype: AMNH 6068 (Fig. 20.14a–d), left and right parts of anterior portion of carapace, each preserving lateral portion of nuchal and adjacent costals, and right hypoplastron, consisting of two broken pieces.

Holotype locality, unit, and age: Ojo Alamo, San Juan County, New Mexico; De-na-zin Member, Kirtland Formation (late Campanian; late Kirtlandian).

Referred specimens: See Appendix for complete list of referred specimens.

Occurrence: San Juan Basin., New Mexico; De-na-zin Member, Kirtland Formation; late Campanian (Kirtlandian).

Remarks: Hay (1908) named three species of "*Aspideretes*"—*A. austerus*, A. *fontanus* and A. *vorax*; all here considered species of the genus *Aspideretoides* Gardner et al. 1995—from the "Ojo Alamo/Laramie beds" of the San Juan Basin, New Mexico. The holotype specimens, AMNH 6068 (*A. austerus*, Fig. 20.14a–d), AMNH 6070 (*A. fontanus*), and AMNH 6140 (*A. vorax*), are incomplete, represent different parts of the carapace and plastron, and have nearly identical sculpturing and shell thickness. Morphological differences are insignificant and are considered by us to be individual variation. We designate *Aspideretoides austerus* as the valid species, because it is the most diagnostic of the three species named. We recognize *A. fontanus* and *A. vorax* as subjective junior synonyms of *A. austerus*. All three specimens have identical preservation, and a bluish-gray color with dark maroon specks or clumps of hematite. Hematite adhering to bluish-gray fossil bone is common only in specimens that come from the shaly De-na-zin Member of the Kirtland Formation and not the overlying Naashoibito Member of the Ojo Alamo Formation. It is the same preservation seen in SMP VP-1717 (Fig. 20.14e, f), which was also collected in the De-na-zin Member of the Kirtland Formation. Therefore, we argue that all the holotype material is from the De-na-zin Member.

Aspideretoides robustus (Gilmore 1919) new combination (Figs. 20.15a, b, d, 20.16)

Synonymies: *Plastomenus* (in part): Cope 1873. *Plastomenus robustus*: Gilmore 1919, p. 53, Figs. 23 and 24, pls. 17 and 18.1. *Plastomenus* sp.: Mateer 1981, p. 70, text-fig. 3.6. "*Plastomenus*" *robustus*: Sullivan and Lucas 2006, p. 18.

Holotype: USNM 8538 (Figs. 20.15a, 20.16c, in part), complete carapace. (The right hyoplastron and hypoplastron are excluded from the holotype and, for reasons discussed below, are considered a referred specimen.)

Holotype locality, unit, and age: South side of Ah-shi-sle-pah Wash, Sec. 4, T22N, R10W, San Juan County, New Mexico; Hunter Wash Member, Kirtland Formation; late Campanian (early Kirtlandian).

Referred specimens: See Appendix for complete list of referred specimens.

Occurrence: San Juan Basin, New Mexico; lower part of Kirtland Formation (Hunter Wash Member) and, possibly, upper part of Fruitland Formation (Fossil Forest Member); late Campanin (Kirtlandian).

Revised diagnosis: Differs from *Aspideretoides splendidus* (Hay 1908) in the xiphipastron having a broader medial contact with the anterior portions arched forward, and a broad contact with the posterior margins of the hypoplastra. Differs from *Aspideretoides austerus* in having a broader hypoplastron; and the posterior margin of the hypoplastron deep, resulting in a more acute margin than in *A. splendidus* and *A. austerus*. Lateral and medial margins of the hypoplastron and hyoplastron directed anteroposteriorly and subparallel.

Remarks: Gilmore (1919) established the species *Plastomenus robustus* based on USNM 8538 (holotype), which according to Gilmore (1919) was collected by J. B. Reeside, Jr. from the southern part of section 2, T22N, R10W, in Ah-shi-sle-pah Wash, in the lower part of the Kirtland Formation (Hunter Wash Member). The fused nature of the plastral elements (hyoplaston + hypoplastron) was the reason Gilmore (1919) assigned USNM 8538 to the genus *Plastomenus*. Our examination of the carapace and the fused right hyoplastron + hypoplaston reveal that the

Fig. 20.15 *Aspideretoides* species. **a**, **b** *Aspideretoides robustus* (new combination), USNM 8538 (holotype), originally described as complete carapace and right hyoplastron + hypoplastron, but we regard only the carapace as the holotype (see text for discussion): **a** carapace, in dorsal view; **b** right hyoplastron + hypoplastron, in ventral view (we interpret these plastal plates as belonging to a different individual than the holotype carapace). **c** *Aspideretoides austerus*, PMU.R31, incomplete plastron consisting of right hypo- and xiphi-plastra and left hyo-, hypo-, and xiphi-plastra, in ventral view. **d** *Aspideretoides robustus* (new combination), PMU.R30, incomplete carapace, in dorsal view. Specimens at different magnifications; bar scales = 10 cm

plastral elements are much larger than would be expected if they had originated from the same individual as the carapace. Thus, we are of the opinion that Gilmore's (1919) holotype actually represents two individuals. Both the carapace and the hyo-hypoplastra have the same external surface sculpturing, so presumably they represent the same species. We have no evidence that they were collected from different localities, so we infer that the stratigraphic horizon and the geographic location are nearly the same for both components. We here restrict the holotype specimen to the carapace and designate the plastral elements (which retain the same institutional catalogue number as the carapace) as a referred specimen of the same species.

Another specimen (PMU.R31), consisting of the right hypoplastron and xiphiplastron, plus the left hyoplastron, hypoplastron and xiphiplastron, was described by Wiman

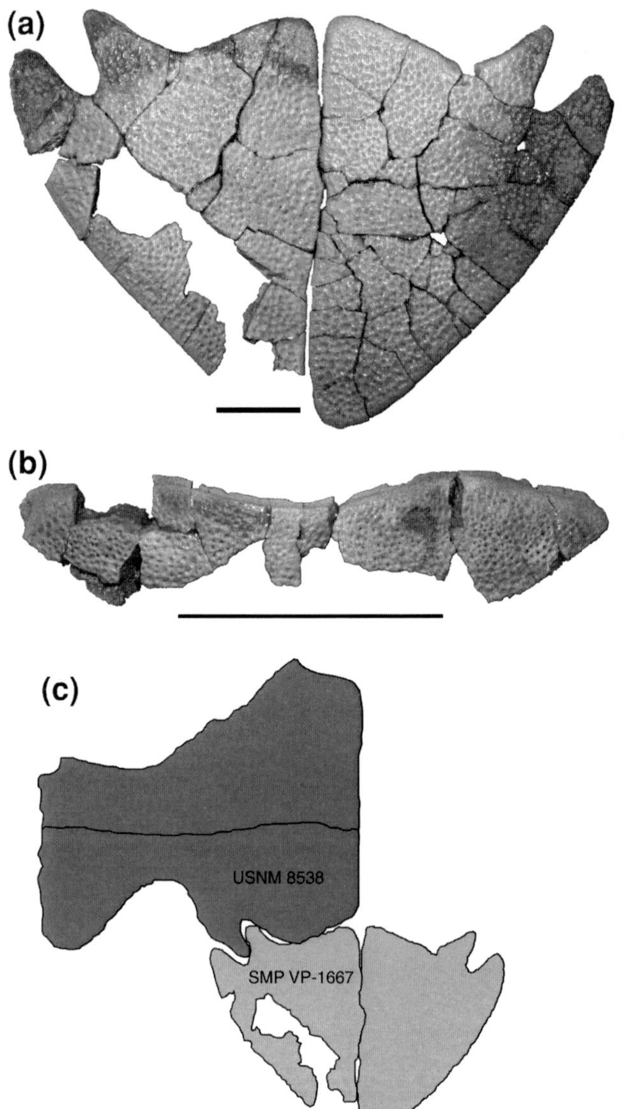

Fig. 20.16 *Aspideretoides robustus* (new combination) **a** SMP VP-1667, incomplete left and right xiphiplastra, in ventral view; **b** incomplete nuchal, in dorsal view; **c** *line* drawing of right hyoplastron + hypoplastron (part of USNM 8538 and originally designated as part of holotype) with right xiphiplastron (SMP VP-1667) showing conformity in the interlocking margins of the respective elements. Specimens at different magnifications; bar scales = 3 cm

(1933, p. 32) and assigned to *Plastomenus* sp. This specimen came from the "Ojo Alamo" beds, but based on its preservation it is almost certainly from the De-na-zin Member of the Kirtland Formation. PMU.R31 (Fig. 20.15c) differs from the hyoplastron + hypoplastron of USNM 8538 (Fig. 20.15b) in the following features: (1) the sutural contact uniting the hyoplastron and hypoplastron is directed anteriorly toward the midline; (2) the femoral notch is more constricted; and (3) the posterior border of the xiphiplastra is more rounded (compared to SMP VP-1667). Based on the morphological differences, and taking into account the different stratigraphic province of these two specimens, we considered PMU.R31 to be referable to *Aspideretoides austerus*. Consequently, *Aspideretoides robustus* is found with certainty only in the lower Kirtland Formation (Hunter Wash Member), whereas *A. austerus* is known only from the upper Kirtland Formation (De-na-zin Member). *Plastomenus* is no longer recognized in the Upper Cretaceous strata of New Mexico.

Confusion regarding the taxonomic status of *Plastomenus* has recently been reviewed by Joyce and Lyson (2010), who concluded that the type species of *Plastomenus* (*P. thomasii* [Cope 1872]) is a nomen dubium. However, in the interest of taxonomic stability, and following previous workers such as Gaffney (1979), they designated AMNH 6018 as the neotype of *Plastomenus thomasii* (Joyce and Lyson 2010). AMNH 6018 is from Grizzly Buttes, Wyoming, and is of Eocene (Bridgerian) age.

The xiphiplastron of AMNH 6018 (*Plastomenus thomasii*) differs from that of USNM 8538 (here referred to *Aspideretoides robustus*) in having a deep, narrow posterolateral margin; reduced lateral projection; and hyoplastron/hypoplastron border projecting anteriorly at the midline rather than nearly straight across (Hay 1908, Fig. 632). The sculpturing of *A. robustus* is similar to that of *A. austerus*, which differs from the sculpturing seen in specimens that are now identified as Plastomeninae (see account below for Plastomeninae Indet. and Appendix). We regard the two aforementioned species as distinct based on the characters of the hypoplastron and xiphiplastron described above, and we synonymize *Plastomenus robustus* with *Aspideretoides*, as the new combination *Aspideretoides robustus*.

Lastly, we note that Armstrong-Ziegler (1980) reported "*Trionyx* sp.", based on plastron and carapace fragments (MNA Pl. 1629), from the Fruitland Formation. Assuming she was correct in the identification of this material as trionychid, we infer this to be referable to *Aspideretoides* sp., and it is most likely referable to *A. robustus*. However, we have not seen the specimen, so we are unable to confirm this identification.

Gilmore (1935) named a new species "*Aspideretes*" *ovatus* and noted the following features: oval carapace with broadest portion anteriorly; posterior portion of carapace broadly pointed; carapace sculpturing "shallow, rounded pits, separated by ridges whose summits are flat topped," outer margins with parallel rows of pits, reduced seventh neural, preneural not evident, eighth pair of costals in contact on median line, seventh pair of costals separated by the seventh neural. Of those features, we regard only the following as possibly significant: (1) oval shape; (2) preneural apparently absent; and (3) seventh pair of costals separated by the seventh neural. However, among the 52 specimens from the Hunter Wash (Kirtland Formation) and Fossil Forest (Fruitland Formation) members, no other specimens of *A. ovatus* have been recognized. We note that "*A. ovatus*" is smaller then many of the coeval *A. robustus* specimens, suggesting that it may be a juvenile of the latter

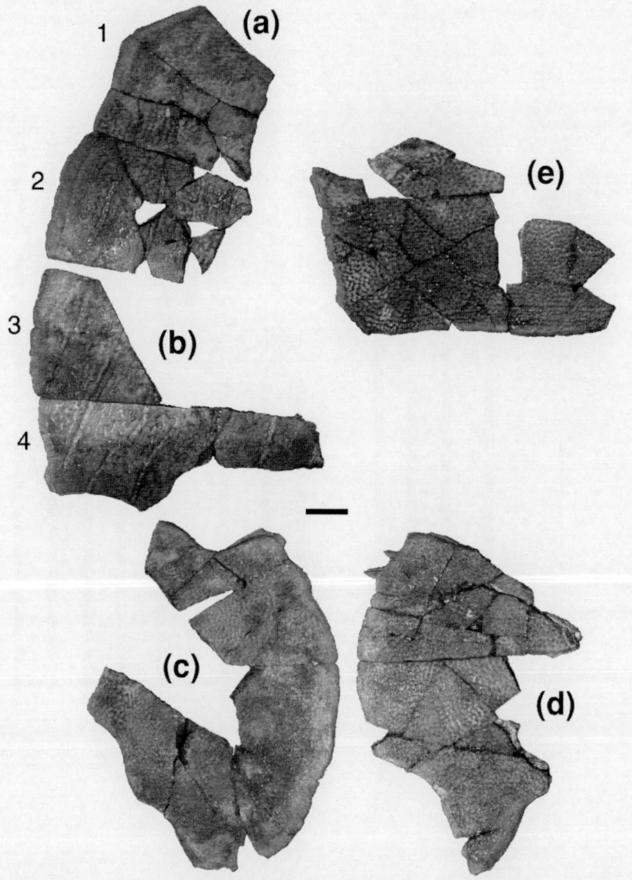

Fig. 20.17 Plastomeninae indeterminate; SMP VP-742, portions of carapace and plastron. **a** lateral portions of left costals *1* and *2*, in dorsal view; **b** lateral portions of left costals *3* and *4*, in dorsal view; **c** medial portions of right hyo- and hypo-plastra, in ventral view; **d** medial portions of left hyo- and hypo-plastra, in ventral view; **e** posterior margin of *left* side of carapace, in dorsal view. Bar scale = 1 cm

species. The apparent absence of the preneural may be the result of aberrant fusion of the first neural. Expansion of the posterior region (costals 6–8) laterally, as a consequence of ontogeny, would result in a more square-shape as seen in adult *A. robustus*. Concomitant with this lateral expansion posteriorly, the seventh pair of costals could contact posteriorly at the midline. Therefore, we do not believe *A. ovatus* is a valid species; instead, we suggest it is best regarded as a juvenile of *A. robustus*.

Aspideretoides sp.

Referred specimens: See Appendix for complete list of referred specimens.

Remarks: There are a number of trionychid specimens that are too fragmentary to be identified to species level. Presumably, based on stratigraphic parsimony, the majority of material from the upper Fruitland and the lower Kirtland formations is *Aspideretoides robustus*, whereas material from the upper Kirtland Formation is *A. austerus*, but this cannot be demonstrated. Many specimens from the Fruitland-Kirtland interval that have been previously assigned to *Plastomenus* are included here.

Plastomeninae indet.
(Fig. 20.17)

Referred specimens: See Appendix for complete list of referred specimens.

Occurrence: Within San Juan Basin, New Mexico, limited to the Hunter Wash Member of the Kirtland Formation; late Campanian (early Kirtlandian).

Remarks: SMP VP-742 (Fig. 20.17) consists of fragmentary sections of the carapace and plastron. The surface texture of both is very similar to that of *Compsemys*, consisting of fine tubercles that frequently fused together forming irregular strands. In places, they are enclosed, forming pits as in members of the Trionychidae. Elongated raised bands are prominently present on parts of the carapace sections. The lateral edges of SMP VP-742 lack peripherals, which is the reason for assigning it to the Trionychidae (Joyce et al. 2009). Characters cited by Joyce et al. (2009) that would allow assignment to the Plastomeninae are not readily recognizable. We note, however, that the sculpturing of the carapace is similar to some plastominine trionychids in having widely-spaced ridges that stretch obliquely across the carapace from the anterior midline region, posterolaterally.

Joyce et al. (2009, p. 322) reviewed the characterization of the clade Plastomeninae and concluded that an evenly rounded entoplastron was a better diagnostic character than the "higher degree of ossification of the plastron," a character that has been traditionally used for taxonomic discrimination in this group. However, they noted that the entoplastron is rarely preserved, making most specimens difficult to identify. Plastomenines are considered stem-cyclanorbines (Joyce et al. 2009; Joyce and Lyson 2010).

Biostratigraphy and Taxonomic Diversity

The rocks of the upper part of the Fruitland Formation (Fossil Forest Member) and the Kirtland Formation (Hunter Wash, Farmington and De-na-zin members) span approximately 2.0 my (Sullivan and Lucas 2003, 2006). The time represented by this depositional sequence is called the Kirtlandian land-vertebrate "age," which encompasses approximately 2.2 million years of late Campanian time (Sullivan and Lucas 2003, 2006). This interval coincides with the Bearpaw transgression and fills a gap between the classic Late Cretaceous Judithian and "Edmontonian" land-vertebrate ages. The fossil vertebrates, including turtles,

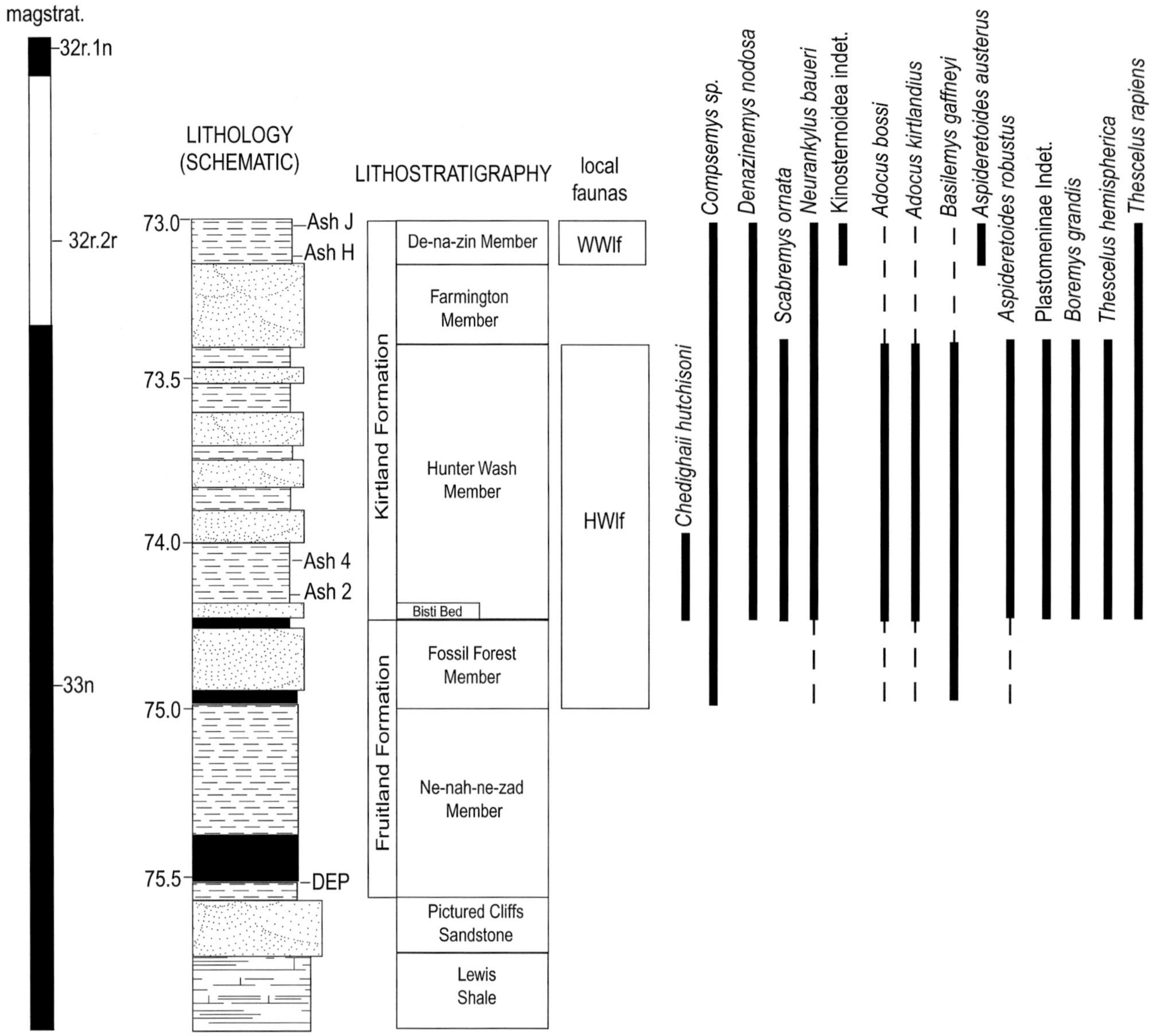

Fig. 20.18 Biostratigraphic distribution of turtles during the late Campanian through the Fruitland and Kirtland formations of the San Juan Basin, New Mexico, USA

which serve to characterize Kirtlandian time were revised by Sullivan and Lucas (2006). Here, we further refine what is known about the testudines from the upper part of the Fruitland and the Kirtland formations.

The turtles from the lower Kirtland Formation (Hunter Wash Member) are very similar to those reported from the Dinosaur Park Formation of Alberta (Brinkman 2005). Five genera are found in common: *Boremys*, *Adocus*, *Basilemys*, *Neurankylus*, and *Aspideretoides*. Both horizons yield an indeterminate plastomenine, but this record is questionable for the Dinosaur Park Formation (Brinkman 2005). The turtle taxa known from the Dinosaur Park Formation that are absent from the Kirtland Formation (both the Hunter Wash and De-na-zin members) include *Plesiobaena antiqua*, *Judithemys sukhanovi* Parham and Hutchison 2003, "*Apalone*" *latus* (Gilmore 1919), and Chelydridae gen. et sp. indet. Turtles that are present in the Fruitland and Kirtland formations, but absent from the Dinosaur Park Formation include: *Chedighaii hutchisoni* (restricted to the Fruitland Formation); *Scabremys ornata* (known with certainty only from the lower Kirtland Formation); *Compsemys* sp. (known throughout both the upper Fruitland and Kirtland formations); Kinosternoidea, gen. et sp. indet. (known only from the upper Kirtland Formation); *Thescelus hemispherica* (restricted to the lower Kirtland Formation); and *T. rapiens* (known throughout the entire Kirtland Formation). Differences in species among the genera shared by the Dinosaur Park Formation and the Fruitland/Kirtland formations are interpreted by us to reflect largely temporal, rather than any latitudinal, differences. A summary of the stratigraphic distribution of the Fruitland-Kirtland turtles is presented in Fig. 20.18.

Our assessment of the stratigraphic distribution of turtles in the Upper Cretaceous rocks of the San Juan Basin differs from that reported by McCord (1996), who largely addressed only the distribution of genera and families which, in large part, is redundant because of the monogeneric nature of many of the turtle families. Thus, McCord (1996) listed only Baenidae, "*Baena*" *nodosa-ornata*, *Compsemys*, *Trionyx* Geoffroy St. Hilaire 1809, and *Adocus* from the Fruitland and the Kirtland formations. He also listed cf. *Basilemys*, *Hoplochelys* and *Plastomenus* as present in the overlying Naashoibito Member of the Ojo Alamo Formation (included in the Kirtland Formation by McCord 1996). McCord (1996, p. 147) then concluded that *Compsemys* is "known only from the Maastrichtian (Lancian) and later sediments" to argue that the Fruitland-Kirtland turtles "suggest a latest Cretaceous age." However, *Compsemys* has records as old as late Turonian (Eaton et al. 1999) and, as noted above, had been reported previously by Armstrong-Zeigler (1978) from the San Juan Basin. Indeed, the Fruitland-Kirtland turtle assemblage well resembles other late Campanian turtle assemblages (see above) and is not indicative of a Maastrichtian (or Lancian) age.

In a recent paper, Gates et al. (2010) provided a list of vertebrate taxa (including turtles) for the undivided Fruitland and Kirtland formations without regard to a more precise (member-level) stratigraphy. A number of taxa listed are in error and their precision with respect to biostratigraphic distribution is also not entirely correct. Thus, Gates et al. (2010) listed as present: *Compsemys victa*, *Boremys pulchra*, *Adocus lineolatus* Cope 1874, and *Basilemys variolosa*. None of these species are present in either the Fruitland or Kirtland formations. The taxon "*Denazinemys*" *ornata* (now *Scabremys ornata*) is known only from the lower Kirtland Formation, not the Fruitland and upper Kirtland formations, as listed by Gates et al. (2010). The taxon "*Plastomenus*" *robustus* (now *Aspideretoides robustus*) is known only from the Fruitland and the lower Kirtland formations, and not the upper Kirtland Formation as indicated by them (Gates et al. 2010). The genus *Thescelus* is clearly present, represented by the species *T. hemispherica*, from the Hunter Wash Member (Kirtland Formation) and *T. rapiens*, from both the Hunter Wash and De-na-zin members of the Kirtland Formation, contrary to its absence reported by Gates et al. (2010). *Neurankylus eximius* is not known from the Fruitland Formation (Sullivan and Lucas 2006) and its identification in this unit is probably due to the synonymy of Gaffney (1972). The "kinosternid n. sp." listed by Gates et al. (2010) for the Kaiparowits and the Fruitland/Kirtland ("KF" of Gates et al. 2010, Appendix 1) are not the same taxon (Hutchison 2010, pers. comm.). These misidentifications and lack of tighter stratigraphic precision based on turtle taxa, call into question the reliability of the rest of the fauna list complied by Gates et al. (2010, Appendix 1).

The biostratigraphic distribution of turtles from the Fruitland and Kirtland formations is relatively useful. Many turtle specimens formerly considered from the Fruitland Formation are now thought to be from the lower Kirtland Formation based on the revised stratigraphy of Brown (1983). Even so, the upper Fruitland (Fossil Forest Member) and lower Kirtland (Hunter Wash Member) formations' vertebrates make up a single local fauna (Hunter Wash local fauna) (Sullivan and Lucas 2006). The vertebrates from the upper Kirtland Formation (De-na-zin Member) are part of the Willow Wash local fauna, and have been shown to be different from those found stratigraphically lower (Sullivan and Lucas 2006). Indeed, turtles from the Willow Wash local fauna differ from those of the Hunter Wash local fauna (Fig. 20.18).

The baenids *Boremys grandis*, *Denazinemys nodosa*, *Neurankylus baueri*, *Scabremys ornata*, *Thescelus hemispherica*, and *T. rapiens* are all present in the Hunter Wash local fauna (HWlf). The bothremydid *Chedighaii hutchisoni*, the adocids *Adocus bossi* and *A. kirtlandius*, the nanhsiungchelyid *Basilemys gaffneyi*, the trionychids *Aspideretoides robustus* and an indeterminate plastomenine are also present in the HWlf. *Compsemys* is known throughout the entire Kirtlandian interval. Previously, *Trionyx* and *Plastomenus* had been reported from the Fruitland and Kirtland formations (Armstrong-Ziegler 1978; Gilmore 1919; McCord 1996), but these taxa most certainly are either *Aspideretoides* or an indeterminate trionychid.

The baenids *Denazinemys nodosa*, *Neurankylus eximius*, and *Thescelus rapiens* are present in the Willow Wash local fauna (WWlf). An indeterminate kinosternoid, probably representing a new genus and species, and similar or identical to the one reported from the Cerro del Pueblo Formation, Mexico, is also present in the WWlf. Only one trionychid, *Aspideretoides austerus*, is present. Turtle genera representing the adocid *Adocus* and the nanhsiungchelyid *Basilemys* are also present, but are presently not identifiable to species level. A summary of turtle distribution for the Hunter Wash and Willow Wash local faunas is presented in Fig. 20.18.

On the face of it, the Hunter Wash local fauna has a higher diversity of turtle taxa, whereas the Willow Wash local fauna has considerably fewer taxa (Fig. 20.18). This apparent decrease in diversity could be attributed to a sampling of different paleoenvironments. The stratum (De-na-zin Member) that contains the Willow Wash local fauna is fossiliferous and has a more prominent terrestrial component represented mostly by dinosaurs. In contrast, the Hunter Wash local fauna, arguably, has a stronger aquatic component. This may be attributed to the fact that fishes and turtles are more common in the HWlf. Furthermore, the HWlf is derived from a much thicker stratigraphic section that is much more widely exposed than the rocks that yield the WWlf—this also may account for the greater diversity of turtles in the HWlf.

Non-marine Late Cretaceous turtle distribution and paleobiogeography have been discussed by Hirayama et al. (2000), Holroyd and Hutchison (2002), and Brinkman (2005). Attempts at assessing the relative abundance of taxa within geographically-limited regions have been discussed in detail by Holroyd and Hutchison (2002). Of particular interest is the idea that different temporal time spans produce different abundances or numbers of taxa. The upper Fruitland and Kirtland formations represent an ideal stratigraphic sequence to study temporal differences. These two formations have been interpreted to represent a relatively continuous depositional sequence, beginning with delta plain deposits (Fruitland-lower Kirtland) overlain by more inland river and floodplain deposits that prograde over the coastal deltas and swamps as the shoreline prograded to the northeast (Fassett and Hinds 1971; Lucas 1981; Lucas and Mateer 1983). This took place during approximately 2 million years of late Campanian time.

With the epicontinental seaway retreating to the east/ northeast, there was a decrease in the near-shore aquatic realm concomitant with an increase in the terrestrial realm. The shift in the near-shore environment eastward no doubt resulted in the reduction of large channels, together with a corresponding increase in terrestrial expanses between riparian regions, a factor noted by Brinkman (2003) in the decrease in taxonomic diversity of turtles during the Late Cretaceous of Alberta. Although terrestrial taxa, such as *Basilemys*, might be expected to have been more common in a more terrestrial-dominated sequence, their relative rarity may reflect dwindling numbers of individuals locally, although Bryant (1989) thought that their rarity was attributed to the presence of habitats away from streams. We note that specimens of the Plastomeninae are rare (as is the case in the Dinosaur Park Formation) and that the genus *Aspideretoides* is common and present throughout the Fruitland and Kirtland formations (Brinkman 2005). Brinkman (2005) noted that *Aspideretoides splendidus* was restricted to the upper part of the Dinosaur Park Formation, and not found in the lower part of the formation or in the Oldman Formation below that. He inferred that *A. splendidus* preferred a more coastal environment (Brinkman 2005). Similar preferences for a coastal environment may be recognized for the stratigraphically younger *A. robustus*, known only from the lower Kirtland Formation (Hunter Wash Member). With respect to the other turtle taxa, there is clearly an apparent decrease in the number taxa from the lower Kirtland, to the upper Kirtland, which may be environment related.

Brinkman (2005) noted the role of latitudinal differences in affecting the geographic distribution and stratigraphic presence of turtles as well as factors such as a low tolerance to salinity and climate. However, we are struck by the generic similarity between the higher latitudinal Dinosaur Park Formation and the lower latitudinal Fruitland/Kirtland formations. Differences among species of the shared genera are more likely to be temporal rather than geographic, assuming similar temperatures and salinity.

In summary, turtle species diversity is greater for the Hunter Wash local fauna than it is for the Willow Wash local fauna. While many of the same genera are found in both local faunas, there are fewer genera known from the WWlf (see Fig. 20.18). This most likely reflects an environmental shift from a delta plain with numerous streams, channels and swamps, to that of a more upland landscape. This decreased turtle diversity during the Late Cretaceous in the San Juan Basin is consistent with patterns of decreasing turtle diversity elsewhere in North America during this time interval (Brinkman 2003).

Acknowledgments This study was made possible by the assistance and help of a number of individuals. For access to collections we thank Eugene S. Gaffney, Mark Norell, and Carl Mehling, (all American Museum of Natural History); John Bolt and William Simpson (both Field Museum of Natural History); Larry Martin and Desui Miao (both University of Kansas); Solwieg Stuenes (Paleontologiska Museet, University of Uppsala); and Michael Brett Surman (United States National Museum, Smithsonian). For discussions and verification of specimen identities we thank Don Brinkman (Royal Tyrrell Museum of Palaeontology); Walter Joyce (University of Tübingen), and Tyler Lyson (Yale University). For help in the field we thank John Burris (San Juan Community College), Mike Burns (University of Alberta), Denver Fowler (Montana State University), Kesler Randall (San Diego Natural History Museum), and Justin A. Spielmann (New Mexico Museum of Natural History and Science). Additional field assistance provided by independent individuals include Arjan Boere, James Hartley, James Nikas, Jim Murphy and Fred Widmann. Kevin Seymour (Royal Ontario Museum) kindly provided the information for ROM 864, and we thank him for these data. Donald Brinkman, Derek W. Larson (University of Toronto), and J. Howard Hutchison reviewed the manuscript, and we thank them for their valuable comments and suggestions. Finally, we thank Pat Hester and Sherrie Landon (Bureau of Land Management, Albuquerque and Farmington offices) for providing the necessary Paleontological Resource Permits for collecting specimens housed in the collections of the NMMNH and the SMP, and for their continuing support of our fieldwork.

Appendix

List of all known previously referred, and newly referred, turtle specimens from the Fruitland and Kirtland formations, as of 2010. Order is alphabetic by (1) family/higher taxonomic category, (2) genus, (3) species, and (4) institution.

Current number	Previous number	Family	Genus	Species	Type status	Locality Number	Formation	Member	Description
USNM 8577		Adocidae	*Adocus*	*bossi*	paratype		Kirtland	Hunter Wash	partial carapace & plastron
USNM 8613		Adocidae	*Adocus*	*bossi*	holotype		Kirtland	Hunter Wash	partial carapace & nearly complete plastron
USNM 11082		Adocidae	*Adocus*	*bossi*			Kirtland	De-na-zin	carapace & plastron
USNM 11326		Adocidae	*Adocus*	*bossi*			Kirtland	Hunter Wash	incomplete carapace & plastron
USNM 12838		Adocidae	*Adocus*	*bossi*			Kirtland	Hunter Wash	carapace & plastron
USNM 12842		Adocidae	*Adocus*	*bossi*			Kirtland	Hunter Wash	carapace & plastron
USNM 12982		Adocidae	*Adocus*	*bossi*			Kirtland	Hunter Wash	carapace & plastron
USNM 12983		Adocidae	*Adocus*	*bossi*			Kirtland	Hunter Wash	carapace, plastron, & pelvis
USNM 8593		Adocidae	*Adocus*	*kirtlandius*	holotype		Kirtland	Hunter Wash	incomplete carapace & plastron
SMP VP-1725		Adocidae	*Adocus*	sp		313a	Kirtland	De-na-zin	carapace fragments
SMP VP-1280		Adocidae	*Adocus*	sp.		313a	Kirtland	De-na-zin	carapace & plastron fragments
SMP VP-1295		Adocidae	*Adocus*	sp.		350	Kirtland	De-na-zin	carapace & plastron fragments
SMP VP-1328		Adocidae	*Adocus*	sp.		386	Kirtland	Hunter Wash	peripheral carapace fragment
SMP VP-1330		Adocidae	*Adocus*	sp.		385	Kirtland	Hunter Wash	part of right carapace with peripherals & fragments
SMP VP-1331		Adocidae	*Adocus*	sp.		385	Kirtland	Hunter Wash	carapace/bridge fragment
SMP VP-1346		Adocidae	*Adocus*	sp.		373	Kirtland	Hunter Wash	carapace fragment with part of bridge
SMP VP-1350		Adocidae	*Adocus*	sp.		373	Kirtland	Hunter Wash	nuchal & anterior carapace fragments
SMP VP-1417		Adocidae	*Adocus*	sp.		388a	Kirtland	De-na-zin	carapace fragments
SMP VP-1482		Adocidae	*Adocus*	sp.		281	Kirtland	Hunter Wash	plastron & other fragments
SMP VP-1502		Adocidae	*Adocus*	sp.		283	Kirtland	Hunter Wash	carapace fragments
SMP VP-1504		Adocidae	*Adocus*	sp.		282	Kirtland	Hunter Wash	costals
SMP VP-1523		Adocidae	*Adocus*	sp.		281	Kirtland	Hunter Wash	peripheral of carapace
SMP VP-1544		Adocidae	*Adocus*	sp.		372	Kirtland	Hunter Wash	carapace fragment
SMP VP-1660		Adocidae	*Adocus*	sp.		283	Kirtland	Hunter Wash	carapace & plastron fragments
SMP VP-2032		Adocidae	*Adocus*	sp.		421	Kirtland	Hunter Wash	part of carapace
SMP VP-2033		Adocidae	*Adocus*	sp.		421	Kirtland	Hunter Wash	carapace fragments

(continued)

(continued)

Current number	Previous number	Family	Genus	Species	Type status	Locality Number	Formation	Member	Description
SMP VP-2361		Adocidae	*Adocus*	sp.		451	Kirtland	Hunter Wash	carapace fragments & incomplete limb bone
SMP VP-2438		Adocidae	*Adocus*	sp.		435	Kirtland	Hunter Wash	carapace fragment
SMP VP-2439		Adocidae	*Adocus*	sp.		435	Kirtland	Hunter Wash	neural & other carapace fragments
SMP VP-2602		Adocidae	*Adocus*	sp.		319a	Kirtland	De-na-zin	two carapace fragments
SMP VP-2616		Adocidae	*Adocus*	sp.		319a	Kirtland	De-na-zin	carapace fragment
SMP VP-2818		Adocidae	*Adocus*	sp.		228	Kirtland	Hunter Wash	carapace neural
SMP VP-3264		Adocidae	*Adocus*	sp.		228	Kirtland	Hunter Wash	carapace fragments
SMP VP-3275		Adocidae	*Adocus*	sp.		463	Fruitland	Fossil Forest	carapace fragments
NMMNH P-22621		Adocidae	*Adocus*	sp.		1636	Kirtland	Hunter Wash	incomplete nuchal
NMMNH P-22700		Adocidae	*Adocus*	sp.		1623	Kirtland/ Fruitland	unknown	6 carapace fragments
NMMNH P-22709		Adocidae	*Adocus*	sp.		1649	Kirtland	Hunter Wash	2 carapace fragments
NMMNH P-22747		Adocidae	*Adocus*	sp.		1674	Fruitland	Fossil Forest	carapace/plastron fragments
NMMNH P-26225		Adocidae	*Adocus*	sp.		3227	Kirtland	De-na-zin	incomplete neural
NMMNH P-26258		Adocidae	*Adocus*	sp.		3243	Kirtland	De-na-zin	carapace fragment
NMMNH P-26262		Adocidae	*Adocus*	sp.		3245	Kirtland	De-na-zin	incomplete costal
NMMNH P-27427		Adocidae	*Adocus*	sp.		1708	Kirtland	Hunter Wash	numerous shell fragments
NMMNH P-27463		Adocidae	*Adocus*	sp.		3875	Kirtland	Hunter Wash	23 carapace/plastron fragments
NMMNH P-27601		Adocidae	*Adocus*	sp.		4015	Fruitland	Fossil Forest	carapace/plastron fragments
NMMNH P-27629		Adocidae	*Adocus*	sp.		4015	Fruitland	Fossil Forest	carapace fragments
NMMNH P-27760		Adocidae	*Adocus*	sp.		4015	Fruitland	Fossil Forest	carapace/plastron fragments
NMMNH P-27774		Adocidae	*Adocus*	sp.		4015	Fruitland	Fossil Forest	carapace fragments & neural
NMMNH P-27926		Adocidae	*Adocus*	sp.		1688	Kirtland	Hunter Wash	shell fragments
NMMNH P-30759		Adocidae	*Adocus*	sp.		3117	Kirtland	Hunter Wash	miscellaneous shell fragments
NMMNH P-32871		Adocidae	*Adocus*	sp.		2387	Fruitland	Fossil Forest	6 shell fragments
NMMNH P-32964		Adocidae	*Adocus*	sp.		2387	Fruitland	Fossil Forest	13 fragments
NMMNH P-38403		Adocidae	*Adocus*	sp.		4010	Fruitland	Fossil Forest	entoplaston
NMMNH P-41022		Adocidae	*Adocus*	sp.		5400	Kirtland	De-na-zin	carapace fragments
NMMNH P-41023		Adocidae	*Adocus*	sp.		5400	Kirtland	De-na-zin	15 carapace/plastron fragments

(continued)

(continued)

Current number	Previous number	Family	Genus	Species	Type status	Locality Number	Formation	Member	Description
NMMNH P-49867	UA 5572	Adocidae	*Adocus*	sp.		6203	Fruitland?	Fossil Forest?	nearly complete plastron
NMMNH P-22537		Adocidae	*Adocus*	sp.		1582	Kirtland	Hunter Wash	shell fragment
NMMNH P-22543		Adocidae	*Adocus*	sp.		1584	Kirtland	Hunter Wash	shell fragment
NMMNH P-22652		Adocidae	*Adocus*	sp.		1672	Fruitland	Fossil Forest	3 shell fragments
NMMNH P-22653		Adocidae	*Adocus*	sp.		1672	Fruitland	Fossil Forest	shell fragment
NMMNH P-22733		Adocidae	*Adocus*	sp.		1668	Kirtland	Hunter Wash	shell fragments
NMMNH P-22740		Adocidae	*Adocus*	sp.		1674	Fruitland	Fossil Forest	carapace fragments
NMMNH P-22750		Adocidae	*Adocus*	sp.		1674	Fruitland	Fossil Forest	incomplete costals
NMMNH P-22822		Adocidae	*Adocus*	sp.		1706	Fruitland	Fossil Forest	2 shell fragments
NMMNH P-27883		Adocidae	*Adocus*	sp.		3227	Kirtland	De-na-zin	costal, plastron, peripheral, & other fragments
NMMNH P-28327		Adocidae	*Adocus*	sp.		1882	Fruitland	Fossil Forest	2 carapace fragments
NMMNH P-30069		Adocidae	*Adocus*	sp.		4267	Fruitland	Fossil Forest	numerous shell fragments
NMMNH P-30361		Adocidae	*Adocus*	sp.		3490	Kirtland	Hunter Wash	numerous carapace/plastron fragments
NMMNH P-32850		Adocidae	*Adocus*	sp.		4544	Fruitland	Fossil Forest	14 costal fragments
NMMNH P-49724	UA 13483	Adocidae	*Adocus*	sp.		6333	Kirtland?	Hunter Wash?	shell fragment
NMMNH P-49725	UA 14413	Adocidae	*Adocus*	sp.		6333	Kirtland?	Hunter Wash?	4 shell fragments
NMMNH P-49826	UA 14404	Adocidae	*Adocus*	sp.		6559	Kirtland?	Hunter Wash?	2 shell fragments
NMMNH P-49831	UA 14411	Adocidae	*Adocus*	sp.		6622	Kirtland?	De-na-zin?	shell fragment
NMMNH P-49849	UA 14402	Adocidae	*Adocus*	sp.		6547	Kirtland?	Hunter Wash?	2 shell fragments
NMMNH P-49851	UA 14405	Adocidae	*Adocus*	sp.		6551	Kirtland?	De-na-zin?	carapace fragment
NMMNH P-49854	UA 14401	Adocidae	*Adocus*	sp.		6552	Kirtland?	De-na-zin?	shell fragment
NMMNH P-49859	UA 14403	Adocidae	*Adocus*	sp.		6558	Kirtland?	Hunter Wash?	2 shell fragments
NMMNH P-49866	UA 5575	Adocidae	*Adocus*	sp.		6203	Fruitland?	Fossil Forest?	plastron & carapace fragments
NMMNH P-49895	UA 14406	Adocidae	*Adocus*	sp.		6542	Kirtland?	Hunter Wash?	shell fragment
NMMNH P-49926	UA 8559	Adocidae	*Adocus*	sp.		6272	Kirtland?	Hunter Wash?	shell fragment

(continued)

(continued)

Current number	Previous number	Family	Genus	Species	Type status	Locality Number	Formation	Member	Description
NMMNH P-49927	UA 8560	Adocidae	*Adocus*	sp.		6272	Kirtland?	Hunter Wash?	shell fragment
NMMNH P-50625	UA 14407	Adocidae	*Adocus*	sp.		6558	Kirtland?	Hunter Wash?	shell fragment
NMMNH P-49846	UA 14426	Adocidae	*Adocus*	sp.		6544	Kirtland?	Hunter Wash?	2 shell fragment
USNM 8654		Adocidae	*Adocus*	sp.			Kirtland	Hunter Wash	partial plastron
SMP VP-1098		Adocidae	cf. *Adocus*	sp.		361	Kirtland	De-na-zin	carapace & plastron fragments
SMP VP-1294		Adocidae	cf. *Adocus*	sp.		372	Kirtland	Hunter Wash	carapace & plastron fragments from multiple individuals
SMP VP-1477		Adocidae	cf. *Adocus*	sp.		365	Kirtland	Hunter Wash	carapace fragment
SMP VP-1666		Adocidae	cf. *Adocus*	sp.		398	Kirtland	Hunter Wash	carapace & plastron fragments
SMP VP-1843		Adocidae	cf. *Adocus*	sp.		319a	Kirtland	De-na-zin	plastron fragment
SMP VP-2096		Adocidae	cf. *Adocus*	sp.		389a	Kirtland	De-na-zin	incomplete plastron
SMP VP-2382		Adocidae	cf. *Adocus*	sp.		382	Kirtland	De-na-zin	carapace fragment
SMP VP-2578		Adocidae	cf. *Adocus*	sp.		450	Fruitland	Fossil Forest	neural
SMP VP-2841		Adocidae	cf. *Adocus*	sp.		450	Fruitland	Fossil Forest	carapace fragment & associated indeterminate bone material
SMP VP-2856		Adocidae	cf. *Adocus*	sp.		450	Fruitland	Fossil Forest	incomplete costal & other carapace fragments
NMMNH P-28322		Adocidae	cf. *Adocus*	sp.		1882	Fruitland	Fossil Forest	numerous carapace fragments
NMMNH P-44902		Adocidae	cf. *Adocus*	sp.		3517	Kirtland	Hunter Wash	carapace fragments
NMMNH P-22756		Adocidae	?*Adocus*	sp.		1678	Fruitland	Fossil Forest	various shell fragments
NMMNH P-27781		Adocidae	?*Adocus*	sp.		4015	Fruitland	Fossil Forest	numerous carapace/plastron fragments
SMP VP-1444		Adocidae	?*Adocus*	sp.		281	Kirtland	Hunter Wash	fragments of carapace
SMP VP-1956		Adocidae	?*Adocus*	sp.		382	Kirtland	De-na-zin	neural
SMP VP-2018		Adocidae	?*Adocus*	sp.		420a	Kirtland	De-na-zin	plastron fragment
SMP VP-2491		Adocidae	?*Adocus*	sp.		389a	Kirtland	De-na-zin	carapace fragment
SMP VP-2802		Adocidae	?*Adocus*	sp.		450	Fruitland	Fossil Forest	costal, marginal, & other carapace fragments

(continued)

(continued)

Current number	Previous number	Family	Genus	Species	Type status	Locality Number	Formation	Member	Description
USNM 12979		Baenidae	*Boremys*	*grandis*	holotype		Kirtland	De-na-zin	carapace & plastron
USNM 12978		Baenidae	*Boremys*	sp.			Kirtland	Hunter Wash?	partial carapace & plastron
SMP VP-1565		Baenidae	cf. *Boremys*	sp.		391	Kirtland	Hunter Wash	plastron fragment
FMNH PR-2250		Baenidae	*Denazinemys*	*nodosa*			Kirtland	Hunter Wash	nearly complete carapace & plastron
NMMNH P-22564		Baenidae	*Denazinemys*	*nodosa*		1602	Kirtland	Farmington	2 carapace fragments
NMMNH P-22716		Baenidae	*Denazinemys*	*nodosa*		1654	Kirtland	Hunter Wash	2 shell fragments
NMMNH P-22790		Baenidae	*Denazinemys*	*nodosa*		1696	Fruitland	Fossil Forest	numerous shell fragments
NMMNH P-22881		Baenidae	*Denazinemys*	*nodosa*		1755	Kirtland	Hunter Wash	numerous carapace & bone fragments
NMMNH P-25075		Baenidae	*Denazinemys*	*nodosa*		1674	Fruitland	Fossil Forest	nearly complete
NMMNH P-25087		Baenidae	*Denazinemys*	*nodosa*		3117	Kirtland	Hunter Wash	complete carapace & plastron
NMMNH P-27732		Baenidae	*Denazinemys*	*nodosa*		4015	Fruitland	Fossil Forest	carapace fragments
NMMNH P-27830		Baenidae	*Denazinemys*	*nodosa*		4015	Fruitland	Fossil Forest	partial carapace & plastron
NMMNH P-32890		Baenidae	*Denazinemys*	*nodosa*		4010	Kirtland	Hunter Wash	incomplete carapace & fragments
NMMNH P-37763		Baenidae	*Denazinemys*	*nodosa*		5163	Kirtland	Hunter Wash	numerous carapace & plastron fragments
NMMNH P-39633		Baenidae	*Denazinemys*	*nodosa*		4275	Kirtland	Hunter Wash	numerous carapace & plastron fragments
NMMNH P-49858	UA 14394	Baenidae	*Denazinemys*	*nodosa*		6558	Kirtland?	Hunter Wash?	shell fragment
NMMNH P-49896	UA 8826	Baenidae	*Denazinemys*	*nodosa*		6539	Kirtland?	Hunter Wash?	carapace fragment
NMMNH P-26254		Baenidae	*Denazinemys*	*nodosa*		3226	Kirtland	De-na-zin	carapace fragment
NMMNH P-26255		Baenidae	*Denazinemys*	*nodosa*		3226	Kirtland	De-na-zin	carapace fragment
NMMNH P-26298		Baenidae	*Denazinemys*	*nodosa*		3169	Kirtland	De-na-zin	numerous carapace fragments
NMMNH P-27278		Baenidae	*Denazinemys*	*nodosa*		3530	Kirtland	De-na-zin	2 carapace fragments
NMMNH P-27439		Baenidae	*Denazinemys*	*nodosa*		3531	Kirtland	De-na-zin	neural part of carapace
NMMNH P-33026		Baenidae	*Denazinemys*	*nodosa*		4275	Kirtland	Hunter Wash	12 carapace fragments
NMMNH P-33028		Baenidae	*Denazinemys*	*nodosa*		4275	Kirtland	Hunter Wash	neural fragment
NMMNH P-41229		Baenidae	*Denazinemys*	*nodosa*		3506	Kirtland	Hunter Wash	nearly complete carapace & plastron

(continued)

(continued)

Current number	Previous number	Family	Genus	Species	Type status	Locality Number	Formation	Member	Description
NMMNH P-22636		Baenidae	*Denazinemys*	*nodosa*		1640	Kirtland	Hunter Wash	carapace fragments
NMMNH P-22643		Baenidae	*Denazinemys*	*nodosa*		1606	Fruitland	Fossil Forest	carapace fragments
NMMNH P-28323		Baenidae	*Denazinemys*	*nodosa*		1882	Fruitland	Fossil Forest	numerous carapace fragments
NMMNH P-32867		Baenidae	*Denazinemys*	*nodosa*		2387	Fruitland	Fossil Forest	5 carapace fragments
PMU.R 11	Exemplar 1	Baenidae	*Denazinemys*	*nodosa*			Kirtland	Hunter Wash	carapace, plastron, pelvis, & other postcrania
PMU.R 12	Exemplar 2	Baenidae	*Denazinemys*	*nodosa*			Kirtland	De-na-zin	nearly complete carapace & plastron
PMU.R 13	Exemplar 3	Baenidae	*Denazinemys*	*nodosa*			Kirtland	Hunter Wash	incomplete carapace & [plastron
PMU.R 14	Exemplar 4	Baenidae	*Denazinemys*	*nodosa*			Kirtland	Hunter Wash	incomplete carapace & plastron
PMU.R unnumbered	Exemplar 5	Baenidae	*Denazinemys*	*nodosa*			unknown	unknown	incomplete carapace & plastron
PMU.R unnumbered	Exemplar 6	Baenidae	*Denazinemys*	*nodosa*			Kirtland	De-na-zin	incomplete carapace & plastron
PMU.R 15	Exemplar 7	Baenidae	*Denazinemys*	*nodosa*			unknown	unknown	incomplete carapace
PMU.R unnumbered	Exemplar 8	Baenidae	*Denazinemys*	*nodosa*			Kirtland	De-na-zin	incomplete plastron & other shell material
PMU.R 16	Exemplar 9	Baenidae	*Denazinemys*	*nodosa*			Kirtland	Hunter Wash	incomplete carapace
PMU.R 17	Exemplar 10	Baenidae	*Denazinemys*	*nodosa*			Kirtland	Hunter Wash	incomplete carapace & nearly complete plastron
PMU.R 18	Exemplar 11	Baenidae	*Denazinemys*	*nodosa*			Kirtland	Hunter Wash	nearly complete carapace & plastron
PMU.R 19	Exemplar 12	Baenidae	*Denazinemys*	*nodosa*			Kirtland	Hunter Wash	incomplete plastron
PMU.R 20	Exemplar 13	Baenidae	*Denazinemys*	*nodosa*			unknown	unknown	incomplete carapace & plastron
PMU.R unnumbered	Exemplar 14	Baenidae	*Denazinemys*	*nodosa*			unknown	unknown	shell fragments
PMU.R unnumbered	Exemplar 15	Baenidae	*Denazinemys*	*nodosa*			Kirtland	Hunter Wash	carapace & plastron fragments
PMU.R unnumbered	Exemplar 16	Baenidae	*Denazinemys*	*nodosa*			Kirtland	Hunter Wash	carapace fragments
PMU.R unnumbered	Exemplar 17	Baenidae	*Denazinemys*	*nodosa*			Kirtland	Hunter Wash	shell fragments
SMP VP-1284		Baenidae	*Denazinemys*	*nodosa*		350	Kirtland	De-na-zin	?carapace fragment
SMP VP-1287		Baenidae	*Denazinemys*	*nodosa*		350	Kirtland	De-na-zin	numerous plastron & carapace fragments

(continued)

(continued)

Current number	Previous number	Family	Genus	Species	Type status	Locality Number	Formation	Member	Description
SMP VP-1290		Baenidae	*Denazinemys*	*nodosa*		360a	Kirtland	De-na-zin	carapace fragments
SMP VP-1323		Baenidae	*Denazinemys*	*nodosa*		319a	Kirtland	De-na-zin	carapace fragment
SMP VP-1403		Baenidae	*Denazinemys*	*nodosa*		388a	Kirtland	De-na-zin	carapace & associated fragments
SMP VP-1869		Baenidae	*Denazinemys*	*nodosa*		382	Kirtland	De-na-zin	complete carapace & plastron
SMP VP-2072		Baenidae	*Denazinemys*	*nodosa*		376a	Kirtland	De-na-zin	nearly complete carapace & plaston
SMP VP-2495		Baenidae	*Denazinemys*	*nodosa*		389a	Kirtland	De-na-zin	carapace fragment
SMP VP-830		Baenidae	*Denazinemys*	*nodosa*		316	Kirtland	Hunter Wash	incomplete plastron with left bridge & numerous carapace fragments
SMP VP-1348		Baenidae	*Denazinemys*	*nodosa*		373	Kirtland	Hunter Wash	carapace fragments
SMP VP-1469		Baenidae	*Denazinemys*	*nodosa*		228	Kirtland	Hunter Wash	anterior part of plastron (intergular, gular, & humeral)
SMP VP-1480		Baenidae	*Denazinemys*	*nodosa*		281	Kirtland	Hunter Wash	carapace fragment
SMP VP-1506		Baenidae	*Denazinemys*	*nodosa*		228	Kirtland	Hunter Wash	right plastron fragment (intergular, gular, & humeral)
SMP VP-1524		Baenidae	*Denazinemys*	*nodosa*		281	Kirtland	Hunter Wash	carapace fragment
SMP VP-1665		Baenidae	*Denazinemys*	*nodosa*		398	Kirtland	Hunter Wash	medial part of carapace
SMP VP-1668		Baenidae	*Denazinemys*	*nodosa*		398	Kirtland	Hunter Wash	complete plastron, partial carapace, & fragments
SMP VP-1948		Baenidae	*Denazinemys*	*nodosa*		228	Kirtland	Hunter Wash	four carapace fragments
SMP VP-1986		Baenidae	*Denazinemys*	*nodosa*		419	Kirtland	Hunter Wash	incomplete plastron & carapace
SMP VP-2529		Baenidae	*Denazinemys*	*nodosa*		452	Kirtland	Hunter Wash	carapace fragment
SMP VP-2828		Baenidae	*Denazinemys*	*nodosa*		349	Kirtland	Hunter Wash	incomplete plastron
SMP VP-3255		Baenidae	*Denazinemys*	*nodosa*		398	Kirtland	Hunter Wash	plastron fragment, including part of xiphiplastron & bridge
USNM 8345		Baenidae	*Denazinemys*	*nodosa*	holotype		Kirtland	De-na-zin	carapace & plastron
USNM 8603		Baenidae	*Denazinemys*	*nodosa*			Kirtland	Hunter Wash	median part of carapace
USNM 8622		Baenidae	*Denazinemys*	*nodosa*			Fruitland	Fossil Forest	median part of carapace
USNM 11323		Baenidae	*Denazinemys*	*nodosa*			Kirtland	Hunter Wash	carapace & plastron

(continued)

(continued)

Current number	Previous number	Family	Genus	Species	Type status	Locality Number	Formation	Member	Description
USNM 11327		Baenidae	*Denazinemys*	*nodosa*			Fruitland	Fossil Forest	crushed shell
USNM 12819		Baenidae	*Denazinemys*	*nodosa*			Kirtland	Hunter Wash	carapace & incomplete plastron
USNM 12834		Baenidae	*Denazinemys*	*nodosa*			Kirtland	De-na-zin	incomplete carapace & plastron
NMMNH P-49945	UA 5573	Baenidae	cf. *Denazinemys*	*nodosa*		6205	Kirtland?	Hunter Wash?	10 plastron & associated fragments
SMP VP-1704		Baenidae	cf. *Denazinemys*	*nodosa*		409	Kirtland	Hunter Wash	partial carapace & plastron
SMP VP-2853		Baenidae	cf. *Denazinemys*	*nodosa*		450	Fruitland	Fossil Forest	carapace fragments
SMP VP-3273		Baenidae	cf. *Denazinemys*	*nodosa*		372	Kirtland	Hunter Wash	carapace fragments
SMP VP-1610		Baenidae	?*Denazinemys*	*nodosa*		398	Kirtland	Hunter Wash	numerous carapace & plastron fragments
SMP VP-2751		Baenidae	?*Denazinemys*	*nodosa*		228	Kirtland	Hunter Wash	carapace fragments
NMMNH P-22698		Baenidae	*Neurankylus*	*baueri*		1616	Kirtland	Hunter Wash	carapace fragments
NMMNH P-27988		Baenidae	*Neurankylus*	*baueri*		3117	Kirtland	Hunter Wash	carapace & plastron sections
NMMNH P-22745		Baenidae	*Neurankylus*	*baueri*		1674	Fruitland	Fossil Forest	various shell fragments
NMMNH P-27788		Baenidae	*Neurankylus*	*baueri*		4015	Fruitland	Fossil Forest	numerous shell fragments
PMU.R 24	Exemplar 1	Baenidae	*Neurankylus*	*baueri*			Kirtland	Hunter Wash	nearly complete carapace, plastron, axial, & appendicular skeleton
PMU.R 25	Exemplar 2	Baenidae	*Neurankylus*	*baueri*			Kirtland	Hunter Wash	nearly complete carapace & incomplete plastron
PMU.R 26	Exemplar 3	Baenidae	*Neurankylus*	*baueri*			Kirtland	Hunter Wash?	incomplete carapace, plastron, & pelvic girdle
PMU.R 27	Exemplar 4	Baenidae	*Neurankylus*	*baueri*			Kirtland	Hunter Wash	incomplete carapace
SMP VP-1503		Baenidae	*Neurankylus*	*baueri*		283	Kirtland	Hunter Wash	plastron fragments
SMP VP-2379		Baenidae	*Neurankylus*	*baueri*		389a	Kirtland	De-na-zin	carapace with associated fragments & plastron
USNM 8344		Baenidae	*Neurankylus*	*baueri*	holotype		Kirtland	Hunter Wash	nearly complete carapace & plastron
USNM 8531		Baenidae	*Neurankylus*	*baueri*	hypotype		Kirtland	Hunter Wash	carapace & plastron
NMMNH P-22749		Baenidae	*Neurankylus*	cf. *N. baueri*		1674	Fruitland	Fossil Forest	shell fragment
NMMNH P-27776		Baenidae	*Neurankylus*	cf. *N. baueri*		4015	Fruitland	Fossil Forest	numerous shell fragments
NMMNH P-22655		Baenidae	*Neurankylus*	cf. *N. baueri*		1672	Fruitland	Fossil Forest	3 shell fragments
NMMNH P-22713		Baenidae	*Neurankylus*	cf. *N. baueri*		1652	Kirtland	Hunter Wash	various shell & bone fragments

(continued)

(continued)

Current number	Previous number	Family	Genus	Species	Type status	Locality Number	Formation	Member	Description
NMMNH P-22792		Baenidae	*Neurankylus*	cf. *N. baueri*		1697	Kirtland	Hunter Wash	incomplete plastron & various fragments
NMMNH P-28850		Baenidae	?*Neurankylus*	cf. *N. baueri*		3117	Kirtland	Hunter Wash	carapace sections
PMU.R 28		Baenidae	*Neurankylus*	cf. *N. baueri*			Kirtland	De-na-zin?	incomplete carapace & plastron
ROM 864		Baenidae	*Neurankylus*	cf. *N. baueri*			Kirtland	Hunter Wash	incomplete carapace & plastron
SMP VP-1854		Baenidae	*Neurankylus*	cf. *N. baueri*		283	Kirtland	Hunter Wash	portion of plastron
SMP VP-1931		Baenidae	*Neurankylus*	cf. *N. baueri*		228	Kirtland	Hunter Wash	partial carapace
SMP VP-2102		Baenidae	*Neurankylus*	cf. *N. baueri*		421	Kirtland	Hunter Wash	incomplete carapace & nearly complete plastron
USNM 8621		Baenidae	*Neurankylus*	cf. *N. baueri*			Kirtland	Hunter Wash	posterior carapace
USNM 13228		Baenidae	*Neurankylus*	cf. *N. baueri*			Kirtland	Hunter Wash	carapace & plastron
USNM 11083		Baenidae	*Scabremys*	*ornata*			Kirtland	De-na-zin	carapace & plastron
USNM 12821		Baenidae	*Scabremys*	*ornata*			Kirtland	Hunter Wash	partial carapace & plastron
USNM 13229		Baenidae	*Scabremys*	*ornata*	holotype		Kirtland	Hunter Wash	carapace & plastron
AMNH 6066		Baenidae	*Thescelus*	*rapiens*	holotype		Kirtland	De-na-zin	incomplete carapace & plastron
SMP VP-2100		Baenidae	*Thescelus*	*rapiens*		421	Kirtland	Hunter Wash	plastron & carapace
PMU.R 22		Baenidae	*Thescelus*	cf. *T. rapiens*			Kirtland	Hunter Wash	incomplete carapace & complete plastron
SMP VP-1553		Baenidae	*Thescelus*	cf. *T. rapiens*		393	Kirtland	De-na-zin	carapace fragments & associated fragments
USNM 12818		Baenidae	*Thescelus*	*hemispherica*	holotype		Kirtland	Hunter Wash	carapace & plastron
PMU.R 23		Baenidae	*Thescelus*	cf. *T. hemispherica*			Kirtland	Hunter Wash	incomplete carapace & plastron
SMP VP-2168		Baenidae	*Thescelus*	sp.		421	Kirtland	Hunter Wash	nearly complete carapace & plastron
NMMNH P-49940	UA 5586	Baenidae	Baenidae indet.	Baenidae indet.		6205	Kirtland?	Hunter Wash?	plastron fragment
SMP VP-1845		Baenidae	Baenidae indet.	Baenidae indet.		319a	Kirtland	De-na-zin	carapace section
SMP VP-2055		Baenidae	Baenidae indet.	Baenidae indet.		421	Kirtland	Hunter Wash	costal fragment
SMP VP-2600		Baenidae	Baenidae indet.	Baenidae indet.		319a	Kirtland	De-na-zin	incomplete plastral bridge
SMP VP-3257		Baenidae	Baenidae indet.	Baenidae indet.		313a	Kirtland	De-na-zin	plastron fragment
SMP VP-3258		Baenidae	Baenidae indet.	Baenidae indet.		350	Kirtland	De-na-zin	carapace fragment

(continued)

(continued)

Current number	Previous number	Family	Genus	Species	Type status	Locality Number	Formation	Member	Description
NMMNH P-22793		Baenidae	Baenidae indet.	Baenidae indet.		1698	Fruitland	Fossil Forest	various shell and bone fragments
NMMNH P-22796		Baenidae	Baenidae indet.	Baenidae indet.		1699	Kirtland	Hunter Wash	shell fragment
NMMNH P-22841		Baenidae	Baenidae indet.	Baenidae indet.		1721	Kirtland	De-na-zin	carapace/plastron fragments
NMMNH P-27632		Baenidae	Baenidae indet.	Baenidae indet.		4015	Fruitland	Fossil Forest	numerous shell fragments
NMMNH P-27759		Baenidae	Baenidae indet.	Baenidae indet.		4015	Fruitland	Fossil Forest	4 carapace fragments
NMMNH P-27762		Baenidae	Baenidae indet.	Baenidae indet.		4015	Fruitland	Fossil Forest	plastron/carapace fragments
NMMNH P-33024		Baenidae	Baenidae indet.	Baenidae indet.		4507	Kirtland	De-na-zin	incomplete innominate
NMMNH P-38401		Baenidae	Baenidae indet.	Baenidae indet.		4010	Kirtland	Hunter Wash	incomplete carapace
SMP VP-1154		Baenidae	Baenidae indet.	Baenidae indet.		372	Kirtland	Hunter Wash	costal plate
SMP VP-1387		Baenidae	Baenidae indet.	Baenidae indet.		391	Kirtland	Hunter Wash	carapace fragments
SMP VP-1399		Baenidae	Baenidae indet.	Baenidae indet.		388a	Kirtland	De-na-zin	right xiphiplastron
SMP VP-1470		Baenidae	Baenidae indet.	Baenidae indet.		281	Kirtland	Hunter Wash	carapace fragment
SMP VP-1507		Baenidae	Baenidae indet.	Baenidae indet.		228	Kirtland	Hunter Wash	part of carapace
SMP VP-1661		Baenidae	Baenidae indet.	Baenidae indet.		398	Kirtland	Hunter Wash	carapace & plastron fragments
SMP VP-1669		Baenidae	Baenidae indet.	Baenidae indet.		398	Kirtland	Hunter Wash	incomplete carapace & associated fragments
SMP VP-1791		Baenidae	Baenidae indet.	Baenidae indet.		281	Kirtland	Hunter Wash	xiphiplastron & other plastron fragments
SMP VP-1798		Baenidae	Baenidae indet.	Baenidae indet.		228	Kirtland	Hunter Wash	dorsal vertebra
SMP VP-1946		Baenidae	Baenidae indet.	Baenidae indet.		228	Kirtland	Hunter Wash	costal
SMP VP-2014		Baenidae	Baenidae indet.	Baenidae indet.		388a	Kirtland	De-na-zin	complete lower jaw
SMP VP-2644		Baenidae	Baenidae indet.	Baenidae indet.		450	Fruitland	Fossil Forest	nearly complete femur
SMP VP-2706		Baenidae	Baenidae indet.	Baenidae indet.		450	Fruitland	Fossil Forest	incomplete costal & plastron fragments
SMP VP-2721		Baenidae	Baenidae indet.	Baenidae indet.		450	Fruitland	Fossil Forest	carapace & plastron fragments, plus associated indet. bone material
USNM 8614		Baenidae	Baenidae indet.	Baenidae indet.			Kirtland	Hunter Wash	anterior half of plastron
USNM 8623		Baenidae	Baenidae indet.	Baenidae indet.			Kirtland	Hunter Wash	anterior half of plastron
USNM 8624		Baenidae	Baenidae indet.	Baenidae indet.			Kirtland	Hunter Wash	partial plastron
NMMNH P-22878		?Baenidae	?Baenidae indet.	?Baenidae indet.		1754	Kirtland	Hunter Wash	bone scrap
NMMNH P-32854		?Baenidae	?Baenidae indet.	?Baenidae indet.		4541	Kirtland	De-na-zin	15 carapace/plastron fragments

(continued)

(continued)

Current number	Previous number	Family	Genus	Species	Type status	Locality Number	Formation	Member	Description
SMP VP-1491		?Baenidae	?Baenidae indet.	?Baenidae indet.		309a	Kirtland	De-na-zin	posterior marginal?
SMP VP-1670		?Baenidae	?Baenidae indet.	?Baenidae indet.		399	Kirtland	Hunter Wash	nuchal & plastron fragments
SMP VP-2358		?Baenidae	?Baenidae indet.	?Baenidae indet.		281	Kirtland	Hunter Wash	plastron fragments
SMP VP-2605		?Baenidae	?Baenidae indet.	?Baenidae indet.		319a	Kirtland	De-na-zin	neural
SMP VP-2820		?Baenidae	?Baenidae indet.	?Baenidae indet.		228	Kirtland	Hunter Wash	carapace neural
NMMNH P-22813		?Baenidae	?Baenidae indet.	?Baenidae indet.		1702	Fruitland	Fossil Forest	?plastron section & various carapace/plastron fragments
SMP VP-3389		?Baenidae indet.	?Baenidae indet.	?Baenidae indet.		461	Kirtland	Hunter Wash	marginal/carapace fragment
SMP VP-1907		Kinosternoidea indet.	Kinosternoidea indet.	Kinosternoidea indet.		350	Kirtland	De-na-zin	part of left hypoplastron
SMP VP-2004		Kinosternoidea indet.	Kinosternoidea indet.	Kinosternoidea indet.		350	Kirtland	De-na-zin	two costal fragments & right hypoplastron
SMP VP-2009		Kinosternoidea indet.	Kinosternoidea indet.	Kinosternoidea indet.		350	Kirtland	De-na-zin	right xiphiplastron
SMP VP-2533		Kinosternoidea indet.	Kinosternoidea indet.	Kinosternoidea indet.		350	Kirtland	De-na-zin	right costal (8th?)
SMP VP-3274		Kinosternoidea indet.	Kinosternoidea indet.	Kinosternoidea indet.		350	Kirtland	De-na-zin	left costal (2nd?)
NMMNH P-49785	UA 14836	Kinosternoidea indet.	Kinosternoidea indet.	Kinosternoidea indet.		6340	Kirtland?	De-na-zin?	peripheral
USNM 11084		Nanhsiungchelyidae	*Basilemys*	*gaffneyi*	holotype		Kirtland	Hunter Wash	carapace & plastron
NMMNH P-27613		Nanhsiungchelyidae	*Basilemys*	cf. *B. gaffneyi*		4015	Fruitland	Fossil Forest	carapace fragments
NMMNH P-27623		Nanhsiungchelyidae	*Basilemys*	cf. *B. gaffneyi*		4015	Fruitland	Fossil Forest	peripheral fragment
NMMNH P-27630		Nanhsiungchelyidae	*Basilemys*	cf. *B. gaffneyi*		4015	Fruitland	Fossil Forest	numerous fragments
NMMNH P-27775		Nanhsiungchelyidae	*Basilemys*	cf. *B. gaffneyi*		4015	Fruitland	Fossil Forest	plastron section & 3 fragments
NMMNH P-28935		Nanhsiungchelyidae	*Basilemys*	cf. *B. gaffneyi*		3510	Kirtland	Hunter Wash	carapace fragments
NMMNH P-28952		Nanhsiungchelyidae	*Basilemys*	cf. *B. gaffneyi*		3496	Kirtland	Hunter Wash	peripheral fragments
NMMNH P-22651		Nanhsiungchelyidae	*Basilemys*	cf. *B. gaffneyi*		1672	Fruitland	Fossil Forest	shell fragment
NMMNH P-22742		Nanhsiungchelyidae	*Basilemys*	cf. *B. gaffneyi*		1674	Fruitland	Fossil Forest	shell fragments
NMMNH P-22744		Nanhsiungchelyidae	*Basilemys*	cf. *B. gaffneyi*		1674	Fruitland	Fossil Forest	various shell fragments
NMMNH P-22871		Nanhsiungchelyidae	*Basilemys*	cf. *B. gaffneyi*		1746	Kirtland	Hunter Wash	various shell & bone fragments
NMMNH P-22935		Nanhsiungchelyidae	*Basilemys*	cf. *B. gaffneyi*		1791	Kirtland	Hunter Wash	carapace fragment
NMMNH P-22947		Nanhsiungchelyidae	*Basilemys*	cf. *B. gaffneyi*		1777	Kirtland	Hunter Wash	limb osteoderm

(continued)

(continued)

Current number	Previous number	Family	Genus	Species	Type status	Locality Number	Formation	Member	Description
NMMNH P-27942		Nanhsiungchelyidae	*Basilemys*	cf. *B. gaffneyi*		1586	Kirtland	Hunter Wash	carapace fragments
NMMNH P-32828		Nanhsiungchelyidae	*Basilemys*	cf. *B. gaffneyi*		4010	Kirtland	Hunter Wash	limb osteoderm
NMMNH P-44517		Nanhsiungchelyidae	*Basilemys*	cf. *B. gaffneyi*		4000	Kirtland?	Hunter Wash?	carapace fragments
PMU.R 29		Nanhsiungchelyidae	*Basilemys*	cf. *B. gaffneyi*			Kirtland	Hunter Wash	nearly complete carapace & plastron
SMP VP-1106		Nanhsiungchelyidae	*Basilemys*	cf. *B. gaffneyi*		316	Kirtland	Hunter Wash	incomplete humerus & femur, plus other indet. fragments
SMP VP-1382		Nanhsiungchelyidae	*Basilemys*	cf. *B. gaffneyi*		361	Kirtland	De-na-zin	carapace fragment
SMP VP-1475		Nanhsiungchelyidae	*Basilemys*	cf. *B. gaffneyi*		365	Kirtland	Hunter Wash	edge of carapace
SMP VP-1510		Nanhsiungchelyidae	*Basilemys*	cf. *B. gaffneyi*		228	Kirtland	Hunter Wash	plastron fragment
SMP VP-1522		Nanhsiungchelyidae	*Basilemys*	cf. *B. gaffneyi*		281	Kirtland	Hunter Wash	carapace fragments
SMP VP-1527		Nanhsiungchelyidae	*Basilemys*	cf. *B. gaffneyi*		281	Kirtland	Hunter Wash	carapace fragments
SMP VP-1566		Nanhsiungchelyidae	*Basilemys*	cf. *B. gaffneyi*		281	Kirtland	Hunter Wash	carapace fragments
SMP VP-1652		Nanhsiungchelyidae	*Basilemys*	cf. *B. gaffneyi*		281	Kirtland	Hunter Wash	carapace fragment
SMP VP-1653		Nanhsiungchelyidae	*Basilemys*	cf. *B. gaffneyi*		281	Kirtland	Hunter Wash	carapace/plastron fragments
SMP VP-1678		Nanhsiungchelyidae	*Basilemys*	cf. *B. gaffneyi*		400	Kirtland	Hunter Wash	partial right humerus
SMP VP-1852		Nanhsiungchelyidae	*Basilemys*	cf. *B. gaffneyi*		228	Kirtland	Hunter Wash	limb osteoderm
SMP VP-2153		Nanhsiungchelyidae	*Basilemys*	cf. *B. gaffneyi*		281	Kirtland	Hunter Wash	carapace fragment
SMP VP-2359		Nanhsiungchelyidae	*Basilemys*	cf. *B. gaffneyi*		281	Kirtland	Hunter Wash	carapace fragment
SMP VP-2797		Nanhsiungchelyidae	*Basilemys*	cf. *B. gaffneyi*		450	Fruitland	Fossil Forest	epiplastron
SMP VP-3366		Nanhsiungchelyidae	*Basilemys*	cf. *B. gaffneyi*		450	Fruitland	Fossil Forest	partial plastron, fragments of femur, pelvic girdle, & carapace/plastron
SMP VP-1146		Nanhsiungchelyidae	cf. *Basilemys*	cf. *B. gaffneyi*		365	Kirtland	Hunter Wash	terminal phalanx
SMP VP-1158		Nanhsiungchelyidae	cf. *Basilemys*	cf. *B. gaffneyi*		228	Kirtland	Hunter Wash	incomplete humerus
NMMNH P-22568		Pleurosternidae	*Compsemys*	sp.		1464	Kirtland	Hunter Wash	2 carapace/plastron fragments
NMMNH P-22741		Pleurosternidae	*Compsemys*	sp.		1674	Fruitland	Fossil Forest	plastron fragment
NMMNH P-22809		Pleurosternidae	*Compsemys*	sp.		1700	Fruitland	Fossil Forest	3 shell fragments
NMMNH P-28326		Pleurosternidae	*Compsemys*	sp.		1882	Fruitland	Fossil Forest	2 shell fragments
NMMNH P-33021		Pleurosternidae	*Compsemys*	sp.		4570	Kirtland	De-na-zin	4 carapace fragments
NMMNH P-40512		Pleurosternidae	*Compsemys*	sp.		3488	Kirtland	De-na-zin	incomplete carapace & plastron

(continued)

(continued)

Current number	Previous number	Family	Genus	Species	Type status	Locality Number	Formation	Member	Description
NMMNH P-49819	UA 14391	Pleurosternidae	*Compsemys*	sp.		6551	Kirtland?	De-na-zin?	carapace fragment
NMMNH P-49821	UA 14388	Pleurosternidae	*Compsemys*	sp.		6554	Kirtland?	De-na-zin?	entoplaston
NMMNH P-49827	UA 14393	Pleurosternidae	*Compsemys*	sp.		6559	Kirtland?	Hunter Wash?	4 shell fragment
NMMNH P-49844	UA 14392	Pleurosternidae	*Compsemys*	sp.		6542	Kirtland?	Hunter Wash?	plastron fragment
SMP VP-1892		Pleurosternidae	*Compsemys*	sp.		350	Kirtland	De-na-zin	peripheral
SMP VP-1904		Pleurosternidae	*Compsemys*	sp.		350	Kirtland	De-na-zin	carapace fragment
SMP VP-1915		Pleurosternidae	*Compsemys*	sp.		350	Kirtland	De-na-zin	five carapace fragments
SMP VP-1916		Pleurosternidae	*Compsemys*	sp.		350	Kirtland	De-na-zin	carapace fragment
SMP VP-2002		Pleurosternidae	*Compsemys*	sp.		350	Kirtland	De-na-zin	costal fragment
SMP VP-2003		Pleurosternidae	*Compsemys*	sp.		350	Kirtland	De-na-zin	five costal fragments
SMP VP-2006		Pleurosternidae	*Compsemys*	sp.		350	Kirtland	De-na-zin	four plastron fragments
SMP VP-2201		Pleurosternidae	*Compsemys*	sp.		319a	Kirtland	De-na-zin	peripheral
SMP VP-2648		Pleurosternidae	*Compsemys*	sp.		350	Kirtland	De-na-zin	carapace fragments
SMP VP-2630		Pleurosternidae	?*Compsemys*	sp.		313a	Kirtland	De-na-zin	incomplete costal
NMMNH P-26230		Trionychidae	*Aspideretoides*	*austerus*		3227	Kirtland	De-na-zin	costal fragments & neural
NMMNH P-27847		Trionychidae	*Aspideretoides*	*austerus*		3226	Kirtland	De-na-zin	incomplete costals, xiphiplastron, hyoplastron, & girdle
SMP VP-811		Trionychidae	*Aspideretoides*	*austerus*		210	Kirtland	De-na-zin	several pieces of plastron and carapace
SMP VP-1364		Trionychidae	*Aspideretoides*	*austerus*		389a	Kirtland	De-na-zin	hypoplastron & other plastral elements
SMP VP-1717		Trionychidae	*Aspideretoides*	*austerus*		319a	Kirtland	De-na-zin	incomplete right hypoplastron & other plastron fragments
SMP VP-1844		Trionychidae	*Aspideretoides*	*austerus*		319a	Kirtland	De-na-zin	carapace fragments
SMP VP-1868		Trionychidae	*Aspideretoides*	*austerus*		382	Kirtland	De-na-zin	costals & nuchal
SMP VP-1959		Trionychidae	*Aspideretoides*	*austerus*		210	Kirtland	De-na-zin	large plastron fragment
SMP VP-2376		Trionychidae	*Aspideretoides*	*austerus*		389a	Kirtland	De-na-zin	costal & carapace fragments
USNM 6550		Trionychidae	*Aspideretoides*	*austerus*	see text		Kirtland	De-na-zin	partial carapace
USNM 12988		Trionychidae	*Aspideretoides*	*austerus*			Kirtland	De-na-zin	anterior part of carapace
USNM 12986		Trionychidae	*Aspideretoides*	*ovatus*	holotype		Kirtland	Hunter Wash?	partial carapace & plastron

(continued)

(continued)

Current number	Previous number	Family	Genus	Species	Type status	Locality Number	Formation	Member	Description
USNM 12987		Trionychidae	*Aspideretoides*	*ovatus*			Kirtland	Hunter Wash?	posterior half of carapace
NMMNH P-27779		Trionychidae	*Aspideretoides*	*robustus*		4015	Fruitland	Fossil Forest	incomplete right hyoplastron
NMMNH P-27826		Trionychidae	*Aspideretoides*	*robustus*		4015	Fruitland	Fossil Forest	incomplete costal & numerous fragments
NMMNH P-27572		Trionychidae	*Aspideretoides*	*robustus*		4015	Fruitland	Fossil Forest	fragments
NMMNH P-27628		Trionychidae	*Aspideretoides*	*robustus*		4015	Fruitland	Fossil Forest	numerous carapace/plastron fragments
NMMNH P-27633		Trionychidae	*Aspideretoides*	*robustus*		4015	Fruitland	Fossil Forest	various carapace/plastron fragments
NMMNH P-27752		Trionychidae	*Aspideretoides*	*robustus*		4015	Fruitland	Fossil Forest	incomplete costal & various fragments
NMMNH P-27768		Trionychidae	*Aspideretoides*	*robustus*		4015	Fruitland	Fossil Forest	2 plastron fragments
NMMNH P-27777		Trionychidae	*Aspideretoides*	*robustus*		4015	Fruitland	Fossil Forest	carapace/plastron fragments
NMMNH P-27941		Trionychidae	*Aspideretoides*	*robustus*		1581	Kirtland	Hunter Wash	incomplete costal
NMMNH P-28321		Trionychidae	*Aspideretoides*	*robustus*		1882	Fruitland	Fossil Forest	costal fragment
NMMNH P-30066		Trionychidae	*Aspideretoides*	*robustus*		4274	Kirtland	Hunter Wash	costal
NMMNH P-30756		Trionychidae	*Aspideretoides*	*robustus*		4104	Fruitland/ Kirtland	unknown	costal fragments
NMMNH P-39632		Trionychidae	*Aspideretoides*	*robustus*		5163	Kirtland	Hunter Wash	2 fused incomplete costals & associated fragments
NMMNH P-49856	UA 11737	Trionychidae	*Aspideretoides*	*robustus*		6548	Kirtland?	Hunter Wash?	incomplete costal
NMMNH P-49862	UA 13319	Trionychidae	*Aspideretoides*	*robustus*		6560	Kirtland?	Hunter Wash?	incomplete costal
PMU.R 30		Trionychidae	*Aspideretoides*	*robustus*			Kirtland	Hunter Wash	nearly complete carapace
PMU.R 31		Trionychidae	*Aspideretoides*	*robustus*			Kirtland	Hunter Wash	right hypoplatron & xiphiplastron, plus left hyoplastron, hypoplastron, & xiphiplastron
SMP VP-1612		Trionychidae	*Aspideretoides*	*robustus*		400	Kirtland	Hunter Wash	carapace fragments
SMP VP-1667		Trionychidae	*Aspideretoides*	*robustus*		398	Kirtland	Hunter Wash	incomplete nuchal, both xiphiplastra, & associated fragments
SMP VP-1679		Trionychidae	*Aspideretoides*	*robustus*		400	Kirtland	Hunter Wash	incomplete costal
SMP VP-1681		Trionychidae	*Aspideretoides*	*robustus*		400	Kirtland	Hunter Wash	costal fragments

(continued)

(continued)

Current number	Previous number	Family	Genus	Species	Type status	Locality Number	Formation	Member	Description
SMP VP-1682		Trionychidae	*Aspideretoides*	*robustus*		400	Kirtland	Hunter Wash	nearly complete costal
SMP VP-1685		Trionychidae	*Aspideretoides*	*robustus*		396	Kirtland	Hunter Wash	incomplete xiphiplastron
SMP VP-1856		Trionychidae	*Aspideretoides*	*robustus*		281	Kirtland	Hunter Wash	carapace fragments
SMP VP-2166		Trionychidae	*Aspideretoides*	*robustus*		228	Kirtland	Hunter Wash	carapace fragments
SMP VP-2698		Trionychidae	*Aspideretoides*	*robustus*		463	Fruitland	Fossil Forest	carapace fragments, including nuchal & neural
USNM 8350		Trionychidae	*Aspideretoides*	*robustus*			Kirtland	Farmington?	left hypoplastron
USNM 8538		Trionychidae	*Aspideretoides*	*robustus*	holotype		Kirtland	Hunter Wash	complete carapace
USNM 12989		Trionychidae	*Aspideretoides*	*robustus*			Kirtland	Hunter Wash	left hyoplastron & hypoplastron
NMMNH P-33914		Trionychidae	*Aspideretoides*	cf. *A. austerus*		4721	Kirtland	De-na-zin	2 incomplete costals
NMMNH P-22751		Trionychidae	*Aspideretoides*	cf. *A. robustus*		1675	Fruitland	Fossil Forest	various carapace/plastron fragments
NMMNH P-22934		Trionychidae	*Aspideretoides*	cf. *A. robustus*		1790	Kirtland	Hunter Wash	incomplete costal
NMMNH P-22949		Trionychidae	*Aspideretoides*	cf. *A. robustus*		1805	Kirtland	Hunter Wash?	2 shell fragments
NMMNH P-27769		Trionychidae	*Aspideretoides*	cf. *A. robustus*		4015	Fruitland	Fossil Forest	plastron fragments
NMMNH P-33916		Trionychidae	*Aspideretoides*	cf. *A. robustus*		4010	Kirtland	Hunter Wash	nearly complete left xiphiplastron
NMMNH P-28320		Trionychidae	*Aspideretoides*	cf. *A. robustus*		1882	Fruitland	Fossil Forest	numerous carapace fragments
NMMNH P-33912		Trionychidae	*Aspideretoides*	cf. *A. robustus*		4717	Fruitland	Fossil Forest	numerous carapace/costal fragments
NMMNH P-49848	UA 14397	Trionychidae	*Aspideretoides*	cf. *A. robustus*		6547	Kirtland?	Hunter Wash?	shell fragment
NMMNH P-49786	UA 14399	Trionychidae	*Aspideretoides*	cf. *A. robustus*		6340	Kirtland?	Hunter Wash?	shell fragment
SMP VP-1525		Trionychidae	*Aspideretoides*	cf. *A. robustus*		281	Kirtland	Hunter Wash	costal & other fragments
NMMNH P-19200		Trionychidae	*Aspideretoides*	sp.		1549	Kirtland	Hunter Wash	incomplete carapace
NMMNH P-22550		Trionychidae	*Aspideretoides*	sp.		1587	Kirtland	Hunter Wash	carapace fragment
NMMNH P-27829		Trionychidae	*Aspideretoides*	sp.		4015	Fruitland	Fossil Forest	numerous carapace fragments
NMMNH P-22729		Trionychidae	*Aspideretoides*	sp.		1665	Kirtland	De-na-zin	incomplete costal
NMMNH P-22735		Trionychidae	*Aspideretoides*	sp.		1673	Kirtland	Hunter Wash	hypoplastron & associated fragment

(continued)

(continued)

Current number	Previous number	Family	Genus	Species	Type status	Locality Number	Formation	Member	Description
NMMNH P-49818	UA 14396	Trionychidae	*Aspideretoides*	sp.		6559	Kirtland?	Hunter Wash?	2 shell fragments
NMMNH P-49822	UA 11741	Trionychidae	*Aspideretoides*	sp.		6559	Kirtland?	Hunter Wash?	8 neurals & other fragments
NMMNH P-49832	UA 14412	Trionychidae	*Aspideretoides*	sp.		6622	Kirtland?	De-na-zin?	costal fragment
SMP VP-1493		Trionychidae	*Aspideretoides*	sp		319a	Kirtland	De-na-zin	carapace fragments
SMP VP-1519		Trionychidae	*Aspideretoides*	sp.		365	Kirtland	Hunter Wash	carapace fragments
SMP VP-2045		Trionychidae	*Aspideretoides*	sp.		388a	Kirtland	De-na-zin	carapace fragments
SMP VP-2601		Trionychidae	*Aspideretoides*	sp.		319a	Kirtland	De-na-zin	carapace fragment
NMMNH P-49853	UA 14400	Trionychidae	cf. *Aspideretoides*	sp.		6552	Kirtland?	De-na-zin?	plastron fragment
SMP VP-1368		Trionychidae	cf. *Aspideretoides*	sp.		349	Kirtland	Hunter Wash	part of carapace
SMP VP-1426		Trionychidae	cf. *Aspideretoides*	sp.		361	Kirtland	De-na-zin	proximal end of left femur
SMP VP-1594		Trionychidae	cf. *Aspideretoides*	sp.		398	Kirtland	Hunter Wash	carapace fragments
SMP VP-1595		Trionychidae	cf. *Aspideretoides*	sp.		398	Kirtland	Hunter Wash	carapace fragments
SMP VP-1810		Trionychidae	cf. *Aspideretoides*	sp.		409	Kirtland	Hunter Wash	carapace fragment
SMP VP-1949		Trionychidae	cf. *Aspideretoides*	sp.		228	Kirtland	Hunter Wash	carapace fragment
SMP VP-2803		Trionychidae	cf. *Aspideretoides*	sp.		450	Fruitland	Fossil Forest	costal, marginal, & other carapace fragments
NMMNH P-22832		Trionychidae	?*Aspideretoides*	sp.		1714	Kirtland	Hunter Wash	various shell fragments
NMMNH P-26204		Trionychidae	?*Aspideretoides*	sp.		3228	Kirtland	De-na-zin	costal fragment
NMMNH P-27782		Trionychidae	?*Aspideretoides*	sp.		4015	Fruitland	Fossil Forest	numerous carapace/plastron fragments
NMMNH P-22627		Trionychidae	Plastomeninae indet.	Plastomeninae indet.		1639	Kirtland	De-na-zin	hypoplastron fragment
NMMNH P-27479		Trionychidae	Plastomeninae indet.	Plastomeninae indet.		4015	Fruitland	Fossil Forest	numerous carapace fragments
SMP VP-742		Trionychidae	Plastomeninae indet.	Plastomeninae indet.		228	Kirtland	Hunter Wash	carapace & plastron fragments
SMP VP-2034		Trionychidae	Plastomeninae indet.	Plastomeninae indet.		421	Kirtland	Hunter Wash	three carapace fragments

(continued)

(continued)

Current number	Previous number	Family	Genus	Species	Type status	Locality Number	Formation	Member	Description
SMP VP-2154		Trionychidae	Trionychidae indet.	Trionychidae indet.		281	Kirtland	Hunter Wash	plastron fragment
SMP VP-2156		Trionychidae	Trionychidae indet.	Trionychidae indet.		281	Kirtland	Hunter Wash	carapace fragment
SMP VP-3348		Trionychidae	Trionychidae indet.	Trionychidae indet.		409	Kirtland	Hunter Wash	incomplete costal
NMMNH P-22583		Trionychidae	Trionychidae indet.	Trionychidae indet.		1613	Kirtland	Hunter Wash	carapace fragment
NMMNH P-22804		Trionychidae	Trionychidae indet.	Trionychidae indet.		1699	Kirtland	Hunter Wash	costal fragment & indet. fragment
NMMNH P-26261		Trionychidae	Trionychidae indet.	Trionychidae indet.		3245	Kirtland	De-na-zin	incomplete humerus
NMMNH P-27903		Trionychidae	Trionychidae indet.	Trionychidae indet.		1625	Fruitland	Fossil Forest	carapace & ?dinosaur bone fragment
NMMNH P-28324		Trionychidae	Trionychidae indet.	Trionychidae indet.		1882	Fruitland	Fossil Forest	numerous carapace fragments
NMMNH P-28909		Trionychidae	Trionychidae indet.	Trionychidae indet.		3964	Kirtland	Hunter Wash	incomplete humerus
NMMNH P-32845		Trionychidae	Trionychidae indet.	Trionychidae indet.		4542	Kirtland	De-na-zin	carapace/plastron fragments
NMMNH P-32851		Trionychidae	Trionychidae indet.	Trionychidae indet.		2387	Fruitland	Fossil Forest	carapace/plastron fragments
NMMNH P-32852		Trionychidae	Trionychidae indet.	Trionychidae indet.		2387	Fruitland	Fossil Forest	carapace/plastron fragments
NMMNH P-49825		Trionychidae	Trionychidae indet.	Trionychidae indet.		6558	Kirtland	Hunter Wash	costal fragment
NMMNH P-49855	UA 14399	Trionychidae	Trionychidae indet.	Trionychidae indet.		6552	Kirtland?	De-na-zin?	2 shell fragments
SMP VP-1476		Trionychidae	Trionychidae indet.	Trionychidae indet.		365	Kirtland	Hunter Wash	plastron fragment
SMP VP-1509		Trionychidae	Trionychidae indet.	Trionychidae indet.		228	Kirtland	Hunter Wash	carapace & plastron fragments
SMP VP-1513		Trionychidae	Trionychidae indet.	Trionychidae indet.		365	Kirtland	Hunter Wash	carapace/plastron fragments
SMP VP- SMP VP-1558		Trionychidae	Trionychidae indet.	Trionychidae indet.		389a	Kirtland	De-na-zin	carapace fragment
SMP VP-1567		Trionychidae	Trionychidae indet.	Trionychidae indet.		228	Kirtland	Hunter Wash	carapace fragments

(continued)

(continued)

Current number	Previous number	Family	Genus	Species	Type status	Locality Number	Formation	Member	Description
SMP VP-1683		Trionychidae	Trionychidae indet.	Trionychidae indet.		400	Kirtland	Hunter Wash	carapace fragments
SMP VP-1690		Trionychidae	Trionychidae indet.	Trionychidae indet.		409	Kirtland	Hunter Wash	carapace fragment
SMP VP-1691		Trionychidae	Trionychidae indet.	Trionychidae indet.		409	Kirtland	Hunter Wash	carapace fragment
SMP VP-1695		Trionychidae	Trionychidae indet.	Trionychidae indet.		409	Kirtland	Hunter Wash	carapace fragments
SMP VP-1817		Trionychidae	Trionychidae indet.	Trionychidae indet.		365	Kirtland	Hunter Wash	two carapace fragments
SMP VP-1857		Trionychidae	Trionychidae indet.	Trionychidae indet.		365	Kirtland	Hunter Wash	incomplete costal
SMP VP-1917		Trionychidae	Trionychidae indet.	Trionychidae indet.		350	Kirtland	De-na-zin	carapace fragments
SMP VP-1947		Trionychidae	Trionychidae indet.	Trionychidae indet.		228	Kirtland	Hunter Wash	incomplete costal
SMP VP-1996		Trionychidae	Trionychidae indet.	Trionychidae indet.		412	Kirtland	Hunter Wash	costal fragment
SMP VP-2005		Trionychidae	Trionychidae indet.	Trionychidae indet.		350	Kirtland	De-na-zin	four costal fragments
SMP VP-2224		Trionychidae	Trionychidae indet.	Trionychidae indet.		433	Kirtland	Hunter Wash	carapace/plastron fragments
SMP VP-2388		Trionychidae	Trionychidae indet.	Trionychidae indet.		450	Fruitland	Fossil Forest	carapace fragments
SMP VP-2403		Trionychidae	Trionychidae indet.	Trionychidae indet.		450	Fruitland	Fossil Forest	carapace & plastron fragments from multiple individuals
SMP VP-2404		Trionychidae	Trionychidae indet.	Trionychidae indet.		450	Fruitland	Fossil Forest	carapace fragments
SMP VP-2423		Trionychidae	Trionychidae indet.	Trionychidae indet.		350	Kirtland	De-na-zin	carapace fragments
SMP VP-2442		Trionychidae	Trionychidae indet.	Trionychidae indet.		450	Fruitland	Fossil Forest	carapace fragments from multiple taxa
SMP VP-2460		Trionychidae	Trionychidae indet.	Trionychidae indet.		451	Kirtland	Hunter Wash	carapace fragments
SMP VP-2487		Trionychidae	Trionychidae indet.	Trionychidae indet.		350	Kirtland	De-na-zin	two carapace fragments
SMP VP-2528		Trionychidae	Trionychidae indet.	Trionychidae indet.		452	Kirtland	Hunter Wash	carapace fragments

(continued)

(continued)

Current number	Previous number	Family	Genus	Species	Type status	Locality Number	Formation	Member	Description
SMP VP-2530		Trionychidae	Trionychidae indet.	Trionychidae indet.		452	Kirtland	Hunter Wash	neural & carapace fragments
SMP VP-2531		Trionychidae	Trionychidae indet.	Trionychidae indet.		452	Kirtland	Hunter Wash	carapace fragments
SMP VP-2532		Trionychidae	Trionychidae indet.	Trionychidae indet.		452	Kirtland	Hunter Wash	carapace fragments
SMP VP-2541		Trionychidae	Trionychidae indet.	Trionychidae indet.		363a	Kirtland	De-na-zin	carapace fragment
SMP VP-2551		Trionychidae	Trionychidae indet.	Trionychidae indet.		450	Fruitland	Fossil Forest	carapace fragment
SMP VP-2586		Trionychidae	Trionychidae indet.	Trionychidae indet.		450	Fruitland	Fossil Forest	carapace fragment
SMP VP-2635		Trionychidae	Trionychidae indet.	Trionychidae indet.		450	Fruitland	Fossil Forest	carapace fragment
SMP VP-2716		Trionychidae	Trionychidae indet.	Trionychidae indet.		228	Kirtland	Hunter Wash	carapace fragment
SMP VP-2728		Trionychidae	Trionychidae indet.	Trionychidae indet.		450	Fruitland	Fossil Forest	carapace fragments
SMP VP-2764		Trionychidae	Trionychidae indet.	Trionychidae indet.		450	Fruitland	Fossil Forest	incomplete costal
SMP VP-2766		Trionychidae	Trionychidae indet.	Trionychidae indet.		450	Fruitland	Fossil Forest	carapace fragments
SMP VP-2772		Trionychidae	Trionychidae indet.	Trionychidae indet.		450	Fruitland	Fossil Forest	incomplete neural
SMP VP-2819		Trionychidae	Trionychidae indet.	Trionychidae indet.		228	Kirtland	Hunter Wash	carapace neural
SMP VP-2839		Trionychidae	Trionychidae indet.	Trionychidae indet.		450	Fruitland	Fossil Forest	carapace fragments
SMP VP-2847		Trionychidae	Trionychidae indet.	Trionychidae indet.		450	Fruitland	Fossil Forest	carapace fragments
SMP VP-2852		Trionychidae	Trionychidae indet.	Trionychidae indet.		450	Fruitland	Fossil Forest	carapace fragments & associated indet. bone material
SMP VP-3271		Trionychidae	Trionychidae indet.	Trionychidae indet.		372	Kirtland	Hunter Wash	incomplete neural
SMP VP-3272		Trionychidae	Trionychidae indet.	Trionychidae indet.		372	Kirtland	Hunter Wash	carapace fragment
SMP VP-1135		Cryptodira indet.	Cryptodira indet.	Cryptodira indet.		373	Kirtland	Hunter Wash	incomplete scapula

(continued)

(continued)

Current number	Previous number	Family	Genus	Species	Type status	Locality Number	Formation	Member	Description
NMMNH P-22743		Testudines indet.	Testudines indet.	Testudines indet.		1674	Fruitland	Fossil Forest	various shell & bone fragments
NMMNH P-22826		Testudines indet.	Testudines indet.	Testudines indet.		1708	Kirtland	Hunter Wash	various bone fragments
NMMNH P-22921		Testudines indet.	Testudines indet.	Testudines indet.		1782	Kirtland	Hunter Wash	distal end of femur
NMMNH P-22939		Testudines indet.	Testudines indet.	Testudines indet.		1795	Kirtland/ Fruitland	unknown	incomplete scapula?
NMMNH P-26246		Testudines indet.	Testudines indet.	Testudines indet.		1708	Kirtland	Hunter Wash	5 pelvic fragments
NMMNH P-26251		Testudines indet.	Testudines indet.	Testudines indet.		3226	Kirtland	De-na-zin	incomplete ?coracoid
NMMNH P-27827		Testudines indet.	Testudines indet.	Testudines indet.		4015	Fruitland	Fossil Forest	incomplete humerus
NMMNH P-22510		Testudines indet.	Testudines indet.	Testudines indet.		1571	Kirtland	Hunter Wash	2 vertebrae
NMMNH P-32823		Testudines indet.	Testudines indet.	Testudines indet.		4010	Kirtland	Hunter Wash	incomplete phalanx
NMMNH P-49847		Testudines indet.	Testudines indet.	Testudines indet.		6546	Kirtland?	Hunter Wash?	shell fragment
NMMNH P-22624		Testudines indet.	Testudines indet.	Testudines indet.		1638	unknown	unknown	pelvis
NMMNH P-37762		Testudines indet.	Testudines indet.	Testudines indet.		5197	Kirtland	De-na-zin	2 carapace fragments
NMMNH P-37787		Testudines indet.	Testudines indet.	Testudines indet.		5197	Kirtland	De-na-zin	14 carapace fragments
NMMNH P-33023		Testudines indet.	Testudines indet.	Testudines indet.		4570	Kirtland	De-na-zin	tibia
SMP VP-817		Testudines indet.	Testudines indet.	Testudines indet.		210	Kirtland	De-na-zin	?posterior part of left ramus
SMP VP-1091		Testudines indet.	Testudines indet.	Testudines indet.		365	Kirtland	Hunter Wash	ungual
SMP VP-1332		Testudines indet.	Testudines indet.	Testudines indet.		385	Kirtland	Hunter Wash	incomplete scapulocoracoid
SMP VP-1420		Testudines indet.	Testudines indet.	Testudines indet.		385	Kirtland	Hunter Wash	girdle & miscellaneous fragments
SMP VP-1471		Testudines indet.	Testudines indet.	Testudines indet.		281	Kirtland	Hunter Wash	pelvic element

(continued)

(continued)

Current number	Previous number	Family	Genus	Species	Type status	Locality Number	Formation	Member	Description
SMP VP-1484		Testudines indet.	Testudines indet.	Testudines indet.		281	Kirtland	Hunter Wash	ungual
SMP VP-1530		Testudines indet.	Testudines indet.	Testudines indet.		281	Kirtland	Hunter Wash	ungual & distal end of phalanx
SMP VP-1531		Testudines indet.	Testudines indet.	Testudines indet.		281	Kirtland	Hunter Wash	girdle
SMP VP-1800		Testudines indet.	Testudines indet.	Testudines indet.		228	Kirtland	Hunter Wash	limb fragments
SMP VP-1903		Testudines indet.	Testudines indet.	Testudines indet.		350	Kirtland	De-na-zin	promimal portion of pelvic element (ilium?)
SMP VP-1919		Testudines indet.	Testudines indet.	Testudines indet.		417	Kirtland	Hunter Wash	skull roof?
SMP VP-2000		Testudines indet.	Testudines indet.	Testudines indet.		350	Kirtland	De-na-zin	skull element
SMP VP-2015		Testudines indet.	Testudines indet.	Testudines indet.		420a	Kirtland	De-na-zin	humerus fragment
SMP VP-2035		Testudines indet.	Testudines indet.	Testudines indet.		421	Kirtland	Hunter Wash	unidentified girdle element
SMP VP-2037		Testudines indet.	Testudines indet.	Testudines indet.		421	Kirtland	Hunter Wash	nearly complete left humerus
SMP VP-2084		Testudines indet.	Testudines indet.	Testudines indet.		421	Kirtland	Hunter Wash	incomplete right humerus
SMP VP-2537		Testudines indet.	Testudines indet.	Testudines indet.		316	Kirtland	Hunter Wash	?humerus
SMP VP-2549		Testudines indet.	Testudines indet.	Testudines indet.		450	Fruitland	Fossil Forest	five carapace fragments
SMP VP-2643		Testudines indet.	Testudines indet.	Testudines indet.		450	Fruitland	Fossil Forest	plastron fragment
SMP VP-2681		Testudines indet.	Testudines indet.	Testudines indet.		228	Kirtland	Hunter Wash	incomplete plastron
SMP VP-2739		Testudines indet.	Testudines indet.	Testudines indet.		228	Kirtland	Hunter Wash	incomplete left femur
SMP VP-2747		Testudines indet.	Testudines indet.	Testudines indet.		228	Kirtland	Hunter Wash	cervical vertebra
SMP VP-2768		Testudines indet.	Testudines indet.	Testudines indet.		450	Fruitland	Fossil Forest	distal end of humerus
SMP VP-2777		Testudines indet.	Testudines indet.	Testudines indet.		450	Fruitland	Fossil Forest	?limb osteoderm

(continued)

(continued)

Current number	Previous number	Family	Genus	Species	Type status	Locality Number	Formation	Member	Description
SMP VP-2795		Testudines indet.	Testudines indet.	Testudines indet.		450	Fruitland	Fossil Forest	incomplete ?humerus
SMP VP-3343		Testudines indet.	Testudines indet.	Testudines indet.		409	Kirtland	Hunter Wash	incomplete plastron
SMP VP-3344		Testudines indet.	Testudines indet.	Testudines indet.		409	Kirtland	Hunter Wash	incomplete plastron
SMP VP-3373		Testudines indet.	Testudines indet.	Testudines indet.		392	Kirtland	De-na-zin	plastron fragment
SMP VP-3377		Testudines indet.	Testudines indet.	Testudines indet.		461	Kirtland	Hunter Wash	incomplete costal & 2 carapace fragments
SMP VP-3382		Testudines indet.	Testudines indet.	Testudines indet.		450	Fruitland	Fossil Forest	2 plastron fragments
SMP VP-3388		Testudines indet.	Testudines indet.	Testudines indet.		461	Kirtland	Hunter Wash	2 carapace fragments
SMP VP-2381		Testudines indet.	Testudines indet.	Testudines indet.		382	Kirtland	De-na-zin	?girdle element & 2 fragments
SMP VP-2443		Testudines indet.	Testudines indet.	Testudines indet.		421	Kirtland	Hunter Wash	distal & proximal ends of a limb bone
SMP VP-3373		?Testudines indet.	?Testudines indet.	?Testudines indet.		461	Kirtland	Hunter Wash	?carapace/plastron fragments
NMMNH P-28984		various taxa	various taxa	various taxa		4098	Kirtland	unknown	numerous bone fragments
NMMNH P-32887		various taxa	various taxa	various taxa		2387	Fruitland	Fossil Forest	numerous carapace, plastron, axial skeleton fragments
NMMNH P-49857	UA 14835	various taxa	various taxa	various taxa		6552	Kirtland?	De-na-zin?	13 shell fragments

References

Armstrong-Ziegler, J. G. (1978). An aniliid snake and associated vertebrates from the Campanian of New Mexico. *Journal of Paleontology*, *52*, 480–483.

Armstrong-Ziegler, J. G. (1980). Amphibia and reptilia from the Campanian of New Mexico. *Fieldiana*: *Geology*, *New Series*, *No. 4*, 1–39.

Bauer, C. M. (1916). Contributions to the geology and paleontology of San Juan County, New Mexico. 1. Stratigraphy of a part of the Chaco River Valley. *United States Geological Survey Professional Paper*, *98*, 271–278.

Baur, G. (1891). Notes on some little known American fossil tortoises. *Proceedings of the Academy of Natural Sciences of Philadelphia*, *43*, 411–430.

Brinkman, D. B. (2003). A review of nonmarine turtles from the Late Cretaceous of Alberta. *Canadian Journal of Earth Sciences*, *40*, 557–571.

Brinkman, D. B. (2005). Turtles: Diversity, paleoecology, and distribution. In P. J. Currie & E. B. Koppelhus (Eds.), *Dinosaur Provincial Park*: *A spectacular ancient ecosystem revealed* (pp. 202–220). Bloomington: Indiana University Press.

Brinkman, D. B., & Nicholls, E. L. (1991). Anatomy and relationships of the turtle *Boremys pulchra* (Testudines: Baenidae). *Journal of Vertebrate Paleontology*, *11*, 302–315.

Brinkman, D., & Nicholls, E. L. (1993). New specimen of *Basilemys praeclara* Hay and its bearing on the relationships of the Nanhsiungchelyidae (Reptilia: Testudines). *Journal of Paleontology*, *67*, 1027–1031.

Brinkman, D., & Peng, J.-H. (1996). A new species of *Zangerlia* (Testudines: Nanhsiungchelyidae) from the Upper Cretaceous redbeds at Bayan Mandahu, Inner Mongolia, and the relationships of the genus. *Canadian Journal of Earth Sciences*, *33*, 526–540.

Brinkman, D., & Rodriguez de la Rosa, R. (2006). Nonmarine turtles from the Cerro del Pueblo Formation (Campanian), Coahuila State, Mexico. *New Mexico Museum of Natural History and Science Bulletin*, *35*, 229–243.

Brown, B. (1910). The Cretaceous Ojo Alamo beds of New Mexico with a description of the new dinosaur genus *Kritosaurus*. *Bulletin of the American Museum of Natural History*, *28*, 267–274.

Brown, J. (1983). Geologic and isopach maps of the Bisti, De-na-zin and Ah-she-sle-pah [sic] wilderness study areas, New Mexico. Department of the Interior, Geological Survey, Reston, VA, Map MF-1508-A, scale 1:50,000, 2 sheets.

Bryant, L. J. (1989). Non-dinosaurian lower vertebrates across the Cretaceous-Tertiary boundary in northeastern Montana. *University of California Publications in Geological Sciences*, *134*, 1–107.

Cope, E. D. (1865). Third contribution to the herpetology of tropical America. *Proceedings of the Academy of Natural Sciences Philadelphia*, *1865*, 185–198.

Cope, E. D. (1868). On the origin of genera. *Proceedings of the Academy of Natural Sciences of Philadelphia*, *1868*, 242–300.

Cope, E. D. (1870). Stated meeting, October 7, 1870. *Proceedings of the Academy of Natural Sciences of Philadelphia*, *11*, 514–515.

Cope, E. D. (1872). Descriptions of some new Vertebrata from the Bridger group of the Eocene. *Proceedings of the American Philosophical Society*, *12*, 460–465.

Cope, E. D. (1873). Some extinct turtles from the Eocene strata of Wyoming. *Proceedings of the Academy of Natural Sciences, Philadelphia*, *1873*, 277–279.

Cope, E. D. (1874). Description of *Adocus lineolatus*. *Bulletin of the United States Geological and Geographical Survey of the Territories* (Vol. 1, pp. 1–30). Bulletin 1, series 2.

Cope, E. D. (1876). Description of some vertebrate remains from the Fort Union beds of Montana. *Proceedings of the Academy of Natural Sciences of Philadelphia*, *24*, 248–261.

Cope, E. D. (1882). Contributions to the history of Vertebrata of the Lower Eocene of Wyoming and New Mexico, made during 1881. *Proceedings of the American Philosophical Society*, *20*, 139–197.

Eaton, J. G., Cifelli, R. L., Hutchison, J. H., Kirkland, J. I., & Parrish, J. M. (1999). Cretaceous vertebrate faunas from the Kapairowits Plateau, south-central Utah. *Utah Survey Miscellaneous Publications*, *no. 99-1*, 345–353.

Fassett, J. E., & Hinds, J. S. (1971). Geology and fuel resources of the Fruitland Formation and Kirtland Shale of the San Juan Basin, New Mexico. *Geological Survey Professional Paper*, *676*, 1–76.

de Fitzinger, L. J. F. J. (1826). *Neue Classification der Reptilien nach ihren natürlichen Verwandtschften*. Vienna, 66 pp.

Gaffney, E. S. (1972). The systematics of the North American family Baenidae (Reptilia, Cryptodira). *Bulletin of the American Museum of Natural History*, *147*, 243–319.

Gaffney, E. S. (1975). A phylogeny and classification of the higher categories of turtles. *Bulletin of the American Museum of Natural History*, *155*, 389–436.

Gaffney, E. S. (1979). Comparative cranial morphology of recent and fossil turtles. *Bulletin of the American Museum of Natural History*, *164*, 67–376.

Gaffney, E. S., & Meylan, P. A. (1988). A phylogeny of turtles, In M. J. Benton (Ed.), *The phylogeny and classification of Tetrapods*: *Amphibians*, *reptiles*, *birds* (Vol. 1, pp. 157–219). Oxford: Systematics Association Special Volume, Clarendon Press.

Gaffney, E. S., Tong, H., & Meylan, P. A. (2006). Evolution of the side-necked turtles: The families Bothremydidae, Euraxemydidae, and Araripemydidae. *American Museum of Natural History Bulletin*, *300*, 1–698.

Geoffroy St. Hilaire, E. F. (1809). Sur les tortues molles, nouveau genre sous le nom de *Trionyx* et sur la formation des carapaces. *Annales du Muséum National d'histoire naturelles*, *Paris*, *14*, 1–20.

Gardner, J. D., Russell, A. P., & Brinkman, D. B. (1995). Systematics and taxonomy of soft-shelled turtles (Family Trionychidae) from the Judith River Group (mid-Campanian) of North America. *Canadian Journal of Earth Sciences*, *32*, 631–643.

Gates, T. A., Sampson, S. D., Zanno, L. E., Roberts, M. E., Eaton, J. G., Nydam, R. L., et al. (2010). Biogeography of terrestrial and fresh-water vertebrates from the Late Cretaceous (Campanian) western interior of North America. *Palaeogeography*, *Paleoclimatology*, *Palaeogeography*, *291*, 371–387.

Gilmore, C. W. (1916). Vertebrate faunas of the Ojo Alamo, Kirtland, and Fruitland formations. *United States Geological Survey Professional Paper*, *98*, 279–302.

Gilmore, C. W. (1919). Reptilian faunas of the Torrejon, Puerco, and underlying Upper Cretaceous formations of the San Juan Basin, New Mexico. *United States Geological Survey Professional Paper*, *19*, 1–68.

Gilmore, C. W. (1935). On the Reptilia of the Kirtland Formation of New Mexico, with descriptions of new species of fossil turtles. *Proceedings of the United States National Museum*, *83*, 159–188.

Hay, O. P. (1902). Bibliography and catalog of fossil vertebrates from North America. *Bulletin of the United States Geological Survey*, *179*, 1–868.

Hay, O. P. (1908). *The fossil turtles of North America*. Washington, DC: Carnegie Institution of Washington.

Hay, O. P. (1910). Descriptions of eight new species of fossil turtles from the west of the one hundredth meridian. *Proceedings of the United States National Museum*, *38*, 307–325.

Hirayama, R., Brinkman, D. B., & Danilov, I. G. (2000). Distribution and biogeography of non-marine Cretaceous turtles. *Russian Journal of Herpetology*, *7*, 181–198.

Hirayama, R., Sakurai, K., Chitoku, T., Kawakami, G., & Kito, N. (2001). *Anomalochelys angulata*, an unusual land turtle of family Nanhsiungchelyidae (Superfamily Trionychoidea; Order Testudines) from the Upper Cretaceous of Hokkaido, North Japan. *Russian Journal of Herpetology*, *8*, 127–138.

Holman, J. A., & Sullivan, R. M. (1981). A small herpetofauna from the type section of the Valentine Formation (Miocene: Barstovian), Cherry County, Nebraska. *Journal of Paleontology*, *55*, 138–144.

Holroyd, P. A., & Hutchison, J. H. (2002). Patterns of geographic variation in latest Cretaceous vertebrates: Evidence from the turtle component. In J. H. Hartman, K. R. Johnson, & D. J. Nichols (Eds.), *The Hell Creek Formation and the Cretaceous-Tertiary boundary in the northern Great Plains: An integrated continental record of the end of the Cretaceous* (pp. 177–190). Boulder: Geological Society of America.

Hunt, A. P., & Lucas, S. G. (1992). Stratigraphy, paleontology and age of the Fruitland and Kirtland formations (Upper Cretaceous), San Juan Basin, New Mexico. In S. G. Lucas, B. S. Kues, T. E. Williamson & A. P. Hunt (Eds.), *San Juan Basin IV* (pp. 217–239). Albuquerque: New Mexico Geological Society.

Hunt, A. P., & Lucas, S. G. (1993). Cretaceous vertebrates of New Mexico. *New Mexico Museum of Natural History and Science Bulletin*, *2*, 77–91.

Hutchison, J. H. (1991). Early Kinosterninae (Reptilia: Testudines) and their phylogenetic significance. *Journal of Vertebrate Paleontology*, *11*, 145–167.

Hutchison, J. H., & Holroyd, P. A. (2003). Late Cretaceous and early Paleocene turtles of the Denver Basin, Colorado. *Rocky Mountain Geology*, *38*, 121–142.

Jasinski, S. E., Sullivan, R. M., & Lucas, S. G. (2011). Revision of the Alamo Wash local fauna, Upper Cretaceous (Maastrichtian) Ojo Alamo Formation (Naashoibito Member), San Juan Basin, New Mexico. *New Mexico Museum of Natural History and Science Bulletin*, *53*, 216–271.

Joyce, W. G. (2007). Phylogenetic relationships of Mesozoic turtles. *Bulletin of the Peabody Museum of Natural History*, *48*, 3–102.

Joyce, W. G., & Lyson, T. R. (2010). A neglected lineage of North American turtles fills a major gap in the fossil record. *Palaeontology*, *53*, 241–248.

Joyce, W. G., & Norell, M. A. (2005). *Zangerlia ukhaachelys*, new species, a nanhsiungchelyid turtle from the Late Cretaceous of Ukhaa Tolgod, Mongolia. *American Museum Novitates*, *3481*, 1–19.

Joyce, W. G., Revan, A., Lyson, T. R., & Danilov, I. G. (2009). Two new plastomenine softshelled turtles from the Paleocene of Montana and Wyoming. *Bulletin of the Peabody Museum of Natural History*, *50*, 307–325.

Lambe, L. M. (1902). New genera and species from the Belly River series (Mid-Cretaceous). *Geological Survey of Canada, Contributions to Canadian Paleontology*, *3*, 25–81.

Lambe, L. M. (1906). *Boremys*, a new chelonian genus from the Cretaceous of Alberta. *Ottawa Naturalist*, *19*, 232–234.

Langston, W. Jr. (1956). The shell of *Basilemys varialosa* (Cope). *Annual Report of the National Museum of Canada, Bulletin*, *142*, 155–163.

Larson, D. W., Longrich, N. R., Evans, D. C., & Ryan, M. J. (2012). A new species of *Neurankylus* from the Milk River Formation (Cretaceous: Santonian) of Alberta, Canada, and a revision of the type species *N. eximius*. In D. B. Brinkman, P. A. Holroyd & J. D. Gardner (Eds.), *Morphology and evolution of turtles*. Dordrecht: Springer.

Lehman, T. M. (1981). The Alamo Wash local fauna: A new look at the old Ojo Alamo fauna. In S. G. Lucas, J. K. Rigby Jr. & B. S. Kues (Eds.), *Advances in San Juan Basin Paleontology* (pp. 189–221). Albuquerque: University of New Mexico Press.

Leidy, J. (1856). Notices of extinct Vertebrata discovered by Dr. F. V. Hayden, during the expedition to the Sioux country under the command of Lieut. G. K. Warren. *Proceedings of the National Academy of Sciences*, *1856*, 311–312.

Leidy, J. (1870). Descriptions of *Emys jeansi*, *E. haydeni*, *Baena arenosa*, and *Saniwa ensidens*. *Academy of Natural Sciences, Philadelphia*, *1870*, 123–124.

Leidy, J. (1872). On a new genus of turtles. *Proceedings of the Academy of Natural Sciences of Philadelphia*, *1872*, 162.

Linneaus, C. (1758). *Systema naturae* (10th ed., Vol. 1, 842 p.). Stockholm: Laurentii Sylvii.

Lucas, S. G. (1981). Dinosaur communities of the San Juan Basin: A case for the lateral variations in the composition of Late Cretaceous dinosaur communities. In S. G. Lucas, J. K. Rigby Jr. & B. S. Kues (Eds.), *Advances in San Juan Basin Paleontology* (pp. 337–393). Albuquerque: University of New Mexico Press.

Lucas, S. G., & Mateer, N. J. (1983). Vertebrate paleoecology of the Late Campanian (Cretaceous) Fruitland Formation, San Juan Basin, New Mexico (USA). *Acta Palaeontologica Polonica*, *28*, 195–204.

Lucas, S. G., & Sullivan, R. M. (2006). *Denazinemys*, a new name for some Late Cretaceous turtles from the Upper Cretaceous of the San Juan Basin, New Mexico. *New Mexico Museum of Natural History and Science Bulletin*, *35*, 223–227.

Mateer, N. J. (1981). The reptilian megafauna from the Kirtland Shale (Late Cretaceous) of the San Juan Basin, New Mexico. In S. G. Lucas, J. K. Rigby, Jr. & B.S. Kues (Eds.), *Advances in San Juan Basin Paleontology* (pp. 49–75). Albuquerque: University of New Mexico Press.

McCord, R. D., II. (1996). Turtle biostratigraphy of Late Cretaceous and early Tertiary continental deposits San Juan Basin, New Mexico. *Fossils of Arizona: Proceedings, Southwest Paleontological Society and Mesa Southwest Museum, Mesa*, 135–153.

Parham, J. F., & Hutchison, J. H. (2003). A new eucrytodiran turtle from the Late Cretaceous of North America (Dinosaur Provincial Park, Alberta, Canada. *Journal of Vertebrate Paleontology*, *23*, 783–798.

Riggs, E. S. (1906). The carapace and plastron of *Basilemys sinuosus*, a new fossil tortoide from the Laramie beds of Montana. *Field Columbian Museum of Natural History, Geology Series*, *2*, 249–256.

Russell, L. S. (1934). Fossil turtles from Saskatchewan and Alberta. *Transactions of the Royal Society of Canada, Series 3*, *148*, 101–110.

Sankey, J. T. (2006). Turtles of the upper Aguja Formation (late Campanian), Big Bend National Park, Texas. *New Mexico Museum of Natural History and Science Bulletin*, *35*, 235–243.

Schneider, J. G. (1783). *Allgemeine Naturgeschichte der Schildkröten, nebst einem Systematischen Verseichnisse der einzelnen Arten*. Leipzig: Müller.

Sukhanov, V. B., & Narmandakh, P. (1975). Turtles of the *Basilemys* group (Chelonia, Dermatemyidae) in Asia. *Trudy Sovmestonoj Sovetsko-Mongol'skoj Paleontologičeskoj Ekspedicii*, *2*, 94–101 (in Russian).

Sukhanov, V. B., & Narmandakh, P. (1977). The shell and limbs of *Basilemys oreintalis* (Chelonia, Dermatemyidae): A contribution to the morphology and evolution of the genus. *Trudy Sovmestonoj Sovetsko-Mongol'skoj Paleontologičeskoj Ekspedicii*, *4*, 57–79 (in Russian).

Sukhanov, V. B., & Narmandakh, P. (2006). New taxa of Mesozoic turtles from Mongolia. In I. G. Danilov & J. F. Parham (Eds.), *Fossil turtle research* (Vol. 1, pp. 119–127). St. Petersburg: Zoological Institute of the Russian Academy of Sciences (*Russian Journal of Herpetology*, *13*, supplement).

Sukhanov, V. B., Danilov, I. G., & Syromyatnikova, E. V. (2008). The description and phylogenetic position of a new nanhsiungchlyid turtle from the Late Cretaceous of Mongolia. *Acta Paleontologia Polonica*, *53*, 601–614.

Sullivan, R. M., & Lucas, S. G., (2003). The Kirtlandian, a new land-vertebrate "age" for the Late Cretaceous of western North America. *New Mexico Geological Society, Guidebook*, *54*, 369–377.

Sullivan, R. M., & Lucas, S. G. (2006). The Kirtlandian land-vertebrate "age"-faunal composition, temporal position and biostratigraphic correlation in the nonmarine Upper Cretaceous of western North America. *New Mexico Museum of Natural History and Science Bulletin*, *35*, 7–29.

Sullivan, R. M., Lucas, S. G., Hunt, A. P., & Fritts, T. H. (1988). Color pattern on the selmacryptodiran turtle *Neurankylus* from the early Paleocene (Puercan) of the San Juan Basin, New Mexico. *Contributions in Science, Natural History Museum of Los Angeles County*, *401*, 1–9.

Swofford, P. L. (2002). PAUP, version 4.0b10. Sunderland: Sinauer Associates.

Tomlinson, S. L. (1997). Late Cretaceous and early Tertiary turtles from the Big Bend region, Brewster County, Texas (Ph.D. Dissertation, Texas Tech University, Lubbock, Texas).

Wiman, C. (1930). Über Ceratopsia aus der Oberen Kreide in New Mexico. *Nova Acta Regiae Societatis Scientiarum Upsaliensis*, *7*, 1–19.

Wiman, C. (1931). *Parasaurolophus tubicen* n. sp. aus der Kreide in New Mexico. *Nova Acta Regiae Societatis Scientiarum Upsaliensis*, *7*, 1–11.

Wiman, C. (1932). *Goniopholis kirtlandicus* n. sp. aus der oberen Kreide in New Mexico. *Bulletin of the Geological Institution of the University of Upsala*, *23*, 181–189.

Wiman, C. (1933). Über Schildkröten aus der Oberen Kreide in New Mexico. *Nova Acta Regiae Societatis Scientiarum Upsaliensis*, *9*, 1–34.

Yeh, H.-Y. (1966). A new Cretaceous turtle of Nanhsiung, northern Kwangtang. *Vertebrata PalAsiatica*, *10*, 191–200.

Chapter 21
A New Species of *Neurankylus* from the Milk River Formation (Cretaceous: Santonian) of Alberta, Canada, and a Revision of the Type Species *N. eximius*

Derek W. Larson, Nicholas R. Longrich, David C. Evans, and Michael J. Ryan

Abstract A new species of *Neurankylus* (*N. lithographicus* sp. nov.) is described on the basis of skull and shell material from the Santonian-aged Milk River Formation, Alberta, Canada. The genus *Neurankylus* is also rediagnosed on the basis of the Milk River material and on new material pertaining to the type species *N. eximius*. *N. eximius* previously was considered to be a long-lived and cosmopolitan taxon. New *Neurankylus* material provides insights into the range of morphological variation present in the genus and indicates that many specimens previously referred to *N. eximius* may belong to different species. The congeners recognized in this chapter have a more restricted geographical and temporal range than has been suggested previously. A new phylogenetic analysis of all known baenid taxa, including all described species of *Neurankylus* and several basal paracryptodiran taxa of uncertain affinities, yields two important results: a monophyletic *Neurankylus* is recovered as a basal radiation within Baenidae and parallel evolution is identified among many features previously regarded as synapomorphies for Baenidae. In light of this study and other recent work on turtle systematics, it is now apparent the biogeography and biostratigraphy of Cretaceous turtles may have been more complex than previously appreciated.

Keywords Baenidae • *Neurankylus* • Phylogeny • Santonian • Testudines

D. W. Larson (✉)
University of Alberta Laboratory for Vertebrate Paleontology, University of Alberta, CW 405 Biological Sciences Building, Edmonton, AB T6G 2E9, Canada
e-mail: derek.larson@utoronto.ca

N. R. Longrich
Department of Geology and Geophysics, Yale University, P.O. Box 208109 New Haven, CT 06520-8109, USA
e-mail: longrich@gmail.com

D. C. Evans
Department of Natural History (Palaeobiology), Royal Ontario Museum, 100 Queen's Park, Toronto, ON M5S 2C6, Canada
e-mail: d.evans@utoronto.ca

M. J. Ryan
Cleveland Museum of Natural History, 1 Wade Oval Drive, University Circle, Cleveland, OH 44106, USA
e-mail: mryan@cmnh.org

Introduction

During the Late Cretaceous, North America hosted a remarkably diverse fauna of turtles. This fauna included a number of taxa that are familiar today, including soft-shelled turtles (Trionychidae) and snapping turtles (Chelydridae). However, it also included a number of extinct families, including the terrestrial Nanhsiungchelyidae and the aquatic Solemydidae, Adocidae, and Baenidae (Holroyd and Hutchison 2002; Brinkman 2003a, 2005). The Baenidae are medium-to-large freshwater turtles characterized by fusion of the elements of the carapace and plastron, thick shells, and adaptation of the jaws for crushing (Gaffney 1972; Hutchison 1984; Gaffney and Meylan 1988; Joyce 2007). Baenids were endemic to North America and formed a major part of North America's non-marine aquatic fauna from the Cretaceous through the Paleogene. Although younger and more derived baenid taxa, the Baenodda (Gaffney 1972), are well-represented from the Campanian through Eocene in North America (Gaffney 1972; Sullivan et al. 1988; Brinkman 2003a, 2005; Holroyd and Hutchison 2002; Hutchison and Holroyd 2003; Lyson and Joyce 2009a, b, 2010), comparatively little is known about the earlier and more basal members of the group.

Neurankylus eximius Lambe 1902, was named from an incomplete carapace of a juvenile individual collected from the Campanian Belly River Group along the Red Deer River of Alberta, Canada. An incomplete skull from the Dinosaur Park Formation, likely the same formation that produced the holotype carapace, was subsequently referred to this species

D. B. Brinkman et al. (eds.), *Morphology and Evolution of Turtles*, Vertebrate Paleobiology and Paleoanthropology,
DOI: 10.1007/978-94-007-4309-0_21,

(Brinkman and Nicholls 1993). Since its initial description, referred material has increased this species' apparent stratigraphic and geographic range, from Campanian through Danian time and from southern Alberta south to New Mexico (Hutchison and Archibald 1986; Sullivan et al. 1988; Holroyd and Hutchison 2002; Hutchison and Holroyd 2003).

Gilmore (1916) erected a second species of *Neurankylus*, *N. baueri* Gilmore 1916, from the Kirtland Formation. Gaffney (1972) subsequently subsumed that name under *N. eximius*. However, Sullivan and Lucus (2006) continued to recognize *N. baueri* as distinct on the basis of its shorter and wider first suprapygal and the coincidence of its posterior costal and posterior neural sutures. Most recently, Sullivan et al. (2012) re-diagnosed *N. baueri* based on well-developed gulars and U-shaped intergulars. Gilmore (1919) erected *N. wyomingensis* Gilmore 1919 on material collected from Wyoming, although this taxon was regarded a nomen dubium by Gaffney (1972). However, it too appears to possess a unique combination of features as revealed from Gilmore's (1919) description and figures, and for that reason, we regard *N. wyomingensis* as a valid species. These three species are currently the only formally named species of *Neurankylus*. In addition to holotype and referred specimens of the above three congeners, we also considered ROM 864, probably from the Kirtland Formation of New Mexico, and YPM 8239, from the Lance Formation. Those two specimens have not been systematically described or studied in detail, but they are of interest because of suspicions that they may represent additional species of *Neurankylus*. Specimens referred to *N. eximius* also have been documented from Paleocene deposits in Montana and New Mexico (Hutchison and Archibald 1986; Sullivan et al. 1988); however, these specimens have neither been illustrated nor described in sufficient detail to permit comparison. Based on the descriptions, referral of those Paleocene specimens to *Neurankylus* is here accepted, although their species designation needs to be confirmed.

Brinkman (2003a) was the first to recognize *Neurankylus* from the Santonian stage on the basis of material recovered from the Milk River Formation of southern Alberta. He regarded these specimens as distinct from *N. eximius* based on smaller shell size and a proportionately longer, more squared posterior lobe of the plastron, but did not name the Milk River species due to the fragmentary nature of the material available at the time. Enough material is now available to reconsider the specific identity of the Milk River *Neurankylus*. Two new specimens, an incomplete shell and an incomplete skull, are pivotal for confirming Brinkman's (2003a) suggestion that the Milk River *Neurankylus* represents a distinct species, which we formally name and describe as *N. lithographicus* sp. nov. In an effort to put this new species into its evolutionary context, we also reassess the phylogeny of baenid turtles.

Institutional Abbreviations used in this chapter are: AMNH, American Museum of Natural History, New York, New York, USA; BYU, Brigham Young University, Provo, Utah, USA; CMN, Canadian Museum of Nature, Ottawa, Ontario, Canada; CPS, Colorado Palaeontographical Society, University Museum, University of Colorado, Boulder, Colorado, USA; DNM, Dinosaur National Monument, Vernal, Utah, USA; DORCM, Dorset County Museum, Dorchester, United Kingdom; FMNH, Field Museum of Natural History, Chicago, Illinois, USA; MCZ, Museum of Comparative Zoology, Harvard, Massachusetts, USA; MRF, Marmarth Research Foundation, Marmarth, North Dakota, USA; PMU, Paleontologiska Museet, Uppsala, Sweden; PU, Princeton University, Princeton, New Jersey, USA; ROM, Royal Ontario Museum, Toronto, Ontario, Canada; TGM, Tate Geological Museum, Casper College, Casper, Wyoming, USA; TMP, Royal Tyrrell Museum of Palaeontology, Drumheller, Alberta, Canada; UALVP, University of Alberta Laboratory for Vertebrate Paleontology, Edmonton, Alberta, Canada; UCMP, University of California Museum of Paleontology, Berkeley, California, USA; UMMP, University of Michigan Museum of Paleontology, Ann Arbor, Michigan, USA; UMZC, University Museum of Zoology, Cambridge, United Kingdom; USNM, United States National Museum, Washington, D.C., USA; YPM, Yale Peabody Museum, New Haven, Connecticut, USA.

Geologic Setting

The Milk River Formation is a latest Santonian siliciclastic wedge of the Western Interior Sedimentary Basin (Payenberg et al. 2002). This formation crops out east of the town of Milk River in southernmost Alberta, Canada. It is equivalent both to the Virgelle and Deadhorse Coulee (middle) members of the Eagle Formation and to the Telegraph Creek Formation, farther south in Montana, USA (Payenberg et al. 2002). The Milk River Formation is composed of three members, the uppermost of which is the terrestrial Deadhorse Coulee Member. This member consists of alternating sandstones and siltstones with intermittent coal, interpreted as fluvial channel and floodplain deposits, respectively. The member is up to 70 m thick (Leahy and Lerbekmo 1995), although no single outcrop exposes the full vertical extent of the unit. Vertebrate fossils are known from the lower 35 m of the Deadhorse Coulee Member and those *Neurankylus lithographicus* sp. nov. specimens with known stratigraphic provenance occur only from 8.0 to 26.5 m above the base of the member. Although the maximum age of the unit is poorly constrained, the 34n-33r magnetochron boundary has been noted at 38 m above the base of the Deadhorse Coulee Member (Leahy and Lerbekmo 1995) and this boundary has been dated to 83.5 Ma by Ogg and Smith (2004). At the time of deposition, the localities of interest were roughly 40 km inland or west of the paleoshoreline during a relatively warm time interval (Brinkman 2003a).

Systematic Palaeontology

Testudines Linnaeus 1758
Paracryptodira Gaffney 1975
Baenoidea Williams 1950
Baenidae Cope 1882
Neurankylinae Williams 1950
Neurankylus Lambe 1902

Type species: *Neurankylus eximius* Lambe 1902.

Referred species: *Neurankylus baueri*, *N. wyomingensis*, and *N. lithographicus* sp. nov.

Occurrence: Late Cretaceous-Paleocene (middle Coniacian–early Danian), Alberta, Canada, and New Mexico and Montana, USA. Although specimens from the Maastrichtian and Danian have been referred to this genus (Hutchison and Archibald 1986; Sullivan et al. 1988), no characters assigning these specimens to a species have been reported.

Revised diagnosis: Large relative to other baenid turtles (carapace averaging 500–600 mm long and skull more than 74 mm long in adult individuals); dorsolateral gutters along the peripherals of the carapace; shell ornamentation essentially smooth, but sometimes consisting of subtle parallel ridges and fine, low-density pitting; foramen stapediotemporale bordered by the opisthotic, quadrate, and prootic, but not the supraoccipital; posteriorly rounded short supraoccipital ridge consisting of only the paired parietals in dorsal view.

Remarks: Some of the characters listed as diagnostic for the genus may, in fact, be diagnostic for a more inclusive Neurankylinae (here defined as all baenids closer to *Neurankylus eximius* than to *Baena arenosa* Leidy 1870 or to *Trinitichelys hiatti* Gaffney 1972), such as the position of the foramen stapediotemporale, also seen in *Arundelemys dardeni* Lipka et al. 2006, but differs from all other known baenids by the exclusion of the supraoccipital and from all except *Stygiochelys estesi* Gaffney and Hiatt 1971 by the inclusion of the opisthotic (Gaffney 1972; Brinkman and Nicholls 1991; Brinkman 2003b; Lyson and Joyce 2009a, b, 2010). This feature suggests a close relationship between *Neurankylus* and *Arundelemys* and supports the inclusion of *Arundelemys* within the Neurankylinae. However, such conclusions are hampered by the incomplete nature of described material (Lipka et al. 2006) and a general paucity of known basal baenids. The Neurankylinae as defined here also excludes *Thescelus insiliens* Hay 1908, which is likely the senior synonym of the baenine *Hayemys latifrons* Gaffney 1972 (Lyson and Joyce 2010). *N. wyomingensis*, known only from the posterior portion of a single shell, may not be referable to the genus *Neurankylus* based on the above diagnosis; although the known material is insufficient for diagnosis of a distinct genus, it is provisionally retained here in *Neurankylus*. The shell ornamentation of *N. wyomingensis*, which consists of small nodes with shallow anastomosing channels, is distinct from other *Neurankylus* species, but the species does share large size and dorsolateral gutters with *Neurankylus*.

Adult size has been shown to be a valid character in baenids, which, unlike many species of turtles, exhibit determinate growth and full fusion of their shells (Hutchison 1984). The dorsolateral gutters noted for the genus by previous authors (Gilmore 1916; Gaffney 1972) also occur in *Glyptops plicatulus* (Cope 1877; Gilmore 1916) but are regarded here as a valid character for defining the genus because they are not known to occur in any other baenid.

In addition to the diagnostic characters, many basal paracryptodire characters can be used for differentiating *Neurankylus* from other Late Cretaceous baenids. *Neurankylus* differs from members of the Baenodda in having primitively wide vertebral scutes bordered posteriorly by marginal scutes, and from *Thescelus* by the lack of deep anterior emargination. Unlike many baenines, the plastron is primitive in the roughly parallel anterior free borders, intergular scutes separating the gulars, and an anal scute that seldom overlaps the hypoplastron. In contrast to most baenodds (except *Gamerabaena sonsalla* Lyson and Joyce 2010) and *Trinitichelys*, the skull has prefrontals with large dorsal lappets.

Phylogenetically, the genus can be defined as a stem-based clade whose members are more closely related to *Neurankylus eximius* than to *Arundelemys dardeni*, *Trinitichelys hiatti*, or *Thescelus insiliens*. Morphologic differences between identified species of *Neurankylus* are subtle (Table 21.1), but among the most useful appears to be the scute morphology on the anterior lobes of the plastra (Fig. 21.1).

Neurankylus eximius Lambe 1902
(Figs. 21.1b, 21.2, 21.3, 21.6b, d, f, h)

Synonymies: *Charitemys captans*: Hay 1908, pp. 98–100, Figs. 93–95; *Baena fluviatilis*: Parks 1933, pp. 19–25, Figs. 1, 2, pl. 7.

Holotype: CMN 1504 (Fig. 21.2), incomplete carapace preserving left costals 3, 6–9, right costals 3, 4, neural 8 (fused to first suprapygal), and both suprapygals. The holotype comes from the Belly River Group (Lambe 1902), probably from the area now encompassed in Dinosaur Provincial Park, but there is no information as to whether it comes from the Dinosaur Park Formation or the underlying

Table 21.1 Comparison of named species of *Neurankylus* (exclusive of *N. wyomingensis*)

Character	*N. eximius*	*N. baueri*	*N. lithographicus*
Intergular lateral edges	Curved	Straight	Straight
Size of intergulars	<gulars	<gulars	>gulars
First–second costal suture orientation	Anteriorly inclined	Perpendicular to midline	Greatly anteriorly inclined
Posterior neurals	Hexagonal	Rectangular	(Unknown)
Eighth neural	Fused to suprapygal	Not fused	(Unknown)
Fifth vertebral	Wide	Narrow	?Wide
Fourth marginal wide	Middle	Posteriorly	Posteriorly
Plastron length	>carapace width	<carapace length	<carapace length
Anterior carapace squared-off	Yes	No	No
Posterior carapace scalloped	No	Yes	No
Dorsal median keel	Yes	Yes	(Unknown)
Skull width	Wide	(Unknown)	Narrow

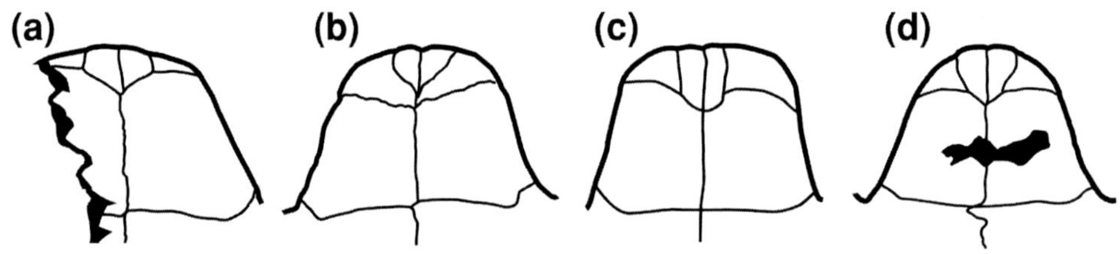

Fig. 21.1 Variation in the ventral surface of the anterior plastral lobe of *Neurankylus*. **a** TMP 2007.035.0045, holotype of *N. lithographicus* sp. nov. **b** TMP 2003.012.0171, *N. eximius*. **c** USNM 8344, *N. baueri* (modified from Gilmore 1916). **d** ROM 864, *Neurankylus* sp. Images drawn to same width

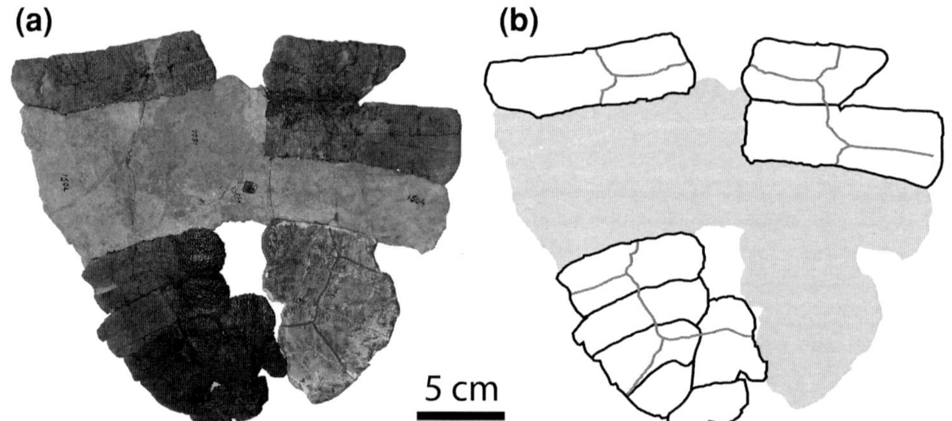

Fig. 21.2 Dorsal view of holotype carapace (CMN 1504) of *Neurankylus eximius*. This is an anteriorly incomplete carapace consisting of left costals 3 and 6–9, right costals 3 and 4, neural 8 (fused to first suprapygal), and both suprapygals, with some missing portions reconstructed in plaster. **a** Photograph. **b** Interpretive drawing, with plaster reconstruction shown in *light grey*. Images at same magnification

Oldman Formation. Given that the Dinosaur Park Formation is more fossiliferous, is exposed over a larger area, and that collecting has tended to focus on this formation, the specimen is probably from the Dinosaur Park Formation.

Referred specimens: Based on synapomorphies, the following specimens are identified as having diagnostic features for the species, and are referable to *Neurankylus eximius*. (Specimens that have previously been referred to *N. eximius*, but which, in our opinion, can only be identified as *Neurankylus* indet. are excluded from this list.) AMNH 6098, holotype of *Charitemys captans*, fragmentary skull, carapace, and plastron from the Judith River Formation of Montana; ROM 854, holotype of *Baena fluviatilis*, incomplete carapace and plastron probably from the Dinosaur Park Formation of Alberta; ROM 1943, complete carapace and plastron from the Dinosaur Park Formation of Alberta; TMP 1989.036.0112, anterior portion of anterior plastral lobe from the Dinosaur Park Formation of Alberta; TMP 1999.055.0134, complete carapace and plastron from the Dinosaur Park Formation of Alberta; TMP 2003.012.0171,

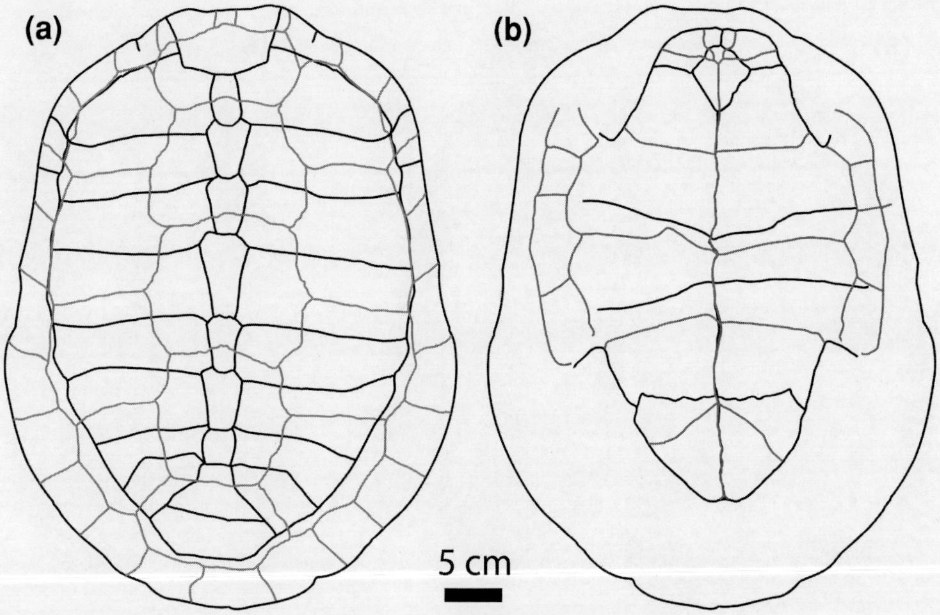

Fig. 21.3 Reconstruction of *Neurankylus eximius* carapace and plastron, based on referred specimen TMP 1999.055.0134 (for photograph see Brinkman 2005, Fig. 11.1c *right*). **a** Dorsal view. **b** Ventral view. *Black lines* denote bone sutures and *grey lines* denote scute sulci. Images at same magnification

complete carapace and plastron from the Dinosaur Park Formation of Alberta; UALVP 30824, incomplete skull preserving much of the roof and braincase from the Dinosaur Park Formation of Alberta. Where specific locality data is known, Albertan specimens are from Dinosaur Provincial Park; specimens without precise locality data are consistent with this location (except AMNH 6098, which is from farther south in Montana).

Occurrence: Dinosaur Park Formation, Dinosaur Provincial Park, Alberta, Canada; Judith River Formation, Montana. The Dinosaur Park Formation in this area is 76.5–74.8 Ma in age (Eberth 2005). The Judith River Formation ranges minimally from 78.5 to 75.4 Ma (Goodwin and Deino 1989; Rogers 1998).

Revised diagnosis: *Neurankylus eximius* is here rediagnosed on the basis of the following autapomorphies: eighth neural and first suprapygal fused even when other carapacial sutures are open; coincidence of the fourth–fifth vertebral sulcus along the aforementioned fused suture; plastron length exceeding carapace width; intergulars only barely separating gulars and having little or no shared sulcus with humerals; and sigmoid intergular–gular sulcus oriented anterolaterally to posteromedially, creating a heart shape (Figs. 21.1b, 21.3). The species differs further from *N. baueri* in that the first–second costal suture is oriented anterolaterally, the posterior neurals are hexagonal with short sides anteriorly, the fifth vertebral is wider than second suprapygal, and the posterior margin of the carapace is smooth and not scalloped. It differs from both *N. baueri* and *N. lithographicus* in that its fourth marginals widen halfway along their length. *N. eximius* differs from *N. baueri*, *N. lithographicus*, and the Lance Formation *Neurankylus* in that the anterior margin of the carapace is squared off, but the carapace possesses a dorsal median keel as in *N. baueri*. The skull differs from that of *N. lithographicus* in being broadly wedge-shaped, with the opisthotic posterior width exceeding 1.5 times the dorsoventral height, the stapedial foramen is more than half of the anteroposterior length of the otic region away from the quadrate-squamosal suture, and the foramen nervi hypoglossi is directed posteriorly.

Remarks: Using the above rediagnosis, *Neurankylus eximius* is recognized in the current study only on the basis of material recovered from middle to upper Campanian strata in Alberta and Montana. The presence of a dorsal median keel on the carapace, although noted to be polymorphic for the species (Gaffney 1972), is present on all known material of *N. eximius* sensu stricto (six specimens) and on specimens of *N. baueri* (five specimens). All four Dinosaur Park Formation specimens that preserve the relevant area lack a scalloped posterior margin of the carapace. Other features thought to be variable within the species, such as the scute morphology on the anterior lobe of the plastron (Gaffney 1972), upon investigation, appear in five of the specimens examined and diagnose the currently recognized species (Fig. 21.1; Table 21.1). Referral of the holotype (AMNH 6098) of *Charitemys captans* to *N. eximius* (Gaffney 1972) is accepted here based on the fourth marginal scute morphology and the orientation of the first–second costal suture. However, the fragmentary nature of the specimen restricts further comparison. The holotype of *Baena fluviatilis* (ROM 854) preserves all of the diagnostic characters for the species except for the fourth marginal morphology. Other material previously referred to *N. eximius*

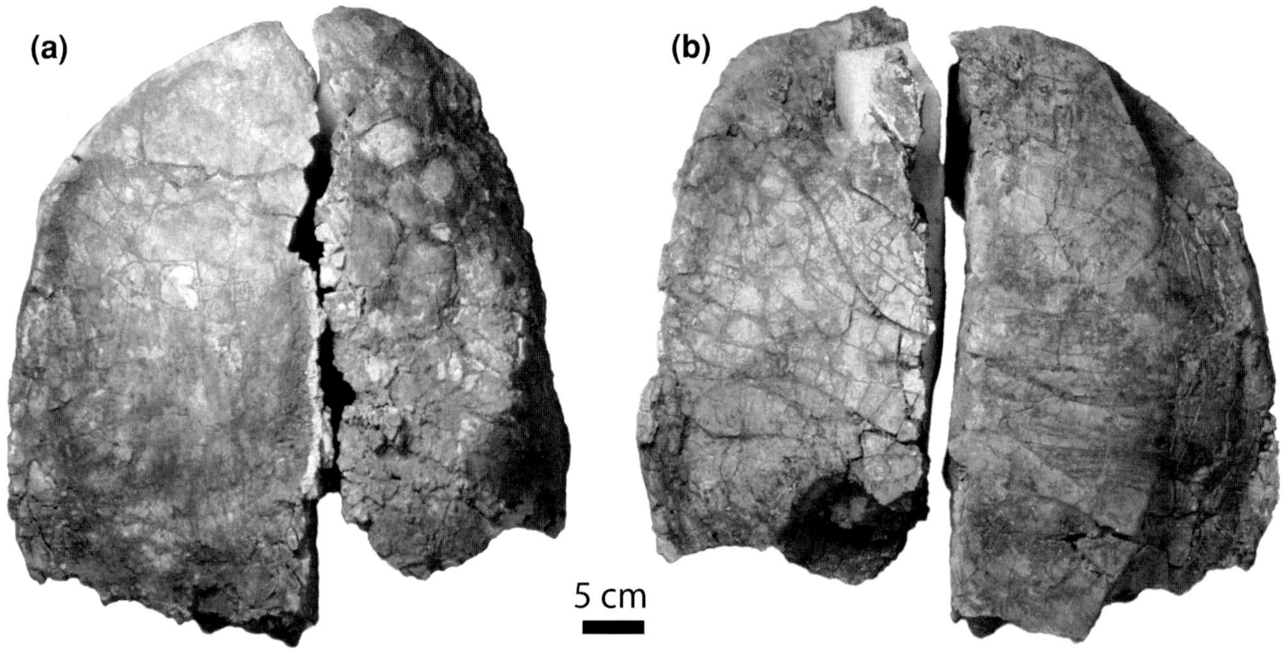

Fig. 21.4 Photographs of holotype (TMP 2007.035.0045) shell of *Neurankylus lithographicus* sp. nov. **a** Dorsal view. **b** Ventral view. Images at same magnification

(Gaffney 1972; Hutchison and Archibald 1986; Sullivan et al. 1988) either does not fit our revised diagnosis or has not been described in enough detail to facilitate comparison.

The diagnostic value of the skull of *Neurankylus eximius*, and thereby the diagnosis of the entire genus, is hampered by the fragmentary preservation of the only described cranial specimens. Although most of the skull roof, braincase, and otic regions are preserved, large proportions of the facial region remain unknown. It is also unknown whether or not a squamosal-parietal contact is present. However, further preparation of UALVP 30824 after its initial description (Brinkman and Nicholls 1993) reveals the true ventral extent of the quadrates and mandibular condyles (Fig. 21.6). The mandibular condyles are comparable in relative size to those of most other baenids, with the exception of the genus *Palatobaena* Gaffney 1972 (Lyson and Joyce 2009a).

Neurankylus lithographicus sp. nov.
(Figs. 21.1a, 21.4, 21.5, 21.6a, c, e, g)

Synonymies: *Neurankylus* sp.: Brinkman 2003a, pp. 560–561.

Holotype: TMP 2007.035.0045 (Fig. 21.4), incomplete carapace and plastron.

Holotype locality, unit, and age: Deadhorse Coulee Member of the Milk River Formation, 26.5 m above the Virgelle Member contact, 1.5 km east of Writing-On-Stone Provincial Park, Alberta, Canada. Detailed locality information on file at the TMP. The Deadhorse Coulee Member is latest Santonian (~ 83.5 Ma) in age (Payernberg et al. 2002).

Etymology: Specific epithet derived from Greek *lithos*, stone, and Greek *graphikos*, of writing, referring to Writing-On-Stone Provincial Park, the closest major landmark to the locality.

Designated paratypes: TMP 1991.113.0001, fragmentary carapace, plastron, caudal vertebrae, limb elements, and possible squamosal; TMP 1991.113.0009, plastron fragment.

Provisionally referred specimens: TMP 1994.377.0001, carapace fragment; TMP 1998.102.0014, carapace fragment and skull fragment including portions of basioccipital, basisphenoid, pterygoids, and palatines; TMP 2007.036.0001, incomplete skull, including portions of pterygoid, basisphenoid, basioccipital, exoccipital, opisthotic, prootic, quadrate, and squamosal.

Occurrence: Deadhorse Coulee Member of the Milk River Formation, around Writing-On-Stone Provincial Park, Alberta, Canada.

Diagnosis: Autapomorphies: intergular scutes larger than gular scutes; distinct first marginal scutes medially narrowing; first–second costal sutures greatly anterolaterally directed such that the angle between the suture and the anterior midline of the shell is less than 70°; and anterior rim of the carapace thickened.

Description: The holotype (TMP 2007.035.0045; Figs. 21.4, 21.5) and TMP 2007.036.0001 (Fig. 21.6a, c, e, g), together with other referred material from the Milk River Formation, preserve a morphology distinct from all published descriptions and illustrations of *Neurankylus* (Table 21.1).

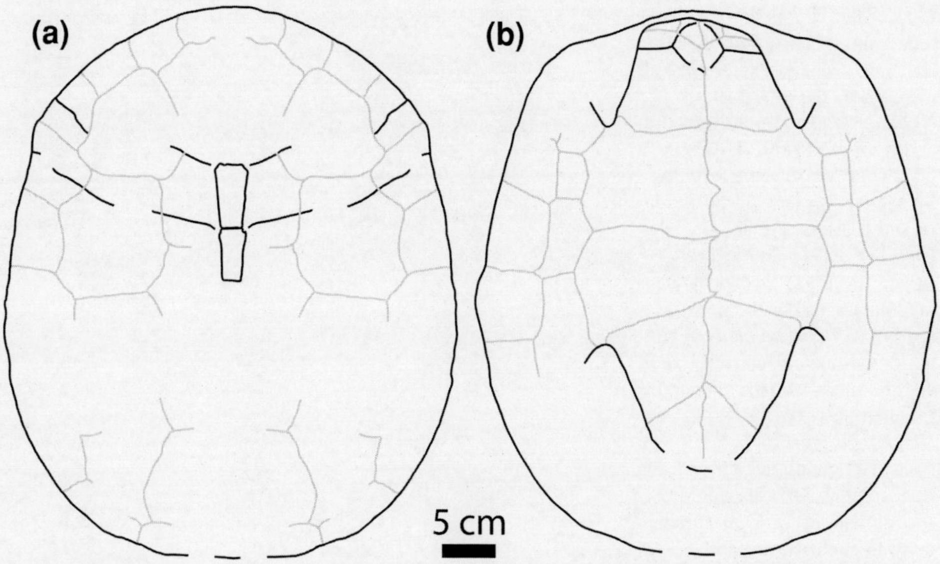

Fig. 21.5 Reconstruction of *Neurankylus lithographicus* sp. nov., carapace and plastron, based on TMP 2007.035.0045 (holotype) and referred specimens TMP 1994.377.0001 and TMP 1991.113.0009. **a** Dorsal view. **b** Ventral view. *Black lines* denote bone sutures and *grey lines* denote scute sulci. Images at same magnification

The paratype specimens are consistent in morphology with that seen in the holotype, although in most respects the former specimens are not as complete.

The holotype shell (Figs. 21.4, 21.5) is referred to the genus *Neurankylus* based on wide vertebral scutes bordered posteriorly by marginal scutes, a lack of deep anterior emargination, dorsolateral gutters along the lateral peripherals, intergular scutes separating the gulars, and an anal scute that does not overlap the hypoplastron. All elements exhibit the large size typical of the genus. *Neurankylus lithographicus* possesses a dorsoventrally thickened anterior rim of the carapace relative to the overall thickness of the carapace. The anterior rim of the carapace margin also appears anteriorly tapered in dorsal view and resembles that of *N. baueri* and the Lance Formation *Neurankylus*, contrasting with the broader more squared-off shell of *N. eximius*. As well, the fourth marginal scute widens posteriorly, again resembling the state in *N. baueri*, rather than *N. eximius* in which the scute widens halfway along its length. The first marginal scute narrows rapidly medially, producing a first vertebral scute with a medial anterior projection, an autapomorphy of the species. The posterior rim of the carapace, from what is preserved, does not appear to be scalloped, similar to *N. eximius* but not *N. baueri*. On the anterior lobe of the plastron, the intergular surface area is larger than the gular surface area, and the intergular–gular sulcus is straight and oriented anteroposteriorly. Like other non-baenine paracryptodires, the gular scutes are completely separated from each other by the intergulars. However, unlike in *N. eximius*, these intergulars are not heart-shaped with restricted contact with the humeral scutes. TMP 1991.113.0009, a potentially juvenile specimen based on size, preserves portions of the right epiplastron, entoplastron, and hyoplastron and has been used to reconstruct these elements in Fig. 21.5a. If indicative of the adult configuration, the most striking thing about these elements is the anterior position of the entoplastron, which seems to be separated by the anterior plastral margin by an epiplastron that is anteroposteriorly shorter than that present in *N. eximius* or *N. baueri*. The holotype (TMP 2007.035.045) carapace has a reconstructed width of 472 mm and an estimated length of 623 mm. The anterior lobe of the plastron has a median length of 115 mm and a bridge length of 204 mm. The plastron is not complete posteriorly; however, plastron length is estimated to be 450 mm (based on the preservation of the femoral-anal sulcus), which is less than the carapace width, unlike *N. eximius*, but similar to other specimens of *Neurankylus* for which measurements are known. Most of the characters mentioned by Brinkman (2003a) to differentiate the species, namely a smaller shell size and a more squared off posterior lobe of the plastron, do not serve to diagnose the species. Compared to shells for the other congeners, TMP 2007.035.0045 has (1) an estimated length comparable to referred shells of *N. eximius* and the type and referred material of *N. baueri* and (2) the taper of the posterior lobe appears similar to that of *N. baueri*, yet more tapered than that of *N. eximius*. However, based on the preserved portion of the posterior lobe, which preserves a complete femoral-anal sulcus, the lobe appears to have been longer relative to the bridge in *N. lithographicus* than that in *N. eximius* or *N. baueri*.

The referred skull (TMP 2007.036.0001; Fig. 21.6a, c, e, g) is identified as *Neurankylus* based on its relatively large size and the foramen stapediotemporale bordered by the opisthotic, quadrate, and prootic, but not the supraoccipital.

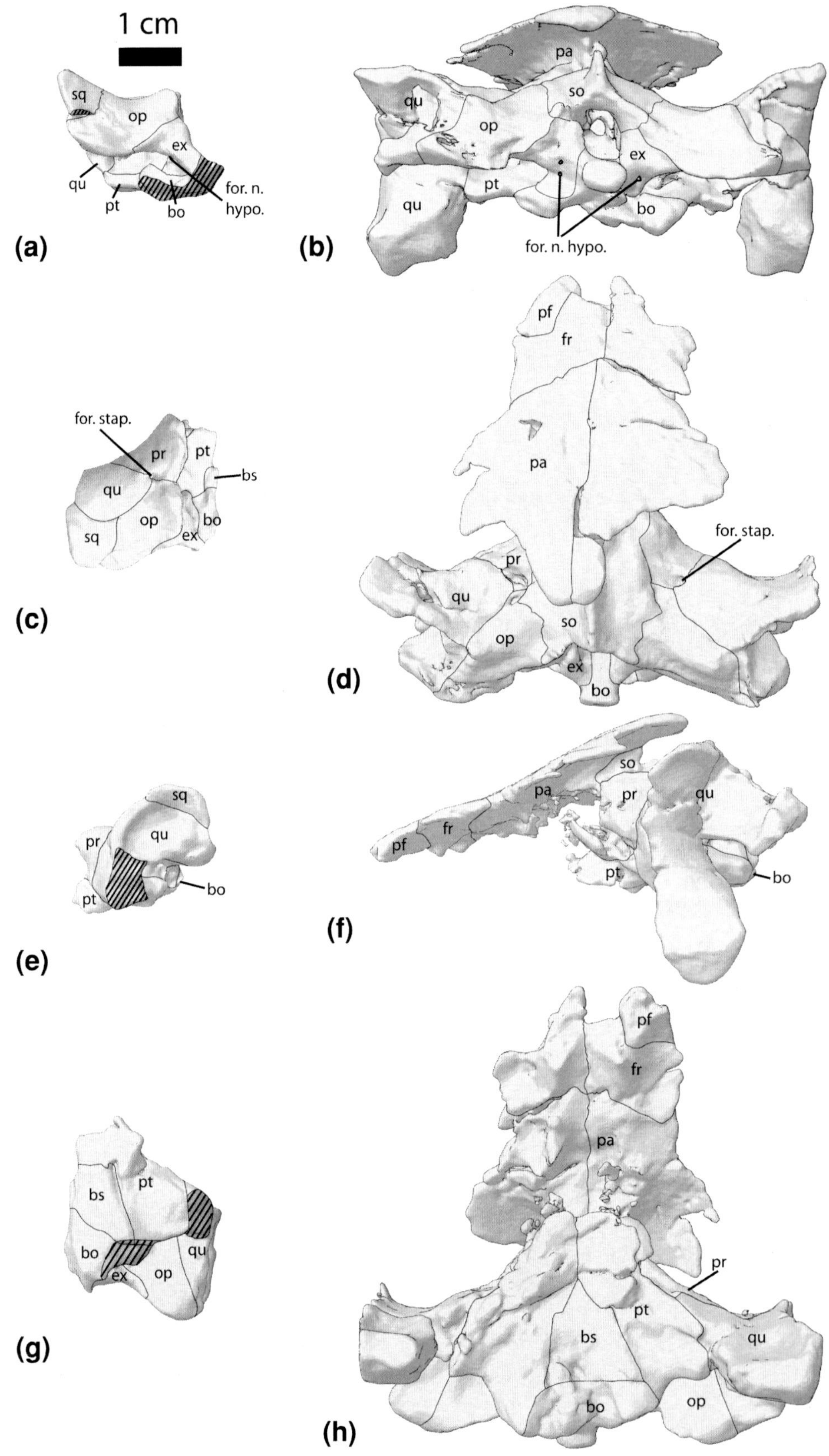

Fig. 21.6 CT volume renderings of two *Neurankylus* skulls. **a**, **c**, **e**, and **g** referred skull (TMP 2007.036.0001) of *N. lithographicus* sp. nov. **b**, **d**, **f**, and **h** referred skull (UALVP 30824) of *Neurankylus eximius*. Specimens depicted in posterior (**a** and **b**), dorsal (**c** and **d**), left lateral (**e** and **f**), and ventral (**g** and **h**) views. Images at same magnification. For the new species *N. lithographicus*, note the proportionally narrower skull and the more laterally directed foramen nervi hypoglossi. *Hatching* represents broken areas. Abbreviations: *bo* basioccipital; *bs* basispenoid; *ex* exoccipital; *for. n. hypo.* foramen nervi hypoglossi; *for. stap.* foramen stapediotemporale; *fr* frontal; *op* opisthotic; *pa* parietal; *pf* prefrontal; *pr* prootic; *pt* pterygoid; *qu* quadrate; *so* supraoccipital; *sq* squamosal

It possesses the baenid feature of the presence of the foramen posterius canalis caroticum internus along the basisphenoid-pterygoid suture. In the skull, the mediolateral width of the opisthotic is less than 1.5 times the dorsoventral height of the element. The foramen stapediotemporale is less than half of the anteroposterior length of the otic region away from the quadrate-squamosal suture. As well, the foramen nervi hypoglossi is directed posterolaterally and is not visible in posterior view. All of the cranial characters indicate a more elongate skull, unlike the wedge-shaped skull typical of most other baenids, including *Neurankylus eximius* (Fig. 21.6b, d, f, h), but more similar to the condition seen in *Arundelemys dardeni* (Lipka et al. 2006). TMP 1991.113.0001 includes an external skull element (based on ornamentation), which is possibly the right sqamosal, although overlap with TMP 2007.036.0001 is too limited to facilitate comparison. If that isolated bone is indeed a squamosal, *N. lithographicus* seems to possess more elongate squamosal processes than other baenid species. No other elements from this specimen have been identified as skull elements. TMP 1998.102.0014 preserves only the posteriormost portions of the palatines, but these are not complete enough to describe any notable morphology. The skull characters mentioned are likely primitive for the genus, but due to a paucity of comparable material, this is not known.

Remarks: The paratype material is referred to *Neurankylus lithographicus* based on its anterior plastral scute morphology. All provisionally referred material have features consistent with the genus *Neurankylus*, and is thought to be *N. lithographicus* largely because of provenance. Both the holotype shell TMP 2007.035.0045 and the referred skull TMP 2007.036.0001 were found at a similar stratigraphic level (roughly 26.5 m above the base of the Deadhorse Coulee Member) 185 m apart. The specimens both possess autapomorphies of *Neurankylus*. TMP 1991.113.0001 is estimated to have been deposited roughly 8.0 m above the base of the Deadhorse Coulee Member in Verdigris Coulee, making it the farthest west and lowest known occurrence of this taxon. Other more fragmentary specimens have less precise stratigraphic data, but all occur within the Deadhorse Coulee Member east of the town of Milk River.

In many shell characters, such as the morphology of the intergular scutes and fourth marginals, the shape of the carapace, and relative size of the plastron, *Neurankylus lithographicus* more closely resembles the shells of later and more southern taxa, such as *N. baueri* and the Lance Formation *Neurankylus*, although neither of these taxa have described skull material. All three of these taxa appear markedly different from the morphology exhibited in the type species *N. eximius*.

Phylogenetic Analysis

To assess the systematic position of species of *Neurankylus*, maximum parsimony and Bayesian phylogenetic analyses were conducted based on a modified baenid character-taxon matrix of Lyson and Joyce (2009b). Thirty-four characters and 11 taxa were added to the matrix, two characters were deleted, and three were modified or rescored (Appendix 1). Fifteen characters represented sequential or continuously variable morphological series and these were treated as ordered in the parsimony analysis. Phylogenetically uninformative characters used in the Bayesian analysis were also removed for the parsimony analysis. For the purposes of this analysis, six operational taxonomic units of *Neurankylus* with the best morphological data were included: *N. eximius*, *N. baueri*, *N. wyomingensis*, *N. lithographicus*, the Lance Formation *Neurankylus*, and an unkeeled lower Kirtland specimen (Appendix 2). The Kirtland specimen (ROM 864) was previously identified (Gaffney 1972) as coming from the Fruitland Formation, but that is now believed to be unlikely (Sullivan et al. 2012). Nonetheless, ROM 864 differs from the type and referred material of *N. baueri* in the differently shaped intergulars (Fig. 21.1c, d), more rounded anterior plastral margin, the lack of dorsal median keel, and the lack of posterior scalloping. While it is possible that the specimen is referable to *N. baueri*, the Kirtland OTU has been analyzed with its own character codings to reflect these differences until it can be demonstrated that these differences are the result of individual variation. Specimens previously identified by Wiman (1933) as coming from the Ojo Alamo Formation were likely collected from the upper Kirtland Formation, and at least one specimen (PMU.R26) may be referable to *N. baueri* (Sullivan et al. 2012). Specimens from the Hell Creek, Tullock, and Nacimiento formations referred to the genus were not examined by us and could not be coded from the literature. The states for OTUs for *Dorsetochelys delairi* Evans and Kemp 1976, and *Thescelus insiliens* were coded using likely shell-skull referrals. This combination of character states is justified by comments from previous authors (Milner 2004; Lyson and Joyce 2010) and the fact that separating these OTU into skull and shell taxa does not result in changes to tree topology.

The maximum parsimony analysis using the modified character matrix (Appendix 3) was conducted in PAUP* 4.0b10 (Swofford 2001). A heuristic search with 1000 random sequence addition replicates and tree-bisection reconnection (TBR) swapping produced 34 most parsimonious trees of 296 steps. The majority rule consensus of these trees is depicted in Fig. 21.7b. Each tree has a consistency index (CI) of 0.416, retention index (RI) of

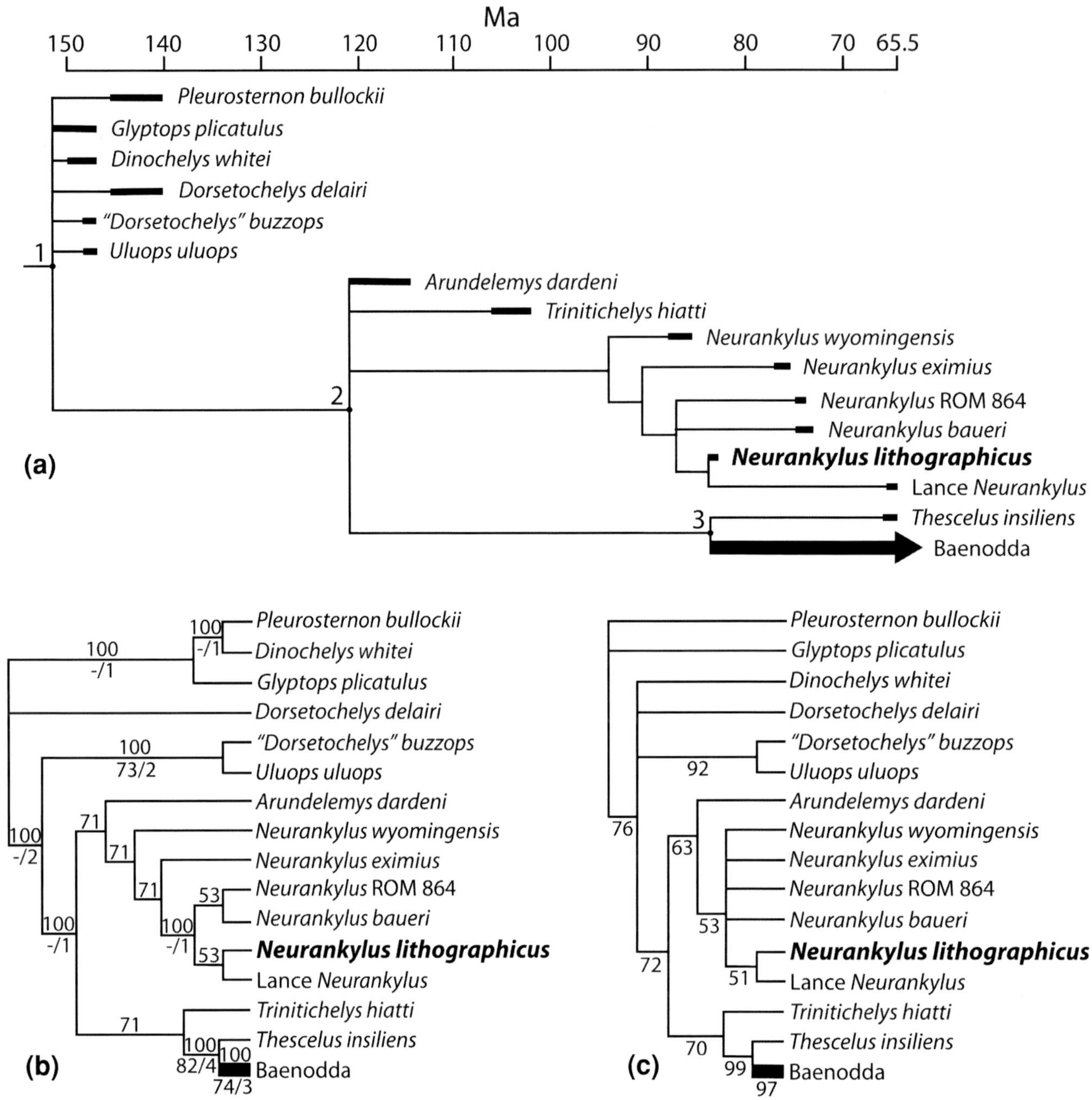

Fig. 21.7 Phylogenetic relationships among basal baenids, including species and specimens of *Neurankylus*, supported by this study. See text and appendices for details. **a** Time-calibrated parsimony and maximum probability Bayesian congruent phylogeny. Numbers correspond to the following clades: *1* Baenoidea; *2* ?Baenidae; *3* Baeninae. **b** Majority-rule consensus parsimony tree. Frequency percentages reported *above* the branch and to *left* of the corresponding clade; support values (where applicable) are *below* the branch, with bootstrap on the *left* and decay on the *right*. **c** Bayesian analysis greatest likelihood tree. Posterior probabilities reported *below* the branch and to *left* of the corresponding clade. For simplicity and to emphasize relationships among less derived taxa, in all three trees the more derived Baenodda are depicted as a single terminal taxon

0.634, and an rescaled consistency index (RC) of 0.264. To evaluate support, decay and bootstrap values were calculated. The bootstrap consisted of 10,000 bootstrap replicates using a 100 random addition sequence replicate heuristic search and TBR swapping.

The Bayesian analysis was conducted in MrBayes 3.1.2 (Ronquist and Huelsenbeck 2003) with default priors. Two runs of four Markov Chains were run for 1×10^6 generations and sampled every 100 generations. Burn-in was determined by noting when cold chain probabilities stopped increasing sharply in both runs. In this case, the first 2000 generations (20 samples) were discarded as burn-in. The results of the analysis are presented in Fig. 21.7c.

Similarities in topologies calculated from different search methods suggest a robust phylogenetic signal in the dataset (Brooks et al. 2007). A congruent topology was constructed using the maximum posterior probability tree and the most topologically similar most-parsimonious tree. The figured topology (Fig. 21.7a) is a strict consensus of the two topologies, and represents the best-available hypothesis of relationships between the species of *Neurankylus* and basal paracryptodires pending the description of more complete material.

Dorsetochelys delairi, which has been placed as a sister taxon to the Paracryptodira (Joyce 2007), is in the parsimony analysis found to be closely associated with pleurosternids. As this species does not possess the single synapomorphy known for Pleurosternidae (namely basisphenoid-palatine contact) inclusion in that family is not supported. However, a similar maximum parsimony analysis including the taxon *Kallokibotion bajazidi* Nopcsa 1923, as an outgroup (not shown) placed *Dorsetochelys delairi* closer to the pleurosternids than to the baenids. In the Bayesian analysis, *Dinochelys whitei* Gaffney 1979 was found in a polytomy with *Dorsetochelys delairi* at the base of Baenidae. The maximum probability Bayesian tree found *Dinochelys whitei* nested within the clade of Jurassic paracryptodires at the base of Baenidae. Closer examination and future discoveries of these basal paracryptodires may resolve the current uncertainty, including polarizing characters for the Baenidae.

The placement of "*Dorsetochelys*" *buzzops* Bakker 1998, with *Uluops uluops* Bakker and Carpenter 1990 as basal baenids with a sister-group relationship to those taxa traditionally known as baenids was unexpected, although this finding is similar to the relationships initially proposed by Carpenter and Bakker (1990) and Bakker (1998). Another novel taxon placement is that of *Arundelemys dardeni* as sister to *Neurankylus* in both the majority-rule consensus parsimony and greatest likelihood trees. Even though such a group was not recovered in the maximum probability Bayesian tree, *A. dardeni* is provisionally considered as belonging to the Neurankylinae. This placement seems to result from the unique configuration of elements around the foramen stapediotemporale, excluding the supraoccipital and including the opisthotic. A previous analysis of this taxon (Lipka et al. 2006) placed it as the sister group to Baenoidea (sensu Williams 1950: combined Baenidae and Pleurosternidae) within a more inclusive Paracryptodira.

In our analysis, *Neurankylus* is identified as a monophyletic clade with low Bayesian posterior probability support and in the maximum probability tree, although this and other basal nodes were better supported when *N. wyomingensis* (a specimen consisting of only the posterior portion of the shell examined only from figures) was pruned from the analysis. However, *N. wyomingensis*, in most analyses, was supported as the basal-most member of the genus. *N. eximius* is also typically placed in a basal position within the genus, second only to *N. wyomingensis*. A clade consisting of *N. lithographicus*, the Lance Formation *Neurankylus*, *N. baueri*, and ROM 864 was found in the maximum parsimony (with a low decay value) and maximum probability analyses. *N. lithographicus*, the second oldest (late Santonian) member of the genus, is consistently placed as the sister taxon to the younger (late Maastrichtian) Lance specimen in all analyses.

Trinitichelys hiatti, in the parsimony and Bayesisan analyses, as in the study by Brinkman and Nicholls (1993) but unlike in the studies by Lyson and Joyce (2009b, 2010), is found to be more closely related to the Baeninae than to *Neurankylus*, although this placement does not have particularly strong support, and is not supported by the maximum probability tree. As in other analyses (Joyce 2007; Lyson and Joyce 2009a, b), the Baeninae and the Baenodda are well-supported in both parsimony and Bayesian analyses. *Thescelus insiliens*, here including *Hayemys latifrons* (Lyson and Joyce 2010), was well-supported as the basal-most baenine, unlike in the classification of Holroyd and Hutchison (2002), which treated *H. latifrons* as a neurankyline. The latter position was not supported by our study, and even when these two taxa were separated in the parsimony analysis, *H. latifrons* and *T. insiliens* occupied the same positions in the cladogram.

Discussion and Implications

A reevaluation of *Neurankylus eximius* reveals numerous characters specific only to the type material and other specimens from the Dinosaur Park Formation of Alberta and the Judith River Formation of Montana, and which are absent in all other known specimens of *Neurankylus*. Thus, all known specimens of *N. eximius* are known exclusively from the middle to late Campanian of Alberta and Montana. From similarly-aged deposits in New Mexico, *N. baueri* and, possibly, an additional species distinct from *N. eximius* are known. *N. lithographicus* is represented by specimens exclusively from the latest Santonian Deadhorse Coulee Member of the Milk River Formation. *Neurankylus wyomingensis*, if valid, may be the earliest known member of this genus (see next paragraph for discussion about the possible age of this species). Yet another distinct form is present in the Lance Formation. The recognition of multiple species in this genus reveals differences in biogeographic and stratigraphic distributions that are invisible when only one species is recognized. These results imply that investigation of other turtle specimens may show complex biogeographic, biostratigraphic, and evolutionary patterns. These patterns may eventually help elucidate distribution pathways and patterns of extinction in the vertebrate fossil record of North America.

Neurankylus lithographicus, along with other specimens that are poorly known (most often regarded as nomina dubia) or whose phylogenetic relationships remain poorly resolved, serves to fill in the ghost lineage between the well-known late Campanian baenids and their purported sister-group, the pleurosternids (Fig. 21.7a). *Trinitichelys hiatti* is known from one specimen whose age can be newly interpolated as about 105.6 Ma (Jacobs et al. 1991; Gradstein et al. 2004). Interpolating the age of "*Glyptops*" *pervicax* Hay 1908, which may be conspecific with *T. hiatti* (Ostrom 1970), from the Cloverly Formation of Montana, yields an age of at least 101.6 Ma. *Neurankylus wyomingensis*, reported from the Colorado shale near Cody, Wyoming, was likely recovered from the Cody Shale of current usage, which in that region contains the biostratigraphic zones of *Scaphites ventricosus* Meek and Hayden 1862, and *Schaphites depressus* Reeside 1927 (Dyman et al. 1997), giving a conservative age estimate of middle to late Coniacian (87.88–85.85 Ma; Gradstein et al. 2004). *Arundelemys dardeni*, here provisionally hypothesized to be a basal neurankyline, occurs in the Arundel Clay facies of the Potomac Group, which was dated to middle Aptian by palynological biostratigraphy (Doyle 1992). The middle Aptian was dated by Ogg et al. (2004) to range from 115 to 121 Ma, pre-dating the age estimate of *T. hiatti*.

The placements of *Neurankylus wyomingensis* as the basal-most *Neurankylus* and of *Trinitichelys hiatti* as the basal-most member of the lineage leading to the Baenodda suggest that both lineages independently lost the coarse shell ornamentation. Also, the probable presence of narrow-skulled species in both lineages (*N. lithographicus* and *T. hiatti*, respectively), as well as the basal occurrence of *D. buzzops* and *U. uluops*, indicates that the evolution of broad skulls within baenoids may have at least three different origins. It also indicates a certain degree of parallelism between the Neurankylinae and the lineage leading to the Baeninae.

Convergent evolution of derived baenid characters within these two lineages has implications for defining Baenidae. Previous analyses have placed the group as the descendants of the node which leads both to *Neurankylus* and *Baena* (Joyce 2007) or to *Trinitichelys* and *Baena* (Gaffney and Meylan 1988). Based on the morphological diagnosis of the family presented in Gaffney (1972), it is more consistent to regard Baenidae as the least inclusive clade containing both *Uluops uluops* and *Baena arenosa*, as has been similarly suggested previously (Bakker 1998). This relationship, although consistent among most-parsimonious trees, was not found in our Bayesian analysis to the exclusion of all pleurosternids and basal paracryptodires, making the higher systematics of these taxa equivocal at this time.

Conclusions

Neurankylus eximius has previously been regarded as the longest-lived baenid species (Gaffney and Haitt 1971; Lyson and Joyce 2009a, b). Description of the new species and a re-examination of the type species have shown its range to be greatly restricted both chronologically and geographically. Comparison of *N. lithographicus* and *N. eximius* allows for diagnostic characters of the latter species to be identified and may serve as guidelines for identifying characters, both in the shells and skulls, of other specimens putatively identified as, but now not referable to, *N. eximius*.

This study provides evidence for previously undocumented morphological variation in the genus *Neurankylus*. Such variation suggests other consistent, although subtle, differences may be overlooked in other taxa. The degree of morphological similarity between species of *Neurankylus* may also indicate that baenid turtles were well adapted for relatively stable ecological niches. Many of the changes that are seen in turtle faunas appear to result from shifting patterns of geographic ranges in response to climate change (Brinkman 2003a) rather than faunal turnover. These patterns of diversification and extinction may elucidate such patterns in other organisms living at the same time. This information may help us better understand the ecological shifts in these ancient faunas and the mechanisms of how those shifts occurred.

Acknowledgments We thank D. Brinkman, P. Currie, M. Wilson, T. Lyson, and the entire UALVP for helpful discussion, resources, and guidance, as well as T. Lyson and W. Joyce for providing the character-taxon matrix for the phylogenetic analysis. Financial support for this research was provided by the Jurassic Foundation and the University of Alberta, Department of Biological Sciences (to DWL), Alberta Ingenuity, National Science Foundation, and Yale Institute for Biospheric Studies (to NRL), National Sciences and Engineering Research Council of Canada Graduate Scholarship and Discovery Grant (to DCE). Specimens were expertly prepared by I. Morrison, J. McCabe, and D. Lloyd. We are grateful to R. Audet for land access permission, the Royal Tyrrell Museum of Palaeontology and the Sloboda family for logistical support, B. Strilisky and T. Lyson for access to specimens, and the 2007 Southern Alberta Dinosaur Research Group field crew for their hard work. This manuscript represents part of DWL's MSc thesis at the University of Alberta. An earlier version of this manuscript was greatly improved by P. Currie, D. Brinkman, M. Wilson, A. Wolfe, and M. Burns. Reviewers R. Sullivan and T. Lyson and editors D. Brinkman and J. Gardner provided excellent constructive criticism for the improvement of this manuscript. Lastly, we salute Eugene Gaffney for his contributions to baenid research, without which the current study would not have been possible.

Appendices

Appendix 1

Modifications to character matrix of Lyson and Joyce (2009b).

(a) Taxa added (see Appendix 2 for list of specimens and sources):

Dinochelys whitei; *Dorsetochelys delairi*; "*Dorsetochelys*" *buzzops*; *Uluops uluops*; *Arundelemys dardeni*; *Thescelus insiliens*; *Neurankylus baueri*; *Neurankylus* ROM 864; *Neurankylus lithographicus*; *Neurankylus wyomingensis*; and Lance Formation *Neurankylus*.

(b) Modified characters:

Character 36: rescored for *Neurankylus eximius*

Character 38: changed and combined with Character 57.

Character 39: scored for *Peckemys brinkman*.

Character 50: rescored for consistency.

Character 53: scored for *Peckemys brinkman*.

Character 57: deleted, because combined with Character 38.

Character 60: deleted, because too variable to be reliable.

(c) New characters:

Character 71. Temporal emargination: 0) does not expose otic cap-sule; (1) stapedial foramen exposed in dorsal view; (2) anterior to an-terior margin of otic capsule.

Character 72. Contribution of supraoccipital to stapedial foramen: (0) enters stapedial foramen; (1) excluded from stapedial foramen.

Character 73. Anterior nuchal projection: (0) absent; (1) present.

Character 74. Skull ornamentation: (0) pustolose (*Glyptops*-like); (1) rugose.

Character 75. Foramen posterius canalis caroticum internus position: (0) in posterior half; (1) in anterior half of pterygoid–basisphenoid suture.

Character 76. Ventral angle between pterygoid occipital process and transverse process of pterygoid: (0) less than 90°; (1) more than 90° or equal.

Character 77. Narrowest part of ventral pterygoid: (0) even or posterior to anteriormost portion of ventral exposure of basisphenoid; (1) anterior to basisphenoid.

Character 78. Opisthotic in posterior view: (0) width less than 1.5 times height; (1) width greater than 1.5 times height.

Character 79. External ventrolateral quadrate-quadratojugal suture: (0) shorter; (1) roughly equal in height or taller than tympanic recess.

Character 80. Foramen nervi hypoglossi directed: (0) posterolaterally not visible from posterior view; (1) posteriorly visible from posterior view.

Character 81. Exoccipital processes of basioccipital: (0) thinner; (1) roughly same thickness as height of condyle.

Character 82. Anterior border of frontals: (0) extends anterior to prefrontals; (1) does not extend past prefrontals.

Character 83. Anterior border of frontals: (0) pointed anteromedially; (1) straight.

Character 84. Shell ornamentation: (0) rugose (*Glyptops*-like); (1) faintly ridged.

Character 85. Dorsal medial keel on posterior vertebrals: (0) absent; (1) present.

Character 86. Dorsolateral gutters on carapace: (0) absent; (1) present.

Character 87. Cervical scute: (0) small and rectangular long edge on carapace margin; (1) small and rectangular short edge on margin; (2) not as above.

Character 88. Anterior edge of nuchal: (0) slightly recessed; (1) anterior or even with first marginals.

Character 89. Nuchal lateral margins: (0) tapers anteriorly; (1) parallel; (2) widens anteriorly.

Character 90. Second and third vertebrals: (0) do not narrow anteriorly; (1) narrow anteriorly.

Character 91. Fifth vertebral: (0) wide anterolateral sulci not present on second suprapygal (same width as fourth vertebral); (1) narrow sulci present on second suprapygal (narrower than fourth vertebral).

Character 92. Fourth marginal resembling asymmetrical trapezoid: (0) no; (1) yes, widest part posterior.

Character 93. Early fusion of first suprapygal to eighth neural: (0) absent; (1) present.

Character 94. Posterior neurals: (0) hexagonal; (1) square.

Character 95. First–second costal suture: (0) slightly posteriorly inclined or perpendicular to long axis; (1) slightly anteriorly inclined; (2) greatly anteriorly inclined.

Character 96. Greatest plastron length: (0) less; (1) equal or greater than greatest carapace width.

Character 97. Carapace width to carapace length: (0) around 0.8; (1) around 0.85.

Character 98. Plastron length to carapace length: (0) less than or equal 0.8; (1) greater than 0.8.

Character 99. Anterior lobe length to posterior lobe length: (0) 0.82 or less; (1) greater than 0.84.

Character 100. Intergular shape: (0) rectangular with large intergular humeral sulcus; (1) heart-shaped with little or no intergular-humeral sulcus; (2) semicircular, no humeral contact.

Character 101. Intergulars: (0) do not overlap entoplastron; (1) overlap entoplastron.

Character 102. Intergular-gular sulci: (0) straight, (1) curved.

Character 103. Inframarginals: (0) narrower; (1) roughly equal to; (2) wider than ventral exposure of marginals.

Character 104. Size: (0) large (carapace width 300 mm or greater, skull basal length greater than 60 mm; *Neurankylus*-sized); (1) small (carapace width less than 300 mm, skull basal length 60 mm or less; *Plesiobaena*-sized).

Appendix 2

Specimens and literature used to score taxa for phylogenetic analysis. Asterisk (*) indicates specimens examined from published material only.

(a) *Neurankylus* taxa scored for all characters:

Neurankylus eximius: CMN 1504* (Hay 1908; Gaffney 1972; probably Dinosaur Park Formation), AMNH 6098 (Judith River Formation), ROM 854 (probably Dinosaur Park Formation), ROM 1943 (probably Dinosaur Park Formation), TMP 1989.036.0112 (Dinosaur Park Formation), TMP 1999.055.0134 (Dinosaur Park Formation), TMP 2003.012.0171 (Dinosaur Park Formation), UALVP 30824 (Dinosaur Park Formation).

Neurankylus lithographicus (all specimens from Milk River Formation): TMP 2007.035.0045, TMP 1991.113.0001, TMP 1991.113.0009, TMP 1994.377.0001, TMP 1998.102.0014, TMP 2007.036.0001.

Neurankylus baueri (all specimens from Kirtland Formation): USNM 8344*(Gilmore 1916), PMU.R24* (Wiman 1933), PMU.R25* (Wiman 1933), PMU.R26* (Wiman 1933), PMU.R27* (Wiman 1933).

Neurankylus wyomingensis: USNM 7581* (Gilmore 1919; Upper Colorado Group).

Other *Neurankylus* specimens: ROM 864 (probably Kirtland Formation), YPM 8239 (Lance Formation).

(b) New taxa scored for all characters:

Arundelemys dardeni: USNM 497740* (Lipka et al. 2006; Arundel clay of Potomac Formation).

Dinochelys whitei: DNM 986* (Gaffney 1979), BYU 13099* (Brinkman et al. 2000); both Morrison Formation.

Dorsetochelys delairi: DORCM G23* (Evans and Kemp 1976; Purbeck Limestone Group).

"*Dorsetochelys*" *buzzops*: TGM 5001* (Bakker 1998; Morrison Formation).

Thescelus insiliens: AMNH 1108* and AMNH 6139* (Gaffney 1972; Hell Creek Formation).

Uluops uluops: CPS 100* (Carpenter and Bakker 1990; Morrison Formation).

(c) Other specimens scored for new characters:

Pleurosternon bullockii: BMNH 28618* (Milner 2004), BMNH R911* (Milner 2004), UMZC T1041* (Evans and Kemp 1975); all Purbeck Limestone Group.

Glyptops plicatulus: AMNH 336* and AMNH 5458* (Gaffney 1979; Morrison Formation).

Trinitichelys hiatti: MCZ 4070* (Gaffney 1972; Trinity Sands of Trinity Group).

Plesiobaena antiqua: TMP 1999.055.0145* and TMP 1985.058.0045* (Brinkman 2003b; Dinosaur Park Formation).

Peckemys brinkman: UMMP 20490* (Lyson and Joyce 2009b; Hell Creek Formation).

Cedrobaena putorius: FMNH PR 2258* (Lyson and Joyce 2009b; Hell Creek Formation).

Boremys pulchra: CMN 2281*, CMN 1130* (Gaffney 1972), TMP 1988.002.0010* (Brinkman and Nicholls 1991); all Dinosaur Park Formation).

Boremys grandis: USNM 12979* (Gilmore 1935; Kirtland Formation).

Eubaena cephalica: YPM 1785* (Gaffney 1972; Lance Formation).

Palatobaena cohen: YPM 57498*, YPM 57498*, MRF 123* (Lyson and Joyce 2009a; Hell Creek Formation).

Palatobaena bairdi: PU 16839* (Gaffney 1972; Fort Union Formation).

Palatobaena gaffneyi: UCMP 114529* (Archibald and Hutchison 1979; Wasatch Formation).

Stygiochelys estesi: AMNH 2601* (Gaffney 1972; Hell Creek Formation).

Baena arenosa: MCZ 4072* (Willwood Formation), AMNH 5977* (Gaffney 1972; Bridger Formation).

Chisternon undatum: AMNH 5961* (Gaffney 1972; Bridger Formation).

Appendix 3

Character-taxon matrix of baenid turtles used for parsimony and Bayesian analyses. Asterisk (*) denotes character retained for Bayesian analysis, but removed for parsimony analysis.

[1] *Pleurosternon bullockii* 0010000000 0000000000 000?0?0000 00?0?00000 0(0*)0000(0*)000 0000000000 00(0*)0000100 0000-00?(0*)? ?001102100 00011?0?10 10?0

[2] *Glyptops plicatulus* 0110010000 0000000000 000?0?0000 00?0000100 0(0*)0000(0*)000 0000000000 00(0*)0000000 0?00-000(0*)0 0000000000 0000101110 1000

[3] *Dinochelys whitei* ??1000???? ?000000?0? 0????????? ????00000 0(0*)0000(0*)000 ?10101000? ??(?*)?0???0? ??00????(?*)? ?001000000 1000110110 00-1

[4] *Dorsetochelys delairi* 001000?001 0000000100 0000?10200 2????10100 1(0*)0000(0*)00? ????0?0000 0?(?*)0001201 0?00110?(0*)? ?000100000 00001?1??? ???0

[5] "*Dorsetochelys*" *buzzops* 110001?0?0 0000000000 00?1?10111 1?00?????? ?(?*)????(?*)??? ???????00 0?(?*)0000201 0??0?001(0*)1 1??0?????? ?????????? ???1

[6] *Uluops uluops* 110000?0?0 0000000000 0011?1021? 1????????? ?(?*)????(?*)??? ???????00 0?(?*)0000201 0??0110?(0*)? ?????????? ?????????? ???1

[7] *Arundelemys dardeni* 000000?000 000011001- 100?000210 21???????? ?(?*)????(?*)??? ???????00 00(?*)0010201 21?11001(?*)1 100??????? ?????????? ???1

[8] *Trinitichelys hiatti* 0000001000 0001000100 10001002?? 21??????00 ?(0*)0000(?*)000 110?000000 00(?*)00000 100 10010001(0*)? 1110000 1?1 ?00?1????0 1001

[9] *Neurankylus eximius* 10???0???? ??00100??? ?101???211 21???00100 0(0*)0000(0*)010 120(01)0100?? ??(?*) 0?100?1 11011101(0*)1 1101111121 0010110111 1110

[10] *Neurankylus baueri* ?????????? ?????????? ?????????? ????10100 0(0*)0000(0*)000 ?2000000?? ??(?*)???? ??? ??0?????(?*)? ???1111101 11(01)10010(01)0 1010

[11] *Neurankylus* ROM 864 ?????????? ?????????? ?????? ???? ??????0100 0(0*)0000(0*)0?0 ?1000100?? ??(?*) ??????? ??0?????(?*)? ???10111?1 01??1010?0 1100

[12] *Neurankylus lithographicus* 0????????? ?????????? ?????0021? 21??????00 0(0*)0000(?*)0?0 ?0000?00?? ??(?*)0?????? ?1011100(?*)0 0??1?1?1?1 01??200010 ?010

[13] *Neurankylus wyomingensis* ?????????? ?????????? ???? ?????? ?????0010? 0(?*)?0??(0*)0?? ???0?0???? ??(?*) ??????? ????????(?*)? ???001???1 0????????? ??10

[14] Lance Formation *Neurankylus* ?????????? ?????????? ?????????? ????10000 0(?*)0000(0*)0?? ???01100?? ??(?*)??????? ????????(?*)? ???10101?1 0????0000? ??10

[15] *Thescelus insiliens* 100??0??0? 0000021100 100??0121? 20???00100 0(0*)1000(0*)001 1211010100 00(0*) 0000?00 1001101?(0*)? ?011000000 00??111112 1120

[16] *Plesiobaena antiqua* 1000001011 01(02)1101010 1000110111 1001011210 1(0*)0000(0*)011 121(01) 010100 00(0*)1000100 2011010?(0*)? ?011000110 00(01)1111102 0121

[17] *Peckemys brinkman* 1000001011 0??1111110 10001 00211 100111?200 1(0*)1000(0*)001 121?0?0000 00(0*) 1200100 10011011(0*)1 1??10001?0 00???1??0 –21

[18] *Cedrobaena putorius* 1000011111 1??1111100 111011 (01)211 1001?????? ?(?*)????(?*)??? ???????00 00(0*) 0001010 10?1111?(0*)? ?????????? ?????????? ???1

[19] *Boremys pulchra* 111000?10? 0101101?00 1000?00211 1010031201 1(0*)1(01)01(1*)101 1110111100 00(0*) 0000100 2001110?(1*)? ?011000100 1001101112 0021

[20] *Boremys grandis* ?????????? ?????????? ?????????? ????31201 1(0*)1011(0*)100 11101111?? ??(?*)???? ??? ??0?????(?*)? ???1001120 0???111112 1120

[21] *Eubaena cephalica* 1110102100 0101121000 1000 300211 10???????? ?(?*)????(?*)??? ???????00 00(0*) 0110100 10?1110?(0*)? ?00??????? ?????????? ???1

[22] *Palatobaena cohen* 2000012111 1122011110 10002 10211 10011?1210 ?(0*)0000(0*)011 1211110111 11(0*)1111110 20111101(0*)1 1111000100 00?? 01??1 –21

[23] *Palatobaena bairdi* 2000012111 1122111210 1110210 211 10011????? ?(?*)????(?*)??? ????????11 11(0*) 0111210 0??11101(0*)1 111??????? ?????????? ???1

[24] *Palatobaena gaffneyi* 2000112111 1122111110 1110 210111 10???????? ?(?*)????(?*)??? ????????11 11(0*) 0111210 ?1?11101(0*)1 101??????? ?????????? ???1

[25] *Stygiochelys estesi* 1101001110 1112101100 100100 1111 11???????? ?(?*)1111(?*)0?0 ?21?1?0100 00(0*) 0100100 20?10101(0*)1 101??????? ?????????? ???1

[26] *Baena arenosa* 1101001110 01?2101201 0111000211 1121021200 1(1*)1101(0*)000 1211110100 00(0*) 0000001 0?11110?(0*)? ?011002110 0011010102 1121

[27] *Chisternon undatum* 1101001110 0112101201 001100 1111 1120021201 1(0*)1111(0*)001 1211110100 00(0*) 0010101 1001100?(0*)? ?011002110 ?0?0001112 1010

[28] *Gammerabaena sonsalla* ?000011110 111102101? ??1??10211 10???????? ?(?*)????(?*)??? ?????0??00 00(1*)?11??10 ????????(?*)? ?????????? ?????????? ????

[29] *Goleremys mckennai* 1000001100 0102101210 10??100211 10???????? ?(?*)????(?*)??? ???????01 00(0*)000??00 ????????(?*)? ?????????? ?????????? ????

References

Bakker, R. T. (1998). Dinosaur mid-life crisis: The Jurassic–Cretaceous transition in Wyoming and Colorado. *New Mexico Museum of Natural History and Science Bulletin, 14*, 67–77.

Brinkman, D. B. (2003a). A review of nonmarine turtles from the Late Cretaceous of Alberta. *Canadian Journal of Earth Sciences, 40*, 557–571.

Brinkman, D. B. (2003b). Anatomy and systematics of *Plesiobaena antiqua* (Testudines: Baenidae) from the mid-Campanian Judith River Group of Alberta, Canada. *Journal of Vertebrate Paleontology, 23*, 146–155.

Brinkman, D. B. (2005). Turtles: Diversity, paleoecology, and distribution. In P. J. Currie & E. B. Koppelhus (Eds.), *Dinosaur Provincial Park: A spectacular ancient ecosystem revealed* (pp. 202–220). Bloomington: Indiana University Press.

Brinkman, D. B., & Nicholls, E. L. (1991). Anatomy and relationships of the turtle *Boremys pulchra* (Testudines: Baenidae). *Journal of Vertebrate Paleontology, 11*, 302–315.

Brinkman, D. B., & Nicholls, E. L. (1993). The skull of *Neurankylus eximius* (Testudines: Baenidae) and a reinterpretation of the relationships of this taxon. *Journal of Vertebrate Paleontology, 13*, 273–281.

Brinkman, D. B., Stadtman, K., & Smith, D. (2000). New material of *Dinochelys whitei* Gaffney, 1979, from the Dry Mesa Quarry (Morrison Formation, Jurassic) of Colorado. *Journal of Vertebrate Paleontology, 20*, 269–274.

Brooks, D. R., Bilewitch, J., Condy, C., Evans, D. C., Folinsbee, K. E., Fröbisch, J., et al. (2007). Quantitative phylogenetic analysis in the 21st century: Progress and prognosis. *Revista Mexicana de Biodiversidad, 78*, 225–252.

Carpenter, K., & Bakker, R. T. (1990). Part II: A new baenid turtle. *Hunteria, 2*(6), 3–4.

Cope, E. D. (1877). On reptilian remains from the Dakota beds of Colorado. *Proceedings of the American Philosophical Society, 17*, 193–196.

Cope, E. D. (1882). Contributions to the history of the Vertebrata of the Lower Eocene of Wyoming and New Mexico, made during 1881. *Proceedings of the American Philosophical Society*, *20*, 139–197.

Doyle, J. A. (1992). Revised palynological correlations of the lower Potomac Group (USA) and the Cocobeach sequence of Gabon (Barremian–Aptian). *Cretaceous Research*, *13*, 337–349.

Dyman, T. S., Tysdal, R. G., Perry, W. J. Jr., Nichols, D. J., & Obradovich, J. D. (1997). Stratigraphy and structural setting of Upper Cretaceous Frontier Formation, western Centennial Mountains, Southwestern Montana and Southeastern Idaho. *Cretaceous Research*, *29*, 237–248.

Eberth, D. A. (2005). The geology. In P. J. Currie & E. B. Koppelhus (Eds.), *Dinosaur Provincial Park: A spectacular ancient ecosystem revealed* (pp. 54–82). Bloomington: Indiana University Press.

Evans, J., & Kemp, T. S. (1975). The cranial morphology of a new Lower Cretaceous turtle from southern England. *Palaeontology*, *18*, 25–40.

Evans, J., & Kemp, T. S. (1976). A new turtle skull from the Purbeckian of England and a note on the early dichotomies of cryptodire turtles. *Palaeontology*, *19*, 317–324.

Gaffney, E. S. (1972). The systematics of the North American family Baenidae (Reptilia, Cryptodira). *Bulletin of the American Museum of Natural History*, *147*, 1–319.

Gaffney, E. S. (1975). A phylogeny and classification of the higher categories of turtles. *Bulletin of the American Museum of Natural History*, *155*, 389–436.

Gaffney, E. S. (1979). The Jurassic turtles of North America. *Bulletin of the American Museum of Natural History*, *162*, 1–135.

Gaffney, E. S., & Hiatt, R. (1971). A new baenid turtle from the Upper Cretaceous of Montana. *American Museum Novitates*, *2443*, 1–9.

Gaffney, E. S., & Meylan, P. A. (1988). A phylogeny of turtles. *The Systematics Association Special Volume*, *35A*, 157–219.

Gilmore, C. W. (1916). Vertebrate faunas of the Ojo Alamo, Kirtland, and Fruitland formations. *U. S. Geological Survey, Professional Paper*, *98Q*, 279–308.

Gilmore, C. W. (1919). New fossil turtles, with notes on two described species. *Proceedings of the U. S. National Museum*, *56*, 113–132.

Gilmore, C. W. (1935). On the Reptilia of the Kirtland Formation of New Mexico, with descriptions of new species of fossil turtles. *Proceedings of the United States National Museum*, *83*, 159–188.

Goodwin, M. B., & Deino, A. L. (1989). The first radiometric ages from the Judith River Formation (Upper Cretaceous), Hill County, Montana. *Canadian Journal of Earth Sciences*, *26*, 1384–1391.

Gradstein, F. M., Ogg, J. G., & Smith, A. G. (2004). Construction and summary of the geologic time scale. In F. M. Gradstein, J. G. Ogg, & A. G. Smith (Eds.), *A geologic time scale* (pp. 455–464). Cambridge: Cambridge University Press.

Hay, O. P. (1908). *The fossil turtles of North America*. Washington, DC: Carnegie Institute.

Holroyd, P. A., & Hutchison, J. H. (2002). Patterns of geographic variation in latest Cretaceous vertebrates: Evidence from the turtle component. In J. H. Hartman, K. R. Johnson & D. J. Nichols (Eds.), *The Hell Creek Formation and Cretaceous-Tertiary Boundary in the Great Plains: An integrated continental record of the end of the Cretaceous* (pp. 177–190). The Geological Society of America, Special Paper 361.

Hutchison, J. H. (1984). Determinate growth in the Baenidae (Testudines): Taxonomic, ecologic, and stratigraphic significance. *Journal of Vertebrate Paleontology*, *3*, 148–151.

Hutchison, J. H., & Archibald, J. D. (1986). Diversity of turtles across the Cretaceous/Tertiary boundary in northeastern Montana. *Palaeogeography, Palaeoclimatology, Palaeoecology*, *55*, 1–22.

Hutchison, J. H., & Holroyd, P. A. (2003). Late Cretaceous and early Paleocene turtles of the Denver Basin, Colorado. *Rocky Mountain Geology*, *38*, 121–142.

Jacobs, L. L., Winkler, D. A., & Murry, P. A. (1991). On the age and correlation of Trinity mammals, Early Cretaceous of Texas, USA. *Newsletter of Stratigraphy*, *24*, 35–43.

Joyce, W. G. (2007). Phylogenetic relationships of Mesozoic turtles. *Bulletin of the Peabody Museum of Natural History*, *48*, 3–102.

Lambe, L. M. (1902). New genera and species from the Belly River Series (mid-Cretaceous). *Contributions to Canadian Palaeontology*, *3*, 23–81.

Leahy, G. D., & Lerbekmo, J. F. (1995). Macrofossil magnetobiostratigraphy for the upper Santonian—lower Campanian interval in the Western Interior of North America: Comparisons with European stage boundaries and planktonic foraminiferal zonal boundaries. *Canadian Journal of Earth Sciences*, *32*, 247–260.

Leidy, J. (1870). Descriptions of *Emys jeanesi*, *E. haydeni*, *Baena arenosa*, and *Saniwa ensidens*. *Proceedings of the Academy of Natural Sciences of Philadelphia*, *1870*, 123–124.

Linnaeus, C. (1758). Systema naturae per regna tria naturae, secundum classes, ordines, genera, species, cum characteribus, differentiis, synonymis, locis, Tomus I, Editio Decima. Stockholm: Holmiae.

Lipka, T. R., Therrien, F., Weishampel, D. B., Jamniczky, H. A., Joyce, W. G., Colbert, M. W., et al. (2006). A new turtle from the Arundel clay facies (Potomac Formation, Early Cretaceous) of Maryland, U.S.A. *Journal of Vertebrate Paleontology*, *26*, 300–307.

Lyson, T. R., & Joyce, W. G. (2009a). A new species of *Palatobaena* (Testudines: Baenidae) and a maximum parsimony and bayesian phylogenetic analysis of Baenidae. *Journal of Paleontology*, *83*, 457–470.

Lyson, T. R., & Joyce, W. G. (2009b). A revision of *Plesiobaena* (Testudines: Baenidae) and an assessment of baenid ecology across the K/T boundary. *Journal of Paleontology*, *83*, 833–853.

Lyson, T. R., & Joyce, W. G. (2010). A new baenid turtle from the Upper Cretaceous (Maastrichtian) Hell Creek Formation of North Dakota and a preliminary taxonomic review of Cretaceous Baenidae. *Journal of Vertebrate Paleontology*, *30*, 394–402.

Meek, F. B., & Hayden, F. V. (1862). Descriptions of new lower Silurian (Primordial), Jurassic, Cretaceous, and Teriery fossils collected in the Nebraska Territory, with some remarks on the rocks from which they were obtained. *Proceedings of the Academy of Natural Sciences of Philadelphia*, *13*, 415–447.

Milner, A. R. (2004). The turtles of the Purbeck Limestone Group of Dorset, Southern England. *Palaeontology*, *47*, 1441–1467.

Nopcsa, F. (1923). On the geological importance of the primitive reptilian fauna of the uppermost Cretaceous of Hungary; with a description of a new tortoise (*Kallokibotion*). *Quarterly Journal of the Geological Society*, *79*, 100–116.

Ogg, J. G., & Smith, A. G. (2004). The geomagnetic polarity time scale. In F. M. Gradstein, J. G. Ogg, & A. G. Smith (Eds.), *A geologic time scale* (pp. 63–86). Cambridge: Cambridge University Press.

Ogg, J. G., Agterberg, F. P., & Gradstein, F. M. (2004). The Cretaceous Period. In F. M. Gradstein, J. G. Ogg, & A. G. Smith (Eds.), *A geologic time scale* (pp. 344–383). Cambridge: Cambridge University Press.

Parks, W. A. (1933). New species of dinosaurs and turtles from the Upper Cretaceous formations of Alberta. *University of Toronto Studies, Geological Series*, *31*, 3–33.

Payenberg, T. H. D., Braman, D. R., Davis, D. W., & Miall, A. D. (2002). Litho- and chronostratigraphic relationships of the Santonian–Campanian Milk River Formation in southern Alberta and Eagle Formation in Montana utilizing stratigraphy, U-Pb geochronology, and palynology. *Canadian Journal of Earth Sciences*, *39*, 1553–1577.

Reeside, J. B. (1927). Cephalopods from the lower part of the Cody Shale of Oregon Basin, Wyoming. *U.S. Geological Survey Professional Paper*, *150A*, 1–19.

Rogers, R. R. (1998). Sequence analysis of the Upper Cretaceous Two Medicine and Judith River formations, Montana: A nonmarine response to the Claggett and Bearpaw marine cycles. *Journal of Sedimentary Research, 68*, 615–631.

Ronquist, F., & Huelsenbeck, J. P. (2003). MrBayes 3: Bayesian phylogenetic inference under mixed models. *Bioinformatics, 19*, 1572–1574.

Sullivan, R. M., & Lucas, S. G. (2006). The Kirtlandian land-vertebrate "age"—faunal composition, temporal position and biostratigraphic correlation in the nonmarine Upper Cretaceous of Western North America. *New Mexico Museum of Natural History and Science Bulletin, 35*, 7–29.

Sullivan, R. M, Lucas, S. G., Hunt, A. P., & Fritts, T. H. (1988). Colour pattern on the selmacryptodiran turtle *Neurankylus* from the early Paleocene (Puercan) of the San Juan Basin, New Mexico. *Natural History Museum of Los Angeles County Contributions in Science, 401*, 1–9.

Sullivan, R. M., Jasinski, S. E., & Lucas, S. G. (2012). Re-assessment of late Campanian (Kirtlandian) turtles from the Upper Cretaceous Fruitland and Kirtland Formations, San Juan Basin, New Mexico, USA. In D. B. Brinkman, P. A. Holroyd, & J. D. Gardner (Eds.), *Morphology and evolution of turtles*. Dordrecht: Springer.

Swofford, D. L. (2001). *PAUP*: Phylogenetic analysis using parsimony (* and other methods) version 4.0b10 edition*. Sunderland: Sinauer Associates.

Williams, E. E. (1950). Variation and selection in the cervical central articulations of living turtles. *Bulletin of the American Museum of Natural History, 94*, 505–562.

Wiman, C. (1933). Über Schildkröten aus der Oberen Kreide in New Mexico. *Nova Acta Regiae Societatis Scientiarum Upsaliensis, 9*, 1–34.

Chapter 22
Redescription of *Zangerlia dzamynchondi* (Testudines: Nanhsiungchelyidae) from the Late Cretaceous of Mongolia, with a Reassessment of the Phylogenetic Position and Relationships of *Zangerlia*

Igor G. Danilov, Vladimir B. Sukhanov, and Elena V. Syromyatnikova

Abstract In this chapter we redescribe *Zangerlia dzamynchondi* Sukhanov and Narmandakh 2006, a poorly described nanhsiungchelyid turtle from the Djadokhta Formation (Late Cretaceous, Campanian), Zamin Khond locality, southern Mongolia. We also present new observations on the holotype of *Z. testudinimorpha*, the type of the genus *Zangerlia*, and reassess some characters of this genus. Inclusion of these new data in a phylogenetic analyses of Adocusia (the clade uniting Adocidae and Nanhsiungchelyidae) allows us to reassess the phylogenetic position and relationships of *Zangerlia*. Our phylogenetic analysis does not support a monophyletic *Zangerlia*. All species of "*Zangerlia*" form a polytomy with *Hanbogdemys orientalis*, *Anomalochelys angulata* + *Nanhsiungchelys wuchingensis*, and *Basilemys*.

Keywords Adocusia • Late Cretaceous • Mongolia • Nanhsiungchelyidae • Turtles • *Zangerlia*

Introduction

Nanhsiungchelyidae Yeh 1966 are an extinct family of cryptodires that are allied with the Adocidae Cope 1870 within the Adocusia Danilov and Parham 2006. The family contains seven or eight genera that are restricted to the Late Cretaceous of Asia (six or seven genera) and North America (one genus), and its members are characterized by a peculiar combination of aquatic and terrestrial features (Brinkman and Peng 1996; Sukhanov 2000; Hutchison 2000; Hirayama et al. 2001; Joyce and Norell 2005; Sukhanov et al. 2008; Danilov and Syromyatnikova 2008, 2009). *Zangerlia* Młynarski 1972 is the most specious Asiatic genus of nanhsiungchelyids, containing four latest Cretaceous species [see Sukhanov et al. (2008) and Danilov and Syromyatnikova (2008) for the most recent reviews of the Asian nanhsiungchelyid record and map of localities]: *Z. testudinimorpha* Młynarski 1972 (type species), Campanian and/or Maastrichtian of Mongolia; *Z. neimongolensis* Brinkman and Peng 1996, Late Cretaceous (probably Campanian) of Inner Mongolia, China; *Z. ukhaachelys* Joyce and Norell 2005, Campanian of Mongolia; and *Z. dzamynchondi* Sukhanov and Narmandakh 2006, Campanian of Mongolia (Młynarski 1972; Brinkman and Peng 1996; Sukhanov 2000; Joyce and Norell 2005; Sukhanov and Narmandakh 2006). According to the latest phylogenetic studies, *Zangerlia* is considered monophyletic and forms a clade with the genera *Anomalochelys* Hirayama et al. 2001, *Basilemys* Hay 1902, *Hanbogdemys* Sukhanov and Narmandakh 2006, and *Nanhsiungchelys* Yeh 1966, although relationships among these genera and their species are unresolved (Joyce and Norell 2005; Sukhanov et al. 2008; Danilov and Syromyatnikova 2009).

Zangerlia dzamynchondi is a poorly known turtle that was described based on a single, incomplete shell from exposures of the Djadokhta Formation, at the Zamin Khond locality in southern Mongolia. The description of this species was very brief and it was assigned to the genus *Zangerlia* without a clear argumentation (Sukhanov 2000; Sukhanov and Narmandakh 2006). As a result, *Z. dzamynchondi* never has been included in any phylogenetic analyses of Nanhsiungchelyidae or Adocusia (Hirayama et al. 2001; Joyce and Norell 2005; Sukhanov et al. 2008; Danilov and Syromyatnikova 2009).

The primary aims of our paper are to give a detailed description of *Zangerlia dzamynchondi* and, for the first time, to include this species in a phylogenetic analysis. In addition, we also present new observations on the holotype

I. G. Danilov (✉) · E. V. Syromyatnikova
Department of Herpetology,
Zoological Institute of the Russian Academy of Sciences,
Universitetskaya Emb. 1, St. Petersburg, Russia 199034
e-mail: igordanilov@gmail.com

E. V. Syromyatnikova
e-mail: esyromyatnikova@gmail.com

V. B. Sukhanov
Paleontological Institute of the Russian Academy of Sciences,
Profsoyuznaya 123, Moscow, Russia 117997
e-mail: sukhanovturtle@yandex.ru

D. B. Brinkman et al. (eds.), *Morphology and Evolution of Turtles*, Vertebrate Paleobiology and Paleoanthropology,
DOI: 10.1007/978-94-007-4309-0_22,

specimen of the type species *Z. testudinimorpha*. Our new observations and data allow us to reassess the phylogenetic position and relationships of *Zangerlia*. Because our study does not support monophyly of *Zangerlia*, below we refer to species of *Zangerlia*, other than *Z. testudinimorpha*, as "*Zangerlia*" spp.

Institutional abbreviations used in this paper are: PIN, Paleontological Institute of the Russian Academy of Sciences, Moscow, Russia; ZPAL, Institute of Paleobiology of the Polish Academy of Sciences, Warsaw, Poland.

Materials and Methods

In addition to the type specimen of "*Zangerlia*" *dzamynchondi* described below, our study also is based on firsthand examination (by IGD and EVS) of the holotype (ZPAL MgCh/72, incomplete shell) of *Z. testudinimorpha* (see Discussion). The phylogenetic analysis of Adocusia by Danilov and Syromyatnikova (2009) was the basis for our analysis. The taxon-character matrix of Danilov and Syromyatnikova (2009) was changed in three ways. First, we decided that a previously used character was too problematic and, for that reason, we eliminated it from our analysis. This problematic character described the form of the posterior part of the carapace [i.e., character 27 of Danilov and Syromyatnikova (2009); = characters 22 and 24 of Hirayama et al. (2001); = character 16 of Joyce and Norell (2005)]. For that character, previous authors had coded all species of "*Zangerlia*" as "1" (steep deflection of post-neural part of carapace present, posterior peripherals shortened), whereas other nanhsiungchelyids were coded as "0" (steep deflection absent, posterior peripherals greatly flared). We identified the following problems with that character: it was not illustrated for any species of "*Zangerlia*"; in "*Z.*" *ukhaachelys* the critical posterior part of the carapace is not preserved; and based on our firsthand examinations of *Z. testudinimorpha* and "*Z.*" *dzamynchondi* we could not differentiate either species from other nanhsiungchelyids based on that character. We did, however, identify a new informative character (length of posterior plastral lobe), which we added to our analysis in place of the older character 27; see Appendix 1 for definition and codings for our replacement character 27. Second, we added "*Z.*" *dzamynchondi* to our analysis; see Appendix 1 for character codings for that species. Third, for *Z. testudinimorpha* we were able to re-score three characters, as follows: character 47 (expansion of ventromedial edge of marginal 6) changed from "0" to "1"; character 66 (participation of pectoral in rim of axillary notch) changed from "?" to "1"; and character 70 (overlapping of scutes onto dorsal surface of plastron) changed from "?" to "1". Our updated matrix was assembled using NDE 0.5.0 (Page 2001) and analyzed using PAUP 4.0b10 (Swofford 2002). Characters were left unordered, considered reversible, and of equal weight. Bremer supports were calculated using Autodecay 4.0.1 (Eriksson 1998).

Systematic Paleontology

Testudines Batsch 1788
Cryptodira Cope 1868
Adocusia Danilov and Parham 2006
Nanhsiungchelyidae Yeh 1966
"*Zangerlia*" *dzamynchondi* Sukhanov and Narmandakh 2006
(Figs. 22.1, 22.2, 22.3)

Synonymy: *Zangerlia dzamynchondi*: Sukhanov 2000, p. 343, Fig. 17.25 (unavailable name); Sukhanov and Narmandakh 2006, p. 124 (type description); Danilov and Syromyatnikova 2008, p. 10.

Holotype: PIN 4698-1, incomplete shell, missing most of central part of carapace and left half of posterior lobe of plastron (Figs. 22.1, 22.2, 22.3).

Holotype locality, unit, and age: Zamin Khond locality, Umnegov Aimag (southern Gobi), southern Mongolia; Djadokhta Formation; Late Cretaceous, Campanian (Sukhanov 2000; Suzuki and Narmandakh 2004).

Occurrence: Known only by the holotype shell from the holotype locality.

Diagnosis: Differs from other nanhsiungchelyids in the following combination of three characters: sulci between pleural 1 and marginals 2 and 3 and between pleural 3 and marginals 7–9 located on peripherals (except *Basilemys* spp.); presence of a knobby protrusion on external surface of suprapygals (except some "*Zangerlia*" spp.); and length of posterior plastral lobe less than distance between posterior ends of plastron and carapace (except some "*Zangerlia*" spp.). Differs further from non-"*Zangerlia*" nanhsiungchelyids, as follows: from *Kharakhutulia kalandadzei* by the presence of a nuchal notch formed by the nuchal and peripherals 1, anterior plastral lobe protrudes farther anteriorly than leading margin of carapace, and broad dorsal extension of gulars onto thickened anterior plastral lip; from *Hanbogdemys orientalis* by longer bridges; from *Anomalochelys angulata* and *Nanhsiungchelys wuchingensis* by a small and trapezoidal nuchal; from *Basilemys* spp. by the presence of a nuchal notch formed by the nuchal and peripherals 1 and by the presence of four pairs of inframarginals. Differs from other species of "*Zangerlia*" in the following characters: wide suprapygal 1 (as wide as suprapygal 2); relatively wide peripherals; pleural-marginal sulcus distant from medial border of peripherals; relatively short contact between epiplastra;

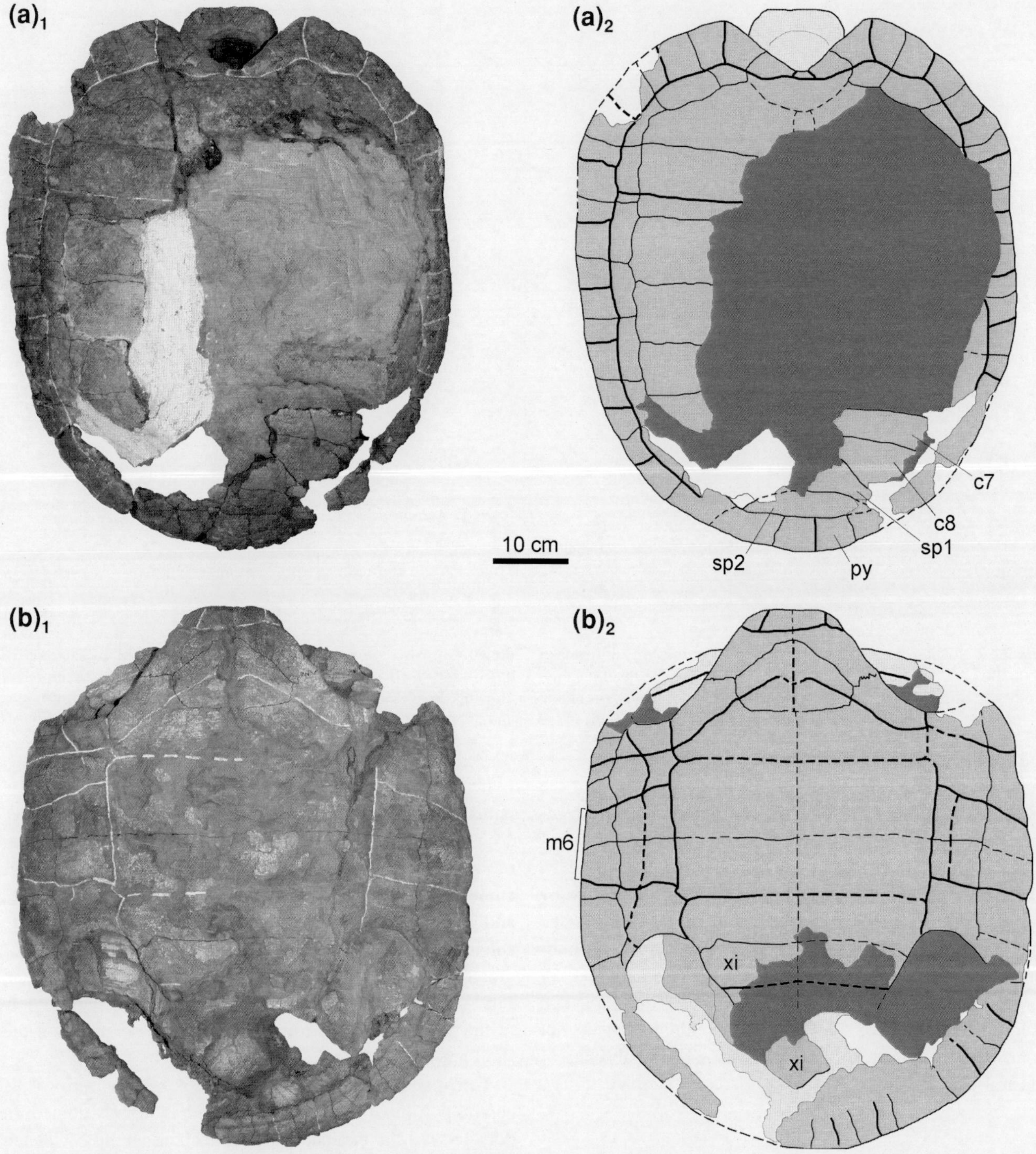

Fig. 22.1 "*Zangerlia*" *dzamynchondi* Sukhanov and Narmandakh 2006, PIN 4698-1 (holotype), incomplete shell; Zamin Khond locality, Umnegov Aimag (southern Gobi), Mongolia; Djadokhta Formation, Campanian. **a** Shell in dorsal view: $\mathbf{a_1}$ photograph; $\mathbf{a_2}$ interpretive drawing. **b** Shell in ventral view: $\mathbf{b_1}$ photograph; $\mathbf{b_2}$ interpretive drawing. Images at same scale. Diagrammatic conventions in interpretive drawings: bone in foreground shaded *medium grey*; bone in background shaded *light grey*; matrix shaded *dark grey*; broken surfaces indicated by *hatched lines*; and reconstructed edges or profiles indicated by *dashed lines*. Abbreviations: *abd* abdominal, *an* anal, *c* costal, *egu* extragular, *ent* entoplastron, *epi* epiplastron, *fe* femoral, *gu* gular, *h* humerus, *hu* humeral, *hyo* hyoplastron, *hypo* hypoplastron, *im* inframarginal, *kp* knobby protrusion, *m* marginal, *pe* pectoral, *py* pygal, *sc* scapula, *sp* suprapygal, *xi* xiphiplastron. Arabic numerals designate element numbers

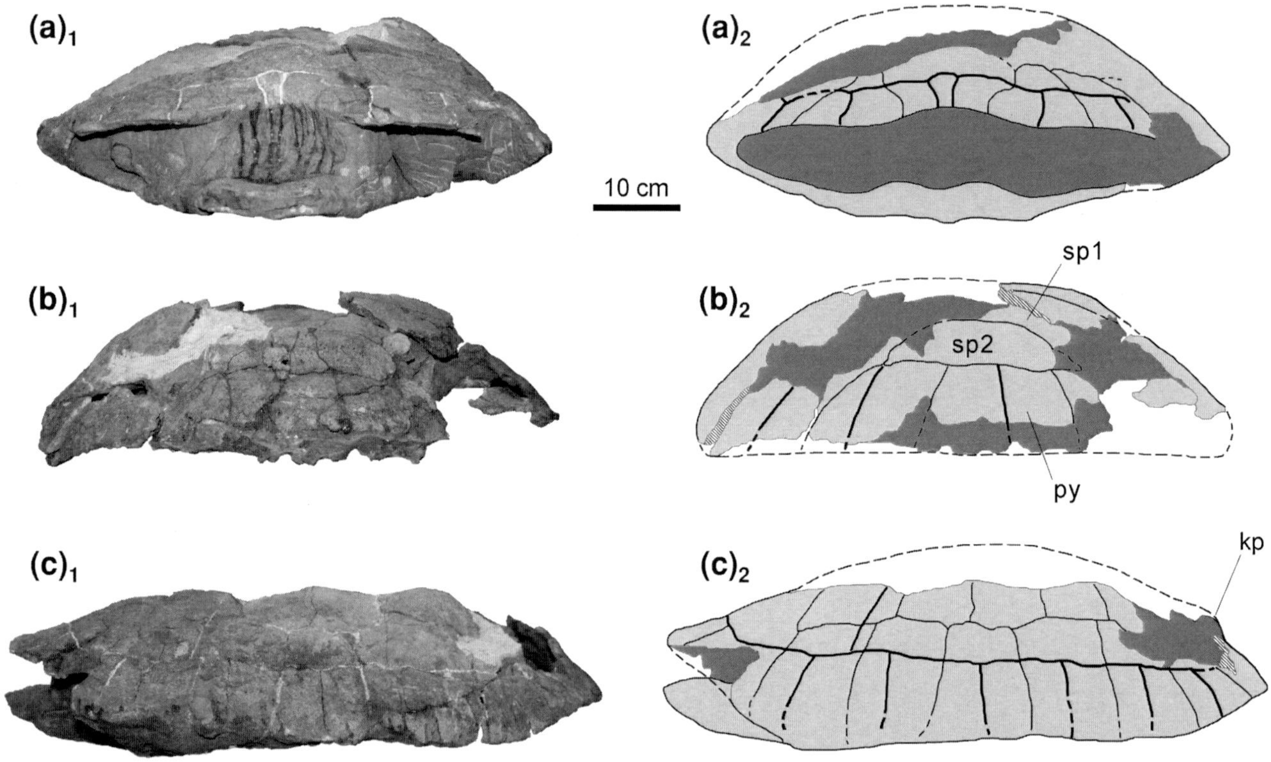

Fig. 22.2 "*Zangerlia*" *dzamynchondi* Sukhanov and Narmandakh 2006, PIN 4698-1 (holotype), incomplete shell; Zamin Khond locality, Umnegov Aimag (southern Gobi), Mongolia; Djadokhta Formation, Campanian. **a** Shell in anterior view: $\mathbf{a_1}$ photograph; $\mathbf{a_2}$ interpretive drawing. **b** Shell in posterior view: $\mathbf{b_1}$, photograph; $\mathbf{b_2}$ interpretive drawing. **c** Shell in left lateral view: $\mathbf{c_1}$ photograph; $\mathbf{c_2}$ interpretive drawing. Images at same scale. See Fig. 22.1 caption for conventions and abbreviations

pentagonal entoplastron that does not reach the level of the axillary notch; presence of a shallow anal notch; short gulars that do not reach the entoplastron; and four pairs of inframarginals that form a continuous row (except "Z." *ukhaachelys*). Differs further from "Z." *neimongolensis* by the presence of small extragulars that do not extend onto the entoplastron and do not meet one another across the midline; from *Z. testudinimorpha* and "Z." *ukhaachelys* by larger size; and from "Z." *ukhaachelys* by the expanded ventromedial edge of marginal 6, abdominals that do not contribute to the rim of the inguinal notch, and absence of plastral fontanelles.

Description

Shell: The length of the shell is about 70 cm and its width is 58 cm, which is comparable with the holotype specimen of "*Zangerlia*" *neimongolensis*, but much bigger than other preserved specimens of "*Zangerlia*". The plastron is estimated to be approximately 85% of the carapace length and projects anteriorly beyond the leading rim of the carapace. Fontanelles are absent in both the carapace and plastron. Among other species of "*Zangerlia*", plastral fontanelles are present only in "Z." *ukhaachelys*. The external surfaces of the shell are covered with a characteristic nanhsiungchelyid type of sculpturing, consisting of irregular grooves and pits ("pockmarks" sensu Joyce and Norell 2005). Scute sulci are deep and distinct. Photographs and interpretive drawings of the shell in multiple views are presented in Figs. 22.1 and 22.2, and a reconstruction of the ventral view of the shell is presented in Fig. 22.3. Measurements of the shell are given in Table 22.1.

Carapace: Although the carapace is missing most of its central portion, it can be reconstructed as slightly domed. In dorsal view, the carapace is oval shaped, slightly widened anteriorly, and has a deep and wide nuchal notch (notch length makes up about 10% of carapace length; notch width makes up about 30% of carapace width) that is formed by the nuchal and peripherals 1. The nuchal is relatively small (compared to *Anomalochelys angulata* and *Nanhsiungchelys wuchingensis*), and it is strongly emarginated and constricted anteriorly (ratio of width across anterior portion of nuchal relative to maximum width of nuchal is 0.59, which is the same as in *Hanbogdemys orientalis*). The length and posterior outline of the nuchal are not clear. The visceral

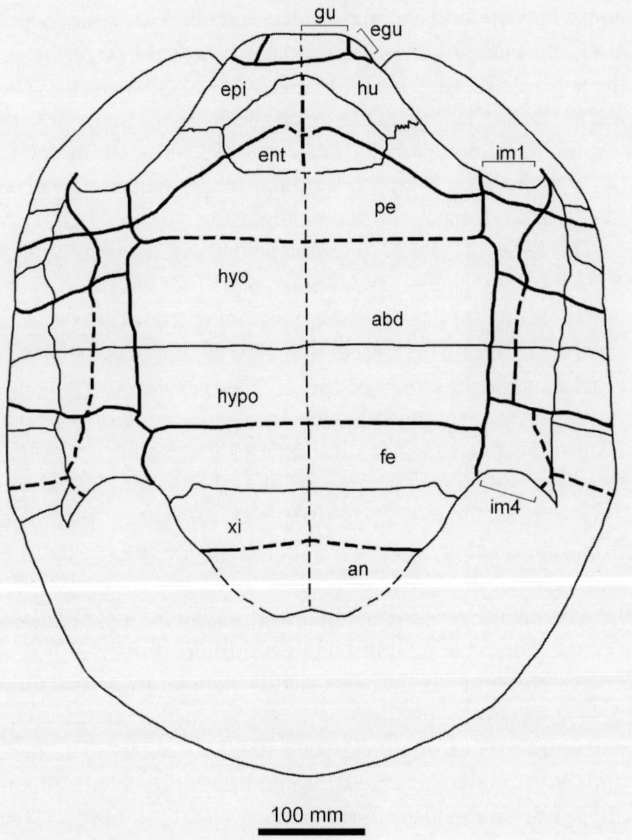

Fig. 22.3 New reconstruction of the shell of "*Zangerlia*" *dzamynchondi* Sukhanov and Narmandakh 2006, based on PIN 4698-1 (holotype), in ventral view. See Fig. 22.1 caption for conventions and abbreviations

Table 22.1 Measurements (in mm) for the holotype shell (PIN 4698-1) of "*Zangerlia*" *dzamynchondi*. Multiple measurements are separated by a forward slash

Bone, scute, or portion of shell and description (in parentheses) of measurement(s)	Measurements
Peripheral 1 (length along free edge/width of dorsal plate across middle)	100/95
Peripheral 2 (as for Peripheral 1)	110/95
Cervical (length/width across anterior edge/width across posterior edge)	25/14/30
Marginal 2 (height/length along free edge)	60/90
Marginal 4 (as for Marginal 2)	115/80
Marginal 5 (as for Marginal 2)	120/75
Marginal 6 (as for Marginal 2)	125/85
Marginal 7 (as for Marginal 2)	70/80
Plastral bridges (minimal length)	270
Anterior lobe of plastron (length/width across base)	150/400
Posterior lobe of plastron (length/width across base)	?/380
Epiplastra (medial length)	68
Entoplastron (length/width)	105/130
Hyoplastra (medial length)	155
Hypoplastra (medial length)	130
Gulars (medial length)	46
Extragulars (medial length)	40
Humerals (medial length)	80
Pectorals (medial length)	105
Abdominals (medial length)	160
Inframarginal 1 (length at contact with plastral scutes/length at contact with marginals/width at contact with anterior scute)	40/55/?
Inframarginal 2 (as for Inframarginal 1)	45/50/45
Inframarginal 3 (as for Inframarginal 1)	125/110/40
Inframarginal 4 (as for Inframarginal 1)	65/?/25

Abbreviation: "?", cannot be measured

surface of the nuchal is covered with matrix; consequently, the presence of costiform processes cannot be determined. The only part of the neural series that is preserved is the anterior part of neural 1, and its precise outline is unclear. There are two suprapygals, but they are preserved only partially. Suprapygal 1 is represented by its right lateral part; it is reconstructed as trapezoidal in outline and as wide as suprapygal 2. Among nanhsiungchelyids, a similarly shaped suprapygal 1 is present in *Basilemys sinuosa* Riggs 1906, whereas other species of "*Zangerlia*" have a narrower suprapygal 1. Suprapygal 2 is wide and has a knobby protrusion midway along its length, similar to those in "*Z*". *neimongolensis* and *Z. testudinimorpha*. The pygal is approximately trapezoidal in outline, wider than long, and slightly widened posteriorly. Its morphology generally corresponds to those of other Nanhsiungchelyidae.

The costals are represented by the lateral parts of costals 1–6 on the left side and fragments of costals 7 and 8 on the right side. Costal 1 is much narrower anteroposteriorly than in other Nanhsiungchelyidae. Costals 2–7 seem to have almost parallel anterior and posterior borders (at least in their lateral parts), whereas costal 8 is strongly widened laterally. Visceral surfaces of the costals are observable on costals 7 and 8 only, both of which bear well-developed rib heads and rib thickenings; costal 8 also bears an attachment for the ilium. The morphology of the costals generally corresponds to those of other Nanhsiungchelyidae.

The peripherals are best preserved on the left side of shell, where they are represented by a complete row of elements. The free edge of peripheral 1 is angled as in some Nanhsiungchelyidae (*Hanbogdemys orientalis* and *Nanhsiungchelys wuchingensis*); this character is unknown in other species of "*Zangerlia*". The other peripherals are approximately as wide as long, and differ from those of other "*Zangerlia*" spp. in being relatively wider. The posterior peripherals are narrowed (shortened).

The following scutes of the carapace are observable: cervical; lateral part of left pleural 1; and marginals 1–10. The cervical is wide and trapezoidal in dorsal view. Its anterior shape is rectangular, but its ventral shape is not clear. Although the cervical is not know for any other species of "*Zangerlia*", in other Nanhsiungchelyidae it is usually wide and either rectangular or trapezoidal in dorsal view. Laterally, the pleurals strongly overlap the corresponding peripherals, similar to *Basilemys* spp. The observable marginals are restricted to the peripherals; the position of the pleural-marginal sulcus is distant from the medial border of the peripherals, whereas in other "*Zangerlia*" spp. this sulcus coincides with the medial border of the peripherals. Marginals 4–7 clearly extend onto the plastron. The ventromedial edge of marginal 6 is expanded, as in other members of Nanhsiungchelyidae (except "Z." *ukhaachelys*).

Plastron: The plastron is almost completely preserved, missing only the left half of the posterior lobe. The contact of the anterior buttress is not observable; however, contact of the posterior buttress with peripheral 8, which is known in other nanhsiungchelyids, is clearly visible. The length of the bridge is about 50% of the plastron length. The anterior lobe of the plastron is wider than long (ratio of midline length of anterior lobe versus width across its base is 0.40), trapezoidal, narrowed anteriorly, has a straight anterior edge, and lacks a gular notch. The length of the anterior lobe makes up 27% of the plastron length. The anterior lobe reaches farther anteriorly than the carapace rim, as in "Z." *neimongolensis*, *Basilemys* spp., and *Hanbogdemys orientalis*. The posterior lobe of the plastron is damaged. Its precise shape and proportions are not clear, but, as reconstructed, are similar to those of the anterior lobe. The length of the posterior plastral lobe is less than the distance between the posterior ends of the plastron and carapace, as "Z." *neimongolensis* and *Z. testudinimorpha*. The posterior lobe is slightly constricted at the femoral-anal sulcus and has a shallow anal notch. Among nanhsiungchelyids, an anal notch is known only in *Kharakhutulia kalandadzei*. Both lobes are thickened along the free edge, as in other Nanhsiungchelyidae. The anterior plastral lip is thickened with a broad dorsal extension of the gulars, as in most other Nanhsiungchelyidae (except *K. kalandadzei*).

The epiplastra have a sutural contact with one another across the midline; this contact is shorter than in other species of "*Zangerlia*" and Nanhsiungchelyidae. The epi-entoplastral suture is very long. The entoplastron is a large pentagonal element, wider than long, and does not reach the level of the axillary notch. Its shape is somewhat similar to *Basilemys* spp. and *Hanbogdemys orientalis,* whereas in other "*Zangerlia*" spp. and *Nanhsiungchelys wuchingensis* the entoplastra are diamond-shaped and reach the level of the axillary notch. The hyoplastron and hypoplastron make equal contributions to the bridge length. The xiphiplastra are represented by the anterior parts from both sides and by the separated, posterior part of the right xiphiplastron. The reconstructed length of the xiphiplastron is about 90% of the midline length of the posterior lobe; this is similar to other "*Zangerlia*" spp., but unlike other Nanhsiungchelyidae, which all have shorter xiphiplastra.

The plastral scutes are represented by a complete set including the gulars, extragulars, humerals, pectorals, abdominals, femorals, and four pairs of inframarginals. The gulars are very short medially and do not reach the entoplastron. As in other species of "*Zangerlia*", it is unclear if the gulars are separated or fused. The gulars of other Nanhsiungchelyidae usually touch or overlap the entoplastron. The gulars extend onto the dorsal anterior plastral lip. The extragulars are small triangular elements, separated from one another by a wide contact of the gulars and humerals. This condition is similar to some *Basilemys* spp., *Hanbogdemys orientalis*, and *Kharakhutulia kalandadzei.* On the other hand, this condition differs from "Z." *neimongolensis*, in which the extragulars are large, extend onto the entoplastron, and meet one another at the midline. The humerals have a relatively long median contact. The pectoral is strongly increased in anteroposterior length medially. The humeral-pectoral sulcus clearly intersects the entoplastron, as in all Nanhsiungchelyidae. The pectoral does not contribute to the anterior plastral rim, as in other Nanhsiungchelyidae, and it has both a long contact with inframarginal 2 and a point contact with inframarginal 1. The abdominal is a rectangular element separated from the inguinal notch by contact of the femoral and inguinal, as in most other Nanhsiungchelyidae. In some species of *Basilemys*, *K. kalandadzei*, and "Z." *ukhaachelys* the abdominal contributes to the rim of the inguinal notch. The femoral contacts inframarginal 4 laterally, as in "Z." *neimongolensis* and *Z. testudinimorpha*, and it laps onto the dorsal surface of the posterior plastral lobe. The inframarginals are represented by a complete row of four scutes that are clearly visible on the right side. They are restricted to the plastron and situated rather distant from the plastron-carapace suture. Inframarginal 1 is tetragonal, variable in length between the right and left sides, contributes to the axillary notch anteriorly, and contacts the humeral medially, inframarginal 2 posteriorly, and marginal 4 laterally. Inframarginal 2 has an irregular pentagonal outline, is variable in length between the right and left sides, and contacts inframarginal 1 anteriorly, the pectoral anteromedially, the abdominal posteromedially, inframarginal 3 posteriorly, and marginal 5 laterally. Inframarginal 3 is tetragonal and elongate (at least three times longer than wide), and contacts inframarginal 2 anteriorly, the abdominal medially, inframarginal 4 posteriorly, and marginal 6 laterally. Inframarginal 4 contacts inframarginal 3

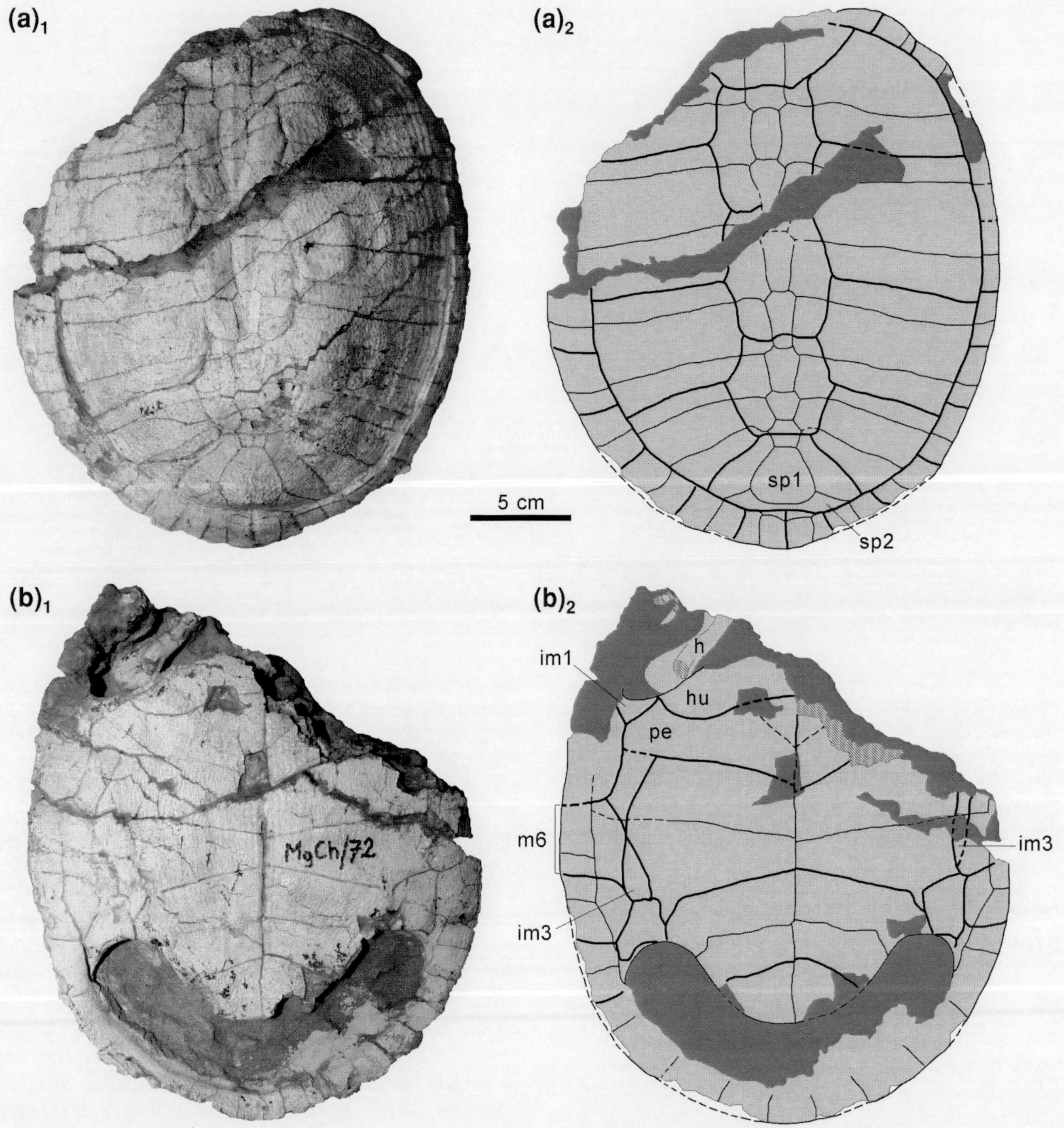

Fig. 22.4 *Zangerlia testudinimorpha* Młynarski 1972, ZPAL MgCh/72 (holotype), incomplete shell; Nemegt locality, Gov-Altai Aimag (Transaltai Gobi), Mongolia; Barungoyot and Nemegt formations, Campanian–Maastrichtian. **a** Shell in dorsal view: $\mathbf{a_1}$ photograph; $\mathbf{a_2}$ interpretive drawing. **b** Shell in ventral view: $\mathbf{b_1}$ photograph; $\mathbf{b_2}$ interpretive drawing. Images at same scale. See Fig. 22.1 caption for conventions and abbreviations

anteriorly, has a short medial contact with the abdominal and a longer contact with the femoral, is slightly widened posteriorly and embraces the inguinal notch, and has a lateral contact with marginal 7. If correctly reconstructed, this pattern is similar to "Z." *ukhaachelys* in the presence of a complete row of four inframarginals, although the shapes and contacts of the scutes are different. Four pairs of inframarginals are also present in *Z. testudinimorpha*, but they do not form continuous rows (see Discussion), whereas "Z." *neimongolensis* was reported to have only two or three pairs of inframarginals (1, 3, and 4 according to Brinkman and Peng 1996). The shape of the midline sulcus is unclear.

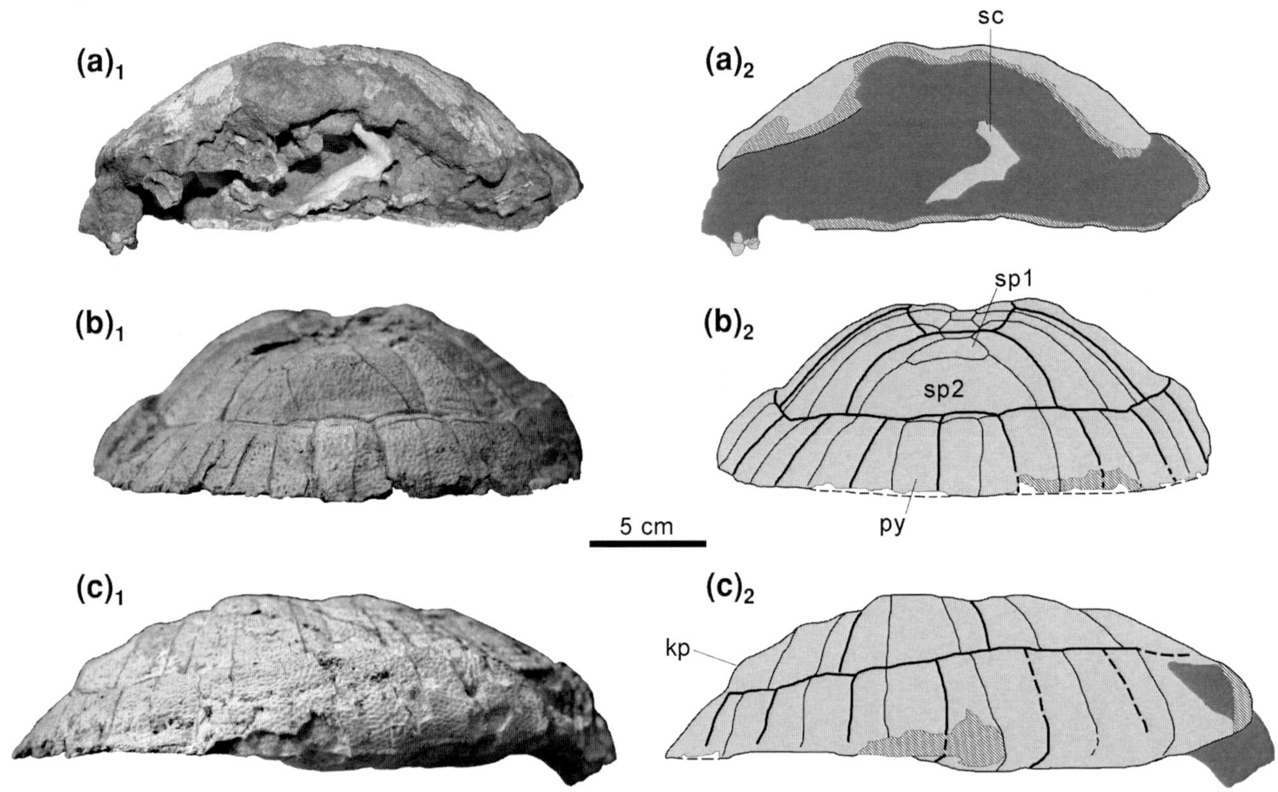

Fig. 22.5 *Zangerlia testudinimorpha* Młynarski 1972, ZPAL MgCh/72 (holotype), incomplete shell; Nemegt locality, Gov-Altai Aimag (Transaltai Gobi), Mongolia; Barungoyot and Nemegt formations, Campanian–Maastrichtian. **a** Shell in anterior view: $\mathbf{a_1}$ photograph; $\mathbf{a_2}$ interpretive drawing. **b** Shell in posterior view: $\mathbf{b_1}$ photograph; $\mathbf{b_2}$ interpretive drawing. **c** Shell in right lateral view: $\mathbf{c_1}$ photograph; $\mathbf{c_2}$ interpretive drawing. Images at same scale. See Fig. 22.1 caption for conventions and abbreviations

Discussion

New Data on the Morphology of "Zangerlia" dzamynchondi and Z. testudinimorpha

Our reexamination of the holotype shell of "*Zangerlia*" *dzamynchondi* allows us to present new images of this specimen (Figs. 22.1, 22.2) and suggest a new reconstruction in ventral view (Fig. 22.3). Our observations and reconstruction differ from previously published ones (e.g., Sukhanov 2000, Fig. 17.25) in the following details: the posterior border of costal 1 is placed farther posteriorly and coincides with the posterior border of peripheral 3; the posterior borders of costals 2–5 do not coincide with the posterior borders of the corresponding peripherals; suprapygal 2 and the pygal are relatively wider; the gulars do not touch the entoplastron; the inframarginals are clearly represented by a complete row of inframarginals 1–4 on the right side and by at least inframarginal 3 on the left side; and the posterior plastral lobe is relatively shorter and narrower posteriorly.

Our observations on the holotype of *Zangerlia testudinimorpha* allow us to present new images for this taxon as well (Figs. 22.4, 22.5) and clarify some characters of this specimen. Among these characters is an expansion of the ventromedial edge of marginal 6. Previously, *Z. testudinimorpha* was considered to be the only member of Nanhsiungchelyidae (in which the morphology of this region was known) that lacked such an expansion (Brinkman and Nicholls 1993; Hirayama et al. 2001; Joyce and Norell 2005; Sukhanov et al. 2008; Danilov and Syromyatnikova 2009). Another important character that can now be recognized for *Z. testudinimorpha* is that the pectoral does not participate in the rim of the axillary notch, as in other Nanhsiungchelyidae. Previously, this character was coded as variable or unknown (Joyce and Norell 2005; Danilov and Syromyatnikova 2009). Although the upper surface of the plastron of *Z. testudinimorpha* is covered with matrix, this species appears to have scutes overlapping onto the dorsal (visceral) surface of the posterior lobe of the plastron, as in other Nanhsiungchelyidae. In addition, we note the presence of an elongate and narrow inframarginal 3 on the left side of the plastron, which is separated from inframarginal 4 by a contact of the abdominal and marginal 6. Previously, inframarginal 3 was shown only on the right side of the plastron, where it was depicted as wider and having a broad contact with inframarginal 4 (Młynarski

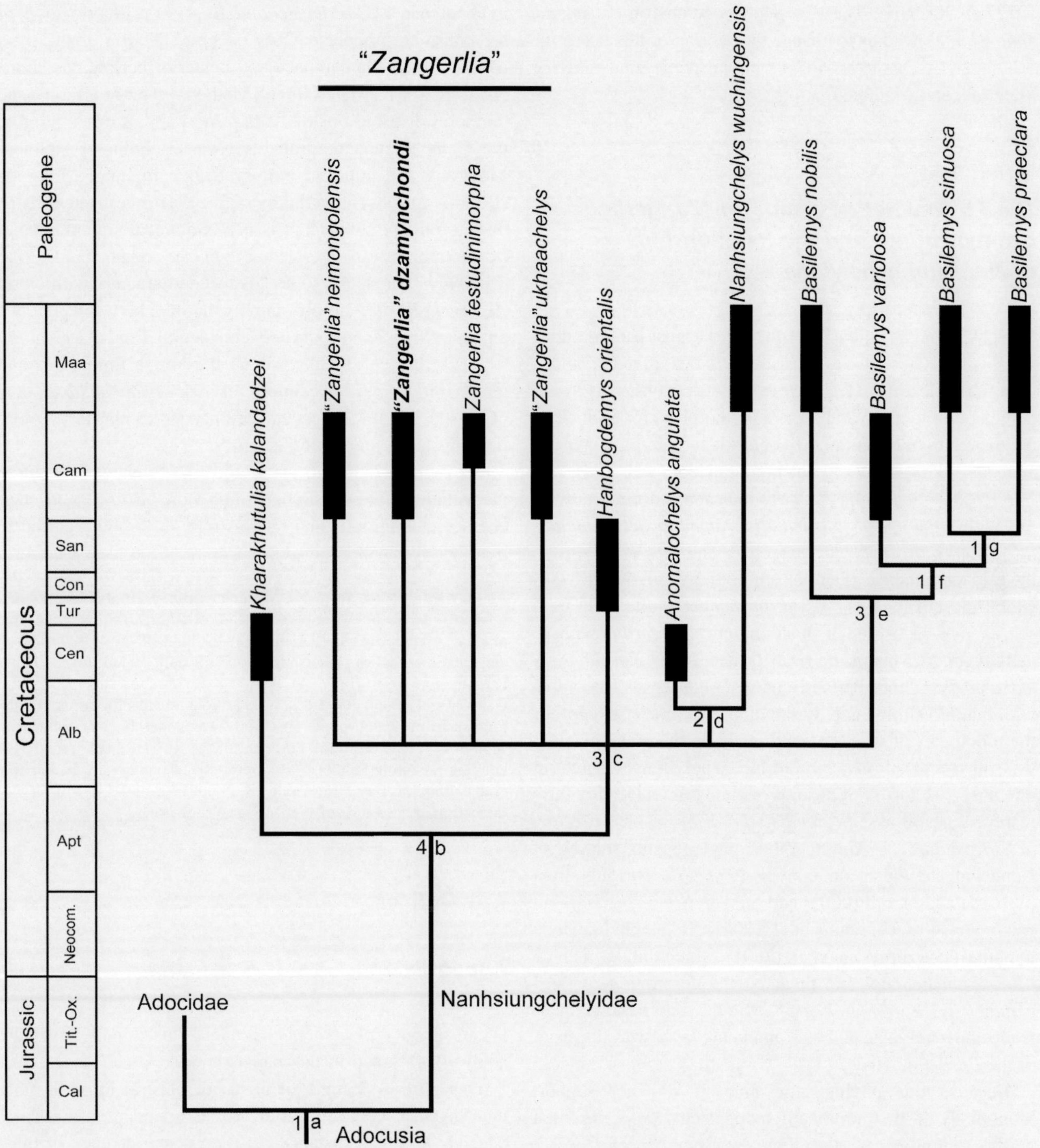

Fig. 22.6 Phylogeny of Adocusia showing the hypothesized position of "*Zangerlia*" *dzamynchondi* (*in bold*) based on a strict consensus of 11 phylogenetic trees resulting from this study. See text (Discussion) for tree description. For diagrammatic simplicity, outgroups and adocid taxa are not shown. *Black boxes* indicate temporal ranges of ingroup taxa. Numbers are Bremer support indices. Letters designate nodes with the following unambiguous synapomorphies: node a (Adocusia): 32(1), 52(1); node b (Nanhsiungchelyidae): 47(1), 54(1), 55(1), 66(1), 70(1), 72(1); node c (unnamed): 29(1), 43(1), 53(1), 60(1); node d (unnamed) 30(1), 43(1); node e (*Basilemys*): 29(0), 46 (1), 48(0), 51(1), 67(1); node f (unnamed): 59(0); and node g (unnamed) 62(1). Abbreviations for geological stages: *Alb* Albian, *Apt* Aptian, *Cal* Callovian, *Cam* Campanian, *Cen* Cenomanian, *Con* Coniacian, *Maa* Maastrichtian, *Neocom* Neocomian, *Ox* Oxfordian, *San* Santonian, *Tit* Tithonian, *Tur* Turonian

1972, pl. XXVIII). These new observations on *Z. testudinimorpha* allowed us to change the scoring of this taxon for a number of characters in our phylogenetic analysis (see Materials and Methods).

The Phylogenetic Position of "Zangerlia" dzamynchondi and the Relationships of Nanhsiungchelyidae

Our phylogenetic analysis resulted in 11 most parsimonious trees with 128 steps, a consistency index of 0.64, and a retention index of 0.81. The resulting strict consensus tree is depicted in Fig. 22.6 and the synapomorphies for each clade are listed in the caption. According to our phylogenetic analysis, "*Zangerlia*" is not monophyletic. All species of "*Zangerlia*", including "*Z.*" *dzamynchondi*, form a polytomy with *Hanbogdemys orientalis*, *Anomalochelys angulata* + *Nanhsiungchelys wuchingensis*, and *Basilemys* spp. In the previous analysis (Danilov and Syromyatnikova 2009)—which did not include "*Z.*" *dzamynchondi*—the strict consensus tree recovered a monophyletic *Zangerlia*, with *Z. neimongolensis* being the sister of *Z. testudinimorpha* + *Z. ukhaachelys*. Other parts of our strict consensus tree agree with that of Danilov and Syromyatnikova (2009), except for the topologies of the *Basilemys* clade. In the previous analysis, all species of *Basilemys* formed a polytomy, whereas in our analysis, the relationships within this clade are fully resolved as follows: *B. nobilis* (*B. variolosa* (*B. praeclara* + *B. sinuosa*)). Among other phylogenetic studies of Nanhsiungchelyidae, monophyly of *Zangerlia* was supported by Joyce and Norell (2005) and Sukhanov et al. (2008), but was rejected by Hirayama et al. (2001). In the phylogenetic hypothesis of Hirayama et al. (2001), which did not include "*Z.*" *dzamynchondi* and predated the description of "*Z.*" *ukhaachelys*, *Z. testudinimorpha* was the most basal taxon of Nanhsiungchelyidae and "*Z.*" *neimongolensis* was the sister of the *Basilemys* clade.

Because our phylogenetic analysis did not support monophyly of the traditionally recognized "*Zangerlia*", the generic affinities of the three non-type species ("*Z.*" *dzamynchondi*, "*Z.*" *neimongolensis*, and "*Z.*" *ukhaachelys*) are unclear. "*Zangerlia*" *dzamynchondi* shares two advanced characters with at least *Z. testudinimorpha* and "*Z.*" *neimongolensis*: presence of a knobby protrusion on suprapygals and the length of the posterior plastral lobe is less than the distance between the posterior ends of the plastron and carapace. Neither character can be determined for "*Z.*" *ukhaachelys*, because the posterior part of the carapace is missing in the holotype and only known specimen. Other characters previously considered diagnostic for "*Zangerlia*" either are not known in "*Z.*" *dzamynchondi* (loss of the lingual ridge) or are considered unclear by us (steep deflection of the postneural part of the carapace and posterior peripherals shortened; see above, Materials and Methods). The geographic and temporal distributions of "*Zangerlia*" spp. are restricted to the Campanian and, probably, Maastrichtian of southern Mongolia and adjacent part of China, in Inner Mongolia (Danilov and Syromyatnikova 2008). This restricted distribution could be considered as an additional argument in favor of monophyly of the genus. On the other hand, "*Z.*" *dzamynchondi* shares some advanced characters with North American *Basilemys* spp., namely the sulci between pleural 1 and marginals 2 and 3 and between pleural 3 and marginals 7–9 are on the peripherals. Given the above limitations and uncertainties, new specimens of "*Z.*" *dzamynchondi* and other species of "*Zangerlia*" are needed to clarify the relationships and content of this genus.

As a final comment, we would like to encourage future nanhsiungchelyid researchers to be more accurate with their character definitions and to illustrate described material in more aspects and better detail.

Acknowledgments We thank Magdalena Borsuk-Białynicka (ZPAL) for access to the holotype of *Zangerlia testudinimorpha* and her hospitality; Andrei Bochkov (Zoological Institute of the Russian Academy of Sciences, St. Petersburg, Russia); and Walter Joyce, James Parham, and Ren Hirayama for their reviews of this paper. Jim Gardner and Walter Joyce deserve additional thanks for correcting the English in our manuscript. This study was undertaken with financial support by grants from the President of the Russian Federation to the Leading Scientific Schools (NSh-6560-2012.4) and from the Russian Foundation for Basic Research 08-05-00557-a and the Ministry of Education and Science of the Russian Federation.

Appendices

Appendix 1

Details about replacement character 27 used in phylogenetic analysis presented in this paper.

Description: Length of posterior plastral lobe: 0, more than distance between posterior ends of plastron and carapace; 1, less than distance between posterior ends of plastron and carapace.

Comments: This replaces character 27 in the previous analysis of Danilov and Syromyatnikova (2009). The primitive state of the new character 27 is found in the outgroups, adocids, and most nanhsiungchelyids. The derived condition is considered diagnostic for *Zangerlia* (Brinkman and Peng 1996; this chapter).

Codings: *Xinjiangchelys levensis*, 0; *X. tianshanensis*, 0; *Carettochelys insculpta*, 0; *Apalone ferox*, -; *Yehguia tatsuensis*, 0; *Adocus aksary*, ?; *A. beatus*/*Adocus* sp., 0;

Ferganemys itemirensis, 0; *F. verzilini*, 0; *Shachemys ancestralis*, 0; *S. baibolatica*, 0; *S. laosiana*, 0; *Kharakhutulia kalandadzei*, 0; *Zangerlia testudinimorpha*, 1; *Z. neimongolensis*, 1; *Z. ukhaachelys*, ?; *Z. dzamynchondi*, 1; *Hanbogdemys orientalis*, 0; *Anomalochelys angulata*, ?; *Nanhsiungchelys wuchingensis*, ?; *Basilemys variolosa*, 0; *B. nobilis*, 0; *B. sinuosa*, 0; *B. praeclara*, 0.

Appendix 2

Characters coded for "*Zangerlia*" *dzamynchondi* and added to the matrix of Danilov and Syromyatnikova (2009).

?????????? ?????????? ??????1110 ???0110?0? 00????1000 11111010?1 00001100?1 010?

References

Batsch, A. J. G. C. (1788). *Versuch einer Anleitung, zur Kenntniß und Geschichte der Thiere und Mineralien*. Jena: Akademische Buchhandlung.

Brinkman, D. B., & Nicholls, E. L. (1993). New specimen of *Basilemys praeclara* Hay and its bearing on the relationship of the Nanhsiungchelyidae (Reptilia: Testudines). *Journal of Paleontology, 67*, 1027–1031.

Brinkman, D. B., & Peng, J. H. (1996). A new species of *Zangerlia* (Testudines: Nanhsiungchelyidae) from the Upper Cretaceous redbeds at Bayan Mandahu, Inner Mongolia, and the relationships of the genus. *Canadian Journal of Earth Sciences, 33*, 526–540.

Cope, E. D. (1868). On the origin of genera. *Proceedings of the Academy of Natural Sciences of Philadelphia, 1868*, 242–300.

Cope, E. D. (1870). On the Adocidae. *Proceedings of the American Philosophical Society, 11*, 547–553.

Danilov, I. G., & Parham, J. F. (2006). A redescription of '*Plesiochelys*' *tatsuensis* from the Late Jurassic of China, with comments on the antiquity of the crown clade Cryptodira. *Journal of Verterbate Paleontology, 26*, 573–580.

Danilov, I. G., & Syromyatnikova, E. V. (2008). New materials on turtles of the family Nanhsiungchelyidae from the Cretaceous of Uzbekistan and Mongolia, with a review of the nanhsiungchelyid record in Asia. *Proceedings of the Zoological Institute of the Russian Academy of Sciences, 312*, 3–25.

Danilov, I. G., & Syromyatnikova, E. V. (2009). Phylogeny of the extinct turtle clade Adocusia. *Gaffney Turtle Symposium (October 17–18, 2009, Royal Tyrrell Museum, Drumheller, Canada). Abstract Volume*, 44–51.

Eriksson, T. (1998). *AutoDecay. 4.0.1*. (Program distributed by author). Stockholm: Department of Botany, Stockholm University.

Hay, O. P. (1902). Bibliography and catalogue of fossil vertebrates of North America. *United States Geological Survey Bulletin, 179*, 1–449.

Hirayama, R., Sakurai, K., Chitoku, T., Kawakami, G., & Kito, N. (2001). *Anomalochelys angulata*, an unusual land turtle of family Nanhsiungchelyidae (Superfamily Trionychoidea; Order Testudines) from the Upper Cretaceous of Hokkaido, North Japan. *Russian Journal of Herpetology, 8*, 127–138.

Hutchison, J. H. (2000). Diversity of Cretaceous turtle faunas of Eastern Asia and their contribution to the turtle faunas of North America. *Paleontological Society of Korea Special Publication, 4*, 27–38.

Joyce, W. G., & Norell, M. A. (2005). *Zangerlia ukhaachelys*, new species, a nanhsiungchelyid turtle from the Late Cretaceous of Ukhaa Tolgod, Mongolia. *American Museum Novitates, 3481*, 1–19.

Młynarski, M. (1972). *Zangerlia testudinimorpha* n. gen., n. sp., a primitive land tortoise from the Upper Cretaceous of Mongolia. *Palaeontologia Polonica, 27*, 85–92.

Page, R. D. M. (2001). *WinClada version 1.00.08*. http://www.cladistics.com.

Riggs, E. S. (1906). The carapace and plastron of *Basilemys sinuosus*, a new fossil tortoise from the Laramie Beds of Montana. *Field Columbian Museum, Geological Series, 2*, 249–256.

Sukhanov, V. B. (2000). Mesozoic turtles of Middle and Central Asia. In M. J. Benton, M. A. Shishkin, D. M. Unwin, & E. N. Kurochkin (Eds.), *The age of dinosaurs in Russia and Mongolia* (pp. 309–367). Cambridge: Cambridge University Press.

Sukhanov, V. B., Danilov, I. G., & Syromyatnikova, E. V. (2008). The description and phylogenetic position of a new nanhsiungchelyid turtle from the Late Cretaceous of Mongolia. *Acta Palaeontologica Polonica, 53*, 601–614.

Sukhanov, V. B., & Narmandakh, P. (2006). New taxa of Mesozoic turtles from Mongolia. In: I. G. Danilov & J. F. Parham (Eds.) *Fossil Turtle Research* (Vol. 1, pp. 119–127). St. Petersburg: Zoological Institute of the Russian Academy of Sciences.

Suzuki, Sh., & Narmandakh, P. (2004). Change of the Cretaceous turtle faunas in Mongolia. *Hayashibara Museum of Natural Sciences Research Bulletin, 2*, 7–14.

Swofford, P. L. (2002). *PAUP, Version 4. 0b10*. Sunderland: Sinauer Associates.

Yeh, H.-K. (1966). A new Cretaceous turtle of Nanhsiung, Northern Kwangtung. *Vertebrata PalAsiatica, 10*, 191–200.

Chapter 23
Cretaceous Trionychids of Asia: An Expanded Review of Their Record and Biogeography

Igor G. Danilov and Natasha S. Vitek

Abstract This chapter presents an expanded review of all described and/or figured specimens and taxa of trionychids from the Cretaceous (Barremian to Maastrichtian) of Asia. Cretaceous occurrences in Asia are limited to indeterminate trionychids and to the subfamily Trionychinae; cyclanorbines are unknown and previous reports of plastomenids are based on misidentified trionychine material. For the Trionychinae, we recognize 13 taxa of uncertain placement within the subfamily ("*Aspideretes*" *maortuensis*, four species of *Khunnuchelys*, "*Paleotrionyx*" *riabinini*, "*Trionyx*" *kansaiensis* and "*T.*" *kyrgyzensis*, and Trionychinae indet. 1–5), seven taxa within the tribe Trionychini ("*Amyda*" *menneri* and "*Am*". *orlovi*, "*Aspideretes*" *alashanensis*, *Aspideretoides riabinini* and *Aspideretoides* sp., and Trionychini indet. 1 and 2), and three taxa within the subtribe Apalonina (Apalonina indet. 1–3). We also recognize *Sinamyda* and four indeterminate taxa as Trionychidae incertae sedis. Although two other tribes (Ulutrionychini and Rafetini) have been proposed to include some Asian Cretaceous trionychines, monophyly of those tribes has yet to be satisfactorily demonstrated. During the Cretaceous, earlier and more basal trionychids (i.e., trionychines with eight neurals) were more broadly distributed across Asia, whereas later and more derived taxa had more restricted ranges. The recognition of *Aspideretoides* spp. and three indeterminate Apalonina in the Cretaceous of Asia provides evidence for a shared history with North American trionychids.

Keywords Asia • Cretaceous • Soft-shelled turtles • Trionychidae

Introduction

The Trionychidae Gray 1825, or soft-shelled turtles, are a group of highly aquatic cryptodires (Meylan 1987). They first appeared in Asia in the Early Cretaceous (Barremian), then in North America in the Late Cretaceous (Cenomanian), and spread to most other continents in the Cenozoic (Nessov 1995a; Hutchison 2000; Hirayama 2002; Brinkman 2003; Danilov 2005). The phylogeny and taxonomy of this group are still not entirely understood (Meylan 1987; Gardner et al. 1995; Karl 1998; Engstrom et al. 2004; Joyce and Lyson 2010). This is especially true for Cretaceous trionychids, which are important for understanding the early diversification and evolution of the family (see Fig. 23.1 for known distributions of Cretaceous trionychids). Their record is poor and includes, besides numerous indeterminate specimens, many taxa that are based entirely on either skulls or shells (Hutchison 2000).

In this chapter we present a review of all described and/or figured trionychid specimens from the Cretaceous of Asia, and provide comments on their morphology, systematic position, and taxonomic status. We also discuss the main suprageneric taxa and biogeography of Asian Cretaceous trionychids. Parts of this review were presented earlier (Danilov and Vitek 2009, Vitek and Danilov 2010). Here we expand upon that work with photographs, illustrations, or reconstructions of the most important specimens and add information for each taxon. Our review does not include undescribed or figured trionychid specimens or occurrences mentioned by Gilmore (1931), Kordikova (1994a), Nessov (1997), Khosatzky (1999), Hirayama et al. (2000), and Lapparent de Broin (2004) or any stem-trionychid or trionychoid material from the Akaiwa Formation

I. G. Danilov (✉)
Department of Herpetology, Zoological Institute of the Russian Academy of Sciences, Universitetskaya Emb. 1, St. Petersburg, Russia 199034
e-mail: igordanilov@gmail.com

N. S. Vitek
Yale University, PO 202411 New Haven, CT 06520, USA
e-mail: nsvitek@utexas.edu

D. B. Brinkman et al. (eds.), *Morphology and Evolution of Turtles*, Vertebrate Paleobiology and Paleoanthropology,
DOI: 10.1007/978-94-007-4309-0_23,

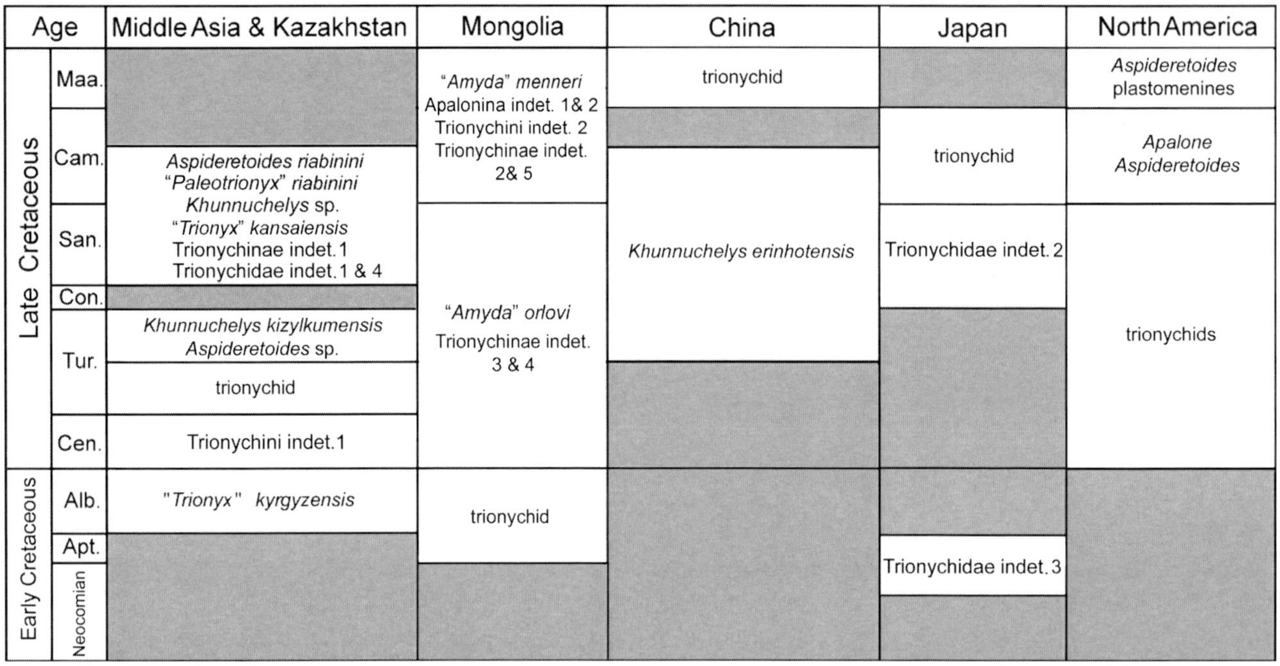

Fig. 23.1 Temporal and geographic distributions of Cretaceous Trionychidae. Gaps in record are filled with *grey*. For Asian columns, "trionychid" denotes undescribed occurrences taken from the following publications: Middle Asia and Kazakhstan: Nessov (1997); Mongolia: Suzuki and Narmandakh (2004); China: Moiseenko et al. (1997); and Japan: Hirayama et al. (2000). Distributions of North American trionychids taken from Gardner et al. (1995), Eaton et al. (1999), Brinkman (2003), and Hutchison and Holroyd (2003). Apalonina indet. 3 (Late Cretaceous of Mongolia), "*Aspideretes*" *alashanensis* and "*A.*" *maortuensis* (both Early or Late Cretaceous of China), *Sinamyda fuchienensis* (?Early Cretaceous of China), and Trionychinae indet. 6 (Early Cretaceous of Mongolia) are not not shown due to uncertainties about their ages. See text for further details

(late Neocomian or Barremian) of Japan (Hirayama 2008; Hirayama et al. 2012).

A substantial amount of the relevant literature on Asian Cretaceous trionychids was published in Russian and in journals, books, and conference abstract volumes that are not readily accessible. Consequently, many of the taxa included in our review originally were diagnosed and described in Russian. For most of those taxa, here we provide the first English translations of the original Russian diagnoses. For the remaining taxa, the diagnoses presented here follow either the original version or the most recently revised version. Because we follow previously published diagnoses and have not made any attempt to provide our own revised or emended versions, the format, presentation, and terminology in those diagnoses varies. Changes (if present) to the original text, spelling, and grammar in those diagnoses are given in square brackets. In Appendix, we provide annotated and translated diagnoses for three supra-specific trionychid taxa (two tribes and one genus) that have been erected for Cretaceous and Cenozoic material from Asia (Kordikova 1994b; Chkhikvadze 1999), but which have not been widely accepted.

The taxonomy of Trionychidae used here generally follows Meylan (1987), who presented the first comprehensive phylogenetic analysis for living trionychids based on skeletal morphology and provided diagnoses for the family and all less inclusive taxa. Because Meylan's (1987) work relied on skeletal characters, it is especially useful for dealing with fossil specimens and taxa. More recent attempts to develop a phylogenetic taxonomy of trionychids (Joyce et al. 2004; Engstrom et al. 2004) are less useful for Cretaceous taxa, because those studies also used non-skeletal features and because few specimens of Cretaceous trionychids have been included in cladistic analyses.

Institutional abbreviations used in this chapter are: CCMGE, Chernyshev's Central Museum of Geological Exploration, St. Petersburg, Russia; FPDM, Fukui Prefectural Dinosaur Museum, Fukui, Japan; HMNS, Hayashibara Museum of Natural Sciences, Okayama, Japan; IPGAS, Institute of Paleobiology, Georgian Academy of Sciences, Tbilisi, Georgia; IVPP, Institute of Vertebrate Paleontology and Paleoanthropology, Beijing, China; IZK, Institute of Zoology, Academy of Sciences of Kazakhstan, Almaty, Kazakhstan; MDM, Mifune Dinosaur Museum, Mifune, Japan; MPS, Mongolian Paleontological Center, Ulan Bator, Mongolia; PIN, Paleontological Institute, Moscow, Russia; ZIN PH (= ZIN PHT), Paleoherpetological collection, Zoological Institute of the Russian Academy of Sciences, St. Petersburg, Russia; ZPAL, Institute of Paleobiology, Polish Academy of Sciences, Warsaw, Poland.

The term "Collection" immediately preceding an institutional catalogue number (e.g., Collection ZIN PH 108) indicates that multiple specimens are accessioned under that number.

Systematic Paleontology

Family Trionychidae Gray 1825
Subfamily Trionychinae Gray 1825

Remarks: Assignment of shell material listed below to Trionychinae is based on the possession of a nuchal bone that is at least three times wider than long, a neural series that always contains at least one reversal in neural orientation, or both of those features (Meylan 1987). Useful skull characters of Trionychinae include (Meylan 1987): dorsal edge of apertura narium externum is slightly to strongly emarginate and foramen jugulare posterius is isolated from the fenestra postotica by ventral arch of the opisthotic. Monophyly of Trionychinae is widely supported by phylogenetic studies (Meylan 1987; Engstrom et al. 2004; Joyce and Lyson 2010). Most of the specimens included in our review can be assigned to the Trionychinae; some of these cannot be identified more precisely, but others can be assigned to lower taxonomic levels.

Tribe Trionychini Gray 1825

Remarks: Assignment of shell material listed below to Trionychini is based on the presence of seven or fewer neurals (Meylan 1987). Useful skull characters of Trionychini include (Meylan 1987): parietal makes up nearly one-quarter of processus trochlearis oticum and dorsal margin of aperture narium externum is, as a rule, strongly emarginate. Although Meylan's (1987) cladistic analysis recovered a monophyletic Trionychini, subsequent phylogenetic analyses have not (Engstrom et al. 2004; Joyce and Lyson 2010). Of the two subtribes (Trionychina and Apalonina) recognized by Meylan (1987), only the later can be identified among specimens included in our review. The remaining Trionychini specimens are indeterminate at the subtribe level, although some of them can be identified to genus or species.

Subtribe Apalonina Meylan 1987

Remarks: Assignment of shell material listed below to Apalonina is based on reduction or absence of costals 8 (Meylan 1987). Useful skull characters of Apalonina include (Meylan 1987): the intermaxillary foramen is about 60% of primary palate in length and the vomer divides the maxillae and reaches the intermaxillary foramen. The Apalonina clade is widely supported by contemporary phylogenetic studies (Meylan 1987; Engstrom et al. 2004; Joyce and Lyson 2010). Below we recognize three indeterminate Apalonina taxa from the latest Cretaceous of Mongolia.

Apalonina indet. 1
(Fig. 23.2a)

Synonymies: Apalonini: Khosatzky 1999, pp. 146–147, Fig. 5. Apalonina indet.: Danilov and Vitek 2009, p. 55.

Referred specimen: ZIN PHT M68-1, incomplete carapace (Fig. 23.2a), from Nogon Tsav, Gov-Altai Aimag, Mongolia, upper part of Barungoyot Formation, Campanian.

Remarks: This specimen originally was reported as Apalonini (Khosatzky 1999) and it is characterized by having a relatively short neural 1, an isometric neural 5, and seven pairs of costals.

Apalonina indet. 2

Synonymies: *Platypeltis* sp.: Merkulova 1978, p. 156. Apalonina indet.: Danilov and Vitek 2009, p. 55.

Referred specimen: PIN unnumbered, nearly complete shell, from Nemegt, Gov-Altai Aimag, Mongolia, Nemegt Formation, Maastrichtian.

Remarks: This specimen, reported as *Platypeltis* sp. based on seven pairs of costals and absence of a preneural (Merkulova 1978), probably, also belongs to Apalonina indet. According to Merkulova (1978, p. 156) it is distinguished from other trionychids by "long processes of hyo- and hypoplastra, medial processes of hypoplastra of equal length and thickness, fontanelles of the regular roundish-triangular shape, and different pattern of sculpturing".

Apalonina indet. 3

Synonymies: Apalonina: Suzuki 2005, p. 27.

Referred specimen: HMNS unnumbered, almost complete skeleton, from unreported locality and formation in Mongolia, Late Cretaceous.

Remarks: According to Suzuki (2005), the following features suggest that this specimen belongs to the Apalonina: dorsal edge of apertura narium externum laterally emarginate strongly; costals 8 reduced; maxilla subdivided by vomer.

Trionychini incertae sedis
"*Amyda*" *menneri* Chkhikvadze in Chkhikvadze and Shuvalov 1988

Synonymies: *Amyda menneri*: Chkhikvadze and Shuvalov 1988, p. 200, Fig. 1; Khosatzky 1999, p. 147; Sukhanov 2000, p. 345. "*Amyda*" *menneri*: Danilov and Vitek 2009, p. 54; Vitek and Danilov 2010, Fig. 1, Table 1.

Holotype: IPGAS 11-5-1, incomplete postcranium of one individual, including nuchal, proximal part of costal 1, fragment of right hyo- and hypo-plastron, fragment of xiphiplastron, and assorted non-shell bones, from Gurilin Tsav, Gov-Altai Aimag, Mongolia, Nemegt Formation, Maastrichtian.

Referred specimens: Paratypes (all the specimens listed below): IPGAS 11-5-2, nuchal fragment, IPGAS 11-5-3, anterior part of carapace; IPGAS 11-5-4, distal part of right

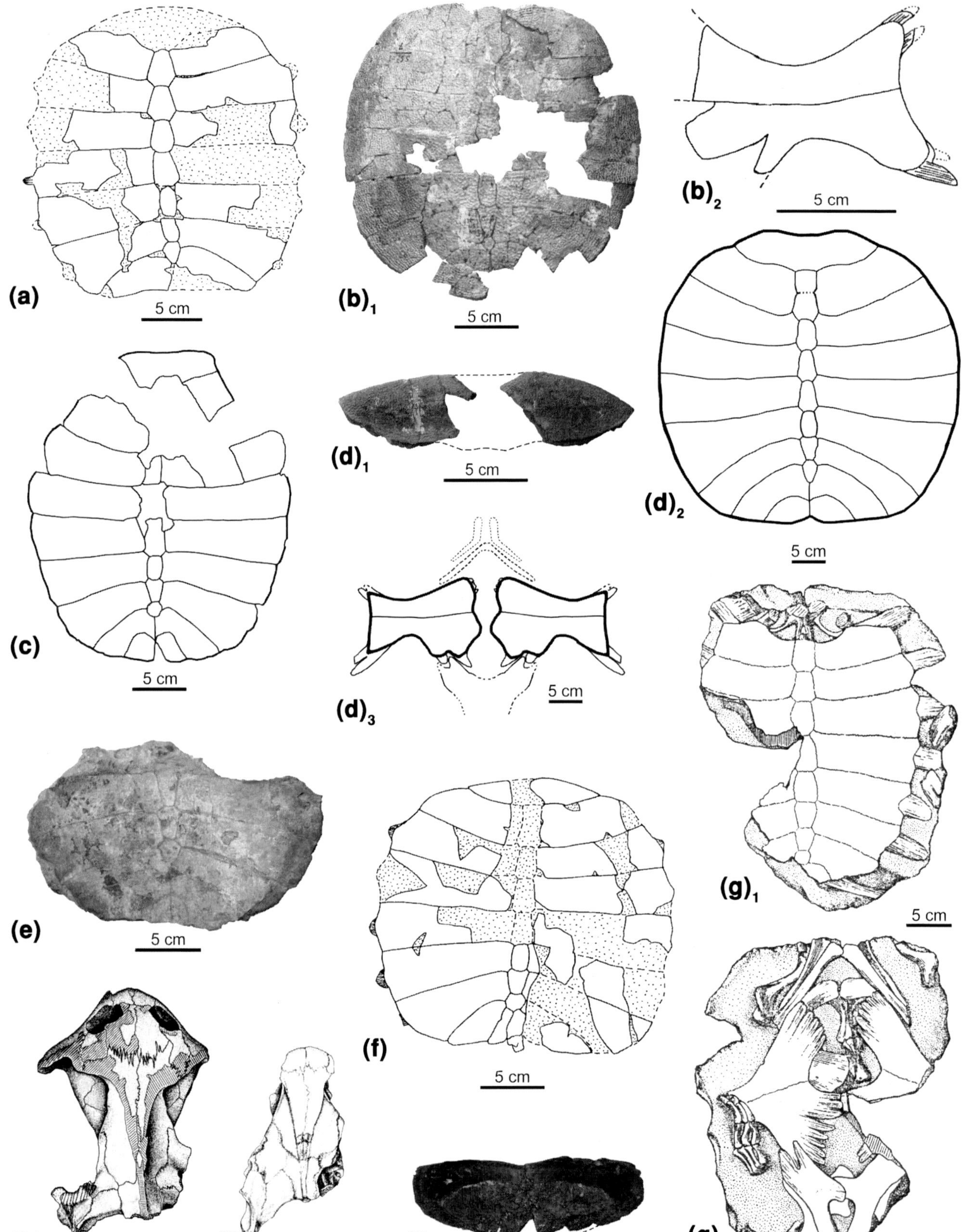
(a)
5 cm
(b)1
5 cm
(b)2
5 cm
(c)
5 cm
(d)1
5 cm
(d)2
5 cm
(d)3
5 cm
(e)
5 cm
(f)
5 cm
(g)1
5 cm
(g)2
(h)
5 cm
(i)
5 cm
(j)
5 cm

◀**Fig. 23.2** Specimens of Cretaceous Trionychidae from Asia. **a** Apalonina indet. 1, ZIN PHT M68-1, incomplete carapace, drawing (after Khosatzky 1999). **b** "*Amyda*" *orlovi*: $\mathbf{b}_1$ PIN 557-1/1 (holotype), incomplete carapace in dorsal view, photograph; $\mathbf{b}_2$ ZIN PHT M46-1, hyo- and hypoplastra in ventral view, drawing (after Khosatzky 1999). **c** "*Aspideretes*" *alashanensis*, IVPP V2865 (holotype), incomplete carapace in dorsal view, drawing (after Yeh 1965). **d** *Aspideretoides riabinini*: $\mathbf{d}_1$ IZK R-3919 (holotype), incomplete nuchal in dorsal view, photograph; $\mathbf{d}_2$ reconstruction of carapace in dorsal view; $\mathbf{d}_3$ reconstruction of plastron in ventral view (both reconstructions after Vitek and Danilov 2010); **e** Trionychini indet. 1, CMGE 7/11659, imprint of posterior part of dorsal surface of carapace, photograph. **f** Trionychini indet. 2, ZIN PHT M67-6, incomplete carapace, drawing (after Khosatzky 1999). **g** "*Aspideretes*" *maortuensis*, IVPP V2864 (holotype), incomplete postcranium, drawings (after Yeh 1965): $\mathbf{g}_1$ dorsal view; $\mathbf{g}_2$ ventral view. **h** *Khunnuchelys erinhotensis*, IVPP V9535 (holotype), incomplete skull in dorsal view, drawing (after Brinkman et al. 1993). **i** *Khunnuchelys kizylkumensis*, CCMGE 8/12458 (holotype), skull roof and braincase in dorsal view, drawing (after Brinkman et al. 1993). **j** "*Paleotrionyx*" *riabinini*, IZK R-3920 (holotype), nearly complete nuchal in dorsal view, photograph. Images at different scales

hyoplastron, IPGAS 11-5-5, right costal 7, and IPGAS 11-5-6, right hyoplastron, all from holotype locality; IPGAS 11-13-11, medial part of right hyoplastron, from Bugin Tsav, Gov-Altai Aimag, Mongolia, Nemegt Formation, Maastrichtian; IPGAS 11-14-2, left posterior part of carapace, IPGAS 11-14-3, medial part of left hypoplastron, and IPGAS 11-14-4, posterior part of carapace, all from Inegeni Khobur, Gov-Altai Aimag, Mongolia, Nemegt Formation, Maastrichtian; IPGAS 11-17-1, left half of carapace, from Bambu Khuduk, Gov-Altai Aimag, Mongolia, Nemegt Formation, Maastrichtian.

Diagnosis (translated from Chkhikvadze and Shuvalov 1988): Length of carapace disc is 20–25 cm. Preneural is absent. There are seven neurals and eight pairs of costals. Postnuchal fontanelles are absent in adults; two round fontanelles in this area are present only in juveniles. Costals 6 are wedge-shaped with expanded distal part. Costals 7 and 8 are small, costals 8 are very small, disappearing. Upper surface of the carapace is completely covered with ornamentation. Posterolateral borders of the nuchal underlie anterior borders of costals 1, but do not reach ends of free ribs of the costals. Two types of neurals: anterior neurals are hexagonal and short-sided posteriorly, followed by tetragonal neural (usually 5), whereas posterior neurals are also hexagonal, but short-sided anteriorly. Ends of free ribs in adults are relatively short and weakly protrude behind the disc of the carapace. Ventral surface of the nuchal bears peculiar rugosities for attachment of the dorsal processes of the scapulae. These rugosities are located near the posterior border of the nuchal; anteromedially to them, there is always a blood vessel opening. There are also lateral ridges of the nuchal (homologies of the nuchal ribs) there, which are more or less straight and reach the pointed lateral ends of the nuchal. The proximal end of the second thoracic rib is located closer to the posterior border of costal 1. The first thoracic rib (place of its attachment to costal 1) is located at a weak angle to the anterior border of costal 1. Hyo- and hypoplastra have paired lateral "horns". The ornamentation of the hyo- and hypoplastra is weakly developed; some, more often medial, parts of these plates are devoid of ornamentation. Medial processes of the hyoplastron form a narrow bundle directed anteromedially. The hypoplastron has anterior process (probably strong) and four smaller medial processes; the anterior of the latter, probably, is perpendicular to the midline, whereas two posterior ones contact corresponding processes of the xiphiplastron.

Remarks: Figures in the published type description are of such low quality that we have found it difficult to meaningfully compare the above-listed specimens with those of other trionychids. The disappearing (i.e., reduced) costals 8, mentioned in the diagnosis, suggest possible affinities with Apalonina, which is known by at least two indeterminate carapaces from the Barungoyot and Nemegt formations of Mongolia (see previous two accounts), but pending further study we conservatively assign "*Amyda*" *menneri* only to the more inclusive Trionychini. Chkhikvadze and Shuvalov (1988) suggested that material they called "*Trionyx* sp. a" (= Trionychinae indet. 2 here; see account below) from a different locality (Tsagan Khushu) in the Nemegt Formation might belong to "*A.*" *menneri*. Some of the materials mentioned by Khosatzky (1999, p. 147) from the Nemegt Formation may also belong to this species.

Trionychini incertae sedis
"*Amyda*" *orlovi* Khosatzky 1976
(Fig. 23.2b)

Synonymies: *Amyda orlovi*: Sochava 1975, p. 115 (nomen nudum); Khosatzky 1976, p. 6; Chkhikvadze and Shuvalov 1988, pp. 200, 202; Sukhanov 2000, p. 345, Fig. 17.27. *Amyda* sp. cf. *A. orlovi*: Khosatzky 1999, Fig. 1. "*Amyda*" *orlovi*: Danilov and Vitek 2009, p. 54; Vitek and Danilov 2010, Fig. 1, Table 1.

Holotype: PIN 557-1/1, incomplete carapace (Fig. 23.2b_1), from Bain Shire, Dornogov Aimag, Mongolia, lower part of Bainshire Formation, Cenomanian-early Turonian.

Referred specimens: ZIN PHT M46-1, hyo- and hypoplastra (Fig. 23.2b_2), from holotype locality.

Diagnosis (translated from Khosatzky 1976): Shell is rounded. Nuchal is wide. Neural 1 is large, twice longer than wide. Neurals 2–4 are hexagonal, widened posteriorly. Neural 5 is tetragonal, elongated. Neural 6 is widened anteriorly, hexagonal. The last neural (7) is small, pentagonal. It wedges between costals 6 and 7 on both sides. Costals 1 are rather wide proximally, sharply narrowed distally. Costals 2–4 are not wedge-shaped. Costals 5 are characterized by moderate widening at distal ends, whereas costals 6 have significant widening at distal ends. Costals 7 embrace the small costals 8.

Remarks: Khosatzky's (1976) published description of this species lacked illustrations; those were not published until much later (Sukhanov 2000, Fig. 17.27). Khosatzky (1999, p. 142, Fig. 1) reported a hyo- and hypoplastra (Fig. 23.2b_2) of *Amyda* sp. cf. *A. orlovi* from the same locality and noted that this specimen had thickened anterior and posterior lateral lobes. In addition, undescribed remains attributed to *Amyda orlovi* (according to identifications by L.I. Khosatzky) were mentioned by Sochava (1975) from the Khara Khutul locality, Dornogov Aimag, Mongolia, Bainshire Formation, Cenomanian-Santonian.

Trionychini incertae sedis
"*Aspideretes*" *alashanensis* Yeh 1965
(Fig. 23.2c)

Synonymies: ?*Aspideretes alashanensis*: Yeh 1965, pp. 51 and 63, Fig. 2, pl. III. *Aspideretes*? *alashanensis*: Yeh 1994, p. 90. '*Aspideretes*'? *alashanensis*: Brinkman et al. 2008, p. 68. "*Aspideretes*" *alashanensis*: Danilov and Vitek 2009, p. 54; Vitek and Danilov 2010, Fig. 1.

Holotype: IVPP V2865, carapace with broken anterior portion but retaining a damaged nuchal (Fig. 23.2c), from Dashukou, Maortu, Alxa, Inner Mongolia, China; age is Late Cretaceous (stage uncertain) according to Brinkman et al. (2008; see below, Remarks).

Diagnosis (after Brinkman et al. 2008): Carapace ovoid; anterior border shallowly concave, posterior margin straight; seven neurals, last one much reduced and rounded; seventh and eighth pairs of costals contacted each other partly and entirely at midline; sculpture of carapace surface weak.

Remarks: The age of "*Aspideretes*" *alashanensis* is uncertain. In the original description (Yeh (1965) the age was given as Early Cretaceous (Aptian–Albian) and this was followed in our previous paper (Danilov and Vitek 2010, Fig. 1). However, other authors have listed the age as Late Cretaceous (Yeh 1994; Brinkman et al. 2008).

Trionychini incertae sedis
Genus *Aspideretoides* Gardner et al. 1995

Synonymies: *Aspideretoides*: Gardner et al. 1995, p. 632 (see also for older synonyms); Vitek and Danilov 2010, p. 383. *Eugenichelys*: Chkhikvadze 2000a, p. 207. *Hummelichelys*: Chkhikvadze 2000a, p. 204. *Aspideretoides* (= *Eugenichelys*; = *Hummelichelys*): Hutchison and Holroyd 2003, p. 135.

Type species: *Aspideretoides foveatus* (Leidy 1856) (= *Aspideretes subquadratus* Lambe 1914; = *A. maturus* Lambe 1914; = *A. rugosus* Parks 1933), middle Campanian and ?Maastrichtian of Canada and USA.

Referred species: *Aspideretoides allani* (Gilmore 1923), middle Campanian of Canada; *A. beecheri* Hay 1905, Maastrichtian of USA; *A. riabinini* (Kuznetsov and Chkhikvadze 1987), early Santonian of Tadzhikistan and Santonian-early Campanian of Kazakhstan; *A. robertemryi* (Chkhikvadze 2008), early Eocene of USA; and *A. splendidus* (Hay 1908) (= *Aspideretes granifer* Hay 1909; = *A. planus* Parks 1933; = *Trionyx planoides* Kuhn 1964), Campanian of Canada and USA.

Diagnosis (after Gardner et al. 1995): Preneural present; nuchal width at least four times greater than length; and plastral bridge length greater than one-half hypoplastral maximum (width) (latter measurement erroneously given as "length" in original diagnosis). Further differs from *Aspideretes* (sensu Meylan 1987, p. 92), the only other valid trionychine genus with a preneural, in primitively retaining short epiplastral projections and four plastral callosities, one pair each on the hyo-, hypo-, and xiphiplastra.

Remarks: The genus *Aspideretoides* was established by Gardner et al. (1995) for Late Cretaceous North American trionychines with a preneural, and originally included three species (*A. foveatus*, *A. allani*, and *A. splendidus*). The genera *Eugenichelys* Chkhikvadze 2000a (type species *Aspideretes allani*) and *Hummelichelys* Chkhikvadze 2000a (type species *Aspideretes beecheri*), both established by Chkhikvadze (2000a), were placed into synonymy of *Aspideretoides* by Hutchison and Holroyd (2003). For this reason, a second species of *Eugenichelys* (*A. robertemryi*; early Eocene of USA) described later by Chkhikvadze (2008) is here tentatively referred to *Aspideretoides* as well. Recently we (Vitek and Danilov 2010) recognized *A. riabinini* (see next account) as the first Asian occurrence for the genus. Phylogenetic analyses have placed *Aspideretoides* within Trionychini as the sister taxon of Apalonina (Gardner et al. 1995; Karl 1998, 1999a).

Aspideretoides riabinini (Kuznetsov and Chkhikvadze 1987)
(Fig. 23.2d)

Synonymies: *Trionyx riabinini*: Kuznetsov and Chkhikvadze 1987 (in part), p. 35, Figs. 3, 4, 6, 7. *Plastomenus riabinini*: Chkhikvadze and Shuvalov 1988, p. 199; Chkhikvadze 1990, pp. 22 and 75; Kordikova 1992a, p. 135. *Paraplastomenus riabinini*: Kordikova 1991, p. 4 (unavailable name); 1992b, p. 648 (unavailable name); 1994a, pp. 343–345; 1994b, p. 8; Nessov 1997, p. 109. *Crassithecachelys riabinini*: Chkhikvadze 2000b, p. 56; 2007, p. 127. *Aspideretoides riabinini*: Danilov and Vitek 2009, p. 54; Vitek and Danilov 2010, p. 383, Table 1, Figs. 1, 3a, b, d, 4, 5.

Holotype: IZK R-3919, incomplete nuchal in two pieces (Fig. 23.2d$_1$), from Shakh–Shakh, northeastern Aral Sea area, Kazakhstan; Bostobe Formation, Santonian-early Campanian.

Referred specimens: Numerous isolated shell plates and fragments from the holotype locality and Kansai, Fergana Depression, Tajikistan, Yalovach Formation, early Santonian; for complete specimen list see Vitek and Danilov (2010, p. 384).

Revised diagnosis (modified from Vitek and Danilov 2010): Largest carapace size approximately 50 cm; can be differentiated from *Aspideretoides foveatus* by bigger size and sculpture pattern; from *A. allani* by bigger size and presence of sculptured plastral callosities at both large and small sizes; from *A. beecheri* by bigger size, narrower posterior notch of the carapace, presence of neural 7, lateral hyoplastral lobe shorter than medial hyoplastron lobe; from *A. robertemryi* by smaller size, smaller axillary and inguinal notches and more complete sculpturing of the plastron; and from *A. splendidus* by having inguinal hypoplastral border not thickened and by shape of xiphiplastra.

Remarks: *Trionyx riabinini* was described based on an isolated, incomplete nuchal (holotype) and fragmentary shell pieces from Shakh–Shakh (Kuznetsov and Chkhikvadze 1987). Later, this species was transferred into the genus *Paraplastomenus* Kordikova 1994a (= *Crassithecachelys* Chkhikvadze 2000b, 2007; Kordikova 1994b). Most recently, we (Vitek and Danilov 2010) described new and abundant shell material of this species from both Shakh–Shakh and Kansai localities, which allowed for a more robust diagnosis and assignment to the genus *Aspideretoides*. Reconstructions of the carapace and plastron are presented in Fig. 23.2d$_2$ and d$_3$, respectively.

Aspideretoides sp.

Synonymies: Undescribed trionychid with slender jaws: Brinkman et al. 1993, p. 2218. *Palaeotrionyx* sp.: Nessov 1997, p. 145, pl. 40, Figs. 3, 4. Trionychini indet.: Danilov 2007, p. 66A. *Aspideretoides* sp.: Danilov and Vitek 2009, p. 54.

Referred specimens: Collection ZIN PH 108, skull and shell material, from Dzharakuduk, Central Kizylkum Desert, Uzbekistan, Bissekty Formation, late Turonian.

Remarks: Skull material of this taxon previously has been mentioned as an "undescribed trionychid with slender jaws" (Brinkman et al. 1993), *Paleotrionyx* sp. (Nessov 1997), and Trionychini indet. (Danilov 2007). Our preliminary study of fragmentary shell material from the same locality (Dzharakuduk) indicates that at least part of the shell collection is diagnostic for *Aspideretoides*. For that reason, we suggest that the above mentioned skull material also probably belongs to *Aspideretoides* (Danilov and Vitek, in prep.).

Trionychini incertae sedis
Trionychini indet. 1
(Fig. 23.2e)

Synonymies: Trionychidae indet.: Nessov 1984, Figs. 6, 7, 9; 1997, p. 137, pl. 34, Fig. 17; pl. 35, Fig. 7; Kordikova 1994a, p. 344. *Trionyx* sp.: Nessov 1985, p. 216. *Trionyx* s. lato: Kordikova 1992a, p. 133. Trionychinae indet.: Nessov 1997, p. 137. Trionychini indet.: Danilov and Vitek 2009, p. 55.

Referred specimens: CCMGE 6/11659, imprint of posterior part of carapace; CCMGE 7/11659, external and visceral imprints of posterior part of carapace of one individual (Fig. 23.2e); CCMGE 8/11659, imprint of plastron fragment in area of inguinal notch; all from Itemir, Central Kizylkum, Uzbekistan, Kulbike Member, Cenomanian.

Remarks: Imprints of the posterior part of the carapace show a reduced pair of costals 8 and the final neural between costals 5 and 6. Detailed description of this material will be presented elsewhere (Danilov and Vitek, in prep.).

Trionychini indet. 2
(Fig. 23.2f)

Synonymies: Trionychidae indet.: Khosatzky 1999, pp. 145–146, Fig. 4. Trionychinae indet. 5: Danilov and Vitek 2009, p. 54.

Referred specimen: ZIN PHT M67-6, incomplete carapace (Fig. 23.2f), from Bugin Tsav, Gov-Altai Aimag, Mongolia, Barungoyot Formation, Campanian.

Remarks: This specimen was reported as a moderately advanced representative of Trionychidae having strongly reduced costals 8, an isometric neural (probably 5), and the last neural contacting costals 6–8 (Khosatzky 1999). Danilov and Vitek (2009) mentioned this specimen as Trionychinae indet. 5. However, further examination of this specimen and the published figure shows that it has only seven neurals. For that reason here we assign this specimen to Trionychini.

Trionychinae incertae sedis
"*Aspideretes*" *maortuensis* Yeh 1965
(Fig. 23.2g)

Synonymies: *Aspideretes maortuensis*: Yeh 1965, p. 48, Fig. 1, pls. I, II; 1994, p. 90, Fig. 67. *Axestemys* (*Axestemys*) *maortuensis*: Kordikova 1994b, p. 7. *Kuhnemys maortuensis*: Chkhikvadze 1999, p. 223. *Dogania maortuensis*: Karl 1999b, Fig. 2:2. '*Aspideretes*' *maortuensis*: Brinkman et al. 2008, p. 68, Fig. 70. "*Aspideretes*" *maortuensis*: Danilov and Vitek 2009, p. 53; Vitek and Danilov 2010, Fig. 1, Table 1.

Holotype: IVPP V2864, incomplete postcranium with parts of carapace, plastron, and girdles, two cervical vertebrae, and right pes (Fig. 23.2g), from Dashukou, Maortu,

Alxa, Inner Mongolia, China; age is Late Cretaceous (stage uncertain) according to Brinkman et al. (2008; see below, Remarks).

Diagnosis (after Brinkman et al. 2008): Shell depressed, longer than broad. Eight neurals, last one much reduced; last pair of costals meeting at midline; distal (end of costals 7 longer than proximal end); a pair of fontanelles present between nuchal, first neural and first pair of costals; distal end of nuchal overlapping rib-end of first costal; surface of carapace ornamented with net-like sculpture but margin smooth; proximal finger-like processes of hyo- and hypoplastra numerous.

Remarks: Originally described by Yeh (1965) as belonging to *Aspideretes* sensu lato, the generic affinities of this species has been contentious. Kordikova (1994b) placed it in the genus *Axestemys* Hay 1899 within the tribe Ulutrionychini Kordikova 1994b, whereas Chkhikvadze (1999) designated it as the type for the new genus *Kuhnemys* Chkhikvadze 1999 within the tribe Rafetini Chkhikvadze 1999. (See Appendix for an annotated version of Kordikova's (1994b) diagnosis of Ulutrionychini and English translations of Chkhikvadze's (1999) diagnoses of *Kuhnemys* and Rafetini.) Karl (1999b) placed the species in the extant genus *Dogania* Gray 1844 within the tribe Pelodiscini (sensu Meylan 1987). We think that all three opinions, although possible, are based on little evidence and, thus, we do not accept any of those arrangements. That is why we place the original generic name in quotes, as "*Aspideretes*" *maortuensis*. Detailed study of the specimen and its inclusion in a phylogenetic analysis are needed to establish its generic affinities. Like "*A.*" *alashanensis* (see account, above), the age of "*A.*" *maortuensis* is uncertain; it has been given as either Aptian–Albian (Yeh 1965; Danilov and Vitek 2010, Fig. 1) or simply Late Cretaceous (Yeh 1994; Brinkman et al. 2008).

Trionychinae incertae sedis
Genus *Khunnuchelys* Brinkman et al. 1993

Synonymies: *Khunnuchelys*: Brinkman et al. 1993, p. 2215; 2008, p. 74; Chkhikvadze 1999, p. 217; 2000a, p. 202; 2000b, p. 56.

Type species: *Khunnuchelys erinhotensis* Brinkman et al. 1993, Late Cretaceous of Inner Mongolia, China.

Referred species: *Khunnuchelys kizylkumensis* Brinkman et al. 1993, late Turonian of Uzbekistan. Also at least one undescribed species, from the Santonian-early Campanian of Kazakhstan.

Diagnosis (after Brinkman et al. 1993): A trionychid in which the external narial opening is located well ventral to the orbits; the orbits face strongly forwards; the suborbital region of the maxilla is deep; the palate is vaulted; the triturating surfaces of the maxillae meet at the midline, forming a midventral ridge; the internal nares are located far posteriorly and are roofed by the palatine and maxilla; the anterolateral corner of the palatine and posterolateral corner of the maxilla are greatly enlarged to form about half of the triturating surface; the maxilla and jugal are swollen to form a broadly convex surface in dorsal view; the postorbital contacts the maxilla, excluding the jugal from the margin of the orbit; the descending flange of the parietal ends well medial to the processus trochlearis oticum; a posteriorly facing occipital surface is bordered dorsally by a rounded ridge formed by the opisthotic and supraoccipital; a concave surface is present on the occiput above the fenestra postotica and lateral to the foramen magnum; the occipital surface of the exoccipital faces posteriorly; nervi hypoglossi exiting via a single foramen in the exoccipital. Similar to *Conchochelys* Hay 1905 (now *Axestemys* according to Hutchison and Holroyd 2003) in the presence of a deeply excavated palate, extensively developed secondary palate with broad contact of the maxilla at the midline, grooves extending backwards from the internal choanal openings, and short basioccipital region. Similar to *Dogania* in that the prefrontal makes a small contribution to the orbit margin between the frontal and maxilla.

Remarks: Although *Khunnuchelys* has a distinctive skull, its relationships within Trionychinae are unclear (Brinkman et al. 1993). Chkhikvadze (1999) placed *Khunnuchelys* in his tribe Rafetini, but for reasons discussed below (see Discussion) that assignment is unproven.

Khunnuchelys erinhotensis Brinkman et al. 1993
(Fig. 23.2h)

Synonymies: *Khunnuchelys erinhotensis*: Brinkman et al. 1993, p. 2215, Figs. 1–3; 2008, p. 74, Fig. 79; Danilov and Vitek 2009, p. 53; Vitek and Danilov 2010, p. 391.

Holotype: IVPP V9535, incomplete skull (Fig. 23.2h), from Erinhot, Inner Mongolia, China, Iren Dabasu Formation; age is Late Cretaceous (stage uncertain) according to Brinkman et al. (1993).

Diagnosis (after Brinkman et al. 1993): A member of the genus *Khunnuchelys*, differing from *K. kizylkumensis* in the presence of a separate foramen jugulare posterius formed by a ventral process of the opisthotic that meets the pterygoid; the absence of a deeply excavated tympanic cavity; that the incisura columellae auris is a large oval opening; the groove formed by the palatines leading posteriorly from the internal narial openings is narrow; the lateral edge of the supraoccipital has a long, spinelike process that extends into a groove in the opisthotic; and the jugal is excluded from the orbit by a contact between the maxilla and postorbital.

Remarks: This species is known only by the distinctive holotype skull.

Khunnuchelys kizylkumensis Brinkman et al. 1993
(Fig. 23.2i)

Synonymies: *Trionyx* sp.: Nessov 1986, pl. I, Fig. 9; Nessov 1987 Figs. 8, 9. Cf. *Eurycephalochelys*: Chkhikvade and Shuvalov 1988, p. 199. *Khunnuchelys kizylkumensis*: Brinkman et al. 1993, p. 2216, Figs. 4–8; Nessov 1997, pp. 144–145, pl. 13, Fig. 18; Chkhikvadze 1999, p. 217; Chkhikvadze 2000b, p. 56; Danilov and Parham 2005, p. 789; Danilov and Vitek 2009, p. 53; Vitek and Danilov 2010, p. 391. *Axestemys* (*Axestemys*) sp.: Kordikova 1994b, p. 7. Cf. *Axestemys riabinini*: Kordikova 1994a, p. 344. "*Trionyx*" sp.: Nessov 1997, pl. 41, Fig. 6. ?*Khunnuchelys* sp.: Nessov 1997, pl. 41, Fig. 7.

Holotype: CCMGE 8/12458, braincase and skull roof (Fig. 23.2i), from Dzharakuduk, Central Kizylkum Desert, Uzbekistan, Bissekty Formation, late Turonian.

Referred specimens: CCMGE 303/12458, maxilla, and CCMGE 8a/12458, braincase; both from holotype locality (see Remarks).

Diagnosis (after Brinkman et al. 1993): A member of the genus *Khunnuchelys* differing from *K. erinhotensis* in that the foramen jugulare posterius is not separated from the fenestra postotica; the incisura columellae auris is small in diameter; the grooves leading posteriorly from the internal narial openings are wide; and the supraoccipital does not have a lateral, spinelike process extending into a groove in the opisthotic.

Remarks: Shell material has also been collected from the holotype locality (Dzharakuduk), and at least two species of trionychids appear to be present there (Brinkman et al. 1993; see above account for *Aspideretoides* sp.). Chkhikvadze (1999) synonymized *Khunnuchelys kizylkumensis* with *Paleotrionyx riabinini*, but without any argumentation (see account below for "*Paleotrionyx*" *riabinini*). We currently are studying the shell material from Dzharakuduk, including some that probably belongs to *K. kizylkumensis* (Danilov and Vitek, in prep.).

Khunnuchelys sp. 1

Synonymies: Cf. *Lophorhothon*: Nessov 1995b, p. 107 (see for previous references). *Khunnuchelys riabinini*: Glinskiy 2008, p. 23. *Khunnuchelys* sp.: Glinskiy and Danilov 2008, p. 19; Danilov and Vitek 2009, p. 53; Vitek and Danilov 2010, p. 390.

Referred specimens: ZIN PH 5/55, incomplete skull, from Baybishe, northeastern Aral Sea area, Kazakhstan, Bostobe Formation; age is Santonian-early Campanian according to Nessov (1997).

Remarks: This specimen originally was interpreted as a dinosaur and identified as cf. *Lophorhothon* by Nessov (1995b). *Lophorhothon* Langston 1960 is a monotypic hadrosaur (duck-billed dinosaur) genus restricted to the Campanian of Alabama and North Carolina, USA (see Horner et al. 2004, Table 20.1). Later, ZIN PH 5/55 was attributed to *Khunnuchelys* based on the following characters (Glinskiy 2008; Glinskiy and Danilov 2008): deep suborbital region of maxilla; vaulted palate; triturating surfaces of maxilla meet along midline; internal nares positioned posteriorly; palatines and maxilla form much of triturating surface; maxilla and jugal form broadly convex surface in dorsal view; postorbital contacts maxilla and excludes jugal from orbital margin; and descending flange of parietal does not participate in formation of processus trochlearis oticum. The specimen differs from other species of *Khunnuchelys* in its large size (the complete skull is estimated to have been 20 cm long) and in that the groove leading posteriorly from the internal narial openings is more strongly narrowed. A detailed description of this specimen will be presented elsewhere (Danilov, in prep.).

Khunnuchelys sp. 2

Synonymies: Trionychid: Khosatzky 1957, p. 18; Rozhdestvensky and Khosatzky 1967, p. 87. Trionychidae indet.: Kordikova 1994a, p. 345. *Khunnuchelys* sp.: Nessov 1997, p. 110.

Referred specimen: ZIN PH 1/146, incomplete skull, from Baykhozha, northeastern Aral Sea area, Kazakhstan, Bostobe Formation; age is Santonian-early Campanian according to Nessov (1997).

Remarks: This specimen was found by the Russian geologist M.E. Voskoboynikov in 1948 and studied by L.I. Khosatzky; although it has been mentioned in print (Khosatzky 1957; Rozhdestvensky and Khosatzky 1967), it has never been described. This skull can be attributed to *Khunnuchelys* because its external narial opening is located well ventral to the orbits and its palate is vaulted. A description of this skull will be presented elsewhere (Danilov, in prep.).

Trionychinae incertae sedis
"*Paleotrionyx*" *riabinini* (Kuznetsov and Chkhikvadze 1987)
(Fig. 23.2j)

Synonymies: *Palaeotrionyx riabinini*: Kuznetsov and Chkhikvadze 1987, p. 35, Fig. 1; Chkhikvadze and Shuvalov 1988, p. 199. "*Palaeotrionyx*" *riabinini*: Kordikova 1992a, p. 133. *Axestemys* (*Axestemys*) *riabinini*: Kordikova 1994b, p. 7. ?*Axestemys riabinini*: Kordikova 1991, p. 3; 1992b, p. 647; 1994a, pp. 343–345; Nessov 1997, p. 109. *Paleotrionyx riabinini* (= *Khunnuchelys kizylkumensis*): Chkhikvadze 1999, p. 217. *Khunnuchelys riabinini*: Chkhikvadze 2000b, p. 56; Glinskiy 2008, p. 23. *Eurycephalochelys riabinini*: Chkhikvadze 2007, p. 127. *Paleotrionyx riabinini*: Danilov and Vitek 2009, p. 53; Vitek and Danilov 2010, p. 391, Fig. 8.

Holotype: IZK R-3920, nearly complete nuchal (Fig. 23.2j), from Shakh–Shakh, northeastern Aral Sea area, Kazakhstan; Bostobe Formation, Santonian-early Campanian.

Diagnosis (translated from Kuznetsov and Chkhikvadze 1987): The width of the nuchal is about 15 cm. The estimated length of the carapace is 30–35 cm. The anterior and lateral areas of the nuchal are devoid of ornamentation. Postnuchal fontanelles are absent, as indicated by the continuous suture along the posterior border of the plate. However, the posterior border of the nuchal is noticeably thinner in the areas where the postnuchal fontanelles usually are located. On the ventral surface of the nuchal, the medial ridge for contact with the first thoracic rib is absent. Lateral borders of the nuchal are free and have no suture with costals 1, whereas posterolateral borders of the nuchal underlie costals 1 and overlap onto the free ribs of costals 1. The body of the nuchal is arched dorsoventrally. Nuchal emargination is weak.

Remarks: Kuznetsov and Chkhikvadze (1987) described this species (as *Palaeotrionyx riabinini*) based on an isolated nuchal. Later, "*Paleotrionyx*" *riabinini* was referred to various other genera: *Axestemys* (= *Conchochelys*; = *Paleotrionyx* Schmidt 1945); *Eurycephalochelys* Moody and Walker 1970; and *Khunnuchelys* (Kordikova 1994a, b; Nessov 1997; Chkhikvadze 1999, 2007; Glinskiy 2008). The nuchal of "*Paleotrionyx*" *riabinini* is comparable in size and degree of nuchal emargination to *Aspideretoides riabinini* (a species known from the same locality; see earlier account), but it has a substantial unsculptured area similar to those in smaller specimens of the latter species. However, the proportions of the nuchal of "*Paleotrionyx*" *riabinini*, being three times wider than long, are different enough from both *Aspideretoides riabinini* and "*Trionyx*" *kansaiensis* (also known from the same locality; see next account) that the nuchal appears to represent a different species. Sculpturing on the holotype nuchal, which could help further diagnose the species, unfortunately is unclear. The systematic position of "*P.*" *riabinini* is uncertain and its attribution to Trionychinae is tentative (Vitek and Danilov 2010).

Trionychinae incertae sedis
"*Trionyx*" *kansaiensis* Vitek and Danilov 2010
(Fig. 23.3a)

Synonymies: *Trionyx riabinini*: Kuznetsov and Chkhikvadze 1987 (in part), p. 35, Figs. 2, 5. *Khunnuchelys* sp.: Vitek and Danilov 2008, p. 17. "*Trionyx*" sp. nov.: Danilov and Vitek 2009, p. 53. "*Trionyx*" *kansaiensis*: Vitek and Danilov 2010, p. 387, Table 1, Figs. 3c, e, 6, 7.

Holotype: ZIN PH 630/64, incomplete nuchal (Fig. 23.3a_1), from Kansai, Fergana Depression, Tajikistan, Yalovach Formation, early Santonian.

Referred specimens: Numerous isolated shell plates and fragments from the holotype locality and from Shakh–Shakh, northeastern Aral Sea area, Kazakhstan, Bostobe Formation, Santonian-early Campanian; for complete specimen list see Vitek and Danilov (2010, p. 387).

Diagnosis (after Vitek and Danilov 2010): A trionychine, which can be differentiated from all other Cretaceous trionychines for which shells are known, by bigger size (up to 75 cm long), strong nuchal emargination, eight neurals (except "*Aspideretes*" *maortuensis* and "*Trionyx*" *kyrgyzensis*), unreduced costals 8 (except "*Trionyx*" *kyrgyzensis*), lateral lobe of hyoplastron longer than its medial lobe (except "*Trionyx*" *kyrgyzensis*), and, probably, absence of the separate anteromedial process of the hypoplastron (except "*Aspideretes*" *maortuensis*); besides that, can be differentiated from species of *Aspideretoides* by absence of a preneural and by sculpture pattern, and from "*Trionyx*" *kyrgyzensis* by presence of sculpture on plastron.

Remarks: Reconstructions of the carapace and plastron are presented in Fig. 23.3a_2 and a_3, respectively. Vitek and Danilov (2008) suggested that "*Trionyx*" *kansaiensis* may belong to the skull-based genus *Khunnuchelys*, based on two lines of evidence, namely that both taxa co-occur in the Bostobe Formation and are known from similarly large specimens. Although this interpretation may have merit, here we follow our more recent publications (Danilov and Vitek 2009; Vitek and Danilov 2010) in refraining from formally synonymizing the two taxa pending new discoveries and descriptions of relevant Cretaceous trionychids from Asia.

Trionychinae incertae sedis
"*Trionyx*" *kyrgyzensis* Nessov 1995
(Fig. 23.3b)

Synonymies: *Trionyx* sp.: Nessov 1977, Fig. 4; 1985, p. 216; 1986, pl. I, Figs. 10–12; Kordikova 1992a, p. 132; 1994a, p. 344. ?*Trionyx* sp.: Nessov and Khosatzky 1978, p. 267. Trionychinae gen. indet.: Kordikova 1991, p. 1. "*Trionyx*" *kyrgyzensis*: Nessov 1995a, p. 137, Figs. 3, 4; 1997, p. 117, pl. 40, Fig. 7; pl. 41, Figs. 3–5, pl. 42; Danilov and Vitek 2009, p. 53; Vitek and Danilov 2010, pp. 391–392, Fig. 1, Table 1. *Kuhnemys kyrgyzensis*: Chkhikvadze 1999, p. 223; 2000b, p. 56; 2001, p. 236. *Dogania maortuensis*: Karl 1999b, Fig. 2.

Holotype: CCMGE 186/12458, incomplete xiphiplastron (Fig. 23.3b_1), from left bank of Sarykungoi Spring, Kylodzhun (the former Klaudzin) settlement, south-eastern Fergana Depression, Kyrgyzstan, Alamyshik Formation, early-middle Albian.

Referred specimens: Collection CCMGE 12458, isolated plates of carapace and plastron; collection ZIN PHT F67, posterior part of the skull, lower jaw, cervical and trunk vertebrae, coracoid, and pubis; all from the type locality.

Diagnosis (after Nessov 1995a): Small trionychid (up to 15 cm long). Alveolar surface of the lower jaw is narrow, only slightly widened anteriorly. Nuchal plate is not long antero-posteriorly. The last neural plate is situated between the posterior costal plates. Distal ends of trunk ribs are relatively long. Bones of plastron without callosities (pits and ridges, which are typical for the trionychid shell relief, are absent on the plastron). Anterior projections of epiplastra are not very long. Xiphiplastra are very long, their posterior projection is narrow, almost straight.

Remarks: Reconstructions of the carapace and plastron are presented in Fig. 23.3b_2 and b_3, respectively. The generic attribution of "*Trionyx*" *kyrgyzensis* varies among authors. Chkhikvadze (1999) placed this species in *Kuhnemys* within the tribe Rafetini (see Discussion), whereas Karl (1999b) synonymized it with the species here identified as "*Aspideretes*" *maortuensis* (see above account for that species) and placed it in *Dogania* under the name *D. maortuensis*.

Trionychinae incertae sedis

Trionychinae indet. 1

(Fig. 23.3c)

Synonymies: *Trionyx zakhidovi*: Khosatzky 1966 (in part), p. 151, Fig. 1. *Paraplastomenus riabinini*: Kordikova 1991, p. 4. Cf. *Paraplastomenus* sp.: Nessov 1997, p. 107. Trionychinae indet. 1: Danilov and Vitek 2009, p. 54. Trionychinae gen. et sp. indet.: Vitek and Danilov 2010, p. 392.

Referred specimen: CCMGE 522/1341, posterior part of carapace (Fig. 23.3c), from area of Kyrkkuduk Well (= Sary-Agach; = Kyrkkuduk I), southern Kazakhstan, either Syuk–Syuk Formation or lower part of Darbaza Formation, Santonian-?middle Campanian (Kordikova 1994a; Nessov 1997).

Remarks: This incomplete carapace originally was identified as *Trionyx zakhidovi* (Khosatzky 1966; see below, account for Trionychidae indet. 1), but later was identified as *Paraplastomenus riabinini* (see above account for *Aspideretoides riabinini*) and cf. *Paraplastomenus* sp. (Kordikova 1991; Nessov 1997). Vitek and Danilov (2010) noted similarities between this incomplete carapace and carapace material of "*Trionyx*" *kansaiensis* in the outline of the posterior border, large size and triangular shape of the posteriormost pair of costals, and, probably, also in the sculpture patterns. The last neural in this trionychid is situated more anteriorly than in "*T.*" *kansaiensis*; however, the significance of that difference is uncertain, because that feature is variable in trionychids (Meylan 1987).

Trionychinae indet. 2

(Fig. 23.3d)

Synonymies: *Trionyx* sp. a: Khosatzky and Młynarski 1971, p. 141, pl. XXIV, Fig. 2; text-fig. 7; Młynarski and Narmandach 1972, p. 100. "*Trionyx*" sp.: Khosatzky 1999, p. 144. Trionychinae indet. 2: Danilov and Vitek 2009, p. 54.

Referred specimens: ZPAL MgCh/52 (Fig. 23.3d) and ZPAL MgCh/76, two nearly complete carapaces from Tsagan Khushu and Nemegt localities, respectively, Gov-Altai Aimag, Mongolia, Nemegt Formation, Maastrichtian; and ZPAL MgCh/60 and unnumbered, isolated costal fragments, from Tsagan Khushu.

Remarks: Khosatzky and Młynarski (1971, text-fig. 7) provided a reconstruction of ZPAL MgCh/52 in which that carapace had no preneural, eight neurals, and eight pairs of costals. However, judging from a photograph of that specimen (Fig. 23.3d) it appears that the areas between costals 1, 7, and 8 are too damaged to establish the morphology of the preneural and neural 8. If ZPAL MgCh/52 has no preneural and only seven neurals, it could be attributed to "*Amyda*" *menneri*, a species that also occurs in the Nemegt Formation (see above account for that species).

Trionychinae indet. 3

(Fig. 23.3e)

Synonymies: Trionychina indet.: Khosatzky 1999, pp. 143–144, Fig. 2. Trionychinae indet. 3: Danilov and Vitek 2009, p. 54.

Referred specimens: ZIN PHT M71-1, incomplete carapace (Fig. 23.3e), from Khongil, Dornogov Aimag, Mongolia, Bainshire Formation, Cenomanian-Santonian.

Remarks: Originally identified as Trionychina indet. (Khosatzky 1999), this carapace is characterized by a wide nuchal, no postnuchal fontanelles, no preneural, eight neurals, a long neural 1, an isometric neural 5, neural 8 located between the last pair (eighth) of costals, and a nearly straight posterior edge.

Trionychinae indet. 4

(Fig. 23.3f)

Synonymies: Trionychidae indet.: Khosatzky 1999, pp. 144–145, Fig. 3. Trionychinae indet. 4: Danilov and Vitek 2009, p. 54.

Referred specimens: ZIN PHT M67-5, anterior part of carapace (Fig. 23.3f) from Unegetu Ula, Dornogov Aimag, Mongolia, Bainshire Formation, Cenomanian-Santonian.

Remarks: The specimen originally was reported as Trionychidae, possibly representing a new species, and it is characterized by having a rather long neural 1 that is widened anteriorly and posteriorly, costals 1 that considerably turn distally to the anterior, and an isometric neural 4 (Khosatzky 1999).

Trionychinae indet. 5

(Fig. 23.3g)

Synonymies: *Trionyx* sp. b: Młynarski and Narmandach 1972, p. 100. "*Plastomenus*"?: Khosatzky 1999, p. 147. "*Plastomenus*": Khosatzky 1999, p. 147, Fig. 6. Trionychinae indet. 6: Danilov and Vitek 2009, p. 54.

Referred specimens: ZIN PHT M46-2, medial part of hyo- and hypoplastra (Fig. 23.3g), from Gilbentu, Gov-Altai Aimag, Mongolia, Nemegt Formation, Maastrichtian; ZPAL

MgCh unnumbered, shell fragments, from Nemegt locality, Gov-Altai Aimag, Mongolia, Nemegt Formation, Maastrichtian (see Młynarski and Narmandach 1972, p. 100).

Remarks: Khosatzky (1999) described this specimen as a relatively large trionychid with tuberculate sculpture and assigned it to "*Plastomenus*". However, we assign this specimen to Trionychinae based on similarities in the shape of the hyo- and hypo-plastra and the sculpture with *Trionyx kansaiensis*, although we would not assign ZIN PHT M46-2 to that species. Trionychinae indet. 5 and "*Trionyx* sp. b," the latter based on shell fragments of a large trionychid that also came from the Nemegt locality (Młynarski and Narmandach 1972), probably are the same taxon.

Trionychinae indet. 6

Synonymies: Trionychidae: Suzuki and Narmandakh 2004, p. 8, pl. 1, pl.-fig. 2. Trionychinae indet. 7: Danilov and Vitek 2009, p. 54.

Referred specimens: HMNS and MPS unnumbered, incomplete nuchal and costals from Dzun Shakhai, Dornogov Aimag, Mongolia, unknown formation, Early Cretaceous.

Remarks: This material was mentioned and figured as Trionychidae by Suzuki and Narmandakh (2004). Here it is assigned to Trionychinae based on the nuchal, which is at least three times wider than long.

Trionychidae incertae sedis
Genus *Sinamyda* Chkhikvadze 2000a

Synonymies: *Sinamyda*: Chkhikvadze 2000a, p. 208.

Type species: *Aspideretes fuchienensis* Yeh 1974.

Referred species: Type species only.

Diagnosis (translated from Chkhikvadze 2000a): The only representative of the family Trionychidae with an unusually elongated carapace (width makes up 60% of length). Bony callosities completely overgrow distal ends of ribs. The whole upper surface of the carapace is ornamentated. Nuchal is segment-shaped. Preneural is present. Free edge of costal 1 is more expanded than that of costal 2. This character is very rare in trionychids. Seven pairs of costals are well preserved. However, it can be supposed, that these turtles also had a reduced pair of costals 8. Costals 3 and 6 are the most wedge-shaped of the costals. The preserved neurals 1–5 are hexagonal and short-sided posteriorly. All of them are relatively big and more elongated (in relation to carapace length) in comparison with other trionychids. For instance, the length of neural 2 is only 6–7 times less than the carapace length. Neural 5 is, most probably, tetragonal. More caudal neurals are not preserved in this specimen, however, it can be supposed that the total number of neurals in *Sinamyda* was eight or at least seven.

Remarks: According to Chkhikvadze (2000a), *Sinamyda* can be differentiated from other trionychids by its very elongate carapace, elongate neurals, and expanded lateral (free) edge of costal 1. The expanded lateral edge of costal 1 is considered to be a primitive character also known in Carettochelyidae Boulenger 1887, Cyclanorbinae Hummel 1929, and Plastomeninae Hay 1902. Chkhikvadze (2000a) also noted that *S. fuchienensis* resembles *Aspideretes ellipticus* Hay 1908 and *A. guttatus* Leidy 1869 (both from the middle Eocene of North America) in the outline of the carapace, nuchal shape, and the orientation and shape of costals 5–7. The two North American species were placed in the genus *Hummelichelys* by Chkhikvadze (2000a). However, Hutchison and Holroyd (2003) considered *Hummelichelys* to be a junior synonym of *Aspideretoides* (see above account for the latter genus). Although the generic status of *Sinamyda* seems to be well founded, its systematic position within Trionychidae remains unclear.

Sinamyda fuchienensis (Yeh 1974)
(Fig. 23.3h)

Synonymies: *Trionyx* (*Aspideretes*) *fuchienensis*: Yeh 1974, p. 190, pl. I. *Aspideretes fuchienensis*: Yeh 1994, p. 91, Fig. 68. *Sinamyda fuchienensis*: Chkhikvadze 2000a, p. 208; Danilov and Vitek 2009, p. 52. '*Aspideretes*' *fuchiensis*: Brinkman et al. 2008, p. 69, Fig. 71.

Holotype: IVPP V4708, incomplete carapace (Fig. 23.3h), from Hekou, Ninghua County, Fujian Province, China, unknown formation; age is ?Early Cretaceous according to Yeh (1974, 1994) and Brinkman et al. (2008).

Diagnosis: Same as for the genus.

Trionychidae incertae sedis
Trionychidae indet. 1
(Fig. 23.3i)

Synonymies: *Trionyx* (*Aspideretes*) *zakhidovi*: Khosatzky 1966, p. 151, Fig. 2; Kuznetsov and Chkhikvadze 1987, p. 34. *Trionyx zakhidovi* = cf. *Paraplastomenus riabinini* or cf. *Axestemys*: Kordikova 1991, p. 5; 1994a, p. 345. "*Trionyx*" *zakhidovi*: Nessov 1997, p. 107; Chkhikvadze 2000b, p. 56. *Trionyx zakhidovi* ? = *Eurycephalochelys riabinini*: Chkhikvadze 2007, p. 127. *Trionyx zakhidovi* (nomen dubium): Danilov and Vitek 2009, p. 52; Vitek and Danilov 2010, p. 392.

Referred specimens: CCMGE 411/1341, femur (holotype of *Trionyx zakhidovi*; Fig. 23.3i), from area of Kyrkkuduk Well (= Sary-Agach; = Kyrkkuduk I), southern Kazakhstan, either Syuk–Syuk Formation or the lower part of the Darbaza Formation, Santonian-?middle Campanian (Kordikova 1994a; Nessov 1997).

Remarks: *Trionyx zakhidovi* Khosatzky 1966 is based on a giant, approximately 20 cm long femur (the holotype) and a referred, posterior part of a large trionychid carapace with an estimated shell length of about 70 cm (see above account for Trionychinae indet. 1) from the area of the Kyrkkuduk Well (Khosatzky 1966). Later, some authors (Kordikova 1994a;

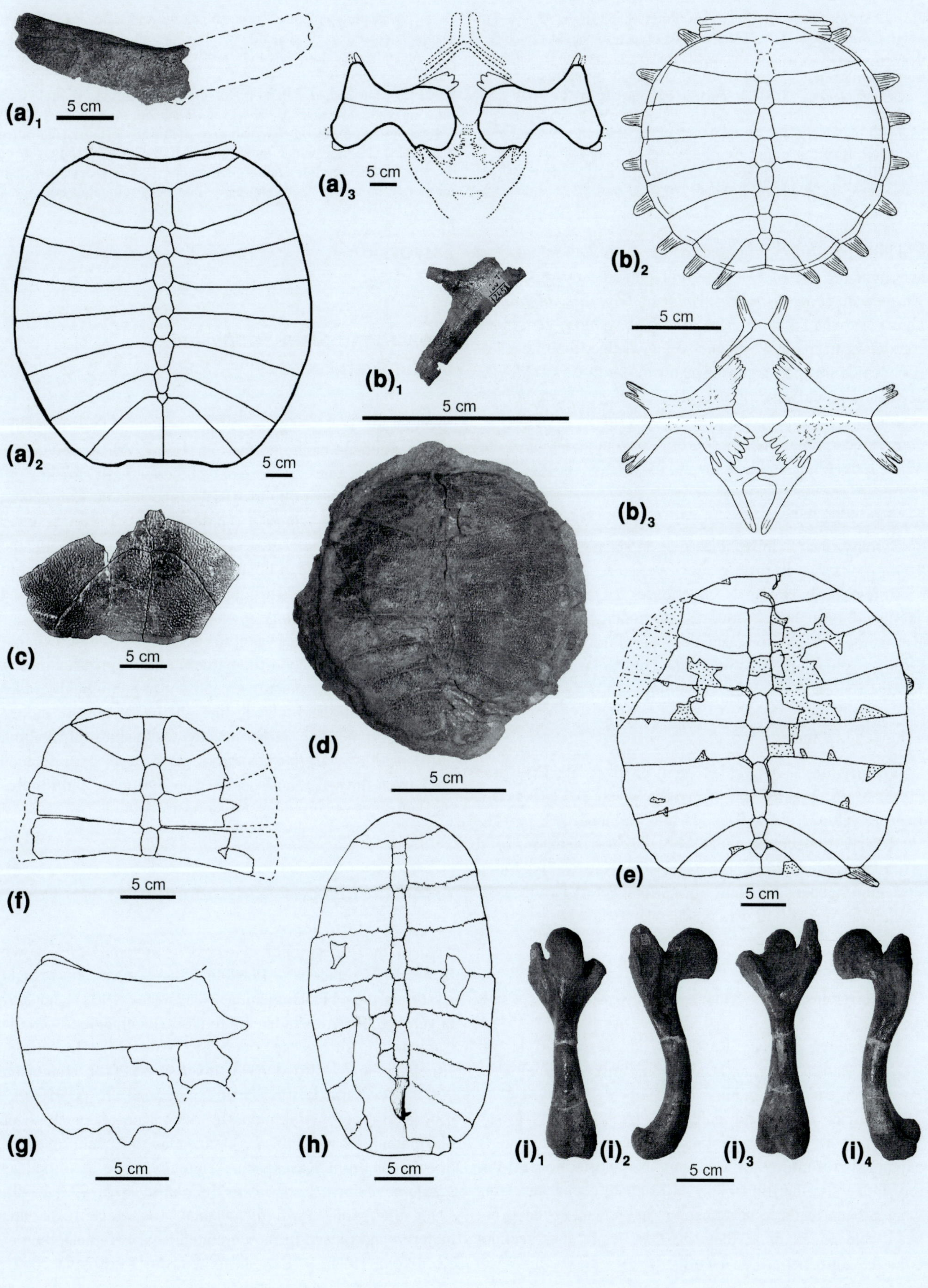
(a)1
5 cm
(a)3
5 cm
(b)2
5 cm
(a)2
5 cm
(b)1
5 cm
(b)3
(c)
5 cm
(d)
5 cm
(e)
5 cm
(f)
5 cm
(g)
5 cm
(h)
5 cm
(i)1
(i)2
(i)3
(i)4
5 cm

◀**Fig. 23.3** Specimens of Cretaceous Trionychidae from Asia. **a** *Trionyx kansaiensis*: $\mathbf{a}_1$ ZIN PH 630/64 (holotype), incomplete nuchal in dorsal view, photograph; $\mathbf{a}_2$ reconstruction of carapace in dorsal view; $\mathbf{a}_3$ reconstruction of plastron in ventral view (both reconstructions after Vitek and Danilov 2010). **b** *Trionyx kyrgyzensis*: $\mathbf{b}_1$ CCMGE 186/12458 (holotype), left xiphiplastron, photograph; $\mathbf{b}_2$ reconstruction of carapace in dorsal view; $\mathbf{b}_3$ reconstruction of plastron in ventral view (both reconstructions after Nessov 1995a). **c** Trionychinae indet. 1, CCMGE 522/1341, posterior part of carapace in dorsal view, photograph. **d** Trionychinae indet. 2, ZPAL MgCh/52, carapace in dorsal view, photograph. **e** Trionychinae indet. 3, ZIN PHT M71-1, incomplete carapace in dorsal view, drawing (after Khosatzky 1999). **f** Trionychinae indet. 4, ZIN PHT M67-5, anterior part of carapace in dorsal view, drawing (after Khosatzky 1999). **g** Trionychinae indet. 5, ZIN PHT M46-2, medial part of hyo- and hypo-plastra in ventral view, drawing (after Khosatzky 1999). **h** *Sinamyda fuchienensis*, IVPP V4708 (holotype), incomplete carapace in dorsal view, drawing (after Yeh 1974). **i** Trionychidae indet. 1, CCMGE 411/1341 (holotype of *Trionyx zakhidovi*), right femur, photographs: $\mathbf{i}_1$ dorsal view; $\mathbf{i}_2$ medial view; $\mathbf{i}_3$ ventral view; $\mathbf{I}_4$ lateral view. Images at different scales

Chkhikvadze 2007) proposed that *T. zakhidovi* might be a synonym for one of two other contemporaneous taxa from Kazakhstan (*Paleotrionyx riabinini* or *Trionyx riabinini*, see above accounts) and/or placed it in Trionychidae gen. indet. According to the current state of knowledge, limb bones of trionychids are non-diagnostic below the family level (Meylan 1987) and should not be used to characterize a species. For this reason, we previously suggested that *Trionyx zakhidovi* be considered a nomen dubium within Trionychidae (Danilov and Vitek 2009; Vitek and Danilov 2010).

Trionychidae indet. 2

Synonymies: Trionychidae gen. et sp. indet.: Hirayama 1998, p. 90, Fig. 4, pl. 2 g–m.

Referred specimens: Collection of MDM, fragmentary shell and limb bones (total 20 specimens) from Mifune, Kumamoto Prefecture, Japan, Upper Formation, Mifune Group, Coniacian-Santonian (see Hirayama 1998; source for age is Ren Hirayama, pers. comm. 2007 to IGD).

Trionychidae indet. 3

Synonymies: Trionychidae gen. et sp. indet.: Hirayama 2002, p. 35, Fig. 5F, G; Hirayama et al. 2012, Fig. 6A. Pantrionychidae indet.: Danilov and Parham 2007, p. 441.

Referred specimens: FPDM-V127, fragmentary costal (Hirayama et al. 2012, Fig. 6A) from Katsuyama, Fukui Prefecture, Japan, Kitadani Formation, Tetori Group, Barremian or Aptian (Hirayama 2002; Hirayama et al. 2012).

Trionychidae indet. 4

Synonymies: ?Trionychidae indet.: Nessov 1997, p. 131, pl. 40, Figs. 5, 6.

Referred specimens: Repository unknown; two dentaries from Kansai, Fergana Depression, Tajikistan, Yalovach Formation, early Santonian.

Remarks: Based on published photographs (Nessov 1997, pl. 40, Figs. 5, 6) these dentaries have strongly expanded triturating surfaces, but their attribution to Trionychidae is questionable. Unfortunately, because the whereabouts of these specimens is unknown, we have not been able to examine them firsthand and, thus, cannot comment further on their identity.

Discussion

Crown- and Stem-Trionychids of the Cretaceous

Joyce et al. (2004) suggested that the age of the trionychid crown extended back into the Late Cretaceous, based on the occurrence of the derived trionychines *Aspideretoides* spp. and *Apalone latus* Gilmore 1919 in the middle Campanian of western North America (Gardner et al. 1995). Scheyer et al. (2007) described plywood-like structures in the external cortex of shell bones in all crown-group trionychid turtles and considered this character synapomorphic for the clade. Preliminary study of shell histology in Cretaceous trionychids from Asia (Nakajima et al. 2009) has shown that this distinctive plywood-like structure first appeared during the late Turonian, whereas it is absent in geologically older trionychids. If the plywood-like shell structure is indeed synapomorphic for crown-group trionychids, Nakajima et al's (2009) work pushes the origin of crown-trionychids back even further, to at least the late Turonian, and implies that all older occurrences are stem-trionychids.

Asian Cretaceous "Plastomenids"

Two of the three major clades of trionychids are known from the Cretaceous—Plastomenidae (sensu Joyce and Lyson 2010; = Plastomeninae sensu Hay 1902) and Trionychinae. Plastomenidae are reliably documented from the Campanian to Eocene of North America (Hay 1908; Hutchison and Holroyd 2003; Joyce et al. 2009; Hutchison 2009; Joyce and Lyson 2010). There also are poorly corroborated reports (Chkhikvadze 1990; Khosatzky 1999) of plastomenids from the Late Cretaceous to Eocene of Asia. Specimens from Kazakhstan originally were identified as *Plastomenus* or "*Plastomenus*" (Chkhikvadze 1973, 1990). Later the genus *Paraplastomenus* was erected for this material and placed in the unrelated tribe Paraplastomenini Kordikova 1994a (= Paraplastomenusini Kordikova 2002), which is characterized by a hyperossification of the shell that is not considered homologous with that seen in true Plastomenidae (= Plastomenusinae; Kordikova 1994a, 2002). Chkhikvadze (1999, 2000a, b, 2008) proposed a different arrangement; he partitioned specimens from Kazakhstan originally identified as "*Plastomenus*" into three genera: *Altaytrionyx* Chkhikvadze 2008 (Paleocene or Eocene); *Crassithecachelys* (junior synonym of *Paraplastomenus*, see Vitek and Danilov 2010; Late Cretaceous to middle Eocene); and *Francedebroinella* Chkhikvadze 1999 (early Oligocene). The content of *Paraplastomenus* (= *Crassithecachelys*) has varied from two to 10 species (Kordikova 1994a; Chkhikvadze 2007) and included *Trionyx riabinini* (as *Paraplastomenus riabinini*) as the only Cretaceous species. More recently, however, *Trionyx riabinini* was reassigned to *Aspideretoides*, a genus within the Trionychinae (Vitek and Danilov 2010). So-called "*Plastomenus*" specimens from the Late Cretaceous of Mongolia (Khosatzky 1999) also are considered to be from a trionychine (see above account for Trionychinae indet. 5). Reassignment of these supposed plastomenids to the Trionychinae means that, at present, no diagnosable Plastomenidae specimens are known from the Cretaceous of Asia. The status of Paleogene "*Plastomenus*" specimens from Kazakhstan, however, remains unresolved.

Tribes of Asian Cretaceous Trionychinae

The subfamily Trionychinae has been subdivided into no fewer than six tribes (Meylan 1987; Kordikova 1994a; Chkhikvadze 1999), all of which include Asian taxa. Four of those tribes have been proposed to include Cretaceous taxa from Asia: Pelodiscini, Trionychini, Ulutrionychini Kordikova 1994a, and Rafetini Chkhikvadze 1999.

The tribe Pelodiscini (sensu Meylan 1987) includes the extant Asian genera *Dogania*, *Palea* Meylan 1987, and *Pelodiscus* Gray 1844. Monophyly of the Pelodiscini has been supported by Meylan (1987) and by Joyce and Lyson (2010), but not by Engstrom et al. (2004). Karl (1999b) recognized *Dogania* in the Early Cretaceous of Asia by placing *Aspideretes maortuensis* (= "*Trionyx*" *kyrgyzensis*) into the genus, based on the presence of eight neurals and absence of the processesus hypoplastralis medialis anterior. However, other authors have assigned the last two species to the tribe Ulutrionychini (or Rafetini; Kordikova 1994a; Chkhikvadze 1999) or have considered them to be outside any known trionychine tribe (Nessov 1995a; this chapter).

The tribe Trionychini (sensu Meylan 1987) includes genera from Asia, Africa, and North America. The genus *Aspideretoides* and the subtribe Apalonina are both recognized in the Cretaceous of Asia. Phylogenetic analyses of osteological characters place *Aspideretoides* as the sister to Apalonina (Gardner et al. 1995; Karl 1998, 1999a). According to Meylan (1987), Apalonina unites two genera: the extant Asian *Rafetus* (Daudin 1802) and the North American *Apalone* Rafinesque 1832, the latter of which includes three extant (Meylan 1987) and one Campanian species (Gardner et al. 1995). Monophyly of Apalonina is supported by modern phylogenetic studies (Meylan 1987; Engstrom et al. 2004; Joyce and Lyson 2010).

The tribe Ulutrionychini (see Appendix for an annotated diagnosis) was established by Kordikova (1994b) for the genera *Axestemys* (Cretaceous-Paleogene of Asia, North America, and Europe), *Ulutrionyx* Kordikova 1994 (Paleogene of Asia), and *Rafetus* (Recent of Asia). *Axestemys* was circumscribed by Kordikova (1994b) in a broad sense to contain three other Paleogene genera (*Conchochelys*, *Paleotrionyx*, *Eurycephalochelys*) and several Cretaceous taxa [*Aspideretes maortuensis*, *Axestemys* (*Axestemys*) sp. (= *Khunnuchelys kizylkumensis*), and *Trionyx riabinini*]. The tribe Ulutrionychini was considered to be a group of paedomorphic trionychids with broad skulls and underdeveloped shells that retained postnuchal fontanelles (Kordikova 1994b). According to the non-cladistic phylogenetic hypothesis of Kordikova (2002), Ulutrionychini was the sister to the clade uniting Trionychini (excluding *Rafetus*), Pelodiscini, and Aspideretini. However, that hypothesis contradicts phylogenetic analyses based on both morphological and molecular data (Meylan 1987; Engstrom et al. 2004; Joyce and Lyson 2010) and has received little support.

The tribe Rafetini (see Appendix for a translation of the original diagnosis) was proposed by Chkhikvadze (1999) for almost the same grouping as Kordikova's (1994b) Ulutrionychini. According to Chkhikvadze (1999), the tribe Rafetini united the genera *Eurycephalochelys*, *Khunnuchelys*, *Kuhnemys*, and *Rafetus*. Chkhikvadze's (1999) arrangement differed from that of Kordikova's (1994b) in two important ways: (1) *Axestemys* and *Ulutrionyx* were considered subgenera of *Rafetus*—*Rafetus* (*Axestemys*) and *Rafetus* (*Ulutrionyx*)—and (2) *Conchochelys* and *Paleotrionyx* were interpreted as being convergently similar, but not related, to members of Rafetini. Among Rafetini, *Khunnuchelys* and *Kuhnemys* are Cretaceous in age. However, the diagnosis of Rafetini included only shell characters and placement of skull genera in this tribe was based on supposed skull-shell associations. For example, *Khunnuchelys* was placed into Rafetini based on the supposed synonymy of *Khunnuchelys kizylkumensis* and *Paleotrionyx riabinini* (Chkhikvadze 1999), which in our opinion is not supported by any existing evidence. More recently, Hutchison and Holroyd (2003) argued that the genus *Axestemys*

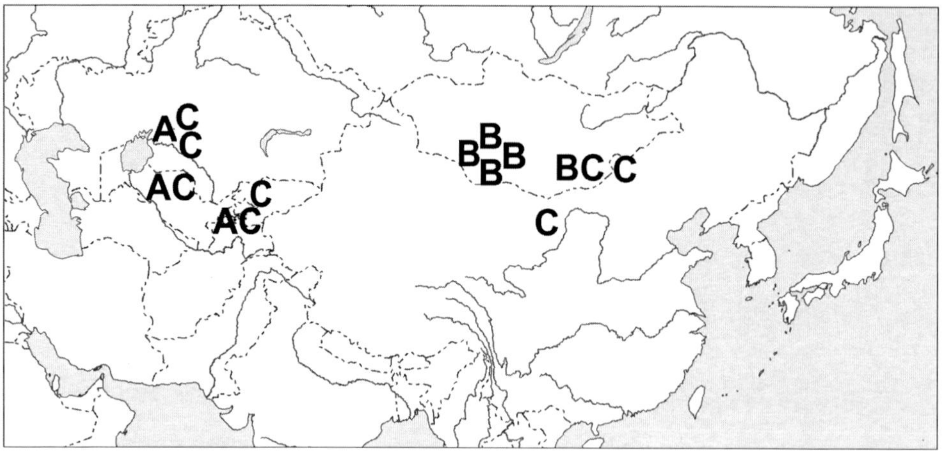

Fig. 23.4 Geographic distributions of Cretaceous trionychids in Asia. Taxa represented by letters, as follows: **A** *Aspideretoides* spp.; **B** Apalonina indet. and other Trionychini without a preneural (i.e., "*Amyda*" *menneri* and "*A.*" *orlovi*); **C** Trionychinae with eight neurals (i.e., "*Aspideretes*" *maortuensis*, *Khunnuchelys* spp., "*Trionyx*" *kansaiensis*, "*T.*" *kyrgyzensis*, and Trionychinae indet. 3). See text for further details

(= *Conchochelys*; = *Paleotrionyx*) is separate from *Rafetus* and placed it outside any known trionychine tribe. In closing, we caution that although the tribes Ulutrionychini or Rafetini may represent natural (i.e., monophyletic) groups, neither name should be used until the monophyly and membership of those clades has been demonstrated by a robust cladistic analysis.

Biogeography of Cretaceous Trionychids

Vitek and Danilov (2010) presented the first reliable evidence of relationships between Cretaceous trionychids of Asia and North America, when they recognized the genus *Aspideretoides*, formerly known only in the Campanian–Maastrichtian of North America (Gardner et al. 1995), in the Santonian-early Campanian of Asia. In this chapter, we expand the distribution of *Aspideretoides* in Asia back to the late Turonian. Other evidence of relationships between Cretaceous trionychids of Asia and North America are based on poor and mostly undescribed specimens, including a record of *Apalone*-like trionychids: Apalonina indet. from the Late Cretaceous of Mongolia (see above) and *Apalone latus* from the Campanian of North America (Gardner et al. 1995). Besides that, *Paleotrionyx riabinini* from the Late Cretaceous of Kazakhstan was previously considered an Asian representative of *Paleotrionyx* (now *Axestemys*), a genus distributed in the Paleogene of North America (Hutchison and Holroyd 2003). However, as shown above, the status and systematic position of *Paleotrionyx riabinini* is unclear and, thus, it should not be used in biogeographic discussions.

This review allows us to identify some broad patterns in the distribution of Cretaceous trionychids in Asia (Fig. 23.4). For instance, representatives of the genus *Aspideretoides* are known only from Middle Asia and Kazakhstan, whereas Apalonina indet. and other Trionychini without a preneural (i.e., "*Amyda*" *menneri*, "*A.*" *orlovi* and Trionychini indet. 2) are recorded only from Mongolia, although the morphology of the preneural is unknown in "*Aspideretes*" *alashanensis* (China) and Trionychini indet. 1 (Uzbekistan). Members of Trionychinae with eight neurals (i.e., "*Aspideretes*" *maortuensis*, *Khunnuchelys* spp., "*Trionyx*" *kansaiensis*, "*T.*" *kyrgyzensis*, and Trionychinae indet. 3) have the broadest geographic distribution, being present in Middle Asia, Kazakhstan, Mongolia, and China. In general, it appears that earlier, basal, forms had wider distributions, whereas later, more advanced forms had more restricted ranges.

It is also clear from our review that the diversity of the Cretaceous trionychids of Asia is poorly understood. The presence of up to seven trionychid taxa in the Santonian-early Campanian interval of Middle Asia and Kazakhstan and of up to six trionychid taxa in the Campanian–Maastrichtian of Mongolia seems to be an overestimation. We predict that at least some of these taxa will be synonymized by future studies.

Acknowledgments We thank T. V. Kurazheva (CCMGE) and V.B. Sukhanov (PIN) for access to collections in their care; E. V. Syromyatnikova (ZIN) for providing a photograph of ZPAL MgCh/52; Ren Hirayama for translating report of Suzuki (2005) from Japanese and J. F. Parham (University of Alabama, Tuscaloosa, USA), C.-F. Zhou (Paleontological Institute, Shenyang Normal University, Shenyang, China), and W. G. Joyce (Institut für Geowissenschaften, University of Tübingen, Tübingen, Germany) for their reviews and useful comments on the submitted version of the manuscript. Finally, we would like to express special thanks to J. D. Gardner (Royal Tyrrell

Musuem of Palaeontology, Drumheller, Canada) for his help and editorial work with our manuscript. Financial support for our study was provided by a grant from the President of the Russian Federation to the Leading Scientific Schools NSh-6560-2012.4 and by the Ministry of Education and Science of the Russian Federation.

Appendix

Diagnoses for three supra-specific trionychid taxa (two tribes and one genus) that have been erected for Cretaceous and Cenozoic material from Asia. The diagnosis for the tribe Ulutrionychini is modified from Kordikova's (1994b) original diagnosis; diagnoses for the tribe Rafetini and genus *Kuhnemys* are translations by us from Chkhikvadze's (1999) original Russian text. Changes to the original text, spelling, and grammar are given in square brackets.

Tribe Ulutrionychini Kordikova 1994b

Type genus: *Ulutrionyx* Kordikova [1994b…].

[**Referred genera**: *Axestemys* Hay 1899; *Rafetus* Daudin 1802; and *Ulutrionyx* Kordikova 1994.]

Diagnosis: The skull is wide. The facial part of it is shortened. The jugal [arches] are slender. The postnuchal fontanelles are present in the definitive age. The nuchal, neural [arches] of vertebrae, ribs and plastron elements [are not] expanded completely: ossifications are absent on lateral processes of the nuchal, on the distal ends of ribs and on the lateral and medial processes of the plastron elements. Sculpture is absent in peripheral parts of the carapace and plastron bones.

Comparison: [Ulutrionychini differ from other tribes] by the considerable fetalization which [is displayed in] the structure of the skeleton—the preservation of the wide skull, the partial expansion of the carapace and plastron bones, the preservation of the postnuchal fontanelles [at] the definitive stage as well as the [peculiarities] of the bone surface of the shell.

Tribe Rafetini Chkhikvadze 1999

[**Type genus**: *Rafetus* Gray 1864.]

Referred genera: [*Eurycephalochelys* Moody and Walker 1970; *Khunnuchelys* Brinkman et al. 1993; *Kuhnemys* Chkhkivadze 1999; and *Rafetus* Gray 1864].

Diagnosis: 1. Carapace and plastron of adults, in most cases, have well expressed features of fetalization. 2. Preneural is present only in Mesozoic and some Paleogene forms. 3. Suprascapular fontanelles, as a rule, are well or strongly developed. 4. Costals 7 and, especially, 8 are strongly reduced. 5. Plastron has ornamentation only on hyo- and hypoplastra. 6. Anteromedial process of hyoplastron is present in advanced forms. 7. Axillary processes of hyoplastra are paired. 8. Ento-hyoplastral fontanelle is always big. 9. Hypo-xiphiplastral fontanelle is always big. 10. Interxiphiplastral fontanelle is always present. 11. Ornamentation is present only on hyo- and hypoplastra (usually only on lateral parts of these plates), whereas ornamentation of xiphiplastra is absent or very weakly developed. 12. Hyo- and hypo-plastra have no tendency to fusion.

Comparison: Rafetini differ from representatives of the tribe Apalonini (sensu stricto) by morphological characters (1–3, 5–7, 11 and 12) and by considerably expressed fetalization of the shell. Besides that, suprascapular fontanelles, usually, are preserved at a definitive stage.

Genus *Kuhnemys* Chkhikvadze 1999

Type species: *Aspideretes maortuensis* Yeh 1965 […].

Etymology: The genus name is in honour of the famous German paleontologists O. Kuhn; this name coincides with Turkic "kun" (= sun).

Referred species: One species *Kuhnemys maortuensis* (Yeh 1965). This genus, undoubtedly, includes also "*Trionyx*" *kyrgyzensis* Nessov 1995. […]

Diagnosis: Preneural, most probably, is present. Suprascapular fontanelles are big. Eight neurals, the last of them contacts costals 8. Neural 6 is tetragonal. Costals 7 and 8 form posterior edge of the carapace disc, however costals 7 are big with weakly expressed wedging, whereas costals 8 are strongly reduced. (1) Ends of ribs are long and wide. Distribution and depth of relief of ornamentation of the carapace and plastron are typical for representatives of the tribe Rafetini. (2) Anteromedial process of hyoplastron is wide, rather long, its distal end is clearly divided into separate processes. (3) Medial row of hypoplastral processes is continuous, weakly reduced. (4) Anteromedial processes of this row are weakly separated, but there is a tendency to formation of the anteromedial process.

Comparison: Differs from other representatives of the tribe Rafetini by characters 1–4 listed in the diagnosis.

References

Boulenger, G. A. (1887). On a new family of pleurodiran turtles. *Annals and Magazine of Natural History, 19*, 170–172.

Brinkman, D. B. (2003). A review of nonmarine turtles from the Late Cretaceous of Alberta. *Canadian Journal of Earth Science, 40*, 557–571.

Brinkman, D. B., Nessov, L. A., & Peng, J.-H. (1993). *Khunnuchelys* gen. nov., a new trionychid (Testudines: Trionychidae) from the Late Cretaceous of Inner Mongolia and Uzbekistan. *Canadian Journal of Earth Sciences, 30*, 2214–2223.

Brinkman, D. B., Li, J., & Ye, X. (2008). Order Testudines. In J. Li, X. Wu & F. Zhang (Eds.), *The Chinese fossil reptiles and their kin.* (2nd ed., pp. 35–102). Beijing: Science Press.

Chkhikvadze, V. M. (1973). *Tertiary turtles of the Zaissan Depression*. Tbilisi: Metsniereba Publishers (In Russian).

Chkhikvadze, V. M. (1990). *Paleogene turtles of USSR*. Tbilisi: Metsniereba Publishers (In Russian).

Chkhikvadze, V. M. (1999). Some fossil soft-shell turtles of Asia (Rafetini trib. nov.). *Trudy Tbilisskogo Gosudarstvennogo Pedagogicheskogo Universiteta, 5*, 215–225 (In Russian).

Chkhikvadze, V. M. (2000a). On the systematic position of some extinct trionychids of North America and Asia. *Trudy Tbilisskogo Gosudarstvennogo Pedagogicheskogo Universiteta, 7*, 199–213 (In Russian).

Chkhikvadze, V. M. (2000b). Fossil trionychid turtles from the territory of the former Soviet Union. *Program and Abstracts, Fourth Asian Herpetological Conference, Chengdu*, 56.

Chkhikvadze, V. M. (2001). About systematic position of some fossil turtles of Asia. *Trudy Tbilisskogo Gosudarstvennogo Pedagogicheskogo Universiteta, 10*, 235–240 (In Russian).

Chkhikvadze, V. M. (2007). A brief catalogue of fossil turtles of the North Eurasia. *Problems of Paleobiology, Tbilisi, 2*, 126–137 (In Russian).

Chkhikvadze, V. M. (2008). Soft-shell turtles (Trionychidae) of Asia and North America: Morphology, phylogeny, systematics, terminology of elements of carapace. *Problems of Paleobiology, Tbilisi, 3*, 85–95 (In Russian).

Chkhikvadze, V. M., & Shuvalov, V. F. (1988). A new species of a trionychid from the Upper Cretaceous deposits of Mongolia. *Izvestiya Akademii Nauk Gruzinskoi SSR, 14*, 198–204 (In Russian).

Danilov, I. G. (2005). Die fossilen Schildkröten Europas. In U. Fritz (Ed.), *Handbuch der Reptilien und Amphibien Europas. Band 3/IIIB: Schildkröten (Testudines) II* (pp. 329–441). Wiebelsheim: Aula.

Danilov, I. G. (2007). New data on soft-shelled turtles (Trionychidae) from the Bissekty Formation (Late Turonian) of Dzharakuduk, Uzbekistan. *Journal of Vertebrate Paleontology, 27* (Suppl 3), 66A.

Danilov, I. G., & Parham, J. F. (2005). A reassessment of the referral of an isolated skull from the Late Cretaceous of Uzbekistan to the stem-testudinoid turtle genus *Lindholmemys*. *Journal of Vertebrate Paleontology, 25*, 784–791.

Danilov, I. G., & Parham, J. F. (2007). The type series of '*Sinemys*' *wuerhoensis*, a problematic turtle from the Lower Cretaceous of China, includes at least three taxa. *Palaeontology, 50*, 431–444.

Danilov I. G., & Vitek N. S. (2009). Cretaceous trionychids of Asia: A review of record and biogeography. *Gaffney Turtle Symposium (October 17–18, 2009, Royal Tyrrell Museum, Drumheller, Canada). Abstract Volume* (pp. 52–58).

Daudin, F. M. (1802). *Histoire Naturelle, Génerale et Particulière des Reptiles* (Vol. 2). Paris: F. Dufart.

Eaton, J. G., Cifelli, R. L., Hutchison, J. H., Kirkland, J. I., & Parrish, J. M. (1999). Cretaceous vertebrate faunas from the Kairparowits Plateau, south-central Utah. In D. D. Gillette (Ed.), *Vertebrate Paleontology in Utah* (Vol. 99, pp. 345–353). Salt Lake City: Utah Geological Survey Miscellaneous Publication.

Engstrom, T. N., Shaffer, H. B., & McCord, W. P. (2004). Multiple data sets, high homoplasy, and the phylogeny of softshell turtles (Testudines: Trionychidae). *Systematic Biology, 53*, 693–710.

Gardner, J. D., Russell, A. P., & Brinkman, D. B. (1995). Systematics and taxonomy of soft-shelled turtles (Family Trionychidae) from the Judith River Group (mid-Campanian) of North America. *Canadian Journal of Earth Sciences, 32*, 631–643.

Gilmore, C. W. (1919). New fossil turtles, with notes on two described species. *Proceedings of the United States National Museum, 56*, 113–132.

Gilmore, C. W. (1923). A new species of *Aspideretes* from the Belly River Cretaceous of Alberta, Canada. *Transactions of the Royal Society of Canada, Section 4, Series 3, 17*, 1–10.

Gilmore, C. W. (1931). Fossil turtles of Mongolia. *Bulletin of the American Museum of Natural History, 59*, 213–257.

Glinskiy, V. N. (2008). About the finding of a skull of a gigantic soft-shelled turtle of the genus *Khunnuchelys* in the Late Cretaceous of Kazakhstan. *Geologiya—Nashe Budushchee. MaterialyMolodezhnoy Nauchnoy Konferentsii (April 29–30, 2008, St. Petersburg State University, St. Petersburg, Russia). Abstracts* (pp. 21–23) (In Russian).

Glinskiy, V. N., & Danilov, I. G. (2008). About the finding of a skull of a gigantic soft-shelled turtle of the genus *Khunnuchelys* in the Late Cretaceous of Kazakhstan. *Modern Paleontology: Classic and Newest Methods. The Fifth All-Russian Scientific School for Young Scientists in Paleontology (October 6–8, 2008, Paleontological Institute of the Russian Academy of Sciences, Moscow, Russia). Abstracts* (pp. 18–19) (In Russian).

Gray, J. E. (1825). A synopsis of the genera of reptiles and amphibia, with a description of some new species. *Annals of Philosophy, 10*, 193–217.

Gray, J. E. (1844). *Catalogue of the tortoises, crocodiles, and amphisbaenians, in the collection of the British Museum*. London: Taylor and Francis.

Hay, O. P. (1899). On the nomenclature of certain American fossil vertebrates. *American Geologist, 24*, 345–349.

Hay, O. P. (1902). Bibliography and catalogue of fossil vertebrates of North America. *United States Geological Survey Bulletin, 179*, 1–449.

Hay, O. P. (1905). On the shell of a new trionychid, *Conchochelys admirabilis*, from the Puerco beds of New Mexico. *Bulletin of the American Museum of Natural History, 21*, 335–338.

Hay, O. P. (1908). The fossil turtles of North America. *Carnegie Institute of Washington Publications, 75*, 1–568.

Hay, O. P. (1909). Descriptions of five new species of North America fossil turtles, four of which are new. *Proceedings of the United States National Museum, 35*, 161–169.

Hirayama, R. (1998). Fossil turtles from the Mifune Group (Late Cretaceous) of Kumamoto Prefecture, Western Japan. In *Report of the research on the distribution of important fossils in Kumamoto Prefecture, "Dinosaurs from the Mifune Group, Kumamoto Prefecture, Japan"* (pp. 85–99). Mifune Town Education Board (In Japanese with English abstract).

Hirayama, R. (2002). Preliminary report of the fossil turtles from the Kitadani Formation (Early Cretaceous) of the Tetori Group of Katsuyama, Fukui Prefecture, Central Japan. *Memoir of the Fukui Prefectural Dinosaur Museum, 1*, 29–40.

Hirayama, R. (2008). Trionychid turtles (Trionychoidea; Testudines) from the Early Cretaceous of Central Japan. *Journal of Vertebrate Paleontology, 28* (Suppl 3), 91A.

Hirayama, R., Brinkman, D. B., & Danilov, I. G. (2000). Distribution and biogeography of non-marine Cretaceous turtles. *Russian Journal of Herpetology, 7*, 181–198.

Hirayama, R., Isaji, S., & Hibino, T. (2012). *Kappachelys okurai* gen. et sp. nov., a new stem soft-shelled turtle from the Early Cretaceous of Japan. In D. B. Brinkman, P. A. Holroyd & J. D. Gardner (Eds.), *Morphology and evolution of turtles*. Dordrecht: Springer.

Horner, J. R., Weishampel, D. B., & Forster, C. A. (2004). Hadrosauridae. In D. B. Weishampel, P. Dodson & H. Osmólska (Eds.), *The Dinosauria* (2nd ed., pp. 438–463). Berkeley: University of California Press.

Hummel, K. (1929). Die fossilen Weichschildkröten (Trionychia). Eine morphologisch-systematische und stammesgeschichtliche studie. *Geologische und Palaeontologische, Abhandlungen, 16*, 359–487.

Hutchison, J. H. (2000). Diversity of Cretaceous turtle faunas of Eastern Asia and their contribution to the turtle faunas of North

America. *Paleontological Society of Korea Special Publication, 4*, 27–38.

Hutchison, J. H. (2009). New soft-shelled turtles (Plastomeninae: Trionychidae: Testudines) from the Late Cretaceous and Paleocene of North America. *Paleobios, 29*, 36–47.

Hutchison, J. H., & Holroyd, P. A. (2003). Late Cretaceous and early Paleocene turtles of the Denver Basin. *Rocky Mountain Geology, 38*, 121–142.

Joyce, W. G., & Lyson, T. R. (2010). A neglected lineage of North America turtles fills a major gap in the fossil record. *Palaeontology, 53*, 241–248.

Joyce, W. G., Parham, J. F., & Gauthier, J. A. (2004). Developing a protocol for the conversion of rank-based taxon names to phylogenetically defined clade names, as exemplified by turtles. *Journal of Paleontology, 78*, 989–1013.

Joyce, W. G., Revan, A., Lyson, T. R., & Danilov, I. G. (2009). Two new plastomenine softshell turtles from the Paleocene of Montana and Wyoming. *Bulletin of the Peabody Museum of Natural History, 50*, 307–325.

Karl, H.-V. (1998). Zur Taxonomie der känozoischen Weichschildkröten Österreichs und Deutschlands (Trionychidae: Trionychinae). *Mitteilungen der Abteilungs für Geologie und Paläontologie der Landesmuseum Joanneum, 56*, 273–328.

Karl, H.-V. (1999a). Die Zoogeographie der känozoischen Weichschildkröte *Trionyx triunguis* Forskäl, 1775. *Joannea Geoogie und Paläontologie, 1*, 27–60.

Karl, H.-V. (1999b). Paleogeography and systematics of the genus *Dogania* Gray, 1844 (Testudines: Trionychidae). *Studia Geologica Salmanticensia, 35*, 3–8.

Khosatzky, L. I. (1957). To the history of trionychid turtles in Kazakhstan. *Izvestiya Akademii Nauk Kazakhskoy SSR. Seriya Bilogicheskaya, 2*, 15–30 (In Russian).

Khosatzky, L. I. (1966). About a gigantic Mesozoic representative of trionychids and several features of the shell of these turtles. In *Pozvonochnye Zhivotnye Sredney Azii* (pp. 150–157). Tashkent: Fan Publishers (In Russian).

Khosatzky, L. I. (1976). A new representative of trionychids from the Late Cretaceous of Mongolia. *Gerpetologiya. Kubanskiy Gosudarstvennyy Universitet. Nauchnye Trudy, 218*, 3–19 (In Russian).

Khosatzky, L. I. (1999). Turtles—trionychids of the Cretaceous of Mongolia. *Voprosy Paleontologii, 11*, 141–149 (In Russian).

Khosatzky, L. I., & Młynarski, M. (1971). Chelonians from the Upper Cretaceous of the Gobi Desert Mongolia. *Palaeontologia Polonica*, 25131–144.

Kordikova, E. G. (1991). *Catalogue of fossil trionychid turtles of the USSR*. Preprint. Tbilisi: L.Sh. Davitashvili Institute of Paleobiology (In Russian).

Kordikova, E. G. (1992a). Review of fossil trionychid turtles of the USSR. *Izvestiya Akademii Nauk Gruzii, 18*, 131–141 (In Russian).

Kordikova, E. G. (1992b). About time and ways of prochoresis of fossil tronychid turtles of Kazakhstan. *Soobshcheniya Akademii Nauk Gruzii, 145*, 647–650 (In Russian).

Kordikova, E. G. (1994a). Review of fossil trionychid localities in the Soviet Union. *Courier Forschungs-Institut Senckenberg, 173*, 341–358.

Kordikova, E. G. (1994b). About systematics of fossil trionychids in Kazakhstan. *Selevinia, 2*, 3–8.

Kordikova, E. G. (2002). Heterochrony in the evolution of the shell of Chelonia. Part 1: Terminology: Cheloniidae: Dermochelyidae: Trionychidae: Cyclanorbidae: Carettochelyidae. *Neues Jahrbuch für Geologie und Paläontologie, Abhandlungen, 226*, 343–417.

Kuhn, O. (1964). Testudines. In F. Westphal (Ed.), *Fossilium catalogus. I: Animalia, Pars 107* (pp. 3–299). S-Gravenhage: Uitgeverij Dr W. Junk.

Kuznetsov, V. V., & Chkhikvadze, V. M. (1987). The Late Cretaceous trionychids from Shakh–Shakh locality in Kazakhstan. *Materialy po Istorii Fauny i Flory Kazakhstana, 9*, 33–39 (In Russian).

Lambe, L. M. (1914). On a new species of *Aspideretes* from the Belly River Formation of Alberta, with further information regarding the carapace of *Boremys pulchrapulchra. Transactions of the Royal Society of Canada,, Section 4, Series 3, 8*, 11–16.

Langston, W., Jr. (1960). The vertebrate fauna of the Selma Formation of Alabama. Part VI. The dinosaurs. *Fieldiana Geology Memoirs, 3*, 313–363.

De Lapparent de Broin, F. (2004). A new Shachemydinae (Chelonii, Cryptodira) from the Lower Cretaceous of Laos: Preliminary data. *Comptes Rendus Palevol, 3*, 387–396.

Leidy, J. (1856). Notices of the remains of extinct reptiles and fishes discovered by Dr. F.V. Hayden in the bad lands of the Judith River, Nebraska Territory. *Proceedings of the Academy of Natural Sciences of Philadelphia, 8*, 72–73.

Leidy, J. (1869). Notice of some extinct vertebrates from Wyoming and Dakota. *Proceedings of the Academy of Natural Sciences of Philadelphia, 20*, 63–67.

Merkulova, N. N. (1978). A new *Trionyx* from the Nemegt (MPR). *Byulleten' Moskovskogo Obshchestva Ispytateley Prirody. Otdel Geologicheskiy, 53*, 156 (In Russian).

Meylan, P. A. (1987). The phylogenetic relationships of soft-shelled turtles (Family Trionychidae). *Bulletin of the American Museum of Natural History, 186*, 1–101.

Młynarski, M., & Narmandach, P. (1972). New turtle remains from the Upper Cretaceous of the Gobi Desert, Mongolia. *Palaeontologia Polonica, 27*, 95–102.

Moiseenko, V. G., Sorokin, A. P., & Bolotsky, Yu. L. (1997). *Fossil reptiles of the Amur region*. Khabarovsk: Amur Scientific Center of the Far East Branch of the Russian Academy of Sciences (In Russian).

Nakajima, Y., Danilov, I. G., & Hirayama, R. (2009). Plywood-like structures in costal bones of Mid- to Late Cretaceous trionychids from Asia. *Gaffney Turtle Symposium (October 17–18, 2009, Royal Tyrrell Museum, Drumheller, Canada). Abstract Volume*, 128.

Nessov, L. A. (1977). Skull morphology of Early Cretaceous turtle of the family AdocidaeAdocidae. *Trudy Zoologicheskogo Instituta AN SSSR, 74*, 75–80 (In Russian).

Nessov, L. A. (1984). Preservation of remains of organisms and conditions of their burial in Kulbike Member of Itemir. In V. G. Ochev (Ed.), *Tafonomiya i Voprosy Paleogeografii* (pp. 62–76). Saratov: Publishers of Saratov University (In Russian).

Nessov, L. A. (1985). Data on Late Mesozoic turtles from the USSR. *Studia Palaeocheloniologica, 1*, 215–223.

Nessov, L. A. (1986). Some Late Mesozoic and Paleocene turtles of Soviet Middle Asia. *Studia Palaeocheloniologica, 2*, 7–22.

Nessov, L. A. (1987). On some Mesozoic turtles of Soviet Union, Mongolia and China, with comments on systematics. *Studia Palaeocheloniologica, 2*, 87–102.

Nessov, L. A. (1995a). On some Mesozoic turtles of the Fergana Depression (Kyrgyzstan) and Dzhungar Alatau Ridge (Kazakhstan). *Russian Journal of Herpetology, 2*, 134–141.

Nessov, L. A. (1995b). *Dinosaurs of Northern Eurasia*. St. Petersburg: St. Petersburg State University (In Russian).

Nessov, L. A. (1997). *Cretaceous Nonmarine Vertebrates of Northern Eurasia*. St. Petersburg: St. Petersburg State University, Institute of Earth's Crust (In Russian).

Nessov, L. A., & Khosatzky, L. I. (1978). Early Cretaceous turtles of Kirghizia. *Ezhegodnik Vsesoyuznogo Paleontologicheskogo Obshchestva, 21*, 267–279 (In Russian).

Parks, W. A. (1933). New species of dinosaurs and turtles from the Upper Cretaceous formations of Alberta. *University of Toronto, Geological Series, 34*, 1–33.

Rafinesque, C. S. (1832). Description of two new genera of soft shell turtles of North America. *Atlantic Journal and Friend of Knowledge, 1*, 64–65.

Rozhdestvensky, A. K., & Khosatzky, L. I. (1967). Late Mesozoic terrestrial vertebrates of the Asiatic part of the USSR. In G. G. Martinson (Ed.), *Stratigrafiya i Paleontologiya Mezozoyskikh i Paleogen-Neogenovykh Kontinentalnykh Otlozheniy Aziatskoy Chasti SSSR* (pp. 82–92). Leningrad: Nauka.

Scheyer, T. M., Sander, P. M., Joyce, W. G., Böhme, W., & Witzel, U. (2007). A plywood structure in the shell of fossil and living soft-shelled turtles (Trionychidae) and its evolutionary implications. *Organisms, Diversity and Evolution, 7*, 136–144.

Schmidt, K. P. (1945). A new turtle from the Paleocene of Colorado. *Fieldiana Geology, 10*, 1–4.

Sochava, A. V. (1975). Stratigraphy and lithology of the Upper Cretaceous deposits of southern Mongolia. *Trudy Sovmestnoy Sovetsko-Mongol'skoy Nauchno-Issledovatelskoy Geologicheskoy Ekspeditsii, 13*, 113–182 (In Russian).

Sukhanov, V. B. (2000). Mesozoic turtles of Middle and Central Asia. In M. J. Benton, M. A. Shishkin, D. M. Unwin & E. N. Kurochkin (Eds.), *The age of Dinosaurs in Russia and Mongolia* (pp. 309–367). Cambridge: Cambridge University Press.

Suzuki, S. (2005). An almost complete skeleton of the trionychid turtle (Trionychidae: Trionychinae) from the Upper Cretaceous of Mongolia. *Abstracts with Programs of the 154th Regular Meeting of the Palaeontological Society of Japan* (p. 27) (In Japanese).

Suzuki, S., & Narmandakh, P. (2004). Change of the Cretaceous turtle faunas in Mongolia. *Hayashibara Museum of Natural Sciences Research Bulletin, 2*, 7–14.

Vitek, N., & Danilov, I. G. (2008). Morphology and systematic position of trionychid turtles from the Late Cretaceous of Kazakhstan and Tadzhikistan. *Modern Paleontology: Classic and Newest Methods. The Fifth All-Russian Scientific School for Young Scientists in Paleontology (October 6–8, 2008, Paleontological Institute of the Russian Academy of Sciences, Moscow, Russia). Abstracts* (pp. 17–18) (In Russian).

Vitek, N. S., & Danilov, I. G. (2010). New material and a reassessment of soft-shelled turtles (Trionychidae) from the Late Cretaceous of Middle Asia and Kazakhstan. *Journal of Vertebrate Paleontology, 30*, 383–393.

Yeh, H.-K. (1965). New materials of fossil turtles of Inner Mongolia. *Vertebrata PalAsiatica, 9*, 47–69.

Yeh, H.-K. (1974). A new fossil *Trionyx* from Fuchien. *Vertebrata PalAsiatica, 12*, 190–192 (In Chinese).

Yeh, H.-K. (1994). *Fossil and recent turtles of China*. Beijing: Science Press.

Chapter 24
Fossil European Sea Turtles: A Historical Perspective

Richard T. J. Moody, Cyril A. Walker, and Sandra D. Chapman

Abstract This chapter presents a historic perspective on the study of marine turtles from the Cretaceous-Tertiary strata of Europe from the Sixteenth Century to the Present Day. It details the historic episodes that have occurred in terms of the first illustrations or the first scientific descriptions. It outlines the backgrounds and societal links of the individuals involved and throws light on the emergence of an intellectual elite in paleontology. The first natural scientists or philosophers came, almost inevitably from middle or upper middle class families, and their interests were those of histomaths or polymaths. Many initially trained as physicians or alchemists with interests in comparative anatomy, zoology and botany; some are famous for the recognition and treatment of 'new' diseases or the discovery of new medicines, others were immensely talented as illustrators; some were subsequently accused of conveying racial dogma or of being at best devious for their own gain. The history of fossil collecting and the acquisition of great personal collections or 'cabinets' is a significant component in the historic description of fossil turtles but the advent of regional and national museums and collections heralds an age when access did not depend on patronage or favour. Universal education, and the mesmeric advances in communication and modes of travel have resulted in a surge of new workers in the field of Testudine paleontology with the work of Gene Gaffney as focal point for a bright future.

Keywords Europe • History • Marine • Testudines

C. A. Walker—Deceased (formerly Department of Palaeontology, The Natural History Museum, Cromwell Road, London SW7 5BD, UK)

R. T. J. Moody (✉)
School of Geology and Earth Sciences, Kingston University, Penrhyn Road, Kingston, Surrey KT1 2EE, UK
e-mail: rtj.moody@virgin.net

S. D. Chapman
Department of Palaeontology, The Natural History Museum, Cromwell Road, London, SW7 5BD, UK
e-mail: s.chapman@nhm.ac.uk

Introduction

The collection of fossils effectively began in the Sixteenth Century with the debate on their origin and meaning conducted by an emergent group of 'naturalists' and 'natural philosophers' (Rudwick 1976). Conrad Gesner (1515–1565) completed his illustrated book *On Fossil Objects* shortly before he died in 1565 and he is considered to be the founding father of paleontology. His distant relative Johannes Gessner (1709–1790) was the first European scientist to figure a fossil sea turtle. A common theme running through the early history of 'fossils' was this relationship between an intellectually elite group of naturalists, many of whom trained initially as doctors and lawyers. A number were appointed as physicians to the courts of European Kings and Queens, whereas others were sponsored by the super rich of the day.

During the Eighteenth Century Holland, Belgium, and southeast England became focal points for the collection and description of new material. It also witnessed the development of comparative anatomy as a true science under the burgeoning influence of Baron Cuvier (1769–1832). During the eighteenth and nineteenth centuries the natural sciences, specifically geology, changed from being encyclopaedic to a practical, interpretive science (Rudwick 1976). In 1824 Cuvier beautifully illustrated a number of living and fossil turtle specimens from several European countries, whereas Richard Owen (1804–1892), the 'Cuvier of Britain', published a paper in 1841 on six

D. B. Brinkman et al. (eds.), *Morphology and Evolution of Turtles*, Vertebrate Paleobiology and Paleoanthropology,
DOI: 10.1007/978-94-007-4309-0_24, © NHM The Natural History Museum 2013

Fig. 24.1 Naturalists of late mediaeval to early modern Europe. **a** Conrad Gesner (1516–1565), the first natural historian to observe, describe, and illustrate fossils and minerals in a scientific manner. (*Sourced from Wikipedia, image in Public Domain*). **b** Johannes Conrad Gessner (Gesner) (1709–1790), a distant relative of Conrad, Johannes is thought to be the first paleontologist to publish an illustration of a European sea turtle. (*Source* From Wikipedia, image in Public Domain)

species of marine turtles and a classification that remained unchallenged for almost 50 years until the release of Lydekker's *Catalogue of Fossil Reptiles & Amphibians for the British Museum (Natural History)* (1889c). This work was greatly influenced by the work of E. D. Cope (1840–1897) and provided an invaluable data set of hundreds of specimens. In the early Twentieth Century studies on European sea turtles were dominated by the description of single specimens exacerbating the need for a more holistic approach to research. Subsequently the main influence came from American workers including Rainer Zangerl and Eugene Gaffney. Zangerl's work greatly influenced the studies of European Mesozoic–Cenozoic sea turtles undertaken by Collins (1970), Moody (1972, 1980a, b, 1993, 1996, 1997), Evans and Kemp (1975), Seago (1979), and Smith (1989). However, the world of fossil turtles went through a seismic change in the 1970s when Gaffney wrote three papers (1972, 1975 and 1979) that literally changed the way we looked at turtles. One outcome of the 'Gaffney' effect was the attraction of more and more comparative anatomists and paleozoologists to turtle research across the globe.

Late Mediaeval to Early Modern Europe

The late Fifteenth Century to early Sixteenth Century marked the transition from the Mediaeval or Middle Ages to the Early Modern Period. It witnessed the voyages of Columbus, the rise of the Elizabethan period and the Protestant Reformation in Northern Europe. English literature was marked by the writings of Bacon, Marlowe, and Shakespeare. The Tudors Henry and Elizabeth ruled England for over 100 years and Spain suffered under the Spanish Inquisition. Elsewhere in Europe, Gutenberg invented moveable type printing and Copernicus developed his heliocentric thesis that moved the Earth from the centre of the universe. Life for the poor remained the same, however, and social enhancement was confined to successful merchants, the aristocracy, and those blessed with genius and financial support. This emergent period is also marked by the growth of the natural sciences.

Conrad Gesner (1516–1565) is recognized as the greatest naturalist of his time. He was a Professor of Greek and Aristotelian Physics and a medical Physician (Fig. 24.1a). His work the *Bibliotheca Universalis* was a compendium of every book published at that time. The last of his 72 published books was entitled '*A Book on Fossils Objects, Chiefly Stones and Gems, their Shapes and Appearances*'. It was completed in 1565 the year he died of a 'bubonic' plaque. The '*De omni rerum fossilium genere*' is a compendium of papers by a host of European natural historians, with Gesner as both contributor and editor. It was believed to be the prelude to a much greater work. His books were among the first to be illustrated by fine woodcuts, with Albrecht Durer among the craftsmen he called on to help. Martin Rudwick (pers. comm.) records that the woodcuts are embedded in letterpress on many different pages and include a Kupferschiefe fish, a crab, and other invertebrates including echinoids, ammonites and brachiopods. But sadly no fossil turtle!

Gesner (1565) includes an essay by Johannes Kentmann (1534–1599), a surgeon who had accumulated a collection of more than thousand mineral specimens. At this time the word 'fossil' was used to describe any object that was dug from the earth and included objects of organic or inorganic origin (Rudwick 1976). By the late Eighteenth Century the number of people involved in the collection and acquisition of fossils and fossil turtle material increased significantly and the use of the term 'Cabinet' referenced the private collection of individuals, including Kings and Queens.

It was also evident that a 'community' of very well educated, affluent natural scientists existed throughout Europe at this time and that there was a web of noted collectors and dealers. Mineralogy was perhaps the preferred study of many naturalists, but a number of intellectuals and the super-rich of the day also took up the search and acquisition of fossils.

Like his ancestor, Conrad, Johannes Gessner (1709–1790) (Fig. 24.1b) was a polymath, born and educated in Zurich who went on to study medicine in Leiden. Once there, he was introduced to the science of botany. Later he studied higher mathematics in Paris. His true calling however, was the study of plants and he wrote prodigiously on systematics. Late in life he turned to geology and saw the link between fossil and living plants. According to Buc'hoz (1781), Gessner (*Tractatus physicus de petrificatis* S84) illustrated the first 'petrification' of a turtle from 'une ardoise de Glaris' in Switzerland in 1758. The specimen in question was housed in the 'Cabinet' of a M. Zoller and was subsequently described as *Glarichelys knorri* by Gray (1831). The specimen was reportedly collected by Georg Wolfgang Knorr (1705–1761) who together with Johann Walch (1725–1778) (see Mlynarski 1959) published several tomes on the natural history of the Canton of Glaris (Glarus); the last three of which were completed by Walch after the death of his co-author. The 1773 volume is devoted to fossils and the quality of the original copper plates made by Knorr is comparable with the work of Durer.

James Parkinson (Letter XVIII, 1811) wrote that:

> *M. Knorr gives the representation of a fossil tortoise from a very valuable specimen in the possession of Dr Gesner, found near Glaris.*

This suggests that this holotype of *Glarichelys* changed hands several times, but Zangerl (1958) provides strong evidence of the link between Johannes Gessner and the study of the Glaris turtle.

Like Gessner, Pierre Joseph Buc'hoz (1731–1807) (Fig. 24.2) was also devoted to the study of plants. He was the son of a lawyer and was born in Metz (northern France). He was initially persuaded by his father to enter the legal profession and duly completed his entrance to the bar in 1750. Law was not his chosen profession, however, and he moved to Nancy to study medicine, a field more closely related to his preferred science—natural history. Having qualified as a doctor in 1763 he was somewhat surprisingly appointed as a physician to King Stanislaus of Poland. His devotion to botany caused him to lose this post and so he devoted his life to the study of botany, medicine, economics, and mineralogy, eventually publishing over 300 volumes on these subjects (Wilson and Schuh 2001). His works were wonderfully illustrated both by himself and numerous brilliant artists and engravers of the day. Today posters and artwork based on the engravings by Buc'hoz are sold throughout the world, some for hundreds of dollars. Buc'hoz's genius encompassed the medical use of plants and plant extracts as cosmetics for ladies. Unfortunately he was somewhat chaotic in his approach and was frequently criticised for his lack of attention to detail. So much so that he failed in his candidature for the Academié des Sciences in 1779 and again in 1800 when Lamark supposedly proposed him along with five other candidates (Vernier and Klein 2005).

Fig. 24.2 Lithograph of Joseph Pierre Buc'hoz at the age of 64. Image created circa 1795. (Image in Public Domain. Original Housed in Musée Lorrain de Nancy, France)

Buc'hoz was also interested in fossils. He published on the 'fossils' of Lorraine and produced a magnificent illustration of an Eocene turtle in 1781. This illustration was in colour and is referred to as figure 5 from: '*Centuries de planches eluminees et non illuminees*' 1778–1781.

The specimen remained unnamed by Buc'hoz, but was thought to be a marine species by Burtin (1784). Later it became the holotype of *Puppigerus camperi* (Gray) by Dollo in 1923. Sadly Buc'hoz was a poor businessman and had to sell off his collections and later died in poverty in Paris in 1807.

François Xavier Burtin (1743–1818) was another individual who represented the close ties that existed between practising physicians and the emergent earth sciences, especially mineralogy and paleontology. He effectively broke the mould however, when as the son of an 'aubergiste' he rose to the highest ranks of the scientific community in Belgium during the Eighteenth Century. He was born in Maastricht, which was then under the rule of France, and became a physician who specialised in diseases associated with syphilis. Practising in Brussels in the reign of King Joseph II, he was to become a member of the sovereign's Privy Council. During his lifetime he was lauded with the membership of many societies and his achievements are documented in the records of the L'Académie impériale et Royale de Bruxelles (Callatay 2009). He was a contemporary of Buc'hoz, Camper and Faujas St Fond, professional men with a love of natural history.

In 1784 Burtin published the luxuriously bound and illustrated: *Oryctographie de Bruxelles, ou Description des Fossiles Tant Naturels qu'accidentels déscouverts jusqu'à ce Jour dans les environs de cette Ville*. This is claimed to be the first detailed or descriptive book on fossils and is justifiably acclaimed for its beautiful coloured plates. Included in this work is a groundbreaking account of the 'Tortues Petrifiées' and included therein, is a copy of the turtle carapace (Fig. 24.3) first illustrated by Buc'hoz in 1781. Burtin was possibly one of the first naturalists to collect and describe numerous fossils from the Bruxelles region; he took part in several excavations with one shaft dug to over 30 m deep. Rudwick (1993) noted that 142 copies of the *Oryctographie* were printed and because Belgium was then part of The Austrian Netherlands many copies went to regional governors, to Kings and Queens, and to noted intellectuals and booksellers whose subscriptions probably helped fund the work. Hand written letters from Burtin to his sponsors give further insight into the ties that prevailed between the rich and the gifted in the Eighteenth Century. According to Rudwick (2005) Burtin was not a geological theorist in the mold of Buffon, but he did register relative ages and a number of '*general revolutions*' in his works.

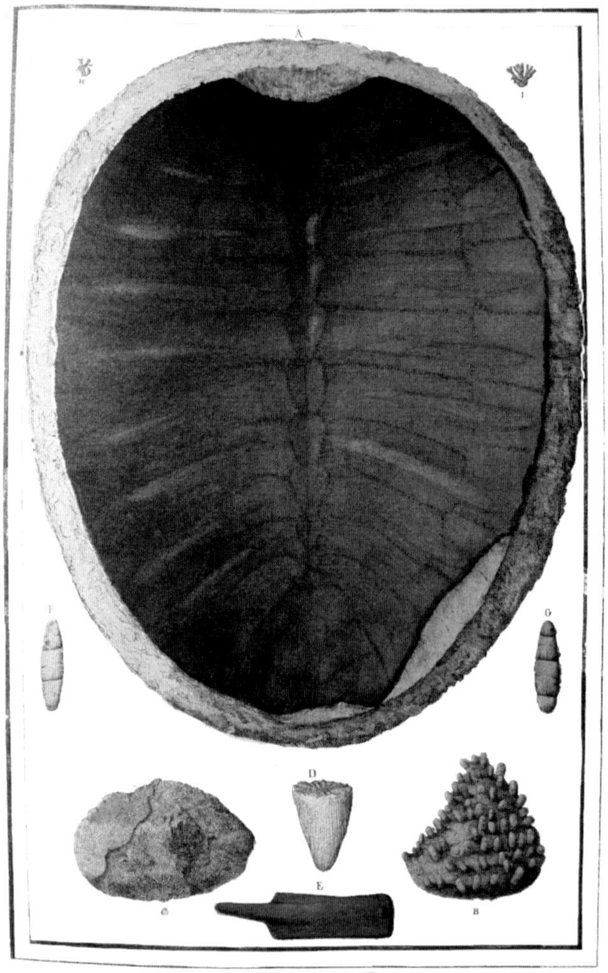

Fig. 24.3 Planche V from Burtin (1784) of a fossil turtle from the Tertiary strata of the Bruxelles region, which was first figured by Buc'hoz (1781). The carapace is a fine example of *Puppigerus camperi* (Gray). (Image supplied R.T.J. Moody)

Petrus' A. G. Camper (1722–1789) (Fig. 24.4a) clearly qualifies as one of the most outstanding polymaths of his era. He was accepted at university at the age of 12 and became a brilliant comparative anthropologist, teacher, and artist. He was born in Leiden where his father and mother lived as people of independent means. His father entertained the rich and famous and introduced Petrus to many famous scientists. Throughout his career Camper specialised in midwifery and invented the vectis lever, a precursor to the obstetrical forceps used today. During his life he was inducted as a member of 20 societies including the Royal Society in London and the Royal Academies of Science in Paris and Berlin. He was wealthy not only by inheritance but also by marriage, and was one of dozens of the elite who built up superb 'Cabinets'. He developed his interest in geology and the natural sciences relatively late in life and believed in a period of old life, thus doubting the soundings of Archbishop James Ussher (1581–1656). He was both social climber and a workaholic. He was deemed by some to be a 'scientific butterfly' who skipped from subject to subject.

Petrus Camper and his son Adriaan Gilles Camper wrote several papers (1786, 1800, 1812) together on the fossilised reptiles of Mont St Pierre and formed a working relationship with an Irish collector who lived in Liege. A specimen collected by Preston was first figured by Camper. In 1786 he published *Translations Philosophique* which figured several incontestable pieces of turtle material housed in the collection of J. L. Hoffman (Mulder 1996).

Frequent reference is made by Camper to the material housed in the collections of other savants—such notes serve to improve our understanding of the growing interest in natural history at this time. The direct purchase or auction of personal collections was fairly frequent. van Veen (2004) and Meijer (1999) noted that Petrus and his son Adriaan travelled some distance to buy material.

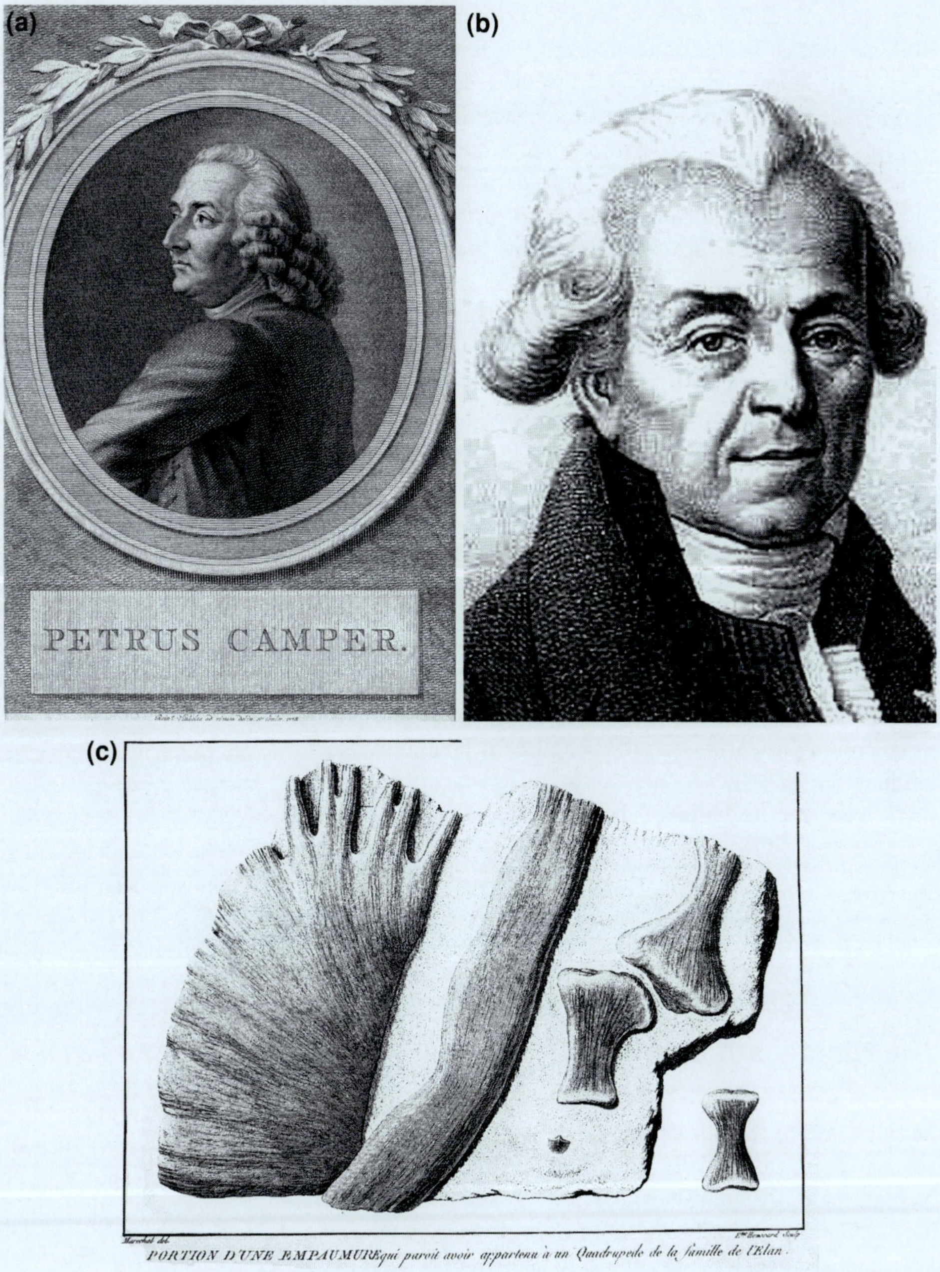

Fig. 24.4 Natural history studies of early modern Europe. **a** Lithograph of Petrus Camper (1722–1789) (Copyright of the Natural History Museum, London). **b** Lithograph of Barthélemy Faujas St Fond (1741–1819) author of *Histoire Naturelle de La Montagne de Saint-Pierre de Maestricht* in 1799. (Sourced from Wikipedia, image in Public Domain). **c** Bones including a plastral plate of *Allopleuron hoffmani* originally defined as a *Portion d'une Empaumure* or the top antlers of deer by St Fond in 1799. (Image provided by R.T.J. Moody)

Camper was a noted student of human anatomy or "menschkunde" and employed the principle of 'metamorphosis' in his studies and was the instigator of the hypothesis of the facial angle. These compared the facial characteristics of European and African hominids and the apes, and belatedly earned him the title of 'one of the founders of racist anthropology'. Miriam Claude Meijer (1999) in her book, entitled *Race and Aesthetics in the Anthropology of Petrus Camper (1722–1789)*, wrote that:

> *Camper's racist reputation originated from his illustrations of animal and human profiles, taken out of its original context. The acute jaw angle became synonymous with "primitivism," "savagery," or "mental deficiency."*

Barthélemy Faujas St Fond (1741–1819) (Fig. 24.4b), like Buc'hoz, first studied law and rose to be President of the Seneschal's Court in Montélimar. However, he too, preferred science and became Royal Commisioner for Mines in France and in 1793 he was appointed as the first Professor of Geology at the Jardin des Plantes in Paris.

As a mining specialist and then an academic versed in geology his appointments heralded a new era where professional geologists were at the forefront of the growth and development of the earth sciences. As a result of war, Holland was essentially dominated by France between 1795–1810; during that interval the Netherlands was known as the Batavian Republic. As a consequence the Dutch were second-rate citizens in their own country and French officials such as Faujas St Fond had considerable powers. Not surprisingly he was personally involved in the removal of the second skull of *Mosasaurus* to Paris after the fall of Maastricht.

Faujas St Fond published his tome the *Histoire Naturelle de La Montagne de Saint-Pierre de Maestricht* in 1799, in which he figured four incomplete specimens, three of which he presumed to be from a sea turtle; the other was part of the antler of a quadruped from the Elan family (Fig. 24.4c). The turtle material was later attributed to the carapace, plastron, and paddle of *Allopleuron hoffmani*.

Faujas St Fond noted the similarity of the turtles of Maastricht with living species that swam off the coast of Nouvelle Holland in the Far East. He also commented on the relative abundance of turtle remains in the 'cabinets' in Paris, London, and Haarlem. Three specimens in Paris at that time came from the collection of a M. Roux of Maastricht.

According to Parkinson (1811) Faujas St Fond noted the unique characteristics of the turtle skeletons from Maastricht were due to the fact that:

> *the hard and osseous covering was extended only along the vertebral column whilst the remaining part of the back was covered with coriaceous or horny covering somewhat resembling T. lyra* Linn.

The Nineteenth Century

During the Nineteenth Century, geology became a major science focussed initially on the arguments between the Neptunists and Plutonists which were gradually subsumed by the findings and theories of an ever growing number of geoscientists. Medicine continued to source great scientists such as Parkinson, Mantell, and Owen, whereas Cuvier worked initially as a private tutor before he became the leading comparative anatomist of his time and, to many, the father of vertebrate paleontology. Professional geologists thrived in the service of the Industrial Revolution and the exploration for new resources, and so-called 'gentlemen scientists' prospered and organised themselves into societies and associations (Veneer 2006). The Nineteenth Century also witnessed an ever-increasing role for women in earth science as collectors, illustrators, patrons, and co-authors (Turner et al. 2010).

James Parkinson (1785–1824) was a founding father of the Geological Society of London and a member of the Dining Club at which the idea of a Society was conceived in 1807. He was a medic and his name is forever associated with the dreaded Parkinson's Disease, not because he was a victim but because he was first to observe and record the decline of patients in '*An Essay on the Shaking Palsy*' 1817.

Parkinson ranks with the Gesners, Campers, and Cuviers of this world, as a talented polymath who included an in-depth knowledge of paleontology in his extensive portfolio. His outstanding contribution to paleontology is forever marked by the three volumes of *Organic Remains of a Former World* published in 1811.

The third of these volumes deals with, starfish, echinoids, bivalves, insects, fishes, amphibia, reptilia, and mammalia. The text is presented as a series of letters, and Letter XVIII is concerned with Amphibiolithi—Tortoise—Crocodile. Therein he refers to the all-embracing genus *Testudo* into which he places both marine and terrestrial species. He was aware of the material figured by Burtin (1784) and noted that Lacépède (1789):

> *thought himself authorized in considering them as belonging to Testudo* marina vulgaris, *of Ray or Testudo* mydas *of Linnaeus*

He further noted that:

> *Camper mentions his possessing the entire back of a tortoise four feet in length and 6 inches in breadth, found in in St Peter's Mountain, Maestricht. He speakes also of other remains of the tortoise fond in the same part, and particularly of a fossil, similar to his own, in the Museum of John Hunter. Philos Trans 1786.*

Parkinson was fully aware of material from the Isle of Sheppey and in 1811 noted that although his own collection is poorly represented in terms of Sheppey 'tortoises' there were two or three finely preserved specimens in the British Museum. He also noted that a Colonel Hawker of the 14th Light Dragoons possessed a very perfect specimen, but that the most complete example was with Mr Francis Crow of Faversham, who had allowed Parkinson (1811) to illustrate it as a 'tortoise' skull from Sheppey as plate XVIII, fig. 3.

There is little doubt that Parkinson was a very able paleontologist with a good eye for detail and a capacity for description. At times, however, he could be a little truculent and in disagreeing with a Mr Johnson as to whether or not one of his specimens was the palate of a fish or 'plates' of *Trionyx aegypticus* of Geoffroy St Hilaire, he wrote:

> *I must not conceal from you, that this ingeneious man who possesses these fossils is disposed to entertain a different opinion, and to believe that they are parts of the jaw or palate of some fish. This must remain to be determined by some more illustrative and analogous specimen: until then I shall hold my opinion with diffidence.*

Based on his own illustration of this material in 1811 (pl. XVIII, fig. 1) we suggest that Parkinson was somewhat overconfident in his own ability. That said, his accounts on turtles, crocodiles, and 'the large fossil animal of Maestricht' make fascinating reading from a time of rapid advancement in the field of vertebrate paleontology.

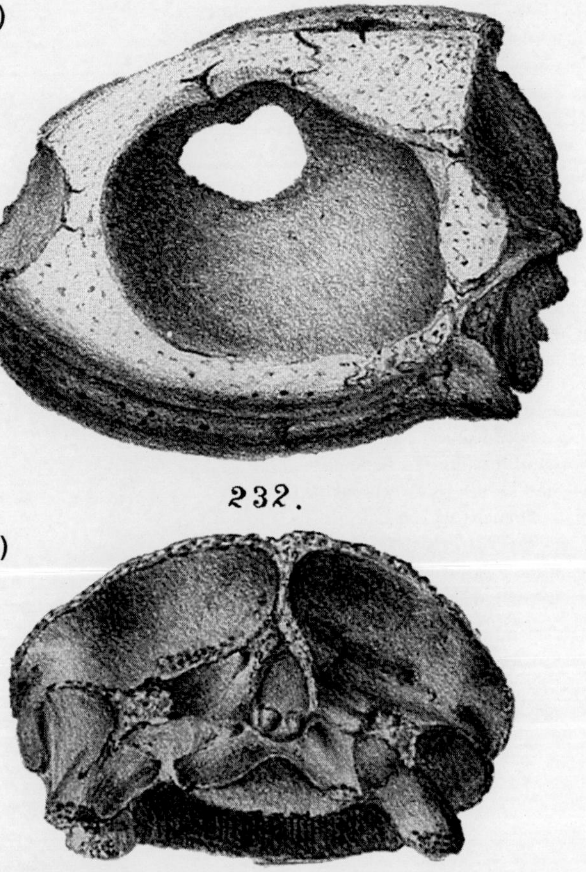

Fig. 24.5 Lateral and posterior views of the skull of *Chelone antiqua* (Pl.XVIII. 232 a and b) as depicted by Konig in *Icones fossilium sectiles* (1820–1825). (Image supplied by Moody)

Carl Dietrich Ebhard 'Charles' Konig (1774–1851) (Fig. 24.6a) was a German paleontologist, born in Brunswick and educated in Gottingen. In 1807, however, he was appointed as assistant Keeper in the Department of Natural History and Modern Curiosities at the British Museum (London); rising to the position of Keeper in 1813. He later became Head of Geology and Mineralogy, retaining that post until he died in 1851.

Konig is also noted for his purchase on behalf of the Museum of the first (claimed) ichthyosaur found by Mary Anning and her brother Richard at an auction in May 1819. The specimen had originally been sold by the Anning family in 1811 for £23 to the Lord of the Manor at Colway, Henry Hoste Chase, but eight years later it was bought on behalf of the British Museum for £45 and five shillings.

Konig described many fossils in the British Museum in a classic work entitled *Icones fossilium sectiles* (1820-1825) and was the first person to allocate specific names to London Clay (Eocene) material with the use of *Chelone (Argillochelys) antiqua* (Pl.XVIII. 232 a and b) (Fig. 24.5).

Georges Léopold Chrétien Frédéric Dagobert Cuvier (1769–1832) (Fig. 24.6b) was thought of as the greatest comparative anatomist of his age, who through sheer brilliance rose from humble beginnings to the title of Baron and a Peer of France.

Rudwick (1997) in his translation and interpretations of the primary texts written by Cuvier reveals that he was fully aware of the various turtle discoveries that had been made in several European countries and indicated that the material could be divided into several species. In his classic work '*Recherches Ossements Fossiles'* (second edition (1824), however, he referred to the Eocene turtles of Sheppey and Bruxelles as:

Les Emydes de l'ile de Sheppey

and

Les Emydes d'environs de Bruxelles

This association with a freshwater environment would influence fellow workers for a long time.

In the *Nouvelle Édition of Recherches sur les Ossemens Fossiles*, Tome Cinquième, IIe Partie, 1824, plate Xl, Cuvier figured the skulls of several living testudines including that of *Chelone (Chelonia) mydas*. He also provided an authoritative description of the cranial characters of each specimen. Plate Xll illustrated the limbs bones, girdles, and hyoid apparatus of the same animals, whereas plate XIII illustrated both carapaces and plastra. These three plates provided emergent paleontologists with the perfect guide to the recognition and classification of extinct species. Plate Xllll included illustrations of a small, almost complete marine turtle from Glaris and various pieces of the skeleton of 'the turtle from Maestricht'. He also illustrated (pl. XV, fig. 13) shell material previously figured by Parkinson (1811) and dorsal and ventral views of the carapaces of turtles that would later be referred to the genus *Puppigerus*. Sadly this work is marred by his association of these fossils with the non-marine emyds [Emydidae].

In the spring of 1830, a great debate raged in Paris involving Cuvier, who divided animals into four distinct, unbridgeable kingdoms or 'embranchements' as founder and leader of the 'functionalists'; believing that anatomical characteristics were exquisitely adapted, by the hand of the Creator, to an organism's "conditions of existence". His opponent was Etienne Geoffroy Saint-Hilaire (1772–1844), a 'formalist' (Waggoner 1996) who considered such adaptations as being of secondary importance when viewed as modifications of a single basic plan, a "unity of type". Cuvier and his British followers including Owen were also described as 'saltationists' (Punctuated Evolution).

It is likely that John Edward Gray (1800–1875) (Fig. 24.6c) was greatly influenced by Cuvier, whom he had met in Paris during his early studies. Gray was born in Walsall, a metalworking town, north of Birmingham in the English Midlands. His father was a pharmacologist and renowned botanist. John Edward studied medicine in

Fig. 24.6 Comparative anatomists and paleontologists of the early Nineteenth Century. **a** 'Charles' Konig (1774–1851) was the first person to allocate specific names to London Clay turtle material. (Sourced from Wikipedia, image in Public Domain). **b** Georges Léopold Chrétien Frédéric Dagobert Cuvier (1769–1832) was regarded as the finest comparative anatomist of his era. (Sourced from Wikipedia, image in Public Domain). **c** John Edward Gray (1800–1875). (Copyright Natural History Museum, London). **d** Gideon Mantell (1790–1852) who named both *Rhinochelys benstedi* and *Cimochelys* (*Chelone*) *benstedi* from the Cambridge Greensand. (Copyright of the Natural History Museum, London)

London, but joined the staff of the Zoology Department at the British Museum in 1824 and was appointed Keeper of Zoology in 1840. He was a contemporary of Thomas Bell (1792–1880), James de Carle Sowerby (1787–1871), and Edward Lear (1812–1888). All four were involved in the writing, illustration, and production of *A Monograph of Testudinata* (Bell 1832), which was finalised by Sowerby et al., in 1872. Gray is reported to have written over a thousand papers, but his dedication to science did not save him from being 'blackballed' by the members of the Linnean Society.

In his publication *Synopsis Reptilium* of 1831, Gray figured material of *Emys camperi* and *Emys parkinsonii*, but sadly his reference to numerous specimens figured by Parkinson (1811) and Cuvier (1824) were not accompanied by detailed descriptions.

According to Donald Smith (pers. comm.), Gray made reference to the families Sphargidae and Cheloniidae in 1825, but strangely gave up this separate grouping in 1831 when he introduced the term 'Cheloniadae' enclosing both *Sphargis* and *Chelonia.*

Education in France in the early 1800s was greatly influenced by the rule of Napolean Bonaparte and the establishment of the Lycee system of secondary schooling (10–16-year-olds) in 1802; with a state curriculum including elementary mathematics and chemistry. In England at this time a call for two years of state education for all, was rejected due to cost and education remained a family matter.

As a result most children of the poorer classes went without, unless they were of the Anglican faith or went to either a parish or a dame school.

Gideon Algernon Mantell (1790–1852) (Fig. 24.6d), the son of a shoemaker, was initially taught in a dame school by an old lady in his home-town of Lewes in Sussex, Southeast England. He was then taught by John Button, a member of the radical section of the Whig Party that supported parliamentary reform, greater access to education, and was against the slave trade. At the age of fifteen Mantell was apprenticed and by great endeavour rose to become a medical practitioner and a member of a group of elite paleontologists and geologists during the first half of the Nineteenth Century. Sadly he was to lead a troubled life and a detailed account of his travails is set out by Deborah Cadbury in her book *The Dinosaur Hunters: A True Story of Scientific Rivalry and the Discovery of the Prehistoric World* (2000). Mantell collected and purchased thousands of specimens for his personal collection and first published in 1814. His work on turtles was limited, but he published *On the Fossil Remains of Turtles, discovered in the Chalk Formation of the South-East of England* in 1841. This was the year that he was both effectively crippled in a carriage accident and when his great antagonist, Richard Owen (1851), criticised his poor attention to detail.

In his 1841 paper, Mantell named *Rhinochelys benstedi* and *Cimochelys* (*Chelone*) *benstedi* from the Chalk (Cambridge Greensand), taxa that have both been subsequently referred to the Chelospharginae of Zangerl (1953).

The vast majority of vertebrate paleontologists are aware of Richard Owen (1804–1892) (Fig. 24.7a) and his skills as an anatomist, comparative anatomist, and vertebrate paleontologist. He was born in Lancaster, England, and attended the local grammar school until he was apprenticed to Leonard Dickson a local surgeon between 1820–1822. After Dickson's death Owen studied under Joseph Seed and Stockdale Harrison until he left for Edinburgh in 1824 to study medicine. He became a member of the Royal College of Surgeons (RCS) in 1826 (Anon 1962) and between 1826–1830 catalogued the collection of John Hunter (Hunterian Collection) at the RCS; during this interval, Owen focused his career on comparative anatomy rather than medicine. Owen was a social climber and in later life a grumpy old man, who prospered for much of his life through nasty, arrogant, and vindictive behaviour.

He made many enemies and was, to many, a stranger to the truth. He lived in a dynamic period, however, when geology and paleontology grew as sciences and when communication and travel became much easier. He first met Cuvier in 1830 and travelled to Paris to attend the debates between Cuvier and Étienne Geoffroy St. Hilaire.

As Hunterian Lecturer at the RCS, Owen talked to a wide audience including royalty and Charles Darwin. In 1834 he became Professor of Comparative Anatomy at St Bartholomews, London and a Fellow of the Royal Society. In 1856 he became Superintendent of the Natural History Department of the British Museum and oversaw the transfer of the collections to South Kensington where he 'ruled' until he retired in 1884. He formed a working relationship with George Bell, but according to one account (Anon 1962) Huxley commented that:

> *The heartburnings and jealousies about this matter are beyond all conception. Owen is both feared and hated, and it is predicted that if Gray and he come to be officers of the same institution, in a year or two the total result will be a caudal vertebra of each remaining after the manner of Kilkenny cats.*

Owen on his own and collaboratively with Bell published a number of major papers on the Cretaceous-Tertiary turtles of England. Skilled observations, coupled with clear text and well crafted illustrations made these publications the definitive source of information on those fossils for more than a 165 years.

In 1841, Owen and Bell drastically changed the adopted status of British Eocene turtles by stating that Cuvier had extended the association of the Sheppey and Harwich material to the Emydes too far and insisted the majority belonged to the group Marina. They also increased the number of species by describing *Chelone longiceps*, *C. breviceps*, *C. convexa*, *C. subcristata*, *C. latiscultata*, and *C. planimentum*. In 1849 they added another five species of sea turtle to those already described from the Isle of Sheppey and Bracklesham, namely *Chelone subcarinata*, *C. crassicostata*, *C. declivis*, *C. cuneiceps*, and C. *trigoniceps*. Van Poelman (1868) continued to follow Cuvier in the assignment of the Eocene turtles to the Emyidae in questioning the separate status of *Emys camperi* and *Emys parkinsonii* and effectively came to the same wrongful conclusion that they were conspecific with *E. parkinsonii* as the senior synonym.

Seago (1979) noted that Owen and Bell (1849) recorded the first material belonging to a giant sea turtle from Sheppey and that by 1861 they had enough material, including a skull to erect the species *Chelone gigas*.

It is worthwhile noting that the detailed publications of Owen and Bell are the only record of material once housed in the Hunterian Collection of The Royal College of Surgeons that was destroyed by bombing in World War II.

In 1889, Owen decided that *Chelone gigas* was closer to the cheloniids than to the dermochelyids due to the shell structure and the fact that skulls of *C. gigas* lacked the descending processes of the premaxillae seen in *Dermochelys*.

The classic essay by Edward Drinker Cope (1840–1897) (Fig. 24.7b) on the *Extinct Batrachia and Reptilia of North America* published in 1870 represented the next major step in the history of Eocene turtle research, for in that publication he erected several generic names that became very important in the synonymy of the cheloniids worldwide.

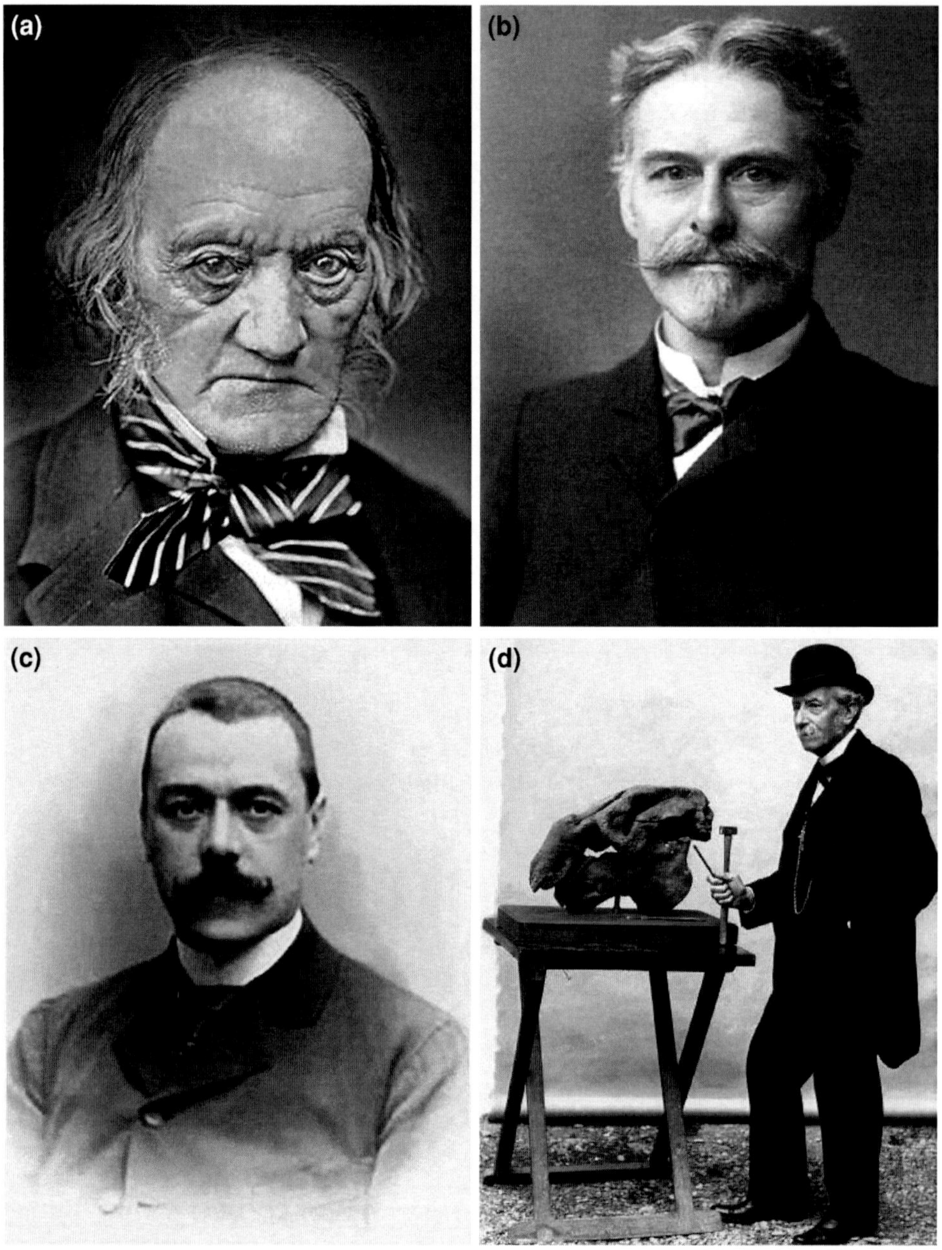

Fig. 24.7 Comparative anatomists and paleontologists of the mid-late Nineteenth Century. **a** Richard Owen (1804–1892) regarded by many as one of the greatest comparative anatomists of his day. This portrait captures him as an old man, possibly aware of the ill-deeds he perpetrated on others. (Copyright of the Natural History Museum, London). **b** Edward Drinker Cope (1840–1897). (*Source* From Wikipedia, image in public domain). **c** Louis Antoine Marie Joseph Dollo (1857–1931). (*Source* From Wikipedia Commons, image in Public Domain). **d** Giovanni Capellini (1833–1922) was a brilliant, but somewhat arrogant geologist, paleontologist, curator, and teacher. (Image sourced from Federico Fanti)

The names *Lytoloma*, *Puppigerus* and the earlier defined *Euclastes* were all used extensively by authors such as Dollo and Lydekker to receive the species originally erected by Owen and Bell. In 1886, Cope wrote a paper critiquing Dollo's work on extinct tortoises, and provided further data on Cretaceous turtles of New Jersey in 1871.

Tiberius Cornelius Winkler (1822–1897) was also someone who could be described as a polymath. He was born into relatively humble family. His father apprenticed him to a grain merchant but Tiberius was driven to succeed—he learnt several languages and after he married in 1844 he studied medicine and graduated as a surgeon in 1852 from Haarlem. His professional interests were in anatomy and zoology, but he was to become a renowned natural historian, paleontologist, and geologist. He accepted the post of Curator of the Teyler Museum fossil and mineral collections ('Cabinets') and the year before his death in 1897 he had catalogued over 15,400 fossil specimens and the museum mineral collection using a numerical system. He wrote many articles popularising natural history and also published a detailed description of *Allopleuron hoffmani* in 1869. Winkler was also attributed with the task of translating Darwin's *Origin of Species* into Dutch by 1860.

Johan Casimir Ubaghs (1829–1894) a contemporary of Winkler also wrote several papers on the species *Chelone hoffmani* including a description of the skull in 1888.

He was born in Aachen in Germany, across the border from Maastricht, and although linked with one of Europe's great families he is poorly known historically and is simply described as a malacologist who referred the teeth of a mosasaur to *Megalosaurus bredai*.

Louis Antoine Marie Joseph Dollo (1857–1931) (Fig. 24.7c) was born in France where he graduated as a Civil Engineer in 1877. Then he emigrated to Belgium where, in 1878, he oversaw the collection, preparation, and articulation of the Bernissart iguanodonts. Dollo was highly regarded for his prolific publication record, and for the fact that he arranged and labelled the unique collection of fossil vertebrates in the then new museum on Parc Leopold in 1905.

Dollo wrote a number of definitive papers on the Eocene turtles of Belgium in 1886, 1887a, 1887b, 1903, 1907, 1909, and 1923 referencing genera such as *Pachyrhynchus* (later *Erquelinnesia*), *Euclastes*, *Eochelone*, *Lytoloma, Puppigerus*, and *Psephophorus*. In 1893, he also established Dollo's Law, also known as the *Law of Irreversible Evolution.*

Dollo's acknowledgement of the value and validity of Cope's contribution to the study of fossil turtles is an indication of the growing status of vertebrate paleontology and the spread of the written word through the publication of an ever growing number of journals and periodicals.

One of the first paleontologists to fully exploit the expansion of geological knowledge and the growth of the earth sciences in universities and museums was Giovanni Capellini (1833–1922) (Fig. 24.7d). He was born into a lower middle class family in La Spezia (Liguria, North West Italy) and was initially encouraged to follow a career as a musician and then to take up a monastic life. He was forced, however, to leave the priory where he worshiped in 1854, at the age of 21, on the death of his father. From then on he had to support himself and his family, and he worked variously as a bookbinder, a tutor at a college in La Spezia, and as a repairer of mechanical devices. He was brilliant and devoted to the natural sciences, and attended university in Pisa where he had the opportunity to demonstrate both his academic ability and personal skills (Fanti 2010). To some Capellini was an agreeable person, who rapidly gained recognition, funding, letters of introduction, and promotion and honours from various institutions and academies. To others he appeared arrogant and somewhat devious, an intellectual who became a major player in the politics associated with the development of the Italian Geological Survey (Corsi 2003). He was the first intellectual from his social background to enter the realm of the natural sciences in Italy, his contemporaries being from the rich and famous of the day (Vai 2003).

After obtaining his first degree in Pisa in 1857, Capellini began his career as a field geologist. Throughout his career he sought to establish long-lasting personal and scientific relationships with the leading scientists of the day, including geologists, zoologists, and archaeologists, at home and abroad (Fanti 2010). At the age of 27, he was honoured by a visit from Charles Lyell (1797–1875). One of his long-standing friendships was with Louis Agassiz (1807–1873), who greatly influenced his career as an outstanding paleontologist, curator, and teacher.

Fig. 24.8 Richard Lydekker (1849–1915). Geologist, naturalist and author numerous books and catalogues, including a joint paper with G. A. Boulenger in 1887 and his *Catalogue of Fossil Reptilia and Amphibia* (1889c). (Copyright of the Natural History Museum, London)

Capellini wrote numerous papers on fossil mammals, fossil birds, invertebrates, and plants. He published two definitive papers in 1884 and 1897 on *Protosphargis veronensis* (Capellini) from the Late Cretaceous near Sant'Anna di Alfaedo in Valpolicella.

Richard Lydekker (1849–1915) (Fig. 24.8) was born in Harpenden in England and attended Trinity College in Cambridge; graduating in 1871 in second place with a first class degree in The Natural History Tripos. The early part of his career was spent with the Geological Survey of India (1874–1882) and his beautifully illustrated publications of that time have continued to increase in value over the decades.

He was a remarkable geologist, naturalist, and a prolific author who collaborated with Sir Henry Flower, who had succeeded his sworn enemy Richard Owen as Director of the Natural History Departments at the British Museum in 1884.

In 1889 Lydekker wrote several papers on Eocene and Cretaceous chelonia and the skull of *Lytoloma* in 1889a, b, as well as the *Catalogue of Fossil Reptilia and Amphibia* (1889c) which is fundamental reading for anyone working on the collections of the NHM, nevertheless up to 30% of his material was wrongly referred to given genera and species (R. T. J. Moody, unpublished data). Lydekker's 1889 Catalogue was to change the list of species for years to come, including:

Emys camperi
Argillochelys antiqua
Argillochelys cuneiceps
Argillochelys subcristata
Argillochelys convexa
Lytoloma longiceps
Lytoloma trigoniceps
Lytoloma crassicostatum
Lytoloma planimentum

George Reber Wieland (1865–1953) was also a paleontologist of great renown. He was born in Boalsburg, Pennsylvania, USA, and studied at Penn State College, the University of Göttingen, the University of Pennsylvania, and then at Yale where he obtained his PhD in 1890. Although he originally wanted to study vertebrate paleontology, he is best known for his outstanding research on fossil cycads and the discovery of the petrified cycad forest of South Dakota. He was obviously a formidable field geologist who found and described several turtles, including the giant *Archelon* (1900, 1902), *Toxochelys* (1902), *Lytoloma* (1904a), *Osteopygis* (1904b), and *Protostega* in 1906 and 1909. His work on *Lytoloma* supported the findings of both Lydekker (1889c) and Dollo (1903), referring this genus to the Cheloniidae. However, unlike Dollo he referred *Lytoloma* to the sub-family Cheloninae. The famous photograph of Wieland and *Archelon* (Fig. 24.9) is simply a masterpiece; a calling card for all nascent paleontologists. The global stature of Wieland was acknowledged when he received a medal and gave a dedicatory address at the new Aldovandrus (now Capellini) Geological Museum in Bolgna in 1907 at the invitation of Giovanni Capellini.

The Twentieth Century: A Revolution in Vertebrate (Turtle) Paleontology

Rainer Zangerl (1912–2004) was born in Winterhur in Switzerland and did his doctoral research under Bernhard Peyer (1885–1963) at the University of Zurich. In 1937, at the age of 25 he emigrated to the USA. Most of his working life was spent at The Field Museum of Natural History in Chicago where he specialised in sharks and turtles. He was truly meticulous in his presentation of data and introduced new methodologies to the study of comparative anatomy (1948). His work on the Protostegidae and Toxochelyidae (Zangerl 1953a, b) set the parameters for workers outside America and his note (Zangerl 1953b, p. 146) that:

> *The various species from the London Clay, referred to Lytoloma Cope by Lydekker (1889), will have to be reassigned, but to do this from the literature would I fear, merely result in a further complication of the already confused taxonomic status. A restudy of the European material seems necessary to determine the generis allocation of these species.*

was inspirational to generations that followed.

From a European perspective, Zangerl's work on the turtles of the Erquelinnes sands (Landenian) of the Hainault region of Belgium published in 1971, was extremely relevant. Moody (1970) was greatly influenced by the work of Zangerl and had concluded that the material referred to *Erquelinnesia gosseleti* and *Lytoloma planimentum* were essentially conspecific and diagnostic of the toxochelyids. Zangerl had studied the Bruxelles material in the mid 1960s and discussed the phylogenetic relationships of Cretaceous-Paleogene sea turtles with Moody and Cyril A. Walker (1939–2009) during a visit to the BMNH in the late 1960s. Zangerl's (1980) paper on *Patterns of Phylogenetic Differentiation in the Toxochelyid and Cheloniid Sea Turtles* proposed that the cheloniids of the Cretaceous, Paleogene, Neogene, and modern day formed distinctive stratigraphical groupings with the taxa present in each grouping exhibiting a greater similarity with each other than with genera of older or younger timeframes. The suggestion being that this phylogenetic differentiation was considered to be the result of repeated episodes of specialization from a conservative ancestral stock.

After Lydekker and Zangerl, the marine turtle collections of Europe gathered dust or, in the case of the Belgian protostegid material, rotted in the humid tropical basement of Institut Royal in Bruxelles, until Dick Moody began to publish on Eocene cheloniids in 1968. Moody was based at the then Kingston College of Technology, which subsequently morphed into Kingston Polytechnic and lastly Kingston University London. He is a geologist-paleontologist by training and his interest in marine turtles began with a fortuitous discovery of well preserved specimen during a student fieldtrip to the Isle of Sheppey. Seeking more information Moody visited the Natural History Museum, or BMNH as it was called then, where he met up with a young Cyril Walker. Walker was a mine of information on living and fossil turtles and a life long friendship was launched in a local hostellery. Over the next 3 years Moody visited

Fig. 24.9 *Archelon ischyros* Wieland, 1896 [holotype] YPM 3000. George R. Wieland standing beside *Archelon* skeleton (mount finished in 1907) in the old Peabody Museum building, 03 November 1914. (Photograph by F. H. Simonds, Nove, Courtesy of the Peabody Museum of Natural History, Yale University, New Haven, Connecticut, USA)

museums throughout Europe, but most of his studies were undertaken in London and Bruxelles. Fortunately the Eocene turtles of Belgium are mostly found in sand-rich sediments that lack a pyrite cement and the fossils are beautifully preserved. As noted above, the same could not be said for protostegid material from the Oligo-Miocene clays, as the destructive oxidation of pyrite cement was to take a terrible toll on those fossils

The Belgium material included a number of almost complete, articulated skeletons, as well as excellent isolated skulls, shells, and limbs. In contrast, the UK material collected from Sheppey, Bracklesham, and the Harwich area of southeast England was comprised of a large number of isolated skulls and shells often preserved in a fine grained, indurate, calc-siltstone characteristic of London Clay septaria. The preparation of this material using air abrasive tools and acid became an integral and essential part of this study.

In 1968 Moody referred the specimen discovered near Warden Point on the Isle of Sheppey to the new genus *Eochelys* and the species *crassicostata* (Owen). The new genus reflected the author's lack of confidence in the taxonomic validity of the genera *Lytoloma* and *Puppigerus*. This uncertainty was also evident in his 1970 doctoral thesis (unpublished) on Eocene marine turtles, in which he referred *Emys parkinsonii* (Gray), *Lytoloma bruxelliensis* (Dollo), *L. wemelliensis* (Dollo), *Chelone longiceps* (Owen), and *C. trigoniceps* (Owen) all to the species *Eochelys camperi* (Gray). By 1974 *Eochelys* together with the subfamily Eochelyinae also containing *Argillochelys* and *Eochelone* were discarded following Moody's detailed review of the taxonomy and morphology of *Puppigerus camperi*.

Moody was able to give a very detailed account of the skull and braincase of *Puppigerus camperi* based on the type specimen IRSNB I.G.8402. This in conjunction with other postcranial characteristics proved that *P. camperi* and *P. crassicostata* were indisputable representatives of the Cheloniidae and together with *Argillochelys cuneiceps* and *A. antiqua* were considered to be ancestral to modern day genera.

Initially the general belief was that the various cheloniid and toxochelyid species present in the northwest European Anglo-Belgium Basin were very restricted in terms of their paleogeographical distribution. Moody (1980a,b, 1993, 1996, 1997) inferred that the late Maastrichtian seas of Holland provided ideal feeding sites for hordes of *Allopleuron hoffmani* and that the Bruxelles and Sheppey areas, during the Eocene, were ideal for the likes of *Puppigerus camperi*, *P crassicostata*, *Argillochelys*

Table 24.1 List of dermochelyid turtles identified by Seago (1979) in his review of fossils from the family Dermochelyidae

Genus and species	Date	Distribution
Psephophorus polygonus Von Meyer	1847	Miocene of Austria
Psephophorus pseudostracion (Gervais)	1848	Miocene of France
Psephophorus scaldii (Van Beneden)	1871	Pliocene and Miocene Belgium
Psephophorus oregonensis Packard	1940	Miocene of North America
Psephophorus rupeliensis (Van Beneden)	1883	Oligocene of Belgium
Psephophorus eocaenus Andrews	1901	Eocene of Egypt
Psephophorus calvertensis Palmer	1909	Miocene of' North America
Cosmochelys dolloi Andrews	1919	Eocene of Nigeria
Eosphargis gigas (Owen)	1861	Eocene of England and Belgium
Eosphargis breineri Nielsen	1959	Eocene of Denmark
Protosphargis veronensis Capellini	1884	Cretaceous of Italy

cuneiceps, *A. antiqua*, *A.* athersuchii (Moody 1980a), *Eochelone brabantica* (see Casier 1968), and the toxochelyid *Erquelinnesia gosseleti*.

Since then Averianov (2005) and Tong (2009) have described new occurrences of *Puppigerus* from the Eocene of Uzbekistan and Morocco. Tong and Hirayama (2008) also recently published on a new species of *Argillochelys*, namely *A. africana*, from Morocco and on the 'first' Mesozoic African dermochelyid in 2004. Moody (1972, 1997) had indicated that the skull of *Thalassochelys testei* Bergounioux (1952, 1959) from the Eocene phosphates of Tunisia was closely related to *Eosphargis*, thus demonstrating that the paleogeographic distribution of Eocene turtles was far greater than previously defined.

Unfortunately neither Seago (1979) nor Smith (1989) published any of the results documented in their respective doctoral theses. However, copies of their theses have been available for consultation in the Paleontology Department at the NHM London and other museums worldwide. In addition brief summaries of their findings were presented by Moody (1997). Publishers accept the use of unpublished theses in reference to original research and copies of both Seago and Smith's theses can be obtained as pdf's from the University library at Kingston University London. Their doctoral theses were examined by peer review and both have made significant contributions to our science. Therefore their original work is rightly documented in this publication that above all is devoted to the presentation of a historical perspective.

It is unlikely that all the people working on turtles, tortoises and terrapins will truly make it into history, but many will be referenced through an ever-growing digital media. After the Second World War, people worldwide had the right of access to both primary and secondary education and post 1950 the percentage of families with their very first university graduate grew dramatically. Patronage has been replaced by grants and scholarships and the number of paleontologists from middle and working class families (UK class system) is testament to the gift of education and that the number of students entering university annually has changed from 2 to 50%. Suffice to say that Richard Moody's father was a train driver of the Great Western Railway—a company that Gene Gaffney, with his lifelong interest in trains, would be well aware of!

Alan Seago (1979) carefully documented the history of the fossil dermochelyids and in his thesis recognised the taxa listed in Table 24.1 prior to his revision of known species.

The majority of the European dermochelyid material is from the Paleogene-Neogene deposits of Belgium, although the first records were somewhat unscientific in that De Serres et al. (1839) attributed the first material of a fossil dermochelyid to the genus *Ostracion* which is an acanthopterygian fish having a box-like body covered in polygonal plates. In 1872 Gervais referred these specimens to *Sphargis pseudostracion*. Von Meyer (1846) saw only disarticulated plates of the shell of *Psephophorus polygonus* (Fig. 24.10) when he initially named this taxon from the Miocene sands of Neudorfl in Czechoslovakia. Subsequently in 1847 he received detailed drawings of the specimen from a Custos Partsch and published a note in the Jahrbuch in 1847 (p. 579). A review of the type material was later undertaken by Seeley (1880).

With regard to Seago's (1979) revision of *Eosphargis*, Owen and Bell (1850) referred the proximal end of a large femur (plate 28, figure 5) from the London Clay of Sheppey to *Chelonia mydas* and in 1889 Owen assigned a skull, scapula, coracoid, eroded humerus, and fragments of the carapace (presented to the British Museum Natural History in 1858), to the new species *Chelone gigas*. A new skull and postcranial material was described by Owen (1880), but it was Lyddeker (1889c) who erected the genus *Eosphargis* to receive the specimens referred by Owen and by Owen and Bell to *C. gigas*. Dollo (1907) reported the first nearly

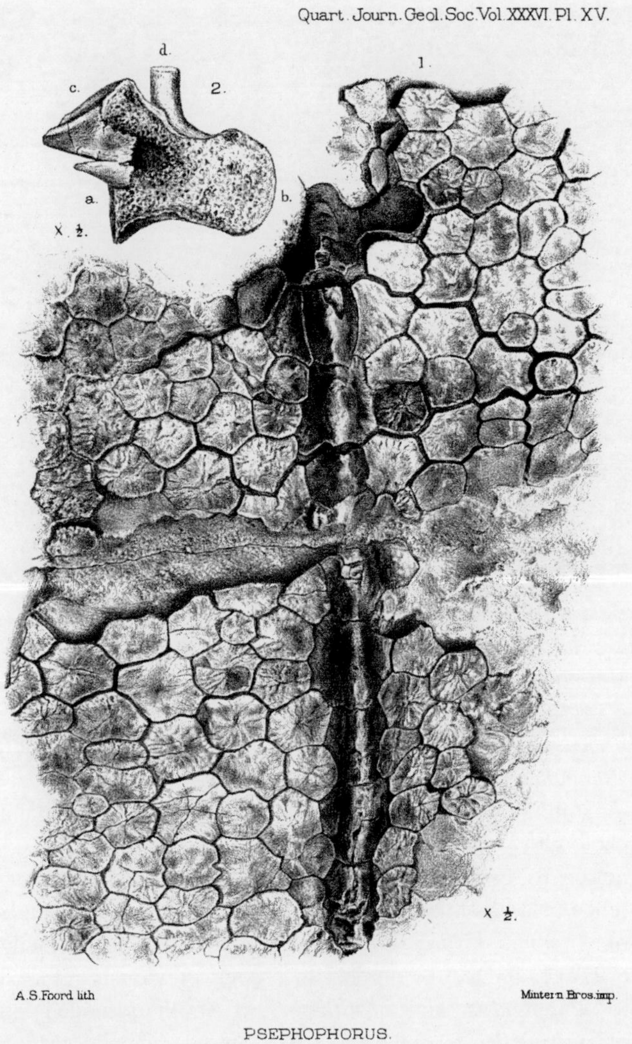

Fig. 24.10 Detailed drawing of the shell of *Psephophorus polygonous* Meyer drawn by A.S. Foord in Seeley (1880, plate XV). (Image printed with permission of the Geological Society, London)

complete skeleton of *Eosphargis* from the lower Eocene clays of Quenast, but the specimen has never been fully described although Nielsen (1958, 1963) briefly compared it with the beautiful example of *Eosphargis breineri* he uncovered in 1957, from the Mo Clay (Ypresian) of Für in Demark. A brief note on the Belgian *Eosphargis* also was presented by Quintart and Plisnier-Ladame (1968).

The upper Oligocene (Rupelian) deposits of Belgium were associated with the discovery of material thought to belong to the genus *Sphargis* (Luth), which is a subjective synonym of *Dermochelys.* The material was attributed by Van Beneden (1883a) to a creature "of which the living form is remarkable in its organisation and world-wide distribution". Also in 1883, Van Beneden (1883b) described additional remains which he referred to the new species *Sphargis rupeliensis*. Van Beneden's original material is sadly missing, but new material from The Argile de Boom at Steendorp was described by Dollo (1886), who in the next year (Dollo 1887a, b) referred the species to the genus *Psephophorus*; Seago (1979) referred the same species to *Cardiochelys*.

Seago's (1979) thesis included a complete study of the osteology of the extant *Dermochelys coriacea* and recognised the subdivision of the Dermochelyidae into the Eospharginae and Dermochelyinae. The main reasons for this were that the eosphargines had a partial primary palate, a slightly emarginated skull, thecal dermal armour, and a primitive cheloniid flipper and hind limb.

The Eosphanginae was thought to include *Eosphargis gigas*, *Eosphargis breineri*, and possibly *Thalassochelys testei* (see Moody 1997).

Seago (1979) referred *Cosmochelys dolloi*, *Cardiochelys rupeliensis*, *Cardiochelys ?eocaenus*, *Psephophorus polygonus*, *P. ?scaldii*, *P. ? oregonensis*, and *Miochelys fermini* to the Dermochelyinae.

Based on a study of the cranial arterial system Seago (1979) recognised that there was a link between the cheloniids and dermochelyids. He noted that there was a strong resemblance between the 'primitive' flippers of the eosphargids and *Desmatochelys*, suggesting that the former had branched off from a cheloniid lineage during the Cretaceous.

Seago (1979) also noted that much of the basiphenoid in the dermochelyids remain unossified, highlighted distinct similarities between *Desmatochelys*, *Corsochelys*, and *Protosphargis veronensis*, and considered the last species to be a specialised cheloniid. More recently, Hirayama (1994) referred *Corsochelys* to the Dermochelyidae claiming that like *Dermochelys*, *Corsochelys* retained the ossification of the *rostrum basiphenoidale*. He also erected the Dermochelyoidae to house both the Dermochelyidae and Protostegidae.

The shell characteristics of the dermochelyids are fascinating and Seago (1979) gave considerable space to the study and evolution of an 'abnormal condition'! The epithecal armour of the Dermochelyines is comprised of a mosaic of hundreds or even thousands of small to relatively large plates or osteoderms (Rieppel 2001). The development of this extreme aquatic specialisation and the presence of 'blubber' in the extant *Dermochelys coriacea* has enabled the animal to dive to great depths and roam the oceans from ice sheets to the Equator. In *Dermochelys* the epithecal carapace is characterised by the presence of seven keel structures that also exist in *Cosmochelys* and *Psephophorus* (Seago 1979).

Deraniyagala (1932, 1936) studied the embryology and growth of *Dermochelys* and recorded the development of epithecal ossifications in 157-day old animals. By that stage of development, epithecal ossifications occur in rows that

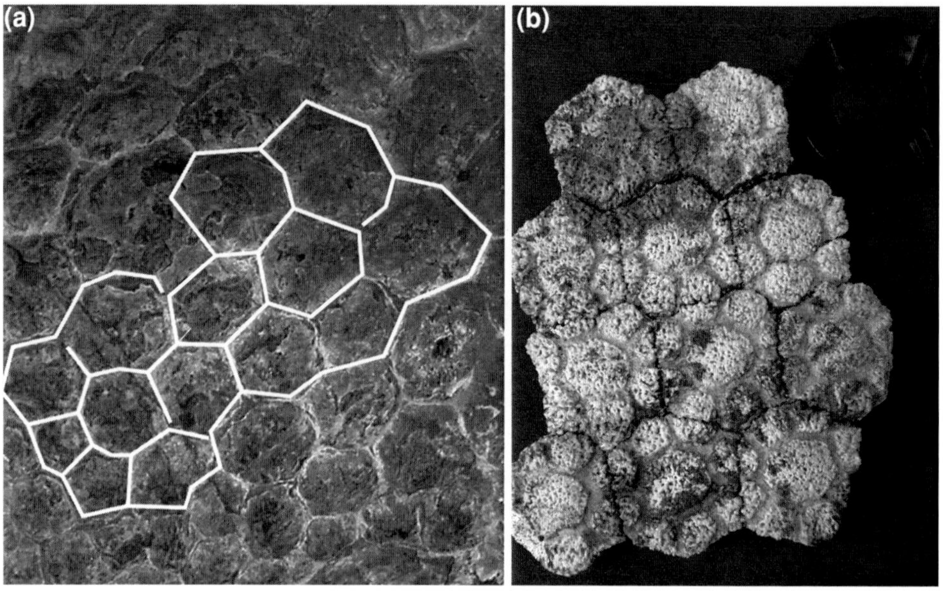

Fig. 24.11 The rosette-style pattern of the epithecal carapace in *Psephophorous* is every similar to those found in **a** the skin plate pattern of the dinosaur *Pelorosaurus* and **b** the armour of the mammal *Glyptodon*. (Images by R. T. J. Moody)

correspond with the keels in older animals. *Dermochelys* and *Psephophorus* exhibit a distinctive arrangement of the interlocking, epithecal plates that form the carapace. This was discussed in detail by Wood et al. (1996) and Wood and Moody (1998), and was used to argue that '*Psephophorus rupeliensis*' constituted a new genus. The predominant pattern or arrangement between the keels is for five, six-sided plates encircling a similar sized central plate. The keels, however, frequently exhibit a larger, multifaceted central plate surrounded by up to ten or eleven smaller plates. It is also common to observe several elongate, central plates forming a feature similar in style to a 'neural' ridge. This rosette development of the epithecal carapace is very similar to those found in the skin plate pattern of the dinosaur *Pelorosaurus*, the armour of the mammal *Glyptodon*, and the scales covering the legs of the tortoise *Geochelone gigantea*, the body of stump-tailed skink (*Tiliqua rugosa*), and the limbs of the giant komodo dragon (*Varanus komodoensis*)! Surprisingly it is also observed on the thick skins of the bread fruit (*Artocarpus altilis*) and durian (*Durio zibethinus*). One suspects that these are analogies rather than homologies; facilitating the solution of a common problem with size and growth. Seeley was aware of this growth phenomenon when he undertook his review of *Psephophorus polygonous* in 1880 (Fig. 24.11).

Smith (1989) set out to identify and evaluate the cranial characteristics of three families of sea turtles: the Cheloniidae and Dermochelyidae (of Gray 1825), and the Desmatochelyidae (of Williston 1894). The genera and species included in his study were the living *Chelonia mydas mydas*, *C. mydas japonica*, *Lepidochelys olivacea olivacea*, *L. olivacea kempi*, *Eretmochelys imbricata*, *Caretta caretta*, *Natator depressa*, and *Dermochelys coriacea*. Fossil taxa included *Puppigerus camperi*; *Eosphargis gigas*, *E. breineri*, *"Miochelys" fermini*, *Cardiochelys rupeliensis*, *Psephophorus oregenensis*; *Desmatochelys lowi*, *Rhinochelys pulchriceps*, *R. elegans*, *R.cantabrigensis*, *Notochelone costata*; *Toxochelys moorevillensis*, *Eochelone brabantica*, *Erquelinnesia gosseleti*, *Ctenochelys tenuitesta*, and *C. arcris*.

Smith's (1989) bone-by-bone description of living cheloniids revealed a large number of diagnostic, taxonomically useful characters. He commented on the occurrence of a great deal of intraspecific variation in the skulls of *Chelonia mydas* a factor that earlier had led Moody (1974) to reduce the number of species previously referred to *Puppigerus* and *Argillochelys*. Smith provided an exhaustive description of the basisphenoid and cautioned on the use of specific characters that are modified by ageing. In 1989, Smith concluded that there was little cranial evidence to separate the living Cheloniidae into the Cheloniinae and Carettinae and that the cranial characteristics of the dermochelyids suggested that they are an extremely conservative group. He noted from his study of the arterial system that the Desmatochelyidae differed from the cheloniids and dermochelyids in that the palatine artery is always smaller than the carotid, the opposite being evident in the other two families. From this observation, he concluded that the Desmatochelyidae were a stable grouping inclusive of *Desmatochelys*, *Rhinochelys*, and *Notochelone*. Seago (1980) additionally discussed the status of *Protosphargis veronensis* and noted several similarities with *Desmatochelys lowi*,but remained noncommittal about its relationship because of its greatly reduced shell. Collins (1970), like Zangerl and Sloane (1960), placed *Rhinochelys* within the Cheloniidae. Smith (1989), however, proposed that the presence of well developed nasals and a small posterior palatine foramen and the absence of a secondary palate were characteristics that place it alongside *Desmatochelys* and

Notochelone. Hirayama (1994, 1997, 1998) subsequently referred *Rhinochelys*, *Notochelone*, and *Desmatochelys* to the family Protostegidae and inferred it was closely related to the earliest known sea turtle *Santanachelys*.

Other Contributions

It is always foolish to judge the contribution of individuals on the number rather than the quality of their publications and it would be remiss of us to exclude certain recent works from this overview. The compilation of papers on Late Cretaceous tetrapods of the Maastricht area of Holland by Mulder (1996, 2004) provides a definitive description and history of *Allopleuron hoffmani*. This work is an excellent data source; providing a valuable photographic record of the quantity and quality of skeletal material referred to this magnificent example of a species that was adapted to a life in the open sea and fully equipped to 'fly beneath the surface'. Bardet and Jagt (1996) have also added to our knowledge of the material found in the Maastricht chalks over the past 200 years.

Europe and Russia have produced a number of dedicated turtle, tortoise, and terrapin researchers during the past 50 years or so, but mostly their works have been dedicated to non-marine taxa. France Lapparent de Broin, Emiliano Jiménez Fuentes, and Lev Aleksandrovich Nessov (1947–1995) have written voluminous descriptions of many testudinates and have contributed greatly to our understanding of turtle evolution and the paleogeographic distribution of given genera. Nessov (1987) also described the Paleogene sea turtles of southern Kazakhstan and detailed the phylogenetic differences between the constituent toxochelyids and cheloniids. Jérémy Anquetin, Andrew Milner, Walter Joyce, and Vladimir B. Sukhanov have written significant papers on stem, crown group and archaic testudinates. Alexander (Sasha) Averianov and Haiyan Tong (see Tong and Hirayama 2004) have expanded on the link between European and North African faunas with descriptions of new species such as *Puppigerus nessovi* (2005) and *Argillochelys africana* (2008) from Uzbekistan and North Africa, whereas Tong et al. (in press) have recognized the presence of *Puppigerus camperi* in Morocco. Haiyan Tong has interests in *Rhinochelys*, the argillochelyids, and dermochelyids and appears to have adopted the mantle of research into Euro-African marine turtles. Most recently Danilov et al. (2010) described the new genus and species *Itilochelys rasstrigin* as a cheloniid from the early Paleocene of the Volograd Province of Russia. Chesi et al. (2007) recorded the presence of *Psephophorus polygonus* Meyer and *Trachyaspis lardyi* Meyer from the Miocene of Pietra Leccese (late Burdigalian-early Messinian), southern Italy. They also made comparisons between the sea turtle fauna of southern Italy with that of the Calvert Formation of North America, adding yet another data point for mapping the paleodistributions of marine turtles.

No review of the history of turtle research would be complete without reference to Eugene Gaffney, without whom the 'movement' into turtle research would have probably been delayed for decades. Gene obtained his doctorate from Colombia University in 1969. In 1972 he wrote an illustrated glossary of turtle nomenclature and by 1975 he had published his classic paper on '*A phylogeny and classification of the higher categories of turtles*' and set new standards in the research into a group of almost forgotten reptiles. In those days people refused to believe that grown-ups could study fossil turtles for a living, let alone get paid for it. With time Gene became a driving force who greatly influenced and inspired a small population of turtle workers and helped pave the way for the introduction of cladistics to paleontology on a global scale. His paper on '*Comparative cranial morphology of Recent and fossil turtles*' (1976) was an outstanding contribution to our science. During the last 40 years Gene has fostered links with workers all over the world and has published several influential papers with Peter Meylan (e.g., Gaffney and Meylan 1988). In turn Meylan, has taken our science a step further with work on molecular phylogeny (Meylan and Shaffer 2001; Shaffer et al. 1997). The influence of Gene Gaffney will last decades, with turtle science having gathered sufficient members to afford one or two opposing groups. The future looks bright.

Acknowledgments The extant authors would like to thank Cyril Walker for his companionship and dedication over several decades. His contribution to vertebrate research has been noted in several obituaries and by the organization of a symposium at the Bristol joint meeting of the SVP/SVPCA in 2009. He was always considerate and very helpful to visitors to the reptile collections in Store Room 41 in the old basement complex of the BMNH/NHM. He was a great friend and companion. Thanks also to Alan Seago for his contribution to this work and to Ross Sandman for his comments and suggestions that have helped improve this manuscript. We are indebted to Ren Hirayama, Benjamin Kear, and Andrew Milner for their review and constructive commentary on this chapter and the Permissions Section of the Yale/Peabody Museum for supplying the image of George Wieland and *Archelon*. Wendell Wilson of the Mineralogical Record provided valuable information on copyright related to the illustrations by Buc'hoz. Lastly we would like to thank Don Brinkman for his encouragement with this paper—especially as it presents a geohistory perspective on our research and thus differs markedly from the many papers written in honor of Gene Gaffney and a career devoted to our science.

References

Anon, (1962). The conservators of the Hunterian Museum: II. Richard Owen. *Annals Royal College of Surgeons England, 30*, 117–124.

Averianov, A. O. (2005). A new sea turtle (Testudines: Chelonidae) from the Middle Eocene of Uzbekistan. *Paleontological Journal, 39*, 646–651.

Bardet, N., & Jagt, J. W. M. (1996). *Mosasaurus hoffmanni*, le "Grand Animal fossile des Carrières de Maestricht": Deux siècles d'histoire. *Bulletin du Muséum national d'Histoire naturelle Paris, 4*, 569–593.

Bell, T. (1832–1836). *A monograph of the Testudinata* (8 parts). Samuel Highley: London.

Bergounioux, F. M. (1952). Cheloniens fossiles des phosphates de Gafsa. *Notes et Memoires de Service des Mines et le Carte Geologie du Maroc, 92*, 375–396.

Bergounioux, F. M. (1959). Cheloniens fossiles des phosphates de Gafsa. *Annales des Mines et de la Geologie de Tunisie, 15*.

Boulenger, G. A., & Lydekker, R. (1887). Notes on Chelonia. *Geological Magazine, 3*, 270–275.

Buc'hoz, P. J. (1781). *Centuries de planches eluminees et non illumines, 1778–1781*. Paris: Chez Lacombes.

Burtin, F. X. (1784). *Oryctographie de Bruxelles, ou Description des Fossiles Tant Naturels qu'accidentels déscouverts jusqu'à ce Jour dans les environs de cette Ville*. Bruxelles: L'imprimerie de Le Maire.

Cadbury, D. (2000). *The dinosaur hunters: A true story of scientific rivalry and the discovery of the prehistoric world*. London: Fourth Estate.

Camper, P. (1786). Conjectures relative to the petrifactions found in St. Peter's Mountain near Maestricht. *Philosophical Transactions, of the Royal Society of London, 76*, 443–456.

Camper, P. (1800). Lettre de A. G. Camper à G. Cuvier sur les ossemens fossiles de la montagnede St. Pierre, à Maëstricht. *Journal de Physique, de chimie et d'histoire naturelle et des arts, 51*, 278–291.

Camper, P. (1812). Mémoire sur quelques parties moins connues du squelette des sauriens fossiles de Maëstricht. *Annales du Muséum d'Histoire Naturelle, Paris, 19*, 215–241.

Capellini, G. (1884). Il chelonio veronese (*Protosphargis veronensis*, Cap.) scoperto nel 1852 nel Cretaceo superiore presso Sant'Anna di Alfaedo in Valpolicella. *Memoire della Classe di Scienze Fisiche, Matematichie, e Naturali Roma, 3*, 291–320.

Capellini, G. (1897). La piastre marginali della *Protosphargis veronensis*. *Rendiconto delle Sessioni Ordinarie dell'Accademia delle Scienze dell'Istituto di Bologna, 2*, 97–113.

Casier, E. (1968). Le squelette cephalique de *Eochelone brabantica* L.Dollo, du Bruxellien (Lutetien Inferieur) de Belgique, et sa comparaison avec celui de *Chelone mydas* Linne. *Bulletin De l'Insititut Royal Des Sciences Naturelles De* Belgique, *44*, 1–22.

Chesi F., Delfino M., Varola A., & Rook L. (2007). Fossil sea turtles (Chelonii, Dermochelyidae and Cheloniidae) from the Miocene of Pietra Leccese (late Burdigalian-early Messinian), Southern Italy. *Geodiversitas, 29*, 2, 321–333.

Collins, J. I. (1970). The chelonian *Rhinochelys* Seeley, from the Upper Cretaceous of England and France. *Palaeontology, 13*, 355–378.

Cope, E. D. (1870). Synopsis of the extinct batrachia, reptilia and aves of North America. *Transactions American Philosophical Society*, N. S., *14*, 1–252.

Cope, E. D. (1871). Extinct tortoises of the Cretaceous of New Jersey. *American Naturalist, 5*, 562–564.

Cope, E. D. (1886). Dollo on extinct tortoises. *American Naturalist, 20*, 967–968.

Corsi, P. (2003). The Italian geological survey: The early history of a divided community. In G. B. Vai & W. Cavazza (Eds.), *Four centuries of the word 'Geology": Ulisse Aldrovandi 1603 in Bologna* (pp. 255–279). Bologna, Italy: Minerva Edizioni.

Cuvier, G. (1824). *Recherches sur les ossements fossiles*, 2nd ed. Paris: Dufour & D'Occagne,

Danilov, I. G., Averianov, A. O., & Yarkov, A. A. (2010). *Itilochelys rasstrigin* gen. et sp. nov, a new hard-shelled sea turtle (Cheloniidae sensu lato) from the Lower Paleocene of Volgograd Province, Russia. *Proceedings of the Zoological Institute of Russian Academy of Sciences, 314*, 24–41.

de Callatay, F. (2009). A bibliography of François Burtin. In H. Hasquin (Ed.), *L'Académie impériale et royale de Bruxelles. Ses académiciens et leurs réseaux intellectuels au XVIIIe siècle, Bruxelles* (pp. 156–161). L'Académie impériale et royale de Bruxelles.

de Carle Sowerby, J. , Lear, E., & Gray, J. E. (1872). *Tortoises, terrapins and turtles drawn from life by James de Carle Sowerby and Edward Lear; (with introduction and explanatory text by J. E. Gray)*. London: Henry Sotheran, Joseph Baer & Co.

Deraniyagala, P. E. P. (1932). Notes on the development of the leathery turtle, *Dermochelys coriacea*. *Spolia Zeylanica, 17*, 73–102.

Deraniyagala, P. E. P. (1936). Some postnatal changes in the leathery turtle, *Dermochelys coriacea*. *Spolia Zeylanica, 19*, 225–239.

Dollo, L. (1886). Premier note sur les chéloniens landeniens (Éocène infèrieur) de la Belgique. *Musée royal d'Histoire naturelle de Belgique, 4*, 129–141.

Dollo, L. (1887a). On some Belgian fossil reptiles. *Geological Magazine, 3*, 392–396.

Dollo, L. (1887b). *Psephophorus*. *Annales de la Societe Scientifique de Bruxelles, 2*, 139–176.

Dollo, L. (1903). *Eochelone brabantica*, tortue marine nouvelle du Bruxellien (Éocène moyen) de la Belgique. *Bulletin de l'Academie royale des Sciences de Belgique*, 792–850.

Dollo, L. (1907). Nouvelle note sur les reptiles de L'Eócène infèrieur la Belgique et des regionséé voisines (*Eosuchus lecheri, Eosphargis gigas*). *Bulletin de la Société Belge de Géologie, 21*, 81–85.

Dollo, L. (1909). The fossil vertebrates of Belgium. *Annals of the New York Academy of Science, 19*, 99–119.

Dollo, L. (1923). L'*Emys camperi* est une Tortue Marine. *Bulletin de l'Academie royale des Sciences de Belgique, 9*, 416–427.

Dumèril, A. M. C., & Bibron, G. (1835). Erpétologie générale ou histoire naturèlle complete des reptiles. Paris: Roret. 2, 1–68C.

Evans, J., & Kemp, T. S. (1975). The cranial morphology of a new Lower Cretaceous turtle from Southern England. *Palaeontology, 18*, 25–40.

Fanti, F. (2010). Life and ideas of Giovanni Capellini (1833–1922): A palaeontological revolution in Italy. In R. T. J. Moody, E. Buffetaut, D. Naish & D. M. Martill (Eds.), *Dinosaurs and other extinct saurians: A historical perspective* (pp. 79–87). Geological Society, London. (Special Publications, 343)

Faujas Saint Fond, B. (1799). *Histoire Naturelle de La Montagne de Saint-Pierre de Maestricht*. Paris: H. J. Hansen.

Fitzinger, L. J. (1826). *Neue Classification der Reptilien, nach ihren Natürlichen Verwandtschaften nebst einer Verwandtschafts-Tafel und einem Verzeichnisse der Reptilien-Sammlung des k.k. ZoologischenMuseum zu Wien*. Wien, Austria: J. G. Hübner Verlagen.

Gaffney, E. S. (1972). An illustrated glossary of turtle nomenclature. *American Museum Noviatates, 2486*, 1–33.

Gaffney, E. S. (1975). A phylogeny and classification of the higher categories turtles. *Bulletin of American Museum of Natural History, 155*, 387–436.

Gaffney, E. S. (1979). Comparative cranial morphology of recent and fossil turtles. *Bulletin of American Museum of Natural History, 164*, 65–376.

Gaffney, E. S., & Meylan, P. A. (1988). A phylogeny of turtles. In M. J. Benton (Ed.), *The phylogeny and classification of the Tetrapods, 1: Amphibians, reptiles and birds* (pp. 157–219). Oxford, UK: Systematics Association, Clarendon Press.

Gervais, M. P. (1872). Osteologie du Sphargis Luth (*Sphargis coriacea*). *Nouvelles Archives du Muséum d'Histoire naturelle de Paris, 8,* 199–228.

Gesner, G. (1545). *Bibliotheca Universalis, sive Catalogus omnium scriptorum locupletissimus, in tribus linguis, Latina, Graeca, et Hebraica.* Tiguri: Apud Christophorum Froschoverum.

Gesner, G. (1565). *A book on fossils objects, chiefly stones and gems, their shapes and appearance* (De Omni Rerum Fossilium).

Gessner, J. C. (1758). *Tractatus physicus de petrificatus in duas partes distinctus.* usw, 84.

Gray, J. E. (1825). A synopsis of the genera of reptiles and amphibia, with a description of some new species. *Annals of Philosophy, New Series, 10,* 193–217.

Gray, J. E. (1831). *Synopsis Reptilium or short descriptions of the species of reptiles. Part I: Cataphracta, tortoises, crocodiles, and enaliosaurians.* London: Treuttel, Wurz & Co.

Hay, O. P. (1898). On *Protostega*, the systematic position of *Dermochelys*, and the morphogeny of the chelonian carapace and plastron. *American Naturalist, 32,* 929–948.

Hay, O. P. (1908). *The fossil turtles of North America.* Washington: Carnegie Institute.

Hirayama, R. (1994). Phylogenetic systematics of chelonioid sea turtles. *The Island Arc, 3,* 270–284.

Hirayama, R. (1997). Distribution and diversity of Cretaceous chelonioids. In J. M. Calloway & E. L. Nicholls (Eds.), *Ancient Marine Reptiles.* San Diego, CA: Academic Press/Harcourt Brace & Company.

Hirayama, R. (1998). Oldest known sea turtle. *Nature, 392,* 705–708.

Howe, S. R., Sharpe, T., & Torrens, H. S. (1981). *Ichthyosaurs: A history of fossil 'sea-dragons'.* Cardiff: National Museum Wales.

Knorr, G. W., & Walch, J. E. I. (1773). *Die Naturgeschichte der Versteinerungen zur Erläuterung der Knorrischen.* Nuremberg: Sammlung von Merkwurdigkeiten der Natur, 1–2.

Koenig, C. D. E. (1825). *Icones fossilium sectiles.* London: G. B. Sowerby.

Lacépède, B. C. (1789). *Histoire des quadrupèdes, ovipares et des serpents.* Paris: De Thou.

Lyddeker, R. (1889a). On remains of Eocene and Mesozoic Chelonia and the tooth *Ornithopsis. Quarterly Journal of the Geological Society of London, 45,* 227–246.

Lyddeker, R. (1889b). On the skull of the chelonian genus *Lytoloma. Proceedings of the Zoological Society London* (pp. 60–66).

Lyddeker, R. (1889c). *Catalogue of Fossil Reptilia and Amphibia in the British Museum. Part III. Containing the order Chelonia.* London: Printed by order of the Trustees of the British Museum.

Mantell, G. A. (1841). On the fossil remains of turtles, discovered in the chalk formation of the South-East of England. *Philosophical Transactions of the Royal Society of London, 131,* 153–158.

Meijer, M. C. (1999). *Race and aesthetics in the anthropology of Petrus Camper (1722–1789).* Amsterdam and Atlanta, GA: Rodopi.

Meylan, P. A., & Shaffer, H. B. (2001). Molecular evidence for higher relationships among turtles. *Tree of Life web project.* www.tolweb.org/articles.

Moody, R. T. J. (1968). A turtle, *Eochelys crassicostata* (Owen) from the London Clay of the Isle of Sheppey. *Proceedings of the Geologists' Association, 79,* 129–140.

Moody, R. T. J. (1970). A revision of the taxonomy and morphology of certain Eocene Cheloniidae. Ph.D. dissertation, University of London.

Moody, R. T. J. (1972). The turtle fauna of the Eocene phosphates of Metlaoui, Tunisia. *Proceedings of the Geologists' Association. 83,* 327–336.

Moody, R. T. J. (1974). The taxonomy and morphology of *Puppigerus camperi* (Gray), an Eocene sea-turtle from Northern Europe. *Bulletin of the British Museum (Natural History), Geology, 25,* 153–186.

Moody, R. T. J. (1980a). Notes on some European Palaeogene turtles. *Tertiary Research, 2,* 161–168.

Moody, R. T. J. (1980b). The distribution of turtles in the British Palaeogene. In J. J. Hooker, A. N. Insole, R. T. J. Moody, C. A. Walker, & D. J. Ward (Eds.). *The distribution of cartilaginous fish, turtles, birds and mammals in the British Palaeogene.* Tertiary Research, *3,* 1–45.

Moody, R. T. J. (1993). Cretaceous-Tertiary marine turtles of northwest Europe. *Revue de Paléobiologie, volume special, 7,* 151–160.

Moody, R. T. J. (1996). *Allopleuron hoffmani*—problematic or simply misunderstood? *Journal of Vertebrate Paleontology, 16* (Suppl. 3), 54A.

Moody, R. T. J. (1997). The paleogeography of marine and coastal turtles of the North Atlantic and Trans-Saharan regions, In: J. M. Callaway & E. Nicholls (Eds.), *Ancient marine reptiles* (pp 259–280). Academic Press: San Diego.

Mulder, E. W. A. (1996). Maastricht Cretaceous finds and Dutch pioneers in vertebrate palaeontology. *Journal of Vertebrate Paleontology, 16,* 673–687.

Mulder, E. W. A. (2004). Maastricht Cretaceous finds and Dutch pioneers in vertebrate palaeontology. In J. L. R. Touret & R. P. W. Visser (Eds.), *Dutch pioneers of the earth sciences* (pp. 165–176). Amsterdam: Royal Netherlands Academy of Arts and Sciences (KNAW).

Mlynarski, M. (1959). *Glarichelys Knorri* (Gray)—a cheloniid from the Carpathian Menilitic Shales (Poland). *Acta Palaeontologica Polonia, 5,* 177–198.

Nielsen, E. (1959). Eocene turtles from Denmark. *Meddelelser fra Dansk Geologisk Forening, 14,* 96–114.

Nielsen, E. (1963). On the post-cranial skeleton of *Eosphargis breineri* Nielsen. *Meddelelser fra Dansk Geologisk Forening, 15,* 281–328.

Nessov, L. A. (1987). The Paleogene sea turtles of Southern Kazakhstan and the phylogenetic relationships between Toxochelyidae and Cheloniidae. *Paleontologicheskii zhurnal, 4,* 76–87.

Owen, R. (1850). Monograph on the fossil Reptilia of the London Clay. Part I: Chelonia, Supp. 1. *Palaeontographical Society Monograph, 3,* 1–4.

Owen, R. (1851). A monograph on the fossil Reptilia of Cretaceous Formations. *Paleontographical. Society Monograph, 5,* 1–118.

Owen, R. (1880). Fossil reptiles of the London Clay. Vol. II. Part I. *Palaeontograpical Society. Monograph 34,* 1–4.

Owen, R., & Bell, T. (1841). Description of the remains of six species of marine turtles (*Chelones*) from the London Clay of Sheppey and Harwich. *Proceedings of the Geological Society of London 3,* 570–578.

Owen, R., & Bell, T. (1849). Monograph on the fossil Reptilia of the London Clay, and on the Bracklesham and other Tertiary beds. Part 1, Chelonia. *Palaeontographical Society Monograph, 2,* 1–76.

Owen, R., & Bell, T. (1850). Description of the remains of the fossil reptiles from the tertiary deposits of Bracklesham and Bognor, in the Museum of Frederick Dixon, Esq., or Figured in the present work. In F. Dixon (Ed.) (1850). *Geology of Sussex.* London: Longman, Brown and Green.

Parkinson, J. (1811). *The organic remains of a former world. An examination of the mineralized remains of the vegetables and animals of the antediluvian world.* London: Sherwood, Neely & Jones, Paternoster Row.

Parkinson, J. (1817). *An essay on the shaking palsy.* London: Sherwood, Neely, and Jones, Paternoster Row.

Quintart, A., & Plisnier-Ladame, F. (1968). *Eosphargis gigas* (Owen, 1861). A propos d'une tortue géante récemment exposée dans les salles de l'Institut Royal des Sciences Naturelles de Belgique. *Les Naturalistes belges, 49,* 366–381.

Rieppel, O. (2001). Turtles as hopeful monsters. *BioEssays 23,* 987–991.

Rudwick, M. J. S. (1976). *The meaning of fossils: Episodes in the history of palaeontology*. Chicago: University of Chicago Press.

Rudwick, M. J. S. (1993). *The meaning of fossils: Episodes in the History of Palaeontology* (1st ed.). London: Macdonald.

Rudwick, M. J. S. (1997). *Georges Cuvier, fossil bones, and geological catastrophes: New translations and interpretations of the primary texts*. Chicago: University of Chicago Press.

Rudwick, M. J. S. (2005). *Bursting the limits of time: The reconstruction of geohistory in the age of revolution*. Chicago: University of Chicago Press.

Seago, A. K. J. (1979). A review of the fossil turtles of the family Dermochelyidae. Ph.D. Dissertation, University of London.

Seeley, H. G. (1880). Note on *Psephophorus polygonus*, v. Meyer, a new type of Chelonian reptile allied to the leathery turtle. *Quarterly Journal of the Geological Society, 361,* 406–413.

Serres de, M., Dubreuil, J. M., & Jeanjean, B. (1839). *Recherches sur les ossemens humatiles des cavernes de Lunel-Viel*. Montpellier.

Shaffer, H. B., Meylan, P. A., & McKnight, M. L. (1997). Tests of turtle phylogeny: Molecular, morphological, and paleontological approaches. *Systematic Biology, 46,* 235–268.

Smith, D. T. J. (1989). *The cranial morphology of fossil and living sea turtles (Cheloniidae, Dermochelyidae, and Desmatochelyidae)*. Ph.D., Council For National Academic Awards (United Kingdom).

Tong, H., & Hirayama, R. (2004). First Cretaceous dermochelyid turtle from Africa. *Revue de Paléobiologie*, spécial *9*, 55–59.

Tong, H., & Hirayama, R. (2008). A new species of *Argillochelys* (Testudines: Cryptodira: Cheloniidae) from the Ouled Abdoun phosphate basin, Morocco. *Bulletin de la Société géologique de France, 179,* 623–630.

Tong, H., Hirayama, R., & Tabouelle, J. (in press). *Puppigerus camperi* (Testudines: Cryptodira: Cheloniidae) from the Ypresian (Early Eocene) of Ouled Abdoun Basin, Morocco. *Bulletin Société Géologigue de France.*

Turner, S., Burek, C. V., & Moody, R. T. J. (2010). Forgotten women in an extinct saurian (man's) world. In R. T. J. Moody, E. Buffetaut , D. Naish & D. M. Martill (Eds.) *Dinosaurs and other extinct saurians: A historical perspective*. Geological Society, London, Special Publications, *343,* 111–153.

Ubaghs, J. C. (1888). Le crane de *Chelone hoffmanni. Bulletin de Société Belge Geologié, Paleontologié, Hydrologié, 2,* 383–392.

Vai, G. B. (2003). Giovanni Capellini and the origin of the International Geological Congress. In G. B. Vai & W. Cavazza (Eds.), *Four centuries of the word geology* (pp. 301–315). Ulisse Aldrovandi 1603 in Bologna. Minerva Edizioni.

Van Poelman, C. A. C. (1868). *Catalogue des Collections d'Anatomie comparee, y compris les ossements fossiles, de l'Universite de Gand*. Universite de L'Etat de Gand, *8,* 120.

Waggoner, B. (1996). Étienne Geoffroy St. Hilaire (1772–1844). Retrieved February 10, 2011, from http://www.ucmp.berkeley.edu/history/hilaire.html.

Wieland, G. R. (1900). The skull, pelvis, and probable relationships of the huge turtles of the genus *Archelon* from the Fort Pierre Cretaceous of South Dakota. *American Journal of Science, 4th Series, 9,* 237–251.

Wieland, G. R. (1902). Notes on the Cretaceous turtles *Toxochelys* and *Archelon*, with a classification of the marine Testudinata. *American Journal Science, 4th Series, 14,* 95–108.

Wieland, G. R. (1904a). Structure of the Cretaceous turtles of New Jersey: *Adocus, Osteopygis*, and *Propleura. American Journal Science, 4th Series, 17,* 112–132.

Wieland, G. R. (1904b). Structure of the Upper Cretaceous turtles of New Jersey: *Lytoloma. American Journal Science, 4th Series, 18,* 183–196.

Wieland, G. R. (1906). The osteology of *Protostega. Memoirs of the Carnegie Museum, 2,* 279–305.

Wieland, G. R. (1909). Revision of the Protostegidae. *American Journal of Science, 4th Series, 27,* 101–130.

Winkler, T. C. (1860). *Ch. Darwin, Het ontstaan der soorten van dieren en planten door middel van de natuurkeus of het bewaard blijven van bevoorregte rassen in de strijd des levens*. Haarlem.

Wieland, G. R. (1869). *Des Tortues Fossiles conservées dans le Musée Teyler et dans quelques autres musées*. Haarlem: Loosjes.

Wilson, W. E., & Schuh, C.,Buc'hoz, P. J. (2001). The marvellous gifts of nature, diversely colored, in the mineral kingdom (1782). *Mineralogical Record, 32,* 5–11.

Wood, R. C., Johnson-Gove, J., Gaffney, E. S., & Maley, K. F. (1996). Evolution and phylogeny of leatherback turtles (Dermochelyidae), with descriptions of new fossil taxa. *Chelonian Conservation and Biology, 2,* 266–286.

Wood, R. C., & Moody, R. T. J. (1998). Fossil (dermochelyid) leatherback turtles from the Tertiary of Belgium. *Journal of Vertebrate Palaeontology, 18*(Suppl. 3), 88A.

Van Beneden, P. J. (1883a). Sur quelques formes nouvelles de terrains tertiaires du Pays. *Bulletin Academie Royal Belge des Science, des Lettres et des Beaux-Arts de Belgique, 3,* 132.

Van Beneden, P. J. (1883b). Notes sur des ossements de *Sphargis* trouvés dans la Terre à brique du pays de Waas. *Bulletin Academie Royal Belge des Science, des Lettres et des Beaux-Arts de Belgique, 3,* 665–684.

Van Veen, J. C. (2004). The history of the Palaeontological-Mineralogical Cabinet of the Teylers Museum, Haarlem, The Netherlands. *The PalArch Foundation Newsletter, 1,* 7–21.

Von Meyer, H. (1846). Mittheilungen. *Neues Jahrbuch, Mineralogie und Geologie. Geol Petrefact,* 462–476.

Von Meyer, H. (1847). Mittheilungen an Professor Bronn gerichtet. *Neues Jahrbuch, Mineralogie und Geologie,* 572–880.

Vernier, F., & Klein, J. P. (2005). Buc'hoz, médecin, botaniste de Stanislas Roi de Pologne et Duc de Lorraine. *Médiathèque de Metz,* 1–32.

Veneer, L. (2006). Provincial geology and the Industrial Revolution. *Endeavour, 30,* 72–80.

Zangerl, R. (1948). The methods of comparative anatomy and its contributions to the study of evolution. *Evolution, 2,* 351–374.

Zangerl, R. (1953a). The vertebrate fauna of the Selma formation of Alabama. Part IV. The turtles of the family Protostegidae. *Fieldiana: Geology Memoirs, 3,* 57–135.

Zangerl, R. (1953b). The vertebrate fauna of the Selma formation of Alabama. Part IV. The turtles of the family Toxochelyidae. *Fieldiana: Geology Memoirs, 3,* 136–288.

Zangerl, R. (1958). Die oligozanen Meerschildkroten von Glarus. *Schweizerische Paläontologische Abhandl, 73,* 1–56.

Zangerl, R. (1960). The vertebrate fauna of the Selma formation of Alabama. An advanced cheloniid sea turtle. *Fieldiana: Geology Memoirs, 4,* 281–312.

Zangerl, R. (1969). The turtle shell. In C. Gans, A. d'A. Bellairs & T. Parsons (Eds.), *Biology of the Reptilia, 1* (pp. 311–339). New York: Academic Press.

Zangerl, R. (1971). Two toxochelyid sea turtles from the Landenian Sands of Erquelinnes (Hain ault), of Belgium. *Institut Royal Sciences Naturelles, Belgique, 169,* 1–32.

Zangerl, R. (1980). Patterns of phylogenetic differentiation in the toxochelyid and cheloniid sea turtles. *American Zoology, 20,* 585–596.

Zangerl, R., & Sloan, R. E. (1960). A new specimen of *Desmatochelys lowi* Williston. A primitive cheloniid sea turtle from the Cretaceous of South Dakota. *Fieldiana: Geology Memoirs, 14,* 7–40.

Chapter 25
Fossil Kinosternidae from the Oligocene and Miocene of Florida, USA

Jason R. Bourque

Abstract Kinosternid remains are generally rare through the Oligocene and Miocene. Fossil material from eight Florida localities is presented here. Specimens discussed include the latest and most southeastern occurrence of the genus *Xenochelys* (new species), one of the earliest occurrences of *Kinosternon*, the oldest record of the *Kinosternon subrubrum-baurii* group in Florida, and the presence of Miocene *Kinosternon* species either convergent with or closely related to taxa that occur today in the southwestern United States and Central-South America.

Keywords Arikareean • Kinosternidae • *Kinosternon* • Miocene • *Xenochelys*

Introduction

Extant members of the family Kinosternidae are taxonomically diverse in Florida with between six and eight reported taxa (Carr 1952; Iverson 1978; Conant and Collins 1998; Ernst and Lovich 2009). These taxa include: the loggerhead and stripe-necked musk turtles, *Sternotherus minor minor* (Agassiz 1857) and *Sternotherus minor peltifer* (Smith and Glass 1947); stinkpot musk turtle, *Sternotherus odoratus* (Latreille, *in* Sonnini De Manoncourt and Latreille 1802); striped mud turtle, *Kinosternon baurii* (Garman 1891); eastern and Mississippi mud turtles, *Kinosternon subrubrum subrubrum* (Lacépède 1788) and *Kinosternon subrubrum hippocrepis* (Gray 1856); and the endemic Florida mud turtle, *Kinosternon subrubrum steindachneri* (Siebenrock 1906). These small turtles inhabit almost all types of freshwater habitats throughout the state and occasionally are found in estuarine habitats (e.g., *K. s. subrubrum*). Fossil records for *S. odoratus* (late Pleistocene), *S. minor* (late Pleistocene), *K. subrubrum* (late Pleistocene), and *K. baurii* (early-late Pleistocene) have been reported from Florida (Hulbert 1992; Holman 1995; Hulbert 2001; Ernst and Lovich 2009).

The oldest kinosternids known from the southeastern United States are from the Paleocene of South Carolina (Hutchison and Weems 1998), but the oldest records from Florida date back only to the late Oligocene (Hayes 2000). Oligocene and Miocene kinosternid fossils are rare. As a result, little is known about the evolution of kinosternids during this interval; for example, changes in the plastron, such as complete loss of the entoplastron and development of plastral kinesis, that are characteristic of the modern kinosternines *Sternotherus* and *Kinosternon*. Morphological developments such as these likely occurred during the Oligocene and Miocene, meaning that kinosternine fossils from those intervals could assist in understanding the transitioning from fossil taxa, such as *Baltemys* and *Xenochelys*, to modern kinosternines, such as *Sternotherus* and *Kinosternon* (Hutchison 1991). Kinosternids from this part of the Cenozoic are also significant because the earliest records of the two extant kinosternine genera *Sternotherus* and *Kinosternon* are probably from the Miocene of North America, at approximately 4.5–7 and 17–17.9 Ma respectively (Hutchison 1991; Holman 1998; Bourque 2011).

Oligo-Miocene kinosternid specimens from eight localities in Florida are described in this paper in hope that future discoveries will elaborate on the scanty material discussed here. To date, only isolated and/or fragmentary specimens are known from these intervals in Florida. Material from the late Oligocene (Arikareean) through at least the late Miocene (early Hemphillian) does not appear to represent any of the modern forms found in the state today. Although many specimens apparently represent new taxa, lack of sufficient material has inhibited diagnosable descriptions here, with the exception of Brooksville 2, for which composite material has

J. R. Bourque (✉)
Division of Vertebrate Paleontology, Florida Museum of Natural History, Dickinson Hall, University of Florida, Gainesville, FL 32611, USA
e-mail: jbourque@flmnh.ufl.edu

D. B. Brinkman et al. (eds.), *Morphology and Evolution of Turtles*, Vertebrate Paleobiology and Paleoanthropology,
DOI: 10.1007/978-94-007-4309-0_25,

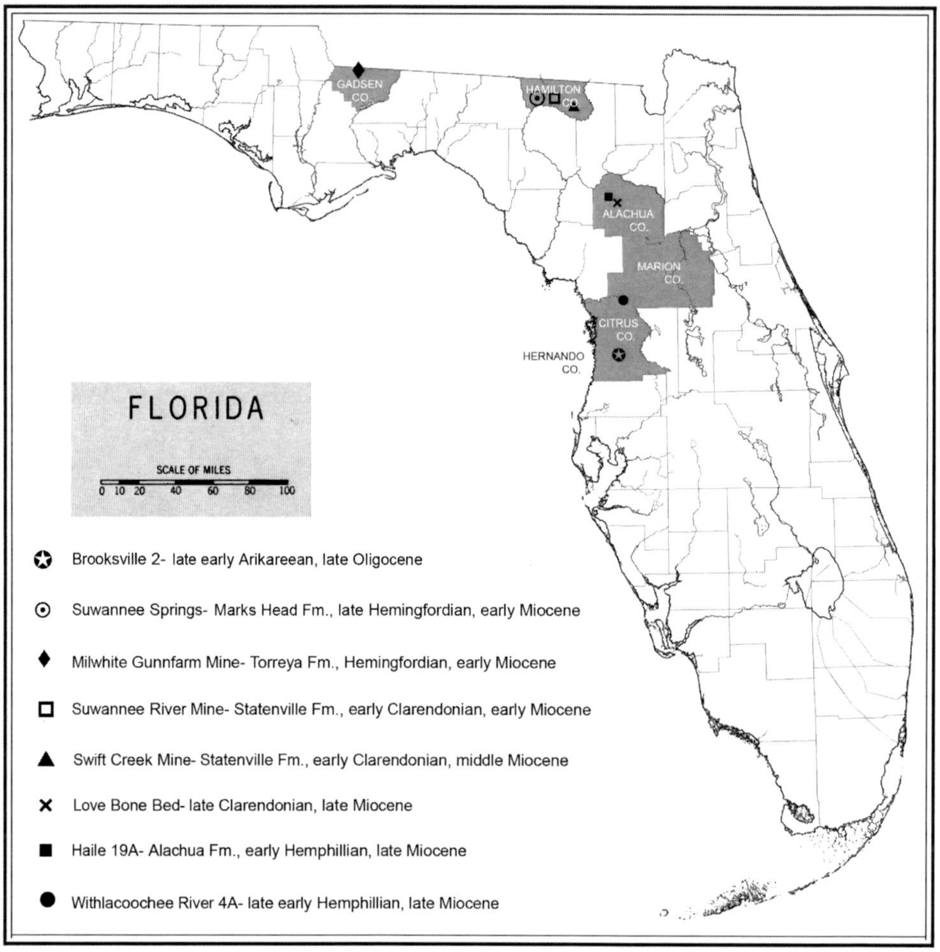

Fig. 25.1 Map of Florida illustrating Oligocene and Miocene localities with fossil kinosternids

been used to describe a new species. All of the fossils represent members of the Kinosterninae sensu Hutchison (1991) and conform to his amended diagnosis for the subfamily.

Institutional Abbreviations used in this paper are: AMNH, American Museum of Natural History, New York, USA; FLMNH, Florida Museum of Natural History, Gainesville, Florida, USA; UCMP, University of California Museum of Paleontology, Berkeley, California, USA; UF, University of Florida, Division of Vertebrate Paleontology, Gainesville, Florida, USA; UF/H, University of Florida Division of Herpetology, Gainesville, Florida, USA; UF/TRO, specimens originally from the collection of John Waldrop now housed at the FLMNH; USNM, National Museum of Natural History, Washington D. C., USA; YPMPU, Yale Peabody Museum of Natural History, New Haven, Connecticut, USA.

Anatomical abbreviations used in this paper are: C, costal; M, marginal scale; P, peripheral; PL, pleural scale; N, neural; V, vertebral scale. The terms scale and scute are used here interchangeably.

The inventory of kinosternid specimens reported here from eight Florida localities is as follows (see next section for locality details and Fig. 25.1 for map): Brooksville 2: UF 150002-150028; Suwannee Springs: UF/TRO 2621; Milwhite Gunn Farm Mine: UF 111725, UF 111753; Suwannee River Mine: UF 225682; Swift Creek Mine: UF 150035; Love Bone Bed: UF 27141, UF 43031, UF 150029-UF 150032; Haile 19A: UF 150034; Withlacoochee River 4A: UF 13893; UF 150033. The following comparative fossil taxa were examined: *Xenochelys lostcabinensis*: UCMP 112341 (holotype); *Xenochelys bridgerensis*: UCMP 121734 (holotype); *Xenochelys formosa*: AMNH 1097 (holotype) and YPMPU 13686 (referred skull); *Baltemys staurogastros*: UCMP 127200 (holotype) and UCMP 121776. Most of the modern skeletal material examined was from the Division of Herpetology collection at the FLMNH.

Overview of Oligocene and Miocene Kinosternid Localities in Florida

The following eight localities have produced fossil kinosternid specimens from the late Oligocene through late Miocene of Florida (Figs. 25.1, 25.2).

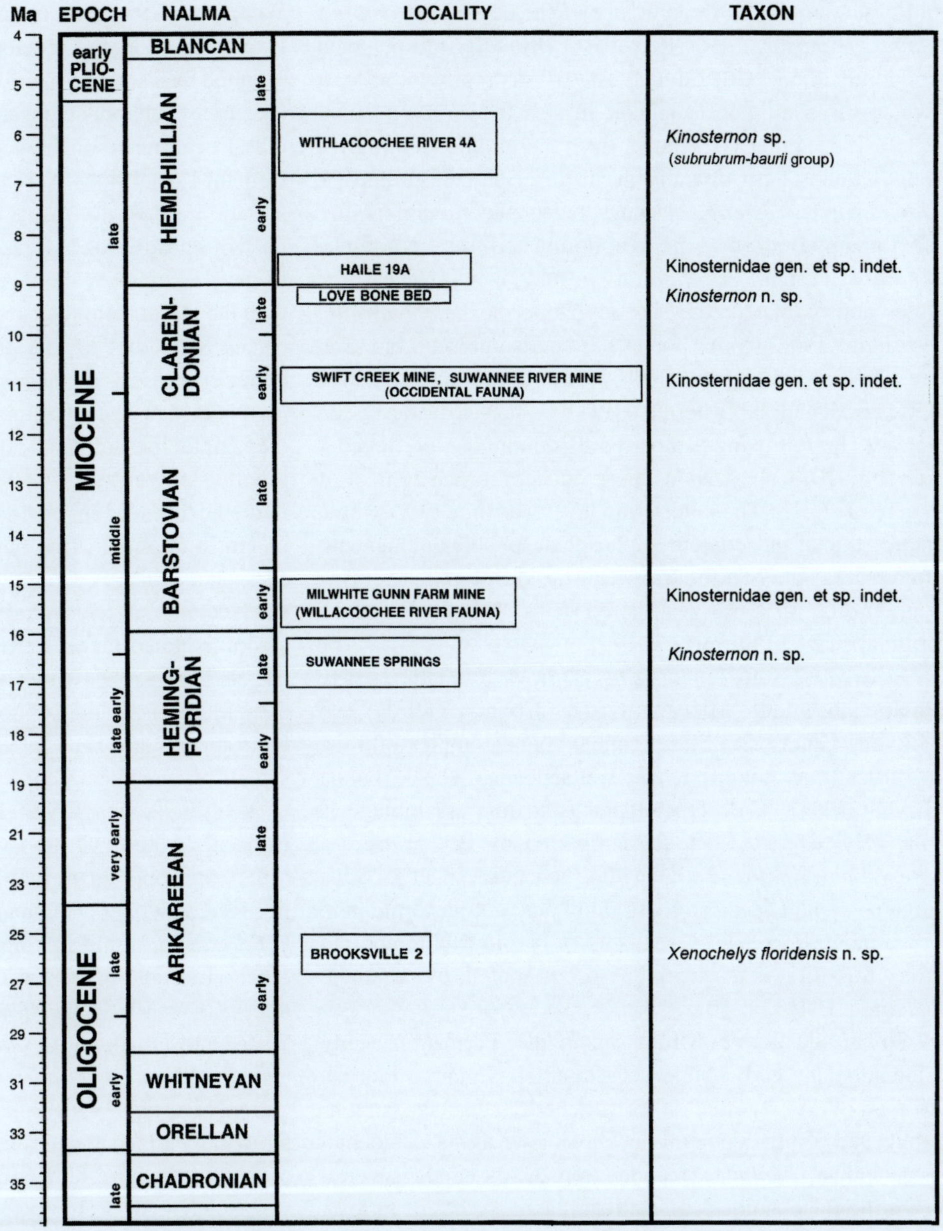

Fig. 25.2 Correlation chart of Oligo-Miocene Kinosternidae in Florida (modified from Hulbert 2001). The presence of *Xenochelys floridensis* sp. nov. at Brooksville 2 represents the latest and most southeastern occurrence of the genus in North America, with congeners known previously from the Eocene of Arkansas, South Dakota, Texas, and Wyoming (Hutchison 1991)

Brooksville 2: Late early Arikareean, late Oligocene, 25–26 Ma (Tedford et al. 2004) or ~25–28 Ma (Hayes 2000); Hernando County, Florida. Hayes (2000) published an article describing the mammals and background of the Brooksville 2 Local Fauna and mentioned Kinosternidae in a faunal list. Brooksville 2 was a complex of at least five clay-filled crevice deposits (sub-localities referred to as 1A-1E) rich in diverse microfossil vertebrate remains.

Chelonians from the site include a small tortoise (from sub-localities 1A and 1B) and a kinosternid (at sub-localities 1A, 1B, 1D, and 1E). Kinosternid specimens that will be discussed from these sub-localities are as follows: 1A: UF 150001-150003; 1B: UF 150005-150009, UF 150011-150013; 1D: UF 150015; and 1E: UF 150016-150023. Sites 1A, 1B, and 1E were collected as matrix and screen washed (Hayes 2000) and also contained the bulk of the kinosternid sample examined here. Other noteworthy members of the Brooksville 2 herpetofauna include a gila monster (Helodermatidae) and dwarf alligatorid, both represented by numerous osteoderms and cranial fragments.

Suwannee Springs: Marks Head Formation (after Scott 1988), late Hemingfordian, early Miocene; Hamilton County, Florida. Little has been published on this roadcut locality on US 129 just north of the Suwannee River (see Frailey 1978; Tedford and Hunter 1984). Material collected in 1973 by John Waldrop was recently donated and is part of the UF/TRO collection now housed at the FLMNH Division of Vertebrate Paleontology. This site is approximately 16–17 million years old based largely on the presence of the horse *Merychippus gunteri* Simpson 1930 (Richard C. Hulbert, Jr., personal communication). Crocodylian osteoderms represent at least two taxa: an alligatorid similar in size to *Alligator olseni* White 1942, otherwise known from the middle Hemingfordian Thomas Farm locality (Tedford et al. 2004), and cf. *Gavialosuchus* (Alex K. Hastings, personal communication). Chelonians from the site are dominated by testudinids, possibly a small as well as a mid-sized species, the latter approximately the size of *Hesperotestudo tedwhitei* (Williams 1953) from Thomas Farm. Also present, but rare, are a pleurodire (represented by a right mesoplastron, UF/TRO 2620) and a kinosternid (UF/TRO 2621) to be discussed below. The following is a personal communication (dated 4 March 2010) from John Waldrop concerning the context of UF/TRO 2621: "The kinosternid layer was thin, two to three inches thick, and contained phosphate pebbles along with fragmented bone. The matrix below the kinosternid layer was a green clay that was considered to be 'Hawthorn' in the original usage of the term."

Milwhite Gunn Farm Mine: Torreya Formation, early Barstovian, middle Miocene; Gadsen County, Florida. The Milwhite Gunn Farm Mine is one within a complex of mine localities that comprise the Willacoochee Creek Fauna (Bryant 1991). The herpetofauna from this assemblage is relatively diverse. Chelonians reported by Bryant were a trionychid (?*Apalone* sp.), emydid (gen. et sp. indet.), both a large and small species of testudinid, and a kinosternid gen. et sp. indet. (to be discussed below). Previously unreported from this site is a pleurodire, represented by a single specimen (UF 111760).

Suwannee River Mine: Statenville Formation, early Clarendonian, early Miocene; Hamilton County, Florida. Morgan (1989) reported on the vertebrate fauna, age, stratigraphy, and correlation of this locality known as the Occidental Local Fauna. Chelonians reported from this assemblage were a sea turtle (Cheloniidae), a small and large species of tortoise (Testudinidae), and an emydid described as a pond turtle cf. *Pseudemys* sp. (Morgan 1989). Previously unreported and minimally represented is a pleurodire (UF 225683).

Swift Creek Mine: Statenville Formation, early Clarendonian, middle Miocene; Hamilton County, Florida, 10 km SE of Jasper.

Love Bone Bed: Late Clarendonian, late Miocene; Alachua County, Florida. Webb et al. (1981) published on the diverse fauna from this locality. Chelonians reported from the site include: a trionychid, *Apalone* cf. *ferox* (Schneider 1783); the emydids *Pseudemys caelata* (Hay 1908; Jackson Jackson 1976), *Deirochelys carri* Jackson 1978, and "*Terrapene* n. sp."; and two testudinids represented by both a giant and a small tortoise species (Webb et al. 1981). The smooth-shelled *Pseudemys williamsi* (Rose and Weaver 1966) is also present at the site. Although *P. williamsi* is missing from the faunal list in Webb et al. (1981), it is mentioned on p. 537 of the same paper. Unmentioned in any previous publication is a chelydrid, represented by a single posterior left peripheral (UF 27141). This specimen probably belongs to a species of *Chelydra* and is the oldest occurrence of the genus from Florida. The trionychid from the Love Bone Bed is a relatively small somewhat gracile species that does not appear to be *A. ferox*, which is endemic to most of Florida today. *A. ferox* tends to be larger and the bones of the shell thicker, more robust, and with larger dermal pits than the taxon from the Love Bone Bed. A kinosternid (to be discussed below) is also present but rare, represented by a perfectly preserved nuchal as well as unassociated bits of peripheral and costal fragments. The status and whereabouts of the single uncatalogued specimen of "*Terrapene* n. sp." reported by Webb et al. (1981) is uncertain, and I was unable to find evidence of *Terrapene* from the Love Bone Bed.

Haile 19A: Alachua Formation, early Hemphillian, late Miocene; Alachua County, Florida. Chelonians from this site include: a trionychid; two emydids consisting of a smooth-shelled taxon (cf. *Pseudemys williamsi*) and a rugose-shelled taxon; two testudinids consisting of a large and a small species; and a kinosternid (to be discussed below). Kinosternids are rare in this assemblage, and only represented by one specimen.

Withlacoochee River 4A: Late early Hemphillian, late Miocene; Marion/Citrus County boundary, Florida. Becker (1985) first reported the family Kinosternidae in a faunal list for this locality. Other chelonians in his list included: a trionychid ("*Trionyx* sp."); the emydids *Terrapene* sp., *Pseudemys concinna* (LeConte 1830) or *Pseudemys williamsi*, *Pseudemys nelsoni* (Carr 1938), *Trachemys inflata* (Weaver and Robertson 1967), and *Trachemys scripta* (Schoepff 1792); and the giant testudinid *Hesperotestudo* ("*Geochelone*") sp. Based on my current investigations, turtle specimens from this fauna may be temporally mixed and in some cases appear to be an admixture of fossil and more temporally modern specimens, the latter suggested by the occurrence of brown-stained, seemingly non-permineralized bone. This may explain the presence of extant riparian taxa such as *P. concinna*, *P. nelsoni*, and *T. scripta* within this Miocene assemblage.

Description of Oligocene Kinosternid Material

Brooksville 2

Remarks: The oldest kinosternids known from Florida are from the Brooksville 2 Local Fauna. The most abundant turtle fossils in this assemblage represent the

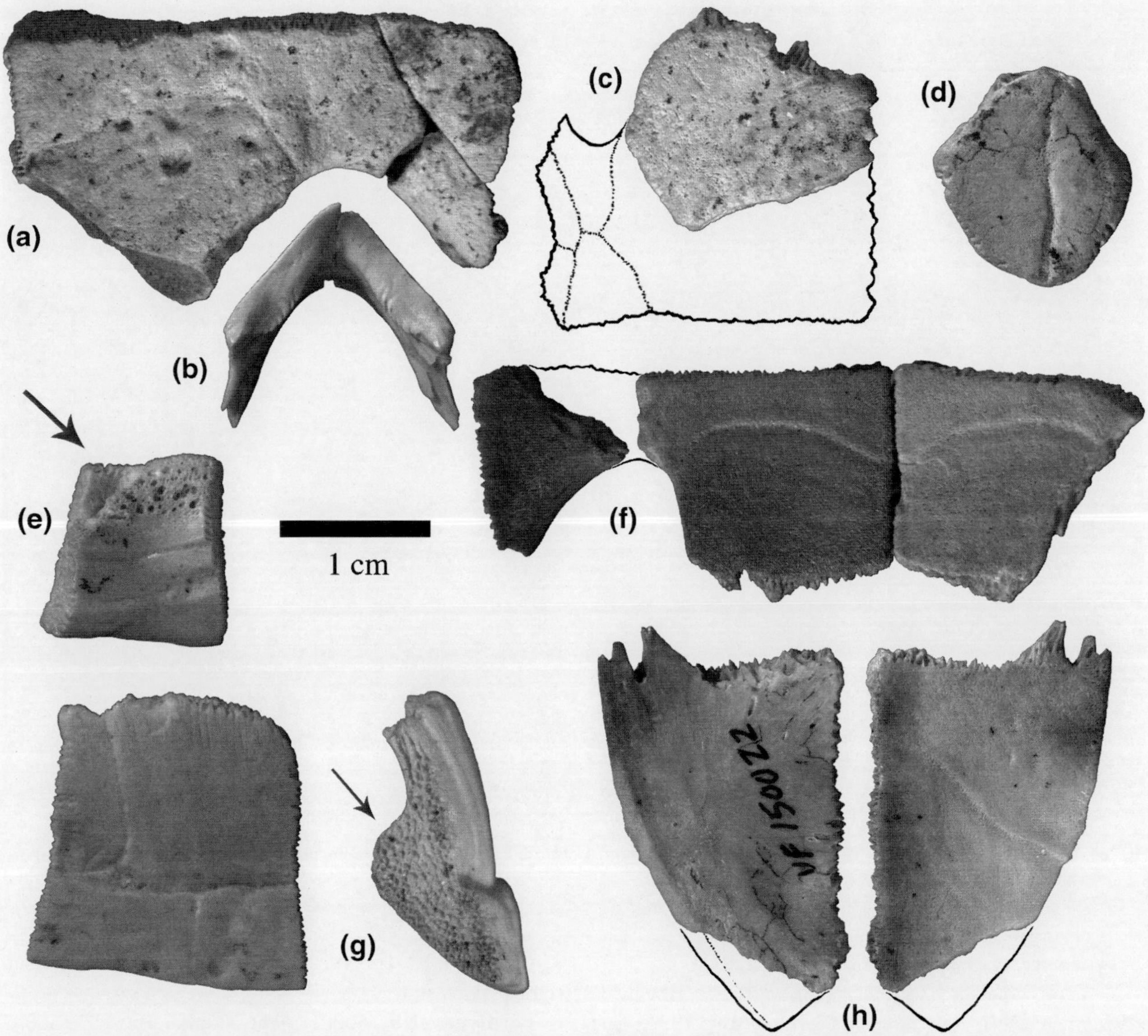

Fig. 25.3 *Xenochelys floridensis* sp. nov. from the late Oligocene (Arikareean) Brooksville 2 locality, Hernando Co., Florida. **a** UF 150005, partial left hypoplastron, in ventral view. **b** UF 150023, mandible, in dorsal (occlusal) view. **c** UF 150008, partial right hyoplastron, in ventral view. **d** UF 150009, entoplastron, in ventral view. **e** UF 150018, left P3, in visceral view. *Arrow* indicates position of sutural cavity for insertion of the C1 rib end that enters the posterior moiety of P3 and terminates in the anterior of P4 or P3-P4 suture in *Xenochelys*. **f** UF 150006-7 (holotype), associated partial left and right hypoplastra, in ventral view. In this specimen the humeral-femoral sulci meet at the midline and form an obtuse "V" shape on the anterior to middle portions of the elements, similar to the condition in *Xenochelys formosa*. **g** UF 150001, right P8, in dorsal (*left*) and posterior sutural (*right*) views. *Arrow* indicates the pronounced step on the visceral scale margin, a synapomorphy with *X. formosa*. **h** UF 150022, partial left xiphiplastron, in dorsal (*left*) and ventral (*right*) views. This element lacks a substantial notch between the femoral and anal scales, which is similar to the condition in *X. formosa*. Additionally, although the visceral scale margin is relatively thick and developed in *X. formosa*, it is very reduced and near to the margin of the xiphiplastron in the new species. All specimens at same magnification

latest and most southeastern occurrence of the genus *Xenochelys* (Fig. 25.3). Prior to this record, *Xenochelys* was known to have persisted from the Wasatchian through Chadronian NALMAs of Arkansas, South Dakota, Texas, and Wyoming, with the most southern and eastern occurrences being from the Uintan of Texas and Arkansas (Hay 1906, 1908; Hutchison 1991) (Figs 25.1, 25.2, 25.3, 25.4, 25.5). The Floridian *Xenochelys* fossils represent a new species that is described below.

Order Testudines Linnaeus 1758
Suborder Cryptodira Cope 1868
Family Kinosternidae Hay1892
Subfamily Kinosterninae Hay 1892

Genus *Xenochelys* Hay 1906

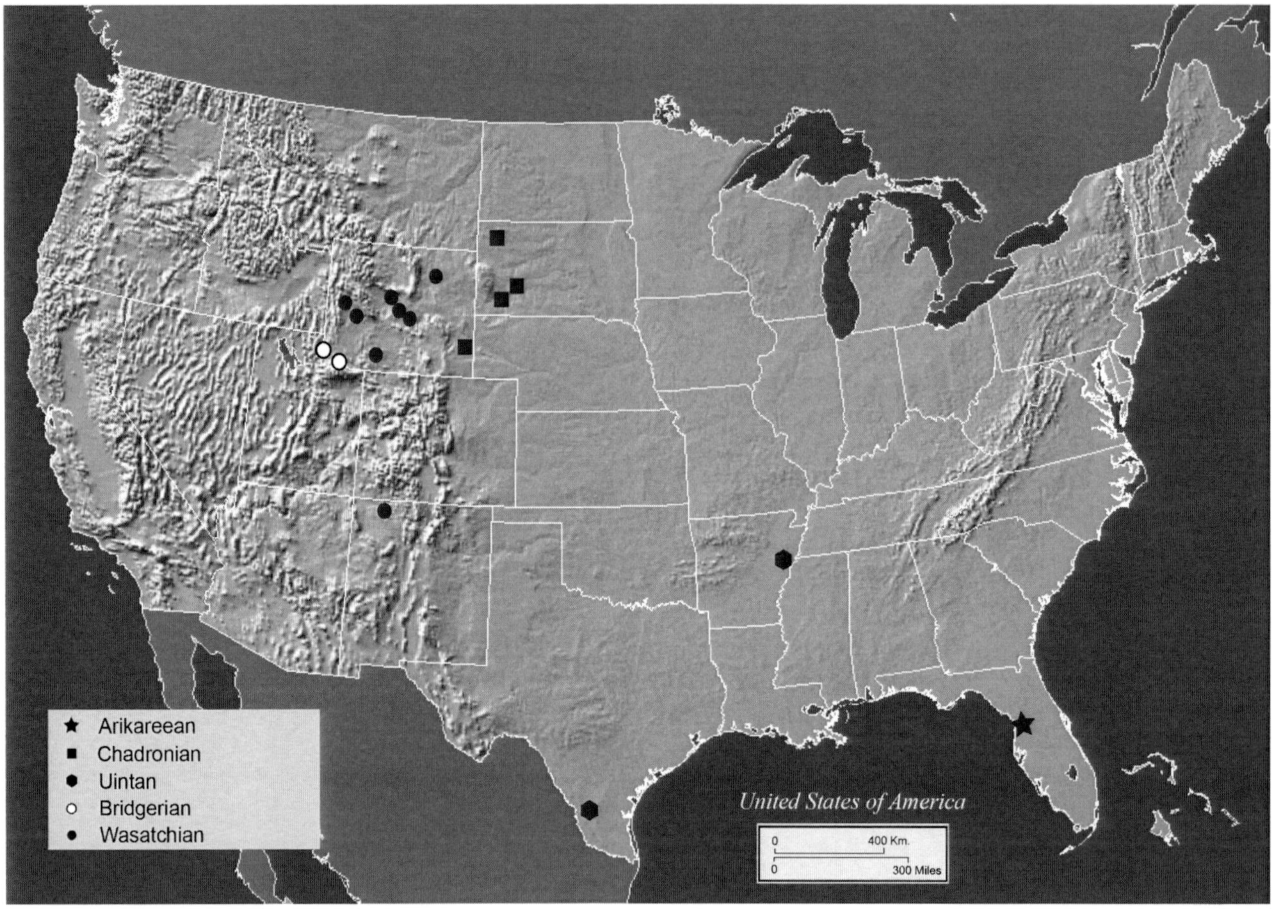

Fig. 25.4 Distribution map of the United States illustrating reported occurrences of the genus *Xenochelys* (Wasatchian through Chadronian localities after Hutchison 1991)

Type species: *Xenochelys formosa* Hay 1906.

Referred species: *Xenochelys bridgerensis* Hutchison 1991; *Xenochelys lostcabinensis* Hutchison 1991; and *Xenochelys floridensis* sp. nov. (described below).

Diagnosis: Plastron lobes moderately broad; plastron length 67–78% of carapace length; entoplastron present; no anterior or posterior lobe kinesis; humeral undivided; scar for the origin of the acromial ligament restricted to the area of the epi-entoplastral suture; epiplastron with anterolaterally, anterodorsally directed projection; humeral-femoral sulcus lies on anterior moiety of hypoplastron; moderate to well-developed caudal notch between xiphiplastra; C2 not contacting P3; carapace distinctly tricarinate to smooth; C1 rib enters middle or posterior moiety of P3 and terminates in anterior P4 or P3-P4 suture; six neurals; perimeter length of nuchal longer to shorter than any peripheral; distinct cusp to no cusp at posterolateral margin of femoral; posterior buttress terminates in anterior two-thirds of P7; gular-humeral sulcus on anterior moiety of entoplastron; M10 distinctly elevated (Hutchison 1991).

Xenochelys floridensis sp. nov.

Holotype: UF 150006-150007, associated partial left and right hypoplastra.

Paratypes: UF 150001, right peripheral 8; UF 150002, left costal 3; UF 150005, left hypoplastron; UF 150008, right hyoplastron fragment; UF 150009, entoplastron; UF 150017, left peripheral 1; UF 150018, left peripheral 3; UF 150019, right peripheral 5; UF 150020, neural 1; UF 150021, neural 2 or 4; UF 150022, left xiphiplastron; UF 150023, mandible. It is possible that UF 150017-UF 150023 are elements from a single individual, because they were found in the same fissure-fill, exhibit the same color and preservation, and appear to be from a similar sized animal.

Holoype locality, unit, and age: Brooksville 2, Hernando Co., Florida, USA; sub-locality 1B; early Arikareean, late Oligocene.

Etymology: Specific epithet refers to the provenance of the type material from the State of Florida.

Diagnosis: Small adult body size (12–14 cm estimated carapace length); lateral sulci for V1 (V1-PL1 sulci) contained on nuchal, not extending onto P1; carapace smooth with no dorsal carinae on neurals or costals (at least on the

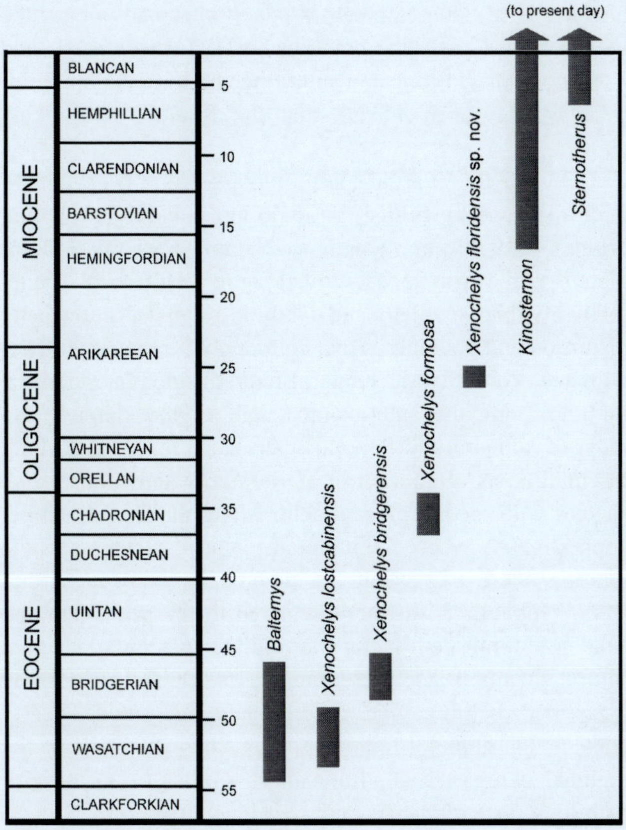

Fig. 25.5 Temporal distribution of occurrences for North American kinosternines. Occurrences of *Baltemys*, *Xenochelys lostcabinensis*, *X. bridgerensis*, *X. formosa*, and *Kinosternon* after Hutchison (1991, 1998), and Holman (1998). Hemphillian occurrence of *Sternotherus* after Bourque (2011)

anterior and mid portions of the carapace; the posterior portion is unknown at this time); costiform processes extend into P2; lack of femoral-anal scale cusp along the posterolateral margin of the xiphiplastron; visceral margin of femoral and anal scales very reduced and narrow; inguinal-humeral sulcus nearly contacting humeral-femoral sulcus at the inguinal buttress of the hypoplastron; humeral-femoral sulci curve posteriorly before contacting at midline creating an obtuse "V" shape; mandible moderately wide with slightly expanded triturating surfaces just at and anterior to the coronoid processes; lingual shelf fades anteriorly, becoming smooth at the dentary symphysis.

Description

Short descriptive accounts are provided below for each of the referred carapace, plastron, and mandible specimens of *Xenochelys floridensis*.

Carapace: UF 150020—Complete N1; length = 14.22 mm, width = 8.44 mm. The N1 is hexagonal in shape and elongate, being widest near the posterior end. This is also typical of the kinosternids *Staurotypus*, *Claudius*, *Baltemys*, *Xenochelys lostcabinensis*, *X. bridgerensis*, *X. formosa*, *Kinosternon*, and *Sternotherus* (character state 5 in Hutchison 1991). A lightly incised V1-2 sulcus is present at approximately half the length of the specimen, and by its shape indicates that V1 on this individual would have been rounded or squared (not pointed) posteriorly. Although the dorsal face of the element shows some relief, no discernible keel is present. The N1-nuchal suture is broad indicating full contact with the nuchal.

UF 150021—Complete N2 or N4; length = 9.71 mm, width = 7.03 mm. This neural is hexagonal, small, thick, and widest near the posterior. No trace of a dorsal keel is present.

UF 150002—Complete left C3; height = 33.58 mm. Element is smooth and acarinate.

UF 150015—Complete left P1; length = 10.09 mm, height = 10.52 mm. The dorsal face lacks any trace of the V1 sulcus. Ventrally, a cavity for the costiform process runs through the entire element, which likely terminated on P2. The element lacks an incised musk duct groove or any trace of the terminus for this groove.

UF 150016—Complete right P1; length = 10.22 mm, height = 11.53 mm (description same as above for UF 150015).

UF 150017—Complete left P1; length = 12.12 mm, height = 13.54 mm (description same as above for UF 150015).

UF 150011—Complete right P3; length = 11.94 mm, height = 12.58 mm. The ventral face exhibits an incised musk duct groove that runs the complete length of the element. This fact, coupled with the complete lack of a musk duct impression on the P1 (as mentioned above), indicates that the anterior musk duct groove terminated on P2. Just distal and adjacent to the musk duct and on the P3-4 suture are two indentations where the axillary buttress would have terminated. The cavity for the C1 rib end is present on the posterior of the element and terminates just anterior of the P3-4 suture. This character is primitive and is shared with *Hoplochelys*, *Staurotypus*, *Claudius*, *Baltemys*, *X. lostcabinensis*, *X. bridgerensis*, and *X. formosa*, but is not seen in *Kinosternon* and *Sternotherus* species (character state 32 in Hutchison 1991).

UF 150018—Complete left P3 (Fig. 25.3e); length = 12.69 mm, height = 12.73 mm. The ventral face exhibits an incised musk duct groove that runs the complete length of the element. Just distal and adjacent to the musk duct and on the P3-4 suture is an indentation where the axillary buttress would have terminated. The cavity for the C1 rib end is present posteriorly and runs through the P3-4 suture, indicating that the rib end would have terminated on P4.

UF 150019—Complete right P5; length = 14.50 mm, height = 21.37 mm. The cavity for the C3 rib end is situated approximately mid-length along the specimen.

UF 150003—Complete right P6; length = 14.97 mm, height = 20.47 mm. The specimen is worn along its edges. The marginal rim is wide with two distinct lateral carinae.

UF 150012—Partial left P7; length = 14.11 mm. The inguinal buttress would have terminated at approximately half the length of the element. This condition is also observed in *Baltemys* and *X. lostcabinensis*, *X. bridgerensis*, and *X. formosa*, and *Sternotherus* (character state 8 in Hutchison 1991). Just distal to this suture is a deep musk pore along the posterior of the M7-inguinal sulcus. The cavity for the C5 rib end enters the posterior of the specimen.

UF 150013—Partial right P7; length = 16.01 mm. This specimen differs from UF150012 in that the inguinal buttress runs the whole length of the specimen and likely terminated on the anterior-most P8. A deep musk pore is present along the M7-inguinal sulcus. Although broken, the very end of the C5 rib cavity is discernible where this rib would have entered at the posterior of the element.

UF 150001—Complete right P8 (Fig. 25.3g); length = 18.91 mm, height = 18.81 mm. The C3-4 sulcus lies on the posterior-most third of the element. The marginal rim is narrow dorsally, but with a wide underlap ventrally. Viscerally, a deeply inset step is formed by thickened bone along the marginals, similar to that seen in *Xenochelys formosa* and *Carettochelys*.

Plastron: UF 150009—Complete entoplastron (Fig. 25.3d); length = 14.42 mm, width = 13.41 mm. The specimen is somewhat tumbled looking, with the sutures worn and the sulci only moderately discernible. The medial humeral sulcus runs approximately 80% of the element, where it terminates and bifurcates anteriorly at the intergular sulci. Viscerally, the entoplastron lacks acromial pits, which implies these instead would have been on the epiplastra. The element is thickened anteriorly and posteriorly. Posteriorly, a nub-like (and very worn) acromial process is present. A slight depression, located along the medial sulcus at approximately the middle of the ventral face of the element, likely represents a scar from a bacterial or fungal infection between the scales.

UF 150008—Partial anterior portion of a right hyoplastron (Fig. 25.3c) that is broken posteriorly and along the axillary buttress. The anterior-most portion of the medial, ento-hyoplastral, and epi-hyoplastral sutures are preserved on the specimen. The piece is quite worn and tumbled, but is ascribed as a kinosternid here based on the quality of the sutures and very narrow visceral scale margin. The epi-hyoplastral suture is rounded as in the holotype of *X. bridgerensis*. The element is also thickened along the medial suture, as in the holotypes of *X. lostcabinensis* and *X. bridgerensis*, Elongated sutural teeth along the epi-ento-hyoplastral suture indicate that the forelobe was likely akinetic.

UF 150005—Partial left hypoplastron (Fig. 25.3a); width (including bridge) = 35.66 mm, width (excluding bridge) = 19.09 mm, length ≈ 21 mm. This fairly complete hypoplastron appears to belong to a large adult, being relatively thick medially and exhibiting signs of ontogenetic distortion and pockmarks that commonly accompany turtles with age. Ventrally, the humeral-femoral sulcus contacts the midline along the anterior-most half of the element. No reduced humeral and femoral scale scars are present along the midline as often seen in *Sternotherus* and *Kinosternon* species with medial plastral skin. Along the medial suture, approximately where the humeral-femoral sulcus meets the midline, there is a vacuity with broadly enlarged sutural teeth. This area is distinct compared to the accompanying smaller sutural teeth along the midline. A similar, distinct sutural vacuity is also present in UF 150006 and UF 150007 (described below). The inguinal-humeral sulcus nearly contacts the humeral-femoral sulcus along the waist of the inguinal buttress, a condition more similar to *X. formosa* than to *X. lostcabinensis* and *X. bridgerensis*.

UF 150006 and 150007 (holotype)—Partial associated left and right hypoplastra (Fig. 25.3f); posterior width = 14.51 mm, length = 14.25 mm (UF 150006). These appear to be the remains of an adult animal, but are slightly smaller and lack pathological wear seen in UF 150005. There are no reduced scale or plastral skin scars present as seen in modern kinosternines with plastral skin. The humeral-femoral sulci curve posteriorly and contact one another medially, just anterior to the center of the hypoplastra. As seen in UF 150005, a broadly-toothed sutural vacuity is present along the midline in both UF 150006 and UF 150007. When both of the hypoplastra are articulated, the medial vacuities form a cavity along the midline that is visible on the dorsal (visceral) faces of the hypoplastra. Isolated hypoplastra of cf. *Baltemys* from the Wasatchian of Wyoming also exhibited a similar morphology along the medial suture. I interpret this feature as having facilitated passive kinesis along the midline, perhaps in housing more connective tissue along the medial suture.

UF 150022—Partial left xiphiplastron (Fig. 25.3h) with posterior-most end missing; approximate length (at midline) = 22.33 mm, width (at hypo-xiphiplastral suture) = 16.54 mm. The element is relatively thick anteriorly, tapering off to very thin posteriorly. The hypo-xiphiplastral suture is straight to slightly concave transversely across. This suture has elongated sutural teeth, with the two longest projections at the distal edge of the element, which implies that the posterior lobe was akinetic. The femoral-anal sulcus is moderately sinuous, not straight as in the types of *Xenochelys lostcabinensis* and *X. bridgerensis*. No reduced femoral or anal scale scars are present along the midline as in most *Sternotherus* with plastral skin. The lateral edge is smooth and un-notched, a condition shared with *X. formosa* (as well as some *Sternotherus* specimens, particularly in older individuals of *S. minor* and *S. odoratus*). A distinct cusp is present at the femoral-anal sulcus in *X. lostcabinensis* and *X. bridgerensis*. The visceral femoral and anal scale margin is very narrow, indicating that the skin would have attached very near to the margin of the element in life, a condition similar to that seen in *Sternotherus*. Conversely, the visceral femoral-anal scale margin is broader in other *Xenochelys* species, particularly so in *X. formosa* for which it is very broad and robust. Although the posterior of the element is broken, *X. floridensis* likely possessed a caudal notch similar in shape to that of *X. formosa* and *Sternotherus*.

Mandible: UF 150023—Partial mandible comprised of the left and right dentaries (Figs. 25.3b, 25.8b); length = 15.40 mm, width (at coronoid processes) = 15.82 mm. This specimen represents the only known mandible for the genus *Xenochelys*. It is moderately wide, with a lingually expanded shelf at the base of and just anterior to the coronoid processes. This shelf tapers off anteriorly, where it terminates just posterior to the dentary symphysis. Though the triturating surfaces are moderately expanded in UF 150023, they appear not as expanded as the broad palate seen in the only known skull of *X. formosa*, YPMPU 13686 sensu Williams (1952) (see Fig. 25.8). The symphyseal hook is moderately curved upward, being relatively low compared to turtles such as *Kinosternon subrubrum* ssp. and *Claudius*. In many respects, the jaw of *X. floridensis* looks similar to *Kinosternon oaxacae* (Fig. 25.8i) and *Sternotherus* species being comparable in overall form to younger pre-megacephalic individuals of *S. minor minor* and *S. m. peltifer*, as well as some older individuals of *S. odoratus*. However, it lacks the lingual shelf at the jaw symphysis seen in *Sternotherus*.

Remarks

Type material of *Xenochelys floridensis* described above is ascribed to the genus *Xenochelys* based on a suite of characters, including: presence of an entoplastron; possession of akinetic plastral lobes; position of the humeral-femoral sulcus lying on the anterior portion of the hypoplastron; smooth carapace with no discernible dorsal carinae on the neurals or costals (although the rear of the carapace is unknown and potentially could possess a slight medial keel

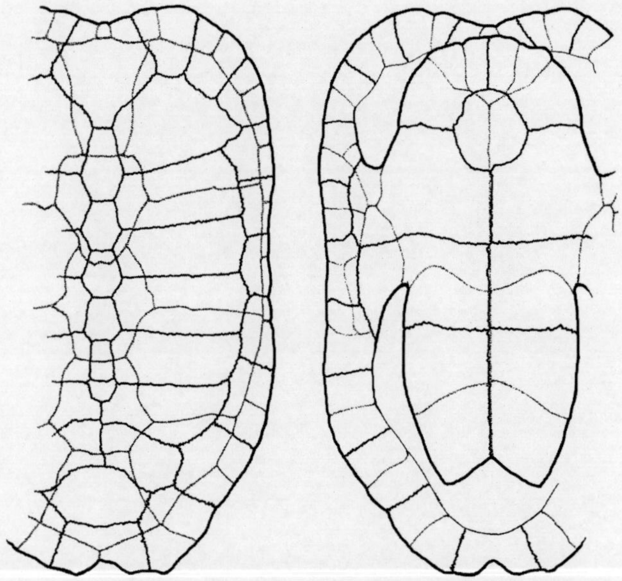

Fig. 25.6 Illustration of shell of *Xenochelys formosa* Hay, in dorsal (*left*) and ventral (*right*) views. Modified from Hay (1906, 1908)

posteriorly); C1 rib entering the posterior portion of P3 and terminating on P4 or just at the P3-P4 suture; no cusp along the posterolateral margin of the femoral-anal scales; inguinal buttress terminates (in one of two available specimens) on anterior two-thirds of P7; and gular-humeral sulcus is contained on anterior portion of the entoplastron. As of yet, no complete associated specimens are known. The description presented here is based on numerous, primarily isolated elements from the Brooksville 2 locality used to create a composite.

Xenochleys floridensis is morphologically and temporally most closely related to *X. formosa*, a relatively large kinosternid (Figs. 25.6, 25.7, 25.8c) from the Chadronian (34–37 Ma) (Prothero and Emry 2004) of South Dakota and Wyoming (Hutchison 1991). Synapomorphies of *X. floridensis* with *X. formosa* include: the lack of a notched femoral-anal scale along the margin of the xiphiplastron; the presence of a deeply undercut step along the visceral margin of the posterior peripherals (similar to that seen in *Carettochelys*); and posteriorly inflected humeral-femoral sulci on the hypoplastra that form a V-shape at the midline, whereas these sulci meet transversely straight at the midline in *X. lostcabinensis* and *X. bridgerensis*. (Fig. 25.7). However, the new species is plesiomorphic in that it attains a smaller adult body size than *X. formosa*; in this feature it is more similar to the Eocene species *X. lostcabinensis*, *X. bridgerensis*, and *Baltemys staurogastros* Hutchison 1991. *X. floridensis* also differs from the type of *X. formosa* (AMNH 1097) in having a highly reduced visceral margin on the xiphiplastron for the femoral and anal scales, and no

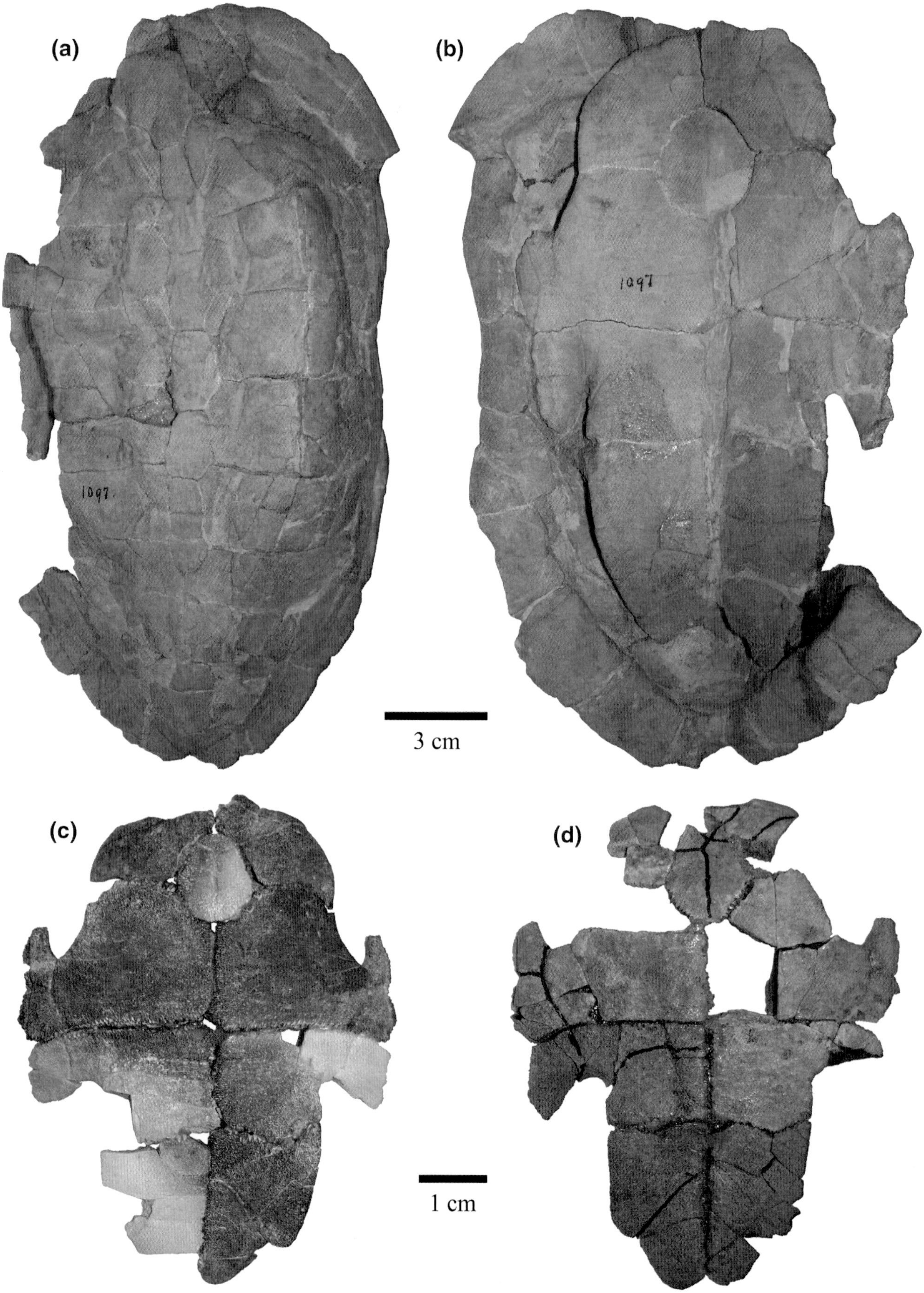
(a)
(b)
1097
1097.
3 cm
(c)
(d)
1 cm

◄**Fig. 25.7** Comparison of previously described *Xenochelys* species. **a**, **b** Shell of *Xenochelys formosa* Hay, AMNH 1097 (holotype), from the Chadronian of South Dakota: **a** dorsal view; **b** ventral view. Although only about half the size, *X. floridensis* sp. nov. is morphologically most similar to this taxon, however the two species are separated temporally by at least 6 million years. **c** Plastron of *Xenochelys bridgerensis* Hutchison, UCMP 121734 (holotype), in ventral view, from the Bridgerian of Wyoming. **d** Plastron of *Xenochelys lostcabinensis* Hutchison, UCMP 112341 (holotype), in ventral view, from the Wasatchian of Wyoming. *Xenochelys floridensis* is similar to *X. bridgerensis* and *X. lostcabinensis* with respect to size, but the new species lacks the distinct notch along the edge of the xiphiplastron between the femoral and anal scales. That character is shared between the early Eocene *Xenochelys* and the genus *Baltemys*. Specimens at different magnifications

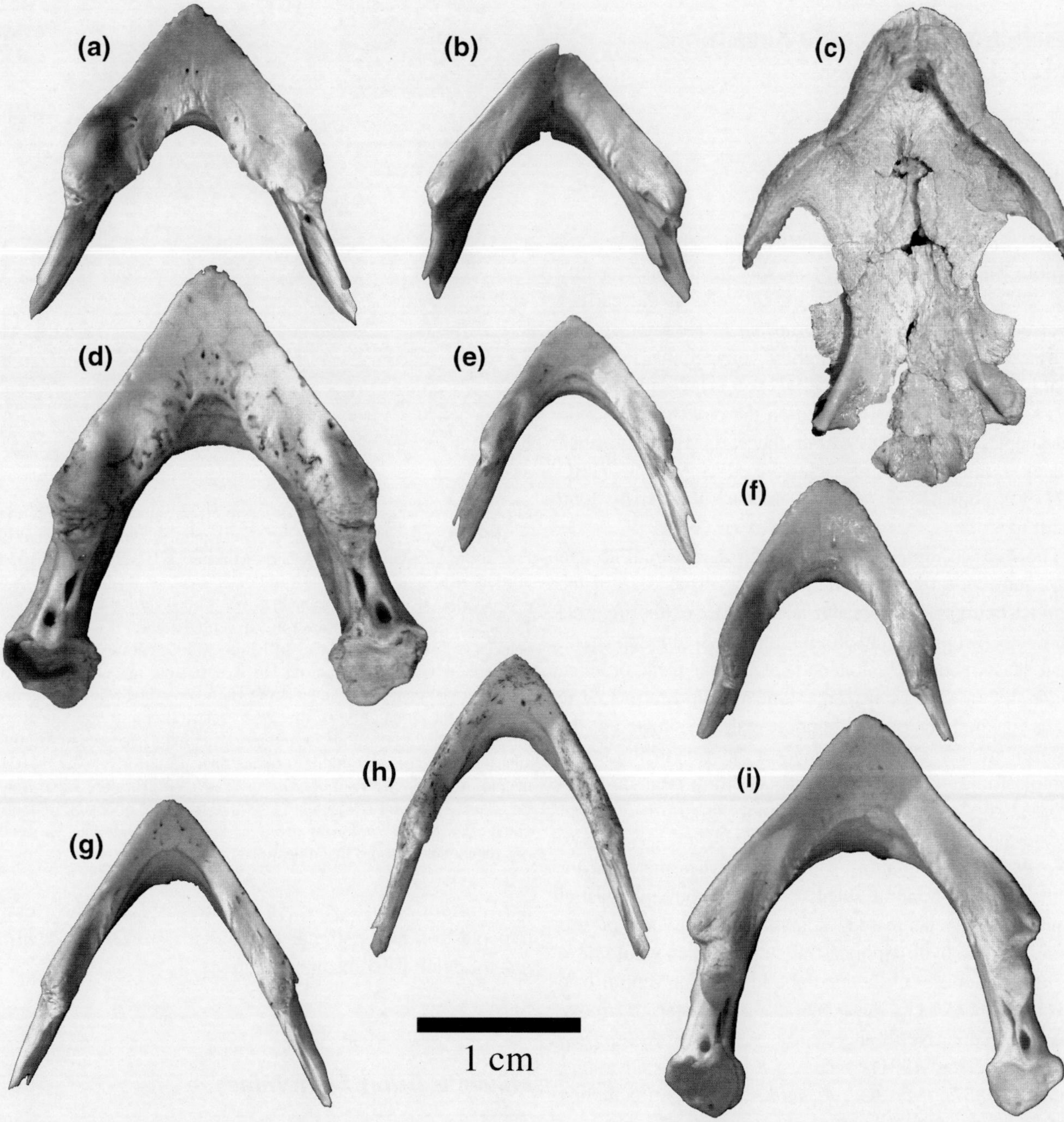

Fig. 25.8 Comparison of jaws among some extant kinosternines and fossil *Xenochelys* (all in occlusal view). **a** UF/H 50058, *Sternotherus m. minor*. **b** UF 150023, *Xenochelys floridensis* sp. nov. **c** YPMPU 13686, palate of cf. *Xenochelys formosa* (Williams 1952). **d** UF/H 150286, *Sternotherus carinatus*. **e** UF/H 47704, *Sternotherus odoratus*. **f** UF/H 18988, *Kinosternon baurii*. **g** UF/H 57914, *Kinosternon herrerai*. **h** UF/H 68599, *Kinosternon subrubrum*. **i** UF/H 57910, *Kinosternon oaxacae*. Image of YPMPU 13686 © Division of Vertebrate Paleontology, Yale Peabody Museum of Natural History, Yale University, New Haven, Connecticut, USA. All specimens at same magnification

signs of carinae on the neurals or costals on at least the anterior portion of the carapace. In AMNH 1097 the visceral margin on the xiphiplastron is relatively thick and wide, and a subtle medial keel is present on the nuchal, N1, and suprapygal, and lateral keels present on C1-C6, being most pronounced on C1-C3 and becoming less so along the anterior portion of C4.

Descriptions of Miocene Kinosternid Material, by Locality

Suwannee Springs

***Kinosternon* n. sp.**: A single left hypoplastron of a kinosternid, UF/TRO 2621, is known from this locality (Fig. 25.9a). This specimen represents one of the earliest occurrences of the genus *Kinosternon* at between 16 and 17 Ma, which is as old or slightly younger than an early record reported by Hutchison (1991) at ~17 Ma (FAMNH 11711). Holman (1998) reported on the oldest *Kinosternon* material (USNM 483389) from the early Hemingfordian Pollack Farm locality of Delaware (17.9 Ma). UF/TRO 2621 was collected and donated to the FLMNH by John Waldrop.

The hypoplastron is 9.11 mm long at the midline and 21.32 mm wide from the midline to the distal edge of the inguinal buttress. The breadth at the waist of the inguinal notch is 6.76 mm. The hypo-xiphiplastral suture appears to be at least moderately kinetic, the sutural teeth oriented diagonally inward as in some modern *Kinosternon*. The sulcus for the humeral and femoral scales is situated at the posterior moiety of the element just anterior to the hypo-xiphiplastral suture, which is the condition seen in *Kinosternon*. This sulcus is positioned more anteriorly in other kinosternids, such as *Sternotherus*, *Xenochelys*, *Baltemys*, *Staurotypus*, and *Claudius*. The area between the posterior of the humeral-femoral sulcus and the hypo-xiphiplastral suture is lightly depressed, indicating the presence of soft tissue over the hypo-xiphiplastral suture, which would have facilitated mobility of the hindlobe in life. The ventral face is very flat and lacks the humeral-inguinal step or groove present in some North and Central American *Kinosternon* species (Iverson 1991), such as *Kinosternon flavescens* (Agassiz 1857), *Kinosternon subrubrum* ssp. (excluding some *steindachneri*), *Kinosternon baurii*, some *Kinosternon sonoriense* LeConte 1854, and *Kinosternon hirtipes* (Wagler 1830) (e.g., UF/H 61938). A relatively flattened hypoplastron such as this is probably primitive for the genus. The inguinal buttress curves upward only slightly at the posterior-most end. In some respects, the fossil hypoplastron looks similar to modern *Kinosternon angustipons* Legler 1965 examined here.

Fig. 25.9 Miocene kinosternid fossils from north-central Florida. **a** UF/TRO 2621, left hypoplastron, in ventral view, of *Kinosternon* n. sp. from Suwannee Springs, early Miocene. **b** UF 225682, proximal portion of a right costal (C4?), in dorsal view, of Kinosternidae gen. et sp. indet. from the Suwannee River Mine, early Miocene. **c** UF 111725, partial right hyoplastron, in ventral view, of Kinosternidae gen. et sp. indet. from the Milwhite Gunn Farm Mine, middle Miocene. **d** UF 150034, partial left hyoplastron, in ventral view, of Kinosternidae gen. et sp. indet. from Haile 19A, late Miocene. In this specimen the orientation of the epi-hyoplastral suture is atypical for modern *Kinosternon* and *Sternotherus* in that it is arched slightly anteriorly towards the midline, rather than posteriorly. **e** UF 43031, nuchal, in dorsal view, of cf. *Kinosternon* n. sp. from the Love Bone Bed, late Miocene. **f** UF 150029, proximal fragment of costal, in dorsal view, of cf. *Kinosternon* n. sp. from the Love Bone Bed, late Miocene. All specimens at same magnification

Milwhite Gunn Farm Mine

Kinosternidae gen. et sp. indet.: Kinosternids are represented in the Willacoochee fauna (Bryant 1991) by shell fragments consisting of a partial right hyoplastron (UF 111725), posterior left costal (UF 111753), partial proximal end of a costal (UF 246135), and a fragment of either a right hyoplastron or left hypoplastron (UF 246136).

The right hyoplastron UF 111725 (Fig. 25.9c) is diagnosed as a kinosternid based on its small size, shallow nature of the sutural teeth, relatively low and shortened axillary bridge buttress, and lack of a transverse sulcus across the element. UF 111725 likely represents a new taxon and looks dissimilar to any currently described fossil or recent taxa that I have examined. Overall the specimen is relatively thin. The axillary buttress is preserved with the sulci for the axillary and inguinal scales, however, the proximal portion of the element is missing. The distal-most portion of the epi-hyoplastral suture is present, but it is speculative as to whether or not this suture was kinetic. The waist of the bridge is wide. The axillary buttress is slightly upwardly curved anteriorly, but almost flat posteriorly. The axillary and inguinal scales broadly contact one another just anterior to centrally on the element, similar to that observed in *Xenochelys*. The ventral-most portion of M5 is present along the bridge-peripheral suture. The hyo-hypoplastral suture projects posteriorly at the inguinal scale, similar to the condition observed in *Xenochelys* species. The costals UF 111753 and 246135 are smooth dorsally and lack any sign of carinae.

Suwannee River Mine

Kinosternidae gen. et sp. indet.: UF 225682, the proximal portion of a right costal (C4?) (Fig. 25.9b), is identified as a kinosternid based on its small size, shallow sutural teeth, and stepped pleural sulcus (where the anterior scale overlaps the adjacent pleural scale). The specimen appears to have come from an animal with an estimated carapace length of between 15 and 17 cm. With respect to size and shape, this specimen looks very similar to a C4 of *Kinosternon oaxacae* Berry and Iverson 1980 examined (e.g., UF/H 57910). It has a flat neural-costal suture anteriorly and a rounded one posteriorly. A very subtle lateral keel is present, at which a slight thickening of the shell and downward curving of the element occurs.

Swift Creek Mine

Kinosternidae gen. et sp. indet.: This questionable record is based on UF 150035, which is a complete left costal that is very worn. This specimen may be from a Recent animal, because the bone is lightweight, porous, and deteriorated. Nevertheless, it is reported here in the chance that it is a fossil. It is relatively small, thin, smooth, and lacks a keel, similar to *Sternotherus odoratus, Kinosternon baurii*, and *K. subrubrum* ssp.

Love Bone Bed

Cf. *Kinosternon* n. sp.: Kinosternids are present, but rare in the Love Bone Bed fauna. Material from the site includes a nuchal (UF 43031), proximal portion of a mid-body costal (UF 150029), a costal fragment (UF 150030), a right P1 (UF 150031), and a right P8 (UF 150032). This material represents an undescribed *Kinosternon* or *Kinosternon*-like turtle with a moderately tricarinate carapace, and it is presumed that these unassociated specimens belong to the same taxon. Each is described in more detail below.

UF 43031—A complete nuchal (Fig. 25.9e), length (at midline) = 20.76 mm, width = 27.04 mm. UF 43031 is ascribed to the Kinosternidae based on its relatively small size, shallow nature of the sutural teeth, and presence of costiform processes. The element is very thick at the anterior margin and highly buttressed viscerally. A distinct medial keel runs the length of the nuchal dorsally, beginning just posterior of the cervical scute. V1 is broad anteriorly, with the lateral sulci not contacting M1 on the surface of the nuchal. This indicates that the V1 sulci either contacted along the nuchal-P1 suture or on P1. The posterior-most suture (pinnacle of the element) is flat and not pointed, indicating broad sutural contact with N1. The cervical scale is longer than broad dorsally and wider than long ventrally, and it is not highly reduced as seen in some *Sternotherus* and *Kinosternon* species. Well-discerned growth annuli are present on the M1 scale set and V1. V1 is slightly raised from scale set PL1. The costiform processes are present but small, being reduced to nub-like teeth that, that if broken, appear to be only slightly so.

The Love Bone Bed nuchal most closely resembles nuchals of *Kinosternon*. Reasons for this include the overall robustness of the element, the wide breadth of V1, and the wide visceral margin of the cervical and first marginal scales. The nature of the keel and extent of growth annuli are similar to the *Kinosternon scorpioides* (Linnaeus 1766) complex. The lack of or reduction of the costiform processes in UF 43031 is seen in *Kinosternon herrerai* Stejneger 1925 (Iverson 1991) such as in UF/H 57914, but a fossil right P1 (see below UF 150031) indicates that these processes were more extensive in some individuals. However, in the absence of diagnostic plastral material from this locality, identification at the generic level remains speculative.

UF 150029 (Fig. 25.9f) and UF 150030 (not figured)—Both specimens are costal fragments that each possess slight lateral carinae accompanied by some creasing.

At these carinae, the elements become slightly angled downward. These along with the nuchal UF 43031 indicate that this animal had a relatively well-defined medial keel with two lesser lateral carinae, similar to some populations of *Kinosternon scorpioides*.

UF 150031—Complete right P1; length = 10.38 mm, height = 10.8 mm. A large pockmark at the nuchal-P1 suture makes it difficult to discern whether or not the V1-PL1 sulcus contacted the face of the element. Two or three well-defined growth annuli are present on the dorsal face. M1-2 are somewhat bulbous. In visceral aspect, a groove for the costiform process runs through the element and appears to terminate at the P1-2 suture, contrary to the lack of extensive costiform processes observed in UF 43031.

UF 150032—Complete right P8; length = 14.03 mm, height = 13.18 mm. The marginals are thick, fairly bulbous, and non-flared. The inguinal bridge of the plastron likely terminated at the P7-8 suture, which is indicated by a small, pitted indentation. This implies that the hyoplastra were posteriorly extensive, and that this species had a relatively broad plastron as in the following (after Iverson 1991): *Kinosternon acutum* (Gray 1831), *Kinosternon creaseri* (Hartweg 1934), *Kinosternon integrum* (LeConte 1854), *Kinosternon leucostomum* (Duméril and Bibron in ® 1851), *Kinosternon oaxacae*, *Kinosternon scorpioides*, and *Kinosternon subrubrum*.

The kinosternid from the Love Bone Bed had a relatively thick carapace with a well-defined medial keel and two lesser lateral carinae, and posteriorly extensive hyoplastra similar to some Central American *Kinosternon* including members of the *K. scorpioides* complex. Notably, Holman (1998) reported the occurrence of a thick-shelled species of *Kinosternon* from the early Miocene (Hemingfordian) Pollack Farm fauna in Delaware that he felt resembled modern *K. scorpioides* or *K. leucostomum*. This and the Love Bone Bed records are evidence for the presence of turtles possibly sister to or convergent with modern Central-South American kinosternids in the eastern United States during the early and late Miocene.

Haile 19A

Kinosternidae gen. et sp. indet.: A single partial left hyoplastron (UF 150034) (Fig. 25.9d) represents the family at this site. The hyoplastron is 14.41 mm long at the midline and ~23.90 mm wide from the midline to the waist of the axillary buttress. It is identified as a kinosternid based on its small size, shallow nature of the sutural teeth, and lack of a transverse sulcus across the element. It is broken at the axillary buttress and appears to be of an aged adult, bearing pockmarks and signs of wear on the ventral face. This specimen likely represents a new taxon, but again insufficient material hinders a diagnosable description here.

The epi-hyoplastral suture appears highly kinetic, with a distinct lip present along the suture on the ventral surface, and inwardly oriented sutural teeth that are reduced in number (~8). This suture is transversely more straight than observed in extant *Kinosternon* and *Sternotherus* species examined. The epi-hyoplastral suture usually curves inward (posteriorly) at the midline in modern members of these genera, whereas in UF 150034 the suture conversely flares slightly anteriorly. No distinct suture for an entoplastron is present. Two or three discernible annuli are present along the midline, and these appear to be growth annuli as opposed to plastral skin scars seen in most *Sternotherus* and some *Kinosternon*. Large sutural teeth are present along the medial suture. The axillary bridge is narrow at the waist and relatively flattened, indicating that the buttress was not highly inclined.

Withlacoochee River 4A

***Kinosternon* sp. (aff. *K. subrubrum-baurii* clade)**: Two kinosternid specimens are known from this locality. The more complete specimen, UF 13893, is a partial carapace consisting of associated right P4, C2, and C4-C5 (Fig. 25.10). P4 is approx. 15.2 mm tall and 11 mm wide, C2 is 26 mm tall and 12 mm wide, C4 is 30.2 mm tall and 12 mm wide, and C5 is 30.3 mm tall and 10.3 mm wide. The overall size of the shell is small, comparable to *Sternotherus odoratus*, *Kinosternon baurii*, and *K. subrubrum* ssp. The costals are smooth, lack any trace of a lateral keel dorsally, and are highly curved proximally. UF 13893 is identified here as *Kinosternon* due to the presence of a well-formed sutural contact for the axillary bridge buttress on the P4 (just above the musk duct groove), which in *Sternotherus* is either not as well-developed or not present at all on P4, or occurs on the P4-5 suture. The condition in the fossil is much like that seen in *K. baurii* and *K. subrubrum* ssp. with a long sutural groove (that would have received a very thin, elongate, anteriorly projecting process of the axillary bridge) present along the visceral face of the element and more developed sutural contact for the axillary bridge at the P4-5 suture. UF 13893 greatly resembles specimens of *K. subrubrum* ssp. and *K. baurii* in terms of size, thickness, and morphology. This specimen likely represents the oldest occurrence of the *K. subrubrum-baurii* group in Florida. UF 13893 is well-mineralized and was noted in the specimen record as having been collected from in situ clay, so it is unlikely a modern contaminant.

Fig. 25.10 UF 13893, partial carapace, consisting of associated right P4, C2, and C4-C5, of *Kinosternon* sp. (aff. *K. subrubrum-baurii* clade) from Withlacoochee 4A, late Miocene, in **a** dorsal and **b** visceral views

A second specimen, UF 150033 (not figured) is the proximal portion of a left costal (C3?). This specimen belongs to a turtle with an approximate carapace length of 9 cm. It is similar to costals of *Sternotherus odoratus, Kinosternon baurii*, and *K. subrubrum* ssp. in its size, shape, in being smooth and thin, and in lacking a keel.

Discussion and Conclusions

The fossils examined here are fragmentary, and as such, taxonomic assignment at the generic level was done conservatively and was based on comparisons with fossil and Recent taxa for which more complete specimens are known. From an evolutionary perspective, the Brooksville 2 kinosternid is important in that it occurs during an interval of time in which almost nothing is known about kinosternid diversity (Fig. 25.5). The Brooksville 2 material is tentatively assigned to the genus *Xenochelys* until more complete specimens become available to say otherwise. Alternatively, future specimens of this taxon could warrant the use of *Xenochelys* to encompass *formosa* and *floridensis*, and a new genus or subgenus possibly erected for the Eocene taxa *bridgerensis* and *lostcabinensis*, due to differences (albeit relatively subtle) in plastral characters such as the orientation of the humeral-femoral and femoral-anal sulci, femoral-anal scale and anal notches, and carapacial characters like the presence or lack of a visceral shelf along the posterior peripherals. Hopefully future discoveries of this small turtle will elaborate on its true significance.

Interestingly, the first representatives of the genus *Kinosternon* from the Miocene of Florida look similar in some respects to modern members of the genus that today live in the southwestern United States and Central-South America, such as *K. angustipons*, *K. hirtipes*, *K. herrerai*, and *K. scorpioides*. This is with the exclusion of the late early Hemphillian Withlacoochee 4A material, which appears to represent the first occurrence of the *K. subrubrum-baurii* group in the state. This late early Hemphillian occurrence is not unlikely, given that a portion of kinosternid material I am currently studying from the late Hemphillian (latest Miocene, early Pliocene) of the Bone Valley Formation (Polk County, Florida), also represents this group. Needless to say, many paleogeographic and evolutionary questions still remain regarding the Kinosternidae, particularly in the eastern United States.

Acknowledgments I wish to thank the following for their assistance during this study: R. Hulbert, Jr., J. Bloch, D. Ehret, K. Krysko, and A. Hastings (FLMNH); P. Holroyd and H. Hutchison (UCMP); W. Joyce, D. Brinkman, and S. Chester (YPM); C. Mehling (AMNH); J. Knight (South Carolina State Museum), J. Waldrop for locality information and donation of his Suwannee Springs material. Additionally, Don Brinkman (Royal Tyrrell Museum), R. Hulbert, Jr. (FLMNH), G. Gaffney (AMNH), G. Morgan (New Mexico Museum of Natural History and Science), and J. Parham (Alabama Museum of Natural History) provided helpful comments and suggestions that greatly improved this manuscript, and Jim Gardner (Royal Tyrrell Museum) did the final editing and formatting. This is University of Florida Contribution to Paleobiology 632.

References

Agassiz, L. (1857). *Contributions to the natural history of the United States of America* (Vols. 1–2). Boston: Little, Brown.

Becker, J. J. (1985). Fossil herons (Aves: Ardeidae) of the late Miocene and early Pliocene of Florida. *Journal of Vertebrate Paleontology*, *5*, 24–31.

Berry, J. F., & Iverson, J. B. (1980). A new species of mud turtle, genus *Kinosternon*, from Oaxaca, Mexico. *Journal of Herpetology*, *14*, 313–320.

Bourque, J. R. (2011). Reassessment of a putative fossil stinkpot (Kinosternidae: *Sternotherus*) from the late Miocene (Clarendonian) of Kansas. *Journal of Herpetology, 45*, 234–237.

Bryant, J. D. (1991). New early Barstovian (middle Miocene) vertebrates from the upper Torreya Formation, eastern Florida panhandle. *Journal of Vertebrate Paleontology, 11*, 472–489.

Carr, A. (1938). *Pseudemys nelsoni*, a new turtle from Florida. *Occasional Papers of the Boston Society of Natural History, 8*, 305–310.

Carr, A. (1952). *Handbook of turtles: The turtles of the United States, Canada, and Baja California*. Ithaca: Comstock Publishing.

Conant, R., & Collins, J. T. (1998). *Reptiles and amphibians of Eastern/Central North America. Peterson Field Guides* (3rd ed.). New York: Houghton Mifflin Co.

Duméril, A. M. C., & Duméril, A. H. A. (1851). *Catalogue Methodoque de la Collection des Reptiles du Museum d'Histoire Naturelle*. Paris: Gide and Boudry.

Ernst, C. H., & Lovich, J. E. (2009). *Turtles of the United States and Canada* (2nd ed.). Baltimore: Johns Hopkins Press.

Frailey, C. D. (1978). An early Miocene (Arikareean) fauna from northcentral Florida (*the SB-1A Local Fauna*). *Occasional Papers, Museum of Natural History, the University of Kansas, 85*, 1–20.

Garman, S. (1891). On a tortoise found in Florida and Cuba, *Cinosternum baurii*. *Bulletin of the Essex Institute, 22*, 1–14.

Gray, J. E. (1831). *Synopsis reptilium or short descriptions of the species of reptiles. Part* I. *Cataphracta, tortoises, crocodiles, and enaliosaurians*. London: Treuttel, Wurz Co.

Gray, J. E. (1856). On some new species of freshwater tortoises from North America, Ceylon, and Australia, in the collection of the British Museum. *Proceedings of the Zoological Society of London, 1855* (pp. 197–202).

Hartweg, N. (1934). Description of a new kinosternid from Yucatan. *Occasional Papers from the Museum of Zoology, University of Michigan, 371*, 1–5.

Hay, O. P. (1906). Descriptions of two new genera (*Echmatemys* and *Xenochelys*) and two new species (*Xenochelys formosa* and *Terrapene putnami*) of fossil turtles. *Bulletin of the American Museum of Natural History, 22*, 27–31.

Hay, O. P. (1908). *The fossil turtles of North America*. Washington: Carnegie Institute, Publication No. 75.

Hayes, F. G. (2000). The Brooksville 2 local fauna (Arikareean, latest Oligocene): Hernando County, Florida. *Bulletin of the Florida Museum of Natural History, 43*, 1–47.

Holman, J. A. (1995). *Pleistocene amphibians and reptiles in North America*. New York: Oxford University Press.

Holman, J. A. (1998). Reptiles of the lower Miocene (Hemingfordian) Pollack Farm fossil site, Delaware. In R. N. Benson (Ed.), *Geology and Paleontology of the lower Miocene Pollack Farm Fossil Site Delaware* (pp. 141–148). Delaware Geological Survey, Special Publication, No. 21. State of Delaware: University of Delaware.

Hulbert, R. C. (1992). *A checklist of the fossil vertebrates of Florida. Papers in Florida Paleontology #6*. Gainesville, Florida: Florida Paleontological Society, Inc.

Hulbert, R. C. (2001). Chapter 3. In R. C. Hulbert (Ed.), *The fossil vertebrates of Florida* (pp. 34–74). Gainesville: University Press of Florida.

Hutchison, J. H. (1991). Early Kinosterninae (Reptilia: Testudines) and their phylogenetic significance. *Journal of Vertebrate Paleontology, 11*, 145–167.

Hutchison, J. H. (1998). Turtles across the Paleocene/Eocene epoch boundary in west-central North America. In M. P. Aubry, S. G. Lucas, & W. A. Berggren (Eds.), *Late Paleocene-Early Eocene climatic and biotic events in the marine and terrestrial records* (pp. 401–408). New York: Columbia University Press.

Hutchison, J. H. & Weems R. E. (1998). Paleocene turtle remains from South Carolina. In A. E. Sanders (Ed.), *Paleobiology of the Williamsburg Formation (Black Mingo Group; Paleocene) of South Carolina, U.S.A.* (pp. 165–195). Transactions of the American Philosophical Society 88: Pt. 4.

Iverson, J. B. (1978). Variation in striped mud turtles, *Kinosternon baurii* (Reptilia, Testudines, Kinosternidae). *Journal of Herpetology, 12*, 135–142.

Iverson, J. B. (1991). Phylogenetic hypotheses for the evolution of modern kinosternine turtles. *Herpetological Monographs, 5*, 1–27.

Jackson, D. R. (1976). The status of the Pliocene turtles *Pseudemys caelata* Hay and *Chrysemys carri* Rose and Weaver. *Copeia, 1976*, 655–659.

Jackson, D. R. (1978). Evolution and fossil record of the chicken turtle *Deirochelys*, with a re-evaluation of the genus. *Tulane Studies in Zoology and Botany, 20*, 35–55.

Lacépède, B. G. E. (1788). *Histoire naturelle des quadrupèdes ovipares et des serpents* (Vol. 1) [*Ovipares*]. Paris: Hôtel de Thou.

LeConte, J. (1830). Description of the species of North American tortoises. *Annales of the Lyceum Natural History (New York), 3*, 91–131.

LeConte, J. (1854). Description of four new species of Kinosternon. *Proceedings of the Academy of Natural Sciences of Philadelphia, 7*, 180–190.

Legler, J. M. (1965). A new species of turtle, genus *Kinosternon* from Central America. *University of Kansas Publications, Museum of Natural History, 15*, 615–625.

Linnaeus, C. (1766). *Systema Naturae* (12th ed.). Sweden: Halae Magdeborgicae.

Morgan, G. S. (1989). Miocene vertebrate faunas from the Suwannee River Basin of north Florida and south Georgia. In G. S. Morgan (Ed.), *Miocene paleontology and stratigraphy of the Suwannee River Basin of north Florida and south Georgia* (pp. 26–53). Tallahassee, Florida: Southeastern Geological Society, Guidebook #30.

Prothero, D. R., & Emry, R. J. (2004). The Chadronian, Orellan, and Whitneyan North American land mammal ages. In M. O. Woodburne (Ed.), *Late Cretaceous and Cenozoic mammals of North America: Biostratigraphy and geochronology* (pp. 156–168). New York: Columbia University Press.

Rose, F. L., & Weaver, W. G., Jr. (1966). Two new species of *Chrysemys* (= *Pseudemys*) from the Florida Pliocene. *Tulane Studies in Geology, 5*, 41–48.

Schneider, J. G. (1783). *Allegemeine naturgeschichte der schildkröten, nebst einem system*. Leipzig: Verzeichnisse der einzelnen Arten.

Schoepff, J. D. (1792–1801). *Historia testudinum iconibus illustrata*. Germany: Palm, Erlangae.

Scott, T. M. (1988). *The lithostratigraphy of the Hawthorn Group (Miocene) of Florida*. Tallahassee: Florida Geological Survey.

Siebenrock, F. (1906). Eine neue *Cinosternum*- Artaus Florida. *Zoologische Anzeigel, 30*, 727–728.

Simpson, G. G. (1930). Tertiary land mammals of Florida. *Bulletin of the American Museum of Natural History, 59*, 149–211.

Smith, H. M., & Glass, B. P. (1947). A new musk turtle from southeastern United States. *Journal of the Washington Academy of Sciences, 37*, 22–24.

Sonnini De Manoncourt, C. S., & Latreille, P. A. (1802). *Histoire naturelle des reptiles avec figures dessinées d'après nature* (Vol. 1). Paris: Chez Deterville.

Stejneger, L. (1925). New species and subspecies of American turtles. *Journal of the Washington Academy of Sciences, 15*, 462–463.

Tedford, R. H., & Hunter, M. E. (1984). Miocene marine-nonmarine correlations, atlantic and gulf coastal plains, North America. *Palaeogeography, Palaeoclimatology, Palaeoecology, 47*, 129–151.

Tedford, R. H., Albright, L. B., III, Barnosky, A. D., Ferrusquia-Villafranca, I., Hunt, R. M., Jr., Storer J. E., et al. (2004). Mammalian biochronology of the Arikareean through Hemphillian interval (late Oligocene through early Pliocene epochs). In M. O. Woodburne (Ed.), *Late Cretaceous and Cenozoic Mammals of North America: Biostratigraphy and geochronology* (pp. 169–231). New York; Columbia University Press.

Wagler, J. (1830). *Natürliches System der Amphibien, mit Vorangehender Classification der Säugthiere und Vögel*. Munich: J. G. Cotta'schen.

Weaver, W. G., Jr, & Robertson, J. S. (1967). A re-evaluation of fossil turtles of the *Chrysemys scripta* group. *Tulane studies in geology*, *5*, 53–66.

Webb, S. D., MacFadden, B. J., & Baskin, J. A. (1981). Geology and paleontology of the love bone bed from the late Miocene of Florida. *American Journal of Science*, *281*, 513–544.

White, T. E. (1942). A new alligator from the Miocene of Florida. *Copeia*, *1942*, 3–7.

Williams, E. E. (1952). A staurotypine skull from the Oligocene of South Dakota (Testudinata, Chelydridae). *Breviora*, *2*, 1–16.

Williams, E. E. (1953). A new fossil tortoise from the Thomas Farm Miocene of Florida. *Bulletin of the Museum of Comparative Zoology*, *107*, 537–554.

Chapter 26
New Turtles from the Paleogene of North America

J. Howard Hutchison

Abstract Formal names, diagnoses, and descriptions are supplied for previously identified, but informally named, cryptodiran turtles from the Paleogene of Montana and Wyoming, USA. Two new early Paleocene (Puercan) turtles are described from Montana: a peculiar trionychid, *Atoposemys entopteros* gen. et. sp. nov., from the Tullock Member of the Fort Union Formation, and a chelydrid, *Tullochelys montanus* gen. et. sp. nov., from both the Tullock Member of the Fort Union Formation and the Hell Creek Formation. From the Wasatchian (early Eocene) Wasatch and Willwood formations of Wyoming, the following taxa are described: the earliest North American emydid, *Psilosemys wyomingensis* gen. et. sp. nov.; a platysternid, *Cardichelyon rogerwoodi* gen. et. sp. nov.; and *Planetochelys dithyros* sp. nov. A new family, Planetochelyidae, within the Trionychoidae, is erected for *Planetochelys.*

Keywords Emydidae • Fort Union Formation • Planetochelyidae • Platysternidae • Tullock Member • Wasatch Formation • Willwood Formation

Introduction

Over the course of many years, I encountered fossils of several new turtle taxa from the Paleogene of Wyoming and Montana, USA, that were distinct enough to warrant scientific names, but I refrained from providing formal names because of procrastination, the hope of finding better material, discussion then in non-systematic works, page limitations, or a combination of these factors. I think now that the time has arrived to provide proper names to some of these taxa to avoid nebulous or inconsistent use of their informal names. While it is desirable to wait for the discovery of complete specimens for purposes of applying scientific names, many taxa are just not found this way. If enough fragmentary specimens and individual elements are found then composite reconstructions are possible. Because of the amount of bone preserved in a turtle shell, turtles are frequently the most common bones found at many sites. Collecting and documenting of such collections leads to a good understanding of the morphology and diversity. Most taxa in an area or formation can be recognized on even very fragmentary material once a critical amount of material has been accumulated and studied relating to size, shape, and surface texture. Although sufficient descriptions of the shell are provided here to characterize each taxon, full descriptions of variation in shells and non-shell elements and more thorough treatments of distributions and relationships will be presented elsewhere.

J. H. Hutchison (✉)
Museum of Paleontology, University of California Berkeley, Berkeley, CA 94720, USA
e-mail: howard.hutchison@gmail.com

Materials and Methods

All measurements are in millimeters. The bone and shield terminology follows that of Zangerl (1969) and Ernst and Barbour (1972). Terminology for the carapacial seams is after Tinkle (1962). Neural formula refers to the position and shape of the neurals, where "(" indicates broad end anterior (normal) and ")" indicated broad end posterior (reversed); e.g., 1)2)3)4(5(6(7(8. "Perimeter length" of the peripheral series refers to the straight-line length of a peripheral element measured from one suture to another where they intersect the perimeter of the shell in dorsal view. Carapace bones (upper case) and scales (lower case) are denoted by a letter and position number starting anteriorly (e.g., C2), as follows: C = costal; N = neural; Nu = Nuchal; P = peripheral plus number (e.g., P4); SPy = suprapygal; Py = pygal; m = marginal plus number (e.g., m4); p = pleural; v = vertebral. Abbreviations

D. B. Brinkman et al. (eds.), *Morphology and Evolution of Turtles,* Vertebrate Paleobiology and Paleoanthropology,
DOI: 10.1007/978-94-007-4309-0_26,

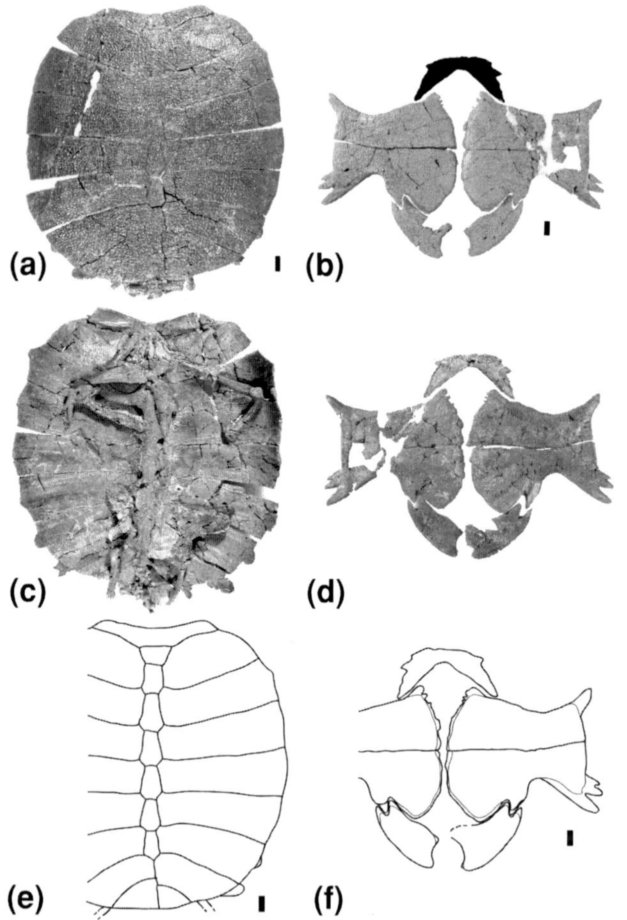

Fig. 26.1 *Atoposemys entopteros* gen. et sp. nov., UCMP 133900 (holotype). **a** Photograph of carapace, in dorsal view. **b** Photograph of incomplete plastron (hyo-, hypo-, and xiphiplastra) prepared separately from rest of specimen, in ventral view, plus infilled silhouette (shaded in *black* and in dorsal view) of entoplastron adhering to underside of carapace to show shape and in-life position of entoplastron. **c** Photograph of carapace, in ventral view, with entoplastron (exposed in dorsal aspect) and assorted elements adhering to underside of carapace. **d** Composite photograph of reconstructed plastron, depicting entoplastron (preserved with carapace) in approximate life articulation with more posterior (hyo-, hypo-, and xiphiplastra) plastral elements, all bones in dorsal view. **e** Line drawing of carapace showing sutures, in dorsal view. **f** Line drawing of reconstructed plastron, in ventral view. Scale bars about 1 cm

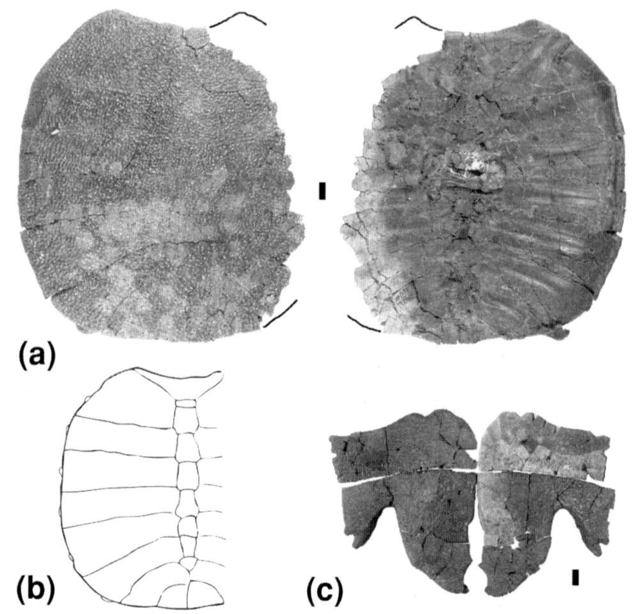

Fig. 26.2 *Atoposemys entopteros* gen. et sp. nov., UCMP 130100 (paratype). **a** Photographs of carapace, plus inferred outlines (*solid lines*) of right anterior and right posterior margins, in dorsal (*left*) and ventral (*right*) views. **b** Line drawing of carapace showing sutures, in dorsal view. **c** Photograph of incomplete plastron (hyo-, hypo-, and xiphiplastra) prepared separately from carapace, in ventral view. Scale bars about 1 cm

for other elements and structures are presented in the relevant figure captions. Arrow heads ("<" and ">") are used in pleural to marginal sulcus contact formulae, as follows: < = anterior (e.g., < m5), > = posterior, < > = middle (e.g., <>m5), thus p1-2 : <>m5.

Geological ages are given in terms of the international units and, where relevant, North American Land Mammal Ages (NALMA) following Berggren et al. (1995) and Woodburne and Swisher (1995). The abbreviation "loc." is used within formal locality numbers. Detailed locality information is on file at the indicated institutions.

Institutional abbreviations used in this paper are: AMNH, American Museum of Natural History, New York, New York, USA; CM, Carnegie Museum of Natural History, Pittsburg, Pennsylvania, USA; TMM, Texas Memorial Museum, University of Texas, Austin, Texas, USA; UCMP, University of California Museum of Paleontology, Berkeley, California, USA; UMMP, University of Michigan Museum of Paleontology, Ann Arbor, Michigan, USA; USGS, United States Geological Survey, Denver, Colorado, USA; USNM, United States Natural History Museum, Washington, D. C., USA; UW, University of Wyoming, Laramie, Wyoming, USA; YPM (PU), Yale Peabody Museum (Princeton collection), New Haven, Connecticut, USA.

Systematic Paleontology

Testudines Batsch 1788
Cryptodira Cope 1869
Trionychoidea Gray 1870
Trionychidae Dumeril and Bibron 1835
Trionychinae? Lydekker 1889
Atoposemys gen. nov.
Atoposemys entopteros sp. nov.
(Figs. 26.1, 26.2)

Synonymy: "Plastomenine" C type: Hutchison and Archibald 1986, p. 5. Plastomenine C: Holroyd and Hutchison 2002, Fig. 2L.

Holotype: UCMP 133900 (Fig. 26.1), incomplete postcranial skeleton consisting of complete carapace, nearly complete plastron except epiplastra, right and left scapula, right coracoid, partial pelvis, and partial tail. UCMP loc. V86094, Garfield County, Montana, USA; Tullock Member, Fort Union Formation; Puercan NALMA, early Paleocene. Collected by J. H. Hutchison, 12 July 1985.

Paratype: UCMP 130100 (Fig. 26.2), carapace lacking right margin, nearly complete plastron except for epiplastra, and pelvis. UCMP loc. V84198, Garfield County, Montana; Tullock Member, Fort Union Formation; Puercan NALMA, early Paleocene:

Referred specimen: UCMP 149216, partial shell. UCMP loc. V92095, Garfield County, Montana; Tullock Member, Fort Union Formation; Puercan NALMA, early Paleocene.

Etymology: *Atoposemys* from the Greek *atopos,* strange, and *emys*, turtle. *Entopteros* from the Greek *ento*, within, and *ptero*, wing, alluding to the wing-shaped entoplastron.

Diagnosis: This is a trionychid based on the absence of epidermal scales, pitted sculpture of the external carapace surfaces, loss of the posteromedial spike of the entoplastron, and detachment of the epiplastra from the hyoplastra. It is placed in the Trionychinae on the basis of the absence of peripheral elements, basic boomerang shape of the entoplastron, and anterolateal orientation of the axillary spikes. It is distinguished from all other Trionychidae on the basis of the entoplastron with two lateral spikes anterior to the terminal spike producing a wing shape.

Description: The carapace (UCMP 133900, Fig. 26.1) of the holotype skeleton is subcircular in shape and convex with eight pairs of costals, six neurals, a preneural, and a nuchal. Carapace length is 191 mm. The entire dorsal surface is covered with a fine and even pattern of pits that do not become aligned into troughs near the margins. The lateral edges of the nuchal form the anterior most points of the carapace and form the lateral margins of concave embayment of the nuchal. The width of the nuchal is eight times its length. The posterior margin is broadly convex. The costal margins are subscalloped and terminate abruptly laterally (profile D of Gardner and Russell 1994). The rib ends are convex and, except for the tenth rib, short, and usually not extending beyond the costal margins. The preneural is trapezoidal with the anterior face longest. The six neurals are all reversed (wide end posterior) with the sixth terminating at the anterior sutures of the seventh costals.

The plastron is nearly complete except for the epiplastra. The entoplastron is appressed to the underside of the carapace and only its dorsal surface is visible. The arms of the entoplastron subtend a 100° angle but are broadened anterolaterally and form one or more weak laterally directed spikes. The anterior margin is nearly transverse. There is no visible evidence a callused ventral surface. The ragged lateral margins of the entoplastron indicate that the epiplastral, when present, did not reach the hyoplastra and the nearly transverse anterior margin suggest the lateral arms of the epiplastra were reduced or absent as in plastomenines. The hyoplastron is roughly rectangular with the broad axillary notch asymmetrically concave between the buttress and entoplastral process. There is no distinct notch in the anterior margin for reception of the entoplastral tip but a series of posterior progressively smaller anteromedially directed spikes for the edge of the anterior fontanelle. The ventral surface is covered with a fine pattern of raised tubercles and short ridges. Only the axillary buttress, entoplastral process, and a narrow band along the medial margin are not patterned. The callused surface forms an underhang between the buttresses laterally. The convex medial margins of the paired hyoplastra closely approach each other but do not form a suture as in plastomenines. The buttress is formed by a single, anterolaterally ribbed spike. The hyoplastron forms a firm but unfused suture with the hypoplastron. The hypoplastron is similarly patterned as the hyoplastron. The inguinal notch is asymmetrically curved with the medial edge nearly longitudinal. The inguinal buttress consists of two posterolaterally directed, fluted spikes with the anterior spike bifurcate. As with the hyoplastra, the convex medial margins of the hypoplastra closely approach but do not form a suture. The posterior margin has a strong xiphiplastral process and notch but is only closely appressed to the xiphiplastron laterally leaving a large midline fontanelle between the four elements. The xiphiplastron is roughly quadratic with the ventral surface fully pattered as with the hypoplastron with only the tips of the processes unsculptured. The medial margin between the tips of the medial processes is concave, thus enclosing a small interxiphiplastral fontanelle.

The paratype (UCMP 130100, Fig. 26.2) has a carapace that generally resembles the type, except the preneural is small and quadratic, the nuchal is somewhat narrower and deeper, and a seventh neural is present. The plastron also differs in details. The anterior margin of the hyoplastron exhibits a small notch for reception of the entoplastron tip and the margin of the anterior fontanelle is smooth and rounded. The axillary buttress of the hypoplastron is not bifurcate. The xiphiplastra contact the hypoplastra all along their mutual border leaving only a narrow midline fontanelle.

Remarks: Differences described above between the carapaces of the holotype and paratype are within expected variation in trionychids. The differences in the plastron may indicate gender rather than just individual variation or specific differences. Trionychids may be dramatically sexually dimorphic, especially in size with the males being

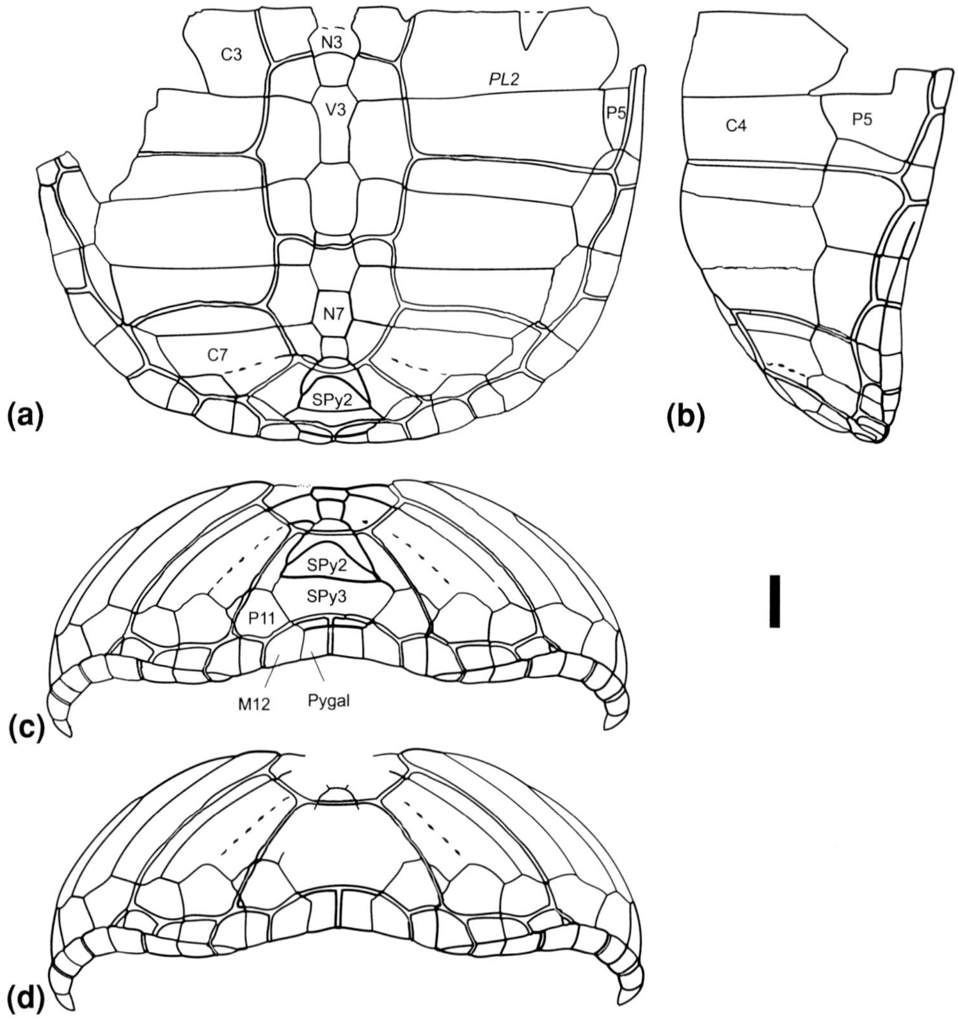

Fig. 26.3 *Planetochelys savoiei* Weems 1988, USNM 412107 (holotype), line drawings of posterior part of carapace (modified after Weems 1988, Fig. 5). **a–c** New interpretation, recognizing eight neurals and three syprapygals: **a** dorsal view; **b** right lateral view, rotated 90 degrees with posterior end pointing to bottom of figure; **c** posterior view. **d** Posterior view showing Weems' (1988) original interpretation of neurals and suprapygals. See "Materials and Methods" in text for abbreviations. Scale bar equals 1 cm

smaller. Although the holotype and paratype are similar in size, the paratype has a more fully callused plastron, more intimate contact of the xiphiplastra and hypoplastra, better developed notch on the hyoplastron for reception of the entoplastron, smaller and smooth margined hyoplastral margin of the anterior fontanelle, and lacks bifurcation of the anterior hypoplastron buttress spur. Lacking evidence to the contrary, I think that the paratype is an adult male and the holotype is a subadult female. The differences are ascribed to the relatively earlier stage of development of the holotype. A test of this hypothesis would be to see if the entoplastron, when found with the paratype plastral morphology, resembles the holotype.

Although the overall morphology resembles that of plastomenines in reduction of the rib ends, extent of plastral callusing, and possible reduction of the posterior arm of the epiplastron, the sharply angled entoplastron, anterolateral orientation and forward position of the hyoplastron buttress spur, open anterior and posterior fontanelles, and persistent midline kinesis resembles *Helopanoplia* Hay 1908 (Holroyd and Hutchison 2002, Fig. 2J) and trionychines to which it is tentatively referred.

Trionychoidea Gray 1870
Planetochelyidae fam. nov.

Type species: *Planetochelys savoiei* Weems 1988, p. 116.

Remarks: Monogeneric family. Included in the Trionychoidea on the basis of the presence of extragular scales, costiform processes of the nuchal, and broad plastron. Distinguished from other trionychoids based features listed below (see Diagnosis) for the type and only known genus. *Planetochelys* Weems 1988

Synonymy: *Planetochelys*, family incertae sedis: Hutchison and Weems 1998, p. 192.

Included species: *Planetochelys savoiei* Weems 1988 (type species) and *P. dithyros* sp. nov.

Occurrence: The oldest records of the genus are late Paleocene Tiffanian NALMA (Virginia and Wyoming) and the youngest are from the middle Eocene Uintan NALMA of California (pers. obs.). It is distributed from Wyoming to Texas and west to southern California (pers. obs.).

Revised diagnosis: Differs from other trionychoids in the following combination of features: moderate size plastron divided by a transverse hinge joint formed along the hyo-hypoplastral suture (entoplastron may also contribute); plastral lobes ligamentally attached to carapace and marked by a low rugose longitudinal ridges on the bridge portions; plastral articular areas on the bridge peripherals (P4-7) truncated and nearly smooth; nuchal with a pair of strong costiform processes that span P1 and insert into P2 viscerally; some distal rib ends hooked; rib ends of C1-5 cross the sutures of two peripherals medially; extragulars may be present; entoplastron distinctly longer than wide; and no plastral caudal notch.

Planetochelys savoiei Weems 1988
(Fig. 26.3)

Synonymy: *Planetochelys savoiei* Weems 1988, pp. 116–118, Figs. 5–8. *Planetochelys*: Hutchison and Weems 1998, Table 1.

Holotype: USNM 412107 (Fig. 26.3), posterior moiety of carapace starting from C3. West bank of Aquia Creek, Virginia, USA; Piscataway Member, Aquia Formation; late Paleocene (Thanetian).

Revised diagnosis: Differs from *Planetochelys dithyros* sp. nov in smaller size (estimated carapace length = 143 mm) and the following carapace features (plastron unknown): C3–C6 weakly wedged-shaped; ventrolateral margins of bridge peripherals rounded; posterior peripherals strongly upturned at margin and without small projections at intermarginal sulci; marginal-pleural sulcus marked by a step or change in plane; v4-5 sulcus on suprapygal 2; v4-5 sulcus on SPy1; no ilial fossa on C8; N3 normal hexagonal; sulci on peripherals depressed; and neural area of carapace concave.

Description: This species is only known from the partial carapace of the holotype specimen (Fig. 26.3). The entire specimen was figured by Weems (1988). The suprapygal area is tightly sutured and partly fused indicating an adult individual. Close examination of the holotype led me to an altered reconstruction of the pygal area—specifically, I interpret the presence of an eighth neural and three suprapygals, instead of the seven neurals and one suprapygal originally reported by Weems (1988; see Fig. 26.3c vs. d). Weems (1988) did not figure or describe the internal (visceral) view of the carapace other than the limit of the ligamental plastral contact. The internal costal-peripheral contacts are generally the same as the external contacts, except that the rib end of C4 crosses the anterodorsal corner of P7 before terminating in P6, as in the other species. The capitular processes of the ribs are narrow near the costal-neural sutures.

Remarks: Weems (1988) assigned *Planetochelys* to the Sinemydidae Yeh 1963 based upon the ligamental attachment of the plastron, the lack of costal-peripheral fontanelles, and misinterpretation of the suprapygal area in USNM 412107. New material and more extensive descriptions (Brinkman and Peng 1993, Sukhanov 2000) of *Sinemys*, the type genus of the Sinemydidae, have documented a suite of diagnostic features for that genus (e.g., low arch of the shell, presence of peripheral spines, and lack of costiform processes) that do not support a close relationship to *Planetochelys* (see below).

Planetochelys dithyros sp. nov.
(Figs. 26.4, 26.5)

Synonymies: Box turtle: Savage et al. 1972, Table 3; Hutchison 1992, p. 455. Box turtle, n. gen.: Hutchison 1980, p. 116. Genus K: Hutchison 1983, p. 65. Emydid B: Bartels 1983, p. 362. *Planetochelys*: Hutchison 1998, p. 402, Fig. 18.2G; Holroyd and Hutchison 2000, p. 75; Holroyd et al. 2001, p. 98, Figs. 2–4, Table 1; Hutchison 2003, p. 23.

Holotype: UCMP 120000 (Fig. 26.4b, 26.5a, b), incomplete skeleton, consisting of: carapace preserving right and left P1-11, nuchal, N1-5, right C2-4, left C2; plastron lacking posterior margin of xiphiplastra; damaged skull; incomplete dentary; fragments of articular and angular; fragments of cervical vertebrae 1–8; two caudal vertebrae; appendicular skeleton bones that are complete or compositely complete (counts indicated in brackets) are scapula (2), coracoid (2), radius (2), ulna (2), humerus (2), metacarpal, innominate (2), femur (2), tibia (2), distal fibula, astragalus-calcaneum (2), tarsale 4(?), tarsal, metatarsal 1 and 5 (2), metatarsal, proximal phalanges (2.5), and ungual phalanges (4). UCMP loc. V77050, Sweetwater County, Wyoming, USA; main body of the Wasatch Formation; Wasatchian NALMA, Graybullian subage, early Eocene. Collected by J. H. Hutchison, 1974.

Paratypes: UCMP 125005, nearly complete shell, limb bones; and UCMP 130896, partial shell. Both from UCMP loc. V81045, Big Horn County, Wyoming, USA; Willwood Formation; Wasatchian NALMA, early Eocene.

Referred specimens: Alkali Creek Tongue Member of Wasatch Formation, Wasatchian NALMA, Sublette County, Wyoming, USA (1 locality): UCMP loc. V99338: UCMP 159173, partial shell. Main Body Member of Wasatch Formation (Wasatchian NALMA), Sweetwater County, Wyoming, USA (1 locality): UCMP loc. V70227: UCMP 117301, suprapygal. Willwood Formation (Wasatchian NALMA), Big Horn County, Wyoming, USA (5 localities): UCMP loc. V85005 (= USGS D1424): UCMP 130937, anterior lobe of plastron, P fragments; UCMP loc. V80007 (= Yale 27): UCMP 124493, right and left epiplastra, entoplastron fragment; UCMP loc. V80010 (= Yale 84):

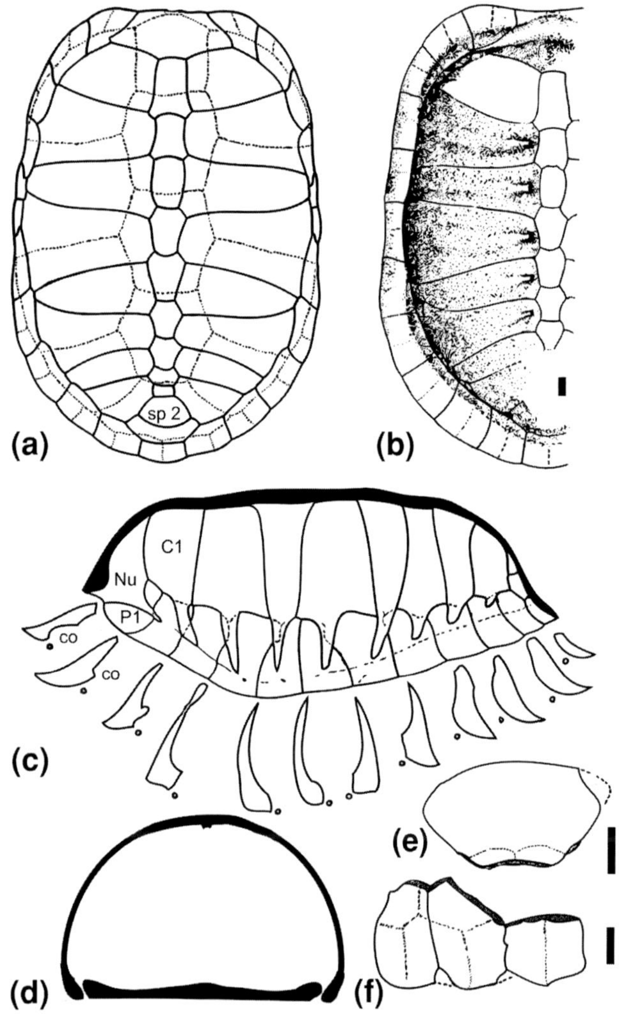

Fig. 26.4 *Planetochelys dithyros* sp. nov., shell. **a** Line drawing of reconstructed carapace, based on UCMP 120000 (holotype) and 125005, in dorsal view, showing sutures (*solid lines*) and scute sulci (*dotted lines*). **b** UCMP 120000 (holotype), drawing of ventral view of carapace. **c** Diagramatic reconstruction of visceral right lateral view of carapace and outlines of the anterior peripheral sutures (*dots* indicate dorsal margin of the marginals viscerally; *co* groove for costiform process of nuchal). **d** Diagramatic reconstruction of cross-section of carapace and plastron at about the level of the peripheral 5–6 suture. **e** UCMP 117301, suprapygal 3, in external view. **f** UCMP 136950, articulated peripherals 10–11 and pygal, in external view. Scale bars equal 1 cm

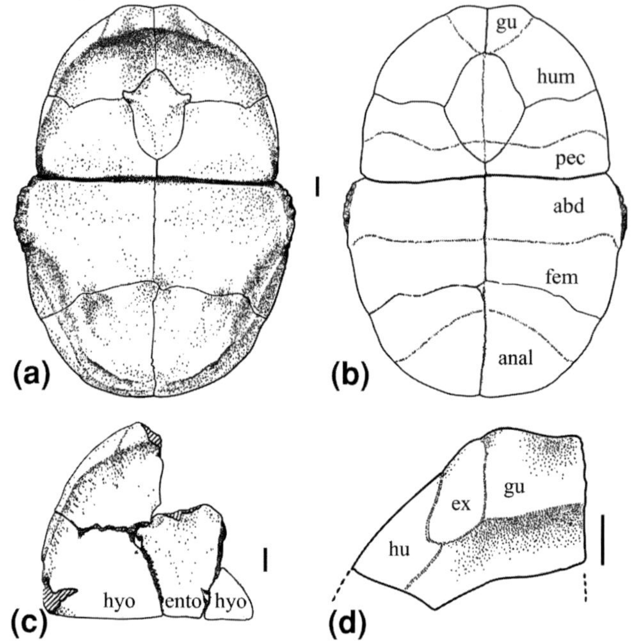

Fig. 26.5 *Planetochelys dithyros* sp. nov., plastron. **a** Stipple drawing of plastron of UCMP 120000 (holotype), in dorsal view. **b** Line drawing of plastron of UCMP 120000 (holotype), in ventral view, showing sutures (*solid lines*) and scute sulci (*dotted lines*). **c** Stipple drawing of UCMP 131008, partial anterior lobe, in dorsal view, showing variant exposure of entoplastron on hinge line. **d** Stipple drawing of UCMP 130909, left epiplastron, in dorsal view. Abbreviations: *abd* abdominal, *ento* entoplastron, *ex* extragular, *fem* femoral, *gu* gular, *hu* humeral, *hyo* hyoplastron, *pec* pectoral. Scale bars equal 1 cm

UCMP 124498, partial shell; UCMP loc. V80012 (= Yale 280): UCMP 124500, partial shell; UCMP 124501, partial plastron, carapace fragments; UCMP loc. V80016: UCMP 124519, epiplastron fragment; UCMP 124521, partial shell. Willwood Formation (Wasatchian NALMA), Waskakie County, Wyoming, USA (6 localities): UCMP loc. V81085: UCMP 130908, partial shell; UCMP loc. V81092 (= Yale 132 plus): UCMP 130861, 130862, 130866, 130867,130868, 130870, 130873, 130874, 130875, 130888, partial shells; UCMP 130918, partial shell, limb bones; UCMP loc. V82198 (= USGS D1342): UCMP 130930, partial shell. UCMP loc.V82300, (= Yale 157 general): UCMP 130934, associated partial shell. UCMP loc. V85065 (= USGS D1401 part): UCMP 130945, anterior plastron; UCMP loc. V85066 (= USGS D1421 low): UCMP 130944, nearly complete plastron, carapace fragments. Lysite Member of Wind River Formation, Wasatchian NALMA, Fremont County, Wyoming, USA (1 locality): UCMP loc. V72174: UCMP 136950, P9-pygal.

Etymology: From the Greek *dithyros*, two-doored or bivalved.

Diagnosis: Differs from *Planetochelys savoiei* in larger size (carapace length 293 cm or longer); C3–C6 strongly wedged-shaped; bridge peripherals without sharp carina marking the transition from dorsal to ventral surfaces in adults (juveniles may show a carina); posterior peripherals weakly upturned at margin and with small projections at intermarginal sulci; marginal-pleural sulcus not marked by a step or change in plane; v4-5 sulcus on N8; N3 quadratic; ilial fossa well developed; sulci on peripherals perched; and neural area of carapace convex.

Description: The description is based principally upon the holotype and UCMP 125005, a nearly complete shell and parts of the skeleton. The shell of the holotype specimen, although less complete than UCMP 125005, is less distorted, permits estimates of shell proportions, has skull material, and has more complete and less distorted appendicular skeleton (to be described elsewhere). Details of the posterior costals, neurals and C1 are based primarily upon UCMP 125005 and referred material.

The shell (Figs. 26.4, 26.5) is highly modified into a box turtle type of specialization. The plastron has no sutural contact with the carapace and lacks the bony pivot seen in *Terrapene* Merrem 1820 and *Cuora* Gary 1855. P4-7 show thickened, flattened, and medially facing surfaces which by analogy with other box turtles are the areas of origin of ligaments that bind the plastron to the carapace. Complementary elongate bosses are present on the adjacent margins of the hyo- and hypoplastra. The longest and strongest boss is on the hypoplastron in this taxon.

The carapace generally resembles that of other box turtles in the pronounced vaulting and high angle of the anterior peripheral arch (Fig. 26.4c). The shell is devoid of distinctive sculpturing and lacks carina. C1 articulates distally only with P1-2 and the nuchal, although the distal tip of the rib is imbedded in P3 internally. Such an anteriorly situated C1-2 suture is unique among Cretaceous and Cenozoic turtles. C2, C4, C6, and C8 are considerably longer ventrally than dorsally (wedge-shaped), whereas C3, C5, and C7 show the reverse condition (Fig. 26.4a–c). The free distal ends of the ribs of C1–C5 contact two peripheral bones, thus P3–P8 are crossed by two ribs. The high angles of approach of the more anterior and posterior ribs are sufficient to bring this about without peculiar modifications of the peripherals and a similar condition can be seen in *Terrapene*. The C3 rib approaches the peripherals nearly perpendicularly. The posterodorsal portion of the P4 is extended posteriorly in such a way as to constrict the dorsal margin of P5, thus allowing the rib to traverse both peripherals along part of the interperipheral suture. The relationship of C4 to P6-7 is a mirror image of the C3 condition. The suture between P5-6 represents the juncture of the anterior and posterior peripheral arches. The ribs are recessed into the internal faces of the peripherals but are well exposed until they reach the thickened inner rim of the peripherals where they insert into pits (Fig. 26.4c). Each rib continues to attenuate within the pit to a sharp point but the rib tips of C1 (?)-3 are hook-shaped, thus preventing the ribs from becoming disengaged, a condition I have seen in no other turtle. The recessed part of the rib ends are thin and strap-like suggesting primarily a tensional function. The rib heads are supported by necks with round cross-sections. They insert on the costals distinctly (but not remarkably) lateral to the neural-costal sutures.

The nuchal has surprisingly long costiform processes that span all of the P1 and one-fifth to two-thirds of P2 (Fig. 26.4b, c). The costiform processes lie ventral to the lateral tips of the nuchal proper as in chelydrids and most kinosternids. The long costiform processes, costal wedging, and double contacts of the rib ends form an elegant truss-and-arch latticework to maintain a relatively light but strong shell.

There are eight neurals and three suprapygal bones. With the exception of N8, all are longer than broad. N1-2 are reversed hexagonal, N3 is quadratic, and N4-8 are normal hexagonal. The suprapygals increase in width posteriorly. Only the posterior suprapygal contacts the peripherals and pygal.

The carapacial sulci are adequately described by the figures (Fig. 26.4a, b, d) The dorsal sulci of m12 lie on SPy3 in UCMP 130950 and 117301 but lie on the pygal only in UCMP 124498. The intermarginal sulci are perched on raised ridges as in many testudinids and terminate in small projections on the free edge of the peripherals as in *Terrapene*. Raised sulci are also present at the juncture of the plastral shields. This type of sulcus indicates a thick epidermal scale. The sulci dorsal to the marginal series are depressed normally into the bone.

The plastron has a nearly symmetrical oval outline, with its greatest width through the posterior part of the hypoplastral bosses. The measurements of the holotype plastron are: length = 202 mm; maximum width of hypoplastron = 145 mm; maximum with of anterior lobe = 125 mm; maximum length of anterior lobe = 89 mm; and maximum length of posterior lobe = 115 mm. The suture between the hyo- and hypoplastra has been transformed into a nearly straight transverse hinge, is only partly and weakly dentate and it is coincident with the posterior margin of the pectoral and anterior margin of the abdominal scales. The contribution of the entoplastron to this hinge is remarkably variable. In the holotype and paratype, the paired hyoplastra meet along the midline and exclude the entoplastron (Fig. 26.5a, b), but in some specimens the entoplastron may form part of the hinge margin (Fig. 26.5c). Both morphotypes may occur within the same localities (e.g., UCMP loc. V77050). The entoplastron in modern box turtles forms no part of the hinge. There also appears to be a kinetic joint between the epiplastra and entoplastron. The epiplastral sutures of the entoplastron form ventral sloping nondentate surfaces that become very finely dentate only along the ventral edge. The epiplastra have a posteromedial flange of bone which overlaps these surfaces. The lateral tips of the entoplastron are formed into transverse processes that have smooth-curved surfaces on the anterior and dorsal sides. These apparently served as rotary joints with the epiplastra. The thinning of the hyo-epiplastral suture (2 mm on the margins of the type) could have permitted some dorsoventral movement via bending of the bones. Such a mechanism would probably function only during extreme retraction when the anterior lobe is elevated. The epiplastra turn gently upward anteriorly, causing the anterior lobe of the plastron to be ventrally arched as in modern box turtles. Extra gular scales may be present on the dorsal side of the gular projections (Fig. 26.5d).

The ventral surface of the posterior lobe of the plastron in all specimens preserving this region is flat. Either all of these are females, which seems unlikely, or the males are not sexually dimorphic as in modern box turtles. Unlike the modern genera, the thickest part of the xiphiplastron does not coincide with the dorsomedial edge of the plastral shields dorsally but lies medial and parallel to this edge and is separated from it by a shallow trough (Fig. 26.5a). This ridge diminishes posteriorly but continues anteriorly to merge with the lateral boss of the hypoplastron. Medial to this ridge and just posterior to the hypo-xiphiplastral suture lies a shallow depression which is probably associated with the seat of the pelvis.

Remarks: The genus shows a mélange of character states that are found in such groups as the Chelydridae, Dermatemydidae, and testudinoids. *Planetochelys* is highly specialized with a heavy overprint of features that are strongly functionally correlated to being a box turtle. The dorsal overlap of the plastral scales, reduction (loss) of inframarginals, and highly domed shell with wedged costals bear striking general similarities to adaptations in some testudinoids (e.g., *Testudo* Linnaeus 1758, *Terrapene, Cuora*). The presence of long costiform processes on the nuchal resembles apomorphies of chelydrids, kinosternids, and dermatemydids. The occasional presence of extragular scales (Fig. 26.5d) is a primitive feature apparently lost early in the chelydrid-testudinoid lineage but retained in some trionychoids (e.g., *Adocus* Cope 1869). The retention of that primitive feature in *Planetochelys* suggests it lies near the adocid grade of trionychoids (sensu Gaffney and Meylan 1988). The development of a bilobe kinetic plastron and terrestriality, however, produces a cluster of character states strikingly similar that those of several testudinoids, such as the broadening of the plastron, overlap of the plastral scales dorsally, and wedging of the costals (also in some chelydrids). The broadening of the plastron and overlap of the scales also occurs in the kinetically lobed kinosternid *Kinosternon* Spix 1824 (Hutchison and Bramble 1981; Bramble et al. 1984), emydid and batagurid box turtles, and the pleurodire *Pelusios* Wagler 1830 (Bramble and Hutchison 1981). These features as well as the ability to withdraw the pectoral girdle in order to accommodate the elevation of the anterior lobe have been derived at least twice within the Emydidae (Bramble 1974), once in the Kinosternidae (Bramble et al. 1984; Hutchison 1991), and once in the Pelomedusidae (Bramble and Hutchison 1981).

That plastral lobes were able to be raised to close the shell openings is indicated by the very close correspondence of the shape of the plastral lobes with the inner rim of the adjacent peripherals and the presence of a shallow trough on the inner surfaces of the peripherals which would allow the

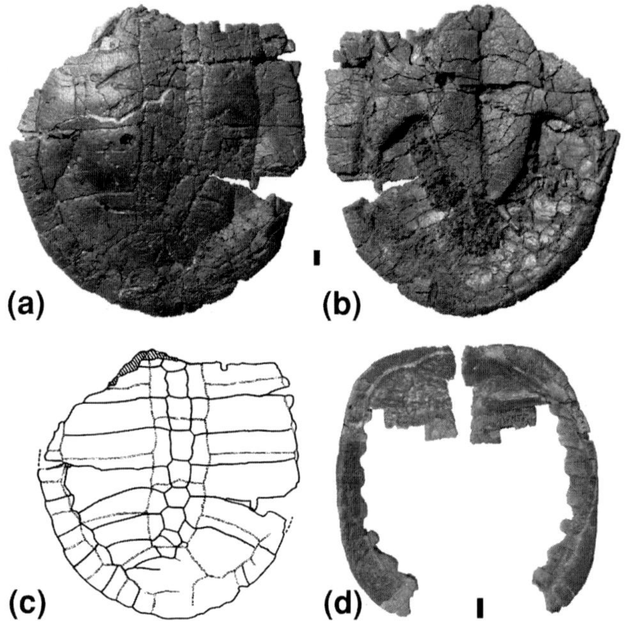

Fig. 26.6 *Tullochelys montana* gen. et sp. nov., shells. **a**, **b** UCMP 136481 (holotype), photographs of anteriorly incomplete shell, in dorsal **a** and ventral **b** views. **c** UCMP 136481 (holotype), line drawing of carapace, in dorsal view, showing sutures (*solid lines*) and scute sulci (*dotted lines*); cross hatching denotes broken surface. **d** UCMP 154567, photographs of left side of juvenile, partial carapace preserving nuchal, P1-10, and C1-2, in dorsal (*left*) and ventral (*right*) views. Scale bars equal 1 cm

plastron to seat tightly against the carapace. Costal wedging is usually associated in living turtles with relative lightening of the shell, an important consideration in terrestrial turtles but also present in some thin-shelled aquatic turtles. The combination of the truss-like mosaic and the smooth high arch of the carapace produce a structurally strong shell and permit a reduction of gross shell weight without loss of strength. Taken together, these lines of evidence clearly indicate that *Planetochelys* was a terrestrial turtle comparable to *Terrapene* or a small testudinid.

The unique combination of primitive (retention of extragular scales) and derived (very long costiform processes and plastral kinesis) characters in *Planetochelys* means that the genus does not fit comfortably in any known family. For that reason, a new family is provisionally erected to contain the genus.

Chelydroidea sensu Gaffney and Meylan 1988
Chelydridae (Agassiz) Gray 1870
Tullochelys gen. nov.
Tullochelys montana sp. nov.
(Figs. 26.6, 26.7)

Synonymies: Chelydridae genus indet.: Hutchison and Archibald 1986, p. 5. *Protochelydra* sp.: Hutchison 2008. p. 18.

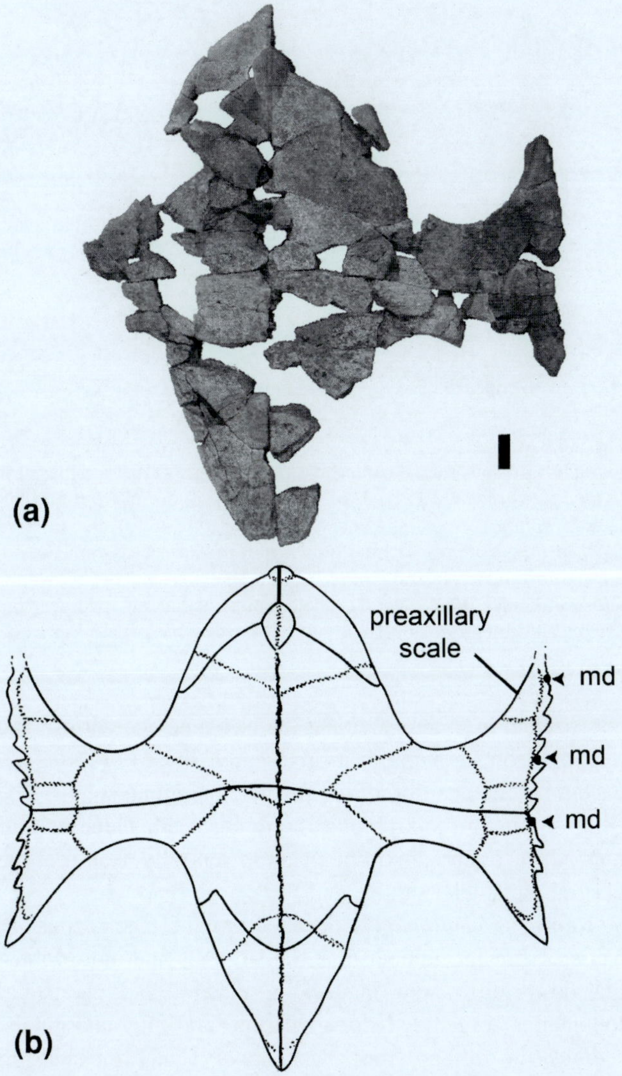

Fig. 26.7 *Tullochelys montana* gen. et sp. nov., plastron. **a** UCMP 129693, photograph of incomplete plastron, in ventral view. **b** Line drawing of reconstructed plastron, based on UCMP 129693, in ventral view, showing sutures (*solid lines*), scute sulci (*dotted lines*), and muck ducts (md). Scale bar equals 1 cm

Holotype: UCMP 136481 (Fig. 26.6a–c), crushed shell anterior portion of carapace and anterior extremities of epiplastron and entoplastron, with corroded and disarticulated limb bones and vertebrae cemented to the shell. UCMP loc. V90001, McCone County, Montana, USA; Hell Creek Formation; early Puercan NALMA, early Paleocene.

Paratypes: Tullock Member of Fort Union Formation, Puercan NALMA, Garfield County, Montana, USA (2 localities): UCMP loc. V83086: UCMP 129189, left P1-P2, peripheral fragments, costal 1 fragment; UCMP loc. V84051: UCMP 129693, fragmentary carapace, most of plastron, limb fragments. Tullock Member of Fort Union Formation, Puercan NALMA, McCone County, Montana, USA (2 localities): UCMP loc. V73023: UCMP 107791 left P1; UCMP 117434 articulated left P11-right P11 and suprapygal; UCMP loc. V85085: UCMP 154567, partial shell of a juvenile lacking the anterior plastral bones, xiphiplastra, majority of the costals and about half of the neurals. Hell Creek Formation, Lancian NALMA, Garfield County, Montana, USA (2 localities): UCMP loc. V93087: UCMP 123314, partial carapace; UCMP loc. V72211: UCMP 158704, P1-P2, partial C1.

Etymology: Genus name refers to the Tullock Member, plus *emys*, turtle. Species name is for the State of Montana.

Diagnosis: This genus is assigned to the Chelydridae on the basis of the narrowing of the plastral, loss of extragulars, retention of ventrally exiting musk ducts, anal-femoral scale crossing part of the hypoplastron, and long costiform processes of the nuchal. It differs from all other known chelydrids in the following combination of character states: epiplastra relatively broad but tapering medially to a short gular beak; gulars small and confined to tip of epiplastra; entoplastron small, elliptical, and firmly sutured to surrounding elements; humeral-pectoral sulcus crossing the posteriormost part of the entoplastron or lying well posterior; hyoplastron and hypoplastron thickened, without central or lateral fontanelles, sutured to peripherals with gomphotic sutures; four inframarginals; v1 on-lapping the nuchal; costiform process of nuchal terminating on middle area of P2; hyoplastral buttress extending to anterior of C1 rib tip; hypoplastral buttress extends to anterior part of P8; carapace without fontanelles; only slight development of serration of the peripherals; pygal unnotched; suprapygal 1 broad with divergent sides; normally nine neurals; and no costal or neural carina.

Description: The carapace in the holotype (Fig. 26.6a–c) and UCMP 123314, although flattened by compression, appears to have been moderately arched as in *Protochelydra zangerli* Erickson 1973. Carapace length is about 290 mm. There is no indication of carina including the low posterior midline carina as in *P. zangerli*. The small serration of the posterior margins seen in adult *P. zangerli* is best indicated in the juvenile of *P. montana* (UCMP 154567) and more weakly developed in adults. The neural contacts vary: UCMP 154567 juvenile-1(2(3(4(5(6–; UCMP type-1) 2(3(4(5(6(7/8/9 (the asymmetrical "N8" is probably an anomaly with "N9" the homolog of "N8"); UCMP 123314 -2(3(4(5(6(7(8|9. The pleural/marginal sulcus formula is: v1-p1: >m1, p2-3: <m5, p2-3: <>m7, p3-4: <>9, p4-5: >m11.

Two suprapygals (only observable in the holotype) follow the neurals. SPy1 is extremely rhomboid, nearly triangular with its posterior suture nearly as wide as SPy2. SPy2 terminates laterally at the anterior moiety of P11. The anterior and lateral marginals are slightly raised above the plane of the pleurals but become flush posteriorly. Aside from a very finely incised vermiform reticulations on the carapace and plastral surfaces, the sculpture on the adult shell ranges from none (holotype) to weak corrugations on the vertebrals and distal pleurals. The juvenile specimen (UCMP 154567, Fig. 26.6d), however, has a distinct sculpture of irregular ridges and corrugations. Sulci are generally well defined. The long nuchal costiform process lies in a groove on P1 and anterior moiety of P2 (anterior of P3 in the juvenile specimen). The axillary buttress terminates in a pit at the posterior extreme of P3 with the tip of the C1 rib inserting in a pit immediately above in the P3–P4 suture. The fluted, peg-like dentations of the hyo and hypoplastra insert into well-defined pits in P4 to anterior P8. The rib tips also insert into well-developed pits in the peripherals. Vertebrals 2–4 are rectangular to weakly hexagonal in adults, and v1 and v5 are rhomboid. A traversing puncture on the upper anterior corner of p3 is probably a tooth puncture. Other small pathological circular, to irregular ovoid, to tract-like grooves mark the surface of the holotype carapace.

The pleural marginal sulci lie distinctly below the costals anteriorly but rise laterally and posteriorly to near or coincident with the costal-peripheral sutures. The twelfth marginals rise onto SPy2. The extent of the marginals is confined to the distal one-third of the peripheral; posteriorly they rise onto the middle portion with a distinct offset from the visceral surface. The cervical scale is wider than long. The p1–v1 sulcus crosses the nuchal as in *Chelydra* Schweigger 1812 and "*Chelydrasia*" *kusnetzovi* Chkhikvadze in Gaiduchenko and Chkhikvadze 1985 (Hutchison 2008, Fig. 2.2d).

The plastron is stout, lacking fontanelles (Fig. 26.7). The individual bones are relatively thick in the holotype and UCMP 129693 with well incised sulci. The epiplastron is relatively broad unlike chelydropsines with a gently arched free margin except for a small gular beak at the extreme anterior end as in *Chelydrasia* Chkhikvadze 1999 and *Protochelydra* Erickson 1973. The gular-humeral sulci lie entirely on the epiplastra as in *Chelydropsis* Peters 1868. The epiplastra are distinctly thickened along their mutual contact forming a keel on the dorsal side. The entoplastron is small, almond-shaped, and firmly attached to the surrounding bones. The posterior sutures are dentate, tongue-and-groove style, whereas the anterior sutures are relatively smooth and abutting surfaces. The humeral-pectoral sulcus crosses the posterior margin of the entoplastron in UCMP 129693 but lies well posterior to the entoplastron in UCMP 136481. The posterior dorsal extension of the entoplastron is broken away or covered but appears to extend in a groove at least as far as the level of the humeral/pectoral sulcus in UCMP 129693. The scale-covered part of the axillary buttress is relatively longer and bears an additional inframarginal scale (preaxillary scale) unlike *P. zangerli*. The lateral and medial suture dentations are very well developed as in *P. zangerli*. The abdominal scale is distinctly withdrawn from the midline but less so than in *P. zangerli* and the Neogene chelydrids. The hypoplastron and xiphiplastron resemble those of *P. zangerli* except for the position of the medial abdominal sulcus. The extreme tips of the xiphiplastra are unscaled. Three or four musk duct foramina exit the plastron ventrally near or at the plastral-carapacial suture.

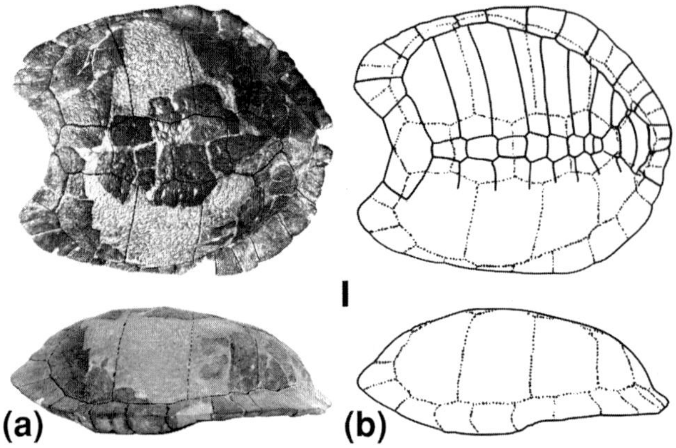

Fig. 26.8 *Cardichelyon rogerwoodi* gen. et sp. nov., carapace. **a** Photographs of reconstructed carapace (original bone is *dark* and plaster is *lighter grey*) of YPM (PU) 14671 (holotype), in dorsal (*top*) and left lateral (*bottom*) views, showing preserved (*solid lines*) and inferred (*dashed lines*) sutures. **b** Line drawing of reconstructed carapace, in dorsal (*top*) and left lateral (*bottom*) views, showing sutures (*solid lines*) and scute sulci (*dotted lines*); shell outline based on holotype and sutures and sulci based on UCMP 126364. Scale bar equals 1 cm

Remarks: *Tullochelys* differs from the Chelydrinae in having a greater number of inframarginals and more robust plastral bridges, and in lacking plastral and carapacial fontanelles, organized dorsal carina, and supramarginals. *Tullochelys* differs from the known Chelydropsinae in having rectangular vertebrals. Within the Protochelydrinae, *Tullochelys* differs from *Protochelydra zangerli* in smaller size, smaller entoplastron, more inframarginals, shorter posterior peripherals, and greater extension of the hyoplastral buttress. *Tullochelys* differs from *Denverus middletoni* Hutchison and Holroyd 2003 in lacking longitudinal dorsal carinae, inflation of the bridge peripherals, and diamond-shaped vertebrals. *Tullochelys* exhibits the most similarities with early chelydropsines and protochelydrines; its detailed relationships will be treated elsewhere.

Testudinoidea sensu Gaffney and Meylan 1988
Platysternidae Williams 1950
Cardichelyon gen. nov.
Cardichelyon rogerwoodi sp. nov.
(Figs. 26.8, 26.9, 26.10, 26.11, 26.12, 26.13)

Synonymies: *Ptychogaster* sp.: Estes 1975, p. 376. Macrocephalic emydid: Bartels 1980, p. 77. Large headed emydid nov. gn.: Hutchison 1980, p. 116. Genus C: Hutchison 1983, p. 65. Emydid A: Bartels 1983, p. 361.

?Emydid C: Hutchison 1992, p. 455; Holroyd and Hutchison 2000, p. 75; Holroyd et al. 2001, p. 98, Figs. 2–4, Table 1. Large-headed emydid: McCord 1996, p. 139. ?Emydid C: Hutchison 1998, p. 402, Fig. 18.2J. Unnamed genus C: Hutchison and Frye 2001, p. 12, Fig. 2.

Holotype: YPM (PU) 14671 (Figs. 26.8, 26.12a, b, 24.13), incomplete skeleton consisting of an incomplete but restored shell, partial skull, both dentaries, parts of cervical vertebrae 1–5, 7, and 8, incomplete humeri and femora, five caudal vertebrae, much of the pelvic girdle on the left side and parts of the right, parts of the scapula and coracoids, and a few phalanges. From the J. P. Reis Quarry (= UMMP loc. FG-10), Park County, Wyoming, USA; Fort Union Formation; Clarkforkian NALMA. Collected by Princeton University field crews under the direction of G. L. Jepsen, circa 1930.

Paratypes: Willwood Formation, Clarkforkian NALMA, Park County, Wyoming, USA (4 localities): UMMP loc. SC-120: UMMP L67563, partial shell; UMMP loc. SC-134: UMMP L67564, partial shell, limb fragments; UMMP loc. SC-234: UMMP L71484, partial shell; UCMP loc. V81169 (= UMMP SC-8): UCMP 126288, 126289, 126515, partial shells. Willwood Formation, Wasatchian NALMA, Big Horn County, Wyoming, USA (3 localities): UCMP loc. V81179 (= Yale 289): UCMP 170581, partial carapace and plastron; UCMP loc. V81223: UCMP 126497, shell; UCMP loc. V81225: UCMP 126498, 126499, shells. Willwood Formation, Wasatchian NALMA, Park County, Wyoming, USA (2 localities): UMMP loc. RB-15: UMMP L77644, partial shell; UMMP loc. SC-146: UMMP L77646, partial shell. Willwood Formation, Wasatchian NALMA, Washakie County, Wyoming, USA (15 localities): UW loc. V73027: UW8316, partial shell; UCMP loc. V810932: UCMP 126337, shell fragments; UCMP loc. V81075: UCMP 126343, shell fragments; UCMP loc. V81090: UCMP 126364, partial shell; UCMP loc. V81092 (= Yale 132 plus): UCMP 126367, 126390, 126395, 126418, shell fragments, UCMP 126432, anterior quadrant of carapace; UCMP loc. V81214: UCMP 126486, shell; UCMP loc. V82184: UCMP 126500, partial shell; UCMP loc. V82299: UCMP 127141, partial shell; UCMP loc. V82300 (= Yale 157): UCMP 127138, 127139, partial shells; UCMP loc. V81075: UCMP 126523, partial shell, ilium; UCMP loc. V81209: UCMP 127130-32, partial shells; UCMP loc. V81214: UCMP 126486, shell; UCMP loc. V81072 (= Upper Yale 157): UCMP 127133-35, partial shells; UCMP loc. V84088: UCMP 169035, partial carapace and plastron; UCMP loc. V96036: UCMP 157437, nuchal, P1, 6–8.

Etymology: From the Greek *kardia*, heart, plus *chelyon*, tortoise-shell. The species is named in honor of the Roger Wood, the neo- and paleochelonologist and conservationist.

Diagnosis: Placed in the Platysternidae on the basis of the wide cephalic emargination of the carapace, extensive roofing over the temporal region of the skull, short plastral bridges, ventrally exiting musk ducts, similarity of the relative midline lengths of the plastral scales (humeral > femoral > anal > pectoral > gular > abdominal), and kinetic contact of the carapace and hypoplastron. Differs from *Platysternon* in the following characters: axillary buttress lies along the costal-peripheral suture and terminates on or just above the dorsal margin of the first peripheral internally; the posterior buttress does not contact the costals; broad and deep U-shaped depression on the dorsal surface of the anterior end of the plastron; cervical scale usually absent or distinctly reduced in width; medial portion of the pectoral-abdominal sulcus converging on the hyoplastral-hypoplastral suture; and sulcus dividing the twelfth marginals frequently absent or incomplete.

Description: The carapace (Fig. 26.8) is moderately domed and about 80% as wide (at P4's) as long (maximum length). There is the normal complement of twelve paired marginal scales (m12's may be partly or completely fused; see Fig. 26.9, Py var. 1), four pleural pairs, and five vertebrals. The marginals all terminate below the costal-peripheral sutures. The vertebrals are generally weakly hexagonal in shape with v1 longer than broad and v2-3 roughly equidimentional. The sulcus between v4-5 is considerably shorter than the other intervertebral sulci and results in an hourglass configuration for this pair. The sulci are usually well defined and finely incised. Modal pleural seam formula: v1-p1: < m1, p1-2: < m5, p2-3: <> or > m7, p3-4: < m9, p4-v5: < m11.

The peripheral series consists of the nuchal, pygal, and usually eleven paired peripheral bones (Fig. 26.9). In general, the peripherals in the axillary and inguinal areas tend to be longer in their perimeter dimensions than those of the bridge and caudal areas. The anterior part of the carapace is deeply notched for reception of the large head. Pl-2 are extended to produce a broad roof over the opening for the forelimbs. The posterior peripherals are only slightly flared lateral to the pygal. The interperipheral sutures are finely and densely dentate. A series of musk duct foramina (Figs. 26.9, 26.10) are probably usually present along each side of the plastron. They lie along or just lateral to the peripheral-plastral suture, one for each bridge peripheral from P4 to P7.

The nuchal (Fig. 26.9) is about half as long (midline) as wide. The free margin (perimeter) length is only about half the maximum width. It is thickened internally by a broad transverse ridge that extends from the lateral apices. The ventral surface of the first marginal terminates just anterior to the internal ridge. The nuchal remains stout toward the perimeter but thins rapidly posteriorly. It reaches its maximum thickness at the lateral apices. The v1-p1 sulcus contacts ml well on the nuchal. The right and left m1's are usually in contact at the midline but there may be a small gap or minute cervical scale. The bridge peripherals

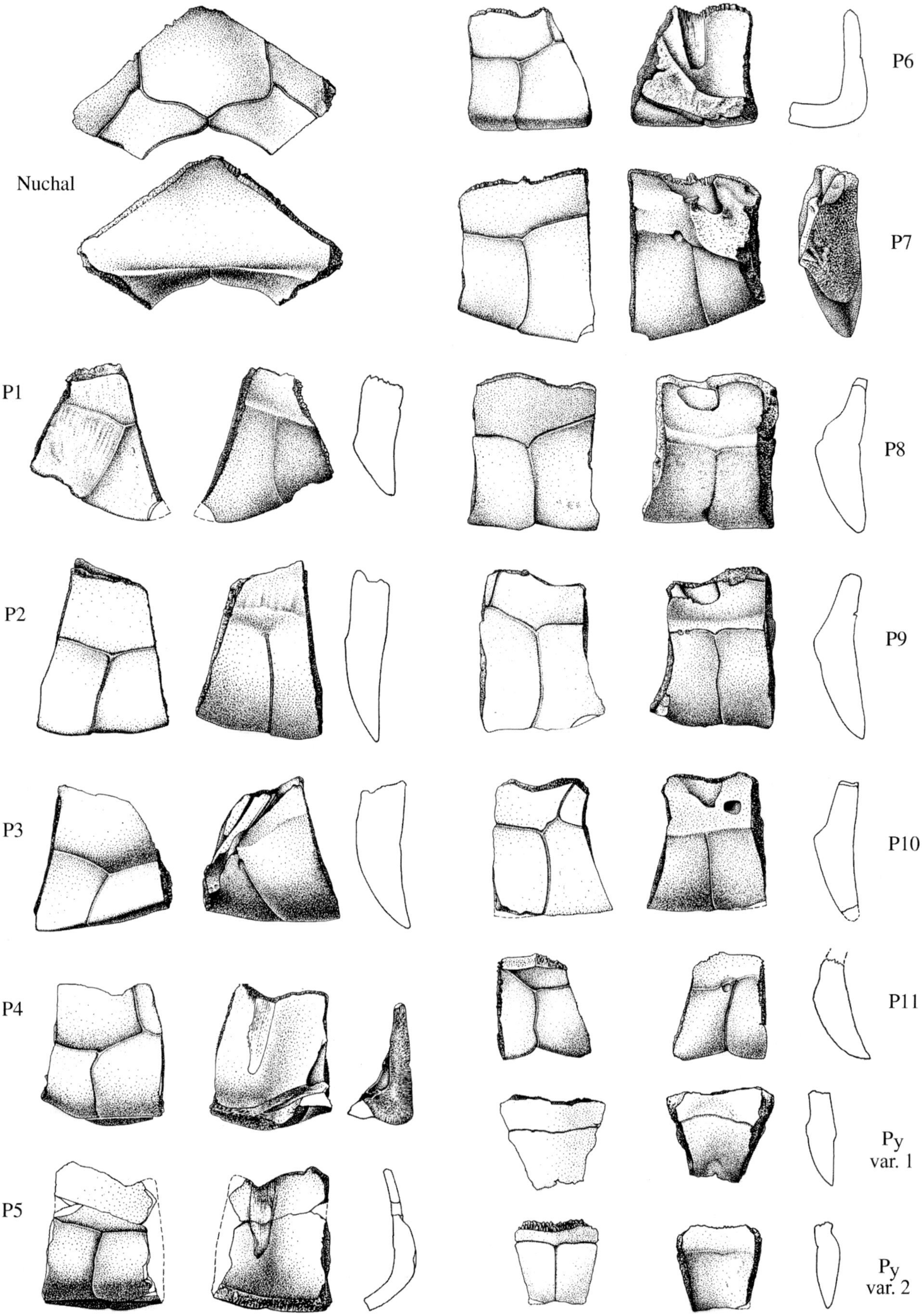
Nuchal
P6
P7
P1
P8
P2
P9
P3
P10
P4
P11
Py
var. 1
P5
Py
var. 2

◀**Fig. 26.9** *Cardichelyon rogerwoodi* gen. et sp. nov., drawings of peripheral elements. Nuchal. UCMP 126289, nuchal plate, in external (*top*) and visceral (*bottom*) views. *P1-Py* left peripherals (P1, 2, 4, 5, and 8 reversed for comparison), each depicted in external view (*left*), visceral view (*middle*), and planar view of cephalad suture (*right*): *P1* UCMP157437; *P2-P3* UCMP 126367; *P4* UCMP 126264; *P5* UCMP 126364; *P6* and *P7* UCMP157437; *P8 P9* and *P10* UCMP 126390, *P11*, 126395; *Py var. 1* UCMP 126418. *Py var. 2*, UCMP 126395. Scales vary

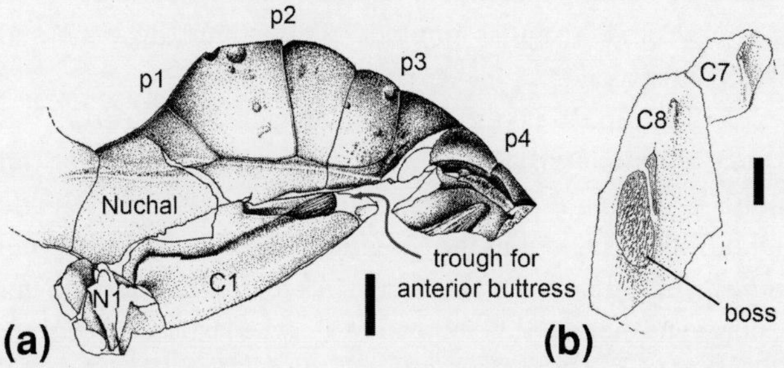

Fig. 26.10 *Cardichelyon rogerwoodi* gen. et sp. nov., details of carapace **a** UCMP 126432, stipple drawing of median and left anterior part of carapace, in visceral view showing trough for anterior buttress and pathological pits. **b** UCMP 126450, stipple drawing of proximal portions of left C8 and dorsal C7, in visceral view, showing boss that probably serves as an area of muscle and ligament attachment for the pelvis. Scale bars equal 1 cm

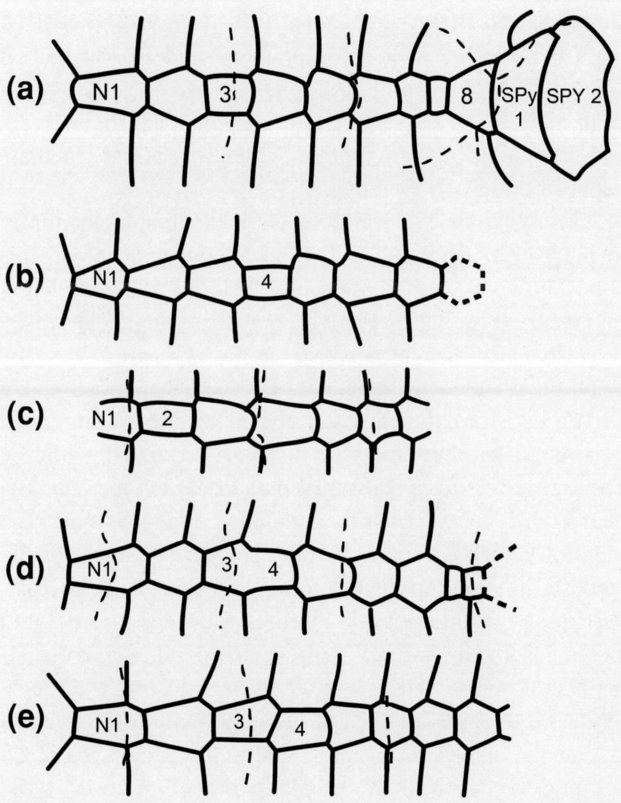

Fig. 26.11 *Cardichelyon rogerwoodi* gen. et sp. nov., diagrams depicting the two dominant neural patterns **a** and **b** and three variants **c–d**. **a** UCMP 126364. **b** UCMP 126480. **c** UCMP 126343. **d** UCMP 126560. **e** UCMP 126337. Scales vary

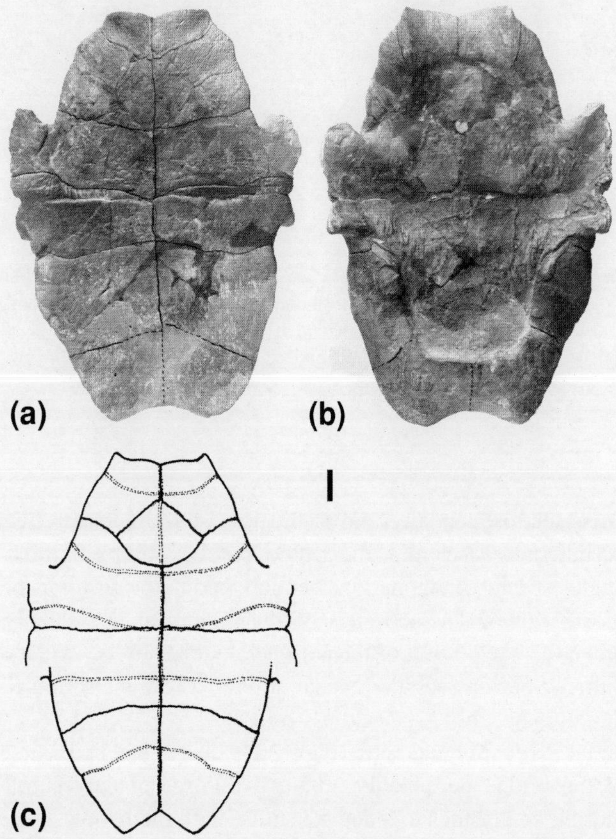

Fig. 26.12 *Cardichelyon rogerwoodi* gen. et sp. nov., plastron. **a**, **b** YPM (PU) 14671 (holotype), photographs in ventral (**a**) and dorsal (**b**) views. **c** UCMP 126288, line drawing in ventral view, showing sutures (*solid lines*) and scute sulci (*dotted lines*). Scale bar equals 1 cm

generally lack a lateral carina and the angulation of the pleural-marginal surfaces is reduced or absent.

P1 is distinctly asymmetrical (Figs. 26.9, 26.10a) with the nuchal suture distinctly shorter than that with the P2. The groove which houses the axillary buttress may lie partly on P1 and adjacent C1 suture or be mostly confined to C1 (Fig. 26.10).

P2 is generally trapezoidal in shape (Figs. 26.9, 26.10a). The caudal suture is about 80% as long as the cephalad suture. The visceral side is extensively covered by the marginals. The suture with C1 internally is coincident with the recessed groove for the axillary buttress (Fig. 26.10).

P3 is also trapezoidal in outline (Figs. 26.9, 26.10a) with the caudal and cranial sutures nearly equal in length. The perimeter margin becomes progressively more rounded in

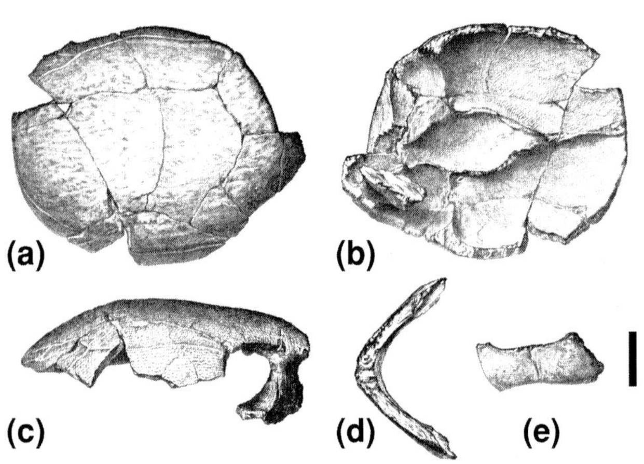

Fig. 26.13 *Cardichelyon rogerwoodi* gen. et sp. nov., YPM PU 14671 (holotype), drawings of incomplete skull. **a–c** Skull roof: **a** dorsal view, with anterior to *right*; **b** ventral view, with anterior to left; **c** right lateral view. **d** Left and right dentaries, in dorsal view. **e** Left dentary, in lateral view. Scale bar equals 1 cm

cross-section caudally and is the last of the anterior free peripherals. Viscerally, the groove for the axillary buttress angles strongly diagonally, essentially along the terminus of the marginal. The terminus of the C1 rib angles steeply down the caudal part of the peripheral lateral to the axillary buttress but crosses the caudal suture to terminate immediately adjacent to the buttress on P4.

P4 is roughly quadratic (Figs. 26.9, 26.10) and is the first of the bridge peripherals. About two-thirds of the ventral margin contributes to a dentate suture with the hyoplastron. The scar for the root of the buttress occupies the anterior half of the plastral suture. The tip of the first rib terminates just within the anteromedial margin. The laterally flattened tip of the second costal rib extends nearly two-thirds the longitudinal height of the peripheral and is recessed into the visceral wall but is not covered. Part of the ventral external margin may be covered by the pectoral scale.

P5 marks the transition between anterior and posterior peripherals and spans the hyo-hypoplastral suture. It is quadratic in lateral outline (Fig. 26.9) and placement of the third rib is similar to that of the second on P4. P5 is taller than long in lateral view.

P6 (Fig. 26.9) is the penultimate bridge peripheral. The plastral suture rises from the ventral surface and thickens as it progresses caudally. The rib of C4 is well exposed viscerally along the caudal moiety before fully penetrating P6. The P5-6 suture forms a right angle bend and thins gradually dorsally, but the P6-7 suture is greatly thickened at and just below the level of the plastral suture. The angle of the ventral and dorsal arms progress caudally from a right angle to an obtuse angle or is unangled. The p2-p3 sulcus lies on or very near the caudal suture. The abdominal scale covers a narrow strip on the cranial moiety of P6 but appears to coincide with the P6-hypoplastral suture caudally. The marginal and pleural scales cover about equal areas of the dorsal ramus of P6. A small musk duct may traverse the suture with the hypoplastron or may lie within the peripheral. The P6-hypoplastral suture is usually covered with fine, low, and irregular pimples. This suture deepens and rises caudally.

P7 (Fig. 26.9) is the last to contact the plastron but retains a complete free marginal edge as well. The plastral suture begins anteriorly about halfway up the visceral side and terminates about at the longitudinal midpoint or beyond and overlaps the pit for the C5 rib. As in P6, the suture is not dentate and nearly flat indicative of an elastic connection. The rib penetrates at about the middle portion just lateral to the plastral suture. There is very little exposure of the rib viscerally and it probably parallels this suture within the body of the peripheral. The P7-8 suture is much narrower than the P7-9 suture and has a narrow half-moon outline. The maximum length and height dimensions of P7 are about equal but the cranial suture is about 70% the length of the caudal.

P8-10 are unremarkable (Fig. 26.9). P11 varies in size and shape from being transversely much shorter than its vertical dimension to longer than this dimension. Sutures on some isolated P10's suggest that P11 may even be wedged out from contact with the suprapygal and C8 by mutual contact of the pygal and P10. In at least one case, the pygal is absent and the opposite P11's are elongated transversely and in contact. The common condition is intermediate between these two extremes.

The pygal is also variable is shape (complementarily to the P11's) (Fig. 26.9). The m12's which meet on this bone show all combinations of fusion from completely separated (Fig. 26.9, Py var. 2) to completely fused (Fig. 26.9, Py var. 1).

Of the eight costal pairs, only C1 and C8 are notable. On C1 (Fig. 26.10a), the precostal rib and head of C1 are joined by a raised step but otherwise distinctly separated medially. The suture for the precostal rib is narrow and extends less than a third of the C1 width. A rounded ridge that marks the trend of C1 rib continues distally beyond the suture of the precostal rib and terminates in the P3-4 (Fig. 26.9, 26.10a). Viscerally the anterodistal margin of the C1 is inserted into a channel for the axillary buttress that parallels the costal-peripheral suture. There is a characteristic fine transverse fluting on the anterior surface of the groove for the buttress just medial to the rib extension. Only a few specimens of C8 are preserved but these usually show (Fig. 26.10b) the presence of a large centrally or anteriorly placed suture for a presacral rib. Where present, it tends to form a prominent boss that may span the C7-8 suture although fragments of the rib indicate that only its costal attachment is enlarged.

The boss probably serves as an area of muscle and ligament attachment for the pelvis.

The neural series consists of eight or nine neurals and three suprapygal bones. The typical neural pattern appears to be about equally divided between 1)2)3(4(5(6(7(8 and 1)2)3)4(5(6(7(8 (Fig. 26.7), although the sample is small (n = 8) (Fig. 26.11b). There is no indication of a mid-dorsal carina on any of the bones of the neural series, even in juvenile individuals. The neurals are longer than wide anteriorly but this may be reversed in the posterior part of the series. Neural 8 is variable in shape and expanded posteriorly with the anterior suture less than twice the length of the posterior suture. In UCMP 126372 a small quadratic neural is inserted between N8 and N7. Neural 8 contacts suprapygal 1 via an anteriorly arched suture; the curvature of the suture is more pronounced on the visceral side. Neural 8 may coossify with suprapygal 1. Ventrally, a thin midline ridge divides the internal surface into two halves. This ridge continues on through SPy2. SPy1 is large and has extensive contact with C8. The triple juncture of v4-5 and p4 (n = 16) lies on N8 or across the N8-SPy suture area. SPy2 is completely covered by v5 except at its extreme lateral tips. It is over two and one-half times as wide as long and completely spans the pygal and p11's. It contacts a small part of C8 and P10 laterally. The anterior suture is upturned at the lateral margins, thus tending to "grip" SPy1.

The plastron is broad, truncated anteriorly, and has a distinct caudal notch (Fig. 26.12). The bridge is short in relation to the lobes and less than one-third the plastral length. Some specimens have a broadly concave ventral surface, probably indicating males. The anterior margin of the plastron protrudes into the space below the cephalic emargination of the carapace.

The epiplastra (Fig. 26.12) are thickened into anteriorly projecting spurs at the lateral sides of the channel for the neck. The neck channel forms a deep U-shaped valley between these spurs. The spurs continue posteriorly as raised ridges on the dorsal surface just medial to the gular-humeral sulci. The dorsal surface is extensively covered by the gular and humeral scales. The posterior tip of the gulars lap onto the entoplastron in 37% (n = 22) of the specimens, just contacts the entoplastron suture in 29% (n = 17), and is confined to the epiplastron in 34% (n = 20) of the Willwood specimens. The posterodorsal margin of the gular scale is marked by a sharp but not large step-down in adult-sized individuals and always slants gradually anteromedially. The dorsal portion of the epiplastron covered by the humeral slopes sharply anterolaterad further accentuating the spur-like projections. Of the three epiplastral sutures, the interepiplastral suture is consistently the shortest with the hyo- and entoplastral sutures about equal. The anterior margin between the spur and the interepiplastral suture is broadly convex; the tip of the spur may or may not actually project anterior to this margin.

The entoplastron (Fig. 26.12) is maybe nearly as long as wide or shorter. The sutures with the hyoplastra form a convex arc, whereas the sutures with the epiplastra are straight and converge to a point anteriorly. The hyoplastron (Fig. 26.12) is shorter than wide with the inter-hyoplastral suture a little more than half the maximum length. A remarkable gently arched and rod-like axillary buttress (Fig. 26.9) arises from the axillary margin and curves toward the nuchal. The buttress parallels the P1-3 and C1 suture and is finely fluted. The channel for the buttress tends to be smooth-walled anteriorly. The buttress is secured by the slightly overlapping rims of the channel. The suture with the entoplastron is longer than the inter-hyoplastral and epiplastral sutures which are about equally long. The minimum length of the bridge part of the hyoplastron is greater than or equal to the interhyoplastral suture length. The humeral-pectoral sulcus lies entirely posterior to the anterior sutures and nearly parallels them, although generally converging with the entoplastral suture medially. The pectoral-abdominal sulcus lies near the hyo-hypoplastral suture laterally and converges on it medially. The dorsal exposure of the scales rapidly broadens anteriorly so that half of the suture with the epiplastron is covered. The hyoplastron gradually thickens anteriorly from its suture with the hypoplastron.

The hypoplastron (Fig. 26.12) is generally rectangular in shape and wider than long. It is thickest where the femoral scale terminates on the dorsal surface, at the bridge, and in the central area near the plastral midline. The suture with the hyoplastron tends to slant posteroventrally along its vertical plane and is about straight transversely, as is the suture with the xiphiplastron. The femoral-abdominal sulcus extends transversely through the middle portion of the hypoplastra with a small anteriad flexure near the free margin. Coincidence of major sutures and sulci are usually indicative of kinesis along the joint (Hutchison and Bramble 1981). The later samples of *Cardichelyon rogerwoodi* show little or no break in outline of the xiphiplastron where the femoral-anal sulcus meets its perimeter of the plastron. Older samples tend to be more variable in this feature. The hypoplastron in a few specimens suggests that the plastron was somewhat concave on the ventral surface, suggesting that they may be males as in *Terrapene* and testudinids.

The xiphiplastron (Fig. 26.12) is about as long as broad in maximum dimensions. The thickest parts are along the dorsal terminations of the anal and femoral scales. The distal ends are separated by a distinct caudal notch. Just anterior to the anal notch and in front of the scale margins on the dorsal surface are symmetrical raised transverse scars. On the ventral surface, the anal-femoral sulci extend posterolaterally from the anterior moiety of midline.

Only the roof of the skull and the descending part of the right prefrontal is preserved in YPM (PU) 14671 (Fig. 26.13a–c). The skull roof is oval in outline without a supraoccipital spine and convex posterior margin. The frontal and prefrontal are co-ossified. A sulcus extends posteriorly from the orbit, follows the contour of the skull and curves along the posterior margin to the midline, thus defining a large unpaired dorsal scale. Two short sulci branch off posteriorly in the temporal region and define two pairs of temporal scales. On the ventral surface of the skull roof, the orbital cavity is separated from the temporal cavity by a transverse ridge. The orbits are separated in the midline by the margins of an amphora-shaped nasal-brain cavity. The dentary (Fig. 26.13d, e) is narrow and deep and diverges about 90 degrees from its mate. The tip is upturned, with a nearly vertical symphysis and a small chin. The triturating surface is composed of two closely spaced, parallel and "untoothed" ridges; the labial ridge is higher and sharper.

Remarks: Estes (1975) referred specimens of this genus to the emydid *Ptychogaster* Pomel 1847, but a number of distinctive character states of *Cardichelyon* clearly distinguish these two superficially similar genera. *Cardichelyon* differs from *Ptychogaster* in having a deep emargination of the nuchal area, wide and short nuchal, neurals 2–8 hexagonal, three suprapygals, posterior plastral lobe only is kinetic, near flat anterior lobe, bridge length more than one-fourth of plastral length, only slight or no overlap of humeral-pectoral sulcus onto entoplastron, angle of axillary buttress to plastron horizontal, axillary buttress terminates at P1-C1 suture and little or no emarginations of the skull roof.

Some general features of the shell (e.g., broad plastron and extent of the dorsal overlap of the plastral scales) of *Cardichelyon* favor emydid relationships. However, the retention of multiple musk ducts and a pair of functionally correlated features (i.e., the roofed over skull and deep emargination of the anterior carapace indicate that the head was not retracted under the carapace) are more convincing of a relationship with *Platysternon* Gray 1831. Detailed comparisons are deferred to the description of another species of *Cardichelyon* with a more complete skull (see below).

Cardichelyon is widely distributed along the eastern and southern slopes of the Rocky Mountains (Montana-Texas, see Hutchison 1998; Holroyd et al. 2001) and as far north as Ellesmere Island (in preparation), Arctic Canada. The genus ranges in age from the Torrejonian through the Wasatchian and, possibly, into the Bridgerian (pers. obs.).

Emydidae Bell 1825
Psilosemys gen. nov.
Psilosemys wyomingensis sp. nov.
(Figs. 26.14, 26.15, 26.16, 26.17, 26.1826.19)

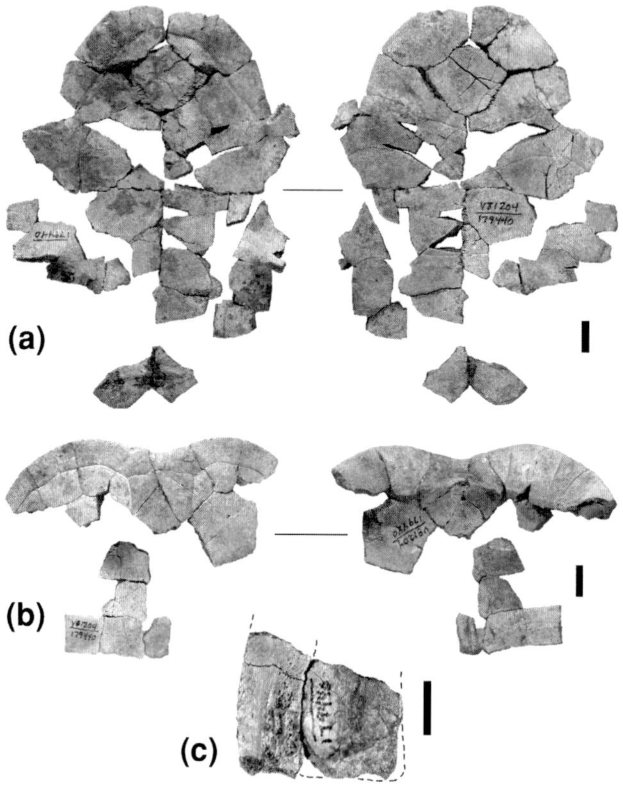

Fig. 26.14 *Psilosemys wyomingensis* gen. et sp. nov., UCMP 179440 (holotype), photographs of shell. **a** Plastron, in dorsal (*left*) and ventral (*right*) views. **b**, **c** Less complete carapace: **b** anterior portion, consisting of nuchal, right P1-3, left C1-3, and N3, in dorsal (*left*) and ventral (*right*) views; **c** distal ends of C5-6, in visceral view. Scale bars equal 1 cm

Synonymies: High-domed emydid: Hutchison 1980, p. 116. Emydid P: Hutchison 1992, p. 455; Hutchison 1998, p. 402; Holroyd and Hutchison 2000, p. 75; Holroyd et al. 2001, p. 98, Fig. 2–4, Table 1.

Holotype: UCMP 179440 (Fig. 26.14), fragmentary shell including anterior lobe of plastron and fragments of posterior lobe, nuchal and peripherals 1–3, 6–8, parts of costals 1–3, 6, and neurals 1 and 6. UCMP loc.V81204, Fifteen Mile Creek, Washakie County, Wyoming, USA; Willwood Formation; Wasatchian NALMA, Graybullian subage. Collected by J. H. Hutchison, 29 July 1980.

Paratypes: Willwood Formation, Wasatchian NALMA, Big Horn County, Wyoming, USA (2 localities): UCMP loc. V97074: UCMP 179484, partial plastron, nuchal and carapace fragments; UCMP loc. V99129: UCMP 179464, crushed central and anterior part of juvenile shell. Willwood Formation, Wasatchian NALMA, Washakie County, Wyoming, USA (3 localities): UCMP loc. V81204: UCMP 179441, crushed but nearly complete shell with crushed posterior part of the skull; UCMP loc. V81067: UCMP 179439, partial plastron and carapace fragments; UCMP

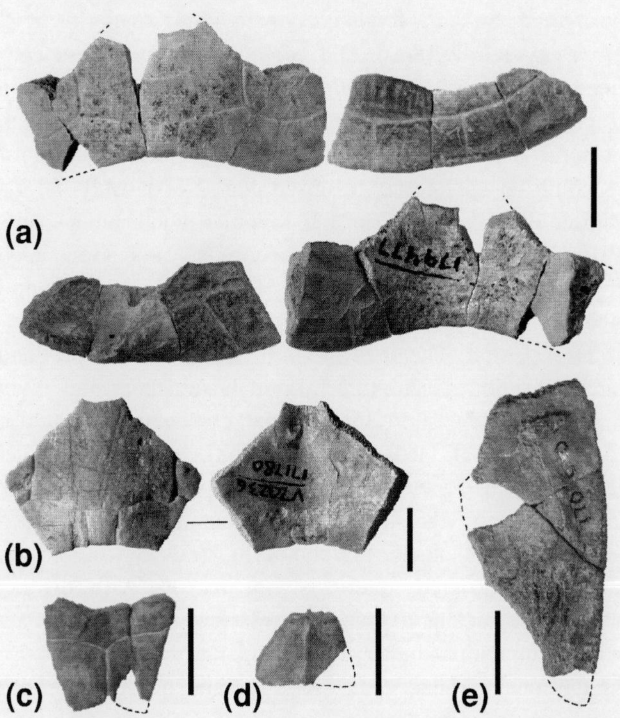

Fig. 26.15 *Psilosemys wyomingensis* gen. et sp. nov., photographs of selected carapace elements. **a** UCMP 179477, nuchal, partial right P1, left P1-2, and partial P3, in exterior (*top*) and visceral (*bottom*) views. **b** UCMP 171780, nuchal, in external (*left*) and visceral (*right*) views. **c** UCMP 179502, pygal, in external view. **d** UCMP 179502, superpygal 1, in visceral view. **e** UCMP 170690, right C1, in visceral view. Scale bars equal 1 cm

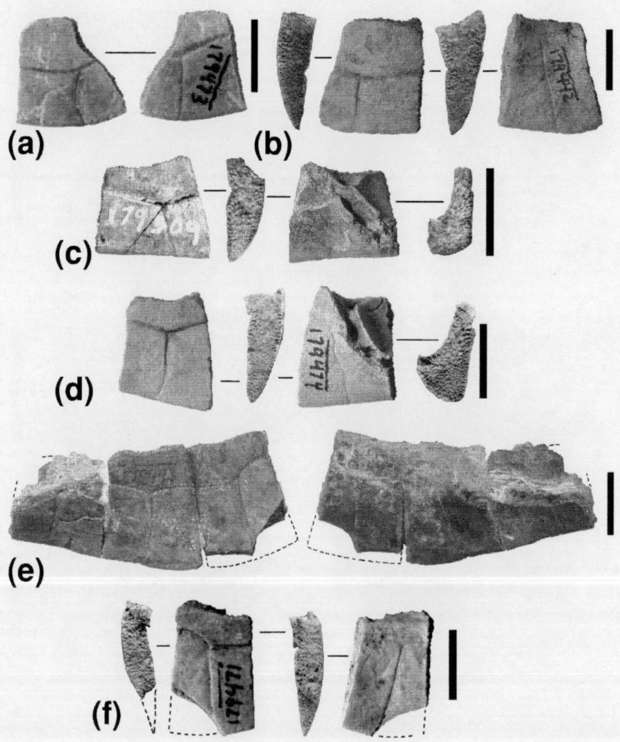

Fig. 26.16 *Psilosemys wyomingensis* gen. et sp. nov., photographs of selected peripherals. **a** UCMP 179473, right P1, in exterior (*left*) and visceral (*right*) views. **b** UCMP 179442, left P2, views from left to right are anterior suture, external, posterior suture, and visceral. **c** UCMP 179509, right P3, views from left to right are external, anterior suture, visceral, and posterior suture. **d** UCMP 179474, left P7, views from left to right are external, posterior suture, visceral, and anterior suture. **e** UCMP 179443, left P6-8, in exterior (*left*) and visceral (*right*) views. **f** UCMP 179471, right P8, views from left to right are anterior suture, external, posterior suture, and visceral. Scale bars equal 1 cm

loc. V81079: UCMP 179442, partial plastron and carapace fragments.

Referred specimens: Willwood Formation, Wasatchian NALMA, Big Horn County, Wyoming, USA (3 localities): UCMP loc. V78106: UCMP 170690, partial shell; UCMP loc. V81083: UCMP 179447, shell fragments; UCMP 179460, epiplastron; UCMP loc. V96248: UCMP 179443, peripherals 2, 3, 7–9. Main Body Member of Wasatch Formation, Wasatchian NALMA, Sweetwater County, Wyoming, USA (8 localities): UCMP loc. V70236: UCMP 171781, epiplastron; UCMP loc. V70272: UCMP 179467, epiplastron; UCMP loc. V73155: UCMP 179509, peripheral 3; UCMP loc. V73157: UCMP 179469, epiplastron; UCMP 179471, shell fragments; UCMP loc. V73161: UCMP 170690, plastron and carapace fragments; UCMP 179474, P7; UCMP loc. V73181: UCMP 179477, partial plastron and carapace fragments; UCMP loc. V74024: UCMP 154883, epiplastron; UCMP loc. V99151: UCMP 179502, partial juvenile shell.

Etymology: From the Greek *psilos*, smooth, plus *emys*, turtle and *wyomingensis*, for the State of Wyoming.

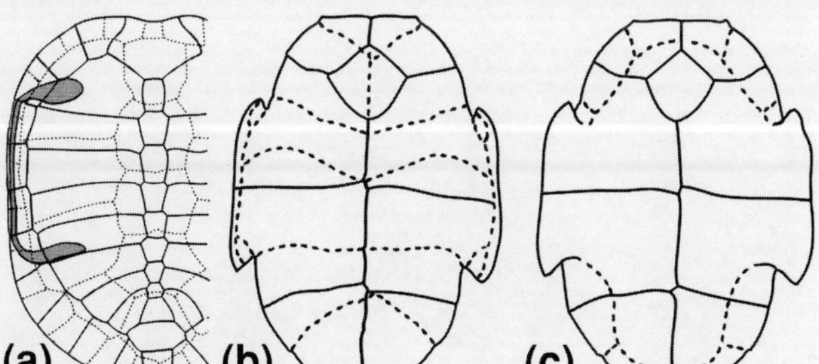

Fig. 26.17 *Psilosemys wyomingensis* gen. et sp. nov., line drawings of reconstructed shells showing sutures (*solid lines*) and scute sulci (*dotted lines*). **a** Carapace in dorsal view, based on UCMP 170690, 179440 (holotype), 179441, and 17947; *shaded area* indicates plastral sutures on visceral surface of carapace. **b, c** Plastron based on UCMP 179440 (holotype) and 179441, in dorsal (**b**) and ventral (**c**) views

Diagnosis: *Psilosemys* is included in the Testudinoidea on the basis of the loss of extragular scales, extensive dorsal overlap of the plastral scales, broad plastron, loss of the inframarginals other than inguinal and axillary, and well

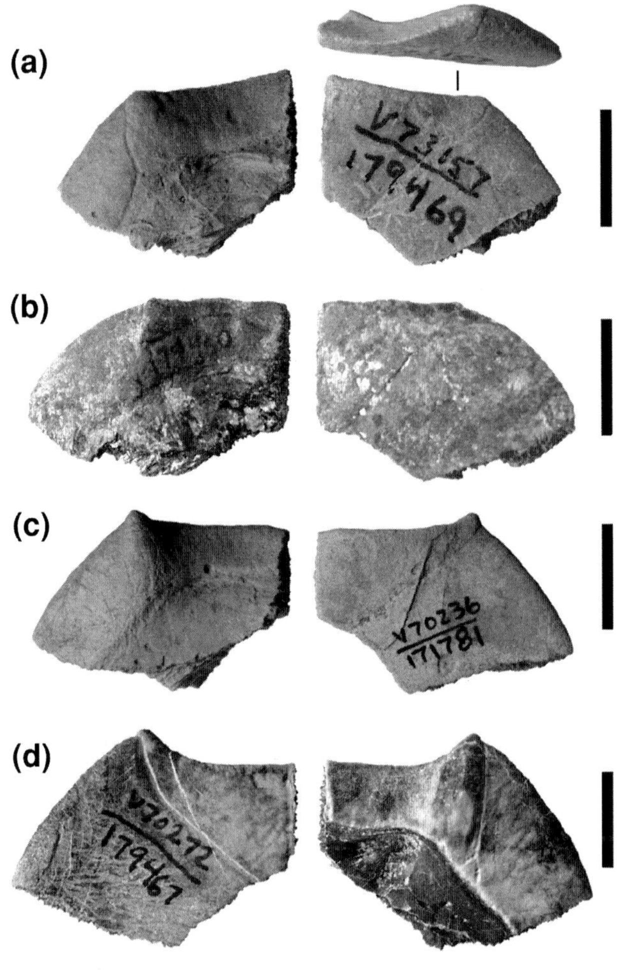

Fig. 26.18 *Psilosemys wyomingensis* gen. et sp. nov., photographs of selected left epiplastra. All specimens shown minimally in dorsal (*left*) and ventral (*right*) views. **a** UCMP 179469, plus anterior view (upper right). **b** UCMP 179460. **c** UCMP 171781. **d** UCMP 179467. Scale bars equal 1 cm

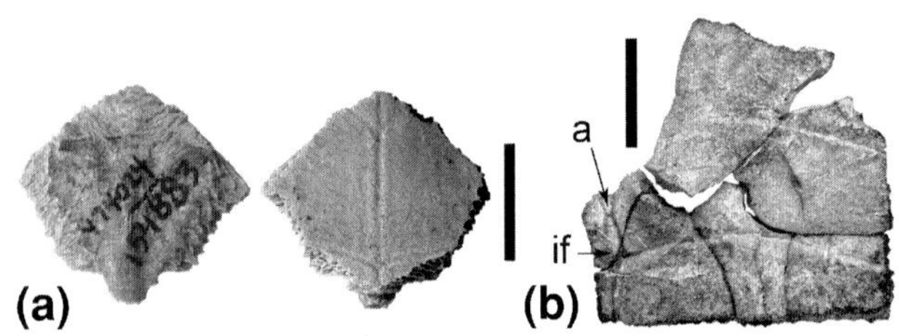

Fig. 26.19 Psilosemys wyomingensis gen. et sp. nov., photographs of selected plastral elements. **a** UCMP 154883, entoplastron, in dorsal (*left*) and ventral (*right*) views. **b** UCMP 179502, *right* hyoplastron showing axillary (**a**) and inframarginal (if) scales, in ventral view. Scale bars equal 1 cm

developed axillary and inguinal buttresses. It is included in the Emydidae because of the absence of musk duct foramina, supracaudal marginals not extending onto the suprapygal (also occurs in a few batagurids; e.g., Joyce and Bell 2004), inframarginals reduced to two, normal hexagonal neurals 2–8 (also occurs in a few batagurids; e.g., *Mauremys* Gray 1869), and costal-inguinal buttress confined to C5. It differs from other emydids in the combination of plastral vertebral scale crossing lateral corners of nuchal, pectoral scale posterior to entoplastron, plastral buttress just reaching costals, carapace unsculptured or weakly so, C5 shorter distally than C6,; and inguinal small and narrow. *Psilosemys* resembles the Deirochelyinae and is excluded from the Emydinae on the basis of distinct lack of pectoral overlap of the entoplastron (one exception).

Description: The shell of adult *Psilosemys* is ovoid and lacks any dorsal sculpture, although subadults and juveniles may show short longitudinal corrugations posterior to the vertebral, pleural (Fig. 26.15a), and plastral sulci (Figs. 26.14b, c, 26.17). The carapace (Figs. 26.14b, 26.17a) shows the following characters: finely incised sulci; no dorsal keels; peripherals not notched or serrate; v1 not constricted anteriorly but narrow with distinct overlap of nuchal; axillary buttress has moderate contact with C1; inguinal buttress has moderate contact with C5; v5-p4 sulcus contacts m11 on P10; and posterior peripherals only weakly flared.

The peripheral series includes the normal complement of eleven peripherals, nuchal, and pygal (Figs. 26.14b, 26.17a). The nuchal is slightly shorter than wide with the anterior margin slightly concave (Figs. 26.14b, 26.15a, b). The margin of the first vertebral lies on the nuchal and reaches the middle part of m1 on the nuchal. Dorsally, the cervical scale is quadratic to slightly trapezoidal with the wide end posterior. The ventral side of the cervical scale is similar but shorter.

The first peripheral (Figs. 26.14b, 26.15a, 26.16a) contacts the nuchal, P21, and C1. The pleural-marginal sulcus lies in the middle and there is no change in plane. The marginal scales extend viscerally to nearly the top of the peripheral. The P1 thickens progressively dorsally with a sharply defined free margin. The P2 (Figs. 26.14b, 26.15a, 26.16b) resembles P1 except for being broader dorsally and has a broad dentate suture with C1. The axillary buttress may just cross the extreme posterodorsal part of P2. P3 (Figs. 26.14b, 26.15a, 26.16a) is transitional to the bridge and is crossed by a narrow but deeply pitted axillary buttress scar. The scale exposures are as for P1-2. The free margin broadens into an obtuse carina in the bridge area. There is no sign of a musk duct foramen. P4-6 (Fig. 26.17a) are bridge peripherals with the obtuse to sharp lateral carina. P7 (Figs. 26.16d, e, 26.17a) is transitional from the bridge to the posterior free margin. The inguinal buttress rises on P7. There is no indication of a musk duct foramen. The lateral carina sharpens to a knife edge at the free margin. The marginal scales rise onto the upper one-third on the visceral side. P8 (Figs. 26.16e, f, 26.17a) is the first free posterior peripheral. The marginal scales occupy over two-thirds of the external and visceral sides. The free margin of this and the remaining peripherals is sharp. On P9, the marginals cover three-fourths of the peripheral externally and more viscerally (UCMP 179443). The p3-4 sulcus contacts the anterior part of m9 on this peripheral. P10 is similar to P9 but with a modest flare (UCMP 170690). P11 is similar to P10 and p5-v5 sulcus contacts the anterior part of M11 (UCMP 170690). The pygal (Fig. 26.15c) is trapezoidal with the m12 confined dorsally to the pygal.

The neural series consist of eight neurals and two suprapygals (Figs. 26.14b, 26.17a). Neurals 1–6 are flat, hexagonal, and coffin-shaped (wide end anterior). SPy1 (Fig. 26.15d) is trapezoidal with the posterior margin broadly concave anteriorly and did not reach the peripherals.

The costals are smooth and broadly arched with finely incised sulci and finely dentate contacts (Figs. 26.14b, c, 26.15e, 26.17a). C1 contacts P1–P3. The thin suture for the first rib extends nearly to the middle of the costal and lies on a ridge that broadens distally. The buttress scar is confined to the distal one-third and is not distinctly raised but forms a rather dentate trough (Fig. 26.15e). The sutural area for P1–P2 is the thickest part of C1. The rib heads lie distinctly lateral to the neural suture. The v1 is constricted at the v1–v2 sulcus. The second costal is unremarkable with a straplike rib head. The costal thickens dorsally but not abruptly. Costals 3–4 resemble C2. Costal 5 (Figs. 26.14c, 26.17a) is distinctly shorter than C6 laterally and bears the inguinal buttress that lies on the distal one-third and occupies about two-thirds of the distal length. The bone is thickened under the buttress scar, which is irregularly pitted but not raised into a ridge. The scar terminates at the C5-6 suture and does not overlap C6. Costal 6 is similar to C5 but the rib head shaft is rounder in cross-section. The anterior margin is thickened to form a buttress for the inguinal buttress but does not contact it directly.

The plastron has broad plastral lobes with a distinct caudal notch, wide gular gutter, small inguinal and axillary scales, and lack musk ducts or foramina (Figs. 26.14a, 26.17a). The plastron length is about 130 mm.

The epiplastra are broader than long (Figs. 26.14a, 26.18). The gulars overlap the anterior point of the entoplastron. The neck gutter is broad and distinctly U-shaped with the lateral edges raised into ridges. Dorsally, the gular scale overlap is relatively short anteroposteriorly and reaches its maximum extent at the lateral gular ridges. The gular-humeral sulcus lies immediately lateral and parallel to the gular ridges. The medial ventral surface of the gular area is in the same plane as the hyoplastron but curves up laterally to produce the gular ridges which terminate in small points or spurs. The hyoplastral suture of the epiplastron is slightly concave, sinusoidal, or slightly convex but always anterolaterally oriented. The median margin of the gular gutter may be slightly convex anteriorly and, thus, equal to or extending slightly beyond the gular spurs (Fig. 26.18a) or concave (Fig. 26.18d).

The entoplastron is roughly fan-shaped to diamond-shaped with the posterior margins convex to nearly straight, angular, or rounded and the anterior margins roughly straight to slightly concave (Figs. 26.14a, 26.19a). It is completely covered ventrally by the humeral scales except for the extreme anterior point. One specimen, UCMP 154883 (Fig. 26.19a) exhibits a slight onlap of the pectoral scale.

The hyoplastron is wider than the midline and maximum lengths (Figs. 26.14a, 26.17b, c, 26.19b). There is a relatively broad overlap of both humeral and pectoral scales onto the dorsal surface. The axillary buttress is anteriorly recumbent and extends forward to the base of C1 contacting P3 and the extreme posterodorsal part of P2. The bone below the dorsal scale overlap region is only slightly thicker than the other parts and not distinctly set off from the dorsal plane except for a shallow scale margin trough. The humeral-pectoral sulcus is relatively straight and confined to the hyoplastron with a distinct gap between it and the entoplastron suture except for UCMP 154883 noted above. The pectoral-abdominal sulcus is also confined to the hyoplastron but is distinctly anteriorly convex and slopes nearly to the hypoplastral suture medially. The marginal scales narrowly overlap the plastron. There is a small axillary scale. The right but not the left hyoplastron of UCMP 179502 (Fig. 26.19b) has a small inframarginal scale posterior to the axillary scale. The suture with the hypoplastron is firm and well dentate and exhibits no indication of kinesis.

The hypoplastron is wider than the midline and maximum lengths (Figs. 26.14a, 26.17c, d). The abdominal-femoral sulcus lies on the posterior half of the hypoplastron. The inguinal buttress is posteriorly recumbent and the base extends just medial to the scale covered area dorsally. The buttress rises on P7 and contacts the lower part of costal 5. The inguinal scale is narrow laterally and does not contact the femoral scale. The abdominal and femoral scales broadly overlap the dorsal surface.

The xiphiplastron is well sutured to the hypoplastron. The femoral-anal sulcus slopes distinctly anteromedially and terminates medially at or just short of the hypoplastron. There is only a moderate to slight notch where the anal-femoral suture meets the lateral margin. The dorsal scale overlap is extensive, occupying more than a third of the dorsal surface with the thickest part along the scale margin. The caudal notch is V-shaped and distinct but not deep.

Remarks: *Psilosemys* is included in the Emydidae because of the absence of musk duct foramina, supracaudal marginals not extending onto the suprapygal and costal-hypoplastral suture confined to C5. Gaffney and Meylan (1988) and Hirayama (1985) use no shell features for diagnosing the genera of the Deirochelyinae but *Psilosemys* resembles the Deirochelyinae and is excluded from the Emydinae on the basis of distinct lack of pectoral overlap of the entoplastron (one exception). The exclusion of the pectoral from the entoplastron appears to be primitive as this is the condition in the Asia outgroups Lindholmemydidae (Sukhanov 2000; Joyce and Bell 2004) and *Pseudochrysemys* Sukhanov and Narmandakh 1976. Relatively narrower plastral lobes, distinct caudal notch on the plastron, and unexpanded distal costal 5 are also primitive in regard to extant emydids. *Psilosemys* thus appears to be a sister taxon to the Emydinae and Deirochelyinae (i.e., crown group Emydidae sensu stricto).

In general, *Psilosemys* lacks derived features of the Emydinae and Deirochelyinae, but most closely resembles *Chrysemys* Gray 1844, among living taxa, in general shape and sculpture. Among fossils, *Psilosemys* most closely resembles unpublished material of *Pseudochrysemys* (Sukhanov 2000; Sukhanov and Danilov, personal communication). The presence of an extra inframarginal in one specimen (UCMP 179502, Fig. 26.19b) is a similarity to *Pseudochrysemys*, where this is the normal state. *Psilosemys* differs from *Pseudochrysemys* Sukhanov and Narmandakh 1976 (Paleocene of Mongolia) in great reduction or lack of sculpture, small size of the axillary and inguinal scales, less excursion of the axillary buttress onto C1, hypoplastral buttress centered near C5-6 suture, and presence of only two inframarginals.

Distribution of *Psilosemys* is currently limited to the early Eocene (Wasatchian NALMA) of Wyoming, specifically the Bighorn (including Clarks Fork), Wind River, and Washakie basins. Within these basins, its stratigraphic distribution is limited to the Graybullian (including "Sandcouleean" of Gingerich 1989) and Lysitian subages (Hutchison 1980, 1992, 1998; Holroyd and Hutchison 2000; Holroyd et al. 2001).

Acknowledgments I thank P. Gingerich and W. Bartels (both UMMP), D. Baird (formerly of Princeton University), D. M. Bramble and John Legler (both University of Utah), J. A. Lillegraven (UW), P. Holroyd (UCMP), T. M. Bown and E. L. Lindsay and B. McCord (University of Arizona Paleontology Laboratory, Tucson), T. Rowe (Texas Memorial Museum, Austin) and E. S. Gaffney (AMNH), for access to collections in their care and loan of specimens. Bob O'Donnell (USGS), B. T. Waters (formerly of UCMP), T. Bown, (formerly with the USGS), K. D. Rose (John Hopkins University), S. L. Wing (USNM), P.A. Holroyd (UCMP), K.T. Smith (formerly UCMP), D. E. Savage (UCMP), W. A. Clemens (UCMP), and D. M. Bramble provided field assistance in collecting and geology. I have received financial support from the following sources: NSF Grant GB 40361 to D. E. Savage; National Science Foundation and National Geographic Society grants to W. A. Clemens; a fellowship from the University of Utah Department of Biology; the Annie M. Alexander Endowment to the University of California Museum of Paleontology; and teaching fellowship (1979–1980) from the Department of Biology, University of Utah. The finely detailed stipple drawings were prepared by A. Lucas-Andreae. R. Wood provided photographs and a Princeton University artist provided the shaded drawings of the holotype of *Cardichelyon*. Other figures are by me. I thank K. Kishi for the expert preparation of many specimens. The BLM provided collecting permits to Federal lands. I thank I. G. Danilov and V. B. Sukhanov for access to unpublished material of *Pseudochrysemys*, and J. F. Parham for facilitating that access. I thank J. Parham, M. J. Ryan and particularly K. T. Smith for provided useful and helpful criticisms, many of which were followed, and P. A. Holroyd and D. Brinkman for their patience.

References

Bartels, W. S. (1980). Early Cenozoic reptiles and birds from the Bighorn Basin, Wyoming. *University of Michigan Museum of Paleontology Papers on Paleontology, 24*, 73–79.

Bartels, W. S. (1983). A transitional Paleocene-Eocene reptile fauna from the Bighorn Basin, Wyoming. *Herpetologica, 39*, 359–374.

Batsch, A. J. (1788). Versuch einer Anleitung, zur Kenntniss und Geschichte der Thiere und Mineralien (p. 528). Jena: Akademische Buchhandlung.

Bell, T. (1825). A monograph of the tortoises having a moveable sternum with remarks on that arrangement and affinities. *Zoological Journal, 2*, 299–310.

Berggren, W. A., & Aubry, M. P. (1995). The Paleocene-Eocene Epoch/Series boundary: Chronostratigraphic framework and estimated geochronology. In M.-P. Aubry, S. Lucas & W. A. Berggren (Eds.), *Late Paleocene-Early Eocene climate and biotic events in the marine and terrestrial records* (pp. 18–36). New York: Columbia University Press.

Bramble, D. M. (1974). Emydid shell kinesis: Biomechanics and evolution. *Copeia, 1974*, 707–727.

Bramble, D. M., & Hutchison. J. H. (1981). A reevaluation of plastral kinesis in African turtles of the genus *Pelusios*. *Herpetologica, 37*, 205–212.

Bramble, D. M., Hutchison, J. H., & Legler, J. M. (1984). Kinosternid shell kinesis: Structure, function and evolution. *Copeia, 1984*, 454–473.

Brinkman, D. B., & Peng. J.-H. (1993). New material of *Sinemys* (Testudines, Sinemydidae) from the Early Cretaceous of China. *Canadian Journal of Earth Sciences, 30*, 2139–2152.

Chkhikvadze V. M. (1999). The history of the development of the Paleogene herpetofauna of the former Soviet Union territory (in Russian). *Problems of Paleobiology, Tbilisi, 'Metsniereba' 1*, 270–279.

Chkhikvadze, V. M., In: Gaiduchenko, L. L., & Chkhikvadze, V. M. (1985). A new species of cayman turtle from Pavlodar Priirtischia (in Russian). *Geologiya i Geofizilca. Novosibirsk, 1*, 116–118.

Cope, E. D. (1869/1870). Seventh contribution to the herpetology of tropical America. *Proceedings of the American Philosophical Society, 11*, 547–553.

Dumeril, A. M. C., & Bibron, G. (1835) Erpétologie Générale ou Histoire Naturelle Complète des Reptiles (Vol. 2). Paris: Librairie Encylopédigue de Roret.

Erickson, B. R. (1973). A new chelydrid turtle *Protochelydra zangerli* from the late Paleocene of North America. *Scientific Publications of the Science Museum of Minnesota, n. s. 2*(2), 1–16.

Ernst, C. H., & Barbour, R. W. (1972). *Turtles of the United States*. Lexington: University of Kentucky Press.

Estes, R. (1975). Lower vertebrates from the Fort Union Formation, Late Paleocene, Big Horn Basin, Wyoming. *Herpetologica, 31*, 365–385.

Gaffney, E. S., & Meylan. P. A. (1988). A phylogeny of turtles. *Systematics Association Special Volume, 35A*, 157–519.

Gardner, J. D., & Russell. A. P. (1994). Carapacial variation among soft-shelled turtles (Testudines: Trionychidae), and its relevance to taxonomic and systematic studies of fossil taxa. *Neues Jahrbuch für Geologie und Paläontogie, Abhandlungen, 193*, 209–244.

Gingerich, P. D. (1989). New earliest Wasatchian mammalian faunas from the Eocene of northwestern Wyoming: Composition and diversity in a rarely sampled high-floodplain assemblage. *University of Michigan Papers in Paleontology, 28*, 1–97.

Gray, J. E. (1831). Characters of a new genus of freshwater tortoise from China. *Proceedings of the Zoological Society of London, 1831*, 106–107.

Gray, J. E. (1844). *Catalogue of tortoises, crocodilians, and amphisbaenians in the collection of the British Museum.* London: British Museum (Natural History), 80 pp.

Gray, J. E. (1855). *Catalogue of shield reptiles in the collection of the British Museum. Part I . Testudinata (tortoises)*. London: Taylor and Francis, 79 pp.

Gray, J. E. (1869). Notes on the families and genera of tortoises (Testudinata), and the characters afforded by the study of their shells. *Proceedings of the Zoological Society of London, 1869*, 165–225.

Gray, J. E. (1870). *Supplement to the catalogue of shield reptiles in the collection of the British Museum. Part I. Testudinata (tortoises)*. London: Taylor and Francis.

Hay, O. P. (1908). The fossil turtles of North America (Vol. 75). Washington, DC: Carnegie Institution of Washington Publication, 568 pp.

Hirayama, R. (1985). Cladistic analysis of batragurine turtles (Batagurinae: Emydidae: Testudinoidea); a preliminary result. *Studia Palaeocheloniologica, 1*, 141–157.

Holroyd, P. A., & Hutchison, J. H. (2000). Proximate causes for changes in vertebrate diversity in the early Paleogene: An example from turtles in the Western Interior of North America. *GFF, 122*, 75–76.

Holroyd, P. A., Hutchison, J. H., & Strait. S. G. (2001). Turtle diversity and abundance through the lower Eocene Willwood Formation of the southern Bighorn Basin. *University of Michigan Papers on Paleontology, 33*, 97–107.

Holroyd, P. A. & Hutchison, J. H. (2002). Patterns of geographic variation in latest Cretaceous vertebrates: Evidence from the turtle component. *Geological Society of America Special Paper, 361*, 177–190.

Hutchison, J. H. (1980). Turtle stratigraphy of the Willwood Formation, Wyoming: Preliminary results. *University of Michigan Museum Paleontology Papers on Paleontology, 24*, 115–118.

Hutchison, J. H. (1983). Two new and unusual genera of Paleocene Emydidae from Wyoming. *Society of the Study Amphibians and Reptiles and Herpetologists' League, Abstracts with Program,* 65.

Hutchison, J. H. (1991). Early Kinosterninae (Reptilia: Testudines) and their phylogenetic significance. *Journal of Vertebrate Paleontology, 1*, 145–167.

Hutchison, J. H. (1992). Western North American reptile and amphibian record across the Eocene/Oligocene boundary and its climatic implications. In D. R. Prothero & W. A. Berggren (Eds.), *Eocene-Oligocene climatic and biotic evolution* (pp. 451–467), Princeton: Princeton University Press.

Hutchison, J. H. (1998). Turtles across the Paleocene/Eocene Epoch boundary in west-central North America. In M.-P. Aubry, S. Lucas & W. A. Berggren (Eds.), *Late Paleocene-Early Eocene climate and biotic events in the marine and terrestrial records* (pp. 401–408). New York: Columbia University Press.

Hutchison, J. H. (2003). *Planetochelys*, the oldest perfect box turtle. *Symposium on Turtle Origins, Evolution and Systematics, Saint-Petersburg, Russia, Program and Abstracts*, p. 23.

Hutchison, J. H. (2008). History of Fossil Chelydridae. In A. C. Steyermark, M. S. Finkler & R. J. Brooks (Eds.), *Biology of the snapping turtle (Chelydra serpentina)* (pp. 14–30). Baltimore: The John Hopkins University Press.

Hutchison, J. H., & Archibald, J. D. (1986). Diversity of turtles across the Cretaceous/Tertiary boundary in northeastern Montana. *Palaeogeography, Palaeoclimatology, Palaeoecology, 55*, 1–22.

Hutchison, J. H., & Bramble, D. M. (1981). Homology of the plastral scales of the Kinosternidae and related turtles. *Herpetologica, 37*, 73–85.

Hutchison, J. H., & Frye, F. L. (2001). Evidence of pathology in early Cenozoic turtles. *PaleoBios, 21*, 12–19.

Hutchison, J. H., & Holroyd, P. A. (2003). Late Cretaceous and early Paleocene turtles of the Denver Basin. *Rocky Mountain Geology, 38*, 1–22.

Hutchison, J. H., & Weems, R. E. (1998). Paleocene turtle remains from South Carolina. *American Philosophical Society Transactions, 88*, 165–195.

Joyce, W. G., & Bell, C. J. (2004). A review of the comparative morphology of testudinoid turtles (Reptilia: Testudines). *Asiatic Herpetological Research, 10*, 53–109.

Lydekker, R. (1889). *Catalogue of the fossil Reptilia and Amphibia in the British Museum (Natural History), Part III (Containing the order Chelonia)*. London: Taylor and Francis.

McCord, R. D. (1996). Turtle biostratigraphy of Late Cretaceous and early Tertiary continental deposits San Juan Basin, New Mexico. *Fossils of Arizona Proceedings, 4*, 135–153.

Merrem, B. (1820). Veruch eines Systems der Amphibien. *Tentamen Systematis Amphibiorum.* Marburg: Krieger, 199 pp.

Peters, K. F. (1868). Zur Kenntniss der Wirbelthiere aus den Miocänschichten von Eibiswald in Steiermark. I. Die Schldkrötenreste (Auszug). *Sitzunsberichte der Akademie der Wissenschaften in Wien, Mathematisch-Naturwissenschaftliche Klasse, 57*, 72–74.

Pomel, A. (1847). Note sur des animaux fossils décoverts dans le départment de l'Allier. *Bulletin de la Société Géologique de France, Ser. 2, 4*, 378–385.

Savage, D. E., Waters, B. T., & Hutchison, J. H. (1972). Northwestern border of the Washakie Basin, Wyoming. In R. M. West (Coordinator), *Guidebook of field conference on tertiary biostratigraphy of southern and western Wyoming* (pp. 32–39). New York: Adelphi University.

Schweigger, F. (1812). Prodromus monographie cheloniorum. *Kongsberg Archive fur Naturwissenschafte und Mathematik, Königsberg, 1*, 271–368, 406–462.

von Spix, J. B. (1824). *Animalia nova, species novae Testudinum et Ranarum in itinere per Brasiliam annis 1817–1820 collegit et descripsit*. München: Hübachmann.

Sukhanov, V. B. (2000). Mesozoic turtles of Middle and Central Asia. In J. M. Benton, M. A. Shishkin, D. M. Unwin & E. N. Kurochkin (Eds.), *Age of dinosaurs in Russian and Mongolia* (pp. 309–367). New York: Cambridge University Press.

Sukhanov, V. B., & Narmandakh, P. (1976). Paleocene turtles from Mongolia (in Russian). *Paleontology and Biostratigraphy of Mongolia, the Joint Soviet-Mongolian Paleontological Expedition, Transactions, 3*, 107–133.

Tinkle, D. W. (1962). Variation in shell morphology of North American turtles I. The carapacial seam arrangements. *Tulane Studies in Zoology, 9*, 331–349.

Wagler, J. G. (1830). *Natürliches System der Amphibien, mit vorangehender Classification der Säugethiere und Vögel.* Münch, Stuttgart and Tübigen. Cotta.

Weems, R. E. (1988). Paleocene turtles from the Aquia and Brightseat Formations, with a discussion of their bearing on sea turtle evolution and phylogeny. *Proceedings of the Biological Society of Washington, 10*, 109–145.

Williams, E. E. (1950). Variation and selection in the cervical central articulations of living turtles. *Bulletin American Museum of Natural History, 94*, 505–562.

Woodburne, M. O., & Swisher, C. C. III. (1995). Land mammal high-resolution geochronology, intercontinental overland dispersals, sea level, climate, and vicariance. *Society for Sedimentary Geology (SEPM), Special Publication, 54*, 335–364.

Yeh, X.-K. (1963). Fossil turtles of China. *Palaeontologica Sinica, 150*, 1–112.

Zangerl, R. (1969). The turtle shell: In C. Gans, A. Bellairs & T. S. Parsons (Eds.), *Biology of the Reptilia. I.* (pp. 311–339). New York: Academic Press.

Part VI
Pathologies, Anomalies, and Variation in Turtle Skeletons

Chapter 27
Osseous and Other Hard Tissue Pathologies in Turtles and Abnormalities of Mineral Deposition

Bruce M. Rothschild, Hans-Peter Schultze, and Rodrigo Pellegrini

Abstract Actualistic interpretation (on the basis of disorders documented in life) of pathology in modern turtles, and across the phylogenetic spectrum, allows confident diagnosis of disease in fossil turtles. Fortunately, those diseases that affect bone do so in a manner that appears independent of phylogeny and even of geologic chronology. To date, most pathologies have been reported as isolated observations. Therefore the latter have been characterized and placed within the framework of recognized diseases and activities. Toward this end, we surveyed the literature, assessing reports in terms of contemporary definitions. Epidemiologic studies of pathology in turtles are rare, with the exception of frequency of anomalies and of avascular necrosis and recognition of diving behavior. Notation of avascular necrosis in 50% of Cretaceous turtles, limitation to marine turtles, and its gradual reduction over geologic time present a unique window into turtle behavior and their development of strategies that minimize its occurrence. Anomalies, typically thought of as representing mutations, whether inherited or de novo, may also provide insight to environmental conditions. While amphibians are often utilized as markers of environmental health, extrapolation to reptiles is also pertinent. Fractures imply behavior, such as falls or mating injuries. Bites and other forms of trauma appear to be the most common cause of bone infections by bacteria and fungi. The most common forms of arthritis are gout and infections, although calcium pyrophosphate crystal disease and osteoarthritis also have been reported. Calcium and urate bladder stones have been reported. Vitamin D deficiency is essentially a phenomenon of captive animals, but can complicate kidney disease. Hyperparathyroidism may occur as part of what is referred to as renal osteodystrophy or it may occur as an isolated phenomenon. Carapace and plastron anomalies may represent normal variation, disease, or possibly even phylogenetic differences. The etiology of shell pitting has been highly controversial, with bites, parasites, mixed bacterial and fungal infections, and even algae invoked as causative agents. While many reports of such alterations, as well as isolated or identified associated organisms, exist, there has not been a delineation of which mechanism is responsible for specific types of shell lesions, or even a working vocabulary for describing such lesions.

Keywords Bone pathology • Extant • Fossil • Sea turtle • Tortoise • Turtle

Introduction

As the study of bone disease in recent and fossil amphibians and reptiles has evolved from observational speculation to analysis of testable hypotheses, so too has recognition of its contribution to our understanding of diseases and organisms (Rothschild and Martin 2006; Rothschild 2009a). Given the development of a 'library' of macroscopic osseous manifestations of a variety of diseases (Rothschild and Martin 2006; Rothschild and Ruhli 2007), the power of such examination of skeletons for identification of the etiology of pathology has greatly reduced the need for destructive analysis. Scientific approaches, a comprehensive database of pathological occurrences, and taxonomically independent pathology recognition form the basis for this review of the current knowledge of hard tissue (e.g., skeleton, urolith) pathology in contemporary and extinct turtles. In superficial examination, shell distortion and swollen bones may be attributed to fracture, neoplasia, osteomyelitis, or fibrous

B. M. Rothschild · H.-P. Schultze (✉)
Biodiversity Research Center and Natural History Museum, The University of Kansas, Lawrence, KS 66045, USA
e-mail: hp1937@ku.edu

B. M. Rothschild
e-mail: brothschild@ku.edu

R. Pellegrini
New Jersey State Museum, PO Box 530 Trenton, NJ 08625-0530, USA
e-mail: Rodrigo.Pellegrini@sos.state.nj.us

D. B. Brinkman et al. (eds.), *Morphology and Evolution of Turtles*, Vertebrate Paleobiology and Paleoanthropology,
DOI: 10.1007/978-94-007-4309-0_27, © Springer Science+Business Media Dordrecht 2013

dysplasia (Boyer 1996c). It is only with careful examination and an actualistic approach that sense can be made of this information.

The character and pattern of the manifestations of specific diseases addressed below have been validated in the contemporary and fossil record across reptilian, mammalian, and avian phylogeny (Rothschild and Panza 2005; Rothschild and Martin 2006; Natarajan et al. 2007; Nunn et al. 2007; Rothschild and Ruhli 2007; Beatty and Rothschild 2008; Rothschild 2009a). Such a data-based approach follows the veterinary medical model, which uses the manifestations of disease as reported in humans, to allow its recognition in the zoological record (McGavin and Zachary 2006; Rothschild 2010a). While variation in histologic appearance of bone is clearly documented across phylogenetic lines (Enlow 1969), macroscopic and radiologic variation in the appearance of various forms of pathology in the osseous skeleton has been remarkably minimal and reproducible across phylogeny and indeed across geologic time (Rothschild and Martin 2006; Rothschild 2009a, 2010a).

Any up to date interdisciplinary review requires consideration of new concepts and semantics. Each field has its own vocabulary, often utilizing terms quite disparate in meaning (Rothschild and Martin 2006). Terminology is not always used consistently, even in a given field. Some pathological terms have been used both as general inclusive concepts and more narrowly for specific diseases. Spondylitis and spondyloarthropathy are classic examples, sometimes used to refer to any pathology of the axial skeleton or, as used herein, to refer to a specific category of disease that often, but not necessarily, affects the vertebral column (Resnick 2002; Rothschild and Martin 2006; Rothschild 2010a). To facilitate clarity, terms are narrowly defined below and elsewhere according to their use in this review.

Focal bony outgrowths (exostoses), where occurring at areas of tendon attachment are recognized as enthesophytes (Resnick 2002), and osteochondroma are exostoses that are covered by a cartilage cap. Osteoarthritis is identified by the presence of bone spurs (osteophytes) at diarthrodial joint margins (Moskowitz et al. 1984; Resnick 2002; Rothschild and Martin 2006). Infection was recognized because of the joint erosions with ill-defined margins, disorganized architecture, and characteristic bone surface (periosteal reaction) (Resnick 2002; Rothschild and Martin 2006). "Punched out" joint erosions are characteristic of gout (Rothschild and Heathcote 1995) and are often covered by an overhanging margin of bone (Resnick 2002; Rothschild and Martin 2006). Calcium pyrophosphate deposition disease (CPPD) is recognized by deposition of calcium on the joint surface or within intervertebral disks (Resnick 2002; Rothschild and Martin 2006; Rothschild and Bruno 2009).

Our recognition of spondyloarthropathy, as related in this review, is based on the presence of zygapophyseal (facet joint) or sacroiliac joint erosion or fusion, asymmetrical pattern of arthritis, reactive new bone formation, or syndesmophytes (Bennett 1991; Resnick 2002; Rothschild and Martin 2006). The last appear as longitudinally directed ossifications bridging (ankylosis) from one vertebral body to the next. This is distinguished from the ligamentous ossification of diffuse idiopathic skeletal hyperostosis (DISH) and the bony spurs (osteophytes) more properly referred to as spondylosis deformans (Ruffer 1921; Resnick 2002; Rothschild 2009c, 2010b). A major source of past confusion has been the mistaken concept that bone spurs can fuse together, rather than recognize that they were caused by a totally different pathogenetic process related to spondyloarthropathy (Resnick 2002; Rothschild and Martin 2006; Rothschild 2010b). The term spondyloarthropathy identifies a category of disease, which may occur in the absence of axial skeleton involvement. Erosions affecting subchondral surfaces (areas originally covered by cartilage) are highly diagnostic, although joint infections occasionally produce similar, but distinguishable damage (Resnick 2002; Martin and Rothschild 2006).

Osteomalacia and rickets are generally considered the result of deficiency in the active form of vitamin D after and before epiphyseal plate fusion, respectively (Resnick 2002), although phosphorus deficiency can produce similar changes. These conditions appear to be a component of what is often described in the veterinary literature as metabolic bone disease or nutritional osteodystrophy (Thompson 2007). The major manifestations relate to failure of mineralization of bone matrix (osteoid) and resulting bone softening. The term osteodystrophy fibrosa is sometimes used to describe the intraosseous cystic phenomenon (related to proliferation of fibrous tissue) that can accompany it. The latter is sometimes claimed to be the result of calcium deficiency (Jarofke and Lange 1993).

Our presentation of turtle pathologies is part of a larger, comprehensive literature survey of amphibian and reptile publications from Aristotle to present, including extant and fossil forms. Scientific names have changed over time. Here we use the most recently accepted names, not a synonym, which may have been used by the cited author. We also use common names for extant taxa.

Fossil and recent turtle specimens were investigated at the following institutions: AMNH, American Museum of Natural History, New York, New York, USA; FMNH, Field Museum of Natural History, Chicago, Illinois, USA; IVPP, Institute of Vertebrate Paleontology and Paleoanthropology, Beijing, China; MSU, Michigan State University Museum, East Lansing, Michigan, USA; UCMP, University of California at Berkeley, Museum of Paleontology, Berkeley, California, USA; UMMZ, University of Michigan Museum of Zoology, Ann Arbor, Michigan, USA; and USNM, United States National Museum, Washington, D.C., USA.

Congenital Versus Environmental Exposure

It is difficult at times to determine whether anomalies observed in the embryo or neonate are the result of genetic alterations or exposure to adverse environmental conditions, or to distinguish anomalies that have been inherited versus those occurring secondarily as the result of a new mutation. One does not necessarily need to invoke an environmental toxin. Environmental conditions may be just as important, either as a direct or indirect effect. Lynn and Ullrich (1950) examined turtle embryos from eggs that were exposed to partial drying during gestation interval days 35–50. Embryo pathologies observed in that study ranged from "almost unrecognizable carapaces and plastrons, distorted limbs and eyeless or jawless heads" (Lynn and Ullrich 1950, p. 250). Newman (1923) suggested temperature effect on twinning, whereas Coker (1910) considered scute anomalies as an effect of egg distension. Packard et al. (1977) noted that prolonged exposure to temperatures several degrees below the optimum produced developmental abnormalities. For the California desert tortoise, *Gopherus agassizii*, incubation of eggs above the optimum temperature of 32°C was associated with upper jaw foreshortening and lack of forelimbs (Frye 1991c), whereas incubation at 33.5°C was associated with anophthalmy, maxillofacial clefts, harelips, forelimb and partial hind limb agenesis, coccygeal hypoplasia, and scute and shell plate duplication (Frye 1989, 1991c). Of course, toxins can produce developmental anomalies. In a study of snapping turtle, *Chelydra serpentine*, populations in the Great Lake region, Bishop et al. (1989) documented how organochlorine compounds (e.g., PCBs) induced missing claws and eyes and resulted in deformed carapaces, tails, limbs and crania (especially in the upper and lower jaws). The same study also showed that frequencies of deformities varied among study areas, and was greatest in areas with the highest organochlorine levels. Whereas most studies represent isolated reports of deformities (e.g., Tables 27.1, 27.2, 27.3, 27.4, 27.5), few have reported statistical results. Yntema (1960) reported that eggs of the common snapping turtle, *Chelydra serpentina*, kept at 15°C during early somite formation and then moved to room temperature developed malformed heads, microphthalmy, and reduced or absent tails, legs, or carapace, and he contrasted that with lack of such abnormalities at 20°C. Bellairs (1981) citing Yntema (1960) referred to 15% of common snapping turtle, *Chelydra serpentina*, individuals examined with an abnormal tail, hind limb, or carapace. Carswell and Lewis (2003) reported 0.32% abnormalities in 14,361 loggerhead turtle, *Caretta caretta,* eggs sampled in 1992 and 2001; these abnormalites were manifested as small size, carapace deformities, folded or foreshortened flippers, misshapened mouthparts, and twinning.

The most complex of congenital anomalies in turtles is perhaps the Siamese, conjoined or parasitic twin (Kuvano 1902; Nakamura 1938; Maddux 1996; Anonymous 2007a). Most examples have been reported as isolated phenomena (Table 27.1). A variety of duplication patterns have been observed: plastron-to-plastron; side-to-side; and posterior-to-posterior. Noting Hildebrand's (1938) tally of 27 cases, Table 27.1 suggests the rarity of this phenomenon. Only 45 cases appear to have been reported to date and no epidemiologic studies are available.

Complete or partial absence of a limb has been reported in isolated occurrences (Table 27.2). Not all are congenital. For example, Jackson (1980) noted one instance in which a foreleg was gnawed off during hibernation by a rat and Frye (1994) reported digit necrosis in a red-eared slider turtle, *Trachemys scripta elegans.*

Head, Leg, and Tail Anomalies

Anomalies of the head (Table 27.3) may be as minor as variation in size and shape of skull, as has been reported in the diamond-back terrapin, *Malaclemys terrapin* (Bangs 1896) and the Florida scooter, *Pseudemys floridana* (Hildebrand 1938). Lack of frontal bones in a bothremydid side-necked turtle, †*Ummulisani rutgersensis*, from the Eocene of Morocco (Gaffney and Tong 2008; Fig. 27.1) is an example of a limited defect. More extensive anomalies include: a protruding lower jaw and eyeless upper jaw (common snapping turtle, *Chelydra serpentina*); cyclopia [red-eared slider, *Trachemys scripta elegans*; loggerhead, *Caretta caretta caretta*; and Florida cooter, *Pseudemys concinna floridana* (Table 27.3)]; an eyeless stump [common snapping turtle, *Chelydra serpentina* (Lynn and Ullrich 1950)]; and lack of a mouth and jaw [box turtle, *Terrapene* sp. (Ewert 1979)]. In the desert tortoise, *Gopherus agassizii*, both shortening of the upper jaw (Frye 1989) and a cleft palate (Bellairs 1981; Frye 1989) have been reported. A cleft palate also has been reported by Ippen (1982) in the pond slider, *Trachemys scripta*, and in the green turtle, *Chelonia mydas*.

Duplications are the most commonly reported anomalies (Fig. 27.2). Edwards (1751) reported partial skull duplication in an eastern box turtle (*Terrapene carolina* = his "Carolina tortoise"). Isolated, sometimes redundant reports of dicephalism in unnamed and named species abound (Table 27.1); consequently, it is difficult to assess the frequency of dicephalism among turtles. Extrapolating from

Table 27.1 Duplications in turtles

Species	Common name	Kind of duplication	References
Cryptodira Cheloniidae			
Caretta caretta	Loggerhead	Dicephalic	Anonymous (2002)
Chelonia mydas	Green turtle	Duplication of head and forelimbs	Haft (1994)
		Second, atrophic head	Coquelet (1983)
Chelydridae			
Chelydra serpentina	Common snapping turtle	Dicephalic	Cederstrom (1931); Canella (1932); Campbell (1967); Feldman (1983); Sims (1989); Anonymous (1995, 2001)
		Twins	Yntema (1970, 1971)
Geoemydidae			
Chinemys reevesii	Reeve's turtle	Dicephalic	Khosatzky (1991)
		Derodymus	Nakamura (1938)
Emydidae			
Chrysemys picta	Northern painted turtle	Dicephalic	Barbour (1888,1896a, b); Anonymous (1889); Girard (1891–1892); Hildebrand (1938)
		Derodymus	Barbour (1896c); Derickson (1927); Townsend (1928)
		Duplication of head and forelimbs	Bishop (1908)
		Siamese twins attached at posterior ends	Hildebrand (1938)
Chrysemys sp.	Painted turtle	Derodymus	Schmidt and Inger (1957); Bellairs (1981)
Emys orbicularis	European pond turtle	Duplication anterior head	Epure and Pogorevici (1940)
Emys sp.	European pond turtle	Dicephalic	Bateson (1894); Przibram (1909); Newman (1923)
Graptemys	*pseudogeographica*	False map terrapin	Dicephalic
Clement (1967); Sims (1989)			
Malaclemys sp.	Diamond-back terrapin	Dicephalic, derodymus	Brogard (1987)
Malaclemys terrapin	Diamond-back terrapin	Dicephalic	Anonymous (1975); Wright (2005)
		Derodymus	Schmidt and Inger (1957); Frye (1991a)
Malaclemys terrapin centrata	Carolina diamond-back terrapin	Duplication of head and forelimbs, anakatamesodidymus	Hildebrand (1938)
Mauremys leprosa	Mediterranean pond turtle	Dicephalic	Martins d'Alte (1937)
Pseudemys sp.	River cooter turtle	Dicephalic	Anonymous (2000)
Pseudemys concinna floridana	Florida cooter	Dicephalic	Hildebrand (1938)
		Siamese twins	Reichenbach-Klinke and Elkan (1965); Reichenbach-Klinke (1977)
		Plastron to plastron	Hildebrand (1938)
Pseudemys concinna hieroglyphica	Hieroglyphic slider turtle	Dicephalic	Wallach (2004)
Pseudemys nelsoni	Florida redbelly turtle	Dicephalic	Ippen (1985)
		Derodymus	Bellairs (1981); Ippen (1982)
Terrapene carolina	Eastern box turtle	Twins	Cohen (1986)
		partial skull duplication	Edwards (1751)
Terrapene carolina triungis	Eastern box turtle	Twins	Crooks and Smith (1958); Cohen (1986)
Terrapene sp.	Box turtle	Dicephalic	Mitchell (1994); Anonymous (2006a)

(continued)

Table 27.1 (continued)

Species	Common name	Kind of duplication	References
Trachemys scripta elegans	Red-eared slider	Dicephalic	Clement (1967); Sims (1989); Frye (1991c); Chapple (1999)
		Derodymus	Frye (1991a); Hasel (1992a, b)
		Duplication of head and forelimbs	Chapple (1999); Anonymous (2007a, b)
		Siamese twins	Anonymous (1967, 2007a); Chapple (1999)
		Siamese twins fused at pelvis	Mader (2006a)
		Thoraco-omphalopagus	Anonymous (1967)
Trachemys scripta scripta	Yellowbelly slider	Dicephalic	Kritzer (2002); Perkins (2002); Reaves (2004); Anonymous (2007c)
		Siamese twins	Reichenbach-Klinke and Elkan (1965); Reichenbach-Klinke (1977)
Trachemys scripta troostii	Cumberland slider	Dicephalic	Byrd (1939)
Trachemys sp.	Slider turtles	Conjoined	Frye (1991a)
Kinosternidae			
Kinosternon flavescens	Yellow mud turtle	Dicephalic	Porras and Beraducci (1980); Frye (1991c)
Testudinidae			
Chersina angulata	Angulate tortoise	Dicephalic	Swarts (2003)
Gopherus agassizii	California desert tortoise	Dicephalic	Frye (1991c)
		Siamese twins attached along caudal portion of shell	Frye (1991a, c); Rothschild (2009a)
Gopherus polyphemus	Gopher tortoise	Twins	Hundsacker (1968)
Testudo graeca	Spur-thighed or Greek tortoise	Dicephalic	Stojanov (2000)
Testudo graeca ibera	Turkish tortoise	Dicephalic	Vellard and Penteado (1931)
		Duplication of head and forelimbs	Caullery (1931)
Testudo hermanni	Hermann's tortoise	Dicephalic	Reichenbach-Klinke and Elkan (1965)
Testudo hermanni boettgeri	Eastern Hermann's tortoise	Siamese twins	Sailer et al. (1997)
Trionychidae			
Apalone ferox	Florida soft-shelled turtle	Siamese twins	Bellairs (1981)
		Siamese twins fused at plastron	Hildebrandt (1938); Ippen (1982)
Unspecified	Turtle/tortoise	Duplication of head and forelimbs	Anonymous (1931)
		Dicephalic	Anonymous (1888a, b, 1895, 1999a, b, c, 2003a, b, c, 2004, 2005, 2006b); Kuvano (1902); Hildebrand (1930); Brogard (1987); Brady (1991); Benest (1994); Eigner (1994); Kleinberg (1994); Meisel (1994); Hamilton (1995); Walker (1995); Maddux (1996); Martin (1996); Wilcox and Perrin (1996); Cisneros (1997); Thomason (1997); Wilkins (1998); Gumbel (1999); Nicholson (1999); Mason (2000); Miller (2002); Reeder (2002); Brown (2003); Hall (2003); Holahan (2003); Montgomery (2003); Smith (2003); Wechsler (2003); Feuer (2005); Goh (2005); Lee (2005); Rothschild (2009a)
		Derodymus	Anonymous (1897)
Pleurodira			
Podocnemididae			
Podocnemis unifilis	Yellow-spotted Amazon river turtle	Derodymus	Ferreira (1923)

Definitions: anakatamesodidymus = two heads, two tails, and rudimentary fifth leg; derodymus = vertebral column bifurcated in the cervical region, double-headed (i.e., two complete heads and necks); dicephalic = bicephalic, having two heads; Siamese twins = conjoined twins; thoraco-omphalopagus = twins fused from upper to lower part of chest region

Table 27.2 Limb abnormalities in turtles

Species	Common name	Abnormality	References
Cryptodira			
Protostegidae			
†*Archelon ischyros*		Shortened fibula and tibia	Tasnádi-Kubacska (1962)
Cheloniidae			
Caretta caretta	Loggerhead	Complete absence of limb	Ewert (1979); Ippen (1982); Brogard (1987)
		Supernumerary limb	Mader (2006a)
		Partial absence of limb	Ewert (1979); Bellairs (1981)
†*Syllomus crispatus aegyptiacus*		Bony ridge on femur between torchanters and caput	Weems (1974)
Chelydridae			
Chelydra serpentina	Common snapping turtle	Hindlimb	Yntema (1960); Bellairs (1981)
		Dwarfism	Bishop et al. (1989)
Emydidae			
Emydidae	Pond turtles	Partial absence of limb	Dutta (1931); Bellairs (1981); Ippen (1982)
Emys orbicularis	European pond turtle	Complete absence of limb	Dürigen (1897)
Malaclemys sp.	Diamond-back terrapin	Supernumerary limb	Brogard (1987)
Malaclemys terrapin centrata	Carolina diamond-back terrapin	Partial absence of limb	Hay (1904)
Terrapene carolina	Eastern box turtle	Partial absence of limb, congenital tarsal absence	Wilkinson et al. (2004)
Trachemys scripta	Slider	Achondroplasy	Frye and Carney (1974); Brogard (1987); Arvy and Fertard (2002)
Trachemys scripta elegans	Red-eared slider	Achondroplasy	Frye and Carney (1974); Frye (1981a, 1991a, c); Marcus (1981); Rothschild (2009a)
Testudinidae			
Gopherus agassizii	Desert tortoise	Syndactyly	Good (1987)
		Achondroplasy	Frye and Carney (1974); Frye (1981b, 1991a, c); Marcus (1981); Brogard (1987); Arvy and Fertard (2002); Rothschild (2009a)
Testudo hermanni	Hermann's tortoise	Oligodactyly	Wermuth (1961); Fojtl (1989)
		Polydactyly	Martínez-Silvestre etal. (1998)
Trionychidae			
Trionyx sp.	Soft-shelled turtle	Complete absence of limb	Dutta (1931); Bellairs (1981); Ippen (1982)
Unspecified	Turtle	Ectromely	Hildebrand (1930); Brogard (1987)
		Achondroplastic dwarfism	Gillespie (1994)

Definitions: achondroplasy = abnormal cartilage conversion to bone, producing short, abnormally limbed dwarfism; ectromely = absence of one or more limbs or incomplete limb with missing distal portion; *oligodactyly* = partial absence of limb; polydactyly = increased number of metatarsals (= hyperdactyly), supernumerary digit(s); syndactyly = fusion or failure of separation of fingers or toes

Table 27.3 Skull defects in turtles

Species	Common name	Defects	References
Cryptodira			
Cheloniidae			
Caretta caretta caretta	Loggerhead	Cyclopia	Adelmann (1936a, b); Ewert (1979); Bellairs (1983)
Chelonia mydas	Green turtle	Cleft palate and shortened upper jaw	Ippen (1982)
Chelydridae			
Chelydra serpentina serpentina	Common snapping turtle	Eyeless	Lynn and Ullrich (1950)
Emydidae			
Chrysemys picta	Northern painted turtle	Cyclopia	Adelmann (1936a, b)
Pseudemys concinna floridana	Florida cooter	Cyclopia	Hildebrand (1938)
Malaclemys terrapin	Diamond-back terrapin	Variation in size and shape of skull	Bangs (1896)
Terrapene sp.	Box turtle	Absent mouth and jaw	Ewert (1979)
Trachemys scripta	Pond slider	Cleft palate	Ippen (1982)
Trachemys scripta elegans	Red-eared slider	Cyclopia	Adelmann (1936a, b); Ewert (1979)
		Shortened upper jaw	Frye and Carney (1974)
Testudinidae			
Gopherus agassizii	Desert tortoise	Cleft palate	Bellairs (1983); Frye (1989)
		Shortened upper jaw	Frye (1991c)
Pleurodira			
Bothremydidae			
†*Ummulisani rutgersensis*		Loss of frontals	Gaffney and Tong (2008)

Definitions: cyclopia = failure of eye separation; cleft palate = failure of fusion of the two halves of the palate

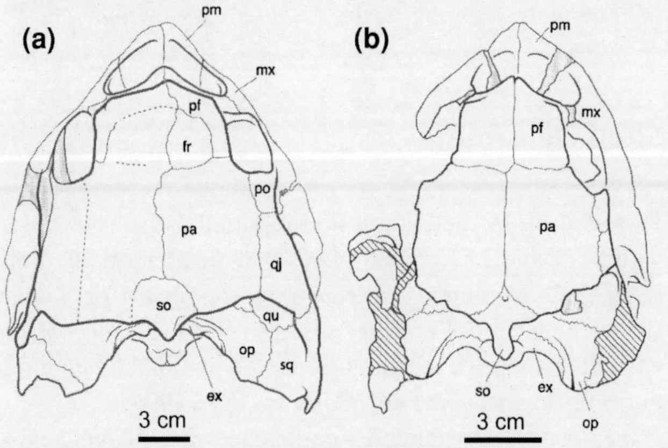

Fig. 27.1 Anomalous absence of skull roofing bones in the fossil side-necked turtle †*Ummulisani rutgersensis* (Bothremydidae), from the Eocene of Morocco. **a** Normal skull, AMNH 30569, in dorsal view and having paired frontals. **b** Anomalous skull, AMNH 30536 (holotype), in dorsal view and lacking frontals. Images reproduced from Gaffney and Tong (2008, Figs. 6A and 3A, respectively) and Courtesy The American Museum of Natural History. Skulls at different scales. Abbreviations: *ex* exoccipital; *fr* frontal; *mx* maxilla; *op* opisthotic; *pa* parietal; *pf* prefrontal; *pm* premaxilla; *po* postorbital; *qj* quadratojugal; *qu* quadrate; *so* supraoccipital; *sq* squamosal

Hildebrand's (1938) findings of only two dicephalic individuals among 100,000 Carolina diamond-back terrapins, *Malaclemys terrapin centrata*, hatched over 25 years, it appears that dicephalism is an extremely uncommon phenomenon. The same appears true for the more complicated form of duplication, called derodymus (Anonymous 1888a, b, 1897, 2007c; Epure and Pogorevici 1940), which involves duplication of the neck as well as the head (Table 27.1). Newman (1923) reported dichotomous fission of a rib in association with such doubling. Duplication of the head and forelimbs (Anonymous 1931) has been reported in the red-eared slider, *Trachemys scripta elegans* (Chapple 1999; Anonymous 2007a, b), the northern painted turtle, *Chrysemys picta* (Bishop 1908), the Carolina diamond-back terrapin, *Malaclemys terrapin centrata* (Hildebrand 1938), and the Greek tortoise, *Testudo graeca ibera* (Caullery 1931). Hildebrand (1938) reported an extreme case of duplication in which a diamond-back terrapin, *Malaclemys terrapin centrata*, had two heads, two tails, and a rudimentary fifth leg. Brogard (1987) reported a bifid tail in *Malaclemys* sp. A duplication of the vertebral column has

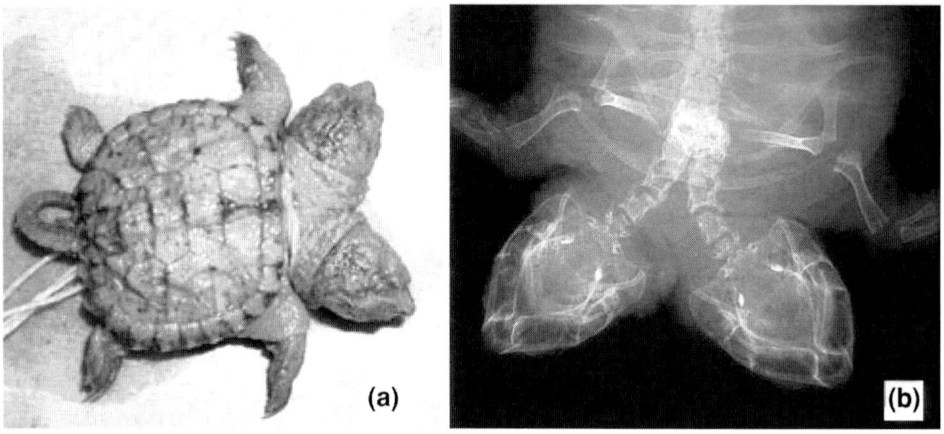

Fig. 27.2 Extreme case of duplication (derodymus = two heads and two necks) in an extant, juvenile, preserved specimen (UMMZ 119985a) of *Chelydra serpentina* (Chelydridae) from the USA. **a** Photograph of entire specimen in dorsal view. **b** X-ray of anterior part of body, enlarged and rotated about 90° clockwise relative to photograph, showing paired skulls and cervicals

been mentioned in a red-eared slider, *Trachemys scripta elegans* (Anonymous 1967).

Reports of limb abnormalities (Table 27.2) include: supernumerary limbs in the diamond-back terrapin, *Malaclemys* sp. (Brogard 1987), and the loggerhead, *Caretta caretta* (Mader 2006a); polydactyly in the Hermann's tortoise, *Testudo hermanni* (Martínez-Silvestre et al. 1998); and syndactyly (fused toes) in the California desert tortoise, *Gopherus agassizii* (Good 1987). Abnormally shortened limbs have been reported in turtles, in the form of achondroplasy (Gillespie 1994). Such long bone shortening, accompanied by cartilage hypertrophy with ossification spicules, has been noted in a dwarf red-eared slider, *Trachemys scripta*, and in an inbred California desert tortoise, *Gopherus agassizii*. Other reported anomalies include a Cretaceous †*Archelon ischyros* with shortened fibula and tibia (Tasnádi-Kubacska 1962) and a Miocene †*Syllomus aegyptiacus* (specimen USNM 24872) with a bony ridge located on the femur between the trochanters and the caput (Weems 1974). Dwarfism, while apparently maintaining normal body proportions, was reported in snapping turtles, *Chelydra serpentina*, by Bishop et al. (1989). Absence of limbs was reported by Dutta (1931), oligodactyly by Fojtl (1989), reduction of toes by Wermuth (1961), and leg anomalies in the European pond turtle, *Emys* sp. by Dürigen (1897).

Vertebral Deformities

Most reports on kyphosis (Table 27.4; see also Shell Deformities) document only isolated occurrences. Frequencies in epidemiologic studies (Rhodin et al. 1984; Stuart 1996) have been significantly under 1%, as follows: 0.2–0.5% in the loggerhead turtle, *Caretta caretta*; 0.08% in the green turtle, *Chelonia mydas*; 0.07% in the leatherback turtle, *Dermochelys coracia*; 0.1–0.4% in the Pacific ridley turtle, *Lepidochelys olivacea*; 0.03% in the western painted turtle, *Chrysemys picta belli*; and 0.04% in the California desert tortoise, *Gopherus agassizii*. Stuart's (1996) report of one instance of kyphosis among 20 Texas spiny soft-shelled turtles, *Apalone spiniferus emory*, may be an artifact of small numbers and likely is not statistically significant. Scoliosis may occur in isolation (Table 27.4; Hildebrand 1930) or it may be associated with kyphosis, as Fox (1941) observed in the ornate box turtle, *Terrapene ornata.*

Caudal agenesis (= failure of formation of all or part of the tail; Table 27.4) has been noted in the California desert tortoise, *Gopherus agassizii* (Frye 1989, 1991a), and the common snapping turtle, *Chelydra serpentina* (Finkler and Claussen 1997). McCallum and Trauth (2000) reported a curly-tail deformity (presented as a tight spiral or coil) in hatchlings of the alligator snapping turtle, *Macroclemys temminckii*, and Feldman (1983) reported a kinked tail in the common snapping turtle, *Chelydra serpentina*.

Shell Deformities

Distorted turtle shells are often reported as isolated observations (Table 27.5). Shell plates that are abnormally broad (e.g., eastern painted turtle, *Chrysemys picta*) or wedge-shaped (e.g., red-footed tortoise, *Geochelone carbonaria*) have been reported, as has 180 degree torsion in a common musk turtle, *Sternotherus odoratus* (Frye 1991a, c).

Few statistically reliable samples are available for gauging the prevalence of shell distortions. Grant (1937) reported that 10% of gopher tortoise specimens (*Gopherus* sp.) had abnormally hard, brittle shells. McEwan (1982) noted that 50% of gopher tortoise specimens, *Gopherus polyphemus*, had supernumerary carapace bones compared with 68% of Texas tortoise specimens, *Gopherus berlandieri*. Cherepanov (1994) reported anomalies in the number of neurals, costals, and peripherals in 121 of 510 individuals of four species:

Table 27.4 Vertebral column deformities in turtles

Species	Common name	Deformity	References
Cryptodira			
Cheloniidae			
Caretta caretta	Loggerhead	Kyphosis in 0.2–0.5%	Coker (1910); Plymale et al. (1978)
Chelonia mydas	Green turtle	Kyphosis in 0.08%, lordosis/scoliosis in 0.03%	Rhodin et al. (1984)
Lepidochelys olivacea	Olive ridley	Kyphosis in 0.1–0.4%, lordosis/scoliosis in 0.3%	Rhodin et al. (1984)
Chelydridae			
Chelydra serpentina	Common snapping turtle	Kyphosis	Chan (1937); Plymale et al. (1978); Wilhoft (1980); Willft (1980)
		Caudal agenesis	Finkler and Claussen (1997)
		Kinked tail	Feldmann (1983)
Macroclemys temminckii	Alligator snapping turtle	Curly-tail deformity	McCallum and Trauth (2000)
Dermatochelyidae			
Dermatochelys coracia	Leatherback	Kyphosis in 0.07%, lordosis/scoliosis in 0.07–0.11%	Rhodin et al. (1984)
Geoemydidae			
Melanochelys trijuga	Indian pond terrapin	Kyphosis	Plymale et al. (1978)
Melanochelys trijuga thermalis	Sri Lanka pond terrapin	Kyphosis	Deraniyagala (1939)
Emydidae			
Chrysemys sp.	Painted turtle	Kyphosis	Wilhoft (1980)
Chrysemys picta	Northern painted turtle	Kyphosis	Ernst (1971); Plymale et al. (1978); Pavalko (1986)
Chrysemys picta marginata	Midland painted turtle	Kyphosis	Werner (1959); Ippen (1982)
Chrysemys picta belli	Western painted turtle	Kyphosis in 1/429	Stuart (1996)
Chrysemys picta bellii x marginata	Painted turtle cross	Kyphosis	Necker (1940); MacCulloch (1981)
Clemmys guttata	Spotted turtle	Kyphosis	Ernst (1971); Plymale et al. (1978)
Deirochelys reticularia	Chicken turtle	Kyphosis	Mertens (1940); Plymale et al. (1978)
Graptemys pseudogeographica	False map turtle	Kyphosis	Plymale et al. (1978)
Graptemys ouachitensis	Ouachita map turtle	Kyphosis	Carpenter (1958)
Malaclemys terrapin centrata	Carolina diamond-back terrapin	Scoliosis	Hildebrand (1930)
		kyphosis	Plymale et al. (1978)
Pseudemys concinna floridana	Florida cooter	Kyphosis	Plymale et al. (1978)
Terrapene carolina	Eastern box turtle	Kyphosis	Plymale et al. (1978)
Terrapene carolina triunguis	Three-toed box turtle	Kyphosis	Black (1976)
Terrapene ornata	Ornate box turtle	Kyphosis	Plymale et al. (1978)
		Kyphoscoliosis	Fox (1941)
Trachemys scripta	Slider	Kyphosis	Plymale et al. (1978)
Trachemys scripta scripta	Yellowbelly slider	Kyphosis	Carr (1952)

(continued)

Table 27.4 (continued)

Species	Common name	Deformity	References
Trachemys scripta yaquia	Yaqui slider	Kyphosis	Plymale et al. (1978)
Kinosternidae			
Sternotherus odoratus	Common musk turtle	Kyphosis	Nixon and Smith (1949); Plymale et al. (1978); Saumure (2001)
Testudinidae			
Gopherus agassizii	Desert tortoise	Kyphosis in <0.04%	Rhodin et al. (1984)
		Caudal agenesis	Frye (1989, 1991a)
Testudo hermanni	Hermann's tortoise	Kyphosis	Wandolleck (1904); Plymale et al. (1978)
Unindentified	Tortoise	Kyphosis	Dämmrich (1967); Jarofke and Lange (1993)
Trionychidae			
Apalone ferox	Florida soft-shelled turtle	Kyphosis	Nixon and Smith (1949); Plymale et al. (1978); Barber (1991)
Apalone mutica	Smooth soft-shelled turtle	Kyphosis	Smith (1947); Plymale et al. (1978)
Apalone spinifera	Spiny soft-shelled turtle	Kyphosis	Chan (1937); Smith (1947); White and Murphy (1972); Burke (1994)
Apalone spinifera emoryi	Texas spiny soft-shelled turtle	Kyphosis	Smith (1947)
		Kyphosis in 1/20	Stuart (1996)
Lissemys sp.	Flap-shelled turtle	Kyphosis	Wilhoft (1980)
Lissemys punctata punctatai	Indian flap-shelled turtle	Kyphosis	Duda and Gupta (1977)
Palea steindachneri	Wattle-necked soft-shelled turtle	Kyphosis	Gressitt (1936, 1937); Smith (1947); Plymale et al. (1978)
Pelodiscus sinensis	Chinese soft-shelled turtle	Kyphosis	Vogt (1922); Gressitt (1936); Plymale et al. (1978)
Trionychidae	Soft-shelled turtles	Kyphosis	Smith (1947); Bellairs (1981)
Trionyx triunguis	African soft-shelled turtle	Kyphosis	Mertens (1940); Plymale et al. (1978); Ippen (1985)
		Scoliosis	Reichenbach-Klinke (1977)
Unidentified		Kyphosis	Williams (1957); Schlumberger (1958)

Definitions: kyphosis = dorsoventral curvature of axial skeleton, with accentuated posterior apical apex, producing hump-like arching of carapace, commonly attributed to metabolic bone disease; kyphoscoliosis = combination of kyphosis and scoliosis; lordosis = curvature of axial skeleton, with accentuated anterior apical apex, usual refers to lumbar region; scoliosis = lateral curvature of vertebral column

Caspian pond turtle, *Mauremys caspica*; European pond turtle, *Emys orbicularis*; Mediterranean spur-thighed tortoise, *Testudo graeca*; and Russian tortoise, *Agrionemys horsfieldi*. Congenital carapace anomalies in captive tortoises were reported by Dollinger (1997).

Carapace and plastron anomalies also are known in fossil sea turtles (Table 27.5), such as the Miocene †*Caretta* (Zangerl and Turnbull 1955), †*Procolpochelys grandaeva* (Zangerl and Turnbull 1955; Weems 1974), and †*Syllomus* (Weems 1974; Fig. 27.3). Such anomalies occur only rarely in the extant ridley *Lepidochelys* (Zangerl and Turnbull 1955).

Frye (1981a) suggested that deformation of the shell was caused by downward tension on the carapace underside at girdle attachments. Shallow concavities have been found in portions of the bony shell overlying limb attachments (Frye 1981a); for example, in the common snapping turtle, *Chelydra serpentina serpentina* (Lynn and Ullrich 1950). Renal rickets has also been proposed as an etiological cause of shell deformation (Frye 1994). Smith (1947) suggested that humps in the carapace could form if adjacent costal plates ankylosed or became tightly bound to one another when the shell was still actively growing.

Hunchback or kyphosis deformity (see also Table 27.4) takes two forms: (1) the more common form consists of deformed, raised plates that allegedly form due to an excess of dietary protein (Rosskopf 1986; McArthur 1996) or (2) thick, porous, raised plates that form because of renal or nutritional osteodystrophy (Jackson 1987; Highfield 1990a,

Table 27.5 Shell anomalies in turtles

Species	Common name	Anomaly	References
Cryptodira			
Cheloniidae			
Caretta caretta	Loggerhead	Lengthwise division of neural plates	Zangerl and Turnbull (1955)
		Carapace deformities	Carswell and Lewis (2003)
Lepidochelys olivacea	Ridley	Division of neural plate into three parts	Zangerl and Turnbull (1955)
†*Procolpochelys grandaeva*	Miocene carettine sea turtle	Supernumerary peripheral carapace element, division of neural plate into three parts	Zangerl and Turnbull (1955); Weems (1974)
†*Syllomus aegyptiacus*		Splitting of first neural, extra postneurals and costals	Weems (1974)
Chelydridae			
Chelydra serpentina serpentine	Common snapping turtle	Shell anomalies, deeply indented carapace	Lynn and Ullrich (1950); Frye (1991a, c)
		Abnormal carapace	Yntema (1960); Bellairs (1981); Bishop et al. (1989)
		Deformed carapace	Hutchinson and Simmonds (1991)
Emydidae			
Chrysemys picta	Northern painted turtle	One or more carapace or plastron anomalies	MacCulloch (1981)
		Fifth neural plate reduced and triangular	Derickson (1927)
		Abnormally broad carapace with extra plates	Hildebrand (1938); Lynn and Ullrich (1950)
		Fusion of plastron pectorals and abdominals, plastron bulges	Lynn and Ullrich (1950)
		Shell abnormalities	Zangerl and Johnson (1957); MacCulloch (1981)
		Extra right costal plate, first vertebral plate divided by suture, and fifth vertebral plate formed by 4 irregular plates	Bateson (1894)
Chrysemys picta marginata	Midland painted turtle	Shell abnormalities	Whillans and Crossman (1977); MacCulloch (1981)
		Changes in neurals and costals	Grant (1936)
		Supernumerary or deficient bony plates	Newman (1906)
Chrysemys picta picta	Eastern painted turtle	Shell anomalies	Frye (1991a, c)
		7 right and 3 left marginals, and unusually broad neurals	Lynn and Ullrich (1950)
Emys orbicularis	European pond turtle	Anomalies related to number of neurals, costals, and peripherals	Cherepanov (1994)
		Double nuchal plate	Epure and Pogorevici (1940)
Glyptemys insculpta	Wood turtle	Irregularities and reduction in number of bony plates in carapace	Parker (1901)
Graptemys geographica	Common map turtle	Supernumerary or deficient bony plates	Newman (1906)
Malaclemys terrapin	Diamond-back terrapin	Supernumerary plates, dwarfed specimen with broad and short shell, twisted to one side	Hay (1904)
Pseudemys concinna	Eastern river cooter	Disfigured carapace and plastron	Jacobson (2007a)
Terrapene carolina carolina	Eastern box turtle	Rigid plastron without hinge	Holman (1984)
Trachemys scripta elegans	Red-eared slider	Carapace anomaly	Bouwhuis (1972)
		Asymmetrical shells, concave plastron	Frye (1973)
		Anomalous vertebral plates, plastral axial deviation	Frye (1991a)
Trachemys scripta troostii	Cumberland slider	Plastron and carapace anomalies	Cagle (1950)
Geoemydidae			
Cuora amboinensis	Southeast Asian box turtle	Shell deformity	Chou (1979)

(continued)

Table 27.5 (continued)

Species	Common name	Anomaly	References
Kinosternidae			
Sternotherus odoratus	Common musk turtle	180° torsion	Frye (1991a)
Testudinidae			
Chelonoidis chilensis	Chaco tortoise	Abnormal carapace	Matz (1977)
Chelonoidis tabulata	South American yellow-footed tortoise	Additional neurals	Grant (1936)
Geochelone carbonaria	South American red-footed tortoise	Asymmetrical shells	Frye (1973)
		Wedge-shaped vertebral plates	Frye (1991a)
		Soft, deformed carapace and plastron, caused by hyperparathyroidism	Frye and Carney (1975); Rivera and Lock (2008)
		Carapace or plastron anomalies	Vieira de Dandrade and Shinya (1993)
Geochelone nigra	Galapagos giant tortoise	Shell anomaly	Hayes and Beaman (1985)
		Congenital carapace anomaly	Dollinger et al. (1997)
Geochelone sulcata	African spurred tortoise	Deformed carapace	Donoghue (2006)
		Carapace asymmetries	Cloudsley-Thompson (1970)
Gopherus agassizii	Desert tortoise	Carapacial and plastral plate abnormalities, Supernumerary vertebral and marginal or reduced carapacial plates	Frye (1989, 1991a)
		Plastral axial deviation	Frye (1991a, c)
		Wavy carapace	Rosskopf et al. (1982)
		Shell hump	Mader (1990a)
Gopherus berlandieri	Texas tortoise	Supernumerary carapace bones	McEwan (1982)
Gopherus polyphemus	Gopher tortoise	Asymmetrical shells	Frye (1973)
		Supernumerary carapace bones	McEwan (1982)
Gopherus sp.		10% abnormal (hard, brittle shells)	Grant (1937)
Mauremys caspica	Caspian terrapin	Anomalies related to number of neurals, costals, and peripherals	Cherepanov (1994)
Testudo graeca ibera	Turkish tortoise	Carapace anomalies	Kabisch (1989)
Testudo graeca	Spur-tighed tortoise	Anomalies related to number of neurals, costals, and peripherals	Cherepanov (1994)
		Distorted carapace	Lynn and Ullrich (1950)
		1 cm high humps on neural and costal plates	Weichmann (1989)
Testudo hermanni	Hermann's tortoise	Humped carapace	Jackson and Cooper (1981)
Testudo horsfieldi	Horsefield's tortoise	Anomalies related to number of neurals, costals, and peripherals	Cherepanov (1994)
Testudo sp.	Tortoise	Carapace hump	Highfield (1990b)
Trionychidae			
Apalone mutica	Florida soft-shelled turtle	Prominent curved hump (5 times normal shell height) change in number of neurals and costals	Smith (1947)
Apalone spinifera	Spiny soft-shelled turtle	Sharp hump, change in number of neurals and costals	Smith (1947)
Apalone spinifera emoryi	Texas spiny soft-shelled turtle	Sharp hump, change in number of neurals and costals	Smith (1947)
Palea steindachneri	Wattle-necked soft-shelled turtle	Carapace hump	Gressitt (1937)
Unidentified			
	Turtles	Carapace deformities	Brogard (1987)
		Necrotic changes in carapace	Ippen (1965)
	Tortoise	Carapace defects	Frank (1976)

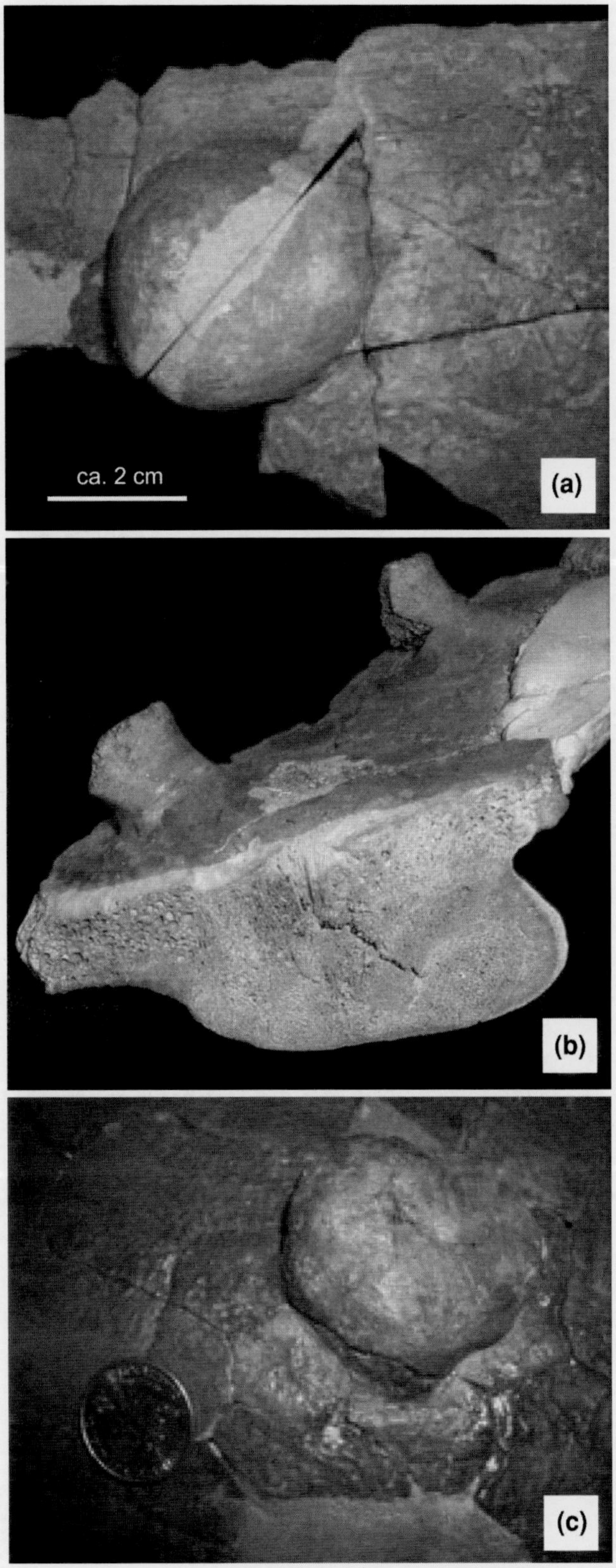

◀**Fig. 27.3** Osteoblastoma (= tumorous bony growths) in shells of the fossil sea turtle †*Syllomus aegyptiacus* (Cheloniidae) from the Miocene of eastern USA. **a**, **b** Fragmentary carapace, USNM 24876: **a** Dorsal view, showing prominently raised, spherical osteoblastoma that has been cut in half (note cut line extending diagonally from *lower left* to *upper right*). **b** Closeup of cut surface of osteoblastoma, in oblique view and orientated with ventral (visceral) surface of bone facing upwards [*Note* orientation here is approximately 180° from how the same piece was depicted by Weems (1974, pl. 2, Fig 2)]. **c** Closeup of the dorsal surface of a more nearly complete carapace, USNM 24872, showing a raised, more irregular-shaped osteoblastoma in upper right of image [*Note* American quarter in lower left for scale and that in this image the specimen has been rotated 90° counterclockwise from how the entire carapace was depicted by Weems (1974, pl. 2, Fig. 1)]

b; Scott 1992) or calcium deficiency (Rosskopf 1986; White 1989). Cagle (1950) proposed that the deformed shell in a Cumberland slider, *Trachemys scripta troostii*, formed because an abnormally large yolk mass causing distension of the carapace. Zwart et al. (1997) reported a thickened (up to 2 cm), loose-textured carapace with large fat-filled spaces in a Hermann's tortoise, *Testudo hermanni*, and slightly elevated rims in another specimen (Zwart et al. 1994). Amorphous carapace lumps were reported in a specimen (USNM 24876) of †*Syllomus aegyptiacus* (Weems 1974).

Plastron bulges with wavy borders (e.g., eastern painted turtle, *Chrysemys picta picta*) have been noted (Lynn and Ullrich 1950), as has a wavy carapace in a California desert tortoise, *Gopherus agassizii* (Rosskopf et al. 1982). Similar features in the Galapagos tortoise, *Geochelone nigra,* have been attributed to calcium deficiency (Hayes and Beaman 1985). A concave plastron was reported in a red-eared slider, *Trachemys scripta elegans* (Frye 1973) and Holman (1984) reported an eastern box turtle, *Terrapene carolina carolina*, with a rigid plastron that lacked a hinge.

Fractures

Published records of bony trauma in turtles are idiosyncratic (Bourdeau 1988b; Devaux 1992; Mautino and Page 1993; Jacobson 1994; Arvy and Fertard 2002; Abou-Madi et al. 2004). Many such injuries are attributed to turtles having been dropped, stepped on, or falling—the last may be more common that previously suspected, considering Vella's (2007) observation that some turtles are avid climbers. Pathologic fractures from nutritional secondary hyperparathyroidism (Avery 1990) and osteomalacia [e.g., Spanish terrapin, *Mauremys leprosa* (Wallis 1927)] must also be considered.

Galois and Ouellet (2007) reported traumatic injuries in spiny soft-shelled turtles, *Apalone spinifera*, without specifying their nature. A "shark's tooth pertaining to a scavenger species related to *Lamna* was found with the type" of the leatherback turtle, *Dermochelys* (Wieland 1909, p. 120). Wieland (1909) also reported an †*Archelon ischyros* with obliquely bitten, healed tibia and fibula. Parker (1901) reported sculptured tortoise marginal carapace fractures associated with scutes that were irregular or reduced in number. Korschelt (1927) reported a 2.5 cm long opening with a bony callus on the inner side of the left costal plates in the carapace of an European pond turtle, *Emys orbicularis*. Frye (1991d) reported a limb fracture in a soft-shelled turtle that apparently was caused by a falling rock.

Healed breaks in the carapace have been reported in the green turtle, *Chelonia mydas*, the Mediterranean pond turtle, *Mauremys leprosa*, the European pond turtle, *Emys orbicularis*, the Mediterranean spur-thighed tortoise, *Testudo graeca*, the Galapagos giant tortoise *Geochelone nigra* (Korschelt 1932), the three-toed box turtle, *Terrapene carolina triunguis* (Barten 1996), the ornate box turtle, *Terrapene ornata* (Bennett 1989), the Gopher tortoise, *Gopherus* sp. (Frye 1973), and in unspecified turtles (Dämmrich 1985). Fractures that cross the dorsal midline of the carapace may be associated with spinal fractures (Isaza and Jacobson 1995). Carrick and Reddacliffe (1980) reported automobile-related, traumatic shell fractures in two species of Australian side-necked turtles: the common snake-neck turtle, *Chelodina longicollis*, and the Australian short-neck turtle, *Emydura macquarii krefftii*. Epidemiologic information is generally lacking, with exception of Meek and Inskeep's (1981) report of 4% frequency of carapace injuries in Hermann's tortoise, *Testudo hermanni*. Cracks in the rear of the carpace in female Galapagos tortoises, *Geochelone nigra*, can occur during mating with the much more massive males (pers. comm. from Margaret Mettler, University of Zürich, Veterinary School).

The literature record also includes reports of a self-amputated hind limb and fractured (shattered) tibia and fibula in an eastern box turtle, *Terrapene carolina carolina* (Boylan 2003), and of a fractured humerus in an Aldabra giant tortoise, *Aldabrachelys gigantea* (Crane et al. 1980). Raidal et al. (2006) reported a healed humeral fracture (encompassing a wire) in a loggerhead turtle, *Caretta caretta*. McArthur (2004) noted rat bite trauma to limbs in post-hybernation turtles, which may explain why Carolina diamond-back terrapin, *Malaclemys terrapin centrata*, individuals are occasionally missing feet and have irregular stumps or irregular growth (Hay 1904); one such individual had lost the posterior half of its body, including one leg.

Arthritis

Arthritis (see also sections below for gout, calcium crystal disease, and infectious arthritis) is a common sickness in mammals including humans, but most reports in non-mammals are from lizards and crocodiles. The condition apparently is not common in turtles and tortoises. Shoulder remodeling and bony proliferation in a loggerhead turtle, *Caretta caretta*, was called osteoarthritis by Raidal et al. (2006), but there seems to be subchondral damage. Given negative cultures, the possibility of a spondyloarthropathy must be entertained (Rothschild 2009a). Dämmrich (1985) has used the terms spondylopathia deformans and spondyloarthropathia deformans, but the meanings are unclear. He did describe ankylosis, lending credence to the possibility of spondyloarthropathy. Fitzgerald and Vera (2006) reported proliferative spinal osteopathy, which could represent an infectious process or spondyloarthropathy. Septic arthritis was reported in a leatherback turtle, *Dermochelys coriacea*, by Ogden et al. (1981).

Metabolic Diseases

Vitamin D Deficiency and Hyperparathyroidism

Metabolic abnormalities in turtles predominantly are the result of urate-related disease, calcium crystal-related disease, or other conditions such as metabolic bone disease, hypoplastic osteoporosis (reduction in quantity and quality of trabecular components of bone), osteomalacia, rickets osteodystrophy, nutritional osteodystrophy, osteodystrophy fibrosa, osteogenesis imperfecta (an inborn error of metabolism), cage paralysis (complete loss of muscle function) and renal osteodystrophy (defective bone formation from compromised metabolism of vitamin D to the active form producing calcium deficiency and secondary hyperparathyroidism) (Jarofke and Lange 1993; Table 27.6). The last condition may be expressed in the skeleton as accentuated 'interplates' and growth lines (Frye 1994).

Nutritional osteodystrophy is most commonly reported in captive animals. For example, in a pair of surveys that considered 144 chelonians reported by the Zoological Society of London, Keymer (1978a, b) documented osteodystrophy in 12 individuals. Barten (1982, 1983) also reported nutritional shell defects. Coquelet (1983) cited Keymer's (1978a) reports of osteodystrophy in 8.3% of terrestial and 9.8% of freshwater aquatic turtles, whereas Glazebrook (1980) noted its extreme rarity in marine turtles.

Frequencies of nutritional and metabolic disorders among captive animals are difficult to assess from the literature, because most reports deal with isolated occurrences (Table 27.6). Epidemiologic or facility reports may also be difficult to interpret. For example, it was reported that 23 (19%) of 122 terrapins autopsied at the Zoological Society of

Table 27.6 Metabolic bone diseases in turtles

Species	Common name	Disease	References
Cryptodira			
Cheloniidae			
Chelonia mydas	Green turtle	Nutritional osteodystrophy	Glazebrook (1980); Coquelet (1983)
Chelydridae			
Chelydra serpentina	Common snapping turtle	Demineralization due to secondary nutritional hyperparathyroidism	Wilkinson et al. (2004)
		Osteopathy	Keymer (1978a)
Geoemydidae			
Heosemys annandalii	Yellow-headed temple turtle	Shell rot	Guirley (2003)
Mauremys caspica rivulata	Western Caspian turtle	Osteopathy and nutritional osteodystrophies	Keymer (1978a)
Mauremys leprosa	Mediterranean pond turtle	Osteopathy and nutritional osteodystrophy	Keymer (1978a)
		Calcioprive osteopathy = osteomalacy + rachitis	Wallis (1927); Dämmrich (1967)
		Gout	Figureres (1997)
Emydidae			
Chrysemys ornata calirostris	Ornate slider	Nutritional osteodystrophy	Keymer (1978a)
Chrysemys picta picta	Eastern painted terrapin	Osteopathy	Keymer (1978a)
Graptemys pseudogeographica ouachitensis	Ouachita map turtle	Osteopathy	Keymer (1978a)
Malaclemys sp.	Terrapin	Nutritional osteodystrophy in 31 of 100	Jackson (1980)
Pseudemys sp.	Scooter	Osteoporotic fracture and misshapened carapace (Osteomalacy)	Frye (1973)
		Calcium pyrophosphate deposition	Brogard (1987)
Terrapene sp.	Box turtle	Ulcerative shell disease, osteoporosis	Bourdeau (1988a)
Terrapene carolina	Eastern box turtle	Congenital tarsal absence	Wilkinson et al. (2004)
Terrapene carolina triunguis	Three-toed box turtle	Bridge between carapace and plastron	Boyer (1996b)
Terrapene ornata	Ornate box turtle	Curling of carapace edges, overgrowth of beak (parrot beak), and broken legs	Boyer (1996b)
Trachemys calirostris	Ornate terrapin	Metabolic bone disease	Keymer (1978a)
Trachemys scripta	Slider	Articular gout	López del Castillo (1998)
		Osteodystrophy with deformed, non-calcified extremities	Bourdeau (1988a)
		Rachitis	Reichenbach-Klinke (1963, 1977)
Trachemys scripta elegans	Red-eared slider	Osteopathy, nutritional osteodystrophy	Keymer (1978a); Jackson and Sainsbury (1992)
		Articular gout	Marcus (1981); McArthur (1996)
		Infected shoulder	McArthur (2004)
		Secondary hyperparathyroidism producing osteodystrophy fibrosa with residual lacy calcification	Gabrisch and Zwart (1992)
		Osteodystrophy fibrosa with residual lacy calcification	Bourdeau (1988a)
		Pseudogout = calcium pyrophosphate deposition	Frye and Dutra (1976); Frye (1991d, 1994); Rothschild and Bruno (2009)
Trachemys scripta scripta	Yellow-bellied slider	Osteopathy	Keymer (1978a)
		Gout	López del Castillo (1998)

(continued)

Table 27.6 (continued)

Species	Common name	Disease	References
Kinosternidae			
Staurotypus triporcatus	Three-keeled terrapin	Osteopathy	Keymer (1978a)
Testudinidae			
Chelonoides carbonaria	South American red-legged tortoise	Soft, deformed carpace and plastron, caused by hyperparathyroidism	Frye and Carney (1975); Rivera and Lock (2008)
Dipsochelys (Aldabrachelys) gigantea	Aldabra tortoise	Osteodystrophia fibrosa and rickets	Hauser et al. (1977)
		Hyperparathyroidism	Gerlach (2004)
Geochelone nigra	Galapagos tortoises	Metabolic bone disease	Dollinger et al. (1997)
Geochelone pardalis pardalis	Leopard tortoise	Secondary nutritional metabolic bone disease	Raiti and Haramati (1997)
		Hyperparathyroidism	Zwart et al. (1994)
Geochelone pardalis babcocki	African leopard tortoise	Ossification of liver and other internal organs	Frye (1991b)
Geochelone radiata	Radiated tortoise	Osteoporosis	Isenbügel and Frank (1985)
		Osteopathy	Keymer (1978b)
		Articular gout	Appleby and Siller (1960)
Geochelone sulcata	African spurred tortoise	Increased mineralization from secondary renal hyperparathyroidism Articular gout	Wilkinson et al. (2004) Appleby and Siller (1960)
Gopherus sp.		Nutritional osteodystrophy	Bourdeau (1988b)
Gopherus agassizii	California desert tortoise	Uncalcified	Boyer (1996b); Mader 1990a, 2006b)
		Gout, osteomalacy	Frye (1984); Homer et al. (1998)
		Osteoclastic resorption, osteopenia, articular gout	Homer et al. (1998)
		Nutritional bone disease	Mader (2006c)
		Calcium pyrophosphate deposition	Rothschild and Bruno (2009)
Kinixys belliana	Bell's hinged tortoise	Osteopathy	Keymer (1978b)
Testudo denticulata	Jaboty or rain forest tortoises	Nutritional osteodystrophy Articular gout	Keymer (1978b) Appleby and Siller (1960)
Testudo elegans	Starred tortoise	Osteopathy and nutritional osteodystrophy	Keymer (1978b); López del Castillo (1998)
Testudo graeca	Mediterrenaean spur-thighed tortoise	Nutritional osteodystrophy	Jackson and Duff Fasal (1981)
		Demineralization due to secondary nutritional hyperpareathyroidism	Wilkinson et al. (2004)
		Osteopathy	Keymer (1978b)
		Articular gout	Appleby and Siller (1960)
		Articular pseudogout = calcium pyrophosphate deposition	López del Castillo (1998); Rothschild and Bruno (2009)
Testudo graeca ibera	Turkish tortoise	Metabolic bone disease	McArthur (2004)
		Shell 'infractions' and osteolysis from suppurated infections	Stojanov (2005)
		Hyperparathyroidism	Gerlach (2004)
Testudo hermanni	Hermann's tortoise	Osteodystrophia fibrosa and rickets	Dämmrich (1967, 1979)
		Short mandible, misshapened limb bones and humped carapace shields of nutritional osteodystrophy	Jackson and Cooper (1981)
		Osteopathy	Keymer (1978b)
		Carapace bone loss from secondary hyperparathyroidism	Wilkinson et al. (2004)
		Elbow osteolysis with osteochondromatosis	Frye (1981a)
		Articular gout	Appleby and Siller (1960); López del Castillo (1998)

(continued)

Table 27.6 (continued)

Species	Common name	Disease	References
Testudo horsfieldi	Horsefield's tortoise	Osteopathy	Keymer (1978b)
Testudo marginata	Marginated tortoise	Hypoplastic osteoporosis	Häfeli and Zwart (2000)
		Osteopathy and osteodystrophy	Keymer (1978b)
		Metabolic bone disease	McArthur (2004)
Trionychidae			
Apalone mutica	Smooth soft-shelled turtle	Carapace curling related to new born mouse diet	Barten (1982, 1983)
Pleurodira			
Chelidae			
Emydura subglobosa	Red-bellied short-neck turtle	Periarticular hydroxyapatite deposition	Wenker et al. (1999)
Pelomedusidae			
Pelusios subniger	Blackish terrapin	Osteopathy	Keymer (1978a)
Unidentified	Turtles	Nutritional osteodystrophy	Cowie (1976); Jackson (1987)
		Osteomalacy	Bourdeau (1988a)
		Gout	Marcus (1968)
	Tortoises	Gout	Frye (1981a); McArthur (1996); Messonnier (1996); Arvy and Fertard (2002); Mader (1996)
		Metabolic bone disease	Mader (1990b)
		Hyperparathyoridism	Frye (1981a); O'Malley (2008)
		Osteoporosis, rachitis, osteomalacy, osteodystrophia fibrosa	Jarofke and Lange (1993)

Definitions: articular pseudogout = calcium pyrophosphate deposition disease; gout = metabolic disorder in which sodium uric acid crystals deposit in joints (referred to as articular gout); hyperparathyroidism = disorder caused by overactivity of the named glands, producing osteitis fibrosa cystica (= fibrous tissue replacement of bone secondary) and other bone changes; osteodystrophy = defective bone formation; osteomalacy = softening of bone due to lack of vitamin D; osteopathy = combination of osteomalacia (= vitamin D deficiency-related softening of bone) and rickets; osteopenia = bone mineral density lower than normal; osteoporosis = reduction in quantity and quality (e.g., thickness) of trabecular components of bone; rickets = failure of bone osteoid to calcify (= rachitis)

London between 1965 and 1975 had "lesions of the skeletal system" (Keymer 1978b, p. 577). This included the category of miscellaneous osteopathies, which encompasses both nutritional and infectious origins of disease (Table 27.6). Although nutritional and metabolic disorders were reported in 19.7% of studied animals (Keymer 1978a), actual nutritional osteodystrophies were listed in 9.8%. This is difficult to reconcile with Keymer's (1978b) report of 'miscellaneous osteopathies' in 4.2% and nutritional osteodystrophies in 0.3% of turtles autopsied at the Zoological Society of London during the same period. Reports of frequency differences are even more pronounced in freshwater turtles that were not raised in the London Zoo. McArthur and Barrows (2004) reported that one-third of freshwater turtles seen in a Leeds/London clinic had nutritional osteodystrophy.

Nutritional osteodystrophy can produce deformities in the shell (e.g., soft, lumps, humps, or abnormally large plastron fontanelles), maxillary region (e.g., overgrown beak), and mandible (e.g., shortened) (Table 27.6), as well as deformed, inadequately calcified, misshapened or swollen extremities, kyphoscoliosis, and pathologic (e.g., greenstick) fractures (Frye 1973; Jackson and Cooper 1981). Pliable mandible or maxilla, fractures of long bones, fibrodysplasia (= replacement of bone tissue by fibrous tissue) of long bones or jaw, kyphosis (= dorsoventral vertebral column curvatures), lordosis (= curvature of axial skeleton, with accentuated anterior apical apex), scoliosis (= lateral curvature of the vertebral column), and widened radiolucent ends of long bones have been described in box turtles, *Terrapene* sp., and desert tortoises, *Gopherus agassizii* (Boyer 1996b; Mader 1990b, 2006d). Widened, calcific rings may occur near epiphyseal junctions (Glazebrook 1980). Overgrowth of the beak (i.e., parrot beak), curling of carapace edges, and net-like porosity of and overgrowth of the bridge between the carapace and plastron have been noted (Boyer 1996b). Thick, spongy bone may also occur in the shell (Gerlach 2004). Highly porous bone is especially prominent at rib ends and in peripheral plates (Stojanov 2005). Paradoxically, increased mineralization from secondary renal hyperparathyroidism (= disorder caused by overactivity of the named glands, producing osteitis fibrosa cystica and other bone changes) can also occur (Wilkinson et al. 2004). Cystic calculi (renal stones) may result (Gillespie 1994; Mader et al. 1999). Osteopenia (reduced bone ossification/density) and metastatic calcification (including joints), however, can also occur from hypervitaminosis (= excess vitamin) D (Gillespie 1994). The term osteodystrophy includes secondary hyperparathyroidism (Avery 1990), which can be caused by dietary calcium or vitamin D deficiency, negative Ca:PO_4 ratio, or by lack of UV-light exposure (Boyer 1996b). It also can occur if vitamin D metabolism is altered secondary to kidney, liver, intestinal, thyroid or parathyroid disease, and hypocalcemia (= presence of low serum calcium levels) (Boyer 1996b).

One component of metabolic bone disease, hyperparathyoridism, can occur as a primary event Frye (1981a), causing a soft, deformed carapace and plastron (Frye and Carney 1975). Parathyroid adenomas (a benign neoplasm of glandular cells) have been reported in the South American red-footed tortoise, *Geochelone carbonaria*, the Mediterannean spur-thighed tortoise, *Testudo graeca*, and the California desert tortoise, *Gopherus agassizii* (Frye and Carney 1975; Frye 1994).

The other component, osteomalacia (also called rickets prior to epiphyseal closure), is caused by vitamin D deficiency. In addition to softened, malformed bones, enlarged joints and soft shells, widened calcitic rings occur near epiphyseal junctions (Wallach 1971; Glazebrook 1980). Osteomalacia has been noted in the painted turtle, *Chrysemys picta* (Reichenbach-Klinke 1977), ornate box turtle, *Terrapene ornata* (Reichenbach-Klinke and Elkan 1965), and unspecified captive turtles (Hunt 1957).

Urate Disease

Although most reptiles excrete nitrogen predominantly as uric acid (Wallach 1978), aquatic turtles primarily excrete urea and sea turtles primarily excrete ammonia (Allen and Oftedal 1994). Urate disease occurs from buildup of uric acid in blood or focal locations. Hyperuricemia (elevated uric acid level) results from water deprivation (e.g., post-hibernation dehydration), excessive consumption of protein or shrimp (both high in uric acid), or renal disease (Frye 1981a; Marcus 1981; McArthur 1996). Vitamin A deficiency produces renal dysfunction, which produces secondary gout (a metabolic disorder in which sodium urate crystals deposit in joints or internal organs). The latter condition was reported in a red-eared slider, *Trachemys scripta elegans* (McArthur 1996).

Uric acid may be deposited in the kidneys (in the form of stones, discussed in section on stones below), in joints (referred to as articular gout), and around internal organs (referred to as visceral gout) (Rothschild 2009b). The last condition is beyond the scope of this review and will not be discussed further.

Articular gout is well recognized in turtles (Table 27.6), either by the classical presence of crystals within synovial fluid that are negatively birefringent (i.e., retard light transmission, as visualized by polarizing microscopy) or by the presence of characteristic bone erosions having sclerotic margins and overhanging edges (Casimire-Etzioni et al. 2004; Rothschild 2009b; Rothschild and Heathcote 1995). Articular gout typically affects joints of the limbs, shoulder, and hip (Homer et al. 1998; López del Castillo 1998; McArthur 2004). Mader (2006c) reported destruction of cervical vertebrae by articular gout.

Calcium Crystal Disease

Calcium pyrophosphate deposition disease (= articular pseudogout) has been reported in turtles, specifically the red-eared slider, *Trachemys scripta elegans* (Frye and Dutra 1976; Frye 1994; Rothschild and Bruno 2009), *Pseudemys* (Brogard 1987), the California desert tortoise, *Gopherus agassizii* (Rothschild and Bruno 2009), and the Mediterranean spur-thighed tortoise, *Testudo graeca* (López del Castillo 1998; Rothschild and Bruno 2009). Hydroxyapatite deposition disease in the red-belly short-necked turtle, *Emydura albertisii*, was recorded by Wenker et al. (1999). Frye (1991b) reported ossification of liver and other internal organs in the African leopard tortoise, *Geochelone pardalis babcocki*.

Stones

Bladder stones are well recognized in Chelonia (Jackson and Cooper 1981). Kölle et al. (2001) reported calculi in 4% of necropsied tortoises. Keymer (1978b) reported a similar value of 4.2% in tortoises. In a survey of bladder stones, Osborne et al. (2008) reported that in tortoises 94% of stones were urate, 3% were calcium carbonate, and 3% were of mixed composition; by contrast, the same study reported that in turtles 58% of stones were urate, 17% were calcium phosphate, 17% were calcium carbonate, and 8% were of mixed composition.

Urate Stones

Salts (ammonium, sodium, calcium, and potassium) of uric acid also deposit in renal tubules or as bladder stones or calculi (Mebs 1965; Kölle et al. 2001). Table 27.7 summarizes isolated reports in turtles, but as with many of the

Table 27.7 Stones in turtles

Species	Common name	Kind of stone	References
Cryptodira			
Cheloniidae			
Chelonia mydas	Green turtle	Calcium carbonate, calcium phosphate, calcium sulfate, and magnesium phosphate	Virchow (1878)
Emydidae			
Trachemys scripta elegans	Red-eared slider	Uric acid	Mebs (1965)
		Calcium phosphate	Mader (2006b)
Testudinidae			
Chersina angulata	South African bowsprit tortoise	Uric acid	Mebs (1965)
Dipsochelys (Aldabrachelys) gigantea	Aldabaran giant tortoise	Not specified	Keymer (1978b)
		Uric acid	Hammerton (1934, 1939); Grünberg (1963/64)
Geochelone elegans	Indian starred tortoise	Not specified	Keymer (1978b)
Geochelone nigra	Galapagos tortoise	Uric acid	Grünberg (1963/64); Wallach (1971)
Geochelone pardalis	Leopard tortoise	Calcium phosphate	Mader (2006b)
Geochelone ?platynota	Burmese star tortoise	Uric acid	Fife (2007)
Geochelone sp.		Uric acid + other substances	Kölle et al. (2001)
Geochelona sulcata	African spurred tortoise	Uric acid	Raiti (2004)
		Calcium phosphate (urolith)	Mader (2006b)
Gopherus agassizii	Desert tortoise	Bladder stone	Long Beach Animal Hospital (2007)
		Uric acid	Frye (1972); Mader et al. (1999); Mangone and Johnson (1998); Homer et al. (1998)
Kinixys belliana	Bell's hingeback tortoise	Uric acid	Appleby and Siller (1960)
Kinixys homeana	Kuhl's tortoise	Not specified	Keymer (1978b)
	Desert tortoise	Bladder stone	Long Beach Animal Hospital (2007)
Testudo graeca	Spur-thighed tortoise	Not specified	Keymer (1978b)
		Uric acid/urate	Grünberg (1963/64); Ebstein (1899)
		Tophi gout	Kölle and Hoffmann (2002)
Testudo hermanni	Hermann's tortoise	Not specified	Keymer (1978b)
Testudo sp.		Uric acid + other substances	Kölle et al. (2001)
Trionychidae			
Apalone spinifera hartwegi	Western spiny soft-shelled turtle	Calcium phosphate, apatite, and struvite	McKown (1998); Mader (2006b)

Definition: tophi = mass of uric acid

previous discussed conditions frequencies are more difficult to assess. Urate stones were found in three of 24 California desert tortoises, *Gopherus agassizii*, studied by Homer et al. (1998). Kölle et al. (2001) reported calculi in 4% of necropsied tortoises and lizards, and noted that the tortoise genera *Testudo* and *Geochelone* were most often affected among chelonians. This is at odds with the claim by Grünberg et al. (1977) that uric acid/urate calculi only occur in carnivorous species. Mangone and Johnson (1998) reported that calculi are frequently encountered in many species of captive tortoises. Kölle and Hoffmann (2002) reported that 64% of European tortoises (especially spur-thighed tortoises, which are more sensitive to a high protein diet than Afghan tortoises) had 'renal alterations' on necropsy. Additionally, 16% had gouty tophi in the kidney, usually occurring as single stones (Kölle and Hoffmann 2002). Mader et al. (1999) reported calculi in 81 individuals of the Californian desert tortoise, *Gopherus agassizii*; of those, 73 individuals had only a single calculi, two individuals each had two calculi, five indviduals each had three calculi, and one individual had five calculi. The calculi, ranged in size from 0.4–14 cm (average = 5.4 cm).

Calcium Stones (Calculi)

Calcium carbonate, calcium phosphate, calcium sulfate, and magnesium phosphate stones have been found in marine turtles (Virchow 1878; Grünberg 1963/1964). Basic calcium phosphate, calcium apatite, and struvite (magnesium ammonium phosphate hexahydrare) crystals (Table 27.7) have been noted. A pseudo-calculus (actually an egg) was reported by Mader (2006b) in a California desert tortoise, *Gopherus agassizii*. A low calcium diet produces calculi (Wallach 1969).

Infection

Osteomyelitis

Direct infection of bone is referred to as osteomyelitis (Hutchison and Frye 2001; Resnick 2002). Note that although some definitions of osteomyelitis also include inflammation of marrow cavity, here we limit the term to infection of the bone. This condition appears as an osteolytic area or an area of reactive new bone formation, often with a filigree or 'aero candy'-like (i.e., microbubbly) surface texture. Osteomyelitis in turtles produces carapace and plastron necrosis. It can be caused by 'ordinary' bacteria (e.g., *Pseudomonas*, *Citrobacter*, or *Klebsiella*), mycobacteria, fungi (especially *Mucor*), and algae (Arvy and Fertard 2002, Barnett 2003; Table 27.8). Shell rot has been attributed to algae, to the fungi *Mucorales*, *Rusarium*, *Geotrichum*, *Trichosporon*, and *Coniothyrium*, and to the bacterium *Beneckea chitinovora* (Garner et al. 1997). The forearm abscess reported by Jacobson (1994) in a Burmese mountain tortoise, *Manouria emys*, was associated with a focal ulcerative lesion of the plastron and linked to a bacterial abscess in the coelomic cavity. Specific shell infections will be discussed separately in the sections below.

Infectious Arthritis

Septic arthritis describes a joint infection. In chelonians, these kinds of infections have been attributed to 'ordinary' bacteria, mycobacteria, and fungi (Rhodin et al. 1990; Table 27.8). Ogden et al. (1981) reported a leatherback turtle, *Dermochelys coriacea*, having erosion of the distal end of the humerus and replacement of the elbow joint by fibrous ankylosis with reactive overgrowth, and sclerosis of the proximal ends of the radius and ulna; this condition suggests infection. Plastron lesions and a swollen left elbow bearing osteolytic radial and ulnar lesions with sclerotic margins in an Atlantic ridley sea turtle, *Lepidochelys kempii*, were attributed to the bacterium *Mycobacterium chelonae* (Greer et al. 2003). Harms et al. (2002) isolated the bacterium *Nocardia* and an unidentified fungus from carpal swelling with proximal first metacarpal radiolucency, obliteration of distal carpal row, ulnare, pisiform and metacarpals II and III in an Atlantic ridley sea turtle, *Lepidochelys kempii*.

Neoplasms

Neoplasms (literally, new abnormal tissue) are rare in chelonians (Arvy and Fertard 2002). They are caused by abnormal growth of cells and may be benign or malignant. Garner et al. (2004) reported neoplasms in 2.7–3.2% of turtles and 1.4% of tortoises referred to his clinic, but did not comment on bony involvement. Those authors did, however, note that metastasis was uncommon. Sykes and Trupkiewicz (2006) reported that neoplasia was found only in turtles (not tortoises) in the Philadelphia zoo from 1901 to 2002. The frequencies ranged from 0.3% in 1901–1967, zero in 1968–1979, 2.4% in 1980–1991, and zero in 1992–2002, for a total of six afflicted among 511 turtles. Zwart and Harshbarger (1991) reported a Hermann's tortoise, *Testudo hermanni*, with squamous cell carcinoma that caused pathologic fracture of one femur and necrosis of the adjoining portion of the carapace. A neurilemmal sarcoma was found in *Testudo hermanni* (Cooper et al. 1983).

Shell neoplasia is extremely rare. One reported case in a California desert tortoise, *Gopherus agassizii*, was classified as a chondroma (Rosskopf 1986). Plastron nodules in the loggerhead, *Caretta caretta*, were caused by a lymphoblastic lymphoma (Orós et al. 2001; Mauldin and Done 2006). Weems (1974, p. 279) described a "tumorous bony growth" in †*Syllomus aegyptiacus* (specimen USNM 24872), which the senior author subsequently recognized as an osteoma or osteoblastoma (Rothschild, unpublished observation). The specimen had been intentionally cut through the affected area, revealing the fine trabecular pattern that is characteristic for that kind of benign tumor (Resnick 2002).

Frye (1981a, 1994) reported osteochondromatosis (= benign neoplasm producing a cartilage cap on an exostosis) and osteolysis (= resorption or destruction of bone) in the elbow of a Hermann's tortoise, *Testudo hermanni*. While this could be the primary disease of numerous osteochondrous nodule production, a more likely explanation is that it occurred secondary to the osteolytic process and actually represents a neuropathic process.

Table 27.8 Infectious diseases (osteomyelitis) in turtles

Species	Common name	Site affected	Organism	References
Cryptodira				
Carettochelyidae				
Carettochelys insculpta	Fly river turtle	Shell	*Paecilomyces lilacinus*	Lafortune et al. (2005); Paré and Jacobson (2007)
Cheloniidae				
Caretta caretta	Loggerhead	Shell	*Fusarium solani*	Rebell et al., (1971); Austwick and Keymer (1981); Rothschild (2009a)
		Shell	barnacles	Jacobson (2007b)
Lepidochelys kempii	Atlantic ridley sea turtle	Plastron, elbow	*Mycobacterium chelonei*	Greer et al. (2003); Jacobson (2007a)
		Carpals, metacarpals	*Nocardia* and unidentified fungus	Harms et al. (2002); Jacobson (2007a); Rothschild (2009a)
Chelydridae				
Chelydra serpentina	Common snapping turtle	Shell (necrosis)	Unidentified	Barten (2006)
		Not listed	*Erysipothrix*	Jacobson (2007a)
		Plastron	fungi	Rothschild (2009a)
Dermochelyidae				
Dermochelys coriacea	Leatherback	Erosion of distal humerus and replacement of elbow with fibrous ankylosis with reactive overgrowth of proximal radius and ulna (arthritis)	Not listed	Ogden et al. (1981); Coquelet (1983); Brogard (1987)
Emydidae				
Chrysemys picta picta	Eastern painted turtle	Shell	*Beneckea chitinovora*	Wallach (1977)
Chrysemys sp.	Painted turtle	Shell	*Beneckea chitinovora*	Brogard (1980); Wallach (1975); Jacobson (2007a); Rothschild (2009a)
Pseudomys concinna	Eastern river cooter	Shell	*Mucorales, Rusarium, Geotrichum, Trichosporon, Coniothyrium*, algae, and *Beneckea chitinovora*	Garner et al. (1997)
Trachemys scripta scripta	Yellow bellied slider	Shell	*Mucorales, Rusarium, Geotrichum, Trichosporon, Coniothyrium*, algae, and *Beneckea chitinovora*	Garner et al. (1997)
Trachemys scripta elegans	Red-eared slider	Loosened plates around sutures	*Beneckea chitinovora*	Wallach (1975, 1977); Jacobson (2007a)
		Carapace	*Trichosporon*	Schildger et al. (1991)
Trachemys sp.	Slider	Loosened plates around sutures	*Beneckea chitinovora*	Wallach (1975); Jacobson (2007a); Rothschild (2009a)
Emydide indet		Shell	*Spirorchis*	Jacobson (2007b)
Geoemydidae				
Heosemys annandalii	Yellow-headed temple turtle	Shell rot	*Beneckea chitinovora*	Guirley (2003)
Kinosternidae				
Sternotherus minor peltifer		Shell	*Beneckea chitinovora*	Wallach (1977)

(continued)

Table 27.8 (continued)

Species	Common name	Site affected	Organism	References
Sternotherus sp.	Musk turtle	Carapace	*Beneckea chitinovora*	Wallach (1975); Brogard (1980); Coquelet (1983); Jacobson (2007a); Rothschild (2009a)
Testudinidae				
Astrochelys radiata	Radiated tortoise	Mandible	Chromomycosis	Keymer (1978b); Jacobson (1994); Rothschild (2009a)
Dipsochelys (Aldabrachelys) gigantea	Aldabra tortoise	Plastron	Fungi	Hammerton (1935); Austwick and Keymer (1981)
		Carapace	*Exophiala oligosperma*	Stringer et al. (2009)
Geochelone elegans	Indian star tortoise	Temporomanidbular joints and knees	Mycotic infection	McArthur (2004)
Gopherus agassizii	Desert tortoise	Leg	Mycobacteria	Frye (1994)
		Shell (plastron)	*Astragalus, Stanleya* and *Xylorrhiza*	Jacobson et al. (1994)
Manouria emys	Asian brown tortoise	Forelimb	Bacterial abscess	Jacobson (1994)
Testudo sp.		Tibia, fibula, and tarsals	Not listed	Jackson and Sainsbury (1992)
Testudo graeca	Spur-thighed tortoise	Shell	*Coniothyrium fuckelianum*	Goodwin (1976); Austwick and Keymer (1981); Rothschild (2009a)
Testudo hermanni	Hermann's tortoise	Carapace	Bacterial infection	Schildger et al. (1991)
Trionychidae				
Trionyx sp.	Soft-shelled turtle turtle	Shell	*Beneckea chitinovora*	Wallach (1975); Brogard (1980); Jacobson (2007a)
Apalone spinifera	Spiny soft-shelled turtle	Shell	*Beneckea chitinovora*	Wallach (1977)
Trionychidae indet.	Soft-shelled turtle	Caseated crateriform ulcers on plastron	*Citrobacter freundii*	Boyer (1996a)
		Shell (necrosis)	*Mucorales*	Jacobson (1980)
Cryptodira indet.		Shell	*Beneckea chitinovora*	Marcus (1980)
Pleurodira **Chelidae**				
Acanthochelys macrocephala	Pantanal swamp turtle	Septic arthritis	Not listed	Rhodin et al. (1990)
Chelodina longicollis	Long-necked terrapin	Plastron	Fungi	Hammerton (1939); Austwick and Keymer (1981); Rothschild (2009a)
Chelus fimbriata	Matamata terrapin	Plastron	Fungi	Hammerton (1934); Austwick and Keymer (1981); Rothschild (2009a)
Emydura subglobosa	Red-bellied short-necked turtle	Hydroxyapatite deposition disease	Not listed	Wenker et al. (1999)
Podocnemidae				
Podocnemis sp.	South American river turtle	Shell	*Beneckea chitinovora*	Brogard (1980); Arvy and Fertard (2002)
Podocnemis unifilis	Yellow-spotted Amazon river turtle	Shell	*Beneckea chitinovora*	Wallach (1977)
Pleurodira indet.		Loosened shell plates around sutures	*Beneckea chitinovora*	Wallach (1975); Rothschild (2009a)

Neuropathic Disease

Frye (1981a, 1994) reported a case of osteochondromatosis in Hermann's tortoise, *Testudo hermanni,* that appeared to be the result of denervation. Such dramatic osteolysis and new bone formation, with formation of multiple osteochondral nodules (correctly referred to as osteochondromatosis) is characteristic of "neuropathic disease" (Resnick 2002; Rothschild 2009a, p. 57: "partial or complete loss of intervertebral space associated with new bone and irregular hyaline cartilage island formation"), not previously recognized in turtles.

Shell Disease

Shell disease, which is a broadly inclusive term used to describe damage to the carapace and plastron, is controversial and its pathologies have been attributed to trauma, infection, and metabolic disease. Trauma may be direct or contributory. As an example of the latter, carapace and plastron injuries sustained by contact with rocks may subsequently become infected (Wright et al. 1977; Balazs 1980). Frye (1981b) reported rostral abrasions, bites (including bites by cage mates), lacerations, and crushing injuries. Puncture wounds associated with thickened bone on posterior and posterodorsal surfaces of the carapace in † *Protochelydra* and a fossil trionychid (Erickson 1984) suggest failed predation attempts.

Although some authors have attributed shell damage to specific infectious agents (e.g., Rosskopf 1986; Jacobson 1994), simple descriptions of the pathology seem inadequate to allow à priori etiologic assignment. It is unclear if better descriptions of pathology would help identify which agent(s) is responsible or if the limited manner in which bone can respond to insults precludes identification of the specific infection from gross examination, unless the organism is seen, components are isolated, or the actual organism is recovered (i.e., cultured). Perhaps attention to details, such as elevated margins and their sharpness or curved nature [as noted by Sowerby and Lear (1872)] and the oval shell lesions reported by Wu (1994) in Chinese soft-shelled turtles, *Pelodiscus sinensis*, might provide clues to specific agents.

Most of the discussion of shell disease revolves around pitting, ulcerative shell disease (Rebell et al. 1971), and necrosis. Hutchison and Frye (1989) described bone sclerosis (associated with osteolysis) in fossil Emydidae, which they interpreted as osteomyelitis. At present, however, descriptions or illustrations are inadequate to further pursue perspectives of shell sclerosis.

Disfiguring shell disease associated with dermal bone remodeling was reported in the eastern river cooter, *Pseudemys concinna*, and the yellow-belly turtle, *Trachemys scripta scripta* (Lovich et al. 1996; Garner et al. 1997; Jacobson 2007a). Jacobson (2007a) described the damage as segmental necrosis and remodeling. Much of the literature simply reports isolated observations. Epidemiologic studies have revealed 6.5% of three-toed box turtle, *Terrapene carolina triunguis*, and Florida box turtle, *Terrapene carolina bauri*, had carapace pits (Carpenter 1956; Schwarz and Schwartz 1974; Dodd et al. 1997; Dodd 2001). Rosskopf (1986) and Bailey (1987) suggested that pitting represented residua of old, healed ostcomyelitis, but that is only one of several possible diagnoses. Rosskopf (1986) also suggested that shell ulcerations in the diamond-back terrapin, *Malaclemys terrapin*, resulted from too much time in fresh water. Jacobson et al. (1994) suggested a possible seasonal relationship of shell necrosis in the California desert tortoise, *Gopherus agassizii,* to its spring foods, specifically the locoweed *Astragalus,* princesplume *Stanleya,* and woody aster *Xylorrhiza,* all of which produce aliphatic nitro compounds. The situation may be even more complex. Garner et al. (1997) reported shell necrosis in turtles with pancreatitis and iron deposition (possible hemochromatosis); inflammation of the pancreas can produce bone necrosis in humans (Resnick 2002). Hutchison and Frye (1989) raised the possibility that the high frequency of pitting in female painted turtles, *Chrysemys picta*, might be related to calcium and phosphate withdrawal for egg laying.

Predation injury has also been proposed for the origins of shell pits. Holes in the shell of the Cretaceous marine turtle † *Protostega gigas* have been matched with mosasaur teeth based on similarities in size, shape, and distribution of the holes (Rothschild and Martin 2006; Rothschild 2009a). Other holes in shells of †*Echmatemys* (specimen UCMP 128283) and †*Hadrianus corsoni* (specimen UCMP 128418) were attributed to crocodilian bites (Hutchison and Frye 1989; Rothschild 2009a). The latter appeared as conical perforations from predation. They were divided pits into circular to ovoid with flat bottoms, versus rounded bottoms, irregular pits with discrete margins, track (linear) and rot, the latter represented by large areas of irregular depression or dead lamellar bone. Scratches may imply failed predation.

Shell disease is a common ailment in at least some extant turtle populations. For example, Lovich et al. (1996) reported that 74% of eastern river cooters, *Pseudemys concinna*, and 35% of sliders, *Trachemys scripta*, from Lake Blackshear, Georgia, had carapace necrosis. Typically, the macroscopic appearance of the pathology is not fully described; only the putative responsible infectious agent.

Carapace necrosis and osteomyelitis have been attributed to various microorganisms (Barnett 2003), specifically the bacterium *Beneckea chitinovora* (Marcus 1980, Table 27.8), ordinary bacteria [e.g., *Aeromonas, Arizona, Bacteroides, Citrobacter, Enterobacter, Escherichia coli* and *Es. freundii, Klebsiella, Pasteurella, Proteus, Providencia, Pseudomonas, Salmonella typhimurium, Sa. regent, Sa. amrina, Serratia, Staphylococcus aureus*, and alpha-hemolytic *Streptococcus* (Rosskopf 1986; Highfield 1990b; López del Castillo 1996; Arvy and Fertard 2002)], acid fast organisms [e.g., *Mycobacterium* and *Nocardia* (López del Castillo 1996)], spirochetes (Jacobson 2007b), fungi [e.g., *Aspergillus, Basidolobus, Candida albicans, Coniothyrium, Dermatophyton, Fusarium solani, Geotrichum, Mucor, Mucorales, Paecilomyces, Penicillium, Rusarium, Trichosporon, Trichoderma,* and *Trichophyton* (Rosskopf 1986; López del Castillo 1996; Garner et al. 1997)], algae (Marcus 1971; Garner et al. 1997), and even barnacles and parasites (Jacobson 2007b; Weems 1974). Barnett (2003) attributed dry forms (i.e., in terrestrial turtles) to fungus and wet forms (i.e., in aquatic turtles) to *Pseudomonas* or the coliform bacilli *Citrobacter* and *Klebsiella*. Shell necrosis in fungal infections especially affects the plastron. In many cases, shell infections are often of mixed bacterial and fungal origin (Lafortune et al. 2005; Schumacher 2003).

Alleged fungal agents reportedly involved in shell damage include *Aspergillus, Basidobolus ranarum, Beauvaria bassiana, Candida albicans, Cladosporium, Fusarium, Geotrichum, Microsporum, Mucor, Paecilomyces, Penicillium,* and *Rhodotolura* (Lafortune et al. 2005; Schumacher 2003). Turtles are more susceptible to these agents when subjected to suboptimal environmental conditions of temperature or humidity, compromised hygiene, or chronic stressors such as overcrowding.

Fifty percent of loggerhead turtles, *Carreta carreta*, and 25% of Kemp's Ridley turtles, *Lepidochelys kempii*, had dermatomycotic scutes, sometimes underlain by ulcerating bone (Duguy et al. 1998; Paré and Jacobson 2007). Irregular lytic carapace lesions were attributed to granulomatous disease (usually used to describe the effects of acid fast or fungal infection) in the eastern long-necked turtle, *Chelodina longicollis* (Gabrisch and Zwart 1992). Although Rosskopf (1986) attributed a hole in the carapace of a Galapagos tortoise, *Geochelone nigra galapagoensis*, to maggots, it is more likely that the presence of maggots was a secondary phenomenon.

The situation is more complex and mixed infections may also be responsible. Rosskopf (1986) suggested that *Beneckea chitinovora*-induced shell disease in fresh water turtles requires intermediary crustacean hosts.

Interestingly, pathology and taxonomy have crossed paths in the form of the species name applied to a pond turtle from Europe, North Africa, and Western Asia. Elkan (1983) identifies Pritchard (1967) as attributing the name *Mauremys leprosa* to shell flaking related to a fungal disease. They thought it mimicked leprosy.

The Paleontological Record

Evidence of genitic malformations and anomalies is rare in the fossil record. Gaffney and Tong (2008) noted lack of frontal bones (and septum orbitotemporale) in a specimen (AMNH 30569) of †*Ummulisani rutgersensis*, a bothremydid side-necked turtle from the Eocene of Morocco. Tasnádi-Kubacska (1962) reported a shortened fibula and tibia in a Cretaceous †*Archelon ischyros*, and Weems (1974) described a specimen (USNM 24872) of †*Syllomus crispatus aegyptiacus* with a bony ridge located between the trochanter and the head of the femur. Zangerl and Turnbull (1955) reported a supernumerary peripheral carapace element in a Miocene †*Procolpochelys grandaeva*.

Trauma has also rarely been documented. Wieland (1909) reported an †*Archelon ischyros* with obliquely bitten, healed tibia and fibula. Hutchison and Frye (1989) reported linear excavations (probable scratch marks, perhaps from failed predation attempts) on shells of †*Baptemys garmanii* and †*Chisternon*. A "shark tooth [from a species related to *Lamna*] was found with the type" of †*Dermochelys* (Wieland 1909, p. 120). Erickson (1984, p. 4) reported puncture wounds associated with thickened bone on the posterior and posterodorsal carapace surfaces of †*Protochelydra*, a trionychid, and "at least two other forms." Shallow, well-rounded pits were noted on outer shell surface of †*Syllomus aegyptiacus* (Weems 1974). Hutchison and Frye (1989) noted circular to irregular shaped pits with well-defined margins and conical perforations, which they attributed to predation, on the shells of †*Echmatemys*, †*Baptemys*, and †*Hadrianus corsoni* from the Eocene of Wyoming, USA. Those authors divided the pits into circular to ovoid with flat bottoms, versus rounded bottoms, irregular pits with discrete margins, track (linear) and rot, represented by large areas of irregular depression or dead lamellar bone. †*Echmatemys euthenta* had more than 40 punctuate, pitting lesions measuring 1–3 mm in diameter. Larger (9 mm diameter) shallow pits and full-thickness partially remodeled pits in a specimen of †*Echmatemys* (UCMP 128283) were attributed to a crocodilian bite. Weems (1974, p. 299) noted that circular depressions on the right third and fourth costals of †*Procolpochelys grandaeva* (USNM 24889) were similar to those seen in †*Balanus concavus*. As already mentioned, Hutchison and Frye (1989) also suggested the possibility that high frequency of pitting in the northern painted turtle, *Chrysemys picta*,

Table 27.9 Known occurrences and prevalence of avascular necrosis in fossil and extant turtle families (data from Rothschild 1991)

Family	Number of specimens evaluated	Avascular necrosis	
		Number	Proportion
Baenidae	14	1	0.07
Protostegidae	6	4	0.67
Toxochelyidae	17	5	0.29
Cheloniidae	170	12	0.07
Desmatochelyidae	6	2	0.33
Dermochelyidae	74	4	0.05
Pleurosternidae	4	1	0.25
Trionychidae	204	2	0.001

females might be related to calcium/phosphate withdrawal for egg laying.

Shell rot has been reported in Eocene (Bridgerian) trionychids, in 10 out of 17 emydids, and in one out of seven †*Dermatemys* (Hutchison and Frye 1989; Williams and Bartels 1994). For 12 specimens of †*Echmatemys*, all five †*E. septaria* had extensive rot, compared with three of seven †*E. wyomingensis.* Emydidae specimen UCMP 128418 had near-half-thickness osteolysis crossing sutures and associated with sclerotic bone. Hutchison and Frye (1989, 2001) interpreted the changes as osteomyelitis.

Amorphous carapace lumps were reported on the carapace of USNM 24876 (Weems 1974). Weems (1974, p. 292) also reported a specimen (USNM 24872) of †*Syllomus aegyptiacus* with a "tumorous bony growth." The senior author recognized it as an osteoblastoma (Rothschild, unpublished observation).

Presence of avascular necrosis (= blocked blood vessel-related connective tissue death from bends or decompression syndrome) allowed recognition of diving behavior in Cretaceous mosasaurs and their prey, turtles (Martin and Rothschild 1989). Avascular necrosis was limited to marine turtles, with no cases in terrestrial and one instance in a fossil Mediterranean †*Trionyx* that lived in a marine habitat (Rothschild 1988). Avascular necrosis has been documented in eight families of aquatic turtles, from the Cretaceous to Recent (Rothschild 1987, 1988, 1991, Table 27.9). Desmatochelyidae, Toxochelyidae, Protostegidae, and Pleurosternidae were especially afflicted in the Cretaceous. The documented frequency of avascular necrosis diminished from 41% in the Cretaceous to 9% in the Eocene, 5% in the Oligocene, and 0.3% in the Holocene (Rothschild 1991). Reduction in the frequency of avascular necrosis in the early Eocene was followed by near disappearance after the Oligocene. Among extant turtles, avascular necrosis is limited to infrequent occurrences in the marine Cheloniidae [Ridley's sea turtle, *Lepidochelys olivacea* (Eckert et al. 1986, 1989)] and in the freshwater Chelydridae (snapping turtles, *Chelydra* and *Macroclemys*) and Kinosternidae (mud turtle, *Kinosternon*) (Rothschild 1987).

Conclusions

Reports of pathologies in turtles generally involve descriptions of isolated occurrence of abnormalities, with only rare epidemiologic studies. An exception in fossil turtles has been avascular necrosis and its reduction in frequency through the Cenozoic.

Anomalies may provide insight to environmental conditions, as has been noted in amphibians. Anomalies vary from as simple as variation in the size and shape of the skull, to more complex conditions, the most commonly reported of which include absence of frontal bones, mouth and jaws, protruding or shortened lower jaws, cyclopia, cleft palates, head (dicephaly) and head and neck (derodymous) duplications, and supernumerary limbs. The most complex of the congenital anomalies in turtles is perhaps the Siamese, conjoined or parasitic twin.

It is difficult at times to determine whether an anomaly observed in the fetus or neonate is the result of genetic alterations or of exposure to adverse environmental conditions. Partial drying and temperature variation (e.g., several degrees below the optimum) during gestation produces major deformities of carapaces and plastrons, distorted limb, eyeless or jawless heads, maxillofacial clefts, forelimb and partial hind limb agenesis, coccygeal hypoplasia, and shell plate duplication. Organochlorine compounds (e.g., PCBs) also induce missing claws and eyes, deformed carapaces, tails, limbs, and crania.

Fractures imply behavior, such as falls or mating injuries. Bites and other forms of trauma appear to be the most common cause of bone infections. Gout, infectious arthritis, calcium pyrophosphate and hydroxyapatite crystal disease, and osteoarthritis have been noted only as isolated occurrences, as have calcium and uric acid bladder stones.

Metabolic disease is essentially a phenomenon of captivity, and is typically referred to as nutritional osteodystrophy or metabolic bone disease. This includes vitamin D deficiency (osteomalacia/rickets) and hyperparathyroidism (usually secondary). Uric acid may be deposited in the kidneys or bladder (in the form of stones) and in joints (referred to as articular gout), in contrast to deposits around internal organs (referred to as visceral gout). Another form of crystalline arthritis is one related to calcium pyrophosphate or hydroxyapatite. These are rare causes of arthritis in turtles. One other form of arthritis has been recognized and so far has been limited to one turtle species, *Caretta caretta.* It predominantly affects vertebrae, producing reactive bone bridges characteristic of the form of inflammatory arthritis referred to as spondyloarthropathy.

Neoplasms are rare in chelonians. In veterinary case studies, neoplasms have been recognized in 2.7–3.2% of turtles and 1.4% of tortoises. Nevertheless bone involvement is only rarely reported, typically recognized because of pathologic fracture, which is also a complication of metabolic bone disease. Shell neoplasia are extremely rare, and those few cases are predominantly benign, not malignant disease.

The most contentious aspect of pathology in turtles, anomalies of the carapace and plastron, may not even be pathologies, but may represent normal variation or disease, or even an attempt at speciation. Shell pitting is perhaps the least understood pathologic phenomenon. Shell disease in contemporary turtles has been a source of confusion, partially fueled by a discordant literature. Bites, parasites, mixed bacterial and fungal infections, and even algae have been invoked as causative agents. Although there are a few reports of such alterations associated with isolation or identification of apparent pathologic organisms, it is unclear which agents may be responsible for which types of shell lesion. Actually, there is not even a standardized vocabulary for their description.

We present some of the questions, the solutions of which should allow progress in identification of shell pathologies. It is suggested that a library of casts of the pathology be assembled, such that the macroscopic appearance can be characterized with associated etiologies (as derived from isolating the causative agent in each case) and to assess which might have specificity. As this technique has worked so well for diseases affecting bone, it is suggested that the same approach should provide insight to shell disease. Examination of the frequency of the various lesions under different environmental conditions could also provide guidance as to potential agents pertinent for investigation. It is only with careful examination and an actualistic approach that sense can be made of this information.

Acknowledgments The senior author thanks Zhonghe Zhou (IVPP), Gregory Schneider (UMMZ), Michael Brett-Surman, Pete Kroehler, and Charyl Ito (all USNM), Laura Abraczinskas (MSU), Gene Gaffney and Christopher J. Raxworthy (AMNH), and Kenneth Krysko and Max Nickerson (FMNH) for access to specimens in their care. We thank Andrew Farke (Alf Museum of Paleontology, Claremont, California, USA), Takuya Konishi (Royal Tyrrell Museum of Palaeontology, Drumheller, Alberta, Canada), and an anonymous reviewer for their reviews and constructive suggestions.

References

Abou-Madi, N., Serivani, P. V., Kollias, G. V., & Hernandez-Divers, S. M. (2004). Diagnosis of skeletal injuries in chelonians using computed tomography. *Journal of Zoo and Wildlife Medicine, 35,* 226–231.

Adelmann, H. B. (1936a). The problem of cyclopia. *Quarterly Review of Biology, 11*(2), 161–182.

Adelmann, H. B. (1936b). The problem of cyclopia. *Quarterly Review of Biology, 11*(3) 284–304.

Allen, M. E., & Oftedal, O. T. (1994). The nutrition of carnivorous reptiles. In J. B. Murphy, K. Adler & J. T. Collins (Eds.), *Captive Management and Conservation of Amphibians and Reptiles* (pp. 71–82). St. Louis: Society for the Study of Amphibians and Reptiles.

Anonymous (1888a, October 20). A turtle with two heads. *Milwaukee Daily Journal,* 1.

Anonymous (1888b, June 6). Two heads that do not agree. *The North American (Philadelphia), 1888,* 1.

Anonymous (1889, June 10). A queer pet. *Milwaukee Daily Journal,* 1.

Anonymous (1895). A turtle with two heads. *The Daily Picayune—New Orleans, 21 May 1895,* 8.

Anonymous (1897, February 7). Two-headed tortoise. *The Daily News (Denver),* 21.

Anonymous (1931, September 28). Two headed turtle. *Time Magazine,* 2.

Anonymous (1967). A case of 'Siamese' twins in the turtle (*Pleudemys scripta elegans*). *Texas Journal of Science, 19,* 232.

Anonymous (1975, November 2). Turtle with a twist. *The Daily Review (Hayward, CA),* 9.

Anonymous (1995, December 4). Snapping turtles thrive in Chesapeake's salt water. *The Washington Times,* C2.

Anonymous (1999a, July 28). Stuff—a piece of the action. *Providence Journal-Bulletin (Rhode Island),* 1.

Anonymous (1999b, 25 July). Asides. *Pittsburgh Post-Gazette (Pennsylvania), Editorial,* 1.

Anonymous (1999c, July 24). Two heads are better than one. *Telegraph Herald (Dubuque, IA),* 1.

Anonymous (2000). A rare find. *Richmond Times Dispatch (Virginia), 14,* 1.

Anonymous (2001, October 20). Strange snapper; two-headed hatchling gets chance at long life with Mr. Turtle. *Grand Rapid Press (Michigan),* A3.

Anonymous (2002, September 13, Friday). Kicker: Seeing twice as well. *Greenwire, KICKER, 10*(9), 1.

Anonymous (2003a, February 6). Life! What's On. *The Straits Times (Singapore), 2003,* 1.

Anonymous (2003b, June 3). Senseless crime in Wisconsin. *Saint Paul Pioneer Press (Minnesota),* E2.

Anonymous (2003c, June 12). Don't lump prairie dogs in with pets or pet peeves. *Pittsburgh Post-Gazette (Pennsylvania),* Lifestyle, 1.

Anonymous (2004, May 16). Carlsband Caverns National Park: Awed, one and all. *Los Angeles Times,* Travel 2.

Anonymous (2005, January 27). SOUTH. *South China Daily,* 1.

Anonymous (2006a, October 27). Tiny two-headed terrapin faces a fight for survival. *Post and Courier,* B8.

Anonymous (2006b, December 20). Two-headed beast from 100 m BC. *London Times,* 31.

Anonymous (2007a, September 28). Turtle with 2 heads. *Youngstown Vindicator,* A4. (Based on Associated Press: Two-headed turtle goes on display in PA).

Anonymous (2007b, October 1). Turtle makes do double take: Pet store's oddity may live 20 years. *Grand Rapid Press,* A4.

Anonymous (2007c, December 21). Turtle racing to top of pet ownership list. *Dayton Daily News,* D7.

Appleby, E. C., & Siller, W. G. (1960). Some cases of gout in reptiles. *Journal of Pathology and Bacteriology, 80,* 427–430.

Arvy, C., & Fertard, B. (2002). Pathologie des tortues: étude synthétique. *Bulletin de la Société Herpétologique de France, 100,* 1–152.

Austwick, P. K., & Keymer, I. F. (1981). Fungal and actinomyctes. In J. E. Cooper & O. F. Jackson (Eds.), *Diseases of the reptiles* (Vol. 1, pp. 193–234). New York: Academic Press.

Avery, B. (1990). Surgery of reptiles. *American Association of Zoo Veterinarians,* 168–171.

Bailey, S. W. (1987). Treatment of a terrapin for shell rot. *Rephibearary, 113,* 6–7.

Balazs, G. H (1980). *Synopsis of biological data on the green turtle in the Hawaiian Islands.* National Oceanic and Atmospheric Administration Technical Memorandum National Marine Fisheries Service-SouthWest, Fisheries Center-7, United States Department of Commerce.

Bangs, O. (1896). An important addition to the fauna of Massachusetts. *Proceedings of the Boston Society of Natural History, 27,* 159–161.

Barber, P. (1991). Kyphosis with lung tissue expansion. *Journal of Biological Photography, 59,* 29–30.

Barbour, E. H. (1888). A young tortoise, *Chrysemys picta,* with two heads. *American Journal of Science, 3*(36), 227–230.

Barbour, E. H. (1896a). (*Chrysemis picta*) bicèphale. *Revue Scientifique, 6*(4, 9), 281.

Barbour, E. H. (1896b). A two-headed tortoise, *Chrysemys pictaScience (N.S.), 4,* 159–160.

Barbour, E. H. (1896c). Eine zweiköpfige Schildkröte (*Chrysemys picta*). *Zoologischer Garten 30—Jahrgang 2,* 61–62.

Barnett, S. (2003). Shell infections: When there are chinks in the armor. *Newsletter Mid-Atlantic Turtle and tortoise Society: Terrapin Tales (MATTS) October,* 1–8.

Barten, S. L. (1982). Recovery from nutritional shell deformity in a softshell turtle, *Trionyx muticus. Bulletin of the Chicago Herpetological Society, 18,* 42.

Barten, S. L. (1983). Nutritional bone disease in a softshell turtle, *Trionyx muticus,* fed pinky mice. *Bulletin of the Chicago Herpetological Society, 17,* 51–53.

Barten, S. L. (1996). Shell damage. In D. R. Mader (Ed.), *Reptile Medicine and Surgery* (pp. 413–417). Philadelphia: Saunders.

Barten, S. L. (2006). Shell damage. In D. R. Mader (Ed.), *Reptile Medicine and Surgery* (pp. 893–899). Philadelphia: Saunders.

Bateson, W. (1894). *Materials for the Study of Variation Treated with Especial Regard to Discontinuity in the Origin of Species.* London: Macmillan & Co.

Beatty, B., & Rothschild, B. M. (2008) Decompression syndrome and the evolution of deep diving physiology in the Cetacea. *Naturwissenschaften, 95,* 793–801.

Bellairs, A. d'A. (1981). Congenital and developmental diseases. In J. E. Cooper & O. F. Jackson (Eds.), *Diseases of the reptilia,* (Vol. 2, pp. 469–485). New York: Academic Press.

Bellairs, A. d'A. (1983). Partial cyclopia and monorhina in turtles. In A. G. J. Rhodin & K. Miyata (Eds.), *Advances in herpetology and evolutionary biology. Essays in honor of Ernest E. Williams* (pp. 150–158). Cambridge: Museum of Comparative Zoology.

Benest, S. (1994). DOUBLE TAKE. *Courier-Mail 16 June 1994,* 1.

Bennett, R. (1989). Reptilian surgery, part II, management of surgical diseases. *Compendium on Continuing Education for the Practicing Veterinarian, 11*(2), 122–133.

Bennett, R. A. (1991) Neurology. In D. R. Mader (Ed.), *Reptile Medicine and Surgery* (pp.141–148). Philadelphia: Saunders.

Bishop, M. (1908). Heart and anterior arteries in monsters of the dicephalic group; a comparative study of Cosmobia. *American Journal of Anatomy, 8,* 441–472.

Bishop, C., Carey, J., & Brooks, R. (1989). Hatchability and deformities in populations of snapping turtles. *Proceedings of the Workshop on Cause-Effect Linkages. March 28–30, 1989,* 14.

Black, J. H. (1976). Kyphosis in the three-toed box turtle. *Chelonia 2,* 2–3.

Bourdeau, P. (1988a). Pathologie des tortues, 1ère partie: Examen clinique et maladies générales. *Point Vétérinaire, 20,* 761–775.

Bourdeau, P. (1988b). Pathologie des tortues, 2ème partie: Affections cutanées et digestives. *Point Vétérinaire, 20,* 871–884.

Bouwhuis, H. (1972). Eeen aangeboren afwijking bij *Pseudemys scripta elegans. Lacerta, 30,* 143.

Boyer, T. H. (1996a). Turtles, tortoises, and terrapins. In D. R. Mader (Ed.), *Reptile medicine and surgery* (pp. 332–336). Philadelphia: Saunders.

Boyer, T. H. (1996b). Metabolic bone disease. In D. R. Mader (Ed.), *Reptile Medicine and Surgery* (pp. 385-392). Philadelphia: Saunders.

Boyer, T. H. (1996c). Turtles, tortoises and terrapins. In D. R. Mader (Ed.), *Reptile medicine and surgery* (pp. 696–704). Philadelphia: Saunders.

Boylan, S. (2003). Prosthesis to protect traumatized hind limbs in box turtles. *Exotic DVM, 5*(5), 12.

Brady, T. J. (1991). The scene in the nation and the World. *The Philadelphia Inquirer, 12 June 1994,* A03.

Brogard, J. J. M. (1980). Les maladies bacteriennes et virales des reptiles, Étude bibliographique. *Thèse pour le Doctorat Vétérinaire Diplôme d'État.* École Nationale Vétérinaire de Toulouse.

Brogard, J. (1987). *Les Maladies des Reptiles.* Maisons-Alfort: Point Vétérinaire.

Brown, T. (2003). The whole damn hoopla. *Herald Sun (Melbourne, Australia), 8 March 2003,* 24.

Burke, R. L. (1994). *Apalone spinifer* (spiny softshell). Extreme kyphosis. *Herpetological Review, 25,* 23.

Byrd, E. E. (1939). Certain aspects of the anatomy of a two headed turtle. *Journal of the Tennessee Academy of Science, 14,* 102–106.

Cagle, F. R. (1950). The life history of the slider turtle, *Pseudemys scripta troostii* (Holbrook). *Ecological Monographs, 20,* 31–54.

Campbell, H. (1967). Two-headed snapping turtle *Chelydra serpentine. International Turtle and Tortoise Society Journal, 1*(4), 24–29.

Canella, M. F. (1932). Transformismo e teratologia. *Rivista di psicologia: Organo della Societá italiana di psicologia e degli istituti universitari di psicologia sperimentale, 28,* 291–313.

Carpenter, C. C. (1956). Carapace pits in the three-toed box turtle, *Terrapene carolina triunguis* (Chelonia-Emydidae). *Southwest Naturalist, 1,* 83–86.

Carpenter, C. C. (1958). An unusual Ouachita map turtle. *Herpetologica, 14,* 116.

Carr, A. (1952). *Handbook of Turtles.* Ithaca: Comstock Publishing Co.

Carrick, T., & Reddacliffe, G. (1980). Repair of traumatic shell fractures in Australian side-necked turtles. *International Zoo Yearbook, 20,* 241–243.

Carswell, L. P., & Lewis, T. E. (2003). Embryo and hatchling abnormalities in loggerhead sea turtles on St. Vincent Island, Florida. *NOAA Technical Memorandum NMFS-SEFSC, 503, August,* 185–186.

Casimire-Etzioni, A. L., Wellehan, J. F., Embury, J. E., Terrell, S. P., & Raskin, R. A. E. (2004). Synovial fluid from an African spur-thighed tortoise (*Geochelone sulcata*). *Veterinary Clinical Pathology, 33*, 43–46.

Caullery, M. (1931). Remarques sur des cas de bicéphalie. *Bulletin de la Société zoologique de France, 56*, 362–363.

Cederstrom, J. A. (1931). A two-headed turtle. *Journal of Heredity, 22*, 137–138.

Chan, A. R. (1937). The turtles of Illinois. *Illinois Biological Monographs, 16*, 1–218.

Chapple, C. (1999). Snake's alive, and it has 2 heads! Collector shows off 'flukes of nature'. *The Times-Picayune, New Orleans, LA, 30 September 1999*, B-3.

Cherepanov, G. O. (1994). Anomalies of bony carapace in turtles. *Zoologicheskii Zhurnal, 73*(6), 68–78.

Chou, L. M. (1979). Shell deformity in a box turtle. *Herpetological Review, 10*, 56.

Cisneros, M. (1997). Museum becomes ghoulish success. *Press Enterprise (Riverside, CA), 28. February 1997*, 1.

Clement, H. (1967). Two heads better than one? *International Turtle & Tortoise Society Journal, 1*(2), 4–5.

Cloudsley-Thompson, J. L. (1970). On the biology of the desert tortoise *Testudo sulcata* in Sudan. *Journal of Zoology, 160*, 17–33.

Cohen, H. (1986). *Terrepine carolina: Twinning. Herpetological Review, 17*, 25.

Coker, R. E. (1910). Diversity in scutes of chelonia. *Journal of Morphology, 21*, 1–75.

Cooper, J. E., Jackson, O. F., & Harshbarger, J. C. (1983). A neurilemmal sarcoma in a tortoise (*Testudo hermanni*). *Journal of Comparative Pathology, 93*, 541.

Coquelet, J. P. (1983). Contribution à l'étude de la pathologie des tortues marines, Observations personnelles dans un élevage de *Chelonia mydas* (Linné). *Thèse Doctoral Médicine Vétérinaire*, Nantes, France.

Cowie, A. F. (1976). *A manual of the care and treatment of children's and exotic pets*. London: British Small Animal Veterinary Association.

Crane, S. W., Curtis, M., Jacobson, E. R., & Webb, A. (1980). Neutralization bone plating repair of a fractured humerus in an Aldabra tortoise. *Journal of the American Veterinary Medical Association, 177*, 945–948.

Crooks, F. D., & Smith, P. W. (1958). An instance of twinning in the box turtle. *Herpetologica, 14*, 170–171.

Dämmrich, K. (1967). Pathomorphologische Befunde am Panzer der Schildkröten. *IX. Internationales Symposium zu Erkrankungen der Zootiere, Prag*, 271–274.

Dämmrich, K. (1979). Zur Pathogenese der Skeletterkrankungen bei Zootieren. *Verhandlungsbericht des Internationalen Symposiums über die Erkrankungen der Zootiere, 21*, 65–71.

Dämmrich, K. (1985). Bewegungsorgane. Knochen, Gelenke und Skelett. In R. Ippen, P. Zwart & H. D. Schröder (Eds.), *Handbuch der Zootierkrankheiten, Band 1, Reptilien* (pp. 215–240). Berlin: Akademie Verlag.

Deraniyagala, P. E. P. (1939). *The Tetrapod Reptiles of Ceylon, Volume I, Testudinates and Crocodilians*. London: Dulau & Co.

Derickson, S. H. (1927). A twin turtle. *The Scientific Monthly, 25*, 562–565.

Devaux, B. (1992). Conférence à Sanibel. *Tortue, 21*, 7–9.

Dodd, C. K., Jr. (2001). *North American box turtles: A natural history*. Norman: University of Oklahoma Press.

Dodd, C. K., Jr., Franz, R., & Johnson, S. A. (1997). Shell injuries and anomalies in an insular population of Florida box turtles (*Terrapene carolina bauri*). *Herpetological Natural History, 5*, 66–72.

Dollinger, P., Pagan, O., Jermann, T., Baumgartner, R., & Honegger, R. E. (1997). Husbandry and pathology of land tortoises (Testudinidae) in Swiss zoos. *Erkrankungen der Zootiere: Verhandlungsbericht des Internationalen Symposiums über die Erkrankungen der Zoo- und Wildtiere, 38*, 7–16.

Donoghue, S. (2006). Nutrition. In D. R. Mader (Ed.), *Reptile medicine and surgery* (pp. 251–298). Philadelphia: Saunders.

Duda, P. L., & Gupta, V. K. (1977). A kyphotic individual of *Lissemys punctata punctata* (Bonnaterre). *Herpetologica, 33*, 427–433.

Duguy, R., Morinieri, P., & Le Milinaire, C. (*1998*). Factors of mortality of marine turtles in the Bay of Biscay. *Oceanologica Acta, 21*, 383–388.

Dürigen, B. (1897). *Deutschlands Amphibien und Reptilien. Eine Beschreibung und Schilderung sämmtlicher in Deutschland und den angrenzenden Gebieten vorkommenden Lurche und Kriechthiere*. Magdeburg: Creutz'sche Verlagsbuchhandlung.

Dutta, S. K. (1931). Congenital absence of limbs in tortoises of the genera *Trionyx* and *Emyda*. *Allahabad University Studies, 8*, 1–8.

Eckert, S. A., Nellis, D. W., Eckert, K. K., & Kooyman, G. L. (1986). Diving patterns of two leatherback sea turtles (*Dermatochelys coriacea*) during internesting intervals at Sandy Point, St. Croix, U.S. Virgin Islands. *Herpetologica, 42*, 381–388.

Eckert, S. A., Eckert, K. K., Ponganis, P., & Kooyman, G. L. (1989). Diving and foraging behavior of leatherback sea turtles (*Dermatochelys coriacea*). *Canadian Journal of Zoology, 67*, 2834–2840.

Edwards, G. (1751). *A natural history of uncommon birds, and of some other rare and undescribed animals, quadrupedes, reptiles, fishes, insects, & c., Exhibited in Two Hundred and Ten Copper-Plates,...coloured after life, in four parts*. London: Printed for the author.

Eigner, J. B. (1994). Public eye. *The San Diego Union-Tribune, 13 June 1994*, E2.

Elkan, E. (1983). Giant-cell arteritis in a chelonian (*Clemmys leprosa*). *Journal of Comparative Pathology, 93*, 613–618.

Enlow, D. H. (1969). The bones of reptiles. In C. Gans & A. Bellairs (Eds.), *Biology of the reptilia. Morphology A* (Vol. 1, pp. 45–80). New York: Academic Press.

Epure, E., & Pogorevici, N. (1940). Über einen seltenen Fall von Opodymus bei *Emys orbicularis* L. *Zoologischer Anzeiger, 131*, 39–44.

Erickson, B. R. (1984). Chelonivorous habits of the Paleocene crocodile *Leidyosuchus formidabilis*. *Scientific Publications of the Science Museum of Minnesota, New Series, 5*(4), 1–9.

Ernst, C. H. (1971). Observations of the painted turtle, *Chrysemys picta*. *Journal of Herpetology, 5*, 216–220.

Ewert, M. A. (1979). The embryo and its egg, development and natural history. In M. Harless & H. Morlock (Eds.), *Turtles: Perspectives and research* (pp. 333–413). New York: John Wiley & Sons.

Feldman, M. L. (1983). Life history notes, Testudines, *Chelydra serpentina* (common snapping turtle), morphology. *Herpetological Review, 14*, 47–48.

Ferreira, J. B. (1923). Trabalhos de Erpetologia do Museu Bocage I, Emydosáurios da Colecção Antiga, provenientes da exploração do Dr. Alexandre Rodrigues Ferreira (1783–1793). *Jornal de Academia de Ciências de Lisboa, Ciências Mathematicas, Physicas e Naturaes, Terceira Série, 14*, 77–89.

Feuer, A. (2005). Pickled piglets and other curiosities, in exile. *The New York Times, 4 June 2005*, B1.

Fife, J. D. (2007). Star tortoise—the natural history, captive care and breeding of *Geochelone elegans* and *Geochelone platynota*. *Turtles of the world, Testudinidae, Number 10*. Ada: Living Art Publishing.

Figureres, J. M. 1997. Treatment of articular gout in a Mediterranean pond turtle, *Mauremys leprosa*. *Proceedings of the Association of Reptilian and Amphibian Veterinary, 7*(4), 5–7.

Finkler, M. S., & Claussen, D. L. (1997). Use of the tail in terrestrial locomotor activities of juvenile *Chelydra serpentina*. *Copeia, 1997*, 884–887.

Fitzgerald, K. T, & Vera, R. (2006). Spinal osteopathy. In D. R. Mader (Ed.), *Reptile medicine and surgery* (pp. 906–912). Philadelphia: Saunders.

Fojtl, J. (1989). Oligodaktýlie u želvy. *Akvárium Terárium, 32*(4), 32–33.

Fox, H. (1941). Kyphoscoliosis in a painted tortoise. *Report of the Penrose Research Laboratory, 1941*, 16.

Frank, W. (1976). Amphibien-Reptilien. In H.-G. Klös & E. M. Lang (Eds.), *Zootierkrankheiten. Krankheiten von Wildtieren im Zoo, Wildpark, Zirkus und in Privathand sowie ihre Therapie* (pp. 290–305). Berlin & Hamburg: Paul Parey.

Frye, F. L. (1972). Surgical removal of a cystic calculus from a desert tortoise. *Journal of the American Veterinary Medical Association, 161*, 600–602.

Frye, F. L. (1973). *Husbandry, medicine and surgery in captive reptiles*. Bonner Springs: VM Publishing, Inc.

Frye, F. L. (1981a). *Biomedical and surgical aspects of captive reptile husbandry* (Vol. 1). Malabar: Krieger Publishing Company.

Frye, F. L. (1981b). Chapter 11, Traumatic and physical diseases. In J. E. Cooper & O. F. Jackson (Eds.), *Diseases of the reptilia* (pp. 387–407). London: Academic Press.

Frye, F. L. (1984). Nutritional disorders in reptiles. In G. L. Hoff, F. L. Frye & E. R. Jacobson (Eds.), *Diseases of amphibians and reptiles* (pp. 633–660). New York: Plenum Press.

Frye, F. L. (1989). Temperature related epigenetic developmental anomalies in the tortoise, *Xenobates agassizi. First World Congress of Herpetology, Abstracts*, S4. University of Kent at Canterbury, United Kingdom.

Frye, F. L. (1991a). Developmental abnormalities. In F. L. Frye (Ed.), *Biomedical and surgery aspects of captive reptile husbandry* (2nd enlarged ed., pp. 393–419). Malabar: Krieger Publishing Co.

Frye, F. L. (1991b). Common pathologic lesions & disease processes. In F. L. Frye (Ed.), *Biomedical and surgery aspects of captive reptile husbandry* (2nd enlarged ed., pp. 530–617). Malabar: Krieger Publishing Co.

Frye, F. L. (1991c). Common pathological lesions and disease processes, neoplasia. In F. L. Frye (Ed.), *Reptile care: An atlas of diseases and treatments* (Vol. 2, pp. 576–609). Neptune City: TFH Publishing.

Frye, F. L. (1991d). *Reptile care: An atlas of diseases and treatments* (Vol. 1). Neptune City: TFH Publishing.

Frye, F. L. (1994). Diagnosis and surgical treatment of reptilian neoplasma with a compilation of cases 1966–1993. *In Vivo, 8*, 885–892.

Frye, F. L., & Carney, J. (1974). Achondroplastic dwarfism in a turtle. *Veterinary Medicine, Small Animal Clinician, 9*, 299–301.

Frye, F. L., & Carney, J. D. (1975). Parathyroid adenoma in a tortoise. *Veterinary Medicine, Small Animal Clinician, 70*, 582–584.

Frye, F. L., & Dutra, F. R. (1976). Articular pseudogout in a turtle (*Chrysemys*: *Trachemys*: *p. elegans*). *Veterinary Medicine, Small Animal Clinician, 71*, 655–659.

Gabrisch, K., & Zwart, P. (1992). Tortues. In K. Gabrisch & P. Zwart (Eds.), *La Consultationdes Nouveaux Animaux de Compagnie* (pp. 243–286). Maisons-Alfort: Editions du Point Veterinaire.

Gaffney, E. S., & Tong, H. (2008). Redescription of the skull of *Ummulisani rutgersensis* Gaffney, Tong, and Meylan, 2006, a bothremydid side-necked turtle from the Eocene of Morocco. *American Museum Novitates, 3615*, 1–20.

Galois, P., & Ouellet, M. (2007). Traumatic injuries in eastern spiny softshell turtles (*Apalone spinifera*) due to recreational activities in the northern Lake Champlain basin. *Chelonian Conservation and Biology, 6*, 288–293.

Garner, M. M., Hernandez-Divers, S. M., & Raymond, J. T. (2004). Reptile neoplasia: A retrospective study of case submissions to a specialty diagnostic service. *Veterinary Clinics: Exotic Animal Practice, 7*, 653–671.

Garner, M. M., Herrington, R., Howerth, E. W., Homer, B. L., Nettles, V. F., Isaza, R., et al. (1997). Shell disease in river cooters (*Pseudemys concinna*) and yellow-bellied turtles (*Trachemys scripta*) in a Georgia (USA) lake. *Journal of Wildlife Diseases, 33*, 78–86.

Gerlach, J. (2004). Effects of diet on the systematic utility of the tortoise carapace. *African Journal of Herpetology, 53*, 77–85.

Gillespie, D. (1994). Reptiles. In S. J. Birchard & R. G. Sherding (Eds.), *Saunders manual of small animal practice* (pp. 1390–1411). Philadelphia: Saunders.

Girard, C. (1891–1892). Une tortue (*Crisemys picta*) bicèphale. Note additionelle. *Le naturaliste, 14 Annèe, 130*, 174.

Glazebrook, J. S. (1980). Diseases in farmed sea turtles. In J. S. Glazebrook (Ed.), *Management of turtle resources* (pp. 42–55). North Queensland: James Cook University.

Goh Lilian (2005). Two-headed turtle sets off alarm bells. *South China Morning Post, 12 December 2005*, 5.

Good, H. M. (1987). Shell anomalies in the desert tortoises (*Gopherus agassizii*) populations of the Beaver Dam Slope, Utah, and Desert Tortoise Natural Area, California. *Desert Tortoise Council, Proceedings of 1984 Symposium*, 95–104.

Goodwin, L. G. (1976). Nuffield Institute of comparative medicine. *Journal of Zoology, 178*, 529–542.

Grant, C. (1936). An extraordinary tortoise shell. *Copeia, 1936*, 231–232.

Grant, C. (1937). Orthogenic variation. *Proceedings of Indiana Academy of Science, 46*, 240–245.

Greer, L. L., Strandberg, J. D., & Whitaker, B. R. (2003). Mycobacterium chelonae osteoarthritis in a Kemp's Ridley sea turtle (*Lepidochelys kempii*). *Journal of Wildlife Diseases, 39*, 736–741.

Gressitt, J. L. (1936). Camel-back turtle. *Zoologica, 39*, 246.

Gressitt, J. L. (1937). Soft-shelled turtle monstrosities from Hainan Island. *Peking Natural History Bulletin, 11*, 413–415.

Grünberg, W. (1963/64). Harnsteine bei Tieren, 1, Mitteilung: Vorkommen, Untersuchungsmaterial und angewandte Methoden. *Zentralblatt für allgemeine Pathologie und pathologische Anatomie, 105*, 256–271.

Grünberg, W., Kovaciny Jelinek, C. L., & Preisinger, A. (1977). Harn und Harnsteine bei Tier und Mensch. *Fortschritte der Urologie und Nephrologie, 9*, 125–138.

Guirley, R. (2003). *Keeping and breeding freshwater turtles*. Ada: Living Art.

Gumbel, A. (1999). American Times: Hollywood, California: The doyenne of death heads for Tinseltown. *The Independent (London), 8 December 1999*, 15.

Häfeli, W., & Zwart, P. (2000). Panzerweiche bei jungen Landschildkröten und deren mögliche Ursachen. *Der Praktische Tierarzt, 81*, 129–132.

Haft, J. (1994). Bemerkungen zu den Suppenschildkröten bei Xcacel, Halbinsel Yucatan, Mexiko. *Salamandra, 30*, 254–259.

Hall, L. (2003). The magic of television comes south. *Richmond Times Dispatch (Virginia), 23 April 2003*, 1.

Hamilton, G. (1995). '94 wasn't without its oddities in Citrus. *St. Petersburg Times (Florida), 2 January 1995*, 1.

Hammerton, A. E. (1934). Report on deaths occurring in the Society's Garden for the year 1933. *Proceedings of the Zoological Society London, 104*, 389–422.

Hammerton, A. E. (1935). Report on deaths occurring in the Society's Garden for the year 1934. *Proceedings of the Zoological Society London, 105*, 443–474.

Hammerton, A. E. (1939). Report on deaths occurring in the Society's Garden for the year 1938. *Proceedings of the Zoological Society London, 109B*, 281–327.

Harms, C. A., Lewbart, G. A., & Beasley, J. 2002. Medical management of mixed nocardial and unidentified fungal osteomyelitis in a Kemp's Ridley sea turtle, *Lepidochelys kempii. Journal of Herpetological Medicine and Surgery, 12*(3), 21–26.

Hasel, S. (1992a). 2-headed, 2 spined turtle's tale. *St. Petersburg Times (Florida), 6 October 1992*, 1.

Hasel, S. (1992b). The tale of a tiny turtle with 2 heads, 2 spines. *St. Petersburg Times (Florida), 7 October 1992*, 2.

Hauser, B., Mettler, F., & Honegger, R. E. (1977). Knochenstoffwechselstörungen bei Seychellen-Riesenschildkröten (*Testudo* [*Geochelone*] *gigantea*). *Verhandlungsbericht des Internationalen Symposiums über die Erkrankungen der Zootiere, 19*, 121–125.

Hay, W. P. (1904). A revision of *Malaclemmys*, a genus of turtles. *Bulletin of the Bureau of Fisheries, 24*, 3–19.

Hayes, F. E., & Beaman, K. R. (1985). Life history notes, *Geochelone elephantopus vanderburghi* (Galapagos giant tortoise: Volcan Alcedo race), morphology. *Herpetological Review, 16*, 81–82.

Highfield, A. C. (1990a). Notes on skeletal and carapace deformity in captive-bred tortoises, (genus *Testudo*) related to diet with observations on the use of vitamins and mineral supplements. *Herptile, 15*, 110–113.

Highfield, A. C. (1990b). *Keeping and breeding tortoises in captivity*. London: R & A Publishing Ltd.

Hildebrand, S. F. (1930). Duplicity and other abnormalities in diamond-back terrapins. *Journal of the Elisha Mitchell Science Society, 46*, 41–53.

Hildebrand, S. F. (1938). Twinning in turtles. *Journal of Heredity, 29*, 243–253.

Holahan, J. (2003). 50 years of spiders, snakes and stars, in 1953 the North Museum opened its doors for generations of kids of all ages, it's revealed the wonders of nature. *Lancaster New Era (Pennsylvania), 9 October 2003*, 1.

Holman, J. A. (1984). Life history notes, Testudines. *Terrapene carolina* (eastern box turtle). Morphology and behavior. *Herpetological Review, 15*, 114.

Homer, B. L., Berry, K. H., Brown, M. B., Ellis, G., & Jacobson, E.R. (1998). Pathology of diseases in wild desert tortoises from California. *Journal of Wildlife Disease, 34*, 508–523.

Hundsacker, D., II. (1968). Twinning in tortoises. *International Turtle and Tortoise Society Journal, 6*, 38.

Hunt, T. J. (1957). Notes on diseases and mortality in testudines. *Herpetologica, 13*, 19–23.

Hutchinson J., & Simmonds, M. (1991). A review of the effects of pollution on marine turtles. *Greenpeace Ecotoxicology Project 1*, 1–28.

Hutchison, J. H., & Frye, F. R. (1989). Pathologies of the shell in Eocene turtles. *Journal of Vertebrate Paleontology, 9 (Supplement to 3)*, 26A.

Hutchison, J. H., & Frye, F. L. (2001). Evidence of pathology in early Cenozoic turtles. *PaleoBios, 21*(3), 12–19.

Ippen, R. (1965). Über Sektionsbefunde bei Reptilien. *Zentralblatt für allgemeine Pathologie, 107*, 520–529.

Ippen, R. (1982). Ein Beitrag zu den angeborenen Mißbildungen der Reptilien. *Verhandlungsbericht des Internationalen Symposiums über die Erkrankungen der Zootiere, 24*, 463–473.

Ippen, R. (1985). Entwicklungsbedingte Anomalien. In R. Ippen, P. Zwart & H. D. Schröder (Eds.), *Handbuch der Zootierkrankheiten, Band 1, Reptilien* (pp. 302–316). Berlin: Akademie Verlag.

Isaza, E., & Jacobson, E. R. (1995). Non-nutritional bone diseases in reptiles. *Current Veterinary Therapy, 12*, 1357–1361.

Isenbügel. E., & Frank, W. (1985). *Heimtierkrankheiten*. Stuttgart: Verlag Eugen Ulmer.

Jackson, O. F. (1980). The sick chelonian. *Proceedings of the European Herpetological Symposium, Oxford* (pp. 1–4). Buford: Cotswold Wild Life Parks and Gardens.

Jackson, O. F. (1987). Carapace and other bone injuries in chelonians. *Testudo, 2*(5), 18–21.

Jackson, O. F., & Cooper, J.E. (1981). Nutritional diseases. In J. E. Cooper & O. F. Jackson (Eds.), *Diseases of the reptilia* (Vol. 2, pp. 409–428). London: Academic Press.

Jackson, O. F., & Duff Fasal, M. (1981). Radiology in tortoises, terrapins and turtles as an aid to diagnosis. *Journal of Small Animal Practice, 22*, 707–716.

Jackson, O. F, & Sainsbury, A. W. (1992). Radiological and related investigations. In P. Beynon, M. P. Lawton & J. E. Cooper (Eds.), *Manual of reptiles* (pp. 63–72). Cheltenham: British Small Animal Veterinary Association.

Jacobson, E. R. (1980). Mycotic diseases of reptiles. In R. J. Montali & G. Migaki (Eds.), *The comparative pathology of zoo animals* (pp. 283–290). Washington: Smithsonian Institution Press.

Jacobson, E. R. (1994). Causes of mortality and diseases in tortoises: A review. *Journal of Zoo and Wildlife Medicine, 25*, 2–17.

Jacobson, E. R. (2007a). Bacterial diseases of reptiles. In E. R. Jacobson (Ed.),. *Infectious diseases and pathology of reptiles: Color atlas and text* (pp. 461–526). Boca Raton: CRC Press.

Jacobson, E. R. (2007b). Parasites and parasitic diseases of reptiles. In E. R. Jacobson (Ed.), *Infectious diseases and pathology of reptiles: Color atlas and text* (pp. 571–605). Boca Raton: CRC Press.

Jacobson, E. R., Wronski, T. J., Schumacher, J., Reggiardo, C., & Berry, K. H. (1994). Cutaneous dyskeratosis in free-ranging desert tortoises, *Gopherus agassizii*, in the Colorado Desert of southern California. *Journal of Zoo and Wildlife Medicine, 25*, 68–81.

Jarofke, D., & Lange, J. 1993. *Reptilien, Krankheiten und Haltung*. Berlin: Parey.

Kabisch, K. (1989). Carapaxanomalien bei vier Jungtieren von *Testudo graeca ibera* Pallas 1814. *Sauria, 11*, 23–25.

Keymer, I. F. (1978a). Disease of chelonians: (2) necropsy survey of terrapins and turtles. *The Veterinary Record, December 16, 1978*, 548–552.

Keymer, I. F. (1978b). Disease of chelonians: (1) necropsy survey of tortoises. *The Veterinary Record, December 23 and 30, 1978*, 577–582.

Khosatzky, L. I. (1991). Double monsters of reptiles *Gerpetologicheski Issledovaniya, 1*, 164–170, 174 (In Russian).

Kleinberg, E. (1994). House of refuge last of its kind. *Palm Beach Post (Florida), 12 June 1994, TRAVEL*, 1.

Kölle, P., & Hoffmann, R. (2002). Incidence of nephropathies in European tortoises. *Proceedings of the Association of Amphibian and Reptile Veterinarians, Annual Conference 2002*, 33–35.

Kölle, P., Hoffman, R., Wolters, M., & Hesse, A. (2001). Cystic calculi in reptiles. *Proceedings of the Association of Amphibian and Reptile Veterinarians, 2001*, 190–192.

Korschelt, E. (1927). Über geheilte Knochenbrüche bei einigen Wirbeltieren. *Sitzungsberichte der Gesellschaft für Naturwissenschaften in Marburg, 62*, 1–28.

Korschelt, E. (1932). Über Frakturen und Skelettanomalien der Wirbeltiere, Zweiter Teil: Vögel, Reptilien, Amphibien und Fische. *Beiträge zur pathologischen Anatomie und zur allgemeinen Pathologie, 89*, 668–717.

Kritzer, J. (2002). Summer fun. *News & Record (Greensboro, NC), 6 June 2002*, 1.

Kuvano, H. (1902). A two-headed tortoise *Dobutsu gakuzasschi*, 14 (In Japanese).

Lafortune, M., Wellehan, J. F. X., Terrell, S. P., Jacobson, E. J., Heard, D. J., & Kimbrough, J. W. (2005). Shell and systemic hyalohyphomycosis in Fly River turtles, *Carettochelys insculpta*, caused by *Paecilomyces lilacinus*. *Journal of Herpetological Medicine and Surgery, 15*(2), 15–19.

Lee, J. (2005). A sword-swallowing collector closes an odd little museum. *The New York Times, 1 January 2005*, B6.

Long Beach Animal Hospital (2007). *Tortoise bladder stone*. Retrieved March 24, 2011 from http://lbah.com/reptile/cdtbladderstone.htm.

López del Castillo, C. (1996). Panzernekrosen bei Schildkröten. *Reptilia (D), 1*, 64–67.

López del Castillo, C. (1998). Metabolic diseases: Gout and articular pseudogout. *Reptilia, 3*, 13–16.

Lovich, J. E., Gotte, S. W., Ernst, C. H., Harshbarger, J. C., Laemmerzahl, A. F., & Gibbons, J. W. (1996). Prevalence and histopathology of shell disease in turtles from Lake Blackshear, Georgia. *Journal of Wildlife Diseases, 32*, 259–265.

Lynn, W. G., & Ullrich, M. C. (1950). Experimental production of shell abnormalities in turtles. *Copeia, 1950*, 250–262.

MacCulloch, R. D. (1981). Variation in the shell of *Chrysemys picta belli* from southern Saskatchewan. *Journal of Herpetology, 15*, 181–185.

Maddux, S. (1996). Unusual turtles survive in spite of physical challenges; natural oddities on display at fireworks stand. *South Bend Tribune (Indiana), 19 July 2004*, 1.

Mader, D. R. (1990a). Metabolic bone disease in captive reptiles. *Vivarium, 2*(3), 12–14.

Mader, D. R. (1990b). Metabolic bone disease in captive reptiles, part 2. *Vivarium, 2*(4), 12–13.

Mader, D. R. (1996). Gout. In D. R. Mader (Ed.), *Reptile medicine and surgery* (pp. 374–379). Philadelphia: Saunders.

Mader, D. R. (2006a). Perinatology. In D. R. Mader (Ed.), *Reptile medicine and surgery* (pp. 365–375). Philadelphia: Saunders.

Mader, D. R. (2006b). Calculi: Urinary. In D. R. Mader (Ed.), *Reptile medicine and surgery* (pp. 763–771). Philadelphia: Saunders.

Mader, D. R. (2006c). Gout. In D. R. Mader (Ed.), *Reptile medicine and surgery* (pp. 793–800). Philadelphia: Saunders.

Mader, D. R. (2006d). Metabolic bone diseases. In D. R. Mader (Ed.), *Reptile medicine and surgery* (pp. 841–851). Philadelphia: Saunders.

Mader, D. R., Ling, G. V., & Ruby. A. L. (1999). Cystic calculi in the California desert tortoise (*Gopherus agassizii*): Evaluation of 100 cases. *Proceedings of the Association of Amphibian and Reptile Veterinarians Annual Conference 1999*, 81–82.

Mangone, B., & Johnson, J. D. (1998). Surgical removal of a cystic calculi via the inguinal fossa and other techniques applicable to the approach in the desert tortoise, *Gopherus agassizii. Proceedings of the Association of Amphibian and Reptile Veterinarians, Annual Conference 1998*, 87–88.

Marcus, L. C. (1968). Diseases of snakes and turtles. In R. W. Kirk (Ed.), *Current veterinary therapy III, small animal practice* (pp. 435–442). Philadelphia: Saunders.

Marcus, L. C. 1971. Infectious diseases of reptiles. *Journal of the American Veterinary Medical Association, 159*, 1626–1631.

Marcus, L. C. (1980). Bacterial infections in reptiles. In J. B. Murphy & J. T. Collins (Eds.), *Reproductive biology and diseases of captive reptiles* (pp. 211–221). Contributions to Herpetology 1. Oxford: Society for the Study of Amphibians and Reptiles.

Marcus, L. C. (1981). *Veterinary biology and medicine of captive amphibians and reptiles.* Philadelphia: Lea & Febiger.

Martin, D. (1996). Blackboard; turtle turns heads, literally. *The New York Times, 31 March 1996, 4A*, 12.

Martin, L. D., & Rothschild, B. M. (1989). Paleopathology and diving mosasaurs. *American Scientist, 77*, 460–467.

Martínez-Silvestre, A., Soler, J., Solé, R., & Sampere, X. (1998). Polidactilia en *Testudo hermanni* y causas teratogénicas en reptiles. *Boletín de la Asociación Herpetológica Española, 9*, 35–38.

Martins d'Alte, J. A. (1937). Repteis bicefálicos. *Anais da Faculdade de Ciências do Pôrto, 22*, 53–56.

Mason, R. (2000). Gentle winter visitors. *St. Petersburg Times (Florida), 7 February 2000,* 3D.

Matz, G. (1977). Sur une anomalie de la carapace chez *Testudo (Chelonoidis) chilensis* Gray (Reptilia, Testudines). *Bulletin de la Société Zoologique de France, 102,* 497–500.

Mauldin, G. N., & Done, L. B. (2006). Oncology. In D. R. Mader (Ed.), *Reptile medicine and surgery* (pp. 299–322). Philadelphia: Saunders.

Mautino, M., & Page, D. (1993). Biology and medicine of turtles and tortoises. *Veterinary Clinics North America Small Animal Practice, 23*, 1251–1270.

McArthur, S. (1996). *Veterinary management of tortoises and turtles.* Cambridge: Blackwell Science.

McArthur, S. (2004). Interpretation of presenting signs. In S. McArthur, R. Wilkinson & J. Meyer (Eds.), *Medicine and surgery of tortoises and turtles* (pp. 273–307). Oxford: Blackwell.

McArthur, S., & Barrows, M. (2004). Nutrition. In S. McArthur, R. Wilkinson & J. Meyer (Eds.), *Medicine and surgery of tortoises and turtles* (pp. 73–85). Oxford: Blackwell.

McCallum, M. L., & Trauth, S. E. (2000). Curly-tail malformity in hatchlings of the alligator snapping turtle *Macroclemys temminckii* (Testudines: Chelydridae), from northeastern Arkansas. *Journal of the Arkansas Academy of Science, 54*, 150–152.

McEwan, B. (1982). Bone anomalies in the shell of *Gopherus polyphemus. Florida Scientist, 45,* 189–195.

McGavin, D., & Zachary, J. F. (2006). *Pathologic basis of veterinary disease* (4th ed.). Amsterdam: Elsevier.

McKown, R. D. (1998). A cystic calculus from a wild western spiny softshell turtle (*Apalone* [*Trionyx*] *spiniferus hartwegi*). *Journal of Zoo and Wildlife Medicine, 29,* 347.

Mebs, D. (1965). Harnsteine bei Schildkröten. *Salamandra, 1,* 47–49.

Meek, R., & Inskeep, R. (1981). Aspects of the field biology of a population of Hermann's tortoise (*Testudo hermanni*) in Southern Yugosalvia. *British Journal of Herpetology, 6,* 159–164.

Meisel, J. (1994). 2-headed turtle lacks direction, otherwise healthy. *Arkansas Democrat- Gazette (Little Rock, AR), 28 June 1994*, 1.

Mertens, R. (1940). Der Knochenpanzer einer kyphotischen Weichschildkröte. *Senkenbergiana, 22,* 236–243.

Messonnier, S. P. (1996). *Common reptile diseases and treatment.* Oxford: Blackwell Science.

Miller, J. (2002). Snake eyes. *News and Observer (Raleigh, NC), 22 September 2002,* 1.

Mitchell, J. C. (1994). *The reptiles of Virginia.* Washington: Smithsonian Institution Press.

Montgomery, D. (2003). Strange attraction; as sideshows vanish from the midway, a film recalls their glory days. *The Washington Post, 24 October 2003*, C01.

Moskowitz, R. W., Howell, D. S., Goldberg, V. M, & Mankin, H. J. (1984). *Osteoarthritis: Diagnosis and management*. Philadelphia: W.B. Saunders Co.

Nakamura, K. (1938). Studies on some specimens of double monsters of snakes and tortoises. *Memoirs of the College of Science, Kyoto Imperial University, series B, 14,* 171–192.

Natarajan, L. C., Mellot, A. L., Rothschild, B. M., & Martin, L. D. (2007). Bone cancer rates in dinosaurs compared with modern vertebrates. *Transactions of the Kansas Academy of Science, 110*, 155–158.

Necker, W. L. (1940). Hump-backed turtles. *Chicago Naturalist, 3*, 62.

Newman, H. H. (1906). Correlated abnormalites in the scutes and bony plates of chelonia. *Science*, *23*, 526.

Newman, H. H. (1923). *The physiology of twinning*. Chicago: University of Chicago Press.

Nicholson, K. (1999). Years of animal magnetism retiree closes cages of popular pet shop. *The Denver Post, 3 June 1999, Denver and the West,* 1.

Nixon, C. W., & Smith, H. M. (1949). The occurrence of kyphosis in turtles. *Turtox News, 27*, 1–2.

Nunn, C., Rothschild, B. M., & Gittleman, J. (2007). Why are some species more commonly afflicted by arthritis than others? A comparative study of spondyloarthropathy in primates and carnivores. *Journal of Evolutionary Biology, 20*, 460–470.

Ogden, J. A., Rhodin, A. G. J., Conlogue, G. J., & Light, T. R. (1981). Pathobiology of septic arthritis contiguous osteomyelitis in a leatherback turtle (*Dermochelys coriacea*). *Journal of Wildlife Diseases, 17*, 277–287.

O'Malley, B. (2008). Nutritional problems in reptiles. *The 33rd Congress of the World Small Animal Veterinary Association Proceedings (Exotics), Dublin, Ireland, 20-24 August 2008.* Retrieved March 24, 2011 from http://www.vin.com/proceedings/

Proceedings.plx?CID=WSAVA2008&Category=3866&PID=23994&O=Generic.

Orós, J., Torrent, A., & de los Monteros, A. E. (2001). Multicentric lymphoblastic lymphoma in a loggerhead sea turtle (*Caretta caretta*). *Veterinary Pathology, 38*, 464–467.

Osborne, C. A., Albasan, H., Lulich, J. P., Nwaokorie, E., Koehler, L. A., & Ulrich, L. K. (2008). Quantitative analysis of 4468 uroliths retrieved from farm animals, exotic species, and wildlife submitted to the Minnesota Urolith Center: 1981 to 2007. *Veterinary Clinics Small Animal, 39*, 65–78.

Packard, G. C., Tracy, C. R., & Roth, J. J. (1977). The physiological ecology of reptilian eggs and embryos, and the evolution of viviparity within the class Reptilia. *Biological Reviews, 52*, 71–105.

Paré, J. A., & Jacobson, E. R. (2007). Mycotic diseases of reptiles. In E. R. Jacobson (Ed.), *Infectious diseases and pathology of reptiles: Color atlas and text* (pp. 527–570). Boca Raton: CRC Press.

Parker, G. H. (1901). Correlated abnormalities in the scutes and bony plates of the carapace of the sculptured tortoise. *American Naturalist, 35*, 17–24.

Pavalko, P. (1986). Shell and scute anomalies in some midwestern turtles. *Bulletin of the Chicago Herpetological Society, 21*, 36–38.

Perkins, A. (2002). Two-for-the-show turtle beating odds: Jam and I is a star attraction at the Natural Science Center. *News & Record (Greensboro, NC), 22 May 2002*, 1.

Plymale, H. H., Jackson, C. G., Jr., & Collier, G. (1978). Kyphosis in *Chrysemys scripta yaquia* (Testudines: Emydidae) and other turtles. *Southwestern Naturalist, 23*, 457–461.

Porras, L., & Beraducci, J. (1980). Dicephalic *Kinosternon*. *Herpetological Review, 11*, 35.

Pritchard, P. C. (1967). *Living turtles of the world.* Neptune City: TFH Publishing.

Przibram, H. (1909). *Experimental-Zoologie. 2. Band: Regeneration: Eine Zusammenfassung der durch Versuche ermittelten Gesetzmässigkeiten tierischer Wieder-Erzeugung (Nachwachsen, Umformung, Missbildung).* Leipzig und Wien: Franz Deuticke.

Raidal, S. R., Shearer, P. L., & Prince, R. (2006). Chronic shoulder osteoarthritis in a loggerhead turtle (*Caretta caretta*). *Australian Veterinary Journal, 84*, 231–234.

Raiti, P. (2004). Endoscopic-assisted retrieval of a cloacal urolith in an African spurred tortoise (*Geochelona sulcata*). *Proceedings of the Association of Amphibian and Reptile Veterinarians, Annual Conference, 2004*, 145.

Raiti, P., & Haramati, N. (1997). Magnetic resonance imaging and computerized tomography of a gravid leopard tortoise (*Geochelone pardalis pardalis*) with metabolic bone disease. *Journal of Zoo and Wildlife Medicine, 28*, 189–197.

Reaves, D.A., (2004). Turtle turns heads—2 of them, Moncks Corner brothers find anomaly of nature behind grandparents' house. *The Post and Currier, Charleston, SC, 15 May 2004*, 1.

Rebell, G., Roth, F. J., Taplan, D., & Wodinsky, J. (1971). Fusariosis in marine turtles. *Bacteriological Proceedings, 71*, 121.

Reeder, J. (2002). Treasure coast boasts a rich, varied history. *Palm Beach Post (Florida), 27 October 2002, Sunday: SPECIAL*, 1.

Reichenbach-Klinke, H.-H. (1963). *Krankheiten der Reptilien.* Stuttgart: Gustav Fischer Verlag.

Reichenbach-Klinke, H.-H. (1977). *Krankheiten der Reptilien. 2. Auflage*. Stuttgart: Gustav Fischer Verlag.

Reichenbach-Klinke, H. (1977). *The principal diseases of lower vertebrates.* New York: Academic Press.

Resnick, D. (2002). *Disorders of bones and joints.* Philadelphia: Saunders.

Rhodin, A. G. J., Anders, G. J., Russell, A., & Mittermeier, R.A. (1990). *Acanthochelys macrocephala. Catalogue of American Amphibians and Reptiles, 481*, 1–2.

Rhodin, A. G. J., Pritchard, P. C., & Mittermeier, R.A. (1984). The incidence of spinal deformities in marine turtles, with notes on the prevalence of kyphosis in Indonesian *Chelonia mydas*. *British Journal of Herpetology, 6*, 369–373.

Rivera, S., & Lock, B. (2008). The reptilian thyroid and parathyroid glands. *Veterinary Clinics: Exotic Animals, 11*, 163–175.

Rosskopf, W. J. (1986). Shell disease in turtles and tortoises. In R. W. Kirk (Ed.), *Current veterinary therapy IX* (pp. 751–759). Philadelphia: WB Saunders.

Rosskopf, W.J., Jr., Woerpel, R.W., & Yanoff, S. (1982). Severe shell deformity caused by a deficient diet in a California desert tortoise. *Veterinary Medicine, Small Animal Clinician, 77*, 593–594.

Rothschild, B. M. (1987). Decompression syndrome in fossil marine turtles. *Annals of Carnegie Museum, 56*, 253–258.

Rothschild, B. M. (1988). Avascular necrosis in diving vertebrates: The turtle test. *Journal of Vertebrate Paleontology, 8 (Supplement to 3)*, 25A.

Rothschild, B. M. (1991). Stratophenetic analysis of avascular necrosis in turtles: Affirmation of the decompression syndrome hypothesis. *Comparative Biochemistry and Physiology, 100A*, 529–535.

Rothschild, B.M. (2009a). Scientifically rigorous reptile and amphibian osseous pathology: Lessons for forensic herpetology from comparative and paleo-pathology. *Applied Herpetology, 6*, 47–79.

Rothschild, B. M. (2009b). Orthopedic surgery for gout. *eMedicine Orthopedic Surgery*. Retrieved March 24, 2011 from http://emedicine.medscape.com/article/1267449-overview.

Rothschild, B. M. (2009c). Diffuse idiopathic skeletal hyperostosis. *eMedicine Rheumatology*. Retrieved March 24, 2011 from http://emedicine.medscape.com/article/1258514-overview.

Rothschild, B. (2010a). Macroscopic recognition of nontraumatic osseous pathology in the postcranial skeletons of crocodilians and lizards. *Journal of Herpetology, 44*, 13–20.

Rothschild, B. M. (2010b). Lumbar spondylosis. *eMedicine Neurosurgery.* Retrieved March 24, 2011 from http://emedicine.medscape.com/article/249036-overview.

Rothschild, B. M., & Bruno, M. A. (2009). Imaging in calcium pyrophosphate deposition disease. *eMedicine Radiology*. Retrieved March 25, 2011 from http://emedicine.medscape.com/article/388348-overview.

Rothschild, B. M., & Heathcote, G. M. (1995). Characterization of gout in a skeletal population sample: Presumptive diagnosis in Micronesian population. *American Journal of Physical Anthropology, 98*, 519–525.

Rothschild, B. M., & Martin, L. D. (2006). Skeletal impact of disease. *New Mexico Museum of Natural History and Science Bulletin, 33*, 1–226.

Rothschild, B. M., & Panza, R. K. (2005). Epidemiologic assessment of trauma-independent skeletal pathology in non-passerine birds from museum collections. *Avian Pathology, 34*, 212–219.

Rothschild, B. M., & Ruhli, F. (2007). Comparative frequency of osseous macroscopic pathology and first report of gout in captive and wild-caught ratites. *Journal of Veterinary Medicine, Series A, 54*, 265–269.

Ruffer, M. A. (1921). A pathological specimen dating from the lower Miocene Period. In Ruffer, M. A. (Ed.), *Studies in the Paleopathology of Egypt* (pp. 184–193). Chicago: University of Chicago Press.

Sailer, A., Pyczak, C., & Hartman, U.-K. (1997). Siamesische Zwillinge bei *Testudo hermanni boettgeri*. *Herpetofauna (Weinstadt), 19*, 12–13.

Saumure, R. A. (2001). Kyphosis in a musk turtle (*Sternotherus odoratus*) from Ontario, Canada. *Chelonian Conservation and Biology, 4*, 159.

Schildger, B. J., Frank, H., Göbel, T., & Weiss, R. (1991). Mycotic infections of the integument and inner organs in reptiles. *Herpetopathologica, 2*, 81–87.

Schlumberger, H. (1958). Krankheiten der Fische, Amphibien und Reptilien. In P. Cohrs, R. Jaffé & H. Meessen (Eds.), *Pathologie der Laboratoriumstiere* (Vol. 2, pp. 714–761). Berlin: Springer Verlag.

Schmidt, K. P., & Inger, R. F. (1957). *Living Reptiles of the World*. London: Hamisch Hamilton.

Schumacher, J. (2003). Fungal diseases of reptiles. *Veterinary Clinics of North America: Exotic Animal Practice, 6*, 327–35.

Schwarz, C. W., & Schwartz, E. R. (1974). The three-toed box turtle in central Missouri: Its population, home range and movements. *Missouri Department of Conservation, Terrestrial Series, 5*, 1–28.

Scott, P. W. (1992). Nutritional diseases. In P. Beynon, M. P. Lawton & J. E. Cooper (Eds.), *Manual of reptiles* (pp. 138–152). Cheltenham: British Small Animal Veterinary Association.

Sims, J. (1989). Herpetology matters—bicephalism. *Aquarist and Pondkeeper, 54*(7), 69.

Smith, H. M. (1947). Kyphosis and other variations in soft-shelled turtles. *University of Kansas Publications, Museum of Natural History, 1*, 117–124.

Smith, Z. N. (2003). Basra? What about Boston? *Chicago Sun-Times, 11 September 2003*, 46.

Sowerby, J. de C., & Lear, E. (1872). *Tortoises, terrapins and turtles drawn from life by James de Carle Sowerby*. London: Henry Sotheran, Joseph Baer & Co. (Series of unnumbered plates).

Stojanov, A. (2000). Hornpanzeranomalien bei den Landschildkröten (*Testudo graeca ibera* and *Testudo hermani boettgeri*) in Bulgarien. *Historia Naturalis Bulgarica, 11*, 97–105.

Stojanov, A. (2005). New data of abnormalities on the bony and horny shell of tortoises from Bulgaria. *Acta Zoologica Bulgarica, 57*, 25–30.

Stringer, E. M., Garner M. M., Proudfoot, J. S., Ramer, J. S., Bowman, M. R., Heng, H. G., et al. (2009). Phaeohyphomycosis of the carapace in an Aldabra Tortoise (*Geochelone gigantea*). *Journal of Zoo and Wildlife Medicine, 40*, 160–167.

Stuart, J. N. (1996). Additional records of kyphosis in freshwater turtles. *Bulletin of the Chicago Herpetological Society, 31*, 60–61.

Swarts, C. (2003). Rare two-headed tortoise found in South Africa. *National Geographic News, 30 May 2003*, 1.

Sykes, J. M., & Trupkiewicz, J. G. (2006). Reptile neoplasia at the Philadelphia Zoological Garden, 1901–2002. *Journal of Zoo and Wildlife Medicine, 37*, 12–29.

Tasnádi-Kubacska, A. (1962). *Paläopathologie—Pathologie der vorzeitlichen Tiere*. Jena: Gustav Fischer Verlag.

Thomason, S. (1997). Helena's turtle man Mike Gugliotta ships all over the world. *Birmingham News (Alabama), 13 August 1997*, 8E.

Thompson, K. (2007). Bones and joints. In M. G. Maxie (Ed.), *Jubb, Kennedy, and Palmer's pathology of domestic animals* (5th ed., pp. 1–184). Oxford: Elsevier.

Townsend, C. H. (1928). A double turtle. *Bulletin of the New York Zoological Society, 31*, 36–37.

Vella, D. (2007). Management of freshwater turtle shell injuries. *Lab Animal, 38*, 13–14.

Vellard, J., & Penteado, J. (1931). Un cas de bicéphalie chez un serpent (*Liophis almadensis*). *Bulletin de la Société de Zooologie de France, 56*, 360–362.

Vieira de Dandrade, D., & Shinya, A. A. (1993). Natural occurrence of shell abnormalities in hatchling red-footed tortoises (*Geochelone carbonaria*). *Herpetological Review, 24*, 89.

Virchow, R. (1878). Ein grosser Blasen-(Kloaken?)-Stein von einer Meeresschildkröte. *Virchows Archiv, 73*, 629–630.

Vogt, T. (1922). Zur Reptilien und Amphibienfauna Südchinas. *Archiv für Naturgeschichte, 88A*, 135–146.

Walker, T. (1995). Two-headed turtle. *Courier-Mail, 4 December 1995*, 1.

Wallach, J. D. (1969). Medical care of reptiles. *Journal of the American Veterinary Medical Association, 155*, 1017–1047.

Wallach, J. D. (1971). Environmental and nutritional diseases of captive reptiles. *Journal of the American Veterinary Medicine Association, 159*, 1632–1643.

Wallach, J. D. (1975). The pathogenesis and etiology of ulcerative shell disease in turtles. *Journal of Zoo Animal Medicine, 6*, 11–13.

Wallach, J. D. (1977). Ulcerative shell disease in turtles: Identification, prophylaxis and treatment. *International Zoo Yearbook, 17*, 170–171.

Wallach, J. D. (1978). Reptiles. Feeding and nutritional diseases. In M. E. Fowler (Ed.), *Zoo and Wild Animal Medicine* (pp. 123–128). Philadelphia: Saunders.

Wallach, V. (2004). Dicephalism in snakes: Two heads are not always better than one! *Reptiles, 12*(2), 60–69.

Wallis, K. (1927) Zur Knochenhistologie und Kallusbildung beim Reptil (*Clemys leprosa* SCHWEIGG). *Zeitschrift für Zellforschung und mikroskopische Anatomie, 6*, 1–26.

Wandolleck, B. (1904). Eine bucklige *Testudo hermanni* L. *Zoologisches Jahrbuch für Systematik, 20*, 151–166.

Wechsler, A. (2003). Doing a double take over a heads-up find; Poestenkill family adopts turtle with two noggins, and it's getting the attention of wildlife experts. *The Times Union (Albany, NY), 20 May 2003*, 1.

Weems, R. E. (1974). Middle Miocene sea turtles (*Syllomus, Procolpochelys, Psephorus*) from the Calvert Formation. *Journal of Paleontology, 48*, 278–303.

Weichmann, U. (1989). Extreme Höcker auf dem Panzer einer Landschildkröte. *Deutsche Aquarien- und Terrarien-Zeitschrift, 42*(10), 1.

Wenker, C. J., Bart, M., Guscetti, F., Hatt, J.-M., & Isenbugel, E. (1999). Periarticular hydroxyapatite deposition disease in two red-bellied short-necked turtles (*Emydura albertisii*). *Annual Proceedings of the American Association of Zoo Veterinarians, 1999*, 23–26.

Wermuth, H. (1961). Anomalien bei einer griechischen Landschildkröte (*Testudo hermanni hermanni*). *Sitzungsberichte der Gesellschaft der naturforschenden Freunde zu Berlin, 1961*, 139–142.

Werner, W. E., Jr. (1959). Amphibians and reptiles of the Thousand Islands region, New York. *Copeia, 1959*, 170–172.

Whillans, T. H., & Crossman, E. J. (1977). Morphological parameters and spring activities in a central Ontario population of midland painted turtle, *Chrysemys picta marginata* (Agassiz). *Canadian Field-Naturalist, 91*, 47–57.

White, J. B., & Murphy, G. G. (1972). A kyphotic eastern spiny softshelled turtle, *Trionyx spinifer spinifer*. *Journal of the Tennessee Academy of Science, 47*, 61.

White, M. (1989). Tortoises and soft shell illness. *Rephiberary, 138*, 6.

Wieland, G. R. (1909). Revision of the Protostegidae. *American Journal of Science, (4), 27*, 101–130.

Wilcox, C., & Perrin, M. (1996). Junior journalists. *St. Petersburg Times (Florida), 18 April 1996*, 3.

Wilhoft, D. C. (1980). Kyphosis in the snapping turtle *Chelydra serpentine*. *Herpetological Bulletin of the New York Herpetological Society, 15*, 15–26.

Wilkins, J. (1998). Little shop horrors; It's not your usual museum; this one takes a gruesome look at the darkside. *The San Diego Union-Tribune, 12 May 1998*, E-1.

Wilkinson, R., Hernandez-Divers, S., LaFortune, M., Calvert, I., Gumpenberger, M., & McArthur, S. (2004). Diagnostic imaging techniques. In S. McArthur, R. Wilkinson & J. Meyer (Eds.), *Medicine and Surgery of Tortoises and Turtles* (pp. 187–238). Oxford: Blackwell.

Willft, D. C. (1980). Kyphosis in the snapping turtle *Chelydra serpintina*. *Herpetologica, 15*, 15–26.

Williams, K. L. (1957). Yolk retraction as a possible cause of kyphosis in turtles. *Herpetologica, 13*, 236.

Williams, J. S., & Bartels, W. S. (1994). Paleoecological implications of ulcerative shell disease in fossil turtles from the Eocene of Southwestern Wyoming. *Abstracts with Programs, Geological Society of America, 26*(5), 68.

Wright, D. (2005). Serpentarium a hot spot. *Post and Courier, 5 November 2005*, B8.

Wright, R. J., Lay, M., & Balazs, G. H. (1977). An investigation of the green turtle in the feeding pastures of Ka'u, Hawii. *Report*

prepared for the Marine Option Program, University of Hawaii at Hilo, 23 pp.
Wu, S. (1994). (The mortal cause and control method on the shell-rot of *Trionyx sinensis.*) *Freshwater Fisheries, 24,* 35–37. (In Chinese).
Yntema, C. L. (1960). Effects of various temperatures on the embryonic development of *Chelydra serpentina. Anatomical Record, 136,* 305–306.
Yntema, C. L. (1970). Twinning in the common snapping turtle, *Chelydra serpentina. Anatomical Record, 166,* 491–498.
Yntema, C. L. (1971). Incidence and survival of twin embryos of the common snapping turtle, *Chelydra serpentine. Copeia, 1971,* 755–758.
Zangerl, R., & Johnson, R.G. (1957). The nature of shield abnormalities in the turtle shell. *Fieldiana: Geology, 10,* 341–362.
Zangerl, R., & Turnbull, W. D. (1955). *Procolpochelys grandaeva* (Leidy), an early carettine sea turtle. *Fieldiana: Zoology, 37,* 345–382.
Zwart, P., & Harshbarger, J.C. (1991). A contribution to tumors in reptiles. Description of new cases. *Proceedings of the International Colloquium of Pathological Medicine of Reptiles and Amphibians, Sept. 27-29, German Veterinary Association, Bad Nauheim, Germany, 4,* 219–224.
Zwart, P., Goedegebuure, S. A., & Lambrechts, L. (1994). Relationship between intestinal tract and mineralization of skeleton in reptiles, with reference to normal development of the carapace in young tortoises. *International Symposium zu Erkrankungen der Zootiere Kristiansand, 36,* 315–319.
Zwart, P., Lambrechts, L., DeBatist, P., Bijnens, B., Claessen, H., Mennes, S. S., et al. (1997). Excessive growth in a Hermann's tortoise (*Testudo hermanni*) and its consequences for the development of the carapace. *Verhandlungen und Berichte über Erkrankungen von Zootieren, Neue Folge, 2,* 61–64.

Chapter 28
Morphological Variation in the Carapace and Plastron of *Terrapene coahuila* Schmidt and Owens 1944

Robert W. Burroughs, Christopher J. Bell, Travis J. LaDuc, and Dean A. Hendrickson

Abstract *Terrapene coahuila* is one of four extant species of North American box turtles. It is restricted in distribution to the Cuatro Ciénegas Basin in Coahuila, México. Results of previous examinations of extant *T. carolina* and *T. ornata* revealed relatively high levels of morphological variation, but morphological studies of *T. coahuila* are rare, and data on skeletal morphology are limited. We examined 214 skeletal specimens of *T. coahuila* and documented variation in 51 mensurative and discrete characters of the carapace and plastron. Overall levels of variation are low, as predicted by previously documented levels of gene flow between the sub-populations of the species. However, significant polymorphism is present in the positions of the anterior and posterior sulci of the fourth vertebral scute and the configuration of neural bones 2, 3, and 7. Additionally, co-ossification of the carapacial bones varies substantially within the sample, but independently of carapace length. Genetic, epigenetic, and environmental controls for those features are not known. In addition to documenting skeletal morphology within *T. coahuila*, we provide new perspectives on patterns of variation within *Terrapene*, and contribute data that should help paleontologists to establish more rigorous criteria for the identification of fossil specimens of North American box turtles. Those data will be especially important for critical evaluation of recently discovered early and middle Tertiary fossils that are yielding new insights into the evolution of box turtles and the modernization of the turtle biota.

R. W. Burroughs (✉) · C. J. Bell
Jackson School of Geosciences, The University of Texas at Austin, Austin TX 78712, USA
e-mail: utexasrburroughs@gmail.com; cjbell@mail.utexas.edu

T. J. LaDuc · D. A. Hendrickson
Texas Natural Science Center, The University of Texas at Austin, Austin TX 78705, USA
e-mail: travieso@mail.utexas.edu; deanhend@mail.utexas.edu

Keywords Box turtle • Cuatro Ciénegas Basin • Emydidae • Polymorphism • Variation

Introduction

> "Terrapene *is one of the plastic genera, and the examination of a great number of specimens from different localities doubtless will show some interesting results*" (Baur 1891, p. 191).

Although they are among the most familiar and well-studied turtles in North America, Emydidae still present numerous opportunities for exploration of evolutionary morphology. Detailed understanding of patterns of morphological variation within most emydids remains an unachieved goal. The elucidation of basic patterns of morphological variation in extant lineages certainly would facilitate a broader understanding of the value and limitations of morphological data in phylogenetic analyses, and is a necessary step towards the rigorous integration of an extensive but problematic fossil record into a holistic understanding of diversity, systematics, and biogeography.

This was exemplified in recent decades by the emergence of alternative hypotheses of relationship among emydid turtles. Those alternatives stimulated renewed interest in the evolution of plastral kinesis within the group (Gaffney and Meylan 1988; Bickham et al. 1996; Burke et al. 1996; Feldman and Parham 2002; Angielczyk et al. 2011; Wiens et al. 2010). Parallel discoveries of fossilized emydids from Tertiary deposits in the central and western United States broadened the scope of that question by providing previously unexpected insights on the complexity of plastral evolution in some lineages (e.g., Hutchison 1981; Holman 1995, 2002; Bever et al. 2003), as well as the antiquity of hinging structures and the early occurrence of functional 'box turtle' lineages in the Paleocene (e.g., *Planetochelys savoiei*; Weems 1988; Hutchison 1998; Holroyd et al. 2001) and Eocene (Hutchison 1992) of North America. Those more

D. B. Brinkman et al. (eds.), *Morphology and Evolution of Turtles*, Vertebrate Paleobiology and Paleoanthropology,
DOI: 10.1007/978-94-007-4309-0_28, © Springer Science+Business Media Dordrecht 2013

ancient lineages that show or suggest plastral kinesis present an important framework for addressing patterns of skeletal variation expressed within extant functional box turtles and their extinct relatives, best exemplified in North America by *Terrapene*.

Reported fossils of *Terrapene* extend back to the Miocene (e.g., Hay 1908; Holman 1975, 1987; Holman and Corner 1985; Parmley 1992). The taxonomic affinities and systematic relationships of those fossils are clouded by an historical tendency to rely upon similarity criteria in the identification of fossil remains, and by incomplete investigations of evolutionary apomorphies that may be present in isolated skeletal elements such as those frequently encountered in the fossil record (Bell et al. 2010). The latter problem is exacerbated by a general paucity of knowledge of the relative importance of sources of morphological variation within extant *Terrapene*, and of the distribution of apomorphic morphology among the constituent lineages of *Terrapene*. The relatively limited taxonomic diversity within *Terrapene* makes this a tractable problem, but adequate samples of skeletal material are rare even for the more common species.

Terrapene carolina, *T. coahuila*, *T. nelsoni*, and *T. ornata* are extant. The extinct *T. longinsulae* was named by Hay (1908), but its taxonomic status needs to be evaluated (Dodd 2001). *T. putnami* also was named by Hay (1906) and is still recognized by some as a valid extinct subspecies of *T. carolina*, but purportedly diagnostic features appear to be restricted to large size and shell thickness (see Dodd 2001), and its validity has long been debated (Blaney 1971; Moodie and Van Devender 1977; Bentley and Knight 1998). The recently described *T. corneri* preserves an interesting distribution of morphological features, some of which suggest affinity with *T. coahuila*, others of which are similar to *T. nelsoni* and *T. ornata* (Holman and Fritz 2005).

Carapacial and plastral morphology of *Terrapene carolina* and *T. ornata* were studied previously in some detail, in part by paleontologists seeking to clarify early taxonomic assignations of fossils (e.g., Barbour and Stetson 1931; Milstead 1956, 1967, 1969; Auffenberg 1958). The results of those efforts indicated that the extant populations of those species often show high levels of morphological variation, echoing the early suggestion by Baur (1891) that large series of specimens would reveal extensive morphological plasticity (quoted in our epigraph). It is clear that fragmentary fossils may be difficult to assign with certainty, and that the taxonomic status of even fairly complete specimens could be controversial (e.g., *T. c. putnami*; Blaney 1971; Moodie and Van Devender 1977; Bentley and Knight 1998).

Terrapene nelsoni and *T. coahuila* are poorly studied in terms of skeletal morphology and patterns of variation. They have no documented fossil record, although some fossils from Miocene sediments in Nebraska (now named *T. corneri*) previously were noted to be at least superficially similar to *T. coahuila* in some characteristics (Holman and Corner 1985; Holman and Fritz 2005). The serendipitous arrival in Austin of an extensive collection of skeletal *T. coahuila* inspired our collaborative effort to explore skeletal morphology of extant *T. coahuila*.

Terrapene coahuila was described by Schmidt and Owens (1944) based on 13 specimens from the Cuatro Ciénegas Basin in Coahuila, México, the only locality from which the species is known. It is the most aquatic species of *Terrapene*; different authors suggested previously that individuals may be agile (Brown 1974) or relatively awkward (Milstead 1967) swimmers, and some submerged individuals are reported to walk along the bottom of ponds and streams and only rarely to swim (Bonin et al. 2006). Individuals are known to seek refuge in water when threatened, and may remain submerged for at least 30 min (Bonin et al. 2006). Natural history data were summarized by Webb et al. (1963), Brown (1974), and Howeth et al. (2008). Morphological studies of *T. coahuila* are rare (e.g., Milstead 1969; Brown 1971) and limited data are available about patterns of variation in the skeleton. The phylogenetic position of the species also remains uncertain. Historical and recent hypotheses alternatively place it as a basal taxon within *Terrapene* (Auffenberg 1958; Legler 1960; Williams et al. 1960; Bickham et al. 1996; Burke et al. 1996), or as a more derived member that is sister to *T. carolina* (Milstead 1960, 1967, 1969; Brown 1971; Minx 1996; Feldman and Parham 2002; Wiens et al. 2010), or nested within *T. carolina* (Spinks et al. 2009). Our specific goals are to increase general knowledge of the species, to provide data that will facilitate comparisons with other extant and extinct *Terrapene*, and to contribute important baseline data that ultimately will improve the rigor with which fossil specimens of *Terrapene* and other 'box' turtles can be assessed and identified.

Institutional abbreviations: *UNAM* Universidad Nacional Autónoma de México, Mexico City, Mexico; *TNHC* Texas Natural History Collections, Austin, Texas, USA; *UF* University of Florida, Gainesville, Florida, USA. For specimens referred to by field numbers, the following abbreviations of locality names are used: *AM* Antiguos Mineros; *CP* Charcos Prietos; *LG* Los Gatos; *L* Las Salinas; *PA* Posas Azules; *TC* Tio Cándido. For specimens lacking locality data, X's are used as place holders for the missing locality information.

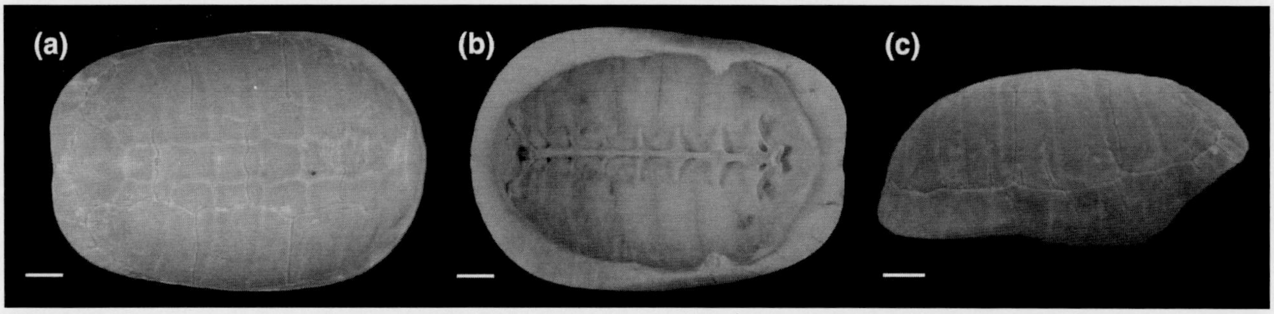

Fig. 28.1 Carapace of *Terrapene coahuila*. **a** Dorsal view, with anterior to left. **b** Ventral (visceral) view, with anterior to right. **c** Right lateral view. Scale bars = 1 cm

Materials and Methods

We examined 214 skeletal specimens of *Terrapene coahuila*, most of which were collected between 2001 and 2005 as part of a field study on the biology of the species. Specimens were salvaged as skeletal remains on the ground surface; discoveries were serendipitous and all collections were made by field workers conducting other types of biological research within the Cuatro Ciénegas Basin. No individuals were sacrificed for our study, and all specimens were collected under SEMARNAT and CITES permits. This represents the largest collection of skeletal material of *T. coahuila*; half of the material is housed at the Texas Natural History Collection (TNHC) at The University of Texas at Austin, the other half at Universidad Nacional Autónoma de México (UNAM). A list of specimens examined is provided in Appendix 1. We gently cleaned the specimens with water to remove dirt, debris, and natural mineral and sediment accumulations. Specimens were air-dried on screens with full circulation to prevent cracking. Specimens with plastral or carapacial scutes intact on the bony surface were cleaned, but the scutes were not removed.

We present morphological data in the form of a character matrix, although ours is specimen-based, not taxon-based. We provide data for 45 morphological features of the carapace and plastron, scored individually for each specimen in our sample (Appendix 2). We include osteological and scute characters previously discussed by Minx [1996; who also included characters described by Legler (1960) and Milstead (1967), (1969)], Joyce and Bell (2004), Holman and Fritz (2005), and Spinks et al. (2009). The majority of our specimens are carapace and plastral elements only, with few associated limb elements, and almost no cranial elements. Our data matrix, therefore, includes only osteological and scute characters from the carapace and plastron. We include several new characters, which assess variation in the carapace and plastron, but whose phylogenetic contexts are not yet rigorously tested.

We gathered mensurative data comparable to those reported by Brown (1971), including carapace length (CL), carapace height, carapace width at bridge, and maximum width of carapace posterior to the bridge. In order to compare our data with those reported by Brown (1971) we took carapace height measurements for all specimens that could be scored (n = 132), and a separate carapace height with plastron intact (n = 109). Our measurements and methods are listed below under "Carapace Characters". In an effort to minimize researcher-induced bias and to maintain consistency within the data set, a single observer recorded all measurements and another observer collected and scored all character data.

Results

Our sample size of *Terrapene coahuila* exceeds those of Schmidt and Owens (1944) (n = 13), Milstead (1969) (n = 59), and Brown (1971) (n = 213). Those studies relied predominantly on whole-body preserved specimens (Schmidt and Owens 1944; Milstead 1969) and live captures and recaptures (Brown 1971), not skeletal specimens. An exemplar of our collection is illustrated in Figs. 28.1 and 28.2, showing dorsal, ventral, and lateral views of the carapace, and ventral and dorsal views of the plastron of *T. coahuila.*

Description and discussion of the morphological characters is provided here, along with annotations on the number of specimens we evaluated, and differences between our observations and data and those previously published by others. For characters 1–5 and 44–48 we provide raw measured values in the matrix; for characters 6–8 and 17–19 we provide the raw counts of the relevant elements. For Figs. 28.1, 28.2, 28.3, 28.4, 28.5, 28.6, 28.7 and 28.8 we provide caption data and the character states shown.

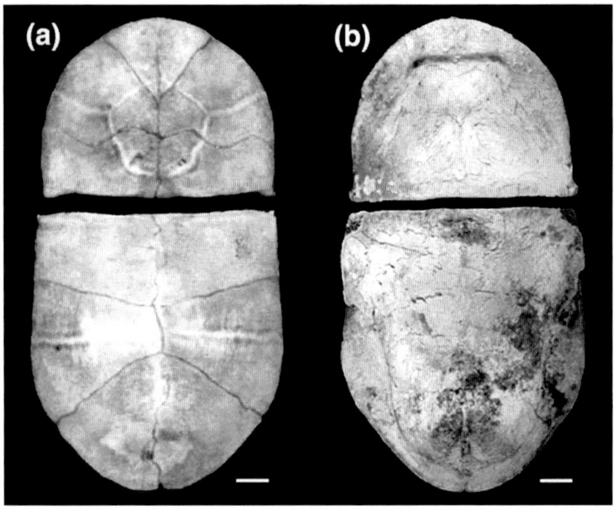

Fig. 28.2 Plastron of *Terrapene coahuila*. **a** Ventral view. **b** Dorsal (visceral) view. Scale bars = 1 cm

Carapace Characters

(1) *Carapace length*

This measurement was taken as described by Brown (1971): it is the straight-line measurement of the carapace from the nuchal to the pygal (Fig. 28.3a). This measurement was taken for 188 specimens, with an average of 133.8 mm, a range of 73.3–182.1 mm and a standard deviation of 22.07 mm. Our largest specimen (TNHC 77202) is 14.1 mm longer than the maximum reported by Milstead (1969), and 27.1 mm longer than the maximum reported by Schmidt and Owens (1944). The average CL from our sample exceeds that reported by Brown (1971) when his males and females are combined.

(2) *Carapace height without plastron*

This measurement was taken as the straight-line distance from the midline of the carapace at its maximum height to the level of the medial projections of the bridge (Fig. 28.4a). This measurement was taken for all specimens that lacked a plastron, as well as those that include a plastron that was not attached by tissue to the carapace. We could reliably measure 132 specimens, which yielded an average of 55.1 mm, a range of 28.5–73.0 mm, and a standard deviation of 7.49 mm.

(3) *Carapace height with plastron*

This measurement was taken as the straight-line distance from the midline of the carapace at its maximum height to the ventral surface of the plastron immediately below (Fig. 28.4b). Indentation of the plastral lobe in many specimens accounts for the lower average score than the average for Character 2. This measurement is most closely comparable to those provided by Brown (1971). One hundred and nine measured specimens yielded an average of 54.8 mm, a range of 30.4–87.2 mm, and a standard deviation of 9 mm.

(4) *Carapace width at the bridge*

This measurement was taken at the level of the medial bridge projections and extends from the lateral edge of the adjacent peripheral bone to the lateral edge of the opposite peripheral bone in a straight line (Fig. 28.3c). One hundred and eighty-nine measured specimens yielded an average of 65.5 mm, a range of 39.0–113.6 mm and a standard deviation of 10.74 mm.

(5) *Maximum carapace width posterior to the bridge*

The location of the maximal width posterior to the bridge varies, but it most commonly occurs near the position of the ninth or tenth marginal scute (Fig. 28.3b). The widest point was established with continuous caliper measurements along the apparent region of maximum width, and the measurement was a direct linear one across the carapace at 90° to the midline. One hundred and ninety-one measured specimens yielded an average of 88.0 mm, a range of 50.8–125.6 mm, and a standard deviation of 15.37 mm.

(6) *Number of neural bones.*

Seven neural bones are present in all 128 specimens we scored.

(7) *Number of costal bones.*

Eight costal bones per side are present in all 134 specimens we scored. In Appendix 2, "7a" corresponds to the number of costal bones on the left side and "7b" to the number of costal bones on the right side.

(8) *Number of peripheral bones.*

Most specimens that could be scored (n = 44) have 11 peripherals per side. One except (TC020629JGH49) has 10 peripherals per side, in this case the first two peripheral bones are fused to form an extra large first peripheral. In Appendix 2, "8a" corresponds to the number of peripheral bones on the left side and "8b" to the number of peripheral bones on the right side.

(9) *Configuration of neural bones based on the number of sutural contacts with surrounding bones: 0—square, 1—pentagonal, 2—hexagonal, 3—heptagonal; modified from* Minx 1996, *character NC;* Joyce and Bell 2004, *characters* 37 *and* 38.

The first through seventh neural bones of different individuals can exhibit various states. Neural configuration was long inferred to be phylogenetically informative for various turtle lineages, and is used as a diagnostic feature for some extinct turtles; these studies typically have small sample sizes and rely on the inferred invariance of neural configuration for an entire series of neurals (e.g., Tong et al. 2010). In the case of *Terrapene coahuila* neural configuration is not stable and most neurals exhibit different states across our sample (Fig. 28.5). Previous workers individuated characters for each neural bone (e.g., Minx 1996; Joyce

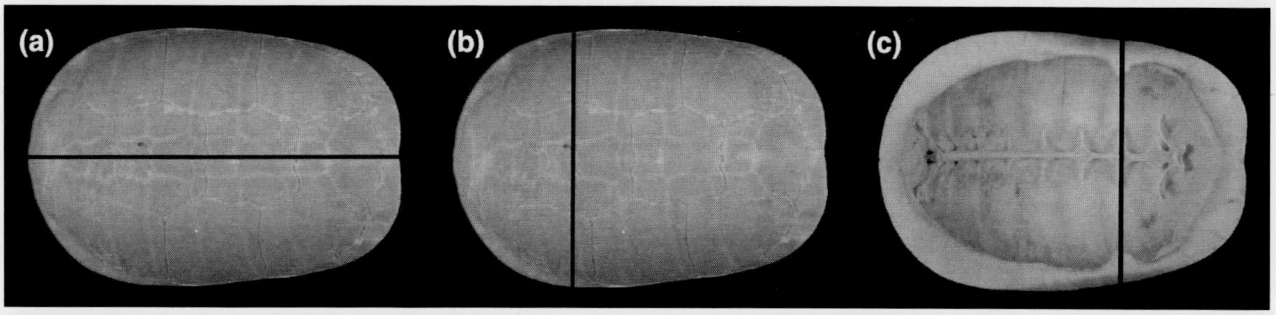

Fig. 28.3 Carapace length and width measurements (*black lines*) used in this study. **a** Carapace in dorsal view, showing length measurement. **b** Carapace in dorsal view, showing maximum width posterior to bridge. **c** Carapace in ventral (visceral) view, showing width at bridge. Anterior to right in all views

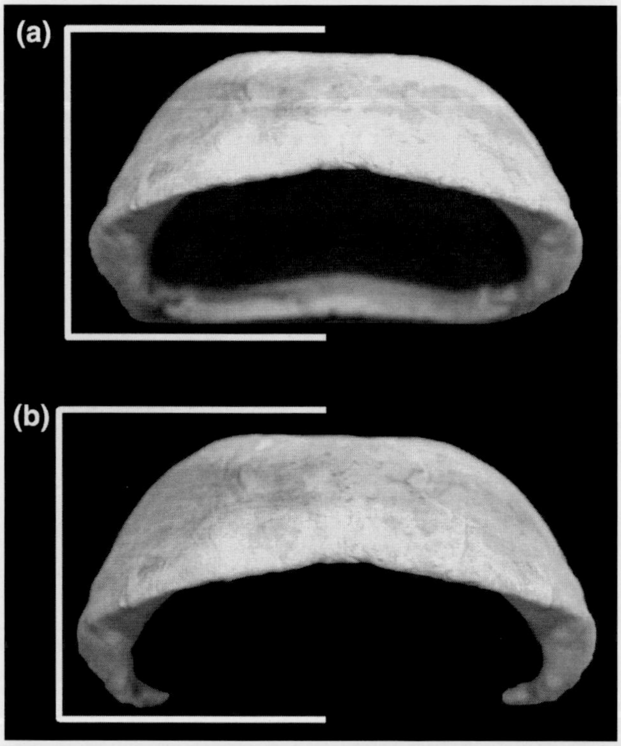

Fig. 28.4 Carapace height measurements used in this study. *White bars* indicate placement of caliper jaws above and below; jaws were then lowered into place to take measurement. **a** Height with posterior lobe of plastron present. **b** Height without plastron

and Bell 2004; Knauss et al. 2011). In our study we present the variation present in our sample and make no assumptions about the phylogenetic affinities of neural configuration as a character or suite of characters. We provide a summary of the relative frequency of various configurations of all neurals in Table 28.1. In Appendix 2 data are provided for each neural bone as a separate column within the appendix, as follows: "9a" corresponds to neural 1; "9b" to neural 2; "9c" to neural 3; "9d" to neural 4; "9e" to neural 5; "9f" for neural 6; and "9g" for neural 7.

(10) *Absence or presence of keels that extend laterally along the carapace: 0—absent, 1—present.*

All specimens that could be scored (n = 198) lack lateral carapacial keels.

(11) *Absence or presence of a mid-dorsal carapacial keel that extends along the midline from vertebral scute two to vertebral scute four: 0—absent, 1—present;* Minx 1996, *character MK.*

Presence of a mid-dorsal keel was evaluated fro 198 specimens. Our findings agree with those presented by Minx (1996); all specimens that could be scored exhibit a weakly present mid-dorsal keel.

(12) *Absence or presence of serrations on the posterior peripherals of the carapace: 0—absent, 1—present;* Joyce and Bell 2004, *character* 35.

We scored 198 specimens; all exhibit some degree of serration, but as noted by Joyce and Bell (2004), the degree of serration is dependent on the presence and shape of intact marginal scutes. All *Terrapene coahuila* have marginal scutes, but posterior marginal scutes often exhibit varying degrees of wear during ontogeny, and often are lost postmortem. In our sample, marginal scutes are not always preserved and definitive determination of serration based on scute morphology was impossible; for those specimens, assessment was based only on the (usually slight) serration present on the peripheral bones. We necessarily simplified this character to a binomial absence or presence.

(13) *Absence or presence of flaring of the posterior peripherals of the carapace: 0—absent, 1—present.*

We scored 203 specimens, and all exhibit a degree of flaring on the posterior peripherals. As with serrations on the posterior peripherals, flaring of the posterior peripherals is variable and degree of flaring is subject to observer bias. Therefore, we rendered our observations of flaring to a simple binomial state.

(14) *Absence or presence of suprapygal: 0—absent, 1—present but fused with pygal, 2—present and not fused with pygal.*

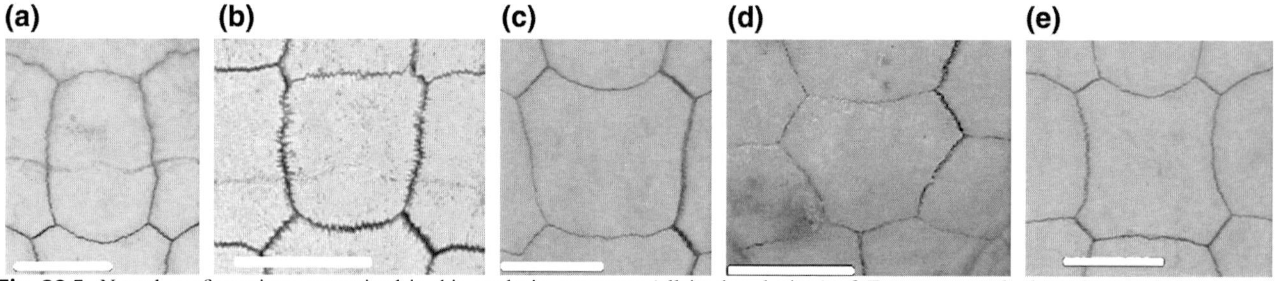

Fig. 28.5 Neural configurations recognized in this study in carapaces (all in dorsal view) of *Terrapene coahuila*. **a** Square. **b** Pentagonal. **c** Hexagonal. **d** Heptagonal. **e** Octagonal. Scale bars = 1 cm

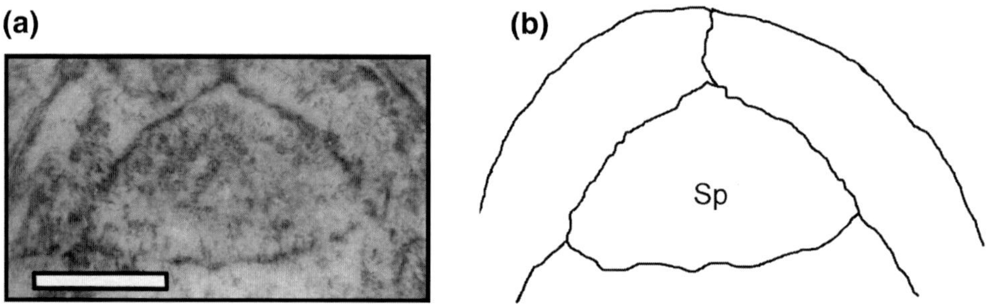

Fig. 28.6 Example of a spade shaped suprapygal (Sp), with apex directed anteriorly and dorsally, in a carapace (dorsal view) of *Terrapene coahuila*. **a** Photograph. **b** Interpretive drawing of same. Both images at same magnification; scale bar = 1 cm

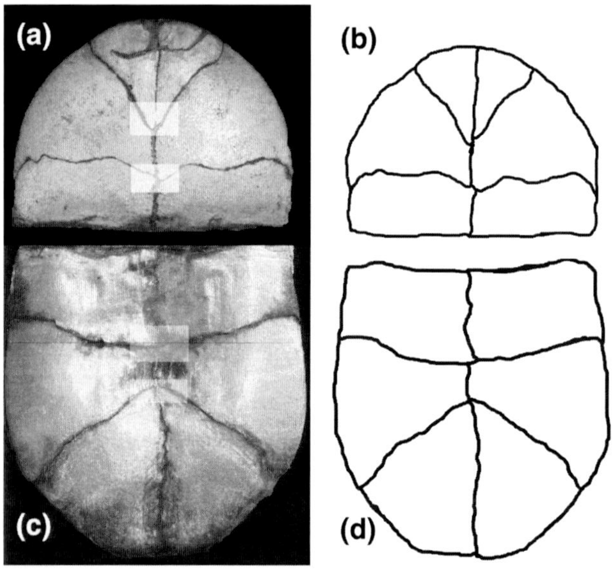

Fig. 28.7 Examples of different patterns of contact between opposite scutes of the plastra (all in ventral view) of *Terrapene coahuila*. **a**, **b** Anterior part of a plastron showing contact between opposite gular and humeral scutes (character 39, state 1) and contact between opposite humeral and pectoral scutes (character 40, state 1): **a** photograph and **b** interpretive *line* drawing. **c**, **d** Posterior part of a plastron showing contact between opposite abdominal and femoral scutes (character 39, state 1) and between opposite femoral and anal scutes (character 43, state 2): **c** photograph and **d** interpretive *line* drawing. *Lighter boxes* in photographs (**a**, **c**) overlie junctions described above between opposite scutes

A suprapygal is present in 82 specimens; 13 specimens exhibit State 1, and 69 exhibit State 2. Although fusion of the suprapygal to the pygal is less common, those specimens that show fusion often exhibit a high degree of carapacial co-ossification. Fusion of the suprapygal to the pygal may therefore be an artifact of co-ossification during ontogeny and may not necessarily be informative with respect to phylogenetic relationships.

(15) *Shape of suprapygal: 0—absent, 1—present with spade-shape, 2—present with triangular shape; modified from* Minx 1996, *character SP.*

We scored 98 specimens; 96 have a spade-shaped suprapygal, and two have a triangular-shaped suprapygal, with the tip of the suprapygal facing anterior and ventral (Fig. 28.6). These results are largely in agreement with the results reported by Minx (1996) who found that *Terrapene coahuila* has a spade-shaped suprapygal, assuming the same orientation that we used. The difference between triangular and spade shape is largely dependent on observer opinion and is not rigorously repeatable. In our approach, if an approximately equilateral triangle was formed by the suprapygal it was scored as triangular in shape, however, if an approximate triangular shape was not found then the specimen was scored as spade-shaped. Because of the difficulty in strictly defining this character, polymorphic character states are likely to be encountered when this character is scored.

Fig. 28.8 Presence of epiplastral beak in plastra (all in ventral view) of *Terrapene coahuila*. **a** Plastron without epiplastral beak (character 49, state 0). **b** and **c** Plastra with epiplastral beak weakly present (character 49, state 1). Scale bars = 1 cm

Table 28.1 Morphological variation in shape of neural bones in *Terrapene coahuila*

Neural number	Total scored	Number oval	Number pentagonal	Number hexagonal	Number heptagonal	Number octagonal
Neural 1	76	69 (91%)	6 (8%)	0	0	1 (1%)
Neural 2	74	2 (3%)	2 (3%)	55 (74%)	6 (8%)	9 (12%)
Neural 3	71	10 (14%)	7 (10%)	54 (76%)	0	0
Neural 4	72	0	0	71 (99%)	1 (1%)	0
Neural 5	68	0	0	68 (100%)	0	0
Neural 6	68	0	0	66 (97%)	2 (3%)	0
Neural 7	66	3 (4%)	9 (13%)	1 (1%)	54 (82%)	0

(16) *Co-ossification of carapacial bones: 0—absent, 1—present in more than 50% of the bones in the carapace; modified from* Minx 1996, *character CO.*

A total of 148 specimens could be scored; 52 lack significant co-ossification of carapacial bones. Co-ossification of 50% or more of the carapacial bones is present in 96 specimens. In most specimens scored as having State 1, the carapace is almost completely fused. Co-ossification of bones was suggested as an apomorphic character for *Terrapene* (Minx 1996), but is variable within *T. coahuila*. Co-ossification might be a better indicator of age or sexual maturity, but it is not clearly tied to carapace length, another possible proxy for maturity and age indication.

(17) *Number of vertebral scutes;* Hirayama 1985, *character P;* Joyce and Bell 2004, *character* 41.

We scored 199 specimens; all have five vertebral scutes with the exception of a single anomalous specimen that exhibits fused third and fourth vertebral scutes. Most testudinoid turtles have five vertebral scutes present (Hirayama 1985; Joyce and Bell 2004).

(18) *Number of pleural scutes.*

We scored 199 specimens; all but one have four pleural scutes per side. The single exception (TNHC 068336) has five pleural scutes on one side, apparently the result of an anomalous division of the fourth pleural. In Appendix 2, "18a" corresponds to the number of pleural scutes on the left side and "18b" to the number of pleural scutes on the right side.

(19) *Number of marginal scutes.*

We scored 185 specimens. Three specimens have less than 12 marginal scutes on the left, and two specimens have less than 12 marginal scutes on the right. In Appendix 2, "19a" corresponds to number of marginal scutes on the left side and "19b" to the number of marginal scutes on the right side.

(20) *Contact between first vertebral and second marginal scutes: 0—absent, 1—present;* Hirayama 1985, *character* O*;* Joyce and Bell 2004, *character 47.*

We scored 195 specimens. Only seven specimens show contact between the first vertebral and the second marginal, 188 lack contact. Although the majority of our specimens lack contact, 3.7% show contact. This was also found to be polymorphic by Joyce and Bell (2004).

(21) *Contact between the third pleural and sixth marginal scutes:* 0—*absent,* 1—*present;* Hirayama 1985, *character B;* Joyce and Bell 2004, *character* 48.

We scored 193 specimens. One specimen (xxxxxxxxxJGH145) exhibits contact between the third pleural and sixth marginal scutes. The remaining specimens lack contact. Our results are concordant with those of Joyce and Bell (2004), who stated that 'emydids' (indicated as a possibly paraphyletic assemblage) lack contact between the third pleural and sixth marginal scutes. The single specimen that has contact is viewed as anomalous.

(22) *Contact between the fifth vertebral and tenth marginal scutes: 0—absent, 1—present;* Hirayama 1985, *character K and* Joyce and Bell 2004, *character* 46.

We scored 190 specimens; 188 specimens lack contact between the fifth vertebral and the tenth marginal. Two specimens, TNHC 069813 and AM020523JGH72, have contact between the fifth vertebral and tenth marginal scutes.

(23) *Twelfth marginal scute: 0—two present, with their common sulcus partially dividing the pygal, 1—two present but their common sulcus fully divides the pygal, 2—twelfth marginal scutes fused along the midline;* Joyce and Bell 2004, *character* 49.

We scored 84 specimens. Shells that do not have scutes preserved were scored based on scute impressions on the bone. In 82 specimens two twelfth marginal scutes share a common sulcus that fully divides the pygal. Two specimens show indications of only 11 marginals; the joint sulci of those fully divide the pygal in both cases. One of those (AM020523JGH72) shows indications that the first two marginal scutes on each side are fused; on the second specimen (TNHC 069813) the eleventh and twelfth marginals are fused. In Appendix 2, "23a" references the number of taxa that could be scored for the presence of twelfth marginal scutes and "23b" provides the data for the number of specimens, which could be scored for pygal division by the twelfth marginal scutes.

(24) *Presence of a cervical scute: 0—present, 1—absent;* Joyce and Bell 2004, *character* 40.

We scored 194 specimens; 190 specimens have a cervical scute, four lack a cervical scute or cervical scute sulci.

(25) *Presence of an axillary scute: 0—absent, 1—present;* Minx 1996, *character AS.*

See notes under character 26.

(26) *Location of the axillary scute with respect to its center: 0—centered on fourth marginal, 1—centered on suture between fourth and fifth marginals, 2—centered on fifth marginal, 3—Axillary scute split with part centered on fourth marginal and separate part centered on fifth marginal; modified from* Minx 1996, *character AS.*

As originally described by Minx (1996) presence/absence and the central location for the scute were scored together. For clarity, we separate the scoring of the presence from the location and define specific character states for the location of the axillary scute with respect to its center. Absence or presence of the axillary scute was scored for 182 specimens, and all have an axillary scute present. Of those, 158 specimens have State 0; the remaining 24 specimens have State 1. Our results are mostly concordant with previous data indicating that *Terrapene coahuila* has an axillary scute present and centered on the fourth marginal (Minx 1996); our sample reveals a higher degree of variation that was previously detected for the position of the scute.

(27) *Anterior margin of the first vertebral scute wider or narrower than posterior margin: 0—narrower, 1—wider, 2—equal width;* Hirayama 1985, *character C; modified from* Joyce and Bell 2004, *character* 44.

We scored 196 specimens. The majority (192 specimens) exhibit State 1. Four specimens exhibit State 2.

(28) *Width of the posterior half of the first vertebral scute relative to the anterior half: 0—wider than anterior half, 1—narrower than anterior half, 2—equal width;* Hirayama 1985, *character R; modified from* Joyce and Bell 2004, *character* 45.

We scored 197 specimens. Of those, 196 specimens exhibit State 1 and a single specimen exhibits State 2.

(29) *Position of the anterior sulcus of the fourth vertebral: 0—sulcus lies on fifth neural, 1—sulcus lies on fourth neural or on the suture between the fourth and fifth neurals, 2—sulcus lies on the sixth neural, or on the suture between the fifth and sixth neurals;* Hirayama 1985, *characters L and M;* Joyce and Bell 2004, *character* 42.

See notes under character 30.

(30) *Position of the posterior sulcus of the fourth vertebral: 0—sulcus lies on the eighth neural or on the homologue of the eighth neural, 1—sulcus lies on the seventh neural or on the suture between the seventh and eighth neurals, 2—eighth neural absent and sulcus overlies costals that meet at midline;* Hirayama 1985, *characters L and M;* Joyce and Bell 2004, *character* 43.

We scored 69 specimens. The location of the anterior sulcus exhibits some variation. Of the total sample scored, 29 specimens have State 0, a single specimen has State 1, and 39 specimens have State 2. With respect to the position of the posterior sulcus, 67 specimens exhibit State 2, a single specimen has State 0, and a single specimen has State 1.

The variation in the position of the anterior sulcus is concordant with the results reported by Joyce and Bell (2004) who found that the locations of these scute sulci were highly variable among taxa. The stability of the posterior location for scute sulci that overlies costals that meet at the midline is potentially informative with respect to the phylogeny of *Terrapene.* A posterior sulcus that overlies the eighth neural was previously reported as the primitive condition (Joyce and Bell 2004). This character should be evaluated across all species of *Terrapene* to clarify whether it is phylogenetically informative.

(31) *Absence of presence of an inguinal scute: 0—absent, 1—present;* Minx 1996, *character IS.*

An inguinal scute is absent in all 184 specimens we scored.

Our results mirror those reported by Minx (1996) who reported absence of the inguinal scute among his sample of *Terrapene coahuila.*

(32) *Absence of presence of an apical scute: 0—absent, 1—present;* Minx 1996, *character AP.*

An apical scute is absent in all 184 specimens we scored. Absence in *Terrapene coahuila* also was reported by Minx (1996).

(33) *Inguinal buttress contact with carapace: 0—presence of a buttress, 1—buttress present but with reduced bony contact between plastron and carapace, 2—buttress weakly present, with virtually no bony contact between plastron and carapace; modified from* Minx 1996, *character IB;* Joyce and Bell 2004, *character* 52.

Within our sample only specimens with a posterior plastral lobe were scored. We limited the scoring in order to compare the level of bony contact between the carapace and the plastron; we did not infer that a buttress was present unless verifiable. All specimens in which the posterior plastral lobe is present (n = 121) exhibit State 1. The level of bony contact between the plastron and carapace is largely uniform, with bony bridge projections meeting distinct contact points on the dorsal surface of the posterior lobe of the plastron. Although there is a reduction in bony contact, plastral lobe/carapace contact is not necessarily reduced. Instead, soft tissues form the majority of the contact between the plastral lobe and the carapace; this was readily observed among specimens that still had soft tissues intact.

(34) *Dark seams between scutes: 0—absent, 1—present in more than 50% of the seams, 2—present in less than 50% of the seams; modified from* Minx 1996, *character DS.*

Seventy specimens had sufficient scute coverage to permit evaluation of the presence of dark seams. Of those, 69 had dark seams present on more than 50% of the seams, and a single specimen showed an absence of dark seams. The presence or absence of dark seams between carapacial scutes was not reported for *Terrapene coahuila* by Minx (1996), who reported the presence of dark seams only in *T. carolina yucatana, T. c. mexicana, and T. c. triunguis.* The presence of those seams was suggested by Minx (1996) to be possibly informative for recognition of those three taxa, but his description was minimal, and we are not certain that we are scoring the same thing. There may also be a difference of degree of darkness in the seams, and interpreter bias could play an important role in scoring. We examined specimens of *T. c. mexicana*, and *T. c. triunguis* from the University of Florida collection (UF 149356, 151623, 151650, 151651, 151656, 151662), but our observations do not permit adequate clarification to confirm Minx's scoring of the character. Adequate documentation with color photography would be desirable.

(35) *Growth annuli: 0—absent, 1—present on more than 50% of scutes, 2—present on less than 50% of scutes; modified from* Minx 1996, *character AN.*

All 68 specimens we scored lack growth annuli. These results are concordant with those reported by Minx (1996), Brown (1974), and Schmidt and Owens (1944), all of whom reported smooth carapacial scutes.

Plastron Characters

(36) *Shape of the posterior lobe of the plastron: 0—rounded, 1—squared;* Minx 1996, *character LR.*

The shape of the posterior lobe of the plastron is variable, and scores may differ based upon whether anal scutes are present or absent. Sometimes the scutes are squared (forming a square-shaped lobe), but the underlying bones do not mimic that shape. For all specimens that were scored (n = 132), only the bony plastron was used to evaluate shape. In all specimens, the shape was rounded.

(37) *Presence of an anal notch at the most posterior portion of the posterior lobe of the plastron: 0—absent, 1—present;* Hirayama 1985, *character W; modified from* Joyce and Bell 2004, *character* 61; Yasukawa et al. 2001, *character* 22.

The presence and depth of the anal notch are highly variable. For our purposes we categorized the notch as simply absent or present, regardless of the depth. We scored 133 specimens. Of those, 118 have a notch present to some depth, and 14 specimens lack a notch. Our results differ from those reported by Joyce and Bell (2004) who reported that the three specimens they scored of *Terrapene coahuila* lacked a notch in the anal portion of the posterior lobe of the plastron.

(38) *Portion of posterior tip of the gular scute extends onto entoplastron: 0—not intruded, 1—intruded.*

Extension of the gular scutes onto the entoplastron was scored only for specimens in which the entoplastron is present, or the bony suture of the entoplastron is present and thus the boundary of the bone was known. Reliable data were collected from 93 specimens; 92 exhibit extension of the gular scute onto the entoplastron, one specimen has State 0.

(39) *Contact between opposite gular and humeral scutes: 0—no contact between opposite scutes, 1—contact present.*

We scored 101 specimens for this character. Of those, 91 have State 1 (Fig. 28.7). The remaining ten specimens lack contact between the opposite gular and humeral scutes.

(40) *Contact between opposite humeral and pectoral scutes: 0—no contact between opposite scutes, 1—left humeral scute contacts right pectoral scute, 2—right humeral scute contacts left pectoral scute.*

An anterior plastral lobe is present in 102 specimens in our sample. Of those, 64 specimens exhibit State 0, 36 exhibit character State 1 (Fig. 28.7), and two specimens exhibit State 2.

(41) *Contact between opposite pectoral and abdominal scutes: 0—no contact between opposite scutes, 1—left pectoral scute contacts right abdominal scute, 2—right pectoral scute contacts left humeral scute.*

We scored 101 specimens. All lack contact between the opposite pectoral and abdominal scutes. The presence of a kinetic plastral hinge means that the pectoral scute (located on the anterior lobe) and the abdominal scute (located on the posterior lobe) do not come in contact with one another in the same manner as found in turtles with akinetic plastra. Instead, these two scutes have a sulcus that fully overlies the hyoplastron/hypoplastron suture, which defines a distinct boundary between the anterior and posterior lobes of the plastron.

(42) *Contact between opposite abdominal and femoral scutes: 0—no contact between opposite scutes, 1—left abdominal scute contacts right femoral scute, 2—right abdominal scute contacts left femoral scute.*

We scored 105 specimens. Of those, 45 have State 0, 31 have State 1, and 29 specimens have State 2 (Figs. 28.7, 28.8).

(43) *Contact between opposite femoral and anal scutes: 0—no contact between opposite scutes, 1—left femoral scute contacts right anal scute, 2—right femoral scute contacts left anal scute.*

We scored 105 specimens. Of those, 74 specimens have State 0, seven specimens have State 1, and 24 specimens have State 2 (Figs. 28.7, 28.8).

(44) *Length of the inter-humeral scute suture.*

Scute suture lengths have been traditionally used in morphometric analyses in a tiered formula (Milstead 1969). We provide raw data on measured lengths of sutures between plastral scutes. The tiered formula based on averages of our measurements is humeral < pectoral < abdominal > femoral < anal. We base this formula on our averages.

This measurement is a straight-line measurement taken along the shared suture of the humeral scutes. The inter-humeral scute suture could be reliably measured for 99 specimens. The average length of the suture was 8.58 mm, a range of 4.5–14.8 mm, and a standard deviation of 1.97 mm.

(45) *Length of the inter-pectoral scute suture.*

Reliable measurement of the length of the inter-pectoral scute suture was made for 100 specimens. Measurements are taken as for character 44. The average length for this suture was 13.34 mm, a range of 6.6–20.7 mm, and a standard deviation of 27.5 mm.

(46) *Length of the inter-abdominal scute suture.*

Measurements for the inter-abdominal scute suture were collected from 118 specimens. Measurements are taken as for character 44. The average length was 24.87 mm, a range of 16.8–36 mm, and a standard deviation of 3.89 mm.

(47) *Length of the inter-femoral scute suture.*

Inter-femoral scute suture length was scored for 116 specimens. Measurements are taken as for character 44. The average length was 6.57 mm, a range of 1.8–14 mm, and a standard deviation of 2.64 mm.

(48) *Length of the inter-anal scute suture.*

We measured 117 specimens. Measurements are taken as for character 44. The average length for this suture was 39.93 mm, a range of 23.4–53.5 mm, and a standard deviation of 5.85 mm.

(49) *Absence or Presence of Epiplastral Beak on the anterior lobe of the plastron—State 0—Absent, State 1—Present; modified from* Holman and Fritz (2005).

Of the 90 specimens scored, 88 do not have an epiplastral beak whereas two specimens (TNHC 068343 and 068956) do have a weakly developed beak (Fig. 28.8). In neither case is the beak as distinct as that illustrated by Holman and Fritz (2005). The epiplastral beak was found by Holman and Fritz (2005) to be variably present in *Terrapene coahuila*, and was hypothesized to be a potentially informative phylogenetic character for *T. corneri*. Our results indicate that the beak is variably present in *T. coahuila* as previously indicated, but the presence of a beak should probably be viewed as an anomalous morphological trait.

(50) *Absence or Presence of notches at the lateral edges of the humeral and pectoral scutes. State 0—Absent, State 1—Present; modified from* Holman and Fritz (2005).

A total of 90 specimens were scored. All lacked any type of notching at the lateral edges of the humeral and pectoral scutes. These results confirm those previously reported by Holman and Fritz (2005). Without examination of *Terrapene* outside of *T. coahuila*, we cannot comment on the phylogenetic affinities of this character as hypothesized by Holman and Fritz (2005) for *T. corneri*.

(51) *Position of the humero-pectoral sulcus with respect to the anterior-posterior mid-line of the entoplastron. State 0—Approximately at the mid-line of the entoplastron, State 1—Sulcus shifted significantly anterior to the mid-line of the entoplastron, State 2—Sulcus shifted significantly posterior to the mid-line of the entoplastron; modified from* Holman and Fritz (2005).

A total of 82 specimens were scored. The humero-pectoral sulcus is located at approximately the mid-line of the entoplastron in 81 specimens; in a single specimen (TNHC 069652) the sulcus was considered to be significantly shifted posterior of the mid-line. These results confirm those reported previously by Holman and Corner (1985) and Holman and Fritz (2005).

Discussion

We evaluated a large sample of specimens of *Terrapene coahuila* for most characters traditionally thought to be informative about relationships among *Terrapene*. However, many previously proposed characters were not clearly defined and may be subject to considerable observer bias (e.g., shell 'robustness' or 'large shape'). Shape often was cited as a major source of variation within *Terrapene* (e.g., Milstead 1956, 1969) but assessment of shape was largely subjective and the various shell shapes within a given species of *Terrapene* may overlap the range of shell shapes in other species. Consequently, many traditionally recognized features associated with shape are difficult to discretely categorize and are ignored in formal cladistic analyses (e.g., Minx 1996). Other features, however, are readily discretized. The presence of a mid-dorsal keel and the degree of flaring and serration found on the posterior rim of the carapace are possible examples (Auffenberg 1958, 1967; Milstead 1967, 1969; Minx 1996). Our observations agree with those of our predecessors that the degree of flaring and serration of the carapace is variable. The development of a mid-dorsal keel is perhaps slightly more informative in that its location may be taxon-specific (Minx 1996). However, expression of the keel is subject to degradation during ontogeny, and may be subject to taphonomic deterioration in fossils. Those considerations led Minx (1996) to suggest that juveniles are the most helpful in evaluating the phylogenetic distribution of a keel. Distinguishing between juveniles and adults often is difficult, and size is sometimes a factor used to assess the level of maturity.

Consideration of size played an important role in the historical development of taxonomy within *Terrapene*. An excellent and persistent example is *T. carolina putnami,* a presumed extinct subspecies, the purported diagnosis of which is based on its large size and robust shell compared to other *Terrapene* (Hay 1908; Dodd 2001). The variation in size range of specimens in our sample is informative in this respect. Previous analyses of *T. coahuila* yielded a range that places the species within the small- to medium-sized box turtle category described by Milstead (1969), but our sample includes turtles smaller than the typical ones examined previously, and specimens as large as an extant *T. c. major*, the subspecies often referred to as the largest *Terrapene* (e.g., Milstead 1969). Increased sample sizes clearly indicate that size criteria may be uninformative for unambiguous determination of species within *Terrapene*. That problem is likely exacerbated for fossils, for which poorly documented patterns of temporal variation and a discontinuous spatial distribution through time provide additional levels of complexity.

We evaluated characters that are readily discretized, as well as standard mensurative features. Those characters form baseline data for *Terrapene coahuila* and can be used in future comparisons with other populations of *Terrapene*, and for phylogenetic analyses of *Terrapene* and relevant fossil specimens of functional box turtles that may, or may not, be closely related to *Terrapene*. Our sample was collected from diverse specific localities within the Cuatro Ciénegas basin, and our data suggest that, at least for most of the features we assessed, levels of variation within *T. coahuila* are relatively low. Such conservative morphology is consistent with predictions derived from documented levels of gene flow between the sub-populations of *T. coahuila* (Howeth et al. 2008). Exceptions include the high degree of variation present in configurations of the neural bones, the level of co-ossification among carapacial bones, and the positions of the anterior and posterior sulci of the fourth vertebral scute. The variation in neural configuration and co-ossification is consistent with results previously reported by Pritchard (1988) and Minx (1996) for species of *Terrapene*, and variation in the sulci of the fourth vertebral also was reported within other testudinoids by Joyce and Bell (2004). Genetic, epigenetic, and/or environmental controls for those features in *T. coahuila* are not yet documented, but provide an interesting area for further research.

Previous considerations of patterns of variation within *Terrapene* included evaluations of the skull and soft tissues (Baur 1891) as well as the shell (e.g., Barbour and Stetson 1931; Milstead 1969; Pritchard 1988), and previous authors routinely noted the variation present within species of *Terrapene*. Levels of variation historically reported for *Terrapene* may be an artifact of the traditional way that morphological features were evaluated. Differences in overall shape patterns, obvious even from casual inspection of small samples of some taxa (e.g., Barbour and Stetson 1931), could reflect a possible level or source of variation that remains undetected in analyses of discrete morphological characters that are readily subjected to phylogenetic analysis. In an interesting departure from traditional perspectives, variation in the shape of the plastron of *Terrapene* was recently documented to be influenced more by phylogeny than by body size (Angielczyk et al. 2011). Those results are consistent with the conservative plastral morphology we found. Aside from features associated with scute sulci, osteological characters scored for the plastron are extremely conserved. If phylogeny is an important factor for the expression of variation among species, then patterns of intraspecific variation would be expected to be relatively low. This provides further impetus to continue the rigorous pursuit to resolve phylogenetic relationships within *Terrapene*.

These considerations raise two additional points. The first is that the suite of characters we used, derived from various sets of characters hypothesized to be phylogenetically informative, are actually a robust set of characters which are not subject to a large amount of intra-specific variation within *Terrapene coahuila*. This lends support to the use of those data in phylogenetic analyses of *Terrapene*, but must be accompanied by a cautionary note. Detailed analyses are needed to assess variation in discrete and continuous characters across all taxa of *Terrapene* in order to develop the most reliable suite of characters for phylogenetic analysis. A second key point is that efforts to elucidate the evolutionary history of *Terrapene* must include rigorous reevaluation of the phylogenetic status of fossils reported to belong to *Terrapene*, as well as other North American functional 'box turtles.'

Evaluation of variation in extant species is a critical consideration in assessing taxonomic affinity of fossils. Classification of a fossil as a functional 'box turtle' often is most reliably made based on assessment of the plastron and the presence of a rudimentary or well-developed hinge, resulting from the approximate or exact alignment of bony sutures with the sutures between the external scutes. In the absence of a preserved plastron, a notable reduction of the bridge is sometimes used to infer the presence of a hinge (e.g., *Planetochelys*; Weems 1988; Hutchison 1998; Holroyd et al. 2001). Beyond an initial diagnosis as a 'box turtle', further taxonomic assignation is best achieved through analysis of a suite of apomorphic characters. A consequent loss of species-level resolution may result (Bell et al. 2010), but isolated shell elements from testudinoid turtles are abundant in the Neogene fossil record, and should not be disregarded; they do sometimes preserve sufficient apomorphic features to permit species-level resolution (e.g., Jass and Bell 2010). Although there is a general acceptance that extant *Terrapene* is represented in Miocene sediments (e.g., Hay 1908; Holman 1975, 1987; Holman and Corner 1985; Parmley 1992; Holman and Fritz 2005), that assertion is not yet rigorously established. The discovery of earlier Tertiary specimens that clearly are 'box turtles' in a functional sense highlights additional uncertainties about the diversity and relationships of Tertiary emydid turtles generally, and of *Terrapene* specifically. Any effort to address the evolution of *Terrapene* in a holistic sense must include not only the evaluation of patterns of variation in previously described characters (such as we did here), but also the elucidation of new characters, particularly in isolated skeletal elements from anatomical systems typically preserved in the fossil record.

Acknowledgments We gratefully acknowledge Jennifer Howeth for her efforts to collect and preserve skeletal specimens of *Terrapene coahuila* during her field research in Cuatro Ciénegas. Her research and, by extension, ours, was conducted under SEMARNAT permits 3054 and 5573, and CITES permits 072019 and 140459. At the time these specimens were collected Susana Moncada Díaz de León was Director of the Area Protegida de Fauna y Flora de Cuatro Ciénegas; she granted local permission to sample within the reserve, greatly facilitated coordination with landowners in the basin, and provided additional staff assistance in the field. We acknowledge Christopher Eckert for his efforts in collecting measurement data for our specimens. This chapter was improved substantially as a result of comments from our colleagues Robert Holmes, Chris Jass, and Walter Joyce; their careful reviews helped clarify our thinking and presentation. Whether he likes it or not, Gene Gaffney helped to inspire us in our continued efforts to understand the evolutionary history of emydid turtles, perhaps one of his least favored groups of Testudines.

Appendices

Appendix 1

Inventory of *Terrapene coahuila* specimens examined. Universidad Nacional Autónoma de México (UNAM) specimens are designated by field numbers only.

Texas Natural History Collections, The University of Texas at Austin (TNHC): TNHC 068333–068362, TNHC 068653, TNHC 068940–068946, TNHC 068950–TNHC 068952, TNHC 068954–068966, TNHC 069624–069645, TNHC 069647–069653, TNHC 069801–069810, TNHC 069812–069825, TNHC 069936, TNHC 069939, TNHC 070241, TNHC 070587, TNHC 070753, TNHC 77201–77207.

Universidad Nacional Autónomo de México (UNAM, designated by field numbers only): AM020523JGH59, AM020523JGH61, AM020523JGH63, AM020523JGH71, AM020523JGH72, AM030111JGH116, AM030111JGH 117, AM030208JGH164, CP030313JGH159, CP030313 JGH161, CP030313JGH162, JGH80, LG020519JGH32, LG020519JGH67, LG020519JGH68, LG020702JGH10, LG020702JGH14, LG020702JGH207, LG020702JGH23, LG020702JGH27, LG020702JGH30, LG020702JGH4, LG020705JGH13, LG020705JGH15, LG020705JGH25, LG020705JGH26, LG020705JGH28, LG020705JGH29, LG020705JGH35, LG020705JGH37, LG020705JGH46, LG020705JGH47, LG020705JGH8, LG020708JGH19, LG020708JGH19, LG020725DAH198, LG020725DAH204, LG020725JGH2, LG020725JGH3, LG020725JGH9, LG020730JGH5, LG020802JGH91, LG020820JGH84, LG020820JGH85, LG020820JGH86, LG020820JGH87, LG020820JGH88, LG020820JGH89, LG020820JGH90, LG020820JGH92, LG020820JGH94, LG020821JGH192, LG020821JGH95, LG020821JGH96, LG020914JGH153, LG020914JGH203, LG020914JGH206, LG021018JGH101, LG021018JGH103, LG030109JGH114, LG030114JGH113, LG030214JGH167, LG030214JGH169, LG030214JGH181, LG030219JGH180, LG030314JGH171, LG030314JGH172, LG030315JGH177, LG030315JH174, LG030719JGH128, LG030728DAH136, LG030728DAH137, LS020730DAH195, LS020730JGH196, LS020730JGH199, LS020730JGH96, PAS030113JGH119, PAS041212JGH124, PAS041212JGH208, TC020629JGH45, TC020629JGH48, TC020629JGH49, TC020629JGH40, XX02JGH98, XX02JGH99, XXX031214JGH129, XXX04XXXXJ GH131, XXX04XXXXJGH132, XXXXXXXXXJGH145.

University of Florida (UF): UF 149356, UF 149511, UF 151623, UF 151650, UF 151651, UF 151656, UF 151662.

Appendix 2

Morphological data for shells of *Terrapene coahuila*.

Data matrices follow the format of the example table below. Specimen number is reported in first row. Measurements and scores for the 51 characters (see descriptions in text) are presented in remaining rows, sequentially from left to right and top to bottom. Value of "?" indicates value could not be determined.

(A) Example matrix:

Specimen									
1	2	3	4	5	6	7a	7b	8a	8b
9a	9b	9c	9d	9e	9f	9g	10	11	12
13	14	15	16	17	18a	18b	19a	19b	20
21	22	23a	23b	24	25	26	27	28	29
30	31	32	33	34	35	36	37	38	39
40	41	42	43	44	45	46	47	48	49
50	51								

(B) Data matrices:

AM020523JGH59								
149.7	58.5	?	74.5	100.2	7	8	8	11
11	0	2	2	2	2	2	3	0
1	1	1	2	0	0	5	4	4
12	12	0	0	0	2	1	?	1
0	1	1	2	2	0	0	?	?
?	?	?	?	?	?	?	?	?
?	?	?	?	?	?	?	?	

AM020523JGH61								
159.7	62.6	?	76.8	107	7	8	8	11
11	?	?	?	?	?	?	?	0
1	1	1	2	0	1	5	5	4
12	12	0	0	0	2	1	1	1
0	1	1	?	?	0	0	?	?
?	?	?	?	?	?	?	?	?
?	?	?	?	?	?	?	?	

AM020523JGH63								
157	65	?	73.4	105	?	8	8	11
11	0	2	2	2	4	4	3	0
1	1	1	2	0	0	5	4	4
12	12	0	0	0	2	1	1	1
0	1	1	2	2	0	0	?	?
?	?	?	?	?	?	?	?	?
?	?	?	?	?	?	?	?	

AM020523JGH70								
153.7	61.4	?	74.9	100	7	8	8	?
?	0	2	2	2	2	?	?	0
1	1	1	?	0	1	5	4	4
12	12	0	0	0	2	1	1	1
0	1	1	0	2	0	0	?	?
?	?	?	?	?	?	?	?	?
?	?	?	?	?	?	?	?	

AM020523JGH71								
171.7	67.8	?	87.9	118.2	7	8	8	11
11	0	2	2	2	2	2	3	0
1	1	1	2	0	1	5	4	4
12	12	0	0	0	2	1	1	1
0	1	1	0	2	0	0	?	?
?	?	?	?	?	?	?	?	?
?	?	?	?	?	?	?	?	

AM020523JGH72								
144.2	60	?	72.7	95.8	7	8	8	?
?	0	2	2	2	2	?	?	0
1	1	1	?	?	1	5	4	4
11	11	0	0	1	0	0	1	1
0	1	1	?	?	0	0	?	?
?	?	?	?	?	?	?	?	?
?	?	?	?	?	?	?	?	

AM030111JGH116								
120	49.3	51.5	63.6	82.9	7	8	8	11
11	0	2	2	2	2	2	3	0
1	1	1	2	0	1	5	4	4
12	12	0	0	0	2	1	1	1
1	1	1	2	2	0	0	1	1
0	0	1	?	?	?	?	?	?
?	?	22.7	4.5	41.4	?	?	?	

AM030111JGH117								
152.8	60.3	?	78.5	107.6	7	8	8	11
11	0	2	2	2	2	2	3	0
1	1	1	2	0	1	5	4	4
12	12	0	0	0	2	1	1	1
0	1	1	2	2	0	0	?	?
?	?	?	?	?	?	?	?	?
?	?	?	?	?	?	?	?	

AM030208JGH164								
119.8	50.6	?	57.6	79	7	8	8	?
?	?	?	?	?	?	?	?	0
1	1	1	2	0	1	5	4	4
12	12	0	0	0	2	1	?	1
0	1	1	?	?	0	0	?	?
?	?	?	?	?	?	?	?	?
?	?	?	?	?	?	?	?	

CP030313JGH159

126.1 51.3 ? 58.8 79.7 7 8 8 11
11 1 2 1 2 2 2 1 0
1 1 1 ? 0 1 5 4 4
12 12 0 0 0 2 1 1 1
0 1 1 2 2 0 0 ? ?
? ? ? ? ? ? ? ? ?
? ? ? ? ? ? ? ?

CP030313JGH161

114.8 48.7 ? 58.6 75.4 7 8 8 ?
? 0 2 2 2 2 2 1 0
1 1 1 ? ? 1 5 4 4
? ? 0 0 0 ? ? ? 1
0 1 1 2 2 0 0 ? ?
? ? ? ? ? ? ? ? ?
? ? ? ? ? ? ? ?

CP030313JGH162

? ? ? ? ? ? ? ? ?
? ? ? ? ? ? ? ? 0
1 1 1 ? ? 1 5 4 4
12 12 0 0 0 2 1 1 1
0 1 1 ? ? 0 0 1 ?
? 0 1 1 1 ? ? ? ?
8.4 12.9 25.1 7.8 42 ? ? ?

JGH80

129 53.6 ? 64.6 84 7 8 8 ?
? 0 2 ? 2 2 2 3 0
1 1 1 2 0 1 5 4 4
12 12 0 0 0 2 1 1 1
0 1 1 2 2 0 0 ? ?
? ? ? ? ? ? ? ? ?
? ? ? ? ? ? ? ?

LG020519JGH32

155.2 58.7 ? 74 101.8 7 8 8 ?
? ? ? ? ? ? ? ? 0
1 1 1 ? 0 1 5 4 4
12 12 0 0 0 2 ? 1 1
0 1 1 ? ? 0 0 ? ?
? ? ? ? ? ? ? ? ?
? ? ? ? ? ? ? ?

LG020519JGH67

133.8 49.6 ? 64.3 86.3 7 8 8 ?
? 0 2 2 2 2 2 3 0
1 1 1 2 0 0 5 4 4
12 12 0 0 0 2 1 1 1
1 1 1 0 2 0 0 ? ?
? ? ? ? ? ? ? ? ?
? ? ? ? ? ? ? ?

LG020519JGH68

124.1 48.6 ? 65.5 83 7 8 8 11
11 0 4 0 2 2 2 3 0
1 1 1 2 0 0 5 4 4
12 12 0 0 0 2 1 1 1
0 1 1 0 2 0 0 ? ?
? ? ? ? ? ? ? ? ?
? ? ? ? ? ? ? ?

LG020702JGH10

143.1 59.8 ? 66.7 90.8 7 8 8 ?
? 0 2 2 2 2 2 3 0
1 1 1 2 0 1 5 4 4
12 12 0 0 0 2 1 1 1
0 1 1 2 2 0 0 ? ?
? ? ? ? ? ? ? ? ?
? ? ? ? ? ? ? ?

LG020702JGH14

158.4 ? ? ? 105.8 7 8 8 ?
? 0 2 2 2 2 2 3 0
1 1 1 2 0 0 5 4 4
? ? 0 0 0 2 1 1 1
0 0 2 0 0 0 0 ? ?
? ? ? ? ? ? ? ? ?
? ? ? ? ? ? ? ?

LG020702JGH207

? ? ? ? ? ? ? ? ?
? ? ? ? ? ? ? ? ?
? ? ? ? ? ? ? ? ?
? ? ? ? ? ? ? ? ?
? ? ? ? ? ? ? ? ?
? ? ? ? ? ? ? 1 ?
? 0 0 ? ? ? ? ? ?
? ? 28.4 10.7 50.7 ? ? ?

LG020702JGH23

112.9 50.2 ? 54.9 74.5 7 8 8 ?
? ? ? ? ? ? ? ? 0
1 1 1 ? 0 1 5 4 4
12 12 0 0 0 2 1 1 1
1 0 1 ? ? 0 0 ? ?
? ? ? ? ? ? ? ? ?
? ? ? ? ? ? ? ?

LG020702JGH27

? 68.2 ? 81.6 122.9 7 8 8 ?
? ? ? ? ? ? ? ? 0
1 ? 1 ? ? 1 5 4 4
? ? 0 0 ? ? ? 1 1
0 1 1 ? ? 0 0 ? ?

? ? ? ? ? ? ? ? ?
? ? ? ? ? ? ? ?

LG020702JGH30
141.2 60.1 ? 70.9 95.5 7 8 8 ?
? 0 2 2 2 2 2 3 0
1 1 1 1 0 1 5 4 4
12 12 0 0 0 ? ? 1 1
0 1 1 0 2 0 0 ? ?
? ? ? ? ? ? ? ? ?
? ? ? ? ? ? ? ?

LG020702JGH4
146.6 57.6 59.8 ? ? 7 8 8 11
11 0 3 1 2 2 2 3 0
1 1 1 2 0 0 5 4 4
? ? 0 ? ? ? ? 1 ?
? 1 1 0 2 ? ? ? ?
? 0 1 ? ? ? ? ? ?
? ? ? ? ? ? ? ?

LG020705JGH13
140.1 54.6 ? 66.8 94.5 7 8 8 11
11 0 2 2 2 2 2 3 0
1 1 1 2 0 1 5 4 4
12 12 0 0 0 2 1 1 1
0 1 1 1 2 0 0 ? ?
? ? ? ? ? ? ? ? ?
? ? ? ? ? ? ? ?

LG020705JGH15
150 60.2 ? 68.6 100.7 7 8 8 ?
? ? ? ? ? ? ? ? 0
1 ? 1 2 0 1 5 4 4
12 12 0 0 0 2 1 1 1
0 1 1 ? ? 0 0 ? ?
? ? ? ? ? ? ? ? ?
? ? ? ? ? ? ? ?

LG020705JGH25
153.2 59.6 ? 73.3 107.3 7 8 8 11
11 0 1 2 2 2 2 3 0
1 1 1 1 0 0 5 4 4
12 12 0 0 0 2 1 1 1
0 1 1 2 2 0 0 ? ?
? ? ? ? ? ? ? ? ?
? ? ? ? ? ? ? ?

LG020705JGH26
133.9 58.4 63.2 68.7 90 7 8 8 ?
? ? ? ? 2 2 2 3 0
1 1 1 1 0 0 5 4 4
12 12 1 0 0 2 ? 0 1

0 1 1 0 2 0 0 1 ?
? 1 1 ? ? ? ? ? ?
? ? 25.5 5.8 44 ? ? ?

LG020705JGH28
? 63.4 ? 80.4 ? 7 8 8 ?
? ? ? ? ? ? ? ? 0
1 1 1 ? 0 1 4 4 4
12 12 0 0 0 2 ? 1 1
0 1 1 ? ? 0 0 ? ?
? ? ? ? ? ? ? ? ?
? ? ? ? ? ? ? ?

LG020705JGH29
164.1 65 ? 79.6 113.1 7 8 8 ?
? ? ? ? ? ? ? ? 0
1 1 1 2 0 1 5 4 4
12 12 0 0 0 2 1 1 1
0 1 1 0 2 0 0 ? ?
? ? ? ? ? ? ? ? ?
? ? ? ? ? ? ? ?

LG020705JGH35
164.8 63.9 ? 79.7 109.3 7 8 8 ?
? ? ? ? ? ? ? ? 0
1 1 1 ? 0 1 5 4 4
12 12 0 0 0 2 ? 1 1
0 1 1 ? ? 0 0 ? ?
? ? ? ? ? ? ? ? ?
? ? ? ? ? ? ? ?

LG020705JGH37
154.2 60.8 ? 74.7 103.8 7 8 8 ?
? ? ? ? ? ? ? ? 0
1 1 1 ? ? ? 5 4 4
12 12 0 0 0 2 ? 1 ?
? 1 1 ? ? ? ? ? ?
? ? ? ? ? ? ? ? ?
? ? ? ? ? ? ? ?

LG020705JGH46
132.9 54.9 ? 67.7 86 7 8 8 ?
? ? ? ? ? ? ? ? 0
1 1 1 ? 0 1 5 4 4
12 12 0 0 0 2 ? 1 1
0 1 1 ? ? 0 0 ? ?
? ? ? ? ? ? ? ? ?
? ? ? ? ? ? ? ?

LG020705JGH47
136.2 55.2 ? 63.9 88.1 7 8 8 ?
? ? ? ? ? ? ? ? 0
1 1 1 ? ? ? 5 4 4

12 12 0 0 0 2 ? 1 ?

? 1 1 ? ? ? ? ? ?

? ? ? ? ? ? ? ? ?

? ? ? ? ? ? ? ?

LG020705JGH8

168 67.1 87.2 113.6 69.7 ? ? ? ?

? ? ? ? ? ? ? ? 0

1 1 1 ? 0 1 5 4 4

12 ? ? ? 0 2 ? ? ?

? ? ? ? ? 0 0 1 ?

? 1 1 1 1 ? ? ? ?

8.9 17.5 36 6.7 53.5 ? ? ?

LG020708JGH19

156.3 61 ? 78.1 107 7 8 8 ?

? 0 2 2 2 2 2 3 0

1 1 1 ? 0 1 5 4 4

12 12 0 0 0 2 1 1 1

0 1 1 0 2 0 0 ? ?

? ? ? ? ? ? ? ? ?

? ? ? ? ? ? ? ?

LG020721JGH96

? ? ? ? ? 7 8 8 11

11 0 2 2 2 2 2 3 0

1 1 1 2 ? 0 5 4 4

12 12 0 0 0 2 1 1 ?

? 1 1 0 2 ? ? ? ?

? 1 1 1 1 ? ? ? ?

8.7 13 27.9 ? 44 0 0 0

LG020725DAH198

? ? ? ? ? ? 8 8 11

11 0 2 ? ? ? 2 3 ?

? 1 1 2 0 0 5 4 4

12 12 0 0 0 2 1 1 1

0 1 1 2 2 0 0 ? ?

? 1 1 1 1 ? ? ? ?

? ? 25.4 7 42.4 0 0 0

LG020725DAH204

128.5 49.2 50.9 60.8 85.7 ? ? ? ?

? 0 2 2 ? ? ? ? ?

? ? ? ? ? ? ? 4 4

? ? 0 0 0 ? ? 1 ?

? 1 1 ? ? ? ? ? ?

? ? ? ? ? ? ? ? ?

? ? ? ? ? ? ? ?

LG020725JGH2

120.3 50.8 56.9 63 82.3 7 8 8 11

11 0 2 2 2 2 2 3 0

1 1 1 2 0 1 5 4 4

12 12 0 0 0 2 1 1 1

1 1 1 0 2 0 0 1 ?

? 1 1 ? ? 0 0 1 2

? ? 23.8 6.3 39.8 ? ? ?

LG020725JGH3

142.7 59.8 ? 67.7 95.3 7 8 8 ?

? ? ? ? ? ? ? ? 0

1 1 1 ? ? 1 5 4 4

12 12 0 0 0 2 ? 1 1

0 0 1 ? ? 0 0 1 ?

? 1 1 ? ? ? ? ? ?

? ? 23.9 3.7 36.8 ? ? ?

LG020725JGH9

134.6 54.9 ? 70.7 89.9 7 8 8 ?

? ? ? ? ? ? ? ? 0

1 1 1 ? ? 1 5 4 4

12 12 0 0 0 2 ? 1 1

0 1 1 ? ? 0 0 ? ?

? ? ? ? ? ? ? ? ?

? ? ? ? ? ? ? ?

LG020730JGH5

128.8 52.7 57.1 64.8 85.4 7 8 8 11

11 0 2 2 2 2 2 3 0

1 1 1 2 0 0 5 4 4

12 12 0 0 0 2 1 1 1

0 1 1 2 2 0 0 ? ?

? ? ? ? ? 0 0 0 1

? ? ? ? ? ? ? ?

LG020802JGH91

129.9 55.2 ? 64.4 86.7 7 8 8 11

11 0 2 2 2 2 2 3 0

1 1 1 2 0 1 5 4 4

12 12 0 0 0 2 1 1 1

0 1 1 2 2 0 0 ? ?

? 1 1 ? ? ? ? ? ?

? ? 25.7 5.9 40.3 ? ? ?

LG020820JGH84

144.5 52.9 ? 69.1 95.1 7 8 8 ?

? ? ? ? ? ? ? ? 0

1 1 1 ? ? 1 5 4 4

12 12 0 0 0 2 ? 1 1

0 1 1 ? ? 0 0 ? ?

? ? ? ? ? ? ? ? ?

? ? ? ? ? ? ? ?

LG020820JGH85

? ? ? ? 84.5 7 8 8 ?

? 0 2 2 2 ? ? 3 0

1 1 1 ? 0 1 5 4 4

12 12 0 0 0 2 ? 1 1

0 1 1 2 2 0 0 1 ?

? 1 1 ? ? ? ? ? ?

? ? ? ? ? ? ? ?

LG020820JGH86

154.9 57.7 ? 72.2 101.7 7 8 8 ?

? 0 3 1 2 2 2 3 0

1 1 1 2 ? 1 5 4 4

12 12 0 0 0 2 1 1 1

0 1 1 0 2 0 0 ? ?

? ? ? ? ? ? ? ? ?

? ? ? ? ? ? ? ?

LG020820JGH87

161.7 65 ? 77.4 110.3 7 8 8 ?

? ? ? ? ? ? ? ? 0

1 1 1 ? ? 1 5 4 4

12 12 0 0 0 2 ? 1 1

0 1 1 ? ? 0 0 ? ?

? ? ? ? ? ? ? ? ?

? ? ? ? ? ? ? ?

LG020820JGH88

127.2 55.2 62.1 66.4 87.6 7 8 8 ?

? ? ? ? ? ? ? ? 0

1 1 1 ? ? 1 5 4 4

12 12 0 0 0 2 ? 1 1

0 1 1 ? ? 0 0 ? ?

? ? ? ? ? ? ? ? ?

? ? ? ? ? ? ? ?

LG020820JGH89

126.1 51 ? 61.4 80.4 7 8 8 11

11 0 2 0 2 2 2 3 0

1 1 1 2 0 1 5 4 4

12 12 0 0 0 2 1 1 1

0 1 1 2 2 0 0 1 ?

? 1 1 ? ? ? ? ? ?

? ? ? ? ? ? ? ?

LG020820JGH90

176.2 65 ? 81.1 115.2 ? ? ? ?

? ? ? ? ? ? ? ? 0

1 1 1 ? ? 1 5 4 4

12 12 0 0 0 2 ? 1 1

0 1 1 ? ? 0 0 ? ?

? ? ? ? ? ? ? ? ?

? ? ? ? ? ? ? ?

LG020820JGH92

119.9 ? ? ? 79.9 ? ? ? ?

? ? ? ? ? ? ? ? 0

1 1 1 ? 0 1 5 4 4

12 12 0 0 0 2 ? 1 1

0 0 1 ? ? 0 0 ? ?

? ? ? ? ? ? ? ? ?

? ? ? ? ? ? ? ?

LG020820JGH94

? ? ? ? ? 7 8 8 11

11 0 2 2 2 2 2 ? 0

1 1 1 ? 0 0 5 4 4

12 12 0 0 0 2 1 1 ?

? 1 1 2 2 0 0 1 ?

? 1 1 ? ? ? ? ? ?

? ? 22.6 6.6 38 ? ? ?

LG020821JGH192

136.4 55.4 ? 64.2 87.9 7 8 8 11

11 0 2 2 2 2 2 3 0

1 1 1 2 ? ? 5 4 4

12 12 0 0 0 2 1 1 ?

? 1 1 2 2 ? ? ? ?

? ? ? ? ? ? ? ? ?

? ? ? ? ? ? ? ?

LG020821JGH95

? ? ? ? ? 7 8 8 ?

? ? ? ? ? ? ? ? 0

1 1 1 ? ? 1 5 4 4

12 12 0 0 0 2 1 1 1

1 1 1 ? ? 0 0 ? ?

? ? ? ? ? ? ? ? ?

? ? ? ? ? ? ? ?

LG020914JGH153

131.5 49.6 ? 62.9 81.8 7 8 8 ?

? ? ? ? ? ? ? ? 0

1 1 1 2 0 1 5 4 4

12 12 0 0 0 2 1 ? 1

0 1 1 ? ? 0 0 ? ?

? ? ? ? ? ? ? ? ?

? ? ? ? ? ? ? ?

LG020914JGH203

135.5 57.7 ? 62.8 92.2 7 8 8 ?

? 0 2 2 2 2 ? ? 0

1 1 1 ? ? 1 5 4 4

12 12 0 0 0 2 1 1 1

0 1 1 ? ? 0 0 ? ?

? ? ? ? ? ? ? ? ?

? ? ? ? ? ? ? ?

LG020914JGH206

125.1 50.4 ? 66.1 85.4 7 8 8 11

11 0 2 2 2 2 2 1 0

1 1 1 1 0 0 5 4 4

12 12 0 0 0 2 1 1 1

0 1 1 2 2 0 0 ? ?

? ? ? 1 1 1 ? ? ?

7.8 13.4 ? ? ? ? ? ?

LG021018JGH101

134.2 60 ? 68.3 90 7 ? ? ?

? ? ? ? ? ? ? ? 0

1 1 1 ? ? 1 5 4 4

12 12 0 0 0 2 ? 1 1

0 1 1 2 2 0 0 ? ?

? ? ? ? ? ? ? ? ?

? ? ? ? ? ? ? ?

LG021018JGH103

141.6 59.5 ? 69.7 92.4 7 8 8 ?

? ? ? ? ? ? ? ? 0

1 1 1 ? ? 1 5 4 4

12 12 0 0 0 2 ? 1 1

0 1 1 ? ? 0 0 ? ?

? ? ? ? ? ? ? ? ?

? ? ? ? ? ? ? ?

LG030109JGH114

142.6 59.4 ? 75.1 99.7 7 8 8 11

11 0 4 0 2 2 2 3 0

1 1 1 1 0 1 5 4 4

12 12 0 0 0 2 1 1 1

0 1 1 2 2 0 0 ? ?

? ? ? ? ? ? ? ? ?

? ? ? ? ? ? ? ?

LG030114JGH113

154.9 63.7 ? 76.7 105.3 ? ? ? ?

? ? ? ? ? ? ? ? 0

1 1 1 ? 0 1 5 4 4

12 12 1 0 0 2 ? 0 1

0 1 1 ? ? 0 0 ? 0

? ? ? ? ? ? ? ? ?

? ? ? ? ? ? ? ?

LG030214JGH167

165.8 62.5 ? 80.8 107.5 ? 8 8 ?

? ? ? ? ? ? ? ? 0

1 1 1 ? ? 1 5 4 4

12 12 0 0 0 2 ? 1 1

1 1 1 ? ? 0 0 ? ?

? ? ? ? ? ? ? ? ?

? ? ? ? ? ? ? ?

LG030214JGH169

150.4 60.4 ? 73.5 99.4 7 8 8 ?

? 0 2 2 2 2 2 3 0

1 1 1 ? 0 0 5 4 4

12 12 0 0 0 2 1 1 1

0 1 1 2 2 0 0 ? ?

? ? ? ? ? ? ? ? ?

? ? ? ? ? ? ? ?

LG030214JGH181

135.5 54 60.5 69.8 91.2 7 8 8 ?

? ? ? ? ? ? ? ? 0

1 1 1 ? 0 1 5 4 4

12 12 0 0 0 2 ? 1 1

0 1 1 ? ? 0 0 ? ?

? 1 1 ? ? ? ? ? ?

? ? 25.6 9.4 42.4 ? ? ?

LG030219JGH180

144.5 61.8 ? 71.5 98.2 7 8 8 11

11 0 2 2 2 2 2 3 0

1 1 1 2 0 1 5 4 4

12 12 0 0 0 2 1 1 1

0 1 1 2 2 0 0 ? ?

? ? ? ? ? ? ? ? ?

? ? ? ? ? ? ? ?

LG030314JGH171

? ? ? ? ? 7 ? ? ?

? 0 ? 2 2 2 2 3 ?

? 1 1 2 ? ? ? 4 4

? ? ? ? ? ? ? 1 ?

? 1 1 ? ? ? ? ? ?

? ? ? ? ? ? ? ? ?

? ? ? ? ? ? ? ?

LG030314JGH172

135 60.3 ? 63.2 90.4 7 8 8 ?

? 0 2 2 2 ? ? ? 0

1 1 1 2 0 1 5 4 4

12 12 1 0 0 2 1 1 1

1 1 1 ? ? 0 0 ? ?

? ? ? ? ? ? ? ? ?

? ? ? ? ? ? ? ?

LG030315JGH177

178.9 70 ? 84.5 105.5 7 8 8 ?

? ? ? ? ? ? ? ? 0

1 1 1 ? 0 1 5 4 4

12 12 0 0 0 2 1 1 1

0 1 1 2 2 0 0 ? ?

? ? ? ? ? ? ? ? ?

? ? ? ? ? ? ? ?

LG030315JH174

163.5 62.3 ? 71.9 105.6 7 8 8 ?
? ? ? ? ? ? ? ? 0
1 1 1 ? ? 1 5 4 4
12 12 0 0 0 2 1 1 1
0 1 1 ? ? 0 0 ? ?
? ? ? ? ? ? ? ? ?
? ? ? ? ?

LG030719JGH128

132.1 52.5 ? 64.5 85.7 7 8 8 ?
? 0 2 2 2 2 2 3 0
1 1 1 2 0 1 5 4 4
12 12 0 0 0 2 1 1 1
0 ? 1 2 2 0 0 ? ?
? ? ? ? ? ? ? ? ?
? ? ? ? ? ? ? ?

LG030728DAH136

? ? ? ? ? ? ? ? ?
? ? ? ? ? ? ? ? ?
? ? 1 ? ? 1 ? ? ?
? ? ? 0 ? ? ? ? 1
0 ? ? ? ? ? ? ? ?
? ? ? 1 1 ? ? ? ?
11.3 20.2 ? ? ? 0 0 0

LG030728DAH137

127.4 51 ? 65.3 82.4 7 8 8 ?
? ? 2 2 2 ? ? ? 0
1 1 1 ? 0 1 5 4 4
12 12 1 0 0 2 ? 1 1
0 1 1 ? ? 0 0 ? ?
? 1 1 ? ? ? ? ? ?
? ? 26.5 4 39.7 ? ? ?

LS020730DAH195

? ? ? ? ? ? ? ? ?
? ? ? ? ? ? ? ? 0
1 ? 1 ? ? 1 ? 4 ?
? ? ? ? ? ? ? ? ?
? ? ? ? ? ? ? ? ?
? ? ? ? ? ? ? ? ?
? ? ? ? ? ? ? ?

LS020730JGH196

? ? ? ? ? ? ? 8 ?
? ? ? ? ? ? ? ? 0
1 1 1 ? ? 1 5 ? 4
? 12 0 0 0 ? ? ? 1
1 ? ? ? ? 0 0 ? ?
? ? ? ? ? ? ? ? ?
? ? ? ? ? ? ? ?

LS020730JGH199

? ? ? ? ? ? 8 ? ?
? ? ? ? ? ? ? ? ?
? ? 1 ? ? 0 5 4 ?
? ? ? ? ? ? ? ? 1
1 ? ? ? ? 0 0 1 ?
? ? ? ? ? ? ? ? ?
? ? ? ? ? ? ? ?

LS020730JGH96

? ? ? ? ? ? ? 8 ?
? ? ? ? ? ? ? ? 0
1 1 1 ? 0 1 5 ? 4
? 12 0 0 0 2 1 ? 1
1 ? ? ? ? 0 0 ? ?
? 1 1 ? ? ? ? ? ?
? ? ? ? ? ? ? ?

PAS030113JGH119

? ? ? ? ? ? 8 ? 11
11 0 ? ? ? ? ? ? ?
? 1 1 2 ? 0 5 4 ?
12 12 1 0 0 2 1 1 1
0 1 1 ? ? 0 0 1 1
? 1 1 1 1 ? ? ? ?
7.5 12.9 24 ? 36.7 0 0 0

PAS041212JGH124

109.6 44.2 ? 56.2 73.6 7 8 8 ?
11 ? ? ? ? ? ? ? 0
1 1 1 2 0 1 5 4 4
12 12 0 0 0 2 1 1 1
0 1 1 ? ? 0 0 0 ?
? ? ? ? ? ? ? ? ?
? ? ? ? ? ? ? ?

PAS041212JGH208

? ? ? ? ? ? ? ? ?
? ? ? ? ? ? ? ? ?
? ? ? ? ? ? ? ? ?
? ? ? ? ? ? ? ? ?
? ? ? ? ? ? ? ? ?
? ? ? 1 1 ? ? ? ?
5.7 6.6 ? ? ? 0 0 0

TC020629JGH45

123 52.2 ? 61.3 81.6 7 8 8 ?
? ? ? ? ? ? ? ? 0
1 1 1 2 0 1 5 4 4
12 12 0 0 0 2 ? 1 1
0 1 1 ? ? 0 0 ? ?
? ? ? ? ? ? ? ? ?
? ? ? ? ?

TC020629JGH48								
178.5	66.6	?	85.1	115.2	7	8	8	?
?	?	?	?	?	?	?	?	0
1	1	1	2	0	1	5	4	4
12	12	0	0	0	2	?	1	1
0	1	1	?	?	0	0	?	?
?	?	?	?	?	?	?	?	?
?	?	?	?	?	?	?	?	

TC020629JGH49								
148	60.6	?	76.7	103.2	7	8	8	10
10	0	2	2	2	2	2	3	0
1	1	1	2	0	0	5	4	4
12	12	0	0	0	2	1	1	1
1	1	1	2	2	0	0	?	?
?	?	?	?	?	?	?	?	?
?	?	?	?	?	?	?	?	

TCO20629JGH40								
142.4	53.3	?	69.3	90.3	7	8	8	11
11	0	2	2	2	2	2	3	0
1	1	1	2	0	1	5	4	4
12	12	0	0	0	2	1	1	1
0	1	1	2	2	0	0	?	?
?	?	?	?	?	?	?	?	?
?	?	?	?	?	?	?	?	

TNHC 068333								
133.7	?	55	64.2	93	?	?	?	?
?	?	?	?	?	?	?	?	0
1	1	1	?	?	?	5	4	4
12	12	0	0	0	2	?	1	1
0	1	1	?	?	0	0	1	1
0	0	1	1	1	1	0	2	2
9	15.3	21.8	9.3	39.5	?	0	0	

TNHC 068334								
154.9	?	59.4	72	106.1	?	?	?	?
?	?	?	?	?	?	?	?	0
1	1	1	?	?	?	5	4	4
12	12	0	0	0	2	?	1	1
1	1	1	?	?	0	0	1	1
0	0	1	1	1	0	0	0	2
7.1	18.8	30.2	8.8	45.1	?	0	0	

TNHC 068335								
149.6	?	56.7	65.7	90.6	?	?	?	?
?	?	?	?	?	?	?	?	0
1	1	1	?	?	1	5	4	4
12	12	0	0	0	2	?	1	1
0	1	1	?	?	0	0	1	?
?	0	1	1	1	0	0	0	0
6.3	14	26.8	6.8	44	0	0	0	

TNHC 068336								
112	?	51.8	56	75.3	?	?	?	?
?	?	?	?	?	?	?	?	0
1	1	1	?	?	?	5	4	5
12	12	0	0	0	2	?	1	1
0	1	1	?	?	0	0	1	1
0	0	1	?	?	?	?	?	?
?	13.7	17.8	10	37.2	?	?	?	

TNHC 068337								
128.2	?	58.6	63	85.6	?	?	?	?
?	?	?	?	?	?	?	?	0
1	1	1	?	?	?	5	4	4
12	12	0	0	0	2	?	1	1
1	1	1	?	?	0	0	1	1
0	0	1	1	1	1	0	2	0
7.3	13.2	24.7	6.7	40.7	0	0	0	

TNHC 068338								
148.7	?	61.6	72	98.7	?	?	?	?
?	?	?	?	?	?	?	?	0
1	1	1	?	?	?	5	4	4
12	12	0	0	0	2	?	1	1
0	1	1	?	?	0	0	1	1
0	0	1	1	1	1	0	2	0
8.2	19.4	25.3	13.9	42.2	0	0	0	

TNHC 068339								
157.9	54.4	58.9	72	101.7	7	8	8	11
11	0	2	2	2	2	2	3	0
1	1	1	2	0	1	5	4	4
12	12	0	0	0	2	1	1	1
0	1	1	0	2	0	0	1	?
?	0	1	1	1	1	0	1	2
9.8	19.3	27.9	14	42.5	0	0	0	

TNHC 068340								
146.8	56	61	70.6	100.6	7	8	8	?
?	?	?	?	?	?	?	?	0
1	1	1	?	?	?	5	4	4
12	12	0	0	0	2	?	1	1
0	1	1	?	?	0	0	1	1
0	0	1	1	1	0	0	1	0
11.5	15.5	27	6.3	45.5	?	0	0	

TNHC 068341								
148.6	54	71.9	92.4	59.2	7	8	8	?
?	?	?	?	?	?	?	?	0
1	1	1	1	0	1	5	4	4
12	12	0	0	0	2	?	1	1
0	1	1	?	?	0	0	1	1
0	0	1	?	?	0	0	1	2
?	?	26.2	11	46.8	?	?	?	

TNHC 068342								
144.9	?	55.1	69.7	96.6	?	?	?	?
?	?	?	?	?	?	?	?	0
1	1	1	1	0	?	5	4	4
12	12	0	0	0	2	1	1	1
0	1	1	?	?	0	0	1	1
0	0	1	1	1	1	0	1	2
8.8	13.6	27.9	9.5	43.1	0	0	0	

TNHC 068343								
175.1	?	68.2	82.3	119.5	7	8	8	?
?	0	2	2	2	2	2	3	0
1	1	1	?	?	0	5	4	4
12	12	0	0	0	2	?	1	1
0	1	1	0	2	0	0	1	1
?	0	1	1	1	0	0	1	0
11.8	20	31.5	10.3	51.3	1	0	0	

TNHC 068344								
130.7	?	54.3	67.5	86.1	?	?	?	?
?	?	?	?	?	?	?	?	0
1	1	1	?	?	?	5	4	4
12	12	0	0	0	2	?	1	1
0	1	1	?	?	0	0	1	1
0	0	1	1	1	2	0	?	?
6.9	12.6	20.6	10.7	42.9	0	0	0	

TNHC 068345								
149.9	55.6	74.6	92.2	59.1	7	8	8	?
?	0	2	2	2	2	2	1	0
1	1	1	1	0	0	5	4	4
12	12	0	0	0	2	1	1	1
0	1	1	0	2	0	0	1	?
?	0	1	1	1	0	0	1	1
8.9	20.2	29.3	11.4	44.4	0	0	0	

TNHC 068346								
152.4	56.1	59.5	75.7	99.7	7	8	8	?
?	?	?	?	?	?	?	?	0
1	1	1	?	?	?	5	4	4
12	12	0	0	0	2	?	0	1
0	1	1	?	?	0	0	1	?
?	0	1	?	1	0	0	2	2
8.5	17	25.7	11.3	46	0	0	0	

TNHC 068347								
?	?	?	?	70.2	?	?	?	?
?	?	?	?	?	?	?	?	0
1	1	1	2	0	0	5	4	4
12	12	0	0	0	2	?	1	1
0	1	1	?	?	0	0	1	?
?	0	1	1	1	0	0	0	0
9.3	9.4	21.3	2.6	32.9	0	0	0	

TNHC 068348								
129.4	53.7	58.7	60.1	82.6	7	8	8	?
?	?	?	?	?	?	?	?	0
1	1	1	2	0	1	5	4	4
12	12	0	0	0	2	?	1	1
0	1	1	?	?	0	0	1	?
?	0	1	1	1	0	0	2	2
9.4	9.5	27	3.8	41.3	?	?	?	

TNHC 068349								
120.1	49	54.3	59.6	79	7	8	8	?
?	?	?	?	?	?	?	?	0
1	1	1	2	0	0	5	?	?
12	12	0	0	0	2	?	1	1
0	1	1	?	?	0	0	1	?
?	0	1	?	?	?	?	0	0
?	?	22	7.5	38.7	?	?	?	

TNHC 068350								
112.5	48.1	54.6	53.6	74.2	?	?	?	?
?	?	?	?	?	?	?	?	0
1	1	1	?	?	1	5	4	4
12	12	0	0	0	2	?	1	1
0	1	1	?	?	0	0	1	?
?	0	1	?	?	?	?	0	0
?	?	24.4	4.2	35	?	?	?	

TNHC 068351								
112.3	?	50.6	56	75.8	7	8	8	?
?	?	?	?	?	?	?	?	0
1	1	1	?	?	?	5	4	4
12	12	0	0	0	2	?	1	1
0	1	1	?	?	0	0	1	1
0	0	1	1	1	0	0	1	0
8.6	11.8	23.3	5.5	37.1	0	0	0	

TNHC 068352								
133.2	?	58.7	62.6	88.6	?	?	?	?
?	?	?	?	?	?	?	?	0
1	1	1	?	0	1	5	4	4
12	12	0	0	0	2	1	1	1
0	1	1	?	?	0	0	1	?
?	0	1	1	1	?	?	?	?
9.7	15.9	26.5	8.2	42.2	?	?	0	

TNHC 068353								
113.2	?	44.3	51.8	73	7	8	8	?
?	4	0	2	2	2	3	0	0
1	1	1	2	0	0	5	4	4
12	12	0	0	0	2	1	1	1
0	1	1	0	2	0	0	1	?
?	?	0	1	0	0	0	0	0
7.1	7.3	20	5.2	33.7	?	?	?	

TNHC 068354								
141.9	55.2	60.5	63.9	91	?	?	?	?
?	?	?	?	?	?	?	?	0
1	1	1	?	?	1	5	4	4
12	12	0	0	0	2	?	1	1
0	1	1	?	?	0	0	1	1
0	0	1	1	1	0	0	2	0
8	16	26.2	11.5	40.2	0	0	0	

TNHC 068355								
112.5	?	44.4	54.7	74.8	?	?	?	?
?	?	?	?	?	?	?	?	0
1	1	1	?	?	?	5	4	4
12	12	0	0	0	2	?	1	1
0	1	1	?	?	0	0	1	1
0	0	1	1	1	0	0	0	0
9	11.1	25.8	5.8	36	0	0	0	

TNHC 068356								
122.6	?	49.2	58.7	82.7	7	8	8	?
?	?	?	?	?	?	?	?	0
1	1	1	?	?	1	5	4	4
12	12	0	0	0	2	?	1	1
0	1	1	?	?	0	0	1	1
0	0	1	?	1	0	0	0	0
9.6	10	21.3	7.1	40.1	0	0	0	

TNHC 068357								
62.5	?	73.9	101.8	?	?	?	?	?
?	?	?	?	?	?	?	0	1
1	1	2	0	1	5	4	4	12
?	?	?	?	?	?	1	1	0
1	1	?	?	0	0	?	?	?
?	?	?	?	?	?	?	?	?
?	?	?	?	?	?	?		

TNHC 068358								
102.9	36.2	40.3	49.9	66.7	?	?	?	?
?	?	?	?	?	?	?	?	0
1	1	1	?	?	?	?	?	?
?	?	?	?	?	?	?	?	?
?	?	?	?	?	?	?	1	?
?	0	1	?	?	?	?	?	?
?	?	?	?	?	?	?	?	

TNHC 068359								
166.8	62.3	69	79.7	108.4	7	8	8	?
?	0	2	2	2	2	2	3	0
1	1	1	1	0	1	5	4	4
12	12	0	0	0	2	1	1	1
0	1	1	?	?	0	0	1	?
?	0	1	?	?	0	0	1	2
?	?	31.8	10.5	51.2	?	?	?	

TNHC 068360								
132	?	54.6	64.3	88.1	7	8	8	?
?	0	4	0	2	2	2	3	0
1	1	1	2	0	0	5	4	4
12	12	0	0	0	2	1	1	1
1	1	1	0	2	0	0	1	?
?	0	1	1	1	0	0	0	0
9.5	15.7	25	7.9	43.2	0	0	0	

TNHC 068361								
138.7	?	57.5	66.6	85	7	8	8	?
?	0	2	2	2	2	2	3	0
1	1	1	2	0	0	5	4	4
12	12	0	0	0	2	1	1	1
0	1	1	0	2	0	0	1	1
0	0	1	1	1	0	0	1	0
9.8	12.2	23.5	8.8	44.1	0	0	0	

TNHC 068362								
173.7	63.3	?	79.3	?	7	8	8	?
?	?	?	?	?	?	?	?	0
1	1	1	?	?	?	5	4	4
?	?	?	?	?	?	?	?	?
?	?	?	?	?	?	?	?	?
?	?	?	?	?	?	?	?	?
?	?	?	?	?	?	?	?	

TNHC 068940								
135.3	55.6	?	64.8	89.5	7	8	8	?
?	?	?	?	?	?	?	?	0
1	1	1	?	?	1	5	4	4
12	12	0	0	0	2	1	1	1
0	1	1	?	?	0	0	?	?
?	?	?	1	0	?	?	?	?
10	15.8	?	?	?	?	?	?	

TNHC 068941								
141.6	57.2	?	67.9	93	7	8	8	?
?	?	?	?	?	?	?	?	0
1	1	1	?	?	1	5	4	4
12	12	0	0	0	2	1	1	1
0	1	1	?	?	0	0	1	?
0	0	1	?	1	0	0	2	0
6.6	14.1	26.6	7.6	42.8	0	0	0	

TNHC 068942								
167.2	60.4	66.4	75.2	105.4	7	8	8	11
11	0	2	2	2	2	2	3	0
1	1	1	2	0	0	5	4	4
12	12	0	0	0	2	1	1	1
0	1	1	0	2	0	0	1	?
?	0	1	?	?	0	0	2	0
?	?	32.3	7.3	49.1	?	?	?	

TNHC 068943

131.4	?	60.2	64.2	86.2	?	?	?	?
?	?	?	?	?	?	?	?	0
1	1	1	?	0	1	5	4	4
12	12	0	0	0	2	?	1	1
0	1	1	?	?	0	0	1	1
0	0	1	1	0	0	0	2	0
6.2	15.1	27.6	5.7	40.8	0	0	0	

TNHC 068944

157.7	63.7	67.8	72.7	103.3	7	8	8	?
?	0	3	1	2	2	2	3	0
1	1	1	2	0	0	5	4	4
?	?	?	?	?	?	?	?	?
?	?	?	0	2	?	?	1	?
?	0	1	?	?	?	?	?	?
?	?	?	?	?	?	?	?	

TNHC 068945

130	?	54.4	62.7	85.7	7	8	8	?
?	0	2	2	2	2	2	3	0
1	1	1	2	?	1	5	4	4
12	12	0	0	0	2	1	1	1
1	1	1	2	2	0	0	1	1
0	0	1	1	1	0	0	1	0
10.4	13.2	26.8	5	40.7	0	0	0	

TNHC 068946

152.9	?	65.5	71.6	98	7	8	8	11
11	1	2	2	2	2	2	2	0
1	1	1	2	0	1	5	4	4
12	12	0	0	0	2	?	1	1
0	1	1	2	2	0	0	1	?
?	0	1	1	0	0	0	0	0
12.2	15.4	32.9	5.8	46.8	0	0	0	

TNHC 068947

111.9	?	48.1	55.6	74.4	7	8	8	11
11	0	4	0	2	2	2	3	0
1	1	1	2	?	?	5	4	4
12	12	0	0	0	2	1	1	?
?	1	1	2	2	?	?	?	?
?	0	1	1	1	0	0	1	0
?	?	?	?	?	0	0	?	

TNHC 068948

152.7	?	65	79.6	106.3	?	?	?	?
?	?	?	?	?	?	?	?	0
1	1	1	?	?	?	5	4	4
12	12	0	0	0	2	?	1	?
?	1	1	?	?	?	?	?	?
?	0	1	1	1	0	0	0	0
?	?	?	?	?	0	0	?	

TNHC 068949

123.4	?	54	60.3	84.9	?	?	?	?
?	?	?	?	?	?	?	?	?
?	1	1	?	?	?	5	4	4
12	12	0	0	0	2	?	0	?
?	1	1	?	?	?	?	?	?
?	0	1	?	?	0	0	1	0
?	?	?	?	?	?	?	?	

TNHC 068950

148.1	?	58.5	70.8	97.3	7	8	8	?
?	1	2	2	2	2	2	3	0
1	1	1	2	0	0	5	4	4
12	12	0	0	0	2	1	1	1
0	1	1	2	2	0	0	1	1
0	0	1	?	1	0	0	0	0
8.6	15	28.3	9.2	44.7	?	?	?	

TNHC 068951

153.1	57.7	61.6	?	101.6	7	8	8	11
?	1	2	2	2	2	2	3	0
1	1	1	2	3	0	5	4	4
12	12	0	0	0	2	1	1	1
0	1	1	2	2	0	0	1	1
0	0	1	1	1	1	0	0	0
8.5	15.7	28.4	8.9	46.5	0	0	0	

TNHC 068952

161.6	61	65.9	72.4	108.4	?	8	8	?
?	?	?	?	?	?	?	?	0
1	1	1	2	0	1	5	4	4
12	12	0	0	0	2	?	1	1
0	1	1	?	?	0	0	1	?
?	0	1	1	1	1	0	1	2
9.8	18.1	29.1	8.5	48.7	0	0	0	

TNHC 068953

163.1	62.3	?	75.7	108.8	7	8	8	?
?	?	?	?	?	?	?	?	0
1	1	1	?	?	1	5	4	4
12	12	0	0	0	2	?	1	1
0	1	1	?	?	0	0	1	?
?	0	1	1	0	0	0	1	0
14.2	15	31.8	8.4	50.6	0	0	0	

TNHC 068954

166.4	?	66.9	72.4	111.7	7	8	8	?
?	?	?	?	?	?	?	?	0

1 1 1 ? ? ? 5 4 4
11 12 0 0 0 2 ? 1 1
0 1 1 ? 1 0 0 1 1
0 0 1 1 ? 1 0 1 2
13 18.7 34.1 4.2 49.7 0 0 0

TNHC 068955
163.5 61.4 65.7 74.2 113.4 7 8 8 ?
? 1 ? 2 2 2 2 3 0
1 1 1 2 0 1 5 4 4
12 12 0 0 0 2 ? 1 1
0 1 1 ? ? 0 0 1 ?
? 0 1 ? 0 0 0 2 2
9.8 14.2 24.5 3.5 42.7 ? ? ?

TNHC 068956
169 62.5 ? 78.1 114.7 7 8 8 ?
? ? ? ? ? ? ? ? 0
1 1 1 ? ? 1 5 4 4
12 12 0 0 0 2 ? 1 1
0 1 1 ? ? 0 0 1 ?
? ? ? 1 1 1 0 0 0
10 20.7 ? ? ? 1 0 0

TNHC 068957
129.3 ? 61.7 64.4 86.2 ? ? ? ?
? ? ? ? ? ? ? ? 0
1 1 1 ? ? 1 5 4 4
12 12 0 0 0 2 ? 1 1
0 1 1 ? ? 0 0 1 1
0 0 1 0 0 1 0 0 2
9.8 14.2 24.5 3.5 42.7 0 0 0

TNHC 068958
150.7 59.4 60.4 75.2 100.5 ? ? ? ?
? ? ? ? ? ? ? ? 0
1 1 1 ? ? 1 5 4 4
12 12 0 0 0 2 ? 1 ?
? 1 1 ? ? ? ? 1 ?
? 0 1 1 ? ? ? ? ?
9.9 17 ? ? ? 0 0 0

TNHC 068959
140.4 54.8 57.1 67.6 93.6 7 8 8 ?
? ? ? ? ? ? ? ? 0
1 1 1 ? ? 1 5 4 4
12 12 0 0 0 2 ? 1 1
0 1 1 ? ? 0 0 1 ?
? 0 1 1 1 1 0 0 0
9.3 14.2 28.5 6.2 42.2 ? ? ?

TNHC 068960
103.9 ? 51.7 57.5 77.2 ? ? ? ?
? ? ? ? ? ? ? ? 0
1 1 1 ? ? ? 5 4 4
12 12 0 0 0 2 ? 1 1
0 1 1 ? ? 0 0 1 1
0 0 1 1 1 1 0 1 2
9.5 7.8 21.9 3.7 37.4 0 0 0

TNHC 068961
127.1 50.2 54.8 58.8 88.5 7 8 8 ?
? ? ? ? ? ? ? ? 0
1 1 1 ? 0 1 5 4 4
12 12 0 0 0 2 ? 1 1
0 1 1 ? ? 0 0 1 ?
? 0 1 1 1 1 0 1 1
6.6 14.2 23.3 5.5 39.2 0 0 0

TNHC 068962
142.6 ? 58 66.7 96.2 7 8 8 11
11 1 2 2 2 2 2 1 0
1 1 1 1 0 0 5 4 4
12 12 0 0 0 2 1 1 1
0 1 1 2 2 0 0 1 1
0 0 1 1 1 1 0 1 1
8 16.7 28.2 6.5 41 0 0 0

TNHC 068963
135 60.1 65.7 64.4 89.8 7 8 8 ?
? ? ? ? ? ? ? ? 0
1 1 1 ? 0 1 5 4 4
12 12 0 0 0 2 1 1 1
0 1 1 ? ? 0 0 1 ?
? 0 1 1 1 1 0 2 0
11 13.9 26.2 6.8 45.1 0 0 0

TNHC 068964
172.3 67.7 68.7 80.6 112.6 7 8 8 ?
? ? ? ? ? ? ? ? 0
1 1 1 ? 0 1 5 4 4
12 12 0 0 0 2 ? 1 1
1 1 1 ? ? 0 0 1 ?
? 0 1 ? ? ? ? 0 0
? ? 31.5 11.6 52.8 ? ? ?

TNHC 068965
141.6 58.5 64.7 69.1 96.2 7 8 8 ?
? ? ? ? ? ? ? ? 0
1 1 1 ? ? 1 5 4 4
12 12 0 0 0 2 ? 1 1

0 1 1 ? ? 0 0 1 ?
? 0 1 ? 1 1 0 2 0
14.8 14.2 28.5 6.4 45 0 0 0

TNHC 068966

124.5 52.1 57.7 61.1 83.9 7 8 8 ?
? ? ? ? ? ? ? ? 0
1 1 1 ? ? 1 5 4 4
12 12 0 0 0 2 1 1 1
0 1 1 ? ? 0 0 1 1
0 0 1 ? 1 0 0 2 2
9.6 12.5 24 4.1 39.7 0 0 ?

TNHC 069624

132 52.3 56.4 64.6 85.7 7 8 8 ?
? ? ? ? ? ? ? ? 0
1 1 1 ? 0 1 5 4 4
12 12 0 0 0 2 1 1 1
0 1 1 ? ? 0 0 1 ?
? 0 1 1 1 0 0 1 0
9.6 14.2 27.7 6 42.1 0 0 0

TNHC 069625

163.2 62.8 67.8 79.3 106.1 7 8 8 ?
? ? ? ? ? ? ? ? 0
1 1 1 ? ? 1 5 4 4
12 12 0 0 0 2 ? 1 1
0 1 1 ? ? 0 0 1 ?
? 0 1 ? 1 1 0 0 0
7.5 20 32.2 8.1 51.5 0 0 0

TNHC 069626

126.2 50.8 ? 62.9 81.9 7 8 8 11
11 0 2 2 2 2 2 1 0
1 1 1 2 0 0 5 4 4
12 12 0 0 0 2 ? 1 1
0 1 1 0 2 0 0 ? ?
? 0 1 ? ? ? ? ? ?
? ? ? ? ? ? ? ?

TNHC 069627

128.4 52 56.9 59 84.8 7 8 8 11
11 0 3 1 2 2 2 3 0
1 1 1 2 0 0 5 4 4
12 12 0 0 0 2 ? 1 1
0 1 1 0 2 0 0 1 ?
? 0 1 1 1 1 0 1 2
9.7 13.3 24.1 9.3 38 0 0 0

TNHC 069628

108.2 46.2 ? 50.6 70.8 ? ? ? ?
? ? ? ? ? ? ? ? 0
1 1 1 ? ? ? 5 4 4
12 12 0 0 0 2 ? 1 ?
? 1 1 ? ? ? ? ? ?
? 0 1 ? ? ? ? ? ?
? ? ? ? ? ? ? ?

TNHC 069629

117.1 44.5 ? 53.4 69.1 7 8 8 11
11 0 1 2 2 2 2 1 0
1 1 1 1 0 0 5 4 4
12 12 0 0 0 2 1 1 1
0 1 1 0 2 0 0 1 ?
? 0 1 ? ? ? ? 0 0
? ? 22.4 7.1 27.5 ? ? ?

TNHC 069630

118.8 49 ? 55.3 74.2 7 8 8 11
11 0 2 2 2 2 2 3 0
1 1 1 2 0 0 5 4 4
12 12 0 0 0 2 1 1 1
1 1 1 2 2 0 0 1 1
0 0 1 1 1 1 0 1 0
7.2 11.1 22 6.5 37.9 0 0 0

TNHC 069631

102.4 45.4 ? 51.5 69.5 ? ? ? ?
? ? ? ? ? ? ? ? 0
1 1 1 ? ? ? 5 4 4
12 12 0 0 0 2 ? 1 1
0 1 1 ? ? 0 0 1 1
0 0 1 1 0 0 0 0 1
6.4 10.7 20.5 4.5 32.8 0 0 0

TNHC 069632

130.3 ? 56.4 64.2 85 ? ? ? ?
? ? ? ? ? ? ? ? 0
1 1 1 ? ? ? 5 4 4
12 12 0 0 0 2 ? 1 1
1 1 1 ? ? 0 0 1 1
0 0 1 1 1 1 0 1 2
8.4 14.4 26 5 40.2 0 0 0

TNHC 069633

112 48.3 54.3 55.2 78 7 8 8 ?
? ? ? ? ? ? ? ? 0
1 1 1 ? 0 1 5 4 4
12 12 0 0 0 2 ? 1 1
0 1 1 ? ? 0 0 1 ?
? 0 1 ? ? 0 0 2 0
? ? 23.6 4.7 34.2 ? ? ?

TNHC 069634

106.8 ? 49.8 53.6 72 ? ? ? ?
? ? ? ? ? ? ? ? 0

1 1 1 ? ? ? 5 4 4
12 12 0 0 0 2 ? 1 1
0 1 1 ? ? 0 0 1 1
0 0 1 1 1 1 0 1 2
6.8 10.9 22.3 3.2 34.7 0 0 0

TNHC 069635
100.9 ? 43.7 49.6 66.5 ? ? ? ?
? ? ? ? ? ? ? ? 0
1 1 1 ? ? ? 5 4 4
12 12 0 0 0 2 ? 1 1
0 1 1 ? ? 0 0 1 1
0 0 1 ? 1 0 0 0 0
6.8 9.6 20.5 7.9 33 0 0 ?

TNHC 069636
112.6 ? 47 59.1 76.4 7 8 8 ?
? 0 2 2 2 2 2 3 0
1 1 1 ? ? ? 5 4 4
12 12 0 0 0 2 ? 1 1
1 1 1 ? ? 0 0 1 1
0 0 1 ? ? 0 0 2 0
? ? 23.3 4.2 37.2 ? ? ?

TNHC 069637
93.5 ? 42.6 50 65 ? ? ? ?
? ? ? ? ? ? ? ? 0
1 1 1 ? ? ? 5 4 4
12 12 0 0 0 2 ? 1 1
0 1 1 ? ? 0 0 1 1
0 0 1 ? 1 0 0 2 0
8.4 7.8 20.4 3.4 31 0 0 ?

TNHC 069638
106.1 ? 48.4 52.2 66.2 ? ? ? ?
? ? ? ? ? ? ? ? 0
1 1 1 ? ? ? 5 4 4
12 12 0 0 0 2 ? 1 1
1 1 1 ? ? 0 0 1 1
0 0 1 1 1 0 0 1 0
9.1 10.7 23.9 5.7 33.5 0 0 0

TNHC 069639
108.9 ? 46.6 54.5 72.3 ? ? ? ?
? ? ? ? ? ? ? ? 0
1 1 1 ? ? ? 5 4 4
12 12 0 0 0 2 ? 1 1
0 1 1 ? ? 0 0 1 1
0 0 1 1 1 1 0 1 1
7.4 11.4 21 4.6 34.9 0 0 0

TNHC 069640
106.3 43.3 48 51.6 71.5 7 8 8 11
11 0 2 2 2 2 2 1 0
1 1 1 2 0 0 5 4 4
12 12 0 0 0 2 1 1 1
0 1 1 2 2 0 0 1 ?
? 0 1 1 1 1 0 2 0
9.3 10 23.7 5.3 36.9 0 0 0

TNHC 069641
100.7 41 45.8 50.8 69.1 ? ? ? ?
? ? ? ? ? ? ? ? 0
1 1 1 ? ? 0 5 4 4
12 12 0 0 0 2 ? 1 1
0 1 1 ? ? 0 0 1 1
0 0 1 ? ? 0 0 2 0
? ? 23.1 3.5 33.4 ? ? ?

TNHC 069642
103.8 ? 45.8 53.8 71.5 ? ? ? ?
? ? ? ? ? ? ? ? 0
1 1 1 ? ? ? 5 4 4
12 12 0 0 0 2 ? 1 1
0 1 1 ? ? 0 0 1 1
0 0 1 1 1 2 0 1 0
8.6 9.7 20.1 5.5 33.7 0 0 0

TNHC 069643
84.9 ? 38.1 43.5 58.2 ? ? ? ?
? ? ? ? ? ? ? ? 0
1 1 1 ? ? ? 5 4 4
12 12 0 0 0 2 ? 1 1
0 1 1 ? ? 0 0 1 1
0 0 1 1 1 1 0 2 0
7.3 9.1 17.6 4.8 28.2 0 0 0

TNHC 069644
81.8 ? 38.2 42.3 57.8 ? ? ? ?
? ? ? ? ? ? ? ? 0
1 1 1 ? ? ? 5 4 4
12 12 0 0 0 2 ? 1 1
0 1 1 ? ? 0 0 1 1
0 0 1 1 1 0 0 0 0
6.5 7.5 17.1 3.8 26.9 0 0 0

TNHC 069645
119.7 ? 49.6 55.8 74.3 ? ? ? ?
? ? ? ? ? ? ? ? 0
1 1 1 ? ? ? 5 4 4
12 12 0 0 0 2 ? 1 1

0 1 1 ? ? 0 0 1 1

0 0 1 1 1 0 0 0 0

7.9 12.9 23 7 39 0 0 0

TNHC 069646

140.7 53.8 ? 71.8 92.4 7 8 8 ?

? ? ? ? ? ? ? ? 0

1 1 1 2 0 1 5 4 4

12 12 0 0 0 2 ? 1 1

0 1 1 ? ? 0 0 ? ?

? ? ? ? ? ? ? ? ?

? ? ? ? ? 0 0 0

TNHC 069647

? ? 58.3 67.6 ? ? ? ? ?

? 0 0 0 2 ? ? ? 0

1 ? ? ? ? 0 ? ? ?

? ? 0 0 ? ? ? 1 1

0 1 1 ? ? 0 0 1 1

0 0 1 1 1 0 0 0 0

9.6 15.3 27.1 7.5 42.1 0 0 0

TNHC 069648

114.2 ? 49.3 56.4 74.4 ? ? ? ?

? ? ? ? ? ? ? ? 0

1 1 1 ? ? ? 5 4 4

12 12 0 0 0 2 ? 1 1

0 1 1 ? ? 0 0 1 1

0 0 0 1 1 0 0 0 0

6 9.3 21.5 6 35.8 0 0 0

TNHC 069649

130.2 ? 53.1 64.5 86.8 ? ? ? ?

? ? ? ? ? ? ? ? 0

1 1 1 ? ? ? 5 4 4

12 12 0 0 0 2 ? 1 1

0 1 1 ? ? 0 0 1 1

0 0 1 1 1 1 0 1 0

5.7 13.7 25.4 7.3 40.4 0 0 0

TNHC 069650

89.4 ? 39.3 ? ? ? ? ? ?

? ? ? ? ? ? ? ? 0

1 1 1 ? ? ? 5 4 4

12 12 0 0 0 2 ? 1 1

0 1 1 ? ? 0 0 1 1

0 0 1 1 1 ? ? ? ?

5 9.7 20.5 2 28.4 0 0 0

TNHC 069651

73.3 ? 30.4 39 51.7 ? ? ? ?

? ? ? ? ? ? ? ? 0

1 1 1 1 0 0 5 4 ?

12 12 0 0 0 2 ? 1 1

0 1 1 ? ? 0 0 1 1

0 0 0 1 0 0 0 0 0

4.5 7.4 16.8 1.8 23.4 0 0 0

TNHC 069652

144.9 ? 58.1 69.2 97.9 ? ? ? ?

? ? ? ? ? ? ? ? 0

1 1 1 ? ? ? 5 4 4

12 12 0 0 0 2 ? 1 1

0 1 1 ? ? 0 0 1 1

0 0 1 1 1 0 0 2 0

10.4 11.4 25.6 10 43.9 0 0 2

TNHC 069653

132.6 55.2 61.8 63.4 85.6 7 8 8 ?

? ? ? ? ? ? ? ? 0

1 1 1 ? 0 1 5 4 4

12 12 0 0 0 2 1 1 1

0 1 1 ? ? 0 0 1 ?

? 0 1 1 1 1 0 0 0

13 13.3 26.6 8.2 43.9 0 0 0

TNHC 069801

113.5 ? 54.7 57.8 74.8 7 8 8 ?

? 0 2 2 2 2 2 3 0

1 1 1 2 ? 1 5 4 4

12 12 0 0 0 2 1 1 1

0 1 1 2 2 0 0 1 1

0 0 1 1 1 0 0 2 0

7.5 10.7 25.8 4.7 38.9 0 0 0

TNHC 069802

109.9 ? 44.6 52.7 74.5 7 ? ? ?

? 0 4 0 2 2 2 ? 0

1 1 1 ? ? 0 5 4 4

12 12 0 0 0 2 ? 1 1

0 1 1 0 ? 0 0 1 1

0 0 1 1 1 0 0 0 0

9.2 11.4 21.7 4.7 36.1 0 0 0

TNHC 069803

117.3 ? 50.1 56.3 72.1 ? ? ? ?

? ? ? ? ? ? ? ? 0

1 1 1 ? ? ? 5 4 4

12 12 0 0 0 2 ? 1 1

0 1 1 ? ? 0 0 1 1

0 0 1 ? ? ? ? ? ?

? ? 24.8 4.7 37.7 ? ? ?

TNHC 069804

116.3 44.1 48.4 57.1 74.9 7 8 8 11

11 0 4 0 3 2 2 3 0

1	1	1	2	0	0	5	4	4
12	12	0	0	0	2	1	1	1
0	1	1	2	2	0	0	1	?
?	0	1	1	0	0	0	0	0
6.9	10.3	21.9	6.8	37.6	0	0	0	
TNHC 069805								
105	42.4	47	54.8	71.6	7	8	8	11
11	0	2	2	2	2	2	3	0
1	1	1	?	?	?	5	4	4
12	12	0	0	0	2	1	1	1
0	1	1	2	2	0	0	1	1
0	0	1	1	1	0	0	0	0
6.7	19.7	24.6	2.6	34.9	0	0	0	
TNHC 069806								
109.4	?	41	42	65.3	?	?	?	?
?	?	?	?	?	?	?	?	0
1	1	1	?	?	0	5	4	4
12	12	0	0	0	2	?	1	1
0	1	1	?	?	0	0	1	1
0	0	1	1	1	0	0	0	0
10.3	11.5	22.5	3.2	36.8	0	0	0	
TNHC 069807								
100.1	?	50.4	56.3	71	7	8	8	11
11	0	3	1	2	2	3	0	0
1	1	1	2	0	0	5	4	4
12	12	0	0	0	2	1	1	1
0	1	1	2	2	0	0	1	1
0	0	1	?	?	0	0	0	0
?	?	20.2	2.4	33.2	?	?	?	
TNHC 069808								
100.4	?	48.4	49.4	64.9	?	?	?	?
?	?	?	?	?	?	?	?	0
1	1	1	?	?	?	5	4	4
12	12	0	0	0	2	?	1	1
0	1	1	?	?	0	0	1	1
0	0	0	1	1	0	0	0	0
5.9	10	18.8	5	34	0	0	0	
TNHC 069809								
134.8	46.6	50	59.9	89.4	?	?	?	?
?	?	?	?	?	?	?	?	0
1	1	1	?	?	0	5	4	4
12	12	0	0	0	2	1	1	1
0	1	1	?	?	0	0	1	1
0	?	?	?	?	?	?	0	0
?	?	23.5	9	?	?	?	?	
TNHC 069810								
102.5	?	43	48.7	66.5	?	?	?	?
?	?	?	?	?	?	?	?	0
1	1	1	?	?	?	5	4	4
12	12	0	0	0	2	?	1	1
0	1	1	?	?	0	0	1	1
0	0	1	1	1	1	0	0	0
5.7	8.2	18.1	5.2	35.1	0	0	0	
TNHC 069811								
110	43.5	47.3	54.3	72.9	7	8	8	11
11	0	4	0	2	2	2	3	0
1	1	1	2	?	?	5	4	4
12	12	0	0	0	2	1	1	?
?	1	1	0	2	?	?	?	?
?	0	1	1	1	0	0	0	0
8.8	10.5	23	5	34.7	0	0	0	
TNHC 069812								
99.9	?	46.7	48.8	66.4	?	?	?	?
?	?	?	?	?	?	?	?	0
1	1	1	?	?	0	5	4	4
12	12	0	0	0	2	?	1	1
0	1	1	?	?	0	0	1	1
0	0	1	1	1	1	0	1	2
5.3	9	19.7	2.9	33.3	0	0	0	
TNHC 069813								
143	?	61.9	71.7	93	?	?	?	?
?	?	?	?	?	?	?	?	0
1	1	1	?	?	?	5	4	4
11	11	0	0	1	0	0	1	1
0	1	1	?	?	0	0	1	1
0	0	1	1	1	1	0	2	2
10	14.7	27.8	7.8	43.1	0	0	0	
TNHC 069814								
109.2	?	43.7	55.5	72.1	?	?	?	?
?	?	?	?	?	?	?	?	0
1	1	1	?	?	?	5	4	4
12	12	0	0	0	2	?	1	1
0	1	1	?	?	0	0	1	1
0	0	1	1	1	0	0	2	2
5.8	9.1	19.3	3.8	34.58	0	0	0	
TNHC 069815								
116.5	?	51.5	56.2	73.2	?	?	?	?
?	?	?	?	?	?	?	?	0
1	1	1	?	?	?	5	4	4
12	12	0	0	0	2	?	1	1

0	1	1	?	?	0	0	1	1
0	0	1	1	1	0	0	0	0
7.5	12.5	20.5	7	36.1	0	0	0	
TNHC 069816								
130.6	?	55.2	62.6	84.5	?	?	?	?
?	?	?	?	?	?	?	?	0
1	1	1	?	?	?	5	4	4
12	12	0	0	0	2	?	1	1
0	1	1	?	?	0	0	1	1
0	0	1	1	1	0	0	1	2
5.8	13.5	27.7	4.5	41.7	0	0	0	
TNHC 069817								
129.3	?	52.2	64.9	81.2	?	?	?	?
?	?	?	?	?	?	?	?	0
1	1	1	?	?	?	5	4	4
12	12	0	0	0	2	?	1	1
0	1	1	?	?	0	0	1	1
0	0	1	1	1	0	0	0	0
7	13.2	22.9	7.4	44	0	0	0	
TNHC 069818								
?	?	?	?	?	?	8	8	?
?	0	2	?	?	?	2	3	0
1	1	1	2	?	?	?	?	?
?	?	?	?	?	?	?	?	?
?	?	?	?	?	?	?	?	?
?	0	1	1	1	1	0	0	1
9.7	10.45	18.65	6.35	34.33	?	?	?	
TNHC 069819								
180.5	65.3	?	87.3	118.6	7	8	8	?
?	?	?	?	?	?	?	?	0
1	1	1	?	?	?	5	4	4
12	12	0	0	0	2	?	1	?
?	1	1	?	?	?	?	?	?
?	?	?	?	?	?	?	?	?
?	?	?	?	?	?	?	?	
TNHC 069820								
141.9	53.5	57.6	73.6	90.7	7	8	8	?
?	?	?	?	?	?	?	?	0
1	1	1	?	?	?	5	4	4
12	12	0	0	0	2	?	1	1
0	1	1	?	?	0	0	1	1
0	0	1	1	1	0	0	0	0
11.2	14.5	24.8	7.5	41.2	0	0	0	
TNHC 069821								
131	56.3	62.5	67	89.1	7	8	8	?
?	?	?	?	?	?	?	?	0
1	1	1	?	?	?	5	4	4

12	12	0	0	0	2	?	1	1
0	1	1	?	?	0	0	1	1
0	0	1	1	1	0	0	0	0
10.8	14.7	29.7	5.9	41.2	0	0	0	
TNHC 069822								
180.9	63.1	?	84.4	122.2	7	8	8	11
11	0	2	2	2	2	2	3	0
1	1	1	?	0	0	5	4	4
12	12	0	0	0	2	1	1	1
0	1	1	2	2	0	0	1	?
?	0	1	1	1	0	0	0	0
9.5	20.6	32.2	13.5	50.1	0	0	0	
TNHC 069823								
107.5	?	44.4	51.1	64.8	?	?	?	?
?	?	?	?	?	?	?	?	0
1	1	1	?	?	?	5	4	4
12	12	0	0	0	2	?	1	1
0	1	1	?	?	0	0	1	1
0	0	1	1	1	0	0	0	0
6.5	10	22.2	2.7	34.8	0	0	0	
TNHC 069824								
139.3	?	61.6	69.3	89.5	?	?	?	?
?	?	?	?	?	?	?	?	0
1	1	1	?	?	?	5	4	4
12	12	0	0	0	2	?	1	1
0	1	1	?	?	0	0	1	1
0	0	1	1	1	0	0	2	0
8.9	14.4	27.2	5.2	42.3	0	0	0	
TNHC 069825								
106.6	41.3	?	54.4	69	?	?	?	?
?	?	?	?	?	?	?	?	0
1	1	1	?	?	?	5	4	4
12	12	0	0	0	2	?	1	1
1	1	1	?	?	0	0	1	1
0	0	0	1	1	?	?	?	?
6.6	9	22.2	4.7	34.4	0	0	0	
TNHC 069936								
137.8	52.9	54.8	64.9	89.7	7	8	8	11
11	0	2	2	2	2	2	3	0
1	1	1	2	0	0	5	4	4
12	12	0	0	0	2	1	1	1
0	1	1	2	2	0	0	1	?
?	0	1	1	1	1	0	0	0
7.7	12.9	26.1	5.8	40	0	0	0	
TNHC 069939								
?	?	?	?	?	?	8	8	?
?	?	?	?	?	?	?	?	?

? ? ? 2 ? ? 5 4 4
? ? 0 ? ? ? ? 1 ?
? 1 1 2 2 ? ? 1 ?
? 0 1 1 1 ? ? ? ?
7.7 8.7 20.3 3.7 29.7 ? ? ?

TNHC 070241
? ? ? ? 112.9 ? ? ? ?
? 0 2 ? ? 2 2 3 0
1 1 1 2 0 0 5 4 4
12 12 0 0 0 2 1 1 1
0 1 1 0 2 0 0 1 ?
? 0 1 1 1 1 0 2 0
7.8 18.5 31.5 11.4 48.3 0 0 0

TNHC 070587
142.1 57.5 ? 75.2 96.3 7 8 8 ?
? ? ? ? ? ? ? ? 0
1 1 1 ? 0 1 5 4 4
12 12 0 0 0 2 1 1 1
0 1 1 ? ? 0 0 ? ?
? ? ? ? ? ? ? ? ?
? ? ? ? ? ? ? ?

TNHC 070753
? ? ? ? ? ? ? ? ?
? 0 2 2 2 ? ? ? 0
1 ? ? ? ? 1 ? ? ?
? ? 0 ? ? ? ? 1 1
0 1 1 ? ? 0 0 ? ?
? ? ? ? ? ? ? ? ?
? ? ? ? ? ? ? ?

TNHC 77201
130.4 53 58.4 62.1 86 7 8 8 11
? 0 2 2 2 2 2 0 0
1 1 1 2 0 0 5 4 4
12 12 0 0 0 2 1 1 1
0 1 1 0 2 0 0 1 1
0 0 1 1 1 1 0 1 0
? ? ? ? ? 0 0 0

TNHC 77202
182.1 73 75.4 80.8 125.6 ? ? ? ?
? ? ? ? ? ? ? ? 0
1 1 1 2 ? ? 5 4 4
12 12 0 0 0 2 ? 1 ?
? 1 1 ? ? ? ? ? ?
? 0 1 ? ? 0 0 2 2
? ? ? ? ? ? ? ?

TNHC 77203
157.7 62.2 ? 72.8 107.7 ? ? ? ?
? ? ? ? ? ? ? ? 0
1 1 1 2 ? 1 5 4 4
12 12 0 0 0 2 ? 1 1
0 1 1 ? ? 0 0 ? ?
? ? ? ? ? ? ? ? ?
? ? ? ? ? ? ? ?

TNHC 77204
127.3 52.3 ? 58.8 85.6 7 8 8 11
11 ? ? ? ? ? ? ? 0
1 1 1 2 0 0 5 4 4
12 12 0 0 0 2 1 1 ?
? 1 1 ? ? 0 0 ? ?
? ? ? ? ? ? ? ? ?
? ? ? ? ? ? ? ?

TNHC 77205
165.3 59.5 ? 72.4 107.3 ? 8 8 11
11 ? ? ? ? ? ? ? 0
1 1 1 2 0 0 5 4 4
12 12 0 0 0 2 1 1 1
0 1 1 ? ? 0 0 ? ?
? ? ? ? ? ? ? ? ?
? ? ? ? ? ? ? ?

TNHC 77206
135.9 ? 56.6 64.5 90.1 7 8 8 ?
? 0 2 2 2 2 2 1 0
1 1 1 2 0 0 5 4 4
12 12 0 0 0 2 1 1 1
1 1 1 0 2 0 0 1 1
0 0 1 ? ? 1 0 2 2
? ? ? ? ? 0 0 0

TNHC 77207
94.2 39.1 41.9 48.1 64.5 ? 8 8 ?
? ? ? ? ? ? ? ? 0
1 1 1 2 3 1 5 4 4
12 12 0 0 0 2 1 1 1
0 1 1 ? ? 0 0 1 ?
? 0 1 ? ? ? ? ? ?
? ? ? ? ? ? ? ?

XX02JGH98
104.5 44 ? 53.2 67 7 8 8 11
11 0 4 0 2 2 2 3 0
1 1 1 2 0 0 5 4 4

12 12 1 0 0 2 1 1 1

0 1 1 2 2 0 0 1 ?

? 0 1 ? ? 0 0 2 0

? ? 23.2 5.4 33.7 ? ? ?

XX02JGH99

159.6 65.6 ? 79.4 107.5 7 8 8 ?

? ? ? ? ? ? ? ? 0

1 1 1 ? 0 1 5 4 4

12 12 1 0 0 2 ? 1 1

1 1 1 ? ? 0 0 ? ?

? ? ? ? ? ? ? ? ?

? ? ? ? ? ? ? ?

XXX031214JGH129

? 44.1 ? 64.3 ? ? 8 8 11

11 0 4 ? ? ? ? ? ?

? 1 1 ? ? 0 5 4 4

12 12 0 0 0 2 1 1 1

0 1 1 0 2 ? ? ? ?

? 0 1 1 1 ? ? ? ?

8.1 12.4 22.5 5.8 37.8 0 0 0

XXX04XXXXJGH131

150.4 53.1 ? 75.8 99.4 7 8 8 ?

? ? ? ? ? ? ? ? 0

1 1 1 ? 0 1 5 4 4

12 12 ? ? ? 2 ? 1 1

0 1 1 ? ? 0 0 0 ?

? 0 1 1 1 1 1 2 0

11.3 16.4 30.3 9.5 41 0 0 0

XXX04XXXXJGH132

127.4 54.9 ? 61 84.1 ? ? ? ?

? ? ? ? ? ? ? ? 0

1 1 1 ? ? 1 5 4 4

12 12 ? ? ? 2 ? 1 ?

? 1 1 ? ? 0 0 ? ?

? ? ? ? ? ? ? ? ?

? ? ? ? ? ? ? ?

xxxxxxxxxJGH145

108.3 41 ? 58 70.5 7 8 8 11

11 0 3 1 2 2 2 3 0

1 1 1 1 0 0 5 4 4

12 12 0 1 0 2 1 1 1

1 1 1 0 2 0 0 ? ?

? ? ? ? ? ? ? ? ?

? ? ? ? ? ? ? ?

uf149511

? 28.5 ? 50.2 50.8 ? ? ? ?

? ? ? ? ? ? ? ? 0

1 1 1 ? ? 0 5 4 4

? ? 0 0 ? ? ? 1 1

0 1 1 ? ? 0 0 ? 1

0 ? ? ? ? ? ? ? ?

? ? ? ? ? ? ? ?

References

Angielczyk, K. D., Feldman, C. R., & Miller, G. R. (2011). Adaptive evolution of plastron shape in emydine turtles. *Evolution, 65*, 377–394.

Auffenberg, W. (1958). Fossil turtles of the genus *Terrapene* in Florida. *Bulletin of the Florida State Museum, Biological Sciences, 3*, 53–92.

Auffenberg, W. (1967). Further notes on fossil box turtles of Florida. *Copeia, 1967*, 319–325.

Barbour, T., & Stetson, H. C. (1931). A revision of the Pleistocene species of *Terrapene* of Florida. *Bulletin of the Museum of Comparative Zoology, 72*, 295–299.

Baur, G. (1891). American box-tortoises. *Science, 17*, 190–191.

Bell, C. J., Gauthier, J. A., & Bever, G. S. (2010). Covert biases, circularity, and apomorphies: A critical look at the North American Quaternary Herpetofuanal Stability Hypothesis. *Quaternary International, 217*, 30–36.

Bentley, C. C., & Knight, J. L. (1998). Turtles (Reptilia: Testudines) of the Ardis local fauna late Pleistocene (Rancholabrean) of South Carolina. *Brimleyana, 25*, 3–33.

Bever, G. S., Bell, C. J., & Hutchison, J. H. (2003). Hinged emydine turtles in North America: Reading the record. *Journal of Vertebrate Paleontology, 23*(Suppl. 3), 34A.

Bickham, J. W., Lamb, T., Minx, P., & Patton, J. C. (1996). Molecular systematics of the genus *Clemmys* and the intergeneric relationships of emydid turtles. *Herpetologica, 52*, 89–97.

Blaney, R. M. (1971). An annotated check list and biogeographic analysis of the insular herpetofauna of the Apalachicola region, Florida. *Herpetologica, 27*, 406–430.

Bonin, F., Devaux, B., & Dupré, A. (2006). Turtles of the World (P. C. H. Pritchard, Trans.). Baltimore, MD: Johns Hopkins University Press.

Brown, W. S. (1971). Morphometrics of *Terrapene coahuila* (Chelonia, Emydidae), with comments on its evolutionary status. *Southwestern Naturalist, 16*, 171–184.

Brown, W. S. (1974). Ecology of the aquatic box turtle, *Terrapene coahuila* (Chelonia, Emydidae) in northern Mexico. *Bulletin of the Florida State Museum Biological Sciences, 19*, 1–65.

Burke, R. L., Leuteritz, T. E., & Wolf, A. J. (1996). Phylogenetic relationships of emydine turtles. *Herpetologica, 52*, 572–584.

Dodd, C. K., Jr. (2001). *North American box turtles: A natural history*. Norman, OK: University of Oklahoma Press.

Feldman, C. R., & Parham, J. F. (2002). Molecular phylogenetics of emydine turtles: Taxonomic revision and the evolution of shell kinesis. *Molecular Phylogenetics and Evolution, 22*, 388–398.

Gaffney, E. S., & Meylan, P. A. (1988). A phylogeny of turtles. In M. J. Benton (Ed.), *The phylogeny and classification of tetrapods, volume 1: Amphibians, reptiles, birds* (pp. 157–219). Oxford: Systematics Association Special Volume, Clarendon Press.

Hay, O. P. (1906). Descriptions of two new genera (*Echmatemys* and *Xenochelys*) and two new species (*Xenochelys formosa* and *Terrapene putnami*) of fossil turtles. *Bulletin of the American Museum of Natural History, 22*, 27–31.

Hay, O. P. (1908). Descriptions of five species of North American fossil turtles, four of which are new. *Proceedings of the United States National Museum, 35*, 161–169 (Plates 26–27).

Hirayama, R. (1985) ("1984"). Cladistic analysis of batagurine turtles (Batagurinae: Emydidae: Testudinoidea); A preliminary result. In F. de Broin & E. Jiménez-Fuentes (Eds.), Stvdia Palaeochelon-iologica I. Communicaciones del I Simposium Internacional sobre Quelonios Fosiles, París, Octubre, 1983 Stvdia Geologica Salmanticensia Volumen Especial I (pp. 141–157). Ediciones Universidad de Salamanca.

Holman, J. A. (1975). Herpetofauna of the Wakeeney local fauna (Lower Pliocene: Clarendonian) of Trego County, Kansas. In G. R. Smith & N. E. Friedland (Eds.), *Studies on Cenozoic paleontology and stratigraphy in honor of Claude W. Hibbard memorial* (Vol. 3, pp. 49–66). The Museum of Paleontology, University of Michigan, Papers on Paleontology, 12.

Holman, J. A. (1987). Herpetofauna of the Egelhoff site (Miocene: Barstovian) of north-central Nebraska. *Journal of Vertebrate Paleontology, 7*, 109–120.

Holman, J. A. (1995). A new species of *Emydoidea* (Reptilia: Testudines) from the late Barstovian (medial Miocene) of Cherry County, Nebraska. *Journal of Herpetology, 29*, 548–553.

Holman, J. A. (2002). Additional specimens of the Miocene turtle *Emydoidea hutchisoni* Holman 1995—new temporal occurrences, taxonomic characters, and phylogenetic inferences. *Journal of Herpetology, 36*, 436–446.

Holman, J. A., & Corner, R. G. (1985). A Miocene *Terrapene* (Testudines: Emydidae) and other Barstovian turtles from south-central Nebraska. *Herpetologica, 41*, 88–93.

Holman, J. A., & Fritz, U. (2005). The box turtle genus *Terrapene* (Testudines: Emydidae) in the Miocene of the USA. *Herpetological Journal, 15*, 81–90.

Holroyd, P. A., Hutchison, J. H., & Strait, S. G. (2001). Turtle diversity and abundance through the lower Eocene Willwood Formation of the southern Bighorn Basin. In P. D. Gingerich (Ed.), *Paleocene–Eocene stratigraphy and biotic change in the Bighorn and Clarks Fork Basins, Wyoming* (pp. 97–107). University of Michigan Papers in Paleontology, 33.

Howeth, J. G., McGaugh, S. E., & Hendrickson, D. A. (2008). Contrasting demographic and genetic estimates of dispersal in the endangered Coahuilan box turtle: A contemporary approach to conservation. *Molecular Ecology, 17*, 4209–4221.

Hutchison, J. H. (1981). *Emydoidea* (Emydidae: Testudines) from the Barstovian (Miocene) of Nebraska. *PaleoBios, 37*, 1–6.

Hutchison, J. H. (1992). Western North American reptile and amphibian record across the Eocene/Oligocene boundary and its climatic implications. In D. R. Protehro & W. A. Berggren (Eds.), *Eocene–Oligocene climatic and biotic evolution* (pp. 451–463). Princeton, NJ: Princeton University Press.

Hutchison, J. H. (1998). Turtles across the Paleocene/Eocene epoch boundary in west-central North America. In M.-P. Aubry, S. G. Lucas, & W. A. Berggren (Eds.), *Late Paleocene–Early Eocene climatic and biotic events in the marine and terrestrial records* (pp. 401–408). New York: Columbia University Press.

Jass, C. N., & Bell, C. J. (2010). Desert tortoises (*Gopherus agassizii*) from Pleistocene sediments in Cathedral Cave, White Pine County, Nevada. *The Southwestern Naturalist, 55*, 558–563.

Joyce, W. G., & Bell, C. J. (2004). A review of the comparative morphology of extant testudinoid turtles (Reptilia: Testudines). *Asiatic Herpetological Research, 10*, 53–109.

Knauss, G. E., Joyce, W. G., Lyson, T. R., & Pearson, D. (2011). A new kinosternoid from the Late Cretaceous Hell Creek Formation of North Dakota and Montana and the origin of the *Dermatemys mawii* lineage. *Paläontologische Zeitschrift, 85*, 125–142.

Legler, J. M. (1960). Natural history of the ornate box turtle, *Terrapene ornata* Agassiz. *University of Kansas Publication of the Museum of Natural History, 11*, 527–669.

Milstead, W. M. (1956). Fossil turtles of Friesenhahn Cave, Texas, with the description of a new species of *Testudo*. *Copeia, 1956*, 162–171.

Milstead, W. M. (1960). Relict species of the Chihuahuan Desert. *Southwestern Naturalist, 5*, 75–88.

Milstead, W. M. (1967). Fossil box turtles (*Terrapene*) from central North America, and box turtles of eastern Mexico. *Copeia, 1967*, 168–179.

Milstead, W. M. (1969). Studies on the evolution of box turtles (genus *Terrapene*). *Bulletin of the Florida State Museum, Biological Sciences, 14*, 1–113.

Minx, P. (1996). Phylogenetic relationships among the box turtles, genus *Terrapene*. *Herpetologica, 52*, 584–597.

Moodie, K. B., & Van Devender, T. R. (1977). Additional late Pleistocene turtles from Jones Spring, Hickory County, Missouri. *Herpetologica, 33*, 87–90.

Parmley, D. (1992). Turtles from the late Hemphillian (latest Miocene) of Knox County, Nebraska. *Texas Journal of Science, 44*, 339–348.

Pritchard, P. C. H. (1988). A survey of neural bone variation among recent chelonian species, with functional interpretations. *Acta Zoologica Cracoviensia, 31*, 625–686.

Schmidt, K. P., & Owens, D. W. (1944). Amphibians and reptiles of northern Coahuila, Mexico. *Zoological Series of Field Museum of Natural History, 29*, 97–115.

Spinks, P. Q., Thomson, R. C., Lovely, G. A., & Shaffer, H. B. (2009). Assessing what is needed to resolve a molecular phylogeny: Simulations and empirical data from emydid turtles. *BMC Evolutionary Biology, 9*, 1–17. doi:10.1186/1471-2148-9-56.

Tong, H., Zhang, J.-Y., & Li, J.-J. (2010). *Anosteria maomingensis* (Testudines: Carettochelyidae) from the Late Eocene of Maoming, Guangdong, southern China: New material and re-description. *Neues Jahrbuch für Geologie und Paläontologie, 256*, 279–290.

Webb, R. G., Minckley, W. L., & Craddock, J. E. (1963). Remarks on the Coahuilan box turtle, *Terrapene coahuila* (Testudines, Emydidae). *Southwestern Naturalist, 8*, 89–99.

Weems, R. E. (1988). Paleocene turtles from the Aquia and Brightseat Formations, with a discussion of their bearing on sea turtle evolution and phylogeny. *Proceedings of the Biological Society of Washington, 101*, 109–145.

Wiens, J. J., Kuczynski, C. A., & Stephens, P. R. (2010). Discordant mitochondrial and nuclear gene phylogenies in emydid turtles: Implications for speciation and conservation. *Biological Journal of the Linnean Society, 99*, 445–461.

Williams, K. L., Smith, H. M., & Chrapliwy, P. S. (1960). Turtles and lizards from northern Mexico. *Transactions of the Illinois Academy of Science, 53*, 36–45.

Yasukawa, Y., Hirayama, R., & Hikida, T. (2001). Phylogenetic relationships of geoemydine turtles (Reptilia: Bataguridae). *Current Herpetology, 20*, 105–133.

Subject Index

D. B. Brinkman et al. (eds.), *Morphology and Evolution of Turtles*, Vertebrate Paleobiology and Paleoanthropology,
DOI: 10.1007/978-94-007-4309-0, © Springer Science+Business Media Dordrecht 2013

Taxonomic Index

DOI: 10.1007/978-94-007-4309-0, © Springer Science+Business Media Dordrecht 2013

MIX
Papier aus verantwortungsvollen Quellen
Paper from responsible sources
FSC® C105338

Printed by Books on Demand, Germany